Emerging Natural and Tailored Nanomaterials for Radioactive Waste Treatment and Environmental Remediation

Interface Science and Technology

Series Editor: Arthur T. Hubbard

In this Series:

Vol. 1: Clay Surfaces: Fundamentals and Applications
Edited by F. Wypych and K.G. Satyanarayana

Vol. 2: Electrokinetics in Microfluids
By Dongqing Li

Vol. 3: Radiotracer Studies of Interfaces
Edited by G. Horányi

Vol. 4: Emulsions: Structure Stability and Interactions
Edited by D.N. Petsev

Vol. 5: Inhaled Particles
By Chiu-sen Wang

Vol. 6: Heavy Metals in the Environment: Origin, Interaction and Remediation
Edited by H.B. Bradl

Vol. 7: Activated Carbon Surfaces in Environmental Remediation
Edited by Teresa J. Bandosz

Vol. 8: Tissue Engineering: Fundamentals and Applications
By Yoshito Ikada

Vol. 9: Particles at Interfaces: Interactions, Deposition, Structure
By Zbigniew Adamczyk

Vol. 10: Interface Science in Drinking Water Treatment: Theory and Applications
Edited by G. Newcombe and D. Dixon

Vol. 11: Surface Complexation Modelling
Edited by Johannes Lützenkirchen

Vol. 12: Theory of Colloid and Interfacial Electric Phenomena
By Hiroyuki Ohshima

Vol. 13: Sorbent Deformation
By A.V. Tvardovskiy

Vol. 14: Advanced Chemistry of Monolayers at Interfaces: Trends in Methodology and Technology
Edited by Toyoko Imae

Vol. 15: Macromolecules in Solution and Brownian Relativity
By Stefano A. Mezzasalma

Vol. 16: The Properties of Water and their Role in Colloidal and Biological Systems
By Carel Jan van Oss

Vol. 17: Chemistry on Modified Oxide and Phosphate Surfaces: Fundamentals and Applications
By Robson Fernandes de Farias

Vol. 18: Interface Science and Composites
By Soo-Jin Park and Min-Kang Seo

Vol. 19: Nanoparticle Technologies: From Lab to Market
By Farid Bensebaa

Vol. 20: Particles at Interfaces
By Zbigniew Adamczyk

Vol. 21: Self-Assembly Processes at Interfaces: Multiscale Phenomena
By Vincent Ball

Vol. 22: Rheology of Emulsions: Electrohydrodynamics Principles
By Aleksandar M. Spasic

Vol. 23: Tailored Thin Coatings for Corrosion Inhibition using a Molecular Approach
By Simo Olavi Pehkonen and Shaojun Yuan

Vol. 24: Charge and Energy Storage in Electrical Double Layers
By Silvia Ahualli and Ángel V. Delgado

Vol. 25: Stimuli Responsive Polymeric Membranes
By Mihir Kumar Purkait, Manish Kumar Sinha, Piyal Mondal, and Randeep Singh

Vol. 26: Theory of Electrophoresis and Diffusiophoresis of Highly Charged Colloidal Particles
By Eric Lee

Vol. 27: Graphene Surfaces: Particles and Catalysts
By Karim Kakaei, Mehdi Esrafili and Ali Ehsani

Vol. 28: An Introduction to Green Nanotechnology
By Mahmoud Nasrollahzadeh, Mohammad Sajadi, Monireh Atarod, Mohadeseh Sajjadi and Zahra Isaabadi

Interface Science and Technology
Volume 29

Emerging Natural and Tailored Nanomaterials for Radioactive Waste Treatment and Environmental Remediation

Principles and Methodologies

Edited by
Changlun Chen
Key Laboratory of Photovoltaic and Energy Conservation Materials, Institute of Plasma Physics, Chinese Academy of Sciences, Hefei, People's Republic of China

ELSEVIER
ACADEMIC PRESS
An imprint of Elsevier

Academic Press is an imprint of Elsevier
125 London Wall, London, EC2Y 5AS, United Kingdom
The Boulevard, Langford Lane, Kidlington, Oxford OX5 1GB, United Kingdom
50 Hampshire Street, 5th Floor, Cambridge, MA 02139, United States
525 B Street, Suite 1650, San Diego, CA 92101, United States

ISBN: 978-0-08-102727-1
ISSN: 1573-4285

Publisher: Susan Dennis
Acquisition Editor: Anita Koch
Editorial Project Manager: Vincent Gabrielle
Production Project Manager: Prem Kumar Kaliamoorthi
Cover Designer: Alan Studholme

Typeset by SPi Global, India

Contents

Contributors

Numbers in Parentheses indicate the pages on which the author's contributions begin.

Changlun Chen (39, 141, 217, 281), Key Laboratory of Photovoltaic and Energy Conservation Materials, Institute of Plasma Physics, Chinese Academy of Sciences, Hefei, People's Republic of China

Wencai Cheng (107), School of National Defence Science and Technology, Southwest University of Science and Technology, Mianyang, People's Republic of China

Congcong Ding (107), Fundamental Science on Nuclear Wastes and Environmental Safety Laboratory, Southwest University of Science and Technology, Mianyang, People's Republic of China

Qiaohui Fan (1), Key Laboratory of Petroleum Resources, Gansu Province/CAS Key Laboratory of Petroleum Resources Research, Institute of Geology and Geophysics, Chinese Academy of Sciences, Lanzhou, People's Republic of China

Jie Li (217), School of Resources and Environment, Anhui Agricultural University, Hefei, People's Republic of China

Ping Li (1), Key Laboratory of Petroleum Resources, Gansu Province/CAS Key Laboratory of Petroleum Resources Research, Institute of Geology and Geophysics, Chinese Academy of Sciences, Lanzhou, People's Republic of China

Songhua Lu (141), Key Laboratory of Photovoltaic and Energy Conservation Materials, Institute of Plasma Physics, Chinese Academy of Sciences, Hefei, People's Republic of China

Huiyang Mei (39), Key Laboratory of Photovoltaic and Energy Conservation Materials, Institute of Plasma Physics, Chinese Academy of Sciences, Hefei, People's Republic of China

Xiaoqin Nie (107), School of National Defence Science and Technology, Southwest University of Science and Technology, Mianyang, People's Republic of China

Duoqiang Pan (1), Radiochemistry Laboratory, School of Nuclear Science and Technology, Lanzhou University, Lanzhou, People's Republic of China

Yubing Sun (141), Key Laboratory of Photovoltaic and Energy Conservation Materials, Institute of Plasma Physics, Chinese Academy of Sciences, Hefei, People's Republic of China

Xiaoli Tan (39), Key Laboratory of Photovoltaic and Energy Conservation Materials, Institute of Plasma Physics, Chinese Academy of Sciences, Hefei, People's Republic of China

Wenling Ye (217), School of Resources and Environment, Anhui Agricultural University, Hefei, People's Republic of China

Kairuo Zhu (281), Key Laboratory of Photovoltaic and Energy Conservation Materials, Institute of Plasma Physics, Chinese Academy of Sciences, Hefei, People's Republic of China

Chapter 1

Radionuclides sorption on typical clay minerals: Modeling and spectroscopies

Qiaohui Fan*, Ping Li* and Duoqiang Pan†
*Key Laboratory of Petroleum Resources, Gansu Province/CAS Key Laboratory of Petroleum Resources Research, Institute of Geology and Geophysics, Chinese Academy of Sciences, Lanzhou, People's Republic of China
†Radiochemistry Laboratory, School of Nuclear Science and Technology, Lanzhou University, Lanzhou, People's Republic of China

1.1 Introduction

1.1.1 Radionuclides (RNs) in the environment

In natural systems, both the primordial and secondary as well as the anthropogenic nuclides have attracted more and more attentions due to the requirements of nuclear weapons and especially, energy. Therefore, this chapter is mainly focused on the radionuclides (RNs) directly related to the production of electricity by nuclear fission. In nuclear power plants, manmade radioactive materials could reach very high specific activities and radiotoxicities, which are the biggest challenges that human beings are facing. For fission reactors, a bimodal fission mass yield distribution occurs with maxima at masses around 90 (e.g., ^{90}Sr, ^{99}Tc, etc.) and 130 (e.g., ^{131}I, ^{133}Ba, ^{135}Cs, ^{137}Cs, etc.). The transuranic elements of Np, Pu, Am, and Cm (and also Cf, Bk, Es, and Fm in very low trace amounts) are generated by neutron capture reactions from U. After several decades of cooling in temporary storage, the fission products of ^{137}Cs and ^{90}Sr will be the main contributors to radioactive levels and heat production. Several hundred years later, the radiotoxicity will be still kept at a significantly high level, which is mainly in terms of the transuranic elements of Pu, Am, and their decay daughters [1]. If the radioactive sources and storage facilities are at or close to the surface (atmospheric weapon tests, accidental releases from nuclear power or

Interface Science and Technology, Vol. 29. https://doi.org/10.1016/B978-0-08-102727-1.00001-7

reprocessing plants and surface storage facilities), RNs will be possibly released into the atmosphere and indirectly (by precipitation) or directly (by effluents) into the soil or the hydrosphere.

Since the last century, nuclear technology has developed very quickly and is widely used within the fields of nuclear weapons, nuclear power, aerospace, and medicine. Unfortunately, a large amount of RNs have been released and entered into the environment due to the nuclear industry, frequent weapons tests, and nuclear accidents such as the Chernobyl and Fukushima incidents [2–6]. From the 1940s to the 1990s, approximately 210,000 m^3 of radioactive byproducts from plutonium processing were deposited in 177 underground tanks on the Hanford Nuclear Reservation in Washington; however, the majority of the storage tanks have exceeded their design lives, resulting in the leaking of plutonium and other RNs [7–9]. As a contaminant, radiocesium (RCs) has been dispersed worldwide in low concentrations following atmospheric nuclear arms testing and the Chernobyl and Fukushima accidents [10, 11]. The total amounts of ^{131}I and ^{137}Cs released into the atmosphere between March 11–31, 2011, were approximately 5.0×10^{17} Bq and 3.3×10^{16} Bq, respectively. ^{239}Pu (and other Pu isotopes) has been detected around the plant and up to several tens of kilometers away, although not as high as in the case of the Chernobyl accident, which is quite similar to the Fukushima case [9]. After the Fukushima accident, the radioactive levels of ^{239}Pu, ^{234}U, and ^{238}U suddenly increased, particular in Hawaii, Alaska, and California [12]. Most of the RNs were transported out over the Pacific Ocean due to the wind orientation. However, about 10%–20% of the total radiocesium and radioiodine emitted from the Daiichi Nuclear Power Plant was still deposited over land in northeastern Japan [13–16]. In the environment, these contaminants of RNs from the anthropogenic activities will have unpredicted impacts on the ecology as well as human beings over the long term, due to their long half-life, radioactivity, and toxicity.

1.1.2 RNs sorption on clay minerals

In general, RNs mobility and bioavailability in the environmental are considered to be determined by the coupled effects of aqueous species, solid/water interface retention, and transport processes (advection or diffusion). Chemical speciation is governed by pH, Eh, ionic strength, the presence of complexing ligands including natural organic matters, etc., in the concerned systems [1]. Chemical speciation is the most important to control RNs reactivity, mobility, bioavailability, and toxicity in the environment. Simultaneously, the speciation of RNs could be easily changed during the presence of microorganisms and biomolecules, which are quite ubiquitous in the environment. The reaction at the solid-water interface mainly contributes to RNs retention in the environment. They can be of either a chemical or physical nature with a broad variability in binding strengths, which depends on the individual RN.

For polyvalent cations such as most actinide ions, the interactions at the solid-water interface prefer the chemical nature; however, the monovalent cations such as Li, Na, K, Rb, and Cs and partial bivalent cations (mainly Ca, Mg, and Sr) prefer the physical nature. The physical nature of RNs at the solid/water interface could cause high mobility and bioavailability in the environment while the chemical nature results in relative immobility and lower bioavailability. Finally, the transport processes can be either the advective movement of water including solutes, or the diffusive transport of solutes driven by concentration gradients. Furthermore, small-sized colloidal particles can also be transported in the aqueous phase.

Clay minerals are ubiquitous in the environment as well as being the most important component of aerosol, soil, sediment, and bedrock. Therefore, the sorption at the clay mineral-water interface is one of the most important reactions controlling geochemical behaviors of RNs, such as speciation, diffusion, migration, transportation, and bioavailability in the environment. RNs sorption at the clay mineral-water interface mainly includes three mechanisms: adsorption, surface precipitation, and polymerization. Adsorption can be defined as the accumulation of an adsorbate at an interface between the solid surface and the batching solution. As the surface coverage or loadings increase, which is affected by the pH, sorption can possibly proceed from mononuclear adsorption to surface precipitation and polymerization (both three-dimensional (3D) phases). There are several thermodynamic driving forces to sorption processes on clay minerals: (1) the surface of clay minerals might lower the energy of nucleation by providing sterically similar sites [17], (2) the activity of the surface precipitates is less than 1.0 [18], and (3) the solubility of the surface precipitate is smaller because the dielectric constant of the solution near the surface is less than that of the bulk solution [19]. During the sorption of RNs at the clay mineral-water interface, several sorption processes can be possibly experienced (it also depends on the individual RN) [20]: (a) adsorption via the formation of an outer-sphere complex (OSC); (b) adsorption via the formation of an inner-sphere complex (ISC) after losing partial or complete hydration water molecules; (c) lattice diffusion and isomorphic substitution with the clay mineral lattice, rapid lateral diffusion and formation either of a surface polymer or adsorption on a ledge (which maximizes the number of bonds to the atom); (d) surface polymers end up embedded in the lattice structure with particle growth; and (f) the adsorbed RN ions can diffuse back in the bulk solution, either as a result of dynamic equilibrium or as a product of surface redox reactions. The sorption mechanism will directly control RNs behaviors of migration, transformation, and bioavailability in the environment. However, in practice, environmental factors such as pH, Eh, ion strength, organic matters, temperature, etc., can have various effects on the sorption mechanism of RNs at the clay mineral-water interface [2, 5, 6, 21–28]. Therefore, the understanding on the sorption mechanism especially at a molecular scale become more important to estimate and predict the environmental behaviors of RNs at clay mineral-water interface.

1.1.3 Aims of this chapter

In this chapter, information on the structure and physicochemical characteristics of typical clay minerals such as micas (muscovite, biotite, and phlogopite), illite, montmorillonite, vermiculite, chlorite, and kaolinite will be summarized in detail. Subsequently, the sorption behaviors of RNs on these typical and important clay minerals will be addressed, and finally the sorption modeling (mainly including surface complexation modeling (SCM) and the general adsorption modeling for Cs (GAM)) and the spectroscopies (X-ray absorption spectroscopy, time-resolved laser-induced fluorescence spectroscopy, X-ray photoelectron spectroscopy, etc.) will be emphatically introduced to show how these powerful approaches explore the sorption mechanism at a molecular scale.

1.2 Clay mineral structure and characteristics

1.2.1 Clay mineral structure

Without question, the clay mineral is the one of the most important phases in controlling the environmental behaviors of RNs. In structure, phyllosilicate clay minerals could be considered as assemblages of tetrahedra and octahedral sheets, where the tetrahedral and octahedral sheets are attributing to the silica sheet and the aluminum oxide sheet, respectively. The distance of the Si—O bond is approximate 1.62 Å, the O—O bond distance is approximate 2.64 Å, the tetrahedra are arranged so that all tips are in the same orientation and the tetrahedra bases are in the same plane, and all the tetrahedra are connected point-to-point (as shown in Fig. 1.1). In the aluminum octahedra sheet, the distances of O-O and OH-OH are 2.67 and 2.94 Å, respectively, and the aluminum octahedra are connected through edge-to-edge (as shown in Fig. 1.2) [20].

When one tetrahedral sheet bounded with one octahedral sheet can form a 1:1 type clay as an example of kaolinite, and then the full-cell chemical formula for the ideal 1:1 type clay should be $Si_4Al_4O_{10}(OH)_8$, where Si and Al are, respectively, in the fourfold and sixfold coordination in the tetrahedra

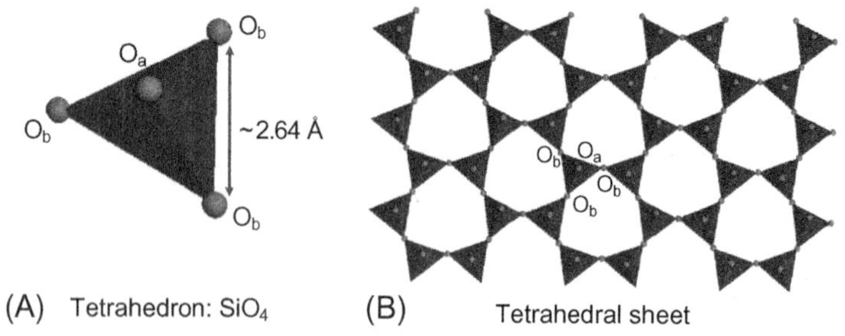

FIG. 1.1 Diagrams of (A) Tetrahedron and (B) Tetrahedra sheet (T). *(Based on F. Bergaya, B. Theng, G. Lagaly, Handbook of Clay Science, Elsevier Science, New York, 2006, pp. 1–1224.)*

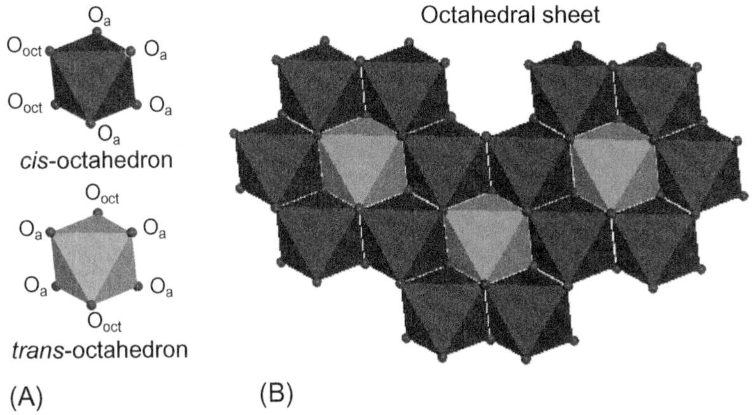

FIG. 1.2 The diagram of (A) *cis-* and *trans*-octahedron and (B) octahedral sheet. *(Based on the reference F. Bergaya, B. Theng, G. Lagaly, Handbook of Clay Science, Elsevier Science, New York, 2006, pp. 1–1224.)*

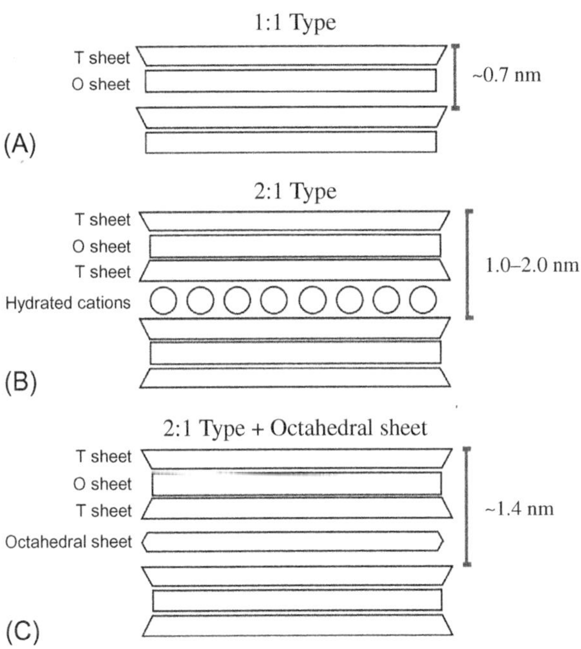

FIG. 1.3 Diagrams of 1:1 and 2:1 type clay minerals, (A) 1:1 type clay with an interlayer distance of ~0.7 nm, (B) 2:1 type clay with an interlayer distance of 1.0–2.0 nm, and (C) 2:1 type + octahedral sheet with an interlayer distance of ~1.4 nm.

and octahedral sheets (as shown in Fig. 1.3A). Two tetrahedral sheets bonded with one octahedral sheet will form a 2:1 type clay, such as illite and montmorillonite, and therefore the full-cell chemical formula is expected to be $Si_8Al_4O_{20}(OH)_4$ for the ideal 2:1 type clay mineral (as shown in Fig. 1.3B).

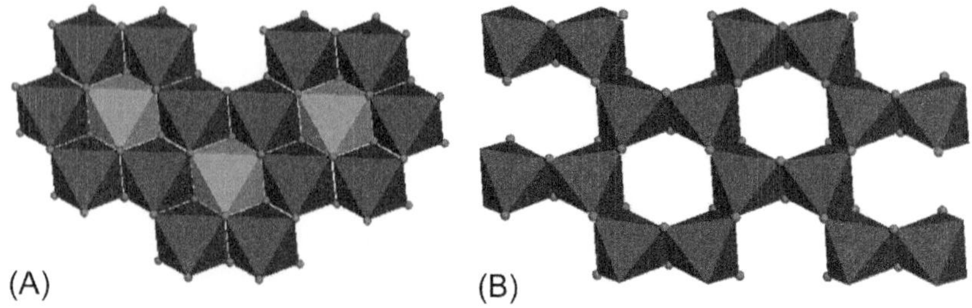

FIG. 1.4 The diagram of (A) tri-octahedral sheet and (B) di-octahedral sheet.

Between the interlayer space, cations might be octahedrally coordinating with the hydroxyls, such as chlorite (as shown in Fig. 1.3C), and they may be present as individual cations, which may or may not be hydrate, as in the micas, vermiculite, and smectite.

Generally, clay minerals can be classified as two species, di-octahedral and trioctahedral, which strictly depend on the number of cation positions in the octahedral sheet. If all three positions are filled, it is the tri-octahedral species (Fig. 1.4A). If two of the three positions are filled, then it is the di-octahedral species, as shown in Fig. 1.4B. For example, if Al^{3+} is present in the octahedral sheet, only two-thirds of the cation positions are filled (di-octahedral species); for every six OH^- anions, only two Al^{3+} satisfy the anionic charge and a formula of $Al_2(OH)_6$ results (Fig. 1.4B). When Mg^{2+} is present, all three cation sites are filled because Mg is divalent and three atoms of Mg^{2+} would be necessary to satisfy the six OH^- ions (tri-octahedral) (Fig. 1.4A). The formula for this sheet would be $Mg_3(OH)_6$. [20].

1.2.2 Isomorphous substitution and structure charge

Isomorphous substitution can normally occur in both tetrahedral and octahedral sheets, which is the most important pathway to produce the permanent negative charge or layer charge for the clay mineral. The size of the cationic radius and the valent normally determine which cation can substitute the Si^{4+} in the tetrahedral sheet and the Al^{3+} in the octahedral sheet. Usually, Al^{3+} can substitute the Si^{4+} in the tetrahedral sheet, whereas in the octahedral sheet, Fe^{2+}, Fe^{3+}, Mg^{2+}, and transition cations such as Zn^{2+}, Ni^{2+}, Mn^{2+}, and Cu^{2+} can replace Al^{3+}.

In the case of isomorphous substitution, a net negative charge in the octahedron and/or tetrahedrons will develop on the surface of clay minerals. For example, if one Si^{4+} is substituted by one Al^{3+} in the ideal 1:1 kaolinite (di-octahedral species), the formula of kaolinite will be $(Si_3Al_1)Al_4O_{10}(OH)_8$ and the net charge on kaolinite is -1, which is balanced by the presence of alkaline cations such as Na^+ and K^+ near the outer surface of kaolinite. In practice, however, the layer charge of 1:1 type phyllosilicate is usually close to zero (Table 1.1). On the other hand, such charge variability is considered to

TABLE 1.1 The layer charge and idealized formula of representative 1:1 and 2:1 clay minerals [29]

Clay minerals	Idealized formulae	Expansibility	Net layer charge per formula unit
Kaolinite	$(Si_2)^{IV}(Al_2)^{VI}O_5(OH)_4$ (Dioctahedral)	Nonexpansible	~0
Serpenite	$(Si_2)^{IV}(Mg_3)^{VI}O_5(OH)_4$ (Trioctahedral)	Nonexpansible	~0
Pyrophyllite	$(Si_4)^{IV}(Al_2)^{VI}O_{10}(OH)_2$(Dioctahedral)	Nonexpansible	~0
Talc	$(Si_4)^{IV}(Mg_3)^{VI}O_{10}(OH)_2$ (Trioctahedral)	Nonexpansible	~0
Muscovite	$K^+(Si_3Al)^{IV}(Al_2)^{VI}O_{10}(OH)_2$	Nonexpansible	0.6–1.0
Biotite	$K^+(Si_3Al)^{IV}(Fe_3)^{VI}O_{10}(OH)_2$	Nonexpansible	0.6–1.0
Phlogopite	$K^+(Si_3Al)^{IV}(Mg_3)^{VI}O_{10}(OH)_2$	Nonexpansible	0.6–1.0
Illite	$(x+y)K^+(Si_{4-x}Al_x)^{IV}(Al_{2-y}Mg_y)^{VI}O_{10}(OH)_2$	Nonexpansible	0.6–1.0
Vermiculite	$(x+y)M^+(Si_{4-x}Al_x)^{IV}(Al_{2-y}Mg_y)^{VI}O_{10}(OH)_2$ (Dioctahedral)	Expansible	0.6–0.9
	$(x+y)/2\,Mg^{2+}(Si_{4-x}Al_x)^{IV}(Mg_{3-y}M_y^{3+})^{VI}O_{10}(OH)_2$ (Trioctahedral)		
Smectite	$yM^+(Si_4)^{IV}(Al_{2-y}Mg_y)^{VI}O_{10}(OH)_2.nH_2O$	Expansible	0.2–0.6
Chlorite	$(M_{3-z}^{2+}M_z^{3+})(Si_{4-x}Al_x)^{IV}(M_{3-y}^{2+}M_y^{3+})^{VI}O_{10}(OH)_2$	Expansible	Variable

be the most important features of 2:1 type phyllosilicates and micas because it could induce occupancy of the interlayer space by exchangeable cations. In the case of 2:1 type phyllosilicates, the negative layer charge is expected to be from 0.2 in montmorillonite and hectorite to 2.0 in true micas, according to the calculation of their structural formula. The layer charge and idealized formula of the typical 1:1 and 2:1 type clay minerals are listed in Table 1.1.

1.2.3 Expandability of clay minerals

Expansion of the interlayer is another important feature that can influence the sorption ability and capacity of clay minerals. The expandability of the representative 2:1 clay minerals is shown in Fig. 1.5. Illite and chlorite are nonexpansible, and vermiculite and smectite are moderately and highly expansible, respectively. In the case of chlorite, the interlayer sites are occupied with a hydroxide sheet as an example of an $Al(OH)_3$ sheet, therefore, chlorite is nonexpansible. As shown in Table 1.1, the charges of vermiculite and smectite are ~0.6–0.9, and ~0.2–0.6, respectively. Illite owns the largest charge (~0.9–1.0), which causes the strong fixation ability of layers. Therefore, the layer charge is the main reason that is controlling the expansible ability of 2:1 clay minerals. On the other hand, the interlayer related to the expandability of clay minerals largely contributes to the cation exchange capacity and the sorption site affinity to RNs. As an example, the sorption species of Cs(I) on illite, vermiculite, smectite, and chlorite were analyzed by an extended X-ray absorption fine structure (EXAFS) approach (Fig. 1.6). It is very clear that the ratios of CN_{IS}/CN_{OS} were following the order of vermiculite > illite > smectite ≈ chlorite, which suggested that vermiculite had the strongest ability to fix Cs(I) and smectite was the weakest. Such phenomena could be attributed to the expandability and charge of clay minerals [2]. Wu et al. [5] have found that there is a good linear relationship between the expansion degree of phlogopite and the sorption of Th(IV), which showed the importance of clay mineral expandability in the retardation and fixation of RNs in the environment.

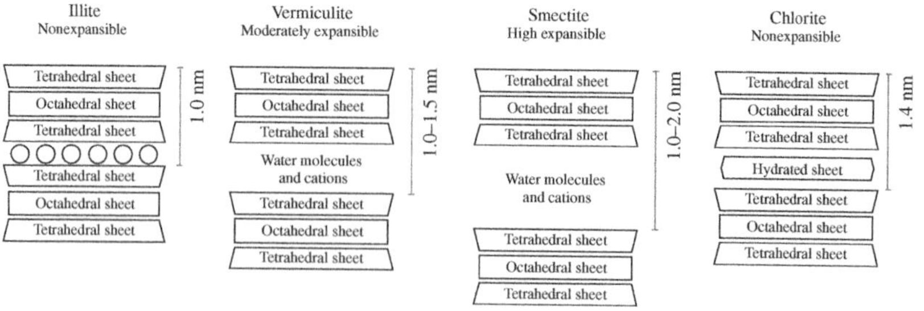

FIG. 1.5 Expandability of typical 2:1 clay minerals (illite, vermiculite, smectite, and chlorite).

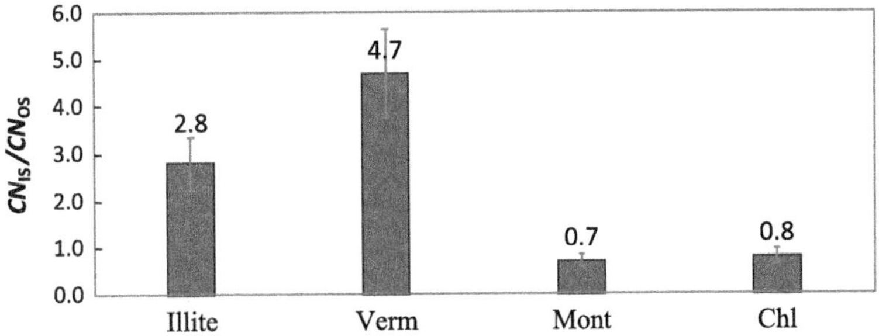

FIG. 1.6 EXAFS analyses of the contributions of ISCs and OSCs to Cs(I) adsorption on clay minerals. CN_{IS}: coordination numbers of the ISCs, and CN_{OS}: coordination numbers of the OSCs. Verm: vermiculite, Mont: montmorillonite, and Chl: chlorite.

1.3 (Semi-)empirical sorption models of RNs

Worldwide activities are focused on the remediation of radioactively contaminated sites. One aim is to provide a more profound chemical base for risk assessment in time and space [30]. To do so in an adequate way requires extending our knowledge about the theoretical basis mechanism of RNs retardation and mobilization phenomena and the underlying basic processes and interactions of RNs at mineral-water interface. Up to now, two theoretical approaches have been commonly used: (1) (semi-)empirical equations of sorption isotherms (Langmuir and Freundlich); and (2) the mechanistic surface complexation model (SCM).

Generally, empirical models for RNs sorption are based on the mathematical relationships between the RN concentrations in the liquid and solid phases at equilibrium and at constant temperature, respectively [31]. These relationships are the so-called sorption isotherms. The most commonly used isotherm is the Langmuir isotherm, which was originally derived for the adsorption of gases on homogeneous plane surfaces by Langmuir [32]. The Langmuir model is based on the assumption that, at equilibrium, the numbers of adsorbed and desorbed molecules are equal in unit time. The lateral interactions and horizontal mobility of adsorbates on mineral surfaces were neglected and the surface was considered to be homogeneous. It is applied for sorption of RNs in the following formation:

$$q_e = q_{max} bC_e / (1 + bC_e) \tag{1.1}$$

where q_e (mol/g) and C_e (mol/L) are the equilibrium concentration of RNs in the solid and aqueous phases, respectively. q_{max} represents the maximum sorption of RNs on the solid phase, and b can be looked upon as a measure of the affinity of the adsorbate for the surface. The relative parameters in the Langmuir model can be calculated from the linear equation of the Langmuir model, as shown in the following:

$$q_e/C_e = -bC_e + bq_{max} \tag{1.2}$$

In this case, q_e/C_e (i.e., distribution coefficient, K_d) can be plotted against C_e. If the sorption of RNs on clay mineral follows the Langmuir model, the plot of q_e/C_e versus C_e should exhibit a highly linear relationship. The Freundlich model is another important empirical equation to describe the sorption behaviors of RNs on clay minerals. The Freundlich model can be expressed by the following equation:

$$q_e = K_F C_e{}^n \tag{1.3}$$

where K_F and n are adjustable positive valued parameters with n ranging from 0 to 1. K_F represents the sorption capacity when the equilibrium concentration of RNs equals 1, and n represents the degree of dependence of sorption. q_e (mol/g) and C_e (mol/L) are the concentrations of RNs in the solid and aqueous phases, respectively.

Until now, both the (derived) Langmuir and Freundlich equations have been widely applied to address the sorption behaviors of RNs at the clay-water interface. The traditional sorption models represented by the (derived) Freundlich and/or Langmuir isotherms are also successful at distinguishing the sorption processes in chemical or physical ways as well as judging the single- or multilayer sorption, predicting the sorption capacity, and sometimes evaluating the surface properties of clay minerals. However, the surface properties of clay minerals are not well described, for example, sorption sites and concentrations on the surface of clay minerals. In addition, a particular set of isotherm parameters is applicable only to the range of conditions where the empirical data have been obtained. This cannot be applied for predictions under a wider range of conditions. Therefore, it is impossible to reveal the internal sorption mechanism of RNs at the clay-water interface.

1.4 Surface complexation models of RNs sorption on clay minerals

The limitations in the (semi-)empirical models mentioned above provide a major impetus for the development of a SCM, which was first proposed by Stumm in the early 1970s. Over the past 50 years, SCM has emerged as a powerful tool for describing sorption processes of metal ions, anions, and RNs at the solid-water interface [31, 33, 34]. In contrast to both the Langmuir and Freundlich models, the strategies of SCM are the decomposition of the distribution coefficient (K_d) into each sorption reaction. For the SCM, there are several basic concepts: (i) the surface is composed of specific functional groups that can react with dissolved solutes to form surface complexes (coordinative complexes) similar to complexation reactions in homogeneous solution; (ii) the equilibria of surface complexation and ionization reactions can be described via the mass law equation, with correction factors applied for variable electrostatic energy using the electric double-layer (EDL) theory; (iii) both surface charge and electrical potential are treated as necessary consequences of chemical reactions of the surface functional groups; and (iv) the apparent binding constants determined for the mass law equations are empirical parameters related to thermodynamic constants via the rational activity coefficients of the surface species. In the mechanism model, the overall free energy of sorption in every reaction could be divided into two parts, mainly including the specific (chemical) sorption energy and the Coulomb static energy, as shown in Eqs. (1.4)–(1.10).

$$\Delta G_{\text{tot}} = \Delta G^0_{\text{chem}} + \Delta G^0_{\text{coul}} \tag{1.4}$$

$$\Delta G^0_{\text{chem}} = -RT \ln K_{\text{int}} \tag{1.5}$$

$$\Delta G^0_{\text{coul}} = \Delta ZF\psi \tag{1.6}$$

$$\Delta G_{\text{tot}} = -RT \ln K_{\text{app}} \tag{1.7}$$

$$-RT \ln K_{\text{app}} = -RT \ln K_{\text{int}} + \Delta ZF\psi \tag{1.8}$$

$$\ln K_{\text{app}} = RT \ln K_{\text{int}} + (-\Delta ZF\psi/RT) \tag{1.9}$$

$$K_{\text{app}} = K_{\text{int}} \exp(-\Delta ZF\psi/RT) \tag{1.10}$$

where ΔG^0_{chem} is the "intrinsic" free energy term and ΔG^0_{coul} is the electrostatic or Coulombic term, reflecting the electrostatic work in transporting ions through the interfacial potential gradient. K_{app} and K_{int} are the apparent and intrinsic equilibrium constants, respectively. F is the Faraday constant, and ΔZ is the change in the charge of the surface species.

However, it is difficult to separate the effects of the two different interactions experimentally, that is, the specific (chemical) sorption and the Coulomb static energies. Chan [35] presented a simple graphical method to analyze the free energy of the formation of the double layer, which demonstrated the role of the chemical and electrostatic interactions. At the solid-water interface, the double layer is considered to be established by the unequal sorption of H^+ and OH^-.

Most models for the electrostatic properties of a surface in contact with an aqueous phase are based on the Gouy-Chapman diffuse layer model [36, 37]. The electric double layer is regarded as consisting of two regions: an outer diffuse region and an inner region. In the outer diffuse region, ions are bound by electrostatic interactions only, and assumed to undergo thermal motion and distribution according to the Boltzmann distribution. In the inner region, ions are relatively immobile. The relationship between the charge in the diffuse layer, the electrolyte concentration in the bulk solution, and the potential is found by solving the Poisson-Boltzmann equation with appropriate boundary conditions. In the current SCMs, the electrostatic interactions are represented by differing numbers of planes and layers in the interfacial regions with ions being adsorbed at different positions. The constant capacitance model (CCM), the diffused layer model (DLM), and the triple-layer model (TLM) are the most frequently and effectively used in SCMs to explore the sorption mechanism of RNs on clay minerals.

1.4.1 Electric double layer at the mineral-water interface

1.4.1.1 Constant capacitance model

In the case of the CCM (Fig. 1.7A), all specifically adsorbed ions are allocated to a single plane and all surface species are at the same potential. Amphoteric surface hydroxyl groups are described as the following reactions (1.11), (1.12):

$$>SOH + H^{+} => SOH_2^{\ +} \tag{1.11}$$

$$>SOH => SO^{-} + H^{+} \tag{1.12}$$

Just one plane is taken into account for the sorption of H^+, OH^-, and specifically adsorbed ions in the interfacial region, where only the ISCs form. Major electrolytes, for example, Na^+, K^+, Cl^-, $NO_3^{\ -}$, are assumed to be inert with respect to the surface and the ionic strength is therefore fixed. The potential is related linearly to the surface charge and can be expressed with Eq. (1.13):

$$\sigma_0(\mathrm{C\,m^{-2}}) = C(\mathrm{Farad\,m^{-2}})\psi_0(\mathrm{V}) \tag{1.13}$$

where C is the capacitance of the mineral-water interface. The surface charge can be calculated from mass-balance equations and the capacitance is fitted to the experiment. Alterations in the capacitance of the surface plane can, to some extent, allow for variations in ionic strength. However, there are no systematic rules for determining the real capacitance for different ionic strengths.

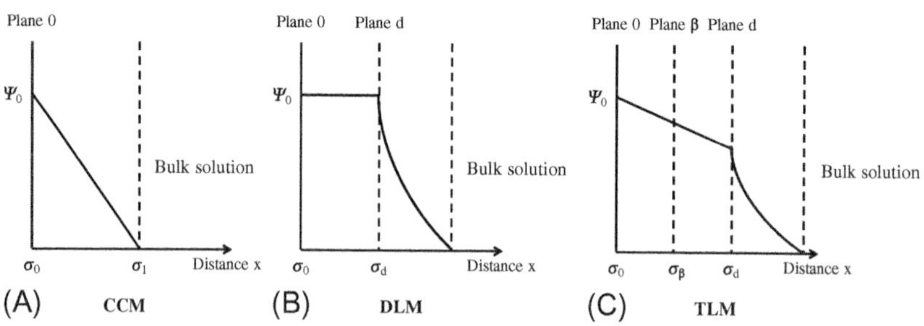

FIG. 1.7 Schematic illustration of the solid-water interfaces according to electric double layer. (A) CCM, (B) DLM, and (C) TLM.

1.4.1.2 Diffuse-layer model

In the DLM, there is charge bound by electrostatic forces only in the diffuse layer (Fig. 1.7B). Therefore, there are two planes in the interfacial regions: (i) a surface plane for the sorption of H^+, OH^-, and all specifically adsorbed solutes; and (ii) a diffuse-layer plane representing the closest distance of approach for all countering ions. In the diffuse layer, the surface charge could be expressed with the Gouy-Chapman theory, as in Eq. (1.14) [38].

$$\sigma_d = -0.1174\sqrt{I}\sinh\frac{ZF\psi_d}{2RT} \tag{1.14}$$

In the case of DLM, the capacitance is fixed according to the theory rather than a variable parameter in the CCM. Similar to the CCM, there is also no surface coordination of electrolyte ions and therefore no explicit dependence on ionic strength.

1.4.1.3 Triple-layer model

CCM and DLM differ in the functions relating σ to ψ, but have the similarities that both models define the sorption of all ions on one sorption plane (o-plane). Therefore, it is hard to distinguish between weak and strong sorption, especially for the cases with different ionic strength. The developed TLM provides a good solution to the above problem (Fig. 1.7C). TLM is closer to the realistic situations because both OSC and ISC reactions are taken into account [39, 40]. In the TLM, the o-plane is valid for the sorption of metal ions and RNs, and the β-plane for the counterions, for example, K^+, Na^+, Cl^-, and NO_3^- (Fig. 1.7C).

The sorption of metal ions and RNs can occur either at the o-plane or β-plane [41]. In the TLM, the surface reactions, mainly including protonation, deprotonation, and specific sorption reactions, are identical to those in the CCM and DLM. The difference is that, for TLM, mass action expressions must have a proper stoichiometry for the electrostatic component representing the β-plane as well as the o-plane. No stoichiometry is needed for the d-plane because no specific adsorption occurs on that plane.

1.4.2 Applications of SCMs in RNs sorption on typical clays

1.4.2.1 RNs sorption on kaolinite (1:1 clay mineral)

As for kaolinite, the surface contains two layers (gibbsite and siloxane layers) and an edge surface, therefore, SCM should contain at least four sets of surface reactions. In practice, the four sets of sorption reactions on the kaolinite surface are too complicated for modeling purposes. Therefore, simplified models are necessary and expected to be more useful and efficient to explore the sorption mechanism of RNs in practice. Kohler et al. [42] found that only the aluminol site (>AlOH) at the edge surface can model and predict the sorption behaviors of actinides on kaolinite. Such simplification was based on the outcome of parallel experiments with gibbsite (which only contains >AlOH sites) and quartz (only >SiOH sites), which showed that actinide sorption on quartz was much weaker. Further, time-resolved fluorescence spectroscopy confirmed that U(VI) binds preferably to >AlOH sites on kaolinite [43].

Tertre et al. [44] investigated the aqueous interfacial chemistry of kaolinite using acid-base continuous titrations and batch experiments. The acid-base continuous titration data could be simulated well only containing two typical edge sites on kaolinite (>SiOH and >AlOH) with the DLM. In agreement with the previous literature, >AlOH sites are amphoteric and >SiOH sites can only be neutral or negative in the focused pH range from 4.0 to 8.0 [45–47]. Therefore, three surface reactions are taken into account, as follows (1.15)–(1.17):

$$> AlOH + H^+ => AlOH_2{}^+ \tag{1.15}$$

$$> AlOH => AlO^- + H^+ \tag{1.16}$$

$$> SiOH => SiO^- + H^+ \tag{1.17}$$

In the above cases, the electrostatic terms are excluded from the mass law equation for surface equilibria and the surface charge density of kaolinite is not considered. Although some previous studies have concluded that there is no net permanent charge in kaolinite, more and more researchers now accept the fact that there is a small cation exchange capacity (about 1.0–8.0 meq/100 g) due to the replacement of Si^{4+} by Al^{3+} in the tetrahedral sheet (isomorphous substitution) and nonideal octahedral occupancy [48]. The negative exchange site can react with both H^+ and Na^+ according to the following reactions (1.18), (1.19):

$$> X^- + Na^+ => XNa \tag{1.18}$$

$$> XNa + H^+ => XH + Na^+ \tag{1.19}$$

In this way, one exchange site (>XNa) beside the two edge sites (>SiOH and >AlOH) was considered in the following models (Fig. 1.8). The main parameters of site capacities and reaction constants (*K*) in SCMs are listed in Table 1.2.

The sorption of U(VI) on kaolinite could be described with three sorption species of $>X_2UO_2$ ($\log K = 1.3$), $>AlOUO_2^+$ ($\log K = -0.8$), and $>AlO(UO_2)_3(OH)_5$ ($\log K = -16.5$) (Fig. 1.9A) [52], and the contribution of >SiOH could be negligible due to its weak reactivity [43, 44]. Based on the similar assumption that the ISCs form preferentially on the edge sites of >AlOH, however, Arda et al. [53] found that only two sorption species of $>X_2UO_2$ ($\log K = -0.9$ to -0.6) and $>AlOUO_2^+$ ($\log K = -3.2$ to -2.0) can simulate the sorption edge in the pH range of 2.0–5.0. Although mono-dentate surface

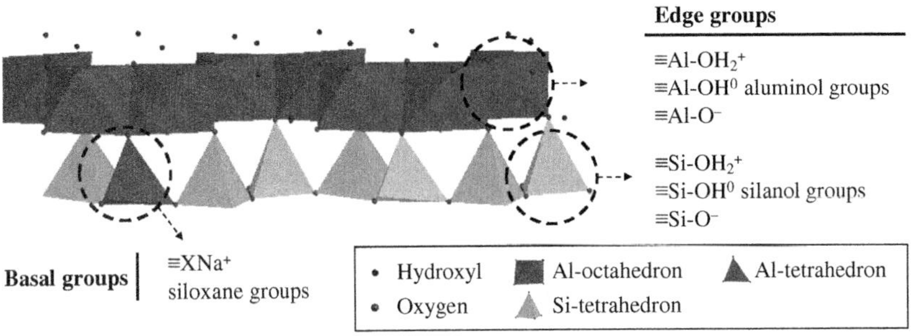

FIG. 1.8 The distribution of sorption sites on kaolonite in the model containing one exchange site and two edge sites [49]. *(Reproduced from X. Gu, L.J. Evans, Surface complexation modelling of Cd (II), Cu (II), Ni (II), Pb (II) and Zn (II) adsorption onto kaolinite, Geochim. Cosmochim. Acta 72 (2008) 267–276. Copyright 2008 Elsevier.)*

TABLE 1.2 Distribution of site density and surface reaction constants for kaolinite [48, 49]

Site types	Site density distribution
>XH	0.014–0.43 mmol/g
>SOH	0.008–0.029 mmol/g
Surface reactions	**Log *K***
$>XH + Na^+ = >XNa + H^+$	−4.07 to −2.02
$> SOH + H^+ => SOH_2{}^+$	4.37–6.6
$>SOH = >SO^- + H^+$	−9.18 to −6.16

Note: >SOH represents for both >AlOH and >SiOH.

complexes have been usually proposed for U(VI) sorption on clay surfaces [54], bi-dentate surface complexes of U(VI) sorption on clay minerals indeed were formed and confirmed by spectroscopies [55]. Dong et al. [56] modeled U(VI) sorption on kaolinite with two bi-dentate surface complexes formed on the amphoteric edge sites (i.e., $(> SOH)_2UO_2{}^+$ and $(> SOH)_2UO_2CO_3{}^-$) without strictly distinguishing >AlOH and >SiOH sites in this study) and a cation exchange species ($>X_2UO_2$).

Moreover, other RNs have also attracted wide attention. Marmier et al. [57] investigated Yb(III) sorption on kaolinite combining CCM. Based on the surface complexation constants determined on single alumina and silica, the model can predict the sorption of Yb(III) on kaolinite well without using any adjustable parameters. It is noticed that no ion-exchange reaction was

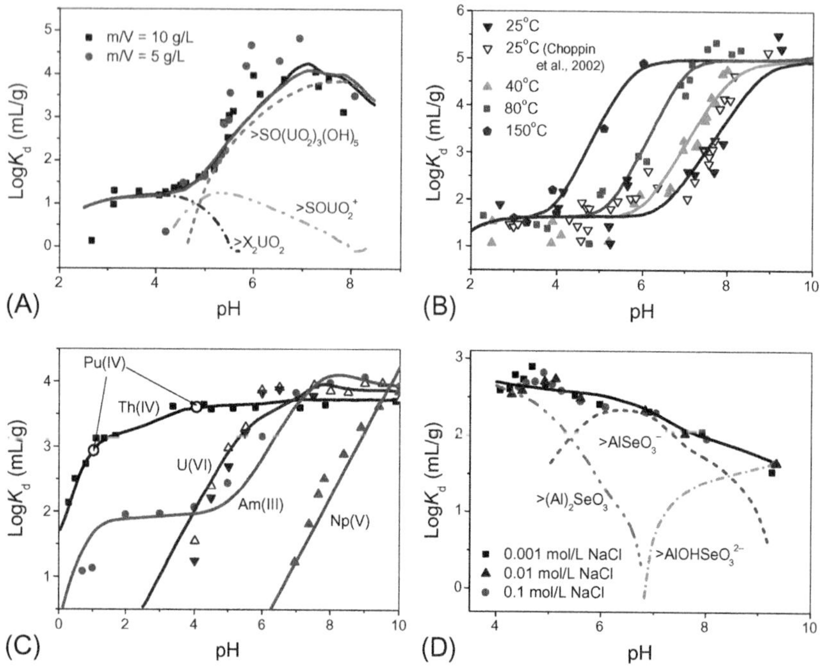

FIG. 1.9 Modeling of the sorption of RNs on kaolinite. (A) U(VI) [52], (B) Eu(III) [50], (C) Am(III), Th(IV), Np(V) and U(VI) [51], and (D) Se(IV) [62]. *((A) Data from L. Gao, Z. Yang, K. Shi, X. Wang, Z. Guo, W. Wu, U (VI) sorption on kaolinite: effects of pH, U (VI) concentration and oxyanions, J. Radioanal. Nucl. Chem. 284 (2010) 519–526; (D) Data from Z.X. Hou, K.L. Shi, X.L. Wang, Y.L. Ye, Z.J. Guo, W.S. Wu, Investigation of Se(IV) sorption on Na-kaolinite: batch experiments and modeling, J. Radioanal. Nucl. Chem. 303 (2015) 25–31; Reproduced from E. Tertre, G. Berger, E. Simoni, S. Castet, E. Giffaut, M. Loubet, H. Catalette, Europium retention onto clay minerals from 25 to 150 °C: experimental measurements, spectroscopic features and sorption modelling, Geochim. Cosmochim. Acta 70 (2006) 4563–4578. Copyright 2008 Elsevier; R. Marsac, N.L. Banik, J. Lützenkirchen, R. A. Buda, J. V. Kratz, C.M. Marquardt, Modeling plutonium sorption to kaolinite: accounting for redox equilibria and the stability of surface species, Chem. Geol. 400 (2015) 1–10. Copyright 2015 Elsevier.)*

considered for Yb(III) sorption on kaolinite. Similar to the models for uranyl sorption, the contribution from >SiOH sites to Eu(III) sorption on kaolinite could be also ignored under the experimental conditions, and ion exchange and surface complexation reactions were considered on >AlOH sites, which was valid in a wide temperature range from 25°C to 150°C (Fig. 1.9B) [50]. By applying the corresponding sorption reaction constant, the presented results for Eu^{3+} could be extrapolated to actinide ions (Ac^{3+}) such as Am^{3+} and Cm^{3+}. Tertre et al. [58] confirmed that the sorption of trivalent lanthanides on kaolinite can be well modeled by the "Generalized Composite Approach" (GCA) [59, 60]. It was not necessary to explicitly treat the acid-base properties of the kaolinite surface, and only one generic surface site (>SOH) was considered. Marsac et al. [51] successfully predicted Pu(III/IV/V/VI) sorption on kaolinite with the GCA model using Am(III), Th(IV), Np(V), and U(VI) as analogues (Fig. 1.9C).

In the environment, RNs existing as anionic species such as I^-, IO_3^-, SeO_3^{2-}, SeO_4^{2-}, and TcO_4^- are highly soluble and mobile and poorly retarded by environmental media. Wang et al. [61] modeled SeO_3^{2-} sorption on kaolinite with DLM. In the model, the formed surface complexes included were $>AlOH_2-HSeO_3^0$, $>SiOH_2-HSeO_3^0$, $>SiO-HSeO_3^{2-}$, and $>AlO-SeO_3^{3-}$. However, in the research of Hou et al. [62], SeO_3^{2-} was considered to form a bi-dentate complex of $(>Al)_2SeO_3$ and two mono-dentate complexes of $>AlSeO_3^-$ and $>AlOHSeO_3^{2-}$ (Fig. 1.9D). The sorption modeling on RNs presented as anion species is still quite limited on clay minerals, and deserves more attention due to high mobility and toxicity.

1.4.2.2 RNs sorption on illite (2:1 clay and nonexpansible)

Du et al. [63] studied the acid-base properties of the illite surface based on the potentiometric titration. In this study, the ion exchange reactions (>XH) were not considered mainly due to the fact that (i) the experiment was conducted under very high background electrolyte concentrations (0.1 and 0.5 mol/L $NaNO_3$ solutions), which were more than 100 times higher than the proton concentrations (~0.001 mol/L at maximum), and (ii) the ion exchange reactions are much weaker than chemisorption. Therefore, only the hydroxyl group (>SOH) was taken into account for describing the surface reactions on illite. Two surface protonation models are proposed: (i) one site—1-pK_a model, where the amphoteric surface hydroxyl groups (>SOH) are treated as homogeneous; and (ii) two sites—2-pK_a model, where the surface was heterogeneous and surface reactions occurred on two kinds of sites with an equivalent density. Both models can give good descriptions of the surface properties of illite under various experimental conditions.

Different from the above studies, a two-site protolysis model with no electrostatic term (2SPNE model) has been developed to describe the titration and sorption edge data for Na-illite, where one strong site ($>S^SOH$) and two weak sites ($>S^{W1}OH$ and $>S^{W2}OH$) were required. $>S^SOH$ and $>S^{W1}OH$ were assumed to have the same protolysis constants and different site densities (as shown in Table 1.3) [64, 65, 68, 69]. Combined with the cation exchange reaction, the Ni(II), Co(II), Eu(III), and Sn(IV) sorption edges were modeled. The $>S^SOH$ sites have a much smaller density but form stronger complexes with metals, especially for trace metal ions and RNs. Therefore, in the cases of Ni(II), Co(II), and Eu(III), the sorption could be described well by one ion exchange reaction and surface complexation reactions only on $>S^SOH$ (Table 1.3). As for Sn(IV), due to the strong hydrolysis of Sn^{4+} at acidic

TABLE 1.3 Summary of site types, site density, protolysis constants, and sorption reaction constants for RNs for montmorillonite and illite [64–67]

	Site density	
Site types	*Montmorillonite*	*Illite*
$>S^SOH$	2.0 mmol/kg	2.0 mmol/kg
$>S^{W1}OH$	4.0 mmol/kg	4.0 mmol/kg
$>S^{W2}OH$	4.0 mmol/kg	4.0 mmol/kg
CEC	0.87 eq./kg	0.225 eq./kg

	Log K_P	
Protolysis reactions	*Montmorillonite*	*Illite*
$>S^SOH/>S^{W1}OH+H^+=>S^SOH_2{}^+/>S^{W1}OH_2{}^+$	4.5	5.5
$>S^SOH/>S^{W1}OH=>S^SO^-/>S^{W1}O^-+H^+$	−7.9	−6.2
$>S^{W2}OH+H^+=>S^{W2}OH_2{}^+$	6.0	9.0
$>S^{W2}OH=>S^{W2}O^-+H^+$	−10.5	−10.5

	Log K	
Surface species	*Montmorillonite*	*Illite*
$>S^SOEu^{2+}$	1.6	1.9
$>S^SOEuOH^+$	−6.4	−4.6
$>S^SOEu(OH)_2{}^0$	−15.7	−12.8
$>S^SOEu(OH)_3{}^-$	–	−24.0
$>S^SOAm^{2+}$	–	1.2
$>S^SOAmOH^+$	–	−5.0
$>S^SOAm(OH)_2{}^0$	–	−13.0
$>S^SOAm(OH)_3{}^-$	–	−24.3
$>S^SOSn(OH)_2{}^+$	–	12.5
$>S^SOSn(OH)_3{}^0$	–	8.5
$>S^SOSn(OH)_4{}^-$	–	2.5
$>S^SOSn(OH)_5{}^{2-}$	–	−5.7
$>S^SOTh^{3+}$	7.2	7.4
$>S^SOThOH^{2+}$	2.7	2.3

TABLE 1.3 Summary of site types, site density, protolysis constants, and sorption reaction constants for RNs for montmorillonite and illite [64–67]—Cont'd

	Log *K*	
Surface species	***Montmorillonite***	***Illite***
$>S^SOTh(OH)_2{}^+$	−2.6	−2.4
$>S^SOTh(OH)_3{}^0$	−9.1	−8.8
$>S^SOTh(OH)_4{}^-$	−16.9	−15.3
$>S^SONpO_2{}^0$	–	−2.0
$>S^SONpO_2OH^-$	–	−10.3
$>S^SOPaO(OH)^+$	–	6.4
$>S^SOPaO_2{}^0$	–	0.4
$>S^SOPaO_2(OH)^-$	–	6.7
$>S^SOUO_2{}^+$	3.1	2.0
$>S^SOUO_2OH^0$	−3.4	−3.5
$>S^SOUO_2(OH)_2{}^-$	−11.0	−10.6
$>S^SOUO_2(OH)_3{}^{2-}$	−20.5	−19.0
	Log K_C	
Cation exchange	***Montmorillonite***	***Illite***
$3Clay\text{-}Na + Eu^{3+} = 3Clay\text{-}Eu + 3Na^+$	1.5	1.9
$3Clay\text{-}Na + Am^{3+} = 3Clay\text{-}Am + 3Na^+$	–	1.9
$Clay\text{-}Na + NpO_2^+ = Clay\text{-}NpO_2 + Na^+$	–	0
$2Clay\text{-}Na + UO_2^{2+} = 2Clay\text{-}UO_2 + 2Na^+$	0.15	0.65

conditions, ion exchange was discarded and only four kinds of complexes formed on $>S^SOH$ sites (the reactions and sorption constants are listed in Table 1.3). Bradbury and Baeyens [66] further described Am(III), Th(IV), Pa(V), Np(V), and U(VI) sorption on illite (Fig 1.10). For U(VI) and Am(III), the sorption edges were modeled using a combination of cation exchange at acidic conditions and surface complexation reactions over a higher pH range. However, in the cases of Th(IV), Pa(V), and Np(V), only surface complexation occurred on the surface hydroxyl group (>SOH). It is noticed that in all cases, the surface complexes were formed only on strong

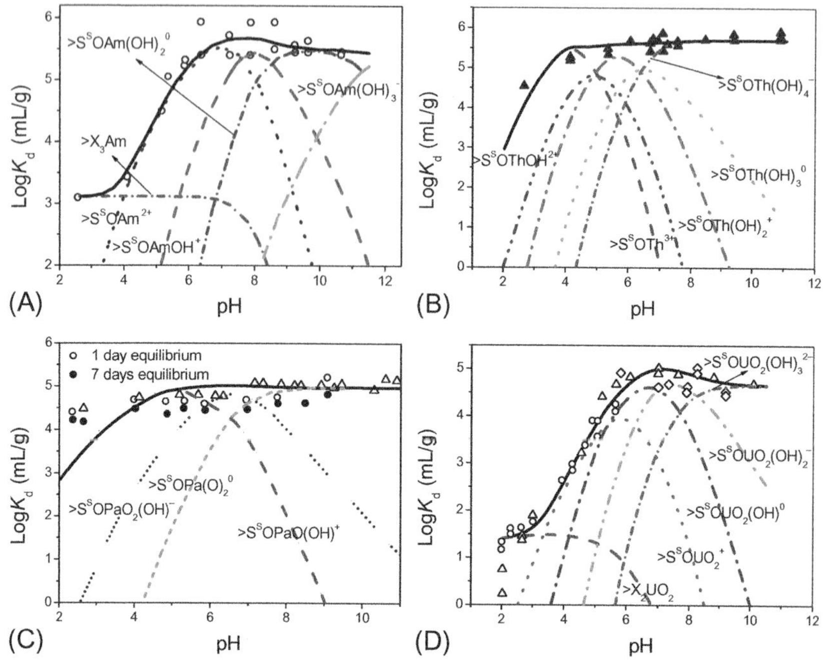

FIG. 1.10 Modeling of the sorption of RNs on illite. (A) Am(III), (B) Th(IV), (C) Pa(V), and (D) U(VI) [66]. *(Reproduced from M.H. Bradbury, B. Baeyens, Sorption modeling on illite Part II: aactinide sorption and linear free energy relationships. Geochim. Cosmochim. Acta 73 (2009) 1004–1013. Copyright 2008 Elsevier.)*

sorption sites, that is, $>S^{S}OH$. When metal ion and RN concentrations are high enough, the use of the strong site alone may not be competent and the surface complexation of metal on weak sites ($>S^{W1}OH$ and $>S^{W2}OH$) would play an instant role. Moreover, the formation of surface precipitation should also be considered [27].

Bradbury and Baeyens [66] took the liner free energy relation (LFER) method to address the sorption data on illite and found a good linear correlation between the logarithms of strong site sorption constants ($\log^{S} K$) and the logarithm of the aqueous hydrolysis constants ($\log^{OH} K$). This is important for those metals for which sorption data are not available. Moreover, by using the same model, similar results were obtained for Eu(III) and Cm(III) sorption on illite [70]. It is noteworthy that, different from Eu(III), TRLFS indicated that a surface species occurs for Cm(III) at high pH, which appears to be related to the interaction of the adsorbed Cm with the dissolved silicate.

1.4.2.3 RNs sorption on montmorillonite/smectite (2:1 clay and high-expansible)

Despite the difference in sorption reactivities [71], illite and montmorillonite have essentially similar T-O-T structures. In essence, this in turn leads to the same structure of the "broken bond" edge sites where metal and RN sorption takes place through surface complexation. Subsequently, the similar steric configurations of corresponding surface species would be anticipated. All the indications are that these two important clay minerals have similar sorption characteristics, which can be described with the same sorption model [64, 67, 69, 72, 73]. Moreover, the LFERs for both illite and montmorillonite reveal a close similarity as well, indicating that the chemical bonding mechanisms of the metal and RN species on both clay minerals must be similar [66]. To date, RNs such as Eu(III), Ga(III), Am(III), Cm(III), Sn(IV), Th(IV) Se(IV), U(VI), Np(V), Pa(V), and sorption on montmorillonite/smectite have been successfully modeled with SCMs in similar ways to illite [67, 73–78]. The sorption species and constants of RNs on both montmorillonite and illite are listed in Table 1.3.

It cannot be ignored that some different models were applied form montmorillonite and smectite in some previous studies. For instance, Tournassat et al. [79] developed a new type of SCM for U(VI) sorption on montmorillonite. In this model, the unique characteristics of the electrostatic surface potentials at the montmorillonite edge sites influenced by neighboring basal plane cation exchange sites were taken into account. The proposed model allowed for a prediction of U(VI) sorption under varying chemical conditions for a number of published datasets without introducing a second site type or including the formation of ternary U(VI)-carbonate surface complexes.

1.5 Generalized adsorption model for cesium

Knowledge about the reactions between Cs^+ and natural media such as clay is therefore very important to predict and evaluate RCs behaviors in the environment due to the high fixation ability of clay to RCs. It has been shown that in most soils, the Cs^+ ion is predominantly adsorbed to specific exchange sites (frayed edge sites) on illitic clay minerals (Fig. 1.11) [2, 3, 22]. It is hard to apply the SCMs to give an accurate prediction of Cs^+ behaviors in such situations. Because Cs^+ exists almost exclusively as the monovalent cation in solution, Freundlich-type sorption implies that there are at least two distinct sorption sites for Cs^+ on illite [2, 80]. If two sites are considered, then they can be broadly classified into high affinity/low capacity and low affinity/high capacity site types (Fig. 1.11). The former is often referred to as FES and is responsible for the adsorption of Cs^+ at low concentrations [2, 81]. The second class of sites is commonly referred to as a "planar site" on the surface of illite arising from isomorphous substitution, for example, Al^{3+} for Si^{4+}. Poinssot et al. [82] also developed a two-site cation exchange model to describe the sorption of Cs^+ over the whole range of experimental conditions. However, the two-site types were termed frayed edge sites (FES, high affinity/low capacity) and type II sites (low affinity/high capacity). According to the nuclear magnetic resonance spectroscopic investigations for Cs^+ adsorbed on illite, except for FESs, there existed at least two

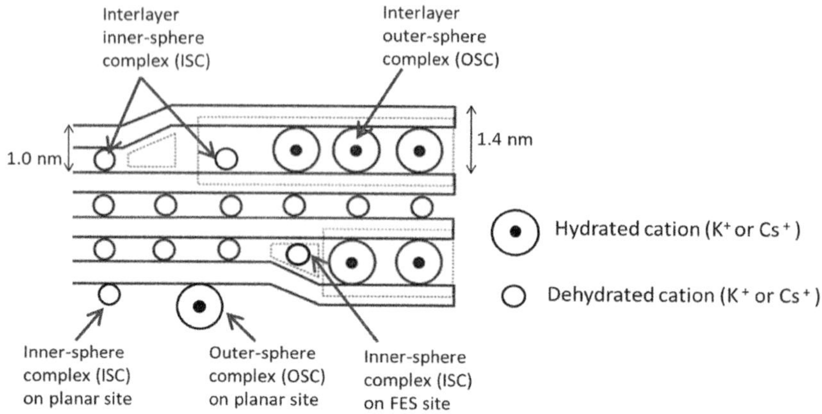

FIG. 1.11 Schematic illustration of the adsorption species and adsorption sites of micaceous minerals [2]. *(Reproduced from Q.H. Fan, M. Tanaka, K. Tanaka, A. Sakaguchi, Y. Takahashi, An EXAFS study on the effects of natural organic matter and the expandability of clay minerals on cesium adsorption and mobility, Geochim. Cosmochim. Acta 135 (2014) 49–65. Copyright 2014 Elsevier.)*

"chemical environments for Cs" at the surface of the illite crystallites [83, 84]. Based on this fact, a GAM for cesium was developed by Bradbury and Baeyens [85] (sometimes also called the generalized cesium sorption model, and in this chapter we use GAM).

The fundamental assumptions for GAM are: (i) that Cs^+ adsorption is mainly contributed to cation exchange on clay, as shown in Eq. (1.20); (ii) that Cs^+ adsorption is envisaged as taking place at three site types with different site capacities and affinities (mainly including the FES, TIIS, and PS); and (iii) that the selectivity coefficients for $Cs^+({}^{Cs}_{M}K_c)$ and the main competing cations are deduced from literature adsorption measurements on illitic minerals based on Eqs. (1.20), (1.21).

$$\text{M-Clay} + Cs^+ \Leftrightarrow \text{Cs-Clay} + M^+ \tag{1.20}$$

$$ {}^{Cs}_{M}K_c = \frac{N_{Cs}}{N_M}\frac{\alpha_M}{\alpha_{Cs}} = \frac{N_{Cs}}{N_M}\frac{\gamma_M[M]}{\gamma_{Cs}[CS]} \tag{1.21}$$

where M represents monovalent cations, such as K^+, Na^+, and $NH_4{}^+$; α and γ are the activities and activity coefficients in the aqueous phase, respectively; and the terms in [·] represent the molar concentration. N_{Cs} and N_M are equivalent fractional occupancies defined as:

$$N_{Cs(\text{or}\,M)} = \frac{\Gamma_{Cs(\text{or}\,M)}}{Q} \tag{1.22}$$

where $\Gamma_{Cs\ (\text{or}\ M)}$ is the quantity of Cs^+ (or M^+) adsorbed (mmol/kg) and Q is the adsorption site capacity (mmol/kg). Only FES, TIIS, and PS were taken

TABLE 1.4 Parameters for three-site ion exchange model used to model Cs adsorption to illite and vermiculite under the experimental conditions [2].

Site type	Site capacity (mmol/kg)[a]	Selectivity coefficient[b]
Illite		
FES	0.0005	$\text{Log}_{\text{Na}}^{\text{Cs}} K=7.0$
		$\text{Log}_{\text{K}}^{\text{Cs}} K=4.6$
Type II site	0.0400	$\text{Log}_{\text{Na}}^{\text{Cs}} K=3.6$
		$\text{Log}_{\text{K}}^{\text{Cs}} K=1.5$
Planar site	0.1600	$\text{Log}_{\text{Na}}^{\text{Cs}} K=1.6$
		$\text{Log}_{\text{K}}^{\text{Cs}} K=0.5$
Vermiculite		
FES	0.0010	$\text{Log}_{\text{Na}}^{\text{Cs}} K=7.0$
		$\text{Log}_{\text{K}}^{\text{Cs}} K=4.6$
Interlayer site	0.4000	$\text{Log}_{\text{Na}}^{\text{Cs}} K=4.7$
		$\text{Log}_{\text{K}}^{\text{Cs}} K=1.7$
Type II site	0.1200	$\text{Log}_{\text{Na}}^{\text{Cs}} K=3.6$
		$\text{Log}_{\text{K}}^{\text{Cs}} K=1.5$
Planar site	0.4800	$\text{Log}_{\text{Na}}^{\text{Cs}} K=1.6$
		$\text{Log}_{\text{K}}^{\text{Cs}} K=0.5$

[a]*For illite, FES: 0.25% of CEC, Type II site: 20% of CEC, and Planar site: 80% of CEC [2, 85]. For vermiculite, FES: 0.1% of CEC, Interlayer site: 40% of CEC; Type II site: 12% of CEC, and planar site: 48% of CEC.*
[b]*The selectivity coefficients of Cs to cations on illite were from Fan et al. [2].*

into account for the sorption of Cs^+ on illite or argillaceous rock [80, 82, 85]. In addition, the ITS was confirmed to be another highly selective adsorption site to the fixation Cs^+ in natural systems [2, 81]. As expected, it will be reasonable to take the ITS into account for Cs^+ adsorption as a modified GAM. The related parameters for GAM are listed in Table 1.4. This model has been widely used to model Cs^+ adsorption on argillaceous rocks and illite under a wide concentration range [2, 80, 85]. In addition, recent investigations have indicated that this three-site model can also be applied to describe Tl^+ adsorption on illite [86].

For vermiculite, the presence of siloxane groups from the ditrigonal cavity of vermiculite (interlayer sites (ITSs)) would form the ISCs with Cs^+.

The adsorption affinity for Cs^+ at the ITS was similar to the FES related to other alkali and alkaline earth ions (Table 1.4). Fan et al. [2, 80] first introduced the ITSs into the GAM to describe the Cs^+ adsorption on vermiculite in addition to the three classic adsorption sites mentioned above. Notably, the adsorption of Cs^+ is closely related to the expandability of the clay minerals. The ISC is one of the important species for Cs^+ adsorption on illite (nonexpansion) and vermiculite (intermediate expansion); however, an almost complete OSC was observed in montmorillonite (large expansion).

Appling the modified GAM, Fan and Takahashi [80] verified that the GAM can precisely predict the distribution of Cs^+ in sediment particles (Fig. 1.12). According to the GAM, the main contributor for Cs^+ adsorption is the FES rather than others due to the very low concentration of Cs^+ in the river-estuary-ocean system. The different solid-water distribution of Cs^+ in the river-estuary-ocean system was dominantly controlled by the salinity in the aqueous phase. Therefore, Cs^+ should be highly reactive with a strong adsorptive character to particulate matter in the river system, whereas a conservative distribution must be dominant in oceans with much weaker affinity to particulate matter because of the high salinity.

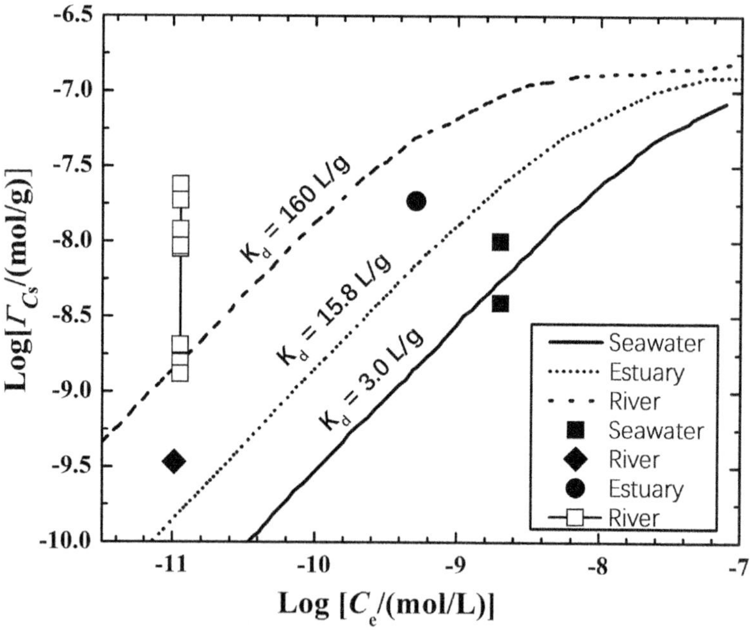

FIG. 1.12 GAM prediction of Cs^+ distribution in the river-estuary-ocean systems. Taking an example using the sediments with the largest *RIP* (5616 mmol/kg) in this summarization [80]. *(Reproduced from Ref. Q.H. Fan, Y. Takahashi, Employment of the generalized adsorption model for the prediction of the solid-water distribution of radiocesium in the river-estuary-ocean system, Appl. Geochem. 79 (2017) 75–84. Copyright 2017 Elsevier.)*

1.6 Applications of spectroscopies in RNs sorption

SCMs and GAM are useful in assuming interaction processes between RNs and minerals; however, the specific surface speciation and chemical environment of target RNs cannot be directly confirmed from wet chemistry and surface complex models. The availability of advanced spectroscopic technologies, such as time-resolved laser fluorescence spectroscopy (TRLFS), EXAFS, and X-ray photoelectron spectroscopy (XPS), can provide a potential approach for investigating the interaction mechanism in great detail by the identification of radionuclide speciation at the mineral/water interface.

1.6.1 Time-resolved laser fluorescence spectroscopy

The excited atoms of some elements, such as Cm, U, Am, Eu, etc., can display fluorescence emission when the electrons deexcite from the lowest excited state to the ground state. The electron transfers to different energy levels of the ground state would generate variable vibronic responses (Fig. 1.13). The fluorescence characteristics are sensitive to the ligand fields, the coordination

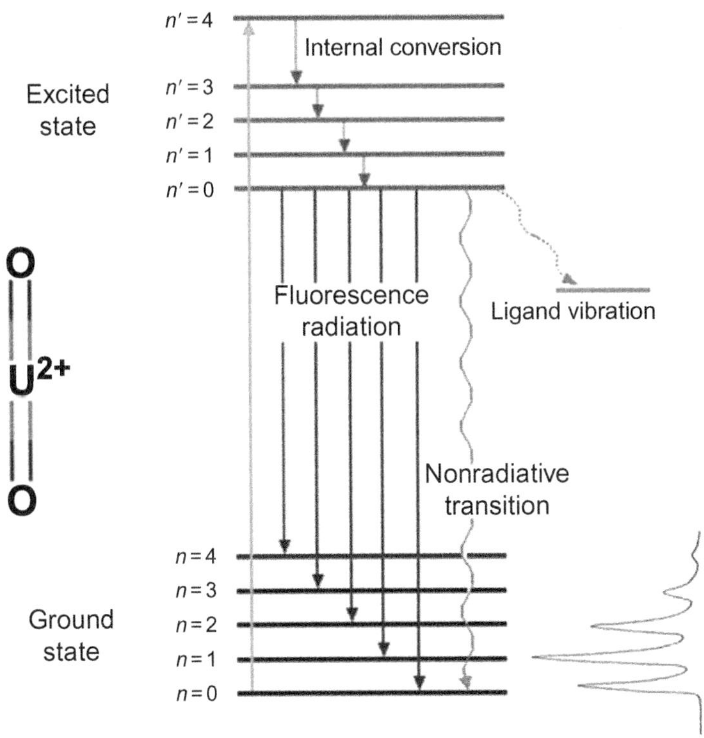

FIG. 1.13 Electronic energy scheme of U.

structure, and the symmetry of the central atom. Therefore, TRLFS affords extremely high sensitivity of radionuclide speciation at the trace concentration.

The fluorescence lifetime, which is the most important parameter in speciation identification fluorescence quench, is controlled by the energy transfer to H_2O/OH^- vibrational modes. Therefore, the presence of H_2O/OH^- around the target atom would lead to a decrease of the fluorescence lifetime [87]. When RNs are adsorbed on mineral surfaces as the ISCs, a portion of the hydration water molecules are displaced in the first coordination sphere, and accordingly the fluorescence lifetimes are elongated. As for Eu, Cm, and Am, the fluorescence lifetime is empirically linearly correlated with the hydration numbers in the first coordination sphere, which enables the determination of the coordination structure in complicated systems [88, 89]. Due to the different quantum characteristics, Eu and Cm display diverse spectroscopic features. In the case of Eu, the vibronic peak positions do not change under different coordination conditions, simply the relative intensities of $^5D_0$-$^7F_1$ (592 nm) and $^5D_0$-$^7F_2$ (616 nm) emission peaks can provide structural information. As for Cm, the emission peaks will shift with the changes of coordination conditions, which offers more precise structural information.

The ISCs, OSCs, and ternary complexes of Eu(III) were identified on smectite and kaolinite surfaces with TRLFS [90]. Generally, Eu(III) is fully hydrated with 8–9 water molecules at low pH, and the OSCs dominated on the mineral surface. The water molecules surrounding Eu(III) gradually decreased with the increase in pH, and the ISCs and surface precipitation prevailed in alkaline conditions [91–94]. TRLFS confirmed that Eu(III), Cm(III), and Am(III) can incorporate into the lattice of hydrotalcite and calcite [70, 95–97].

In similarity, the fluorescence of UO_2^{2+} can be applied for surface speciation. The spectral properties, including the intensity, bandwidth, and spacing of vibronic bands as well as the fluorescence lifetime, are more complicated in comparison with Eu and Cm (Fig. 1.14). At room temperature, the spectra are weak and poorly resolved due to the severe quenching effect, which does not allow the identification of uranium species such as $UO_2(CO_3)_3^{4-}$. However, the spectral intensity and resolution can be greatly increased under cryogenic conditions, especially at liquid helium temperature (LHeT), because of the suppression of fluorescence quenching and spectral broadening effects, which facilitates the identification of uranyl species in environmental samples [98, 99].

Usually, more than one surface species presents at the mineral-water interface not only for laboratory samples but also for natural in situ samples. By controlling the delay time after the excitation pulse, the individual spectrum with a different lifetime can be discriminated. Arnold et al. [100] decomposed two surface complexes and amorphous nanoclusters of uranium on a muscovite surface under neutral conditions by combining TRLFS and high-angle

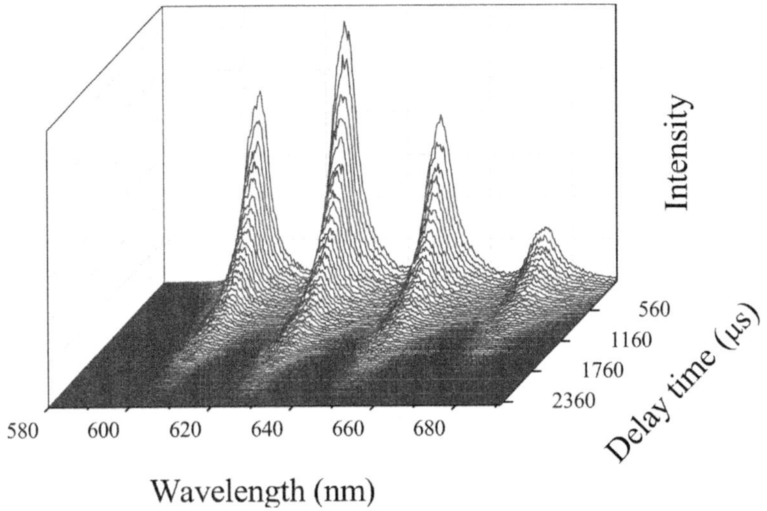

FIG. 1.14 Time-resolved fluorescence spectroscopy of uranium.

annular dark-field scanning transmission electron microscopy (HAADF-STEM). Del Nero et al. [101] confirmed the formation of the ISCs and the uranium phosphate precipitate on the alumina surface by the quantum yield and peak positions as well as the lifetimes of fluorescence spectra. The adsorbed U(VI) on quartz and chlorite displayed individual fluorescence characteristics; therefore, the correlation between the loading amount and the sorption species of U(VI) was revealed by TRLFS [102]. Due to the high sensitivity to trace uranium, TRLFS plays a significant role in speciation identification in in situ environmental samples [98, 99].

In spite of these distinguishing capabilities, it is not easy to estimate the number of species, their contributions, and their structural information from TRLFS data when spectra are overlapped and/or fluorescence lifetimes are similar. To overcome such difficulties, more deconvolution analysis methods have been developed to extract as much information as possible from the TRLFS dataset. Chang et al. [103] used the evolving factor analysis (EFA) combined with multivariate curve resolution to analyze the TRLFS data of adsorbed U(VI) on gibbsite at different conditions, and identified four surface species and their abundant dependence on the conditions. Gabriel et al. [104] applied EFA to resolve the pure spectra and the decay curves of U(VI) species on silica. Saito et al. [105] used parallel factor analysis (PARAFAC) to interpret a series of TRLFS data on Eu(III) complexation with acetate, by which the pure spectra, decay curves, and concentration of each complex were decomposed.

Although TRLFS is a sensitive approach to identify the speciation of RNs at trace concentrations in complicated conditions, the presence of a fluorescence quencher, such as the incorporated iron in the bulk structure of clay minerals, can mitigate or even extinguish the fluorescence emission, which disables the prominent ability of TRLFS. In addition, for those non-fluorescent elements, TRLFS is no longer in force.

1.6.2 X-ray absorption fine structure

The electron waves excited from a certain atom by an X-ray can interfere with scattering waves from neighboring atoms and generate oscillations in the energy absorption coefficient, namely X-ray absorption fine structure (XAFS). This technique is applicable for almost all heavy metals and RNs because the characteristic absorption energy edge of each element is unique and discrete. The photon energy range in the X-ray absorption spectrum is generally divided into two parts: the X-ray absorption near-edge structure (XANES) in the energy range from 20 eV before the absorption edge and 30 eV after absorption edge, and the EXAFS in the energy range from 30 to 1000 eV after the absorption edge. XANES can provide the oxidation state, the coordination number, and the geometry information while EXAFS contains more detailed information on the surrounding chemical environment of the target atom, such as coordination number, neighbor atoms, and bond distances. XAFS is a useful tool in understanding environmental behavior mechanisms, including oxidation/redox reactions, surface complexation, surface precipitation, and structural incorporation.

The XANES region involves transitions on E_0, and the adsorption edge is sensitive to the valence state of the absorbing atom. For uranium oxide, the absorption edge shifts toward higher energies as oxidation state increasing in the order of UO_2, U_3O_8, UO_3. It is possible to determine the oxidation of an unknown sample by comparing with standard samples [106, 107]. Fukushi et al. [108] determined Mg speciation and each contribution in a monohydrocalcite mineral with the aid of XANES, molecular simulation, and geochemical modeling. Stetten et al. [109] revealed that the immobilization of uranium in sediments was correlated with the reduction process derived from the structural Fe(II) in phyllosilicate minerals.

The fine structure information of EXAFS is generated by the interference of photoelectron waves from absorbing atoms and backscattering waves from neighboring atoms. After the extraction of the fine structure oscillation function and the Fourier transform, the oscillation in R space can be induced [Fig. 1.15]. The peak position and intensity in the R space display the position and coordination contribution information of surrounding atoms, respectively. EXAFS can provide metrical parameters describing the short-range environment surrounding the absorbing atom; therefore, it has been widely employed to probe the precise microstructure around central atoms. Based on the structural parameters derived from oscillation paths fitting, both ISCs and OSCs of metals and RNs are identified to be formed on mineral surface [111].

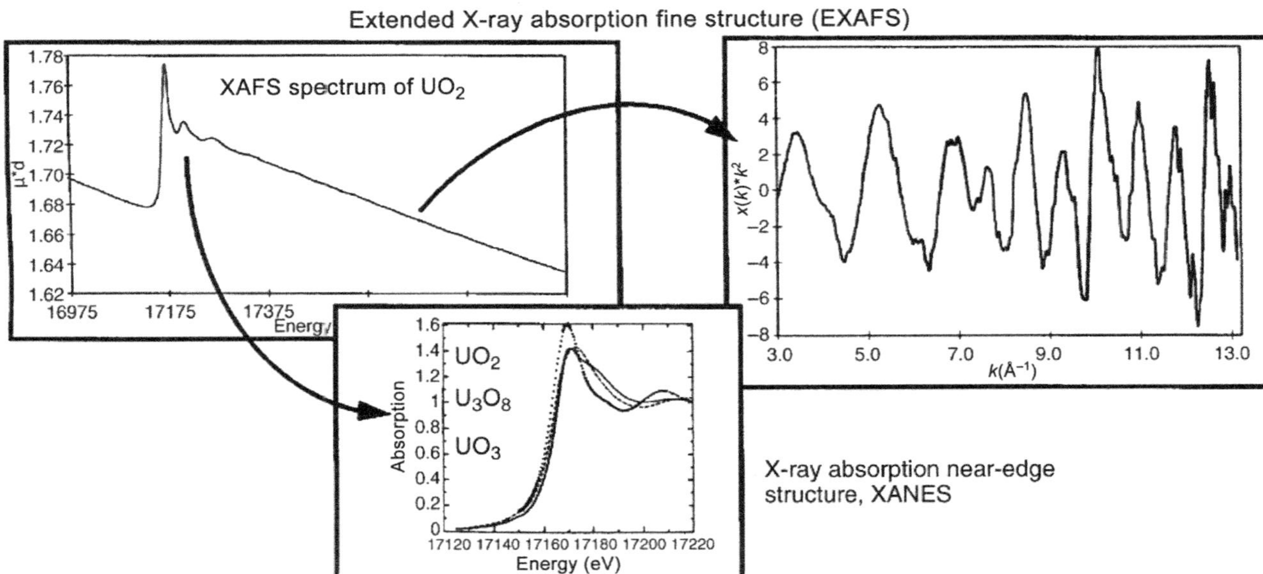

FIG. 1.15 XAFS spectra of uranium [110]. *(Reproduced from M.A. Denecke, Actinide speciation using X-ray absorption fine structure spectroscopy, Coord. Chem. Rev. 250 (2006) 730–754. Copyright 2006 Elsevier.)*

The montmorillonite and vermiculite showed similarity in mineralogical structure; however, their sorption mechanisms toward Cs(I) are varied according to EXAFS analysis. Cs(I) prefers forming the ISCs on vermiculite while the OSCs on montmorillonite [2]. The relationship between the adsorption species of cesium and radiocesium interception potential (*RIP*) in soils and minerals is revealed with the EXAFS approach. The results showed that the adsorption amount of cesium in soil is obviously correlated to *RIP* and *CEC* values, and the fraction of the ISCs ($CN_{IS}/(CN_{OS}+CN_{IS})$) is sensitive to log (*RIP*/*CEC*) (as shown in Fig. 1.16) [81]. Yang et al. [112] found that Eu(III) formed edge-share and corner-share complexes ($R_{Eu\text{-}Al}$ 3.28 and 3.99 Å, respectively) on an alumina surface within 15 days while polynuclear complexes emerged after 15 days. Křepelová et al. [43] measured the L_{III} edge of U adsorption on kaolinite; the U distance between the axis and equatorial O were 1.77 and 2.34 Å, respectively. In the presence of humic acid, uranyl-HA hybrid preferred binding directly on kaolinite surface via uranyl rather than HA. The sorption structure of Eu(III) on muscovite revealed that the ISCs were dominant on the muscovite at neutral range while in the presence of phosphates, the P-bridge type ternary surface complex was the main sorption species (Fig. 1.17) [24].

XAFS can provide an effective method to probe the in situ structural and composition information of samples, taking the advantage of element-specific,

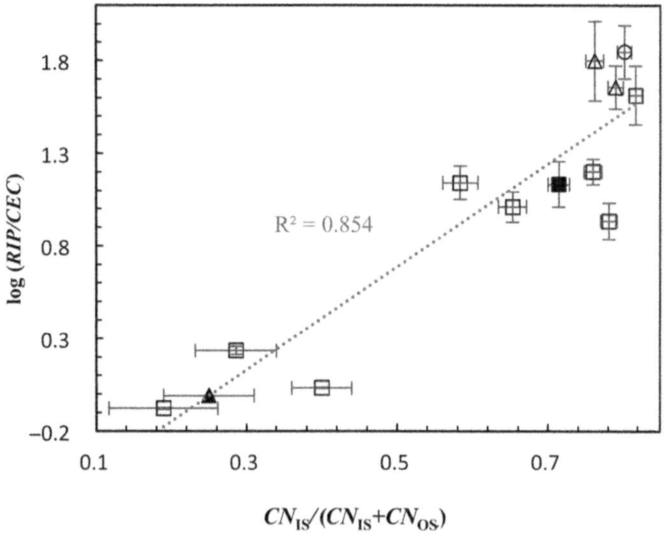

FIG. 1.16 Linear relationship between log (*RIP*/*CEC*) and the ratio of $CN_{IS}/(CN_{IS}+CN_{OS})$ derived from EXAFS fitting [81]. *(Reproduced from Q.H. Fan, N. Yamaguchi, M. Tanaka, H. Tsukada, Y. Takahashi, Relationship between the adsorption species of cesium and radiocesium interception potential in soils and minerals: an EXAFS study, J. Environ. Radioact. 138 (2014) 92–100. Copyright 2014 Elsevier.)*

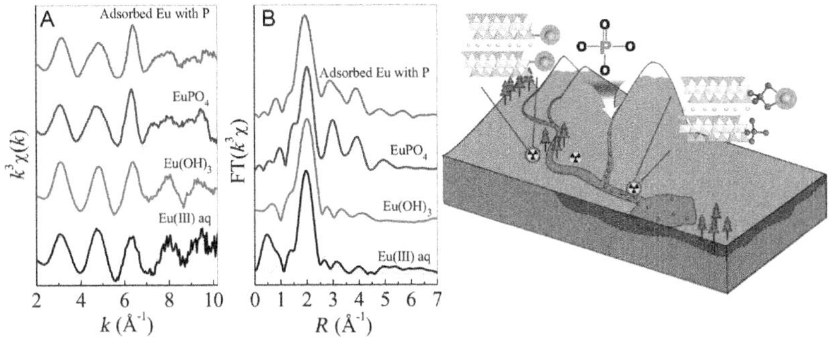

FIG. 1.17 Eu(III) sorption mechanism at muscovite-water interface [24]. *(Reproduced from D. Pan, F. Fan, Y. Wang, P. Li, P. Hu, Q. Fan, W. Wu, Retention of Eu(III) on muscovite environment: batch and spectroscopic studies, Chem. Eng. J. 330 (2017) 559–565. Copyright 2017 Elsevier.)*

bulk-sensitive and nondestructive, and it has been widely employed in microstructure study. However, the drawback that the required concentration of the target element usually exceeds the load under environmental conditions limited its sensitivity, accuracy, and thus applicability on trace level environmental samples. In addition, the structural information stemming from XAFS data is an average set, which usually causes deviation from reality when more than one local structure exists simultaneously.

1.6.3 X-ray photoelectron spectroscopy

X-ray photoelectron spectroscopy (XPS) is a surface-sensitive quantitative spectroscopic technique that is capable of measuring all elements in principle (except hydrogen and helium). The ability makes XPS a unique and valuable tool for understanding the chemistry of surface. XPS spectra are obtained by measuring the photoelectrons emitted from surface atoms when materials are exposed to X-ray irradiation. Because the energy of the X-ray with a certain wavelength is fixed (for Al Kα X-ray, $E_{photon} = 1486.7\,eV$), by measuring the kinetic energy of the emitted electrons, the electron binding energy of each emitted electron can be determined. The binding energy is characteristic of specific elements, thus it can be used to identify the elemental compositions and individual quantities that are present within 1–10nm of the sample surface. In general, the peak position and shape from the XPS spectrum provide information on chemical composition, the oxidation state, and the binding structure while the peak intensity carries information on element content and concentration.

A single element may generate several peaks due to the variation of the local binding environment; therefore, high-energy resolution spectra can provide further structural information by performing data processing such as peak fitting, deconvolution, etc. The binding energy of photoelectrons is correlated with the binding environment and electron configuration of atoms, from which the chemical surrounding and state can be speculated. The sorption processes of RNs and heavy metals and RNs on minerals have been widely investigated by the XPS approach. The species of Eu(III) on feldspar were confirmed by linear correlation fitting of standard spectra [23]. The reduced U(IV) species were identified with the aid of XPS spectra [106, 107]. Additionally, the detailed sorption mechanism of Eu(III) on minerals was revealed by combining XPS and other spectroscopies [21].

1.7 Conclusion and perspectives

Models can be usually used for describing RNs sorption on clays minerals successfully under studied conditions, and can also be used for prediction purposes, to some extent. However, a problem with the SCMs is that it is impossible to identify the exact nature of all the relevant surface complexes. At best an intelligent guess is made as to the surface reactions and this should be remembered when using the models. The models therefore differ in both the form of the electrostatic model chosen and in the species used to describe the surface reactions. In addition, one set of data may be simulated well by models based on different assumptions. Direct molecular structural evidence is thus essential for supporting the formation of the proposed surface complexes. Fortunately, the advanced spectroscopies, such as EXAFS, TRLFS, and XPS, are useful and powerful to obtain the exact surface species of RNs on clay minerals at a molecular scale. With the rapid development of technology, numerous other spectroscopies, such as Fourier Transform infrared spectroscopy (FTIR), second harmonic generation (SHG), secondary ion mass spectrometry (SIMS), nuclear magnetic resonance (NMR), etc., have been attempted to probe the interaction of radionuclides with phyllosilicate minerals. These technologies showed admirable applicability and validity on speciation and structure identification. Although each approach has its unique advantages, the drawbacks are distinct and inevitable as well. Therefore, the SCMs combining advanced spectroscopic approaches can explore the real sorption mechanism of RNs and metal ions on a clay surface.

References

[1] C. Poinssot, H. Geckeis, Radionuclides Behaviour in the Natural Environment Science, Implications and Lessons for the Nuclear Industry, Woodhead Publishing Limited, 2012.

[2] Q.H. Fan, M. Tanaka, K. Tanaka, A. Sakaguchi, Y. Takahashi, An EXAFS study on the effects of natural organic matter and the expandability of clay minerals on cesium adsorption and mobility, Geochim. Cosmochim. Acta 135 (2014) 49–65.

[3] Y. Takahashi, Q. Fan, H. Suga, K. Tanaka, A. Sakaguchi, Y. Takeichi, K. Ono, K. Mase, K. Kato, V.V. Kanivets, Comparison of solid-water partitions of radiocesium in river waters in Fukushima and Chernobyl areas, Sci. Rep. 7 (2017) 12407.

[4] J. Wang, B. He, X. Wei, P. Li, J. Liang, S. Qiang, Q. Fan, W. Wu, Sorption of uranyl ions on TiO_2: effects of pH, contact time, ionic strength, temperature and HA, J. Environ. Sci. 75 (2019) 115–123.
[5] H. Wu, S. Qiang, Q. Fan, X. Zhao, P. Liu, P. Li, J. Liang, W. Wu, Exploring the relationship between Th(IV) adsorption and the structure alteration of phlogopite, Appl. Clay Sci. 152 (2018) 295–302.
[6] C. Xu, M. Athon, Y.F. Ho, H.S. Chang, S. Zhang, D.I. Kaplan, K.A. Schwehr, N. DiDonato, P.G. Hatcher, P.H. Santschi, Plutonium immobilization and remobilization by soil mineral and organic matter in the far-field of the Savannah River site, US, Environ. Sci. Technol. 48 (2014) 3186–3195.
[7] M. Flury, J.B. Mathison, J.B. Harsh, In situ mobilization of colloids and transport of cesium in Hanford sediments, Environ. Sci. Technol. 36 (2002) 5335–5341.
[8] B.Ø. Hansen, P. Kwan, M.M. Benjamin, C.W. Li, G.V. Korshin, Use of iron oxide-coated sand to remove strontium from simulated Hanford tank wastes, Environ. Sci. Technol. 35 (2001) 4905–4909.
[9] J. Zheng, K. Tagami, Y. Watanabe, S. Uchida, T. Aono, N. Ishii, S. Yoshida, Y. Kubota, S. Fuma, S. Ihara, Isotopic evidence of plutonium release into the environment from the Fukushima DNPP accident, Sci. Rep. 2 (2012) 304.
[10] N. Kinoshita, K. Sueki, K. Sasa, J.I. Kitagawa, S. Ikarashi, T. Nishimura, Y.S. Wong, Y. Satou, K. Handa, T. Takahashi, M. Sato, T. Yamagata, Assessment of individual radionuclide distributions from the Fukushima nuclear accident covering central-East Japan, Proc. Natl. Acad. Sci. U. S. A. 108 (2011) 19526–19529.
[11] J.M. Zachara, S.C. Smith, C. Liu, J.P. McKinley, R.J. Serne, P.L. Gassman, Sorption of Cs^+ to micaceous subsurface sediments from the Hanford site, USA, Geochim. Cosmochim. Acta 66 (2002) 193–211.
[12] O. Eiichiro, Hiroshima to Fukushima—Biohazards of Radiation, Springer-Verlag, Berlin Heidelberg, 2014.
[13] M. Chino, H. Nakayama, H. Nagai, H. Terada, G. Katata, H. Yamazawa, Preliminary estimation of release amounts of I-131 and Cs-137 accidentally discharged from the Fukushima Daiichi Nuclear Power Plant into the atmosphere, J. Nucl. Sci. Technol. 48 (2011) 1129–1134.
[14] Y. Morino, T. Ohara, M. Nishizawa, Atmospheric behavior, deposition, and budget of radioactive materials from the Fukushima Daiichi nuclear power plant in March 2011, Geophys. Res. Lett. 38 (2011) L00G11.
[15] T.J. Yasunari, A. Stohl, R.S. Hayano, J.F. Burkhart, S. Eckhardt, T. Yasunari, Cesium-137 deposition and contamination of Japanese soils due to the Fukushima nuclear accident, Proc. Natl. Acad. Sci. U. S. A. 108 (2011) 19530–19534.
[16] N. Yoshida, Y. Takahashi, Land-surface contamination by radionuclides from the Fukushima Daiichi Nuclear Power Plant accident, Elements 8 (2012) 201–206.
[17] M.B. McBride, Processes of heavy and transition metal sorption by soil minerals, in: G.H. Bolt, M.F.D. Boodt, M.H.B. Hayes, M.B. McBride (Eds.), Interactions at the Soil Colloid-Soil Solution Interface, vol. 190, Kluwer Academic Publishers, Dordrecht, 1991.
[18] G. Sposito, Distinguishing adsorption from surface precipitation, in: J.A. Davis, K.F. Hayes (Eds.), Geochemical Processes at Mineral Surface, ACS Symposium Series, 323, American Chemical Society, Washington, DC, 1986, pp. 217–229.
[19] P.A. O'Day, G.E. Brown, G.A. Parks, X-ray-absorption spectroscopy of cobalt(II) multinuclear surface complexes and surface precipitates on kaolinite, J. Colloid Interface Sci. 165 (1994) 269–289.

[20] D.L. Sparks, Environmental Soil Chemistry, second ed., Academic Press, 2003.

[21] Q. Fan, X. Tan, J. Li, X. Wang, W. Wu, G. Montavon, Sorption of Eu(III) on attapulgite studied by batch, XPS, and EXAFS techniques, Environ. Sci. Technol. 43 (2009) 5776–5782.

[22] Q. Fan, K. Tanaka, A. Sakaguchi, H. Kondo, N. Watanabe, Y. Takahashi, Factors controlling radiocesium distribution in river sediments: field and laboratory studies after the Fukushima Dai-ichi Nuclear Power Plant accident, Appl. Geochem. 48 (2014) 93–103.

[23] P. Li, H. Wu, J. Liang, Z. Yin, D. Pan, Q. Fan, D. Xu, W. Wu, Sorption of Eu(III) at feldspar/water interface: effects of pH, organic matter, counter ions, and temperature, Radiochim. Acta 105 (2017) 1049–1058.

[24] D. Pan, F. Fan, Y. Wang, P. Li, P. Hu, Q. Fan, W. Wu, Retention of Eu(III) on muscovite environment: batch and spectroscopic studies, Chem. Eng. J. 330 (2017) 559–565.

[25] B.A. Powell, M.C. Duff, D.I. Kaplan, R.A. Fjeld, M. Newville, D.B. Hunter, P.M. Bertsch, J.T. Coates, P. Eng, M.L. Rivers, S.M. Serkiz, S.R. Sutton, I.R. Triay, D.T. Vaniman, Plutonium oxidation and subsequent reduction by Mn(IV) minerals in Yucca Mountain tuff, Environ. Sci. Technol. 40 (2006) 3508–3514.

[26] K. Schmeide, G. Bernhard, Sorption of Np(V) and Np(IV) onto kaolinite: effects of pH, ionic strength, carbonate and humic acid, Appl. Geochem. 25 (2010) 1238–1247.

[27] X. Zhao, S. Qiang, H. Wu, Y. Yang, D. Shao, L. Fang, J. Liang, P. Li, Q. Fan, Exploring the sorption mechanism of Ni(II) on illite: batch sorption, modelling, EXAFS and extraction investigations, Sci. Rep. 7 (2017) 8495.

[28] X. Zhao, Y. Wang, H. Wu, L. Fang, J. Liang, Q. Fan, P. Li, Insights into the effect of humic acid on Ni(II) sorption mechanism on illite: batch, XPS and EXAFS investigations, J. Mol. Liq. 248 (2017) 1030–1038.

[29] F. Bergaya, B. Theng, G. Lagaly, Handbook of Clay Science, Elsevier Science, New York, 2006, pp. 1–1224.

[30] V. Brendler, T. Arnold, A. Richter, G. Bernhard, Capability of surface complexation models and databases for predicting radionuclide sorption, in: WM'04 Conference, February 29–March 4, 2004, Tucson, AZ, 2004.

[31] H.B. Bradl, Adsorption of heavy metal ions on soils and soils constituents, J. Colloid Interface Sci. 277 (2004) 1–18.

[32] I. Langmuir, The constitution and fundamental properties of solids and liquids. Part I. Solids, J. Am. Chem. Soc. 38 (1916) 2221–2295.

[33] D.A. Kulik, Gibbs energy minimization approach to modeling sorption equilibria at the mineral-water interface: thermodynamic relations for multi-site-surface complexation, Am. J. Sci. 302 (2002) 227–279.

[34] Z. Wang, D.E. Giammar, Mass action expressions for bidentate adsorption in surface complexation modeling: theory and practice, Environ. Sci. Technol. 47 (2013) 3982–3996.

[35] D.Y.C. Chan, Free energies of electrical double layers at the oxide-solution interface, Geochem. Process. Miner. Surfaces. 323 (1987) 99–112.

[36] M. Gouy, Sur la constitution de la charge électrique à la surface d'un électrolyte, J. Phys. Theor. Appl. 9 (1910) 457–468.

[37] D.L. Chapman, A contribution to the theory of electrocapillarity, Lond. Edinb. Dubl. Phil. Mag. J. Sci. 25 (1913) 475–481.

[38] W. Stumm, H. Hohl, F. Dalang, Interaction of metal ions with hydrous oxide surfaces, Croat. Chem. Acta 48 (1976) 491–504.

[39] J.A. Davis, J.O. Leckie, Surface ionization and complexation at the oxide/water interface II. Surface properties of amorphous iron oxyhydroxide and adsorption of metal ions, J. Colloid Interface Sci. 67 (1978) 90–107.

[40] J.A. Davis, R.O. James, J.O. Leckie, Surface ionization and complexation at the oxide/water interface: I. Computation of electrical double layer properties in simple electrolytes, J. Colloid Interf. Sci. 63 (1978) 480–499.

[41] K.F. Hayes, J.O. Leckie, Modeling ionic strength effects on cation adsorption at hydrous oxide/solution interfaces, J. Colloid Interface Sci. 115 (1987) 564–572.

[42] M. Kohler, E. Wieland, J.O. Leckie, Metal–ligand–surface interactions during sorption of uranyl and neptunyl on oxides and silicates, in: Y.K. Kharaka, A.S. Maest (Eds.), Water–Rock Interaction (VII), Balkema, Rotterdam, 1992, pp. 51–54.

[43] A. Křepelová, V. Brendler, S. Sachs, N. Baumann, G. Bernhard, U (VI)-kaolinite surface complexation in absence and presence of humic acid studied by TRLFS, Environ. Sci. Technol. 41 (2007) 6142–6147.

[44] E. Tertre, S. Castet, G. Berger, M. Loubet, E. Giffaut, Surface chemistry of kaolinite and Na-montmorillonite in aqueous electrolyte solutions at 25 and 60°C: experimental and modeling study, Geochim. Cosmochim. Acta 70 (2006) 4579–4599.

[45] P.V. Brady, Alumina surface chemistry at 25, 40, and 60°C, Geochim. Cosmochim. Acta 58 (1994) 1213–1217.

[46] E. Tombacz, M. Szekeres, Colloidal behavior of aqueous montmorillonite suspensions: the specific role of pH in the presence of indifferent electrolytes, Appl. Clay Sci. 27 (2004) 75–94.

[47] D.A. Sverjensky, N. Sahai, Theoretical prediction of single-site enthalpies of surface protonation for oxides and silicates in water, Geochim. Cosmochim. Acta 62 (1998) 3703–3716.

[48] Y.X. Liu, D.S. Alessi, S.L. Flynn, M.S. Alam, W.D. Hao, M. Gingras, H.Z. Zhao, K.O. Konhauser, Acid-base properties of kaolinite, montmorillonite and illite at marine ionic strength, Chem. Geol. 483 (2018) 191–200.

[49] X. Gu, L.J. Evans, Surface complexation modelling of Cd (II), Cu (II), Ni (II), Pb (II) and Zn (II) adsorption onto kaolinite, Geochim. Cosmochim. Acta 72 (2008) 267–276.

[50] E. Tertre, G. Berger, E. Simoni, S. Castet, E. Giffaut, M. Loubet, H. Catalette, Europium retention onto clay minerals from 25 to 150°C: experimental measurements, spectroscopic features and sorption modelling, Geochim. Cosmochim. Acta 70 (2006) 4563–4578.

[51] R. Marsac, N.L. Banik, J. Lützenkirchen, R.A. Buda, J.V. Kratz, C.M. Marquardt, Modeling plutonium sorption to kaolinite: accounting for redox equilibria and the stability of surface species, Chem. Geol. 400 (2015) 1–10.

[52] L. Gao, Z. Yang, K. Shi, X. Wang, Z. Guo, W. Wu, U (VI) sorption on kaolinite: effects of pH, U (VI) concentration and oxyanions, J. Radioanal. Nucl. Chem. 284 (2010) 519–526.

[53] D. Arda, J. Hizal, R. Apak, Surface complexation modelling of uranyl adsorption onto kaolinite based clay minerals using FITEQL 3.2, Radiochim. Acta 94 (2006) 835–844.

[54] R.T. Pabalan, D.R. Turner, Uranium(6^+) sorption onto montmorillonite: experimental and surface complexation modeling, Aquat. Geochem. 2 (1997) 203–226.

[55] A.K. Fouchard, R. Drot, E. Simoni, J.J. Ehrhardt, Use of spectroscopic techniques for uranium(VI)/montmorillonite interaction modeling, Environ. Sci. Technol. 38 (2004) 1399–1407.

[56] W. Dong, K.T. Tokunaga, J.A. Davis, J. Wan, Uranium(VI) adsorption and surface complexation modeling onto background sediments from the F-area Savannah River site, Environ. Sci. Technol. 46 (2012) 1565–1571.

[57] N. Marmier, J. Dumonceau, J. Chupeau, F. Fromage, Modeling of Yb (III) sorption on kaolinite by using single oxide surface complexation models, MRS Proc. 353 (1994) 1085.

[58] E. Tertre, A. Hofmann, G. Berger, Rare earth element sorption by basaltic rock: experimental data and modeling results using the "generalised composite approach", Geochim. Cosmochim. Acta 72 (2008) 1043–1056.

[59] J.A. Davis, J.A. Coston, D.B. Kent, C.C. Fuller, Application of the surface complexation concept to complex mineral assemblages, Environ. Sci. Technol. 32 (1998) 2820–2828.
[60] J.A. Davis, D.E. Meece, M. Kohler, G.P. Curtis, Approaches to surface complexation modeling of uranium (VI) adsorption on aquifer sediments, Geochim. Cosmochim. Acta 68 (2004) 3621–3641.
[61] P.M. Wang, A. Anderko, D.R. Turner, Thermodynamic modeling of the adsorption of radionuclides on selected minerals. II: anions, Ind. Eng. Chem. Res. 40 (2001) 4444–4455.
[62] Z.X. Hou, K.L. Shi, X.L. Wang, Y.L. Ye, Z.J. Guo, W.S. Wu, Investigation of Se(IV) sorption on Na-kaolinite: batch experiments and modeling, J. Radioanal. Nucl. Chem. 303 (2015) 25–31.
[63] Q. Du, Z.X. Sun, W. Forsling, H.X. Tang, Acid-base properties of aqueous illite surfaces, J. Colloid Interface Sci. 187 (1997) 221–231.
[64] M.H. Bradbury, B. Baeyens, A mechanistic description of Ni and Zn sorption on Na-montmorillonite. Part II: modelling, J. Contam. Hydrol. 27 (1997) 223–248.
[65] M.H. Bradbury, B. Baeyens, Sorption modelling on illite Part I: titration measurements and the sorption of Ni, Co, Eu and Sn, Geochim. Cosmochim. Acta 73 (2009) 990–1003.
[66] M.H. Bradbury, B. Baeyens, Sorption modeling on illite Part II: actinide sorption and linear free energy relationships, Geochim. Cosmochim. Acta 73 (2009) 1004–1013.
[67] M.H. Bradbury, B. Baeyens, Modelling the sorption of Mn(II), Co(II), Ni(II), Zn(II), Cd(II), Eu(III), Am(III), Sn(IV), Th(IV), Np(V) and U(VI) on montmorillonite: linear free energy relationships and estimates of surface binding constants for some selected heavy metals and actinides, Geochim. Cosmochim. Acta 69 (2005) 875–892.
[68] M.H. Bradbury, B. Baeyens, Modelling the sorption of Zn and Ni on Ca-montmorillonite, Geochim. Cosmochim. Acta 63 (1999) 325–336.
[69] M.H. Bradbury, B. Baeyens, Sorption of Eu on Na- and Ca-montmorillonites: experimental investigations and modelling with cation exchange and surface complexation, Geochim. Cosmochim. Acta 66 (2002) 2325–2334.
[70] A. Schnurr, R. Marsac, T. Rabung, J. Lützenkirchen, H. Geckeis, Sorption of Cm(III) and Eu(III) onto clay minerals under saline conditions: batch adsorption, laser-fluorescence spectroscopy and modeling, Geochim. Cosmochim. Acta 151 (2015) 192–202.
[71] D.A. Kulik, S.U. Aja, V.A. Sinitsyn, S.A. Wood, Acid–base surface chemistry and sorption of some lanthanides on K^+-saturated Marblehead illite: II. A multisite-surface complexation modeling, Geochim. Cosmochim. Acta 64 (2000) 195–213.
[72] M.H. Bradbury, B. Baeyens, Physico-Chemical Characterization Data and Sorption Measurements of Cs, Ni, Eu, Th, U, Cl, I and Se on MX-80 Bentonite, PSI Bericht 10–09, Paul Scherrer Institut, Villigen, Switzerland and Nagra NTB 09–08, Nagra, Wettingen, Switzerland, 2010.
[73] M.H. Bradbury, B. Baeyens, Predictive sorption modelling of Ni(II), Co(II), Eu(IIII), Th(IV) and U(VI) on MX-80 bentonite and opalinus clay: a "bottom-up" approach, Appl. Clay Sci. 52 (2011) 27–33.
[74] A. Benedicto, C. Degueldre, T. Missana, Gallium sorption on montmorillonite and illite colloids: experimental study and modelling by ionic exchange and surface complexation, Appl. Geochem. 40 (2014) 43–50.
[75] T. Missana, U. Alonso, M. García-Gutiérrez, Experimental study and modelling of selenite sorption onto illite and smectite clays, J. Colloid Interface Sci. 334 (2009) 132–138.
[76] C.J. Chisholm-Brause, J.M. Berg, K.M. Little, R.A. Matzner, D.E. Morris, Uranyl sorption by smectites: spectroscopic assessment of thermodynamic modeling, J. Colloid Interface Sci. 277 (2004) 366–382.
[77] P. Li, Z. Liu, F. Ma, Q.L. Shi, Z.J. Guo, W.S. Wu, Effects of pH, ionic strength and humic acid on the sorption of neptunium(V) to Na-bentonite, J. Mol. Liq. 206 (2015) 285–292.

[78] M.H. Bradbury, B. Baeyens, Modelling sorption data for the actinides Am(III), Np(V) and Pa(V) on montmorillonite, Radiochim. Acta 94 (2006) 619–625.

[79] C. Tournassat, R.M. Tinnacher, S. Grangeon, J.A. Davis, Modeling uranium(VI) adsorption onto montmorillonite under varying carbonate concentrations: a surface complexation model accounting for the spillover effect on surface potential, Geochim. Cosmochim. Acta 220 (2018) 291–308.

[80] Q.H. Fan, Y. Takahashi, Employment of the generalized adsorption model for the prediction of the solid-water distribution of radiocesium in the river-estuary-ocean system, Appl. Geochem. 79 (2017) 75–84.

[81] Q.H. Fan, N. Yamaguchi, M. Tanaka, H. Tsukada, Y. Takahashi, Relationship between the adsorption species of cesium and radiocesium interception potential in soils and minerals: an EXAFS study, J. Environ. Radioact. 138 (2014) 92–100.

[82] C. Poinssot, B. Baeyens, M.H. Bradbury, Experimental and modelling studies of caesium sorption on illite, Geochim. Cosmochim. Acta 63 (1999) 3217–3227.

[83] Y. Kim, R.J. Kirkpatrick, R.T. Cygan, ^{133}Cs NMR study of cesium on the surfaces of kaolinite and illite, Geochim. Cosmochim. Acta 60 (1996) 4059–4074.

[84] Y. Kim, R.J. Kirkpatrick, ^{23}Na and ^{133}Cs NMR study of cation adsorption on mineral surfaces: local environments, dynamics, and effects of mixed cations, Geochim. Cosmochim. Acta 61 (1997) 5199–5208.

[85] M.H. Bradbury, B. Baeyens, A generalized sorption model for the concentration dependent uptake of caesium by argillaceous rocks, J. Contam. Hydrol. 42 (2000) 141–163.

[86] S. Wick, B. Baeyens, M.M. Fernandes, A. Voegelin, Thallium adsorption onto illite, Environ. Sci. Technol. 52 (2017) 571–580.

[87] W.D. Horrocks, D.R. Sudnick, Lanthanide ion probes of structure in biology—laser-induced luminescence decay constants provide a direct measure of the number of metal-coordinated water molecules, J. Am. Chem. Soc. (2) (1979) 334–340.

[88] T. Kimura, G.R. Choppin, Luminescence study on determination of the hydration number of Cm(III), J. Alloys Compd. 213–214 (1994) 313–317.

[89] T. Kimura, Y. Kato, Luminescence study on the inner-sphere hydration number of lanthanide(III) ions in concentrated aqueous salt solutions in fluid and frozen states, J. Alloys Compd. 278 (1998) 92–97.

[90] T. Stumpf, A. Bauer, F. Coppin, T. Fanghänel, J.I. Kim, Inner-sphere, outer-sphere and ternary surface complexes: a TRLFS study of the sorption process of Eu(III) onto smectite and kaolinite, Radiochim. Acta 90 (2002) 345–349.

[91] D. Pan, Q. Fan, F. Fan, Y. Tang, Y. Zhang, W. Wu, Removal of uranium contaminant from aqueous solution by chitosan@attapulgite composite, Sep. Purif. Technol. 177 (2017) 86–93.

[92] T. Stumpf, A. Bauer, F. Coppin, J.I. Kim, Time-resolved laser fluorescence spectroscopy study of the sorption of Cm(III) onto smectite and kaolinite, Environ. Sci. Technol. 35 (2001) 3691–3694.

[93] T. Rabung, M.C. Pierret, A. Bauer, H. Geckeis, M.H. Bradbury, B. Baeyens, Sorption of Eu(III)/Cm(III) on Ca-montmorillonite and Na-illite. Part 1: batch sorption and time-resolved laser fluorescence spectroscopy experiments, Geochim. Cosmochim. Acta 69 (2005) 5393–5402.

[94] T. Jan, S. Thorsten, W. Erich, F. Thomas, Uptake of Cm(III) and Eu(III) by calcium silicate hydrates: a solution chemistry and time-resolved laser fluorescence spectroscopy study, Environ. Sci. Technol. 37 (2003) 3568–3573.

[95] T. Stumph, M. Marques Fernandes, C. Walther, K. Dardenne, T. Fanghänel, Structural characterization of am incorporated into calcite: a TRLFS and EXAFS study, J. Colloid Interface Sci. 302 (2006) 240–245.

[96] T. Stumph, H. Curtius, C. Walther, K. Dardenne, K. Ufer, T. Fanghänel, Incorporation of Eu(III) into hydrotalcite: a TRLFS and EXAFS study, Environ. Sci. Technol. 41 (2007) 3186–3191.

[97] M.M. Fernandes, T. Stumpf, T. Rabung, D. Bosbach, T. Fanghänel, Incorporation of trivalent actinides into calcite: a time resolved laser fluorescence spectroscopy (TRLFS) study, Geochim. Cosmochim. Acta 72 (2008) 464–474.

[98] Z. Wang, J.M. Zachara, W. Yantasee, P.L. Gassman, C. Liu, A.G. Joly, Cryogenic laser induced fluorescence characterization of U(VI) in Hanford vadose zone pore waters, Environ. Sci. Technol. 38 (2004) 5591–5597.

[99] Z. Wang, J.M. Zachara, P.L. Gassman, C. Liu, O. Qafoku, W. Yantasee, J.G. Catalano, Fluorescence spectroscopy of U(VI)-silicates and U(VI)-contaminated Hanford sediment, Geochim. Cosmochim. Acta 69 (2005) 1391–1403.

[100] T. Arnold, S. Utsunomiya, G. Geipel, R.C. Ewing, N. Baumann, V. Brendler, Adsorbed U(VI) surface species on muscovite identified by laser fluorescence spectroscopy and transmission electron microscopy, Environ. Sci. Technol. 40 (2006) 4646–4652.

[101] M.D. Nero, C. Galindo, R. Barillon, B. Madé, TRLFS evidence for precipitation of uranyl phosphate on the surface of alumina: environmental implications, Environ. Sci. Technol. 45 (2011) 3982–3988.

[102] Z. Wang, J.M. Zachara, J. Shang, C. Jeon, J. Liu, C. Liu, Investigation of U(VI) adsorption in quartz–chlorite mineral mixtures, Environ. Sci. Technol. 48 (2014) 7766–7773.

[103] H.-S. Chang, G.V. Korshin, Z. Wang, J.M. Zachara, Adsorption of uranyl on gibbsite: a time-resolved laser-induced fluorescence spectroscopy study, Environ. Sci. Technol. 40 (2006) 1244–1249.

[104] U. Gabriel, L. Charlet, C.W. Schläpfer, J.C. Vial, A. Brachmann, G. Geipel, Uranyl surface speciation on silica particles studied by time-resolved laser-induced fluorescence spectroscopy, J. Colloid Interf. Sci. 239 (2001) 358–368.

[105] T. Saito, H. Sao, K. Ishida, N. Aoyagi, T. Kimura, S. Nagasaki, S. Tanaka, Application of parallel factor analysis for time-resolved laser fluorescence spectroscopy: implication for metal speciation study, Environ. Sci. Technol. 44 (2010) 5055–5060.

[106] C. Ding, W. Cheng, Y. Sun, X. Wang, Effects of Bacillus subtilis on the reduction of U(VI) by nano-Fe^0, Geochim. Cosmochim. Acta 165 (2015) 86–107.

[107] Y. Sun, C. Ding, W. Cheng, X. Wang, Simultaneous adsorption and reduction of U(VI) on reduced graphene oxide-supported nanoscale zerovalent iron, J. Hazard. Mater. 280 (2014) 399–408.

[108] K. Fukushi, Y. Suzuki, J. Kawano, T. Ohno, M. Ogawa, T. Yaji, Y. Takahashi, Speciation of magnesium in monohydrocalcite: XANES, *ab initio* and geochemical modeling, Geochim. Cosmochim. Acta 213 (2017) 457–474.

[109] L. Stetten, A. Mangeret, J. Brest, M. Seder-Colomina, P. Le Pape, M. Ikogou, N. Zeyen, A. Thouvenot, A. Julien, G. Alcalde, J.L. Reyss, B. Bombled, C. Rabouille, L. Olivi, O. Proux, C. Cazala, G. Morin, Geochemical control on the reduction of U(VI) to mononuclear U(IV) species in lacustrine sediments, Geochim. Cosmochim. Acta 222 (2018) 171–186.

[110] M.A. Denecke, Actinide speciation using X-ray absorption fine structure spectroscopy, Coord. Chem. Rev. 250 (2006) 730–754.

[111] D.G. Strawn, D.L. Sparks, The use of XAFS to distinguish between inner- and outer-sphere lead adsorption complexes on montmorillonite, J. Colloid Interf. Sci. 216 (1999) 257–269.

[112] S. Yang, G. Sheng, G. Montavon, Z. Guo, X. Tan, B. Grambow, X. Wang, Investigation of Eu(III) immobilization on γ-Al_2O_3 surfaces by combining batch technique and EXAFS analyses: role of contact time and humic acid, Geochim. Cosmochim. Acta 121 (2013) 84–104.

Chapter 2

Interactions between radionuclides and the oxide-water interfaces in the environment

Huiyang Mei, Xiaoli Tan and Changlun Chen
Key Laboratory of Photovoltaic and Energy Conservation Materials, Institute of Plasma Physics, Chinese Academy of Sciences, Hefei, People's Republic of China

2.1 Introduction

Metal oxides (e.g., iron oxides, alumina oxides) are important natural envir onmental media for the retention of radionuclides (RNs). The interaction between RNs and oxide surfaces, which is closely related to adsorption and precipitation processes, deserves much attention. Immobilization of the RNs depends on not only the chemical forms of the oxide minerals, but also on their surface properties. RNs could be adsorbed on oxide surfaces through different interaction mechanisms, which is greatly relative to the environmental conditions. To better understand and predict the transport and fate of RNs in natural and engineered systems, it is imperative to investigate the molecular interaction mechanism between RNs and oxides. In addition, the selection of the partition coefficient of the theoretical adsorption models also depends on the understanding of the speciation and coordination geometry of RNs on oxide surfaces. Moreover, new efficient treatment technologies for high-level nuclear waste and remediation strategies for sediments, soils, and groundwater contaminated by radioactive waste may emerge during the investigation of interactions between RNs and oxides [1].

This chapter mainly concentrates on the studies involving RNs or their analogues that occupy a great proportion of radioactive waste. The RNs include actinides such as U and Pu isotopes, ^{237}Np, ^{241}Am, and ^{244}Cm; fission products such as ^{99}Tc and ^{79}Se; and isotopes formed in the cladding materials (for instance, ^{60}Co and ^{63}Ni). The use of radioisotope analogues in experiments could avoid the danger of high-level radioactivity during working,

Interface Science and Technology, Vol. 29. https://doi.org/10.1016/B978-0-08-102727-1.00002-9

making the characterization and measurement easier. In addition, the difficulties of limiting the valence state of redox-active RNs could be overcome. Thus, Th(IV) and Eu(III) are employed as analogues of tetravalent and trivalent actinides, respectively, due to their exhibition of a sole oxidation state and low specific activity [2].

Mineral oxide surfaces in contact with radioactive wastewater or the groundwater can influence the migration of RNs through adsorption, redox, transformation, and so on [3]. Thus, the investigation of interactions between RNs and oxides is important and deserves much attention. Trace levels of RNs could be captured and concentrated from aqueous systems by conventional oxides due to their high surface affinity, which endows them with some developmental prospects in exploring resource recovery and environmental remediation.

Fe (hydr)oxides (FHOs) are vital geochemical media influencing the migration and fate of RNs and nuclear wastes in the natural environment. For instance, the host of Atlantic Coastal Plain sediments contains a large mount of hematite (α-Fe_2O_3) and goethite (α-FeOOH) [4, 5] and varying FHOs (such as goethite and ferrihydrite ($Fe_2O_3{\cdot}0.5H_2O$)) are present in the Hanford fine sediments [6, 7]. SiO_2 and Al_2O_3, as representative natural solid particles, can be used to model the layers of silicon oxide or aluminum oxide in clay minerals, which are potential and promising host media for long-term nuclear waste disposal. The adsorption of RNs on oxide minerals is mainly through adsorption reactions with the surface hydroxyl groups. It has been shown that hydroxyl groups on the (hydr)oxide surfaces (e.g., ferrinol (Fe-OH) and aluminol (Al-OH) groups) exert high affinity for RNs [6]. In addition, other reaction processes such as colloid formation, ion exchange, redox processes, and dissolution-precipitation also obviously influence the fate of RNs [3, 8]. In this chapter, advances in the understanding of interactions of RNs on oxide surfaces at the molecular level are presented and the potential challenges are also discussed.

2.2 Surface properties of oxides

In this section, the main focus is the introduction and discussion of the surface properties of several natural oxides. The bulk structures of different oxides and bulk-terminated surfaces with the lowest energy are first presented. We then introduce the properties of several hydroxylated oxide surfaces as well as the metal (oxy)hydroxides. Finally, the dissolution behavior of oxides is explored and elucidated.

2.2.1 Bulk structures of several oxides

Many common oxides are ionic compounds. The oxide crystal structure is usually described by close-packing of oxygen anions, with interstitial sites

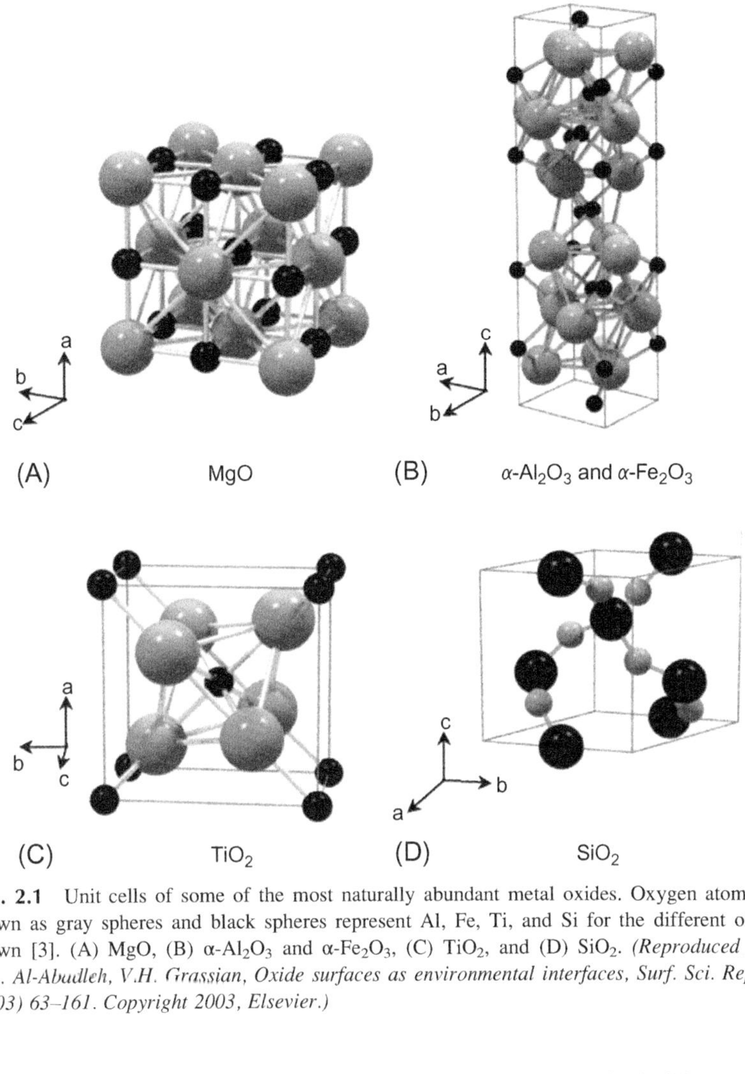

FIG. 2.1 Unit cells of some of the most naturally abundant metal oxides. Oxygen atoms are shown as gray spheres and black spheres represent Al, Fe, Ti, and Si for the different oxides shown [3]. (A) MgO, (B) α-Al_2O_3 and α-Fe_2O_3, (C) TiO_2, and (D) SiO_2. *(Reproduced from H.A. Al-Abadleh, V.H. Grassian, Oxide surfaces as environmental interfaces, Surf. Sci. Rep. 52 (2003) 63–161. Copyright 2003, Elsevier.)*

being occupied by metal cations. Both octahedral and tetrahedral interstitial sites are present in close-packed structures. Fig. 2.1 presents the unit cells of some oxides. Magnesium oxide (MgO) crystallizes in the simplest rock salt cubic structure ($a=8.396$ Å), where all octahedral sites are occupied by Mg [9]. The α-Fe_2O_3 ($a=5.034$ Å and $c=13.75$ Å) exists in the rhombohedral corundum crystalline structure, where Fe^{3+} occupies 2/3 of the octahedral interstices [10]. The α- and γ-aluminum oxide (Al_2O_3) are the most stable phases under ambient conditions. α-Al_2O_3 ($a=4.823$ Å and $c=13.11$ Å) has a similar structure to α-Fe_2O_3 [11]. γ-Al_2O_3 belongs to spinel structure ($a=7.9$ Å) (Fig. 2.2A) [13, 15]. Ferromagnetic Fe_3O_4 crystallizes in the inverse cubic

(A) (B)

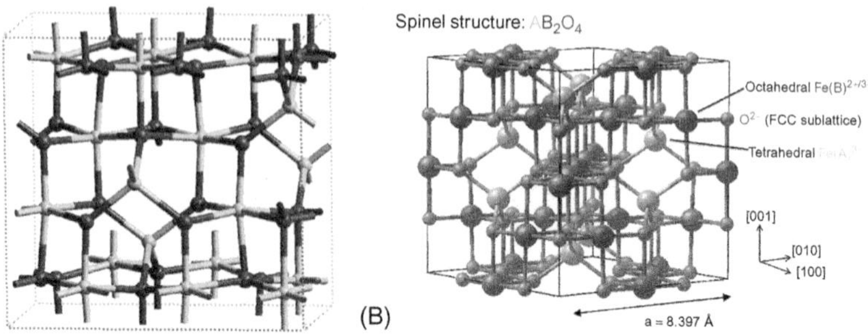

FIG. 2.2 (A) Unit cell of the γ-alumina bulk according to Ref. [12]. *Black* balls, oxygen atoms. *White* balls, aluminum atoms [13]. (B) The inverse spinel structure of Fe_3O_4 is based on an fcc arrangement of O^{2-} cations in which Fe^{3+} cations occupy 1/2 of the tetrahedral interstices, and a 50:50 mix of Fe^{3+} and Fe^{2+} cations occupy 1/8 of the octahedral interstices [14]. *((A) Reproduced from M. Digne, P. Sautet, P. Raybaud, P. Euzen, H. Toulhoat, Use of DFT to achieve a rational understanding of acid–basic properties of γ-alumina surfaces, J. Catal. 226 (2004) 54–68. Copyright 2004, Elsevier; (B) Reproduced from G.S. Parkinson, Iron oxide surfaces, Surf. Sci. Rep. 71 (2016) 272–365. Copyright 2016, Elsevier.)*

spinel structure ($a=8.396$ Å), with Fe^{2+} in octahedral interstices and Fe^{3+} in octahedral/tetrahedral sites (Fig. 2.2B) [16].

Titanium dioxide (α-TiO_2) crystalizes mainly in three phases: tetragonal rutile (α-TiO_2), anatase (β-TiO_2), and rhombohedral brookite (γ-TiO_2) [17]. Their main structural difference depends on the O^{2-} close-packing mode (hexagonal for rutile, cubic for anatase, and mixed for brookite). The tetragonal structures of anatase and rutile are illustrated in Fig. 2.3. The most compact and stable structure among those is rutile under ambient conditions. Silicon dioxide (SiO_2) is a covalent compound and exists mainly in three forms: quartz, tridymite, and cristobalite, where Si atoms are all tetrahedrally coordinated with O in a hexagonal structure. Quartz (hexagonal, $a=4.584$ Å and $c=2.953$ Å) is the densest tetrahedra-packing structure among the three polymorphs. Next, we will present the surface structures of these common oxide minerals.

2.2.2 Bulk-terminated oxide surfaces

The bulk oxide crystalline structure is usually terminated by the lowest energy surface, which is the typical crystal growth direction of natural oxide minerals. The most thermodynamically stable surface has the minimum surface Gibbs energy [19]. When a particular crystallographic plane is cleaved from a crystal, the configuration of surface ions will undergo surface reconstruction or relaxation to minimize the free energy, and thus more or less differs from that of the bulk structure. The too-high energy barrier for reconstruction may lead to the formation of metastable surface structures. The stabilities of surface structures are closely related to the degree of coordination unsaturation and surface polarity.

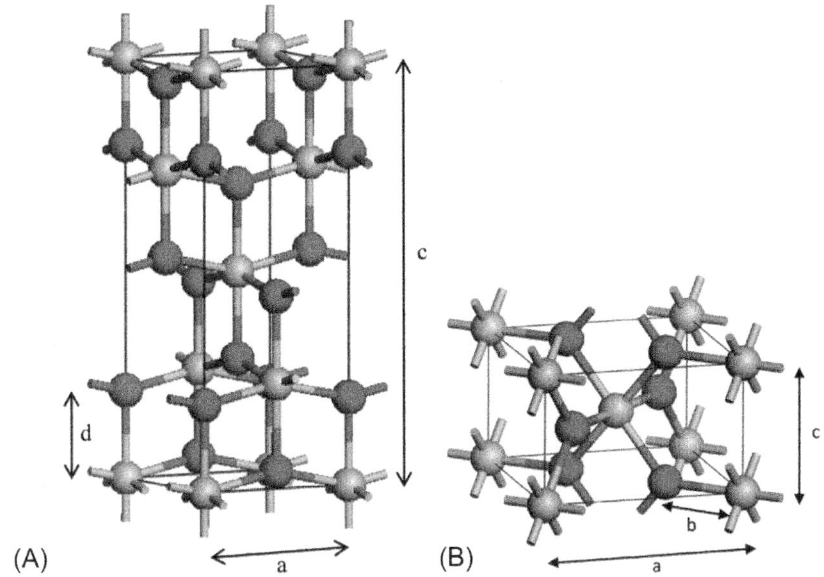

FIG. 2.3 Tetragonal structure of anatase (A) and rutile (B). *Gray* and *black* balls represent, respectively, Ti and O atoms [18]. *(Reproduced from K. Bourikas, C. Kordulis, A. Lycourghiotis, Titanium dioxide (anatase and rutile): surface chemistry, liquid–solid interface chemistry, and scientific synthesis of supported catalysts, Chem. Rev. 114 (2014) 9754–9823. Copyright 2014, American Chemical Society.)*

The most stable surface of MgO is the (1 0 0) surface (Fig. 2.4) [20]. By evaporating Mg in the oxygen, the well-defined MgO(1 0 0) films could grow on substrates [21]. The MgO(1 0 0) face could serve as a support for metal thin film growth due to the strong inductive effect on epitaxial growth [22]. The α-Al_2O_3 and α-Fe_2O_3 exhibit many naturally grown faces, of which the (0 0 0 1) and (1 0 −1 2) surfaces are the two most stable and prevalent growth planes [23, 24]. The neutral bulk terminated α-Al_2O_3 and α-Fe_2O_3(0 0 0 1) surface is represented as the —O_3—M—M—O_3—M—M— stacking sequence. There are three terminations that could be cleaved from the (0 0 0 1) face: M—O_3—M—R, M—M—O_3—R, and O_3—M—M—R, where M denotes the metal and R denotes the subsequent bulk structure [25].

The neutral bulk terminated α-Al_2O_3 and α-Fe_2O_3(1 0 −1 2) surface is represented as the —O_2—M_2—O_2—M_2—O_2— stacking sequence (Fig. 2.5) [26]. After interaction with liquid water, there will be two possible different surface terminations (Fig. 2.6). The full-layer termination is formed by adding an extra O layer to complete the coordination of the upper M site. The upper M site vacant is present in the other termination (half-layer termination). Singly coordinated (>MO) and triply coordinated (>M_3O) groups exist on both terminations while the doubly coordinated ones (>M_2O) only exist in the half-layer termination. It is concluded that the α-Al_2O_3 and

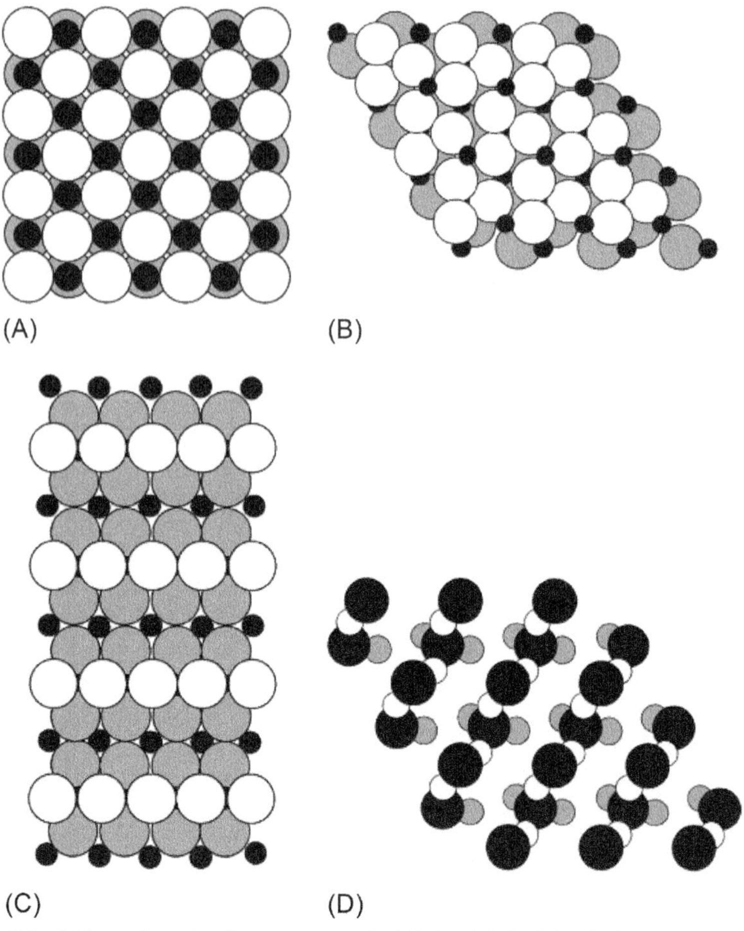

FIG. 2.4 Bulk-terminated surface structures of: (A) MgO(1 0 0), (B) α-Al_2O_3 and α-Fe_2O_3(0 0 0 1), (C) TiO_2(1 1 0), and (D) SiO_2(0 0 0 1). *White* represents the first layer of oxygen atoms whereas the *gray* color represents an underlying layer of oxygen atoms. *Black* spheres represent Al, Fe, Ti, and Si for the different oxides shown [3]. *(Reproduced from H.A. Al-Abadleh, V.H. Grassian, Oxide surfaces as environmental interfaces, Surf. Sci. Rep. 52 (2003) 63–161. Copyright 2003, Elsevier.)*

α-Fe_2O_3(1 0 −1 2) surface may contain two terminations, with the half-layer one being dominant [27].

There are two kinds of titanium (Ti) (five- and sixfold) and oxygen atoms (O) (two- and threefold) on the rutile (110) surface. The surface is represented as the O—Ti_2O_2—O stacking sequence and is the most stable of the crystal faces. The surface O atoms move laterally, which leads to some relaxation [28]. Similar to the rutile (110) surface, the (100) surface also contains two kinds of titanium (Ti) (five- and sixfold) and oxygen atoms (O) (two- and threefold). The surface is represented as the O—Ti—O—O—Ti—O stacking sequence. The surface Ti and O atoms move in opposite directions, which produces a

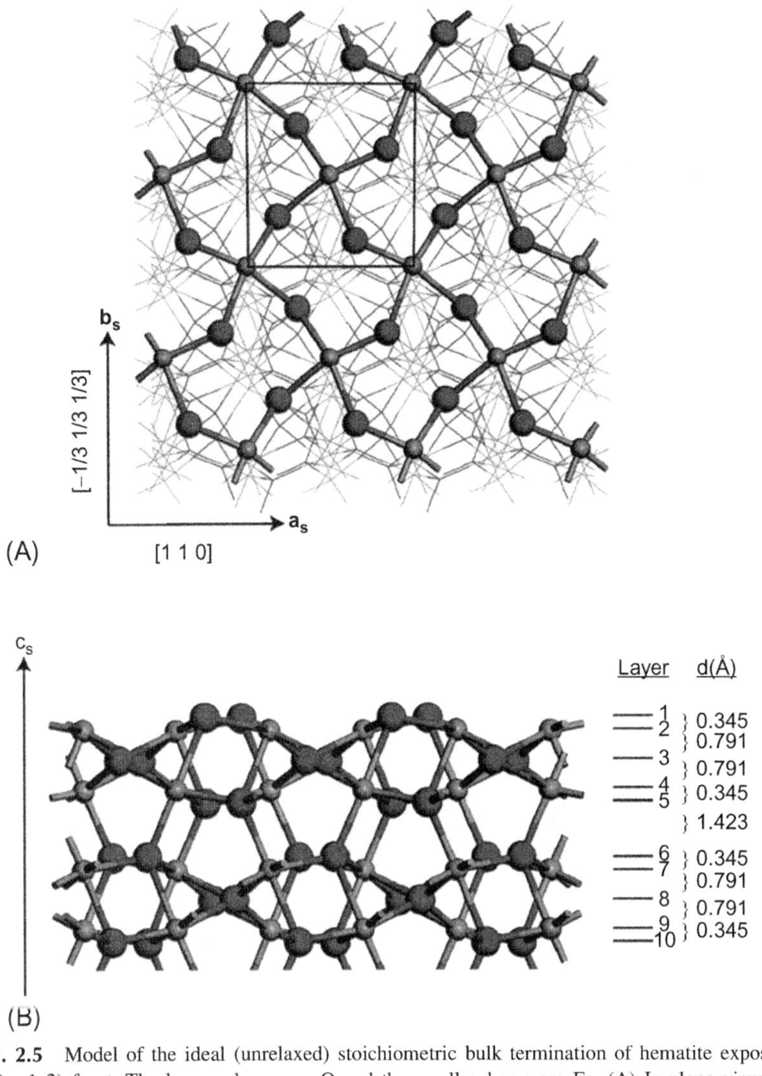

FIG. 2.5 Model of the ideal (unrelaxed) stoichiometric bulk termination of hematite exposed (1 0 −1 2) facet. The large spheres are O and the small spheres are Fe. (A) In-plane view of the (1 0 −1 2) surface showing the surface net. The real-space basis vectors in the surface indexing are shown, along with their corresponding indices in the bulk indexing. (B) Layer stacking sequence and layer spacings along the c_s axis [26]. *(Reproduced from C.S. Lo, K.S. Tanwar, A.M. Chaka, T.P. Trainor, Density functional theory study of the clean and hydrated hematite (1–102) surfaces, Phys. Rev. B 75 (2007) 075425. Copyright 2007, American Physical Society.)*

large relaxation along [0–10] [28, 29]. Unlike the above two surfaces, the Ti atoms on the rutile (001) surface are all fourfold coordinated. In addition, the O atoms on the outermost layer are twofold coordinated. Thus, there are more broken bonds on the rutile (001) surface, leading to a high surface energy [17].

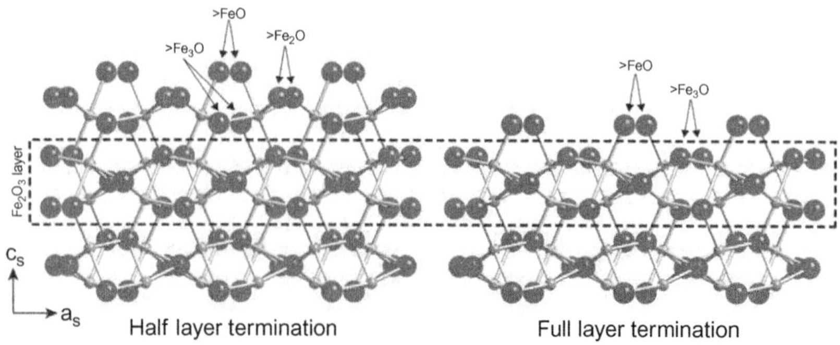

FIG. 2.6 Structural models of two possible terminations of the α-Fe_2O_3(1 0 −1 2) surface in aqueous solution. The small spheres are Fe and the large spheres are O. Also shown are the sites of important surface functional groups (surface oxygens undercoordinated with respect to bulk oxygens) [27]. *(Reproduced from J.G. Catalano, Z. Zhang, C. Park, P. Fenter, M.J. Bedzyk, Bridging arsenate surface complexes on the hematite (012) surface, Geochim. Cosmochim. Acta 71 (2007) 1883–1897. Copyright 2007, Elsevier.)*

As for the anatase (101) surface, the surface O atoms are all twofold and the surface Ti atoms are fivefold. The surface is represented as the O—Ti—O—O—Ti—O stacking sequence. The relaxation of the surface atoms produces tight Ti—O bonds. The fivefold Ti atoms as well as two- and threefold O atoms are present in the outermost plane of the flat anatase (100) surface. The unit cell of the anatase (100) surface is consisted of six atoms (one Ti_2O_4 unit). However, the outward relaxation of O_3 and O_4 and Ti atoms (fivefold Ti inward and sixfold Ti outward) turns the surface to be not flat [28].

The polar (0 0 0 1) surface of SiO_2, which has been found as the lowest energy surface, is presented in Fig. 2.4D. The surface is also predicted to be the most stable by theoretical methods [30, 31]. All the surface silicon (Si) atoms are fourfold coordinated for the relaxed structure. Density functional theory (DFT) results demonstrate that the cleaved unhydrated α-quartz (0 0 0 1) surface is reconstructed, where the Si and O atoms form three- and six-membered rings [32].

2.2.3 Hydroxylated oxide surface

Oxide minerals will be inevitably exposed to aqueous solutions in natural geochemical systems. The interaction between water and the oxides will produce hydrated and hydroxylated surfaces, the properties and reactivities of which are quite different from the metal-terminated surfaces [33, 34]. It has been reported that the hydroxylated oxide surfaces are more stable under aquatic environments. The surface atoms that are unsaturated in coordination may be completed during the hydroxylation process [3].

A fully hydroxylated surface structure ($(HO)_3$—Fe—H_3O_3—Fe—R) of the Fe-terminated surface will be formed after exposure to increasing water

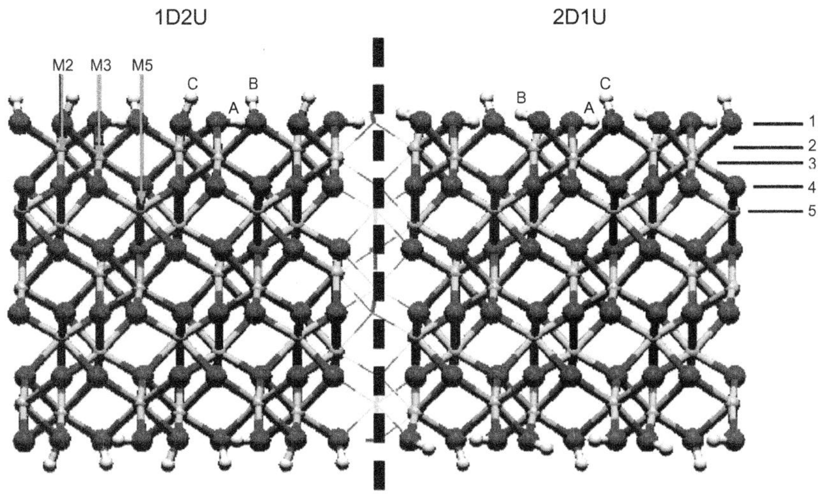

FIG. 2.7 Side views of the oxygen terminated (0001) α-M_2O_3 surfaces with the "one down two up" (1D2U, *left*) and "two down one up" (2D1U, *right*) H-bond networks. The cations that define the centers of the M_2, M_3, and M_5 adsorption sites, and the numeric atomic layers, (both labeling schemes invariant to H-bonding), are indicated on the 1D2U by vertical lines and 2D1U panels by horizontal lines, respectively. Oxygen atoms are shown as large spheres, cations (Al or Fe) are shown by smaller *gray* spheres, and H^+ located at both ends are shown in *white* [11]. *(Reproduced from S.E. Mason, C.R. Iceman, K.S. Tanwar, T.P. Trainor, A.M. Chaka, Pb(II) adsorption on isostructural hydrated alumina and hematite (0001) surfaces: a DFT study, J. Phys. Chem. C 113 (2009) 2159–2170. Copyright 2009, American Chemical Society.)*

partial pressure, but not for the Al-terminated surface [35]. Previous studies support the removal of the surface Al cation and thus the formation of the hydroxylated termination ($(HO)_3$—Al—Al—R) (Fig. 2.7) [33, 36]. A proportion of surface Fe atoms could also be removed for α-Fe_2O_3(0 0 0 1) when exposed to liquid water. Comparison of experiments with the theoretical calculation results implied the dominant termination of the O-terminated structure ($(HO)_3$—Fe—Fe—R) with the remaining structure of Fe-termination ($(HO)_3$—Fe—H_3O_3—Fe—R) [35]. Therefore, the hydroxylated O-termination is the dominant termination for the α-Fe_2O_3(0 0 0 1) surface while the only one for the α-Al_2O_3(0 0 0 1) surface under ambient conditions.

The predicted lowest-energy surface, that is, the most stable surface of α-Fe_2O_3(1 0 −1 2), is in a configuration of the C4 model ($(H_2O)_2$—X—$(HO)_2$—Fe_2—O_2—R), where X represents an absence of the top Fe layer [26]. There are three kinds of O-containing functional groups existing in the hydroxylated surface stoichiometry (i.e., Fe_3—O, Fe_2—OH, and Fe—OH_2). A similar termination is assigned to the hydrated α-Al_2O_3(1 0 −1 2) surface [37].

As for the surface of γ-Al_2O_3, the clean Al-terminated (1 1 0) and (1 0 0) surfaces have been shown to easily produce hydroxyl groups, which are

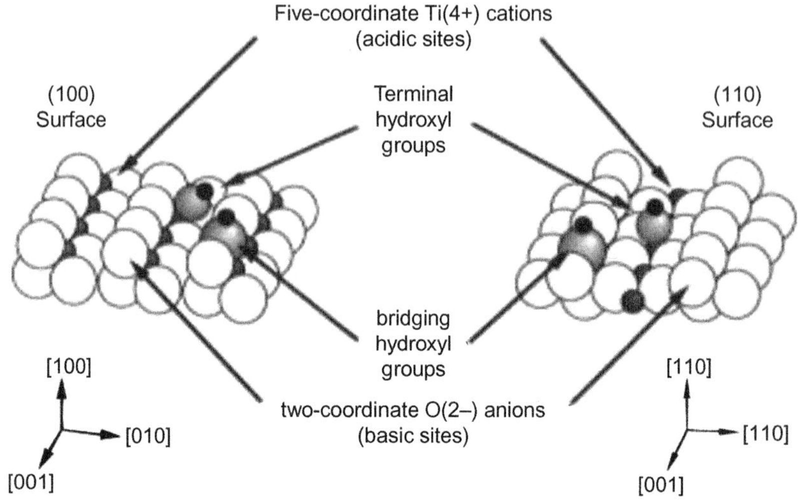

FIG. 2.8 Schematic representations of terminal and bridging hydroxyl groups on the TiO_2(1 1 0) and (1 0 0) surfaces [41]. *(Reproduced from M.A. Henderson, Structural sensitivity in the dissociation of water on TiO_2 single-crystal surfaces, Langmuir 12 (1996) 5093–5098. Copyright 1996, American Chemical Society.)*

created by the dissociation of water [13, 38, 39]. It has been reported that the surface hydroxyl group density of the (1 1 0) surface is higher than that of the (1 0 0) surface [13, 40].

Different hydroxyl groups (e.g., bridging and terminal OH) may form on the TiO_2 (1 0 0) and (1 1 0) surfaces due to the dissociation of water (Fig. 2.8) [41]. The OH group dissociated from a water molecule is coordinated by the terminal Ti atom to form a terminal OH. The bridging OH is formed through the other generated H binding to the bridging O [42]. There are also five-coordinated Ti^{4+} cations (acidic sites) and two-coordinated O^{2-} anions (basic sites) on the two surfaces [43].

The dissociation of water and subsequent adsorption on the SiO_2 surface leads to the formation of silanol (Si—OH) groups (~5 groups/nm^2), which lowers the surface energy significantly [43]. In aquatic environments, the siloxane sites (Si—O—Si) on the silica surfaces will convert to silanol sites (Si—OH) by reacting quickly with water molecules, which changes the hydrophobic surface to a hydrophilic one [3, 32]. The fully hydroxylated α-SiO_2(0 0 1) surface exhibits a herringbone structure, as indicated by previous calculations [30, 44].

2.2.4 Metal (oxy)hydroxides

As a commonly occurring natural iron oxide mineral, α-FeOOH is thermodynamically stable. The Fe octahedra with double chains ($FeO_3(OH)_3$) running

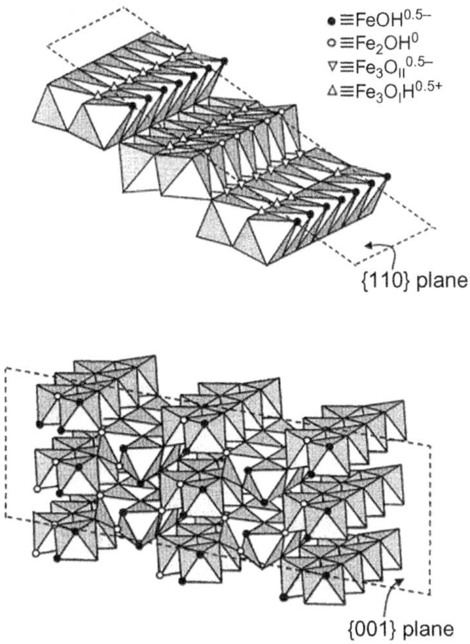

FIG. 2.9 Schematic representation of the {110} and {001} plane of the goethite surface, along with the distribution and density of singly ($\equiv FeOH^{0.5-}$), doubly ($\equiv Fe_2OH^0$), and triply ($\equiv Fe_3O_IH^{0.5+}$ and $\equiv Fe_3O_{II}^{0.5-}$) coordinated surface oxygens with respect to underlying Fe(III) atoms [48]. *(Reproduced from J. Lützenkirchen, J.F. Boily, L. Lövgren, S. Sjöberg, Limitations of the potentiometric titration technique in determining the proton active site density of goethite surfaces, Geochim. Cosmochim. Acta 66 (2002) 3389–3396. Copyright 2002, Elsevier.)*

along the [010] direction is formed by edge sharing (E_2) (space group Pnma). Adjacent corner-sharing (CS) double chains are linked, with one being displaced by b/2 relative to its neighbors [45]. The three Fe-Fe distances are formed with two edge sharing and one CS [46]. The dominant (1 1 0) face leads to an acicular morphology [47]. To interpret RN sorption data on goethite, various reactive sites on the α-FeOOH surface have been noticed and investigated over the years (Fig. 2.9) [48–50]. Different hydroxyl groups present on the surface have been confirmed by infrared spectroscopy data [51, 52].

Lepidocrocite (γ-FeOOH) is another commonly occurring iron mineral (space group of $Cmc2_1$) (Fig. 2.10). It is a less-stable phase compared with goethite [54]. Lepidocrocite exists in nature and has been derived from soils [55, 56] and deep-sea sediments [57]. It can be prepared easily by rapid oxidation and hydrolysis of Fe(II) solutions in laboratory experiments [58, 59]. The dominant exposed face is the (0 1 0) surface.

Similar to Al_2O_3, $Al(OH)_3$ is amphoteric in nature. $Al(OH)_3$ commonly exists with a layered structure ($Al(OH)_6$ octahedra layer), the surficial hydroxyl groups of which are stabilized by hydrogen bond [60]. The different stacking

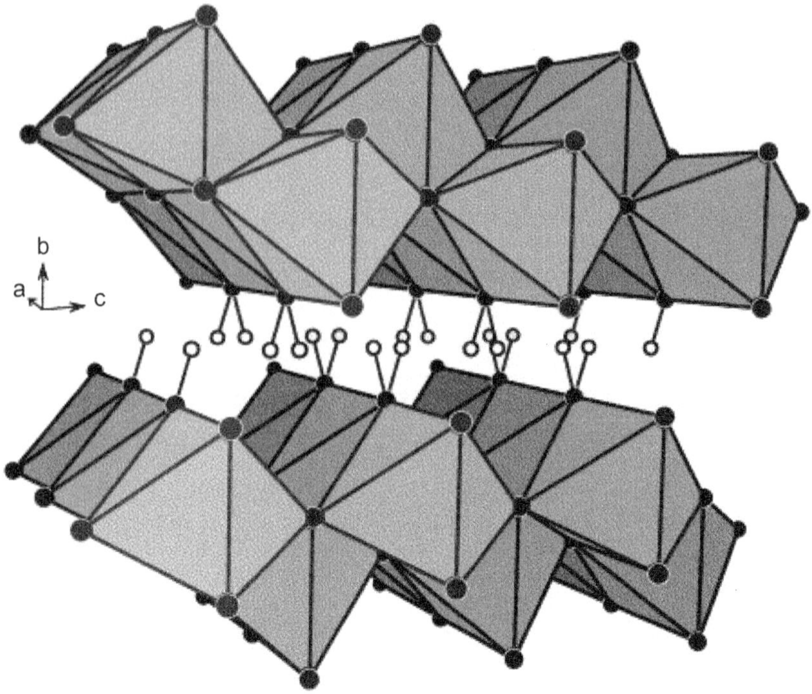

FIG. 2.10 Octahedral layers in the structure of lepidocrocite, bound only by hydrogen bonds. Hydrogen atoms are shown by small circles. The occupancy of these positions is 0.5 [53]. *(Reproduced from J. Majzlan, L. Mazeina, A. Navrotsky, Enthalpy of water adsorption and surface enthalpy of lepidocrocite (γ-FeOOH), Geochim. Cosmochim. Acta 71 (2007) 615–623. Copyright 2006, Elsevier.)*

sequence of the aluminum octahedra layers accounts for the polymorphs of $Al(OH)_3$. Both bayerite and gibbsite have the same space group of P_{21}/n, but with different stacking orders. The bayerite structure consists of these repeated double-layer stacking structures (ABABABAB…) while for gibbsite, these double layers are stacked by reflection of the previous one (ABBAABBA…). It should be noted that only 2/3 of the octahedral geometrical sites are occupied by Al ions in each double layer [61].

2.2.5 Surface charges and zeta potentials

Metal oxides are generally electrically charged in aqueous suspension, which can be shown by electrophoresis experiments. The charged surface is mainly attributed to two mechanisms: dissociation of amphoteric surface hydroxy groups (MOH) and adsorption of M—OH complexes originated from the dissolution of the solid. The pH where the zero net surface charge is zero, (i.e., point of zero charge (PZC) or the isoelectric point (IEP)), could be qualitatively explained by both mechanisms. The IEP corresponds to the PZC if the adsorbed ions are only H^+ and OH^-.

The protolysis behavior of the amphoteric hydroxyl sites on the oxide surfaces can be described by the mass law (M-L) equations [62]:

$$\equiv SOH + H^+ \leftrightarrow \equiv SOH_2^+ \quad K_{app}^+ = \frac{[\equiv SOH_2^+]}{[\equiv SOH] \cdot [H^+]} \tag{2.1}$$

$$\equiv SOH \leftrightarrow \equiv SO^- + H^+ \quad K_{app}^- = \frac{[\equiv SO^-] \cdot [H^+]}{[\equiv SOH]} \tag{2.2}$$

where $\equiv$S is metal oxide. $\equiv$SOH, $\equiv SOH_2^+$, and $\equiv SO^-$ represent uncharged and charged hydroxyl groups on the oxide surface, [ ] denotes concentration, and K_{app}^+ and K_{app}^- are apparent acidity constants.

The surface charge is generally measured through acid-base titration of an oxide suspension in solutions with variable ionic strengths [63]. The site density could also be deduced from the titration data. Surface charge and site density are classically deduced from acid-base titration curves. The transfer of protons at the surface is responsible for the change of surface charge as a function of pH. Surface is negatively charged when pH > PZC. Due to the different surface properties of various oxides, the PZC ranges from 2 to 9.6 [48]. The surface charge may change through adsorption when carbonate or other ions are present [64]. In CO_2-free conditions, the highest PZC value for goethite was obtained. In addition, impurities of the metal oxide may also affect the site density and surface charge [65]. The PZC/IEP of several metal oxides from the recent literature is summarized by Marek Kosmulski and listed in Table 2.1 [66].

2.2.6 Dissolution of the oxides

The dissolution of oxides could be explained by the surface coordination model, where surface functional groups are coordinated with the surface metal atoms [67]. Available thermodynamic data have shown the low solubility of iron oxides ($\log K_{sp} \sim 0$–3.5) [68, 69]. There are three principal reactions for how Fe may be released from solid Fe oxides: protonation, reduction, and complexation. The respective reactions are expressed as follows:

$$FeOOH + H^+ \rightarrow Fe(OH)_2^+ \tag{2.3}$$

$$FeOOH + e^- + 3H^+ \rightarrow Fe^{2+} + 2H_2O \tag{2.4}$$

$$FeOOH + nL^- + 3H^+ \rightarrow [FeL_n]^{3-n} + 2H_2O \ (L = \text{ligand}) \tag{2.5}$$

Connor et al. [70] employed surface titration by the internal reflectance spectroscopy technique to investigate the coordinated species changes on the TiO_2 surface as a function of pH. Both the bending ($\delta(H_2O)$) and stretch ($\nu(OH)$) modes of H_2O are sensitive to the change of pH. At acidic conditions,

TABLE 2.1 pH_{pzc} range of several typical metal (hydr)oxides

Metal (hydr)oxides type	pH_{pzc} range
α-Alumina	5.5–9.5
γ-Alumina	7.4–9.5
Magnetite	3.2–8.0
Hematite	4.2–9.5
Goethite	7.3–10.0
Lepidocrocite	5.6–7.4
Anatase	4.5–7.2
Rutile	2.7–5.6
Boehmite	8.4
Gibbsite	9.0–11.2
Silica	<4.5

Data from M. Kosmulski, The pH dependent surface charging and points of zero charge. VI. Update, J. Colloid Interface Sci. 426 (2014) 209–212.

the complex species (Ti—OH^+—Ti and Ti—OH_2^+) are present at ~3000 cm^{-1}. At pH >6.7, the species of Ti—OH_2 and Ti—OH exist at ~3200 and 3400 cm^{-1}.

Recent studies using X-ray photoelectron spectroscopy (XPS) disclosed the dissolution behavior of SiO_2 [71]. The analysis results showed the O 1s peak position changes with pH, which are closely related to the surface SiOH groups and the oxygen in the bulk structure. As pH increased from 0 to 10, the O 1s peak shifted to the lower energy. After comparing these binding energies to those for standard peaks, it was deduced that the —$SiOH_2^+$ species dominates at low pH while the —SiO^- dominates at the alkaline conditions [3]. The dissolution of the quartz surface in hydrated conditions was studied by Du et al [72]. An OH group is added to the Si atom and a proton to the neighboring O atom after the addition of a water molecule. In the end, the $Si(OH)_4$ species is dissociated from the quartz surface, causing the substitution of four protons for the dissolved silicon atom [72].

2.3 Sorption of RNs on oxide surface

Sorption is, in a broad sense, a mass-transfer process between the fluid and solid phases, resulting in the change of mass of a chemical in the solid phase. The sorption of RNs on oxides surfaces exerts a significant role in controlling the transport and retention of these pollutants in near-surface environments [73–75].

Based on the relative concentrations of RNs and contact methods with oxides, the main sorption mechanisms consist of absorption, including ion exchange and the formation of solid solution through diffusion; adsorption, including inner- and outer-sphere surface complexation; and (co)/(re)precipitation on the surface, leading to the overgrowth through heterogeneous nucleation and subsequent formation of solid solutions, and even topotactic reactions [76, 77]. The initial fast sorption stage is mainly controlled by absorption and adsorption while surface (co)/(re)precipitation and incorporation processes appear to be responsible for the slow stage [78, 79]. These processes usually act simultaneously and it is difficult to distinguish them unless using analytical methods with molecular-scale resolution.

2.3.1 Adsorption

Adsorption is the accumulation of chemicals at the solid/liquid interface and the basis of most surface reaction processes. It influences the distribution of RNs between the aqueous and solid phases and thus affects the transport of RNs in the natural environment. The affinity of oxide surfaces for soluble RNs controls their residence time, their concentration in the water system, and their ultimate fate in the environment. Surface hydroxyl groups that acquire or lose hydrogen ions will result in surface charge and electrostatic effects [80]. Inner- and outer-sphere complexes formed on the surface are two postulated principal mechanisms for RN adsorption on oxide minerals [75]. For inner-sphere complexation, chemical bonds are formed between RNs and the surface O^{2-} or OH groups, which are pH dependent [81]. In the latter case, the adsorption of RNs on the surface is proceeded mainly by electrostatic interaction and possibly hydrogen bonding while the first coordination shell of the adsorbate ion remains intact. A split of the first coordination shell and a short distance between the RN center and the surface are the characteristic features of inner-sphere complexes, whereas for outer-sphere adsorption, the distance is longer and the adsorbed RNs are similar to free hydrated ions. Therefore, inner-sphere complexation as chemisorption will lead to forming stronger bound complexes, which is regarded as the dominant adsorption mode. Adsorption processes are of great significance for the retention of radioactive contaminants [82, 83]. The reactivity of surfaces could be influenced by adsorption, and thereby the rates of subsequent processes such as surface precipitation, dissolution of the oxide surface, and the catalysis of redox processes will also be affected [84].

Metal (hydr)oxide surfaces contain several functional groups, which could be regarded as inorganic polymers. Many interactions are achieved through the coordination of surface functional groups with RNs [75]. The amphoteric behavior of the surface should also be considered. Surface complexation models (SCMs) have been widely applied to modeling the interactions at the solid/liquid interface. Several well-established SCMs for RN adsorption to oxides have been proposed [85–88]. Dzombak and Morel [87] listed some criteria features of SCMs:

(i) Adsorption occurs at specific surface reactive sites.
(ii) Adsorption reactions could be depicted by M-L equations.
(iii) The adsorption reaction itself causes surface charge.
(iv) The surface charge effect could be considered by correcting the mass law constants.

Variable charged hydroxyl groups are responsible for the adsorption of RNs on oxide surfaces [89]. The pH-dependent inner-sphere complex is generally investigated by the batch technique, which was closely related to the deprotonation of surface hydroxyl groups [81, 90, 91]. The adsorption process is expressed by the following equation, where the hydration waters are not considered:

$$\equiv SOH + M^{n+} \leftrightarrow \equiv SOM^{-1+n} + H^{+} \tag{2.6}$$

where $\equiv SOH^{-1+n}$ denotes the formed surface complex. For instance, the $\equiv SOEu(OH)_n^{2-n}$ species are employed to simulate Eu(III) adsorption on oxides [88, 92]. Jakobsson et al. [93] modeled Pm^{3+} adsorption onto TiO_2 by using the 1-pK Basic Stern model approach, which reported that the adsorption is best fitted with an inner-sphere monodentate surface complex.

In addition, spectroscopic characterization is also important to probe the structural information of the surface complexes on a molecular level. Piasecki and Sverjensky [94] investigated the rare earth ion adsorption on oxides by application of the X-ray technique. It was found that the proportion of tetra- and mononuclear surface adsorption species are affected obviously by the environmental factors as well as the oxides types. The monodentate complex of trivalent lanthanides on rutile was inferred relative to hematite or silica. By using time-resolved laser-induced fluorescence spectroscopy (TRLFS), inner-sphere Eu(III) surface complexes are identified by the decrease of the amount of H_2O in the first coordination sphere [94–98].

2.3.1.1 Radionuclide adsorption on iron oxides

Fe(III) (hydr)oxides (FHOs) are important adsorbents of RNs in the environment, owing to their widespread presence in soils and sediments, high surface area, and strong affinity for many trace elements. The sorption of RNs onto FHOs is a fundamental geochemical process that controls the transport and fate of RNs and other nuclear wastes in natural aqueous environments. Meanwhile, understanding the complexation mechanism at the molecular-level is essential to model the behavior in subsurface zones and to provide guidance for the development of new efficient technologies for nuclear waste treatment as well as remediation strategies for contaminated groundwater and soils [1].

A number of studies (experimental and theoretical) have been concentrated on the sorption of RNs on FHOs [99]. The study of Eu(III) adsorption on α-Fe_2O_3 revealed that the edge-shared surface complex (i.e., mononuclear bidentate) was formed [100]. Tamura et al. [101] investigated the adsorption of $^{60}Co^{2+}$ onto spherical magnetite particles. The batch experiments demonstrated that the adsorption process is pH- and temperature-dependent, which can be interpreted quantitatively by taking into consideration the surface complex formation at the neutral site (≡FeOH) with the release of one to two protons [101]. In the batch study of Cs(I) ions interacting with natural magnetite in an aqueous system, the Cs(I) adsorption percentage increased with increasing pH [102]. In addition, the Cs(I) adsorption decreased seriously in the presence of interfering cations.

In the absence of CO_2 or carbonate ligand, the chief hydrolysis species are UO_2^{2+}, UO_2OH^+, and $UO_2(OH)_2^0$ and $UO_2(OH)_3^-$ with increasing pH [103, 104], which may be different in the specific calculated models. The formation of S-type U(VI) adsorption edges on many FHOs is normally similar (Fig. 2.11) [86, 103–115]. Specifically, little sorption could be observed at acidic pH while the increased adsorption could be observed with the pH increase. The adsorption reaches a high plateau from pH 5/6 to 10 [103, 104, 106, 107, 115]. Compared with natural hematite, the much higher sorption capacities of smaller-size hematite, ferrihydrite, and goethite for U(VI) are probably caused by the difference in the property variation of the different surfaces [103, 106] while the ion strength seems not to significantly affect U adsorption onto FHOs [103]. Similarly, the presence of 1 mM Ca and Mg ions also has little influence on uranyl adsorption [104]. Moreover, the adsorption kinetics indicate that the fast reaction generally reached a steady state within 2 days [86, 115], and there is little difference when extending to 3 months for goethite [103].

Sherman et al. [112] investigated the interaction between uranyl and α-FeOOH and concluded that the edge-sharing U(VI) complex could only exist on the α-FeOOH(2 1 0) or (0 1 0) surface (Fig. 2.12). In addition, the

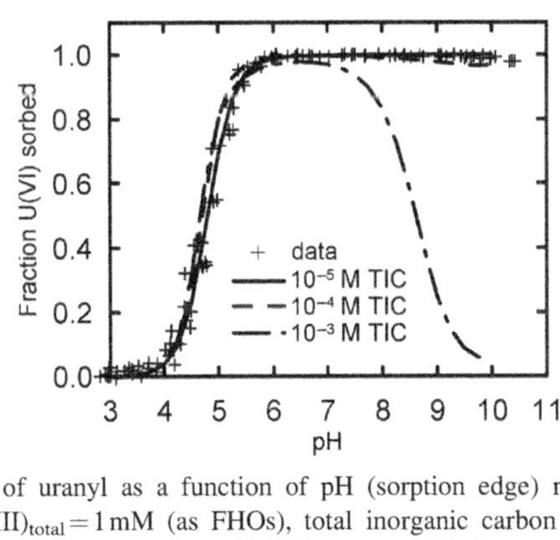

FIG. 2.11 Sorption of uranyl as a function of pH (sorption edge) measured in this study: $U(VI)_{total} = 1\,\mu M$, $Fe(III)_{total} = 1\,mM$ (as FHOs), total inorganic carbon (TIC) = 10–60 μM, and $I = 0.1$ [105]. *(Reproduced from J.H. Jang, B.A. Dempsey, W.D. Burgos, A model-based evaluation of sorptive reactivities of hydrous ferric oxide and hematite for U(VI), Environ. Sci. Technol. 41 (2007) 4305–4310. Copyright 2007, American Chemical Society.)*

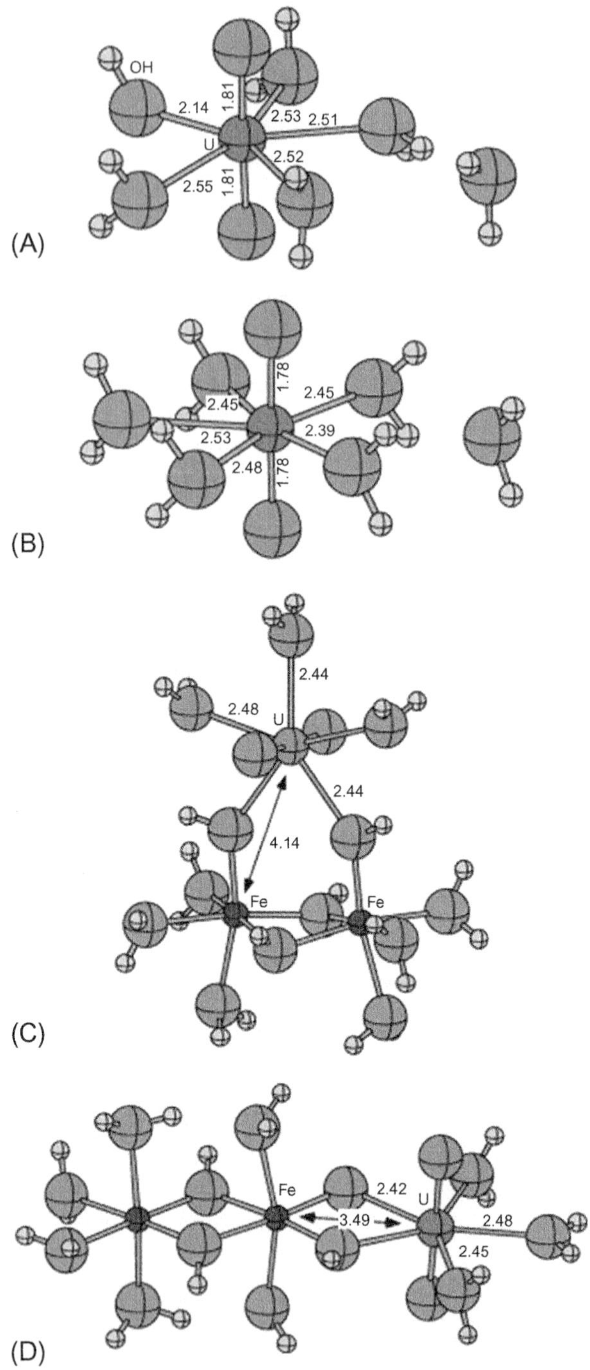

FIG. 2.12 U(VI) ab initio molecular geometry clusters. (A) $UO_2(H_2O)_6^{2+}$ complex, (B) $UO_2(OH)(H_2O)_5^+$ complex, (C) bidentate corner-sharing mononuclear cluster on (predominantly) (101) and (D) bidentate edge-sharing mononuclear cluster on (210) or (010). Bond lengths shown in Å [112]. *(Reproduced from D.M. Sherman, C.L. Peacock, C.G. Hubbard, Surface complexation of U(VI) on goethite (α-FeOOH), Geochim. Cosmochim. Acta 72 (2008) 298–310. Copyright 2007, Elsevier.)*

U-Fe distance in the complex can be well fitted by multiple scattering, which is used to cover the bidentate CS $\equiv Fe(OH)_2UO_2(H_2O)_3$ complex (C_2 complex). Hence, as for the dominant goethite (1 0 1) surface, the main formed complex without carbonate is the C_2 complex.

The presence of carbonate intensely influences the sorption of U onto FHOs. In systems containing carbonate, U(VI)-carbonate complexes begin to form at pH ~4 and occupy the dominant proportion at higher pH [104, 112, 115, 116]. To fit the uranyl sorption profiles in the presence of carbonate, the intrinsic constants of the carbonate complexes need modifying slightly [104]. In the presence of carbonate, it is generally considered that U(VI) sorption profiles on FHOs are in this manner: (i) little adsorption at acidic conditions, (ii) increase sharply when pH increases, (iii) reach adsorption plateau at near-neutral pH, (iv) decrease sharply when pH increases, and (v) little adsorption at alkaline conditions [1, 105] (Fig. 2.11). It is different for Pu adsorption on goethite, where the adsorption is still drastic at a carbonate alkalinity of ~100 meq/L. [117] When increasing the concentration of ferrihydrite suspensions, a left shift of the adsorption edge could be observed. The U(VI) sorption outlines are similar for different initial concentrations (1×10^{-8}–1×10^{-4} M), except for the shift of the adsorption edge. In addition, the change of CO_2 partial pressure will also lead to the adsorption edge [86, 115].

The dissolved uranyl carbonate complexes (e.g., $UO_2CO_3{}^0$, $UO_2(CO_3)_2{}^{2-}$, and $UO_2(CO_3)_3{}^{4-}$) obviously hinder U(VI) sorption onto FHOs [86, 115]. Hiemstra et al. [86] ascribed the increase of U(VI) in solutions to desorption at alkaline conditions. The binding of the uranyl-carbonate complex with only one carbonate ligand is favored from the batch experiment results and SCM. However, the complexes with more than one carbonate ligand may not be favored, which is due to the electrostatic repulsion between the surfaces and aqueous complexes with the same negative charge at high pH [115, 118].

Although the $\equiv FeO_{surface}$-U(VI)-carbonate complexes are difficult to be adsorbed onto FHOs, the ternary complexes has been demonstrated to be adsorbed on hematite by Bargar et al. [108, 119] through electrophoresis and X-ray absorption fine structure (EXAFS) analysis. U(VI) acting as a bridge binds simultaneously to surface FeO_6 octahedra as well as carbonate ligands in bidentate coordination (hematite-U(VI)-carbonate). These results indicate that U(VI)-carbonate complexes are significant transport-limiting species under natural systems.

The adsorption structure of U(VI) on ferrihydrite was identified by employing EXAFS and iterative transformation factor analysis [86, 120, 121]. It was proposed that the spectra data could only be quantified and fitted by two surface complexes: (1) the E_2 complex [115, 122], and (2) a uranyl surface complex containing three carbonate ions $\left(\equiv (UO_2)(CO_3)_3{}^{4-}\right)$, where one acts as a bridge (Fig. 2.13). However, uranyl will polymerize at the

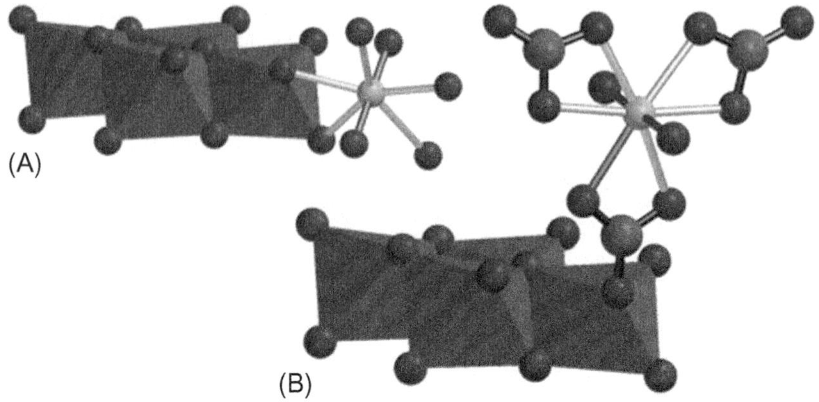

FIG. 2.13 (A) Uranyl bound by two singly coordinated surface groups presented at a free edge. For this geometry with d(Fe—U) = 3.45 Å and d(U—O_{edge}) = 2.49 Å, $d(O—O)_{edge}$ = 2.87 Å. The outer ligand of the uranyl surface complex may be OH, OH_2, or CO_3. (B) A uranyl tris-carbonate complex that is singly coordinated to a Fe ion in the solid via a carbonate group [86]. *(Reproduced from T. Hiemstra, W.H. Van Riemsdijk, A. Rossberg, K.U. Ulrich, A surface structural model for ferrihydrite II: adsorption of uranyl and carbonate, Geochim. Cosmochim. Acta 73 (2009) 4437–4451. Copyright 2009, Elsevier.)*

ferrihydrite surface, giving tris-uranyl surface complexes at high uranyl concentrations [123].

The presence of Ca ions has been identified as noticeably affecting the transport of U(VI) at alkaline conditions due to the large formation constants of Ca-containing complexes [124, 125], which is consistent with previous thermodynamic calculations [126, 127]. Fox et al. [128] explored U(VI) adsorption on ferrihydrite in the presence of Ca(II) and found that U adsorption decreased from 83% without Ca to 57% with 1.8 mM Ca, owing to the $Ca_2UO_2(CO_3)_3^0$ complex being the predominated species.

The effect of HA on U(VI) adsorption on iron (oxy)hydroxides was studied by Li et al. [129] HA facilitated U(VI) adsorption obviously at low pH while the adsorption was obstructed by HA at high pH. Lenhart and Honeyman [130] also observed a similar effect of HA (Fig. 2.14). More HA was considered to adsorb to the surface at acidic conditions, which decreased the electrostatic repulsion of uranyl with the α-Fe_2O_3 surface, thus increasing the adsorption tendency for U(VI). At alkaline pH value, the formative dissolved U(VI)-HA complexes were hard to adsorb. As a result, the effect of HA must be taken into consideration in modeling the natural systems with dissolved organic matter.

Different from the uranyl ion, Am has only one oxidation state, +3. In the absence of CO_2, a typical S-type sorption profile was observed for Am(III) adsorption on α-Fe_2O_3 [131]. It was also found that ionic strength exerted little effect on Am(III) adsorption. HA considerably increased Am(III) adsorption at acidic conditions, independent of ion strength [131], while at alkaline conditions, Am(III) adsorption decreased gradually with the decrease of ion strength.

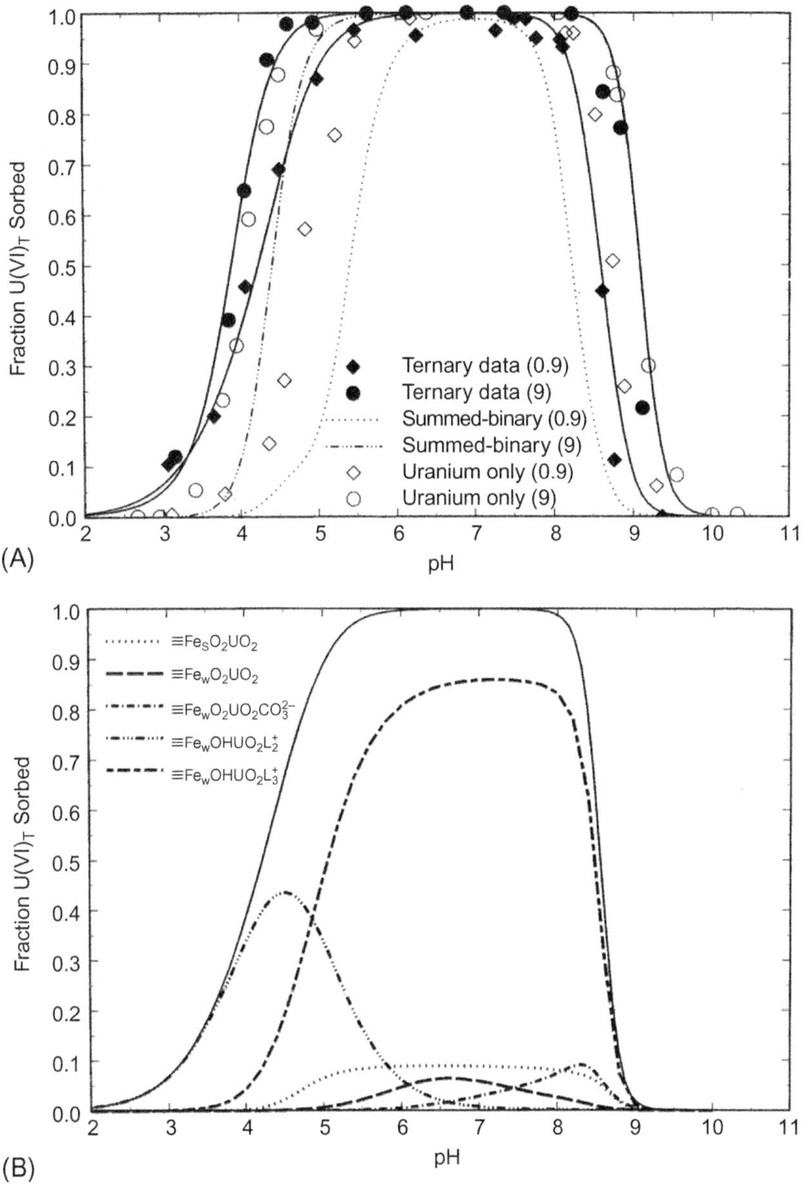

FIG. 2.14 (A) Experimental results and FITEQL simulation of the fraction of U(VI) sorbed versus pH in the presence of HA with postulated ternary complexes ($\equiv Fe_wOH-UO_2L_2^+$ and $\equiv Fe_wOH-UO_2L_3^+$) and (B) surface speciation diagram for the 0.9 g/L ternary system [130]. *(Reproduced from J.J. Lenhart, B.D. Honeyman, Uranium(VI) sorption to hematite in the presence of humic acid, Geochim. Cosmochim. Acta 63 (1999) 2891–2901. Copyright 1999, Elsevier.)*

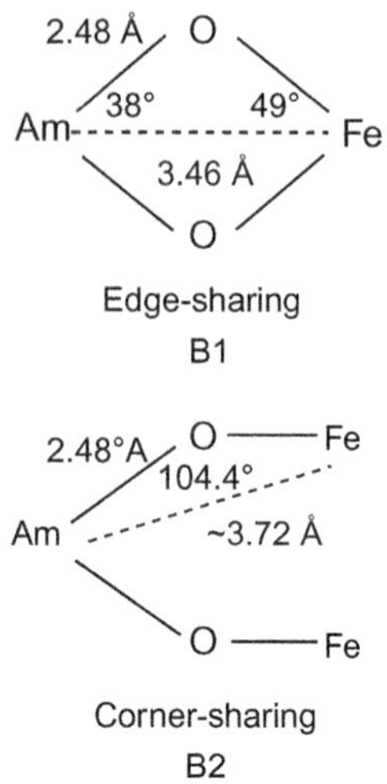

FIG. 2.15 Models of the bidentate bonding of Am(III) onto the ferrihydrite surface [132]. *(Reproduced from S. Stumpf, T. Stumpf, K. Dardenne, C. Hennig, H. Foerstendorf, R. Klenze, T. Fanghanel, Sorption of Am(III) onto 6-line-ferrihydrite and its alteration products: investigations by EXAFS, Environ. Sci. Technol. 40 (2006) 3522–3528. Copyright 2006, American Chemical Society.)*

Am(III) adsorbed onto ferrihydrite surfaces was investigated by Stumpf et al. [132] The coordination of Am(III) was confirmed as a bidentate CS species, which implies that the bonding is more robust than that of a monodentate system (Fig. 2.15). Compared with the EXAFS fitting data, the Am-Fe distances yielding from the other bidentate site B1 are too short. Ferrihydrite gradually transformed to goethite or hematite upon heating, during which the adsorbed Am(III) species may be released at acidic conditions while incorporation may occur at alkaline conditions.

The adsorption of Pu(IV)/(V) on α-FeOOH presented distinct variances due to their different hydrolytic characters. The adsorption edge of Pu(IV) occurred at the left side of Pu(V), which shifted to left gradually with time, owing to the redox transformations from Pu(V) to Pu(IV) [117]. Khasanova et al. [133] validated that HA could distinctly increase Np(V) sorption at acidic conditions, but little impact was observed at pH 5.5–10. However, under an aerobic atmosphere, little effect of HA was observed at acidic conditions, although it showed an obstructing effect at neutral to alkaline conditions [134]. As for the anaerobic atmosphere, HA enhanced Np(IV) sorption at low pH values but exerted little impact at neutral to alkaline conditions.

2.3.1.2 Radionuclide adsorption on aluminum oxides

Spectroscopic investigations could not only provide direct structural coordination of the surface complexes but also reveal the relevant binding sites

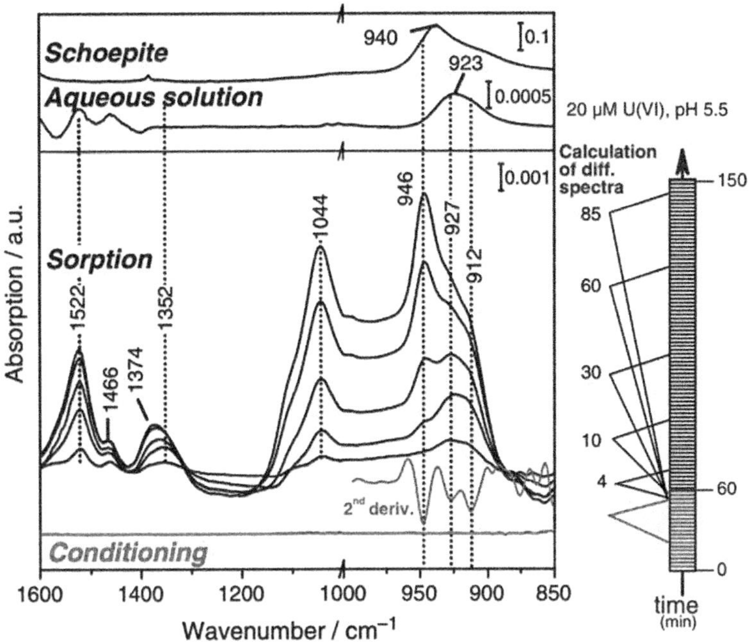

FIG. 2.16 In situ time-resolved mid-IR spectra of U(VI) sorption on γ-Al_2O_3 (20 μM initial U(VI), 0.1 M NaCl, pH6, air, 0.1 mg/cm^3 Al_2O_3, flow rate 0.2 mL/min). The spectra of the sorption process are recorded at different times after induction as given [139]. *(Reproduced from K. Müller, H. Foerstendorf, V. Brendler, A. Rossberg, K. Stolze, A. Gröschel, The surface reactions of U(VI) on γ-Al_2O_3—in situ spectroscopic evaluation of the transition from sorption complexation to surface precipitation, Chem. Geol. 357 (2013) 75–84. Copyright 2013, Elsevier.)*

participating in the sorption process. Several complexes have been demonstrated to form during U(VI) sorption onto aluminum (hydr)oxide [135–138]. The formation of bidentate U(VI) complexes with carbonate coordinated is predicted by the IR spectra at low U(VI) surface loadings on γ-Al_2O_3 (Fig. 2.16) [139].

To obtain a more detailed understanding of U(VI) adsorption on Al-hydroxide surfaces at the molecular level, EXAFS was employed to probe the interaction of gibbsite and U(VI). A monomeric binary surface complex is confirmed through EXAFS data, regardless of the atmospheric condition and surface loading. In addition, a ternary complex containing a bidentate coordinated carbonate was ascertained as an inner-sphere dimeric uranyl species [138].

A series of active sites for adsorption (e.g., surface oxygen or hydroxyl groups) is located at the edges of the oxide surfaces. RNs can directly bond to these active sites in corner-, edge-, or face-shared binding modes [140]. It is generally accepted that inner-sphere complexes are more stable than outer-sphere ones because they lead to substantial microstructure alterations. In addition, surface polynuclear complexes may form through the aggregation of inner-sphere surface complexes. It has been demonstrated that the thermodynamic stability of the surface polynuclear complexes was enhanced by the formation of extra bonds [140]. They surmised that the relative fractions of polynuclear surface complexes affected the adsorption reversibility.

Chen and Wang [141] studied the adsorption of Th(IV) on γ-Al_2O_3 in the absence/presence of soil organic matter by using the batch technique. The obvious effect of organic matter on Th(IV) adsorption at acidic conditions was attributed to the formation of ternary complexes such as $\equiv SO-O-HA-Th$ or $\equiv SO-O-FA-Th$.

The adsorption mechanism of Eu(III) with alumina was mainly ion exchange at acidic conditions, and surface complexation was responsible for it at alkaline conditions [142]. Tan et al. [98] probed the influence of HA/FA on the interaction of Eu(III) with alumina and confirmed the strong impacting effect of HA/FA by using time-resolved laser fluorescence spectroscopy (Fig. 2.17). Li et al. [143] studied the addition order of FA and Co(II) to γ-Al_2O_3 suspension and concluded that the addition order has almost no impact on Co(II) adsorption.

2.3.1.3 Radionuclide adsorption on titanium oxides

Titanium dioxide (TiO_2) is an efficient adsorbent for RN removal from water because of its chemical stability and strong affinity for RNs [144]. Several studies have been devoted to exploring U(VI) sorption on TiO_2, combining classical batch experiments (e.g., Kd, sorption isotherms, and SCM) [144–146] with molecular characterization techniques (e.g., attenuated total reflection-infrared spectroscopy (ATR-IR), X-ray, TRLFS, EXAFS, and theoretical modeling) [147–149].

Ridley et al. [97] examined Nd(III) adsorption on rutile surfaces through potentiometric titration and SCM. An increase of the Nd(III) adsorption was noted with elevated temperature. The formation of a unique tetradentate Nd(III) complex was proposed to rationalize the experimental results. In the study of Pu(VI) adsorption onto TiO_2, no apparent ionic strength dependence was detected, which is an indication of the formation of inner-sphere complexes [80]. In the pH range of 2–7, the adsorption data could be well fitted by considering two surface complexes in a 1-pK Stern model.

Lefèvre et al. [148] probed the adsorption of U(VI) onto TiO_2 by employing ATR-IR spectroscopy. A trimer surface complex was suggested to interpret U(VI) adsorption. The influence of carbonate on the adsorption of U(VI) on TiO_2 was studied by Wazne et al. [146]. The adsorption percentage decreased noticeably with increasing inorganic carbonate concentrations at neutral and alkaline conditions, which was assigned to the formation of strong U(VI)-carbonate aqueous complexes.

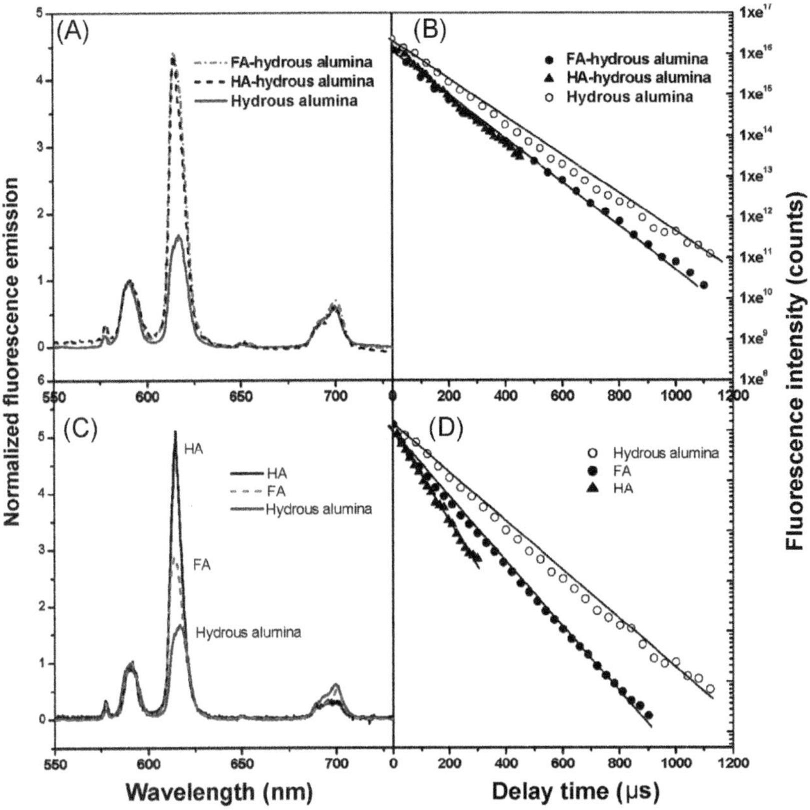

FIG. 2.17 The fluorescence emission spectra (A and C) and time dependence of fluorescence emission decay results (B and D) for Eu(III) bound to HA, FA, hydrous alumina, HA-hydrous alumina, and FA-hydrous alumina hybrids. C(hydrous alumina) = 4.4 g/L, C(Eu(III))$_{(initial)}$ = 4.310–5 mol/L, C(FA/HA)$_{(initial)}$ = 10 mg/L, C(KNO_3) = 0.1 mol/L, pH = 6.3 ± 0.1, T = 20 ± 1°C [98]. *(Reproduced from X.L. Tan, X.K. Wang, H. Geckeis, T.H. Rabung, Sorption of Eu(III) on humic acid or fulvic acid bound to hydrous alumina studied by SEM-EDS, XPS, TRLFS, and batch techniques, Environ. Sci. Technol. 42 (2008) 6532–6537. Copyright 2008, American Chemical Society.)*

Tan et al. [91] studied the adsorption of Eu(III) on anatase and rutile and confirmed the evident influence of HA on Eu(III) adsorption. The EXAFS results denoted that the adsorbed Eu(III) was bound to 7–8 O atoms with a distance of ~2.40 Å (Fig. 2.18 and Table 2.2). The spectral profiles for the sorption samples and reference sample $Eu(OH)_3$(s) are apparently different, and analysis of the EXAFS data implies the formation of polymeric Eu surface complexes at low pH while the coprecipitation phases were formed at high pH. For the sample at pH 9.0, the EXAFS data cannot be well described solely by surface complexes, indicating the formation of a coprecipitated component. The Eu-Ti distance close to 4.41 Å verifies the formation of

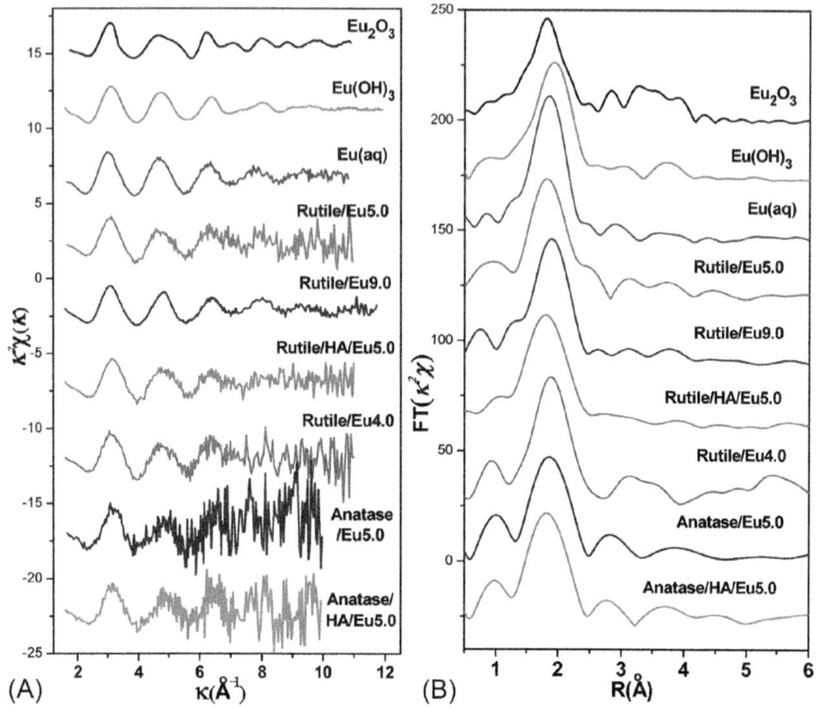

FIG. 2.18 k^2-weighted EXAFS spectra (A) and the corresponding Fourier transforms (B) of the following reference and sorption samples: Eu(III) adsorbed onto rutile at pH5.0 (rutile/Eu5.0), at pH4.0 (rutile/Eu4.0), and at pH9.0 (rutile/Eu9.0); the presence of HA at pH5.0 (rutile/HA/Eu5.0 and anatase/HA/Eu5.0); and Eu(III) adsorbed onto anatase at pH 5.0 (anatase/Eu5.0). $C(TiO_2)=0.3\,g/L$, $C[Eu(III)]_{initial}=3.0\times10^{-5}\,mol/L$, $C(NaClO_4)=0.01\,mol/L$, $C(HA)_{initial}=7.5\,mg/L$, pH 4.0 ± 0.1 or 5.0 ± 1 or 9.0 ± 0.2, $T=25\pm1°C$ [91]. *(Reproduced from X.L. Tan, Q.H. Fan, X.K. Wang, B. Grambow, Eu(III) sorption to TiO_2 (anatase and rutile): batch, XPS, and EXAFS studies, Environ. Sci. Technol. 43 (2009) 3115–3121. Copyright 2009, American Chemical Society.)*

monodentate complexes. As for the Eu/HA/TiO_2 system, the Eu—C single scattering and the Eu—O—C scattering were observed around 2.7 Å, which is assigned to the Eu-HA complexes [150]. Debnath et al. [151] utilized hydrous titanium oxide to adsorb Ni(II) from wastewaters, and found that the presence of Fe(III) ion inhibits Ni(II) adsorption. The mean free energy of reaction revealed the columbic binding of RNs with the solid particles.

2.3.1.4 Radionuclide adsorption on silicon oxides

The adsorption of Yb(III) and Cs(I) on the silica surface was investigated by using SCM [152]. It was shown that the binding of Yb(III) on the silica surface is stronger than that of Cs(I), which is reasonable considering the natural difference of the two cations. Indeed, Yb(III) could form complexes with oxide surfaces, this is not the case for Cs(I). This rationalizes the low affinity

TABLE 2.2 EXAFS results of reference samples and sorption samples at Eu L_{III}-edge [91]

Sample[a]	Shell	R^b (Å)	N^c	σ^{2d} (Å^2)	R_f^e
Eu_2O_3	Eu—O	2.345(6)	5.7(8)	0.0081(13)	0.040
	Eu—Eu	3.59(2)	6.4(2)	0.0077(2)	0.030
$Eu(OH)_3$	Eu—O	2.428(7)	8.0(4)	0.0095(1)	0.020
	Eu—Eu	3.61(6)	2.2(3)	0.001(9)	0.065
Eu(aq)	Eu—O	2.427(5)	8.8(5)	0.0075(4)	0.023
Rutile/Eu5.0	Eu—O	2.39(3)	8.3(2.2)	0.0103(8)	0.075
	Eu—Ti	4.40(2)	1.1(4)	0.0064(1)	0.048
Rutile/HA/Eu5.0	Eu—O	2.40(3)	8.5(1.6)	0.0106(4)	0.034
	Eu—C	2.67(2)	1.4(6)	0.0083(3)	0.134
	Eu—Ti	4.42(3)	1.3(4)	0.0064(4)	0.138
Rutile/Eu5.0	Eu—O	2.41(2)	9.1(1.1)	0.0099(2)	0.041
	Eu—Ti	4.41(3)	1.2(3)	0.0064(1)	0.245
Anatase/Eu5.0	Eu—O	2.40(2)	6.7(4)	0.0064(11)	0.020
	Eu—Ti	4.40(1)	1.0(6)	0.0086(4)	0.102
Anatase/HA/Eu5.0	Eu—O	2.39(2)	7.1(1.5)	0.0065(8)	0.033
	Eu—C	2.70(2)	2.0(6)	0.0064(21)	0.124
	Eu—Ti	4.40(2)	1.1(2)	0.0066(1)	0.114
Rutile/Eu9.0	Eu—O	2.415(7)	8.0(5)	0.0080(4)	0.021
	Eu—Eu	3.65(2)	2.35(9)	0.0026(16)	0.164
	Eu—Ti	4.51	1.3(6)	0.0025(9)	0.137

[a] *Eu_2O_3, $Eu(OH)_3$, and Eu(aq) are named as reference samples, whereas the other samples of anatase or rutile with adsorbed Eu(III) in the presence or absence of HA are named as sorption samples.*
[b] *R is the interatomic distance.*
[c] *N is the number of neighbor oxygens.*
[d] *σ^2 is the Debye-Waller factor.*
[e] *R_f is the residual factor. $R_f = \sum_k (k^3x - k^3x_{calc}) / \sum_k (k^3x_{calc})$, which measures the quality of the model Fourier-filtered contribution (x_{calc}) with respect to the experimental contribution (x_{exp}).*
Reproduced from X.L. Tan, Q.H. Fan, X.K. Wang, B. Grambow, Eu(III) sorption to TiO_2 (anatase and rutile): batch, XPS, and EXAFS studies, Environ. Sci. Technol. 43 (2009) 3115–3121.

of Cs(I) for oxide surfaces, which are thought to present only surface complex sites. The surface charge of silica is negative for pH over 2–3, which facilitates the binding of cesium on the silica surface through electrostatic attraction. This binding strength is weaker compared with that of complexation.

In addition, the electrolyte cation (i.e., Na^+) may compete with Cs(I). However, the competitive effect of Na^+ is insignificant for Yb(III) adsorption.

Chen and Wang [153] researched the removal of Th(IV) by silica with HA/FA through batch experiments. It was found that the presence of HA/FA had a significantly positive impact on Th(IV) adsorption at pH from 2 to 4, which was ascribed to strong binding of HA/FA on the silica surface and consequently the formation of surface complexes (e.g., $\equiv MO-O-HA-Th$ or $\equiv MO-O-FA-Th$).

The adsorption of U(VI) on the silica surface has been examined by Sylwester et al [154]. An inner-sphere bidentate complex appeared to form at pH3.14 as the equatorial oxygens were split into two individual shells, with no near-neighbor Si or U being detected (Fig. 2.19). At near-neutral pH, a polynuclear surface complex or precipitate was surmised to form, with each U(VI) coordinating with a single surface Si in bidentate mode and complexing with other nearby U(VI).

In addition, the interaction of the ternary system of U(VI)-phosphate-silica was thoroughly investigated by Comarmond et al [155]. Two binary U(VI) surface species were formed in the absence of phosphate, which was shown by the luminescence spectroscopy (Fig. 2.20). Whereas with prolonged adsorption with phosphates, the formation of an autunite-like precipitate is convincingly

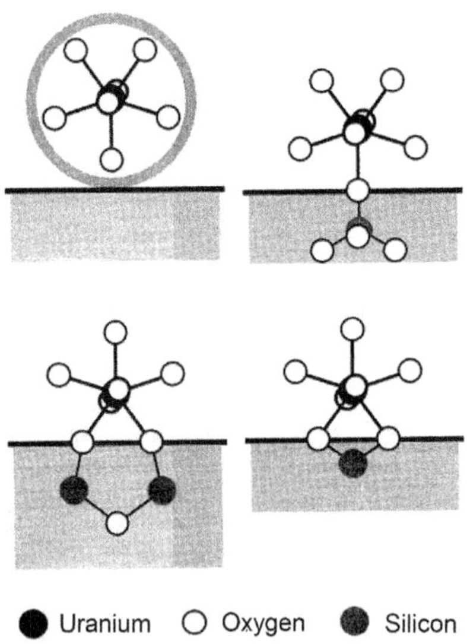

FIG. 2.19 Illustration of possible uranyl-silica surface interactions. *Top left*; outer-sphere adsorption, *top right*; monodentate adsorption, *bottom left*; bidentate adsorption to two silicon atoms, *bottom right*; bidentate adsorption to a single silicon atom. Bond angles and distances are not to scale [154]. *(Reproduced from E. Sylwester, E. Hudson, P. Allen, The structure of uranium(VI) sorption complexes on silica, alumina, and montmorillonite, Geochim. Cosmochim. Acta 64 (2000) 2431–2438. Copyright 2000, Elsevier.)*

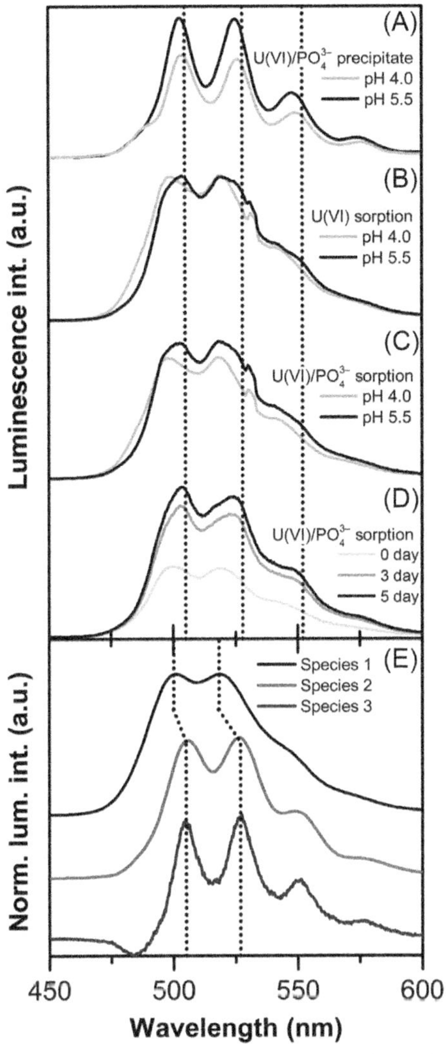

FIG. 2.20 Luminescence spectra of the U(VI)-phosphate precipitate (A) and the batch sorption samples (B–D). The samples were adjusted to pH 4 (*gray* trace) and 5.5 (*black* trace) prior to measurements (A–C). Spectra of the batch samples of the binary U(VI)/SiO_2 (B) and ternary U(VI)/PO_4^{3-}/SiO_2 (C) sorption systems. Spectra of the ternary U(VI)/PO_4^{3-}/SiO_2 sorption sample (pH 5.5) collected at different incubation times (0, 3, and 5 days, from below to *top*) (D). Decomposed luminescence spectra (species 1, 2) obtained from spectra of sorption samples (B and C) at pH 4 and 5.5 (E). Species 3 represents the second species obtained from spectra of aged sorption sample (D) [155]. *(Reproduced from M.J. Comarmond, R. Steudtner, M. Stockmann, K. Heim, K. Müller, V. Brendler, T.E. Payne, H. Foerstendorf, The sorption processes of U(VI) onto SiO_2 in the presence of phosphate: from binary surface species to precipitation, Environ. Sci. Technol. 50 (2016) 11610–11618. Copyright 2016, American Chemical Society.)*

recommended, denoting that the U(VI) adsorption initially forms binary surface complexing species and transforms toward surface precipitation with a longer reaction time.

2.3.2 Surface precipitation

Surface precipitation needs to be taken into consideration when the concentration of RNs goes beyond its solubility by one-tenth or the total surface site concentration by one half [87]. Surface precipitation could be divided into several types, for example, homogenous and heterogeneous precipitation as well as coprecipitation (Fig. 2.21) [156]. Two categories of formation mechanisms have been proposed: (1) the oxide surface induces some changes in RN properties, and (2) the vicinity of the interface alters the composition of the solution. They might act simultaneously and the chemical properties of

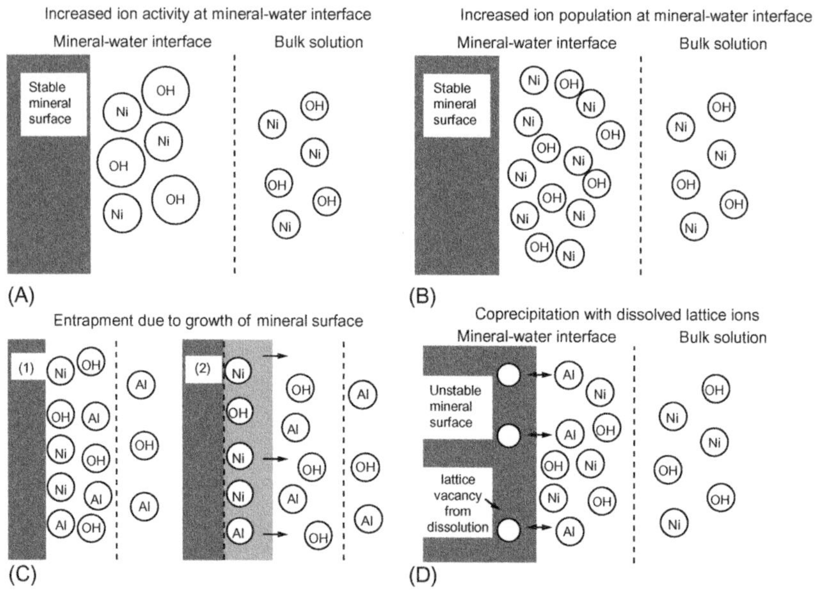

FIG. 2.21 Schematic diagram depicting four mechanisms for the formation of surface precipitate. (A) Increase of ion activities due to a reduction in the dielectric constant of water at the mineral-water interface. (B) Oversaturation within a finite volume of fluid at the mineral-water interface due to ion sorption. (C) Incorporation of a sorption complex within the sorbent structure due to particle growth. This would be considered an epitaxial growth phenomenon if the sorption complex acts as a nucleation site for surface precipitation growth. (D) Formation of a coprecipitate at the mineral-water interface incorporating the sorbate and dissolved sorbent lattice ions. A structural relationship between the sorbent and the coprecipitate is not required as reference reported [156]. *(Reproduced from R.G. Ford, A.C. Scheinost, D.L. Sparks, Frontiers in metal sorption/precipitation mechanisms on soil mineral surfaces, Adv. Agron. 74 (2001) 41–62. Copyright 2001, Elsevier.)*

RNs as well as the physicochemical properties of the oxide surface will determine the dominant mechanism.

Fig. 2.21A depicts the increased activities of RNs, owing to the dielectric property of the bulk solution changing after bathing the solid. The inducement of this precipitation is not ascribed to the excess of the ion population for nucleation of the solid phase, but because of that the increase of the ion activity product (IAP) exceeds the K_{sp} of the homogeneous precipitation. The increased ion populations near the charged oxide-water interface may lead to an increase in the specific IAP, which is similar to the "salting-out" effect in the vicinity of the solid-liquid interface. Generally, surface precipitates could form rapidly even when the surface loadings of RNs are quite low, and in pH where homogeneous precipitates will not form based on the K_{sp} [157–159]. O'Day et al. [159] proposed that the RN precipitate was formed on a quartz surface, which was attributed to the alterations of the dielectric constant of the interface. The changes of the local charge of oxide surfaces will accelerate precipitation more effectively [160]. Besides silica, precipitates and carbonate complexation were also expected to occur at the alumina surface [139, 154, 161–163]. A surface precipitate obtained at neutral pH, the spectral feature of which is similar to schoepite, has been demonstrated by the IR spectroscopic data (Fig. 2.16) [139].

Surface complexation dominates the RN retention process when the concentration of adsorbed RNs on the oxide surface is low. Adsorption of RNs will terminate and the sorption isotherm will reach a plateau when the available sites are saturated with sorbate. Nucleation and subsequently surface precipitate will occur with the increase of surface complex concentration (Fig. 2.21B) [139, 158, 164, 165]. The macroscopic adsorption isotherm could provide evidence for the formation of surface precipitation (Fig. 2.22), which is based on an assumption that there is a limited population of reactive sites for sorption on a given mass of oxide sorbents. Surface precipitation could be regarded as an intermediate product transition from surface complexation to bulk homogeneous precipitation of the RNs [158, 167]. Recent studies have revealed the occurrence of surface-induced precipitation on the oxide surfaces with high loading of sorbate [158, 167, 168].

Fig. 2.21C clarifies the entrapment of RNs into the oxide mineral structure by continued particle growth, which represents that a new phase is produced. Its crystal structure is similar to that of the original oxides. In this situation, although ion is released into solution, the oxide surface is stable and the surface growth can continue with the addition of RNs to solution, which can be deemed the epitaxial growth of new solid phases in a broad sense [169, 170]. It should be noted that only RNs with similar radii as the components in the oxide bulk structures could incorporate into the structure [171].

Fig. 2.21D depicts an unstable surface due to the release of structural ions into solution. The subsequent precipitation and recrystallization process leads to the formation of new phases [165, 172–174]. When the released ions and

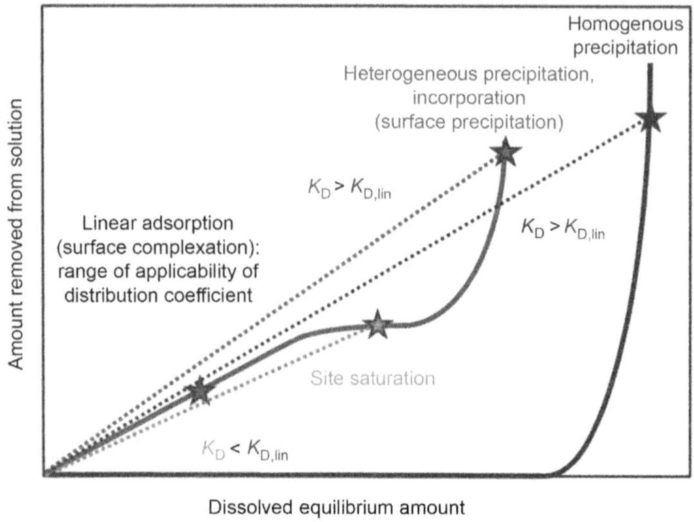

FIG. 2.22 Illustrative sorption isotherm for an actinide on an oxidic sorbent at constant pH and ionic strength *(middle solid curve)*. The stars show various possible measurement results that may be on the equilibrium isotherm [166]. *(Reproduced from H. Geckeis, J. Lützenkirchen, R. Polly, T. Rabung, M. Schmidt, Mineral–water interface reactions of actinides, Chem. Rev. 113 (2013) 1016–1062. Copyright 2013, American Chemical Society.)*

RNs in solution are simultaneously contained in the newly formed solid phase, it is considered that a coprecipitate emerges [172]. Thus, the composition of the coprecipitate is closely related to the composition of the oxides as well as the RN species in bulk solution.

Takahashi et al. [175] employed TRLFS techniques to investigate the interaction between SiO_2 and Eu(III) and identified that the number of water molecules bound to Eu(III) decreased with increased aging time. Almost no water molecules were left around Eu(III) after 280 days in alkaline conditions, implying the incorporation of adsorbed Eu(III) into the bulk structure with time. They assumed that Eu(III) contributes to the hydrolysis of the Si—O—Si bonds and speeds up the dissolution of SiO_2, followed by hydrolysis of silanol on the surface.

Missana et al. [176] studied the retention of selenite (SeO_3^{2-}) by magnetite, the processes of which includes adsorption, reduction, and dissolution/coprecipitation. Adsorption experiments of SeO_3^{2-} on goethite under similar conditions were also conducted for comparison. The sorption edge profiles of the two oxides were different, that is, the sorption data for goethite could be well modeled just assuming an inner-sphere complex with surface sites (≡SOH). The sorption data for magnetite at low pH could be reasonably modeled only by considering the dissolution of oxides and the subsequent

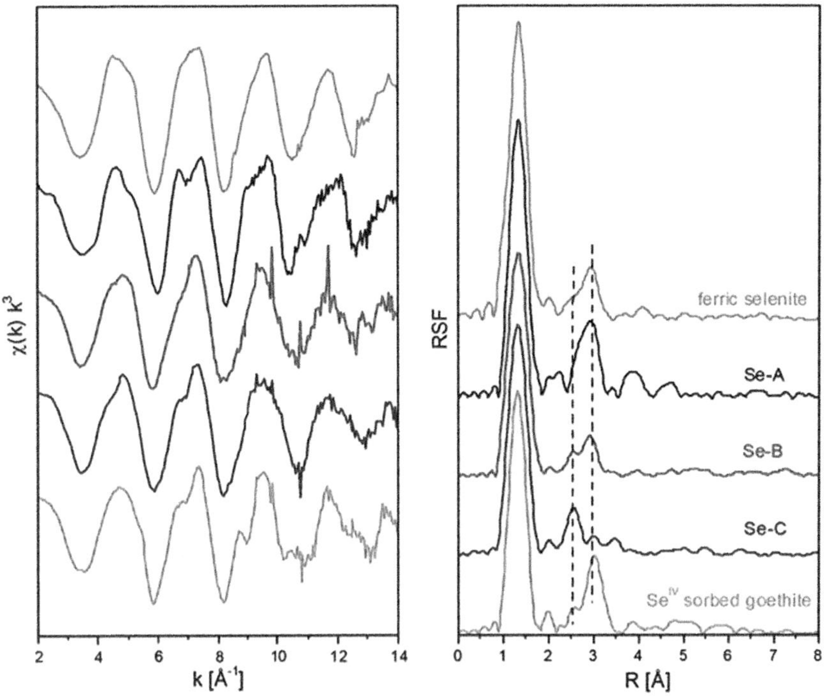

FIG. 2.23 Se-*K*-edge EXAFS spectra and corresponding Fourier transform magnitude of Se^{IV}-reacted magnetite samples (Se-A, Se-B, and Se-C) in comparison to ferric selenite (*top*) and to Se^{IV}-sorbed goethite (*bottom*) [176]. *(Reproduced from T. Missana, U. Alonso, A.C. Scheinost, N. Granizo, M. Garcia-Gutierrez, Selenite retention by nanocrystalline magnetite: Role of adsorption, reduction and dissolution/co-precipitation processes, Geochim. Cosmochim. Acta 73 (2009) 6205–6217. Copyright 2009, Elsevier.)*

formation of surface complexation of the Fe-Se species. Under high SeO_3^{2-} concentrations in solution, the formation of ferric selenite precipitation was the dominant retention mechanism at pH <5, which was confirmed by the analysis of EXAFS results (Fig. 2.23 and Table 2.3). The dissolution and coprecipitation processes exert a great influence on the SeO_3^{2-} retention by magnetite under acidic reaction conditions [176].

2.3.3 Incorporation

RNs can be incorporated into oxide minerals, leading to the formation of new solid phases containing substitutional impurities. These "cocrystallization" processes not only decide oxide growth but they also limit the mobility of RNs in the environment. The incorporation mechanism can be developed as a remediation strategy (e.g., the design of an engineered barrier) to remove

TABLE 2.3 Se-*K* EXAFS fit results ($S_0^2 = 0.9$) [176]

Sample	Coordination shell			Further shells			ΔE_0 (eV)	$\chi^2_{res\%}$
	CN[a]	R[b] (Å)	σ^{2c} (Å^2)	*CN*	R (Å)	σ^2 (Å^2)		
Se-A magnetite	2.8 O	1.70	0.0012	3.1 Fe	3.29	0.0082	16.0	14.2
				1.5 Se	4.19	0.0065		
Se-B magnetite	3.1 O	1.71	0.0029	0.9 Fe	2.97	0.0098	17.5	15.8
				1.2 Fe	3.42	0.0084		
Se-C magnetite	2.9 O	1.70	0.0026	1.0 Fe	2.91	0.0071	16.5	12.3
Se goethite	2.6 O	1.70	0.0020	1.8 Fe	3.35	0.0045	13.7	16.2
Ferric selenite	3.1 O	1.70	0.0015	0.5 O	2.88	0.0032	14.0	18.2
				2.9 Fe	3.30	0.0076		

[a] *CN: coordination number, error ±25%.*
[b] *R: radial distance, error ±0.01 Å.*
[c] *σ^2: Debye-Waller factor, error ±0.0005 Å^2.*

Reproduced from T. Missana, U. Alonso, A.C. Scheinost, N. Granizo, M. Garcia–Gutierrez, Selenite retention by nanocrystalline magnetite: Role of adsorption, reduction and dissolution/co-precipitation processes, Geochim. Cosmochim. Acta 73 (2009) 6205–6217.

RNs from wastewaters. The effectiveness of this remediation tool depends on various environmental factors such as operation at different scales in space, time, and temperature [177]. Iron oxides and oxyhydroxides can act as primary hosts that RNs could incorporate directly into their bulk structures [178–182], or temporary carriers that increase RN loading and stability in other solid waste forms (e.g., glasses) [182–185].

Incorporation of uranium into FHOs provides a new insight into the long-term immobilization of RNs. It has been reported that various elements (e.g., Si, Ti, Mn, Ni) are able to be accommodated into the structures of goethite and hematite as impurities [186, 187]. Specifically, U may be incorporated into goethite in a hexavalent or even a pentavalent state [188–190], and U(VI) incorporation into hematite through coprecipitation has been identified [191–193]. Remarkably, Duff et al. [192] investigated the ferrihydrite coprecipitate and induced hematite from a Fe(III) solution mixed with U(VI). The incorporation environment of U(VI) into the hematite bulk structure was proposed as uranate species without the axial U—O bonds (Fig. 2.24). Ilton et al. [193] drew a similar structure and suggested that the incorporated U(VI) was reduced to U(V) while the adsorbed U(VI) to U(IV), which is confirmed by XPS analysis.

Similarly, the EXAFS data analyzed by Marshall et al. [195] revealed the incorporation of U(VI) into the hematite structure by occupying the octahedrally coordinated Fe(III) site. However, fitting of the EXAFS denoted the presence of axial U—O bonds (~1.81–1.87 Å), contrary to previous studies that the incorporation was accompanied with the loss of axial O atoms. Kerisit et al. simulated the local U incorporation environment in the goethite bulk structure by using ab initio molecular dynamics (AIMD) [196]. The calculated results indicated that U(V) is coordinated in octahedral uranate form by substituting the structural Fe(III) (Fig. 2.25). McBriarty et al. [197] then presented a new structure of the incorporation of U(V) into lepidocrocite. Their AIMD analysis demonstrated that although the formation of excess lepidocrocite was favored under the reaction conditions, the incorporation of U is preferred into goethite over lepidocrocite.

Kerisit et al. [198] also characterized the incorporation environments of U in magnetite structures. The tetrahedral site shares corners with 12 nearest-neighbor FeO_6 octahedra whereas it shares with six for the edge-sharing octahedral site, both of which were considered for U incorporation (Fig. 2.26). The average U—O bond distances are a little longer than the average Fe—O distance, indicating that there is minimal distortion of magnetite after U incorporation. For U substitution for Fe(III) in the tetrahedral site, U(IV) maintains fourfold coordination for all charge compensation schemes, and U(V) changes to sixfold coordination only for the creation of iron vacancies, whereas U(VI) bonds to two extra oxygen atoms.

Technetium-99 (Tc) has received much attention due to its volatility during the generation of solid phases (e.g., glass) for long-term RN immobilization.

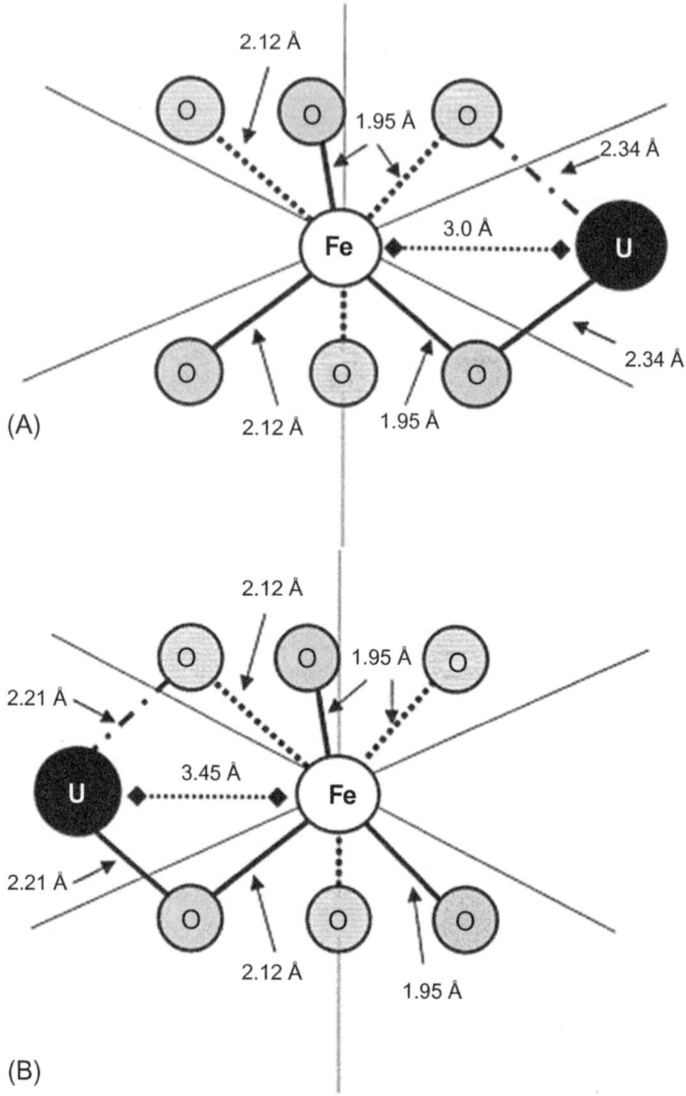

FIG. 2.24 Molecular models of the association of U^{6+} with distorted Fe octahedra from the hematite structure. (A) Uranium bound via the unshared faces of distorted hematite Fe octahedra at a radial U-Fe distance as short as 3.0 Å and (B) adsorbed U on edges of Fe octahedra typically resides at radial U-Fe distance ranging from 3.4 to ~3.5 Å [108]. The hematite Fe octahedra exhibit face and edge sharing, which results in substantial structural distortion and straining. Unshared faces have the Fe—O bond lengths of 1.945 Å, and shared faces have Fe—O bond distances of 2.116 Å [192, 194]. *(Reproduced from M. Duff, J. Coughlin, D. Hunter, Uranium co-precipitation with iron oxide minerals, Geochim. Cosmochim. Acta 66 (2002) 3533–3547. Copyright 2002, Elsevier.)*

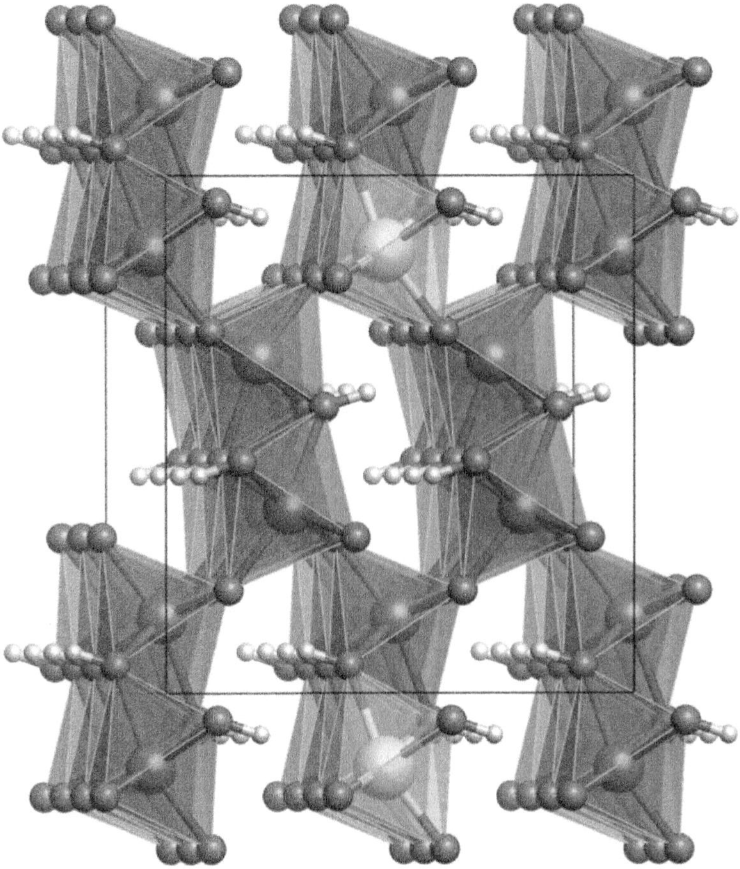

FIG. 2.25 Polyhedral model of the goethite supercell (viewed down the [001] axis with [100] as the horizontal axis) used in this work with uranium substitution at an octahedral site. The largest spheres are uranium atoms, the smaller spheres at the octahedral centers are iron atoms, the spheres at the octahedral apices are oxygen atoms, and the smallest spheres are hydrogen atoms [196]. *(Reproduced from S. Kerisit, E.J. Bylaska, M.S. Massey, M.E. McBriarty, E.S. Ilton, Ab initio molecular dynamics of uranium incorporated in goethite (α–FeOOH): interpretation of X-ray absorption spectroscopy of trace polyvalent metals, Inorg. Chem. 55 (2016) 11736–11746. Copyright 2016, American Chemical Society.)*

Smith et al. [199] evaluated the energy of five scenarios of Tc(IV) incorporation into goethite with different charge compensation mechanisms by using quantum-mechanical calculations. Relative to gas-phase ions, the two most stable scenarios are proceeded by direct substitution of Tc(IV) onto the octahedral Fe(III) lattice site. The charge balance was achieved by removing one nearby H^+ to generate an H vacancy or replacing 3 Tc(IV) for 4 Fe(III) to create an Fe(III) vacancy, with the former being more favored. The Fe(III)-vacancy substitution mechanism becomes more energetically favorable when including the

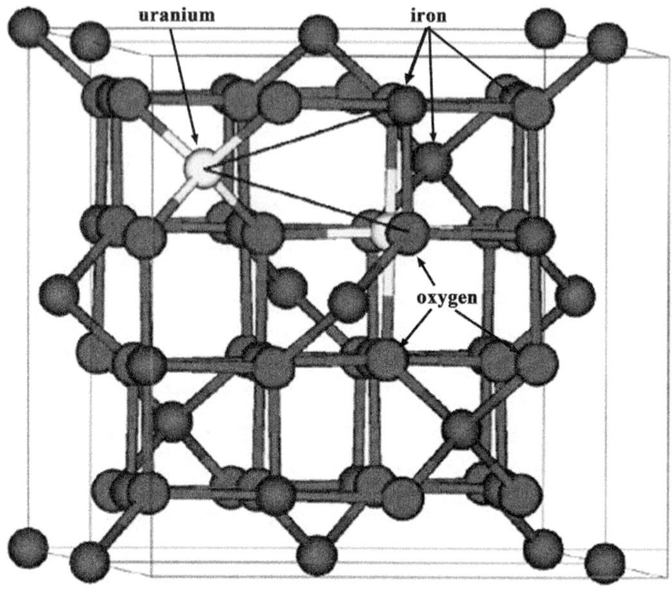

FIG. 2.26 Uranium incorporation into magnetite. Iron atoms are shown in *dark gray*, oxygen atoms are in *gray*. The uranium incorporation sites are shown in *white*. The additional uranium oxygen bonds formed upon relaxation in some of the simulations are shown with solid lines [198]. *(Modified from S. Kerisit, A.R. Felmy, E.S. Ilton, Atomistic simulations of uranium incorporation into iron (hydr)oxides, Environ. Sci. Technol. 45 (2011) 2770–2776. Copyright 2011, American Chemical Society.)*

extra hydration energy. It could be inferred that the dominant charge compensation mechanisms appear to depend largely on the valence state of the incorporated RNs as well as the structure of the host phase. Discerning the coordination environment of RNs associated with the major host phases has significant implications for the transport, mobility, redox activity, and ultimate fate of RNs in the environment.

2.3.4 Surface-induced redox reaction analysis

The $Fe(II)_{aq}$ and Fe(III) (oxyhydr)oxides commonly coexist at redox interfaces (e.g., the weathering of host rocks containing Fe(II), acid mine drainage, and the upwelling of anoxic underground water). It has been demonstrated that Fe(II) undergoes interfacial electron transfer between $Fe(II)_{aq}$ with the Fe(III) in the solid. Frierdich et al. [200] reported the cycle of Ni through hematite and goethite and incorporation into the bulk structure progressively during $Fe(II)_{aq}$ redox-driven recrystallization. Surface-adsorbed Ni becomes incorporated into bulk structures through overgrowth in the Fe(II) oxidative growth regions while existing incorporated Ni is released from regions of reductive dissolution. The rates of Ni incorporation and release are dependent on the rate of recrystallization, to some extent.

Surface-catalyzed U(VI) reduction and incorporation into the host phases by $Fe(II)_{aq}$ are significant processes for uranium sequestration in a variety of redox-stratified environments [109]. The long-term stability of incorporated U is closely related to the U coordination mode, the oxidation state (IV, V, or VI), and the corresponding charge compensation scheme [197]. It should be noted that adsorbed Fe(II) on FHOs can transform to Fe(III) through the surface electron transfer process, even without a soluble oxidant (e.g., O_2) [201–204]. The anaerobic remineralization of ferrihydrite induced by $Fe(II)_{aq}$ with aqueous or absorbed U(VI) was studied by Nico et al. [188] The majority of the U substituted into magnetite or goethite was ascertained to be oxidized U. The EXAFS spectra of the solid sample with no addition of Fe(II) could be well fitted by the *Adsorbed* model (Fig. 2.27), which consists of two axial O atoms, six equatorial O atoms splitting to two shells, a carbonate shell with two C atoms, and an Fe shell with an Fe atom. While in the presence of 3 mM $Fe(II)_{aq}$, the U EXAFS was best represented by the *Adsorbed and Incorporated* model 1, which takes into account two environments of solid state U: the adsorbed U and the incorporated U that takes up the octahedral sites of FHOs. The EXAFS of extraction samples with 30 mM $KHCO_3$ was simulated best with the *Adsorbed and Incorporated* model 2, where the equatorial oxygen shell is not split. The coordination numbers of axial U—O decreased from 2 to 0.7, suggesting that the adsorbed U phase only occupies ~35% of the total

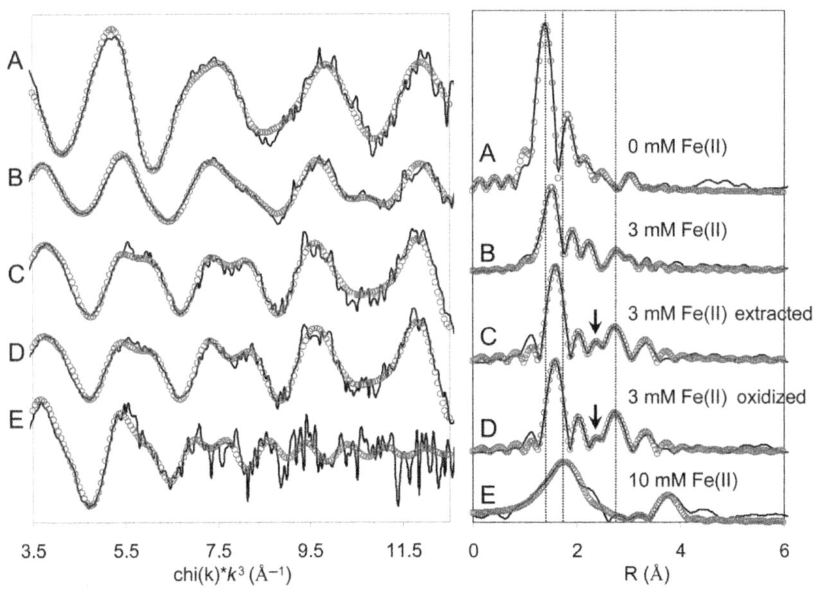

FIG. 2.27 Uranium EXAFS data and fits for chi(k) and chi(R) functions. *Solid lines* represent experimental data and open circles the associated fit values [188]. *(Reproduced from P. Nico, B. Stewart, S. Fendorf, Incorporation of oxidized uranium into Fe (hydr)oxides during Fe(II) catalyzed remineralization, Environ. Sci. Technol. 43 (2009) 7391–7396. Copyright 2009, American Chemical Society.)*

solid U, with the remaining as the incorporated U phase. In addition, no evident alteration was observed after 5 days of aeration, implying that the changing redox conditions rarely affected the coordination mode of the incorporated U. Lastly, the sample datum in the 10 mM Fe(II) systems was best depicted by the Reduced model representing $UO_{2(s)}$.

This research results identified that substantial quantities of U could be retained by incorporation into host phases during the transformation of amorphous FHOs to more crystalline phases. It has been demonstrated that incorporated U is more resistant to both $KHCO_3$ extraction and changing redox conditions than adsorbed ones. The incorporation process is helpful to understand the strong and persistent interactions of U with FHOs in natural environments.

Massey et al. also studied the Fe(II)-induced ferrihydrite transformation experiment and identified incorporation as one of uranium retention pathways using several spectroscopic techniques. The incorporation process occurs mainly in four steps: (1) uranyl is adsorbed onto the ferrihydrite as a mononuclear bidentate complex; (2) binding and electron transfer from Fe(II) binds to the adsorbed uranyl and leads to the formation of U(V) and Fe(II) through electron transfer; (3) the created U(V) shifts to the octahedral site; and, (4) the goethite lattice continues to grow around the U(V) (Fig. 2.28) [205].

Similarly, Stewart et al. [189] examined the sorption and release of uranium interacted with FHOs under redox conditions. U is incorporated into the FHOs during phase transformation catalyzed by Fe(II), and remains stable after exposed to oxic conditions (Fig. 2.29). However, when increasing Fe(II) concentration to 10 mM, the initially adsorbed U(VI) becomes incorporated into the FHOs and is reduced to form precipitation of $UO_{2(s)}$. These solids seem to be less stable in oxic conditions, giving rise to the dissolution of $UO_{2(s)}$, the oxidation of Fe(II)-containing minerals, and the final release of U(VI) into solution. Hence, a stable phase for sequestering U might be preferred when the Fe(II) concentration is at a level that favors formation to goethite or lepidocrocite.

Boland et al. [190] studied the fate of U(VI) during the Fe(II)-induced transformations of ferrihydrite with and without silicate by using EXAFS. Ferrihydrite transformed to goethite in the presence of $Fe(II)_{aq}$, whereas ferrihydrite with silicate transformed into a phase resembling ferrihydrite synthesized without silicate. Analysis of EXAFS data showed that a mononuclear bidentate inner-sphere surface complex was initially created for both systems. During the transformation process, U(VI) was the most likely reduced product for the system without silicate while in the presence of silicate, U(VI) remained similar to the adsorbed state [190].

Missana et al. [110] noticed that the redox potential of the magnetite-solution equilibrium is slightly positive (50–100 mV) under anoxic conditions, providing the thermodynamic possibility of uranium reduction within the liquid phase. The $UO_{2(s)}$ precipitate is likely to form during the reduction

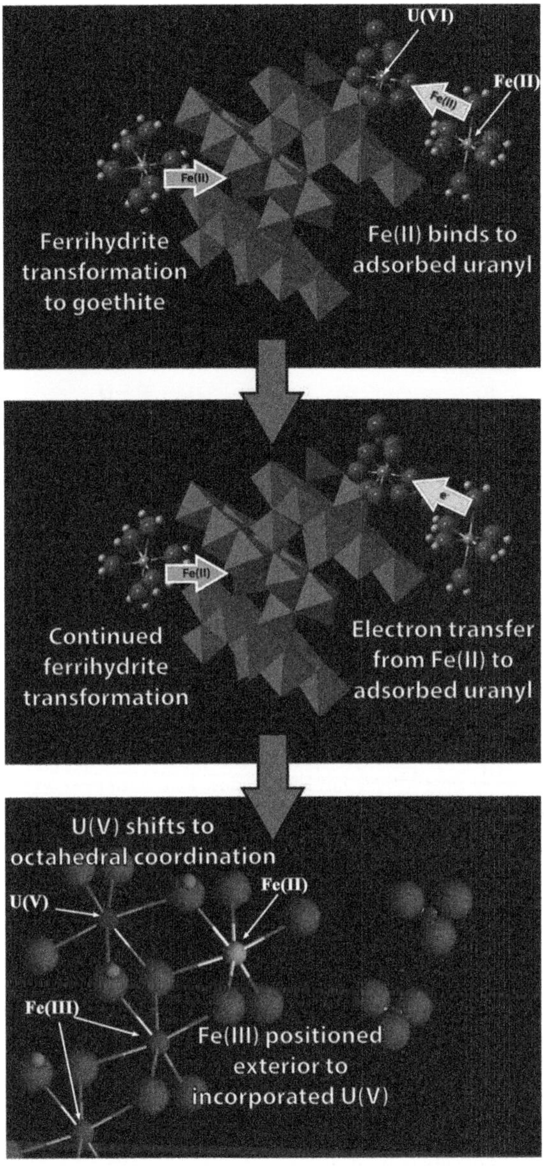

FIG. 2.28 Uranium incorporation mechanism. The addition of Fe(II) drives ferrihydrite transformation to goethite, and Fe(II) binding to adsorbed uranyl initiates incorporation (top). U(V) incorporation occurs via electron transfer from Fe(II) to U(VI) (*middle*), creating a Fe(III) atom exterior to an incorporated U(V) (*bottom*), and enabling continued crystal growth of goethite [205]. *(Modified from M. Massey, J. Lezama-Pacheco, M. Jones, E. Ilton, J. Cerrato, J. Bargar, S. Fendorf, Competing retention pathways of uranium upon reaction with Fe(II), Geochim. Cosmochim. Acta 142 (2014) 166–185. Copyright 2014, Elsevier.)*

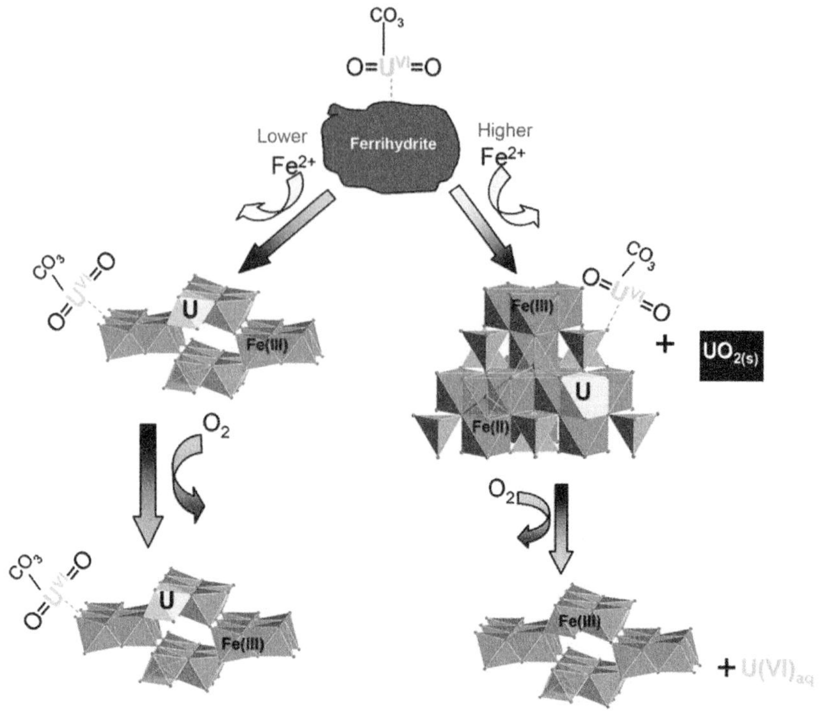

FIG. 2.29 Schematic diagram depicting dominant mechanisms of the U(VI)-ferrihydrite system reacted with Fe(II) [189]. *(Reproduced from B.D. Stewart, P.S. Nico, S. Fendorf, Stability of uranium incorporated into Fe (hydr)oxides under fluctuating redox conditions, Environ. Sci. Technol. 43 (2009) 4922–4927. Copyright 2009, American Chemical Society.)*

process induced by magnetite. However, Ilton et al. [203] examined the anaerobic system containing U(VI) and magnetite and found the partial reduction of U(VI) to U(V) with no observation of U(IV). The incorporation of U(V) and U(VI) into secondary phases is proposed to explain the experimental results. Aamrani et al. [206] explored the reaction of U(VI) on the magnetite surface by means of EXAFS. The spectrum of the UO_2 sample was best modeled with a coordination number (CN) of 8 for the equational oxygen shell (EOS) at a distance of 2.35 Å (Fig. 2.30 and Table 2.4). As for schoepite, the best fit was obtained with CN = 3.3 for the first EOS at 2.31 Å and CN = 0.8 at 2.55 Å for the second EOS. In addition, an EOS of CN = 7.0 at 2.30 Å and CN = 1.7 for the axial oxygen shell at 1.77 Å was derived to simulate the spectrum of the U-magnetite sample well. This observation corroborated the ability of magnetite to reduce U(VI) to U(IV).

Kirsch et al. employed X-ray absorption near-edge structure (XANES) and EXAFS spectra to investigate Pu(III) and Pu(V) interaction with magnetite [207]. Pu(V) was reduced to Pu(III) by magnetite, which is confirmed by

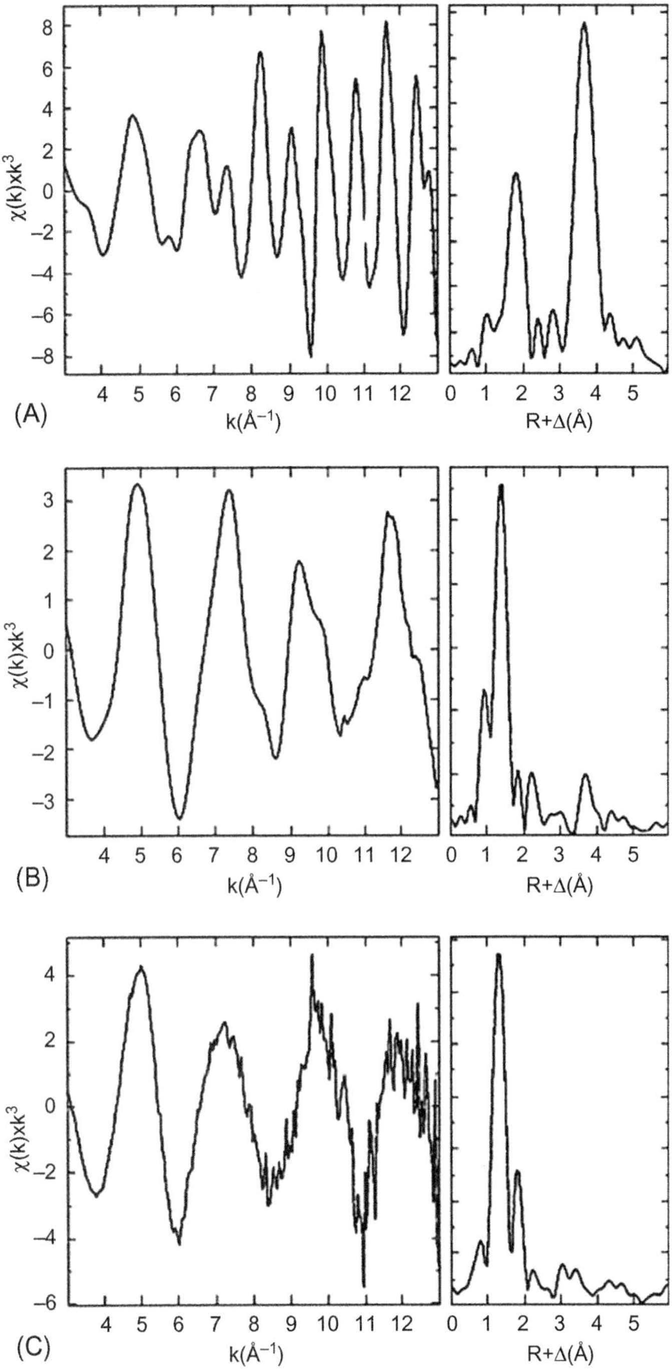

FIG. 2.30 k^3-weighted U-L_{III} edge EXAFS spectra (*left*) and the corresponding Fourier transforms (*right*), (A) UO_2(s), (B) schoepite, and (C) uranium-coated magnetite [206]. *(Reproduced from S. El Aamrani, J. Giménez, M. Rovira, F. Seco, M. Grive, J. Bruno, L. Duro, J. De Pablo, A spectroscopic study of uranium(VI) interaction with magnetite, Appl. Surf. Sci. 253 (2007) 8794–8797. Copyright 2007, Elsevier.)*

TABLE 2.4 EXAFS structural parameters for the three samples [206]

Sample	Shell	R (Å)[a]	N[b]	σ^2 (Å^2)	ΔE_0
UO_2	U–O_{ax}	1.81	1.8	0.00232	−0.56
Schoepite	U–O_{eq}	2.35	8.1	0.0085	−10.0
	U–O_{eq}	2.31	3.3	0001563	−0.56
	U–O_{eq}	2.55	0.8	0.00170	−0.56
U-magnetite	U–O_{ax}	1.77	1.7	0.00086	−10.8
	U–O_{eq}	2.30	7.0	0.01826	−10.8
	U–O_{eq}	2.30	7.0	0.01826	−10.8

[a]*Errors in distances are ±0.02 Å.*
[b]*Errors in the coordination number are ±25%; σ^2 is the Debye-Waller factor, which accounts for the effects of thermal and statistic disorder.*
Reproduced from S. El Aamrani, J. Giménez, M. Rovira, F. Seco, M. Grive, J. Bruno, L. Duro, J. De Pablo, A spectroscopic study of uranium(VI) interaction with magnetite, Appl. Surf. Sci. 253 (2007) 8794–8797.

XANES (Fig. 2.31A). The EXAFS data were best simulated with two shells: the oxygen shell corresponding to 9 O atoms at 2.49 Å and the iron shell with 3 Fe atoms at 3.54 Å (Fig. 2.31B). Refinement led to the creation of Pu—O and Pu—Fe radial pair distribution functions (Fig. 2.31C). It was identified that there is only one specific position for complexation, which is situated on the (111) surface with octahedral geometry (Fig. 2.31D). A trinuclear, tridentate, triple edge-sharing Pu(III) surface was formed on the specific coordination sites (Fig. 2.31E). These results also signified that under anoxic conditions, Pu(III) species other than PuO_2 also needed to be judged for environmental risk assessment.

More Np was adsorbed on magnetite under anaerobic conditions compared with that under aerobic conditions [208]. In the system containing Np(V) and $Fe(II)_{aq}$, the reduction extent of Np(V) to Np(IV) after 7 days was just 6% at pH 4 and 6, which was surmised to take place mainly on the surface of magnetite. For systems with $Fe(II)_{aq}$, the bound Fe(II) on FHOs surfaces could reduce Tc(VII) to Tc(IV) [209]. Um et al. [178] also reported that Tc(VII) was efficiently catalyzed and reduced by Fe(II) and reduced to Tc(IV). The interaction of Tc(VII) and $Fe(II)_{aq}$ in Al (hydr)oxide suspensions was investigated by Peretyazhko et al [210]. They observed complete reduction of Tc(VII) by adsorbed Fe(II). The final redox product on corundum that is analogous to $TcO_2{\cdot}nH_2O$ was revealed by Tc-EXAFS data (Fig. 2.32) [210].

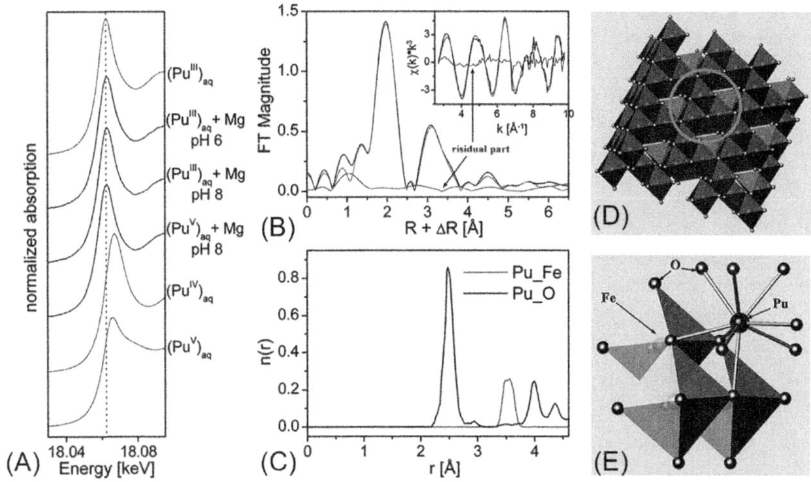

FIG. 2.31 (A) Experimental Pu-L$_{III}$ XANES spectra of Pu reacted with magnetite and (Pu^{III})$_{aq}$, (Pu^{IV})$_{aq}$, and (Pu^{V})$_{aq}$ for reference. (B) Experimental Pu-L$_{III}$ EXAFS spectrum of (Pu^{V})$_{aq}$+Mg, pH 8 sample (*thick black*), and theoretical spectrum (*gray*) resulting from radial refinement by MC simulation, FT, and chi-spectrum (inset). (C) Pu—O and Pu—Fe radial pair distribution functions resulting from radial refinement. (D) Octahedrally terminated (111) face of magnetite with one out of seven possible positions for Pu(III) sorption marked by a circle. (E) Sorption complex structure of Pu(III) on edge-sharing FeO_6-octahedra. *(Modified from R. Kirsch, D. Fellhauer, M. Altmaier, V. Neck, A. Rossberg, T. Fanghänel, L. Charlet, A.C. Scheinost, Oxidation state and local structure of plutonium reacted with magnetite, mackinawite, and chukanovite, Environ. Sci. Technol. 45 (2011) 7267–7274. Copyright 2011, American Chemical Society.)*

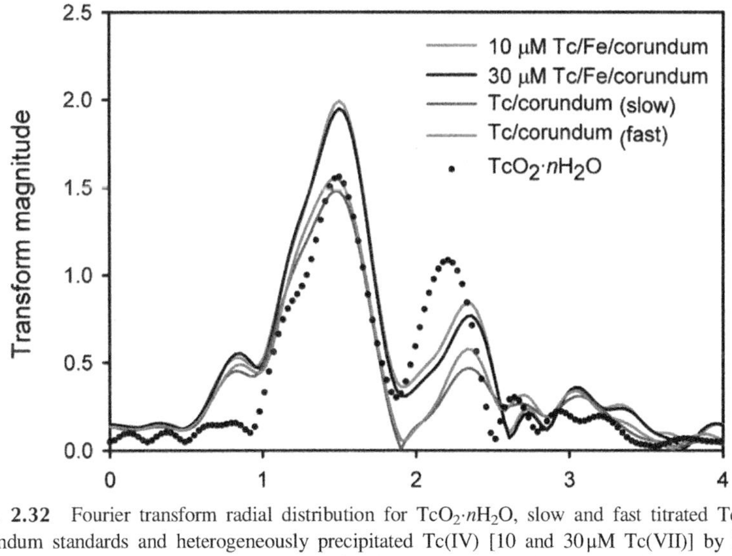

FIG. 2.32 Fourier transform radial distribution for $TcO_2{\cdot}nH_2O$, slow and fast titrated Tc(IV)-corundum standards and heterogeneously precipitated Tc(IV) [10 and 30 μM Tc(VII)] by Fe(II) adsorbed on corundum at pH 7 [210]. *(Reproduced from T. Peretyazhko, J.M. Zachara, S.M. Heald, R.K. Kukkadapu, C. Liu, A.E. Plymale, C.T. Resch, Reduction of Tc(VII) by Fe(II) sorbed on Al (hydr)oxides, Environ. Sci. Technol. 42 (2008) 5499–5506. Copyright 2008, American Chemical Society.)*

2.3.5 Radionuclide sorption on exposed facets of oxides

Generally speaking, it is complicated to interpret the EXAFS data unless based on assumptions that the surfaces of oxides in aqueous solutions have perfect terminations (i.e., no relaxation, defect, or reconstruction), and that the outermost spheres of surface cations are coordinated by water molecules or hydroxyl ions. Powdered substrates with various surface terminations and surface defects, which could produce a quantity of different surface reactive sites for aqueous ion binding, have been used widely in many studies. In such systems, it is often impractical to directly determine the adsorption coordination geometry, and thus challenging to fundamentally understand the complex mechanisms [161]. The surface microstructure and property of oxides (e.g., atom arrangement, unsaturated dangling bonds) affect the adsorption process greatly as adsorption occurs mainly on the surface. Therefore, to explore the reaction mechanism more precisely, it is essential to investigate the adsorption of RNs on different exposed facets of oxides. Even for single-crystal surfaces with only limited potential adsorption sites, it is still hard to ascertain the binding mechanisms of RNs unless the surface structure is clearly known [161]. It has been recognized that different facets of a monocrystalline oxide present distinct geometric and electronic structures, which endow them with different properties [211].

The (0 0 0 1) and (1 0 −1 2) surfaces are two common, energetically stable and well-characterized terminations of α-Al_2O_3 and α-Fe_2O_3 [45, 212–215]. In addition, they are generally taken as model analogues for naturally occurring soil phases (e.g., goethite, gibbsite), and some clay minerals such as kaolinite and smectite, as their exposed surface sites display similar local structure and reactivity [216]. Adsorption of U(VI) on the hydroxylated α-Al_2O_3(0 0 0 1) surface was investigated by means of DFT [163]. Based on the estimation of adsorption energies, the formation of an inner-sphere complex species (i.e., $\equiv S(OH)_2—UO_2(H_2O)_3$) is not likely to form. In contrast, the formation of an outer-sphere complexation with nearly zero enthalpy of adsorption was favored [163].

Catalano et al. [161] determined the adsorption surface complexation of U(VI) on the (1 0 −1 2) surfaces of α-Al_2O_3 and α-Fe_2O_3 by using crystal truncation rod (CTR) diffraction and EXAFS. Ternary uranyl-carbonate complexes to surface oxygens are formed on both surfaces, which is demonstrated by the EXAFS data (Fig. 2.33). On the α-Al_2O_3(1 0 −1 2) surface, a mononuclear monodentate complex was put forward, whereas the formed binuclear bidentate complexes on the α-Fe_2O_3(1 0 −1 2) surface are coordinated to two adjacent oxygen sites. Both XPS and CTR diffraction revealed higher U(VI) coverage density on the α-Fe_2O_3(1 0 −1 2) surface compared with the α-Al_2O_3(1 0 −1 2) surface, implying the intrinsically higher affinity of the α-Fe_2O_3(1 0 −1 2) surface for U(VI) [161].

Yuan et al. [217] investigated the adsorption of Eu^{3+}, Pu^{3+}, Am^{3+}, and Cm^{3+} on α-Fe_2O_3(0 0 0 1) surfaces by using DFT calculations to compare

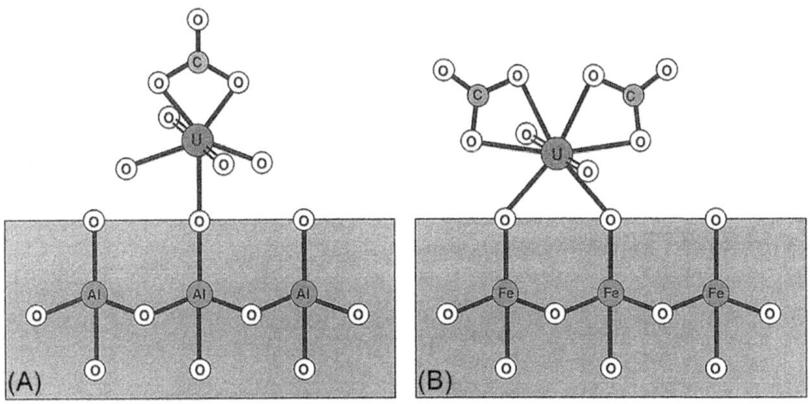

FIG. 2.33 Ball-and-stick representation of hypothetical U(VI) surface complexes consistent with CTR and EXAFS data for (A) α-Al_2O_3 and (B) α-Fe_2O_3(1 0 −1 2) surfaces. Linear uranyl unit is likely aligned parallel to the surface due to steric effects; it is likely perpendicular to plane of page in (B) but may rotate freely in (A) due to the single bond to surface. Unbonded, uncomplexed equatorial oxygen atoms in the α-Al_2O_3 model are likely water and/or hydroxyls [161]. *(Reproduced from J.G. Catalano, T.P. Trainor, P.J. Eng, G.A. Waychunas, G.E. Brown, CTR diffraction and grazing-incidence EXAFS study of U(VI) adsorption onto α-Al_2O_3 and α-Fe_2O_3 (1–102) surfaces, Geochim. Cosmochim. Acta 69 (2005) 3555–3572. Copyright 2005, Elsevier.)*

the difference of electronic properties, coordination structures, and adsorption energies between trivalent actinides An^{3+} and their analogue Eu^{3+}. A tridentate binuclear Eu^{3+} surface complex on the α-Fe_2O_3(0 0 0 1) surface (CN=6) with the lowest adsorption energy was proposed (Fig. 2.34). Under similar adsorption configurations, the distances between An^{3+} cations and the surface are ~0.2 Å closer compared to Eu^{3+}, indicating the tighter binding of An^{3+} on the surface than that of Eu^{3+}. The $[FeO]_6$ hexagon and adjacent hexagonal rings were distorted after An^{3+} adsorption onto the hexagonal cavity.

The local atomic geometries of U(VI) adsorbed onto the rutile (1 1 0) face have been presented by Perron et al [218]. Three possible adsorption sites were put forward in the modeling of the adsorption process (Fig. 2.35): one U atom linked to two terminal O atoms (tt), one U atom linked to two bridging O atoms (bb), and one U atom linked to one bridging and one terminal O atom (bt) Concerning the relative adsorption energies of U(VI), it appeared that the latter two structures were more stable compared with the first one, which is in line with previous experimental results that only two surface complexes were observed [219, 220].

Dossot et al. [221] examined the adsorption of U(VI) on a rutile (0 0 1) single crystal by means of the surface second-harmonic generation. The possible geometry of U(VI) adsorption complexes on rutile is constrained by the nonlinear susceptibility tensor for each sample, which leads to the identification of two sorption sites with different O atoms on the rutile surface. The first adsorption site, noted as the "bb site," denotes a link between the U atom and

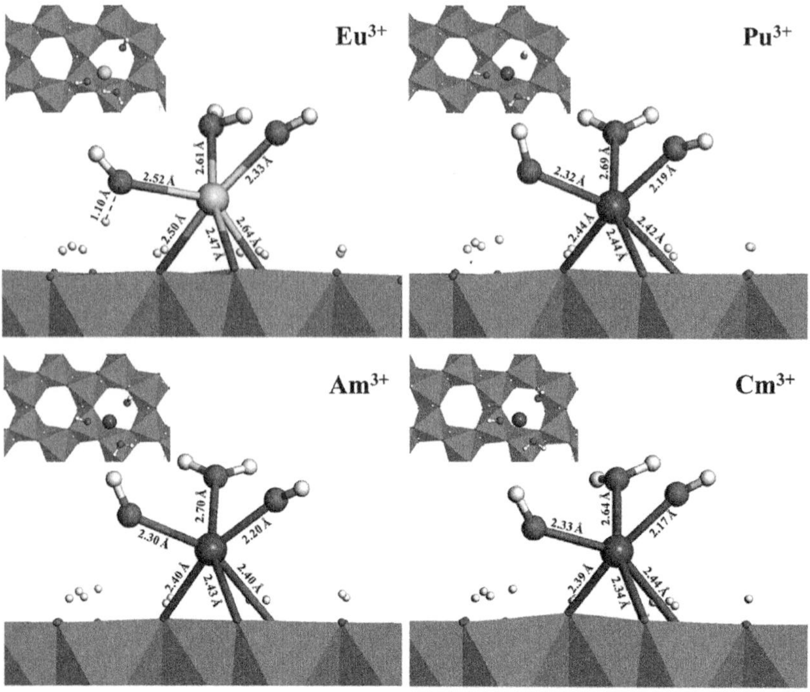

FIG. 2.34 Ground state configuration of the $[Eu/An(OH)_2H_2O]^+$ complexes adsorbed on the α-Fe_2O_3(0 0 0 1) surface, viewed along the [110] direction. Adsorption geometry with the lowest adsorption energy for each cation is shown. For clarity, only the top surface layer of the hematite slab is shown. Inserted graphs are the top views along the [001] direction [217]. *(Reproduced from K. Yuan, S.D. Taylor, B.A. Powell, U. Becker, An ab initio study of the adsorption of* Eu^{3+}*,* Pu^{3+}*,* Am^{3+}*, and* Cm^{3+} *hydroxide complexes on hematite (001) surface: role of magnetism on adsorption, Surf. Sci. 664, (2017), 120–128. Copyright 2017, Elsevier.)*

two bridging O atoms. The second proposed adsorption site, labeled as the "bt site," corresponds to the linking of the U atom to one bridging and one terminal O atom, where the U atom acts as a bridge between the two O atoms [221].

The adsorption of U(VI) on the hydroxylated α-SiO_2(0 0 1) surface was studied by Wang et al. [222] by employing DFT and AIMDs simulation. Six bidentate inner-sphere as well as two monodentate and one outer-sphere complexes were considered to investigate distinct cases. The results showed that the bidentate complex at the double deprotonated site (SiO—SiO) is most favored. The binding strengths of the inner-sphere complexes are similar and become weaker with the increasing number of protons, implying the enhancement effect of higher pH on the adsorption strength within a certain range. Strong chemical interaction exerts great influence on the inner-sphere complexes. The Bader charge analysis demonstrated that the charge-charge interaction between U(VI) and α-SiO_2(0 0 1) surface is the weakest in the SiOHOH and strongest in the SiOO configuration [222].

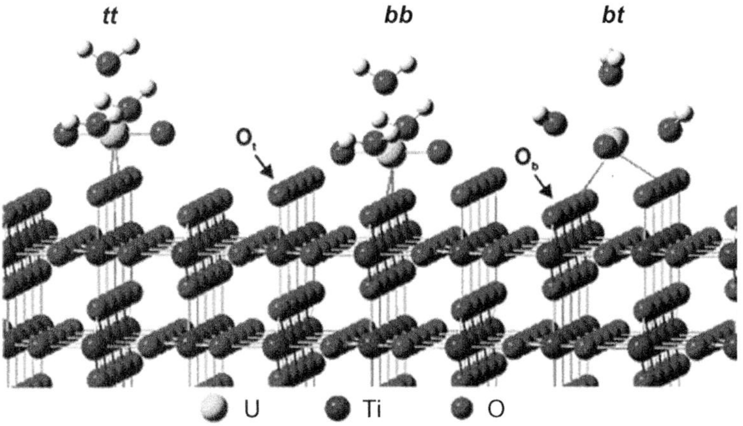

FIG. 2.35 The three possible bidentate sorption sites for the uranyl ion on the TiO_2 rutile (1 1 0) face (from *left* to *right*): terminal-terminal (tt), bridging-bridging (bb), and bridging-terminal (bt). H atoms used to saturate the surface (on terminal and bridging O atoms) are not represented here for clarity [218]. *(Reproduced from H. Perron, C. Domain, J. Roques, R. Drot, E. Simoni, H. Catalette, Periodic density functional theory investigation of the uranyl ion sorption on the TiO_2 rutile (110) face, Inorg. Chem. 45 (2006) 6568–6570. Copyright 2006, American Chemical Society.)*

Veilly et al. [223] researched the adsorption of U(VI) on the gibbsite (0 0 1) face by combining spectroscopies with DFT computations. It was confirmed that there are three possible sites for U(VI) adsorption and the most stable one is depicted in Fig. 2.36. The optimized structures displayed a bidentate interaction of uranyl with two surface oxygen atoms and was stabilized by a hydrogen-bond network. The U-surface bond strength relies on the protonation process of the surface O atoms [223].

2.3.6 Radionuclide adsorption on oxide nanoparticles

Adsorption of RNs may be affected by the nanoscale size of oxides due to their different specific surface areas, the density of surface reactive sites, and surface modification processes (e.g., surface contraction, surface disorder, and quantum confinement) [224]. The surface structures and energetics of oxide nanoparticles can vary substantially to those of larger particles, which can influence the affinity as well as the capacity for RN adsorption.

The nanoscale size effect on U(VI) adsorption to α-Fe_2O_3 particles was explored by Zeng et al [106]. For the aerosol-synthesized particles, the adsorption edge shifted left gradually with increased particle size from 12 to 125 nm. A two-site SCM with strong sites occupying 5% of the total sites was put forward to evaluate the adsorption affinity and capacity of U(VI). The multiple U-Fe distances in EXAFS fitting results also suggested the existence of different types of surface sites (Fig. 2.37). In addition, it is

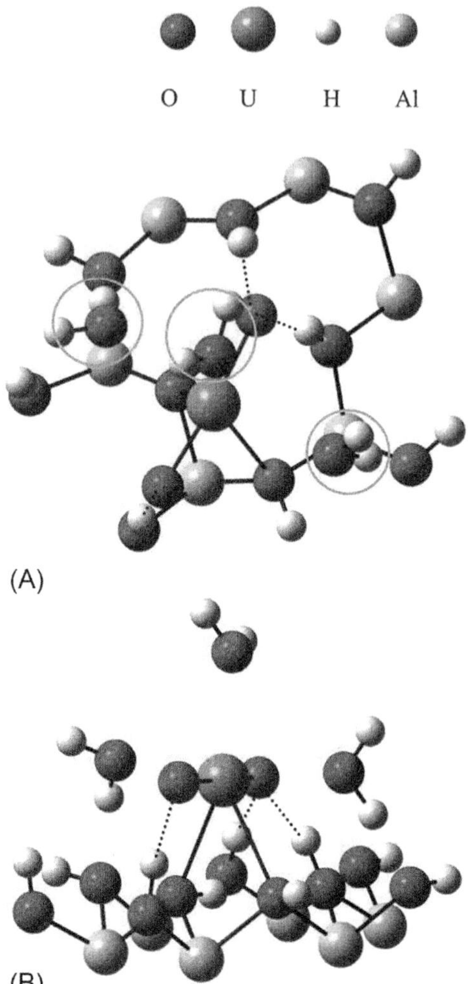

FIG. 2.36 The most stable calculated complex. Hydrogen bonds are in *black dash lines*; water molecules are in circles. (A) Top view and (B) side view [223]. *(Reproduced from E. Veilly, J. Roques, M.C. Jodin-Caumon, B. Humbert, R. Drot, E. Simoni, Uranyl interaction with the hydrated (001) basal face of gibbsite: a combined theoretical and spectroscopic study, J. Chem. Phys. 129 (2008) 244704. Copyright 2008, AIP.)*

concluded that the variance in site concentrations is responsible for the affinity and capacity differences [106].

Nanoscale size can also affect the redox reactions of RNs at oxide-water interfaces. Thus, Zeng et al. [225] then investigated U(VI) reduction in

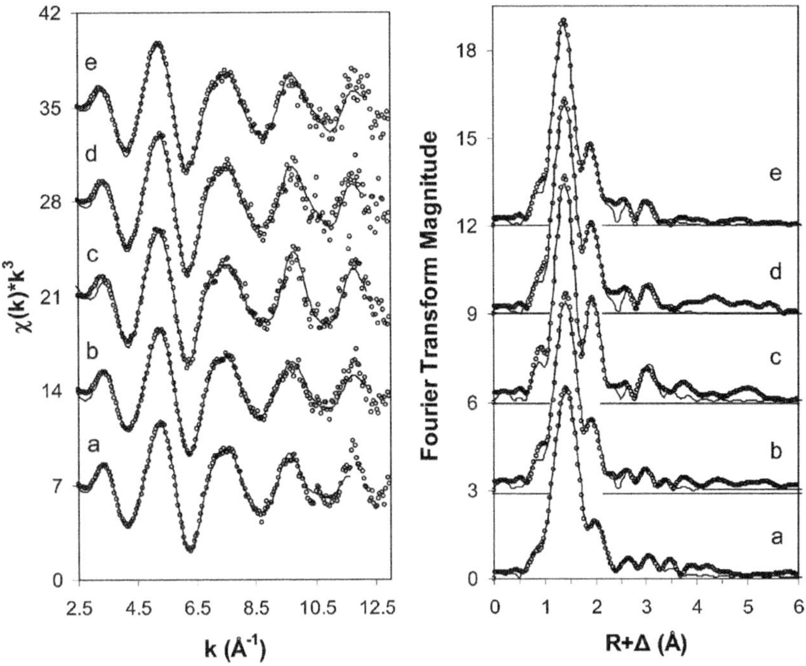

FIG. 2.37 U L_{III}-edge EXAFS spectra (*left*) and Fourier transforms (*right*) of uranyl-sorbed hematite samples for different particle sizes and U surface coverage: (a) 12 nm_1 μmol of U/g (pH 6.54); (b) 12 nm_10 μmol of U/g (pH 6.22); (c) 30 nm_10 μmol of U/g (pH 7.96); (d) 70 nm_1 μmol of U/g (pH 7.01); (e) 70 nm_10 μmol of U/g (pH 6.82). Dots represent the data, and lines represent the least-squares fits to the data [106]. *(Reproduced from H. Zeng, A. Singh, S. Basak, K.U. Ulrich, M. Sahu, P. Biswas, J.G. Catalano, D.E. Giammar, Nanoscale size effects on uranium(VI) adsorption to hematite, Environ. Sci. Technol. 43 (2009) 1373–1378. Copyright 2009, American Chemical Society.)*

hematite suspensions with different sizes, which were preequilibrated with $Fe(II)_{aq}$. The rate constants were similar for aerosol-synthesized hematite particles (30, 50, and 125 nm), but higher than for aqueous-synthesized ones. In addition, the reduction for the 12 nm particles was the fastest among the hematite samples, which is probably due to the faster electron conduction from Fe(II) to U(VI) through hematite and the subsequent formation of a more reactive solid (e.g., magnetite) as well as the quantum confinement effects [225].

The effect of nanoscale size on adsorption-desorption of U(VI) with nano goethite and alumina was studied by Qian et al. [226] The adsorption rate and extent of U(VI) on nano alumina and goethite are higher than that of the non-nanophases as the site densities of nanoparticles are determined to be higher. The batch desorption results denoted that U(VI) binds more strongly on nanoparticles relative to nonnanoparticles, which was in accordance with the

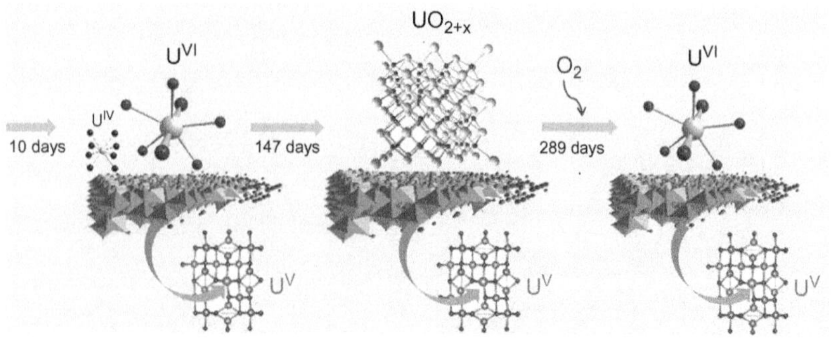

FIG. 2.38 The proposed conceptual model of uranium redox transformations after U(VI) coprecipitation with magnetite nanoparticles [227]. *(Reproduced from I. Pidchenko, K.O. Kvashnina, T. Yokosawa, N. Finck, S. Bahl, D. Schild, R. Polly, E. Bohnert, A. Rossberg, J. Göttlicher, Uranium redox transformations after U(VI) coprecipitation with magnetite nanoparticles, Environ. Sci. Technol. 51 (2017) 2217–2225. Copyright 2017, American Chemical Society.)*

results of SCM, assuming both weak and strong binding sites for nanoparticles while only weak sites for the nonnanophases [226].

The redox states and the speciation of uranium after coprecipitation with nano magnetite were investigated by Pidchenko et al [227]. They proposed a conceptual model to depict the reaction process (Fig. 2.38). Most of the initially added U(VI) was adsorbed on ferrihydrite, which later recrystallizes to magnetite. A proportion of U(VI) was still adsorbed on the magnetite surface after aging for 10 days. It was supposed that the U(V) species were incorporated into the bulk structure of magnetite during the transformation process, as indicated by the more symmetric coordination environment relative to uranyl and coordination to Fe. Minor UO_2/UO_{2+x} or adsorbed U(IV) may also form on the magnetite surface. After aging for 147 days, the adsorbed U(VI) transformed and was reduced to UO_2/UO_{2+x}, whereas most of the incorporated U(V) species was well preserved. When exposed to oxygen for an extra 142 days, U(IV) oxidized back to U(VI) while U(V) remained incorporated into the octahedral sites of magnetite, as the surrounding matrix could prevent U(V) from oxidation. Scott el al [228]. investigated the interaction of natural magnetite with U(VI) and confirmed the presence of $UO_{2(s)}$ at the magnetite surface by XPS and secondary ion mass spectrometry while the reduction and incorporation of U(V) was not observed.

The effect of nanoscale size on the adsorption of Sr(II) to nano anatase samples (diameters of 4, 20, and 40nm) has been evaluated by utilizing macroscopic experimental studies [229]. Adsorption curves of Sr(II) on the anatase surfaces of the 20 and 40nm samples were similar while the curves were shallower for the 4nm samples, meaning that less Sr(II) was adsorbed on the smallest particles than the larger particles. At the molecular scale, DFT calculations were employed to explore the most stable Sr(II) surface complex on the

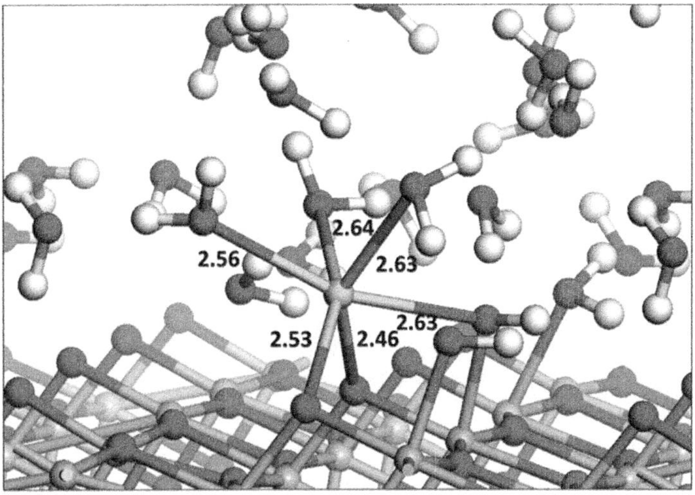

FIG. 2.39 Structural schematic of anatase showing the (101) surface and a representation of the coordination geometry for the inner-sphere Sr^{2+} tridentate species based on the DFT calculations. For clarity, the schematic has been centered on the Sr^{2+} ion and shows only a portion of the cell used in the DFT simulations. All bond lengths are in angstroms [229]. *(Reproduced from M.K. Ridley, M.L. Machesky, J.D. Kubicki, Experimental study of strontium adsorption on anatase nanoparticles as a function of size with a density functional theory and CD model interpretation, Langmuir 31 (2015) 703–713. Copyright 2015, American Chemical Society.)*

predominant anatase (101) surface, which was recognized to be an inner-sphere tridentate Sr(II) surface species (Fig. 2.39). Compared with that for the 20 and 40nm samples, the charge distribution value for the 4nm anatase sample was smaller, reflecting the shallower Sr(II) adsorption curves. This observation can be explained by the fact that water is more highly structured near the smaller anatase particles; moreover, the most stable Sr tridentate surface complex may require more ordered surface terraces than those on the 4nm anatase samples.

Nanosized metal oxides (NMOs) exhibit many advantages (e.g., fast kinetics, high capacity, and adsorption selectivity) for RN removal from wastewater. Nevertheless, there are still some technical barriers to be solved to widen the practical application of NMOs for water purification. For example, the adsorption capacity of NMOs is lost in aqueous solution due to their intense aggregation into large-size particles. Moreover, efficient and cost-effective separation of the NMOs from wastewater still remains interesting but challenging. Fabricating new NMO-based nanocomposite adsorbents seems effective to overcome the above technical problems. However, how to reduce the cost of the new composites and keep the stability during the application are still big problems to solve [230]. In addition, the potential risk of new nanocomposites to the environment and human health also need to be evaluated.

2.4 Conclusion and perspectives

As a conclusion, let us restate the information and content of this chapter. At first, the bulk structures, bulk-terminated surfaces, and hydroxylated surfaces of several common oxides under environmentally relevant conditions were introduced and discussed. The surface properties and dissolution behavior of these oxides were also presented. The interactions between RNs and oxides cover a wide range of topics in dealing with nuclear waste disposal. Then, examples of immobilization or retardation of RNs by common oxides through various mechanisms, for example, adsorption, (co)precipitation, incorporation, and surface-induced redox reaction, have been illustrated. Finally, RN adsorption on specifically exposed facets of oxides and oxide nanoparticles was presented to obtain the specific coordination geometries of RNs on oxide surfaces.

To evaluate the migration, diffusion, adsorption, desorption, precipitation, incorporation, and redox transformation of RNs in the natural environment, it is essential to thoroughly identify the structures and species of RNs. The batch technique studies are necessary to explore the effects of various environmentally relevant conditions. The theoretical modeling and spectroscopy technique analysis of the physicochemical properties of RNs are also imperative to understand their interactions with oxides. Combination of these methods and techniques could draw extraordinarily coherent pictures and provide excellent insights into the understanding of the exact interaction mechanisms of RNs with oxides. In this chapter, the development of interaction mechanisms derived from a range of batch results, spectroscopic techniques, theoretical modeling, and calculations has been highlighted. These data are helpful for risk assessment of the behaviors of the RNs in underground and terrestrial environments. They could also be used to evaluate the RN geochemical behaviors under nuclear waste tank extreme conditions.

Although the understanding of RN interaction with oxides has made great strides, there still exist several ambiguous matters to be solved. Despite the rapid development of modern spectroscopic and computational techniques, it is suggested that they be further improved, as more definitive proof for specific coordination geometries is needed. To make real progress, the theoretical models and spectroscopic techniques should be more closely linked for identifying surface species and their coordination structures. In addition, these models have poor performance in a system containing mixed oxides, which were predicted simply from individual oxide studies [231]. Although the real radioactive wastes are multicomponent systems containing different RNs, there is also a need to study the behavior of targeted RNs in the presence of other ones [232]. Moreover, the long-time performance of the interaction mechanisms is the largest obstacle of the RN adsorption experiments on oxides, as most studies conducted in the laboratory were in a relatively short period of time. We hope to directly and precisely determine the behaviors on the RN-oxide interfaces, eventually casting results into theoretical models with predictive capabilities.

References

[1] D. Li, D.I. Kaplan, Sorption coefficients and molecular mechanisms of Pu, U, Np, Am and Tc to Fe (hydr) oxides: a review, J. Hazard. Mater. 243 (2012) 1–18.

[2] M.J. Keith-Roach, The speciation, stability, solubility and biodegradation of organic co-contaminant radionuclide complexes: a review, Sci. Total Environ. 396 (2008) 1–11.

[3] H.A. Al-Abadleh, V.H. Grassian, Oxide surfaces as environmental interfaces, Surf. Sci. Rep. 52 (2003) 63–161.

[4] W. Dong, T.K. Tokunaga, J.A. Davis, J. Wan, Uranium(VI) adsorption and surface complexation modeling onto background sediments from the F-Area Savannah River site, Environ. Sci. Technol. 46 (2012) 1565–1571.

[5] J.C. Seaman, P.M. Bertsch, R.N. Strom, Characterization of colloids mobilized from southeastern coastal plain sediments, Environ. Sci. Technol. 31 (1997) 2782–2790.

[6] W. Um, R.J. Serne, C.F. Brown, K.A. Rod, Uranium(VI) sorption on iron oxides in Hanford Site sediment: application of a surface complexation model, Appl. Geochem. 23 (2008) 2649–2657.

[7] M.E. Bishop, H. Dong, R.K. Kukkadapu, C. Liu, R.E. Edelmann, Bioreduction of Fe-bearing clay minerals and their reactivity toward pertechnetate (Tc-99), Geochim. Cosmochim. Acta 75 (2011) 5229–5246.

[8] J.M. Zachara, J. Serne, M. Freshley, F. Mann, F. Anderson, M. Wood, T. Jones, D. Myers, Geochemical processes controlling migration of tank wastes in Hanford's vadose zone, Vadose Zone J. 6 (2007) 985–1003.

[9] D. Linde, CRC Handbook of Chemistry and Physics, CRC Press, Boca Raton, Ann Arbor, Boston, 1991 (1992).

[10] L. Pauling, S.B. Hendricks, The crystal structures of hematite and corundum, J. Am. Chem. Soc. 47 (1925) 781–790.

[11] S.E. Mason, C.R. Iceman, K.S. Tanwar, T.P. Trainor, A.M. Chaka, Pb(II) adsorption on isostructural hydrated alumina and hematite (0001) surfaces: a DFT study, J. Phys. Chem. C 113 (2009) 2159–2170.

[12] X. Krokidis, P. Raybaud, A.E. Gobichon, B. Rebours, P. Euzen, H. Toulhoat, Theoretical study of the dehydration process of boehmite to γ-alumina, J. Phys. Chem. B 105 (2001) 5121–5130.

[13] M. Digne, P. Sautet, P. Raybaud, P. Euzen, H. Toulhoat, Use of DFT to achieve a rational understanding of acid-basic properties of γ-alumina surfaces, J. Catal. 226 (2004) 54–68.

[14] G.S. Parkinson, Iron oxide surfaces, Surf. Sci. Rep. 71 (2016) 272–365.

[15] M. Kuemmel, D. Grosso, C. Boissière, B. Smarsly, T. Brezesinski, P.A. Albouy, H. Amenitsch, C. Sanchez, Thermally stable nanocrystalline γ-alumina layers with highly ordered 3D mesoporosity, Angew. Chem. Int. Ed. 44 (2005) 4589–4592.

[16] A. Barbieri, W. Weiss, M. Van Hove, G. Somorjai, Magnetite Fe_3O_4 (111): surface structure by LEED crystallography and energetics, Surf. Sci. 302 (1994) 259–279.

[17] U. Diebold, The surface science of titanium dioxide, Surf. Sci. Rep. 48 (2003) 53–229.

[18] K. Bourikas, C. Kordulis, A. Lycourghiotis, Titanium dioxide (anatase and rutile): surface chemistry, liquid-solid interface chemistry, and scientific synthesis of supported catalysts, Chem. Rev. 114 (2014) 9754–9823.

[19] H.H. Kung, Transition Metal Oxides: Surface Chemistry and Catalysis, Elsevier, Amsterdam, 1989.

[20] V. Henrich, P. Cox, The Surface Science of Metal Oxides, Cambridge University Press, Cambridge, 1994. Oxford.

[21] S.A. Chambers, Epitaxial growth and properties of thin film oxides, Surf. Sci. Rep. 39 (2000) 105–180.
[22] S.L. Hemmingson, G.M. Feeley, N.J. Miyake, C.T. Campbell, Energetics of 2D and 3D gold nanoparticles on MgO(100): influence of particle size and defects on gold adsorption and adhesion energies, ACS Catal. 7 (2017) 2151–2163.
[23] L. Zhang, C. Tian, G.A. Waychunas, Y.R. Shen, Structures and charging of α-alumina (0001)/water interfaces studied by sum-frequency vibrational spectroscopy, J. Am. Chem. Soc. 130 (2008) 7686–7694.
[24] S. Yin, X. Ma, D. Ellis, Initial stages of H_2O adsorption and hydroxylation of Fe-terminated α-Fe_2O_3 (0001) surface, Surf. Sci. 601 (2007) 2426–2437.
[25] K.S. Tanwar, S.C. Petitto, S.K. Ghose, P.J. Eng, T.P. Trainor, Fe(II) adsorption on hematite (0001), Geochim. Cosmochim. Acta 73 (2009) 4346–4365.
[26] C.S. Lo, K.S. Tanwar, A.M. Chaka, T.P. Trainor, Density functional theory study of the clean and hydrated hematite (1-102) surfaces, Phys. Rev. B 75 (2007) 075425.
[27] J.G. Catalano, Z. Zhang, C. Park, P. Fenter, M.J. Bedzyk, Bridging arsenate surface complexes on the hematite (012) surface, Geochim. Cosmochim. Acta 71 (2007) 1883–1897.
[28] F. Labat, P. Baranek, C. Adamo, Structural and electronic properties of selected rutile and anatase TiO_2 surfaces: an ab initio investigation, J. Chem. Theory Comput. 4 (2008) 341–352.
[29] P. Reinhardt, B. Hess, Electronic and geometrical structure of rutile surfaces, Phys. Rev. B 50 (1994) 12015.
[30] G.M. Rignanese, A. De Vita, J.C. Charlier, X. Gonze, R. Car, First-principles molecular-dynamics study of the (0001) α-quartz surface, Phys. Rev. B 61 (2000) 13250.
[31] N.H. de Leeuw, F.M. Higgins, S.C. Parker, Modeling the surface structure and stability of α-quartz, J. Phys. Chem. B 103 (1999) 1270–1277.
[32] T. Goumans, A. Wander, W.A. Brown, C.R.A. Catlow, Structure and stability of the (001) α-quartz surface, Phys. Chem. Chem. Phys. 9 (2007) 2146–2152.
[33] P.J. Eng, T.P. Trainor, G.E. Brown Jr., G.A. Waychunas, M. Newville, S.R. Sutton, M.L. Rivers, Structure of the hydrated α-Al_2O_3 (0001) surface, Science 288 (2000) 1029–1033.
[34] K. Hass, W. Schneider, A. Curioni, W. Andreoni, First-principles molecular dynamics simulations of H_2O on α-Al_2O_3 (0001), J. Phys. Chem. B 104 (2000) 5527–5540.
[35] T.P. Trainor, A.M. Chaka, P.J. Eng, M. Newville, G.A. Waychunas, J.G. Catalano, G.E. Brown, Structure and reactivity of the hydrated hematite (0001) surface, Surf. Sci. 573 (2004) 204–224.
[36] X.G. Wang, A. Chaka, M. Scheffler, Effect of the environment on α-Al_2O_3 (0001) surface structures, Phys. Rev. Lett. 84 (2000) 3650.
[37] T.P. Trainor, P.J. Eng, G.E. Brown, I.K. Robinson, M. De Santis, Crystal truncation rod diffraction study of the α-Al_2O_3 (1-102) surface, Surf. Sci. 496 (2002) 238–250.
[38] A. Ionescu, A. Allouche, J.P. Aycard, M. Rajzmann, F. Hutschka, Study of γ-alumina surface reactivity: adsorption of water and hydrogen sulfide on octahedral aluminum sites, J. Phys. Chem. B 106 (2002) 9359–9366.
[39] E. DeCanio, J. Edwards, J. Bruno, Solid-state 1H MAS NMR characterization of γ-alumina and modified γ-aluminas, J. Catal. 148 (1994) 76–83.
[40] X.L. Tan, X.M. Ren, J.X. Li, X.K. Wang, Theoretical investigation of uranyl ion adsorption on hydroxylated γ-Al_2O_3 surfaces, RSC Adv. 3 (2013) 19551–19559.
[41] M.A. Henderson, Structural sensitivity in the dissociation of water on TiO_2 single-crystal surfaces, Langmuir 12 (1996) 5093–5098.

[42] M. Předota, A. Bandura, P. Cummings, J. Kubicki, D. Wesolowski, A. Chialvo, M. Machesky, Electric double layer at the rutile (110) surface. 1. structure of surfaces and interfacial water from molecular dynamics by use of ab initio potentials, J. Phys. Chem. B 108 (2004) 12049–12060.
[43] A. Fahmi, C. Minot, A theoretical investigation of water adsorption on titanium dioxide surfaces, Surf. Sci. 304 (1994) 343–359.
[44] G.M. Rignanese, J.C. Charlier, X. Gonze, First-principles molecular-dynamics investigation of the hydration mechanisms of the (0001) α-quartz surface, Phys. Chem. Chem. Phys. 6 (2004) 1920–1925.
[45] R.M. Cornell, U. Schwertmann, The Iron Oxides: Structure, Properties, Reactions, Occurrences and Uses, John Wiley & Sons, 2003.
[46] A. Manceau, V. Drits, Local structure of ferrihydrite and feroxyhite by EXAFS spectroscopy, Clay Miner. 28 (1993) 165.
[47] P. Weidler, T. Schwinn, H. Gaub, Vicinal facies on synthetic goethite observed by atomic force microscopy, Clay. Clay Miner. 44 (1996) 437–442.
[48] J. Lützenkirchen, J.F. Boily, L. Lövgren, S. Sjöberg, Limitations of the potentiometric titration technique in determining the proton active site density of goethite surfaces, Geochim. Cosmochim. Acta 66 (2002) 3389–3396.
[49] T. Hiemstra, P. Venema, W.H. Van Riemsdijk, Intrinsic proton affinity of reactive surface groups of metal (hydr)oxides: the bond valence principle, J. Colloid Interface Sci. 184 (1996) 680–692.
[50] T. Hiemstra, W.H. Van Riemsdijk, A surface structural model for ferrihydrite I: sites related to primary charge, molar mass, and mass density, Geochim. Cosmochim. Acta 73 (2009) 4423–4436.
[51] R. Parfitt, J. Russell, Adsorption on hydrous oxides. IV. Mechanisms of adsorption of various ions on goethite, Eur. J. Soil Sci. 28 (1977) 297–305.
[52] N. Kaur, M. Gräfe, B. Singh, B. Kennedy, Simultaneous incorporation of Cr, Zn, Cd, and Pb in the goethite structure, Clay. Clay Miner. 57 (2009) 234–250.
[53] J. Majzlan, L. Mazeina, A. Navrotsky, Enthalpy of water adsorption and surface enthalpy of lepidocrocite (γ-FeOOH), Geochim. Cosmochim. Acta 71 (2007) 615–623.
[54] J. Majzlan, K.D. Grevel, A. Navrotsky, Thermodynamics of Fe oxides: part II. Enthalpies of formation and relative stability of goethite (α-FeOOH), lepidocrocite (γ-FeOOH), and maghemite (γ-Fe2O3), Am. Mineral. 88 (2003) 855–859.
[55] R.W. Fitzpatrick, R. Taylor, U. Schwertmann, C. Childs, Occurrence and properties of lepidocrocite in some soils of New Zealand, South Africa and Australia, Aust. J. Soil Res. 23 (1985) 543–567.
[56] L.H.C. Dos Anjos, D. Franzmeier, D. Schulze, Formation of soils with plinthite on a toposequence in Maranhao State, Brazil, Geoderma 64 (1995) 257–279.
[57] N.T. Goldman, C.B. Koch, A. Singer, Lepidocrocite in hydrothermal sedimentof the Atlantis II and Thetis Deeps, Red Sea, Clay. Clay Miner. 50 (2002) 186–197.
[58] U. Schwertmann, A. Fechter, The formation of green rust and its transformation to lepidocrocite, Clay Miner. 29 (1994) 87–92.
[59] L. Carlson, U. Schwertmann, The effect of CO_2 and oxidation rate on the formation of goethite versus lepidocrocite from an Fe(II) system at pH 6 and 7, Clay Miner. 25 (1990) 65–71.
[60] G.Y. Chao, J. Baker, A.P. Sabina, A.C. Roberts, Doyleite, a new polymorph of $Al(OH)_3$, and its relationship to bayerite, gibbsite and nordstrandite, Can. Mineral. 23 (1985) 21–28.

[61] J.D. Gale, A.L. Rohl, V. Milman, M.C. Warren, An ab initio study of the structure and properties of aluminum hydroxide: gibbsite and bayerite, J. Phys. Chem. B 105 (2001) 10236–10242.
[62] M.M. Fernandes, B. Baeyens, C. Beaucaire, Radionuclide retention at mineral-water interfaces in the natural environment, in: Radionuclide Behaviour in the Natural Environment, Elsevier, 2012, pp. 261–301.
[63] T. Hiemstra, W. Van Riemsdijk, Physical chemical interpretation of primary charging behaviour of metal (hydr)oxides, Colloids Surf. 59 (1991) 7–25.
[64] M. Villalobos, J.O. Leckie, Carbonate adsorption on goethite under closed and open CO_2 conditions, Geochim. Cosmochim. Acta 64 (2000) 3787–3802.
[65] G. Lefevre, S. Noinville, M. Fédoroff, Study of uranyl sorption onto hematite by in situ attenuated total reflection-infrared spectroscopy, J. Colloid Interface Sci. 296 (2006) 608–613.
[66] M. Kosmulski, The pH dependent surface charging and points of zero charge. VI. Update, J. Colloid Interface Sci. 426 (2014) 209–212.
[67] W. Stumm, Reactivity at the mineral-water interface: dissolution and inhibition, Colloids Surf. A 120 (1997) 143–166.
[68] U. Schwertmann, Solubility and dissolution of iron oxides, Plant and Soil 130 (1991) 1–25.
[69] W.L. Lindsay, Solubility and redox equilibria of iron compounds in soils, in: Iron in Soils and Clay Minerals, Springer, New York, 1988, pp. 37–62.
[70] P.A. Connor, K.D. Dobson, A.J. McQuillan, Infrared spectroscopy of the TiO_2/aqueous solution interface, Langmuir 15 (1999) 2402–2408.
[71] Y. Duval, J. Mielczarski, O. Pokrovsky, E. Mielczarski, J. Ehrhardt, Evidence of the existence of three types of species at the quartz-aqueous solution interface at pH $0-10$: XPS surface group quantification and surface complexation modeling, J. Phys. Chem. B 106 (2002) 2937–2945.
[72] Z. Du, N.H. de Leeuw, Molecular dynamics simulations of hydration, dissolution and nucleation processes at the α-quartz (0001) surface in liquid water, Dalton Trans. (2006) 2623–2634.
[73] G.E. Brown, How minerals react with water, Science 294 (2001) 67–69.
[74] P.A. Maurice, Environmental Surfaces and Interfaces From the Nanoscale to the Global Scale, Wiley, 2009.
[75] W. Stumm, Chemistry of the Solid-Water Interface: Processes at the Mineral-Water and Particle-Water Interface in Natural Systems, John Wiley & Son Inc., 1992.
[76] G. Sposito, Distinguishing Adsorption From Surface Precipitation, ACS Publications, 1986.
[77] A. Godelitsas, J.M. Astilleros, Dissolution, sorption/(re) precipitation, formation of solid solutions and crystal growth phenomena on mineral surfaces: implications for the removal of toxic metals from the environment, in: EMU Notes in Mineralogy, vol. 10, 2010, pp. 289–324.
[78] E.J. Elzinga, D.L. Sparks, Nickel sorption mechanisms in a pyrophyllite-montmorillonite mixture, J. Colloid Interface Sci. 213 (1999) 506–512.
[79] A.M. Scheidegger, D.G. Strawn, G.M. Lamble, D.L. Sparks, The kinetics of mixed Ni-Al hydroxide formation on clay and aluminum oxide minerals: a time-resolved XAFS study, Geochim. Cosmochim. Acta 62 (1998) 2233–2245.
[80] M. Olsson, A.M. Jakobsson, Y. Albinsson, Sorption of Pu(VI) onto TiO_2, J. Colloid Interface Sci. 266 (2003) 269–275.
[81] X.L. Tan, M. Fang, J.X. Li, Y. Lu, X.K. Wang, Adsorption of Eu(III) onto TiO_2: effect of pH, concentration, ionic strength and soil fulvic acid, J. Hazard. Mater. 168 (2009) 458–465.

[82] P. Jardine, G. Jacobs, J. O'Dell, Unsaturated transport processes in undisturbed heterogeneous porous media: II. Co-contaminants, Soil Sci. Soc. Am. J. 57 (1993) 954–962.
[83] A. Tessier, R. Carignan, B. Dubreuil, F. Rapin, Partitioning of zinc between the water column and the oxic sediments in lakes, Geochim. Cosmochim. Acta 53 (1989) 1511–1522.
[84] J. Rosenqvist, P. Persson, S. Sjöberg, Protonation and charging of nanosized gibbsite (α-$Al(OH)_3$) particles in aqueous suspension, Langmuir 18 (2002) 4598–4604.
[85] X.M. Ren, S.T. Yang, X.L. Tan, C.L. Chen, X.K. Wang, Investigation of radionuclide $^{60}Co(II)$ binding to TiO_2 by batch technique, surface complexation model and DFT calculations, Sci. China Chem. 55 (2012) 1752–1759.
[86] T. Hiemstra, W.H. Van Riemsdijk, A. Rossberg, K.U. Ulrich, A surface structural model for ferrihydrite II: adsorption of uranyl and carbonate, Geochim. Cosmochim. Acta 73 (2009) 4437–4451.
[87] D.A. Dzombak, F.M. Morel, Surface Complexation Modeling: Hydrous Ferric Oxide, John Wiley & Sons, 1990.
[88] T. Rabung, H. Geckeis, J.I. Kim, H.P. Beck, Sorption of Eu(III) on a natural hematite: application of a surface complexation model, J. Colloid Interface Sci. 208 (1998) 153–161.
[89] X.L. Tan, X.M. Ren, C.L. Chen, X.K. Wang, Analytical approaches to the speciation of lanthanides at solid-water interfaces, Trends Anal. Chem. 61 (2014) 107–132.
[90] Q.H. Fan, X.L. Tan, J.X. Li, X.K. Wang, W.S. Wu, G. Montavon, Sorption of Eu(III) on attapulgite studied by batch, XPS, and EXAFS techniques, Environ. Sci. Technol. 43 (2009) 5776–5782.
[91] X.L. Tan, Q.H. Fan, X.K. Wang, B. Grambow, Eu(III) sorption to TiO_2 (anatase and rutile): batch, XPS, and EXAFS studies, Environ. Sci. Technol. 43 (2009) 3115–3121.
[92] D.D. Shao, Q.H. Fan, J.X. Li, Z.W. Niu, W.S. Wu, Y.X. Chen, X.K. Wang, Removal of Eu(III) from aqueous solution using ZSM-5 zeolite, Microporous Mesoporous Mater. 123 (2009) 1–9.
[93] A.M. Jakobsson, Y. Albinsson, R.S. Rundberg, Measurements and modelling of Pm sorption onto TiO_2 and goethite, Radiochim. Acta 92 (2004) 683–689.
[94] W. Piasecki, D.A. Sverjensky, Speciation of adsorbed yttrium and rare earth elements on oxide surfaces, Geochim. Cosmochim. Acta 72 (2008) 3964–3979.
[95] J. Kuta, M.C. Wander, Z. Wang, S. Jiang, N.A. Wall, A.E. Clark, Trends in Ln(III) sorption to quartz assessed by molecular dynamics simulations and laser-induced fluorescence studies, J. Phys. Chem. C 115 (2011) 21120–21127.
[96] T. Rabung, T. Stumpf, H. Geckeis, R. Klenze, J. Kim, Sorption of Am(III) and Eu(III) onto γ-alumina: experiment and modelling, Radiochim. Acta 88 (2000) 711–716.
[97] M.K. Ridley, M.L. Machesky, D.J. Wesolowski, D.A. Palmer, Surface complexation of neodymium at the rutile-water interface: a potentiometric and modeling study in NaCl media to 250°C, Geochim. Cosmochim. Acta 69 (2005) 63–81.
[98] X.L. Tan, X.K. Wang, H. Geckeis, T.H. Rabung, Sorption of Eu(III) on humic acid or fulvic acid bound to hydrous alumina studied by SEM-EDS, XPS, TRLFS, and batch techniques, Environ. Sci. Technol. 42 (2008) 6532–6537.
[99] A. Richter, V. Brendler, C. Nebelung, Blind prediction of Cu(II) sorption onto goethite: current capabilities of diffuse double layer model, Geochim. Cosmochim. Acta 69 (2005) 2725–2734.
[100] S.L. Estes, Y. Arai, U. Becker, S. Fernando, K. Yuan, R.C. Ewing, J. Zhang, T. Shibata, B.A. Powell, A self-consistent model describing the thermodynamics of Eu(III) adsorption onto hematite, Geochim. Cosmochim. Acta 122 (2013) 430–447.
[101] H. Tamura, E. Matijević, L. Meites, Adsorption of Co^{2+} ions on spherical magnetite particles, J. Colloid Interface Sci. 92 (1983) 303–314.

[102] R. Sheha, E. Metwally, Equilibrium isotherm modeling of cesium adsorption onto magnetic materials, J. Hazard. Mater. 143 (2007) 354–361.
[103] T. Missana, M. Garcı´a-Gutiérrez, C. Maffiotte, Experimental and modeling study of the uranium(VI) sorption on goethite, J. Colloid Interface Sci. 260 (2003) 291–301.
[104] C.D. Hsi, D. Langmuir, Adsorption of uranyl onto ferric oxyhydroxides: application of the surface complexation site-binding model, Geochim. Cosmochim. Acta 49 (1985) 1931–1941.
[105] J.H. Jang, B.A. Dempsey, W.D. Burgos, A model-based evaluation of sorptive reactivities of hydrous ferric oxide and hematite for U(VI), Environ. Sci. Technol. 41 (2007) 4305–4310.
[106] H. Zeng, A. Singh, S. Basak, K.U. Ulrich, M. Sahu, P. Biswas, J.G. Catalano, D.E. Giammar, Nanoscale size effects on uranium(VI) adsorption to hematite, Environ. Sci. Technol. 43 (2009) 1373–1378.
[107] Z.Y. Tao, T.W. Chu, W.J. Li, J.Z. Du, X.X. Dai, Y.J. Gu, Cation adsorption of NpO^{2+}, UO_2^{2+}, Zn^{2+}, Sr^{2+}, Yb^{3+}, and Am^{3+} onto oxides of Al, Si, and Fe from aqueous solution: ionic strength effect, Colloids Surf., A 242 (2004) 39–45.
[108] J.R. Bargar, R. Reitmeyer, J.J. Lenhart, J.A. Davis, Characterization of U(VI)-carbonato ternary complexes on hematite: EXAFS and electrophoretic mobility measurements, Geochim. Cosmochim. Acta 64 (2000) 2737–2749.
[109] E. Liger, L. Charlet, P. Van Cappellen, Surface catalysis of uranium(VI) reduction by iron(II), Geochim. Cosmochim. Acta 63 (1999) 2939–2955.
[110] T. Missana, C. Maffiotte, M. Garcia-Gutiérrez, Surface reactions kinetics between nanocrystalline magnetite and uranyl, J. Colloid Interface Sci. 261 (2003) 154–160.
[111] T. Missana, M. Garcia-Gutierrez, V. Fernńdez, Uranium(VI) sorption on colloidal magnetite under anoxic environment: experimental study and surface complexation modelling, Geochim. Cosmochim. Acta 67 (2003) 2543–2550.
[112] D.M. Sherman, C.L. Peacock, C.G. Hubbard, Surface complexation of U(VI) on goethite (α-FeOOH), Geochim. Cosmochim. Acta 72 (2008) 298–310.
[113] L.N. Moyes, R.H. Parkman, J.M. Charnock, D.J. Vaughan, F.R. Livens, C.R. Hughes, A. Braithwaite, Uranium uptake from aqueous solution by interaction with goethite, lepidocrocite, muscovite, and mackinawite: an X-ray absorption spectroscopy study, Environ. Sci. Technol. 34 (2000) 1062–1068.
[114] S.D. Yusan, S.A. Erenturk, Sorption behaviors of uranium(VI) ions on α-FeOOH, Desalination 269 (2011) 58–66.
[115] T. Waite, J. Davis, T. Payne, G. Waychunas, N. Xu, Uranium(VI) adsorption to ferrihydrite: application of a surface complexation model, Geochim. Cosmochim. Acta 58 (1994) 5465–5478.
[116] T.E. Payne, V. Brendler, M.J. Comarmond, C. Nebelung, Assessment of surface area normalisation for interpreting distribution coefficients (Kd) for uranium sorption, J. Environ. Radioact. 102 (2011) 888–895.
[117] A.L. Sanchez, J.W. Murray, T.H. Sibley, The adsorption of plutonium IV and V on goethite, Geochim. Cosmochim. Acta 49 (1985) 2297–2307.
[118] M. Duff, C. Amrhein, Uranium(VI) adsorption on goethite and soil in carbonate solutions, Soil Sci. Soc. Am. J. 60 (1996) 1393–1400.
[119] J.R. Bargar, R. Reitmeyer, J.A. Davis, Spectroscopic confirmation of uranium(VI)-carbonato adsorption complexes on hematite, Environ. Sci. Technol. 33 (1999) 2481–2484.
[120] K.U. Ulrich, A. Rossberg, H. Foerstendorf, H. Zänker, A.C. Scheinost, Molecular characterization of uranium(VI) sorption complexes on iron(III)-rich acid mine water colloids, Geochim. Cosmochim. Acta 70 (2006) 5469–5487.

[121] A. Rossberg, K.U. Ulrich, S. Weiss, S. Tsushima, T. Hiemstra, A.C. Scheinost, Identification of uranyl surface complexes on ferrihydrite: advanced EXAFS data analysis and CD-MUSIC modeling, Environ. Sci. Technol. 43 (2009) 1400–1406.

[122] T. Reich, H. Moll, T. Arnold, M. Denecke, C. Hennig, G. Geipel, G. Bernhard, H. Nitsche, P. Allen, J. Bucher, An EXAFS study of uranium(VI) sorption onto silica gel and ferrihydrite, J. Electron. Spectrosc. Relat. Phenom. 96 (1998) 237–243.

[123] A.Y. Romanchuk, S.N. Kalmykov, R.A. Aliev, Plutonium sorption onto hematite colloids at femto-and nanomolar concentrations, Radiochim. Acta 99 (2011) 137–144.

[124] G. Bernhard, G. Geipel, T. Reich, V. Brendler, S. Amayri, H. Nitsche, Uranyl(VI) carbonate complex formation: validation of the $Ca_2UO_2(CO_3)_3$(aq) species, Radiochim. Acta 89 (2001) 511–518.

[125] S.N. Kalmykov, G.R. Choppin, Mixed $Ca^{2+}/UO_2^{2+}/CO_2^{2-}$ complex formation at different ionic strengths, Radiochim. Acta 88 (2000) 603–608.

[126] S.C. Brooks, J.K. Fredrickson, S.L. Carroll, D.W. Kennedy, J.M. Zachara, A.E. Plymale, S.D. Kelly, K.M. Kemner, S. Fendorf, Inhibition of bacterial U(VI) reduction by calcium, Environ. Sci. Technol. 37 (2003) 1850–1858.

[127] J.A. Davis, D.E. Meece, M. Kohler, G.P. Curtis, Approaches to surface complexation modeling of uranium(VI) adsorption on aquifer sediments, Geochim. Cosmochim. Acta 68 (2004) 3621–3641.

[128] P.M. Fox, J.A. Davis, J.M. Zachara, The effect of calcium on aqueous uranium(VI) speciation and adsorption to ferrihydrite and quartz, Geochim. Cosmochim. Acta 70 (2006) 1379–1387.

[129] P. Li, Z.X. Yin, J.F. Lin, Q. Jin, Y.F. Du, Q.H. Fan, W.S. Wu, The immobilization of U(VI) on iron oxyhydroxides under various physicochemical conditions, Environ. Sci.: Processes Impacts 16 (2014) 2278–2287.

[130] J.J. Lenhart, B.D. Honeyman, Uranium(VI) sorption to hematite in the presence of humic acid, Geochim. Cosmochim. Acta 63 (1999) 2891–2901.

[131] T. Sakuragi, S. Sato, T. Kozaki, T. Mitsugashira, M. Hara, Y. Suzuki, Am(III) and Eu(III) uptake on hematite in the presence of humic acid, Radiochim. Acta 92 (2004) 697–702.

[132] S. Stumpf, T. Stumpf, K. Dardenne, C. Hennig, H. Foerstendorf, R. Klenze, T. Fanghanel, Sorption of Am(III) onto 6-line-ferrihydrite and its alteration products: investigations by EXAFS, Environ. Sci. Technol. 40 (2006) 3522–3528.

[133] A. Khasanova, S.N. Kalmykov, I. Perminova, S. Clark, Neptunium redox behavior and sorption onto goethite and hematite in the presence of humic acids with different hydroquinone content, J. Alloys Compd. 444 (2007) 491–494.

[134] A. Jain, N. Rawat, S. Kumar, B.S. Tomar, V. Manchanda, S. Ramanathan, Effect of humic acid on sorption of neptunium on hematite colloids, Radiochim. Acta 95 (2007) 501–506.

[135] N. Baumann, V. Brendler, T. Arnold, G. Geipel, G. Bernhard, Uranyl sorption onto gibbsite studied by time-resolved laser-induced fluorescence spectroscopy (TRLFS), J. Colloid Interface Sci. 290 (2005) 318–324.

[136] H.S. Chang, G.V. Korshin, Z.M. Wang, J.M. Zachara, Adsorption of uranyl on gibbsite: a time-resolved laser-induced fluorescence spectroscopy study, Environ. Sci. Technol. 40 (2006) 1244–1249.

[137] T. Hattori, T. Saito, K. Ishida, A.C. Scheinost, T. Tsuneda, S. Nagasaki, S. Tanaka, The structure of monomeric and dimeric uranyl adsorption complexes on gibbsite: a combined DFT and EXAFS study, Geochim. Cosmochim. Acta 73 (2009) 5975–5988.

[138] K. Gückel, A. Rossberg, V. Brendler, H. Foerstendorf, Binary and ternary surface complexes of U(VI) on the gibbsite/water interface studied by vibrational and EXAFS spectroscopy, Chem. Geol. 326 (2012) 27–35.

[139] K. Müller, H. Foerstendorf, V. Brendler, A. Rossberg, K. Stolze, A. Gröschel, The surface reactions of U(VI) on γ-Al_2O_3—in situ spectroscopic evaluation of the transition from sorption complexation to surface precipitation, Chem. Geol. 357 (2013) 75–84.

[140] S.T. Yang, G.D. Sheng, G. Montavon, Z.Q. Guo, X.L. Tan, B. Grambow, X.K. Wang, Investigation of Eu(III) immobilization on γ-Al_2O_3 surfaces by combining batch technique and EXAFS analyses: role of contact time and humic acid, Geochim. Cosmochim. Acta 121 (2013) 84–104.

[141] C.L. Chen, X.K. Wang, Influence of pH, soil humic/fulvic acid, ionic strength and foreign ions on sorption of thorium(IV) onto gamma-Al_2O_3, Appl. Geochem. 22 (2007) 436–445.

[142] X.K. Wang, D. Xu, L. Chen, X.L. Tan, X. Zhou, A. Ren, C.L. Chen, Sorption and complexation of Eu(III) on alumina: effects of pH, ionic strength, humic acid and chelating resin on kinetic dissociation study, Appl. Radiat. Isot. 64 (2006) 414–421.

[143] X.L. Li, C.L. Chen, P.P. Chang, S.M. Yu, W.S. Wu, X.K. Wang, Comparative studies of cobalt sorption and desorption on bentonite, alumina and silica: effect of pH and fulvic acid, Desalination 244 (2009) 283–292.

[144] M.J. Comarmond, T.E. Payne, J.J. Harrison, S. Thiruvoth, H.K. Wong, R.D. Aughterson, G.R. Lumpkin, K. Müller, H. Foerstendorf, Uranium sorption on various forms of titanium dioxide–influence of surface area, surface charge, and impurities, Environ. Sci. Technol. 45 (2011) 5536–5542.

[145] Z.J. Guo, Z.Y. Yan, Z.Y. Tao, Sorption of uranyl ions on TiO_2: effects of contact time, ionic strength, concentration and humic substance, J. Radioanal. Nucl. Chem. 261 (2004) 157–162.

[146] M. Wazne, X. Meng, G.P. Korfiatis, C. Christodoulatos, Carbonate effects on hexavalent uranium removal from water by nanocrystalline titanium dioxide, J. Hazard. Mater. 136 (2006) 47–52.

[147] J. Vandenborre, R. Drot, E. Simoni, Interaction mechanisms between uranium(VI) and rutile titanium dioxide: from single crystal to powder, Inorg. Chem. 46 (2007) 1291–1296.

[148] G. Lefèvre, J. Kneppers, M. Fédoroff, Sorption of uranyl ions on titanium oxide studied by ATR-IR spectroscopy, J. Colloid Interface Sci. 327 (2008) 15–20.

[149] C. Den Auwer, R. Drot, E. Simoni, S. Conradson, M. Gailhanou, J.M. De Leon, Grazing incidence XAFS spectroscopy of uranyl sorbed onto TiO_2 rutile surfaces, New J. Chem. 27 (2003) 648–655.

[150] G. Montavon, C. Hennig, P. Janvier, B. Grambow, Comparison of complexed species of Eu in alumina-bound and free polyacrylic acid: a spectroscopic study, J. Colloid Interface Sci. 300 (2006) 482–490.

[151] S. Debnath, U.C. Ghosh, Nanostructured hydrous titanium(IV) oxide: synthesis, characterization and Ni(II) adsorption behavior, Chem. Eng. J. 152 (2009) 480–491.

[152] N. Marmier, A. Delisée, F. Fromage, Surface complexation modeling of Yb(III) and Cs(I) sorption on silica, J. Colloid Interface Sci. 212 (1999) 228–233.

[153] C.L. Chen, X.K. Wang, Sorption of Th(IV) to silica as a function of pH, humic/fulvic acid, ionic strength, electrolyte type, Appl. Radiat. Isot. 65 (2007) 155–163.

[154] E. Sylwester, E. Hudson, P. Allen, The structure of uranium(VI) sorption complexes on silica, alumina, and montmorillonite, Geochim. Cosmochim. Acta 64 (2000) 2431–2438.

[155] M.J. Comarmond, R. Steudtner, M. Stockmann, K. Heim, K. Müller, V. Brendler, T.E. Payne, H. Foerstendorf, The sorption processes of U(VI) onto SiO_2 in the presence of phosphate: from binary surface species to precipitation, Environ. Sci. Technol. 50 (2016) 11610–11618.

[156] R.G. Ford, A.C. Scheinost, D.L. Sparks, Frontiers in metal sorption/precipitation mechanisms on soil mineral surfaces, Adv. Agron. 74 (2001) 41–62.
[157] S.E. Fendorf, G.M. Lamble, M.G. Stapleton, M.J. Kelley, D.L. Sparks, Mechanisms of chromium(III) sorption on silica. 1. Chromium(III) surface structure derived by extended x-ray absorption fine structure spectroscopy, Environ. Sci. Technol. 28 (1994) 284–289.
[158] G. Waychunas, C. Fuller, J. Davis, Surface complexation and precipitate geometry for aqueous Zn(II) sorption on ferrihydrite I: X-ray absorption extended fine structure spectroscopy analysis, Geochim. Cosmochim. Acta 66 (2002) 1119–1137.
[159] P.A. O'Day, C.J. Chisholm-Brause, S.N. Towle, G.A. Parks, G.E. Brown Jr., X-ray absorption spectroscopy of Co(II) sorption complexes on quartz (α-SiO_2) and rutile (TiO_2), Geochim. Cosmochim. Acta 60 (1996) 2515–2532.
[160] M.N. Nguyen, S. Dultz, J. Kasbohm, D. Le, Clay dispersion and its relation to surface charge in a paddy soil of the Red River Delta, Vietnam, J. Plant Nutr. Soil Sci. 172 (2009) 477–486.
[161] J.G. Catalano, T.P. Trainor, P.J. Eng, G.A. Waychunas, G.E. Brown, CTR diffraction and grazing-incidence EXAFS study of U(VI) adsorption onto α-Al_2O_3 and α-Fe_2O_3 (1-102) surfaces, Geochim. Cosmochim. Acta 69 (2005) 3555–3572.
[162] A. Froideval, M. Del Nero, C. Gaillard, R. Barillon, I. Rossini, J.L. Hazemann, Uranyl sorption species at low coverage on Al-hydroxide: TRLFS and XAFS studies, Geochim. Cosmochim. Acta 70 (2006) 5270–5284.
[163] L.V. Moskaleva, V.A. Nasluzov, N. Rösch, Modeling adsorption of the uranyl dication on the hydroxylated α-Al_2O_3(0001) surface in an aqueous medium. Density functional study, Langmuir 22 (2006) 2141–2145.
[164] S. Kumar, A.S. Kar, B.S. Tomar, D. Bhattacharyya, X-ray absorption fine structure spectroscopy study of Eu(III) sorption products onto amorphous silica and γ-alumina: effect of pH and substrate, Polyhedron 33 (2012) 33–40.
[165] S.N. Towle, J.R. Bargar, G.E. Brown Jr., G.A. Parks, Surface precipitation of Co(II)(aq) on Al_2O_3, J. Colloid Interface Sci. 187 (1997) 62–82.
[166] H. Geckeis, J. Lützenkirchen, R. Polly, T. Rabung, M. Schmidt, Mineral-water interface reactions of actinides, Chem. Rev. 113 (2013) 1016–1062.
[167] L.E. Katz, K.F. Hayes, Surface complexation modeling: II. Strategy for modeling polymer and precipitation reactions at high surface coverage, J. Colloid Interface Sci. 170 (1995) 491–501.
[168] Y.J. Lee, E.J. Elzinga, R.J. Reeder, Sorption mechanisms of zinc on hydroxyapatite: systematic uptake studies and EXAFS spectroscopy analysis, Environ. Sci. Technol. 39 (2005) 4042–4048.
[169] M.L. Schlegel, A. Manceau, L. Charlet, D. Chateigner, J.L. Hazemann, Sorption of metal ions on clay minerals. III. Nucleation and epitaxial growth of Zn phyllosilicate on the edges of hectorite, Geochim. Cosmochim. Acta 65 (2001) 4155–4170.
[170] M.L. Schlegel, A. Manceau, Evidence for the nucleation and epitaxial growth of Zn phyllosilicate on montmorillonite, Geochim. Cosmochim. Acta 70 (2006) 901–917.
[171] X.F. Liang, Y.B. Zang, Y.M. Xu, X. Tan, W.G. Hou, L. Wang, Y.B. Sun, Sorption of metal cations on layered double hydroxides, Colloids Surf. A 433 (2013) 122–131.
[172] R. Dähn, A.M. Scheidegger, A. Manceau, M.L. Schlegel, B. Baeyens, M.H. Bradbury, M. Morales, Neoformation of Ni phyllosilicate upon Ni uptake on montmorillonite: a kinetics study by powder and polarized extended X-ray absorption fine structure spectroscopy, Geochim. Cosmochim. Acta 66 (2002) 2335–2347.

[173] A.C. Scheinost, R.G. Ford, D.L. Sparks, The role of Al in the formation of secondary Ni precipitates on pyrophyllite, gibbsite, talc, and amorphous silica: a DRS study, Geochim. Cosmochim. Acta 63 (1999) 3193–3203.
[174] J.B. d'Espinose de la Caillerie, M. Kermarec, O. Clause, Impregnation of γ-alumina with Ni(II) or Co(II) ions at neutral pH: hydrotalcite-type coprecipitate formation and characterization, J. Am. Chem. Soc. 117 (1995) 11471–11481.
[175] Y. Takahashi, M. Murata, T. Kimura, Interaction of Eu(III) ion and non-porous silica: irreversible sorption of Eu(III) on silica and hydrolysis of silica promoted by Eu(III), J. Alloys Compd. 408 (2006) 1246–1251.
[176] T. Missana, U. Alonso, A.C. Scheinost, N. Granizo, M. Garcia-Gutierrez, Selenite retention by nanocrystalline magnetite: role of adsorption, reduction and dissolution/co-precipitation processes, Geochim. Cosmochim. Acta 73 (2009) 6205–6217.
[177] M. Prieto, J.M. Astilleros, L. Fernández-Díaz, Environmental remediation by crystallization of solid solutions, Elements 9 (2013) 195–201.
[178] W. Um, H.S. Chang, J.P. Icenhower, W.W. Lukens, R.J. Serne, N.P. Qafoku, J.H. Westsik Jr., E.C. Buck, S.C. Smith, Immobilization of 99-technetium(VII) by Fe(II)-goethite and limited reoxidation, Environ. Sci. Technol. 45 (2011) 4904–4913.
[179] W. Um, H. Chang, J.P. Icenhower, W.W. Lukens, R.J. Serne, N. Qafoku, R.K. Kukkadapu, J.H. Westsik Jr., Iron oxide waste form for stabilizing ^{99}Tc, J. Nucl. Mater. 429 (2012) 201–209.
[180] J. McBeth, J. Lloyd, G. Law, F. Livens, I. Burke, K. Morris, Redox interactions of technetium with iron-bearing minerals, Mineral. Mag. 75 (2011) 2419–2430.
[181] T.A. Marshall, K. Morris, G.T. Law, J.F.W. Mosselmans, P. Bots, S.A. Parry, S. Shaw, Incorporation and retention of 99-Tc(IV) in magnetite under high pH conditions, Environ. Sci. Technol. 48 (2014) 11853–11862.
[182] S.A. Luksic, B.J. Riley, M. Schweiger, P. Hrma, Incorporating technetium in minerals and other solids: a review, J. Nucl. Mater. 466 (2015) 526–538.
[183] P. Hrma, J.V. Crum, P. Bredt, L.R. Greenwood, B. Arey, H. Smith, Vitrification and testing of a Hanford high-level waste sample. Part 2: phase identification and waste form leachability, J. Nucl. Mater. 345 (2005) 31–40.
[184] A. Akatov, B. Nikonov, B. Omel'yanenko, S. Stefanovsky, J. Marra, Structure of borosilicate glassy materials with high concentrations of sodium, iron, and aluminum oxides, Glass Phys. Chem 35 (2009) 245–259.
[185] A. Akatov, B. Nikonov, B. Omel'yanenko, O. Stefanovskaya, S. Stefanovsky, D.Y. Suntsov, J. Marra, Influence of the content of a surrogate of iron aluminate high-level wastes on the phase composition and structure of glassy materials for their immobilization, Glass Phys. Chem 36 (2010) 45–52.
[186] B. Singh, D. Sherman, R. Gilkes, M. Wells, J. Mosselmans, Structural chemistry of Fe, Mn, and Ni in synthetic hematites as determined by extended X-ray absorption fine structure spectroscopy, Clay. Clay Miner. 48 (2000) 521–527.
[187] J. Liu, C.H. Liang, H.M. Zhang, Z.F. Tian, S.Y. Zhang, General strategy for doping impurities (Ge, Si, Mn, Sn, Ti) in hematite nanocrystals, J. Phys. Chem. C 116 (2012) 4986–4992.
[188] P. Nico, B. Stewart, S. Fendorf, Incorporation of oxidized uranium into Fe (hydr)oxides during Fe(II) catalyzed remineralization, Environ. Sci. Technol. 43 (2009) 7391–7396.
[189] B.D. Stewart, P.S. Nico, S. Fendorf, Stability of uranium incorporated into Fe (hydr)oxides under fluctuating redox conditions, Environ. Sci. Technol. 43 (2009) 4922–4927.
[190] D.D. Boland, R.N. Collins, T.E. Payne, T.D. Waite, Effect of amorphous Fe(III) oxide transformation on the Fe(II)-mediated reduction of U(VI), Environ. Sci. Technol. 45 (2011) 1327–1333.

[191] T. Payne, J. Davis, T. Waite, Uranium retention by weathered schists—the role of iron minerals, Radiochim. Acta 66 (1994) 297–304.

[192] M. Duff, J. Coughlin, D. Hunter, Uranium co-precipitation with iron oxide minerals, Geochim. Cosmochim. Acta 66 (2002) 3533–3547.

[193] E. Ilton, J. Pacheco, J. Bargar, Z. Shi, J. Liu, L. Kovarik, M. Engelhard, A. Felmy, Reduction of U(VI) incorporated in the structure of hematite, Environ. Sci. Technol. 46 (2012) 9428–9436.

[194] R. Blake, R. Hessevick, T. Zoltai, L.W. Finger, Refinement of hematite structure, Am. Mineral. 51 (1966) 123–129.

[195] T. Marshall, K. Morris, G. Law, F. Livens, J. Mosselmans, P. Bots, S. Shaw, Incorporation of uranium into hematite during crystallization from ferrihydrite, Environ. Sci. Technol. 48 (2014) 3724–3731.

[196] S. Kerisit, E.J. Bylaska, M.S. Massey, M.E. McBriarty, E.S. Ilton, Ab initio molecular dynamics of uranium incorporated in goethite (α-FeOOH): interpretation of X-ray absorption spectroscopy of trace polyvalent metals, Inorg. Chem. 55 (2016) 11736–11746.

[197] M.E. McBriarty, J.A. Soltis, S. Kerisit, O. Qafoku, M.E. Bowden, E.J. Bylaska, J.J. De Yoreo, E.S. Ilton, Trace uranium partitioning in a multiphase nano-FeOOH system, Environ. Sci. Technol. 51 (2017) 4970–4977.

[198] S. Kerisit, A.R. Felmy, E.S. Ilton, Atomistic simulations of uranium incorporation into iron (hydr)oxides, Environ. Sci. Technol. 45 (2011) 2770–2776.

[199] F.N. Smith, C.D. Taylor, W. Um, A.A. Kruger, Technetium incorporation into goethite (α-FeOOH): an atomic-scale investigation, Environ. Sci. Technol. 49 (2015) 13699–13707.

[200] A.J. Frierdich, Y. Luo, J.G. Catalano, Trace element cycling through iron oxide minerals during redox-driven dynamic recrystallization, Geology 39 (2011) 1083–1086.

[201] B.H. Jeon, B.A. Dempsey, W.D. Burgos, M.O. Barnett, E.E. Roden, Chemical reduction of U(VI) by Fe(II) at the solid-water interface using natural and synthetic Fe(III) oxides, Environ. Sci. Technol. 39 (2005) 5642–5649.

[202] R. Crane, M. Dickinson, I. Popescu, T. Scott, Magnetite and zero-valent iron nanoparticles for the remediation of uranium contaminated environmental water, Water Res. 45 (2011) 2931–2942.

[203] E.S. Ilton, J.F. Boily, E.C. Buck, F.N. Skomurski, K.M. Rosso, C.L. Cahill, J.R. Bargar, A.R. Felmy, Influence of dynamical conditions on the reduction of U(VI) at the magnetite solution interface, Environ. Sci. Technol. 44 (2010) 170–176.

[204] R. Nakada, Y. Takahashi, M. Tanimizu, Isotopic and speciation study on cerium during its solid-water distribution with implication for Ce stable isotope as a paleo-redox proxy, Geochim. Cosmochim. Acta 103 (2013) 49–62.

[205] M. Massey, J. Lezama-Pacheco, M. Jones, E. Ilton, J. Cerrato, J. Bargar, S. Fendorf, Competing retention pathways of uranium upon reaction with Fe(II), Geochim. Cosmochim. Acta 142 (2014) 166–185.

[206] S. El Aamrani, J. Giménez, M. Rovira, F. Seco, M. Grive, J. Bruno, L. Duro, J. De Pablo, A spectroscopic study of uranium(VI) interaction with magnetite, Appl. Surf. Sci. 253 (2007) 8794–8797.

[207] R. Kirsch, D. Fellhauer, M. Altmaier, V. Neck, A. Rossberg, T. Fanghänel, L. Charlet, A.C. Scheinost, Oxidation state and local structure of plutonium reacted with magnetite, mackinawite, and chukanovite, Environ. Sci. Technol. 45 (2011) 7267–7274.

[208] K. Nakata, S. Nagasaki, S. Tanaka, Y. Sakamoto, T. Tanaka, H. Ogawa, Sorption and reduction of neptunium(V) on the surface of iron oxides, Radiochim. Acta 90 (2002) 665–669.

[209] T. Peretyazhko, J.M. Zachara, S.M. Heald, B.H. Jeon, R.K. Kukkadapu, C. Liu, D. Moore, C.T. Resch, Heterogeneous reduction of Tc(VII) by Fe(II) at the solid-water interface, Geochim. Cosmochim. Acta 72 (2008) 1521–1539.

[210] T. Peretyazhko, J.M. Zachara, S.M. Heald, R.K. Kukkadapu, C. Liu, A.E. Plymale, C.T. Resch, Reduction of Tc(VII) by Fe(II) sorbed on Al (hydr)oxides, Environ. Sci. Technol. 42 (2008) 5499–5506.

[211] J. Jiang, K. Zhao, X.Y. Xiao, L.Z. Zhang, Synthesis and facet-dependent photoreactivity of BiOCl single-crystalline nanosheets, J. Am. Chem. Soc. 134 (2012) 4473–4476.

[212] J.H. Choi, D.Y. Kim, B.J. Hockey, S.M. Wiederhorn, C.A. Handwerker, J.E. Blendell, W.C. Carter, A.R. Roosen, Equilibrium shape of internal cavities in sapphire, J. Am. Ceram. Soc. 80 (1997) 62–68.

[213] M. Kitayama, T. Narushima, A.M. Glaeser, The Wulff shape of alumina: II, experimental measurements of pore shape evolution rates, J. Am. Ceram. Soc. 83 (2000) 2572–2583.

[214] M. Kitayama, A.M. Glaeser, The Wulff shape of alumina: III, undoped alumina, J. Am. Ceram. Soc. 85 (2002) 611–622.

[215] A. Marmier, S:C. Parker, Ab initio morphology and surface thermodynamics of α-Al_2O_3, Phys. Rev. B 69 (2004) 115409.

[216] G.E. Brown, V.E. Henrich, W.H. Casey, D.L. Clark, C. Eggleston, A. Felmy, D.W. Goodman, M. Grätzel, G. Maciel, M.I. McCarthy, Metal oxide surfaces and their interactions with aqueous solutions and microbial organisms, Chem. Rev. 99 (1999) 77–174.

[217] K. Yuan, S.D. Taylor, B.A. Powell, U. Becker, An ab initio study of the adsorption of Eu^{3+}, Pu^{3+}, Am^{3+}, and Cm^{3+} hydroxide complexes on hematite (001) surface: role of magnetism on adsorption, Surf. Sci. 664 (2017) 120–128.

[218] H. Perron, C. Domain, J. Roques, R. Drot, E. Simoni, H. Catalette, Periodic density functional theory investigation of the uranyl ion sorption on the TiO_2 rutile (110) face, Inorg. Chem. 45 (2006) 6568–6570.

[219] H. Perron, C. Domain, J. Roques, R. Drot, E. Simoni, H. Catalette, Theoretical first step towards an understanding of the uranyl ion sorption on the rutile TiO_2 (110) face: a DFT periodic and cluster study, Radiochim. Acta 94 (2006) 601–607.

[220] R. Drot, J. Roques, E. Simoni, Molecular approach of the uranyl/mineral interfacial phenomena, C.R. Chim. 10 (2007) 1078–1091.

[221] M. Dossot, S. Cremel, J. Vandenborre, J. Grausem, B. Humbert, R. Drot, E. Simoni, Sorption of uranyl cations on a rutile (001) single crystal monitored by surface second-harmonic generation, Langmuir 22 (2006) 140–147.

[222] H. Wang, Z.F. Chai, D.Q. Wang, Adsorption of uranyl on hydroxylated α-SiO_2 (001): a first-principle study, Dalton Trans. 44 (2015) 1646–1654.

[223] E. Veilly, J. Roques, M.C. Jodin-Caumon, B. Humbert, R. Drot, E. Simoni, Uranyl interaction with the hydrated (001) basal face of gibbsite: a combined theoretical and spectroscopic study, J. Chem. Phys. 129 (2008) 244704.

[224] G.A. Waychunas, C.S. Kim, J.F. Banfield, Nanoparticulate iron oxide minerals in soils and sediments: unique properties and contaminant scavenging mechanisms, J. Nanopart. Res. 7 (2005) 409–433.

[225] H. Zeng, D.E. Giammar, U(VI) reduction by Fe(II) on hematite nanoparticles, J. Nanopart. Res. 13 (2011) 3741–3754.

[226] L.P. Qian, M.H. Ma, D.H. Cheng, Adsorption and desorption of uranium on nano goethite and nano alumina, J. Radioanal. Nucl. Chem. 303 (2015) 161–170.

[227] I. Pidchenko, K.O. Kvashnina, T. Yokosawa, N. Finck, S. Bahl, D. Schild, R. Polly, E. Bohnert, A. Rossberg, J. Göttlicher, Uranium redox transformations after U(VI) coprecipitation with magnetite nanoparticles, Environ. Sci. Technol. 51 (2017) 2217–2225.
[228] T. Scott, G. Allen, P. Heard, M. Randell, Reduction of U(VI) to U(IV) on the surface of magnetite, Geochim. Cosmochim. Acta 69 (2005) 5639–5646.
[229] M.K. Ridley, M.L. Machesky, J.D. Kubicki, Experimental study of strontium adsorption on anatase nanoparticles as a function of size with a density functional theory and CD model interpretation, Langmuir 31 (2015) 703–713.
[230] M. Hua, S. Zhang, B. Pan, W. Zhang, L. Lv, Q. Zhang, Heavy metal removal from water/wastewater by nanosized metal oxides: a review, J. Hazard. Mater. 211 (2012) 317–331.
[231] M.H. Bradbury, B. Baeyens, Predictive sorption modelling of Ni(II), Co(II), Eu(IIII), Th(IV) and U(VI) on MX-80 bentonite and opalinus clay: a "bottom-up" approach, Appl. Clay Sci. 52 (2011) 27–33.
[232] R.O. Rahman, H.A. Ibrahium, Y.T. Hung, Liquid radioactive wastes treatment: a review, Water 3 (2011) 551–565.

Chapter 3

Microorganisms and radionuclides

Congcong Ding*, Wencai Cheng† and Xiaoqin Nie†
*Fundamental Science on Nuclear Wastes and Environmental Safety Laboratory, Southwest University of Science and Technology, Mianyang, People's Republic of China
†School of National Defence Science and Technology, Southwest University of Science and Technology, Mianyang, People's Republic of China

3.1 Introduction

In natural systems, the chemical behavior, bioavailability, mobility, and transformation of radionuclides are greatly dependent on interactions with environmental components. Such components include clay and other minerals, soil colloidal materials, humic substances, and living organisms. Microbes are one of the major biotic components, and they possess the following advantages: naturally occurring and ubiquitous presence in the environment, high surface area/volume ratio, abundant functional groups, and strong tolerance to extreme environments. Despite the radiation and toxicity of radionuclides, many microorganisms grow and even flourish in radionuclide-containing locations. Therefore, the potentially positive or negative role of microbial communities may be underestimated in affecting the biogeochemical behavior of radionuclides in the environment.

All kinds of microbes, including prokaryotes and eukaryotes and their symbiotic consortia with each other and "higher organisms," play a key role in the biogeochemical cycling of radionuclides. Microbes and their activities are able to influence radionuclide speciation, toxicity, mobility, and radionuclide and mineral transformation (e.g., mineral formation, dissolution/corrosion, or deterioration). The series of interactions is a critical component of natural biogeochemical cycles for radionuclides and associated elements in organic matter, soil, rocks, and minerals, that is, sulfur, phosphorus, metalloids, and other metals. In addition to the significance of biogeochemical cycling, microbe-radionuclide interaction can result in beneficial or detrimental impacts on human society. One of the advantages is to use microbes to reclaim, immobilize, and detoxify radionuclides or their pollutants.

Interface Science and Technology, Vol. 29. https://doi.org/10.1016/B978-0-08-102727-1.00003-0

Bioremediation demonstrates the ecofriendliness, efficiency, and cost- effectiveness in the clean-up of radionuclide pollution. Biosorption of radionuclides and the nucleation, precipitation, and biomineral formation or reduction by microbes play a more significant role in retarding radionuclide species mobility in the environment than previously supposed. Radionuclides or other heavy metal ions can be deposited by microbes in nanoparticle, crystalline, or colloidal forms that possess catalytic or other properties. Thus, biomineralization contributes to the development of novel biomaterials for structural, technological, environmental, and antimicrobial purposes [1]. But at the same time, microorganisms can also cause the the release of radionuclides into the environment via destruction, corrosion, and dissolution of natural and synthetic materials such as concrete and mine.

This chapter focuses on: (1) identifying the microbial communities and structures in radionuclide-related environments and their physiology types to provide baseline information that can act as an indicator to reflect natural or anthropogenic environment perturbation; (2) summarizing the microbe-radionuclide interaction types and the responding mechanisms as well as impacting factors and research approaches to develop effective, ecofriendly, and long-term treatments of radionuclides; and (3) elucidating the potential for microbial communities to affect geochemical processes relevant to radioactive waste disposal. The information provides insights on the development of future in situ bioremediation techniques, and on the assessment of microbial roles in both potentially positive and negative radioactive waste disposal processes. Concurrently, this information can serve as a guide in the construction of concrete for radioactive waste storage. The "right" order of priority of goals is to have microorganisms as the core of a radionuclide depollution strategy via optimizing microbe-radionuclide interactions by various actions. Microbe-based techniques can secure the requirement of the ecofriendly aspect while not paying the price for saving the environment by replacing one kind of pollution with another.

3.2 Microorganisms in radionuclide-related environments

As witnesses and survivors of various extreme environments on the early Earth, microbes—especially bacteria and archaea—have demonstrated their ability to survive in the radionuclide-related environment. The resistance mechanisms include direct methods by changing the radionuclide species (i.e., decreased or increased mobility) or indirect methods by altering the physicochemical nature of the environment, which impacts the toxicity of radionuclides. Specifically, these ways may be microbe-mediated redox transformation, production of metal-binding peptides and proteins, radionuclide precipitation, active transport, efflux, and intracellular compartmentalization. Apart from the above approaches, cell walls and other structural components have significant binding abilities with radionuclides.

The inhabitant microbes in a radionuclide-related environment are well adapted to the local harsh conditions and play significant roles in the biogeochemical cycling of radionuclide elements, including their mobility and toxicity. Generally, those indigenous microorganisms exhibit high immobilization ability toward radionuclides and most heavy metal ions via precipitating or accumulating processes. Therefore, identification and assessment of radionuclide-resistant populations from radioactive environments not only enables us to understand the role of microbes in radionuclide biogeochemistry, but also helps to develop promising remediation strategies for cleaning up contaminated environments.

3.2.1 Radiation environment

Natural radioactivity primarily results from ^{4}K and the members of the ^{238}U decay chain. A report from Bulgaria found that the presence of radionuclides (U, Th, Pu) and some heavy metals (Cu, Cd, Pb) in terrestrial agricultural lands can be attributed to the commercial acid leaching of radionuclides from mines [2]. In India, naturally occurring radionuclides are ^{238}U, ^{232}Th, ^{4}K, and ^{226}Ra. Levels of radiation for ^{238}U, ^{232}Th, and ^{4}K in South Konkan can reach 68.08×10^{-9} Sv/h [3]. Chronic radiation in an oligotrophic environment would put selection pressure on microbes, thus altering the diversity of the microorganisms in the environment. During the uranium mining and milling processes in the northeastern Indian state of Jharkhand, the emitted alpha particles changed the diversity and structure of the local microbial community [4]. In addition to the natural radioactivity and anthropogenic radioactivity from radionuclide mining and milling processes, accidents from nuclear power plants such as Fukushima and Chernobyl are also radiation sources.

The hazards of gamma rays and other ionizing radiations are mainly because they engage in the alteration of microbial DNA, thus inhibiting the reproduction of microbes. The stress from ionizing radiations can cause lesions and degradation of nuclear materials by direct (i.e., ejection of an electron from DNA) and indirect (radiolysis process) mechanisms [5]. In the radiation environment, some bacteria are able to accept and undergo stable genetic mutations, forming new species of recombinant mutants. Those mutants, possessing the characteristic self-sustaining capability, can be used for the remediation of radioactive waste. The most radiation-resistant bacteria, *Deinococcus radiodurans*, was found to survive ionizing radiation levels up to 10×10^{3} Sv [6]. Few species of *Deinococcus* and *Rubrobacter* have been reported to survive gamma radiation doses of $>25 \times 10^{3}$ Sv. Some species of Chroococcidiopsis have been reported to survive up to 15×10^{3} Sv [7].

It is challenging to study the tolerant mechanisms of microorganisms against radionuclides due to the limitations of scientific technology and our finite knowledge from organisms. Misra et al. [8] found that the genetic basis of radiation resistance was an exonuclease repair pathway. ^{241}Am was observed to accumulate in the DNA of bacteria that exhibit resistance to desiccation, ultraviolet radiation, and oxidative agents [9]. The presence of multiple pathways for amino acid synthesis has been identified to explain the immense resistance of bacteria [7]. Daly et al. [10] determined that the accumulation of Mn(II) can enhance the radiation resistance of bacteria. The tolerance of *Escherichia coli* to nickel and cobalt can be attributed to the synthesis of the YohM protein *rcnA*, a first that does not rely on the traditional NiCoT (nickel cobalt transporter) family. S-layer proteins were also reported to provide resistance against gamma radiation [7]. After the accident at the Chernobyl nuclear power plant, researchers found an interesting phenomenon, in that the radiation had a positive effect on fungal spore germination. Those Chernobyl-associated fungi can use radiation as a source of energy. For many U-resistant bacteria, bioprecipitation and biomineralization processes that limit the toxic effects of U have been identified as being responsible for their resistance characteristics [11]. Overall, the available research on the tolerant mechanisms varies from different levels, that is, the DNA and RNA level, the enzyme and protein or other metabolite level, or just the behavior of cellular levels such as the distribution of radionuclides in cells. No unified and specific mechanisms are defined. Thus, there are still many gaps in our understanding of the survival mechanisms of microbes against a radiation environment.

3.2.2 Deep subsurface environment

Understanding the microbial community structure and its metabolic potential in the deep subsurface is useful to assess the radioactive waste repository function in the long term. The relevant information can also be a guide for future repository materials. Geological sequestration in deep underground repositories is still the prevailing proposal for long-lived radioactive waste disposal. The existence of microorganisms in deep terrestrial ecosystems is a common occurrence due to the presence of buried organic carbon, mineral-associated electron donors and acceptors, and biologically or abiotically formed hydrogen and methane [12]. The indigenous microorganisms can influence radioactive waste disposal in the subsurface via direct immobilization ways or indirect alteration of the mineralogical and geochemical features of the environment (i.e., the biodeterioration of the repository structure). Generally, the corrosion of metals associated with repository materials in the deep subsurface could produce high levels of H_2, which stimulate H_2-oxidizing bacteria and archaea [13]. The activities of these microorganisms via oxidative [NiFe] hydrogenases and electron-bifurcating [FeFe] hydrogenases may reduce the risks associated with H_2 accumulation [12]. H_2 is an important energy currency in the subsurface.

Hernsdorf et al. [12] explored the microbial community structure and metabolic functions of a sediment-hosted ecosystem at the Horonobe Underground Research Laboratory (URL), Hokkaido, Japan. They selected groundwater samples from three boreholes at the URL from a depth range of between 140 and 250 m in two consecutive years. The analysis of the genome taxonomic assignment showed that 15 were from archaea, 195 were from bacteria, and 19 were from phage or other mobile elements. Interestingly, most genomes were evaluated to be a potential for H_2 metabolism via identification of the cofactors (i.e., [NiFe], [FeFe], or [Fe]). Therefore, the majority of these organisms can metabolize H_2. [FeFe] hydrogenases included three major groups (A, B, and C) by sequence identity, and Group A was further subdivided to Group A1, A2, and A3 based on operon structures [14]. Surprisingly, ~80% of the [FeFe] Group A hydrogenases were identified as heterotrimeric/heterotetrameric Group A3 [FeFe] hydrogenases. Group A3 [FeFe] hydrogenases can reversibly bifurcate electrons from H_2 to ferredoxin and NAD. Another 15% were monomeric composition (Group A1), whereas the remaining 5% could not be accurately identified as A1 or A3 due to the location of the genes at the end of the scaffold. The occurrence of [FeFe] hydrogenases in metagenomic surveys has been taken as an indication of H_2 production [15]. Definitely, we cannot ensure the direction of bifurcating hydrogenases solely relying on the sequence because they may function in the reverse direction as confurcating hydrogenases [16]. However, the cooccurrence of both bifurcating [FeFe] (Group A3) and Group A1 indicated an H_2 metabolism but not H_2 production. >70% of the genomes in the Horonobe subsurface were identified as bifurcating [FeFe] hydrogenases. Concurrently, the few monomeric H_2-evolving [FeFe] hydrogenases (Group A1) were accompanied by Group A3 hydrogenases. This result suggested that evolved H_2 may also depend on uptake hydrogenases (i.e., A3 hydrogenases) to fuel the metabolism in times of substrate limitation. It is a favorable strategy for microorganisms in the deep subsurface of a radionuclide-related environment to oxidize H_2 for generating reduced cofactors such as ferrodoxin and NAD. Bifurcating [FeFe] hydrogenases can allow the direct coupling of hydrogen oxidation to generate a proton gradient for ATP synthase.

The stimulation of hydrogenotrophic microbes due to the elevated H_2 is likely to result in the oxidation of FeS_2 minerals and the liberation of sulfate into the groundwater. The microbial production of sulfide would accelerate the corrosion of radioactive waste containers and cause the release of radionuclides. On the other side, hydrogenotrophic iron-reducing organisms may also retard radionuclide migration. Laboratory and field investigations have demonstrated that U(VI) and Tc(VII) can be reduced by biogenic-reduced iron such as Fe(II) and the enzymatic activity of [NiFe] hydrogenases [12], thus preventing their subsurface migration.

3.2.3 Radionuclide ore environment

Understanding the bacterial community from uranium ore might provide some effective approaches for bioremediation of the generated wastes during uranium mining and the milling process. The waste generation is expected to be in bulk, which requires the development of green chemicals and ecofriendly mining processes to minimize environmental deterioration. For example, the Domiasiat ore in India has an average of about 0.1% U_3O_8. This means that, to obtain 1 kg of U, a 1000 kg of ore extracted from the Domiasiat deposit has to be processed at a U milling plant [17]. There are significant populations of both microbial planktonic cells and biofilms at the uranium ore. In order to grow and survive, those microbes have adapted to the environment via different microbial activities to protect against toxic elements [18]. Hence, one of the promising strategies for remediation of the contaminated subsurface is the precipitation or immobilization of radionuclides in situ by microorganisms.

Metal-resistant *Bacillus* and *Pseudomonas* strains were isolated early from U ore deposits. They showed high minimum inhibitory concentration (MIC) values for metals and a larger spectrum of antibiotic resistance [19]. The MIC, which means the lowest concentration of metals or other antibiotics that completely inhibits the growth of microorganisms on the plate, was determined for the two bacteria. *Serratia marcescens* was another representative species found in uranium ore. Kumar et al. [20] identified five *S. marcescens* isolates by phylogenetic analyses of 16S rRNA gene sequences. The MICs for U(VI) ranged from 3.5 to 4.0 mM for these natural isolates. Concurrently, they presented high U(VI) adsorption capacity, such as 90%–92% and 60%–70% U(VI) removal capacity when exposed to 23.8 and 476 mg/L uranyl nitrate solutions, respectively, at pH 3.5 within 10 min. The tolerance of these *S. marcescens* isolates was also confirmed by viability tests at pH 3.5.

In addition to the exploration of specific functional bacteria species for bioremediation, the knowledge of the overall structure and diversity of the microbial community is also essential for accurately predicting the potential risk of radionuclide mining. Recently, an investigation on the cultured bacteria communities of Domiasiat was reported. Among the isolated species, there are 130U-tolerant aerobic chemoheterotrophic bacteria, comprised of 76 gram-positive and 54 gram-negative bacteria [21]. Based on the molecular analysis of the 16S rRNA gene of each isolate, 62 of the 130 isolates were identified to the species level based on high similarity percentage (>98.7%). Among the identified U-tolerant species, *Firmicutes* (51%) and *Actinobacteria* (11%) were the dominant gram-positive bacteria, whereas *Gammaproteobacteria* (26%), *Bacteroidetes* (10%), and *Betaproteobacteria* (2%) were the predominant gram-negative bacteria. Those isolates not only presented higher uranium tolerance but also exhibited higher tolerance toward copper, zinc, lead, and cadmium, which cooccur in uranium deposits. One of the interesting findings is that 76% of the U-tolerate species possess constitutive phosphatase activity, and 53% have metal transporting P_{IB}-type ATPase genes. This suggests the great role of phosphates in the detoxification of uranium, which might be a possible survival strategy in such a harsh environment.

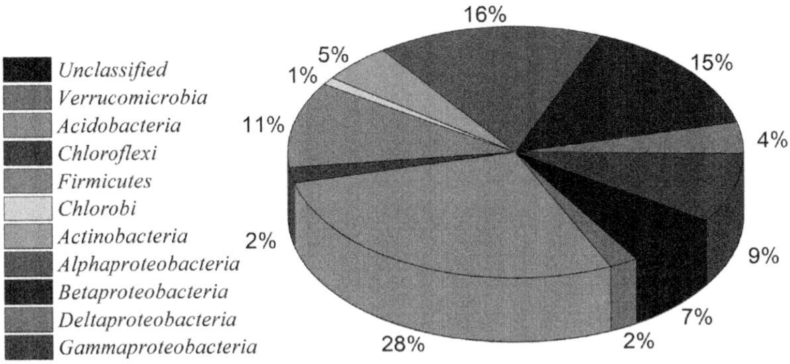

FIG. 3.1 Diversity of the microbial community from the clone library at the Domiasiat U deposit. *(Data from R. Kumar, M. Nongkhlaw, C. Acharya, S.R. Joshi, Bacterial community structure from the perspective of the uranium ore deposits of domiasiat in India, Proc. Natl. Acad. Sci. India Sect. B Biol. Sci. 83 (2013) 485–497.)*

Considering that many bacteria cannot yet be successfully cultured, culture-independent molecular approaches are adopted to analyze the diversity of the microbial community based on PCR amplification, cloning, and sequencing of 16S rRNA genes. Kumar et al. [17] constructed a clone library from the total DNA of the soil in uranium ore. As shown by Fig. 3.1, the clones belonged to 10 different bacteria groups. Of these species, 7% were novel bacterial lineages. Over the studied samples, *Proteobacteria* were the most abundant, among which *Alphaproteobacteria* were observed to be predominant. *Acidobacteria* followed the proteobacterial group that would have been the most abundant if four proteobacterial groups were considered separately.

It is known that the microbial community of a particular microhabitat is dependent on its geochemical matrix composition. Commonly, the diversity and distribution of the microbial community is specific to the sites and varies with the change in geochemical conditions [22]. Any changes in the environment such as anthropogenic activities can alter the structure and composition of the natural microbial community [23]. Therefore, the information of the microbial community, sensitive to microhabitat changes, is increasingly being regarded as an effective tool to determine the impact of environmental contamination. Hence, the investigation of the microbial community of the premined uranium ore is expected to provide baseline information on the microbial structure and diversity of the site. This baseline information can function as an indicator to predict the environmental perturbation that will happen in the U mining process, thus helping to estimate the impact of the mining via comparing the microbial community before and after mining. Relative to the extensive studies on copper and gold mines or U-contaminated sites, the microbial community in premined U ore is less frequently reported.

Extensive investigation of the taxonomic position of microorganisms surviving in the ore deposits and their interaction with U will be useful for developing effective, ecofriendly, and long-term treatments of radionuclides as well as for predicting the microbial impacts on the performance of the radioactive waste repository. While numerous microbes exhibit a high ability toward a particular radionuclide or heavy metal, their sensitivity to the metals might impede the application in situ. Hence, checking the tolerance of the indigenous bacteria to different heavy metals occurring simultaneously at the site is the basis for the bioremediation strategy. Adapted microbial populations in uranium ore are prone to exhibit higher resistance to heavy metals relative to other populations in noncontaminated sites [24]. The reason for many indigenous bacteria to grow in an environment with a high concentration of radionuclides and heavy metals is to develop a homeostasis within cells that keeps the reactive heavy metal ions at an optimal subtoxic level. The reported adaptive mechanisms include biotransformation and alteration of metal species via oxidation or reduction of radionuclides; mobilization of radionuclides by chelation siderophores and other microbial compounds; biosorption of radionuclides by cell surface polymers; uptake of radionuclides into the cells; induction of metal precipitation; and generation of minerals [17].

3.3 Radionuclide-microbe interactions

The microbial-mediated transformation provides the potential for bioremediation of radionuclides. Microbial cells can function as a radionuclide sink, either by sorption to the cell surface, intracellular accumulation, precipitation of radionuclides on or in the cells, or reduction of radionuclides to the more insoluble species. Most microbial materials can be effective radionuclide biosorbents. Herein, we seek to outline the ways that microbes immobilize radionuclides. It is focused on the interaction mechanisms and their influence on the toxicity and transport of radionuclides, their importance in the geological and environmental processes, and their applied significance within the concept of bioremediation.

3.3.1 Biosorption

Biosorption can be described as the sequestration of radionuclides to the outer surfaces or the cell membranes of microorganisms. Biosorption occurs by the electrostatic attraction between radionuclide cations and anionic cell wall functional groups. Additionally, the complexation of radionuclides with ligands such as carboxyl, amine, hydroxyl, and phosphate in an extracellular polymeric substance (EPS), S-layer, or capsule on bacteria also contributes to the biosorption process. Thus, electrostatic attraction, chemical interaction of functional groups, and metal ions all influence biosorption. Generally, biosorption is determined as a passive uptake process. Many studies observed that dead cells presented better sorption capacity relative to living ones.

Xie et al. [25] determined that the sorption process induced by microbes was dependent on the functional groups in the cell wall rather than microbial activity. They identified that carboxyl in the cell wall of *Citrobacter freundii* acted as one of the active sites and played a key role for biosorption capacity.

Overall, the biosorption process is species-specified and is dependent on environmental conditions (i.e., pH, temperature, and aeration), radionuclide species (negative or positive charged species), and the characteristics of microorganisms (i.e., growth phase of cells, production of exopolymers and metabolites, component on the surface cells). One of the shortcomings for biosorption is the relatively low selectivity. Target and other than desired cations both can be quickly adsorbed on the biomass, which leads to competitive desorption. Therefore, biosorption technology is generally used for the removal of low-concentration radionuclides.

Biosorption is usually the first step for further microbe interactions such as bioaccumulation, biomineralization, and bioreduction [26]. The precipitation of uranyl phosphate is initiated by the electrostatic attraction with phosphate groups on the surface of *Citrobacter*. This interaction provides nucleation sites for mineral formation and protects an outer membrane phosphatase. Similarly, the reduction of Pu(VI) to Pu(V) by microbes is caused because Pu(VI) was first bound to the phosphate groups on the cell surface within hours. This process is more effective than direct bioreduction.

3.3.2 Bioaccumulation

Cell surface sequestration (biosorption) of radionuclides may be a protective mechanism against chemical or radioactive toxicity within microorganisms (i.e., the oxidative effect caused by UO_2^{2+}). However, ^{238}U was observed within extra- and intrahyphal structures [27], suggesting that U(VI) was not only adsorbed on the surface of fungi, but also bioaccumulated into the cytoplasm. Bioaccumulation is an enrichment of substance within organisms. Herein, bioaccumulation is defined as intracellular accumulation to distinguish from biosorption.

Bioaccumulation can be either an active or a passive process. The active process is relatively slower and requires more energy, whereas passive bioaccumulation is faster and requires less energy. Passive bioaccumulation is prone to occur under conditions where the nutrient is limited [7]. For active accumulation, the radionuclides are usually essential elements for the metabolisms of microbes, or their physicochemical properties are similar to those of essential elements. Although some radionuclides are essential for life, they will exert toxicity when present above certain threshold concentrations.

Accumulation of iodide on the soil *Bacillus subtilis* cell wall increased with the addition of glucose [28]. Decomposition of organic matter, ranging from simple compounds such as sugar and amino acids to complex molecules such as cellulose and starch, is one of the most important microbial activities. During the degradation process, component elements will be redistributed between microorganisms and environmental compartments. Therefore, it is guessed that all decomposition and degradative microbial activities are linked to the cycling of radionuclides accumulated from natural or anthropogenic sources.

Other radionuclides such as Cs and U have no known essential metabolic functions but will still be accumulated. The formation of U-phosphate complexes was prevalently found in many microorganisms, which might be one reason for ^{238}U fungal uptake [11, 29]. Francis et al. [30] found that intracellular accumulation of U(VI) in *Citrobacter* sp. and *Halomonas* sp. was largely in the form of phosphates such as hydroxophosphate or polyphosphates, or as uranium hydrogen phosphate. However, U-phosphate is not the only U species assimilated by fungal hypha because at higher pH values where uranyl sulfate and uranyl carbonate species are predominant, U is still efficiently assimilated by fungi [31]. Several factors, including organic matter concentration; Ca^{2+}, K^+, and PO_4^{3-} concentrations; and soil pH and type, can directly affect the chemical speciation of radionuclides and consequently their bioavailability for the uptake into inside of microorganisms. U is a nonessential element for fungi. So far, it is not well understood for its biological interactions at the cellular level. For example, how is U transported and partitioned into the fungi? There is evidence that fungi can enhance the bioaccumulation of nonessential elements such as Pb, Cd, and ^{137}Cs [32]. Despite these experiments, there are still many gaps in our understanding of radionuclide transfer into microbes, both mechanistically and in an ecological context.

3.3.3 Biomineralization

The term biomineralization represents the collective processes by which microorganisms forms minerals, as shown in Fig. 3.2. Microorganisms can provide nucleation sites for the precipitation of radionuclide ions to insoluble minerals [33]. This process is called biomineralization, where microbes can serve as a platform for mineral formation. Various microbial groups including bacteria, cyanobacteria, microalgae, protozoa, and fungi have important roles in the transformation of radionuclides to minerals. For example, cyanobacterial and microalgal processes are generally of greater significance in aquatic environments while free-living and symbiotic fungi play a critical role in the terrestrial environment. Biomineralization can be categorized into biological-induced mineralization (BIM) and biological-controlled mineralization (BCM). BIM refers to the process where a microbe changes its local microenvironment to facilitate the extracellular chemical precipitation of the mineral phase, whereas BCM emphasizes the control role of microbes during the biomineralization process. BIM is a common process which can occur by redox reaction with microbes, complexation with metabolite excretion such as

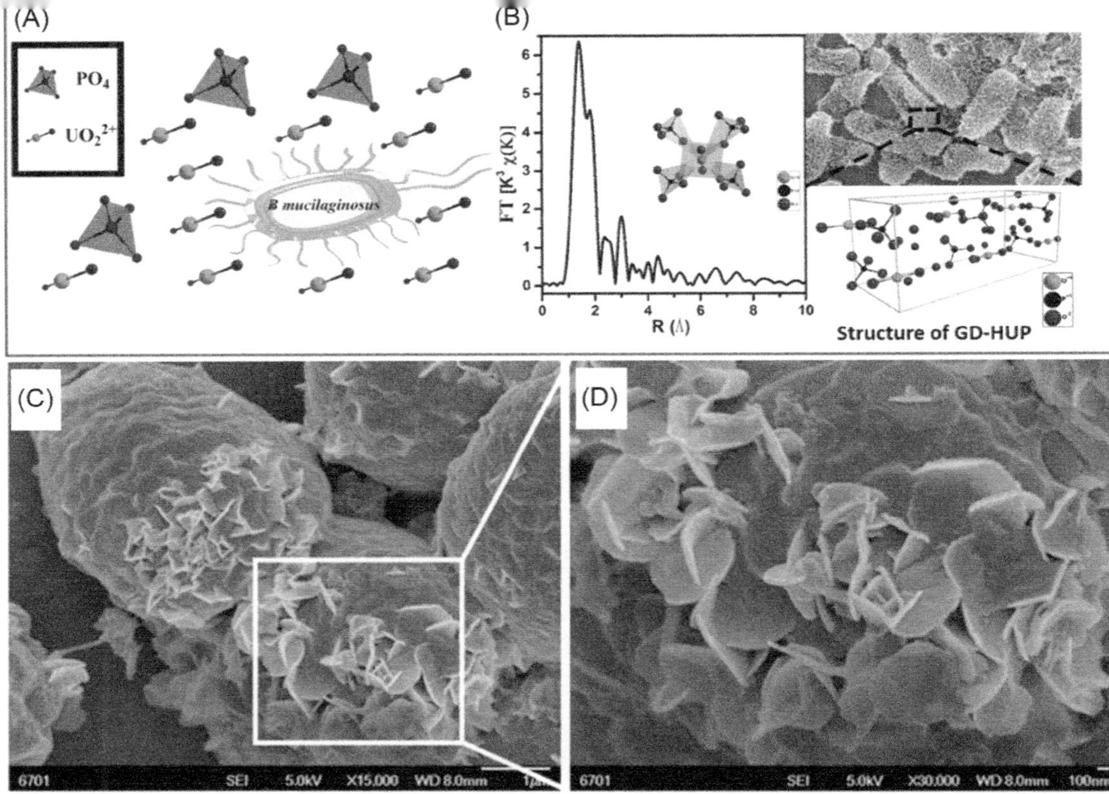

FIG. 3.2 Some examples of uranium-related mineral deposition by microbes. (A) Interaction schematic between U(VI) and *Bacillus mucilaginosus*. (B) GD-HUP-like minerals formed on active *Bacillus mucilaginosus*. (C) SEM micrographs of *S. cerevisiae* after biosorption. (D) Correspond to details the square, U(VI) = 400 mg/L, $I = 0.01$ mol/L $NaNO_3$, pH 5.40 ± 0.10, and 30°C, 250 rpm, $t = 2$ h. *((A and B) From W. Huang, W. Cheng, X. Nie, F. Dong, C. Ding, M. Liu, Z. Li, T. Hayat, N.S. Alharbi, Microscopic and spectroscopic insights into uranium phosphate mineral precipitated by Bacillus* Mucilaginosus, *ACS Earth Space Chem. 1 (2017) 483–492 copyright 2017, American Chemical Society; (B and C) From X.-Y. Zheng, X.-Y. Wang, Y.-H. Shen, X. Lu, T.-S. Wang, Biosorption and biomineralization of uranium(VI) by* Saccharomyces cerevisiae*-crystal formation of chernikovite, Chemosphere 175 (2017) 161–169 copyright 2017, Elsevier.)*

phosphate, or the sorption by the reactive sites on the cell surface, consequently inducing the nucleation and formation of radionuclide mineral precipitates. An active cell surface provides a favorable location for the mineralization process, such as the array of ionizable carboxylate and phosphate ligands in the LPS, or the binding ability of radionuclide ions and inorganic anions to form a mineral phase. The nature of the resultant minerals may rely on the nature of the cell surface, the cellular microenvironment, and the presence of reactive anions. Note that biomineralization can also occur independent of microbial activity, considering that reactive anions such as ionizable phosphate ligands are present on dead microorganisms. Radionuclide-related minerals mediated by microbes may be vulnerable to environmental conditions and undergo further changes in crystallinity and structure with time, such as mineral deposition or dissolution and the incorporation of other metals to their structure.

Biomineralization can be induced by radionuclide binding to biofilms produced by microorganisms. Polycrystalline $NaUO_2PO_4$ deposits were observed inside and around the cell wall of *Citrobacter* sp. Both uranium sorption to LPS and the activity of acid phosphatase in the outer membrane of *Citrobacter* sp. contributed to the biomineralization [34]. Mineral formation is driven by two gradients in the outer membranes: an incoming UO_2^{2+} and an outgoing PO_4^{3-} [26]. The presence of organic agents such as citric acid, hydroxyl-acetic acid, oxalic acid, tartaric acid, etc., is useful for the bioprecipitation of most metal ions. Those organic acids, especially citric acid, can serve as chelating agents and form highly soluble metal-citrate complexes. The metal chelates can be degraded by microorganisms, resulting in subsequent reprecipitation of metals.

Biomineralization of radiostrontium (^{89}Sr and ^{90}Sr) via microbial-generated ligands is also a popular research topic. The concentration of ^{90}Sr in marine environments can be up to 1 PBq (1×10^{15} Bq), which is mainly released from nuclear fission. As one kind of carbonate-producing microorganism, calcified algae attract a great deal of attention in the biomineralization of Sr^{2+} due to their key role in the global carbon cycle [35]. Coccolithophores are one unicellular calcified algae, enclosed by coccoliths formed from calcite crystal plates. The dominant academic view determined that the mineralization of coccoliths occurred on a specialized intracellular Golgi-derived vesicle, and then the coccoliths were extracellularly assembled into a coccosphere constituting the exoskeleton [36]. Sr^{2+} can be simultaneously partly substituted for Ca^{2+} and incorporated in coccoliths during calcification.

There are at least three stages that will impact the immobilization, as coccoliths are formed intracellularly. The first stage is channel or carrier-mediated transport for Ca and Sr ions from the outside to the inside of the cell. The second stage is related to the cations partitioning during coccolith precipitation in internal coccolith-associated vesicles [37]. The second stage relies on kinetics, thermodynamics, and biological impacts. When the Sr^{2+}

concentration increases in a bulk medium, its pumping rate across the cellular membranes will accordingly increase. The selectivity of Sr^{2+} in the channel or carrier may be enhanced relative to Ca^{2+}, thus increasing the chance of Sr^{2+} being incorporated into the coccoliths [38]. The final stage involves Sr^{2+} precipitation outside the cell after forming the coccosphere. During this stage, EPSs, especially surface-active polysaccharides (SAPs), will be an influence factor for Sr precipitation. SAP has been demonstrated to relate to the formation and attachment of coccoliths to the coccosphere [39]. Previous reports found that Sr^{2+} is more strongly bound to SAPs than Ca^{2+} [38]. Thus, SAP may provide nucleation sites and templates for Sr^{2+} precipitation during calcification. It has been a promising bioremediation strategy to precipitate Sr^{2+} with biogenic carbonate due to the common existence of microbial carbonate, especially in marine environments.

3.3.4 Bioreduction

Redox reactions by microbes can alter the solubility of radionuclides, thus forming oxides, coprecipitates, ionic, and organic or inorganic complexes of radionuclides. Microorganisms are able to utilize an oxidized form of radionuclides such as U and Tc as the terminal acceptor. Those oxidized radionuclide species such as U(VI) and Tc(VII) are usually water-soluble, mobile, and biologically available in the environment. When transformed to their reduction state under reducing conditions, those radionuclides become insoluble, thus decreasing their bioavailability. Researchers from South Africa isolated and purified four bacteria strains from local mining sites where the uranium concentration in soil (168 mg/kg) was much higher than that in natural sites (0.30–11.7 mg/kg) [40]. These four bacterial strains were identified as *Pantoea* sp., *Pseudomonas* sp., *Enterobacter* sp., and *Klebsiella* sp. An interesting result was that U(VI) was reduced and reoxidized intermittently by the four bacteria in a consortium. Further experiments by individual species showed that *Klebsiella* sp. was responsible for the remobilization of U(IV) to the U(VI), and the other microbes were able to reduce U(VI) to U(IV) ($U(OH)_4$ [aq]) under pH 5–6. Complete U(VI) reduction can be achieved at concentrations up to 200 mg/L. Even for 400 mg/L U(VI), the reduction percentage of those bacteria can reach 88%. Reduction of Tc has been reported under varying pH conditions [41]. *Desulfovibrio desulfuricans* has been identified as having the ability to reduce Np(V) [42].

Bioreduction of radionuclides can be realized via direct and indirect ways [7]. For the direct enzymatic reduction, bacteria can directly use the electrons provided by the substrate (electron donor) to reduce the oxidized species of U(VI) and Tc(VII) into the reduced U(IV) and Tc(IV). Thus, the addition of electron donors such as lactic acid, acetic acid, ethanol, hydrogen, etc., can enhance the reduction ability of the *Desulfovibrio* and *Shewanella* species. In addition to their function as an electron donor, some organic acids can chelate with radionuclide ions. Research reveals that uranium-citrate complexes are readily available for bacteria as electron acceptors, despite their inability to metabolize the organic ligands of radionuclides [26]. A c-type cytochrome in the periplasm of bacteria was determined to be involved with U(VI) reduction. Vivo studies also found that cytochrome *c*3 played an important role for U(VI) reduction by *D. desulfuricans* (G20) [26]. Similarly, the role of homologous cytochrome (PpcA), a trihaem periplasmic cytochrome *c*7, was also confirmed in U(VI) reduction by the Fe(III)-reducing bacterium *Geobacter sulfurreducens* [43]. Available research found that the formate dehydrogenase component was required when formate was supplied as an electron donor for Tc(VII) reduction. The formate hydrogenase catalyzed the transfer of electrons from dihydrogen to Tc(VII) [26].

Some lithotroph bacteria have no direct enzymatic reduction capacity of radionuclides. However, they can couple the oxidation of organic compounds or hydrogen to the reduction of iron Fe(III) or sulfur S(VI) in the form of sulfate. The iron Fe(III) and S(VI) can be reduced to Fe(II) and S(II) or H_2S, respectively. Those reduced products can further reduce the radionuclides into low valent species. This process is called an indirect mechanism. Thus, the reported microorganisms possessing radionuclide reduction ability are those capable of dissimilatory Fe(II) or sulfate-reducing bacteria (SRB), for example *G. sulfurreducens* strain PCA, *Anaeromyxobacter dehalogenans* strain K, and *Shewanella putrefaciens* strain CN-32 [7]. Dissimilatory reduction of radionuclides by microbes is usually accompanied by the energy metabolism. Thus, substances are needed as electron acceptors for the generation of energy. Lloyd and Lovley [44] found that the electron acceptor was reduced but not incorporated into organic matter during the biosynthetic anabolism process.

3.4 Methods for studying radionuclide-microbe interactions

Traditional methods have been developed for environmental bioremediation and ecological management, such as environmental sample collection and processing, microscopic and spectroscopic techniques, culture methods and physiological methods, etc. Despite papers demonstrating the uptake of microbes toward radionuclides, there are still many gaps in our understanding of radionuclide transfer into microbes, both mechanistically and in an ecological context. Those gaps have not been addressed since these earlier studies 10 years ago. Fortunately, the appearance of some technical advances might hopefully help us understand the interaction mechanisms, such as synchrotron-based spectroscopy imaging methods or systems biology approaches. Herein, we will focus on approaches that could further advance our knowledge of radionuclide transfer from aqueous solutions or soils to microorganisms. This knowledge is necessary for accurately modeling the chemical behavior of radionuclides and predicting their potential risk.

Synchrotron-based X-ray absorption (XAS) spectroscopy has been an important tool in environmental research due to its element-specific, bulk-sensitive, and nondestructive characteristics. The XAS technique includes X-ray absorption near edge spectroscopy (XANES) and extended X-ray absorption fine structure (EXAFS) spectroscopy. The high-resolution XAS technique can give the structural and compositional information of radionuclides sequestered by microorganisms at a cellular level, even at low concentrations. XANES spectroscopy can exactly identify the species of radionuclides on microbes, that is, the oxidation state of radionuclides. Based on the analysis of XANES, Ding et al. [45] found that uranium was dominantly present as a U(IV) species on the surface of nanoscale zero valent iron (NZVI), whereas the addition of *B. subtilis* decreased the reduction ability of NZVI with more U(VI) in the complexed system. EXAFS spectroscopy provides the local atomic structure information of radionuclides (i.e., complexation type, bond distance, and coordination number of near neighbors surrounding a specific radionuclides). XANES and EXAFS technology can be used to elucidate the biomineralization process of uranium on microorganisms. As shown in Fig. 3.3, the molecular structure of uranium on *Bacillus mucilaginosus* was similar to the HUP (a chernikovite-like structure), and the crystal U(VI)-phosphate complexes were predominant with the increased contact time [11].

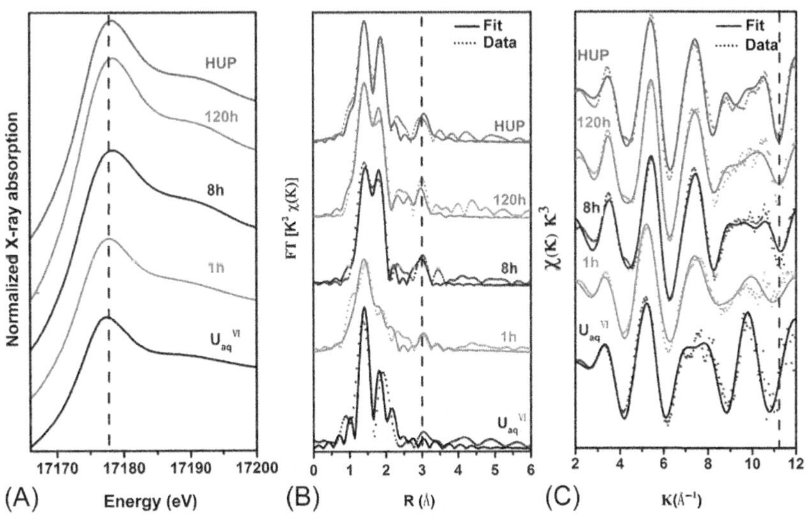

FIG. 3.3 U L_{III}-edge XANES spectra (A), Fourier transform (FT) of the R space (B) and K space (C) for HUP, U(VI) in aqueous solution and U(VI)-loaded samples over a reaction time of 1, 8, and 120h in 100mg/L U(VI) at pH5.0. *(From W. Huang, W. Cheng, X. Nie, F. Dong, C. Ding, M. Liu, Z. Li, T. Hayat, N.S. Alharbi, Microscopic and spectroscopic insights into uranium phosphate mineral precipitated by Bacillus* Mucilaginosus, *ACS Earth Space Chem. 1 (2017) 483–492 copyright 2017, American Chemical Society.)*

Time-resolved laser fluorescence spectroscopy (TRLFS) is applicable to radionuclide-microbe systems due to its high sensitivity and selectivity to targeted ions. TRLFS spectroscopy is determined to identify the coordination chemistry of lanthanides such as Eu(III) and actinides such as U(VI), Am(III), and Cm(III) in aqueous solutions and on the water/solid interfaces [46]. Alteration of emission spectra is able to reflect the complexation type (inner- or outer-sphere complexation) of radionuclides due to variations in the first coordination sphere of the fluorescent metal ion. The fluorescence lifetime can analyze the number of fluorescence-quenching entities present. Relative to EXAFS spectroscopy, the advantage of TRLFS is the enhanced sensitivity and the combined information on concentrations (based on intensities) and coordination (based on emission wave numbers and fluorescence lifetimes). However, the TRLFS technique can only analyze radionuclides that are fluorescent such as uranium. Additionally, the measurements are greatly affected by apparatus properties and temperature. Therefore, TRLFS is not a universal method.

Synchrotron-based X-ray fluorescence (XRF) imaging provides spatially resolved elemental mapping data, which identifies the location of radionuclides immobilized by microbes. A detailed radionuclide and other element mapping can be given by XRF, which identifies the accumulation location of radionuclides in the microbial structure. XRF is helpful to analyze the relationships between different elements. For example, Nayuki et al. [47] observed a positive correlation between Cd and P distribution. One can use XRF imaging for the analysis of uranium or other radionuclides in mycorrhizal plants, such as determining how many radionuclides are accumulated by fungal cells and how many are translocated into roots and shoots [48]. An XRF imaging combination with EXAFS spectroscopy is capable of tracking the chemical speciation of radionuclides and their complexation with microbial components along the biological translocation pathway. The challenge with XRF is the ability to produce and preserve the structure of microbial samples for microscopy or synchrotron analysis. It is easily to destroy the clarity of the intracellular structure and tissues during sample preparation and analysis. Thus, improved sample preparation for XRF imaging is required.

Bioremediation has historically been regarded as a "black box" in terms of fundamental understanding because we seldom give a complete understanding of why. Recently, a systems biology approach has been applied to investigate the microbial community structure and stress response from radionuclides in environments at all levels, which is expected to make bioremediation breakthroughs and illuminate the "black box" [49]. Systems biology is an integrated approach (including modern tools of genomics, transcriptomics, proteomics, metabolomics, phenomics, lipidomics, etc.) to study the interactions and networks of complex biology systems from molecular, cellular, community, and ecosystem levels (Fig. 3.4). Systems biology can characterize the change of

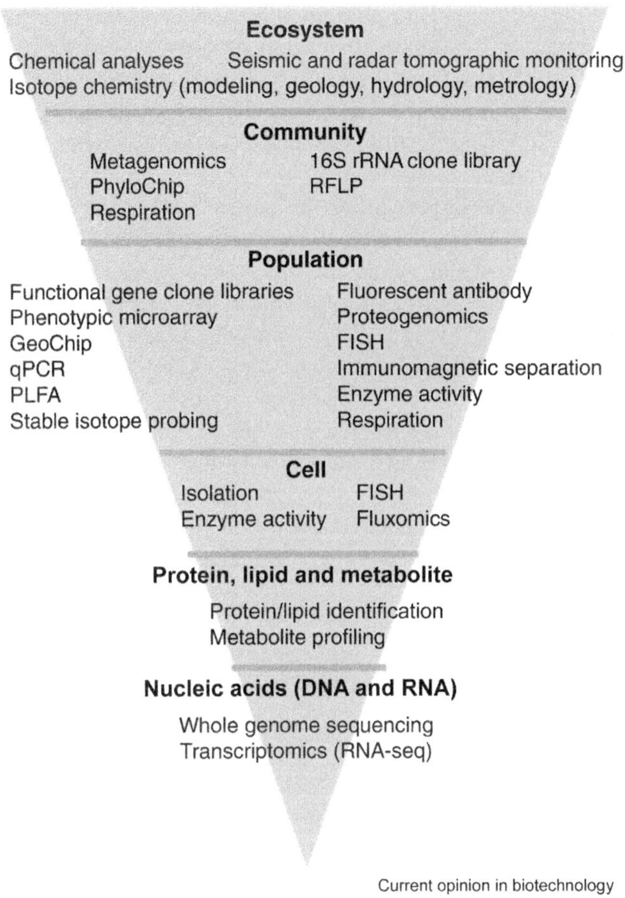

FIG. 3.4 Systems biology from molecules to ecosystems. *(From R. Chakraborty, C.H. Wu, T.C. Hazen, Systems biology approach to bioremediation, Curr. Opin. Biotechnol. 23 (2012) 483–490, copyright 2017, Elsevier.)*

the microbial community composition and molecular activity caused by the presence of radionuclides. As shown by Fig. 3.4, an ecosystem is composed of communities, populations, cells, protein, RNA, and DNA. We can analyze DNA, RNA, and proteins at the cellular level to study the impact of radionuclides on microorganisms, thus understanding how bioremediation functions. Tools such as GeoChip, RNA-seq, and other mass spectrometries can be used to identify DNA, RNA, and proteins. We can analyze populations and communities to investigate the impacts of radionuclides on structure/function relationships. DNA-based "-omics" tools such as 16S rRNA clone library, PhyloChip, or sequencing can be used to identify and track the microbial community composition.

There are numerous unculturable microorganisms in the natural environment. For example, it is estimated that normally only 1%–10% of soil microflora are culturable. Therefore, the traditional plate colony-counting method exhibits obvious limitations when used to analyze the microbial communities in mine tailings, mine wastes, mining-impacted sites, and other radionuclide-contaminated subsurface environments. Advances in molecular biology have led to new tools that are dependent on the type and sequence of nucleic acids (including DNA or RNA) to examine environmental microorganisms. Many studies have described the total microbial community in radionuclide-contaminated sites via amplifying ribosomal DNA restriction analysis and sequencing of 16S rRNA or 18S rRNA genes [17]. In brief, DNA is directly extracted and purified from the contaminated sites. The obtained mixture of DNA as community DNA is ideally representative of all populations within the samples. Amplify all the 16S rRNA genes in the community DNA, clone the PCR products, and subsequently analyze the sequence obtained in the clones.

The systems biology approach has been successfully applied in the Area 3 FRC site at Oak Ridge (contaminated with uranium up to 0.2 mmol/L) to explore the role of microorganisms in the subsurface. An analysis of the 16S rRNA clone library at this site has determined that *Desulfovibrio*, *Geobacter*, *Anaeromyxobacter*, *Desulfosporosinus*, *Acidovorax*, and *Geothrix* spp. are related to U(VI) reduction [50]. Clone libraries of functional gene makers such as *dsrAB*, *nirK*, *nirS*, *amoA*, and *pmoA* revealed the high microbial diversity in functional genes [51]. Via GeoChip analysis, *dsr* genes were observed to decrease during the reoxidation phase, whereas the microbial functional diversity seems to not be altered [52]. This means that the microbial community was able to recover and continue to reduce U(VI) in the postoxidation phase.

In spite of the great promise of the systems biology approach in understanding microbial functions in radionuclide-contaminated sites, the wide application of this approach is still limited due to funding, expertise, and resources. Generally, it needs the high cost associated with sample processing equipment, materials, and reagents for the analysis of large amounts of samples.

Current research has mostly used one model bacterium in the monoxenic cultures method, which enables specific variables to be determined for the interaction between radionuclides and microorganisms. However, this method is not applicable to ecological systems. Undoubtedly, it is extremely rare for the environment to be colonized by a single microbial species. Furthermore, the microbial community and structure are not random in one specific radionuclide-contaminated environment. Therefore, most available studies on radionuclide-microbe interactions lack ecological relevance.

To improve the ecological relevance, more microcosm and mesocosm methods using natural sources of radionuclides and soils from actual

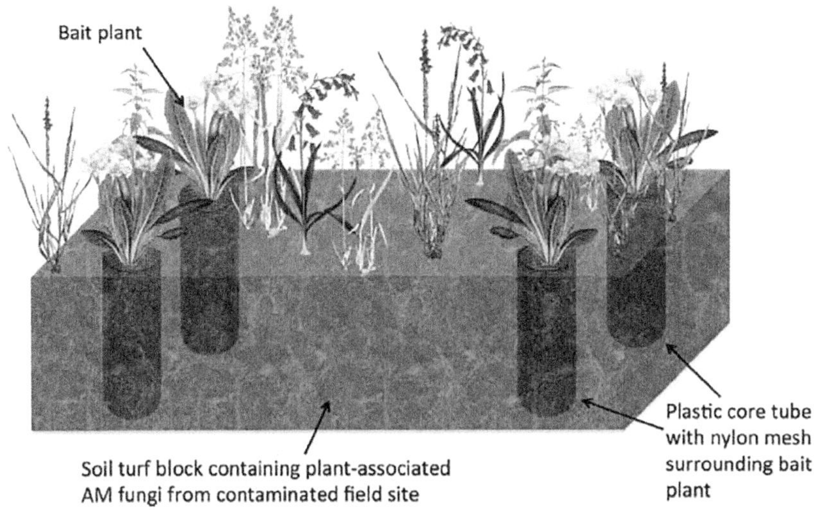

FIG. 3.5 Diagrammatic representation of an intact turf-based experiment to examine the effects of AM fungal colonization on radionuclide accumulation in plant tissue. *(From H.S. Davies, F. Cox, C.H. Robinson, J.K. Pittman, Radioactivity and the environment: technical approaches to understand the role of arbuscular mycorrhizal plants in radionuclide bioaccumulation, Front. Plant Sci. 6 (2015) 580, open-access article.)*

contaminated sites are needed, which are more likely to correspond to what would actually happen in field. A good example is from Johnson et al. [53] for studying radionuclide accumulation by arbuscular mycorrhizal (AM) in a grassland system. As described in Fig. 3.5, a soil turf containing undisturbed vegetation and associated AM fungi from ^{238}U-rich sites was removed. Cores were extracted and added with plastic tubes that contained sterilized soil and the "bait" plants. Part of the tubes was cut away and covered by 35 μm pore-sized nylon mesh to prevent root growth outside the core but allow hypha to penetrate the core and infect the bait plant root. This experiment setup allows a comparison between fungal-associated and nonfungal-associated bait plants without causing drastic changes to the source of AM fungi and the physical-chemical properties of the soils. Similar microcosm and mesocosm methods are helpful to more clearly understand the impacts of radioactivity on the whole ecosystem, thus providing implications on the design characteristics of geological disposal facilities.

3.5 Factors affecting microbial remediation

Various factors from the radionuclide species, microbes, and environmental aspects influence the removal of radionuclides by microbes. The factors of radionuclide species include ionic size, atomic weight, and the concentration or reduction potential of radionuclide ions. Biotic factors from microbes are microbial tolerance, structural components, metabolic activity, etc. Temperature, pH, salinity, hardness, organic or inorganic ligands, competing ions, and the required nutrition for microorganisms can be assigned to the environmental impact factors. It is worth noting that those factors commonly affect each other and exert their impact in concert. For example, the solubility of radionuclide species is greatly influenced by pH.

Biosorption of radionuclides is determined by their speciation and charge in solution, which in turn is dependent on pH and the presence of complexing agents in the environment. Generally, radionuclide-contaminated sites contain more than one radionuclide. The combined effect of all the constituents is often not equivalent to the summation of the individual effect. Three possible joint actions can be summarized: the synergism effect, the antagonism effect, and the noninteractive effect. The synergism effect means that the combined sorption amount is greater than the sum of the individual sorption amount. On the contrary, the antagonism effect refers to the fact that the combined sorption of the mixture is less than the sum of the individual one. For the noninteractive effect, the combined sorption amount is identical to the sum of the individual one. Cheng et al. [54] found the synergism sorption of U(VI) and Sr(II) on GOs, where precipitation of U(VI) at higher concentrations provided additional sorption sites for Sr(II).

The immobilization of radionuclides involves rapid interactions with anionic groups present in components of the cell surface. Therefore, the microbial characteristics, such as the content of lipids and polysaccharides and the cell surface area, greatly influence the remediation. It was reported that the bioaccumulation process was affected by the capsule, slime, and S-layers of bacteria [7]. Jiang et al. [55] determined that the surface characteristics of the bacteria wall greatly influenced biosorption properties. For example, the different cell wall compositions of gram-positive (peptidoglycan) and gram-negative (lipopolysaccharide) bacteria exhibit significantly different sorption capacity of radionuclides. Thus, radionuclide immobilization by microbes is species-specific. Even with the same species, the secretion or production of exopolymers and metabolites is different, consequently affecting the electrostatic attraction or the chemical interaction of metal ions with those functional groups. Additionally, the metabolism ability of microorganisms can impact the solubility of radionuclides via the direct or indirect enzymatic redox reaction, alteration of environmental pH, and Eh (i.e., production of acid metabolites and activity of electrons). The secreted organic substances such as phosphates or carbonates can cause the precipitation with radionuclides such as the formation of $(UO_2)(HPO_4)\cdot 4H_2O$ and $SrCO_3$ precipitants. Those microbial activities eventually affect the bioremediation ability.

Radionuclide-contaminated sites are normally poor in microbe concentration due to the higher toxicity of radionuclides. The captured radionuclides could more or less destroy the microorganisms. Particularly, those radionuclides transported into the cytoplasm of the cells can cause the inhibition of a series of enzymatic actions, thus decreasing microbial growth and reactivity. The end result is that the removal ability of microorganisms toward radionuclides decreases. To improve the bioremediation efficiency, recombinant DNA technology may be considered to construct genetically engineered (GE) microorganisms with high radiation resistance. In terms of liability and ethical issues, the release of GE microorganisms is a tedious task that requires site-dependence compatibility and comparability with other organisms onsite. Another strategy is the introduction of a microbial consortium to enhance the tolerance of microorganisms. In nature, the biological process is usually found to combined or simultaneous enzymatic actions of bacteria, microfungi, and microalgae. One interesting consortium of microorganism periphyton (bioassociation of algae, diatoms, fungi, bacteria, and protozoa) has demonstrated a relatively robust nature, which tends to resist change under variable conditions [56]. Such properties make periphyton communities easily contrived and/or incorporated into bioreactors, resulting in efficiently operating and running bioremediation-based processes.

Among various environmental factors, pH is perhaps one of the most important parameters influencing radionuclide immobilization by microbes. It influences both radionuclide chemistry in solution and cell surface binding sites. Most of the time, it also influences the tolerance of microbes. pH-dependent sorption behavior is related to the radionuclide species as well as the acid-base properties of functional groups on the microbial surface. At low pH, the functional groups are associated with H^+, which restricts the binding of positively charged radionuclides. With the increase of pH, the deprotonation of those functional groups increases, and they are consequently more negatively charged. Generally, it is a common phenomenon that higher pH facilitates negative charged radionuclide sorption on microbes. However, the uranium species is predominantly negatively charged at higher pH (i.e., $UO_2(CO_3)_2^{2-}$ and $UO_2(CO_3)_3^{4-}$), which decreases U(VI) sorption at higher pH [57, 58]. Precipitation of most radionuclides occurs at a higher pH level, contributing to the removal of radionuclides. Temperature is another important impact factor that affects the radionuclide species, the ligand and ligand complex, and the solubility of radionuclide ions [59]. Most reaction rates are sensitive to temperature changes. From a thermodynamic view, biosorption is favored at higher temperatures if the binding is endothermic but weakened if it is exothermic. Other factors such as various electron donors, acceptors, and nutrients in the environment can affect the microbial activity during the biotransformation of radionuclides.

To be honest, a single water treatment technique is rarely able to properly remediate an environmental threat on its own. For example, biosorption is a well-known technique to effectively remove radionuclides from aqueous solutions and concentrate them on the solid phase. However, those radionuclides are just transferred to another media. It still needs to be resolved how to permanently immobilize those radionuclides or transform them to stable and insoluble species. Therefore, a multicomponent water treatment process is considered, involving the combination of different biological processes such as biosorption/biomineralization or biosorption/reduction. The combination strategy is to give fully play to advantages of single treatment technology, as it is said that "even weak men have strength in unit". In theory, biosorption should benefit from biomineralization to permanently immobilize the concentrated radionuclides or from reduction to enzymatically reduce radioactive metals. Hexavalent uranium (U(VI)) is highly soluble, more mobile, and therefore biologically more available in the environment. However, under reducing conditions, the radionuclide is transformed to the less-soluble U(IV) form, hence decreasing its bioavailability [60]. On the other hand, biomineralization or reduction can benefit from the concentrated radionuclides via the biosorption process.

Despite the ecofriendliness, efficiency, and cost effectiveness of microbial decontamination technology, some intrinsic limitations such as relatively low sorption ability and weak selectivity still impede its widespread application. Recently, the application of novel nanoscale adsorbents (i.e., graphene oxide, carbon nanofiber, nano-scale zerovalent iron, and other nanoparticles) with high removal efficiency has definitely become the hot topic for environmental remediation [61]. However, the great debate regarding nanoparticle application is the ecofriendliness and cost effectiveness. Most reported nanoparticles are expensive and not safe (to say the least) for the environment. A perfect approach is to benefit from both to gain the efficiency of nanoparticles as well as the ecofriendliness and cost effectiveness of microorganisms. The possible scenarios are (1) to produce biodegradable biomass-derived NPs, (2) simultaneous actions, or (3) combined via associating inorganic NPs with microorganisms to develop biocomposites and hybrid materials [62, 63].

Overall, having microorganisms as the core of the radionuclide depollution strategy can achieve the "right" order of priority of goals, that is, (1) the ecofriendliness, (2) efficiency, and (3) cost effectiveness. Obviously, we will not save the environment by replacing one kind of pollution with another. As a naturally occurring component, the microbial-based technique can secure the requirement of the ecofriendly aspect. The treatment can be optimized via modifying or transforming the biomass, extracting biochemicals (i.e., biopolymers and enzymes), and combining the diverse actions (i.e., biosorption/reduction) of different microbes. As a renewable and inexpensive biomaterial, microorganisms are very advantageous when considering the last criterion of cost. So far, microbe-based decontamination technology is still in its infancy because most of the studies are performed in laboratories or limited to single radionuclide aqueous solutions. Few studies have focused on the application of microorganisms for the treatment of real radionuclide wastewater or other complex real-case scenarios. Therefore, microbial remediation technology needs to be further proved in large pilot-scale systems.

3.6 Microbially induced corrosion

Recent research has demonstrated the important role of microbes in mineral or concrete corrosion and deterioration, resulting in the release of radionuclides into the environment. Concrete is used in radionuclide repository materials in the deep subsurface. Low-level radioactive wastes generated from nuclear operations are often contained in concrete structures underground for hundreds of years. Similarly, spent fuel from nuclear reactors is also contained in concrete for interim storage aboveground, prior to long-term disposal. However, the interim storage time is uncertain, varying from 60 to 300 years. Thus, understanding the long-term structural integrity of concrete is necessary for the assessment of concrete to contain contaminants over time. Noticeably, microbial degradation combined with extreme natural phenomena may result in jeopardizing the mineral or concrete matrix, causing a serious problem in the geological repository of radionuclides. Therefore, attempts at anthropogenic intervention and efforts should be made to ensure the right direction. Herein, a special focus is on the biogenic deterioration and its protection. Relevant corrosion and deterioration mechanisms induced by microbes, impacting factors, existing control measures, and recent progress have been reviewed for the protection of concrete structures against biogenic deterioration. This review may also serve as a guide in the construction of concrete in the future for radioactive waste storage.

3.6.1 Radionuclide mineral corrosion

Numerous microorganisms in various radionuclide-contaminated environments have the capacity to induce or facilitate the mineralization of radionuclides, thus immobilizing them. Concurrently, there exist a certain number of bacteria or fungi in the environment that dissolve solid minerals into soluble and extractable radionuclide elements. This biohydrometallurgical process is usually called bioleaching, and uses microorganisms to extract radionuclides from their ore materials.

Microbial leaching of radionuclides from minerals is primarily based upon three synchronous steps [64]. Acidolysis is the principal mechanism during the corrosion process of minerals. Fungi produce organic acids such as citric, oxalic, gluconic, and malic because of microbial activity. Those organic acids dissolve radionuclides from minerals by displacement of metal chelates. Concurrently, complexolysis occurs where metal ions complex with various organic acids, thus solubilizing radionuclides such as iron chelate with oxalic acid. On the one hand, complexolysis facilitates the dissolution of radionuclides or metal ions from mineral ore and destabilizes the bonds between the surface metal and bulk leaching materials. On the other hand, this process often reduces the adverse effect of heavy metals toward fungi through chelation or complexation. Redoxolysis, the reduction of metal ions, also happens during the mineral corrosion process, such as in the case of ferric ions and/or manganese under the influence of oxalic acid. Generally, the production of a series of organic acids by a fungal metabolism causes the following acidolysis complex and chelate formation. The decrease of pH and the increase of redox potential were observed in the aqueous leach solutions by Lee et al. [65] when *Acidithiobacillus ferrooxidans* were initially incubated in uranium-bearing black shale.

A wide variety of microorganisms is capable of contributing to the corrosion of minerals. Chemolithoautotrophic bacteria are the most-studied species for their bioleachability of metals from sulfidic ores. Those bacteria such as *Acidithiobacillus* spp., *Ferrimicrobium* spp., and *Leptospirillum* spp. use ferrous iron and/or reduced sulfur sources as the electron donor to produce sulfuric acid and ferric iron, which subsequently dissolve minerals [66]. Generally, this kind of bacteria is acidophilic and adapted to the environmental pH range of 1.5–2.0.

Some heterotrophic bacteria also enable to dissolve minerals when organic carbon sources are provided as energy sources for their growth. These bacteria secrete acidic metabolites, as mentioned above. The produced organic acids play an important role in the dissolution of minerals or facilitating the mineral dissolution via forming stable soluble metal-organic complexes in solution.

Some metal-reducing bacteria are other alternatives to mediate mineral dissolution and metal ion release. For example, strains of *S. putrefaciens*, *Shewanella oneidensis*, or *Geobacter metallireducens* oxidize organic carbon (e.g., acetate) to carbon dioxide coupled with the reduction of Fe(III) or Mn(IV) oxide under anaerobic conditions. The produced soluble Fe(II) and Mn(II) increase their mineral dissolution [66]. *Aeromonas hydrophila* can reduce different biogenic jarosites followed by a release of Fe(II). The cells were attached on the surface of the mineral via pili, which acts as the electron transfer [67]. *G. metallireducens* GS-15 and *S. oneidensis* MR-1 demonstrated their ability of enzymatic reduction of plutonium(IV) to more soluble plutonium(III) under anaerobic conditions [68]. The solubilization processes are mostly accompanied by secondary mineralization, leading to the formation of siderite, vivianite, or magnetite.

The bioleaching processes are influenced by several factors that are mainly from microorganisms, minerals, and the physicochemical parameters during the incubation. Microbial concentration, activity, and tolerance to toxic metals greatly affect the bioleaching efficiency. Acidic metabolites as a consequence of microbial activity play a key role in radionuclide bioleaching. Accordingly, different mineral components of tested ore possess different binding chelation abilities with different organic acids. Thus, the bioleaching efficiency is greatly determined by the types and concentrations of produced organic acids as a consequence of microbial activity. It is reported that citric, oxalic, galic, malonic, and ascorbic are effective for uranium solubility [64]. Microbial tolerance to toxic heavy metals is also a major factor to influence the bioleaching process. The application of indigenous bacteria adapted to the environment instead of well-known strains from lab collections may alleviate the adverse effect from the toxic heavy metals. Additionally, physicochemical parameters such as temperature, pH, oxygen supply, stirring rate, contact time, composition of media, and properties of the material to be leached such as pulp density, grain size, mineral composition, and surface properties are the impacting factors. For example, the pH of the incubation medium is important for the microorganism growth, the biosynthesis of the types of metabolites, and the surface binding between cells and minerals [64].

Bioleaching approaches have been widely applied for radionuclides or metal extraction from ores. Recently, this approach also demonstrated its potential in other fields, such as the recovery of metals from printed circuit boards from computers and mobile phones, spent catalysts, batteries, and others. Microbial-mediated mineral corrosion causes the release of radionuclides from their repository into the environment, increasing the potential hazards toward humans and the biosphere. Therefore, attempts at anthropogenic intervention and efforts should be made to ensure the right direction.

3.6.2 Microbially induced concrete corrosion

Over the last two decades, the biodeterioration of concrete has become the focus of attention due to its significant impact on radionuclide repositories in the deep subsurface. Deterioration of concrete is mainly caused by the corrosion process, including loss of the effective steel cross-section. The production of acids such as sulfuric acid by microbes lowers the local environmental pH, which facilitates the penetration of corrosion-causing agents (i.e., carbon dioxide, chlorides ions, and moisture) to the interface of the concrete/steel. The corroding surface can act as a mixed electrode consisting of anodes and cathodes (electrically connected through the body of steel), whereas the pore water in concrete can function as an electrolyte. Under the appropriate conditions (dependent on the availability of oxygen and water, the pH of the electrolyte, relative humidity, and temperature), a series of anodic and cathodic reactions occurs, thus leading to the dissolution of metal.

In addition to the modification of local environment, the microorganism itself can impact the corrosion of concrete. For example, the enzymes and other acidic metabolites as a result of microbial activity react with the concrete surface to form expansive low-strength products [69], thus weakening the structural integrity of concrete. On the one hand, microbial activity can alter the local environment. On the other hand, the changed environment such as lowered pH, elevated relative humidity, and high concentration of carbon dioxide, chloride ion, sulfates or other salts, can facilitate the growth of microorganisms on the concrete surface, further speeding up the corrosion rate of concrete [70].

In conclusion, the existing mechanisms regarding the biogenic corrosion of concrete include physical corrosion and chemical corrosion [71]. In the former aspect, concrete corrosion occurs due to the physical or mechanical breaking of the structure, which is caused by the microbial growth or movement. In the latter aspect, it is mainly metabolites or other substances secreted by microbes that react with the surface component of concrete, adversely affecting the structural properties such as increased porosity and an expansive low-strength matrix. It is worth noting that the corrosion process is incapable of being realized by a pure culture of bacteria, algae, or fungi. It requires a series of biochemical interactions from the synergistic interactions of multiple species in the biofilm [72], which makes the reproduction of the microbial corrosion of concrete in the lab difficult.

A specific corrosion process involves the alteration of pH and substances as well as the succession of microorganism types in the local environment. Fig. 3.6 reveals the pH-dependent ecological succession of microbes during the concrete corrosion process, which is related with sulfur cycle [73]. The first colonization species is *Thiobacillus thioparus* when pH is decreased to 9.0. The pH is further lowered to neutral due to the oxidation of thiosulfate to polythionic acids by *T. thioparus*. However, *T. thioparus* disappears with the decrease of pH, whereas *Thiobacillus novellus*, *Thiobacillus neopolitanous*, and *Thiobacillus* are sequentially initiated under moderate or weakly acidic conditions. The microbial activity of these species further lowers the pH. Conversion of thiosulfate to elemental sulfur stimulates the growth of *Acidithiobacillus thiooxidans*, which oxides sulfur to H_2SO_4. H_2SO_4 induces a series of chemical reactions with the surface component of concrete.

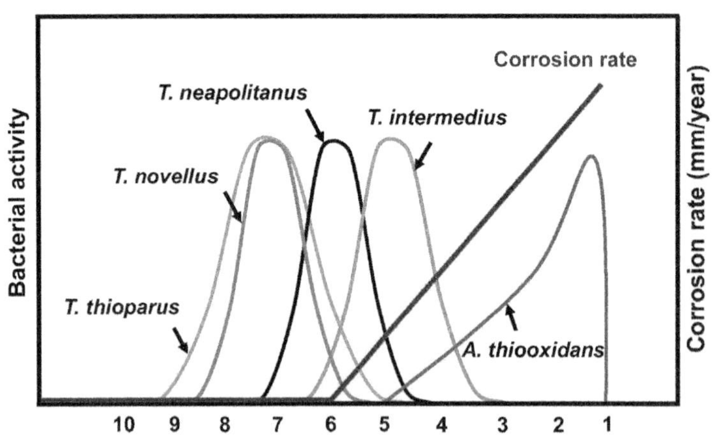

FIG. 3.6 Ecological succession of *Thiobacillus* sp. on the surface of fresh concrete exposed to hydrogen sulfide. *(From T. Noeiaghaei, A. Mukherjee, N. Dhami, S.-R. Chae, Biogenic deterioration of concrete and its mitigation technologies, Constr. Build. Mater. 149 (2017) 575–586, copyright 2017, Elsevier.)*

As a result, more expansive calcium-containing products are formed on the surface, such as gypsum ($CaSO_4 \cdot 2H_2O$), ettringite ($CaO_3 \cdot Al_2O_3 \cdot (CaSO_4)_3 \cdot 32H_2O$), or thaumasite ($Ca_3Si(CO_3)(SO_4)(OH)_6 \cdot 12H_2O$) [74], resulting in internal cracking and pitting of the concrete. The modification of concrete and the local environment is favored for further microbial corrosion. At last, pH is reduced to 1.5, an unfavorable environmental even for *A. thiooxidans*. Although *A. thiooxidans* was observed as the dominant species involved in the concrete corrosion process, the composition of the microorganism community varies with different types of groundwater and soils. As shown in Fig. 3.6, the corrosion rate of concrete is negligible in the initial period until the surface pH of concrete is decreased to 6.0. At this point, the concrete corrosion rate increases linearly with time, accompanied by the proliferation of *A. thiooxidans* at pH <5.0. The needed time to reach this pH is dependent on the local environment and varies from a few months to a few years, but it is <2 years. The observed corrosion rate is related to the temperature, relative humidity, the dissolved oxygen hydrogen level, and the H_2S concentration.

The rate and degree of concrete corrosion is dependent on numerous physical, chemical, and biological parameters [75]. Physical and chemical parameters (i.e., pH, Eh, carbon source type and concentration, moisture level, terminal electron acceptor type, concentration, and bioavailability) control, to a large extent, the distribution and microbial activity of the subsurface environment. In return, microbial activity results in the alteration of physical and chemical conditions. Biological or physiological aspects of microorganisms mainly control concrete biodegradation. For example, a certain range of pH and osmotic pressure is required for the growth of microbes. Similarly, the availability and abundance of energy sources are also affecting parameters because these components are involved in sulfur, iron, and nitrogen oxidation. Carbon flow and availability are also factors to consider in predicting concrete biodegradation.

So far, ways to mitigate biogenic corrosion can be classified into two main classes: (1) controlling the causative environment and (2) improving the resistance of concrete characteristics against microbial corrosion. Many researchers focused on decreasing sulfide and H_2S via dosing of chemicals such as oxygen, sodium hydroxide, magnesium hydroxide, iron salts, and nitrate into the deep surface or dosing sulfide-oxidizing bacteria [76, 77]. These strategies can reduce the sulfide level in the liquid phase, but not work for the gas phase sulfide. From the practical viewpoint, this method still has some problems to be addressed such as high operating and maintenance costs over the service life of the system and avoiding unwanted byproducts.

Due to the uncontrollability of the local environment most of the time, a common protection approach is to isolate the concrete from contact with surrounding environments. Therefore, protective coatings and sealers have been used for new concrete structures to extend their longevity. Epoxy was coated on the concrete surface against *T. ferrooxidans* trough laboratory and field exposure tests in a cooling tower basin [78]. The results showed that epoxy-coated concrete exhibited a good protection performance in that no biofilm was formed on the surface of the coated specimen after a 60-day exposure period. Although epoxy resins have excellent resistance against microbes, they are vulnerable to brittleness and low fracture toughness. In order to improve the structural properties of epoxy coatings such as tensile strength and modulus, several types of fillers such as glass fibers have been added to the coating system [79]. Recently, polymers (i.e., styrene-acrylic ester and acrylic) were introduced to the concrete to decrease its permeability [80]. Both microbiological and chemical tests demonstrated that the polymer-modified concrete can effectively decelerate the penetration of H_2S, CO_2, and the ingress of microorganisms and their metabolites, due to the decreased permeability via bridging of microcracks, reduction of pore size, and blocking of the pores by polymer particles. While decreased permeability was observed, the compatibility of the polymer and the concrete component can negatively influence the behavior of polymer-modified concrete. It was reported that high polymer/cement ratios decreased the compressive strength of concrete [76].

The addition of an antimicrobial agent in the coating is also a strategy to improve the resistance of concrete against microbial corrosion. A silver-bearing zeolite coating for the protection of mortar specimens from attack by *Acidithiobacillus thiooxidans* was studied by examining Ca^{2+} and Si^{4+} from the cementitious [81]. The leaching of Ca^{2+} and Si^{4+} from coated samples was 3.5-fold and 2.6-fold less than the uncoated samples, respectively. Heavy metal oxides (i.e., cuprous oxide and silver oxide) were also reported to mix with epoxy or spray on the concrete surface to reduce the proliferation of SRB [82]. Although the addition of antimicrobial agents can significantly protect concrete from microbial attack, their effectiveness is temporary. In addition to the high economic and environmental costs, another impediment to their use is that only a few of these agents have been trialled for a relatively short term. Therefore, further research is required to explore the ecofriendly and long-lasting antibacterial treatment of concrete.

Although the protective coatings are widely used, their functional properties including strength, elasticity, and adhesion might be influenced by microbial activity, especially the formation of biofilm. Microorganisms might secret aggressive metabolites and enzymes that attack these coatings and disrupt their structure. It was reported that coating defects and the debonded area usually acted as habitats of anaerobic microorganisms such as SRB [83]. Recently, some scientists have turned to nanotechnology for the improvement of concrete characteristics, which might be a promising research field. Considering the novel functionalities and excellent antimicrobial properties of nanomaterials, nano-incorporated concrete and nanocoatings are capable of being a viable solution for improving biocorrosion resistance and prolonging the durability of concrete. However, the application of nanotechnology in concrete is still in its research phase. More research is required for the filed application of nanomaterials modified concrete with low or negligible toxicity.

References

[1] G.M. Gadd, Microbial Roles in Mineral Transformations and Metal Cycling in the Earth's Critical Zone, Springer, The Netherlands, 2013, pp. 115–165.

[2] S. Groudev, P. Georgiev, I. Spasova, Bioremediation of a soil contaminated with radioactive elements, Hydrometallurgy 59 (2001) 311–318.

[3] S.J. Dhawal, M.R. Phadatare, N.D. Thorat, G.S. Kulkarni, S.H. Pawar, Natural radioactivity study in soil samples of South Konkan, Maharashtra, India, Radiat. Prot. Dosim. 157 (2013) 225–233.

[4] P.K. Dhal, P. Sar, Microbial communities in uranium mine tailings and mine water sediment from Jaduguda U mine, India: a culture independent analysis, J. Environ. Sci. Health A 49 (2014) 694–709.

[5] J.L. Ravanat, T. Douki, UV and ionizing radiations induced DNA damage, differences and similarities, Radiat. Phys. Chem. 128 (2016) 92–102.

[6] M.M. Cox, J.R. Battista, *Deinococcus radiodurans*—the consummate survivor, Nat. Rev. Microbiol. 3 (2005) 882–892.

[7] A. Shukla, P. Parmar, M. Saraf, Radiation, radionuclides and bacteria: an in-perspective review, J. Environ. Radioact. 180 (2017) 27–35.

[8] H.S. Misra, N.P. Khairnar, S. Kota, S. Shrivastava, V.P. Joshi, S.K. Apte, An exonuclease I-sensitive DNA repair pathway in *Deinococcus radiodurans*: a major determinant of radiation resistance, Mol. Microbiol. 59 (2006) 1308–1316.

[9] M. Alexandrova, T. Rozhko, G. Vydryakova, N. Kudryasheva, Effect of americium-241 on luminous bacteria. Role of peroxides, J. Environ. Radioact. 102 (2011) 407–411.

[10] M.J. Daly, E.K. Gaidamakova, V.Y. Matrosova, A. Vasilenko, M. Zhai, A. Venkateswaran, M. Hess, M.V. Omelchenko, H.M. Kostandarithes, K.S. Makarova, L.P. Wackett, J.K. Fredrickson, D. Ghosal, Accumulation of Mn(II) in *Deinococcus radiodurans* facilitates gamma-radiation resistance, Science 306 (2004) 1025–1028.

[11] W. Huang, W. Cheng, X. Nie, F. Dong, C. Ding, M. Liu, Z. Li, T. Hayat, N.S. Alharbi, Microscopic and spectroscopic insights into uranium phosphate mineral precipitated by Bacillus *Mucilaginosus*, ACS Earth Space Chem. 1 (2017) 483–492.

[12] A.W. Hernsdorf, Y. Amano, K. Miyakawa, K. Ise, Y. Suzuki, K. Anantharaman, A. Probst, D. Burstein, B.C. Thomas, J.F. Banfield, Potential for microbial H-2 and metal transformations associated with novel bacteria and archaea in deep terrestrial subsurface sediments, ISME J. 11 (2017) 1915–1929.

[13] A. Bagnoud, K. Chourey, R.L. Hettich, I. de Bruijn, A.F. Andersson, O.X. Leupin, B. Schwyn, R. Bernier-Latmani, Reconstructing a hydrogen-driven microbial metabolic network in Opalinus Clay rock, Nat. Commun. 7 (2016) 12770.

[14] C. Greening, A. Biswas, C.R. Carere, C.J. Jackson, M.C. Taylor, M.B. Stott, G.M. Cook, S.E. Morales, Genomic and metagenomic surveys of hydrogenase distribution indicate H2 is a widely utilised energy source for microbial growth and survival, ISME J. 10 (2016) 761–777.

[15] W.J. Brazelton, B. Nelson, M.O. Schrenk, Metagenomic evidence for H2 oxidation and H2 production by serpentinite-hosted subsurface microbial communities, Front. Microbiol. 2 (268) (2011) 1–16.

[16] J.R. Sieber, M.J. McInerney, R.P. Gunsalus, Genomic insights into syntrophy: the paradigm for anaerobic metabolic cooperation, S. Gottesman, C.S. Harwood, O. Schneewind (Eds.), Annu. Rev. Microbiol. 66 (2012) 429–452.

[17] R. Kumar, M. Nongkhlaw, C. Acharya, S.R. Joshi, Bacterial community structure from the perspective of the uranium ore deposits of domiasiat in India, Proc. Natl. Acad. Sci. India Sect. B Biol. Sci. 83 (2013) 485–497.

[18] M.L. Merroun, S. Selenska-Pobell, Bacterial interactions with uranium: an environmental perspective, J. Contam. Hydrol. 102 (2008) 285–295.

[19] K. Rakshak, J.S. Ram, A. Celin, Metal tolerant Bacillus and Pseudomonas from uranium rich soils of Meghalaya, Res. J. Biotechnol. (2008) 345–350.

[20] R. Kumar, C. Acharya, S.R. Joshi, Isolation and analyses of uranium tolerant *serratia marcescens* strains and their utilization for aerobic uranium U(VI) bioadsorption, J. Microbiol. 49 (2011) 568–574.

[21] R. Kumar, M. Nongkhlaw, C. Acharya, S.R. Joshi, Uranium (U)-tolerant bacterial diversity from U ore deposit of domiasiat in north-East India and its prospective utilisation in bioremediation, Microbes Environ. 28 (2013) 33–41.

[22] P.G. Falkowski, T. Fenchel, E.F. Delong, The microbial engines that drive Earth's biogeochemical cycles, Science 320 (2008) 1034–1039.

[23] C. Desai, R.Y. Parikh, T. Vaishnav, Y.S. Shouche, D. Madamwar, Tracking the influence of long-term chromium pollution on soil bacterial community structures by comparative analyses of 16S rRNA gene phylotypes, Res. Microbiol. 160 (2009) 1–9.

[24] G. Haferburg, M. Sineriz, D. Merten, Heavy metal resistance mechanisms in actinobacteria for survival in AMD contaminated soils, Chem. Erde-Geochem. 65 (2005) 131–144.

[25] S. Xie, J. Yang, C. Chen, X. Zhang, Q. Wang, C. Zhang, Study on biosorption kinetics and thermodynamics of uranium by *Citrobacter freudii*, J. Environ. Radioact. 99 (2008) 126–133.

[26] D. Prakash, P. Gabani, A.K. Chandel, Z. Ronen, O.V. Singh, Bioremediation: a genuine technology to remediate radionuclides from the environment, Microb. Biotechnol. 6 (2013) 349–360.

[27] G. Rufyikiri, S. Declerck, Y. Thiry, Comparison of 233U and 33P uptake and translocation by the arbuscular mycorrhizal fungus Glomus intraradices in root organ culture conditions, Mycorrhiza 14 (2004) 203–207.

[28] S. Amachi, T. Fujii, H. Shinoyama, Y. Muramatsu, Microbial influences on the mobility and transformation of radioactive iodine in the environment, J. Nucl. Radiochem. Sci. 6 (2005) 21–24.

[29] W. Huang, X. Nie, F. Dong, C. Ding, R. Huang, Y. Qin, M. Liu, S. Sun, Kinetics and pH-dependent uranium bioprecipitation by *Shewanella putrefaciens* under aerobic conditions, J. Radioanal. Nucl. Chem. 312 (2017) 531–541.

[30] A.J. Francis, Microbial mobilization and immobilization of plutonium, J. Alloys Compd. 444 (2007) 500–505.

[31] G. Rufyikiri, Y.L. Thiry, B. Delvaux, S. Declerck, Uranium uptake and translocation by the arbuscular mycorrhizal fungus, Glomus intraradices, under root-organ culture conditions, New Phytol. 156 (2002) 275–281.

[32] A. Heggo, J.S. Angle, R.L. Chaney, Effects of vesicular-arbuscular mycorrhizal fungi on heavy metal uptake by soybeans, Soil Biol. Biochem. 22 (1990) 865–869.

[33] X.-Y. Zheng, X.-Y. Wang, Y.-H. Shen, X. Lu, T.-S. Wang, Biosorption and biomineralization of uranium(VI) by *Saccharomyces cerevisiae*-crystal formation of chernikovite, Chemosphere 175 (2017) 161–169.

[34] J.D. Keasling, S.J.V. Dien, P. Trelstad, N. Renninger, K. Mcmahon, Application of polyphosphate metabolism to environmental and biotechnological problems (review), Biochem. Mosc. 65 (2000) 324–331.

[35] M.E. Marsh, Regulation of CaCO(3) formation in coccolithophores, Comp. Biochem. Physiol. B 136 (2003) 743–754.

[36] J.R. Young, Biomineralization within vesicle: the calcite of coccoliths, Rev. Mineral. Geochem. 54 (2003) 189–215.

[37] E.I. Stevenson, M. Hermoso, R.E.M. Rickaby, J.J. Tyler, F. Minoletti, I.J. Parkinson, F. Mokadem, K.W. Burton, Controls on stable strontium isotope fractionation in coccolithophores with implications for the marine Sr cycle, Geochim. Cosmochim. Acta 128 (2014) 225–235.

[38] G. Langer, N. Gussone, G. Nehrke, U. Riebesell, A. Eisenhauer, H. Kuhnert, B. Rost, S. Trimborn, S. Thoms, Coccolith strontium to calcium ratios in *Emiliania huxleyi*: the dependence on seawater strontium and calcium concentrations, Limnol. Oceanogr. 51 (2006) 310–320.

[39] J.S. Chow, C. Lee, A. Engel, The influence of extracellular polysaccharides, growth rate, and free coccoliths on the coagulation efficiency of *Emiliania huxleyi*, Mar. Chem. 175 (2015) 5–17.

[40] S. Chabalala, E.M.N. Chirwa, Uranium(VI) reduction and removal by high performing purified anaerobic cultures from mine soil, Chemosphere 78 (2010) 52–55.

[41] T. Jabbar, G. Wallner, Biotransformation of Radionuclides: Trends and Challenges, Springer International Publishing, 2015.

[42] G.A. Icopini, H. Boukhalfa, M.P. Neu, Biological reduction of Np(V) and Np(V) citrate by metal-reducing bacteria, Environ. Sci. Technol. 41 (2007) 2764–2769.

[43] J.R. Lloyd, C. Leang, A.L. Hodges Myerson, M.V. Coppi, S. Cuifo, B. Methe, S.J. Sandler, D.R. Lovley, Biochemical and genetic characterization of PpcA, a periplasmic *c*-type cytochrome in *Geobacter sulfurreducens*, Biochem. J. 369 (2003) 153–161.

[44] J.R. Lloyd, D.R. Lovley, Microbial detoxification of metals and radionuclides, Curr. Opin. Biotechnol. 12 (2001) 248–253.

[45] C. Ding, W. Cheng, Y. Sun, X. Wang, Effects of *Bacillus subtilis* on the reduction of U(VI) by nano-Fe-0, Geochim. Cosmochim. Acta 165 (2015) 86–107.

[46] X. Tan, M. Fang, X. Wang, Sorption speciation of lanthanides/actinides on minerals by TRLFS, EXAFS and DFT studies: a review, Molecules 15 (2010) 8431–8468.

[47] K. Nayuki, B. Chen, R. Ohtomo, Y. Kuga, Cellular imaging of cadmium in resin sections of arbuscular mycorrhizas using synchrotron micro X-ray fluorescence, Microbes Environ. 29 (2014) 60–66.

[48] H.S. Davies, F. Cox, C.H. Robinson, J.K. Pittman, Radioactivity and the environment: technical approaches to understand the role of arbuscular mycorrhizal plants in radionuclide bioaccumulation, Front. Plant Sci. 6 (2015) 580.

[49] R. Chakraborty, C.H. Wu, T.C. Hazen, Systems biology approach to bioremediation, Curr. Opin. Biotechnol. 23 (2012) 483–490.

[50] E. Cardenas, W.M. Wu, M.B. Leigh, J. Carley, S. Carroll, T. Gentry, J. Luo, D. Watson, B. Gu, M. Ginder-Vogel, Microbial communities in contaminated sediments, associated with bioremediation of uranium to submicromolar levels, Appl. Environ. Microbiol. 74 (2008) 3718–3729.

[51] M.W. Fields, C.E. Bagwell, S.L. Carroll, T. Yan, X. Liu, D.B. Watson, P.M. Jardine, C.S. Criddle, T.C. Hazen, J. Zhou, Phylogenetic and functional biomakers as indicators of bacterial community responses to mixed-waste contamination, Environ. Sci. Technol. 40 (2006) 2601–2607.

[52] J.D.V. Nostrand, W.M. Wu, L. Wu, Y. Deng, J. Carley, S. Carroll, Z. He, B. Gu, J. Luo, C.S. Criddle, GeoChip-based analysis of functional microbial communities during the reoxidation of a bioreduced uranium-contaminated aquifer, Environ. Microbiol. 11 (2009) 2611–2626.

[53] D. Johnson, J.R. Leake, D.J. Read, Novel in-growth core system enables functional studies of grassland mycorrhizal mycelial networks, New Phytol. 152 (2001) 555–562.
[54] W. Cheng, C. Ding, Q. Wu, X. Wang, Y. Sun, W. Shi, T. Hayat, A. Alsaedi, Z. Chai, X. Wang, Mutual effect of U(VI) and Sr(II) on graphene oxides: evidence from EXAFS and theoretical calculations, Environ. Sci.: Nano 4 (2017) 1124–1131.
[55] W. Jiang, A. Saxena, B. Song, B.B. Ward, T.J. Beveridge, S.C. Myneni, Elucidation of functional groups on gram-positive and gram-negative bacterial surfaces using infrared spectroscopy, Langmuir 20 (2004) 11433–11442.
[56] S.T. Larned, P. Silver, A.D. Steinman, I. Polls, A prospectus for periphyton: recent and future ecological research, J. N. Am. Benthol. Soc. 29 (2010) 182–206.
[57] C. Ding, W. Cheng, X. Nie, F. Yi, Synergistic mechanism of U(VI) sequestration by magnetite-graphene oxide composites: evidence from spectroscopic and theoretical calculation, Chem. Eng. J. 324 (2017) 113–121.
[58] W. Cheng, C. Ding, X. Nie, T. Duan, R. Ding, Fabrication of 3D macroscopic graphene oxide composites supported by montmorillonite for efficient U(VI) wastewater purification, ACS Sustain. Chem. Eng. 5 (2017) 5503–5511.
[59] K.S. Kumar, H.U. Dahms, E.J. Won, J.S. Lee, K.H. Shin, Microalgae—a promising tool for heavy metal remediation, Ecotoxicol. Environ. Saf. 113 (2015) 329–352.
[60] L. Newsome, K. Morris, J.R. Lloyd, The biogeochemistry and bioremediation of uranium and other priority radionuclides, Chem. Geol. 363 (2014) 164–184.
[61] S. Yu, X. Wang, S. Yang, G. Sheng, A. Alsaedi, T. Hayat, X. Wang, Interaction of radionuclides with natural and manmade materials using XAFS technique, Sci. China: Chem. 60 (2017) 170–187.
[62] W. Cheng, C. Ding, Y. Sun, X. Wang, Fabrication of fungus/attapulgite composites and their removal of U(VI) from aqueous solution, Chem. Eng. J. 269 (2015) 1–8.
[63] C. Ding, W. Cheng, X. Nie, F. Yi, S. Xiang, A.M. Asiri, H.M. Marwani, Reactivity of carbonized fungi supported nanoscale zero-valent iron toward U(VI) influenced by naturally occurring ions, J. Ind. Eng. Chem. 61 (2018) 236–243.
[64] M.M. Amin, I.E. Elaassy, M.G. El-Feky, A.S. Sallam, M.S. Talaat, N.A. Kawady, Effect of mineral constituents in the bioleaching of uranium from uraniferous sedimentary rock samples, Southwestern Sinai, Egypt, J. Environ. Radioact. 134 (2014) 76–82.
[65] J.U. Lee, S.M. Kim, K.W. Kim, I.S. Kim, Microbial removal of uranium in uranium-bearing black shale, Chemosphere 59 (2005) 147–154.
[66] K. Pollmann, S. Kutschke, S. Matys, J. Raff, G. Hlawacek, F.L. Lederer, Bio-recycling of metals: recycling of technical products using biological applications, Biotechnol. Adv. 4 (2018) 1048–1062.
[67] L. Castro, M.L. Blazquez, F. Gonzalez, J.A. Munoz, A. Ballester, Reductive leaching of jarosites by Aeromonas hydrophila, Miner. Eng. 95 (2016) 21–28.
[68] G.A. Icopini, J.G. Lack, L.E. Hersman, M.P. Neu, H. Boukhalfa, Plutonium(V/VI) reduction by the metal-reducing bacteria *Geobacter metallireducens* GS-15 and *Shewanella oneidensis* MR-1, Appl. Environ. Microbiol. 75 (2009) 3641–3647.
[69] N. Kip, J.A. van Veen, The dual role of microbes in corrosion, ISME J. 9 (2015) 542–551.
[70] S. Wei, Z. Jiang, H. Liu, D. Zhou, M. Sanchez-Silva, Microbiologically induced deterioration of concrete—a review, Braz. J. Microbiol. 44 (2013) 1001–1007.
[71] C. Gaylarde, M.R. Silva, T. Warscheid, Microbial impact on building materials: an overview, Mater. Struct. 36 (2003) 342–352.
[72] J.W.S. Domingo, R.P. Revetta, B. Iker, V. Gomez-Alvarez, J. Garcia, J. Sullivan, J. Weast, Molecular survey of concrete sewer biofilm microbial communities, Biofouling 27 (2011) 993–1001.

[73] H. Satoh, M. Odagiri, T. Ito, S. Okabe, Microbial community structures and in situ sulfate-reducing and sulfur-oxidizing activities in biofilms developed on mortar specimens in a corroded sewer system, Water Res. 43 (2009) 4729–4739.

[74] E. Kirsten, Thaumasite attack on concrete at Marbjerg waterworks, Cem. Concr. Compos. 25 (2003) 1147–1150.

[75] C.E. Turick, C.J. Berry, Review of concrete biodeterioration in relation to nuclear waste, J. Environ. Radioact. 151 (2016) 12–21.

[76] T. Noeiaghaei, A. Mukherjee, N. Dhami, S.-R. Chae, Biogenic deterioration of concrete and its mitigation technologies, Constr. Build. Mater. 149 (2017) 575–586.

[77] M. Nemati, G.E. Jenneman, G. Voordouw, Impact of nitrate-mediated microbial control of souring in oil reservoirs on the extent of corrosion, Biotechnol. Prog. 17 (2001) 852–859.

[78] M.L. Berndt, Evaluation of coatings, mortars and mix design for protection of concrete against sulphur oxidising bacteria, Constr. Build. Mater. 25 (2011) 3893–3902.

[79] C. Vipulanandan, J. Liu, Glass-fiber mat-reinforced epoxy coating for concrete in sulfuric acid environment, Cem. Concr. Res. 32 (2002) 205–210.

[80] A. Beeldens, J. Monteny, E. Vincke, N.D. Belie, D.V. Gemert, L. Taerwe, W. Verstraete, Resistance to biogenic sulfuric acid corrosion of polymer-modified mortars, Cem. Concr. Compos. 23 (2001) 47–56.

[81] T. Halle, G. Nakhla, E. Allouche, Evaluation of the resistance of mortars coated with silver bearing zeolite to bacterial-induced corrosion, Corros. Sci. 50 (2008) 713–720.

[82] E.H. Hewayde, G.F. Nakhla, E.N. Allouche, P.K. Mohan, Beneficial impact of coatings on biological generation of sulfide in concrete sewer pipes, Struct. Infrastruct. Eng. 3 (2007) 267–277.

[83] S.Y. Li, Y.G. Kim, K.S. Jeon, Y.T. Kho, Microbiologically influenced corrosion of underground pipelines under the disbonded coatings, Met. Mater. 6 (2000) 281–286.

Chapter 4

Adsorption of radionuclides on carbon-based nanomaterials

Songhua Lu, Yubing Sun and Changlun Chen
Key Laboratory of Photovoltaic and Energy Conservation Materials, Institute of Plasma Physics, Chinese Academy of Sciences, Hefei, People's Republic of China

4.1 Introduction

The nuclear industry generates large amounts of radioactive wastewater that must be effectively treated before discharge into the environment. For the last few decades, there has been increasing global attention on public health issues due to environmental pollution caused by radionuclides [1, 2]. Therefore, owing to ever-growing concern over the safety of nuclear power, the efficient removal and cleanup of radioactive waste remain fundamental challenges in nuclear science. Many radioactive elements, namely, uranium (U(VI)), thorium (Th(IV)), europium (Eu(III)), trontium (Sr(II)), cesium (Cs(I)), etc., are highly hazardous and toxic to human health [3]. As a representative of the hexavalent actinide element, uranium is the main component of nuclear waste and plays an important role in the nuclear fuel cycle [4]. Thorium, a representative tetravalent actinide element, has a toxic nature, even at trace levels [5]. Europium reveals similar adsorption properties to both trivalent lanthanides and actinides [6]. Due to its long half life ($T_{1/2} = 28.8$ years) and abundance in nuclear fission products, strontium causes a huge hazard to human safety [7]. Cesium has a long half-life of 30.1 years and is a biological hazard [8]. Based on the above information, it can be concluded that radionuclide contaminants cause a serious environmental problem that has attracted many researchers to investigate the removal of radionuclides.

For the past few decades, a variety of technologies have been investigated and developed for the removal and recovery of radionuclides from mine tailings, nuclear fuel effluents, seawater, and other waste sources [9–11]. The adsorption process has numerous advantages, including ease of operation, the possibility of regeneration and reuse as well as low cost, being suitable both for batch and continuous processes, and applicability at very low concentrations [12]. The solid-liquid separation technique based on carbonaceous

Interface Science and Technology, Vol. 29. https://doi.org/10.1016/B978-0-08-102727-1.00004-2

materials, including activated carbon [13–16], graphene oxides (GOs) [3, 17], carbon nanotubes (CNTs) [18, 19], and carbon fibers (CNFs) [20, 21], has been gradually applied to the adsorption field. The carbonaceous materials have been chosen for application in the adsorption field due to their better acid-base stability than common inorganic sorbents and higher radiation and thermal resistance than organic exchanger resins [13, 16]. Activated carbons are a kind of unique adsorbent due to their extended surface area and microporous structure. In the early stages, the pristine activated carbons were first used to adsorption radionuclides with low adsorption capacity, which was attributed to the short of functional groups on the surface of activated carbons [10, 22]. The mainly governed adsorption mechanism of activated carbons toward metal ions is electrostatic interaction (electrostatic attraction or repulsion). Subsequently, the modified activated carbons with different functional groups were utilized to remove radionuclides with enhanced adsorption capacity compared to pristine activated carbons [16, 23]. CNTs were first discovered in 1991 by Iijima as a new member of the carbon family [24]. Afterward, CNTs have attracted extensive attention and research due to their unique structural, electronic, optoelectronic, semiconductor, mechanical, chemical, and physical properties. CNTs have proven to have great potential as superior adsorbents for removing many kinds of organic and inorganic pollutants [25–28]. The chemical interaction between the metal ions and the surface functional groups of CNTs is commonly believed to be the major sorption mechanism [29–31] (Fig. 4.1).

Simultaneously, the pristine and modified CNTs were used in the remediation of radionuclide contaminants [32, 33]. Tan et al. observed that the adsorption of Eu(III) was irreversible and difficult to be desorbed from CNTs [34]. Moreover, the large adsorption capacity makes CNTs a possible enrichment material for radionuclides. Chen et al. found that the interaction mechanism between oxidized MWCNTs and Sr(II) or Eu(III) was largely governed

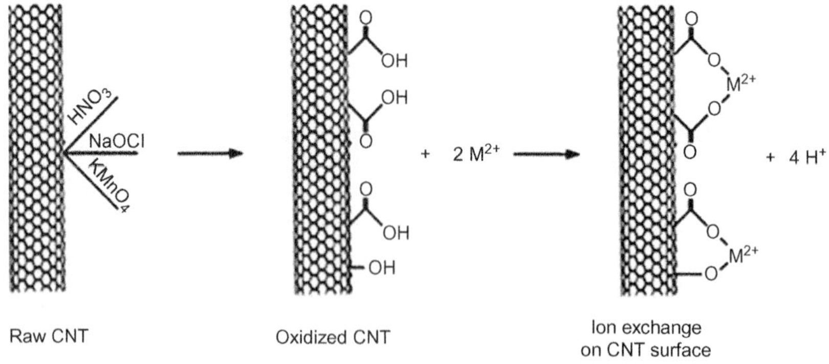

FIG. 4.1 Schematic diagram of the major mechanism for sorption of divalent metal ions onto CNT surface [31]. *(Data from G.P. Rao, C. Lu, F. Su, Sorption of divalent metal ions from aqueous solution by carbon nanotubes: a review, Sep. Purif. Technol. 58 (2007) 224–231. Copyright 2007, Elsevier.)*

by forming complexes with carboxyl groups (low pH) and deprotonated hydroxyl groups (high pH) [35]. A novel amido-amine functionalized MWCNT was fabricated by Singh Deb et al., which revealed an excellent adsorption performance toward radionuclides ^{233}U (20.66 mg/g) and $^{152+154}Eu$ (16.1 mg/g) [33]. As one of the most important graphene derivatives, GO was generated from the oxidation of graphite. The oxidation process of graphite to GO can introduce abundant oxygen functional groups (carboxyl, hydroxyl, and epoxy groups) on its edges and its basal plane. These functional groups can be used as anchoring sites to bind radionuclide ions by electrostatic and/or complexation interactions [6, 36]. Additionally, these introduced oxygen groups also make GO hydrophilic, implying that it can be applied in aquatic systems. However, the exfoliated graphene nanosheets tend to aggregate, which causes a low specific surface area in aqueous systems and is unfavorable for the adsorption process. In addition, the nanosized particle and hydrophilic property of GO make it hard to be separated from the adsorption suspension. To overcome these disadvantages, the decoration of inorganic nanoparticles on graphene nanosheets to form inorganic GO-based materials is an effective and feasible approach [37, 38]. Besides, owing to the high reactivity of some functional groups (i.e., amino, amidoxime) toward radionuclide ions, the organic GO-based materials are obtained by grafting an organic substance [3, 39]. Generally, the obtained inorganic/organic GO-based materials with larger surface area or abundant functional groups show superior adsorption performance, compared with pristine GO. CNFs as a kind of new breed and promising nanomaterials are used to investigate the enrichment behaviors toward radionuclides [21, 40]. CNTs with an interconnected three-dimensional framework and abundant oxygen functional groups are potentially valuable and important adsorbents for application in the removal of aquatic contaminants. Because of their versatile properties, CNFs have been widely utilized in various fields [41, 42], Yu's groups found that CNFs possess unique properties (e.g., huge specific surface area, high porosity, and low density) [43, 44]. Due to the high adsorption performance and excellent recyclability, CNFs were usually used to remove heavy metal ions [45, 46] and organic contaminants [45, 47, 48] for environmental cleanup. Liang et al. observed that CNFs (diameter ~50 nm) exhibit excellent adsorption performance toward Pb(II) ions [45]. Besides, as another category of CNFs, activated carbon fibers (ACFs) have been applied in gas adsorption/storage and water treatment due to their high specific surface area resulting from pore-creating surface modification [49, 50]. The high porosity and large number of oxygen functional groups on CNFs could be assigned to their excellent adsorption performance [43]. Their adsorption may also affect the fate, transformation, and transfer of radionuclides in the environment. Here, we focused on a review of carbon family materials in the removal of radionuclides (e.g., U(VI), Th(IV), Eu(III), Sr(II), Cs(I)) from diverse aqueous phases. Therefore, this chapter summarizes the surface properties and modification of carbon family materials, and their adsorption behavior and mechanism toward radionuclides.

4.2 Surface properties of carbon family materials

4.2.1 Activated carbon

Generally, a large number of various carbonaceous materials (e.g., lignin, coal, wood, and coconut shells) are utilized to produce active carbon [51]. Moreover, to decrease production costs, industrial or municipal and agricultural wastes are used as alternatives to produce active carbon [52]. Various techniques were used to modify the activated carbon, including chemical, physical, and biological modification methods. These modification methods made some functional groups form on the surface of active carbon. These groups include heteroatoms of hydrogen and oxygen and may also include sulfur, nitrogen, and chlorine. The nature of these groups depends on activation conditions, and it is correlated to the specific interactions with surface adsorbed compounds and the acid/base properties of the surface [53]. It has been proved that the concentration level of surface oxygen functional groups (phenol, carboxyl, lactol, lactone, and quinone) is an important factor related to the removal mechanism and adsorption capacity [54–57]. Other important properties of active carbon that influence the adsorption capacity are pore size distribution [58–60], ash content [61, 62], and the pH of point of zero charge (pH_{PZC}, as an indicator of the surface chemistry of active carbon) [63].

4.2.2 Carbon nanotubes

CNTs have attracted considerable research because of their unique physicochemical properties. CNTs are tubular objects with unique structural, mechanical, and electronic properties [64, 65]. CNTs can be classified into two types, according to their (super)structures (Fig. 5.2): (i) multiwalled carbon nanotubes (MWCNTs) [66], and (ii) single-walled carbon nanotubes (SWCNTs) [67, 68]. Multiwalled carbon nanotubes (MWCNTs) were the first observed CNTs (Fig. 4.2A). They can be considered elongated fullerenes [71] due to their being constituted of up to several tens of graphitic shells with large length/diameter ratio, diameter of ~ 1 nm, and adjacent shell separation of ~ 0.34 nm [72]. Later, Iijima and Ichihashi [67] and Bethune et al. [68] independently discovered single-walled carbon nanotubes (SWCNTs) (Fig. 4.2B) using the arc-discharge method and metal (iron and cobalt) as catalysts. CNTs have been extensively investigated by both theoretical and experimental methods because of their unique properties such as their morphology or nanostructure, the length and diameter of the tubes, and the functional mechanical, optoelectronic, thermal, and electrical properties that rely on the atomic arrangement [73, 74]. As a consequence, the investigations of the application and properties of CNTs in multidisciplinary areas increase exponentially.

The surface properties of MWCNTs are fundamental and important for understanding the sorption behaviors of radionuclide ions. The XRD patterns of pristine and oxidized MWCNTs are shown in Fig. 4.2C [70]. The 2θ degree

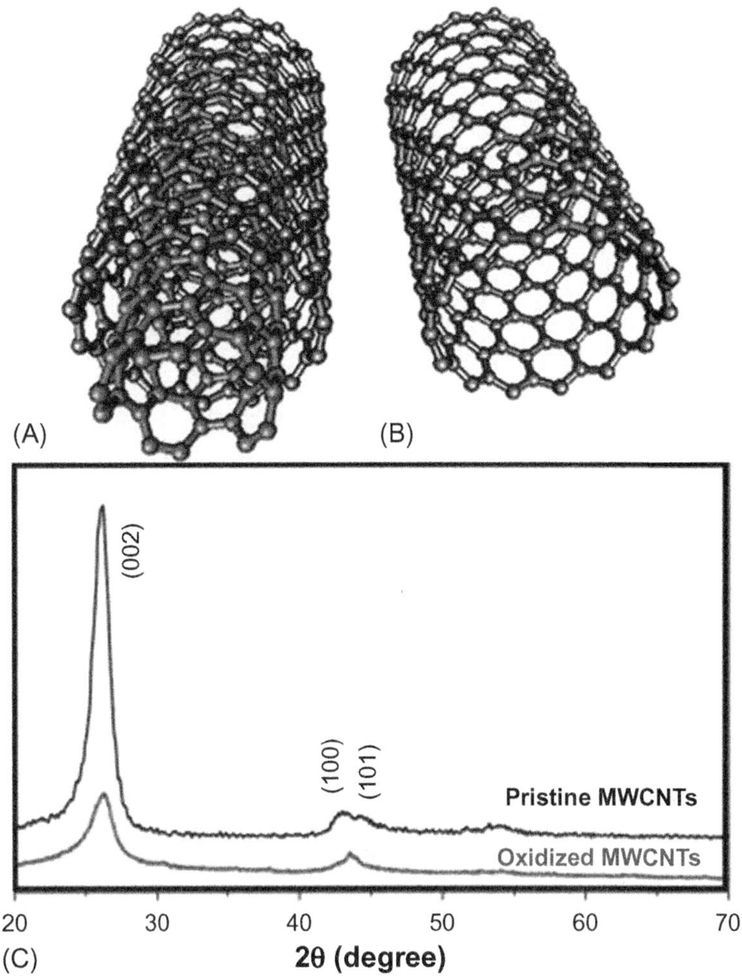

FIG. 4.2 (Super)structure representations of (A) a MWCNT and (B) a SWCNT,[69] the XRD patterns of pristine and oxidized MWCNTs samples (C) [70]. *(Data from Y.L. Zhao, J.F. Stoddart, Noncovalent functionalization of single-walled carbon nanotubes, Acc. Chem. Res. 42 (2009) 1161–1171 and Y.B. Sun, S.T. Yang, G.D. Sheng, Z.Q. Guo, X.K. Wang, The removal of U(VI) from aqueous solution by oxidized multiwalled carbon nanotubes, J. Environ. Radioact. 105 (2012) 40–47. Copyright 2009 and 2012, American Chemical Society, Elsevier.)*

at 26 degrees, 43 degrees, and 54 degrees of XRD patterns is assigned to the (002), (100), and (004) diffraction of MWCNTs, and the broad peak of 2θ degree at 44 degrees belongs to the (101) diffraction of solid carbon materials of pristine MWCNTs. For the oxidized MWCNTs, the disappearance of a sharper peak of 2θ degree at 44 degrees suggests that solid carbon materials are eliminated by oxidation. Also, the lower intensity of oxidized MWCNT peaks compared with pristine MWCNTs implies the poor crystallinity of oxidized MWCNTs.

4.2.3 Graphene oxides

GO with abundant oxygen groups (carboxyl, hydroxyl, epoxy groups) [75, 76] can form complexes with organic pollutants and metal ions through electrostatic interaction, hydrogen bonding, and coordination, which is a benefit to its potential application in environmental pollution remediation and control [77–79]. Generally, GO was prepared by the oxidation of graphite with various oxidants in acidic media [80, 81]. The Hummer method was usually used and considered one of the most efficient methods for synthesizing GO [82, 83]. Subsequently, GO can be chemically reduced to reduced graphene oxides (RGO) by various inorganic and organic reducing agents (Fig. 4.3) [84, 85].

GO-based materials support a low density compared to low-cost adsorbents (e.g., oxides, clay minerals) [86], and with a high surface area and abundant surface functional groups (e.g., carboxyls, hydroxyls, and epoxides). Therefore, these superiorities (low density, high surface area, and abundant surface functional groups) let GO-based materials with excellent adsorption capacity for contaminants. Fig. 4.4A shows the Raman spectra of graphene and GO. Two dominant peaks at 1349 and 1582 cm^{-1} are assigned to the D band and the G band, respectively. Generally, the D band and the G band are derived from

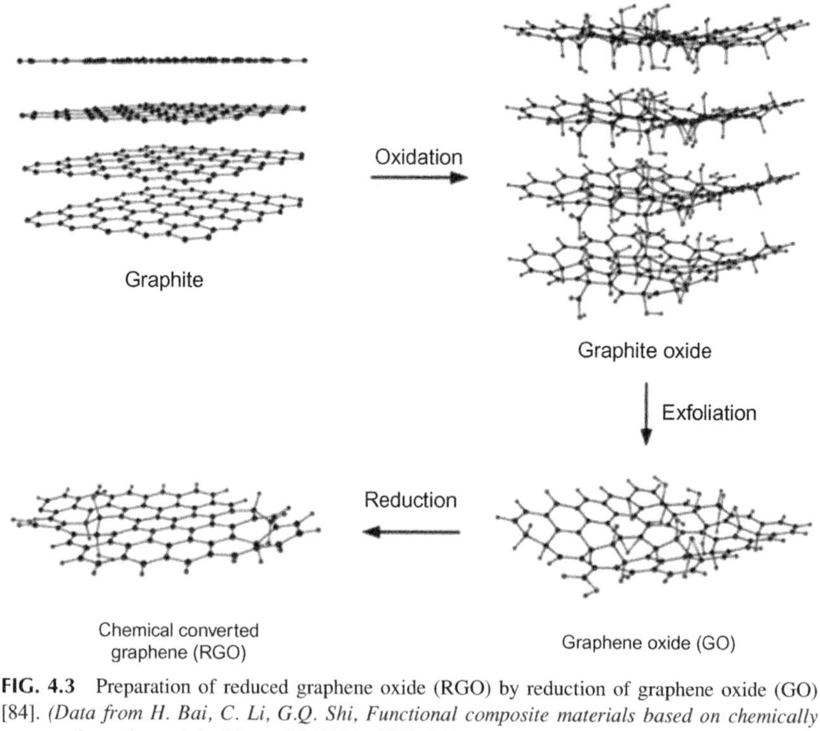

FIG. 4.3 Preparation of reduced graphene oxide (RGO) by reduction of graphene oxide (GO) [84]. *(Data from H. Bai, C. Li, G.Q. Shi, Functional composite materials based on chemically converted graphene, Adv. Mater. 23 (2011) 1089–1115. Copyright 2011, Wiley.)*

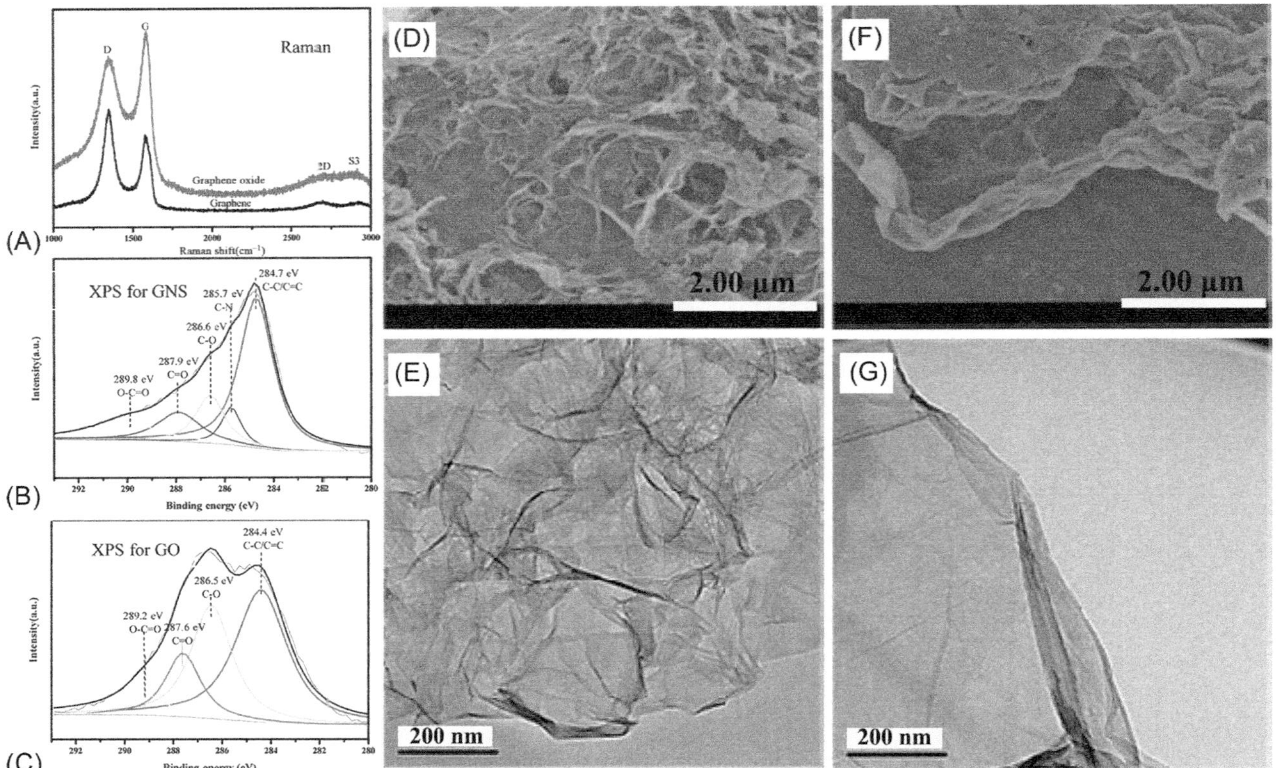

FIG. 4.4 Structural characterization of graphene (GNS) and graphene oxide (GO), (A) Raman spectra of GNS and GO, XPS spectra of GNS (B) and GO (C) for C 1s, the SEM and TEM images of graphene (D) and (E),and graphene oxide (GO) (F) and (G) [87]. *(Data from J. Wang, Z.M. Chen, B.L. Chen, Adsorption of polycyclic aromatic hydrocarbons by graphene and graphene oxide nanosheets, Environ. Sci. Technol. 48 (2014) 4817–4825. Copyright 2014, American Chemical Society.)*

the stretching vibration of sp^3 and sp^2 carbon atoms, respectively. Fig. 4.4B and C show the XPS high-resolution C 1s peak of graphene and GO. The main peak at 284.7 eV of graphene was due to the existence of C=C/C—O [88]. The peaks at 289.8, 287.9, and 286.6 eV correspond to the O—C=O, t C=O, and C—O groups, respectively (Fig. 4.4B). Additionally, the contents of the oxygen group moieties (O—C=O (289.2 eV), C=O (287.6 eV), C—O (286.5 eV)) of GO obviously increased compared with graphene, implying that the surface of GO was more hydrophilic than that of graphene (Fig. 4.4C).

These oxygen-containing functional groups can supply abundant adsorption sites for the removal of radionuclides. Further, these oxygen functional groups of GO made the more hydrophilic property of GO compared with graphene, which may also make the formation of stable complexes between GO and contaminants. Numerous research articles have focused on the removal of contaminants by GO, including small molecules [89–91], heavy metal ions [92–94], radionuclides [95–97], and organic chemicals [98–100]. Wang's group has performed much meaningful work to deeply investigate and develop GO-based materials (e.g., functionalized GO [17, 101, 102] and GO-based composites [103–105]). They utilized these GO-based materials for environmental remediation and systematically investigated their interaction mechanisms. Chai's group combined theoretical calculations with spectroscopy technology to study the interaction mechanisms of radionuclides on GO-based materials [97, 106]. The GO-based materials showed excellent removal ability for contaminants in these reports. The macro/microstructure of GO was obtained through various characterization methods. Fig. 4.4D–G show the SEM and TEM images of graphene and GO nanosheets. From Fig. 4.4D and F, the surface morphology of graphene and GO present a typical ripple state. Both the graphene and GO were revealed in the TEM images (Fig. 4.4E and G). The wrinkles were primarily distributed on the edge of the GO and formed scrolls while the wrinkles were loosely located on the basal planes of graphene formed groove regions. The potential adsorption sites of GO could be divided into groove regions and flat surfaces due to the relatively flat surface of GO compared with graphene.

The colloidal behavior of GO was influenced by environmental cations. Recently, the morphological transformation of GO under heavy metal ions [107] has been investigated. Yang et al. found that two-dimensional (2D) GO nanosheets could roll into one-dimensional (1D) carbon nanotube-like material, a 2D multiplied structure, and three-dimensional (3D) sphere bulk induced by heavy metal ions, which might be helpful in understanding the microstructure manipulation and self-assembly of GO. They proposed interactions of heavy metal cations with GO nanosheets in three steps: the adsorption of heavy metals with GO nanosheets (process I), the formation of 1D-, 2D-, and 3D-like structures (process II), and the aggregation of sphere-like flocculate (process III).

4.2.4 Carbon nanofibers

CNFs are important for industrial and technological applications due to their excellent chemical resistance, thermal conductivity, superior electrical property, and very high strength-to-weight ratio. Fig. 4.5A and B show the SEM and TEM images of CNFs, which were synthesized via a template-directed hydrothermal carbonization procedure [40]. It reveals that CNFs exhibit a morphology with an interconnected 3D framework, and they show a surprisingly high uniformity and narrow diameter size distribution (~100 nm). Further, Liang et al. observed that the diameter of CNFs was related to their adsorption capacity, and the small diameter was more beneficial to the adsorption [45]. The smooth surface morphology of CNFs was observed from a high-resolution TEM image (Fig. 4.5B), and it was not found obvious hollow morphology on the surface of CNFs. The high-resolution C 1s XPS

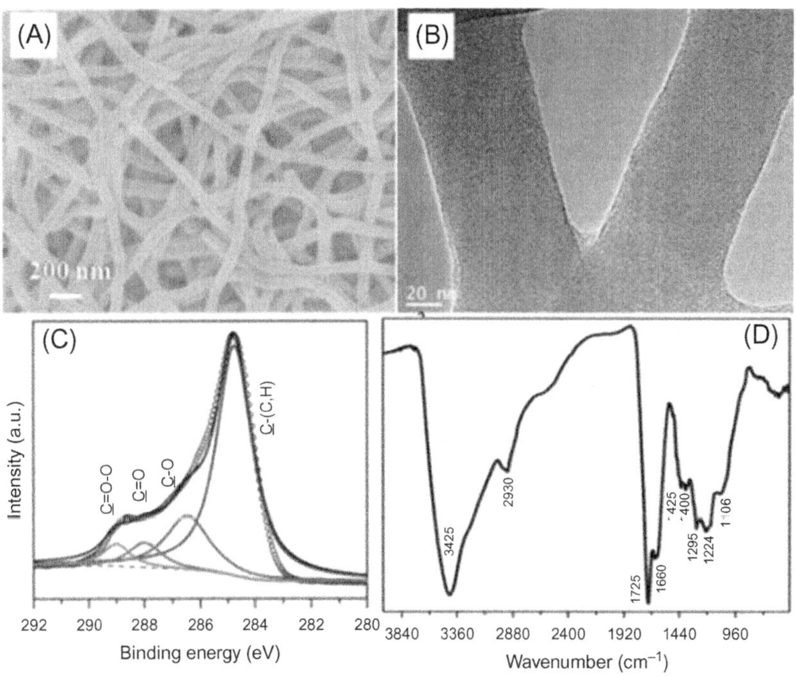

FIG. 4.5 The characterization of CNFs, (A) and (B), SEM and high resolution TEM images, respectively; (C), high resolution of C 1s XPS spectrum [108]; (D), FTIR spectrum [40]. *(Data from Y.B. Sun, X.X. Wang, C.C. Ding, W.C. Cheng, C.L. Chen, T. Hayat, A. Alsaedi, J. Hu, X.K. Wang, Direct synthesis of bacteria-derived carbonaceous nanofibers as a highly efficient material for radionuclides elimination, ACS Sustain. Chem. Eng. 4 (2016) 4608–4616; C.C. Ding, W.C. Cheng, X.X. Wang, Z.Y. Wu, Y.B. Sun, C.L. Chen, X.K. Wang, S.H. Yu, Competitive sorption of Pb(II), Cu(II) and Ni(II) on carbonaceous nanofibers: a spectroscopic and modeling approach, J. Hazard. Mater. 313 (2016) 253–261. Copyright 2016, American Chemical Society, Elsevier.)*

spectrum of CNFs is shown in Fig. 4.5C. It can be deconvoluted into four peaks at ~289.0, 287.8, 286.4, 284.7 eV, corresponding to the O—C=O, C=O/O—C—O, C—O, and C—C/C—H groups, respectively [109]. The IR spectra of CNFs show the characteristic functional groups of CNFs (Fig. 4.5D). The absorption peaks at 1205 and 1296 cm^{-1} can be attributed to the vibration of the C—OH and C—O—C groups, respectively [110]. The absorption peak at ~1382 cm^{-1} was assigned to the bending vibration of C–H groups [111], whereas the other absorption bands at 1087, 1425, 1662, and 2929 cm^{-1} belonged to the stretching vibrations of the C—O (ν-C—O), C—OH (ν-C—OH), C=C (ν-C=C), and C—H (ν-C—H) groups, respectively [112–114]. IR and XPS analysis revealed that CNFs possessed abundant oxygen functional groups (e.g., —OH, —COOH).

4.3 Adsorption of radionuclides on carbon family materials

4.3.1 Adsorption of radionuclides on activated carbon

Activated carbon has been proven to be an effective adsorbent for the removal of a wide variety of organic and inorganic pollutants dissolved in aqueous media. The active carbon is extensively utilized in the treatment of wastewaters because of its possession of a wide spectrum of surface functional groups, well-developed internal microporosity structure, and exceptionally high surface areas (500–1500 m^2/g) [115]. As an inert porous carrier material, it is capable of distributing chemicals on its large hydrophobic internal surface, thus making it accessible to reactants [116]. Preparation of activated carbon from several agricultural byproducts has been given in the literature to obtain less-expensive activated carbon from large amounts of renewable materials, especially for adsorption treatment of wastewater. Researchers have studied the production of activated carbon from bamboo [117], waste bagasse [118], fruit stones and nutshells [119], olive stones [22, 120], plum kernels [121], palm-tree cobs [122], coconut husks [123], and palm kernel shells [124].

Separation and purification processes based on the adsorption technique are also important in nuclear waste and the nuclear industry, where activated carbon is often used for the separation of metal ions from solutions due to its selective adsorption, high radiation stability, and high purity [125]. In past decades, the active carbon and functionalized active carbon were utilized to remediate the uranium contaminant [13, 106, 126–128]. Mellah et al. reported that the pure active carbon showed an excellent adsorption performance for U(VI) removal at acidic conditions. Kütahyalı et al. discovered that the chemical-activated active carbon had selective adsorption toward uranium. Chen et al. found that the graphene oxide-activated carbon composite revealed a high adsorption capacity of U(VI) (298 mg/g) at pH 5.5 (Fig. 4.6A).

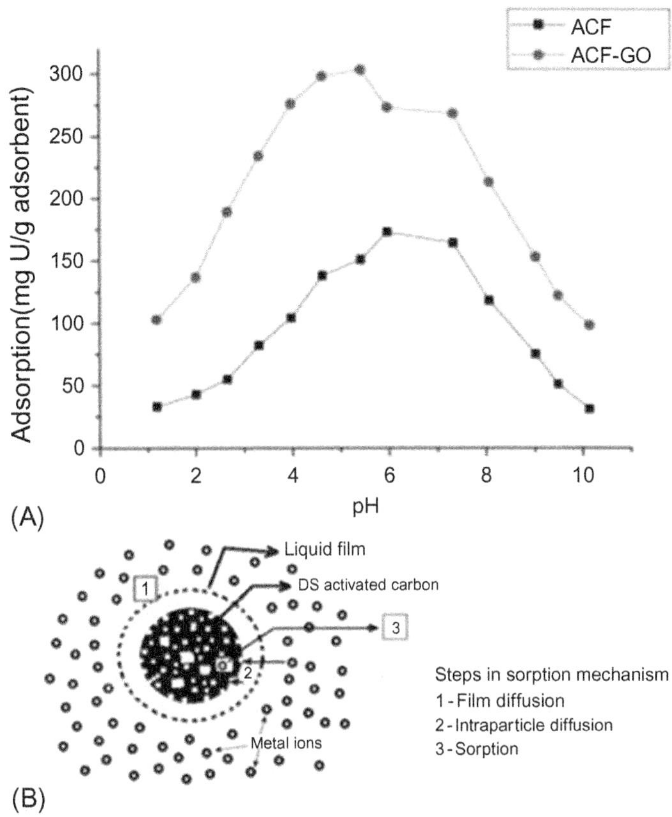

FIG. 4.6 The effect of solution equilibrium pH on the U(VI) adsorption on ACF and ACF-GO (A),[126] schematic diagram of steps in adsorption mechanism for the sorption of Eu(III) (B) [129]. *(Data from S.P. Chen, J.X. Hong, H.X. Yang, J.Z. Yang, Adsorption of uranium(VI) from aqueous solution using a novel graphene oxide-activated carbon felt composite, J. Environ. Radioact. 126 (2013) 253–258; M.M. Hamed, M. Ali, M. Holiel, Preparation of activated carbon from doum stone and its application on adsorption of 60Co and 152+154Eu: equilibrium, kinetic and thermodynamic studies, J. Environ. Radioact. 164 (2016) 113–124. Copyright 2013 and 2016, Elsevier.)*

Using activated carbon for the adsorption of other radionuclides (i.e., Eu(III), Sr(II), Cs(I)) was also studied [129–131]. Hamed et al. used doum stone (DS) to prepare activated carbon as an excellent adsorbent to remove Eu(III) from aqueous solutions. They used Elovich and intraparticle diffusion models to examine the mechanism of adsorption of Eu(III) onto the DS-activated carbon, including three successive steps: (i) film diffusion (external diffusion); (ii) intraparticle diffusion (internal diffusion); and (iii) adsorption (Fig. 4.6B). They found that the adsorption process is controlled by intraparticle diffusion.

4.3.2 Adsorption of radionuclides on carbon nanotubes

4.3.2.1 U(VI) and Th(IV) adsorption

Because CNTs showed encouraging promise in the capability of radionuclide retention, many researchers focused on the interaction of radionuclides on CNTs. Density functional theory (DFT) calculations have been extensively utilized in actinide stoichiometry studies because they are a viable method to investigate molecule systems such as actinide containing tens or more atoms. Xin et al. utilized the DFT method to calculate the interactions between U atoms and the inner/outer surfaces of SWCNT [132]. They found that in the U-CNT (inner and outer) system, the single occupied molecular orbitals were four orbitals (HOMO to HOMO-3). For the U-CNT (inner) system, the contributions of the U atom were 17.84%, 92.61%, 62.52%, and 74.67% of the HOMO, HOMO-1, HOMO-2, and HOMO-3 orbitals, respectively. It primarily revealed the orbital hybridization between the U atom and adjacent C atoms of HOMO-2 and HOMO-3 orbitals, suggesting strong covalent interactions between the U atom and the CNT wall. However, based on the analysis of the orbital component of the U-CNT (outer) system, these four orbitals (HOMO, HOMO-1, HOMO-2, and HOMO-3) were mainly from the U atom's contribution (about 89%–99%), whereas the CNT contribution was not significant. The results implied that the electrostatic attraction between the U and CNT wall was the dominant interaction. The results of the DFT calculations showed that the ground electronic states were all quintet with each CNT surface adsorbed U atoms. Besides the surface (inner or outer), bridge sites on the inner surface and hole sites on the outer surface of CNT could be attributed to the differential adsorption modes of U atoms on CNT. Singh Deb et al. utilized the DFT theoretical studies to calculate the band distance parameters of Th^{4+} ions on pristine CNT in the presence of NO_3 ions and H_2O [133]. The $Th(NO_3)_4(H_2O)_2$ species were observed lying outside the side surface and the open end of the CNT from the optimized structures of complexes of Th^{4+} ions with pristine CNT.

The impact of HA/FA and pH on the U(VI) sorption on oxidized MWCNTs was investigated by Sun et al. [70]. They found that the existence of HA/FA facilitated U(VI) sorption at pH 1.0–6.0 due to the complexation of U(VI) with HA/FA bound on oxidized MWCNTs. They also explained that the sorption of U(VI) declined with increasing pH at pH $>$ 6.0 assigned to the formation of soluble HA/FA-U complexes in aqueous solution. Additionally, the influence of FA on U(VI) sorption on oxidized MWCNT was enhanced more obviously than that of HA at low pH, which might be due to the higher surface site density of FA than that of HA.

4.3.2.2 Eu(III) and Sr(II) adsorption

The main interaction mechanisms of $^{152+154}$Eu(III) on MWCNTs were considered to be the strong surface complexation and/or chemisorption [34]. Tan et al. reported that the interaction of radionuclides on MWCNTs lied mainly in two pathways: (i) fast interaction on the outer surfaces, and

(ii) slow interaction in the inner channel of MWCNTs [34]. Therefore, they proposed a possible adsorption process of radionuclides on MWCNTs: the radionuclides were fast bound to the outer surfaces of MWCNTs, and then with increasing time, more radionuclides were attached to the inner channel of MWCNTs. Further, they also found that the radionuclides on the outer surfaces of MWCNTs were easily desorbed, whereas that was difficult for the radionuclides in the inner channel of MWCNTs. Therefore, compared to other adsorbents, the irreversible interaction of radionuclides on MWCNTs was significant in the disposal of radionuclides.

MWCNTs as promising adsorbents were also investigated to remove other radionuclides. Chen et al. investigated the adsorption performance of oxidized MWCNTs toward Sr(II)/Eu(III), and they found that the adsorption of Sr(II)/Eu(III) decreased with increasing ionic strength and increased with increasing pH [35]. Fig. 4.7A–C show the fractional distributions of functional groups of oxidized MWCNTs as a function of pH at different ionic strengths; they were calculated with the aid of FITEQL 3.2. They assumed that hydroxyl group sites (WOH) and carboxyl group sites (≡SOH) were the most significant surface active sites of oxidized MWCNTs. The positively charged surface of oxidized MWCNTs was attributed to the governance of $\equiv WOH_2{}^+$ functional groups at pH < 4.5, whereas the negatively charged surface of oxidized MWCNTs was assigned to the dominance of $\equiv WO^-$ and $\equiv SO^-$ functional groups at pH > 5. They studied the adsorption performance of oxidized MWCNTs toward Sr(II) as a function of pH at three different ionic strengths, and they found that the adsorption was dependent on the pH and the ionic strength Fig. 4.7D. The adsorption of Sr(II) displayed a weak affinity on oxidized MWCNTs at low pH, indicating that the protonated oxidized MWCNT functional groups were not easy to form strong complexation with Sr(II) ions. As pH increased, the formation of negative oxidized MWCNT charge was due to the progressive deprotonating of functional groups. The attractive forces between cationic metal ions and anionic surface sites of oxidized MWCNT led to the formation of metal-ligand oxidized MWCNT complexes. They measured the lower final values of pH at adsorption equilibrium compared with the initial pH, which might be attributed to the deprotonation of oxidized MWCNTs, and the deprotonation led to more metal ion complexation with oxidized MWCNTs. They also proposed that the interaction of Sr(II)/Eu(III) on oxidized MWCNTs was primarily due to the complexation with carboxyl groups and deprotonated hydroxyl groups at low and high pH conditions, respectively. Additionally, with increasing solution pH, the degree of deprotonation of MWCNTs increased, leading to the increase of adsorption capacity.

4.3.3 Adsorption of radionuclides on graphene oxides

4.3.3.1 U(VI) adsorption

For removal of radionuclides, GO materials have been investigated. In the early investigation stage, GO nanosheets were recognized as a suitable material

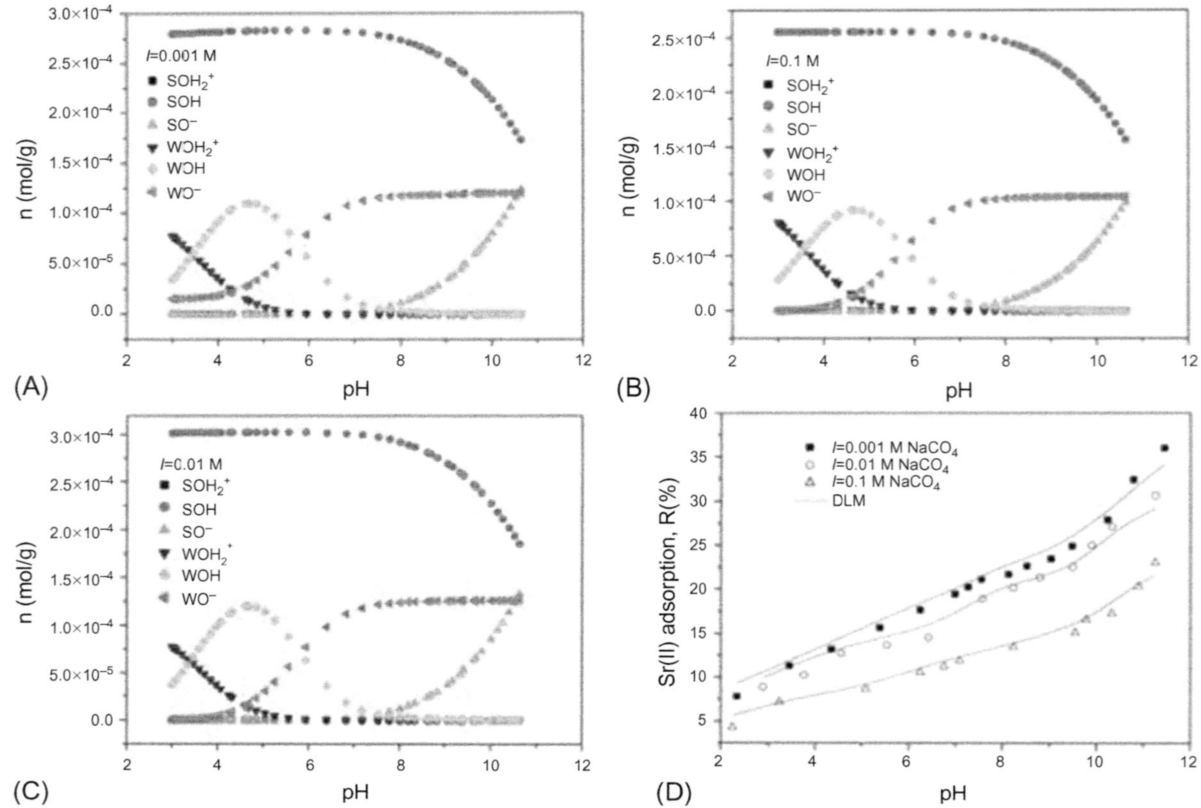

FIG. 4.7 Surface speciation repartition diagram of oxidized MWCNT surface sites as a function of pH at three different ionic strengths (A–C), experimental data and fit curves for Sr(II) adsorption onto oxidized MWCNTs with the aid of FITEQL 3.2 at different ionic strength (initial $C_{Sr(II)} = 2 \times 10^{-7}$ mol/L, $m/V = 0.6$ g/L) (D) [35]. *(Data from C.L. Chen, J. Hu, D. Xu, X.L. Tan, Y.D. Meng, X.K. Wang, Surface complexation modeling of Sr(II) and Eu(III) adsorption onto oxidized multiwall carbon nanotubes, J. Colloid Interface Sci. 323 (2008) 33–41. Copyright 2008, Elsevier.)*

for the removal of U(VI) ions from polluted wastewater due to their high specific surface area and surface oxygen-functional groups [36, 106]. With the development of research, studies on the removal of U(VI) on GO are gradually improving. Sun et al. used a sonochemical approach to introduce the abundant carboxyl and hydroxyl groups on GO, and the treated GO presented an efficient adsorption performance toward U(VI) [134]. From the analysis of the EXAFS spectra, they determined that the adsorption mechanism between U(VI) and GO was inner-sphere surface complexation. Sun et al. also studied the adsorption and desorption behaviors of U(VI) on GOs, HOOC-GOs (carboxylated GOs), and rGOs (reduced GOs) [135]. They found that the adsorption amounts of U(VI) followed in the order of GOs > HOOC-GOs > rGOs, indicating that GOs exhibited the higher adsorption performance toward U(VI) as compared to HOOC-GOs and rGOs (Fig. 4.8A). The higher adsorption performance of GOs toward U(VI) might be due to their excellent dispersive properties and their abundant oxygenated functional groups, which provided more available adsorption sites. The amount of desorbed U(VI) from GOs, HOOC-GOs, and rGO was in the order of rGOs > GOs > HOOC-GOs, which implied that it was harder to extract U(VI) from HOOC-GOs than that from GOs (Fig. 4.8B).

Ding et al. investigated the adsorption and desorption behaviors of U(VI) on GO nanosheets, and the desorption induced by two methods: lowering the solution pH and replacing the radionuclide supernatant [95]. The almost identical trend of desorption edge and adsorption edge of U(VI) on GO nanosheets, indicating that the adsorbed U(VI) on GO nanosheets is completely desorbed when desorption is induced by lowering solution pH. They explained that the desorbed amounts of U(VI) increase with decreasing pH because the pH of solution exhibits a significant impact on the dissolution of surface complexes: $SOM^{n+} + H^{+} = M^{(n+1)+} + SOH$. They conducted the desorption experiments of U(VI) from GO nanosheets by using the replacement radionuclide supernatant method. The desorption and adsorption isotherms of U(VI) from GO nanosheets exhibited an adsorption-desorption hysteresis phenomenon, which was assigned to the preferable binding affinity of GO nanosheets for U(VI).

Recently, Liu et al. studied the adsorption and desorption performances of different-sized GO toward U(VI) [136]. They found that the smaller GO (SGO) (100–300 nm) has a stronger interaction force with U(VI) than that of U(VI) on larger GO (LGO) (2–5 μm). The different interaction forces of U(VI) on SGO and LGO were demonstrated by using adsorption and desorption processes. Besides, they also analyzed the different adsorption kinetics of U(VI) on SGO and LGO by using different adsorption kinetic models to fit the adsorption data. Fig. 4.8C–E show the pseudo-first-order rate model, the pseudo-second-order rate model, and the intraparticle diffusion model, respectively. A faster chemisorption process of U(VI) was found for SGO, which might be attributed to the smaller particle size of SGO and the more

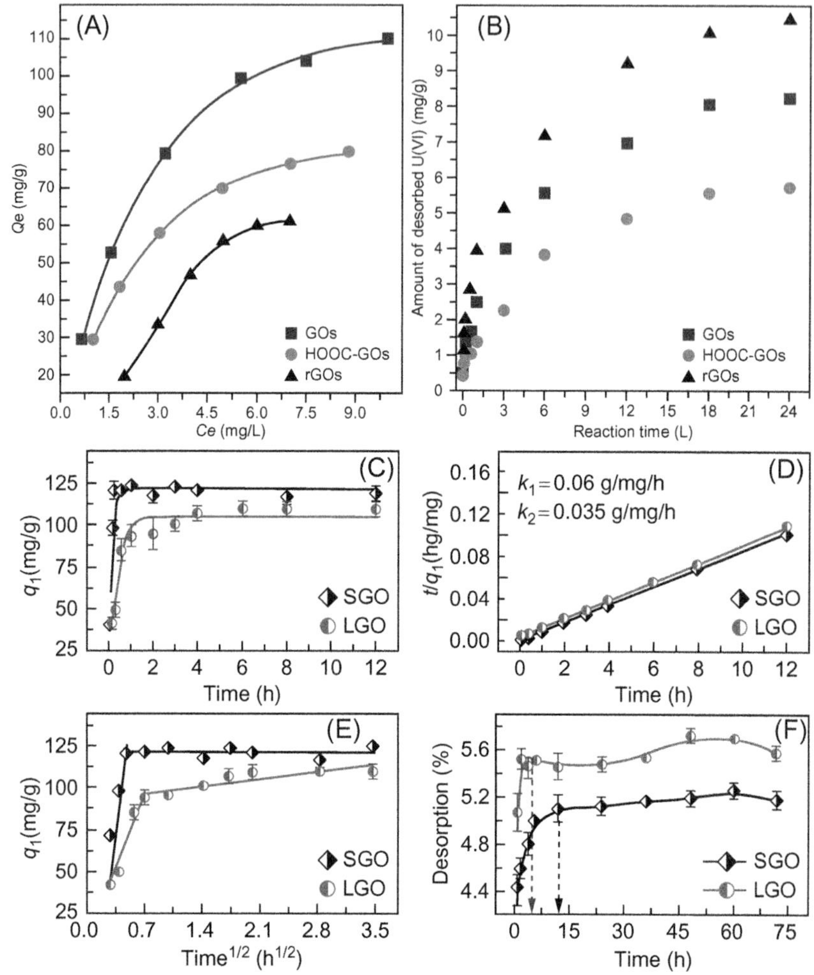

FIG. 4.8 The adsorption isotherms of U(VI) on GOs, HOOC-GOs, and rGOs (A), pH = 4.0, m/V = 0.25 g/L, T = 293 K, I = 0.01 mol/L $NaClO_4$, the desorption kinetics of U(VI) on GOs, HOOC-GOs, and rGOs (B), pH = 4.0, $C_{[NaClO4]}$ = 0.01 mol/L, m/V = 0.25 g/L, T = 293 K [135], adsorption of U(VI) on LGO and SGO at pH = 5, I = 0.01 M NaCl, m/V [GO] = 20 mg/L and C_0 [U(VI)] = 6 mg/L. Pseudo-first-order (C) pseudo-second-order rate models (D) intraparticle diffusion (E). Desorption by lowering pH from 5.0 to 2, m/V [GO] = 50 mg/L, C_0 [U(VI)]= 20 mg/L, I = 0.01 M NaCl (F) [136]. *(Data from Y.B. Sun, S.B. Yang, Y. Chen, C.C. Ding, W.C. Cheng, X.K. Wang, Adsorption and desorption of U(VI) on functionalized graphene oxides: a combined experimental and theoretical study, Environ. Sci. Technol. 49 (2015) 4255–4262; X. Liu, J. Sun, X. Xu, A. Alsaedi, T. Hayat, J. Li, Adsorption and desorption of U(VI) on different-size graphene oxide, Chem. Eng. J. (2018). Copyright 2015 and 2018, American Chemical Society, Elsevier.)*

active oxygen-containing groups of SGO due to its having more edges than LGO. In addition, from the analysis of the intraparticle diffusion model, the diffusion rate of U(VI) from the solution to the SGO surface was faster than that to LGO. Fig. 4.8F shows the desorption processes of U(VI) from SGO and LGO, which revealed that the desorption rate of U(VI) from SGO was lower than that from LGO. The result implied that the complexation interaction of U(VI) on SGO was stronger than that on LGO.

For the past few decades, surface complexation models have been applied as useful tools to describe sorption processes on solid-water interfaces [137, 138]. Owing to the fact that simple inner-sphere surface complexes ($SOUO_2^+$) can't estimate the adsorption edge very well, Hu et al. used two inner-sphere surface complexes ($SOUO_2^+$) and ($(SO)_2UO_2(OH)_2^{2-}$) to simulate the adsorption of U(VI) on GO [139]. They fitted the adsorption data of U(VI) on GO by using three surface complexation models (SCMs), that is, the constant-capacitance model (CCM), the diffuse-layer model (DLM), and the triple-layer model (TLM) with the aid of FITEQL v. 4.0 codes (Fig. 4.9); the optimized parameters

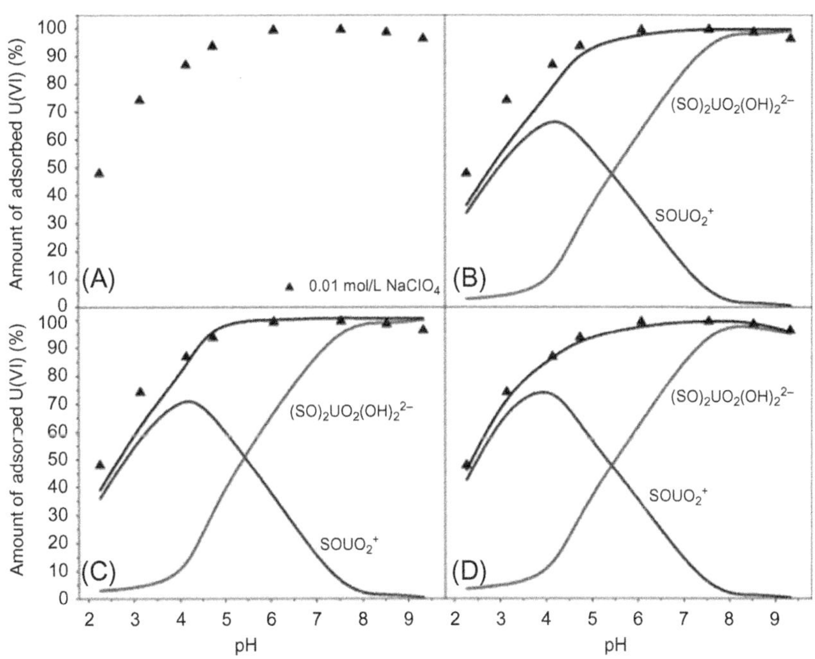

FIG. 4.9 The fitted results of surface complexation modeling for pH-dependent adsorption of U(VI) on GO, (A) original data; (B) CCM; (C) DLM; (D) TLM [139]. *(Data from T. Hu, S.J. Ding, H.J. Deng, Application of three surface complexation models on U(VI) adsorption onto graphene oxide, Chem. Eng. J. 289 (2016) 270–276. Copyright 2016, Elsevier.)*

TABLE 4.1 Optimized parameters of CCM, DLM, and TLM for U(VI) adsorption for GO [139]

Model	[a]Log K^+, Log K^-	[b]Log K_{Cat}, Log K_{An}	C_1, C_2 (F/m^2)	Log K1, Log K2	WSOS/DF
CCM	3.19, −4.26		1.0, 0	2.24, −7.26	11.42
DLM	3.19, −4.26			2.11, −7.47	15.75
TLM	3.19, −4.26	−4.15, 4.15	[c]1.2, 0.2	2.08, −8.11	19.43

[a]*Equilibrium constants of protonation and deprotonation reactions.*
[b]*Equilibrium constants of cation and anion surface complexation reactions.*
[c]*Data cited from Torner and Sparks [140].*
Data from T. Hu, S.J. Ding, H.J. Deng, Application of three surface complexation models on U(VI) adsorption onto graphene oxide, Chem. Eng. J. 289 (2016) 270–276, Copyright 2016, Elsevier.

of simulated adsorption data are summarized in Table 4.1. From Fig. 4.9B–D, the $SOUO_2{}^+$ and $(SO)_2UO_2(OH)_2{}^{2-}$ are the main adsorbed U(VI) species at pH <5.0 and >6.0, respectively. Further, the fitted results showed that using TLM is better than that of DLM and CCM, which might be attributed to more parameters used in TLM.

The strong surface complexation of U(VI) on GOs was also determined by using theoretical DFT calculations [141]. Wang et al. considered that the adducts $[GOs—OH_2\cdots UO_2]^{3+}$ and $[GOs—OH—OH_2\cdots UO_2]^{3+}$ are represented as the protonation situation of —OH group at low pH values, and the adducts $[GOs—O\cdots UO_2]^+$ and $[GOs—OH—O\cdots UO_2]^+$ assigned to the deprotonation of —OH group at high pH (Fig. 4.10). The adducts $[GOs—\oplus COOH_2\cdots UO_2]^{3+}$ and $[GOs—COO\cdots UO_2]^+$ are represented as the situation of the —COOH group at low and high pH values, respectively. They calculated that the binding energy (E_{bd}) value of $[GOs—OH_2\cdots UO_2]^{3+}$ (14.4kcal/mol) (low pH) is much lower than that of $[GOs—O\cdots UO_2]^+$ (67.6kcal/mol) (high pH). The geometries in Fig. 4.10 show that the U(VI) species is away from the GO nanosheets together with the protonated —OH group in $[GOs-—OH_2\cdots UO_2]^{3+}$ at low pH values compared to the species at high pH values. Similarly, the low E_{bd} value of $[GOs—OH—OH_2\cdots UO_2]^{3+}$ (13.8kcal/mol) at low pH value exhibits a longer U···O bond distance of 2.914Å compared to the shorter U···O— bond distance of 2.204Å in $[GOs—OH—O\cdots UO_2]^+$ (84.2kcal/mol) at high pH value. The high positive E_{bd} value of $[GOs—\oplus COO\cdots UO_2]^+$ (50.5kcal/mol) suggests the strong interaction between the uranyl species and the Gos while the negative E_{bd} value of $[GOs—COOH_2\cdots UO_2]^{3+}$ (−4.2kcal/mol) further proves the instability of the interaction between the positive charged carboxyl group and the uranyl cation species.

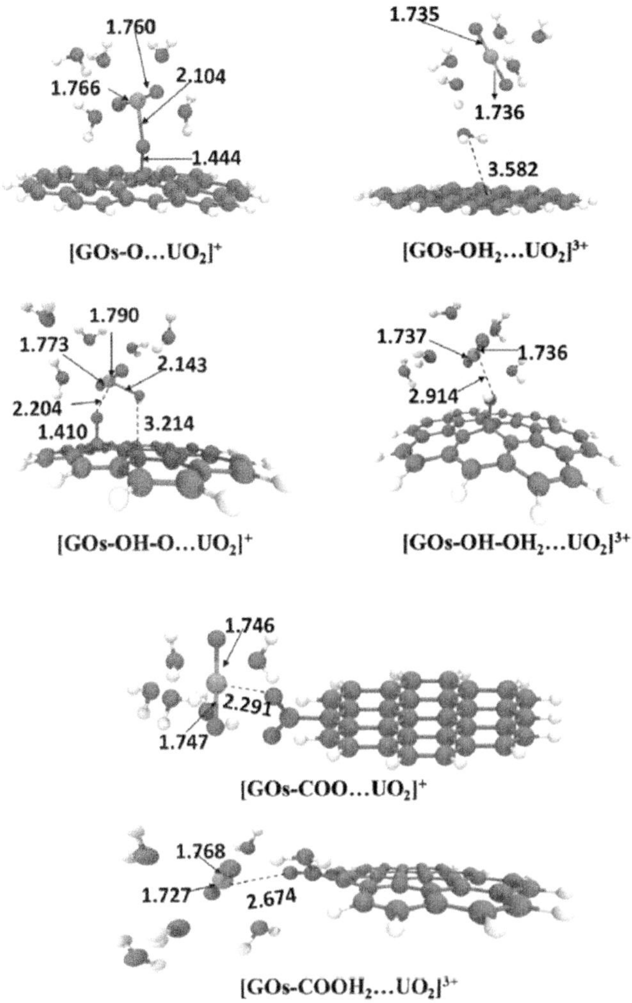

FIG. 4.10 The optimized geometrical parameters for GOs_uranyl complexes at high and low pH values. *Green*: uranium atoms, *red*: oxygen atoms, *blue*: carbon atoms, *white*: hydrogen atoms. (For interpretation of the references to color in this figure legend, the reader is referred to the web version of this article) [141]. *(Data from X.X. Wang, Q.H. Fan, S.J. Yu, Z.S. Chen, Y.J. Ai, Y.B. Sun, A. Hobiny, A. Alsaedi, X.K. Wang, High sorption of U (VI) on graphene oxides studied by batch experimental and theoretical calculations, Chem. Eng. J. 287 (2016) 448–455. Copyright 2016, Elsevier.)*

4.3.3.2 Th(IV) adsorption

Experiments on the performance of GOs toward Th(IV) immobilization were conducted by Xu et al. They found that the whole adsorption process was dependent on pH and independent of ionic strength [142]. These results indicated that surface complexation was the dominant adsorption mechanism. Besides, they fitted the adsorption data of the pH effect at different ionic strengths by using the diffusion-layer model (DLM) (Fig. 4.11). They proposed that possible reactions for Th(IV) retention with the functional groups on GOs were as follow (Eqs. 4.1–4.3):

$$\equiv XOH_2{}^+ + Th^{4+} \rightarrow \equiv XOHTh^{4+} + H^+ \quad (4.1)$$

$$\equiv XOH + ThOH^{3+} \rightarrow \equiv XOThOH^{2+} + H^+ \quad (4.2)$$

$$\equiv YOH + ThOH^{3+} \rightarrow \equiv YOThOH^{2+} + H^+ \quad (4.3)$$

where ≡XOH was utilized to represent strong acidic groups and ≡YOH represented weak acidic groups. In the different ionic strengths (0.001, 0.01, 0.1M), $\equiv XOHTh^{4+}$ and $\equiv XOTh(OH)^{2+}$ are the controllable surface species at low pH and high pH values, respectively. The $\equiv YOTh(OH)_2{}^+$ surface species is at a relatively low level in the whole pH, indicating that the strong acidic groups (≡XOH) provided more contribution in Th(IV) adsorption on

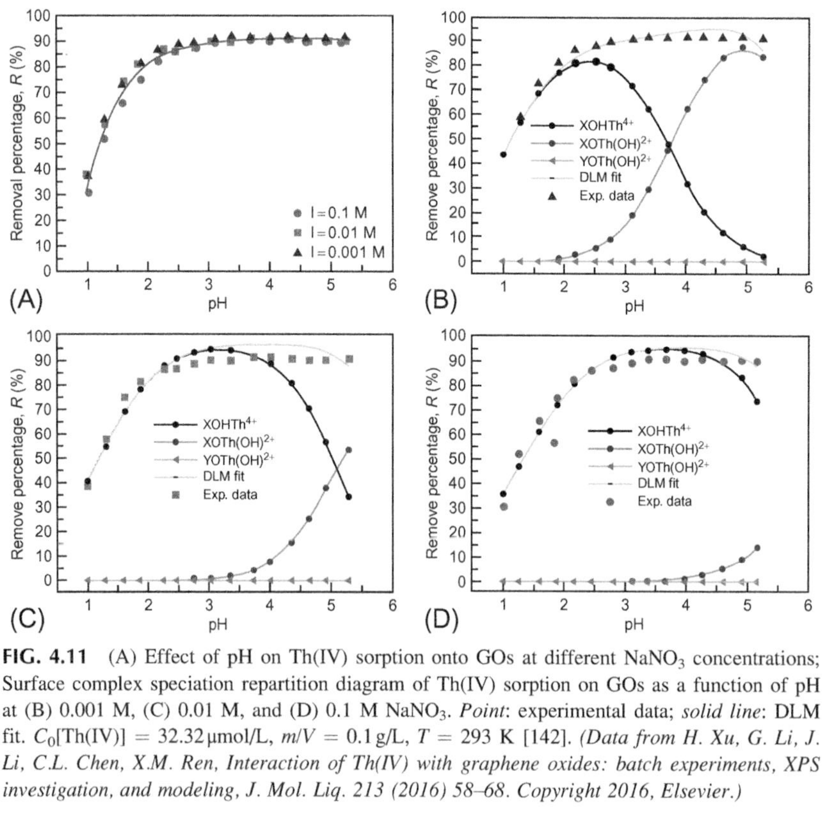

FIG. 4.11 (A) Effect of pH on Th(IV) sorption onto GOs at different $NaNO_3$ concentrations; Surface complex speciation repartition diagram of Th(IV) sorption on GOs as a function of pH at (B) 0.001 M, (C) 0.01 M, and (D) 0.1 M $NaNO_3$. *Point*: experimental data; *solid line*: DLM fit. C_0[Th(IV)] = 32.32 μmol/L, *m/V* = 0.1 g/L, *T* = 293 K [142]. *(Data from H. Xu, G. Li, J. Li, C.L. Chen, X.M. Ren, Interaction of Th(IV) with graphene oxides: batch experiments, XPS investigation, and modeling, J. Mol. Liq. 213 (2016) 58–68. Copyright 2016, Elsevier.)*

GOs than the weak acidic groups (≡YOH). They also deduced that the ionic strength influenced the species of the adsorbed Th(IV) on GOs because the relative ratio of $\equiv XOTh(OH)^{2+}$ and $\equiv XOHTh^{4+}$ was not the same with each other at different ionic strengths.

The sorption of Th(IV) onto GO was researched by Bai et al., who observed that the high sorption capacity for Th(IV) was ascribed to the sufficient exposed surface abundant oxygen-containing functional groups of GO [143]. Apart from the extended X-ray absorption fine structure (EXAFS) and IR analysis, they used DFT calculations to further confirm the rationality of the experimental and the EXAFS results. At the B3LYP/6-31G(d)/RECP level of theory, seven structures of Th(IV) coordinated to hydroxyl or carboxyl oxygen atoms on the middle and edge sites of GO were optimized to determine the different binding motifs formed between hydroxyl or carboxyl oxygen atoms and Th atoms. They found that three of the NO_3^- ions act as bidentate ligands and the other NO_3^- ions coordinate as monodentate to Th^{4+} for three structures, whereas all the NO_3^- ions are coordinated as bidentate ligands for another four structures. Additionally, the optimized structure results showed that all the lengths of the Th—O bond for the carboxyl oxygen atoms are longer than those for the hydroxyl oxygen atoms, indicating that the carboxyl oxygen atoms have stronger coordination ability than hydroxyl oxygen atoms.

4.3.3.3 Eu(III) adsorption

Due to the chemical homologues of europium (Eu) and actinide radionuclides, the adsorption behavior of Eu(III) on the solid-liquid interface can be as an assist reference to evaluate the risks of the exposure of radioactive trivalent actinides [144, 145]. However, few studies focused on the treatment of Eu(III) by GO- based materials in nuclear waste management. Graphene oxide nanosheets (GONS) are considered as applicable materials for the disposal of heavy metal contaminants, and it they were was prepared and utilized for the adsorption of Eu(III) [6, 95]. In addition, Ding et al. observed that the adsorption of Eu(III) on GO was related to the surface properties of the GO and the relative distribution of the Eu(III) species in solution [95]. The adsorption of Eu(III) increased from pH 2.0 to 6.0 ascribed to electrostatic attraction between positive Eu(III) species and negatively charged surface of GO, and then adsorption kept high level at pH 6.0–9.0 attributed to surface complexation between Eu(III) species and functional groups of GO.

Xie et al. conducted batch experiments for Eu(III) adsorption on GO as a function of pH (Fig. 4.12A) [146]. The Eu(III) adsorption was achieved at 97% between pH 2.74 and 7.25, and the adsorption edge occurred at pH 2.7. The batch experiments data as a function of pH were modeled using the DLM-SCM model, assuming Eu(III) adsorption to both the carboxylate ($[\equiv COO^-]$) and sulfonate sites ($[\equiv SO_3^-]$) (Fig. 4.12A). The dominant adsorption species

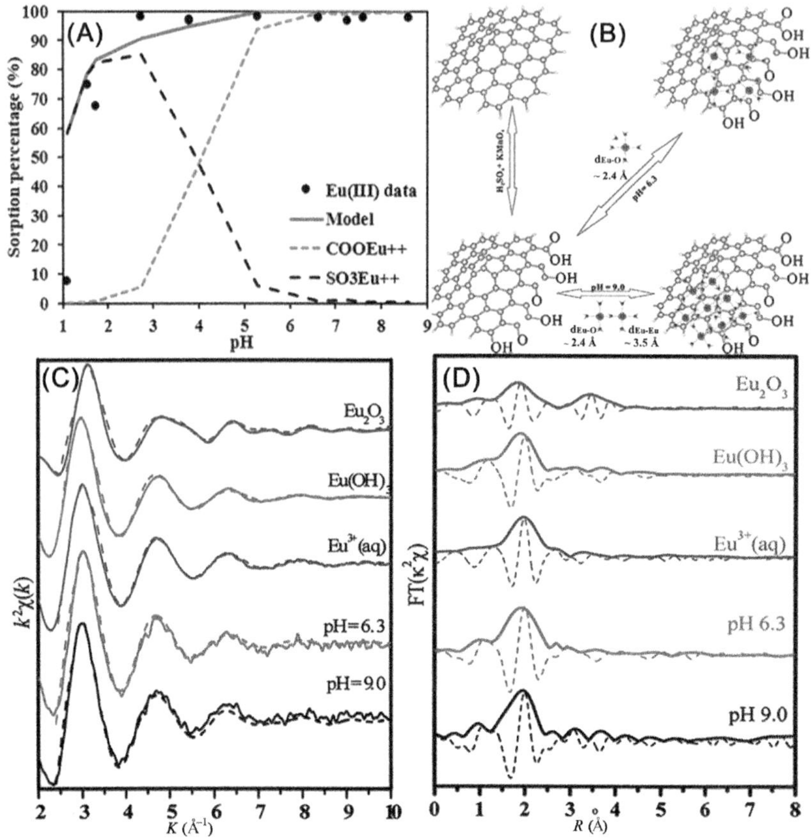

FIG. 4.12 Experimental data *(dots)* and SCM fit *(lines)* of Eu(III) sorption on GO as a function of pH (A),[146] (B) schematic diagram of preparation of GONS and its interaction toward Eu(III) at pH 6.3 and 9.0, (C) k^2-weighted Eu L_{III}-edge EXAFS spectra and (D) corresponding Fourier transforms of the reference samples and selected adsorption samples at pH 6.3 and 9.0. $m/V = 0.2\,g/L$, $C_{Eu(III)} = 3.35 \times 10^{-4}$ mol/L, $I = 0.01$ mol/L $NaClO_4$, and $T = 298$ K [6]. *(Data from Y.B. Sun, Q. Wang, C.L. Chen, X.L. Tan, X.K. Wang, Interaction between Eu(III) and graphene oxide nanosheets investigated by batch and extended X-ray absorption fine structure spectroscopy and by modeling techniques, Environ. Sci. Technol. 46 (2012) 6020–6027; Y. Xie, E.M. Helvenston, L.C. Shuller-Nickles, B.A. Powell, Surface complexation modeling of Eu(III) and U(VI) interactions with graphene oxide, Environ. Sci. Technol. 50 (2016) 1821–1827. Copyright 2012 and 2016, American Chemical Society.)*

was [$\equiv SO_3Eu^{2+}$] at low pH value while [$\equiv COOEu^{2+}$] dominated at high pH condition. The SCM model demonstrates that Eu(III) adsorption to GO occurs mostly via sulfonate group sites at low pH and through complexation with carboxylate sites at high pH. The removal of Eu(III) in the existence of phosphate on GO was investigated by Xu et al. They observed that the existence of phosphate could enhance the adsorption amount of Eu(III) [147]. Further, they

TABLE 4.2 Surface reactions and complexation constants describing Eu(III) sorption on GO [146]

Reaction	Log K_s
$\equiv COOH \rightarrow \equiv COO^- + H^+$	-4^a
$\equiv SO_3H \rightarrow \equiv SO_3^- + H^+$	1.7^a
$\equiv COO^- + Eu^{3+} \rightarrow \equiv COOEu^{2+}$	0.2 ± 0.02^b
$\equiv SO_3^- + Eu^{3+} \rightarrow \equiv SO_3Eu^{2+}$	-1.95 ± 0.02^b

[a]*Constants fixed when modeling surface site densities and Eu(III) or U(VI) sorption.*
[b]*Uncertianty reported at a 95% confidence level based on FITEQL output.*
Data from Y. Xie, E.M. Helvenston, L.C. Shuller-Nickles, B.A. Powell, Surface complexation modeling of Eu(III) and U(VI) interactions with graphene oxide, Environ. Sci. Technol. 50 (2016) 1821–1827. Copyright 2016, American Chemical Society.

observed that the sorption edge of Eu(III) shifted to lower pH values in the presence of phosphate, indicating that the phosphate had a positive effect on Eu(III) sorption. They proposed three possible reasons for the enhanced Eu(III) removal in the presence of phosphate: (i) the Eu(III)-phosphate complex may have higher affinity to GO; (ii) phosphate can be captured and act as the new active site for removal of Eu(III); and (iii) the Eu(III)-phosphate complex can precipitate from solution (Table 4.2).

In the past decade, spectroscopic techniques experienced a boom due to the growing challenges for radionuclide sorption investigations [20, 148]. The structure information of the surface complexes formed between surface sites and radionuclide ions can be directly detected by using surface spectroscopic techniques. X-ray absorption fine structure (XAFS), X-ray absorption near-edge structure (XANES) spectroscopy, EXAFS spectroscopy, and X-ray photoelectron spectroscopy (XPS) have all been utilized as powerful characterization tools in the analysis of the solid surface structure information. The EXAFS technique is quite useful and meaningful to depict the coordination mechanism due to the fact that it can supply valuable coordination information of nearby atoms around a specific metal ion, even on the interface of two phases [149]. Sun et al. utilized the EXAFS technique to study the interaction mechanism of Eu(III) on graphene oxide nanosheets (GONS) [6]. Fig. 4.12B shows the schematic diagram of the preparation of GONS and its interaction toward Eu(III) at pH 6.3 and 9.0 conditions. They studied the k^2-weighted Eu L_{III}-edge EXAFS spectra and corresponding Fourier transforms of Eu(III) loaded on GONS at pH 6.3 and 9.0 (Fig. 4.12C). From the analysis results of the EXAFS spectra, they concluded that the presence of inner-sphere surface complexes in the Eu(III) loaded on GONS (pH 6.3 and 9.0) attributed to the existence of broadened oscillation at $k \sim 6.5 \text{Å}^{-1}$ (Fig. 4.12D). The analysis results of Fourier transforms are in quite good agreement with the results of the interaction of Eu(III) on calcium silicate hydrates [150]. From the results of the EXAFS spectra and Fourier transforms, they observed that a difference between aqueous Eu(III) ions and Eu(III) loaded on GONS and Eu(III) loaded on GONS at pH 9.0 exhibited some similarity with the crystalline $Eu(OH)_3$. The analysis results of the EXAFS data indicate the multinuclear surface complexes on the surface of GONS occurred at pH 9.0 [150]. Further, using EXAFS spectroscopy and surface complexation modeling, they determined that the interaction mechanism of Eu(III) on GONS transformed mononuclear monodentate complexes to binuclear bidentate complexes with increasing pH.

4.3.3.4 Cs(I) and Sr(II) adsorption

The investigations of the adsorption of Cs(I) and Sr(II) on GO were relatively fewer than other high-valence radionuclides (e.g., U(VI), Th(IV), Eu(III)). Romantic et al. found that the interaction mechanism of Cs(I) and Sr(II) with GO was different at high pH values (pH > 6) [151]. Fig. 4.13A and B show the pH dependence of Cs(I) adsorption on GO at different concentrations of $NaClO_4$ (0.1 and 0.0001 M) while the adsorption of Sr(II) in 0.0001 M and 0.1 M $NaClO_4$ was similar and almost quantitative at high pH conditions (pH > 6). Therefore, they assumed that Cs(I) was adsorbed via the ion-exchange mechanism, whereas Sr(II) was sorbed via both ion-exchange and inner-sphere complex formation, and the second mechanism dominates under neutral pH conditions (Fig. 4.13). The two mechanisms (ion-exchange and inner-sphere complex) were expressed by Reactions (4.4), (4.5):

$$\equiv XNa + Cs^+ \rightleftharpoons \equiv XCs + Na^+ \tag{4.4}$$

where $\equiv X^-$ stands for the ion exchange sites of GO.

$$\equiv SOH + Sr^{2+} \rightleftharpoons \equiv SOSr^+ + H^+ \tag{4.5}$$

where ≡SOH stands for the hypothetical surface group of GO.

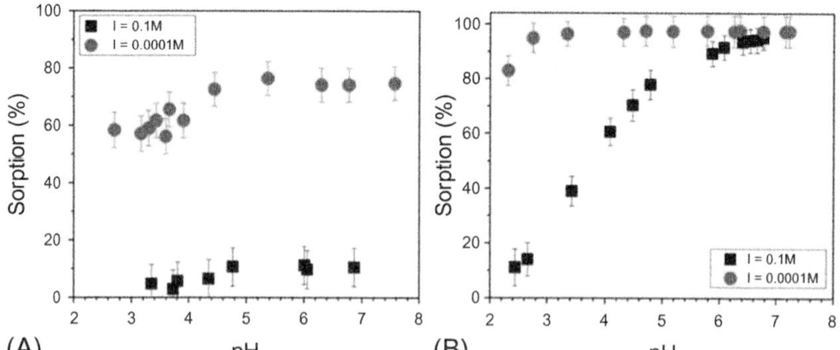

FIG. 4.13 Cs(I) (A) and Sr(II) (B) sorption onto GO at different pHs and ionic strengths; ([Cs] = 5×10^{-11} M, [Sr] = 2×10^{-10} M, [GO] = 1.5 g/L) [151]. *(Data from A. Romanchuk, A. Kuzenkova, A. Slesarev, J. Tour, S. Kalmykov, Cs (I) and Sr (II) sorption onto graphene oxide, Solvent Extr. Ion Exc. 34 (2016) 594–602. Copyright 2016, Taylor & Francis.)*

4.3.3.5 Competitive adsorption

Apart from the investigation of the interaction mechanisms of individual radionuclides on GOs in single-metal systems, many reports focused on the competitive sorption mechanisms of radionuclides on GOs in multimetal systems. From the sorption isotherm results of U(VI) and Sr(II) on GOs, the sorption of U(VI) was little affected by the presence of Sr(II) while the sorption of Sr(II) was obviously different in the single-metal and binary-metal systems [152]. The sorption of Sr(II) was gradually increased in the single-metal system while it was rapidly increased in the binary-metal system. Moreover, when the initial concentration of U(VI) was below 0.2 mmol/L, the sorption of Sr(II) in the binary-metal system was obviously lower than that in the single-metal system, whereas the facilitated sorption of Sr(II) on GOs was observed at U(VI) concentrations above 0.2 mmol/L. They proposed that the significantly enhanced sorption amount of Sr(II) at a higher U(VI) concentration might be due to the formation of U(VI) surface precipitates (schoepite) above an U(VI) initial concentration of 0.2 mmol/L. The formation of polymeric or precipitated U(VI) species might supply additional sorption sites, containing further sorption of Sr(II) on the U(VI) precipitate similar with many metal oxy-hydroxides [153, 154]. In the single-metal and binary-metal systems, U(VI) sorption onto GOs showed a two-segment sorption trend: (i) well fitted by the Langmuir model (0.1–0.6 mmol/L C_0[U(VI)]); (ii) further increased U(VI) sorption at C_0[U(VI)] $>$ 0.6 mmol/L, indirectly indicating the formation of U(VI) precipitate with increasing initial U(VI) concentrations.

Wang et al. found the differences of the competitive and simultaneous sorption of U(VI), $^{152+154}$Eu(III), $^{85+89}$Sr(II), and ^{134}Cs(I) from solution on GOs [155]. The sorption of U(VI) and Eu(III) was not significantly influenced by the existence of Sr(II) and Cs(I), whereas the sorption of Sr(II) and Cs(I) was obviously decreased by the existence of U(VI) and Eu(III). They assigned the reason for the decreased sorption of Sr(II) and Cs(I) ions to the ion exchange mechanism of Sr(II) and Cs(I) ions on GOs, which could be affected by Na^+ ions in solution. Besides, the competitive sorption also existed between Sr(II) ions and Cs(I) ions, which could result in their lower sorption.

The separation of Th(IV)/U(VI) by GO was reported by Jiang et al., and the results revealed that GO showed high selectivity toward Th(IV) [156]. The separation factor of Th(IV)/U(VI) (SFTh(IV)/U(VI)) enhanced with increasing pH from 2.0 to 3.8, and then subsequently declined with increasing pH from 3.8 to 5.0 (Fig. 4.14A). They deduced that the GO likely prefers to adsorb the hydrolysate species. According to the XRD analysis, the addition of U(VI) caused a more pronounced decrease of the interlayer distance of GO while the addition of Th(IV) induced a negligible decrease. The result indicated that the interlayer distance of GO naturally fitted the adsorption of Th(IV). Moreover, after the Th(IV)/U(VI) loaded on GO, the interlayer distance was similar to that of Th((IV) on the treated GO, which also helped

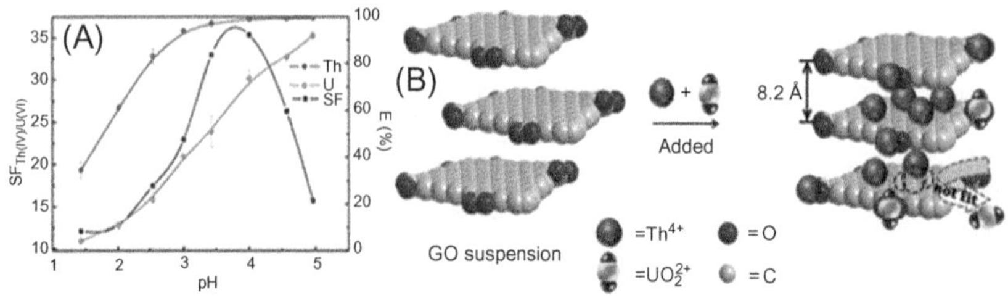

FIG. 4.14 The dependence of the separation factor of Th(IV)/U(VI) *(squares)* and the removal efficiencies of Th(IV) *(circles)* and U(VI) *(diamonds)* on the pH values in the Th(IV)/U(VI) mixture solutions, C_0(Th(IV)) ≈ 0.1mM, C_0(U(VI)) ≈ 0.1mM, GO concentration 0.2mg/mL, $T = 298 \pm 1$ K, $t = 2$h (A), separation mechanism of Th(IV)/U(VI) by GO (B) [156]. *(Data from D. Jiang, L. Liu, N. Pan, F.Q. Yang, S.K. Li, R.B. Wang, I.W. Wyman, Y.D. Jin, C.Q. Xia, The separation of Th(IV)/U(VI) via selective complexation with graphene oxide, Chem. Eng. J. 271 (2015) 147–154. Copyright 2015, Elsevier.)*

to demonstrate that Th(IV) competed strongly with U(VI). They deduced that the spatial selectivity of GO was attributed to the selective complexation of GO with Th(IV) over U(VI) (Fig. 4.14B).

Zhao et al. studied the competitive adsorption of U(VI) and Sr(II) on GO. They observed that the GO exhibited higher affinity performance toward U(VI) than Sr(II) in single and binary-solute systems [157]. Besides, they used the ETLM model of surface complexation modeling to fit the adsorption of U(VI) and Sr(II) on GO (Fig. 4.15), and summarized the optimized parameters (Table 4.3). The fitted results showed that the two inner-sphere surface complexes (>SOM^+ and (>$SO)_2MOH^-$ species) could be satisfactorily fitted with the adsorption data at single- and binary-solute systems. Besides, in binary-solute systems, the adsorption rate and amount of U(VI) and Sr(II) on GO was obviously lower than that of U(VI) and Sr(II) adsorption in single-solute systems, which was attributed to competitive ions occupying the abundant surface active sites.

4.3.4 Adsorption of radionuclides on carbon nanofibers

Carbonaceous nanofiber (CNF) materials were used to investigate the adsorption performance and mechanisms toward U(VI) and Eu(III) [158]. The effect of pH on the adsorption of U(VI) and Eu(III) on the CNFs was shown in Fig. 4.16A. The adsorption performance of CNFs toward U(VI) and Eu(III) was obviously enhanced from pH 3.0 to 6.0, which was assigned to the strong surface complexation between the surface oxygen groups of the CNFs and the radionuclide species. The high adsorption performance of CNFs toward U(VI) and Eu(III) at neutral condition could be attributed to the strong complexation between U(VI) and Eu(III) and the carboxylic groups of CNFs. The decreased adsorption at high pH was due to the electrostatic repulsion interaction between negatively charged radionuclide species and the negatively charged

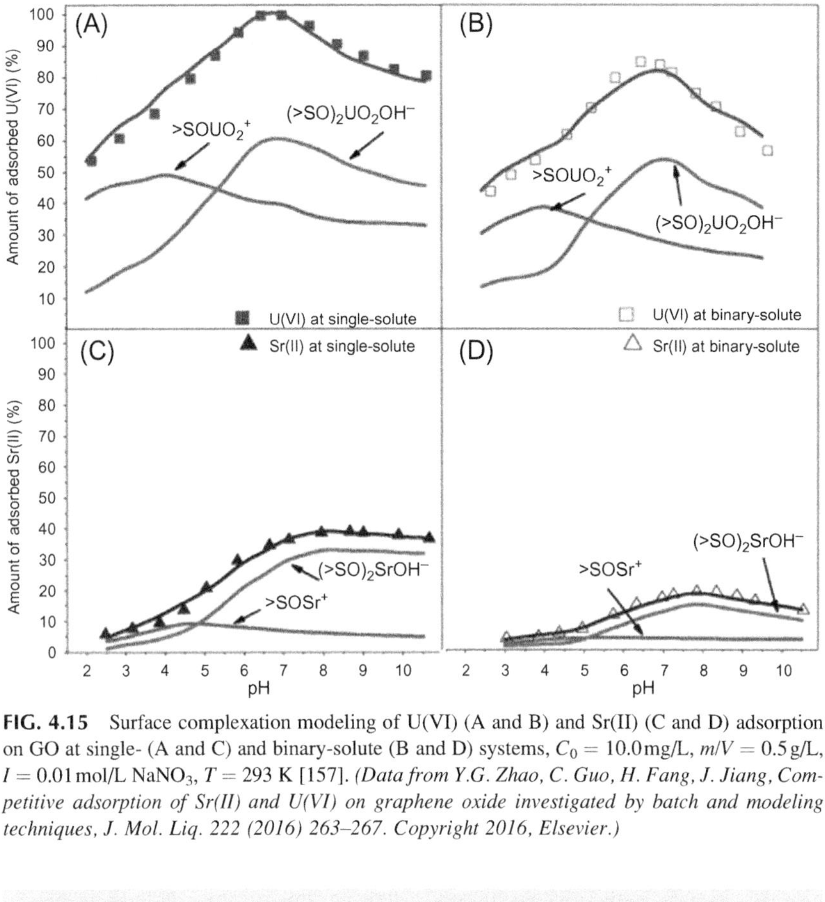

FIG. 4.15 Surface complexation modeling of U(VI) (A and B) and Sr(II) (C and D) adsorption on GO at single- (A and C) and binary-solute (B and D) systems, $C_0 = 10.0\,mg/L$, $m/V = 0.5\,g/L$, $I = 0.01\,mol/L$ $NaNO_3$, $T = 293$ K [157]. *(Data from Y.G. Zhao, C. Guo, H. Fang, J. Jiang, Competitive adsorption of Sr(II) and U(VI) on graphene oxide investigated by batch and modeling techniques, J. Mol. Liq. 222 (2016) 263–267. Copyright 2016, Elsevier.)*

TABLE 4.3 The ETLM parameters for U(VI) and Sr(II) on graphene oxide [157]

Reaction type	Reaction	Log *K*
[a]$\text{Log } K_1^0$	$>SOH + H^+ => SOH_2^+$	2.43
[a]$\text{Log } K_2^0$	$>SO^- + H^+ = >SOH$	6.03
$\text{Log } K_{SOUO2}^{+0}$	$>SOH + UO_2^{2+} => SOUO_2^+ + H^+$	−5.42
$\text{Log } K_{(SO)2UO2OH}^{2-0}$	$2>SOH + UO_2^{2+} + H_2O = (>SO)_2UO_2OH^- + 3H^+$	−4.86
$\text{Log } K_{SOSr}^{+0}$	$>SOH + Sr^{2+} = >SOSr^+ + H^+$	−3.85
$\text{Log } K_{(SO)2SrOH}^{-0}$	$2>SOH + Sr^{2+} + H_2O = (>SO)_2SrOH^- + 3H^+$	−2.13

[a]*The protonation and deprotonation constants from the fitting the data of acid-base titration.*
Data from Y.G. Zhao, C. Guo, H. Fang, J. Jiang, Competitive adsorption of Sr(II) and U(VI) on graphene oxide investigated by batch and modeling techniques, J. Mol. Liq. 222 (2016) 263–267. Copyright 2016, Elsevier.

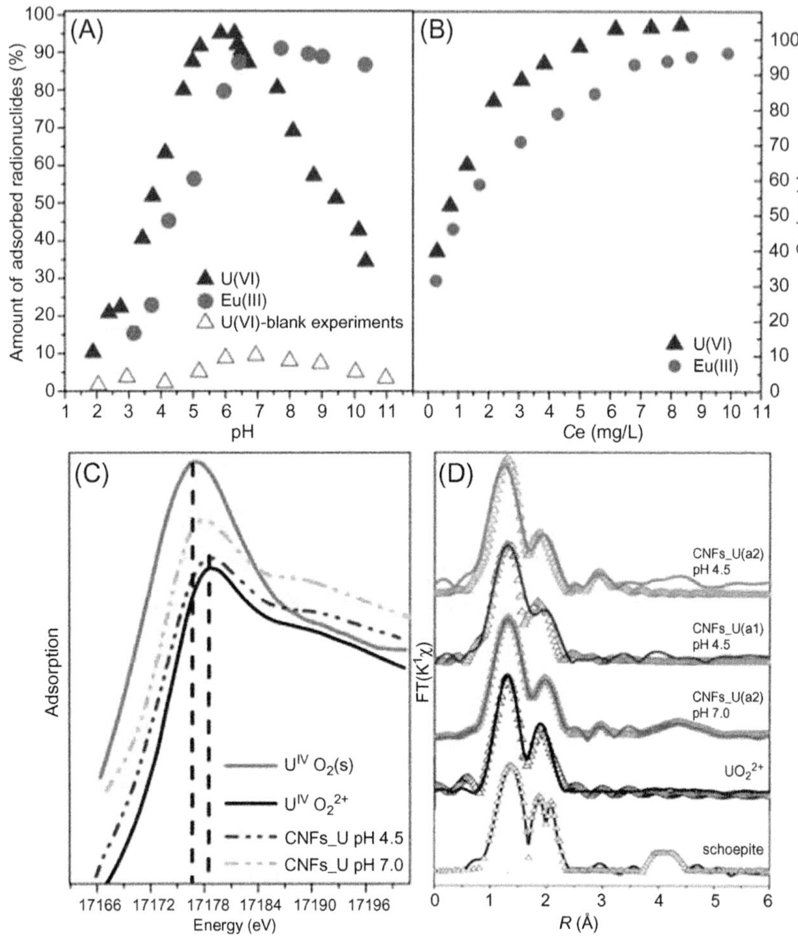

FIG. 4.16 Effect of pH (A) and initial concentration (B) on adsorption of U(VI) and Eu(III) on the CNFs, A: C_0 = 10 mg/L, I = 0.01 mol/L $NaClO_4$, m/V = 0.16 g/L, T = 298 K. B: pH 4.5, I = 0.01 mol/L $NaClO_4$, m/V = 0.16 g/L, T = 298 K; the XANES spectra of reference and uranium adsorption samples (C) and the k^2-weighted U L_{III}-edge EXAFS spectra of samples after adsorption (D), $C_{U(VI)}$ = 60 and 10 mg/L for the CNFs_U(a1) and CNFs_U(a2), respectively, m/V = 0.016 g/L, I = 0.01 mol/L $NaClO_4$, T = 298 K, *solid lines*: experimental data; *hollow triangles*: fitted data [158]. *(Data from Y.B. Sun, Z.Y. Wu, X.X. Wang, C.C. Ding, W.C. Cheng, S.H. Yu, X.K. Wang, Macroscopic and microscopic investigation of U(VI) and Eu(III) adsorption on carbonaceous nanofibers, Environ. Sci. Technol. 50 (2016) 4459–4467. Copyright 2016, American Chemical Society.)*

surface of the CNFs. An experiment of a single U(VI) (CNF-free control) under different pH conditions was also studied (Fig. 4.16A). Approximately 10% of U(VI) was reduced at pH 6.0, which might be attributed to the formation of U(VI) precipitation (e.g., schoepite). Therefore, the precipitation process could be a reason for the high-level removal of U(VI) on CNFs at neutral conditions. The CNFs exhibited excellent adsorption performance toward both U(VI) and Eu(III) at pH 4.5 and $T=298\,K$ (Fig. 4.16B). The maximum adsorption capacities of U(VI) and Eu(III) on CNFs were calculated at 125 and 91 mg/g, respectively. Further, the XANES and EXAFS spectra of U(VI) adsorption at pH 4.5 and 7.0 were shown in Fig. 4.16C and D. The results of EXAFS showed that the inner-sphere surface complexation contributed to the interaction of U(VI) on the CNFs at pH 4.5, whereas the surface coprecipitation was observed at pH 7.0. Therefore, the enrichment of U(VI) and Eu(III) on CNFs was due to the sufficient active sites (e.g., —COOH and —OH groups) on the surface of CNFs.

Hu et al. also investigated the adsorption mechanism of U(VI) on CNFs through EXAFS analysis [159]. From the analysis results of Fig. 4.17A and B, they concluded that a splitting of U-Oeq shell in the CNFs-U(VI) system could be attributed to be inner-sphere surface complexes. The existence of U-C and U-U shells in the sample of CNF_U pH 7.0 indicated that the adsorption was due to the inner-sphere/surface coprecipitate mechanism. For the interaction mechanism of CNFs toward U(VI), they concluded that the outer-sphere surface complexation controlled the adsorption at pH 4.0, whereas inner-sphere/surface coprecipitate governed the adsorption at pH 7.0. Additionally, the bacteria-derived CNFs were used to investigate the adsorption of Cs(I) and Sr(II) [40]. They observed that the adsorption of Sr(II) and Cs(I) on CNFs was dependent of pH and ionic strength. The results showed that the adsorption enhances significantly from pH 2.0 to 6.0,which was assigned to the strong surface complexation between the surface oxygen groups of the CNFs and Sr^{2+} and Cs^{+} species. The adsorption of CNFs toward Sr(II) and Cs(I) slightly declined with increasing ionic strength at $pH < 5.0$, which could be attributed to the outer-sphere surface complexation. However, no effect of ionic strength on adsorption was observed at $pH > 6.0$, which was mainly ascribed to the inner-sphere surface complexation at $pH > 6.0$. According to the ionic strength-dependent adsorption experiments, the adsorption of Sr(II) and Cs(I) on CNFs was simulated by using outer-sphere ($SOHM^{n+}$) and inner-sphere ($SOM^{(n-1)+}$) complexes under different pH conditions. Two surface complexation reactions are applied as Eqs. (4.6), (4.7):

$$SOH + M^{n+} = SOHM^{n+} \tag{4.6}$$

$$SOH + M^{n+} = SOHM^{(n-1)+} + H^{+} \tag{4.7}$$

where M^{n+} represents Sr(II) or Cs(I) ions. Fig. 4.17C and D shows the fitting results of the removal of Sr(II) and Cs(I) on CNFs, and the corresponding equilibrium constants of that are summarized in Table 4.4. The adsorption data can be desirably fitted by using the $SOHSr^{2+}$, $SOSr^{+}$ and $SOHCs^{+}$, SOCs species, respectively (Fig. 4.17C and D). They concluded that the adsorption mechanism is mainly attributed to the outer-sphere surface complexation ($pH < 4.0$) and the inner-sphere surface complexation ($pH > 5.0$), according to the analysis of surface complexation modeling.

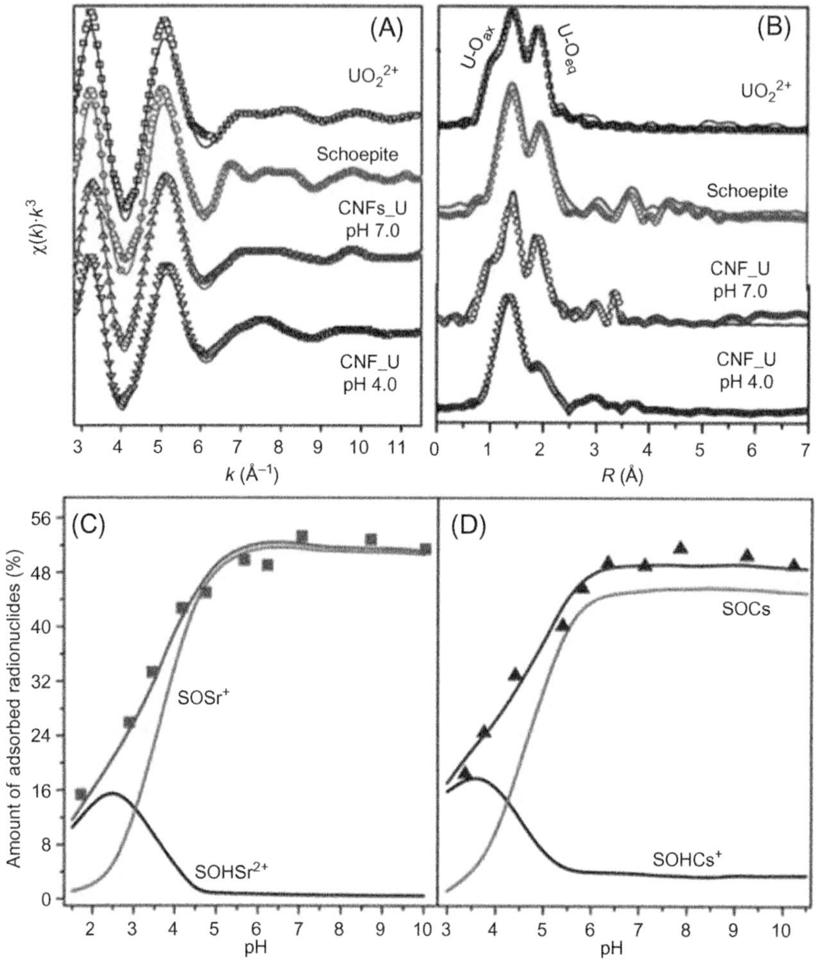

FIG. 4.17 The k^2-weighted U L_{III}-edge EXAFS spectra (A) and the corresponding Fourier transforms (B) of standards (UO_2^{2+} and schoepite) and samples of CNF_U at pH 4.0 and 7.0. C_0 = 10 mg/L, m/V = 0.5 g/L, I = 0.01 mol/L NaCl, T = 293 K, *solid lines*: experimental data; *hollow triangles*: fitted data [159], surface complexation modeling simulation of Sr(II) (C) and Cs(I) (D) adsorption on CNFs, C_0 = 10 mg/L, I = 0.01 mol/L $NaClO_4$, m/V = 0.16 g/L, T = 293 K [40]. *(Data from Y.B. Sun, X.X. Wang, C.C. Ding, W.C. Cheng, C.L. Chen, T. Hayat, A. Alsaedi, J. Hu, X.K. Wang, Direct synthesis of bacteria-derived carbonaceous nanofibers as a highly efficient material for radionuclides elimination, ACS Sustain. Chem. Eng. 4 (2016) 4608–4616; B.W. Hu, Q.Y. Hu, D. Xu, C.G. Chen, The adsorption of U(VI) on carbonaceous nanofibers: a combined batch, EXAFS and modeling techniques, Sep. Purif. Technol. 175 (2017) 140–146. Copyright 2016 and 2017, American Chemical Society, Elsevier.)*

Zhang et al. studied the adsorption capacity and mechanism of CNFs toward U(VI) [21]. They preliminarily explored some research on the application of environmental soil remediation by CNFs. Montmorillonite was chosen as the substitute of soil to investigate the migration behavior of U(VI).

TABLE 4.4 Optimized parameters for surface complexation modeling of Sr(II) and Cs(I) sorption on the CNFs [40]

Adsorbates	Surface complexation reactions	Log *K*
Sr(II)	$SOH + Sr^{2+} = SOHSr^{2+}$	0.98
	$SOH + Sr^{2+} = SOSr^{+} + H^{+}$	−2.15
Cs(I)	$SOH + Cs^{+} = SOHCs^{+}$	0.47
	$SOH + Cs^{+} = SOCs + H^{+}$	−3.47

Data from Y.B. Sun, X.X. Wang, C.C. Ding, W.C. Cheng, C.L. Chen, T. Hayat, A. Alsaedi, J. Hu, X.K. Wang, Direct synthesis of bacteria-derived carbonaceous nanofibers as a highly efficient material for radionuclides elimination, ACS Sustain. Chem. Eng. 4 (2016) 4608–4616. Copyright 2016, American Chemical Society.

They concluded that CNFs could extract the sorbed uranium onto their surface by a pseudo-second-order kinetic process when it coexisted with montmorillonite. The schematic and mechanism diagrams were shown in Fig. 4.18A and B. They found that the CNFs had favorable stability and reusability from desorption and recycling experiments. In addition, amidoxime-grafted activated carbon fibers (ACFs-AO) were tested for the efficient removal of uranium from aqueous solution by Lu et al. [160]. They observed that a favorable adsorption performance of ACFs-AO toward U(VI) (about 191.6 mg/g) was mainly due to surface complexation of U(VI) with the amidoxime groups on ACFs-AO (Fig. 4.18C). Moreover, they found that ACFs-AO selectively adsorbed U(VI) in the existence of competitive ions with efficient regeneration and reusability.

4.4 Adsorption of radionuclides on carbon nanotube-based composites

4.4.1 Adsorption of radionuclides on inorganic carbon nanotube-based composites

4.4.1.1 U(VI) adsorption

The Ca/Al-layered, double-hydroxide decorated carbon nanotube (Ca/Al-LDH@CNTs) was synthesized by Chen et al. through the coprecipitation method and hydrothermal aged treatment [161]. They carried out detailed batch experiments for U(VI) adsorption on CNTs and Ca/Al-LDH@CNTs to investigate the interaction mechanisms and behaviors of U(VI) adsorption on them. Experiments on the influence of pH and ionic strength on U(VI) adsorption were conducted to study the adsorption mechanism of U(VI) on CNTs and Ca/Al-LDH@CNTs (Fig. 4.19A and B). The adsorption process could be composed of two processes: (process I) the adsorption enhanced rapidly with increasing pH ($pH < 6.0$); (process II) the adsorption declined with

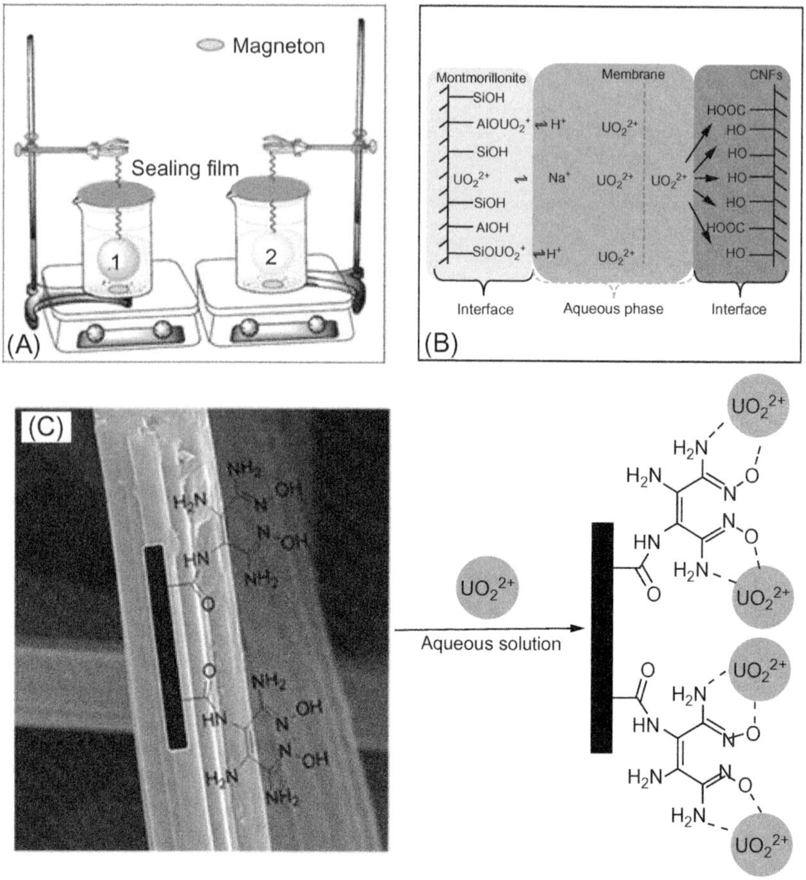

FIG. 4.18 (A) The schematic diagram of the application of CNFs by contrast experiments, (B) The mechanism diagram of bag 2 [21], (C) Schematic of the adsorption of U(VI) on ACFs-AO [160]. *(Data from R. Zhang, C.L. Chen, J. Li, X.K. Wang, Investigation of interaction between U(VI) and carbonaceous nanofibers by batch experiments and modeling study, J. Colloid Interface Sci. 460 (2015) 237–246; X. Lu, D.X. Zhang, A.T. Reda, C. Liu, Z. Yang, S.S. Guo, S.T. Xiao, Y.G. Ouyang, Synthesis of amidoxime-grafted activated carbon fibers for efficient recovery of uranium(VI) from aqueous solution, Ind. Eng. Chem. Res. 56 (2017) 11936–11947. Copyright 2015 and 2017, Elsevier, American Chemical Society.)*

increasing pH (pH > 6.0). Process I was due to the strong surface complexes with cationic U(VI) species and oxygen groups of CNTs and Ca/Al-LDH-@CNTs while process II was attributed to the electrostatic repulsion interaction between anionic U(VI) species and the negatively charged surface of CNTs and Ca/Al-LDH@CNTs. The significant enhanced adsorption performance of Ca/Al-LDH@CNTs toward U(VI) might be assigned to the coordination of U(VI) with the abundant surface active sites (e.g., Ca-O, Al-O groups) of Ca/Al-LDH@CNTs. Besides, the influence of ionic strength on

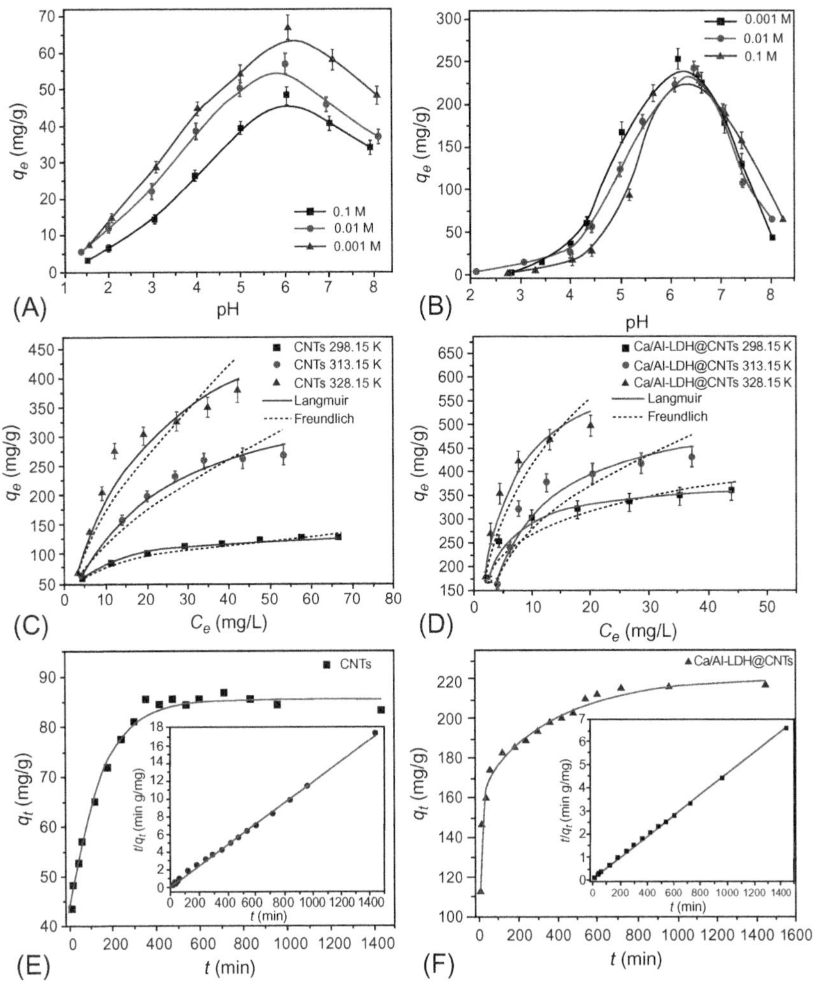

FIG. 4.19 Effect of pH on U(VI) adsorption on CNTs (A) Ca/Al-LDH@CNTs (B) at different $NaNO_3$ concentrations. $m/V = 0.1$ g/L, $T = 298.15$ K, $C_{[U(VI)]initial} = 30$ mg/L. Adsorption isotherms of U(VI) on CNTs (C) and Ca/Al-LDH@CNTs (D), pH $= 6.0 \pm 0.1$, $I = 0.01$ M $NaNO_3$. Adsorption kinetics of U(VI) on CNTs (E) and Ca/Al-LDH@CNTs (F), pH $= 6.0 \pm 0.1$, $I = 0.01$ M $NaNO_3$, $C_{[U(VI)\ initial]} = 30$ mg/L, $T = 298.15$ K [161]. *(Data from H.J. Chen, Z. Chen, G.X. Zhao, Z.B. Zhang, C. Xu, Y.H. Liu, J. Chen, L. Zhuang, T. Haya, X.K. Wang, Enhanced adsorption of U(VI) and 241Am(III) from wastewater using Ca/Al layered double hydroxide@carbon nanotube composites, J. Hazard. Mater. 347 (2018) 67–77. Copyright 2018, Elsevier.)*

U(VI) adsorption on CNTs was observed more distinctly than that of Ca/Al-LDH@CNTs, which might be due to more oxygen functional groups of Ca/Al-LDH@CNTs resulting in the formation of much stronger surface complexes with U(VI). Fig. 4.19C and D show the isotherms of U(VI) adsorption on CTNs and Ca/Al-LDH@CNTs. For both CTNs and Ca/Al-LDH@CNTs, the adsorption capacities of U(VI) were obviously enhanced with the increase of temperature, and the maximum adsorption capacity was 137.80 mg/g on CNTs and 382.89 mg/g on Ca/Al-LDH@CNTs at 298.15 K.

The relative parameters calculated from the Langmuir and Freundlich models are listed in Table 4.5. For the adsorption of U(VI) on CNTs and Ca/Al-LDH@CNTs, the well-simulated data of isotherms by the Langmuir model implied that the adsorption was mainly dominated by monolayer coverage. Further, Fig. 4.19E and F show that the adsorption kinetic experiments of U(VI) were conducted to clarify the adsorption dynamic process. The adsorption first increased rapidly at the initial stage, which was due to the abundant available active and binding sites on the surface of CNTs and Ca/Al-LDH@CNTs. The equilibrium contact time of the adsorption of U(VI) on CNTs was shorter than that of Ca/Al-LDH@CNTs, indicating that the diffusion of U(VI) species into the interlayer of Ca/Al-LDH@CNTs would take a relatively longer contact time to achieve equilibrium.

4.4.1.2 Other radionuclide adsorption

In the absence and presence of poly(acrylic acid) (PAA), Chen et al. studied the interaction of Eu(III) on a MWCNT/iron oxide magnetic composite [162]. The positive ferrous and ferric ions are attached on oxidized MWCNTs, which is attributed to the coordination reaction between negative carboxylic and hydroxylic groups on MWCNTs and ferrous and ferric ions. Chen et al. observed that the existence of PAA increased the adsorption of Eu(III) at pH < 4.5, whereas the decreased Eu(III) adsorption at pH > 5.0 was due to the soluble Eu-PAA complexes. Additionally, they investigated the adsorption of Eu(III) for the three different addition sequences: (Eu – PAA)–magnetic composite, (Eu – magnetic composite)–PAA, and (PAA – magnetic composite)–Eu. They observed that the different addition sequences have no impact on Eu(III) adsorption at pH < 3.5, whereas differences inhibiting the adsorption of Eu(III) were observed at pH > 4.0 with the three different addition sequences. They found that the interaction of Eu(III) with PAA was stronger than the interaction of Eu(III) with the magnetic composite was attributed to the inhibited adsorption in the (Eu – magnetic composite)–PAA and (PAA – magnetic composite)–Eu systems.

Liu et al. studied the adsorption of Th(IV) on magnetic MWCNTs, which was significantly dependent on pH values [163]. The adsorption and desorption isotherms show that the adsorption of Th(IV) on magnetic MWCNTs was endothermic and irreversible. Additionally, the application of the MWCNTs-Fe_3O_4 nanocomposite to the preconcentration of thorium was investigated by Zolfonoun et al.; this shows an improved separation rate due to magnetic separation [164]. MWCNT)/iron oxide magnetic composites were synthesized by Chen et al., and utilized to remove Sr(II) ions [165].

TABLE 4.5 Parameters of U(VI) adsorption on CNTs and Ca/Al-LDH@CNTs at three temperatures [161]

		CNTs			Ca/Al-LDH@CNTs		
Models	**Parameters**	*298.15 K*	*313.15 K*	*328.15 K*	*298.15 K*	*313.15 K*	*328.15 K*
Langmuir	K_L (L/mg)	0.15	0.043	0.045	0.33	0.12	0.20
	q_{max}	137.8	313.44	407.48	382.89	553.10	658.12
	R^2	0.988	0.991	0.962	0.955	0.929	0.942
Freundlich	K_F ($mg^{1-n}L^n/g$)	38.23	28.72	38.11	150.54	109.34	153.24
	$1/n$	0.30	0.60	0.65	0.44	0.41	0.24
	R^2	0.976	0.948	0.906	0.857	0.829	0.843

Data from H.J. Chen, Z. Chen, G.X. Zhao, Z.B. Zhang, C. Xu, Y.H. Liu, J. Chen, L. Zhuang, T. Haya, X.K. Wang, Enhanced adsorption of U(VI) and 241Am(III) from wastewater using Ca/Al layered double hydroxide@carbon nanotube composites, J. Hazard. Mater. 347 (2018) 67–77.

The adsorption results revealed that the adsorption of Sr(II) on the magnetic composites strongly relied on pH and ionic strength. Further, the adsorption capacity of magnetic composites toward Sr(II) was significant higher than that of MWCNTs and iron oxides, which might be the synergistic effect between oxidized MWCNTs and iron oxides.

4.4.2 Adsorption of radionuclides on organic carbon nanotube-based composites

4.4.2.1 U(VI) adsorption

Song et al. synthesized multicomponent composite (CNTs-PDA-PGMA-EDA) by integrating the mussel-inspired PDA (polydopamine) chemistry and surface-initiated activators (ARGET ATRP) for controlled growth of poly(glycidyl methacrylate) (PGMA) brushes on CNTs, and then facile covalently grafted the ethylenediamine (EDA) ligands to the surface of CNTs-PDA-PGMA [166] (Fig. 4.20A). The obtained CNT composite exhibited superior adsorption ability to U(VI) (192.9 mg/g), almost 15-fold higher than that of pristine CNT, which provided a potential use for uranium enrichment.

As the plasma-grafting technique was considered as an environmentally friendly and convenient method to modify the surface properties of materials, an amidoxime-grafted MWCNT composite (AO-g-MWCNTs) has been fabricated utilizing the plasma technique to selectively separate uranium [167]. The excellent adsorption capacity for U(VI) (145 mg/g) was observed at pH 4.5. In addition, AO-g-MWCNTs showed a significantly enhanced adsorption performance toward U(VI) compared to oxidized MWCNTs (Fig. 4.20B). Therefore, AO-g-MWCNT composites exhibited a great potential for applications in remediation and recovery of U(VI) from aqueous solutions. Further, Shao et al. also used plasma techniques to graft carboxymethyl cellulose (CMC) on MWCNTs and obtained the MWCNT-g-CMC composite [19]. In contrast with raw MWCNT, the MWCNT-g-CMC has obviously more favorable adsorption performance toward U(VI). Fig. 4.20C exhibited the adsorption of U(VI) on raw MWCNT, MWCNT treated, and MWCNT-g-CMC. Additionally, the MWCNT-g-CMC revealed a higher adsorption capacity toward U(VI) with eight times that of raw MWCNT, which was due to the strong complexes between U(VI) and abundant functional groups on the surfaces of MWCNT-g-CMC. Considering the economic application of MWCNT or MWCNT-g-CMC, the effect of the sorbent content on the adsorption of U(VI) was studied (Fig. 4.20D). The adsorption percentage of U(VI) increased from ~23% to ~98% with increasing MWCNT-g-CMC content from 0.10 to 1.0 g/L. Further, the adsorption percentage of U(VI) on MWCNT-g-CMC was obviously more favorable than that on raw MWCNT and MWCNT-treated in the whole content range. It also demonstrated the more efficient removal of U(VI) ability of MWCNT-g-CMC compared to raw MWCNT and MWCNT-treated. Mishra et al. studied the sorption behavior of uranium on tributyl phosphate-grafted CNT composites (CNT-TBP) [168]. From the XPS and IR

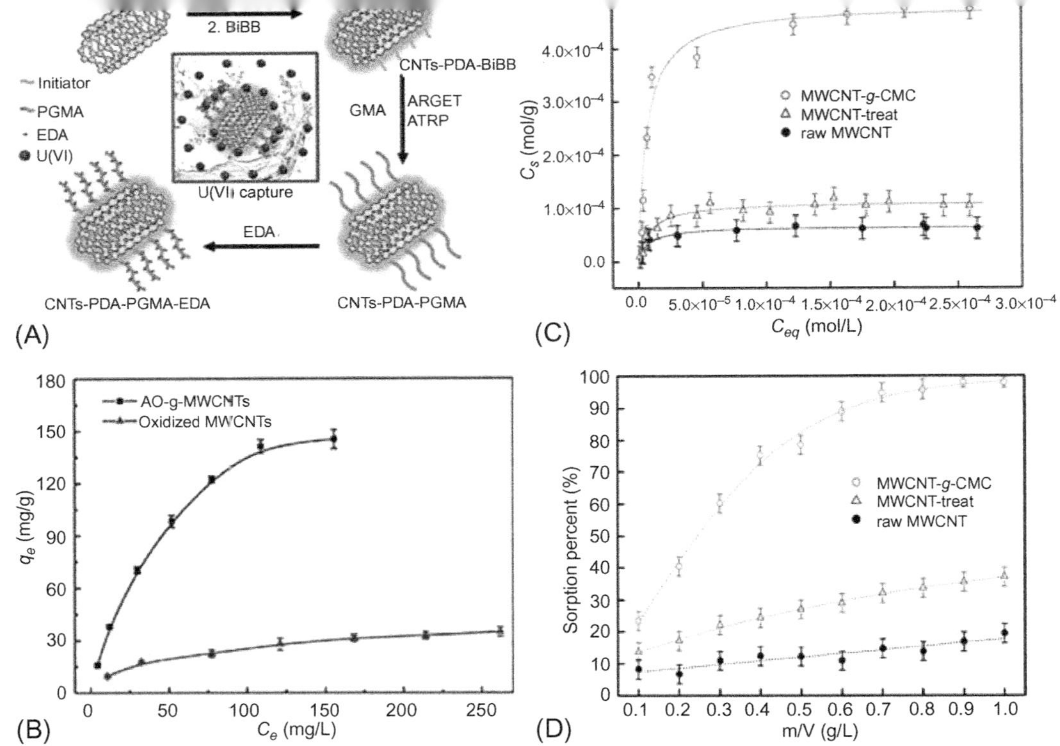

FIG. 4.20 CNT platform for U(VI) capture via anchoring ATRP initiators and PGMA functionalization (A),[166] equilibrium isotherm for the U(VI) sorption on oxidized MWCNTs and AO-g-MWCNTs (B), pH 4.5, $m = 20$ mg, $t = 240$ min, $T = 298.15$ K, and $V = 20$ mL [167], sorption isotherms (C) and effect of sorbent content (D) on the removal of UO_2^{2+} from solution on raw MWCNT, on MWCNT-treat, and on MWCNT-g-CMC. $T = 25 \pm 2°C$, equilibrium time 24 h, pH $= 5.0 \pm 0.1$, $C[NaClO_4] = 1.0 \times 10^{-2}$ mol/L; (C) $m/V = 0.4$ g/L, (D) $C[UO_2^{2+}]_{(initial)} = 2.00 \times 10^{-4}$ mol/L [19]. *(Data from D.D. Shao, Z.Q. Jiang, X.K. Wang, J.X. Li, Y.D. Meng, Plasma induced grafting carboxymethyl cellulose on multiwalled carbon nanotubes for the removal of UO22+ from aqueous solution, J. Phys. Chem. B 113 (2009) 860–864; Y. Song, G. Ye, Y.X. Lu, J. Chen, J.C. Wang, K. Matyjaszewski, Surface-initiated ARGET ATRP of poly(glycidyl methacrylate) from carbon nanotubes via bioinspired catechol chemistry for efficient adsorption of uranium ions, ACS Macro Lett. 5 (2016) 382–386; Y. Wang, Z.X. Gu, J.J. Yang, J.L. Liao, Y.Y. Yang, N. Liu, J. Tang, Amidoxime-grafted multiwalled carbon nanotubes by plasma techniques for efficient removal of uranium(VI), Appl. Surf. Sci. 320 (2014) 10–20. Copyright 2009, 2016, 2014, American Chemical Society, American Chemical Society, Elsevier.)*

analysis, the uranyl ions were complex with the phosphoryl group of the tributyl phosphate ligand. Efficient complexation of U(VI) and the phosphoryl group of CNT-TBP resulted in an excellent sorption capacity (166.6mg/g).

An amide-functionalized CNT (CNT-DHA) was synthesized by Sengupta et al. and used to remove tetravalent (Th^{4+}) and hexavalent (UO_2^{2+}) actinides [169]. The adsorption capacities of CNT-DHA toward UO_2^{2+} and Th^{4+} were 32 and 47 mg/g, respectively. They investigated the effect of feed acidity on the K_d values of uranium and thorium ions, and the K_d value was calculated, as in Eq. (4.8). The K_d values of the UO_2^{2+} and Th^{4+} ions were increased with increasing feed acidity from 0 to 3 M HNO_3, and then maintained a plateau. The initial increase stage of K_d values was assigned to the law of mass action considering Eqs. (4.9a), (4.9b). The relative constant stage of K_d values was due to the sufficient concentration of H^+ ions competing with UO_2^{2+} or Th^{4+}ions (Eq. 4.10).

$$K_d = \frac{(C_0 - C_e)}{C_e}\frac{\nu}{w} \tag{4.8}$$

$$UO_2^{2+} + \text{CNT-DHA} + 2NO_3^- = UO_2(NO_3)_2 \cdot \text{CNT-DHA} \tag{4.9a}$$

$$Th^{4+} + \text{CNT-DHA} + 2NO_3^- = Th(NO_3)_4 \cdot \text{CNT-DHA} \tag{4.9b}$$

$$H^+ + \text{CNT-DHA} + NO_3^- = HNO_3 \cdot \text{CNT-DHA} \tag{4.10}$$

The greater K_d values for uranyl ions was observed at lower feed acidity compared to the thorium ions, which was due to the lower concentration of NO_3^- at the lower feed acidity being insufficient for effective complexation of Th^{4+} as $Th(NO_3)_4$. The NO_3^- concentration was found to be sufficient for both UO_2^{2+} and Th^{4+} ions at moderate feed acidity, which led to a crossover in the K_d values near 1 M HNO_3. Beyond the feed acidity of 3 M HNO_3, the K_d values of Th^{4+} surpassed UO_2^{2+}.

The DFT calculations were used to understand the complexation information between CNT-DHA and the Th^{4+} or UO_2^{2+} ions. They also studied the optimized structure of a single CNT-DHA ligand and its complexes with UO_2^{2+} and Th^{4+} ions in the existence of nitrate ions. They found that the stretched C=O bond (1.289 Å) and shortened C—C bond (1.423 Å) compared to the free CNT-DHA ligand was attributed to the interaction of UO_2^{2+} ions with the amide O atom. For the CNT-DHA-$Th(NO_3)_4$-H_2O complex, the elongated C=O bond (1.305 Å) compared to that of CNT-DHA-$UO_2(NO_3)_2$ indicated stronger interaction compared to the UO_2^{2+} ion.

Singh Deb et al. prepared amido-amine-functionalized CNTs (CNT-AA) to remove U(VI), calculated by DFT calculations [33]. The U(VI) and CNT-AA presents as a 1:1 complex in the existence of nitrate ions, and the complexation reaction mentioned in the following Eq. (4.11).

$$Mn^+ + nNO_3^- + \text{CNT-AA} \rightarrow M(NO_3)_n\text{CNT-AA} \tag{4.11}$$

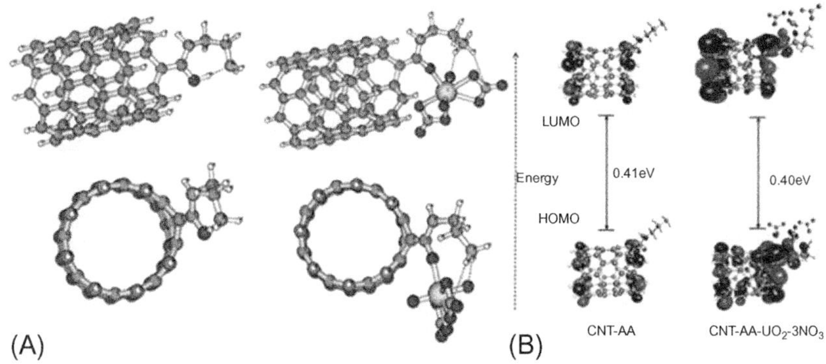

FIG. 4.21 (A) geometric optimized molecular structures of the (a) CNT-AA, (b) $UO_2(NO_3)_2$CNT-AA complex, (B) The distribution of the HOMO and LUMO orbitals of CNT-AA, and its complexes with UO_2^{2+} ions in the presence on nitrate anions, at isosurface value of 0.011e.bohr^{-3} [33]. *(Data from A.S. Deb, S. Pahan, K. Dasgupta, S. Panja, A. Debnath, P. Dhami, S.M. Ali, C. Kaushik, J. Yadav, Carbon nano tubes functionalized with novel functional group-amido-amine for sorption of actinides, J. Hazard. Mater. 345 (2018) 63–75. Copyright 2018, Elsevier.)*

In the optimized structures of the CNT-AA (Fig. 4.21A(a)), it showed that the projected outward of amido-amine unit, which was attached with a carbon atom at the open edge of the CNT. Fig. 4.21A(b) shows the optimized complex structures of the U(VI) with CNT-AA in the existence of NO_3^- ions, which exhibited the amidoamine unit of the CNT coordinates in the bidentate mode. Two NO_3^- coordinates to the central U in the bidentate mode resulted in the formation of an eight-coordinated complex containing two coordinations from two oxo groups. The less positive charge on U(VI) (1.736) suggested a strong complexation between the U(VI) and CNT-AA. The low charge on oxygen and nitrogen coordinating sites of the amidoamine unit of the CNT-AA for U(VI) indicated more charge transfer from these atoms to the metal acceptor center, which in turn strengthened the coordination with U(VI) (Fig. 4.21B). Table 4.6 lists the optimized structures of the bond distances between the donor atom (O and N) and the metal coordination center. The shorter bond length of M—O(C=O) than that of M—O(NO_3) suggested that amide carbonyl groups had more favorable binding affinity than the nitrates. Similarly, the shorter bond distance between the metal and carbonyl oxygen than that between the metal and nitrogen indicated that carbonyl oxygen had more strong binding force than the amine nitrogen. They obtained negative ΔG (Gibbs free energy change) values for the U(VI) ions in the gas and aqueous phases, which implied a favorable adsorption process of CNT-AA toward U(VI). They proposed that the solvation/hydration of the metal ions played an important role in the complexation process assigned to the distinctly decreased ΔE (binding energy change) or ΔG values from the gas phase to the solution phase.

TABLE 4.6 Calculated M-O and M-N distances in Å for the CNT-AA and its complexes with UO_2^{2+} in presence of nitrate ions [33]

Bonds	$UO_2(NO_3)_2$CNT-AA	CNT-AA
M—O(C=O)	2.378	/
M—N(NH_2)	2.689	/
M—O(NO_3)[a]	2.503	/
C=O(Amide)	1.267	1.230
C—N(Amine)	1.468	1.454

[a]*Averages of all the M—O distances between M and* NO_3^-.
Data from A.S. Deb, S. Pahan, K. Dasgupta, S. Panja, A. Debnath, P. Dhami, S.M. Ali, C. Kaushik, J. Yadav, Carbon nano tubes functionalized with novel functional group-amido-amine for sorption of actinides, J. Hazard. Mater. 345 (2018) 63–75. Copyright 2018, Elsevier.

4.4.2.2 Other radionuclide adsorption

Singh Deb et al. utilized the DFT calculation to assist the analysis the experimental observations of the Th^{4+} ion on oxidized CNTs (CNT-COOH) and diglycolamic acid functionalized CNTs (CNT-DGA) [133]. In the existence of NO_3^- ions, they investigated the optimized structures of the complex of Th^{4+} ion with functionalized CNT-COOH and CNT-DGA. For the CNT-COOH complex, the Th^{4+} ion only coordinated to the carbonyl O of the carboxyl group and the length was 2.36 Å in the existence of the NO_3^- ion. Nevertheless, in the absence of the NO_3^- ions, both the O centers of the carboxyl group were bonded to Th^{4+} ions. The Th^{4+} ion coordinated to four NO_3^- ions in the bidentate mode, and the lengthened distance of the Th-O bond (O of NO_3) was attributed to the strong coordination from the carbonyl O atom of CNT-COOH. For the CNT-DGA complex, the Th^{4+} ion was observed to be coordinated to the six donor O atoms of the four NO_3^- ions (two bidentate and two monodentate mode), the three donor O atoms of the DGA moiety, and one from aqua leading to deca coordination. Moreover, the shorter Th—O bond (2.36 Å) of the CNT-COOH than that of the CNT-DGA (2.62 Å) clearly suggested the stronger affinity of Th^{4+} ions on CNT-COOH than that on CNT-DGA. Wang et al. found that the —COOH group of CNTs was mainly corresponding to Eu(III) sorption at low pH values. With the increase of pH the sorption was obviously enhanced, which was attributed to the deprotonation of —COOH and then the —COO— group contributed mainly to the sorption of Eu(III) on CNTs. The fluorescence intensity of the TRLFS spectra of Eu(III) in the CNT suspension at different pH values decreased with increasing pH values (Fig. 4.22A), implying the existence of static quenching and further indicating the formation of nonfluorescent Eu(III)-CNT complexes. Fig. 4.22B shows that the fluorescence intensity of Eu(III) in the CNT

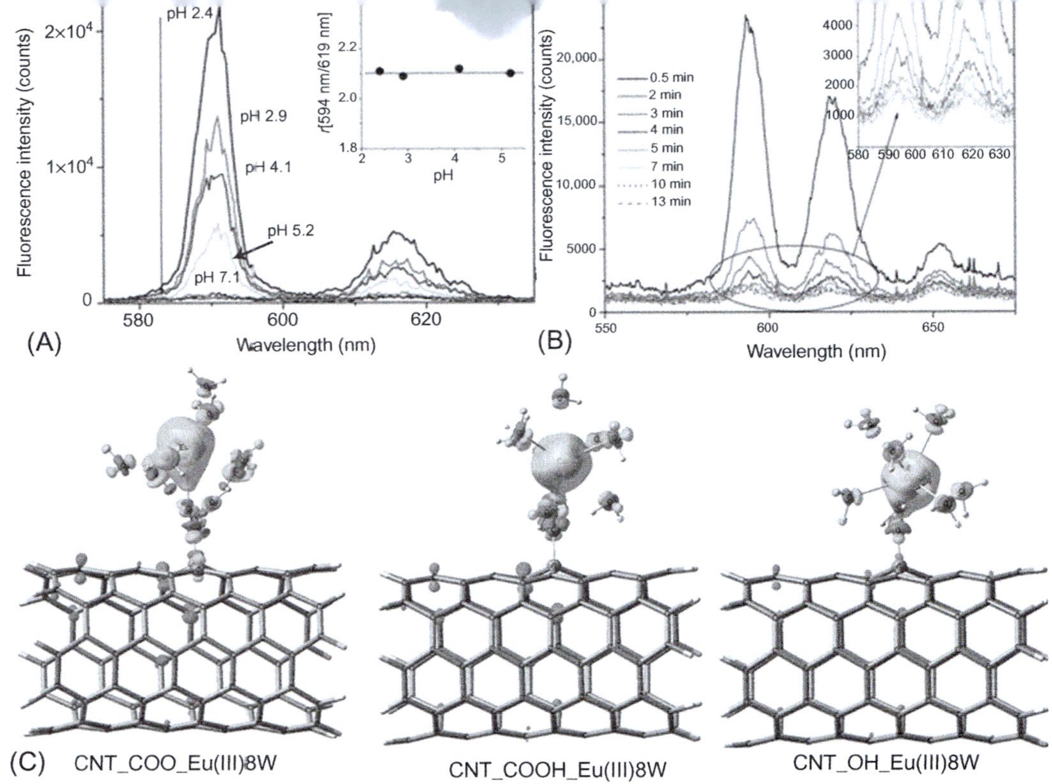

FIG. 4.22 (A) TRLFS spectra of Eu(III) in the presence of carbon nanotubes. C[Eu(III)] $= 1.9 \times 10^{-6}$ mol/L, $I = 0.1$ M $NaClO_4$, $m/V = 0.233$ g/L. (B) TRLFS study of the time dependent Eu(III) sorption to carbon nanotubes at pH 6.0. C[Eu(III)] $= 1.9 \times 10^{-6}$ mol/L, $I = 0.1$ M $NaClO_4$, $m/V = 0.233$ g/L. (C) difference density between the oxidized CNTs_Eu(III)8W and its two fragments (oxidized CNTs and Eu(III)8W). The *light color* and *dark color* represent the increase and decease of electron density, respectively. Both the colors indicated that the charge transfer interaction existed. The significant increase in electron density indicates the strong charge transfer interaction [170]. *(Data from X.X. Wang, S.B. Yang, W.Q. Shi, J.X. Li, T. Hayat, X.K. Wang, Different interaction mechanisms of Eu(III) and 243Am(III) with carbon nanotubes studied by batch, spectroscopy technique and theoretical calculation, Environ. Sci. Technol. 49 (2015) 11721–11728. Copyright 2015, American Chemical Society.)*

suspension decreased quickly with increasing contact time. Therefore, according to the fluorescence intensity variation, they concluded the uptake of Eu(III) by CNTs occurred very rapidly. They also noted that the electron density of Eu(III) increased significantly (Fig. 4.22C), indicating the existence of strong charge transfer interactions between Eu(III) and CNT_COO—, CNT_COOH, and CNT_OH.

4.5 Adsorption of radionuclides on graphene oxide-based composites

4.5.1 Adsorption of radionuclides on inorganic graphene oxide-based composites

4.5.1.1 U(VI) adsorption

Compared with the tedious solid-liquid separation procedure of centrifugation or filtration, the magnetic separation may be an effective and easy separation way at an applied magnetic field condition. Generally, magnetic nanomaterials usually have a selectivity, high adsorption efficiency, and capacity toward radionuclides. Therefore, magnetic GO composites have been widely investigated. Zong et al. prepared Fe_3O_4/GO and utilized it as an adsorbent to adsorb U(VI) ions, which shows a high maximum sorption capacity 69.49 mg/g toward U(VI) [38]. The magnetite-graphene oxide composite (M-GO) was synthesized by Ding et al. and the maximum removal capacity was about 179 mg/g at pH 4.5 and $T=303$ K [171]. Additionally, due to the high reactivity of the amidoxime toward U(VI), Zhao et al. prepared an amidoximated magnetite/graphene oxide (AOMGO) composite in two steps: (1) simple coprecipitation of magnetic GO; and (2) amidation of GO with diaminomaleonitrile followed by treatment with hydroxylamine hydrochloride to obtain the final products [39]. The obtained AOMGO composite exhibited a high maximum adsorption capacity toward U(VI) (284.9 mg/g) at pH 5 assigned to the formation of stable complexes between U(VI) ions and amidoxime/other oxygen-containing functional groups on the surfaces of AOMGO. A magnetic cucurbit[6]uril/graphene oxide (CB[6]/GO/Fe_3O_4) composite was prepared by linking the CB[6] on the GO surface through hydrogen bonding and then introducing numerous well-dispersed Fe_3O_4 nanoparticles on the CB[6]/GO surface by the coprecipitation approach [172]. The CB[6]/GO/Fe_3O_4 composite was utilized to study the removal of U(VI) and exhibited excellent adsorption performance and acceptable reus ability due to its strong complexation ability, high efficiency of magnetic separability, and good stability. Moreover, the composite could be regenerated in acidic solution for 24 h, with only minute iron leaching.

Other inorganic GO composites were also synthesized and utilized to remove U(VI) [37, 173–175]. The 3D macroscopic composites of agar-MMT-GO were obtained by intercalated montmorillonite (MMT) into a sheet of GO and then cross-linked by agar [173]. A depiction of the synthesis process and the schematic structure of agar-MMT-GO composites is shown in Fig. 4.23A. The MMT and GO were first dealt with by ultrasound treatment, and then MMT was proposed to embed into the GO nanosheets. Lastly, the mixture of MMT and GO nanosheets was cross-linked by agar, which formed a porous scaffold structure.

Linghu et al. utilized Mg-Al-layered double hydroxide (LDH) and its supported GO composites (LDH/GO) to uptake U(VI) (Fig. 4.23B) [174].

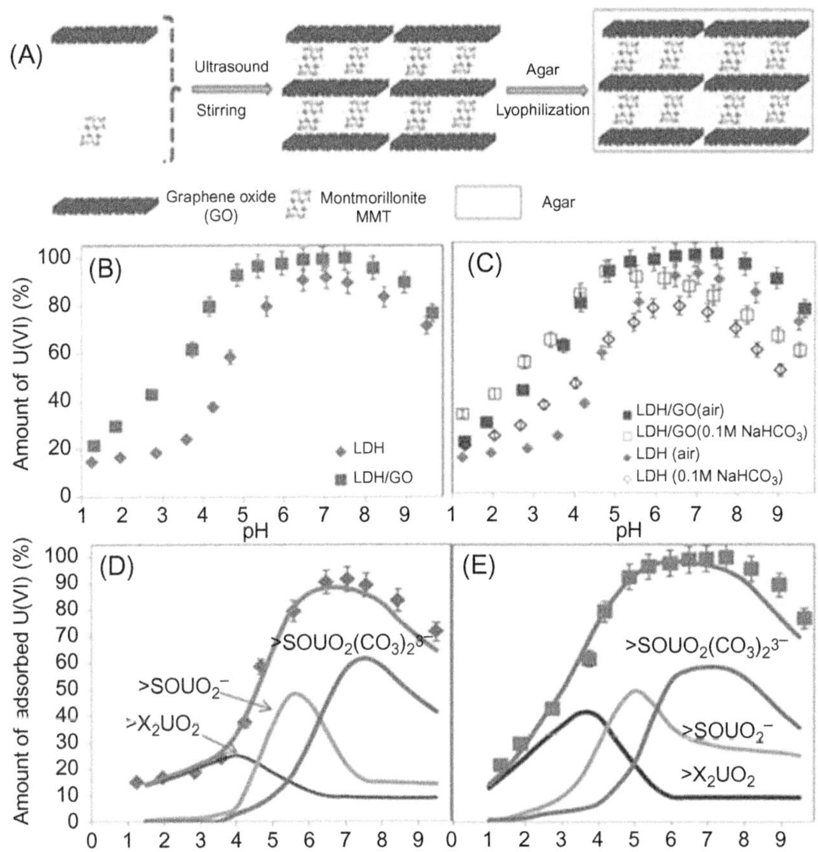

FIG. 4.23 (A) Principle diagram of synthesis process of agar-MMT-GO composites [173], effect of pH (B) and carbonate concentration (C) on U(VI) removal on LDH and LDH/GO composites, Surface complexation modeling of U(VI) on LDH (D) and LDH/GO (E) under different pH conditions: $C_0 = 10\,mg/L$, $m/V = 1.5\,g/L$, $I = 0.01\,mol/L$ $NaClO_4$, $T = 293$ K [174]. *(Data from W.C. Cheng, C.C. Ding, X.Q. Nie, T. Duan, R.R. Ding, Fabrication of 3D macroscopic graphene oxide composites supported by montmorillonite for efficient U(VI) wastewater purification, ACS Sustain. Chem. Eng. 5 (2017) 5503–5511; W.S. Linghu, H. Yang, Y.X. Sun, G.D. Sheng, Y.Y. Huang, One-pot synthesis of LDH/GO composites as highly effective adsorbents for decontamination of U(VI), ACS Sustain. Chem. Eng. 5 (2017) 5608–5616. Copyright 2017, American Chemical Society.)*

The uptake of U(VI) on LDH was gradually increased at pH < 6.5 and then maintained high-level uptake at pH 6.5–7.5, whereas it distinctly declined at pH > 8.0. Compared to LDH, the LDH/GO composite showed enhanced uptake performance toward U(VI) in the whole pH, suggesting that GO improved the uptake of U(VI). The influence of carbonate concentration on the uptake performance of LDH and the LDH/GO composite toward U(VI) is shown in Fig. 4.23C. Compared to the absence of carbonate, the promoted and inhibited uptake of U(VI) was observed in the absence of carbonate at pH < 5.0 and pH > 7.0, respectively. Linghu et al. explained that the electrostatic attraction between the negative charge of the carbonato-uranyl species and the positive charge of the LDH and LDH/GO composites caused the inhibited uptake performance of the LDH and LDH/GO composites toward U(VI) at pH > 7.0. The surface complexation and/or surface coprecipitation reaction of U(VI) on LDH and LDH/GO composites could be assigned to the high-level adsorption at pH 6.0–7.0. Further, they used a double-layer model to study the surface complexation modeling of U(VI) on the LDH and LDH/GO composites (Fig. 4.23D and E). Table 4.7 lists the optimized parameters of surface complexation modeling. Fig. 4.23D and E show that the adsorption data of the LDH and LDH/GO composites toward U(VI) can be well simulated by the double-layer model, which revealed that the dominated sites were cation exchange sites (X_2UO_2) at low pH condition and two inner-sphere complexation sites ($SOUO_2^+$ and $SOUO_2(CO_3)_2^{3-}$ species) at neutral and alkaline conditions. The results of surface complexation modeling implied that cation exchange and inner-sphere surface complexation mainly dominated the adsorption mechanism of U(VI) on the LDH and LDH/GO composite at pH <4.0 and >5.0, respectively.

In recent years, photocatalysis has attracted extensive attention for the reduction of heavy metals and radionuclides, such as Cr(VI) and U(VI), to less harmful oxidation states. Therefore, the photocatalysis technique has great potential application in environmental purification and solar energy conversion.

TABLE 4.7 Optimized parameters of U(VI) adsorption on LDH/GO composites by surface complexation modeling [174]

Reactions	Log *K*
$2{>}XNa + UO_2^{2+} = ({>}X)_2UO_2 + 2Na^+$	3.2
${>}SOH + UO_2^{2+} => SOUO_2^+ + H^+$	−2.7
${>}SOH + 2H_2CO_3 + UO_2^{2+} => SOUO_2(CO_3)_2^{3-} + 5H^+$	−18.9

Data from W.S. Linghu, H. Yang, Y.X. Sun, G.D. Sheng, Y.Y. Huang, One-pot synthesis of LDH/GO composites as highly effective adsorbents for decontamination of U(VI), ACS Sustain. Chem. Eng. 5 (2017) 5608–5616. Copyright 2017, American Chemical Society.

Guo et al. evaluated the photocatalytic reduction efficiencies of the Fe_2O_3-GO composite toward U(VI) [176]. The photocatalytic efficiencies of the Fe_2O_3-GO composite were evaluated by the reduction of U(VI). They suggested that the promoted removal efficiency of U(VI) on the Fe_2O_3-GO composite was due to the improved removal properties of U(VI) on GO, which was beneficial to the photocatalytic removal of U(VI). They also proposed a mechanism to explain the enhanced photocatalytic reduction activity of the Fe_2O_3-GO composite for U(VI) under visible light irradiation. The first path for the transfer of photoinduced carriers can be described in Eqs. (4.12), (4.13). Visible light irradiation induces intramolecular electron transfer from Fe_2O_3 to U(IV), leading to the reduction of U(VI). The redox recycling between Fe^{3+} and Fe^{2+} on the surface of the Fe_2O_3-GO composite was the rate controlling step for U(VI) reduction (Eqs. 4.14, 4.15).

$$Fe_2O_3 + h\nu \rightarrow h^+ + e^- \tag{4.12}$$

$$2UO_2{}^{2+} + 2e^- \rightarrow UO_2 \tag{4.13}$$

$$Fe^{3+} + 2e^- \rightarrow Fe^{2+} \tag{4.14}$$

$$2Fe^{2+} + UO_2{}^{2+} \rightarrow 2Fe^{3+} + UO_2 \tag{4.15}$$

4.5.1.2 Eu(III) adsorption

As is well known, GO has some drawbacks when it is applied to the disposal of large-scale wastewaters. The high cost and agglomeration after drying are the two main problems of GO. Moreover, the agglomerates of GO after drying can reduce its specific surface area and its hydrophilicity, which affects the sorption efficacy. In order to overcome these problems, Wang et al. fabricated a hierarchical structure of a GO@MgSi composite through synthesizing the talc-phase $Mg_3Si_4O_9(OH)_{10}$ on the surface of GO nanosheets [177]. The GO@MgSi composite showed efficient preconcentration of Eu(III) with high adsorption capacity (344.82 mg/L). In addition, the spectroscopy techniques (e.g., IR, XPS) confirmed the structural transformations in the Si-O tetrahedral and Mg-OH octahedral sheet in the talc-phase $Mg_3Si_4O_9(OH)_{10}$, indicating that the talc-phase $Mg_3Si_4O_9(OH)_{10}$ lamella played an important role during the adsorption process. In addition, Li et al. prepared titanium phosphate (TiP)-modified GO (GTiP) composites to remove Eu(III) [178]. GTiP exhibited excellent sorption capacity (64.33 mg/g) for Eu(III). This was assigned to the synergistic effect of the composite material, where TiP promoted the mechanical and hydrophilicity properties of GTiP and GO increased the specific surface area of GTiP. The strategy of Eu^{3+} preconcentration or desorption on GTiP nanocomposites was also studied. They found that the GTiP could be recycled and reused with 1 M HCl solution.

The influence of humic acid (HA) or fulvic acid (FA) on Eu(III) adsorption on magnetite-decorated GOs was investigated by Li et al. [179].

They observed that a positive effect of HA or FA on Eu(III) adsorption was observed at low pH values, whereas a negative effect of HA or FA on Eu(III) adsorption was found at high pH values. The results could be attributed to surface-adsorbed HA or FA, altering surface charges, or changing available sites of magnetite-decorated GOs for Eu(III) uptake. Li et al. studied the sorption of Eu(III) on magnetic GOs by using the batch technique and surface complexation modeling [180]. The results showed that the ionic strength had no effect on Eu(III) sorption, indicating that inner-sphere surface complexation dominated the sorption of Eu(III) on magnetic GOs over a wide range of pH. From the batch sorption, they deduced that the monodentate inner-sphere complexation dominated at pH < 4.0, and the multidentate inner-sphere surface complexation was governed at pH > 5.0. Fig. 4.24 shows the fitted results of the sorption of Eu(III) on GOs and magnetic GOs revealed that the sorption can be well simulated by the DLM model with XNa and SOH sites. The maximum sorption capacity of Eu(III) on magnetic GOs was 70.15 mg/g (pH 4.5), implying the application of magnetic GOs for the immobilization and enrichment of radionuclides from environmental cleanup.

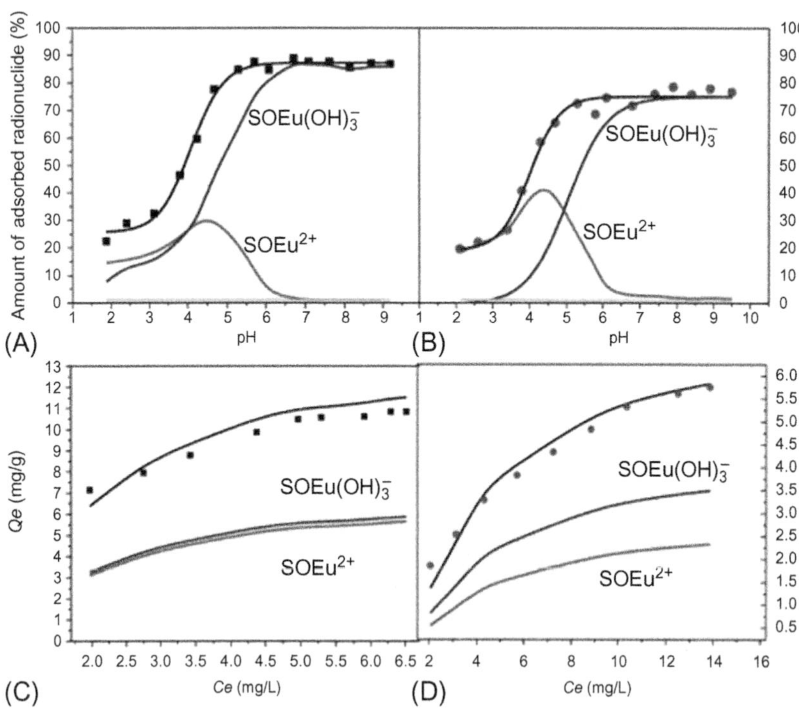

FIG. 4.24 The surface complexation modeling of adsorption of Eu(III) on pH-edge (A and B) and adsorption isotherms (C and D) on GOs (A and C) and magnetic GOs (B and D) [180]. *(Data from D.M. Li, B. Zhang, F.Q. Xuan, The sorption of Eu(III) from aqueous solutions by magnetic graphene oxides: a combined experimental and modeling studies, J. Mol. Liq. 211 (2015) 203–209. Copyright 2015, Elsevier.)*

4.5.1.3 Cs(I) and Sr(II) adsorption

Tayyebi et al. used magnetite nanoparticles loaded on the surface of a GO-obtained M-GO hybrid for the adsorption of Sr^{2+} ions [181]. The adsorption of Sr(II) ions on the M-GO hybrid is quite sensitive to the variation of ionic strength, which might be due to the adsorption of Sr(II) through an ion-exchange mechanism. For the synthesis of the M-GO hybrid, they proposed a poly condensation mechanism as follows: the first layer Fe(II) and Fe(III) ions were linked to oxygen groups of GO via a covalent dative bond of C—O—Fe, and then the next Fe layers were connected to the preceding ones by repeated formation of the Fe—OH—OH—Fe bridge, followed by immediate dehydratation to the Fe—O—Fe structure (Fig. 4.25A). The O1s peak of GO was deconvoluted to two compositions assigned to C=O (carbonyl and carboxyl) (530eV) and C—O (epoxy and hydroxyl) (532eV) groups (Fig. 4.25B). The O1s spectrum of M-GO could be divided into three components at 528, 529.5, and 531.7eV that corresponded to the Fe_3O_4, the bonding between Fe_3O_4 and the graphene, and the C—O groups, respectively (Fig. 4.25C).

4.5.1.4 Competitive adsorption

Pan et al. fabricated a manganese dioxide-decorated GO (α-GOM_2) composite through the fixation of crystallographic α-MnO_2 on the surface of GO [37]. The GOM composite was applied to the simultaneous removal of thorium/uranium ions from aqueous solutions. Compared with the excellent single adsorption capacities of Th(IV) (163.1mg/g) and U(VI) (77.7mg/g), the lower simultaneous adsorption capacity of Th(IV)/U(VI) on α-GOM_2 was calculated at about 152.1 and 51.3mg/g, respectively (Fig. 4.26A). In Fig. 4.26B, the maximum adsorption capacities of single Th(IV) and single U(VI) on α-MnO_2, GO, and α-GOM_2 followed the order α-GOM_2 > GO > α-MnO_2, which might be due to the high specific surface area of α-GOM_2 (2001 m^2/g). According to the XPS analysis, the radionuclides bonded with the O—H groups of α-GOM_2 to form the surface complexes C—O—M (M = Th or U) of the α-GOM_2-Th or α-GOM_2-U (Fig. 4.26C). Four types of molecular interactions (coordination, electrostatic interaction, cation-π interaction, and Lewis acid-base interaction) were considered as the probable interaction mechanism between Th(IV) or U(VI) ions and the surface oxygen groups of α-GOM_2.

4.5.2 Adsorption of radionuclides on organic graphene oxide-based composites

4.5.2.1 U(VI) and Th(IV) adsorption

Polymers have attracted increasing attention because of their facile synthesis, relatively low cost, nontoxicity, thermal stability, and unique redox and

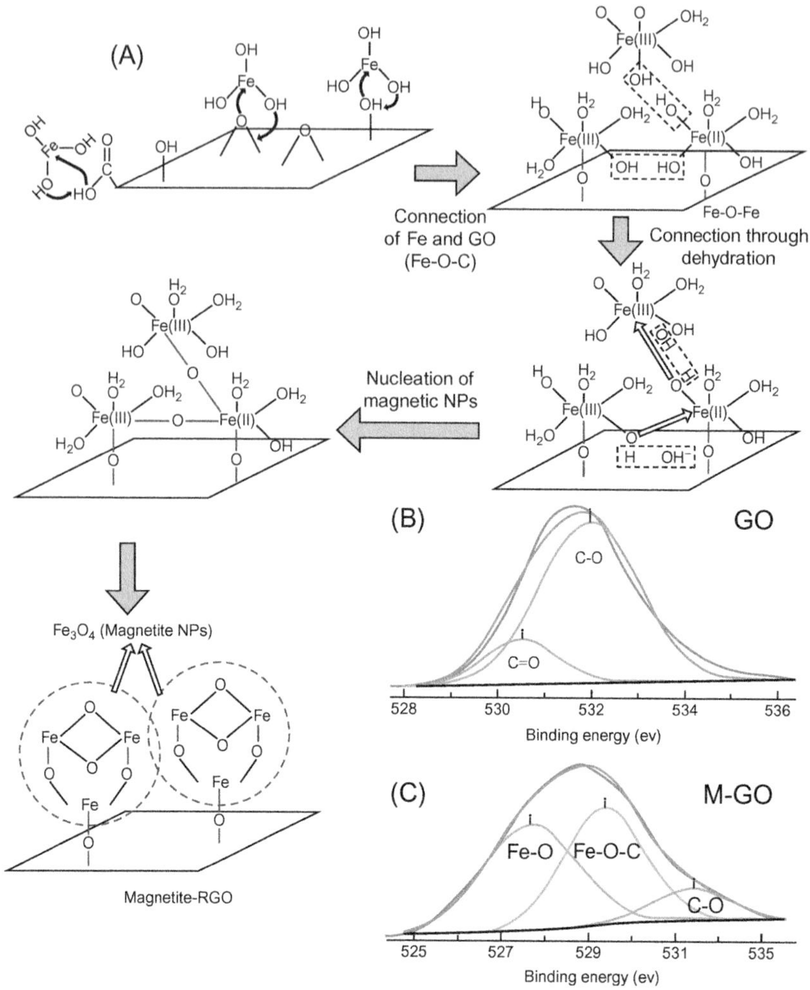

FIG. 4.25 (a) Formation mechanism of M-GO, XPS O 1s high resolution spectra of (b) GO and (c) M-GO [181]. *(Data from A. Tayyebi, M. Outokesh, S. Moradi and A. Doram, Synthesis and characterization of ultrasound assisted "graphene oxide-magnetite" hybrid, and investigation of its adsorption properties for Sr(II) and Co(II) ions, Appl. Surf. Sci. 353 (2015) 350–362. Copyright 2015, Elsevier.)*

electronic properties. The interactions of U(VI) with various GO organic composites were extensively studied [3, 4, 39, 182]. Hu et al. prepared a graphene oxide/polypyrrole (GO/PPy) composite and used it to remove U(VI) from aquatic systems, which showed a rapid sorption rate and high sorption capacity toward U(VI) [4]. The fast sorption rate indicated that the main mechanism for the sorption of U(VI) on GO/PPy composites was assigned to the strong chemical sorption or surface complexation. The sorption percentage of U(VI) increased with the rise of the content of the GO/PPy composite, which

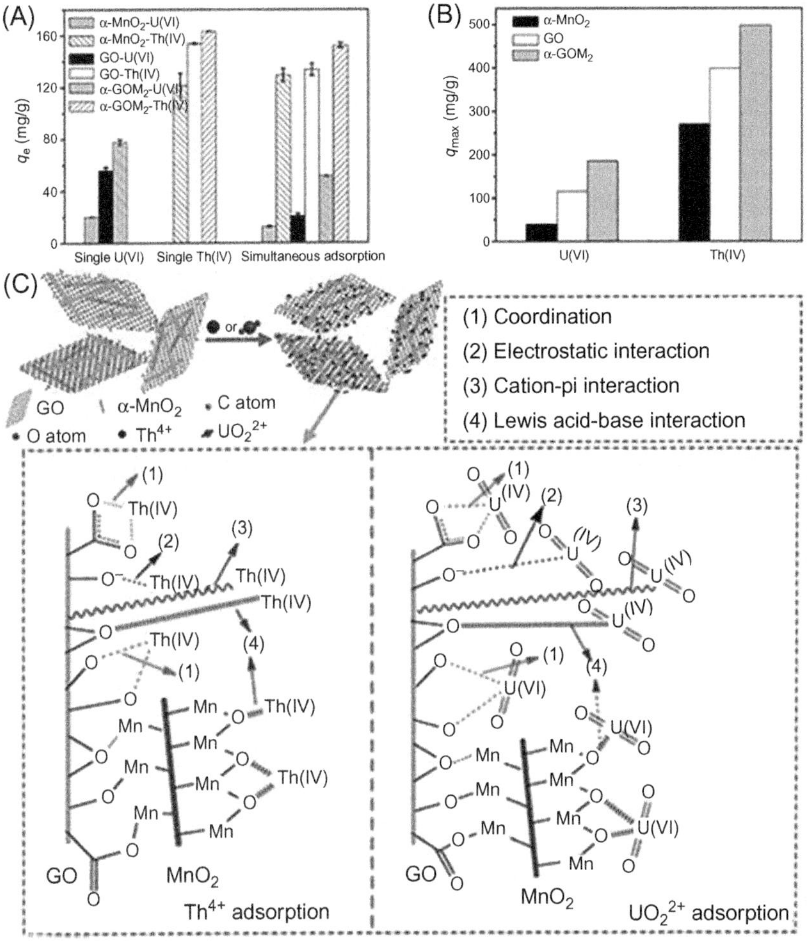

FIG. 4.26 (A) Individual and simultaneous adsorption of Th(IV)/U(VI) on α-MnO_2, GO, and α-GOM_2 under the same conditions; (B) the maximum adsorption capacity of single Th(IV) and single U(VI) ion on α-MnO_2, GO, and α-GOM_2. (A) $C[Th(IV)]_{initial} = 40$ mg/L, $C[U(VI)]_{initial} = 20$ mg/L, $m/V = 0.25$ g/L, (b) $m/V = 0.1$ g/L, pH = 3.8 ± 0.1, $T = 298$ K, contact time = 20 min, (C) schematic diagram of possible interaction between Th(IV) or U(VI) and α-GOM_2 [37]. *(Data from N. Pan, L. Li, J. Ding, S.K. Li, R.B. Wang, Y.D. Jin, X.K. Wang, C.Q. Xia, Preparation of graphene oxide-manganese dioxide for highly efficient adsorption and separation of Th(IV)/U(VI), J. Hazard. Mater. 309 (2016) 107–115. Copyright 2016, Elsevier.)*

was attributed to the fact that more sorption sites were available with high solid content. Further, the solution pH has a great impact on the U(VI) species and the protonation degree of functional groups on the surfaces of GO-based materials. In addition, the sorption of U(VI) on the GO/PPy composite was dependent on pH, which was due to the aqueous U(VI) species and the surface properties of GO/PPy composites were different under various pH conditions. At acidic and neutral pH conditions, the species of U(VI) are positively charged while the species of U(VI) are negatively charged at alkaline conditions. The increase of the sorption percentage of U(VI) at pH < 6.5 was assigned to more surface functional groups of GO/PPy that are deprotonated to provide more available sorption sites for U(VI) ions, whereas the decrease of sorption in the pH range of 6.5–11.0 was due to the electrostatic repulsion between negatively charged U(VI) species and negatively charged surfaces of GO/PPy composites. Additionally, the sorption of U(VI) was also strongly dependent on ionic strength and the sorption percentage of U(VI) decreased with increasing ionic strength, indicating the existence of an outer-sphere surface complexation mechanism. Shao et al. synthesized a poly(amidoxime)-reduced GO (PAO/RGO) composite by using a situ polymerization method following an amidoximation treatment, and PAO/RGO exhibited a high adsorption performance toward U(VI) (872 mg/g) [183]. Polyaniline decorated on a GO (PANI@GO) composite was fabricated by chemical oxidation and showed a maximum sorption capacity (1.03 mmol/g) at pH 3.0 and $T=298$ K for U(VI) [3].

Yang et al. synthesized GO-COOH/UiO-66 composites by using a rapid and effective covalent cross-linking method [184]. The results of the experiment indicated that the introduction of GO-COOH provided more active adsorption sites and achieved high adsorption capacity (1002 mg/g) at pH = 8. They concluded that the adsorption mechanisms included the coordination or π-π interaction between U(VI) and functional groups (—COOH and —OH) on the surface of GO-COOH/UiO-66 composites, and little ion-exchange and chelation between U(VI) and COOH groups.

The FT-IR and XPS spectroscopy techniques have been used to investigate the interaction mechanisms between radionuclides and GO-based materials. The adsorption interaction mechanisms between U(VI) and PANI/GO were determined using FT-IR and XPS spectroscopy techniques (Fig. 4.27) [185]. In Fig. 4.27A, the peaks at ~946 cm^{-1} were assigned to the vibration of O=U(VI)=O, and the U4f 5/2 and U4f 7/2 peaks at 393.0 and 382.1 eV belonged to the U(VI) (Fig. 4.27B). Based on FT-IR and XPS characterizations, they reported that the carboxyl (—COOH) groups were effective active sites for the adsorption of U(VI) ions, and amino (—NH_2) groups could also coordinate with U(VI) ions through electrostatic and hydrogen bond interactions.

To investigate the interaction mechanisms between U(VI) and graphene oxide-chitosan aerogel (GO-CS), Huang et al. utilized EXAFS spectroscopy to analyze the local atomic structure of adsorbed U(VI) on the surface of GO-CS (Fig. 4.27C and D) [182]. As can be seen from Fig. 4.27C, with increasing pH and U(VI) concentration, a slight phase shift toward a low k value and a merge of 6.8 and 7.8 $Å^{-1}$ features can be observed of the GO-CS spectra, which possibly indicated a slight coordination environment change of the surface-adsorbed U(VI). Curve fitting results carried out in

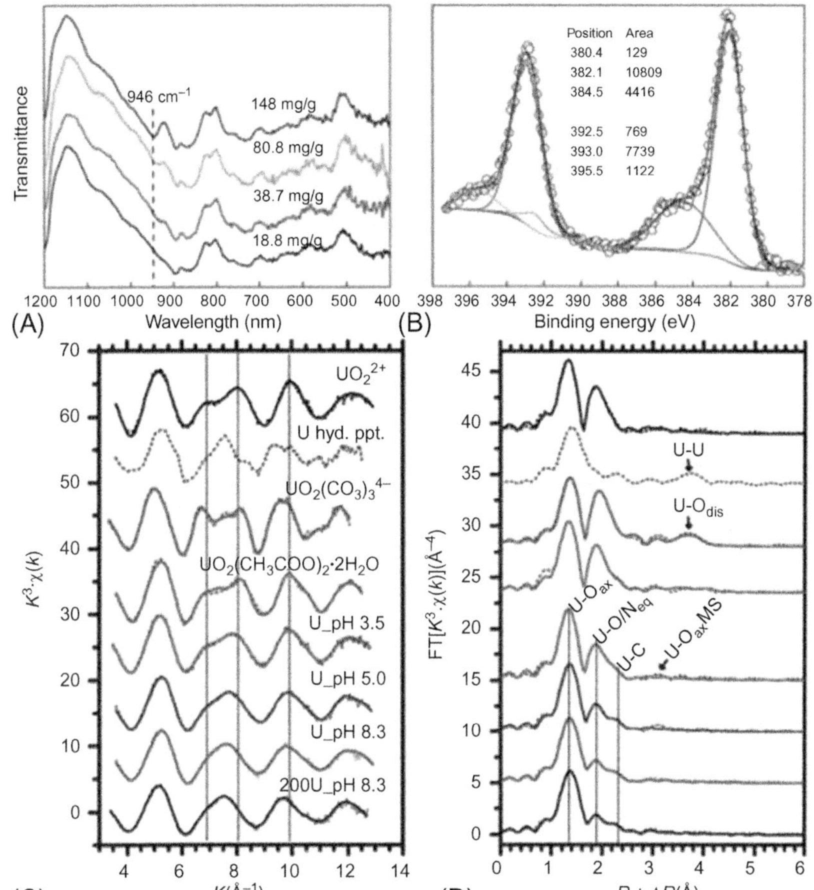

FIG. 4.27 (A) FT-IR spectra of 1.0PANI/GO after adsorbed different amount of U(VI), (B) XPS U4f 5/2 and U4f 7/2 peaks of 1.0PANI/GO after adsorbed U(VI) [185], uranium L_{III}-edge k^3-weighted EXAFS spectra (C) and the corresponding FTs (D) of U_pH 3.5, U_pH 5.0, U_pH 8.3, and 200U_pH 8.3 and the references UO_2^{2+} (aq), uranium hydrolysis precipitate (s), $UO_2(CO_3)_3^{4-}$ (aq), and $UO_2(CH_3COO)_2{\cdot}2H_2O$ (s). *Dotted* and *solid lines* represent experimental and fitting results, respectively [182]. *(Data from Z.W. Huang, Z.J. Li, L.R. Zheng, L.M. Zhou, Z.F. Chai, X.L. Wang, W.Q. Shi, Interaction mechanism of uranium(VI) with three-dimensional graphene oxide-chitosan composite: insights from batch experiments, IR, XPS, and EXAFS spectroscopy, Chem. Eng. J. 328 (2017) 1066–1074; D.D. Shao, G.S. Hou, J.X. Li, T. Wen, X.M. Ren, X.K. Wang, PANI/GO as a super adsorbent for the selective adsorption of uranium(VI), Chem. Eng. J. 255 (2014) 604–612. Copyright 2017, 2014, Elsevier.)*

the $R + \Delta R$ range from 1.1 to 3.4 Å are shown in Fig. 4.27D, and the simulated coordination parameters are summarized in Table 4.8. Based on the EXAFS analyses, they concluded that the inner-sphere surface complexation mechanism was dominated for the U(VI) adsorption on the GO-CS,

TABLE 4.8 Structural parameters around uranium for the reference and GO-CS samples derived from EXAFS analyses. For the GO-CS samples, FT^{-1} ranges from 3.5 to 12.8 $Å^{-1}$ and FT from 1.1 to 3.4 Å [182]

Sample	Shell	CN	R (Å)	σ^2 ($Å^2$)	ΔE_0 (eV)^a	Rf, Red-χ^2
UO_2^{2+} (aq)	U-Oax	2^b	1.76(3)	0.002(0)	6.4(8)	1.2×10^{-4}
	U-Oeq	5	2.41(1)	0.008(1)		14.5
$UO_2(CO_3)_3^{4-}$ (aq)	U-Oax	2^b	1.80(0)	0.002(0)	5.2(1.0)	1.6×10^{-4}
	U-Oeq	6	2.45(1)	0.006(1)		12.3
	U-C	3	2.90(0)	0.001(1)		
	U-Odis	3	4.14			
$UO_2(CH_3COO)_2{\cdot}2H_2O$(s)	U-Oax	2^b	1.76(1)	0.002(0)	4.5(2.7)	1.3×10^{-4}
	U-Oeq1	3	2.37(1)	0.006(1)		51.9
	U-Oeq2	2	2.47(3)	0.001(2)		
	U-C	1	2.91			
U_pH 3.5	U-Oax	2^b	1.77(1)	0.0026(2)	4.4(1.8)	6.5×10^{-5}
	U-O/Neq1	1.5(3)^c	2.26(3)	0.005(1)^d		10.0
	U-O/Neq2	3.5^c	2.41(2)			
	U-C	1.4(6)	2.90(2)	0.004^b		
U_pH 5.0	U-Oax	2^b	1.78(1)	0.003(0)	5.5(2.4)	1.4×10^{-4}
	U-O/Neq1	1.5(8)^c	2.29(6)	0.009(3)^d		18.8
	U-O/Neq2	3.5^c	2.46(4)			
	U-C	1.2(6)	2.92(3)	0.004^b		

U_pH 8.3	U-Oax	2^b	1.78(1)	0.003(0)	3.5(2.4)	1.1×10^{-4}
	U-O/Neq1	$1.9(2)^c$	2.26(3)	$0.007(2)^d$		19.2
	U-O/Neq2	3.1^c	2.45(3)			
	U-C	1.1(6)	2.92(3)	0.004^b		
200U_pH 8.3	U-Oax	2^b	1.79(1)	0.003(0)	3.0(2.2)	1.2×10^{-4}
	U-O/Neq1	$1.8(2)^c$	2.28(3)	$0.008(2)^d$		7.0
	U-O/Neq2	3.2^c	2.48(3)			
	U-C	0.8(5)	2.90(4)	0.004^b		

Note: CN: coordination number; R: bond distance; σ^2: Debye-Waller factor; Rf, Red-χ^2: quality of fit as defined in the literature [186]; Number in parentheses presents the error in the last digit.
[a] *ΔE_0 was allowed to vary but linked for all paths.*
[b] *Value fixed during the fit.*
[c] *The CN sum of equatorial oxygen/nitrogen atoms was constrained to 5.*
[d] *σ^2 was allowed to vary but linked for two equatcrial O/N subshells.*
Data from Z.W. Huang, Z.J. Li, L.R. Zheng, L.M. Zhou, Z.F. Chai, X.L. Wang, W.Q. Shi, Interaction mechanism of uranium(VI) with three-dimensional graphene oxide-chitosan composite: insights from batch experiments, IR, XPS, and EXAFS spectroscopy, Chem. Eng. J. 328 (2017) 1066–1074. Copyright 2017, Elsevier.

and they proposed that the —COO⁻, —OH, and —NH_2 groups of the GO-CS were capable of complexing the uranium.

Surface complexation modeling has enabled predictions of the effect of solution chemistry (e.g., pH and ionic strength) on the binding of inorganic aqueous solutes to solid surfaces. Therefore, Sun et al. used surface complexation modeling to simulate the pH-dependent sorption of Th(IV) on amidoxime/GO composites with the aid of Visual MINTEQ 3.0 [187]. The two surface protonation and deprotonation reactions are expressed in (4.16), (4.17):

$$SOH + H^+ = SOH_2{}^+ \tag{4.16}$$

$$SOH = SO^- + H^+ \tag{4.17}$$

where SOH represents the amphoteric hydroxide functional groups of amidoxime/graphene oxide composites. Two surface complexation reactions and a surface precipitate were described by Eqs. (4.18)–(4.20):

$$SOH + Th^{4+} = SOTh^{3+} + H^+ \tag{4.18}$$

$$2SOH + Th^{4+} + 3H_2O = (SO)_2Th(OH)_3{}^- + 5H^+ \tag{4.19}$$

$$Th^{4+} + 4H_2O = Th(OH)_4(s) + 4H^+ \tag{4.20}$$

They observed that the $SOTh^{3+}$, $(SO)_2Th(OH)_3{}^-$, and $Th(OH)_4(s)$ species were the main sorption species of Th(IV) on amidoxime/GO composites at pH <4.5, 4.5–6.0, and >6.0, respectively. According to the above fitting results, the adsorption mechanism of Th(IV) on amidoxime/GO composites was mainly due to inner-sphere surface complexation at pH < 6.0, and the surface precipitate (i.e., $Th(OH)_4(s)$) dominated the Th(IV) adsorption on amidoxime/GO composites at pH > 6.0. The GO-PDA composite was fabricated through functionalization of GO with phenanthroline diamide, which exhibited excellent maximum sorption capacities of U(VI) (718 mg/g) and Th(IV) (703 mg/g) [188]. The adsorption isotherms were carried out at a constant pH by varying the initial concentration of U(VI) and Th(IV) to investigate the adsorption mechanisms. The strong complexation between U(VI) or Th(IV) and PDA groups on the surface of GO-PDA might play an important role in the high adsorption of U(VI) or Th(IV) on GO-PDA.

A Schiff base/quaternary ammonium salt bifunctional GO (GO-S) composite was prepared by covalently grafting a novel primary amine on GO [189]. The GO-S composite was utilized for the efficient removal of Th(IV)/U(VI), and the results exhibited that the imine group of the Schiff base moiety on GO-S could form strong complexes with U(VI), while the hydroxyl group and imine on GO-S could coordinate with Th(IV). The adsorption isotherm data of GO-S toward Th(IV) or U(VI) were shown in Fig. 4.28A, and the fitting results of adsorption isotherms data were given in Fig. 4.28B–D. The calculated

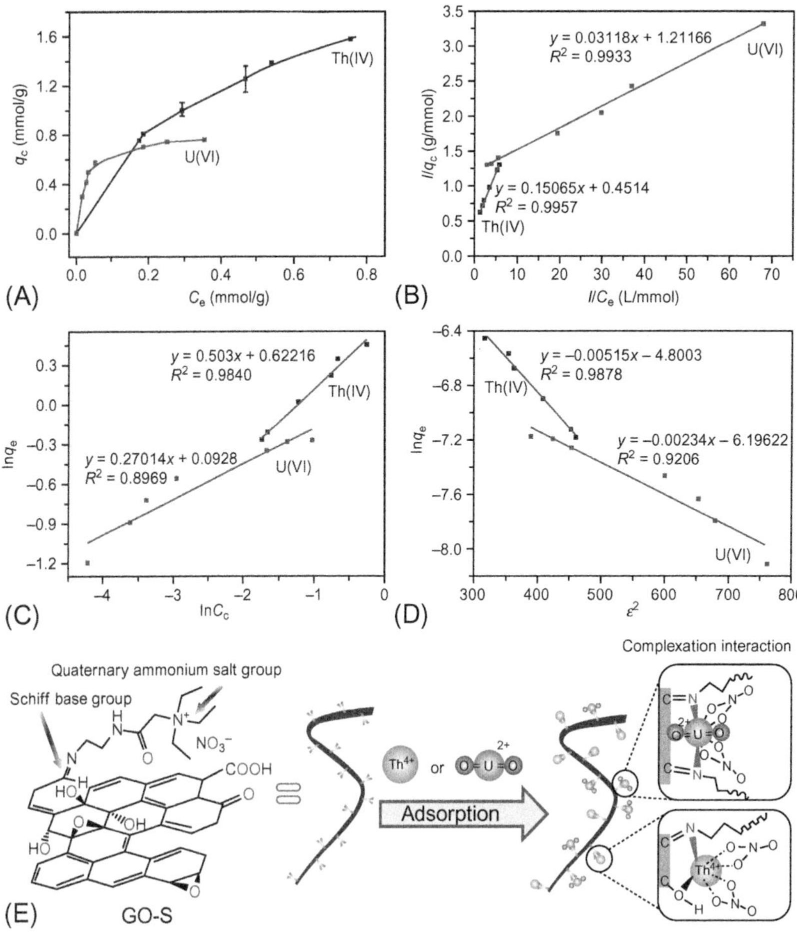

FIG. 4.28 Adsorption isotherms of Th(IV) or U(VI) by GO-S (A) and fitting results using Langmuir (B), Freundlich (C) and D-R models (D), pH = 3.0 ± 0.1, m/V = 0.25 g/L, t = 2 h, T = 298 K, schematic of complexation of Th(IV) or U(VI) with GO-S (E) [189]. *(Data from N. Pan, L. Li, J. Ding, R.B. Wang, Y.D. Jin, C.Q. Xia, A Schiff base/quaternary ammonium salt bifunctional graphene oxide as an efficient adsorbent for removal of Th(IV)/U(VI), J. Colloid Interface Sci. 508 (2017) 303–312. Copyright 2017, Elsevier.)*

results revealed that the maximum adsorption capacity of GO-S toward Th(IV) and U(VI) was about 2.22 and 0.83 mmol/g, respectively, indicating that the GO-S could be a potentially valuable and practical adsorbent for the treatment of contaminants from aquatic systems. Based on the FT-IR and XPS analysis, the interaction mechanism of Th(IV) or U(VI) on GO-S is mainly assigned to the coordination of Th(IV) or U(VI) with hydroxyl and Schiff base groups (Fig. 4.28E).

4.5.2.2 Cs(I) and Sr(II) adsorption

A magnetic polyaniline/GO (PANI/GO) composite was explored to study the interaction of Sr(II) by batch, surface complexation modeling, and EXAFS techniques [190]. The results revealed that the removal of Sr(II) on a magnetic PANI/GO composite was dominated by the outer-sphere surface complexation at pH < 4.0, whereas the removal of Sr(II) at pH > 5.0 was mainly controlled by the inner-sphere complexation attributed to the occurrence of the Sr-C shell. Additionally, the outer-sphere complex ($SOHSr^{2+}$ species) and two inner-sphere complexes (i.e., $(SO)_2Sr(OH)^-$ and $SOSr^+$ species) were dominant surface complexation species, according to the analysis of surface complexation modeling (Fig. 4.29A). These species can be utilized to fit the adsorption isotherms of Sr(II), and the $SOHSr^{2+}$ species was the main uptake species of the adsorption isotherm (Fig. 4.29B). Fig. 4.29C and D showed the Sr K-edge EXAFS spectra and the corresponding Fourier transformed spectra. The optimized parameters of Sr(II) reference and Sr(II) removal under different pH conditions were listed in Table 4.9. The fitted results showed that the removal of Sr(II) at pH 3.0 was similar to those of the aqueous Sr^{2+} species and could not be fitted with Sr-C and/or the Sr-Sr shell, implying that the outer-sphere complexation governed the removal of Sr(II) at pH 3.0. In addition, the existence of an Sr-C shell of the sample at pH 7.0 suggested that Sr was loaded on the magnetic PANI/GO composite as a monodentate and mononuclear inner-sphere surface complex at pH 7.0. These results are significant for the potential utilization of magnetic PANI/GO as a valuable adsorbent for the removal of radionuclides in environmental remediation.

4.5.2.3 Selective adsorption

Amidoxime-functionalized GO nanoribbons (AOGONRs) were synthesized by a chemical grafting technology [191]. The AOGONR composite was used to remove U(VI) and showed an excellent adsorption ability for U(VI), attributed to the strong chelation of U(VI) to amidoxime groups. The simulated adsorption experiment of 12 main sensible nuclides on GO nanoribbons (GONRs) and AOGONRs was conducted in order to evaluate the sorption selectivity of both the GONRs and AOGONRs.

A distinct higher adsorption capacity of U(VI) on AOGONRs was obviously noticed. The total adsorption capacity of all cations on AOGONRs reached 2.27 mmol/g, which was much higher than that of the GONRs (1.46 mmol/g). Moreover, AOGONRs had distinctly preferable affinity toward U(VI) ions compared to other cations, which could be clarified by distribution coefficients (K_d). A high K_d value of AOGONRs was up to nearly 6000 mL/g for U(VI), which was higher than that of other coexistent ions (<500 mL/g). The results indicated that the AOGONR composite showed a satisfactory selectivity toward

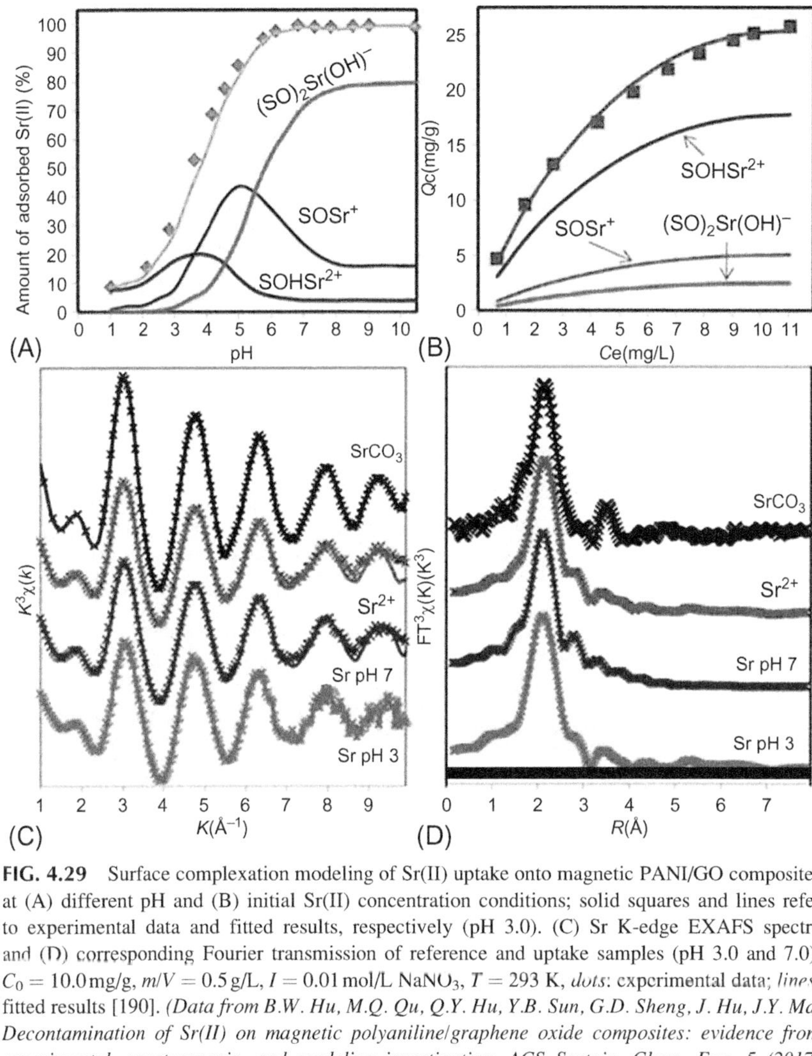

FIG. 4.29 Surface complexation modeling of Sr(II) uptake onto magnetic PANI/GO composites at (A) different pH and (B) initial Sr(II) concentration conditions; solid squares and lines refer to experimental data and fitted results, respectively (pH 3.0). (C) Sr K-edge EXAFS spectra and (D) corresponding Fourier transmission of reference and uptake samples (pH 3.0 and 7.0). $C_0 = 10.0\,mg/g$, $m/V = 0.5\,g/L$, $I = 0.01\,mol/L$ $NaNO_3$, $T = 293$ K, *dots*: experimental data; *lines*: fitted results [190]. *(Data from B.W. Hu, M.Q. Qu, Q.Y. Hu, Y.B. Sun, G.D. Sheng, J. Hu, J.Y. Ma, Decontamination of Sr(II) on magnetic polyaniline/graphene oxide composites: evidence from experimental, spectroscopic, and modeling investigation, ACS Sustain. Chem. Eng. 5 (2017) 6924–6931. Copyright 2017, American Chemical Society.)*

U(VI) ions over a range of competing metal ions. Therefore, AOGONRs had either strong affinity or selectivity toward U(VI) ions, even in a multiion system. Besides, the repeated sorption-desorption experiments showed that the AOGONR composite could be efficiently used for regeneration and reuse for U(VI) adsorption, indicating that the AOGONR composite may be a potential and suitable candidate for the separation of U(VI) from various uranium-containing water samples.

TABLE 4.9 Structural parameters obtained from Sr K-edge EXAFS data analysis for reference and uptake samples [190]

Sample	Shell	*R* (Å)	CN	σ^2 (Å^2)
Sr^{2+}_{aq}	Sr—O	2.62	7.5	0.007
$SrCO_3$	Sr—O	2.57	6.8	0.0075
	Sr—C	3.14	2.2	0.0089
	Sr—Sr	3.95	5.2	0.0425
mPANI/GO_pH 3.0	Sr—O	2.59	7.5	0.0065
mPANI/GO_pH 7.0	Sr—O	2.58	7.1	0.0052
	Sr—C	3.12	2.1	0.0094

Data from B.W. Hu, M.Q. Qu, Q.Y. Hu, Y.B. Sun, G.D. Sheng, J. Hu, J.Y. Ma, Decontamination of Sr(II) on magnetic polyaniline/graphene oxide composites: evidence from experimental, spectroscopic, and modeling investigation, ACS Sustain. Chem. Eng. 5 (2017) 6924–6931.

4.5.3 Adsorption of radionuclides on reduced graphene oxide-based composites

The GO-based inverse spinel nickel ferrite (GONF) and rGO-based inverse spinel nickel ferrite (rGONF) nanocomposite were prepared by coprecipitation of GO with nickel and iron salts in one pot, and this was used to remove U(VI) and Th(IV) [192]. Lingamdinne et al. studied the effect of time on the adsorption of U(VI) and Th(IV) on GONF and rGONF at pH 3.5 and 293 ± 2 K. The initial very fast adsorption of U(VI) and Th(IV) that occurred was attributed to more freely reactive sites on the surface of GONF and rGONF at initial adsorption. The maximum adsorption capacity of U(VI) and Th(IV) on rGONF and GONF was calculated to be (200.00 ± 0.26, 135.13 ± 0.46 mg/g) and (126.58 ± 0.30, 88.49 ± 0.27 mg/g), respectively. They concluded that rGONF was an efficient, promising, and easily recoverable adsorbent for the removal of radioactive ions from wastewater.

In past years, the combination of magnetite nanoparticles with graphene-based materials has been investigated and developed as a novel adsorbent with the advantages of easy separation, high adsorption, and reclamation [193, 194]. Besides, as the magnetite composed of Fe(II) and Fe(III), magnetite was demonstrated to reduce U(VI) to U(IV) by the oxidation of Fe(II) [96, 195]. Many reports stated that the interaction of U(VI) on adsorbents included both adsorption and reduction reactions [171, 196]. Cheng et al. reported the impact of the reaction time on the removal and reduction of U(VI) on magnetite-reduced GO (M-rGO) composites at pH 5.0 [196]. The removal kinetics of U(VI) on M1-rGO, M5-rGO, and M10-rGO (the number representative the mass ratio of Fe_3O_4 to rGO) were fast and could achieve equilibrium during 1 h. The highest removal percentage of U(VI) was observed on M10-rGO (79%), which can be assigned to the increased reduction of U(VI) by Fe_3O_4 nanoparticles. They observed that the reduction of U(VI) to U(IV) by M-rGO enhanced with increasing content of Fe_3O_4 (14% for M10-rGO at 30 min). Further, many reports demonstrated that the reduction of U(VI) to U(IV) by Fe_3O_4 accompanied with the oxidation of ferrous ions in the Fe_3O_4 crystal structure [197, 198].

The kinetics results showed that the removal rates of U(VI) on M-rGO were faster than the reduction rates, which might be due to the fact that the U(VI) adsorption on the surface of magnetite nanoparticles previously occurred with regard to the U(VI) reduction by structural Fe(II) [199]. Additionally, they used a kinetic model to simulate the kinetics experimental data, which was adopted from Yan et al. [200]. The model hypothesized that the initial sorption (Eq. 4.21) and subsequent reduction (Eq. 4.22) reactions constitute the removal of U(VI) on M-rGO.

$$\equiv S + U^{(VI)}O_2{}^{2+} \underset{k_{-1}}{\overset{k_1}{\rightleftharpoons}} \equiv SU^{(VI)}O_2{}^{2+} \tag{4.21}$$

$$\equiv SU^{(VI)}O_2{}^{2+} \xrightarrow{k_2} \equiv SU^{(IV)}O_2(s) \tag{4.22}$$

where $\equiv$S represented the species associated with the surfaces of M-rGO, which could be reviewed as a constant during the modeling attributed to the large specific surface area of M-rGO [200]. k_1 and k_{-1} are the sorption (forward) and desorption (backward) rate constants, respectively. k_2 is the reduction rate constant. They also used Eqs. (4.23)–(4.25) to express the relation between aqueous U(VI) depletion, U(VI) sorption, and U(IV) production and time.

$$\frac{d[U(VI)]}{dt} = -k_1\left[U^{(VI)}O_2{}^{2+}\right] + k_{-1}\left[\equiv SU^{(VI)}O_2{}^{2+}\right] \tag{4.23}$$

$$\frac{d\left[\equiv SU^{(VI)}O_2{}^{2+}\right]}{dt} = k_1\left[U^{(VI)}O_2{}^{2+}\right] - (k_{-1} + k_2)\left[\equiv SU^{(VI)}O_2{}^{2+}\right] \tag{4.24}$$

$$\frac{d[U(IV)]}{dt} = k_2\left[\equiv SU^{(VI)}O_2{}^{2+}\right] \tag{4.25}$$

These kinetics experimental data were well fitted by the kinetics model, indicating that the reduction of U(VI) was induced by the sorption of U(VI) on M-rGO. They also found that the value of k_1 increased with the increasing content of Fe_3O_4 on M-rGO, whereas the k_{-1} value decreased. The result indicated that the M-rGO with higher Fe_3O_4 content is more efficient to remove U(VI), which might be due to the fact that the Fe_3O_4 incorporated in the M-rGO increased surface sorption sites and surface reactivity.

The reduction behaviors of U(VI) on magnetic GO or rGO can be determined by spectroscopic techniques. Sun et al. discovered the reduction information of

U(VI) by the nanoscale zerovalent iron (nZVI) and nanoscale zerovalent iron loaded reduced graphene oxide (nZVI/rGO) composite from the XRD patterns and XPS analysis [96]. Fig. 4.30A shows the XRD patterns of U(VI)-reacted nZVI and nZVI/rGO composites after 2, 16, and 60 days (represented as nZVI_2

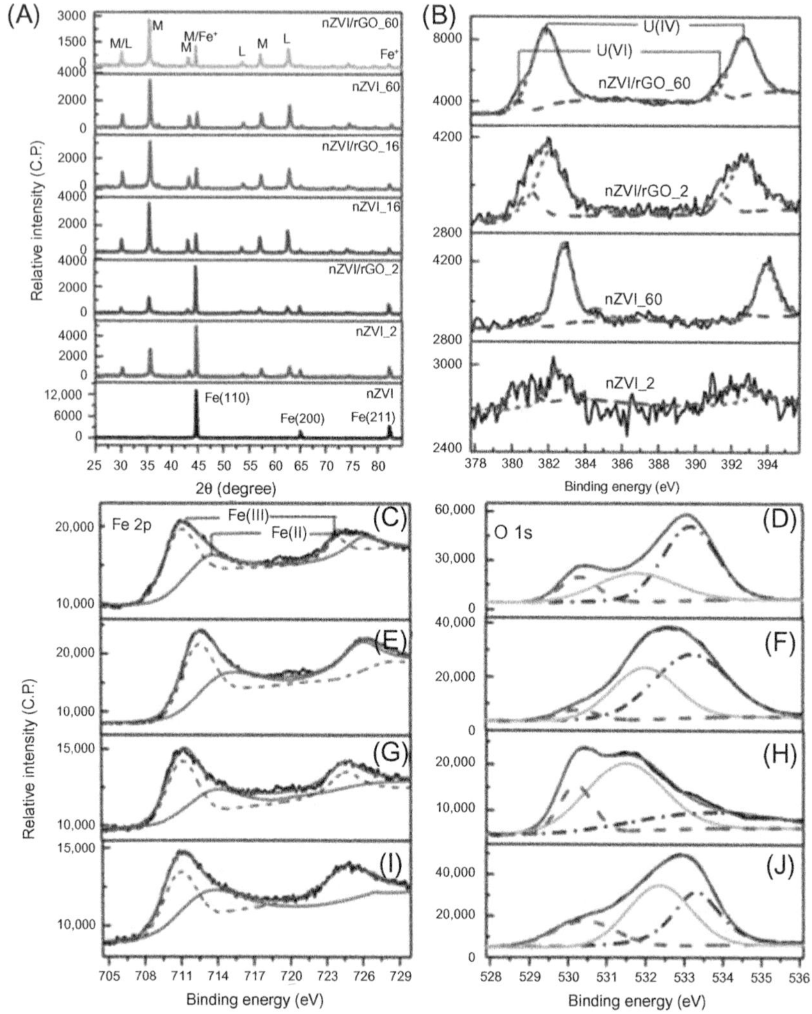

FIG. 4.30 (A) XRD patterns of corrosion products of nZVI and nZVI/rGO composites, M (magnetite/maghemite), L (lepidocrocite), and Fe^0 (nZVI); (B) XPS spectra of U 4f peaks, *dash lines (blue)*: fitted U(IV) components. *Short dash lines (red)*: fitted U(VI) components, XPS of curve-fitted Fe 2p and O 1s for NZVI and NZVI/GO, (C) and (D) NZVI/GO_60; (E) and (F) NZVI/GO_2; (G) and (H) NZVI_60; (I) and (J) NZVI_2, $T = 25 \pm 1\,^{\circ}C$, pH 5.0, $I = 0.01$ mol/L $NaClO_4$ [96]. *(Data from Y.B. Sun, C.C. Ding, W.C. Cheng, X.K. Wang, Simultaneous adsorption and reduction of U(VI) on reduced graphene oxide-supported nanoscale zerovalent iron, J. Hazard. Mater. 280 (2014) 399–408. Copyright 2014, Elsevier.)*

or nZVI/rGO_2, nZVI_16 or nZVI/rGO_16, and nZVI_60 or nZVI/rGO_60). The relative intensities of ZVI peaks of these samples obviously were weakened with increasing contact time while the relative intensities of magnetite/maghemite and lepidocrocite peaks were obviously strengthened with prolonged contact time. Therefore, the ZVI phase of nZVI and the nZVI/rGO composite might be transformed to the magnetite/maghemite and lepidocrocite phases during the adsorption process. The high resolution of U 4f peaks can be quantitatively deconvoluted into U(IV) (at ~380.8 eV) and U(VI) (at ~382.0 eV) components (Fig. 4.30B) [201]. The relative intensity of U(IV) on nZVI_2 or nZVI/rGO_2 was obviously weaker than that of U(IV) on nZVI_60 or nZVI/rGO_60, which indicated that the primary reduction of U(VI) to U(IV) occurred at the longer contact time. Moreover, the relative intensity of U(VI) on nZVI/rGO_60 was slightly stronger than that of U(VI) on the nZVI_60 sample, implying that the U(VI) was more dominant on the nZVI/rGO composites.

The mechanism for the distinct enhancement of U(VI) uptake by the nZVI/rGO composites was confirmed by XPS analysis (Fig. 4.30C–J) [96]. They observed that the relative intensity of Fe(II) for nZVI/rGO_2 and nZVI/rGO_60 is stronger than that of Fe(II) for nZVI_2 and nZVI_60, respectively (Fig. 4.30C, E, G, and I). These results implied that the existence of rGO in the nZVI slurry can improve the enrichment of Fe(II) on the surface of nZVI/rGO, which was more efficient to improve the reduction of U(VI) to U(IV). Fig. 4.30D and H show that the relative intensity of chemisorbed OH^- groups for the nZVI_60 sample at 533 eV was obviously weaker than that of the nZVI/rGO_60 sample, suggesting that the chemisorbed OH^- groups were greatly significant in the uptake of U(VI) on the nZVI/rGO_60 sample. Therefore, according to the XPS analysis, the distinct enhancement uptake of U(VI) on the nZVI/rGO composite was assigned to the abundant chemisorbed OH— groups on rGO and the enrichment of Fe^{2+} ions on the surface of nZVI/rGO.

Zhu et al. successfully synthesized the reduced graphene oxide-supported nanoscale zerovalent iron (NZVI@rGO) composite [202]. They utilized the NZVI@rGO composite to uptake Eu(III), which showed that the adsorption was an endothermic and spontaneous process. Fig. 4.31A showed the impact of ionic strength and pH on Eu(III) removal by NZVI@rGO, and the results revealed that the removal was obviously dependent on ionic strength and pH at $pH < 5.0$. The higher removal percentage at $pH > 7.0$ might be assigned to the functional groups of NZVI@rGO resulting in the inner-sphere surface complexation mechanism, whereas the lower removal at higher ionic strength at $pH < 5.0$ was due to the outer-sphere surface complexation mechanism. Besides, according to the results of surface complexation modeling, the adsorption behavior of Eu(III) on the NZVI@rGO composite can be well fitted by using the diffuse-layer model with cation exchange sites (XNa) and surface complexation sites (SOH) (Fig. 4.31B). The surface complexation modeling fitted results showed that X_3Eu species (cation exchange complexes) dominated at $pH < 4.5$ while $SOEu^{2+}$ species (mononuclear and monodentate complexes)

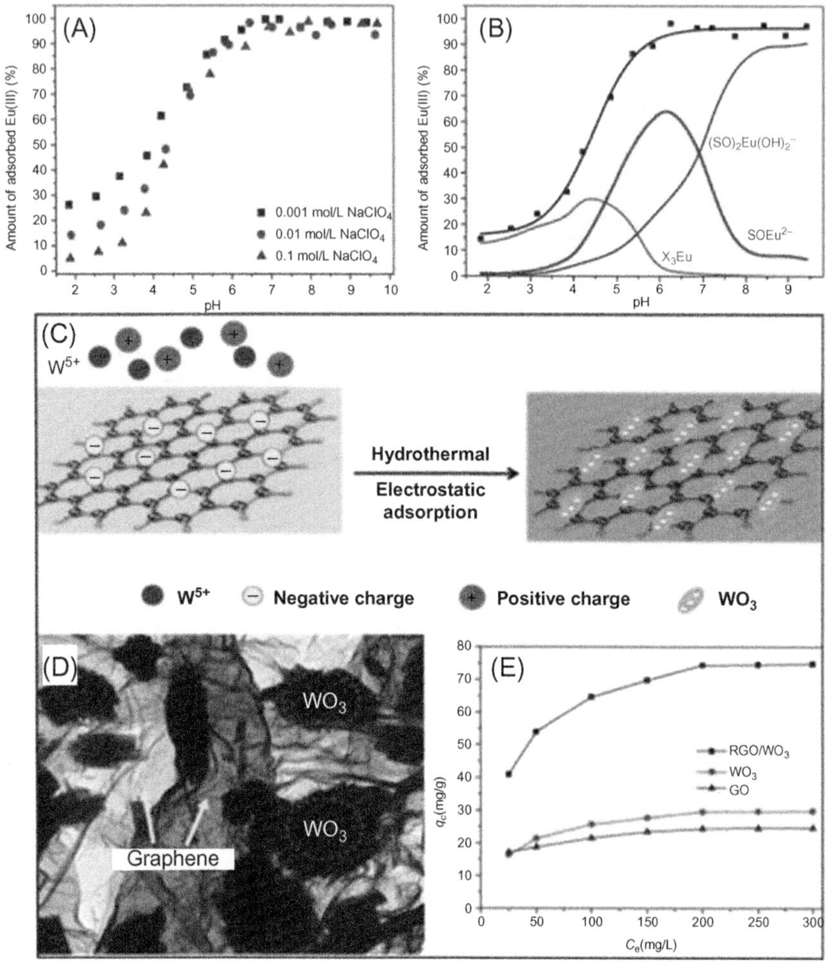

FIG. 4.31 (A) Effect of pH on the uptake of Eu(III) on NZVI@rGO composites at different ionic strength (I = 0.001–0.1 mol/L $NaClO_4$). (B) The fitting of surface complexation modeling of Eu(III) uptake on NZVI@rGO composites at different pH conditions [202]. T = 293 K, C_0 = 10 mg/L, m/V = 1.2 g/L, I = 0.01 M $NaClO_4$. (C) Schematic illustration of WO_3 nanorods anchored to reduced graphene oxide. (D) TEM image of RGO/WO_3 composite. (E) Adsorption isotherm of Sr^{2+} on GO, neat WO_3 and RGO@WO_3 [203]. *(Data from H.J. Zhu, H.Y. Gao, X.M. Huang, W.F. Kong, X.M. Yan, The uptake of europium by reduced graphene oxide-supported nanoscale zerovalent iron investigated by batch and modeling techniques, J. Environ. Chem. Eng. 3 (2015) 2974–2980; W.J. Mu, Q.H. Yu, R. Hu, X.L. Li, H.Y. Wei, Y. Jian, Porous three-dimensional reduced graphene oxide merged with WO3 for efficient removal of radioactive strontium, Appl. Surf. Sci. 423 (2017) 1203–1211. Copyright 2015 and 2017, Elsevier.)*

governed at the neutral condition, and the $(SO)_2Eu(OH)_2^-$ species (mononuclear and bidentate complexes) mainly at high pH condition, respectively. Therefore, they proposed that the removal mechanism of Eu(III) on NZVI@rGO was assigned to cation exchange at low pH, and inner-sphere surface complexation dominated the Eu(III) removal at high pH values. An RGO/WO_3 composite with a 3D nanostructure was synthesized by a simple hydrothermal method (Fig. 4.31C) [203]. In addition, the results of characterization analysis indicated that WO_3 nanorods were uniformly attached on the surface of the RGO with a mesoporous 3D structure (Fig. 4.31D). The large surface of RGO was beneficial to load WO_3 and prevent the agglomeration of WO_3 nanoparticles. The RGO/WO_3 composite showed an excellent adsorption capacity for Sr^{2+} over a wide pH from 4.0 to 11.0. Besides, the RGO/WO_3 composite exhibited a preferable adsorption for Sr^{2+} than WO_3 and GO, which was due to combining the benefits of RGO and WO_3 (Fig. 4.31E).

4.6 Conclusions and prospect

The application of carbon-based materials on the removal of radionuclides is summarized in this chapter. This paper reviews the adsorption behaviors and interaction mechanisms of radionuclides on carbon-based materials. Various spectroscopy techniques involving FT-IR, XPS, Raman, and EXAFS as well as surface complexation modeling and DFT calculations have been used and combined to explore the adsorption mechanisms. The carbon-based materials exhibited excellent removal capacity toward radionuclides due to the high specific surface area and abundant oxygen-containing groups. Further, many modified carbon-based materials were investigated due to the fact that carbon materials have some shortcomings such as easy aggregation and the high cost of GO. Therefore, abundant inorganic and organic modified carbon-based materials were introduced in this chapter, and their adsorption performances and mechanisms were compared with the raw carbon materials in this chapter. These obtained results in our review give an understanding and a guide for the removal of radionuclides by carbon-based materials. Additionally, the results of our review are also useful to design and prepare the more efficient and practical carbon-based adsorbents.

References

[1] E.D. Hwang, K.W. Lee, K.H. Choo, S.J. Choi, S.H. Kim, C.H. Yoon, C.H. Lee, Effect of precipitation and complexation on nanofiltration of strontium-containing nuclear wastewater, Desalination 147 (2002) 289–294.

[2] D.M. Taylor, S.K. Taylor, Environmental uranium and human health, Rev. Environ. Health 12 (1997) 147–158.

[3] Y.B. Sun, D.D. Shao, C.L. Chen, S.B. Yang, X.K. Wang, Highly efficient enrichment of radionuclides on graphene oxide-supported polyaniline, Environ. Sci. Technol. 47 (2013) 9904–9910.

[4] R. Hu, D.D. Shao, X.K. Wang, Graphene oxide/polypyrrole composites for highly selective enrichment of U(VI) from aqueous solutions, Polym. Chem. 5 (2014) 6207–6215.

[5] P. Ilaiyaraja, A.K.S. Deb, K. Sivasubramanian, D. Ponraju, B. Venkatraman, Removal of thorium from aqueous solution by adsorption using PAMAM dendron-functionalized styrene divinyl benzene, J. Radioanal. Nucl. Chem. 297 (2013) 59–69.

[6] Y.B. Sun, Q. Wang, C.L. Chen, X.L. Tan, X.K. Wang, Interaction between Eu(III) and graphene oxide nanosheets investigated by batch and extended X-ray absorption fine structure spectroscopy and by modeling techniques, Environ. Sci. Technol. 46 (2012) 6020–6027.
[7] S.J. Yu, H.Y. Mei, X. Chen, X.L. Tan, B. Ahmad, A. Alsaedi, T. Hayat, X.K. Wang, Impact of environmental conditions on the sorption behavior of radionuclide ^{90}Sr(II) on Na-montmorillonite, J. Mol. Liq. 203 (2015) 39–46.
[8] H. Kim, M. Kim, W. Lee, S. Kim, Rapid removal of radioactive cesium by polyacrylonitrile nanofibers containing Prussian blue, J. Hazard. Mater. 347 (2018) 106–113.
[9] B.H. Gu, Y.K. Ku, P.M. Jardine, Sorption and binary exchange of nitrate, sulfate, and uranium on an anion-exchange resin, Environ. Sci. Technol. 38 (2004) 3184–3188.
[10] W. Abbasi, M. Streat, Adsorption of uranium from aqueous solutions using activated carbon, Sep. Sci. Technol. 29 (1994) 1217–1230.
[11] O. Raff, R.-D. Wilken, Removal of dissolved uranium by nanofiltration, Desalination 122 (1999) 147–150.
[12] K. Mohanty, D. Das, M. Biswas, Preparation and characterization of activated carbons from Sterculia alata nutshell by chemical activation with zinc chloride to remove phenol from wastewater, Adsorption 12 (2006) 119–132.
[13] A. Mellah, S. Chegrouche, M. Barkat, The removal of uranium(VI) from aqueous solutions onto activated carbon: kinetic and thermodynamic investigations, J. Colloid Interface Sci. 296 (2006) 434–441.
[14] S.J. Coleman, P.R. Coronado, R.S. Maxwell, J.G. Reynolds, Granulated activated carbon modified with hydrophobic silica aerogel-potential composite materials for the removal of uranium from aqueous solutions, Environ. Sci. Technol. 37 (2003) 2286–2290.
[15] A. Starvin, T.P. Rao, Solid phase extractive preconcentration of uranium(VI) onto diarylazobisphenol modified activated carbon, Talanta 63 (2004) 225–232.
[16] Y.S. Zhao, C.X. Liu, M. Feng, Z. Chen, S.Q. Li, G. Tian, L. Wang, J.B. Huang, S.J. Li, Solid phase extraction of uranium(VI) onto benzoylthiourea-anchored activated carbon, J. Hazard. Mater. 176 (2010) 119–124.
[17] G.X. Zhao, J.X. Li, X.M. Ren, C.L. Chen, X.K. Wang, Few-layered graphene oxide nanosheets as superior sorbents for heavy metal ion pollution management, Environ. Sci. Technol. 45 (2011) 10454–10462.
[18] A. Schierz, H. Zänker, Aqueous suspensions of carbon nanotubes: surface oxidation, colloidal stability and uranium sorption, Environ. Pollut. 157 (2009) 1088–1094.
[19] D.D. Shao, Z.Q. Jiang, X.K. Wang, J.X. Li, Y.D. Meng, Plasma induced grafting carboxymethyl cellulose on multiwalled carbon nanotubes for the removal of UO_2^{2+} from aqueous solution, J. Phys. Chem. B 113 (2009) 860–864.
[20] Y.B. Sun, S.H. Lu, X.X. Wang, C. Xu, J. Li, C.L. Chen, J. Chen, T. Hayat, A. Alsaedi, N.S. Alharbi, X.K. Wang, Plasma-facilitated synthesis of amidoxime/carbon nanofiber hybrids for effective enrichment of ^{238}U(VI) and ^{241}Am(III), Environ. Sci. Technol. 51 (2017) 12274–12282.
[21] R. Zhang, C.L. Chen, J. Li, X.K. Wang, Investigation of interaction between U(VI) and carbonaceous nanofibers by batch experiments and modeling study, J. Colloid Interface Sci. 460 (2015) 237–246.
[22] C. Kütahyalı, M. Eral, Sorption studies of uranium and thorium on activated carbon prepared from olive stones: kinetic and thermodynamic aspects, J. Nucl. Mater. 396 (2010) 251–256.
[23] G.I. Park, H.S. Park, S.I. Woo, Influence of pH on the adsorption of uranium ions by oxidized activated carbon and chitosan, Sep. Sci. Technol. 34 (1999) 833–854.

[24] S. Iijima, Helical microtubules of graphitic carbon, Nature 354 (1991) 56–58.
[25] X.J. Peng, Y.H. Li, Z.K. Luan, Z.C. Di, H.Y. Wang, B.H. Tian, Z.P. Jia, Adsorption of 1,2-dichlorobenzene from water to carbon nanotubes, Chem. Phys. Lett. 376 (2003) 154–158.
[26] C.S. Lu, Y.L. Chung, K.F. Chang, Adsorption of trihalomethanes from water with carbon nanotubes, Water Res. 39 (2005) 1183–1189.
[27] Y.H. Li, S.G. Wang, Z.K. Luan, J. Ding, C.L. Xu, D.H. Wu, Adsorption of cadmium(II) from aqueous solution by surface oxidized carbon nanotubes, Carbon 41 (2003) 1057–1062.
[28] Y.H. Li, Z.C. Di, J. Ding, D.H. Wu, Z.K. Luan, Y.Q. Zhu, Adsorption thermodynamic, kinetic and desorption studies of Pb^{2+} on carbon nanotubes, Water Res. 39 (2005) 605–609.
[29] C. Lu, C. Liu, Removal of nickel(II) from aqueous solution by carbon nanotubes, J. Chem. Technol. Biotechnol. 81 (2006) 1932–1940.
[30] C.S. Lu, H. Chiu, C.T. Liu, Removal of zinc(II) from aqueous solution by purified carbon nanotubes: kinetics and equilibrium studies, Ind. Eng. Chem. Res. 45 (2006) 2850–2855.
[31] G.P. Rao, C. Lu, F. Su, Sorption of divalent metal ions from aqueous solution by carbon nanotubes: a review, Sep. Purif. Technol. 58 (2007) 224–231.
[32] Q.H. Fan, D.D. Shao, J. Hu, C.L. Chen, W.S. Wu, X.K. Wang, Adsorption of humic acid and Eu(III) to multi-walled carbon nanotubes: Effect of pH, ionic strength and counterion effect, Radiochim. Acta 97 (2009) 141–148.
[33] A.S. Deb, S. Pahan, K. Dasgupta, S. Panja, A. Debnath, P. Dhami, S.M. Ali, C. Kaushik, J. Yadav, Carbon nano tubes functionalized with novel functional group-amido-amine for sorption of actinides, J. Hazard. Mater. 345 (2018) 63–75.
[34] X.L. Tan, D. Xu, C.L. Chen, X.K. Wang, W.P. Hu, Adsorption and kinetic desorption study of $^{152+154}Eu$ (III) on multiwall carbon nanotubes from aqueous solution by using chelating resin and XPS methods, Radiochim. Acta 96 (2008) 23–29.
[35] C.L. Chen, J. Hu, D. Xu, X.L. Tan, Y.D. Meng, X.K. Wang, Surface complexation modeling of Sr(II) and Eu(III) adsorption onto oxidized multiwall carbon nanotubes, J. Colloid Interface Sci. 323 (2008) 33–41.
[36] G.X. Zhao, T. Wen, X. Yang, S.B. Yang, J.L. Liao, J. Hu, D.D. Shao, X.K. Wang, Preconcentration of U(VI) ions on few-layered graphene oxide nanosheets from aqueous solutions, Dalton Trans. 41 (2012) 6182–6188.
[37] N. Pan, L. Li, J. Ding, S.K. Li, R.B. Wang, Y.D. Jin, X.K. Wang, C.Q. Xia, Preparation of graphene oxide-manganese dioxide for highly efficient adsorption and separation of Th(IV)/U(VI), J. Hazard. Mater. 309 (2016) 107–115.
[38] P.F. Zong, S.F. Wang, Y.L. Zhao, H. Wang, H. Pan, C.H. He, Synthesis and application of magnetic graphene/iron oxides composite for the removal of U (VI) from aqueous solutions, Chem. Eng. J. 220 (2013) 45–52.
[39] Y.G. Zhao, J.X. Li, S.W. Zhang, H. Chen, D.D. Shao, Efficient enrichment of uranium(VI) on amidoximated magnetite/graphene oxide composites, RSC Adv. 3 (2013) 18952–18959.
[40] Y.B. Sun, X.X. Wang, C.C. Ding, W.C. Cheng, C.L. Chen, T. Hayat, A. Alsaedi, J. Hu, X.K. Wang, Direct synthesis of bacteria-derived carbonaceous nanofibers as a highly efficient material for radionuclides elimination, ACS Sustain. Chem. Eng. 4 (2016) 4608–4616.
[41] Z.Y. Wu, C. Li, H.W. Liang, J.F. Chen, S.H. Yu, Ultralight, flexible, and fire-resistant carbon nanofiber aerogels from bacterial cellulose, Angew. Chem. 125 (2013) 2997–3001.
[42] L.F. Chen, Z.H. Huang, H.W. Liang, Q.F. Guan, S.H. Yu, Bacterial-cellulose-derived carbon nanofiber@MnO_2 and nitrogen-doped carbon nanofiber electrode materials: an asymmetric supercapacitor with high energy and power density, Adv. Mater. 25 (2013) 4746–4752.

[43] H.W. Liang, L. Wang, P.Y. Chen, H.T. Lin, L.F. Chen, D. He, S.H. Yu, Carbonaceous nanofiber membranes for selective filtration and separation of nanoparticles, Adv. Mater. 22 (2010) 4691–4695.

[44] H.W. Liang, Q.F. Guan, L.F. Chen, Z. Zhu, W.J. Zhang, S.H. Yu, Macroscopic-scale template synthesis of robust carbonaceous nanofiber hydrogels and aerogels and their applications, Angew. Chem. Int. Ed. 51 (2012) 5101–5105.

[45] H.W. Liang, X. Cao, W.J. Zhang, H.T. Lin, F. Zhou, L.F. Chen, S.H. Yu, Robust and highly efficient free-standing carbonaceous nanofiber membranes for water purification, Adv. Funct. Mater. 21 (2011) 3851–3858.

[46] F.A. Al-Khaldi, B. Abu-Sharkh, A.M. Abulkibash, M.A. Atieh, Cadmium removal by activated carbon, carbon nanotubes, carbon nanofibers, and carbon fly ash: a comparative study, Desalin. Water Treat. 53 (2015) 1417–1429.

[47] Y. Patiño, E. Díaz, S. Ordóñez, Performance of different carbonaceous materials for emerging pollutants adsorption, Chemosphere 119 (2015) S124–S130.

[48] M.R. Cuervo, E. Asedegbega-Nieto, E. Díaz, A. Vega, S. Ordóñez, E. Castillejos-López, I. Rodríguez-Ramos, Effect of carbon nanofiber functionalization on the adsorption properties of volatile organic compounds, J. Chromatogr. A 1188 (2008) 264–273.

[49] N. Yusof, A. Ismail, Post spinning and pyrolysis processes of polyacrylonitrile (PAN)-based carbon fiber and activated carbon fiber: a review, J. Anal. Appl. Pyrolysis 93 (2012) 1–13.

[50] J. Wu, D. Chung, Increasing the electromagnetic interference shielding effectiveness of carbon fiber polymer–matrix composite by using activated carbon fibers, Carbon 40 (2002) 445–447.

[51] T. Karanfil, J.E. Kilduff, Role of granular activated carbon surface chemistry on the adsorption of organic compounds. 1. Priority pollutants, Environ. Sci. Technol. 33 (1999) 3217–3224.

[52] J.M. Dias, M.C. Alvim-Ferraz, M.F. Almeida, J. Rivera-Utrilla, M. Sánchez-Polo, Waste materials for activated carbon preparation and its use in aqueous-phase treatment: a review, J. Environ. Manag. 85 (2007) 833–846.

[53] J.E. Kilduff, T. Karanfil, Y.P. Chin, W.J. Weber, Adsorption of natural organic polyelectrolytes by activated carbon: a size-exclusion chromatography study, Environ. Sci. Technol. 30 (1996) 1336–1343.

[54] K.J. Choi, S.G. Kim, C.W. Kim, S.H. Kim, Effects of activated carbon types and service life on removal of endocrine disrupting chemicals: amitrol, nonylphenol, and bisphenol-A, Chemosphere 58 (2005) 1535–1545.

[55] A. Dąbrowski, P. Podkościelny, Z. Hubicki, M. Barczak, Adsorption of phenolic compounds by activated carbon—a critical review, Chemosphere 58 (2005) 1049–1070.

[56] L. Li, P.A. Quinlivan, D.R. Knappe, Effects of activated carbon surface chemistry and pore structure on the adsorption of organic contaminants from aqueous solution, Carbon 40 (2002) 2085–2100.

[57] B. Ruiz, I. Cabrita, A.S. Mestre, J.B. Parra, J. Pires, A.P. Carvalho, C.O. Ania, Surface heterogeneity effects of activated carbons on the kinetics of paracetamol removal from aqueous solution, Appl. Surf. Sci. 256 (2010) 5171–5175.

[58] C. Donati, M. Drikas, R. Hayes, G. Newcombe, Microcystin-LR adsorption by powdered activated carbon, Water Res. 28 (1994) 1735–1742.

[59] J.E. Kilduff, A. Wigton, Sorption of TCE by humic-preloaded activated carbon: Incorporating size-exclusion and pore blockage phenomena in a competitive adsorption model, Environ. Sci. Technol. 33 (1999) 250–256.

[60] G. Newcombe, M. Drikas, R. Hayes, Influence of characterised natural organic material on activated carbon adsorption: II. Effect on pore volume distribution and adsorption of 2-methylisoborneol, Water Res. 31 (1997) 1065–1073.
[61] J.Y. Chen, W. Chen, D.Q. Zhu, Adsorption of nonionic aromatic compounds to single-walled carbon nanotubes: effects of aqueous solution chemistry, Environ. Sci. Technol. 42 (2008) 7225–7230.
[62] A. Mestre, J. Pires, J. Nogueira, A. Carvalho, Activated carbons for the adsorption of ibuprofen, Carbon 45 (2007) 1979–1988.
[63] A.S. Mestre, J. Pires, J.M. Nogueira, J.B. Parra, A.P. Carvalho, C.O. Ania, Waste-derived activated carbons for removal of ibuprofen from solution: role of surface chemistry and pore structure, Bioresour. Technol. 100 (2009) 1720–1726.
[64] D. Tasis, N. Tagmatarchis, A. Bianco, M. Prato, Chemistry of carbon nanotubes, Chem. Rev. 106 (2006) 1105–1136.
[65] H. Dai, Carbon nanotubes: synthesis, integration, and properties, Acc. Chem. Res. 35 (2002) 1035–1044.
[66] R. Andrews, D. Jacques, D. Qian, T. Rantell, Multiwall carbon nanotubes: synthesis and application, Acc. Chem. Res. 35 (2002) 1008–1017.
[67] S. Iijima, T. Ichihashi, Single-shell carbon nanotubes of 1-nm diameter, Nature 363 (1993) 603.
[68] D. Bethune, C.H. Kiang, M. De Vries, G. Gorman, R. Savoy, J. Vazquez, R. Beyers, Cobalt-catalysed growth of carbon nanotubes with single-atomic-layer walls, Nature 363 (1993) 605.
[69] Y.L. Zhao, J.F. Stoddart, Noncovalent functionalization of single-walled carbon nanotubes, Acc. Chem. Res. 42 (2009) 1161–1171.
[70] Y.B. Sun, S.T. Yang, G.D. Sheng, Z.Q. Guo, X.K. Wang, The removal of U(VI) from aqueous solution by oxidized multiwalled carbon nanotubes, J. Environ. Radioact. 105 (2012) 40–47.
[71] M. Endo, H. Kroto, Formation of carbon nanofibers, J. Phys. Chem. 96 (1992) 6941–6944.
[72] V.N. Popov, Carbon nanotubes: properties and application, Mater. Sci. Eng. R 43 (2004) 61–102.
[73] P. Avouris, Molecular electronics with carbon nanotubes, Acc. Chem. Res. 35 (2002) 1026–1034.
[74] E.T. Thostenson, Z. Ren, T.-W. Chou, Advances in the science and technology of carbon nanotubes and their composites: a review, Compos. Sci. Technol. 61 (2001) 1899–1912.
[75] S. Pei, H.-M. Cheng, The reduction of graphene oxide, Carbon 50 (2012) 3210–3228.
[76] G.X. Zhao, X.M. Ren, X. Gao, X.L. Tan, J.X. Li, C.L. Chen, Y.Y. Huang, X.K. Wang, Removal of Pb(II) ions from aqueous solutions on few-layered graphene oxide nanosheets, Dalton Trans. 40 (2011) 10945–10952.
[77] L. Chai, T. Wang, L. Zhang, H. Wang, W. Yang, S. Dai, Y. Meng, X. Li, A Cu–m-phenylenediamine complex induced route to fabricate poly (m-phenylenediamine)/reduced graphene oxide hydrogel and its adsorption application, Carbon 81 (2015) 748–757.
[78] X. Zhou, W.Y. Huang, J. Shi, Z.X. Zhao, Q.B. Xia, Y.W. Li, H.H. Wang, Z. Li, A novel MOF/graphene oxide composite GrO@ MIL-101 with high adsorption capacity for acetone, J. Mater. Chem. A 2 (2014) 4722–4730.
[79] L. Liu, Y.R. Zhang, Y.J. He, Y.F. Xie, L.H. Huang, S.Z. Tan, X. Cai, Preparation of montmorillonite-pillared graphene oxide with increased single-and co-adsorption toward lead ions and methylene blue, RSC Adv. 5 (2015) 3965–3973.
[80] J. Chen, B.W. Yao, C. Li, G.Q. Shi, An improved Hummers method for eco-friendly synthesis of graphene oxide, Carbon 64 (2013) 225–229.

[81] G.X. Zhao, L. Jiang, Y.D. He, J.X. Li, H.L. Dong, X.K. Wang, W.P. Hu, Sulfonated graphene for persistent aromatic pollutant management, Adv. Mater. 23 (2011) 3959–3963.

[82] W.S. Hummers, R.E. Offeman, Preparation of graphitic oxide, J. Am. Chem. Soc. 80 (1958) 1339.

[83] D.C. Marcano, D.V. Kosynkin, J.M. Berlin, A. Sinitskii, Z. Sun, A. Slesarev, L.B. Alemany, W. Lu, J.M. Tour, Improved Synthesis of Graphene Oxide, ACS Nano 4 (2010) 4806–4814.

[84] H. Bai, C. Li, G.Q. Shi, Functional composite materials based on chemically converted graphene, Adv. Mater. 23 (2011) 1089–1115.

[85] J.L. Zhang, H.J. Yang, G.X. Shen, P. Cheng, J.Y. Zhang, S.W. Guo, Reduction of graphene oxide via L-ascorbic acid, Chem. Commun. 46 (2010) 1112–1114.

[86] Y. Shen, Q. Fang, B.L. Chen, Environmental applications of three-dimensional graphene-based macrostructures: adsorption, transformation, and detection, Environ. Sci. Technol. 49 (2014) 67–84.

[87] J. Wang, Z.M. Chen, B.L. Chen, Adsorption of polycyclic aromatic hydrocarbons by graphene and graphene oxide nanosheets, Environ. Sci. Technol. 48 (2014) 4817–4825.

[88] Z.G. Pei, L.Y. Li, L.X. Sun, S.Z. Zhang, X.Q. Shan, S. Yang, B. Wen, Adsorption characteristics of 1,2,4-trichlorobenzene, 2,4,6-trichlorophenol, 2-naphthol and naphthalene on graphene and graphene oxide, Carbon 51 (2013) 156–163.

[89] A.H. Hung, R.J. Holbrook, M.W. Rotz, C.J. Glasscock, N.D. Mansukhani, K.W. MacRenaris, L.M. Manus, M.C. Duch, K.T. Dam, M.C. Hersam, Graphene oxide enhances cellular delivery of hydrophilic small molecules by co-incubation, ACS Nano 8 (2014) 10168–10177.

[90] L. Kong, A. Enders, T.S. Rahman, P.A. Dowben, Molecular adsorption on graphene, J. Phys. Condens. Matter 26 (2014)443001.

[91] X.Y. Liu, J.M. Zhang, Formaldehyde molecule adsorbed on doped graphene: a first-principles study, Appl. Surf. Sci. 293 (2014) 216–219.

[92] J.P. Sun, Q.L. Liang, Q. Han, X.Q. Zhang, M.Y. Ding, One-step synthesis of magnetic graphene oxide nanocomposite and its application in magnetic solid phase extraction of heavy metal ions from biological samples, Talanta 132 (2015) 557–563.

[93] L.M. Cui, X.Y. Guo, Q. Wei, Y.G. Wang, L. Gao, L.G. Yan, T. Yan, B. Du, Removal of mercury and methylene blue from aqueous solution by xanthate functionalized magnetic graphene oxide: sorption kinetic and uptake mechanism, J. Colloid Interface Sci. 439 (2015) 112–120.

[94] S. Yang, L.Y. Li, Z.G. Pei, C.M. Li, X.Q. Shan, B. Wen, S.Z. Zhang, L.R. Zheng, J. Zhang, Y.N. Xie, Effects of humic acid on copper adsorption onto few-layer reduced graphene oxide and few-layer graphene oxide, Carbon 75 (2014) 227–235.

[95] C.C. Ding, W.C. Cheng, Y.B. Sun, X.K. Wang, Determination of chemical affinity of graphene oxide nanosheets with radionuclides investigated by macroscopic, spectroscopic and modeling techniques, Dalton Trans. 43 (2014) 3888–3896.

[96] Y.B. Sun, C.C. Ding, W.C. Cheng, X.K. Wang, Simultaneous adsorption and reduction of U(VI) on reduced graphene oxide-supported nanoscale zerovalent iron, J. Hazard. Mater. 280 (2014) 399–408.

[97] Q.Y. Wu, J.H. Lan, C.Z. Wang, Y.L. Zhao, Z.F. Chai, W.Q. Shi, Understanding the interactions of neptunium and plutonium ions with graphene oxide: scalar-relativistic DFT investigations, J. Phys. Chem. A 118 (2014) 10273–10280.

[98] C. Lattao, X.Y. Cao, J.D. Mao, K. Schmidt-Rohr, J.J. Pignatello, Influence of molecular structure and adsorbent properties on sorption of organic compounds to a temperature series of wood chars, Environ. Sci. Technol. 48 (2014) 4790–4798.

[99] X.Y. Guo, Q. Wei, B. Du, Y.K. Zhang, X.D. Xin, L.G. Yan, H.Q. Yu, Removal of Metanil Yellow from water environment by amino functionalized graphenes (NH_2-G)–influence of surface chemistry of NH_2-G, Appl. Surf. Sci. 284 (2013) 862–869.

[100] X.B. Wang, S.S. Huang, L.H. Zhu, X.L. Tian, S.H. Li, H.Q. Tang, Correlation between the adsorption ability and reduction degree of graphene oxide and tuning of adsorption of phenolic compounds, Carbon 69 (2014) 101–112.

[101] Y.B. Sun, S.B. Yang, G.X. Zhao, Q. Wang, X.K. Wang, Adsorption of polycyclic aromatic hydrocarbons on graphene oxides and reduced graphene oxides, Chem. Asian J. 8 (2013) 2755–2761.

[102] G.X. Zhao, J.X. Li, X.K. Wang, Kinetic and thermodynamic study of 1-naphthol adsorption from aqueous solution to sulfonated graphene nanosheets, Chem. Eng. J. 173 (2011) 185–190.

[103] T. Wen, X.L. Wu, X.L. Tan, X.K. Wang, A.W. Xu, One-pot synthesis of water-swellable Mg–Al layered double hydroxides and graphene oxide nanocomposites for efficient removal of As (V) from aqueous solutions, ACS Appl. Mater. Interfaces 5 (2013) 3304–3311.

[104] X.L. Wu, L. Wang, C.L. Chen, A.W. Xu, X.K. Wang, Water-dispersible magnetite-graphene-LDH composites for efficient arsenate removal, J. Mater. Chem. 21 (2011) 17353–17359.

[105] M.C. Liu, C.L. Chen, J. Hu, X.L. Wu, X.K. Wang, Synthesis of magnetite/graphene oxide composite and application for cobalt(II) removal, J. Phys. Chem. C 115 (2011) 25234–25240.

[106] Z.J. Li, F. Chen, L.Y. Yuan, Y.L. Liu, Y.L. Zhao, Z.F. Chai, W.Q. Shi, Uranium(VI) adsorption on graphene oxide nanosheets from aqueous solutions, Chem. Eng. J. 210 (2012) 539–546.

[107] K.J. Yang, B.L. Chen, X.Y. Zhu, B.S. Xing, Aggregation, adsorption, and morphological transformation of graphene oxide in aqueous solutions containing different metal cations, Environ. Sci. Technol. 50 (2016) 11066–11075.

[108] C.C. Ding, W.C. Cheng, X.X. Wang, Z.Y. Wu, Y.B. Sun, C.L. Chen, X.K. Wang, S.H. Yu, Competitive sorption of Pb(II), Cu(II) and Ni(II) on carbonaceous nanofibers: a spectroscopic and modeling approach, J. Hazard. Mater. 313 (2016) 253–261.

[109] A. Omoike, J. Chorover, Spectroscopic study of extracellular polymeric substances from Bacillus subtilis: aqueous chemistry and adsorption effects, Biomacromolecules 5 (2004) 1219–1230.

[110] P. Chen, H.W. Liang, X.H. Lv, H.Z. Zhu, H.B. Yao, S.H. Yu, Carbonaceous nanofiber membrane functionalized by beta-cyclodextrins for molecular filtration, ACS Nano 5 (2011) 5928–5935.

[111] Y.M. Ren, Q. Xu, J.M. Zhang, H.X. Yang, B. Wang, D.Y. Yang, J.H. Hu, Z.M. Liu, Functionalization of biomass carbonaceous aerogels: selective preparation of MnO_2@CA composites for supercapacitors, ACS Appl. Mater. Interfaces 6 (2014) 9689–9697.

[112] X.L. Wu, T. Wen, H.L. Guo, S.B. Yang, X.K. Wang, A.W. Xu, Biomass-derived sponge-like carbonaceous hydrogels and aerogels for supercapacitors, ACS Nano 7 (2013) 3589–3597.

[113] R.J. White, V. Budarin, R. Luque, J.H. Clark, D.J. Macquarrie, Tuneable porous carbonaceous materials from renewable resources, Chem. Soc. Rev. 38 (2009) 3401–3418.

[114] S. Stankovich, D.A. Dikin, R.D. Piner, K.A. Kohlhaas, A. Kleinhammes, Y. Jia, Y. Wu, S.T. Nguyen, R.S. Ruoff, Synthesis of graphene-based nanosheets via chemical reduction of exfoliated graphite oxide, Carbon 45 (2007) 1558–1565.
[115] P. Chingombe, B. Saha, R.J. Wakeman, Surface modification and characterisation of a coal-based activated carbon, Carbon 43 (2005) 3132–3143.
[116] L. Monser, N. Adhoum, Modified activated carbon for the removal of copper, zinc, chromium and cyanide from wastewater, Sep. Purif. Technol. 26 (2002) 137–146.
[117] B.H. Hameed, A.T.M. Din, A.L. Ahmad, Adsorption of methylene blue onto bamboo-based activated carbon: kinetics and equilibrium studies, J. Hazard. Mater. 141 (2007) 819–825.
[118] W.T. Tsai, C.Y. Chang, M.C. Lin, S.F. Chien, H.F. Sun, M.F. Hsieh, Adsorption of acid dye onto activated carbons prepared from agricultural waste bagasse by $ZnCl_2$ activation, Chemosphere 45 (2001) 51–58.
[119] A. Aygun, S. Yenisoy-Karakas, I. Duman, Production of granular activated carbon from fruit stones and nutshells and evaluation of their physical, chemical and adsorption properties, Microporous Mesoporous Mater. 66 (2003) 189–195.
[120] A.H. El-Sheikh, A.P. Newman, H.K. Al-Daffaee, S. Phull, N. Cresswell, Characterization of activated carbon prepared from a single cultivar of Jordanian Olive stones by chemical and physicochemical techniques, J. Anal. Appl. Pyrolysis 71 (2004) 151–164.
[121] F.C. Wu, R.L. Tseng, R.S. Juang, Pore structure and adsorption performance of the activated carbons prepared from plum kernels, J. Hazard. Mater. 69 (1999) 287–302.
[122] J. Avom, J.K. Mbadcam, C. Noubactep, P. Germain, Adsorption of methylene blue from an aqueous solution on to activated carbons from palm-tree cobs, Carbon 35 (1997) 365–369.
[123] Z.J. Yi, J. Yao, Y.F. Kuang, H.L. Chen, F. Wang, J.S. Xu, Uptake of hexavalent uranium from aqueous solutions using coconut husk activated carbon, Desalin. Water Treat. 57 (2016) 1749–1755.
[124] Z.J. Yi, J. Yao, J.S. Xu, M.S. Chen, W. Li, H.L. Chen, F. Wang, Removal of uranium from aqueous solution by using activated palm kernel shell carbon: adsorption equilibrium and kinetics, J. Radioanal. Nucl. Chem. 301 (2014) 695–701.
[125] R. Qadeer, J. Hanif, M. Saleem, M. Afzal, Effect of alkali metals, alkaline earth metals and lanthanides on the adsorption of uranium on activated charcoal from aqueous solutions, J. Radioanal. Nucl. Chem. 165 (1992) 243–253.
[126] S.P. Chen, J.X. Hong, H.X. Yang, J.Z. Yang, Adsorption of uranium(VI) from aqueous solution using a novel graphene oxide-activated carbon felt composite, J. Environ. Radioact. 126 (2013) 253–258.
[127] R. Donat, K.E. Erden, Adsorption of U (VI) ions from aqueous solutions by activated carbon prepared from Antep pistachio (Pistacia vera L.) shells, Radiochim. Acta 105 (2017) 359–367.
[128] C. Kütahyalı, M. Eral, Selective adsorption of uranium from aqueous solutions using activated carbon prepared from charcoal by chemical activation, Sep. Purif. Technol. 40 (2004) 109–114.
[129] M.M. Hamed, M. Ali, M. Holiel, Preparation of activated carbon from doum stone and its application on adsorption of 60Co and 152+ 154Eu: equilibrium, kinetic and thermodynamic studies, J. Environ. Radioact. 164 (2016) 113–124.
[130] H. Omar, H. Moloukhia, Use of activated carbon in removal of some radioisotopes from their waste solutions, J. Hazard. Mater. 157 (2008) 242–246.
[131] H. Moloukhia, W. Hegazy, E. Abdel-Galil, S. Mahrous, Removal of Eu3+, Ce3+, Sr2+, and Cs+ ions from radioactive waste solutions by modified activated carbon prepared from coconut shells, Chem. Ecol. 32 (2016) 324–345.

[132] M.S. Xin, X. Dai, J. Han, M.X. Jin, C.A. Jimenez-Cruz, D. Ding, Z.G. Wang, R.H. Zhou, Carbon nanotubes adsorb U atoms differently in their inner and outer surfaces, RSC Adv. 4 (2014) 30074–30080.

[133] A.K.S. Deb, S.M. Ali, K.T. Shenoy, Unanticipated favoured adsorption affinity of Th(IV) ions toward bidentate carboxylate functionalized carbon nanotubes (CNT-COOH) over tridentate diglycolamic acid functionalized CNT: density functional theoretical investigation, RSC Adv. 5 (2015) 80076–80088.

[134] Y.B. Sun, S.B. Yang, C.C. Ding, Z.X. Jin, W.C. Cheng, Tuning the chemistry of graphene oxides by a sonochemical approach: application of adsorption properties, RSC Adv. 5 (2015) 24886–24892.

[135] Y.B. Sun, S.B. Yang, Y. Chen, C.C. Ding, W.C. Cheng, X.K. Wang, Adsorption and desorption of U(VI) on functionalized graphene oxides: a combined experimental and theoretical study, Environ. Sci. Technol. 49 (2015) 4255–4262.

[136] X. Liu, J. Sun, X. Xu, A. Alsaedi, T. Hayat, J. Li, Adsorption and desorption of U(VI) on different-size graphene oxide, Chem. Eng. J. (2018).

[137] Z. Wang, D.E. Giammar, Mass action expressions for bidentate adsorption in surface complexation modeling: theory and practice, Environ. Sci. Technol. 47 (2013) 3982–3996.

[138] C. Koretsky, The significance of surface complexation reactions in hydrologic systems: a geochemist's perspective, J. Hydrol. 230 (2000) 127–171.

[139] T. Hu, S.J. Ding, H.J. Deng, Application of three surface complexation models on U(VI) adsorption onto graphene oxide, Chem. Eng. J. 289 (2016) 270–276.

[140] C. Toner, D. Sparks, Chemical relaxation and double layer model analysis of boron adsorption on alumina, Soil Sci. Soc. Am. J. 59 (1995) 395–404.

[141] X.X. Wang, Q.H. Fan, S.J. Yu, Z.S. Chen, Y.J. Ai, Y.B. Sun, A. Hobiny, A. Alsaedi, X.K. Wang, High sorption of U (VI) on graphene oxides studied by batch experimental and theoretical calculations, Chem. Eng. J. 287 (2016) 448–455.

[142] H. Xu, G. Li, J. Li, C.L. Chen, X.M. Ren, Interaction of Th(IV) with graphene oxides: batch experiments, XPS investigation, and modeling, J. Mol. Liq. 213 (2016) 58–68.

[143] Z.Q. Bai, Z.J. Li, C.Z. Wang, L.Y. Yuan, Z.R. Liu, J. Zhang, L.R. Zheng, Y.L. Zhao, Z.F. Chai, W.Q. Shi, Interactions between Th(IV) and graphene oxide: experimental and density functional theoretical investigations, RSC Adv. 4 (2014) 3340–3347.

[144] T. Rabung, H. Geckeis, J.I. Kim, H.P. Beck, The influence of anionic ligands on the sorption behaviour of Eu(III) on natural hematite, Radiochim. Acta 82 (1998) 243–248.

[145] Q.H. Fan, X.L. Tan, J.X. Li, X.K. Wang, W.S. Wu, G. Montavon, Sorption of Eu(III) on attapulgite studied by batch, XPS, and EXAFS techniques, Environ. Sci. Technol. 43 (2009) 5776–5782.

[146] Y. Xie, E.M. Helvenston, L.C. Shuller-Nickles, B.A. Powell, Surface complexation modeling of Eu(III) and U(VI) interactions with graphene oxide, Environ. Sci. Technol. 50 (2016) 1821–1827.

[147] H. Xu, J. Hu, X.M. Ren, G. Li, Macroscopic and microscopic insight into the mutual effects of europium(III) and phosphate on their interaction with graphene oxide, RSC Adv. 6 (2016) 85046–85057.

[148] X.L. Tan, X.K. Wang, H. Geckeis, T.H. Rabung, Sorption of Eu(III) on humic acid or fulvic acid bound to hydrous alumina studied by SEM-EDS, XPS, TRLFS, and batch techniques, Environ. Sci. Technol. 42 (2008) 6532–6537.

[149] S.J. Yu, X.X. Wang, S.T. Yang, G.D. Sheng, A. Alsaedi, T. Hayat, X.K. Wang, Interaction of radionuclides with natural and manmade materials using XAFS technique, Sci. China Chem. 60 (2017) 170–187.

[150] M.L. Schlegel, I. Pointeau, N. Coreau, P. Reiller, Mechanism of europium retention by calcium silicate hydrates: an EXAFS study, Environ. Sci. Technol. 38 (2004) 4423–4431.

[151] A. Romanchuk, A. Kuzenkova, A. Slesarev, J. Tour, S. Kalmykov, Cs (I) and Sr (II) sorption onto graphene oxide, Solvent Extr. Ion Exc. 34 (2016) 594–602.

[152] W.C. Cheng, C.C. Ding, Q.Y. Wu, X.X. Wang, Y.B. Sun, W.Q. Shi, T. Hayat, A. Alsaedi, Z.F. Chai, X.K. Wang, Mutual effect of U (VI) and Sr (II) on graphene oxides: evidence from EXAFS and theoretical calculations, Environ. Sci.: Nano 4 (2017) 1124–1131.

[153] R.M. Tinnacher, J.D. Begg, H. Mason, J. Ranville, B.A. Powell, J.C. Wong, A.B. Kersting, M. Zavarin, Effect of fulvic acid surface coatings on plutonium sorption and desorption kinetics on goethite, Environ. Sci. Technol. 49 (2015) 2776–2785.

[154] Y.B. Sun, S.T. Yang, G. Sheng, Q. Wang, Z.Q. Guo, X.K. Wang, Removal of U(VI) from aqueous solutions by the nano-iron oxyhydroxides, Radiochim. Acta 100 (2012) 779–784.

[155] X.X. Wang, Z.S. Chen, X.K. Wang, Graphene oxides for simultaneous highly efficient removal of trace level radionuclides from aqueous solutions, Sci. China Chem. 58 (2015) 1766–1773.

[156] D. Jiang, L. Liu, N. Pan, F.Q. Yang, S.K. Li, R.B. Wang, I.W. Wyman, Y.D. Jin, C.Q. Xia, The separation of Th(IV)/U(VI) via selective complexation with graphene oxide, Chem. Eng. J. 271 (2015) 147–154.

[157] Y.G. Zhao, C. Guo, H. Fang, J. Jiang, Competitive adsorption of Sr(II) and U(VI) on graphene oxide investigated by batch and modeling techniques, J. Mol. Liq. 222 (2016) 263–267.

[158] Y.B. Sun, Z.Y. Wu, X.X. Wang, C.C. Ding, W.C. Cheng, S.H. Yu, X.K. Wang, Macroscopic and microscopic investigation of U(VI) and Eu(III) adsorption on carbonaceous nanofibers, Environ. Sci. Technol. 50 (2016) 4459–4467.

[159] B.W. Hu, Q.Y. Hu, D. Xu, C.G. Chen, The adsorption of U(VI) on carbonaceous nanofibers: a combined batch, EXAFS and modeling techniques, Sep. Purif. Technol. 175 (2017) 140–146.

[160] X. Lu, D.X. Zhang, A.T. Reda, C. Liu, Z. Yang, S.S. Guo, S.T. Xiao, Y.G. Ouyang, Synthesis of amidoxime-grafted activated carbon fibers for efficient recovery of uranium(VI) from aqueous solution, Ind. Eng. Chem. Res. 56 (2017) 11936–11947.

[161] H.J. Chen, Z. Chen, G.X. Zhao, Z.B. Zhang, C. Xu, Y.H. Liu, J. Chen, L. Zhuang, T. Haya, X.K. Wang, Enhanced adsorption of U(VI) and ^{241}Am(III) from wastewater using Ca/Al layered double hydroxide@carbon nanotube composites, J. Hazard. Mater. 347 (2018) 67–77.

[162] C.L. Chen, X.K. Wang, M. Nagatsu, Europium adsorption on multiwall carbon nanotube/iron oxide magnetic composite in the presence of polyacrylic acid, Environ. Sci. Technol. 43 (2009) 2362–2367.

[163] P. Liu, W. Qi, Y.F. Du, Z. Li, J. Wang, J.J. Bi, W.S. Wu, Adsorption of thorium(IV) on magnetic multi-walled carbon nanotubes, Sci. China Chem. 57 (2014) 1483–1490.

[164] E. Zolfonoun, S.R. Yousefi, Sorption and preconcentration of uranium and thorium from aqueous solutions using multi-walled carbon nanotubes decorated with magnetic nanoparticles, Radiochim. Acta 103 (2015) 835–841.

[165] C.L. Chen, J. Hu, D.D. Shao, J.X. Li, X.K. Wang, Adsorption behavior of multiwall carbon nanotube/iron oxide magnetic composites for Ni(II) and Sr(II), J. Hazard. Mater. 164 (2009) 923–928.

[166] Y. Song, G. Ye, Y.X. Lu, J. Chen, J.C. Wang, K. Matyjaszewski, Surface-initiated ARGET ATRP of poly(glycidyl methacrylate) from carbon nanotubes via bioinspired catechol chemistry for efficient adsorption of uranium ions, ACS Macro Lett. 5 (2016) 382–386.

[167] Y. Wang, Z.X. Gu, J.J. Yang, J.L. Liao, Y.Y. Yang, N. Liu, J. Tang, Amidoxime-grafted multiwalled carbon nanotubes by plasma techniques for efficient removal of uranium(VI), Appl. Surf. Sci. 320 (2014) 10–20.

[168] S. Mishra, J. Dwivedi, A. Kumar, N. Sankararamakrishnan, The synthesis and characterization of tributyl phosphate grafted carbon nanotubes by the floating catalytic chemical vapor deposition method and their sorption behavior toward uranium, New J. Chem. 40 (2016) 1213–1221.

[169] A. Sengupta, J. Sk, A. Boda, S.M. Ali, An amide functionalized task specific carbon nanotube for the sorption of tetra and hexa valent actinides: experimental and theoretical insight, RSC Adv. 6 (2016) 39553–39562.

[170] X.X. Wang, S.B. Yang, W.Q. Shi, J.X. Li, T. Hayat, X.K. Wang, Different interaction mechanisms of Eu(III) and 243Am(III) with carbon nanotubes studied by batch, spectroscopy technique and theoretical calculation, Environ. Sci. Technol. 49 (2015) 11721–11728.

[171] C. Ding, W. Cheng, X. Nie, F. Yi, Synergistic mechanism of U (VI) sequestration by magnetite-graphene oxide composites: evidence from spectroscopic and theoretical calculation, Chem. Eng. J. 324 (2017) 113–121.

[172] L. Shao, X.F. Wang, Y.M. Ren, S.F. Wang, J.R. Zhong, M.F. Chu, H. Tang, L.Z. Luo, D.H. Xie, Facile fabrication of magnetic cucurbit[6] uril/graphene oxide composite and application for uranium removal, Chem. Eng. J. 286 (2016) 311–319.

[173] W.C. Cheng, C.C. Ding, X.Q. Nie, T. Duan, R.R. Ding, Fabrication of 3D macroscopic graphene oxide composites supported by montmorillonite for efficient U(VI) wastewater purification, ACS Sustain. Chem. Eng. 5 (2017) 5503–5511.

[174] W.S. Linghu, H. Yang, Y.X. Sun, G.D. Sheng, Y.Y. Huang, One-pot synthesis of LDH/GO composites as highly effective adsorbents for decontamination of U(VI), ACS Sustain. Chem. Eng. 5 (2017) 5608–5616.

[175] S.J. Yu, J. Wang, S. Song, K.Y. Sun, J. Li, X.X. Wang, Z.S. Chen, X.K. Wang, One-pot synthesis of graphene oxide and Ni-Al layered double hydroxides nanocomposites for the efficient removal of U(VI) from wastewater, Sci. China Chem. 60 (2017) 415–422.

[176] Y.D. Guo, Y.Q. Guo, X.G. Wang, P. Li, L.W. Kong, G.H. Wang, X.M. Li, Y.H. Liu, Enhanced photocatalytic reduction activity of uranium(vi) from aqueous solution using the Fe_2O_3–graphene oxide nanocomposite, Dalton Trans. 46 (2017) 14762–14770.

[177] Y.H. Wang, B. Chi, C.L. Xiao, D.G. Chen, Y.J. Wang, M.H. Liu, Graphene oxide@Mg_3-$Si_4O_9(OH)_{10}$: a hierarchical layered silicate nanocomposite with superior adsorption capacity for enriching Eu(III), Chem. Eng. J. 338 (2018) 628–635.

[178] C.R. Li, Y. Huang, Z. Lin, Fabrication of titanium phosphate@graphene oxide nanocomposite and its super performance on Eu^{3+} recycling, J. Mater. Chem. A 2 (2014) 14979–14985.

[179] Y. Li, G.D. Sheng, J. Sheng, Magnetite decorated graphene oxide for the highly efficient immobilization of Eu(III) from aqueous solution, J. Mol. Liq. 199 (2014) 474–480.

[180] D.M. Li, B. Zhang, F.Q. Xuan, The sorption of Eu(III) from aqueous solutions by magnetic graphene oxides: a combined experimental and modeling studies, J. Mol. Liq. 211 (2015) 203–209.

[181] A. Tayyebi, M. Outokesh, S. Moradi, A. Doram, Synthesis and characterization of ultrasound assisted "graphene oxide-magnetite" hybrid, and investigation of its adsorption properties for Sr(II) and Co(II) ions, Appl. Surf. Sci. 353 (2015) 350–362.

[182] Z.W. Huang, Z.J. Li, L.R. Zheng, L.M. Zhou, Z.F. Chai, X.L. Wang, W.Q. Shi, Interaction mechanism of uranium(VI) with three-dimensional graphene oxide-chitosan composite: insights from batch experiments, IR, XPS, and EXAFS spectroscopy, Chem. Eng. J. 328 (2017) 1066–1074.

[183] D.D. Shao, J.X. Li, X.K. Wang, Poly (amidoxime)-reduced graphene oxide composites as adsorbents for the enrichment of uranium from seawater, Sci. China Chem. 57 (2014) 1449–1458.

[184] P.P. Yang, Q. Liu, J.Y. Liu, H.S. Zhang, Z.S. Li, R.M. Li, L.H. Liu, J. Wang, Interfacial growth of a metal-organic framework (UiO-66) on functionalized graphene oxide (GO) as a suitable seawater adsorbent for extraction of uranium(VI), J. Mater. Chem. A 5 (2017) 17933–17942.

[185] D.D. Shao, G.S. Hou, J.X. Li, T. Wen, X.M. Ren, X.K. Wang, PANI/GO as a super adsorbent for the selective adsorption of uranium(VI), Chem. Eng. J. 255 (2014) 604–612.

[186] J.G. Catalano, G.E. Brown, Uranyl adsorption onto montmorillonite: evaluation of binding sites and carbonate complexation, Geochim. Cosmochim. Acta 69 (2005) 2995–3005.

[187] Y.B. Sun, X.X. Wang, W.C. Song, S.H. Lu, C.L. Chen, X.K. Wang, Mechanistic insights into the decontamination of Th(IV) on graphene oxide-based composites by EXAFS and modeling techniques, Environ. Sci.: Nano 4 (2017) 222–232.

[188] F.H. Li, Z. Yang, H.Q. Weng, G. Chen, M.Z. Lin, C. Zhao, High efficient separation of U(VI) and Th(IV) from rare earth elements in strong acidic solution by selective sorption on phenanthroline diamide functionalized graphene oxide, Chem. Eng. J. 332 (2018) 340–350.

[189] N. Pan, L. Li, J. Ding, R.B. Wang, Y.D. Jin, C.Q. Xia, A Schiff base/quaternary ammonium salt bifunctional graphene oxide as an efficient adsorbent for removal of Th(IV)/U(VI), J. Colloid Interface Sci. 508 (2017) 303–312.

[190] B.W. Hu, M.Q. Qu, Q.Y. Hu, Y.B. Sun, G.D. Sheng, J. Hu, J.Y. Ma, Decontamination of Sr(II) on magnetic polyaniline/graphene oxide composites: evidence from experimental, spectroscopic, and modeling investigation, ACS Sustain. Chem. Eng. 5 (2017) 6924–6931.

[191] Y. Wang, Z.S. Wang, R. Ang, J.J. Yang, N. Liu, J.L. Liao, Y.Y. Yang, J. Tang, Synthesis of amidoximated graphene oxide nanoribbons from unzipping of multiwalled carbon nanotubes for selective separation of uranium(vi), RSC Adv. 5 (2015) 89309–89318.

[192] L.P. Lingamdinne, Y.-L. Choi, I.-S. Kim, J.-K. Yang, J.R. Koduru, Y.-Y. Chang, Preparation and characterization of porous reduced graphene oxide based inverse spinel nickel ferrite nanocomposite for adsorption removal of radionuclides, J. Hazard. Mater. 326 (2017) 145–156.

[193] Y. Yoon, W.K. Park, T.-M. Hwang, D.H. Yoon, W.S. Yang, J.-W. Kang, Comparative evaluation of magnetite-graphene oxide and magnetite-reduced graphene oxide composite for As(III) and As(V) removal, J. Hazard. Mater. 304 (2016) 196–204.

[194] G. Bharath, E. Alhseinat, N. Ponpandian, M.A. Khan, M.R. Siddiqui, F. Ahmed, E.H. Alsharaeh, Development of adsorption and electrosorption techniques for removal of organic and inorganic pollutants from wastewater using novel magnetite/porous graphene-based nanocomposites, Sep. Purif. Technol. 188 (2017) 206–218.

[195] C.C. Ding, W.C. Cheng, Y.B. Sun, X.K. Wang, Effects of Bacillus subtilis on the reduction of U(VI) by nano-Fe^0, Geochim. Cosmochim. Acta 165 (2015) 86–107.

[196] W.C. Cheng, Z.X. Jin, C.C. Ding, M.L. Wang, Simultaneous sorption and reduction of U(VI) on magnetite-reduced graphene oxide composites investigated by macroscopic, spectroscopic and modeling techniques, RSC Adv. 5 (2015) 59677–59685.

[197] D.E. Latta, C.A. Gorski, M.I. Boyanov, E.J. O'Loughlin, K.M. Kemner, M.M. Scherer, Influence of magnetite stoichiometry on U-VI reduction, Environ. Sci. Technol. 46 (2012) 778–786.

[198] D.M. Singer, S.M. Chatman, E.S. Ilton, K.M. Rosso, J.F. Banfield, G.A. Waychunas, U(VI) sorption and reduction kinetics on the magnetite(111) surface, Environ. Sci. Technol. 46 (2012) 3821–3830.

[199] D.M. Singer, S.M. Chatman, E.S. Ilton, K.M. Rosso, J.F. Banfield, G.A. Waychunas, Identification of simultaneous U(VI) sorption complexes and U(IV) nanoprecipitates on the magnetite(111) surface, Environ. Sci. Technol. 46 (2012) 3811–3820.

[200] S. Yan, B. Hua, Z.Y. Bao, J. Yang, C.X. Liu, B.L. Deng, Uranium(VI) removal by nanoscale zerovalent iron in anoxic batch systems, Environ. Sci. Technol. 44 (2010) 7783–7789.

[201] P. Wersin, M.F. Hochella Jr., P. Persson, G. Redden, J.O. Leckie, D.W. Harris, Interaction between aqueous uranium(VI) and sulfide minerals: spectroscopic evidence for sorption and reduction, Geochim. Cosmochim. Acta 58 (1994) 2829–2843.

[202] H.J. Zhu, H.Y. Gao, X.M. Huang, W.F. Kong, X.M. Yan, The uptake of europium by reduced graphene oxide-supported nanoscale zerovalent iron investigated by batch and modeling techniques, J. Environ. Chem. Eng. 3 (2015) 2974–2980.

[203] W.J. Mu, Q.H. Yu, R. Hu, X.L. Li, H.Y. Wei, Y. Jian, Porous three-dimensional reduced graphene oxide merged with WO_3 for efficient removal of radioactive strontium, Appl. Surf. Sci. 423 (2017) 1203–1211.

Chapter 5

Removal of toxic/radioactive metal ions by metal-organic framework-based materials

Jie Li*, Wenling Ye* and Changlun Chen†
*School of Resources and Environment, Anhui Agricultural University, Hefei, People's Republic of China
†Key Laboratory of Photovoltaic and Energy Conservation Materials, Institute of Plasma Physics, Chinese Academy of Sciences, Hefei, People's Republic of China

5.1 Introduction

5.1.1 MOFs and MOF-based materials as adsorbents

Metal-organic frameworks (MOFs) are composed of secondary building units (SBUs) (e.g., metal ions or metal clusters) and organic ligands, which can be connected into three-dimensional (3D) lattices [1]. MOFs derived from polyhedral or small-cluster coordination polymers were later extended into one-dimensional (1D) chains, two-dimensional (2D) layers, and finally 3D frameworks. It is well established that MOFs are one of the most widely investigated materials of the 21st century, owing to their structural tailorability, controlled porosity, and high crystallinity [2, 3]. MOFs and their composites have been applied in a wide spectrum of applications, including drug delivery, gas storage, catalysis, and luminescence sensing [4–11]. MOFs and their composites acting as a class of adsorbent materials have emerged within the past three decades. MOFs have numerous advantages when one considers their application as adsorbents. One advantage is the high surface area and porosity of MOFs, which can aid with adsorption site accessibility and the diffusion of toxic or radioactive metal ions through the framework. The surface areas of typical MOFs range from 1000 to 10,000 m^2/g [12], which can exceed those of traditional porous materials such as zeolites and carbons. The pores of MOFs are highly ordered due to their crystalline structure. In addition, the size of MOF pores as well as the MOF shape can be modulated by selecting the organic linkers as well as the connectivity of the metal ions. Another advantage

Interface Science and Technology, Vol. 29. https://doi.org/10.1016/B978-0-08-102727-1.00005-4

is that MOFs can be fabricated on a large scale via low-cost and simple synthetic methods [13–17]. Also, many MOFs have displayed superior stability under harsh conditions [18–24]. Moreover, MOFs can be functionalized through postsynthetic modification (PSM), which can customize the ways in which they interact with toxic or radioactive metal ions. These advantages as well as their ordered porous structures and tunable physiochemical properties make MOFs highly promising candidates for the sequestration of toxic or radioactive metal ions from aqueous solutions. Over the last three decades, many researchers have used MOFs and MOF-based composites as potential scavengers for the decontamination of various toxic and/or radioactive metal ions [20, 21, 25–37] from aqueous solutions.

5.1.2 Approaches to improving the removal performance of MOFs

The subsequent modification or functionalization of MOFs is highly convenient due to their simple and efficient synthesis methods, which offer facile approaches to obtain MOF-based materials with structural complexity, composition uniformity, and reproducibility. The removal properties of toxic/radioactive metal ions by MOFs and MOF-based materials, including reaction kinetics/thermodynamics, equilibrium time, removal capacity, reaction pH range, and interference ability, are mainly dominated by the highly ordered framework structures, the hydrophilicity, the surface area, the functionality, and the pore size and distribution of MOFs and MOF-based materials. In recent years, researchers have achieved great progress in improving the removal performance of MOFs, which covers five ways (Fig. 5.1): (a) MOFs prepared by longer organic ligands. The aperture of MOFs can expand this approach [18, 20], which is very powerful and can facilitate the diffusion of metal ions through the framework of MOFs [20]. However, this method depends on the commercial availability of the wanted organic ligands, which are often costly; (b) The design of defective MOFs. That is, creating defects in MOFs by adding modulators to the precursor mixture. This method can expand the aperture size and introduce additional reaction sites into MOFs [33, 38–40]. Moreover, generally defective MOFs will become more hydrophilic [41]. For instance, the creation of defects was observed if UiO-66 is prepared by adding monocarboxylic acid modulators under widely used conditions [40]. This method is easier and more economical than method (a); (c) MOFs prepared from organic ligands with functional groups. Functionality can be incorporated into MOFs by using organic ligands with additional functional groups, such as —NH_2 [20, 42–44], —OH [20, 45, 46], and —SH [21]. MOFs prepared by this method have additional specific functional groups, which can provide extra reaction sites and enhance the selectivity of pristine MOFs; (d) MOFs prepared from metal clusters with extra functional groups. Specifically, MOFs synthetized by metal clusters that are functionalized with specific molecules [47, 48]. For example, Liang et al. [47] designed sulfur-modified MOFs by using $Co(NCS)_4^{2-}$ as the metal precursor, in

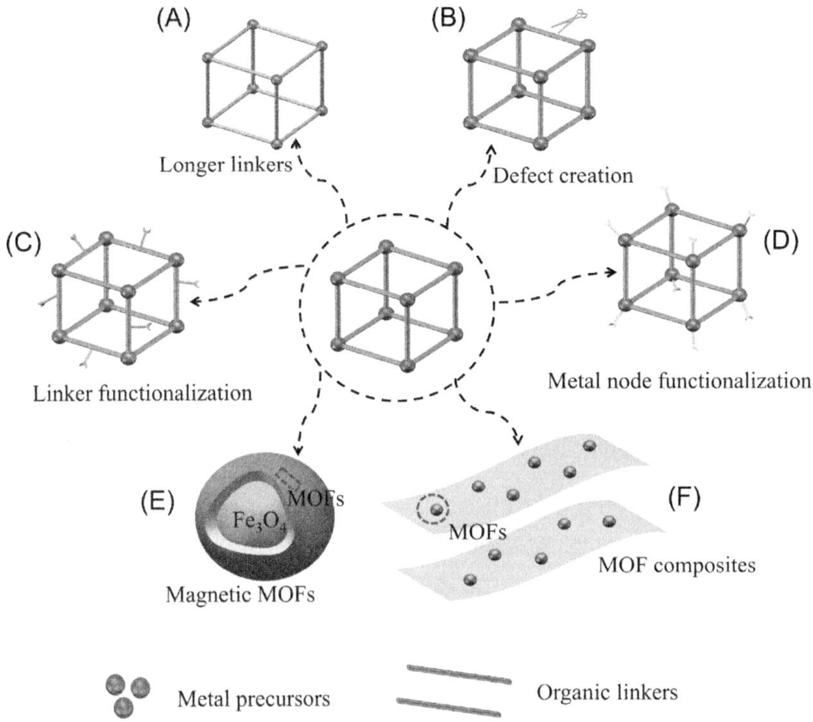

FIG. 5.1 Approaches to improving the adsorption performance of MOFs.

which the chemically soft S atom is available for the capture of Hg(II) ions. Postsynthetic ligand exchange and PSM methods are commonly used to introduce functional groups into MOFs [49–53]. MOFs modified by method (d) are generally more active than those functionalized with method (c) due to the steric hindrance; (e) MOF-based composites, that is, the hybridization of MOFs with other functional materials, such as magnetic materials (e.g., Fe_3O_4) [54, 55]. Magnetic MOFs have the advantage of easy recovery from solutions by using a magnetic field, or with other adsorbents, such as MOF-GO composites [56], showing improved removal performance compared with their individual counterparts due to the synergistic effects of the components [56, 57]. Usually, MOFs are small particles (50–100 nm) that are prone to form fine water suspensions that will move through the columns or lead to column clogging; these disadvantages can be conquered by using method (f) to synthesize MOF-based materials [13, 58].

5.1.3 The aim of this chapter

Very recently, MOFs and their composites have been shown to be extremely powerful porous materials in the capture of toxic and radioactive metal ions. This chapter aims to give insight on the latest advances achieved on the application of MOFs and their composites as adsorbents for water purification. To this end, we will critically provide some typical examples focused on the removal of various metal ions from wastewater. The adsorption properties are evaluated among MOF-based materials and compared with those of other widely used adsorbent materials. Moreover, some future trends of this area are provided. We think that summarizing the applications of MOF-based materials in the decontamination of toxic and radioactive metal ions is very important for future research toward wastewater purification. Such a summarization is anticipated to help researchers get thorough knowledge of approaches to the design and synthesis of high-performance MOF-based materials for wastewater decontamination as well as the interplay mechanism between MOFs and metal ions.

5.2 Removal of radioactive metal ions by MOF-based materials

5.2.1 Uranium

Up to now, a lot of MOFs and MOF-based materials have shown effective and efficient removal performance for U(VI) in aqueous solutions. In the following subsections, we pursued some typical examples of MOFs and MOF-based materials for the preconcentration of U(VI) ions from aqueous solutions. In addition, the adsorption properties of U(VI) by reported MOFs and MOF-based materials are evaluated and compared with those by other adsorptive materials.

5.2.1.1 Pristine MOFs for the removal of U(VI)

The first application of MOFs as novel adsorbents to extract U(VI) from aquatic environments was reported by Lin's group [59]. Two different phosphorylurea groups were covalently grafted onto **UiO-68** MOFs and the resulting materials were efficient in separating U(VI) with the maximum adsorption capacity of 217 mg/g. DFT calculations demonstrated that two phosphorylurea ligands bind to one uranyl ion. Recently, Wang's group reported a family of zirconium phosphonate MOFs and applied them for the removal of U(VI) [60]. Among these MOFs, **SZ-3** displayed high stability in various solutions and even in fuming acids, which is never reported in other reported MOFs. SZ-3 showed a moderate adsorption capacity for U(VI) species and was active in a broad pH range with adsorption efficiency of 46.4% and 97.4% at pH 1 and 4.5, respectively. In addition, SZ-3 showed an adsorption rate of 80.2% for U(VI) in 60 min at pH 4.5. The superior adsorption performance was not affected by large amounts of interfering metal ions, including Li(I), Mg(II), Cu(II), Pb(II), Al(III), Ho(III), and Mo(VI), which are usually present in nuclear wastewaters. However, information about the recyclability of this MOF is not provided, which is crucial for an adsorbent in the practical extraction of fissionable radionuclides from spent nuclear fuel. Except the excellent adsorption performance for U(VI) species, this work gives important hints for the transferability of treasured knowledge in this field. In addition, this study overcame common problems and obtained highly crystalline zirconium phosphonate MOFs by using the synthetic approaches, which would lead to the development of phosphonate-based MOFs and broaden the application range of MOFs.

5.2.1.2 Functionalized MOFs for the removal of U(VI)

Micro- and mesoporous **MOF-74(Zn)** was functionalized with different amounts of coumarin molecules by the postsynthetic method and further used for the effective removal and photo-triggered storage/release of U(VI) ions [61]. Maximum adsorption capacity of 360 mg/g was obtained by micro- and mesoporous MOF-74(Zn) modified with ca. 11.7 wt% of coumarin molecules. As identified by the FTIR technique, coumarin molecules were indeed the preferential active site for U(VI) capture, but higher loadings of coumarin (−20.6, 28.2, 35.7 wt%) led to a decrease in the adsorption performance due to the steric constraints. Although practical application requires conducting at more acidic solution and some regeneration/reusability tests are not provided, this study gives important hints for future developments on the photo-switched storage/release of U(VI).

Very recently, Sun et al. [62] reported amidoxime-modified MOF for the removal of U(VI) ions. They found that the maximum U(VI) adsorption capacity of this modified MOF at pH 7.0 was 454.55 mg/g (Fig. 5.2A), which was much higher than that of heavy metals (e.g., 54, 62, 25, and 68 mg/g for Ca(II), Mg(II), Sr(II), and Na(I), respectively) (Fig. 5.2B). Furthermore, this modified MOF could also extract 2.74 mg/g U(VI) from Huanghai seawater within 90 min. In another study reported by Bai and coworkers [63], they prepared three kinds of **MIL-101(Cr)** amino derivatives by covalent grafting to the open metal sites of MIL-101(Cr), including MIL-101-NH_2(Cr), MIL-101 ED(Cr), (ED = ethanediamine) and MIL-101-DETA(Cr) (DETA = diethylentriamine). These MIL-101(Cr) amino derivatives showed high thermal and acid stability, thus these MOFs were investigated to the effect of different kinds of amino groups on the capture of U(VI) from aqueous solution. The capture of U(VI) was strongly dependent on solution pH. No obvious adsorption performance was observed by any MOF at $pH < 4$; it reached the optimal adsorption performance at $pH = 5.5$. The highest adsorption capacity of 315 mg/g was reached by MIL-101-DETA(Cr), which also showed relatively fast kinetic adsorption. As evidenced by the FTIR and EXAFS results, the adsorbed U(VI) ions were coordinated to the amine groups, thus MIL-101-DETA(Cr) was reasonable to show the best adsorption capacity due to its higher loading of aliphatic amines. Although the coexisting of other metal ions did not exert an obvious influence on the adsorption capacity, the reusability of this MOF still needs improvements because 30% of its adsorption capacity was lost in the second cycle.

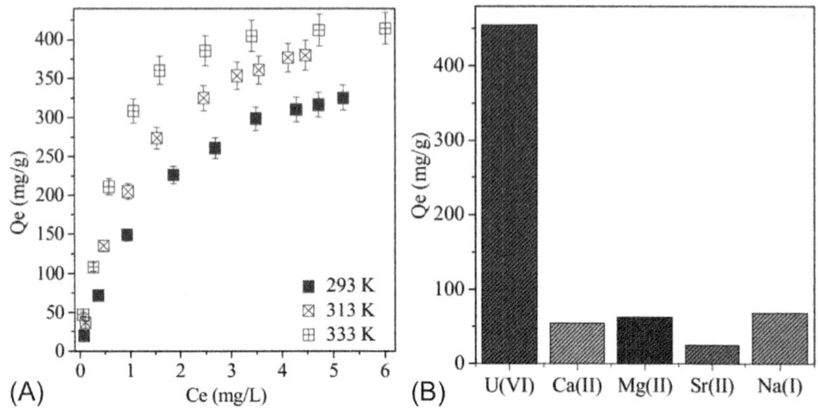

FIG. 5.2 Adsorption isotherms of U(VI) on AO/MOF composites under different temperatures (A); comparison of maximum adsorption capacities of AO/MOF composites for U(VI) and other heavy metals (B) [62]. *(Reprinted with permission from F. Li, X. Li, P. Cui, Y. Sun, Plasma-grafting amidoxime/metal-organic framework composites for the selective sequestration of U(VI), Environ. Sci. Nano (2018), doi:10.1039/C1038EN00583D. Copyright 2018 RSC.)*

5.2.1.3 MOF composites for the removal of U(VI)

MOFs combined with other functional materials can achieve considerably enhanced adsorption performance compared with that of individual counterparts through synergistic effects among components with rationally designed properties [55, 56]. In particular, a composite consisting of UIO-66(Zr) and graphene oxide (GO) (**GO—COOH/UiO-66**) showed a remarkably high adsorption capacity for U(VI) at ppm and ppb levels from seawater, which surpassed each of the components separately [56]. The adsorption performance of this MOF composite was strongly dependent on solution pH, with its optimal adsorption rate achieved at 7.5–8.4, which is the common seawater pH condition. The influence of initial U(VI) concentration and contact time on the adsorption performance was also studied in detail. They found that the maximum adsorption capacity was 1012 mg/g, but the adsorption kinetics were moderate, that is, 99% of U(VI) was removed in 230 min from 500 ppm aqueous solutions. Importantly, the adsorption performance was not affected by the coexisting competing ions, including Ca(II), Mn(II), Pb(II), Fe(III), K(I), Cu(II), Al(III), Ba(II), Na(I), Zn(II), Ni(II), Mg(II), and Sr(II). In addition, this composite displayed excellent recyclability with only a small decrease in the adsorption amount after five adsorption-desorption cycles, which was traceable to the adsorbent mass loss between cycles and to the impossibility of regenerating all the adsorptive sites. Different eluents, including H_2O, 0.1 M HNO_3, 0.1 M $NaHCO_3$, and 0.1 M NaOH, were tried to regenerate this composite. They observed that the best eluent was a diluted HNO_3 solution. As shown in Fig. 5.3, FTIR and XPS results strongly indicate that the superior adsorption performance for U(VI) was driven by the coordination

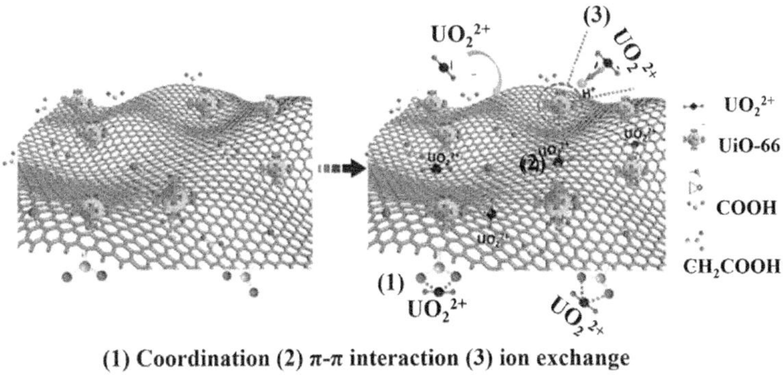

FIG. 5.3 Proposed mechanisms for the adsorption of U(VI) onto GO—COOH/UiO-66 composites [56]. *(Reprinted with permission from P. Yang, Q. Liu, J. Liu, H. Zhang, Z. Li, R. Li, L. Liu, J. Wang, Interfacial growth of a metal-organic framework (UiO-66) on functionalized graphene oxide (GO) as a suitable seawater adsorbent for extraction of uranium(VI), J. Mater. Chem. A 5 (2017) 17933–17942. Copyright 2017 RSC.)*

or the π-π interaction between U(VI) and the carboxylate/hydroxyl groups from the surface of the composite. Interestingly, the superior adsorption performance was maintained at ppb concentration in simulated seawaters.

5.2.1.4 Comparison between MOF-based materials and with other adsorbents

Apart from the above-mentioned typical MOFs and MOF-based materials reported for the removal of U(VI) from aqueous solutions, there are many other MOFs and MOF-based materials showing excellent adsorption efficiency for U(VI); these are shown in Table 5.1. Table 5.1 reveals that the removal of U(VI) by pristine MOFs was mainly via a mechanism of ion exchange due to the highly ordered structures of MOFs [59–61, 63–66]. In terms of functionalized MOFs and MOF-based composites, coordination with functional groups or the synergic effect between different components is another alternative interplay mechanism for the removal of U(VI) from aqueous solutions [55, 67–71].

Table 5.1 reveals that among these MOFs and MOF-based materials, GO—COOH/UiO-66 displayed the highest adsorption capacity for U(VI) [56]; moreover, GO—COOH/UiO-66 showed superior selective properties for the removal of U(VI). That is, GO—COOH/UiO-66 showed excellent adsorption efficiency, even under the interference of various metal ions with different valence states. It is worth noting that MOF Co-SLUG-35 was applied without any pretreatments [65], which is unlike other MOFs where degassing was primarily needed. Relatively few MOFs and MOF-based materials, including MIL-101-DETA [63], GO—COOH/UiO-66 [56], and Fe_3O_4@ZIF-8 [55], have had their detailed removal properties reported. These MOFs and MOF-based materials displayed relatively fast adsorption kinetics and excellent selectivity, but limited reusability. And the modified MOF MIL-101-DETA has a narrow

TABLE 5.1 Uranium adsorption data for reported MOF-based materials

MOF-based materials	Adsorption capacity (mg/g)	Equilibrium time	Selectivity	Reusability	Mechanism	Ref.
Fe_3O_4@ZIF-8	523.5	2h	vs various actinides	Reusable	Coordination, hydrogen bonding	[55]
GO—COOH/UiO-66	1012	230min	vs various cations	Reusable	Coordination, ion-exchange	[56]
MOF-2	217	NA	–	–	Ion-exchange	[59]
MOF-3	109	1h	–	–	Ion-exchange	[59]
SZ-2/SZ-3	58.18/58.18	5h/7h	vs various cations	–	Ion-exchange	[60]
A1	200	5h	–	–	Ion-exchange	[61]
A2	300	5h	–	–	Ion-exchange	[61]
A3	360	5h	–	–	Ion-exchange	[61]
MIL-101-ED	200	1.5h	–	–	Ion-exchange	[63]
MIL-101-DETA	350	1.5h	vs Co^{2+}, Ni^{2+}, Sr^{2+}, etc.	Reusable	Ion-exchange	[63]
MOF-76	300	5h	vs various cations	–	Ion-exchange	[64]
Co-SLUG-35	118	4h	vs $CrO_4{}^{2-}$, Cl^-	Limited	Ion-exchange	[65]
UiO-66	109.9	5h	vs various cations	–	Ion-exchange	[66]
UiO-66-NH_2	114.9	5h	vs various cations	–	Ion-exchange	[66]

Carboxyl-functionalized MIL-101	314	2h	vs various cations	Reusable	Binding with carboxyl groups	[67]
ED grafting MIL-101(Cr)	109	NA	–	Weak	Coordination with amine groups	[68]
CMPO@MIL-101	–	30min	vs various rare earth metals	Reusable	Coordination	[69]
Fe_3O_4@AMCA-MIL53(Al)	227.3	90min	–	Weak	Electrostatic interactions and coordination	[70]
UiO-66-AO	2.68 (real seawater)	120min	–	Reusable	Coordination	[71]
HKUST	787.4	60min	–	Weak	Coordination, electrostatic interactions	[72]
1-NO_2	165.29	60min	–	–	Chemisorption	[73]
Azo-MOFs	312.32	140min	–	–	Chemisorption	[74]

active pH range (pH4.5–5.5). In addition, the information about the radiation stability of these MOFs and MOF-based materials was not provided, which is very important for further evaluation of their feasibility in real-world applications. It is also a crucial aspect for applications in radioactive waste management, especially after a high preconcentration of radioactive U(VI) ions.

The comparison of U(VI) adsorption properties by MOFs and MOF-based materials with other widely used adsorptive materials can clearly show the advantages of using MOFs and MOF-based materials for the removal of U(VI). Table 5.1 reveals that several MOFs and MOF-based materials have displayed ultrahigh adsorption capacities (up to 787.4mg/g) [55, 61, 63, 64, 67, 72], which were superior to other widely used adsorptive materials, including AC (only a few dozen mg/g) [75], carbonaceous materials (125mg/g) [76], clay minerals (only a few dozen mg/g) [77], metal oxides (<100mg/g) [78, 79], LDHs (26–330mg/g) [80, 81], NZVI (215mg/g) [82, 83], GOs (100–299mg/g) [56, 84, 85], and CNTs (19–212mg/g) [86, 87]. They were also comparable to metal sulfide ion exchangers (MSIEs) (380mg/g for KMS-1) [88], chalcogenide (338mg/g) [89], and mesoporous silica (409) [90–93]. However, traditional adsorptive materials such as AC, clay minerals, carbonaceous materials, and LDHs usually showed poorer and slower adsorption kinetics and recyclability compared with MOFs and MOF-based materials [59, 72, 73]. It is reported that metal oxides, LDH, and mesoporous silica were found to be unstable in extreme alkaline or acidic conditions of nuclear waste, which was due to the immediate dissolution of the aluminium/silicon ions [94]. Widely used NZVI particles also have some obvious disadvantages covering low selectivity, a narrow working pH range, and rapid deactivation [95, 96]. Traditional CNTs displayed a low selective adsorption property for U(VI) due to the lack of specific functional groups. Comparing MOFs and MOF-based materials with GOs, the low-cost fabrication method of MOFs is superior to the multistep synthesis of Gos, which also needs toxic organic reagents and solvents [97]. In comparison of MOFs and MOF-based materials with MSIEs, many MSIEs showed low regeneration abilities and reusability after the load of U(VI) species while some MOFs still have remarkable reusability even after several adsorption-desorption cycles [55, 63, 67].

Nowadays, the development of novel adsorptive materials applicable in seawater has attracted considerable interest for the selective uptake of radioactive U(VI) from seawater. UiO-68-type MOFs [59, 71], Co-SLUG-35 [65], and GO—COOH/UiO-66 [56] have shown excellent extraction efficiency for U(VI) from simulated seawater, indicating promising alternatives for U(VI) uptake. In addition, several MOF-based materials [55, 63, 66, 67] displayed superior selective uptake efficiency for U(VI) in the presence of lanthanides, including La(III), Ce(III), Sm(III), Gd(III), Yb(III), Nd(III), Eu(III), Dy(III), and Er(III). To date, the efficient extraction of actinides from lanthanides is still a difficult task in nuclear waste management and is a crucial aspect in spent fuel reprocessing. MOFs and MOF-based materials show great promise in the selective separation of actinides from lanthanides, given that proper structures are constructed and specific functional groups are incorporated into MOFs and MOF-based materials.

5.2.2 Cs(I) and Sr(II)

Cs(I) and Sr(II) are largely produced in the process of nuclear fuel cycle and are difficult to form complex with organic functional groups. Thus, adsorption including ion exchange becomes an efficient and applicable technique to separate them from aqueous solution.

Nowadays, few MOF materials have been investigated to sequester ^{137}Cs and ^{90}Sr because it is challenging to design anionic MOFs with weak-complexing cations in the pores. The selection of SBUs and ligands is crucial in the construction of anionic frameworks, especially in the stability of MOFs. In 2015, Wang et al. reported a rare example of a *3D uranyl organic framework* material consisting of a 3,5-di(4′-carboxylphenyl) benzoic acid and $UO_2(NO_3)_2 \cdot 6H_2O$ by solvothermal reaction under *N,N*-dimethylformamide (DMF), and water [29]. This MOF has three sets of graphene-like layers with exchangeable $[NH_2(CH_3)_2]^+$ cations in the pores. Batch experiments disclosed that Cs(I) adsorption proceeded fast and the maximum adsorption capacity was 432 mg/g. Moreover, the adsorption efficiency of Cs(I) was up to 72%–94%, even in the interference of 20 mass equivalences of coexisting Li(I), Na(I), K(I), Rb(I), Mg(II), or Ca(II) cations. Importantly, no structural or crystal degradation was observed even after 200 kGy ^{60}Co γ irradiation and 200 kGy β irradiation, suggesting excellent resistance to radiation. In another study, Aguila et al. [98] modified the ultrastable **MIL-101** with sulfonic acid groups and obtained excellent adsorbents toward Cs(I) and Sr(II). As shown in Fig. 5.4A and B, this modified MOF showed high adsorption performance toward Cs(I) and Sr(II) at different solution pH and in the presence of the interference of coexisting ions. A total of 99.9% and 98.92% of Cs(I) and Sr(II) is removed at a molar ratio of 4:1 adsorbent to Cs(I) and Sr(II) solution, leading to remarkably high K_d values (Fig. 5.4C). However, the adsorption kinetics were quite slow, requiring 24 h to reach the adsorption equilibrium. MOFs can offer ample pores as well as high specific surface area to accommodate various functional groups. Naeimi et al. [57] modified **HKUST-1** potassium nickel hexacyanoferrate and found that this modified MOF showed fast adsorption kinetics (30 min), high adsorption capacity (153 mg/g), high adsorption affinity ($K_d = 1.5 \times 10^3$ mL/g), and excellent selectivity toward Cs(I).

As shown in Table 5.2, nowadays the 3D uranyl organic framework reported by Wang et al. [29] showed the highest adsorption capacity for Cs(I) (432 mg/g), which represents one of the highest values for the removal

pH	% removal	q_e (mg/g)	K_d (mL/g)
3	88.35	23.43	1896.04
6	100.00	36.47	—
10	79.26	29.17	955.37

(A)

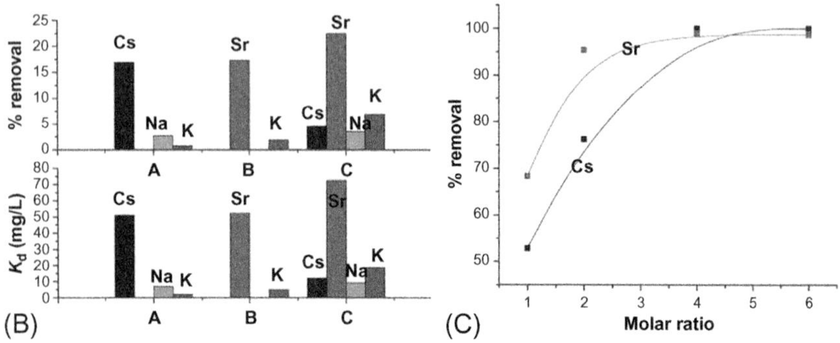

FIG. 5.4 Effect of solution pH on Cs removal (A); Percent removal and K_d values of all ions in A–C experiments (B); percent removal of Cs and Sr with increasing amounts of sorbent (time = 24 hours) (C) [98]. *(Reprinted with permission from B. Aguila, D. Banerjee, Z. Nie, Y. Shin, S. Ma, P.K. Thallapally, Selective removal of cesium and strontium using porous frameworks from high level nuclear waste, Chem. Commun. 52 (2016) 5940–5942. Copyright 2016 RSC.)*

of Cs(I) via solid extraction. In comparison, other adsorptive materials also have been reported for the removal of Cs(I), but with relatively low adsorption capacities (e.g., <20 mg/g for clay minerals [115, 116], <185 mg/g for GO-based materials [117], and 280 mg/g for MSIE (KMS-2) [88, 118]). In addition, metal oxides such as manganese oxides displayed low selectivity for Cs(I) under the interference of salt with high concentrations [119].

5.2.3 Selenium oxyanions

Until now, very few examples have focused on the removal of selenium oxyanions by MOF-based materials, with only the reports from Hupp and Farha's group [20] and Wang's group [33]. For example, researchers in Hupp and Farha's group used a series of stable Zr-based MOFs, including **UiO-66**, **UiO-66-NH$_2$**, **UiO-66-(NH$_2$)$_2$**, **UiO-66-(OH)$_2$**, **UiO-67**, **NU-1000**, and **NU-1000BA**, for the capture of selenite and selenite from water. They found that NU-1000 showed the highest adsorption capacity and fastest adsorption rates for both selenite and selenate among all these MOFs studied (Fig. 5.5A). The maximum adsorption capacity of NU-1000 for selenite and selenate is 102 and 62 mg/g, respectively. X-ray total scattering data and diffuse reflectance infrared Fourier transform spectroscopy analyses clearly showed that selenite and selenate ions were exclusively bound to the zirconium cluster nodes in a

TABLE 5.2 Other radioactive metal ion adsorption on reported MOF-based materials

Targets	MOF-based materials	Adsorption capacity (mg/g)	Equilibrium time	Selectivity	Reusability	Mechanism	Ref.
SeO_3^{2-}/SeO_4^{2-}	UiO-66	27/17	72 h	–	–	Ion-exchange	[20]
SeO_3^{2-}/SeO_4^{2-}	UiO-66-$(OH)_2$	15.5/–	72 h	–	–	Ion-exchange	[20]
SeO_3^{2-}/SeO_4^{2-}	UiO-66-$(NH)_2$	45/38.5	72 h	–	–	Ion-exchange	[20]
SeO_3^{2-}/SeO_4^{2-}	NU-1000	102/62	<1 min	–	–	Ion-exchange	[20]
Cs^+	3D uranyl organic framework	432	20 min	vs halide, Ca^{2+}, Mg^{2+}	–	Ion-exchange	[29]
TcO_4^-	NDTB-1	NA	>36 h	vs various oxyanions	–	Ion-exchange	[30]
Co^{2+}	$[Mn_3(L)_2]^{2-} \cdot [NH_2H_3)_2]^+ \cdot 9DMF$	93	24 h	vs Li^+, Na^+	–	Ion-exchange	[31]
ReO_4^-	SCU-100	541	2 h	vs SO_4^{2-}, CO_3^{2-}, PO_4^{3-}	Reusable	Ion-exchange	[32]
SeO_4^{2-}	UiO-66-HCl	86.8	5 h	–	–	Ion-exchange	[33]
Ba^{2+}	MIL-101-Cr-SO_3H	70.5	<250 min	vs various cations	Reusable	Chelate with sulfonic group	[52]
Ba^{2+}	MOF-808-SO_4	131.1	<250 min	vs various cations	Reusable	Binding with sulfate group	[52]
Cs^+	MOF/KNiFC	153	45 min	vs K^+, Na^+	Reusable	Ion-exchange chemisorption	[57]
ReO_4^-	SLUG-21	602	NA	–	–	Ion-exchange	[99]
MnO_4^-	SLUG-21	283	>48 h	vs NO_3^-, CO_3^{2-}	–	Ion-exchange	[99]
ReO_4^-	UiO-66-NH_2	159	24 h	vs NO_3^-, SO_4^{2-}, PO_4^{3-}	–	Ion-exchange	[100]
ReO_4^-	SBN	786	2 h	vs various anions	–	Ion-exchange	[101]

Continued

TABLE 5.2 Other radioactive metal ion adsorption on reported MOF-based materials—Cont'd

Targets	MOF-based materials	Adsorption capacity (mg/g)	Equilibrium time	Selectivity	Reusability	Mechanism	Ref.
ReO_4^-	SCU-101	217	10 min	vs various anions	Reusable	Ion exchange	[102]
Th^{4+}	UiO-66-$(COOH)_2$	350	30 min	vs various ions	Low	Coordination	[103]
Pd^{2+}	UiO-66-NH_2	167	>3 h	vs Co^{2+}, Ni^{2+}, Cu^{2+}, Zn^{2+}	Reusable	Electrostatic attraction	[104]
Co^{2+}	PCN-100	1.61 formula	NA	–	–	Chelating effect	[105]
Eu^{3+}	HKUST-1@$H_3PW_{12}O_{40}$	14.58	120 min	vs Ca^{2+}, Mg^{2+}, Hg^{2+}, Zn^{2+}	Reusable	Coordination with ligands	[106]
ReO_4^-	SCU-7	NA	>2.5 h	–	–	Ion-exchange	[107]
MnO_4^-	Ag(btr)·PF_6·0.5CH_3CN	0.56 mol/mol	24 h	vs NO_3^-, BF_4^-, ClO_4^-	–	Ion-exchange	[108]
Ba^{2+}	Zr-BDC-NH_2-SO_4	181.8	420 min	vs various cations	–	Chelation with sulfate group	[109]
Co^{2+}	UiO-66-Schiff base	256	5 h	vs various ions	Reusable	Coordination	[110]
Th^{4+}	SCU-3	NA	NA	–	–	Coordination ion-exchange	[111]
Th^{4+}	MIL-100(Al)	167	15 min	vs UO_2^{2+}, Nd^{3+}	–	Chemisorption	[112]
Th^{4+}	dMn-MOF	46.345	–	vs La^{3+}, Eu^{3+}, Lu^{3+}	–	–	[113]
Pd^{2+}	ASUiO-66	45.4	NA	vs various cations	Reusable	Acting with sulfur and alkene	[114]

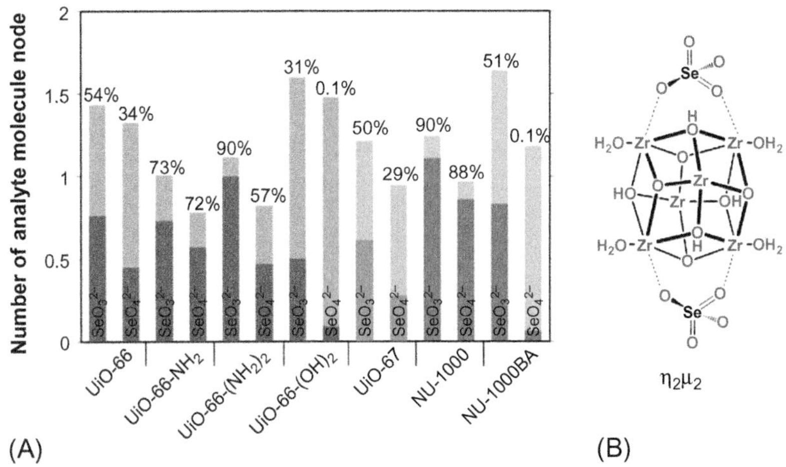

FIG. 5.5 Bar graph illustrating the number of selenate or selenite molecules adsorbed per node in a series of Zr-based MOFs (A); potential binding modes of selenate (or selenite) to the node of NU-1000 (B) [20]. *(Reprinted with permission from A.J. Howarth, M.J. Katz, T.C. Wang, A.E. Platero-Prats, K.W. Chapman, J.T. Hupp, O.K. Farha, High efficiency adsorption and removal of selenate and selenite from water using metal–organic frameworks, J. Am. Chem. Soc. 137 (2015) 7488–7494. Copyright 2015 ACS.)*

$\eta_2\mu_2$ fashion (Fig. 5.5B). Wang's group modified **Zr_6 MOFs** with various defects via an easy modulator synthetic technique and applied them in the adsorption of selenate ions [33]. They found that increasing the concentrations of missing linker defects or missing cluster defects led to more efficient and faster uptake of selenate oxyanions. Selenate oxyanions could substitute the weak-binding compensating ligands, including Cl^-, $O_2C-CH_3^-$, $O_2C-CF_3^-$ or $O_2C-CHF_2^-$, HO^-, and/or H_2O, on Zr_6 nodes via anion-exchange, which can be shown from the XPS results. Moreover, experimental and theoretical results revealed that smaller "capping" ligands, including HO^-, Cl^-, or H_2O, were more easily replaced than the larger "capping" ligands, including $O_2C-CH_3^-$, $O_2C-CF_3^-$, or $O_2C-CHF_2^-$. Importantly, **UiO-66-HCl** showed a higher adsorption capacity than NU-1000, suggesting that the pore size of MOFs can be efficiently expanded by the used modulator synthetic technique. The authors showed that the capture of selenate oxyanions decreased with the increase of solution pH and ionic strength while no information about the regeneration or reusability of UiO-66-HCl is provided.

Nowadays, a lot of materials have been applied in the removal of SeO_3^{2-} and SeO_4^{2-} anions from aqueous solutions, including MOFs and MOF-based materials [20, 33], clay materials [120, 121], metal oxides and their composites [122–124], LDH and functionalized LDH [125], modified silica materials and their resins [126–128], and GO-based materials [129]. Among these materials, NU-1000 showed a comparable adsorption capacity for SeO_3^{2-} anions

(95 mg/g) to that of MgO nanosheets (103.52 mg/g) [122] while Gly-LDH had a much larger adsorption capacity for SeO_4^{2-} anions (208.6 mg/g) [125]. Importantly, NU-1000 displayed the highest reaction kinetics for SeO_3^{2-} and SeO_4^{2-} anions (<1 min). In addition, NU-1000 displayed a quite equal rate for both SeO_3^{2-} and SeO_4^{2-} anions, which can be observed by few materials. As shown in Table 5.2, UiO-66-HCl displayed a relatively larger adsorption capacity for SeO_4^{2-} anions, which represents a promising option for real-world applications due to the low cost and easy preparation method of UiO-66-HCl. Information about the reusability, selectivity, and column studies of NU-1000 and UiO-66-HCl are needed for practical applications.

5.2.4 ReO_4^- and TcO_4^-

Few studies were carried out directly with radioactive $^{99}TcO_4^-$ in the investigation of adsorption process. Generally, ReO_4^- is used as a surrogate for TcO_4^- due to the similar chemical properties and ionic size between ReO_4^- and TcO_4^-. Fei et al. [99] documented the use of a cationic layered MOF (**SLUG-21**) for the capture of ReO_4^- from aqueous solutions. They found that the adsorption capacity of this MOF was up to 602 mg/g, but the ethanedisulfonate anions were exchanged out, which could be a challenge in the practical application. In addition, SLUG-21 has a relatively high solubility in aquatic environments. Afterwards, a stable *cationic Zr-based MOF* was reported by Banerjee et al. [100] and used for the adsorption of ReO_4^-. But the adsorption kinetics, adsorption capacity, and selectivity of this MOF need improvement.

For the adsorption of ReO_4^- and TcO_4^- by MOF-based adsorbents, Wang's groups contributed a lot to this area [32, 101]. For example, researchers in his group prepared a novel 3D cationic MOF (**SCU-100**) by using Ag^+ cations and a tetradentate neutral nitrogen-donor ligand (tetrakis[4-(1-imidazol-yl)phenyl]methane) [32]. They found that this MOF contained 1D channels with exchangeable NO_3^- anions and was stable in aqueous solution over a broad pH range (1–13). Furthermore, this MOF was stable under high β and γ radiation. They directly conducted the adsorption experiments with TcO_4^- and found that this MOF could remove all available TcO_4^- within 30 min, which is a remarkable finding. The total adsorption capacity of SCU-100 for ReO_4^- was 553 mg/g with a highest distribution coefficient of 1.9×10^5 mL/g. Importantly, the adsorption performance of this MOF for TcO_4^- could not be affected by the presence of large amounts of interfering anions, including NO_3^-, SO_4^{2-}, CO_3^{2-}, and PO_4^{3-}. A total of 87% of TcO_4^- could be selectively extracted from simulated Hanford low-level waste by using SCU-100. As revealed by the single crystal X-ray diffraction, the adsorption process of ReO_4^- by SCU-100 occurred in a SC-SC transformation from an eightfold interpenetrated framework with disordered NO_3^- anions to a fourfold interpenetrated framework with ordered ReO_4^- anions. The adsorbed ReO_4^- anions were bound to the open Ag^+ sites by the formation of Ag—O—Re bonds. In addition, the high adsorption capacity and superior selectivity of this MOF could also be ascribed to the dense hydrogen bonding network and the highly hydrophobic nature of the ligand. In 2017, researchers in this group also reported another stable cationic MOF (**Ag(4,4′-bipyridine)NO$_3$**) with a much higher adsorption capacity for ReO_4^- (786mg/g) [101], which is higher than a lot of other reported scavenger materials. They found that the adsorbed ReO_4^- ions were stabilized and immobilized in the solid phase, thus a molar excess of 1000 times NO_3^- could not exert any effect on the removal of ReO_4^- ions. The material after irreversible SC-SC transformation was the least soluble ReO_4^-/TcO_4^- salt, which could be considered a promising candidate for the direct immobilization of TcO_4^-. DFT calculations disclosed that open soft Ag^+ sites contributed to the selective adsorption of ReO_4^-/TcO_4^- rather than hydrogen bonding, which offered crucial information for designing more efficient cationic MOFs for the capture of TcO_4^- ions.

As shown in Table 5.2, MOF SBN showed the largest adsorption capacity for ReO_4^-/TcO_4^- anions (786mg/g) [101] while MOF SCU-101 showed the most selective property for the removal of ReO_4^-/TcO_4^- anions [102], which could selectively remove ReO_4^- in the interference of various coexisting (e.g., NO_3^-,SO_4^{2-},ClO_4^-,CO_3^{2-},and PO_4^{3-}) with a K_d value $>7.5\times10^5$ mL/g. Table 5.2 shows that the adsorption capacities of SBN [101], SLUG-21 [99] and SCU-100 [32] are higher than those of anion exchange resins (252mg/g) [130], GO-based materials (<44mg/g) [131], NZVI (28.73mg/g) [132], and other cost-effective materials, including LDHs [133–136]. However, the adsorption capacity of anion exchange resins might lose in highly alkaline solutions or under high radiation [137]. In addition, further research on improving the adsorption capacity, reaction kinetics, and selectivity of materials is needed due to the noncomplexing nature, high water solubility, and great stability of ReO_4^-/TcO_4^- anions.

5.2.5 Other radioactive metal ions

Other radioactive nuclides, such as Th(IV), Eu(III), and Co(II), are highly toxic and hazardous to human health and the environment [138]. In recent years, MOF-based materials have shown great potential in the enrichment of radioactive metal ions from radioactive wastewater [52, 103, 104]. In the following section, we summarize some typical examples of radioactive metal ion adsorption onto MOF-based adsorbents and compared their adsorption performances with those of other widely used adsorbents.

MIL-101 functionalized with sulfate and sulfonic acid groups was studied for the capture of radioactive Ba(II) ions [52]. The adsorption kinetics of Ba(II) ions onto modified MIL-101 were well fitted to the pseudo-second-order kinetic model with high k_2 values. Based on the Langmuir model, the adsorption capacity of **MOF-808-SO$_4$** (131mg/g) was higher than that of **MIL-101-SO$_3$H(Cr)** (70.05mg/g). One reasonable explanation for this result is that the sulfate groups grafted to the metal cluster were more accessible than sulfonic groups introduced into the ligands. The results of FTIR, XPS, and DFT analysis indicated a stronger binding interaction between the sulfate groups and Ba(II) ions, which could explain the higher selectivity of MOF-808-SO_4 than that of MIL-101-SO_3H(Cr). Additionally, the adsorption of Ba(II) ions by these MOFs was irreversible, which avoided secondary pollution. It is not yet known if the frameworks of these MOFs are damaged by the effects of radiation. This is a crucial requirement for real-world applications under the radiation conditions.

UiO-66(Zr) and its two carboxylate derivatives, including **COOH-UiO-66 (Zr)** and **$(COOH)_2$-UiO-66(Zr),** were used to try and extract Th(IV) from acidic wastewaters [103]. The authors found that the adsorption of Th(IV) was enhanced due to the carboxylate functionalization. Thus, $(COOH)_2$-UiO-66(Zr) was reasonable to show the highest adsorption amount of 360 mg/g for Th(IV). The main interplay mechanism was the coordination between Th(IV) ions and carboxylate groups, which was shown by the EXAFS and FTIR analysis. They observed that the adsorption capacity was independent of the ionic strength and pH-dependent. The coexisting of interfering metal ions, including Zn(II), Co(II), Ni(II), Sr(II), Yb(III), Nd(III), Sm(III), Gd(III), and La(III), did not exert an obvious influence on the adsorption of Th(IV). However, the regeneration of this MOF was difficult and needed harsh acidic conditions due to the strong interaction between the carboxylates and Th(IV) ions. This MOF was partially destroyed in acid solution and about half its adsorption amount was lost.

Table 5.2 tabulates the removal properties of other radioactive metal ions by MOFs and MOF-based materials that have been reported in the literature. As tabulated in Table 5.2, MOFs and MOF-based materials reported for the removal of radioactive metal ions are limited for a relatively few radioactive metal ions. Fewer MOFs and MOF-based materials have been applied for the removal of radioactive metal ions such as Am(III), Np(IV), and Pu(IV). In addition, few studies of MOFs and MOF-based materials reported for the removal of radioactive metal ions have been carried out and explained in detail. Moreover, the extent of radiation damage to the frameworks of most MOFs and MOF-based materials needs further investigation.

Table 5.2 shows that UiO-66-NH_2 displayed an adsorption capacity of 167 mg/g for Pd(II) ions [104], which was superior to that of macrocyclic ligand-modified silica (83 mg/g) [139] and functionalized chitosan (<138 mg/g) [140–142], and comparable to that of calixcrown functionalized microsized silica (160 mg/g) [143]. In addition, the recovery of Pd(II) is often treated by strong HNO_3; thus UiO-66-NH_2 has more promise for the recovery of Pd(II) from extremely acidic high-level liquid wastes due to the large adsorption capacity, facile synthesis, and strong acid-resistance property of UiO-66-NH_2. As shown in Table 5.2, MOF-808-SO_4 showed a higher adsorption capacity for Ba(II) than various adsorptive materials, including AC [144], clays [145], and titanate nanotubes [146]. Titanate nanofibers displayed larger adsorption capacity for Ba(II) than MIL-101-SO_3H(Cr), but MIL-101-SO_3H (Cr) displayed ultrafast adsorption kinetics, which is a promising property for practical applications.

Table 5.2 shows that among MOFs and MOF-based materials reported for the removal of Th^{4+} from aqueous solutions, UiO-66-$(COOH)_2$ showed the largest adsorption capacity (350mg/g), which was much higher than other adsorptive materials, including CNTs (13.22mg/g) [147], chrysotile nanotubes (40.39mg/g) [148], impregnated resin (198.5mg/g) [149], and GO (<214mg/g) [138, 150]. Nowadays, thorium has drawn considerable interest because thorium can act as a potential alternative to uranium as a fertile material for producing nuclear fuel [103] while thorium and uranium are usually present together in minerals [151]. Therefore, the separation of thorium from uranium is very important in future research.

As shown in Table 5.2, MOFs and MOF-based materials reported for the removal of Co(II) [31, 105] and Eu(III) [106] displayed low adsorption capacities, indicating that developing advanced MOFs and MOF-based materials for the removal of Co(II)/Eu(III) and/or other radionuclides (e.g., Am(III), Np(IV), and Pu(IV)) from nuclear waste is highly desirable.

5.3 Removal of heavy metal ions by MOF-based materials

5.3.1 Chromium ions

In the following subsections, we summarize and assess different kinds of MOFs (3D MOFs, layered MOFs, 1D MOFs, and MOF composites) with high adsorption efficiency for the efficient elimination of chromium ions.

5.3.1.1 3D MOFs for the removal of chromium ions

The first 3D MOF applied in the capture of CrO_4^{2-} ions was the heterometallic 3D framework **1-ClO_4**, which was synthesized via a solvothermal reaction of metal salts and H_2BPDC ligands in EtOH-H_2O [152]. Fig. 5.6A and B show the 3D framework of 1-ClO_4 and the location of ClO_4^- in the channel of 1-ClO_4. This framework has strong alkali resistance and was able to remove ~70% of CrO_4^{2-} (initially at ~580ppm) within 30min and 85% of CrO_4^{2-} at equilibrium (after ~6h) via a mechanism of ion exchange with ClO_4^-. The adsorption capacity of 1-ClO_4 reached ~63mg/g. In addition, Fig. 5.6C shows that 1-ClO_4 was selective for CrO_4^{2-} from a solution containing NO_3^-,Cl^-,Br^-,I^-,and CrO_4^{2-}. The exchange of CrO_4^{2-} in the 1-ClO_4 was clearly observed by the UV-vis spectra (Fig. 5.6D). The mechanisms of selectivity and reusability require further study. A silver-triazolato MOF (**1-NO_3**) was reported in 2016 by Li et al. [153] Its chromium(VI) adsorption was moderate via an ion exchange mechanism with an equilibrium time of 4h. The MOF was reusable after four consecutive adsorption-release

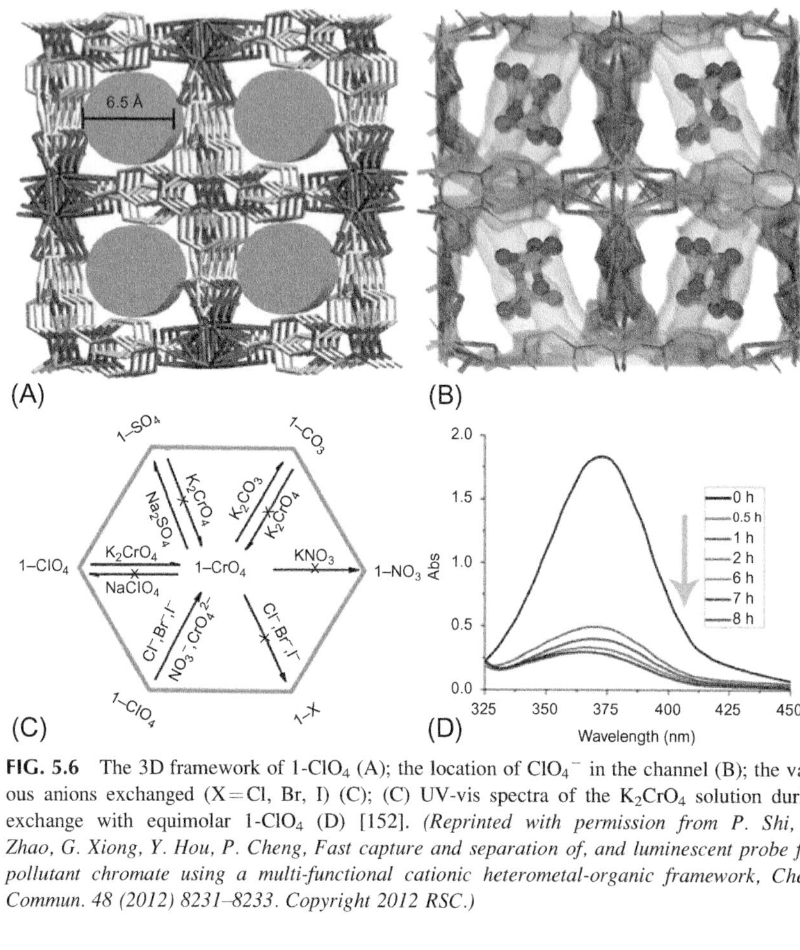

FIG. 5.6 The 3D framework of 1-ClO_4 (A); the location of ClO_4^- in the channel (B); the various anions exchanged (X=Cl, Br, I) (C); (C) UV-vis spectra of the K_2CrO_4 solution during exchange with equimolar 1-ClO_4 (D) [152]. *(Reprinted with permission from P. Shi, B. Zhao, G. Xiong, Y. Hou, P. Cheng, Fast capture and separation of, and luminescent probe for, pollutant chromate using a multi-functional cationic heterometal-organic framework, Chem. Commun. 48 (2012) 8231–8233. Copyright 2012 RSC.)*

cycles. In addition, 1-NO_3 showed a different selectivity performance for $HCrO_4^-$ over common interfering anions such as NO_3^-, SO_4^{2-}, Cl^-, and ClO_4^- with different concentrations due to the Hofmeister bias [153].

For another kind of chromium oxyanions $\left(Cr_2O_7^{2-}\right)$, several 3D MOFs have been applied for $Cr_2O_7^{2-}$ adsorption [25, 154–156]. Among these studies, $Cr_2O_7^{2-}$ adsorption onto 3D MOFs via anion-exchanged single-crystal-to-single-crystal (SC-SC) transformation is notable. The first case involved 3D cationic ABT·2ClO_4, which was isolated by the slow evaporation of an $AgClO_4$-btr solution in water-acetonitrile. This MOF contains two types of cages (i.e., octahedral cages and tetrahedral cavities). The interconnection of these two cages resulted in the formation of a 3-D cationic framework [154]. Observation from UV-vis spectroscopy disclosed that the SC-SC exchange of $Cr_2O_7^{2-}$ ions occurred slowly and the adsorption equilibrium required 48h. The total $Cr_2O_7^{2-}$ ion-exchange capacity was 213–271 mg/g. Due to the severe disorder of ABT·2ClO$_4$ after the exchange with $Cr_2O_7^{2-}$, the exact position of the trapped $Cr_2O_7^{2-}$ was not determined exactly by using the SCXRD technique. Selectivity studies have revealed that ABT·2ClO$_4$ preferentially adsorbed $Cr_2O_7^{2-}$ over NO_3^- and BF_4^-, owing to the stronger affinity of $Cr_2O_7^{2-}$ for the cationic framework than other interfering anions. In 2017, Li et al. reported a new 3D MOF showing a relatively high adsorption capacity for $Cr_2O_7^{2-}$ (207 mg/g) via ion-exchange in an SC-SC fashion [156]. The $Cr_2O_7^{2-}$ ions were adsorbed into the 1D channels by the formation of multiple weak C—H···O interactions with the host framework. Additionally, this MOF showed excellent selectivity and the adsorption efficiency remained high in the presence of various anions (e.g., F^-, Cl^-, Br^-, OAc^-, SO_4^{2-}, NO_3^-, BF_4^-, $CH_3SO_3^-$, $CF_3SO_3^-$, ClO_4^-, and $Cr_2O_7^{2-}$). However, the selectivity and reusability remain unexplained; the chemical stability of this MOF should be improved because its single crystallinity was destroyed after the release of $Cr_2O_7^{2-}$ ions in recycling processes.

5.3.1.2 Layered MOFs for the removal of chromium ions

Two layered MOFs (**FIR-53** [155] and **Ag-3** [157]) have been reported for $Cr_2O_7^{2-}$ adsorption through ion exchange. A typical example of a layered MOF ion exchanger is Ag-3 synthesized by reacting $AgBF_4$ and 4-(5-methyl-3-pyridine)-1,2,4-triazole in CH_3CN at room temperature. The Ag-3 framework displays a two-dimensional structure based on dinuclear Ag_2 units interconnected via L ligands. BF_4^- anions are in the 1D channels of the cationic layers, which can be exchanged by $Cr_2O_7^{2-}$ anions. The total $Cr_2O_7^{2-}$ exchange capacity of Ag-3 was 167 mg/g. Furthermore, batch experiments in the presence of anionic species (SiF_6^{2-}, SO_4^{2-}, CO_3^{2-}, and $Cr_2O_7^{2-}$) disclosed that only $Cr_2O_7^{2-}$ anions can be exchanged with the BF_4^- species. Another example of a 2D MOF ion exchanger is FIR-53. The Zn(II) ions in FIR-53 are tetrahedrally coordinated with one u_2-OH$^-$ group and three N atoms. The latter bridges two Zn-TIPA (6,3) layers, leading to the formation of a double layer. The packing of the double layers via π-π interactions results in the formation of relatively large nanotubular channels. FIR-53 was able to remove 95% $Cr_2O_7^{2-}$ from 4 mL aqueous solutions (2 mM) within only 10 min. The regeneration ability of this MOF was quite high, with 87% $Cr_2O_7^{2-}$ release efficiency after five consecutive cycles. Unfortunately, the selectivity of FIR-53 decreased drastically under the interference of other dianionic species. The SC-SC transformation proceeded during $Cr_2O_7^{2-}$ adsorption as evidenced by the SCXRD data, which showed that the adsorbed $Cr_2O_7^{2-}$ anions were fixed into the double layers of FIR-53 through the formation of weak C—H···O interactions.

5.3.1.3 1D and hydrogen-bonded MOFs for the removal of chromium ions

Some 1D and hydrogen-bonded MOFs (e.g., 1D cationic SLUG-21 [26], Co-SLUG-35, Zn-SLUG-35, and MOFs prepared from mixed Zn(II)/Co(II) metal ions [158], MONT-1 [159], TMU-30 [160], and 1D Fe-gallic acid MOFs [161]) have been applied to the adsorption of CrO_4^{2-} or $Cr_2O_7^{2-}$ ions. For example, Aboutorabi et al. in 2016 prepared **TMU-30**—an isonicotinate N-oxide based MOF consisting of 1D rhombic tunnels for effective CrO_4^{2-} removal from aqueous solutions. This framework was highly efficient for CrO_4^{2-} removal (145 mg/g) within a short time (equilibrium reached within 10 min) and over an extremely wide solution pH range (between 2 and 9) [160]. PXRD analysis disclosed that adsorption at such a wide pH range was found to have little effect on the crystalline structure of the MOF, with the first signs of structural change occurring at pH above 9. Adsorption selectivity tests showed that TMU-30 adsorbed CrO_4^{2-} over other ions (with rare exceptions such as MoO_4^{2-} or WO_4^{2-}). Using FTIR, EDS, and XPS techniques, the authors proposed that the excellent adsorption performance was attributed to the electrostatic interaction between the positively charged nitrogen from N-oxide groups and oxygen atoms from CrO_4^{2-} ions. Thus, TMU-30 can only differentiate into metal ions and oxyanions, but it cannot distinguish between different oxyanions. In addition, the authors pointed out that the selection of lead(II) to build the MOF is not the most convenient toward real-world applications [160]. However, the information from this study will be highly valuable to construct new MOFs with improved capacity and selectivity. In 2017, Niu et al. documented the application of an exceptional **1D Fe-gallic acid MOF** for the capture of $Cr_2O_7^{2-}$ ions [161]. This MOF not only showed an ultrahigh adsorption capacity for $Cr_2O_7^{2-}$ ions (1709.2 mg/g) but also displayed an excellent selective adsorption property under the interference of several coexisting anions (Cl^-, PO_4^{3-}, SO_4^{2-}, or NO_3^-) and metal cations (Cu(II), Ni(II), Zn(II), Cd(II), Ba(II), and Mn(II)). This excellent selectivity could be traceable to the ion exchange ability of the interfering ions. The crystallinity and pore size of this MOF remained unchanged in strong acidic (pH 2) and alkaline (pH 11) solutions, resulting in excellent adsorption performance over a broad pH range (3.0–9.0). FTIR and XPS results showed that $Cr_2O_7^{2-}$ capture by Fe-gallic acid MOFs could be attributed to the combined mechanisms of reduction, metal substitution, and coprecipitation. If the regeneration problems of this MOF are addressed, this material shows great potential for practical applications.

5.3.1.4 MOF composites for the removal of chromium ions

In 2016, Yang et al. used Fe_3O_4 crystal seeds to grow the MIL-100(Fe) MOF in situ via the hydrothermal reaction and obtained **Fe_3O_4@MIL-100(Fe)** magnetic microspheres for the capture of $HCrO_4^-$ from aqueous solutions [162]. At an optimal pH of 2, the maximum adsorption capacity was 18 mg/g due to the electrostatic interactions between the Fe_3O^+ and $HCrO_4^-$ ions. The adsorption efficiency of this MOF composite was decreased with increasing pH. The adsorption equilibrium needed only 2 h [162]. A **chitosan-UiO-66** composite synthesized by grinding and microwave irradiation was able to uptake 94 mg/g of $Cr_2O_7^{2-}$, which was higher than chitosan [163]. Such high adsorption capacity was due to the strong electrostatic attraction between high oxidation state metal ions and oxygen atoms or the -NH_2 groups of the linkers. The adsorption performance was prevalent at lower pH values while increasing the basicity of the solution resulted in decreased adsorption efficiency. In 2017, researchers in our group synthesized a kind of well-defined, core-double-shell structured magnetic polydopamine@ZIF-8 (**MP@ZIF-8**) hybrid through a facile and green approach, which displayed an uptake capacity of 136.56 mg/g for $Cr_2O_7^{2-}$ via synergistic reduction and adsorption mechanisms [164]. The adsorption kinetics of $Cr_2O_7^{2-}$ by this composite followed the pseudo-second-order model, indicating a chemical reaction process. Additionally, the uptake of $Cr_2O_7^{2-}$ by the MP@ZIF-8 was highly dependent on solution pH and the capacity remained 42.59 mg/g at pH 9.05. XPS results revealed that, owing to the reductive properties of the nitrogen-containing groups on ZIF-8 and PDA, Cr(VI) was easily converted into Cr(III) with a lower toxicity, which was then immobilized on the MP@ZIF-8. Optimization of the MOF composite for metal ion uptake and the preparation of new MOF-based adsorbents are currently under way in our group.

Rapti et al. found that MOFs combined with alginic acid (MOR-1-HA [13, 58] and MOR-2-HA [165]) were excellent adsorbents for $Cr_2O_7^{2-}$ ions and showed high adsorption capacities of 162–280 mg/g via a mechanism of ion exchange. **MOR-2** showed high adsorption efficiency in a broad pH range (2–9) and superior selectivity for $Cr_2O_7^{2-}$ and CrO_4^{2-} over common interfering anions, including Cl^-, Br^-, NO_3^-, and SO_4^{2-}. Such high selectivity has also been observed in potable and industrial water systems. The adsorption of $Cr_2O_7^{2-}$ and CrO_4^{2-} by MOR-2 was completed within 1 min. The combination of theoretical calculations and experimental results showed that the excellent adsorption ability of the MOR-2 was determined by the bulky pyridyl-methyl-ammonium functional groups. In the structure of $MOR-2@Cr_2O_7^{2-}$ and $MOR-2@CrO_4^{2-}$, $Cr_2O_7^{2-}$ or CrO_4^{2-} ions were coordinated to the Zr_6 clusters, which acted as ligands in an n^1: n^1: μ_2-fashion.

In addition to the batch experimental technique, the adsorption efficiency of these MOF-based composites was also tested with the use of continuous flow columns [13, 58, 165]. **MOR-2-HA** was mixed with silica sand as an inert material to ensure a stable flow rate of the aqueous solutions through the column. Adsorption of $Cr_2O_7^{2-}$ was clearly observed. Importantly, the column could be easily regenerated through treatment with 4 M HCl solution and remained at a high adsorption capacity after five cycles of the column. Eleven bed volumes of the effluent samples showed a total Cr content $\leq$10 ppb, which was below the acceptable levels. In addition, the MOR-2-HA/sand column was also effective for treating chrome-plating wastewater samples, which indicates potential for real wastewater decontamination. The MOR-1-HA/sand column also showed excellent adsorption efficiency [13, 58]. The high adsorption performance of these MOF-based composites indicates promise for applications in the treatment of Cr-contaminated wastewater.

5.3.1.5 Comparison between MOF-based materials and other adsorbents

To date, more and more MOFs and MOF-based materials have been applied for the removal of chromium species from aqueous solutions. The removal properties of chromium species by MOFs and MOF-based materials is tabulated in Table 5.3. Table 5.3 reveals that the removal of chromium species by MOFs and MOF-based materials was mainly via a mechanism of ion exchange. As shown in Table 5.3, only a few MOFs and MOF-based materials have been applied for the removal of CrO_4^{2-} anions, which contains 1-ClO_4 [152], SLUG-21 [26], Zn-Co-SLUG-35 [158], 1-NO_3 [153], and ZIF-67 [166]. It is worth noting that the adsorption capacity of 1-NO_3 was relatively low while it can be fabricated in a large scale [153]. No additional information about the adsorption capacity under a different solution pH was provided. The adsorption capacity of ZIF-67 was only 5.88–13.34 mg/g, but it displayed excellent reusable ability [166].

Increasing numbers of MOFs and MOF-based materials have been applied for the removal of $Cr_2O_7^{2-}$ anions, which is shown in Table 5.3. Several MOFs and MOF-based materials have a broad active pH range, which includes MOR-1-HA [13, 58], MOR-2-HA [165], and Fe-gallic acid MOFs [161]. Only MOR-1-HA and MOR-2-HA composites have carried out column chromium adsorption experiments [13, 58, 165]. Several MOFs and MOF-based materials have high adsorption capacities (>200 mg/g) [13, 58, 154, 156, 167, 168], in which Fe-gallic acid MOF showed the largest adsorption capacity (up to 1709.2 mg/g) [161]. In addition, the removal performance of Fe-gallic acid MOF is superior to other traditional adsorptive materials, which includes bioadsorbents (333 mg/g) [174, 175], metal/metal oxide-based materials (4–473 mg/g) [176, 177], LDHs (90–183 mg/g) [178–181], organic resins (110–140 mg/g) [182], carbon or GO-based adsorbents (23–581 mg/g) [183, 184], and hybrid polymeric adsorbents (90–377 mg/g) [185, 186].

In addition to the adsorption capacity, the comparison of chromium removal properties of MOFs and MOF-based materials to those of other widely used adsorptive materials is also very necessary. This includes the reaction kinetics, selectivity, and regeneration of these materials. Traditional

TABLE 5.3 Chromium adsorption data for reported MOF-based materials

Targets	MOF-based materials	Adsorption capacity (mg/g)	Equilibrium time	Selectivity	Reusability	Mechanism	Ref.
$Cr_2O_7^{2-}$	MOR-1 (protonated)	247–321	NA	–	–	Ion-exchange	[13, 58]
$Cr_2O_7^{2-}$	MOR-1-HA	242–280	3 min	vs Cl^-, Br^-, NO_3^-,SO_4^{2-}	Reusable	Ion-exchange	[13, 58]
$Cr_2O_7^{2-}$	MOR-1 (nonprotonated)	267	>1 h	–	–	Ion-exchange	[13, 58]
$Cr_2O_7^{2-}$	1-SO_4	166	72 h	vs NO_3^-,BF_4^-,$CF_3SO_3^-$,ClO_4^-	Reusable	Ion-exchange	[25]
CrO_4^{2-}	SLUG-21	60	48 h	vs NO_3^-,CO_3^{2-}	–	Ion-exchange	[26]
$Cr_2O_7^{2-}$	Co-SLUG-35	119	4 h	vs SO_4^{2-},NO_3^-,Cl^-	Limited	Ion-exchange	[65]
CrO_4^{2-}	1-ClO_4	62.9	6 h	vs halide anions	–	Ion-exchange	[152]
CrO_4^{2-}	1-NO_3	37	NA	vs NO_3^-,CO_3^{2-}	–	Ion-exchange	[153]
$Cr_2O_7^{2-}$	ABT·2ClO_4	213–271	48 h	vs NO_3^-,BF_4^-,$CF_3SO_3^-$	–	Ion-exchange	[154]
$Cr_2O_7^{2-}$	FIR-54	103	30 min	–	Reusable	Ion-exchange	[155]
$Cr_2O_7^{2-}$	FIR-53	74	10 min	vs Cl^-,Br^-,NO_3^-	Reusable	Ion-exchange	[155]
$Cr_2O_7^{2-}$	$[Ag(L^{243})](CF_3CO_2)(H_2O)$	207	320 min	vs various anions	Limited	Anion-exchange	[156]
$Cr_2O_7^{2-}$	Ag-3	167	24 h	vs SiF_6^{2-},SO_4^{2-},CO_3^{2-}	–	Ion-exchange	[157]
CrO_4^{2-}	Zn-Co-SLUG-35	68.5	2 h	vs NO_3^-,SO_4^{2-}	–	Ion-exchange	[158]
$Cr_2O_7^{2-}$	MONT-1	211.8	24 h	–	Reusable	Ion-exchange	[159]

Continued

TABLE 5.3 Chromium adsorption data for reported MOF-based materials—Cont'd

Targets	MOF-based materials	Adsorption capacity (mg/g)	Equilibrium time	Selectivity	Reusability	Mechanism	Ref.
$Cr_2O_7^{2-}$	Fe-gallic acid MOFs	1709.2	72h	vs various ions	Limited	Multiple	[161]
$Cr_2O_7^{2-}$	MP@ZIF-8	136.56	16h	–	–	Adsorption Reduction	[164]
$Cr_2O_7^{2-}$	MOR-2	194	1 min	–	Reusable	Anion-exchange	[165]
CrO_4^{2-}	ZIF-67	5.88–13.34	20–60min	–	Reusable	Multiple	[166]
$Cr_2O_7^{2-}$	IL-MIL-100(Fe)	285.7	5h	–	–	Chemisorption	[167]
$Cr_2O_7^{2-}$	ZJU-101	245	10min	vs halide, NO_3^-, SO_4^{2-}	–	Ion-exchange	[168]
$Cr_2O_7^{2-}$	MOF-867	53	>12h	–	–	Ion-exchange	[168]
Cr^{3+}	Fe_3O_4-En-MIL-101	173	NA	vs Cd^{2+}, Pb^{2+}, Zn^{2+}	Reusable	Electrostatic adsorption	[169]
$Cr_2O_7^{2-}$	Cu-BTC	48	NA	–	Reusable	Physisorption	[170]
$Cr_2O_7^{2-}$	γ-Fe_2O_3/C@HKUST-1	101.4	>5h	–	Reusable	Ion-exchange	[171]
$Cr_2O_7^{2-}$	Cd(II) MOF	228	–	vs various anions	–	Ion-exchange	[172]
Cr^{3+}	U-H_4btec	NA	30min	–	–	Chemisorption, physisorption	[173]

adsorptive materials used for the removal of chromium species include cost-effective LDHs, metal oxide-based materials, and organic resins. Cost-effective LDHs usually showed slow adsorption kinetics, that is, the adsorption equilibrium required several hours [178–180]. However, a few MOFs and MOF-based materials can achieve the chromium adsorption equilibrium in 3 min [13, 58, 165]. Moreover, LDHs usually showed inferior selective removal properties under the interference of common competitive anions such as SO_4^{2-} [178] while a few MOFs and MOF-based materials displayed superior selectivity under interfering ions (e.g., halide, NO_3^-, and SO_4^{2-}) [25, 155]. For the removal of chromium species, metal oxide-based adsorbents have a very narrow active pH range [187] while a few MOFs and MOF-based materials displayed high adsorption efficiency over a wide pH range [13, 58, 161]. Generally, organic resins can be oxidized by Cr(VI), thus they cannot be recycled [188], whereas a few MOFs and MOF-based materials have superior regeneration ability and reusability even after several adsorption-desorption cycles [13, 25, 58, 155, 159, 166, 169–171, 189]. Moreover, the fabrication method of organic resins, GO-based adsorbents, and hybrid polymeric adsorbents is usually complex, some of which requires toxic organic reagents and solvents [182] whereas a few MOFs and MOF-based materials can be fabricated through a one-step hydrothermal synthetic method [15, 153]. In comparison to the low yield of GO and CNTs, the preparation method of 1-NO_3 is fast, facile, and can be obtained on a large scale [153]. Based on these comparisons, we can conclude that MOFs and MOF-based materials show great prospects in the removal of chromium species from chromium-contaminated wastewater.

5.3.2 Arsenic ions

Recently, MOF-based materials have drawn considerable interest as adsorbents for arsenic removal. In the following subsections, we assess the adsorption characteristics of As(V), As(III), and organoarsenic species by MOF-based materials, respectively.

5.3.2.1 MOF-based materials for the removal of As(V)

The first MOF used as an adsorbent for the adsorption of arsenic species was **Fe-BTC**, which was obtained through a facile solvothermal method [190]. As shown in Fig. 5.7, a relatively fast uptake of As(V) occurred and then reached the equilibrium value within 10 min for Fe-BTC polymer and bulk Fe_2O_3 powders while the equilibrium was reached within 60 min for the Fe_2O_3 nanoparticles. Experimental data analyzed by the kinetic model suggested that the preconcentration of As(V) by these three materials fitted well to the pseudo-second-order rate equation. The adsorption capacity of Fe-BTC for As(V) was only 12.3 mg/g; however, this value was still more than six times as high as that of Fe_2O_3 nanoparticles and 36 times higher than that of

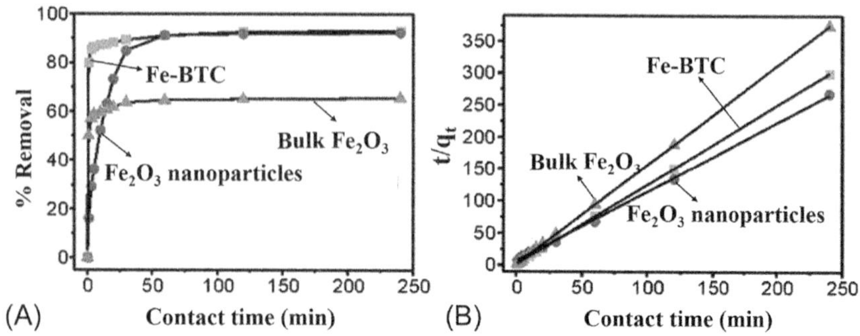

FIG. 5.7 Effect of contact time on absorption rate (A) and (B) pseudo-second-order kinetic plots for As(V) adsorption by Fe-BTC polymer gel, Fe_2O_3 nanoparticles, and bulky Fe_2O_3 powders (B) [190]. *(Reprinted with permission from B. Zhu, X. Yu, Y. Jia, F. Peng, B. Sun, M. Zhang, T. Luo, J. Liu, X. Huang, Iron and 1,3,5-benzenetricarboxylic metal-organic coordination polymers prepared by solvothermal method and their application in efficient As(V) removal from aqueous solutions, J. Phys. Chem. C 116 (2012) 8601–8607. Copyright 2012 ACS.)*

commercial Fe_2O_3 powders. The observed Fe—O—As groups in the IR spectra of As(V)-loaded Fe-BTC confirmed the adsorption of As(V) ions. Interestingly, arsenium elements were not detected by XPS in this study, suggesting that adsorbed As(V) ions were trapped in the interior of the Fe-BTC rather than at the outer surface. A disadvantage of this MOF was its instability under basic conditions. However, this work demonstrates the potential for applications of MOFs to the capture of arsenic species [190].

The MIL families are another example of MOFs applied to As(V) adsorption, such as MIL-53(Fe) [191], MIL-53(Al) [192], and MIL-100(Fe) [193]. In particular, **MIL-53(Al)** displayed an ultrahigh adsorption capacity of 105.6 mg/g for As(V) pH = 8, and it could extract 15.4 mg/g of As(V) at the initial concentration of 10 ppb. The presence of interfering anions including Cl^-, F^-, NO_3^-, and SO_4^{2-} did not exert any obvious effect on its adsorption behavior while it decreased in the presence of PO_4^{3-}. FTIR and XPS analysis disclosed that the adsorption of As(V) by MIL-53(Al) was driven by the electrostatic interactions and hydrogen bonds. But we think that these were not the only dominating underlying causes because the adsorption amount of As(V) by MIL-53(Al) and MIL-53(Fe) was not the same. The adsorption amount of As(V) by MIL-53(Al) is fivefold that of MIl-53(Fe).

5.3.2.2 MOF-based materials for the removal of both As(V) and As(III)

Magnetic MOF composite **Fe_2O_3@MIL-101** [194] was studied in the simultaneous removal of As(III) and As(V). In this study, the advantages of MIL-101(Cr) MOF with great robustness and hematite nanoparticles with high affinity toward arsenic species were combined together. This MOF composite has two aims, including: (i) immobilization of hematite nanoparticles and protecting them from agglomeration, thus improving their adsorption performance; (ii) acting as scavenger materials for toxic arsenic species. This composite showed enhanced adsorption performance for arsenic species due to the synergic effect than the components separately. As high as 121.5 and 80.0 mg/g of As(III) and As(V) were removed by this MOF composite at pH = 7, which could also be maintained even under the interference of large amounts of competing ions. Some leaching of Cr^{3+} and Fe^{2+}/Fe^{3+} was observed. A desired recyclability was therefore not reached even when several different conditions were tested due to the strong interaction between the arsenic species and the MOF composite. We have to acknowledge that strong interaction is desired, but this is also a challenge for the regeneration and reusability of an adsorbent. This is indeed a crucial issue to be considered when it is desirable to regenerate the adsorbent in several adsorption-desorption times.

5.3.2.3 MOF-based materials for the removal of As(III)

Recently, a study from Huo et al. [195] reported the efficient preconcentration of As(III) from an aquatic environment by using a magnetic core-shell composite (**Fe_3O_4-ZIF-8**). The adsorption capacity of this composite was 100 mg/g, which was due to the strong affinity. The adsorption efficiency was affected under interfering CO_3^{2-} and PO_4^{3-}, but other anions such as SO_4^{2-}, Cl^-, and NO_3^- did not exert any obvious effect. Importantly, this MOF composite can be efficiently separated from aquatic environments by using an external magnetic field. In addition, these authors also conducted thorough experiments to ensure the reusability of this composite. The adsorbed As(III) species could be easily desorbed from the composite by using NaOH solutions. This study demonstrated the application prospects of magnetic MOFs for the effective preconcentration of contaminants from aqueous solutions.

5.3.2.4 MOF-based materials for the removal of organoarsenic

Jhung's group has reported extensively on the adsorption of organoarsenic species by MOF-based materials [196–198]. In 2015, researchers in this group compared the p-arsanilic acid (ASA) adsorption capacities among AC, zeolite Y, and various MOFs including **ZIF-8, mesoporous ZIF-8**, **MIL-101(Cr),** and **MIL-53(Cr)** [196]. The authors found that mesoporous ZIF-8 and ZIF-8 were more efficient than the AC. The adsorption difference among these materials was not solely dependent on their structural properties. The main mechanism of ASA adsorption onto mesoporous ZIF-8 and ZIF-8 was electrostatic interaction at pH < 9, and no reason was given for mesoporous ZIF-8 being more efficient than ZIF-8. Another study from this group assessed the adsorption of ASA and ROX over different porous materials, including **MIL-53(Cr)**, **MIL-100(Al)**, **MIL-100(Cr)**, **MIL-100(Fe)**, **MIL-101(Cr)**,

AC, zeolite Y, and goethite [197]. There were two major findings: (i) MIL-100(Fe) showed the highest adsorption capacity; and (ii) MIL-100(Al) and MIL-100(Cr) displayed similar adsorption capacities. MIL-100(Al), MIL-100(Cr), and MIL-100(Fe) were further studied to understand the influence of the central metal ion of MIL-100 on the adsorption performance. The authors found that the adsorption difference among MIL-100 analogues can be traceable to the water adsorption energy and the organic molecular adsorption energy. Water was shown to bind much weaker to MIL-100(Fe) than to other MIL-100 analogues. Additionally, ASA and ROX coordinated more strongly to MIL-100(Fe) than to other MIL-100 analogues. The authors suggested that the anionic species of ASA and ROX were bound to coordinatively unsaturated sites (CUSs) of the MIL-100(Fe). In 2017, researchers in this group compared the adsorption efficiency of phenylarsonic acid (PAA) and ASA onto functionalized MIL-101 (**MIL-101(OH)** and **MIL-101(OH)$_3$**) with pristine MIL-101 and commercial AC [198]. The adsorption capacities of PAA and ASA decreased in the following order: MIL-101(OH)$_3$ > AC > MIL-101(OH) > MIL-101. Hydrogen bonding and electrostatic interactions were proposed to contribute to the adsorption of PAA and ASA onto these materials.

5.3.2.5 Comparison between MOF-based materials and other adsorbents

More and more MOFs and MOF-based materials have been applied for the removal of arsenic species from aqueous solutions. Table 5.4 shows the removal properties of these materials for different arsenic species. We conclude that the removal of arsenic species by MOFs was mainly through the mechanisms of ion exchange and coordination to the MOFs' metal nodes. As shown in Table 5.4, nowadays MOFs and MOF-based materials applied in the removal of arsenic species from aqueous solutions are mainly limited to different forms of ZIFs and MILs, several of which showed high removal efficiency for both As(V) and As(III) [194, 205, 206] while few MOF-based materials displayed reusable properties [39, 193, 196, 197, 201].

Up to now, a lot of cost-effective adsorptive materials, such as AC, metal oxides, LDHs, and laterite soil, have been applied for the removal of As(V) from aqueous solutions [208–217]. Among these adsorptive materials, the yttrium-manganese binary composite showed the largest adsorption capacity (279.9 mg/g) for As(V) at pH 7 [212], but needed 25 h to reach adsorption equilibrium and could be easily regenerated. The preparation process was expensive and complicated, and the yttrium-manganese binary composite has a limited active pH range (pH 4–7). In comparison, Zr-based, ZIF, and MIL family MOFs can be prepared via a facile and low-cost approach. Moreover, UiO-66 displayed superior adsorption capacity (303.4 mg/g) to the yttrium-manganese binary composite [202], but showed low reaction kinetics,

TABLE 5.4 Arsenic adsorption properties of MOF-based materials

Targets	MOF-based materials	Adsorption capacity (mg/g)	Equilibrium time	Selectivity	Reusability	Mechanism	Ref.
As(III)	β-MnO_2@ZIF-8	147.28	24 h	–	–	Oxidation	[28]
ROX	UiO-66-20	730	NA	vs Cl^-, OAc^-, SO_4^{2-}	Reusable	Ion-exchange	[39]
As(V)	Fe-BTC	12.3	10 min	–	–	Ion-exchange	[190]
As(V)	ZIF-8	76.5	7 h	–	–	Inner-sphere complex	[199]
As(V)	Hierarchical ZIF-8	90.93	10 h	–	–	Ion-exchange	[200]
As(V)	MIL-53(Al)	105.6	11 h	vs CO_3^{2-}, Cl^-, NO_3^-, SO_4^{2-}	–	Electrostatic attraction, hydrogen bond	[192]
As(V)	MIL-53(Fe)	21.27	90 min	–	–	Ion-exchange	[191]
As(V)	MIL-100(Fe)	110	1 h	vs Cl^-, SO_4^{2-}, NO_3^-, CO_3^{2-}	Reusable	Coordination	[193]
As(V)/As(III)	Fe_3O_4@MIL-101(Cr)	80.0/121.5	24 h/24 h	vs various oxyanions, CR, HA	–	Ion-exchange	[194]
ASA	ZIF-8	729.9	24 h	–	–	Ion-exchange	[196]
ASA	Mesoporous ZIF-8	791.1	6 h	vs Na^+, Cl^-	Reusable	Ion-exchange	[196]
ROX	MIL-100(Fe)	387	3 h	–	Reusable	Ion-exchange	[197]
ASA	MIL-100(Fe)	366	3 h	–	Reusable	Ion-exchange	[197]
ASA	MIL-101(OH)	163	6 h	–	–	Hydrogen bonding, electrostatic interaction	[198]

Continued

TABLE 5.4 Arsenic adsorption properties of MOF-based materials—Cont'd

Targets	MOF-based materials	Adsorption capacity (mg/g)	Equilibrium time	Selectivity	Reusability	Mechanism	Ref.
ASA	MIL-101$(OH)_3$	238	6h	–	Reusable		[198]
PAA	MIL-101	57.1	6h	–	–	Hydrogen bonding, electrostatic interaction	[198]
PAA	MIL-101(OH)	84.1	6h	–	–		[198]
PAA	MIL-101$(OH)_3$	139	6h	–	Reusable		[198]
As(V)	MOF-808	24.83	30min	–	Reusable	Ion-exchange	[201]
As(V)	UiO-66	303.4	48h	vs CO_3^{2-}, Cl^-, NO_3^-, SO_4^{2-}	–	Ion-exchange	[202]
As(V)	NH_2-MIL-88(Fe)	125	60min	–	–	–	[203]
As(V)	AUBM-1	103.1	3h	–	Reusable	–	[204]
As(V)/As(III)	ZIF-8nps	60.03/49.49	7h/13h	vs NO_3^-, SO_4^{2-}	–	Ion-exchange	[205]
As(V)/As(III)	$CoFe_2O_4$@MIL-100 (Fe)	114.8/143.6	2min/2min	vs various oxyanions, CR, HA	–	Ion-exchange/ hydrogen bonding	[206]
As(III)	Cubic ZIF-8	122.6	10h	–	–	Ion-exchange	[207]
As(III)	Leaf-shaped ZIF-8	108.1	10h	–	–	Ion-exchange	[207]
As(III)	Dodecahedral ZIF-8	117.5	10h	–	–	Ion-exchange	[207]

which might hamper its real-world application. Fe-BTC, MIL-53(Fe), and MOF-808 exhibited fast reaction kinetics (10–30 min), and have a wide active pH range [190, 191, 205]. Nevertheless, these MOFs displayed relatively low adsorption capacities, and the information about the regeneration/reusability of these MOFs was not provided.

In addition, MOFs and MOF-based materials have also been applied for the removal of organoarsenic species [39, 196, 197]. These materials displayed excellent uptake performance in a broad pH range and can be easily recycled several times. For the uptake efficiency, defective UiO-66 showed great promise for the adsorption of organoarsenic species with respect to MIL-100(Fe). It is worth noting that MIL-100(Fe) displayed superior uptake efficiency in both batch experiments and column studies. Some MOFs and MOF-based materials have been applied for the removal of As(III) from aqueous solutions with adsorption capacities ranging from 49.5 to 147.28 mg/g [28, 194, 205–207]. Some other widely used adsorptive materials have also been investigated in the extraction of As(III) from aqueous solutions [217–220]. The adsorption amounts of these materials were very low ($\leq$20 mg/g). Low-cost Fe—Cu binary oxides (122.3 mg/g) [221] and $Mg_{0.27}Fe_{2.50}O_4$ (127.4 mg/g) [222] displayed comparable adsorption capacities.

Except for the adsorption capacity and adsorption kinetics, the influence of antimony on the removal of arsenic species is also very important for assessing the efficiency of materials in real-world applications because arsenic species usually occur together with antimony in water around old mining and smelting areas; they also share many similarities in their chemical properties and toxicities [223, 224]. Table 5.4 shows several MOF-based materials that display excellent selectivity for arsenic species in the presence of other inorganic anions [39, 192, 193, 202, 206]. As a well-known competitor for the adsorption of As(V), the effects of PO_4^- required further investigation.

5.3.3 Mercury ions

To date, MOFs and MOF-based materials have shown great promise for the extraction of mercury species from aquatic environments. In the following subsection, we pursued some typical examples of MOFs and MOF-based materials for the removal of mercury species through coordination and ion-exchange mechanisms, respectively.

5.3.3.1 Binding with sulfur atoms

Owing to the strong binding ability between sulfur atoms and a wide range of metal ions, the incorporation of highly reactive sulfur atoms and sulfur-containing functional groups into MOFs has been widely investigated for the efficient elimination of metal ions [21, 225–228], particularly for Hg(II) adsorption. To date, several thiol-functionalized or thioether-modified MOFs have been investigated for elimination of Hg(II) ions [21, 48, 225, 229, 230]. Yee and coworkers [21] prepared a thiol-laced Zr- and Al-based MOF with the UiO-66 and CAU-1 topology based on the hard-and-soft synthetic strategy by using the 2,5-dimercapto-1,4-benzenedicarboxylic acid (DMBD) ligand (Fig. 5.8A). Among these two MOFs, **Zr-DMBD MOF** performed efficiently for the uptake of mercury in both aqueous solutions and the gas phase. In addition, it could also uptake 99.9% of Hg(II) from 10 mL of a 10 ppm Hg(II) aqueous solution with a maximum adsorption capacity of 1.82 mmol per unit formula of MOF. Spectroscopic analysis strongly disclosed that the uptake of mercury species was driven by the strong binding ability of thiol groups in the MOF (Fig.5.8B and C). PXRD patterns revealed the MOF stability and the adsorbed mercury species in the MOF channels (Fig. 5.8D). Its adsorption property might be improved if the formation of sulfur-sulfur bonds were prevented. The authors proposed that thio-ether groups could also be incorporated into MOFs by using the postsynthetic methods. However, the synthesis method of sulfur-containing ligands used in this study was very complex. Another study reported by He et al. [225] modified the **MOF-5272** channels with thio-ether side chains, which was able to selectively extract $HgCl_2$ from ethanol solutions at very low concentrations. But the performance could not be measured in water because it decomposed rapidly in water.

5.3.3.2 *Binding with other functional groups and ion exchange*

Except for binding with sulfur, Hg(II) can also bind with carbonyl, acylamide, and hydroxyl groups in the MOFs [45, 231]. For example, *acylamide and hydroxyl-functionalized MOF* showed an adsorption capacity of ~250 mg/g toward Hg(II), according to the Langmuir model [45]. In addition, kinetic data fitted well with the pseudo-second-order model and revealed that Hg(II) adsorption by this adsorbent required only 1 h to reach equilibrium, even at an initial Hg concentration as low as 100 μg/L. A combination of characterization techniques including FTIR, XPS, XRD, SEM, and EDX was applied to probe the possible adsorption mechanisms, which revealed that Hg(II) was inserted as $HgCl_2$. Such adsorption properties were determined from the striking structural characteristics of its 1D regular hexagonal channel with a pore size of ~5.4 Å and numerous acylamide and hydroxyl groups. However, this MOF was efficient in a narrow pH range (4–7). Information about the selectivity or reusability properties was not reported for this material. Recently, thymine-functionalized MIL-101 (**MIL-101-Thymine**) also showed adsorption ability for Hg(II) [44]. The isotherm data could be described by the Langmuir model with a maximum adsorption capacity of 51.27 mg/g. The adsorption kinetic data were in good agreement with the pseudo-second-order model. It was found that MIL-101-Thymine exhibited excellent selectivity toward Hg(II) over other cations, owing to the highly selective interactions of T-Hg(II)-T in MIL-101-Thymine and cations. XPS and TGA results disclosed that one Hg atom was coordinated with two N atoms in thymine. Furthermore, MIL-101-Thymine was also able to remove trace amounts of Hg(II) from real water samples, and satisfactory recoveries were received. However,

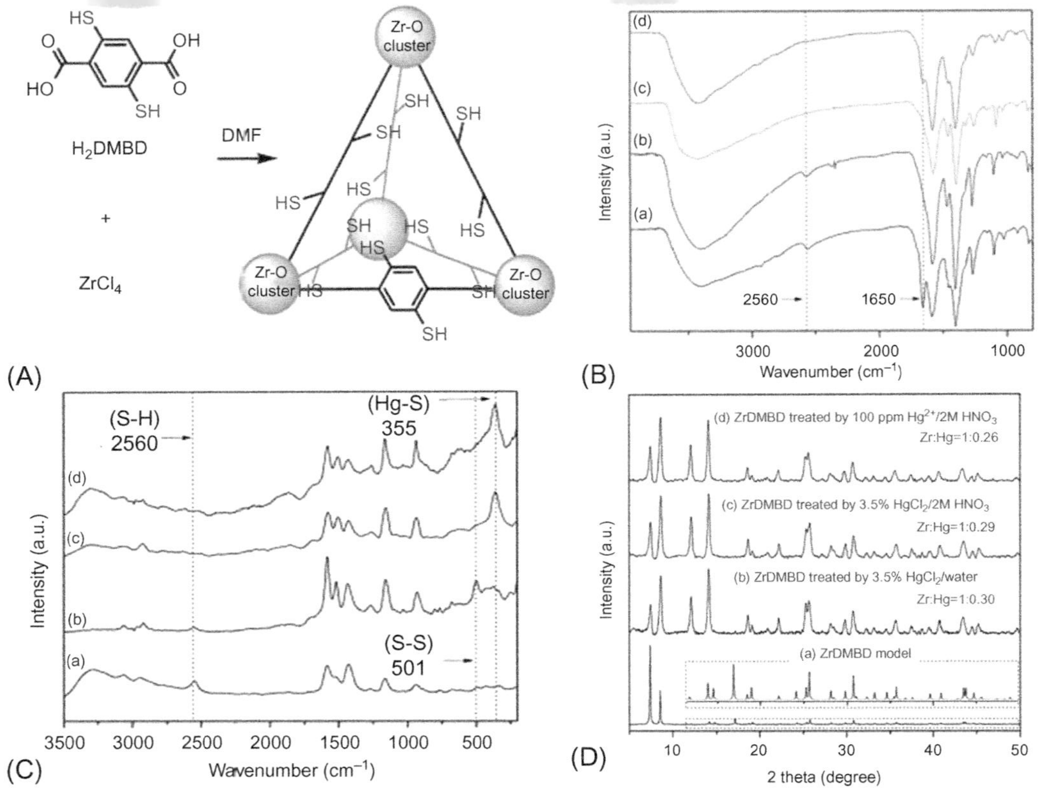

FIG. 5.8 Synthetic scheme for the Zr-DMBD network (A); The topology is the same as for UiO-66 and is shown in simplified form as a tetrahedral cage. The IR spectra of (a) an as-made sample of Zr-DMBD, (b) activated sample of Zr-DMBD, (c) $HgCl_2$ water solution treated Zr-DMBD sample, and (d) Hg vapor treated Zr-DMBD sample (B); the Raman spectra of (a) an as-made sample of Zr-DMBD, (b) activated sample of Zr-DMBD, (c) $HgCl_2$ water solution treated Zr-DMBD sample, and (d) Hg vapor treated Zr-DMBD sample (C); X-ray powder patterns of $HgCl_2$-treated Zr-DMBD and related samples (D) [21]. *(Reprinted with permission from K. Yee, N. Reimer, J. Liu, S. Cheng, S. Yiu, J. Weber, N. Stock, Z. Xu, Effective mercury sorption by thiol-laced metal-organic frameworks: in strong acid and the vapor phase, J. Am. Chem. Soc. 135 (2013) 7795–7798. Copyright 2013 ACS.)*

the working pH range was limited, and the maximum adsorption capacity was only 51.27 mg/g. Furthermore, 34% of the adsorption capacity was lost after only three recycling processes [44].

Hg(II) adsorption via ion exchange was reported for only one MOF (**AMOF-1′**) [232]. The maximum adsorption capacity for Hg(II) was 78 mg/g, owing to the anionic features of this MOF. Inductively coupled plasma analysis revealed that the number of Hg(II) ions per formula unit was 0.31 after ion exchange for 24 h, which increased to 0.48 after 7 days. Nearly complete removal (98.7%) of Hg(II) at an initial concentration of 1 mg/L was achieved within 24 h, which is crucial for real applications.

5.3.3.3 Comparison between MOF-based materials and other adsorbents

The adsorption properties of Hg(II) by MOFs and MOF-based materials are tabulated in Table 5.5. The effective and efficient removal performance for Hg(II) is collectively dominated by the highly porous structure and well-distributed functionalities of MOF-based materials. It is worth noting that ultrahigh adsorption capacities for Hg(II) were observed by using UiO-66-NHC(S)NHMe (769 mg/g) [50], $[Cu_3(BTC)_2]_n$ (714 mg/g) [48], and 3D bio-MOFs (900 mg/g) [27]. These high adsorption capacities are comparable to those by thiol-functionalized mesoporous silica (635 mg/g) [240], and functionalized covalent organic frameworks (COFs) (734 mg/g for TAPB-BMTTPA-COF [241]), and superior to those by MSIEs (377 mg/g for KMS-1) [88], resins (363 mg/g for Duolite GT-73) [242], and mesoporous silica (179 mg/g) [243]. Other materials, including a thiol-modified organic polymer (1014 mg/g for PAF-1-SH [244] and 1216 mg/g for POP-SH [245]) and postsynthetically modified COF (1350 mg/g) [246] displayed relatively high adsorption capacities. While the fabrication of COFs needs expensive organic linkers, the preparation of PAF-SH-1 and POP-SH need multistep reactions, including the modification of porous organic polymers. Many MOF-based materials displayed superior selectivity and reusability [49, 51, 233, 234], which are very crucial properties for real-world applications. In addition, MOF **1** and FJI-H12 are environmentally friendly and can be easily fabricated on a large scale at low cost, indicating realistic possibilities for commercialization and real-world applications [230]. Moreover, mercury species are rarely found in ionic forms and usually form complexes in real wastewater systems. Thus, studies on the removal performance of MOF-based materials for mercury species from practical industrial wastewater are a high priority.

5.3.4 Other heavy metal ions

5.3.4.1 MOF-based materials for the removal of Pb(II)

In 2013, Zou et al. reported on an efficient synthesis method for HKUST-1, and its subsequent functionalization with a polyoxometalate to give **HKUST-1-MW@**$\mathbf{H_3PW_{12}O_{40}}$—a framework with promising selectivity and uptake for

TABLE 5.5 Mercury adsorption data for reported MOF-based materials

MOF-based materials	Adsorption capacity (mg/g)	Equilibrium time	Selectivity	Reusability	Mechanism	Ref.
Zr-DBMD	197	NA	–	–	Binding with thiol	[21]
3D bioMOFs	900	<200 min	vs CH_3Hg^+	–	Binding with S	[27]
MIL-101-Thymine	59.28	300 min	vs various cations	Limited	Coordination with N	[44]
$[Zn(hip)(L)]\cdot(DMF)(H_2O)$	250	1 h	–	–	Contact with C=O	[45]
FJI-H12	439.8	<50 min	vs Mn^{2+}, Ba^{2+}, Ni^{2+}, Cd^{2+}	Reusable	Binding with SCN^-	[47]
$[Cu_3(BTC)_2]_n$	714.29	120 min	–	–	Binding with thiol	[48]
MFC-S	282	20 min	vs various cations	Reusable	Chelating with S	[49]
UiO-66-NHC(S)NHMe	769	NA	–	–	Chelating with S	[50]
Thiol-functionalized $Fe_3O_4@Cu_3(btc)_2$	348.43	120 min	vs various ions	Reusable	Binding with thiol	[51]
PCN-100	364.7	>120 min	–	–	Coordination	[105]
$Zn_4O(L)_3$	102.8	NA	–	–	Binding with S	[225]
Pb-TMBD	361.3	NA	–	–	Chelating with S	[229]
$\{Cu^{II}_4[(S,S)\text{-methox}]_2\cdot 5H_2O\}$	$HgCl_2$@1	15 min	vs various cations	–	Binding with thiol	[230]

Continued

TABLE 5.5 Mercury adsorption data for reported MOF-based materials—Cont'd

MOF-based materials	Adsorption capacity (mg/g)	Equilibrium time	Selectivity	Reusability	Mechanism	Ref.
MOF-74-Zn	74.85	1.5 h	–	–	Coordination	[231]
AMOF-1′	78	24 h	vs Pb^{2+}, Cd^{2+}	–	Ion-exchange	[232]
Magnetic MOF	254	NA	vs. various ions	Reusable	Coordination	[233]
UiO-66-$(SH)_2$	236.4	30 min	vs various ions	Reusable	Binding with thiol	[234]
Magnetic Bi-I-HKUST-1	264	10 min	vs various cations	Weak	Chelating with S	[235]
$[Ni(3\text{-}bpd)_2(NCS)_2]_n$	713	2 h	–	–	Binding with S	[236]
LMOF-263	380	30 min	vs light metals	Limited	Binding with S	[237]
SH@SiO_2/$Cu_3(BTC)_2$	210	15 min	vs various cations	Reusable	Binding with thiol	[238]
ZIF-90-SH	22.45	24 h	–	Reusable	Binding with thiol	[239]

lead(II) ions. Its maximum adsorption was recorded at 98 mg/g, with only a short time required to reach equilibrium (10 min) [247]. Magnetic composites based on **amino-functionalized MIL-53(Al)** reported by Ricco et al. displayed a remarkably higher adsorption capacity (up to 492.4 mg/g) for Pb(II) ions than that of pristine MIL-53(Al), which was mainly attributed to the nitrogen-based functional groups [54]. The authors found that the adsorption capacity could be elegantly tuned by changing the concentration of NH_2 groups in this composite (Fig. 5.9A). The regeneration of this composite could be partially realized by magnetic heating (Fig. 5.9B), although studies on the recycling ability of the regenerated adsorbent were not carried out. The adsorption energy (2.6 eV) calculated by DFT calculations revealed the strong interactions between Pb(II) atoms and NH_2 groups (Fig. 5.9C). PXRD results revealed that Pb(II) ions were sequestered into the pores, which induced the formation of an open form phase

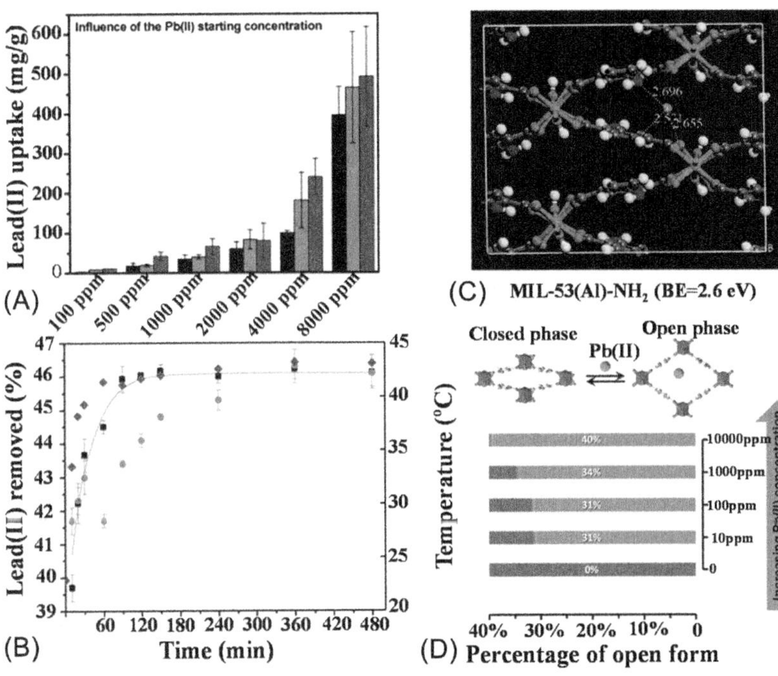

FIG. 5.9 Lead(II) uptake in water by MFCs with different amounts of amino groups. In each Pb(II) starting concentration, the left, middle, and right column represent 0%, 50%, and 100% of NH_2, respectively (A); release of 8000 ppm loaded NH_2-MIL-53(Al) MFC with and without magnetic heating (B). The round, square, and rhombus dots represent control, test, and temperature experiments, respectively. The *solid line* represent the fitting results; DFT-optimized locations and binding energies of lead atom in NH_2-MIL-53(Al) calculated using the DFTD2 method (C); the distances shown are in angstroms; and Lead(II) induced the formation of an open form phase increasing with the ion concentration (D). *(Reprinted with permission from R. Ricco, K. Konstas, M.J. Styles, J.J. Richardson, R. Babarao, K. Suzuki, P. Scopece, P. Falcaro, Lead(II) uptake by aluminium based magnetic framework composites (MFCs) in water, J. Mater. Chem. A 3 (2015) 19822–19831. Copyright 2015 RSC.)*

that enhanced with the increase of Pb(II) concentration (Fig. 5.9D). A thorough understanding via both the experimental results and the DFT calculations for the adsorption mechanism is required to understand the interaction mechanism in more detail at the molecular and atomic levels.

Up to now, many MOFs and MOF-based materials have been applied in the removal of Pb(II) from aqueous solutions, shown in Table 5.6, while few studies have provided detailed information by experimental approaches or computational calculations. As shown in Table 5.6, MOF-based composite MCNC@Zn-BTC displayed the highest adsorption capacity for Pb(II) (558.66 mg/g) [248], which was superior to traditional adsorbents such as MSIEs (319 mg/g for KMS-1) [88], resins (238 mg/g for Duolite GT-73) [255], modified nanocellulose (53 mg/g) [256], and clays and functionalized clays (365 mg/g for Mg-MTMS [257], 19 mg/g for PCB [258]). In addition, magnetic thiol-functionalized $Cu_3(btc)_2$ showed excellent selectivity ($K_d = 1.23 \times 10^4$ mL/g) [51]. In comparison, the traditional adsorbent NZVI displayed an ultrahigh adsorption capacity for Pb(II) (1667 mg Pb(II)/g Fe) [259] while NZVI particles were usually separated as small particles, which tend to aggregate. Thus, this characteristic may restrict their direct application and further restrict their direct use for field-scale applications.

5.3.4.2 *MOF-based materials for the removal of Cd(II)*

Typically, the sulfonic acid-functionalized HKUST-1 (**$Cu_3(BTC)_2$—SO_3H**) showed a relatively low adsorption capacity (88.7 mg/g) [249] but fast kinetics with a rate constant of 0.6818 g min/mg (i.e., 1–3 orders of magnitude higher than those of other adsorbents) (Fig. 5.10). Cd(II) adsorption by $Cu_3(BTC)_2$—SO_3H was not affected by interfering cations (Na(I), Mg(II), Ca(II), Pb(II), Cu(II), and Ni(II)), mainly because of the functionalization of —SO_3H. In addition, this —SO_3H-modified MOF was active in a broad pH range (3–8) and could be recycled more than six times without any obvious loss in adsorption capacity [249]. Such excellent adsorption performance was attributed to the chelation between Cd(II) ions and the —SO_3H groups. Additionally, regenerative **FJI-H9** was studied for Cd(II) adsorption and showed a relatively high adsorption capacity of 225 mg/g and excellent selectivity in the presence of various background metal ions (Cd(II), Hg(II), Ca(II), Mg(II), Co(II), Ni(II), Mn(II), Zn(II), Fe(II), and Pb(II) ions) [250]. Analysis of PXRD and EXAFS data showed that such a high adsorption performance could be attributed to the unusual synergy of active sites and the confined cavity. Additionally, in situ reconstruction of used FJI-H9 was successfully achieved by immersing the used sample in KSCN solutions at ambient temperature.

As shown in Table 5.6, relatively few MOFs and MOF-based materials have been applied in the removal of Cd(II). The adsorption capacity ranges from 49 to 225 mg/g, which is lower than those of KMS-1 (329 mg/g) and CMS (515 mg/g) [88]. It's worth noting that among these MOF-based materials, $Cu_3(BTC)_2$—SO_3H has a wide active pH range and showed fast reaction kinetics,

TABLE 5.6 Other heavy metal adsorption data for reported MOF-based materials

Targets	MOF-based materials	Adsorption capacity (mg/g)	Equilibrium time	Selectivity	Reusability	Mechanism	Ref.
Pb(II)	ED-MIL-101	81.09	30 min	vs various cations	Limited	Specific binding	[43]
Cd(II)	Alkylothiourea-UiO-66	49	NA	–	–	Ion-exchange	[50]
Pb(II)	Thiol-functionalized Fe_3O_4@$Cu_3(btc)_2$	215.05	120 min	vs various cations	Reusable	Binding with thiol	[51]
Pb(II)	MIL-53(Al) MFC	492.4	120 min	–	–	Ion-exchange	[54]
Pb(II)	Fe_3O_4-En-MIL-101	198	NA	–	Reusable	Electrostatic adsorption	[169]
Cd(II)	Fe_3O_4-En-MIL-101	198	NA	–	Reusable	Electrostatic adsorption	[169]
Pb(II)	AMOF-1′	71	24 h	vs Hg^{2+}, Cd^{2+}	–	Ion-exchange	[232]
Cd(II)	AMOF-1′	41	24 h	vs Pb^{2+}, Hg^{2+}	–	Ion-exchange	[232]
Pb(II)	MCNC@Zn-BTC	558.66	30 min	vs Cu^{2+}, Zn^{2+}, Cd^{2+}	Reusable	–	[248]
Cd(II)	$(Cu_3(BTC)_2)$—SO_3H	88	10 min	vs various cations	Reusable	Ion-exchange	[249]
Cd(II)	FJI-H9	225	NA	vs various cations	Reusable	Cd—O interactions	[250]

Continued

TABLE 5.6 Other heavy metal adsorption data for reported MOF-based materials—Cont'd

Targets	MOF-based materials	Adsorption capacity (mg/g)	Equilibrium time	Selectivity	Reusability	Mechanism	Ref.
Cu(II)	Ca-MOF	68	100 min	–	Reusable	Ion-exchange	[251]
Cu(II)	ZIF-8	800	20 min	–	–	Ion exchange coordination	[252]
Pb(II)	NH_2-Zr-MOFs	92.18	120 min	–	–	Coordination interaction	[253]
Cd(II)	NH_2-Zr-MOFs	177.35	120 min	–	–	Coordination interaction	[253]
Pb(II)	TMU-5	251	5 min	–	–	Lewis acid-base interaction	[254]
Cd(II)	TMU-5	43	5 min	–	–	Lewis acid-base interaction	[254]
Cu(II)	TMU-5	57	5 min	–	–	Lewis acid-base interaction	[254]

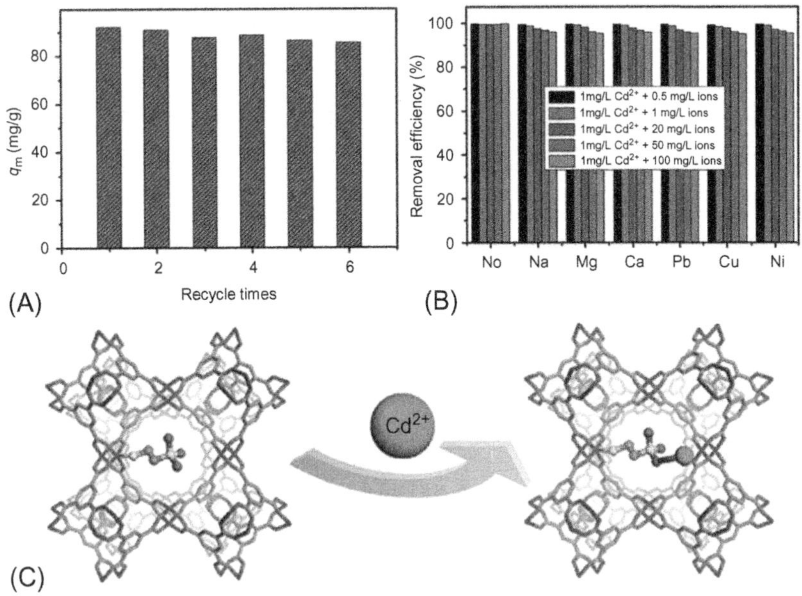

FIG. 5.10 Reusability of the $Cu_3(BTC)_2$—SO_3H for cadmium adsorption (A); effect of coexisting ions on the removal of cadmium by $Cu_3(BTC)_2$—SO_3H (B); proposed mechanism for Cd(II) adsorption in $Cu_3(BTC)_2$—SO_3H (C) [249]. *(Reprinted with permission from Y. Wang, G. Ye, H. Chen, X. Hu, Z. Niu, S. Ma, Functionalized metal-organic framework as a new platform for efficient and selective removal of Cadmium(II) from aqueous solution, J. Mater. Chem. A 3 (2015) 15292–15298. Copyright 2015 RSC.)*

superior selectivity, and reusability. Hence, it is still worthwhile to explore new types of MOF-based materials with excellent removal performance for Cd(II).

5.3.4.3 MOF-based materials for the removal of Cu(II)

The first MOF used for Cu(II) adsorption was **ZIF-8**, which showed fast kinetics and an ultrahigh adsorption capacity (~800 mg/g). ZIF-8 was efficient for Cu(II) adsorption over a pH range of 3–6 at low and high Cu(II) concentrations. The adsorption mechanism included: (i) ion exchange at a low initial Cu(II) concentrations; and (ii) coordination between Cu(II) and the nitrogen groups on 2-methylimidazole at high Cu(II) concentrations. Recently, Margariti et al. prepared a new **Ca-MOF** and discovered that Cu(II) was adsorbed via exchange with the Ca^{2+} ions [251]. As shown in Fig. 5.11, the adsorption kinetics and isotherm data were well described by the pseudo-second-order and Langmuir models. The maximum adsorption capacity was determined to be only 68 mg/g by the batch technique, and a similar adsorption capacity was achieved in column experiments. Furthermore, the adsorption performance could be maintained over a wide pH range (3–8) [251]. This study represents the first case where the constituent metal ions of an MOF could be exchanged by adsorbate ions.

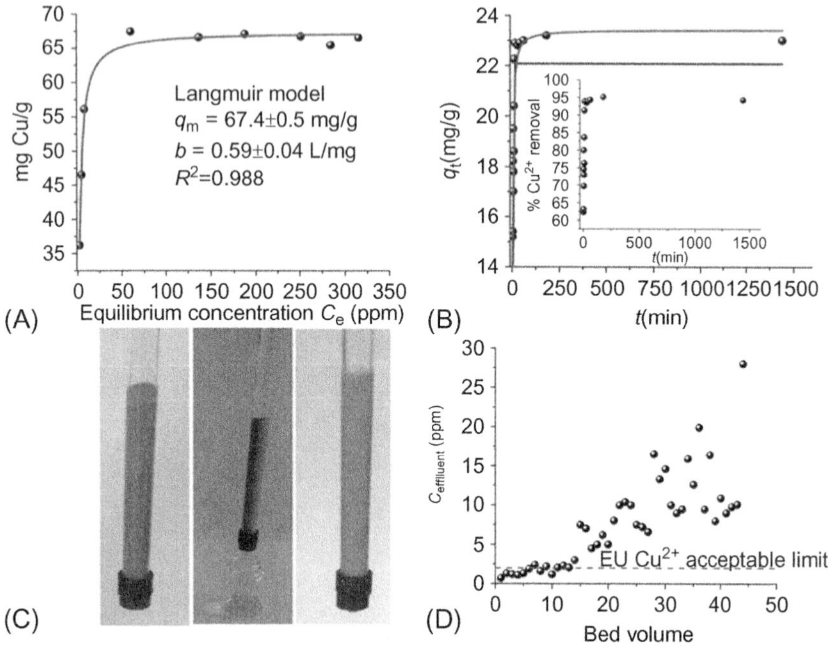

FIG. 5.11 Isotherm data for Cu(II) exchange by the Ca-MOF. The *red line* (gray line in print versions) represents the fitting of the data with the Langmuir model (A); kinetic data for the sorption of Cu(II) by the Ca-MOF (initial Cu(II) concentration was 24.4 ppm, pH~7). The *blue and red solid lines* (gray line in print versions) represent the fitting of the data with Lagergren's first-order (B); column study for the adsorption and desorption of Cu(II) exchange by the Ca-MOF (C); column sorption data (D) [251]. *(Reprinted with permission from A. Margariti, S. Rapti, A. Katsenis, T. Friščić, Y. Georgiou, M. Manos, G. Papaefstathiou, Cu^{2+} sorption from aqueous media by a recyclable Ca^{2+} framework, Inorg. Chem. Front. 4 (2017) 773–781. Copyright 2017 RSC.)*

As shown in Table 5.6, only a few MOFs and MOF-based materials have been applied in the removal of Cu(II) from aqueous solutions. Among these materials, the highest adsorption capacity for Cu(II) was observed by ZIF-8 (~800 mg/g) [252], which is much higher than by AC [260], natural zeolite [261], GO-based materials (18 mg/g for GO/Fe_3O_4 [262] and 24.4 mg/g for PD/GO [263]), organic resins (13–60 mg/g) [264], and LDH composites (75.01 mg/g) [265]. In comparison, Ca-MOF has a relatively low adsorption capacity (68 mg/g), but it has superior reusability and a wide active pH range with respect to the above-mentioned materials.

5.3.4.4 MOF-based materials for the simultaneous removal of various cations

To date, several MOF-based materials have been applied to simultaneously remove different cations [50, 169, 253, 254, 266]. In particular, **UiO-66-NH$_2$(Zr)** was functionalized with thiourea, isothiocyanate, and isocyanate molecules by postsynthetic methods and further applied for the preconcentration of various metal ions, including Pb(II), Cr(III), Cd(II), and Hg(II) from aqueous solutions [50]. The modified MOFs showed a more enhanced adsorption performance for all four metal ions than the original MOFs, due to the strong coordination ability of sulfur functionalities toward metal ions. Among these modified MOFs, the adsorption capacity of **UiO-66-NHC(S)NHMe** for Pb(II) and Hg(II) was up to 232 and 769 mg/g, respectively. Importantly, the Zr-based MOFs showed excellent thermal and chemical stabilities, which could assure the integrity of the framework. This characteristic could prevent the adsorbent from becoming a secondary contamination. Nevertheless, no information about the selectivity and recyclability of these modified MOFs was provided; this needs a deep study.

Mechanosynthesized azine-(TMU-4 and 5) and imine-modified (TMU-6) Zn-based MOFs were tried to extract trace amounts of various heavy metal ions, including Cd(II), Co(II), Cr(III), Cu(II), and Pb(II), from aqueous solutions [254]. All these MOFs were stable in water in a wide pH range. Among these MOFs, **TMU-5** showed superior uptake efficiency and capacity for Pb(II) ions (251 mg/g). TMU-4 and 5 showed higher adsorption performance than TMU-6 because the Lewis basicity of the azine groups contributed to the adsorption. As shown in Fig. 5.12, a solution of ethylenediaminetetraacetic acid (EDTA) (0.4 M) was able to desorb the adsorbed metal ions. The authors focused on the optimization of different parameters, including solution pH, contact time, and the adsorbent dosage. Further study on the recyclability is needed.

The adsorption properties of other heavy metal ions by MOFs and MOF-base materials are tabulated in Table 5.6. Most of these studies are limited to the removal of Cd(II), Pb(II), and Cu(II), with little information on the removal of other heavy metal ions. Information about the competitive removal of coexisting various heavy metal ions is of great importance, while most studies were not provided.

5.4 Conclusions and perspectives

All in all, MOF-based materials for the capture of toxic and radioactive metal ions outperformed current techniques in adsorption capacity, adsorption kinetics, and recyclability. Although great progress has been achieved in this area, other key parameters, including low cost, high processability, and long lifetime of an adsorbent, should be taken seriously to build bridges between theoretical studies and practical applications. Relevant studies in some of these aspects have been carried out to mature MOF technologies during the last few years. For example, MOFs possessing excellent stability under practical environmental conditions were one of the crucial disadvantages of MOFs a few years ago, hampering their application in real water remediation. Now, the development of various high water-stable MOFs has been realized by using various synthetic approaches. Moreover, great progress has been carried out in improving MOFs and incorporating them

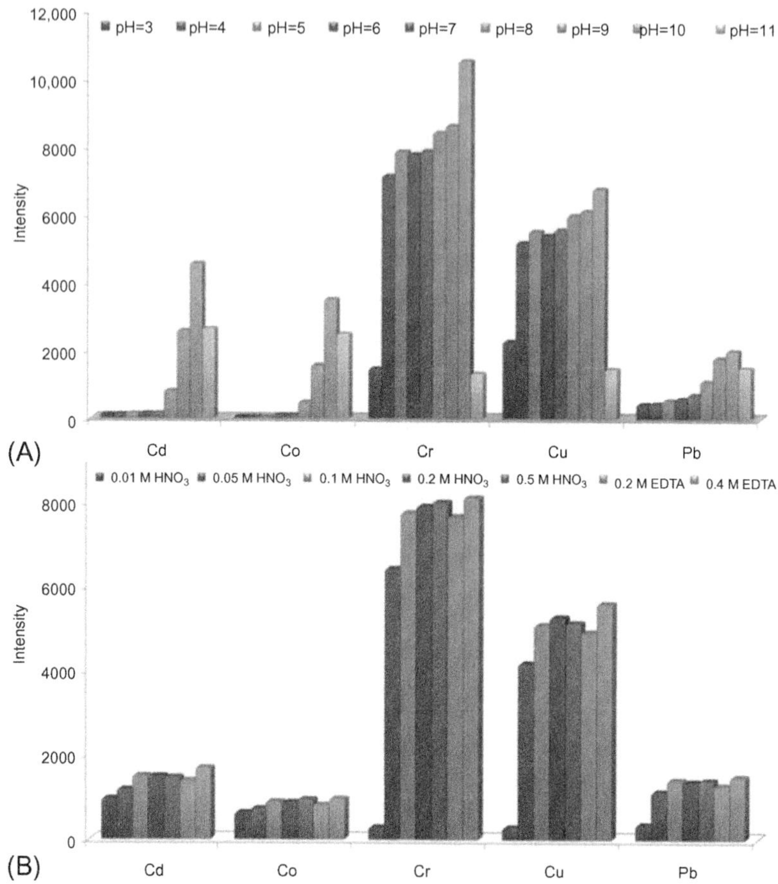

FIG. 5.12 Effect of eluent type on extraction efficiency (A); effect of sample pH on extraction efficiency (B) [254]. *(Reprinted with permission from E. Tahmasebi, M.Y. Masoomi, Y. Yamini, A. Morsali, Application of mechanosynthesized azine-decorated Zinc(II) metal-organic frameworks for highly efficient removal and extraction of some heavy-metal ions from aqueous samples: a comparative study, Inorg. Chem. 54 (2014) 425–433. Copyright 2014 ACS.)*

with other functional adsorbents, which will enhance the processability and open a fruitful area of research in materials science.

Despite these significant developments, MOFs applied in the decontamination of wastewater are still scarce, and are mainly limited to various MILs, ZIFs, UiOs, and their derivatives. Thus, the design of new water-stable MOFs available for wastewater treatment should still be taken into consideration for future interests. Expanding the spectrum of organic linkers beyond carboxylate and imidazolate ligands applied to synthesize MOFs should be the research direction for those working in MOF fields. To achieve this goal, organic chemistry would provide almost endless possibilities for the coordinating chemists to construct functional and robust MOFs. Moreover, the advantage of multivariate and metalloligand synthetic techniques as well as a more synergic link to supramolecular chemistry should be taken into consideration. Each approach or its combination would take MOFs to the next level.

The recent research progress of MOF-based materials used as adsorbents for the decontamination of toxic and radioactive metal ions has been summarized critically in this chapter. More detailed studies focused on the adsorption behaviors and mechanisms of MOFs and MOF-based materials are still required in future research.

References

[1] H. Li, M. Eddaoudi, M. O'Keeffe, O.M. Yaghi, Design and synthesis of an exceptionally stable and highly porous metal-organic framework, Nature 402 (1999) 276–279.

[2] G. Maurin, C. Serre, A. Cooper, G. Ferey, The new age of MOFs and of their porous-related solids, Chem. Soc. Rev. 46 (2017) 3104–3107.

[3] D.J. Tranchemontagne, J.L. Mendoza-Cortes, M. O'Keeffe, O.M. Yaghi, Secondary building units, nets and bonding in the chemistry of metal-organic frameworks, Chem. Soc. Rev. 38 (2009) 1257–1283.

[4] A. Douvali, A.C. Tsipis, S.V. Eliseeva, S. Petoud, G.S. Papaefstathiou, C.D. Malliakas, I. Papadas, G.S. Armatas, I. Margiolaki, M.G. Kanatzidis, Turn-on luminescence sensing and real-time detection of traces of water in organic solvents by a flexible metal-organic framework, Angew. Chem. Int. Ed. 127 (2015) 1671–1676.

[5] S. Horike, S. Shimomura, S. Kitagawa, Soft porous crystals, Nat. Chem. 1 (2009) 695–704.

[6] D. Bradshaw, J. Claridge, E. Cussen, T. Prior, M. Rosseinsky, Design, chirality, and flexibility in nanoporous molecule-based materials, Acc. Chem. Res. 38 (2005) 273–282.

[7] O.K. Farha, J.T. Hupp, Rational design, synthesis, purification, and activation of metal-organic framework materials, Acc. Chem. Res. 43 (2010) 1166–1175.

[8] S. Zheng, T. Wu, J. Zhang, M. Chow, R.A. Nieto, P. Feng, X. Bu, Porous metal carboxylate boron imidazolate frameworks, Angew. Chem. Int. Ed. 49 (2010) 5362–5366.

[9] J.M. Taylor, T. Komatsu, S. Dekura, K. Otsubo, M. Takata, H. Kitagawa, The role of a three dimensionally ordered defect sublattice on the acidity of a sulfonated metal–organic framework, J. Am. Chem. Soc. 137 (2015) 11498–11506.

[10] P. Kumar, A. Deep, K. Kim, Metal organic frameworks for sensing applications, TrAC, Trends Anal. Chem. 73 (2015) 39–53.

[11] R. Kaur, K. Kim, A. Paul, A. Deep, Recent advances in the photovoltaic applications of coordination polymers and metal organic frameworks, J. Mater. Chem. A 4 (2016) 3991–4002.

[12] H. Furukawa, K.E. Cordova, M. O'Keeffe, O.M. Yaghi, The chemistry and applications of metal-organic frameworks, Science 341 (2013) 1230444.

[13] S. Rapti, A. Pournara, D. Sarma, I.T. Papadas, G.S. Armatas, Y.S. Hassan, M.H. Alkordi, M.G. Kanatzidis, M.J. Manos, Rapid, green and inexpensive synthesis of high quality UiO-66 amino-functionalized materials with exceptional capability for removal of hexavalent chromium from industrial waste, Inorg. Chem. Front. 3 (2016) 635–644.

[14] Z. Hu, Y. Peng, Z. Kang, Y. Qian, D. Zhao, A modulated hydrothermal (MHT) approach for the facile synthesis of UiO-66-type MOFs, Inorg. Chem. 54 (2015) 4862–4868.

[15] Z. Hu, D. Zhao, De facto methodologies toward the synthesis and scale-up production of UiO-66-type metal–organic frameworks and membrane materials, Dalton Trans. 44 (2015) 19018–19040.

[16] T. Friščić, Metal-organic frameworks: mechanochemical synthesis strategies, in: Encyclopedia of Inorganic and Bioinorganic Chemistry, John Wiley & Sons, New York, 2014, pp. 1–9.

[17] S. Duan, J. Li, X. Liu, Y. Wang, S. Zeng, D. Shao, T. Hayat, HF-free synthesis of nanoscale metal–organic framework NMIL-100(Fe) as an efficient dye adsorbent, ACS Sustain. Chem. Eng. 4 (2016) 3368–3378.

[18] J.H. Cavka, S. Jakobsen, U. Olsbye, N. Guillou, C. Lamberti, S. Bordiga, K.P. Lillerud, A new zirconium inorganic building brick forming metal organic frameworks with exceptional stability, J. Am. Chem. Soc. 130 (2008) 13850–13851.

[19] M. Kandiah, M.H. Nilsen, S. Usseglio, S. Jakobsen, U. Olsbye, M. Tilset, C. Larabi, E.A. Quadrelli, F. Bonino, K.P. Lillerud, Synthesis and stability of tagged UiO-66 Zr-MOFs, Chem. Mater. 22 (2010) 6632–6640.

[20] A.J. Howarth, M.J. Katz, T.C. Wang, A.E. Platero-Prats, K.W. Chapman, J.T. Hupp, O.K. Farha, High efficiency adsorption and removal of selenate and selenite from water using metal–organic frameworks, J. Am. Chem. Soc. 137 (2015) 7488–7494.

[21] K. Yee, N. Reimer, J. Liu, S. Cheng, S. Yiu, J. Weber, N. Stock, Z. Xu, Effective mercury sorption by thiol-laced metal–organic frameworks: in strong acid and the vapor phase, J. Am. Chem. Soc. 135 (2013) 7795–7798.

[22] C. He, K. Lu, W. Lin, Nanoscale metal–organic frameworks for real-time intracellular pH sensing in live cells, J. Am. Chem. Soc. 136 (2014) 12253–12256.

[23] L. Zhu, D. Zhang, M. Xue, H. Li, S. Qiu, Direct observations of the MOF (UiO-66) structure by transmission electron microscopy, CrystEngComm 15 (2013) 9356–9359.

[24] W. Morris, W.E. Briley, E. Auyeung, M.D. Cabezas, C.A. Mirkin, Nucleic acid–metal organic framework (MOF) nanoparticle conjugates, J. Am. Chem. Soc. 136 (2014) 7261–7264.

[25] A.V. Desai, B. Manna, A. Karmakar, A. Sahu, S.K. Ghosh, A water-stable cationic metal-organic framework as a dual adsorbent of oxoanion pollutants, Angew. Chem. Int. Ed. 55 (2016) 7811–7815.

[26] H. Fei, M.R. Bresler, S.R. Oliver, A new paradigm for anion trapping in high capacity and selectivity: crystal-to-crystal transformation of cationic materials, J. Am. Chem. Soc. 133 (2011) 11110–11113.

[27] M. Mon, F. Lloret, J. Ferrando-Soria, C. Marti-Gastaldo, D. Armentano, E. Pardo, Selective and efficient removal of mercury from aqueous media with the highly flexible arms of a bioMOF, Angew. Chem. Int. Ed. 128 (2016) 11333–11338.

[28] M. Jian, H. Wang, R. Liu, J. Qu, H. Wang, X. Zhang, Self-assembled one-dimensional MnO_2@zeolitic imidazolate framework-8 nanostructures for highly efficient arsenite removal, Environ. Sci. Nano 3 (2016) 1186–1194.

[29] Y. Wang, Z. Liu, Y. Li, Z. Bai, W. Liu, Y. Wang, X. Xu, C. Xiao, D. Sheng, J. Diwu, Umbellate distortions of the uranyl coordination environment result in a stable and porous polycatenated framework that can effectively remove cesium from aqueous solutions, J. Am. Chem. Soc. 137 (2015) 6144–6147.

[30] S. Wang, E.V. Alekseev, J. Diwu, W.H. Casey, B.L. Phillips, W. Depmeier, T.E. Albrecht-Schmitt, NDTB-1: a supertetrahedral cationic framework that removes TcO_4^- from solution, Angew. Chem. Int. Ed. 49 (2010) 1057–1060.

[31] J. Tian, L.V. Saraf, B. Schwenzer, S.M. Taylor, E.K. Brechin, J. Liu, S.J. Dalgarno, P.K. Thallapally, Selective metal cation capture by soft anionic metal–organic frameworks via drastic single-crystal-to-single-crystal transformations, J. Am. Chem. Soc. 134 (2012) 9581–9584.

[32] D. Sheng, L. Zhu, C. Xu, C. Xiao, Y. Wang, Y. Wang, L. Chen, J. Diwu, J. Chen, Z. Chai, S. Wang, Efficient and selective uptake of TcO_4^- by a cationic metal-organic framework material with open Ag^+ sites, Environ. Sci. Technol. 51 (2017) 3471–3479.

[33] J. Li, Y. Liu, X. Wang, G. Zhao, Y. Ai, B. Han, T. Wen, T. Hayat, A. Alsaedi, X. Wang, Experimental and theoretical study on selenate uptake to zirconium metal-organic frameworks: effect of defects and ligands, Chem. Eng. J. 330 (2017) 1012–1021.

[34] J. Li, X. Wang, G. Zhao, C. Chen, Z. Chai, A. Alsaedi, T. Hayat, X. Wang, Metal-organic framework-based materials: superior adsorbents for the capture of toxic and radioactive metal ions, Chem. Soc. Rev. 47 (2018) 2322–2356.

[35] M. Mon, R. Bruno, J. Ferrando-Soria, D. Armentano, E. Pardo, Metal-organic framework technologies for water remediation: towards a sustainable ecosystem, J. Mater. Chem. A 6 (2018) 4912–4947.

[36] P.A. Kobielska, A.J. Howarth, O.K. Farha, S. Nayak, Metal-organic frameworks for heavy metal removal from water, Coord. Chem. Rev. 358 (2018) 92–107.

[37] X. Zhao, Y. Wang, D.-S. Li, X. Bu, P. Feng, Metal-organic frameworks for separation, Adv. Mater. (2018)1705189

[38] C.O. Audu, H.G.T. Nguyen, C. Chang, M.J. Katz, L. Mao, O.K. Farha, J.T. Hupp, S.T. Nguyen, The dual capture of As^V and As^{III} by UiO-66 and analogues, Chem. Sci. 7 (2016) 6492–6498.

[39] B. Li, X. Zhu, K. Hu, Y. Li, J. Feng, J. Shi, J. Gu, Defect creation in metal-organic frameworks for rapid and controllable decontamination of roxarsone from aqueous solution, J. Hazard. Mater. 302 (2016) 57–64.

[40] G.C. Shearer, S. Chavan, S. Bordiga, S. Svelle, U. Olsbye, K.P. Lillerud, Defect engineering: tuning the porosity and composition of the metal-organic framework UiO-66 via modulated synthesis, Chem. Mater. 28 (2016) 3749–3761.

[41] P. Ghosh, Y.J. Colon, R.Q. Snurr, Water adsorption in UiO-66: the importance of defects, Chem. Commun. 50 (2014) 11329–11331.

[42] X. He, X. Min, X. Luo, Efficient removal of antimony (III, V) from contaminated water by amino modification of a zirconium metal-organic framework with mechanism study, J. Chem. Eng. Data 62 (2017) 1519–1529.

[43] X. Luo, L. Ding, J. Luo, Adsorptive removal of Pb(II) ions from aqueous samples with amino-functionalization of metal-organic frameworks MIL-101(Cr), J. Chem. Eng. Data 60 (2015) 1732–1743.

[44] X. Luo, T. Shen, L. Ding, W. Zhong, J. Luo, S. Luo, Novel thymine-functionalized MIL-101 prepared by post-synthesis and enhanced removal of Hg^{2+} from water, J. Hazard. Mater. 306 (2016) 313–322.

[45] F. Luo, J.L. Chen, L.L. Dang, W.N. Zhou, H.L. Lin, J.Q. Li, S.J. Liu, M.B. Luo, High-performance Hg^{2+} removal from ultra-low-concentration aqueous solution using both acylamide-and hydroxyl-functionalized metal-organic framework, J. Mater. Chem. A 3 (2015) 9616–9620.

[46] J. Li, X. Li, T. Hayat, A. Alsaedi, C. Chen, Screening of zirconium-based metal-organic frameworks for efficient simultaneous removal of antimonite (Sb(III)) and antimonate (Sb(V)) from aqueous solution, ACS Sustain. Chem. Eng. 5 (2017) 11496–11503.

[47] L. Liang, Q. Chen, F. Jiang, D. Yuan, J. Qian, G. Lv, H. Xue, L. Liu, H. Jiang, M. Hong, In situ large-scale construction of sulfur-functionalized metal-organic framework and its efficient removal of Hg(II) from water, J. Mater. Chem. A 4 (2016) 15370–15374.

[48] F. Ke, L. Qiu, Y. Yuan, F. Peng, X. Jiang, A. Xie, Y. Shen, J. Zhu, Thiol-functionalization of metal-organic framework by a facile coordination-based postsynthetic strategy and enhanced removal of Hg^{2+} from water, J. Hazard. Mater. 196 (2011) 36–43.

[49] L. Huang, M. He, B. Chen, B. Hu, A mercapto functionalized magnetic Zr-MOF by solvent-assisted ligand exchange for Hg^{2+} removal from water, J. Mater. Chem. A 4 (2016) 5159–5166.

[50] H. Saleem, U. Rafique, R.P. Davies, Investigations on post-synthetically modified UiO-66-NH_2 for the adsorptive removal of heavy metal ions from aqueous solution, Microporous Mesoporous Mater. 221 (2016) 238–244.

[51] F. Ke, J. Jiang, Y. Li, J. Liang, X. Wan, S. Ko, Highly selective removal of Hg^{2+} and Pb^{2+} by thiol-functionalized Fe_3O_4@ metal-organic framework core-shell magnetic microspheres, Appl. Surf. Sci. 413 (2017) 266–274.

[52] Y. Peng, H. Huang, D. Liu, C. Zhong, Radioactive barium ion trap based on metal-organic framework for efficient and irreversible removal of barium from nuclear wastewater, ACS Appl. Mater. Interfaces 8 (2016) 8527–8535.

[53] T. Liu, J. Che, Y. Hu, X. Dong, X. Liu, C. Che, Alkenyl/thiol-derived metal-organic frameworks (MOFs) by means of postsynthetic modification for effective mercury adsorption, Chem. Eur. J. 20 (2014) 14090–14095.

[54] R. Ricco, K. Konstas, M.J. Styles, J.J. Richardson, R. Babarao, K. Suzuki, P. Scopece, P. Falcaro, Lead(II) uptake by aluminium based magnetic framework composites (MFCs) in water, J. Mater. Chem. A 3 (2015) 19822–19831.

[55] X. Min, W. Yang, Y. Hui, C. Gao, S. Dang, Z. Sun, Fe_3O_4@ZIF-8: a magnetic nanocomposite for highly efficient UO_2^{2+} adsorption and selective UO_2^{2+}/Ln^{3+} separation, Chem. Commun. 53 (2017) 4199–4202.

[56] P. Yang, Q. Liu, J. Liu, H. Zhang, Z. Li, R. Li, L. Liu, J. Wang, Interfacial growth of a metal-organic framework (UiO-66) on functionalized graphene oxide (GO) as a suitable seawater adsorbent for extraction of uranium(VI), J. Mater. Chem. A 5 (2017) 17933–17942.

[57] S. Naeimi, H. Faghihian, Performance of novel adsorbent prepared by magnetic metal-organic framework (MOF) modified by potassium nickel hexacyanoferrate for removal of Cs^+ from aqueous solution, Sep. Purif. Technol. 175 (2017) 255–265.

[58] S. Rapti, A. Pournara, D. Sarma, I.T. Papadas, G.S. Armatas, A.C. Tsipis, T. Lazarides, M.G. Kanatzidis, M.J. Manos, Selective capture of hexavalent chromium from an anion-exchange column of metal organic resin–alginic acid composite, Chem. Sci. 7 (2016) 2427–2436.

[59] M. Carboni, C.W. Abney, S. Liu, W. Lin, Highly porous and stable metal-organic frameworks for uranium extraction, Chem. Sci. 4 (2013) 2396–2402.

[60] T. Zheng, Z. Yang, D. Gui, Z. Liu, X. Wang, X. Dai, S. Liu, L. Zhang, Y. Gao, L. Chen, D. Sheng, Y. Wang, J. Diwu, J. Wang, R. Zhou, Z. Chai, T.E. Albrecht-Schmitt, S. Wang, Overcoming the crystallization and designability issues in the ultrastable zirconium phosphonate framework system, Nat. Commun. 8 (2017)15369.

[61] L. Zhang, L.L. Wang, L. Le Gong, X.F. Feng, M.B. Luo, F. Luo, Coumarin-modified microporous-mesoporous Zn-MOF-74 showing ultra-high uptake capacity and photo-switched storage/release of U^{VI} ions, J. Hazard. Mater. 311 (2016) 30–36.

[62] F. Li, X. Li, P. Cui, Y. Sun, Plasma-grafting amidoxime/metal-organic framework composites for the selective sequestration of U(VI). Environ. Sci. Nano (2018), https://doi.org/10.1039/C1038EN00583D.

[63] Z. Bai, L. Yuan, L. Zhu, Z. Liu, S. Chu, L. Zheng, J. Zhang, Z. Chai, W. Shi, Introduction of amino groups into acid-resistant MOFs for enhanced U(VI) sorption, J. Mater. Chem. A 3 (2015) 525–534.

[64] W. Yang, Z. Bai, W. Shi, L. Yuan, T. Tian, Z. Chai, H. Wang, Z. Sun, MOF-76: from a luminescent probe to highly efficient U^{VI} sorption material, Chem. Commun. 49 (2013) 10415–10417.

[65] J.Q. Li, L. Le Gong, X.F. Feng, L. Zhang, H.Q. Wu, C.S. Yan, Y.Y. Xiong, H.Y. Gao, F. Luo, Direct extraction of U(VI) from alkaline solution and seawater via anion exchange by metal-organic framework, Chem. Eng. J. 316 (2017) 154–159.

[66] B. Luo, L. Yuan, Z. Chai, W. Shi, Q. Tang, U(VI) capture from aqueous solution by highly porous and stable MOFs: UiO-66 and its amine derivative, J. Radioanal. Nucl. Chem. 307 (2016) 269–276.

[67] L. Li, W. Ma, S. Shen, H. Huang, Y. Bai, H. Liu, A combined experimental and theoretical study on the extraction of uranium by amino-derived metal-organic frameworks through post-synthetic strategy, ACS Appl. Mater. Interfaces 8 (2016) 31032–31041.

[68] J. Zhang, N. Zhang, L. Zhang, Y. Fang, W. Deng, M. Yu, Z. Wang, L. Li, X. Liu, J. Li, Adsorption of uranyl ions on amine-functionalization of MIL-10(Cr) nanoparticles by a facile coordination-based post-synthetic strategy and X-ray absorption spectroscopy studies, Sci. Rep. 5 (2015) 13514.

[69] J. De Decker, J. Rochette, J. De Clercq, J. Florek, P. Van Der Voort, Carbamoylmethylphosphine oxide-functionalized MIL-101(Cr) as highly selective uranium adsorbent, Anal. Chem. 89 (2017) 5678–5682.

[70] A.A. Alqadami, M. Naushad, Z.A. Alothman, A.A. Ghfar, Novel metal–organic framework (MOF) based composite material for the sequestration of U(VI) and Th(IV) metal ions from aqueous environment, ACS Appl. Mater. Interfaces 9 (2017) 36026–36037.

[71] L. Chen, Z. Bai, L. Zhu, L. Zhang, Y. Cai, Y. Li, W. Liu, Y. Wang, L. Chen, J. Diwu, J. Wang, Z. Chai, S. Wang, Ultrafast and efficient extraction of uranium from seawater using an amidoxime appended metal-organic framework, ACS Appl. Mater. Interfaces 9 (2017) 32446–32451.

[72] Y. Feng, H. Jiang, S. Li, J. Wang, X. Jing, Y. Wang, M. Chen, Metal-organic frameworks HKUST-1 for liquid-phase adsorption of uranium, Colloids Surf. A 431 (2013) 87–92.

[73] L.L. Song, C. Chen, F. Luo, S.Y. Huang, L.L. Wang, N. Zhang, Isoreticular MOFs functionalized in the pore wall by different organic groups for high-performance removal of uranyl ions, J. Radioanal. Nucl. Chem. 310 (2016) 317–327.

[74] S. Liu, M. Luo, J. Li, F. Luo, L. Ke, J. Ma, Adsorption equilibrium and kinetics of uranium onto porous azo-metal-organic frameworks, J. Radioanal. Nucl. Chem. 310 (2016) 353–362.

[75] M.E. Mahmoud, M.A. Khalifa, Y.M. El Wakeel, M.S. Header, T.M. Abdel-Fattah, Engineered nano-magnetic iron oxide-urea-activated carbon nanolayer sorbent for potential removal of uranium (VI) from aqueous solution, J. Nucl. Mater. 487 (2017) 13–22.

[76] Y. Sun, Z. Wu, X. Wang, C. Ding, W. Cheng, S. Yu, X. Wang, Macroscopic and microscopic investigation of U(VI) and Eu(III) adsorption on carbonaceous nanofibers, Environ. Sci. Technol. 50 (2016) 4459–4467.

[77] R. Zhang, C. Chen, J. Li, X. Wang, Preparation of montmorillonite@ carbon composite and its application for U(VI) removal from aqueous solution, Appl. Surf. Sci. 349 (2015) 129–137.

[78] S. Song, S. Huang, R. Zhang, Z. Chen, T. Wen, S. Wang, T. Hayat, A. Alsaedi, X. Wang, Simultaneous removal of U(VI) and humic acid on defective TiO_{2-x} investigated by batch and spectroscopy techniques, Chem. Eng. J. 325 (2017) 576–587.

[79] H. Mei, X. Tan, S. Yu, X. Ren, C. Chen, X. Wang, Effect of silicate on U(VI) sorption to γ-Al_2O_3: batch and EXAFS studies, Chem. Eng. J. 269 (2015) 371–378.

[80] L. Yan, K. Yang, R. Shan, T. Yan, J. Wei, S. Yu, H. Yu, B. Du, Kinetic, isotherm and thermodynamic investigations of phosphate adsorption onto core-shell Fe_3O_4@LDHs composites with easy magnetic separation assistance, J. Colloid Interface Sci. 448 (2015) 508–516.

[81] S. Ma, L. Huang, L. Ma, Y. Shim, S.M. Islam, P. Wang, L. Zhao, S. Wang, G. Sun, X. Yang, Efficient uranium capture by polysulfide/layered double hydroxide composites, J. Am. Chem. Soc. 137 (2015) 3670–3677.

[82] H. Liu, M. Li, T. Chen, C. Chen, N.S. Alharbi, T. Hayat, D. Chen, Q. Zhang, Y. Sun, New synthesis of nZVI/C composites as an efficient adsorbent for the uptake of U(VI) from aqueous solutions, Environ. Sci. Technol. 51 (2017) 9227–9234.

[83] C. Ding, W. Cheng, Y. Sun, X. Wang, Effects of *Bacillus subtilis* on the reduction of U(VI) by nano-Fe^0, Geochim. Cosmochim. Acta 165 (2015) 86–107.

[84] W. Linghu, H. Yang, Y. Sun, G. Sheng, Y. Huang, One-pot synthesis of LDH/GO composites as highly effective adsorbents for decontamination of U(VI), ACS Sustain. Chem. Eng. 5 (2017) 5608–5616.

[85] Z. Li, F. Chen, L. Yuan, Y. Liu, Y. Zhao, Z. Chai, W. Shi, Uranium(VI) adsorption on graphene oxide nanosheets from aqueous solutions, Chem. Eng. J. 210 (2012) 539–546.

[86] L. Tan, Q. Liu, X. Jing, J. Liu, D. Song, S. Hu, L. Liu, J. Wang, Removal of uranium(VI) ions from aqueous solution by magnetic cobalt ferrite/multiwalled carbon nanotubes composites, Chem. Eng. J. 273 (2015) 307–315.

[87] H. Alijani, M.H. Beyki, S.N. Mirzababaei, Adsorption of UO_2^{2+} ions from aqueous solution using amine functionalized MWCNT: kinetic, thermodynamic and isotherm study, J. Radioanal. Nucl. Chem. 306 (2015) 165–173.

[88] M.J. Manos, M.G. Kanatzidis, Metal sulfide ion exchangers: superior sorbents for the capture of toxic and nuclear waste-related metal ions, Chem. Sci. 7 (2016) 4804–4824.

[89] M. Feng, D. Sarma, X. Qi, K. Du, X. Huang, M.G. Kanatzidis, Efficient removal and recovery of uranium by a layered organic-inorganic hybrid thiostannate, J. Am. Chem. Soc. 138 (2016) 12578–12585.

[90] Y. Liu, L. Yuan, Y. Yuan, J. Lan, Z. Li, Y. Feng, Y. Zhao, Z. Chai, W. Shi, A high efficient sorption of U(VI) from aqueous solution using amino-functionalized SBA-15, J. Radioanal. Nucl. Chem. 292 (2012) 803–810.

[91] X. Wang, L. Yuan, Y. Wang, Z. Li, J. Lan, Y. Liu, Y. Feng, Y. Zhao, Z. Chai, W. Shi, Mesoporous silica SBA-15 functionalized with phosphonate and amino groups for uranium uptake, Sci. China Chem. 55 (2012) 1705–1711.

[92] L. Yuan, Y. Liu, W. Shi, Z. Li, J. Lan, Y. Feng, Y. Zhao, Y. Yuan, Z. Chai, A novel mesoporous material for uranium extraction, dihydroimidazole functionalized SBA-15, J. Mater. Chem. 22 (2012) 17019–17026.

[93] L. Yuan, Z. Bai, R. Zhao, Y. Liu, Z. Li, S. Chu, L. Zheng, J. Zhang, Y. Zhao, Z. Chai, W. Shi, Introduction of bifunctional groups into mesoporous silica for enhancing uptake of thorium(IV) from aqueous solution, ACS Appl. Mater. Interfaces 6 (2014) 4786–4796.

[94] A. Braun, D. Evans, P. Gandhi, M. Garamszeghy, H. Geiser, E. Hooper, F. Kaufmann, P. Luycx, B. Martynoy, L. Mezga, Application of Ion Exchange Processes for the Treatment of Radioactive Waste and Management of Spent Ion Exchangers, International Atomic Energy Agency, Vienna, 2002.

[95] T.L. Kirschling, K.B. Gregory, J.E.G. Minkley, G.V. Lowry, R.D. Tilton, Impact of nanoscale zero valent iron on geochemistry and microbial populations in trichloroethylene contaminated aquifer materials, Environ. Sci. Technol. 44 (2010) 3474–3480.

[96] K.D. Grieger, A. Fjordbøge, N.B. Hartmann, E. Eriksson, P.L. Bjerg, A. Baun, Environmental benefits and risks of zero-valent iron nanoparticles (nZVI) for in situ remediation: risk mitigation or trade-off? J. Contam. Hydrol. 118 (2010) 165–183.

[97] D.R. Dreyer, S. Park, C.W. Bielawski, R.S. Ruoff, The chemistry of graphene oxide, Chem. Soc. Rev. 39 (2010) 228–240.

[98] B. Aguila, D. Banerjee, Z. Nie, Y. Shin, S. Ma, P.K. Thallapally, Selective removal of cesium and strontium using porous frameworks from high level nuclear waste, Chem. Commun. 52 (2016) 5940–5942.

[99] H. Fei, D.L. Rogow, S.R. Oliver, Reversible anion exchange and catalytic properties of two cationic metal-organic frameworks based on Cu(I) and Ag(I), J. Am. Chem. Soc. 132 (2010) 7202–7209.

[100] D. Banerjee, W. Xu, Z. Nie, L.E. Johnson, C. Coghlan, M.L. Sushko, D. Kim, M.J. Schweiger, A.A. Kruger, C.J. Doonan, Zirconium-based metal-organic framework for removal of perrhenate from water, Inorg. Chem. 55 (2016) 8241–8243.

[101] L. Zhu, C. Xiao, X. Dai, J. Li, D. Gui, D. Sheng, L. Chen, R. Zhou, Z. Chai, T.E. Albrecht-Schmitt, S. Wang, Exceptional perrhenate/pertechnetate uptake and subsequent immobilization by a low-dimensional cationic coordination polymer: overcoming the Hofmeister bias selectivity, Environ. Sci. Technol. Lett. 4 (2017) 316–322.

[102] L. Zhu, D. Sheng, C. Xu, X. Dai, M.A. Silver, J. Li, P. Li, Y. Wang, Y. Wang, L. Chen, C. Xiao, J. Chen, R. Zhou, C. Zhang, O.K. Farha, Z. Chai, T.E. Albrecht-Schmitt, S. Wang, Identifying the recognition site for selective trapping of $^{99}TcO_4^-$ in a hydrolytically stable and radiation resistant cationic metal-organic framework, J. Am. Chem. Soc. 139 (2017) 14873–14876.

[103] N. Zhang, L. Yuan, W. Guo, S. Luo, Z. Chai, W. Shi, Extending the use of highly porous and functionalized MOFs to Th(IV) capture, ACS Appl. Mater. Interfaces 9 (2017) 25216–25224.

[104] S. Lin, J.K. Bediako, M. Song, W. Wei, J.A. Kim, Y. Yun, Effective adsorption of Pd(II), Pt(IV) and Au(III) by Zr(IV)-based metal-organic frameworks from strongly acidic solution, J. Mater. Chem. A 5 (2017) 13557–13564.

[105] Q. Fang, D. Yuan, J. Sculley, J. Li, Z. Han, H. Zhou, Functional mesoporous metal-organic frameworks for the capture of heavy metal ions and size-selective catalysis, Inorg. Chem. 49 (2010) 11637–11642.

[106] H. Zhang, J. Xue, N. Hu, J. Sun, D. Ding, Y. Wang, L. Li, Selective removal of U(VI) from low concentration wastewater by functionalized HKUST-1@$H_3PW_{12}O_{40}$, J. Radioanal. Nucl. Chem. 308 (2016) 865–875.

[107] Z. Bai, Y. Wang, Y. Li, W. Liu, L. Chen, D. Sheng, J. Diwu, Z. Chai, T.E. Albrecht-Schmitt, S. Wang, First cationic uranyl-organic framework with anion-exchange capabilities, Inorg. Chem. 55 (2016) 6358–6360.

[108] X. Li, Y. Gong, H. Zhao, R. Wang, Anion-directed assemblies of cationic metal-organic frameworks based on 4, 4'-bis (1,2,4-triazole): syntheses, structures, luminescent and anion exchange properties, Inorg. Chem. 53 (2014) 12127–12134.

[109] C. Kang, Y. Peng, Y. Tang, H. Huang, C. Zhong, Sulfate-rich metal-organic framework for high efficiency and selective removal of barium from nuclear wastewater, Ind. Eng. Chem. Res. 56 (2017) 13866–13873.

[110] G. Yuan, Y. Tian, J. Liu, H. Tu, J. Liao, J. Yang, Y. Yang, D. Wang, N. Liu, Schiff base anchored on metal-organic framework for Co(II) removal from aqueous solution, Chem. Eng. J. 326 (2017) 691–699.

[111] Y. Wang, Y. Li, Z. Bai, C. Xiao, Z. Liu, W. Liu, L. Chen, W. He, J. Diwu, Z. Chai, T.E. Albrecht-Schmitt, S. Wang, Design and synthesis of a chiral uranium-based microporous metal organic framework with high SHG efficiency and sequestration potential for low-valent actinides, Dalton Trans. 44 (2015) 18810–18814.

[112] C. Falaise, C. Volkringer, R. Giovine, B. Prelot, M. Huve, T. Loiseau, Capture of actinides (Th^{4+}, $[UO_2]^{2+}$) and surrogating lanthanide (Nd^{3+}) in porous metal-organic framework MIL-100(Al) from water: selectivity and imaging of embedded nanoparticles, Dalton Trans. 46 (2017) 12010–12014.

[113] X.-G. Guo, S. Qiu, X. Chen, Y. Gong, X. Sun, Postsynthesis modification of a metallosalen-containing metal-organic framework for selective Th(IV)/Ln(III) separation, Inorg. Chem. 56 (2017) 12357–12361.

[114] M. Zha, J. Liu, Y. Wong, Z. Xu, Extraction of palladium from nuclear waste-like acidic solutions by a metal-organic framework with sulfur and alkene functions, J. Mater. Chem. A 3 (2015) 3928–3934.

[115] A. Ararem, O. Bouras, F. Arbaoui, Adsorption of caesium from aqueous solution on binary mixture of iron pillared layered montmorillonite and goethite, Chem. Eng. J. 172 (2011) 230–236.

[116] M.R. Mahmoud, G.M. Rashad, E. Metwally, E.A. Saad, A.M. Elewa, Adsorptive removal of $^{134}Cs^{+}$, $^{60}Co^{2+}$ and $^{152+154}Eu^{3+}$ radionuclides from aqueous solutions using sepiolite: single and multi-component systems, Appl. Clay Sci. 141 (2017) 72–80.

[117] Y. Sun, D. Shao, C. Chen, S. Yang, X. Wang, Highly efficient enrichment of radionuclides on graphene oxide-supported polyaniline, Environ. Sci. Technol. 47 (2013) 9904–9910.

[118] D. Sarma, C.D. Malliakas, K.S. Subrahmanyam, S.M. Islam, M.G. Kanatzidis, $K_{2x}Sn_{4-x}S_{8-x}$ (x = 0.65-1): a new metal sulfide for rapid and selective removal of Cs^{+}, Sr^{2+} and UO_2^{2+} ions, Chem. Sci. 7 (2016) 1121–1132.

[119] A. Dyer, M. Pillinger, R. Harjula, S. Amin, Sorption characteristics of radionuclides on synthetic birnessite-type layered manganese oxides, J. Mater. Chem. 10 (2000) 1867–1874.

[120] N. Bleiman, Y.G. Mishael, Selenium removal from drinking water by adsorption to chitosan–clay composites and oxides: batch and columns tests, J. Hazard. Mater. 183 (2010) 590–595.

[121] A.W. Lounsbury, J.S. Yamani, C.P. Johnston, P. Larese-Casanova, J.B. Zimmerman, The role of counter ions in nano-hematite synthesis: implications for surface area and selenium adsorption capacity, J. Hazard. Mater. 310 (2016) 117–124.

[122] W. Cui, P. Li, Z. Wang, S. Zheng, Y. Zhang, Adsorption study of selenium ions from aqueous solutions using MgO nanosheets synthesized by ultrasonic method, J. Hazard. Mater. 341 (2018) 268–276.

[123] Y.T. Chan, W.H. Kuan, T.Y. Chen, M.K. Wang, Adsorption mechanism of selenate and selenite on the binary oxide systems, Water Res. 43 (2009) 4412–4420.

[124] J.S. Yamani, A.W. Lounsbury, J.B. Zimmerman, Adsorption of selenite and selenate by nanocrystalline aluminum oxide, neat and impregnated in chitosan beads, Water Res. 50 (2014) 373–381.

[125] H. Asiabi, Y. Yamini, M. Shamsayei, Highly selective and efficient removal of arsenic(V), chromium(VI) and selenium(VI) oxyanions by layered double hydroxide intercalated with zwitterionic glycine, J. Hazard. Mater. 339 (2017) 239–247.

[126] T. Nishimura, H. Hashimoto, M. Nakayama, Removal of selenium(VI) from aqueous solution with polyamine-type weakly basic ion exchange resin, Sep. Sci. Technol. 42 (2007) 3155–3167.

[127] T.M. Suzuki, D.A. Pacheco Tanaka, M.A. Llosa Tanco, M. Kanesato, T. Yokoyama, Adsorption and removal of oxo-anions of arsenic and selenium on the zirconium(IV) loaded polymer resin functionalized with diethylenetriamine-N,N,N′,N′-polyacetic acid, J. Environ. Monit. 2 (2000) 550–555.
[128] T. Yokoi, T. Tatsumi, H. Yoshitake, Selective selenate adsorption on cationated amino-functionalized MCM-41, Bull. Chem. Soc. Jpn. 76 (2003) 2225–2232.
[129] W. Xiao, B. Yan, H. Zeng, Q. Liu, Dendrimer functionalized graphene oxide for selenium removal, Carbon 105 (2016) 655–664.
[130] J. Zu, M. Ye, P. Wang, F. Tang, L. He, Design of a strong-base anion exchanger and its adsorption and elution behavior for rhenium(VII), RSC Adv. 6 (2016) 18868–18873.
[131] Y. Gao, K. Chen, X. Tan, X. Wang, A. Alsaedi, T. Hayat, C. Chen, Interaction mechanism of Re(VII) with zirconium dioxide nanoparticles archored onto reduced graphene oxides, ACS Sustain. Chem. Eng. 5 (2017) 2163–2171.
[132] J. Li, C. Chen, R. Zhang, X. Wang, Reductive immobilization of Re(VII) by graphene modified nanoscale zero-valent iron particles using a plasma technique, Sci. China Chem. 59 (2016) 150–158.
[133] L. Zhang, X.Q. Jiang, T.C. Xu, L.J. Yang, Y.Y. Zhang, H.J. Jin, Sorption characteristics and separation of rhenium ions from aqueous solutions using modified nano-Al_2O_3, Ind. Eng. Chem. Res. 51 (2012) 5577–5584.
[134] Y. Gao, C. Chen, H. Chen, R. Zhang, X. Wang, Synthesis of a novel organic-inorganic hybrid of polyaniline/titanium phosphate for Re(VII) removal, Dalton Trans. 44 (2015) 8917–8925.
[135] H. Hu, B. Jiang, J. Zhang, X. Chen, Adsorption of perrhenate ion by bio-char produced from *Acidosasa edulis* shoot shell in aqueous solution, RSC Adv. 5 (2015) 104769–104778.
[136] Y. Wang, H. Gao, Compositional and structural control on anion sorption capability of layered double hydroxides (LDHs), J. Colloid Interface Sci. 301 (2006) 19–26.
[137] K. Pillay, A review of the radiation stability of ion exchange materials, J. Radioanal. Nucl. Chem. 102 (1986) 247–268.
[138] S. Yu, X. Wang, X. Tan, X. Wang, Sorption of radionuclides from aqueous systems onto graphene oxide-based materials: a review, Inorg. Chem. Front. 2 (2015) 593–612.
[139] F. Bai, G. Ye, G. Chen, J. Wei, J. Wang, J. Chen, Highly selective recovery of palladium by a new silica-based adsorbent functionalized with macrocyclic ligand, Sep. Purif. Technol. 106 (2013) 38–46.
[140] L. Zhou, J. Liu, Z. Liu, Adsorption of platinum(IV) and palladium(II) from aqueous solution by thiourea-modified chitosan microspheres, J. Hazard. Mater. 172 (2009) 439–446.
[141] L. Zhou, J. Xu, X. Liang, Z. Liu, Adsorption of platinum(IV) and palladium(II) from aqueous solution by magnetic cross-linking chitosan nanoparticles modified with ethylenediamine, J. Hazard. Mater. 182 (2010) 518–524.
[142] A. Ramesh, H. Hasegawa, W. Sugimoto, T. Maki, K. Ueda, Adsorption of gold(III), platinum(IV) and palladium(II) onto glycine modified crosslinked chitosan resin, Bioresour. Technol. 99 (2008) 3801–3809.
[143] Y. Leng, J. Xu, J. Wei, G. Ye, Amino-bearing calixcrown receptor grafted to micro-sized silica particles for highly selective enrichment of palladium in HNO_3 media, Chem. Eng. J. 232 (2013) 319–326.
[144] H. Gad, Y. Lasheen, T. El-Zakla, Efficiency of locally prepared activated carbon in the preconcentration of barium-133 and radium-226 radionuclides in single and binary systems, Radiochemistry 55 (2013) 589–595.

[145] M. Chávez, L. De Pablo, T. García, Adsorption of Ba^{2+} by Ca-exchange clinoptilolite tuff and montmorillonite clay, J. Hazard. Mater. 175 (2010) 216–223.

[146] D.J. Yang, Z.F. Zheng, H.Y. Zhu, H.W. Liu, X.P. Gao, Titanate nanofibers as intelligent absorbents for the removal of radioactive ions from water, Adv. Mater. 20 (2008) 2777–2781.

[147] C. Chen, X. Li, D. Zhao, X. Tan, X. Wang, Adsorption kinetic, thermodynamic and desorption studies of Th(IV) on oxidized multi-wall carbon nanotubes, Colloids Surf. A 302 (2007) 449–454.

[148] L. Cheng, L. Zhai, W. Liao, X. Huang, B. Niu, S. Yu, An investigation on the behaviors of thorium(IV) adsorption onto chrysotile nanotubes, J. Environ. Chem. Eng. 2 (2014) 1236–1242.

[149] A. Rahmani-Sani, A. Hosseini-Bandegharaei, S. Hosseini, K. Kharghani, H. Zarei, A. Rastegar, Kinetic, equilibrium and thermodynamic studies on sorption of uranium and thorium from aqueous solutions by a selective impregnated resin containing carminic acid, J. Hazard. Mater. 286 (2015) 152–163.

[150] L.P. Lingamdinne, Y. Choi, I. Kim, J. Yang, J.R. Koduru, Y. Chang, Preparation and characterization of porous reduced graphene oxide based inverse spinel nickel ferrite nanocomposite for adsorption removal of radionuclides, J. Hazard. Mater. 326 (2017) 145–156.

[151] M.E. Nasab, Solvent extraction separation of uranium(VI) and thorium(IV) with neutral organophosphorus and amine ligands, Fuel 116 (2014) 595–600.

[152] P. Shi, B. Zhao, G. Xiong, Y. Hou, P. Cheng, Fast capture and separation of, and luminescent probe for, pollutant chromate using a multi-functional cationic heterometal-organic framework, Chem. Commun. 48 (2012) 8231–8233.

[153] L. Li, X. Feng, R. Han, S. Zang, G. Yang, Cr(VI) removal via anion exchange on a silver-triazolate MOF, J. Hazard. Mater. 321 (2017) 622–628.

[154] X. Li, H. Xu, F. Kong, R. Wang, A cationic metal-organic framework consisting of nanoscale cages: capture, separation, and luminescent probing of $Cr_2O_7^{2-}$ through a single-crystal to single-crystal process, Angew. Chem. Int. Ed. 52 (2013) 13769–13773.

[155] H. Fu, Z. Xu, J. Zhang, Water-stable metal-organic frameworks for fast and high dichromate trapping via single-crystal-to-single-crystal ion exchange, Chem. Mater. 27 (2014) 205–210.

[156] C. Li, H. Zhou, S. Wang, J. Chen, Z. Wang, M. Du, Highly efficient $Cr_2O_7^{2-}$ removal of a 3D metal-organic framework fabricated by tandem single-crystal to single-crystal transformations from a 1D coordination array, Chem. Commun. 53 (2017) 9206–9209.

[157] B. Ding, J.Z. Huo, Y.Y. Liu, X. Wang, X. Su, X.X. Wu, Z.Z. Zhu, J. Xia, Triazole based Ag coordination clusters: synthesis, structural diversity and anion exchange properties, RSC Adv. 5 (2015) 83415–83426.

[158] H. Fei, C.S. Han, J.C. Robins, S.R. Oliver, A cationic metal-organic solid solution based on Co(II) and Zn(II) for chromate trapping, Chem. Mater. 25 (2013) 647–652.

[159] B. Ding, C. Guo, S.X. Liu, Y. Cheng, X.X. Wu, X.M. Su, Y.Y. Liu, Y. Li, A unique multi-functional cationic luminescent metal-organic nanotube for highly sensitive detection of dichromate and selective high capacity adsorption of Congo red, RSC Adv. 6 (2016) 33888–33900.

[160] L. Aboutorabi, A. Morsali, E. Tahmasebi, O. Buyukgungor, Metal-organic framework based on isonicotinate N-oxide for fast and highly efficient aqueous phase Cr(VI) adsorption, Inorg. Chem. 55 (2016) 5507–5513.

[161] H. Niu, Y. Zheng, S. Wang, S. He, Y. Cai, Stable hierarchical microspheres of 1D Fe–Gallic acid MOFs for fast and efficient Cr(VI) elimination by a combination of reduction, metal substitution and coprecipitation, J. Mater. Chem. A 5 (2017) 16600–16604.

[162] Q. Yang, Q. Zhao, S. Ren, Q. Lu, X. Guo, Z. Chen, Fabrication of core-shell Fe_3O_4@MIL-100(Fe) magnetic microspheres for the removal of Cr(VI) in aqueous solution, J. Solid State Chem. 244 (2016) 25–30.

[163] K. Wang, X. Tao, J. Xu, N. Yin, Novel chitosan–MOF composite adsorbent for the removal of heavy metal ions, Chem. Lett. 45 (2016) 1365–1368.

[164] K. Zhu, C. Chen, H. Xu, Y. Gao, X. Tan, A. Alsaedi, T. Hayat, Cr(VI) reduction and immobilization by core-double-shell structured magnetic polydopamine@zeolitic idazolate frameworks-8 microspheres, ACS Sustain. Chem. Eng. 5 (2017) 6795–6802.

[165] S. Rapti, D. Sarma, S. Diamantis, E. Skliri, G.S. Armatas, A. Tsipis, Y.S. Hassan, M.H. Alkordi, C.D. Malliakas, M. Kanatzidis, All in one porous material: exceptional sorption and selective sensing of hexavalent chromium by a Zr^{4+} MOF, J. Mater. Chem. A 5 (2017) 14707–14719.

[166] X. Li, X. Gao, L. Ai, J. Jiang, Mechanistic insight into the interaction and adsorption of Cr(VI) with zeolitic imidazolate framework-67 microcrystals from aqueous solution, Chem. Eng. J. 274 (2015) 238–246.

[167] A. Nasrollahpour, S. Moradi, Hexavalent chromium removal from water by ionic liquid modified metal-organic frameworks adsorbent, Microporous Mesoporous Mater. 243 (2017) 47–55.

[168] Q. Zhang, J. Yu, J. Cai, L. Zhang, Y. Cui, Y. Yang, B. Chen, G. Qian, A porous Zr-cluster-based cationic metal-organic framework for highly efficient $Cr_2O_7^{2-}$ removal from water, Chem. Commun. 51 (2015) 14732–14734.

[169] M. Babazadeh, R. Hosseinzadeh-Khanmiri, J. Abolhasani, E. Ghorbani-Kalhor, A. Hassanpour, Solid phase extraction of heavy metal ions from agricultural samples with the aid of a novel functionalized magnetic metal-organic framework, RSC Adv. 5 (2015) 19884–19892.

[170] A. Maleki, B. Hayati, M. Naghizadeh, S.W. Joo, Adsorption of hexavalent chromium by metal organic frameworks from aqueous solution, J. Ind. Eng. Chem. 28 (2015) 211–216.

[171] Y. Xiong, F. Ye, C. Zhang, S. Shen, L. Su, S. Zhao, Synthesis of magnetic porous γ-Fe_2O_3/C@HKUST-1 composites for efficient removal of dyes and heavy metal ions from aqueous solution, RSC Adv. 5 (2015) 5164–5172.

[172] M. Guo, H. Guo, S. Liu, Y. Sun, X. Guo, A microporous cationic metal-organic framework for the efficient removal of dichromate and the selective adsorption of dyes from water, RSC Adv. 7 (2017) 51021–51026.

[173] R.M. Vala, D. Wankasi, E.D. Dikio, Morphology and adsorption of chromium ion on uranium 1,2,4,5-benzenetetracarboxylic acid metal organic framework (MOF), Hem. Ind. 70 (2016) 565–572.

[174] M. Bhaumik, H.J. Choi, M.P. Seopela, R.I. McCrindle, A. Maity, Highly effective removal of toxic Cr(VI) from wastewater using sulfuric acid-modified avocado seed, Ind. Eng. Chem. Res. 53 (2014) 1214–1224.

[175] S.K. Prabhakaran, K. Vijayaraghavan, R. Balasubramanian, Removal of Cr(VI) ions by spent tea and coffee dusts: reduction to Cr(III) and biosorption, Ind. Eng. Chem. Res. 48 (2009) 2113–2117.

[176] W. Liu, J. Ni, X. Yin, Synergy of photocatalysis and adsorption for simultaneous removal of Cr(VI) and Cr(III) with TiO_2 and titanate nanotubes, Water Res. 53 (2014) 12–25.

[177] M. Gheju, I. Balcu, G. Mosoarca, Removal of Cr(VI) from aqueous solutions by adsorption on MnO_2, J. Hazard. Mater. 310 (2016) 270–277.

[178] S. Kaneko, M. Ogawa, Effective concentration of dichromate anions using layered double hydroxides from acidic solutions, Appl. Clay Sci. 75–76 (2013) 109–113.

[179] F. Zhang, N. Du, H. Li, X. Liang, W. Hou, Sorption of Cr(VI) on Mg-Al-Fe layered double hydroxides synthesized by a mechanochemical method, RSC Adv. 4 (2014) 46823–46830.
[180] X. Yuan, Y. Wang, J. Wang, C. Zhou, Q. Tang, X. Rao, Calcined graphene/MgAl-layered double hydroxides for enhanced Cr(VI) removal, Chem. Eng. J. 221 (2013) 204–213.
[181] C. Lei, X. Zhu, B. Zhu, C. Jiang, Y. Le, J. Yu, Superb adsorption capacity of hierarchical calcined Ni/Mg/Al layered double hydroxides for Congo red and Cr(VI) ions, J. Hazard. Mater. 321 (2017) 801–811.
[182] A.A. Bhatti, S. Memon, N. Memon, Dichromate extraction by calix[4]arene appended amberlite XAD-4 resin, Sep. Sci. Technol. 49 (2014) 664–672.
[183] D. Dinda, A. Gupta, S.K. Saha, Removal of toxic Cr(VI) by UV-active functionalized graphene oxide for water purification, J. Mater. Chem. A 1 (2013) 11221–11228.
[184] Y. Sun, Q. Yue, Y. Mao, B. Gao, Y. Gao, L. Huang, Enhanced adsorption of chromium onto activated carbon by microwave-assisted H_3PO_4 mixed with Fe/Al/Mn activation, J. Hazard. Mater. 265 (2014) 191–200.
[185] J. Zhao, X. Zhang, X. He, M. Xiao, W. Zhang, C. Lu, A super biosorbent from dendrimer poly(amidoamine)-grafted cellulose nanofibril aerogels for effective removal of Cr(VI), J. Mater. Chem. A 3 (2015) 14703–14711.
[186] D. Xu, K. Zhu, X. Zheng, R. Xiao, Poly (ethylene-co-vinyl alcohol) functional nanofiber membranes for the removal of Cr(VI) from water, Ind. Eng. Chem. Res. 54 (2015) 6836–6844.
[187] Y.T. He, S.J. Traina, Cr(VI) reduction and immobilization by magnetite under alkaline pH conditions: the role of passivation, Environ. Sci. Technol. 39 (2005) 4499–4504.
[188] M. McGuire, N. Blute, L. Fong, D. Froelich, P. Kavounas, G. Qin, Hexavalent Chromium Removal Using Anion Exchange and Reduction With Coagulation and Filtration, Water Environment Research Foundation, USA, 2008.
[189] A.D. Lemly, Aquatic selenium pollution is a global environmental safety issue, Ecotoxicol. Environ. Saf. 59 (2004) 44–56.
[190] B. Zhu, X. Yu, Y. Jia, F. Peng, B. Sun, M. Zhang, T. Luo, J. Liu, X. Huang, Iron and 1,3,5-benzenetricarboxylic metal-organic coordination polymers prepared by solvothermal method and their application in efficient As(V) removal from aqueous solutions, J. Phys. Chem. C 116 (2012) 8601–8607.
[191] T.A. Vu, G.H. Le, C.D. Dao, L.Q. Dang, K.T. Nguyen, Q.K. Nguyen, P.T. Dang, H.T. Tran, Q.T. Duong, T.V. Nguyen, Arsenic removal from aqueous solutions by adsorption using novel MIL-53(Fe) as a highly efficient adsorbent, RSC Adv. 5 (2015) 5261–5268.
[192] J. Li, Y. Wu, Z. Li, M. Zhu, F. Li, Characteristics of arsenate removal from water by metal-organic frameworks (MOFs), Water Sci. Technol. 70 (2014) 1391–1397.
[193] J. Cai, X. Wang, Y. Zhou, L. Jiang, C. Wang, Selective adsorption of arsenate and the reversible structure transformation of the mesoporous metal-organic framework MIL-100 (Fe), Phys. Chem. Chem. Phys. 18 (2016) 10864–10867.
[194] K. Folens, K. Leus, N.R. Nicomel, M. Meledina, S. Turner, G. Van Tendeloo, G.D. Laing, P. Van Der Voort, Fe_3O_4@MIL-101–A selective and regenerable adsorbent for the removal of As species from water, Eur. J. Inorg. Chem. 2016 (2016) 4395–4401.
[195] J.-B. Huo, L. Xu, J.-C.E. Yang, H.-J. Cui, B. Yuan, M.-L. Fu, Magnetic responsive Fe_3O_4-ZIF-8 core-shell composites for efficient removal of As(III) from water, Colloids Surf. A 539 (2018) 59–68.
[196] B.K. Jung, J.W. Jun, Z. Hasan, S.H. Jhung, Adsorptive removal of p-arsanilic acid from water using mesoporous zeolitic imidazolate framework-8, Chem. Eng. J. 267 (2015) 9–15.

[197] J.W. Jun, M. Tong, B.K. Jung, Z. Hasan, C. Zhong, S.H. Jhung, Effect of central metal ions of analogous metal-organic frameworks on adsorption of organoarsenic compounds from water: plausible mechanism of adsorption and water purification, Chem. Eur. J. 21 (2015) 347–354.

[198] M. Sarker, J.Y. Song, S.H. Jhung, Adsorption of organic arsenic acids from water over functionalized metal-organic frameworks, J. Hazard. Mater. 335 (2017) 162–169.

[199] J. Li, Y. Wu, Z. Li, B. Zhang, M. Zhu, X. Hu, Y. Zhang, F. Li, Zeolitic imidazolate framework-8 with high efficiency in trace arsenate adsorption and removal from water, J. Phys. Chem. C 118 (2014) 27382–27387.

[200] Y. Wu, M. Zhou, B. Zhang, B. Wu, J. Li, J. Qiao, X. Guan, F. Li, Amino acid assisted templating synthesis of hierarchical zeolitic imidazolate framework-8 for efficient arsenate removal, Nanoscale 6 (2014) 1105–1112.

[201] Z. Li, J. Yang, K. Sui, N. Yin, Facile synthesis of metal-organic framework MOF-808 for arsenic removal, Mater. Lett. 160 (2015) 412–414.

[202] C. Wang, X. Liu, J.P. Chen, K. Li, Superior removal of arsenic from water with zirconium metal-organic framework UiO-66, Sci. Rep. 5 (2015) 16613.

[203] D. Xie, Y. Ma, Y. Gu, H. Zhou, H. Zhang, G. Wang, Y. Zhang, H. Zhao, Bifunctional NH_2-MIL-88(Fe) metal-organic framework nanooctahedra for highly sensitive detection and efficient removal of arsenate in aqueous media, J. Mater. Chem. A 5 (2017) 23794–23804.

[204] H. Atallah, M. Mahmoud, A. Jelle, A. Lough, M. Hmadeh, Highly stable indium based metal organic framework for efficient arsenic removal from water, Dalton Trans. 47 (2017) 799–806.

[205] M. Jian, B. Liu, G. Zhang, R. Liu, X. Zhang, Adsorptive removal of arsenic from aqueous solution by zeolitic imidazolate framework-8 (ZIF-8) nanoparticles, Colloids Surf. A 465 (2015) 67–76.

[206] J. Yang, X. Yin, $CoFe_2O_4$@MIL-100(Fe) hybrid magnetic nanoparticles exhibit fast and selective adsorption of arsenic with high adsorption capacity, Sci. Rep. 7 (2017) 40955.

[207] B. Liu, M. Jian, R. Liu, J. Yao, X. Zhang, Highly efficient removal of arsenic(III) from aqueous solution by zeolitic imidazolate frameworks with different morphology, Colloids Surf. A 481 (2015) 358–366.

[208] G. Ungureanu, S. Santos, R. Boaventura, C. Botelho, Arsenic and antimony in water and wastewater: overview of removal techniques with special reference to latest advances in adsorption, J. Environ. Manag. 151 (2015) 326–342.

[209] J. Zhu, S.A. Baig, T. Sheng, Z. Lou, Z. Wang, X. Xu, Fe_3O_4 and MnO_2 assembled on honeycomb briquette cinders (HBC) for arsenic removal from aqueous solutions, J. Hazard. Mater. 286 (2015) 220–228.

[210] Z. Wen, Y. Zhang, C. Dai, B. Chen, S. Guo, H. Yu, D. Wu, Synthesis of ordered mesoporous iron manganese bimetal oxides for arsenic removal from aqueous solutions, Microporous Mesoporous Mater. 200 (2014) 235–244.

[211] D. Mohan, C.U. Pittman, Arsenic removal from water/wastewater using adsorbents—a critical review, J. Hazard. Mater. 142 (2007) 1–53.

[212] Y. Yu, L. Yu, J.P. Chen, Introduction of an yttrium–manganese binary composite that has extremely high adsorption capacity for arsenate uptake in different water conditions, Ind. Eng. Chem. Res. 54 (2015) 3000–3008.

[213] J. Yang, H. Zhang, M. Yu, I. Emmanuelawati, J. Zou, Z. Yuan, C. Yu, High-content, well-dispersed γ-Fe_2O_3 nanoparticles encapsulated in macroporous silica with superior arsenic removal performance, Adv. Funct. Mater. 24 (2014) 1354–1363.

[214] M.A. Sabur, S. Goldberg, A. Gale, N. Kabengi, H.A. Al-Abadleh, Temperature-dependent infrared and calorimetric studies on arsenicals adsorption from solution to hematite nanoparticles, Langmuir 31 (2015) 2749–2760.

[215] L. Poon, S. Younus, L.D. Wilson, Adsorption study of an organo-arsenical with chitosan-based sorbents, J. Colloid Interface Sci. 420 (2014) 136–144.

[216] G. Zhang, F. Liu, H. Liu, J. Qu, R. Liu, Respective role of Fe and Mn oxide contents for arsenic sorption in iron and manganese binary oxide: an X-ray absorption spectroscopy investigation, Environ. Sci. Technol. 48 (2014) 10316–10322.

[217] L. Ma, S.M. Islam, H. Liu, J. Zhao, G. Sun, H. Li, S. Ma, M.G. Kanatzidis, Selective and efficient removal of toxic oxoanions of As(III), As(V), and Cr(VI) by layered double hydroxide intercalated with MoS_4^{2-}, Chem. Mater. 29 (2017) 3274–3284.

[218] S.M. Maliyekkal, L. Philip, T. Pradeep, As(III) removal from drinking water using manganese oxide-coated-alumina: performance evaluation and mechanistic details of surface binding, Chem. Eng. J. 153 (2009) 101–107.

[219] A. Gupta, V.S. Chauhan, N. Sankararamakrishnan, Preparation and evaluation of iron–chitosan composites for removal of As(III) and As(V) from arsenic contaminated real life groundwater, Water Res. 43 (2009) 3862–3870.

[220] B. Chen, Z. Zhu, J. Ma, Y. Qiu, J. Chen, Surfactant assisted Ce–Fe mixed oxide decorated multiwalled carbon nanotubes and their arsenic adsorption performance, J. Mater. Chem. A 1 (2013) 11355–11367.

[221] G. Zhang, Z. Ren, X. Zhang, J. Chen, Nanostructured iron(III)-copper(II) binary oxide: a novel adsorbent for enhanced arsenic removal from aqueous solutions, Water Res. 47 (2013) 4022–4031.

[222] W. Tang, Y. Su, Q. Li, S. Gao, J.K. Shang, Superparamagnetic magnesium ferrite nanoadsorbent for effective arsenic(III,V) removal and easy magnetic separation, Water Res. 47 (2013) 3624–3634.

[223] P. Qi, T. Pichler, Competitive adsorption of As(III), As(V), Sb(III) and Sb(V) onto ferrihydrite in multi-component systems: Implications for mobility and distribution, J. Hazard. Mater. 330 (2017) 142–148.

[224] G. Okkenhaug, Y. Zhu, J. He, X. Li, L. Luo, J. Mulder, Antimony (Sb) and arsenic (As) in Sb mining impacted paddy soil from Xikuangshan, China: differences in mechanisms controlling soil sequestration and uptake in rice, Environ. Sci. Technol. 46 (2012) 3155–3162.

[225] J. He, K. Yee, Z. Xu, M. Zeller, A.D. Hunter, S.S. Chui, C. Che, Thioether side chains improve the stability, fluorescence, and metal uptake of a metal-organic framework, Chem. Mater. 23 (2011) 2940–2947.

[226] J. He, M. Zha, J. Cui, M. Zeller, A.D. Hunter, S. Yiu, S. Lee, Z. Xu, Convenient detection of Pd(II) by a metal-organic framework with sulfur and olefin functions, J. Am. Chem. Soc. 135 (2013) 7807–7810.

[227] J. Cui, Y. Wong, M. Zeller, A.D. Hunter, Z. Xu, Pd uptake and H_2S sensing by an amphoteric metal-organic framework with a soft core and rigid side arms, Angew. Chem. Int. Ed. 53 (2014) 14438–14442.

[228] G. Yang, Z. Lang, H. Zang, Y. Lan, W. He, X. Zhao, L. Yan, X. Wang, Z. Su, Control of interpenetration in S-containing metal-organic frameworks for selective separation of transition metal ions, Chem. Commun. 49 (2013) 1088–1090.

[229] X. Zhou, Z. Xu, M. Zeller, A.D. Hunter, Reversible uptake of $HgCl_2$ in a porous coordination polymer based on the dual functions of carboxylate and thioether, Chem. Commun. (2009) 5439–5441.

[230] M. Mon, X. Qu, J. Ferrando-Soria, I. Pellicer-Carreño, A. Sepulveda-Escribano, E.V.R. Fernandez, J.C. Jansen, D. Armentano, E. Pardo, Fine-tuning of the confined space in microporous metal-organic frameworks for efficient mercury removal, J. Mater. Chem. A 5 (2017) 20120–20125.

[231] Y.Y. Xiong, J.Q. Li, L.L. Gong, X.F. Feng, L.N. Meng, L. Zhang, P.P. Meng, M.B. Luo, F. Luo, Using MOF-74 for Hg^{2+} removal from ultra-low concentration aqueous solution, J. Solid State Chem. 246 (2017) 16–22.

[232] A. Chakraborty, S. Bhattacharyya, A. Hazra, A.C. Ghosh, T.K. Maji, Post-synthetic metalation in an anionic MOF for efficient catalytic activity and removal of heavy metal ions from aqueous solution, Chem. Commun. 52 (2016) 2831–2834.

[233] A. Tadjarodi, A. Abbaszadeh, A magnetic nanocomposite prepared from chelator-modified magnetite (Fe_3O_4) and HKUST-1 (MOF-199) for separation and preconcentration of mercury(II), Microchim. Acta 183 (2016) 1391–1399.

[234] K. Leus, J.P.H. Perez, K. Folens, M. Meledina, G. Van Tendeloo, G. Du Laing, P. Van Der Voort, UiO-66-$(SH)_2$ as stable, selective and regenerable adsorbent for the removal of mercury from water under environmentally-relevant conditions, Faraday Discuss. 201 (2017) 146–161.

[235] L. Huang, M. He, B. Chen, B. Hu, A designable magnetic MOF composite and facile coordination-based post-synthetic strategy for the enhanced removal of Hg^{2+} from water, J. Mater. Chem. A 3 (2015) 11587–11595.

[236] S. Halder, J. Mondal, J. Ortega-Castro, A. Frontera, P. Roy, A Ni-based MOF for selective detection and removal of Hg^{2+} in aqueous medium: a facile strategy, Dalton Trans. 46 (2017) 1943–1950.

[237] N.D. Rudd, H. Wang, E.M.A. Fuentes-Fernandez, S.J. Teat, F. Chen, G. Hall, Y.J. Chabal, J. Li, Highly efficient luminescent metal-organic framework for the simultaneous detection and removal of heavy metals from water, ACS Appl. Mater. Interfaces 8 (2016) 30294–30303.

[238] M.R. Sohrabi, Preconcentration of mercury(II) using a thiol-functionalized metal-organic framework nanocomposite as a sorbent, Microchim. Acta 181 (2014) 435–444.

[239] S. Bhattacharjee, Y. Lee, W. Ahn, Post-synthesis functionalization of a zeolitic imidazolate structure ZIF-90: a study on removal of Hg(II) from water and epoxidation of alkenes, CrystEngComm 17 (2015) 2575–2582.

[240] X. Chen, X. Feng, J. Liu, G.E. Fryxell, M. Gong, Mercury separation and immobilization using self-assembled monolayers on mesoporous supports (SAMMS), Sep. Sci. Technol. 34 (1999) 1121–1132.

[241] N. Huang, L. Zhai, H. Xu, D. Jiang, Stable covalent organic frameworks for exceptional mercury removal from aqueous solutions, J. Am. Chem. Soc. 139 (2017) 2428–2434.

[242] S. Chiarle, M. Ratto, M. Rovatti, Mercury removal from water by ion exchange resins adsorption, Water Res. 34 (2000) 2971–2978.

[243] M.R. Awual, Novel nanocomposite materials for efficient and selective mercury ions capturing from wastewater, Chem. Eng. J. 307 (2017) 456–465.

[244] B. Li, Y. Zhang, D. Ma, Z. Shi, S. Ma, Mercury nano-trap for effective and efficient removal of mercury(II) from aqueous solution, Nat. Commun. 5 (2014) 5537.

[245] B. Aguila, Q. Sun, J.A. Perman, L.D. Earl, C.W. Abney, R. Elzein, R. Schlaf, S. Ma, Efficient mercury capture using functionalized porous organic polymer, Adv. Mater. 29 (2017) 1700665.

[246] Q. Sun, B. Aguila, J. Perman, L.D. Earl, C.W. Abney, Y. Cheng, H. Wei, N. Nguyen, L. Wojtas, S. Ma, Postsynthetically modified covalent organic frameworks for efficient and effective mercury removal, J. Am. Chem. Soc. 139 (2017) 2786–2793.

[247] F. Zou, R. Yu, R. Li, W. Li, Microwave-assisted synthesis of HKUST-1 and functionalized HKUST-1-@$H_3PW_{12}O_{40}$: selective adsorption of heavy metal ions in water analyzed with synchrotron radiation, ChemPhysChem 14 (2013) 2825–2832.

[248] N. Wang, X. Ouyang, L. Yang, A.M. Omer, Fabrication of a magnetic cellulose nanocrystal/metal-organic framework composite for removal of Pb(II) from water, ACS Sustain. Chem. Eng. 5 (2017) 10447–10458.

[249] Y. Wang, G. Ye, H. Chen, X. Hu, Z. Niu, S. Ma, Functionalized metal-organic framework as a new platform for efficient and selective removal of cadmium(II) from aqueous solution, J. Mater. Chem. A 3 (2015) 15292–15298.

[250] H. Xue, Q. Chen, F. Jiang, D. Yuan, G. Lv, L. Liang, L. Liu, M. Hong, A regenerative metal-organic framework for reversible uptake of Cd(II): from effective adsorption to in situ detection, Chem. Sci. 7 (2016) 5983–5988.

[251] A. Margariti, S. Rapti, A. Katsenis, T. Friščić, Y. Georgiou, M. Manos, G. Papaefstathiou, Cu^{2+} sorption from aqueous media by a recyclable Ca^{2+} framework, Inorg. Chem. Front. 4 (2017) 773–781.

[252] Y. Zhang, Z. Xie, Z. Wang, X. Feng, Y. Wang, A. Wu, Unveiling the adsorption mechanism of zeolitic imidazolate framework-8 with high efficiency for removal of copper ions from aqueous solutions, Dalton Trans. 45 (2016) 12653–12660.

[253] K. Wang, J. Gu, N. Yin, Efficient removal of Pb(II) and Cd(II) using NH_2-functionalized Zr-MOFs via rapid microwave-promoted synthesis, Ind. Eng. Chem. Res. 56 (2017) 1880–1887.

[254] E. Tahmasebi, M.Y. Masoomi, Y. Yamini, A. Morsali, Application of mechanosynthesized azine-decorated zinc(II) metal-organic frameworks for highly efficient removal and extraction of some heavy-metal ions from aqueous samples: a comparative study, Inorg. Chem. 54 (2014) 425–433.

[255] S. Chamarthy, C.W. Seo, W.E. Marshall, Adsorption of selected toxic metals by modified peanut shells, J. Chem. Technol. Biotechnol. 76 (2001) 593–597.

[256] Q. Xu, Y. Wang, L. Jin, Y. Wang, M. Qin, Adsorption of Cu(II), Pb(II) and Cr(VI) from aqueous solutions using black wattle tannin-immobilized nanocellulose, J. Hazard. Mater. 339 (2017) 91–99.

[257] I.L. Lagadic, M.K. Mitchell, B.D. Payne, Highly effective adsorption of heavy metal ions by a thiol-functionalized magnesium phyllosilicate clay, Environ. Sci. Technol. 35 (2001) 984–990.

[258] B.I. Olu-Owolabi, A.H. Alabi, E.I. Unuabonah, P.N. Diagboya, L. Böhm, R. Düring, Calcined biomass-modified bentonite clay for removal of aqueous metal ions, J. Environ. Chem. Eng. 4 (2016) 1376–1382.

[259] Y. Zhang, Y. Su, X. Zhou, C. Dai, A.A. Keller, A new insight on the core-shell structure of zerovalent iron nanoparticles and its application for Pb(II) sequestration, J. Hazard. Mater. 263 (2013) 685–693.

[260] M.L. Godino-Salido, A. Santiago-Medina, R. López-Garzón, M.D. Gutiérrez-Valero, P. Arranz-Mascarós, M.D.L. de la Torre, M. Domingo-García, F.J. López-Garzón, Preparation and characterization of trihydroxamic acid functionalized carbon materials for the removal of Cu(II) ions from aqueous solution, Appl. Surf. Sci. 387 (2016) 128–138.

[261] T. Batjargal, J. Yang, D. Kim, K. Baek, Removal characteristics of Cd(II), Cu(II), Pb(II), and Zn(II) by natural mongolian zeolite through batch and column experiments, Sep. Sci. Technol. 46 (2011) 1313–1320.

[262] J. Li, S. Zhang, C. Chen, G. Zhao, X. Yang, J. Li, X. Wang, Removal of Cu(II) and fulvic acid by graphene oxide nanosheets decorated with Fe_3O_4 nanoparticles, ACS Appl. Mater. Interfaces 4 (2012) 4991–5000.

[263] Z. Dong, D. Wang, X. Liu, X. Pei, L. Chen, J. Jin, Bio-inspired surface-functionalization of graphene oxide for the adsorption of organic dyes and heavy metal ions with a superhigh capacity, J. Mater. Chem. A 2 (2014) 5034–5040.

[264] L. Lin, R. Juang, Ion-exchange equilibria of Cu(II) and Zn(II) from aqueous solutions with Chelex 100 and Amberlite IRC 748 resins, Chem. Eng. J. 112 (2005) 211–218.

[265] J. Li, Q. Fan, Y. Wu, X. Wang, C. Chen, Z. Tang, X. Wang, Magnetic polydopamine decorated with Mg–Al LDH nanoflakes as a novel bio-based adsorbent for simultaneous removal of potentially toxic metals and anionic dyes, J. Mater. Chem. A 4 (2016) 1737–1746.

[266] A. Abbasi, T. Moradpour, K. Van Hecke, A new 3D cobalt(II) metal-organic framework nanostructure for heavy metal adsorption, Inorg. Chim. Acta 430 (2015) 261–267.

Chapter 6

Application of nZVI and its composites into the treatment of toxic/radioactive metal ions

Kairuo Zhu and Changlun Chen
Key Laboratory of Photovoltaic and Energy Conservation Materials, Institute of Plasma Physics, Chinese Academy of Sciences, Hefei, People's Republic of China

6.1 Introduction

Heavy metal wastewater has attracted wide interest because of the serious threat it represents to aquatic environments and the health of organisms. Heavy metals such as Pb, Hg, and Cd are widely considered harmful for the natural environment because of their toxicological criteria [1–7]. Cr, As, or Se are also toxic [8–10]. Generally, Pb, Hg, Cd, Cr, and As are usually identified as the primary heavy metal ions because of their high toxicity, prevalence, and lack of degradation in the aquatic environment. Ni is widely used [11, 12]. The usual metal ions, Zn and Cu, are vital elements for all kinds of lives; however, they are still toxic in high concentrations [13, 14]. Precious metals such as Au, Ag, Pd, and Pt always attract enhanced attention, primarily to recover them from precious metal-containing water even though they would not pose a threat to the aquatic environment. A number of heavy metals only present as anions, including Cr(VI), Mo(VI), Se(V), and As(V), with diverse properties from those of the common metal cations. Radionuclides are the third significant category of metals, in light of their dangerous environmental effects and wide interest for nuclear facilities, nuclear weapons testing, or nuclear power plants. Among those radionuclides, ^{238}U, ^{99}Tc, ^{90}Sr, ^{137}Cs, and ^{60}Co are important, owing to their environmental impact on human health [15–17]. Uranium is a vital nuclear fuel resource and one of the dominating radioactive elements in radioactive liquid wastes, although uranium is not particularly a problem in the natural environment. ^{99}Tc, the coproduct of ^{235}U fission in nuclear reactors, presents radiological hazards to internal organs and the skin. Long-lived radionuclides, including ^{137}Cs, ^{90}Sr, and ^{60}Co, are supposed to be the most harmful to human health because

Interface Science and Technology, Vol. 29. https://doi.org/10.1016/B978-0-08-102727-1.00006-6

of their high transferability, high solubility, and easy assimilation in living organisms. Owing to the chemical similarity with potassium (K) for Cs or calcium (Ca) for Sr, ^{137}Cs and ^{90}Sr can easily transfer to living organisms, so that they exert persistent radiological impacts on human beings. Therefore, it is a very meaningful and challenging project for environmental pollution management to effectively separate and enrich toxic and radioactive metal ions from water toward maintaining ecological stability and public safety.

Various methods such as chemical precipitation [18, 19], coagulation/filtration [20], ion exchange [21, 22], photocatalysis [23, 24], adsorption [3, 25, 26], and reverse osmosis membrane separation [27, 28] have been applied for toxic and radioactive metal ion contaminant removal from aqueous solutions. However, most of these techniques have many disadvantages, such as high cost, difficult access, and inconvenient application [29, 30]. For example, although precipitation and biological treatments are cost effective, they are unable to sufficiently reduce the metal ion concentration to below the legal limited levels; they also tend to produce large amounts of sludge [31]. Photocatalysis often generates unwanted byproducts (e.g., U(IV) ions produced from U(VI) reduction also need a secondary treatment) [32, 33]. Among these methods, adsorption is seen as the most effective one with traits of simplicity, high efficiency, low cost, etc.

As far as we know, many studies have been performed to investigate CNTs [34, 35], clay minerals [36, 37], LDHs [38], nanosized metal oxides [39, 40], activated carbon [41], and GO [42, 43] as highly efficient adsorbents to remove toxic and radioactive metal ions from aquatic environments. To date, nanoscale zero-valent iron (nZVI) has received significant attention as a functional adsorbent for the treatment of environmental contamination. Detoxification and oxidation of nZVI leads to the generation of a variety of iron oxides, such as FeO, Fe_2O_3, and Fe_3O_4, on the nZVI surface. Both nZVI and the oxide shell are important for reduction, adsorption, precipitation, and coprecipitation of water pollutants (Fig. 6.1) [45–47]. Notably, nZVI has shown a remarkable reductive capacity for treating a variety of metal ions in aqueous environments (e.g., U(VI), Tc(VII), Co(II), Cr(VI), Se(VI), As(V), Cd(II) etc.) with the advantage of its small size and high surface area. Unfortunately, the tiny particle size and the solid powder state of nZVI easily reacts with surrounded media and tends to agglomerate in the preparation process, which reduces the reactivity of nZVI and leads to low mobility and difficultly transform nZVI into the contaminated position, thus, limiting the wide application of nZVI. [48]. To solve the problem, nZVI-based materials such as modified nZVI, porous material-supported nZVI, and inorganic clay mineral-supported nZVI have been successfully prepared and extensively used as an adsorbent to remove metal ions from aquatic environments. However, there have been few summaries that have focused on toxic/radioactive metal ion adsorption onto nZVI-based materials [49, 50], and many related articles only briefly discussed the adsorption behavior of nZVI-based materials and simple reduction immobilization mechanisms. It is still necessary to

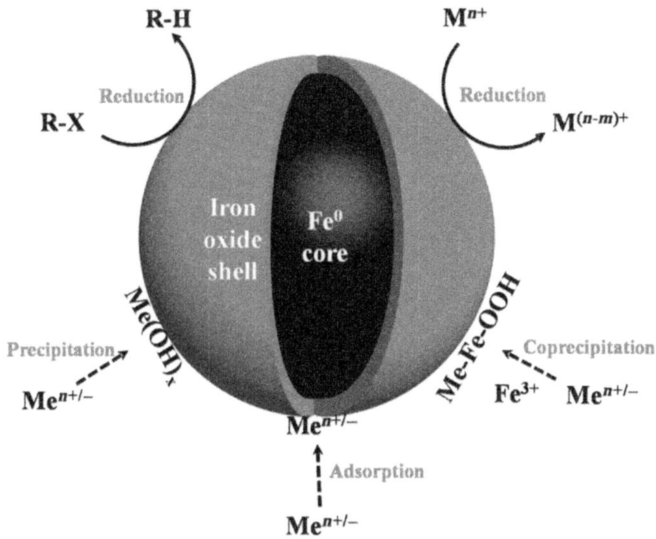

FIG. 6.1 Illustration of oxide shell-mediated contaminant removal by nZVI. R-X represents chlorinated organic compounds. M^{n+} represents inorganic/organic contaminants, and $Me^{n+/-}$ represents metal cations [44]. *(Data from Y. Mu, F. Jia, Z.H. Ai, L.Z. Zhang, Iron oxide shell mediated environmental remediation properties of nano zero-valent iron, Environ. Sci. Nano 4 (2017) 27–45. Copyright 2017, Royal Society of Chemistry.)*

obtain a comprehensive summarization to the specific toxic and radioactive metal ion decontamination.

In the chapter, the aim is to provide an up-to-date account of the radioactive/heavy metal ion removal properties onto nZVI and its composites via the analysis of macroscopic batch experimental performance and microscopic characterization techniques. We put emphasis not only on the potential of nZVI-based composites in realistic applications but also on the methods for characterizing their adsorption behaviors/mechanisms. Our purpose is to present the latest and most objective insight into the existing and emerging applications of nZVI-based materials to the decontamination of toxic/radioactive metal ions. This will help researchers obtain a deep understanding of the reactions between nZVI-based composites and metal ions, and the strategic design and fabrication of nZVI-based materials with high efficiency for environmental remediation.

6.2 nZVI—A brief overview

6.2.1 History

When Gillham et al. [51] started the study of nZVI technology, they found that bulk ZVI could decrease the group of halogenated aliphatic compounds in underground water. That property of ZVI resulted in the development of iron permeable-reactive barriers (Fe-PRBs) [52]. More than 100 PRB structures have been designed in the United States in recent years to various degrees of success [53–55]. The crucial concern for the long-term effect of the PRB installations is the reduction of the permeability because of iron oxidation deposition and biofilm growth [52]. Furthermore, high installation costs limit the application of PRBs for the treatment of deep aquifers or sites. Iron nanoparticles or nZVI are identified as supplementary treatment options to Fe-PRBs [56]. Iron nanoparticles can be injected straight into groundwater via gravity or a pressurized feed owing to their small size, which proposes a highly flexible treatment technology. Based on the reviews by Karn et al. and Mueller et al. [56, 57], in the last 10 years more than 50 pilot or full-scale uses of nZVI for in situ groundwater and soil remediation have been found.

6.2.2 Properties and structure

The nZVI particles prepared by different ways show differences in size, morphology, surface area, and crystallinity, which all hold the core-shell structure. Representatively, the comparison between the gray edge and the dark center seen in TEM images shows that the iron cores of nZVI particles are coated with the superficially amorphous shell (Fig. 6.2A–C). The amorphous shell comprised of iron oxides is generated by the corrosion and oxidation of the metallic core by water and oxygen in the nZVI preparation process [44, 59, 60]. When fresh iron is exposed in aerobic conditions at room temperature, the corrosion and oxidation process is very fast and just needs 0.2 fs to form an original 1 nm thickness of oxide layer, based on the Cabrera-Mott model [61], and the oxide phases (e.g., FeO, Fe_3O_4, and Fe_2O_3) are formed with the exposure distance (Eqs. 6.1–6.8) [60, 62].

$$2Fe^0{}_{(s)} + 4H^+{}_{(aq)} + O_{2(aq)} \rightarrow 2Fe^{2+} + 2H_2O_{(l)}\ E^0 = +1.67\,V \tag{6.1}$$

$$2Fe^0{}_{(s)} + 4H_2O_{(l)} \rightarrow 2Fe^{2+} + 2H_{2(g)} + 2OH^-{}_{(aq)}\ E^0 = -0.39\,V \tag{6.2}$$

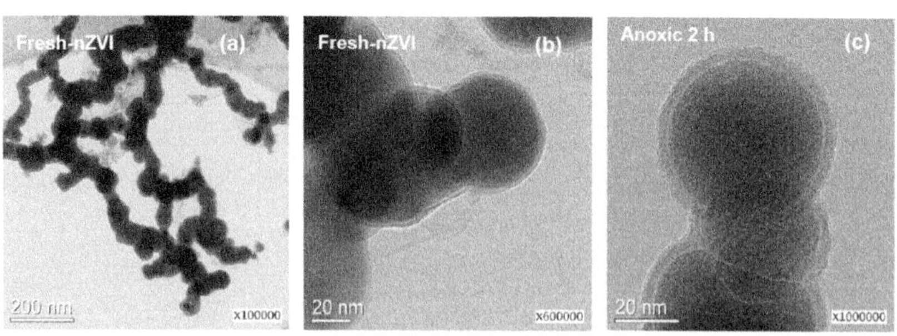

FIG. 6.2 Transmission electron microscopy (TEM) images of (A, B) fresh-nZVI, and (C) nZVI eroded in anoxic water for 2 h [58]. *(Data from A. Liu, J. Liu, J. Han, W.X. Zhang, Evolution of nanoscale zero-valent iron (nZVI) in water: microscopic and spectroscopic evidence on the formation of nano- and micro-structured iron oxides, J. Hazard. Mater. 322 (2017) 129–135. Copyright 2017, Elsevier.)*

$$2Fe^{2+}{}_{(s)} + 2H^{+}{}_{(aq)} + 0.5O_{2(aq)} \rightarrow 2Fe^{3+} + 2H_2O_{(l)}\ E^0 = +0.46\,V \quad (6.3)$$

$$2Fe^{2+}{}_{(s)} + 2H_2O_{(l)} \rightarrow 2Fe^{2+} + H_{2(g)} + 2OH^{-}{}_{(aq)}\ E^0 = -1.60\,V \quad (6.4)$$

$$Fe^{3+} + 3OH^{-} \rightarrow Fe(OH)_3 \downarrow \quad (6.5)$$

$$Fe(OH)_3 \rightarrow FeOOH \downarrow + H_2O \quad (6.6)$$

$$2Fe^{0}{}_{(s)} + O_2 \rightarrow 2FeO \quad (6.7)$$

$$6FeO + O_2 \rightarrow 2Fe_3O_4 \quad (6.8)$$

$$4Fe_3O_4 + O_2 \rightarrow 6Fe_2O_3 \quad (6.9)$$

The aforementioned corrosion and oxidation reactions were commonly found in the nZVI synthesis process. The rapid corrosion reactions led to the unceasing formation of the oxide phases on the surface of the metallic core, even under the exceedingly controlled conditions [63, 64]. Then, the core shell-structured nZVI is assembled, along with the unceasing corrosion of the Fe^0 core and the formation of iron oxide/hydroxide. As a result, the obtained nZVI has already held the oxide shell before the use of nZVI in the environmental contaminant remediation [63, 65, 66]. nZVI owns the magnetic property [67], which leads to the fresh nanoparticles forming chain-like structure and agglomeration (Fig. 6.2A). The magnetic properties of nZVI-based composites have been applied in several recovery procedures, including the magnetic separation of nZVI particles from aquatic solution.

6.2.3 Immobilized or surface-stabilized nZVI

For unmodified nZVI, high surface energy and strong magnetic interactions accentuated the self-aggregation phenomenon, which becomes the main difficulty in synthesizing the stabilized, well-dispersed nZVI suspensions that limits the practical applications of nZVI in a variety of environmental systems. To date, many strategies such as coating the nZVI surface with polyelectrolytes or nonionic surfactants, the combination of the nZVI with the solid support, and the dispersion of the nZVI in oil-water emulsions have been used to improve the stability of nZVI. The coats that have been used can be organic or inorganic molecules, which function as protection shells to stabilize the nZVI and act as functionalization. The nZVI supporting not only efficiently prevents aggregation of the nZVI, but also protects its' further leaching in the harsh environment. Another solution proposed is to combine the nZVI with other metals (e.g., Pd or Ni) to form bimetallic materials.

Surface coating of nZVI changes the charge and steric stabilization to enhance the particle stability. In this connection, some anionic polymers (e.g., polyacrylic acid, polystyrene sulfonate, and carboxymethyl cellulose) are regarded as attractive candidates due to their large molecular weights and the abundance of charged functional groups [68–70]. Among the above polymers, the negative surface charges of the anionic polymers are prior in these applications, because they can reduce the affinity between nZVI and the natural soil media [71].

Immobilizing nZVI on the solid support enhances the dispersion of particles and promotes the practical application of nZVI. With the use of -COOH, -OH, or $-NH_2$ as chelating sites for the nZVI (or the precursor of Fe(0)), nZVI has been anchored on naturally clay minerals such as zeolite, bentonite, and montmorillonite [72–74]. The intrinsic ion-exchange capability of the above minerals permits facile incorporation of Fe(0). Intercalating Fe(0) in the pillared layers of bentonite produces interlayer mesopores or micropores, providing the channels of pollutant transport and plenty of reactive surface area.

There is no denying that the improvement of the stability and transport capability of nZVI-based materials via their encapsulated or supported are at the cost of reducing the chemical reactivity because of the obstacle of reactive sites [75, 76] and the additive diffusive barriers resulted by the encapsulation materials [69]. Evaluating the whole effect of surface-engineered nZVI under field-relevant conditions would be essential for assessment.

6.2.4 The drawbacks of nZVI use

When taking into account the optimum size range, nZVI potentially represents a highly efficient alternative to universal materials used for water clean-up [77]. nZVI has been completely investigated for environmental remediation goals, and it has been shown to remediate various contaminants and at much greater reaction rates than bulk iron.

However, despite the unquestionable effectiveness of nZVI and the simplicity of delivery, there are multiple disadvantages of using "free" nanoparticles for remediation. It is well recognized that the dispersion of nZVI in an aquifer will be limited by multiple steps such as geochemical/geophysical characteristics, mineral adsorption, microbiological activity, and the aggregation and formation of voluminous corrosion products. These variables are hard to predict and nZVI would need to be distinctly tailored for every situation [78, 79]. Further, variation in the groundwater system may also induce contaminants adsorbed to the nZVI surface and the resulting surface adaptations may be reversed or become redundant [80] when considering the operational difficulty of reclaiming nZVI after injection.

Multiple studies have developed ways for solving these problems by altering the nZVI particles to restrict or reduce interparticle attractions. Surfactants [81] or polymers [82] can be affixed to the surfaces of nZVI to hearten steric hindrance or hinder electrostatic attraction, or nZVI can be merged into other mobile structures. However, these treatments are an additional cost and some may be regarded as possible contaminants.

Moreover, investigation about the long-term ecotoxicological effects of monodispersed nZVI in the environment is few [83], the similar properties offer the remediative qualities of nZVI, which also make them hazardous for living organisms. For the time to come, nZVI itself, may be seen as an inacceptable risk, contractors that have employed nZVI should subsequently be responsible for their clean-up.

Because of some of the uncertainties outlined above, the United Kingdom is taking preventative measures for introducing engineered nanoparticles into the natural environment [84, 85], and has called for further research into nanotoxicology and nanoparticle behavior in subsurface environmental systems.

6.3 Identification and quantification techniques of toxic/radioactive metal ion adsorption behavior

The fast development in the use of nZVI-based materials as adsorbents has generated interest in characterizing the interactions between nZVI-based materials and toxic/radioactive metal ions. The main interaction mechanisms such as adsorption, redox, ion exchange, aggregation, hydroxylation, and succedent precipitation have been proposed to explain the adsorption behaviors between nZVI-based materials and toxic/radioactive metal ions. This part is devoted to explaining a variety of traditional and advanced techniques that are frequently employed to identify the species and microstructures of metal ions adsorbed on the surface of nZVI-based materials at the molecular or atomic level. We have focused on batch experiments (such as kinetic and thermodynamic analyses); appropriate spectroscopic analytical tools such as Mössbauer spectroscopy, XPS, EXAFS, XANES, and magnetic measurements; and imaging techniques such as STEM-XEDS. The combination of these powerful techniques can offer precise quantitative measurement of particle size, mineralogy, morphology, and stoichiometry.

6.3.1 Batch experimental techniques

Batch experimental techniques are one of the most widely used methods, and are known to be extremely convenient and efficient. This approach can provide valuable information about the adsorption equilibrium time, maximum adsorption capacity, and selectivity.

6.3.1.1 Kinetic analysis

Kinetic analyses can give information about the contact time required for achieving adsorption completion. A investigation by Xiang et al. [86] reported that the U(VI) adsorption efficiency achieved ~85% in 3 min and achieved equilibrium (almost 97%) after 30 min by nZVI-Al bimetal, which was higher than the single nZVI and nZV-Al particle or the total of the two particles. Such fast adsorption kinetics might be attributed to the nZV-Al core as the electron donor, which retarded the surface oxidation of Fe(0) and held the strong reduction ability of Fe(0), hence giving the unceasing and efficient

U(VI) removal performance. The adsorption equilibrium of many nZVI-based materials for metal ion adsorption requires only several minutes, which is superior to many other widely used adsorbents. Although plentiful reactions occur in the nZVI-metal ion system, many apparent rate constants are lack of the meaning of mechanism and can't be illustrated to quantify the impact of these reactions [87].

Kinetic analyses can also be employed to explore the underlying rate-controlling steps such as mass transfer and chemical reaction processes. For instance, the groups of Hu [88] found that Cr(VI) adsorption by nZVI@Mg $(OH)_2$ consisted of an initial slow adsorption process followed by a more rapid adsorption process. These findings indicated that the removal process was initially governed by Cr(VI) chemisorption to the readily accessible binding sites near the surfaces of $Mg(OH)_2$ and became diffusion-limited over time as Cr(VI) anions migrated into the nZVI core. Generally, adsorption reactions between nZVI-based materials and toxic/radioactive metal ions proceed rapidly at the original stages and gradually retard upon achieving the adsorption equilibrium. The time required for adsorption completion is dependent on the solution conditions, the initial metal ion concentration, and the characteristics of the metal ions and nZVI-based materials.

To study the adsorption kinetics precisely, numerous kinetic models have been used to simulate the adsorption data, including the pseudo-first-order model, the pseudo-second order model, the intraparticle diffusion model, and the Lagergren, Zeldowitsch, and Elovich kinetic models. A study by Zhu et al. [15] applied multiple kinetic models such as pseudo-first-order, pseudo-second order, and intraparticle diffusion models to simulate the U(VI) adsorption data on Fe/N-C-700 (Fig. 6.3). Based on the correlation coefficients and calculated Q_e values, the pseudo-second order model was noticed to be more appropriate for simulating the adsorption kinetic experimental data, indicative that chemical reactions occurred between the U(VI) and adsorption sites of Fe/N-C-700 in the adsorption process. Generally, the adsorption behavior of most metal ions onto nZVI-based materials fits the pseudo-second-order model well, suggesting that chemical adsorption is dominant rather than physical adsorption.

6.3.1.2 Thermodynamic analysis

Adsorption isotherms under specific solution conditions can provide valuable information for evaluating the mobility or the retention of metal ions from the aquatic environments to the solid phase. The interaction mechanisms, surface properties, and the affinity degree of adsorbents can be obtained from physico-chemical parameters and underlying thermodynamic assumptions. Numerous adsorption isotherm models, such as Langmuir, Freundlich, Brunauer-Emmett-Teller, Redlich-Peterson, Dubinin-Radushkevich (D-R), Temkin, Toth, Koble-Corrigan (K-C), Sips, Flory-Huggins, and Radke-Prausnitz, have been used to

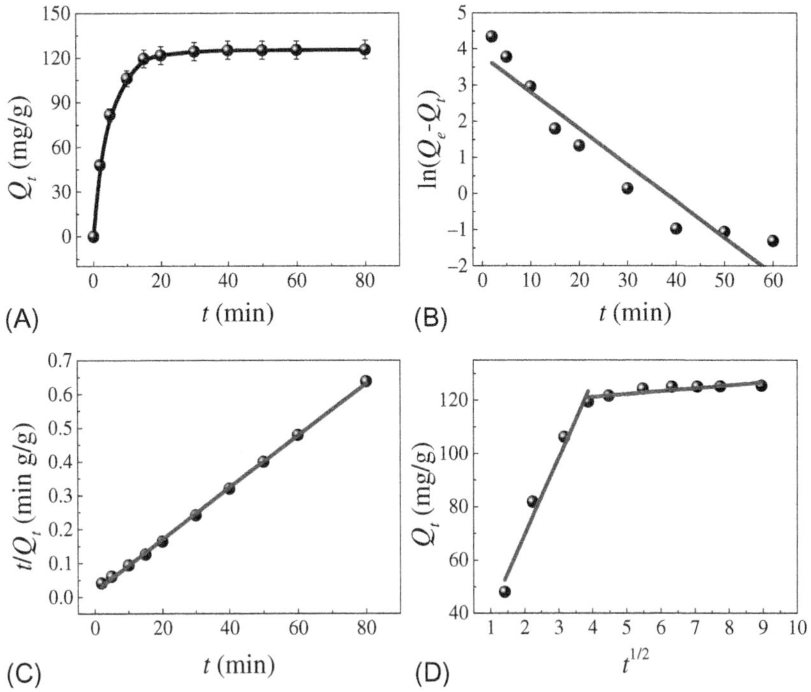

FIG. 6.3 (A) Effect of contact time on U(VI) removal onto Fe/N-C-700, (B) simulation of kinetics data by pseudo-first-order model, (C) pseudo-second-order model, and (D) intraparticle diffusion model [15]. *(Data from K.R. Zhu, C.L. Chen, M.W.C. Xu, K. Chen, X.L. Tan, M. Wakeel, N.S. Alharbi, In situ carbothermal reduction synthesis of Fe nanocrystals embedded into N-doped carbon nanospheres for highly efficient U(VI) adsorption and reduction, Chem. Eng. J. 331 (2018) 395–405. Copyright 2017, Elsevier.)*

simulate the adsorption isotherms. Generally, the behaviors of metal ion adsorption onto nZVI-based materials followed the Langmuir model well, indicating that the adsorption behavior is of the monolayer type. Vilardi et al. [89] used Langmuir, Freundlich, Sips, and D-R models to simulate the isotherm data of Cr(VI) adsorption onto CNTs-nZVI nanocomposites. They found that the adsorption equilibrium capacities of these adsorbents fast increased in the low concentration ranges, then increased slowly and finally reached the highest adsorption equilibrium capacity. The Sips model was suitable to describe the adsorption process, proving to be more flexible and suitable for metal adsorption modeling onto complex surfaces.

In addition to the adsorption capacity and adsorption mechanisms, thermodynamic parameters (i.e., ΔG, ΔH, and ΔS) can also be obtained from the temperature-dependent adsorption isotherms. These parameters can further provide insight into the interaction behaviors between metal ions and nZVI-based composites. Table 6.1 lists the reported thermodynamic parameters

TABLE 6.1 Thermodynamic parameters for metal ion adsorption onto nZVI-based materials at 298 K

Targets	nZVI-based materials	T(k)	ΔG (kJ/mol)	ΔH (kJ/mol)	ΔS (J/(mol·K))	Ref
U(VI)	nZVI/CNF	293	−22.69	9.82	110.78	[90]
U(VI)	N-doped Fe/Fe_3C@C-800	298	−3.18	23.95	3.89	[91]
Cr(VI)	nZVI/Fe_3O_4@C	298	−169.406	61.763	776.190	[92]
Cr(VI)	nZVI/Ni	298	−5.1973	419.57	1.4350	[93]
As(V)	GnIP	293	−16.71	−27.41	−29.51	[94]
Sb(V)	nZVI	298	−12.4	180.1	−0.646	[95]
As(III)	CS-P-NZVI	298	−9.45	−115.4	1.29	[96]
Sb(III)	nZVI/Fe_3O_4	298	−9.34	−16.69	−24.69	[97]
Pb(II)	nZVI-WSBC	293	−8.840	8.637	57.6	[98]
Cd(II)	nZVI/rGOs	293	−27.789	132	13.527	[99]
Tl(I)	Fe@Fe_2O_3	303	−4.45	9.65	46.54	[100]

for metal ion adsorption onto nZVI-based materials. As displayed in Table 6.1, the negative ΔG values suggested that the adsorption process is a spontaneous endothermic reaction. The thermodynamic parameters for adsorption of different metal ions onto different nZVI-based materials are quite different, and are governed by the solution conditions, the properties of the metal ions and the nZVI composites, and experimental conditions.

6.3.2 X-ray photoelectron spectroscopy

XPS analysis can provide the element oxidation states, element species and bonding relationships for all kinds of surface elements (except for H). To date, many studies have used XPS techniques to characterize the adsorption properties of nZVI-based materials for toxic/radioactive metal ions [101, 102].

Researchers in our group used XPS to explore the interaction mechanism between Re(VII) oxyanions and nZVI/rGOs [103]. After the adsorption of Re(VII) oxyanions, the peak of ReO_2 on the nZVI/rGOs-Re was stronger than that on nZVI-Re, attributable to the stronger reduction ability of nZVI/rGOs than nZVI. Under the existence of nZVI, the Fe phase produced atomic hydrogen that reduced ReO_4^- to the low oxidation state of Re. XPS analysis of Ag(I) capture by nZVI revealed the occurrence of Ag_2O reduction to Ag^0 during the capture process [104]. This phenomenon can be demonstrated by the upward shift and downtrend of the binding energy of Ag 1s with the rise of the reaction time. XPS results also revealed that the active FeOOH layer makes function in charge transport in the reduction reaction. XPS analysis of Cr(VI) adsorption by Fe/Al bimetallic nanoparticles revealed the occurrence of Cr_2O_3 and $Cr(OH)_3$ after reaction. Meantime, the XPS analysis of Fe 2p and O 1s also showed that the Cr(VI) ions were immobilized by the formation of $Cr(OH)_3$ and Cr(III)-Fe(III) hydroxide/oxyhydroxide on the Fe/Al bimetal surface [105]. In the report of Sapsford et al. [106], XPS analysis recorded that the outer surface was constituted by a mixed valent of iron oxide. In view of the mean free path of Fe, thus the detection in the XPS analysis suggests that the oxide shell thickness of the nZVI isn't exceeding about 5 nm, which verifies the above nZVI oxide thickness measured *via* the use of HRTEM. XPS analysis is quantitative and has high surface sensitivity, which limits the detection in the top 3–30 atomic layers (10–100 Å) in solid power samples. Hence, XPS is a powerful tool for characterizing the adsorption reactions between toxic/radioactive metal ions and nZVI and its composites.

6.3.3 Mössbauer spectroscopy

Since its discovery in 1958 [107], Mössbauer spectroscopy has established itself as an important technique for determining the mineral composition, oxidation state, and even the crystallinity of Fe minerals. The technique is becoming more and more attractive with the growing number of studies on wastewater decontamination [108].

The existence of both Fe(II) and Fe(III) in nZVI-based composites results in its suitability for Mössbauer spectroscopy, as both sites are visible as distinct doublets. Many works have even tried to investigate the iron composition within different nZVI-based composites. Vinod et al. [109] used Mössbauer spectroscopy technology to understand the adsorption mechanism and coordination environment of arsenic on an AN/MA/IA/nZVI membrane (Fig. 6.4). Mössbauer spectroscopy data show that these impurities arise from partial surface oxidation of Fe(0) or represent some minor iron oxides produced during the immobilization of nZVI on the AN/MA/IA membrane. After the arsenic removal, the Mössbauer spectrum of A the N/MA/IA/nZVI membrane shows a significant increase of ferric oxide content (19%), demonstrating the Fe(0) oxidation during the process of arsenic removal. Similarly, Mössbauer spectroscopy was employed to describe the Cr(VI) redox reaction on the surface of the hybrid nZVI@MCM-41 [110]. Based on analysis of the ^{57}Fe

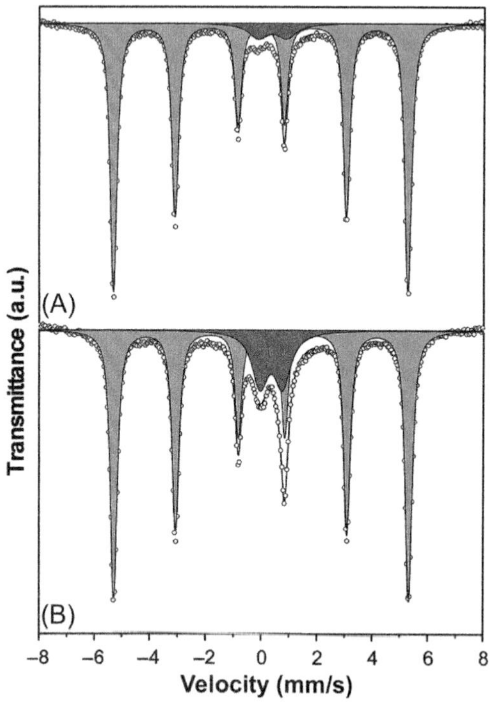

FIG. 6.4 ^{57}Fe Mössbauer spectra of the AN/MA/IA/nZVI membrane. (A) As-prepared AN/MA/IA/nZVI membrane, (B) AN/MA/IA/nZVI membrane after arsenic removal. *Dark grey* component-metallic iron (α-Fe), *orange* component-ultrasmall (superparamagnetic) ferric oxides/hydroxides [109]. *(Data from V.T.P. Vinod, J. Filip, S.K. Ittara, S. Wacławek, M. Cernik, Electrospun membrane composed of poly [acrylonitrile-co-(methyl acrylate)-co-(itaconic acid)] terpolymer and ZVI nanoparticles and its application for the removal of arsenic from water, RSC Adv. 6 (2016) 110288–110300. Copyright 2016, Royal Society of Chemistry.)*

Mössbauer spectra of nZVI@MCM-41 and after reacting with Cr(VI) recorded at 77 K, the diffusion of Cr(VI) anions through the ZVI nanoparticle outer shell is reduced into the Cr(III) species by oxidation of the Fe(0) core into Fe(III) and Fe(II) cations, thus increasing the thickness of the outer shell. Sheng et al. [111] used Mössbauer spectroscopy to decipher the stability of nZVI/LDH during the U(VI) removal reaction. After reaction with U(VI) cations, the Fe(0) content for nZVI/LDH reduced from 96.8% to 88.6% while the Fe(0) content for nZVI reduced from 90.0% into 78.7%. This result shows the higher stability of nZVI/LDH than nZVI. Mössbauer spectroscopy was also conducted to explore the dynamic characteristics of the reaction between nZVI and As(V) with the increasing reaction time and the various forms of Fe existed in As(V)-treated nZVI and have verified the alteration over time of the percentage of Fe(0) [112].

6.3.4 Synchrotron X-ray absorption

Recently, synchrotron-based analytical methods have been widely employed in the geosciences-related fields for the analysis of metal ion species on the surfaces of solid particles. XAFS, including XANES and EXAFS, is a well-established and powerful tool in microstructure characterization.

Results obtained from XAFS analysis, such as the type of near neighbors surrounding targeted ions, the coordination number, and the bond distances between atoms, can give detailed information about the main adsorption species and the interaction mechanisms. For example, Qian et al. [113] utilized XAFS techniques to examine the iron corrosion in the Mo(VI)-nZVI system at pH 6 and 9. As shown in Fig. 6.5A and B, at pH 6, the content of Fe(0) and iron oxide were 45.5% and 54.5%, respectively, while at pH 9, the content of Fe(0) and iron oxide became 69.7% and 30.3%. Meanwhile, the Fe(II) concentration in the supernatant at pH 6 was also more than at pH 9. Moreover, a 15% amorphous Fe-O bond was found at pH 6; however, only 1.7% was measured at pH 9. The above results indicated that iron corrosion is much faster at pH 6 than that at pH 9. Moreover, the Mo(VI) reduction products also were analyzed *via* XANES [113]. Through the linear fitting for Mo K-edge XANES spectra (Fig. 6.5C and D), one could be found that high pH supports for Mo(VI) reduction and H^+ is reliable for the hollow-out structure. Sheng et al. [114] used the XANES spectrum to reveal differences from Se(IV) removal by nZVI/CNT and pristine nZVI, CNT in Fig. 6.6. After Se(IV) sequestered on CNT, the apparent likeness of the XANES edge feature with that of Se(IV) indicates that the oxidation state of Se(IV) didn't change after reaction along with CNT, leading to the adsorptive interaction. However, for Se(IV) removal on nZVI/CNT, the XANES spectrum showed that Se(0)/Se(−II) oxidation states were primary, indicative of complete transformation of Se(IV) into Se(0)/Se(−II) *via* the reductive interaction. Nevertheless, for Se(IV) removal on nZVI, the XANES spectrum is obviously different from

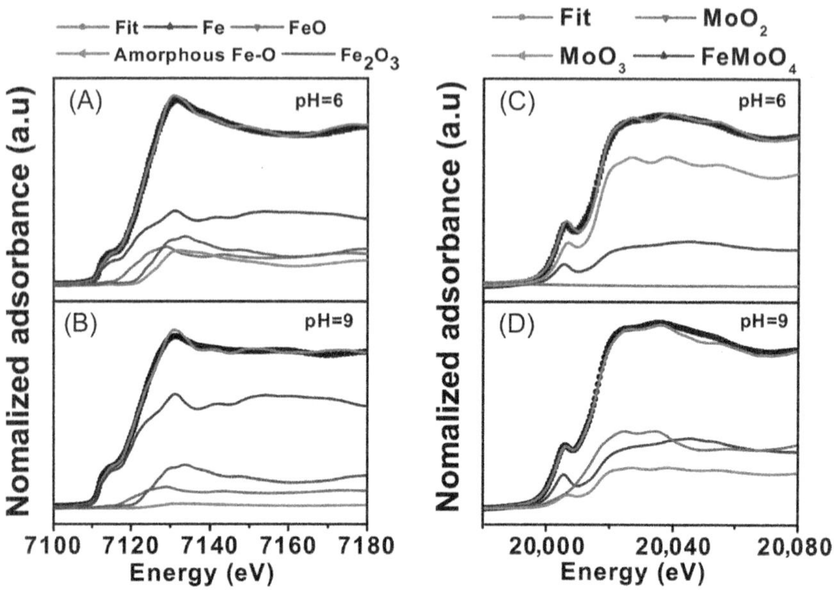

FIG. 6.5 K-edge XANES spectra of Fe (A, B: pH = 6, 9) and Mo (C, D: pH = 6, 9) [113]. *(Data from D.X. Qian, Y.M. Su, Y.X. Huang, H.Q. Chu, X.F. Zhou, Y.L. Zhang, Simultaneous molybdate (Mo(VI)) recovery and hazardous ions immobilization via nanoscale zerovalent iron, J. Hazard. Mater. 344 (2018) 698–706. Copyright 2018, Elsevier.)*

other samples, which appears to be combined with that species of Se(IV) and Se(0)/Se(−II). This result suggests that Se(IV) isn't completely reduced by nZVI. The use of the XAFS tools will be realized to understand the speciation and microstructures of toxic/radioactive metal ions adsorbed onto the surfaces of nZVI-based materials with the increasing demand for molecular-scale metal ion-speciation information and the increasing sophistication of the XAFS tools.

6.3.5 Scanning transmission electron microscopy and X-ray energy dispersive spectroscopy

Visualization of the reaction of nZVI and its composites is routinely examined by STEM-XEDS. This technique provides accurate and simultaneous near-atomic resolution visualization of nZVI structural and compositional changes. Zhang et al. [115] used STEM-XEDS technology to offer unexampled details on the Fe(0)-Se(IV) reactions in a single nZVI nanoparticle (Fig. 6.7). Visualizations of solid-phase transsituation and conversion of Se(IV) such as diffusion, reduction, and deposition as well as the impact of surface defects in Fe(0) nanoparticles were displayed by the STEM-XEDS method. Se(IV) was reduced to Se(−II) and Se(0), forming a 0.5 nm layer of selenium at the interface of iron oxide-Fe(0).

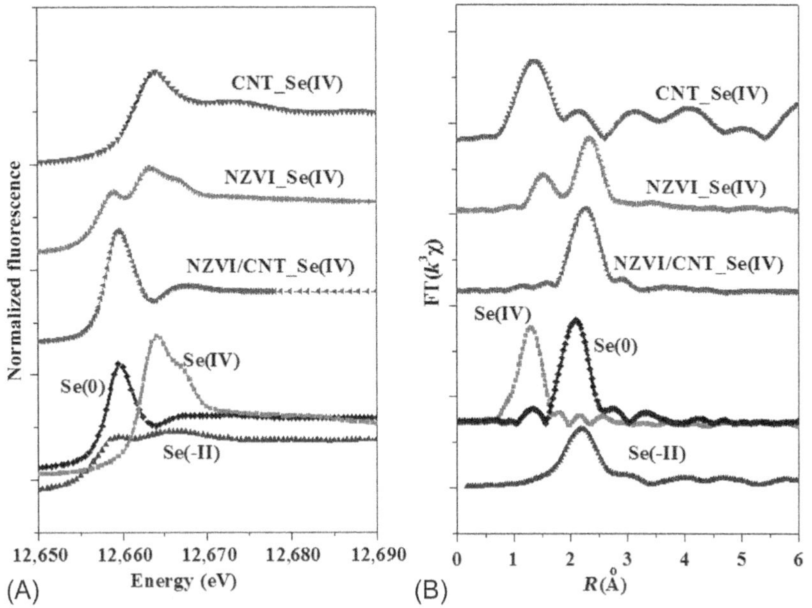

FIG. 6.6 (A) X-ray absorption near-edge structure (XANES), and (B) extended X-ray absorption fine structure (EXAFS) spectra of reference samples and reacted samples (at pH ~6.0 after reaction for 2h) of Se(IV) sequestered on the reactive materials [114]. *(Data from G.D. Sheng, A. Alsaedi, W. Shammakh, S. Monaquel, J. Sheng, X.K. Wang, H. Li, Y.Y. Huang, Enhanced sequestration of selenite in water by nanoscale zero valent iron immobilization on carbon nanotubes by a combined batch, XPS and XAFS investigation, Carbon 99 (2016) 123–130. Copyright 2016, Elsevier.)*

Similarly, Ling et al. [116] extended the investigation about the single nZVI reaction with other metal ions (such as Cr, As, Ni, Ag, Cs, and Zn) using STEM-XEDS technology (Fig. 6.8). Moreover, the removal mechanism is up to the standard reduction potential E^0 (Table 6.2). For Cr(VI) anions, it is diffusion and encapsulation. For As(V) anions, it's adsorption at the surface, following with diffusion and encapsulation into the particle between the core and the shell. For Cs(I) and Zn(II) cations, it's adsorption or surface-complex formation. For Ag(I) cations, it's rapid reduction on the surface of nZVI. At the same time, for Ni(II) ions, it can be immobilized at the nanoparticle surface by adsorption and reduction Adsorption, reduction, and encapsulation mechanisms of nZVI make a synergetic impact in leading to rapid and high-efficient reactions for treatment and immobilization of metal ions.

The former investigations have reported that rapid and almost complete uranium enrichment and encapsulation from water can be reached with nZVI [117, 118]. Direct visualization and chemical mapping of the Fe(0)-U(VI) reactions are acquired with STEM-XEDS (Fig. 6.9). After 1h reacting with U(VI), the profile of the uranium area is less than both iron and oxygen,

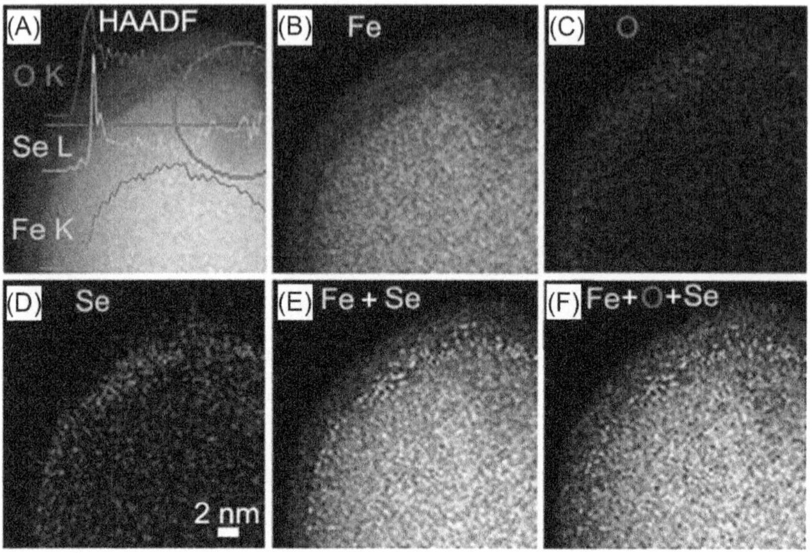

FIG. 6.7 XEDS-STEM elemental mappings of the Fe-Se reactions: (A) HAADF image with XEDS-STEM elemental line scans of the Fe, Se, and O over a defect area, (B) Fe, (C) O, (D) Se, (E) Fe + Se, (F) Fe + O + Se. Signals collected from nZVI after 48h reactions with 1.3mM selenite [115]. *(Data from L. Ling, B. Pan, W.X. Zhang, Removal of selenium from water with nanoscale zero-valent iron: mechanisms of intraparticle reduction of Se(IV), Water Res. 71 (2015) 274–281. Copyright 2015, Elsevier.)*

demonstrating that uranium has penetrated the surface oxide layer. After 24h reacting with U(VI), the uranium is highly concentrated near the particle center to form the unique egg yolk-like structure. The results verify that the most profitable destination for the reduction and immobilization of uranium is right at the particle center where the concentration of Fe(0) is at the maximum.

6.3.6 Other analysis techniques

Other analysis techniques such as XRD, FT-IR, Raman spectroscopy, and magnetic measurement have also been used to characterize the adsorption behaviors of nZVI-based materials. For example, Li et al. [103] investigated ReO_4^- adsorption onto nZVI/go with the use of XRD and found that ReO_4^- was reduced to ReO_2 in the reduction reaction in which nZVI particles served as reductants, indicating that the insoluble ReO_2 formation was the main mechanism for the removal of ReO_4^- by the nZVI/rGOs. Arshadi et al. [119] used FT-IR to investigate the PO_4^{3-} adsorption onto S-NaOH-nZVI (Fig. 6.10), finding that the new peaks appearing at 437 and 497 cm^{-1} were due to the stretching vibration of metal oxide. This indicates the existence of Fe—O—P bonds between phosphate ions and the active sites of nZVI. Similarly, a new Raman band at 1044 cm^{-1} appeared in a used

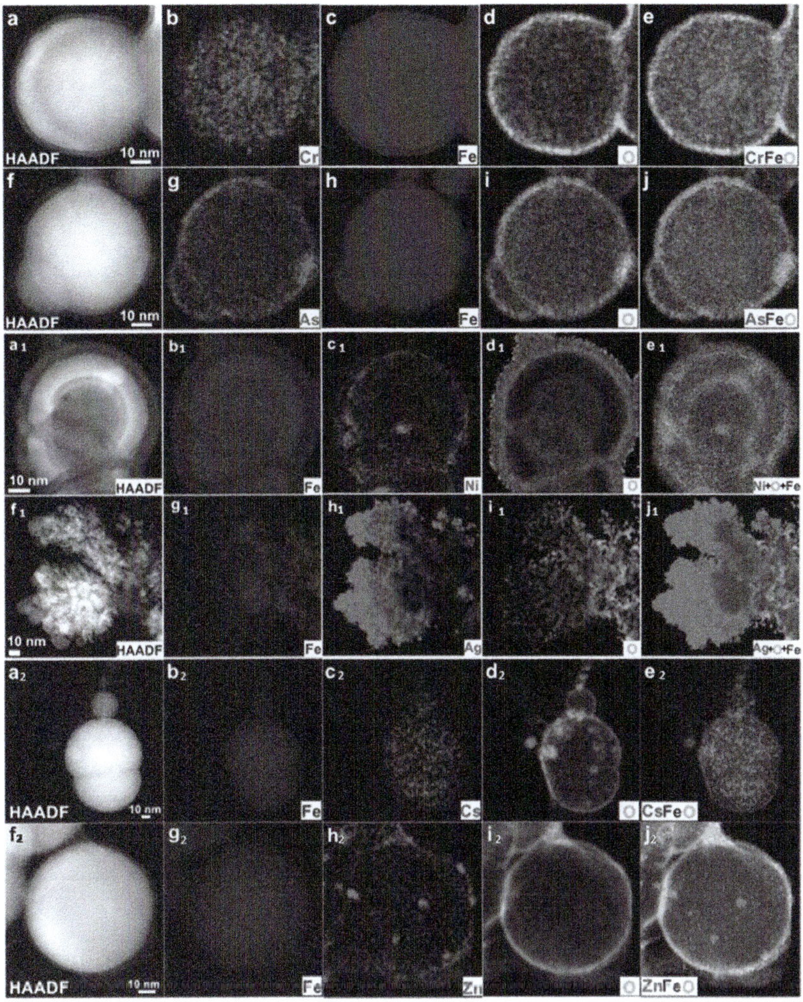

FIG. 6.8 High-resolution XEDS mapping of a single iron nanoparticle with Cr(VI), As(V), Ni(II), Ag(I), Cs(I), and Zn(II) [116]. *(Data from L. Ling, X.Y. Huang, M. Li, W.X. Zhang, Mapping the reactions in a single zero-valent iron nanoparticle, Environ. Sci. Technol. 51 (2017) 14293–14300. Copyright 2016, American Chemical Society.)*

graphene-nZVI sample during the phosphorus-nZVI reaction process, which could also be attributed to the appearance of the Fe—O—P group [120]. Cao et al. [121] reported that magnetic measurements were employed to study the saturation magnetizations of nitrogen-doped magnetic carbons after use, which found that the puny reduction of adsorbent magnetization could be attributed to the expense of the nZVI, generating the iron oxide and $Cr_2O_3/Cr(OH)_3$.

TABLE 6.2 Removal of various metals with nZVI [116]

Metal	Half-reactions	E^0 (V)[a]	E (V)[b]	Reaction mechanisms
Cr	$Cr_2O_7^{2-} + 14H^+ + 6e^- \leftrightarrow 2Cr^{3+} + 7H_2O$	1.33	0.58	Adsorption, reduction, precipitation
As	$H_3AsO_4 + 2H^+ + 2e^- \leftrightarrow HAsO_2 + 2H_2O$	0.56	0.04	Adsorption, reduction, precipitation
	$HAsO_2 + 3H^+ + 3e^- \leftrightarrow As + 2H_2O$	0.25	−0.18	–
Ag	$Ag^+ + e^- \leftrightarrow Ag$	0.88	0.77	Reduction
Fe	$Fe^{2+} + 2e^- \leftrightarrow Fe$	−0.44	−0.48	–
Ni	$Ni^{2+} + 2e^- \leftrightarrow Ni$	−0.25	−0.33	Adsorption, reduction
Zn	$Zn^{2+} + 2e^- \leftrightarrow Zn$	−0.76	−0.82	Adsorption
Cs	$Cs^+ + e^- \leftrightarrow Cs$	−3.03	−3.03	Adsorption

[a] *Standard potentials (E^0) refers to Lange's handbook of chemistry [116].*
[b] *Electrode potentials (E) at pH 7.*

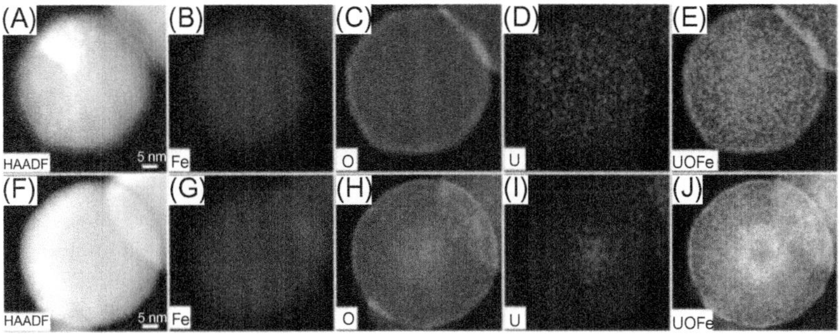

FIG. 6.9 Enrichment and encapsulation of uranium from wastewater. STEM-XEDS mappings of uranium reactions with nZVI: (A–E) after 1 h; (F–J) after 24 h [60]. *(Date from L. Ling, X.Y. Huang, W.X. Zhang, Enrichment of precious metals from wastewater with core-shell nanoparticles of iron, Adv. Mater. 30 (2018) 1705703. Copyright 2018, Wiley.)*

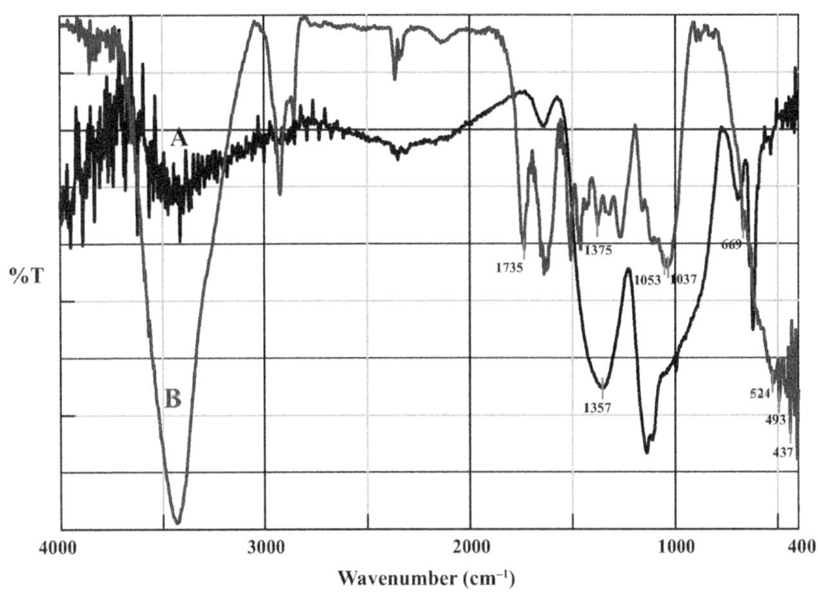

FIG. 6.10 FTIR spectra of S-NaOH-nZVI before (A) and after (B) adsorption of P(V) from aqueous solution [119]. *(Data from M. Arshadi, S. Foroughifard, J.E. Gholtash, A. Abbaspourrad, Preparation of iron nanoparticles-loaded Spondias purpurea seed waste as an excellent adsorbent for removal of phosphate from synthetic and natural waters, J. Colloid Interface Sci. 452 (2015) 69–77. Copyright 2015, Elsevier.)*

6.4 nZVI-based materials for radioactive metal ion decontamination

6.4.1 Uranium

Uranium primarily consists of two dominating states in the aquatic environment, including the high mobility of U(VI) and the low chemical toxicity of U(IV). Many workers have concentrated on preparing functional materials that convert U(VI) into U(IV) [122–124]. Among the reported materials, nZVI is seen as the favorable material, owing to it being highly effective and inexpensive while having a strong reduction property.

Since the report by Cantrell et al. [125], ZVI has been seen as a hopeful technology to remove U(VI) from contaminated sites due to its low cost and high availability [126]. Following that, the adsorption behaviors and mechanisms between ZVI and U(VI) have been generally studied [127, 128]. To date, great attention has been paid to the use of nZVI in the treatment of uranium wastewater because of the high reactivity of nZVI [129–131]. ZVI is the strong reducing agent that can be oxidized to Fe(II) and Fe(III). This procedure generates the poor crystallinity of corrosion products, which can function as the potential adsorption surfaces for removing cations. The interaction of nZVI and uranium ions occurs via any of three mechanisms:

(a) Fe(0) was considered as an electron donor to chemically reduce the aqueous, mobile U(VI) species to a less soluble U(IV) precipitate (i.e., $UO_2{\cdot}xH_2O$) [132]:

$$Fe^0 + 1.5\,UO_2{}^{2+} + 6H+ = Fe^{3+} + 1.5U^{4+} + 3H_2O \tag{6.10}$$

$$U^{4+} + 2H_2O \rightarrow UO_2 + 4H^+ \tag{6.11}$$

(b) Adsorption onto corrosion products of ZVI by ion exchange with hydroxyl sites, and ZVI is oxidized:

$$Fe^0 \rightarrow Fe^{2+} + 2e^- \tag{6.12}$$

$$Fe^{2+} \rightarrow Fe^{3+} + e^- \tag{6.13}$$

In the aqueous phase, Fe(III) reacts with OH^- or H_2O to form hydroxide or oxyhydroxide:

$$Fe^{3+} + 3OH^- \rightarrow Fe(OH)_3 \tag{6.14}$$

Subsequent dehydration of $Fe(OH)_3$ generates α-FeOOH:

$$Fe(OH)_3 + 3H^+ \rightarrow FOOH + H_2O \tag{6.15}$$

FeOOH adsorption properties in aquatic systems have been widely verified [133, 134]. In water, U(VI) is incorporated into the FeOOH layer and can form $(U_xFe_{1-x})(OH)_3$ or $U_xFe_{1-x}OOH$. The U(VI) removal mechanism by nZVI is involved in rather complex since different processes are involved.

Singh et al. [135] explored the molecular-scale U(VI) immobilization mechanisms on goethite by EXAFS spectroscopy. EXAFS analysis showed that the precipitation of U(VI) had a structure that is accordant with the meta-autunite materials. The adsorbed U(VI), in the lack of phosphate at pH 4–7, formed $\equiv Fe(OH)_2UO_2$ and $(\equiv FeOH)_2UO_2$.

(c) The incorporation of adsorption and reduction/precipitation to a new precipitation of iron oxides simultaneously occurred. Fiedor et al. [136] applied XPS to study the soluble U(VI) removal mechanism from groundwater by ZVI. They reported that U(VI) was adsorbed onto hydrous ferric oxide under aerobic conditions; otherwise, U(VI) was reduced into U(IV) under anaerobic conditions.

As mentioned above, nZVI holds the merits of high reactivity because of its large specific surface area. Nevertheless, because nZVI shows high reactivity, it is prone to oxidize and form aggregates in the aquatic environment, which adversely affects its large application. To reduce the tendency to particle aggregation and rapid sedimentation, great attention has been paid to improve nZVI performance [81, 137–139].

For instance, Sun et al. [140] synthesized an nZVI/rGO composite by chemical deposition; the existence of rGO on nZVI nanoparticles significantly improved the U(VI) reaction rate and removal capacity. nZVI-diatomite was prepared and applied for U(VI) enhanced sequestration [74]. nZVI-diatomite showed a distinct synergistic effect between diatomite adsorption and nZVI reduction. XPS results suggested that reducing the high toxicity of UO_2^{2+} into less-toxic UO_2 could be enhanced by the use of diatomite-supported nZVI (Fig. 6.11). Therefore, diatomite could serve as a scavenger for UO_2, and thus more reactive sites could be applied for the reduction of U(VI). Illite was also used to support nZVI (I-nZVI) to remove U(VI) in aqueous solutions [141]. I-nZVI shows higher stability and removal efficiency that nZVI. The investigation shows that illite serves as the good dispersant and stabilizer to decrease the nZVI aggregation and improve the Fe(0) activity. Porous carbon materials are broadly applied as the support for stabilizing nZVI. For instance, new nZVI/C composites were synthesized by heating the mixture of natural hematite and pine sawdust at high temperature under N_2 gas conditions [142]. Simultaneous adsorption and reduction were related in the U(VI) adsorption process based on the analysis of XPS and XANES. The U(VI) adsorption behaviors by the nZVI/C were well fitted by surface complexation modeling. The U(VI) highest adsorption capacity by the nZVI/C was 186.92 mg/g at pH 4.0 and 328 K. The above observations suggested that the nZVI/C can be seen as a potential

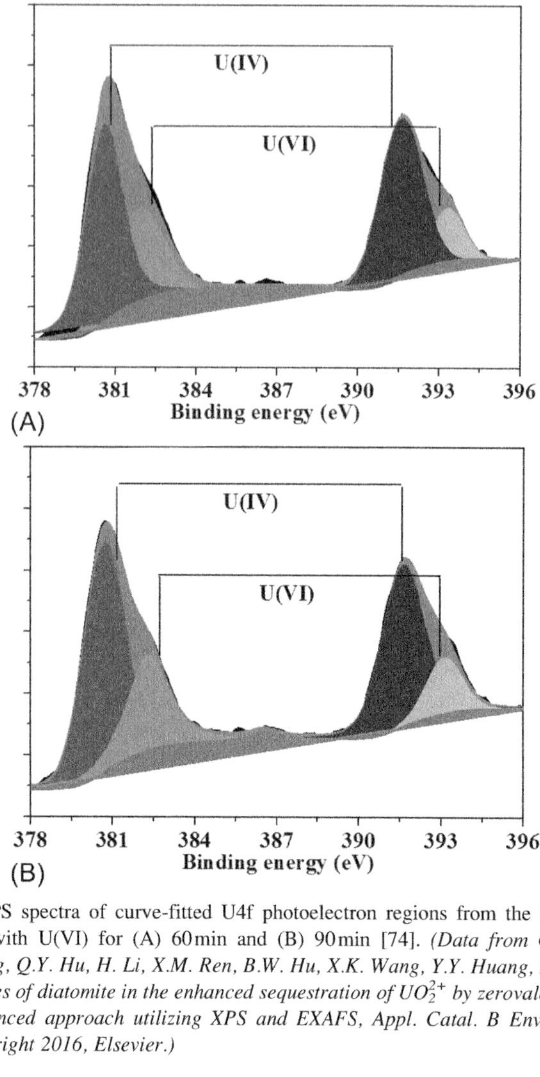

FIG. 6.11 XPS spectra of curve-fitted U4f photoelectron regions from the NZVI-D samples after reaction with U(VI) for (A) 60 min and (B) 90 min [74]. *(Data from G.D. Sheng, P.J. Yang, Y.N. Tang, Q.Y. Hu, H. Li, X.M. Ren, B.W. Hu, X.K. Wang, Y.Y. Huang, New insights into the primary roles of diatomite in the enhanced sequestration of* UO_2^{2+} *by zerovalent iron nanoparticles: an advanced approach utilizing XPS and EXAFS, Appl. Catal. B Environ. 193 (2016) 189–197. Copyright 2016, Elsevier.)*

adsorbent to eliminate radionuclide ions for environmental remediation. The nZVI/CNF composites were synthesized to remove U(VI) and the U(VI) decontamination mechanism was studied via batch, XPS, EXAFS, and modeling techniques [90]. The experimental results suggested that the U(VI) removal amount was obviously higher than that of nZVI. The diffuse layer model showed better fits to U(VI) removal by nZVI/CNFs, whereas underestimation of the removal of U(VI) was noticed at pH $>$ 5.0 because of the existence of reductive precipitation. The excellent adsorption performance and the magnetic separation property, making nZVI-based materials seen as the promising adsorbent for the treatment of U(VI)-containing wastewater, but application of this composite is currently limited to laboratory. Thus, the realization of the fast and simple separation of U(VI) based on nZVI-based composites is still a great challenge.

6.4.2 Technetium

Similar to uranium, immobilizing technetium can be reached *via* reducing the TcO_4^- into the insoluble $TcO_2{\cdot}nH_2O$ [46, 143]. In an acidic solution, reducing Tc(VII) into Tc(IV) occurs *via* [144]:

$$TcO_4^- + 4H^+ + 3e^- \rightarrow TcO_2 + 2H_2O \tag{6.16}$$

Liang et al. [145] studied the Tc(VII) removal on iron filings (size of 420 mm). The Tc removal was due to the incorporation of the reductive precipitation and adsorption mechanisms. Darab et al. [146] synthesized zirconia-supported nZVI and applied it to treat typical Hanford nuclear wastewater. Presently, immobilizing Tc(VII) ions with the use of modified nZVI has been widely reported. For instance, Fan et al. [46] showed that the TcS_2 formation is favored over TcO_2 under the presence of sulfidic stimulated by nZVI. Fan et al. [147] further studied the Tc(IV) sulfide stability by the analysis of solution chemistry, solid-phase characterization, and XPS. Tc dissolution data showed that the initial Tc(IV) dissolution inhibition is attributed to the redox buffer property of FeS. Because no obvious Tc dissolution happened in the process of oxidation, the speciation transformation takes a secondary part in obstructing Tc oxidation.

Based on the current study, few studies were directly applied with radioactive $^{99}TcO_4^-$. Because of the familiar ionic size and chemical properties between ReO_4^- and TcO_4^-, ReO_4^- is commonly seen as the best surrogate in the TcO_4^- adsorption study [148, 149]. For example, Sheng et al. [111] synthesized the nZVI supported on layered double hydroxide (nZVI/LDH) and applied it for immobilizing Tc(VII) by studying its chemical analogue Re(VII) (Fig. 6.12). Compared to pristine nZVI, the obtained nZVI/LDH showed high-efficient Re(VII) immobilization because of the better synergistic effect between sorption and reduction. The analysis of phrase and morphology suggests that nZVI was homodispersed on the surface of LDH, which obviously reduces the nZVI's aggregation and improves the reactivity of nZVI. ^{57}Fe Mössbauer spectra indicate that nZVI/LDH showed excellent antioxidation performance, resulting in enhanced stability in the reaction. XAFS results show that LDH could facilitate completely reducing Re(VII) to Re(IV) by nZVI while adsorption and reduction simultaneously happened on pristine nZVI. Furthermore, LDH not only serves as the pH buffering agent but also as the scavenger for the sparingly soluble products, enhancing the reactivity of nZVI.

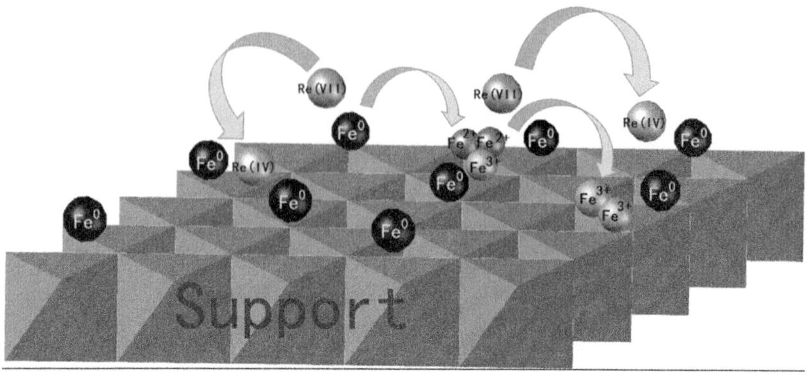

FIG. 6.12 A conceptual model for Re(VII) removal with nZVI/LDH [111]. *(Data from G.D. Sheng, Y.N. Tang, W.S. Linghu, L.J. Wang, J.X. Li, H. Li, X.K. Wang, Y.Y. Huang, Enhanced immobilization of* ReO_4^- *by nanoscale zerovalent iron supported on layered double hydroxide via an advanced XAFS approach: implications for* TcO_4^- *sequestration, Appl. Catal. B Environ. 192 (2016) 268–276. Copyright 2016, Elsevier.)*

6.4.3 Other radioactive metal ions

Other radioactive nuclides (^{60}Co, $^{152+154}Eu$) have high degrees of toxicity and pose hazards to human beings and the environment [150–154]. Recently, nZVI-based materials have displayed wonderful potential in the enrichment of radioactive Co(II) and Eu(III) ions from radioactive wastewater. In the following section, we summarize some typical examples of Co(II) and Eu(III) adsorption onto nZVI-based adsorbents.

Shahwan et al. [155] first reported the adsorption behavior of Co(II) on nZVI under various experimental conditions. Iron nanoparticles verified the rapid uptake and large capacity for Co(II) removal. Co(II) uptake increased with the increase of pH. The structural evolution of nZVI in anoxic Co(II) solution was further studied by Zhang et al [156]. Meanwhile, nZVI/GO composites also were applied to remove Co(II) [157, 158]. The detailed information of the interaction between nZVI/GO and Co(II) at different pH values was displayed by the analysis of XAFS and other characterization technologies (Fig. 6.13). At pH 3.0 and 6.0, the Co-substituted iron oxides converted into a $CoFe_2O_4$-like structure; however, at pH 9.0, they formed green rust-like phases. The Co(II) adsorption mechanism was due to innersphere complexation and the dissolution/reprecipitation of the substituted metal oxides. Eu(III) was the typical trivalent radionuclide. Zhu et al. [159] reported that nZVI@rGO composites were prepared toward the Eu(III) uptake in environmental contamination treatment. The uptake kinetics suggested that the Eu(III) uptake by nZVI@rGO can be well fitted to the pseudo-second-order kinetic model. The Eu(III) uptake by nZVI@rGO was up to ionic strength at pH > 5.0, suggesting that innersphere surface complexation governs the Eu(III) uptake Eu(III) by nZVI@rGO. The highest capacity of nZVI@rGO

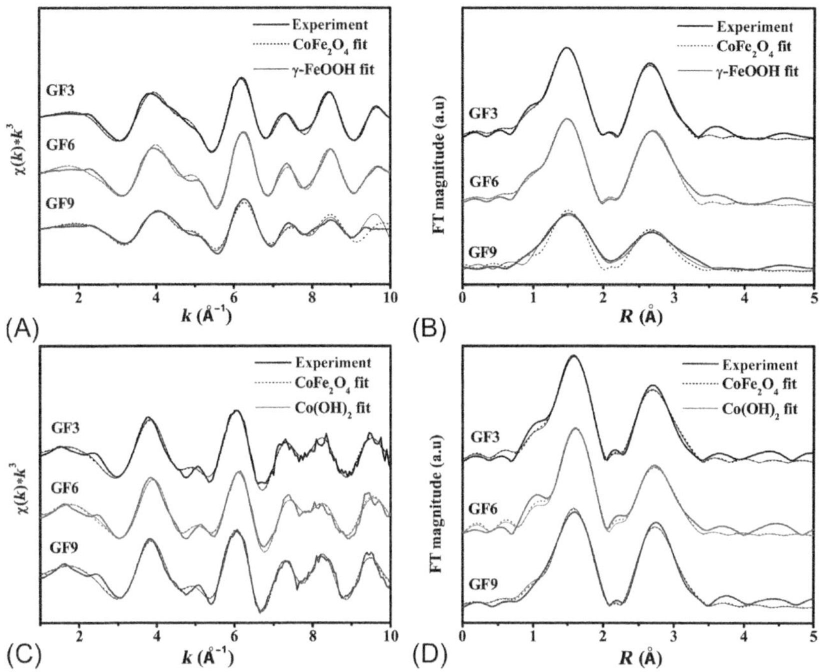

FIG. 6.13 Experimental (*thick lines*) and fitted (*thin lines and dash lines*) data of k^3-weighted and Fourier-transformed EXAFS spectra for nZVI/GO reaction at different initial pH values. (A) and (B) for Fe; (C) and (D) for Co [158]. *(Data from M. Xing, L.J. Xu, J.L. Wang, Mechanism of Co(II) adsorption by zero valent iron/graphene nanocomposite, J. Hazard. Mater. 301 (2016) 286–296. Copyright 2016, Elsevier.)*

composites toward the uptake of Eu(III) computed from Langmuir models was 96.675 mg/g. The thermodynamic parameter suggested that the Eu(III) uptake by nZVI@rGO was the endothermic and spontaneous process. The above results suggested that nZVI@rGO can be seen as a promising adsorbent for trivalent radionuclide uptake from wastewaters.

6.5 nZVI-based materials for heavy metal decontamination

Many investigations of using nZVI and nZVI-based composites are concentrated on the removal of heavy metals. Heavy metals are difficult to biodegrade and hold a strong tendency to accumulate in living organisms. Heavy metals and metalloids such as Cr, Ni, As, Cu, Zn, Pb, Cd, and Hg are of particular concern in wastewater cleanup because of their high degree of toxicity and possible carcinogenicity. nZVI and its composites have displayed high removal efficiency for all kinds of metallic ions, such as Cr(VI) [160], Ni(II) [161], As(III) [72], Pb(II) [48], Zn(II) [162], Cd(II) [163], Cu(II), and Co(II) [164]. However, the most studied is the Cr(VI) removal from water [165, 166].

6.5.1 Chromium

The most common sources of chromium pollution are usually discharged from the dye, photographic, and leather tanning industries into groundwater and surface water [167–171]. It is toxic and carcinogenic as well as having high mobility that could pose a severe threat to the natural environment [160, 172, 173]. In China, it is seen as one of the environmental pollutants. Hence, the remediation of chromium is important.

nZVI and nZVI-based materials have been demonstrated to be highly efficient materials for Cr(VI) removal. Currently, the Cr(VI) removal mechanism is based on the Cr(VI) adsorption onto the nZVI (Fe(0)) surface and then reducing Cr(VI) into Cr(III) with the subsequent oxidation of Fe(0) to Fe(III) (Eq. 6.10). Next, Cr(III) precipitates partially form as coprecipitation of $Cr_x\ Fe_{(1\text{-}x)}\ (OH)_3$ (Eq. 6.11) [174, 175] (Fig. 6.14).

$$Cr_2O_7^{2-}(aq) + 2Fe^0(s) + 14H^+(aq) \rightarrow 2Cr^{3+}(aq) + 2Fe^{3+}(aq) + 7H_2O \quad (6.17)$$

$$xCr^{3+}(aq) + (1\text{-}x)Fe^{3+}(aq) + 3H_2O \rightarrow Cr_xFe_{(1\text{-}x)}(OH)_3 \downarrow + 3H^+ \quad (6.18)$$

To date, many clay-supported nZVI composites have been employed to treat Cr(VI)-contaminated water [166, 177–181]. Typical studies on the application of clay-nZVI composites for the treatment of Cr(VI)-polluted water are shown as the following:

Al-bent-supported nZVI was synthesized by Li et al. [182] and applied to remove Cr(VI) from water. The results of XANES offered evidence that nZVI/Al-bent could entirely reduce Cr(VI) into Cr(III). The structure change of the sample nZVI/Al-bent loaded Cr(VI) displayed the formation of the Cr-Al/Si bond, indicating that the insoluble Cr(III) species can be transferred into the Al-bent surface. Therefore, the precipitates on the iron surface can be obviously decreased. Chen et al. [178] prepared nZVI-supported zeolite MCM-41 (nZVI/MCM) and applied it for Cr(VI) immobilization from water via a batch experiment. They evaluated that the obtained nZVI/MCM

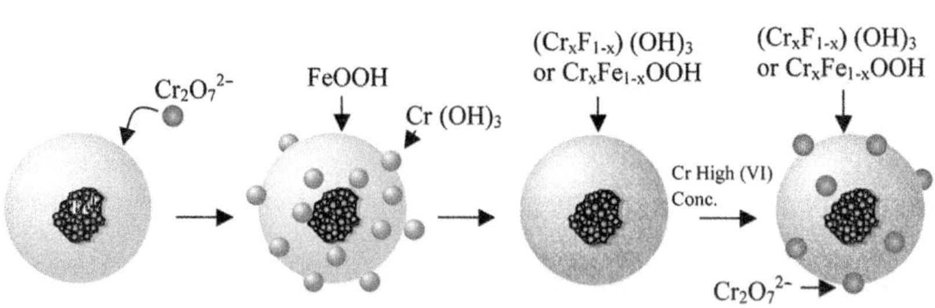

FIG. 6.14 A conceptual model for chromium immobilization with nZVI [176]. *(Data from X.Q. Li, J.S. Cao, W.X. Zhang, Stoichiometry of Cr(VI) immobilization using nanoscale zerovalent iron (nZVI): a study with high-resolution X-ray photoelectron spectroscopy (HR-XPS), Ind. Eng. Chem. Res. 47 (2008) 2131–2139. Copyright 2008, American Chemical Society.)*

displayed higher efficiency on Cr(VI) immobilization because of the synergistic impact between Cr(VI) reduction by nZVI and adsorption by MCM-41. The report on the analysis of the morphology and visualization suggested that nZVI was uniformly distributed on the MCM-41 surfaces, which decreased the aggregation of nZVI and thereby improved the activity of the reduction reaction (Fig. 6.15). The analysis of XPS and EXAFS concluded that the reaction mechanism, which showed that nZVI could facilitate reducing Cr(VI) into Cr(III), and the adsorption and reduction occurred on the surface of bare nZVI. MCM-41 served as a scavenger for the $Cr_2O_3/Cr(OH)_3$ product, which can increase the nZVI reduction activity.

Carbon-based material, including MWCNT, GO, nitrogen-doped carbon, and fluorine-doped carbon, also were applied as the nZVI support for Cr(VI) removal. Lv et al. [183] first reported that nZVI/MWCNT nanocomposites were applied to remove Cr(VI) from water. As a comparison with bare nZVI, the nZVI/MWCNT showed around 36% higher efficiency on the removal of Cr(VI). Li et al. [101] prepared the nZVI/rGOs composites with the use of the H_2/Ar plasma reduction method and utilized this to remove Cr(VI), which showed excellent water treatment performance. In the study by Wang et al [184], MIL-100(Fe) was successfully developed as a precursor for the preparation of nZVI-functionalized porous carbon (denoted as nZVI@C) for removing Cr(VI). The organic ligands in MIL-100(Fe) were converted into the porous carbon matrix. Meanwhile, the Fe-O clusters were reduced in situ to nZVI because of the impact of the carbothermal reduction. The Cr(VI) adsorption capacity of the obtained nZVI@C achieved 206 mg/g under optimal conditions. Moreover, the good magnetic property of nZVI@C makes it easy for Cr(VI) cleanup through simple magnetic separation.

6.5.2 Arsenium

Arsenic usually exists in the form of As (III) and As(V) in water. Arsenic was made famous by the WHO as the first priority contaminant. It is proposed that the concentration of arsenic is under 10 μg/L in groundwater [185–187]. Recently, to meet drinking water standards, the application of nZVI-based materials for remediation of arsenic-contaminated water has been investigated [188–193].

The arsenic removal mechanism by nZVI is rather complex, as multiple reaction processes are involved. It is worthy to note that the main arsenic removal mechanism of nZVI is involved in sorption, reduction, surface precipitation, and coprecipitation (Fig. 6.16). As far as we know, As(III) is more mobile and more toxic than As(V). Nevertheless, removing As(III) is more difficult than removing As(V). Katsoyiannis et al. [195] studied the process of As(III) oxidation and removal by nZVI in the aerated water, showing that As(III) was oxidized by the Fenton reaction and removed by adsorption on hydrous ferric oxides.

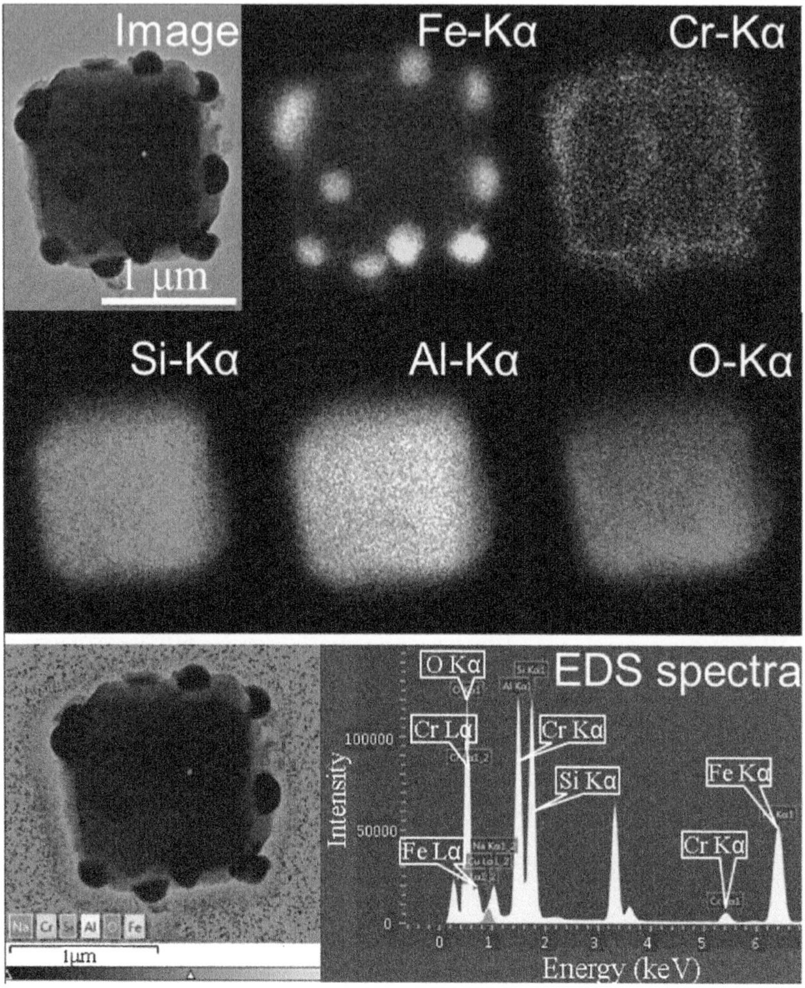

FIG. 6.15 TEM image of nZVI/MCM, and TEM-EDS elemental mapping image of Fe(Kα), Cr (Kα), Si(Kα), Al(Kα), and O(Kα) on this area. TEM-EDX elemental mapping images of Fe, Cr, Si, Al, and O on this area [178]. *(Data from Z.S. Chen, D.L. Wei, Q. Li, X.X. Wang, S.J. Yu, L. Liu, B. Liu, S.Y. Xie, J. Wang, D.Y. Chen, Macroscopic and microscopic investigation of Cr(VI) immobilization by nanoscaled zero-valent iron supported zeolite MCM-41 via batch, visual, XPS and EXAFS techniques, J. Clean. Prod. 181 (2018) 745–752. Copyright 2018, Elsevier.)*

nZVI-based composites were employed for removing As(III) and As(V). Wu et al. [9] synthesized sulfide-modified nZVI and employed it to remove aqueous As(III). The sulfidation of nZVI could enhance the As(III) removal rate, and the enhancement was up to the S/Fe molar ratio. The characterization analysis verified that S-nZVI showed the unique core-shell structure. Incorporation of sulfur into the shell of S-nZVI resulted in the thickness of

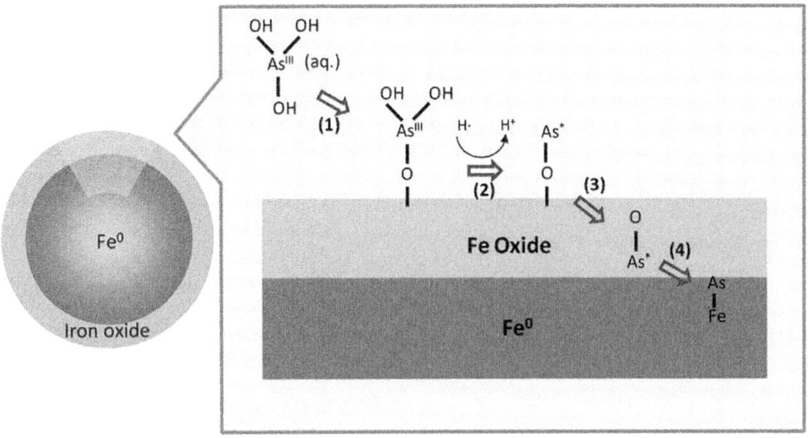

FIG. 6.16 Proposed mechanism of As(III)-nZVI reactions [194]. *(Data from W.L. Yan, R. Vasic, A.I. Frenkel, B.E. Koel, Intraparticle reduction of arsenite (As(III)) by nanoscale zerovalent iron (nZVI) investigated with in situ X-ray absorption spectroscopy, Environ. Sci. Technol. 46 (2012) 7018–7026. Copyright 2012, American Chemical Society.)*

the surface layer increasing from 5 nm to about 30 nm, suggesting that more As(III) ions could be sequestered by the S-nZVI. Sulfidation facilitated the precipitation process and inhibited outersphere complexation, which resulted in improved As removal. An nZVI@$Mg(OH)_2$ composite was also prepared and applied to remove As(V) [196]. The results showed that the variation of corrosion products of nZVI in the existence of $Mg(OH)_2$ and Mg(II) is the vital factor for enhancing As(V) adsorption. XRD analysis suggested that the impotently basic condition induced by $Mg(OH)_2$ reduces the content of γ-FeOOH and increases the content of Fe_3O_4/γ-Fe_2O_3 in the corrosion products of nZVI, and Fe_3O_4/γ-Fe_2O_3 has better As(V) adsorption affinity (Fig. 6.17). In addition, EXAFS suggested that the coordination between arsenic and iron minerals is impacted by the Mg(II) ions, resulting in the potential formation of $MgFe_2O_4$.

6.5.3 Selenium

The toxicity and solubility of Se is up to the redox conditions. The formation of soluble SeO_4^{2-} and SeO_3^{2-} is approved in oxidizing conditions, and the formation of insoluble Se(0) and Se(−II) is approved in reducing conditions [197, 198]. Reducing Se(VI) into Se(−II) can be described as [199]:

$$Se_4^{2-} \rightarrow SeO_3^{2-} \rightarrow Se^0 \rightarrow Se^{2-} \tag{6.19}$$

Reducing Se(VI) into insoluble Se(0) with the use of nZVI is thermodynamically approved, but further reducing to Se(−II) is not. Se(VI) removal

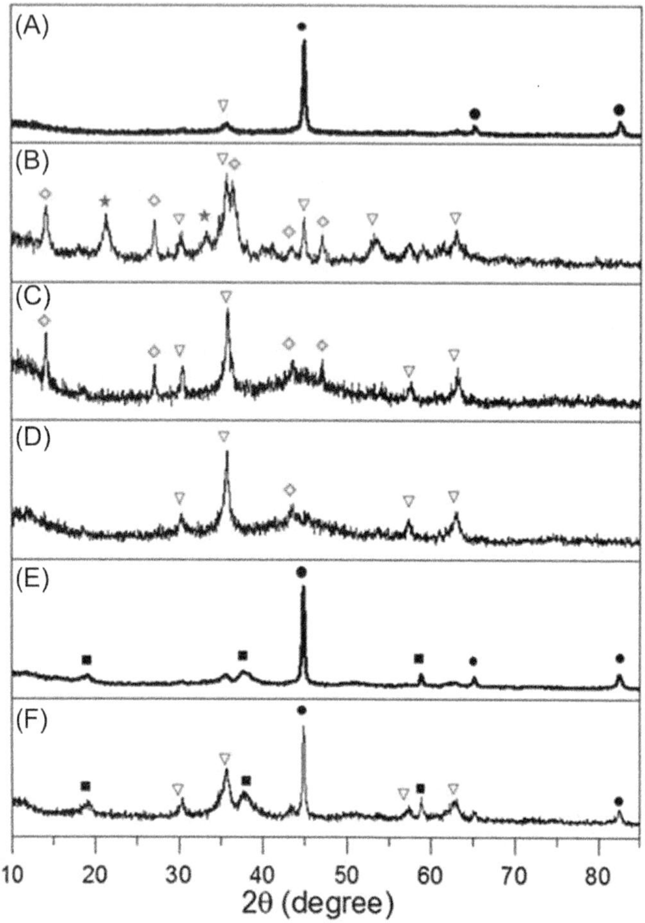

FIG. 6.17 XRD patterns of (a) nZVI before As(V) adsorption, (b) As(V)-loaded nZVI at initial pH 4, (c) As(V)-loaded nZVI at initial pH 7, (d) As(V)-loaded nZVI at initial pH 10, (e) nZVI@$Mg(OH)_2$ before As(V) adsorption, and (f) As(V)-loaded nZVI@$Mg(OH)_2$. ●, Fe(0); ■, $Mg(OH)_2$; ▽, maghemite/magnetite, γ-Fe_2O_3/Fe_3O_4; ◇, lepidocrocite, γ-FeOOH; ★, goethite, α-FeOOH [196]. *(Data from C. Wu, W.Z. Liu, J. Zhang, S.Q. Chu, Z.Q. Shi, Z. Dang, Z. Lin, Mechanisms of synergistic removal of low concentration As(V) by nZVI@Mg (OH)2 nanocomposite, J. Phys. Chem. C 121 (2017) 21411–21419. Copyright 2017, American Chemical Society.)*

is very complex with all kinds of possible removal mechanisms (i.e., reduction, complex formation, adsorption, and reoxidation). The above removal mechanisms have been shown to happen *via* investigation by SEM-EDX, XANES, and EXAFS [10, 200].

nZVI-based composites were employed to remove Se(VI) and Se(IV). Li et al. [201] used the Al-bent as the carrier of nZVI for facilitating the anionic Se(VI) removal from water. The results suggest that the Al-bent can facilitate the reduction of Se(VI) into less soluble Se(−II). Meanwhile, the Al-bent also served as the pH buffering agent, and attributed to the increasing stability and reusability of nZVI in this composite. Moreover, based on EXAFS studies, Al-bent could easily transfer the insoluble products away from the surface of Fe(0). In particular, Al-bent can activate Fe(II) produced in the reaction process, which offers nZVI the additional reduction capacity. Sheng et al. [114] prepared the nZVI/CNT composite and employed it for the Se(IV) sequestration in water. The results from XAFS analysis showed that Se(IV) could be nearly completely reduced into Se(0)/Se(−II) with nZVI/CNT. In addition, the role of pH in the sequestration of Se(IV) by nZVI/CNT was lower than that of nZVI because of the buffering impact of CNT. The remarkable performance of nZVI/CNT provides a promising material for Se(IV) detoxification from wastewater (Fig. 6.18).

The coexisting anions significantly inhibited the trace Se(VI) removal in water. To solve the problem, the millimeter-sized nZVI@D201 composite

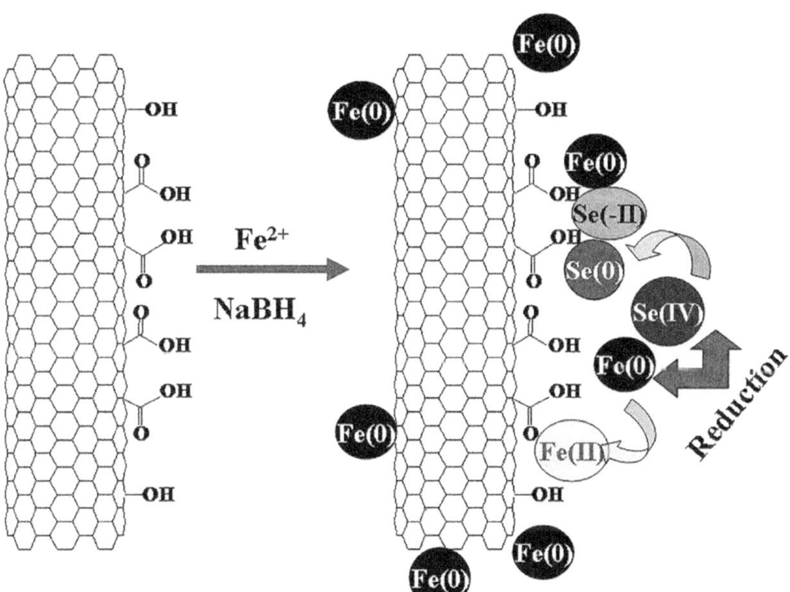

FIG. 6.18 Proposed surface complex models of Se(V) sequestrated on nZVI@CNT [114]. *(Data from G.D. Sheng, A. Alsaedi, W. Shammakh, S. Monaquel, J. Sheng, X.K. Wang, H. Li, Y.Y. Huang, Enhanced sequestration of selenite in water by nanoscale zero valent iron immobilization on carbon nanotubes by a combined batch, XPS and XAFS investigation, Carbon 99 (2016) 123–130. Copyright 2016, Elsevier.)*

was prepared by confining nZVI into the pore channels of the polymeric anion exchanger (D201) [202]. The trace Se(VI) removal by nZVI@D201 was indifferently affected by pH (3–10), dissolved oxygen concentration, coexisting anion levels, and HA levels. The analysis of the XPS spectrum showed that the Se species immobilized in nZVI@D201 was primarily Se(IV) (84.9%), indicative of the synergistic removal mechanism involved in ion exchange, adsorption, and reduction. Through several regeneration experiments, nZVI@D201 could be sustainably applied for removing Se(VI). Moreover, the column experiments showed that nZVI@D201 exhibited excellent potential in practically tracing the removal of Se(VI) in fixed-bed systems.

6.5.4 Nickel and lead

Nickel (Ni(II)) and lead (Pb(II)) can be removed by nZVI through reduction to Ni(0) and Pb(0) as well as adsorption of Ni(II) and Pb(II) [203–208]. When reacting with nZVI, Pb(II) also forms precipitation such as $Pb(OH)_2$ and oxides such as α-PbO_2, as confirmed by the analysis of XRD [48]. Detailed analysis of XPS shows that nZVI can remove Ni(II) through adsorbing Ni(II) on the surface of nZVI and then reducing it to low-toxic Ni(0) as presented here [209]:

$$\equiv FeOH + Ni^{2+} \rightarrow \equiv FeO\text{-}Ni^{+} + H^{+} \tag{6.20}$$

$$\equiv FeO\text{-}Ni^{+} + H_2O \rightarrow \equiv FeONi\text{-}OH + H^{+} \tag{6.21}$$

$$\equiv FeO\text{-}Ni^{+} + Fe^{0} + H^{+} \rightarrow \equiv FeOH\text{-}Ni + Fe^{2+} \tag{6.22}$$

nZVI-based composites were used for Ni(II) and Pb(II) removal. Fan et al. [210] reported that nZVI/rGO was fabricated *via* chemical deposition and used to remove Pb(II). The optimal reaction temperature, pH, initial Pb(II) concentration, and contact time for removing Pb(II) in this experiment were 21°C, 5, 400 mg/L, and 1 h, respectively. Moreover, removing Pb(II) on nZVI/rGO was quantitatively assessed by the use of adsorption isotherms, including the Langmuir and Freundlich isotherm models. Among them, the Langmuir isotherm showed the better correlation, and the highest adsorption capacity was calculated at 910 mg/g. The Pb(II) removal process was completed in 50 min, which could be fitted well by the pseudo-second-order kinetic model. A functionalized carbonaceous material was employed to prepare the hybrid material (Ni/Fe-HC) to achieve Pb(II) removal from water [211]. Near-complete Pb(II) removal (>99.5%) was reached in 1.5 h with the use of Ni/Fe-HC; however, beyond 85% of Pb(II) stayed in the reaction solution after 10 h using pristine HC. Hence, the adsorptive precipitation and specific adsorption governed the removal of Pb(II) by Ni/Fe-HC, regardless of partial reduction of Pb(II) to Pb(0) (Fig. 6.19). But reducing Pb(II) into

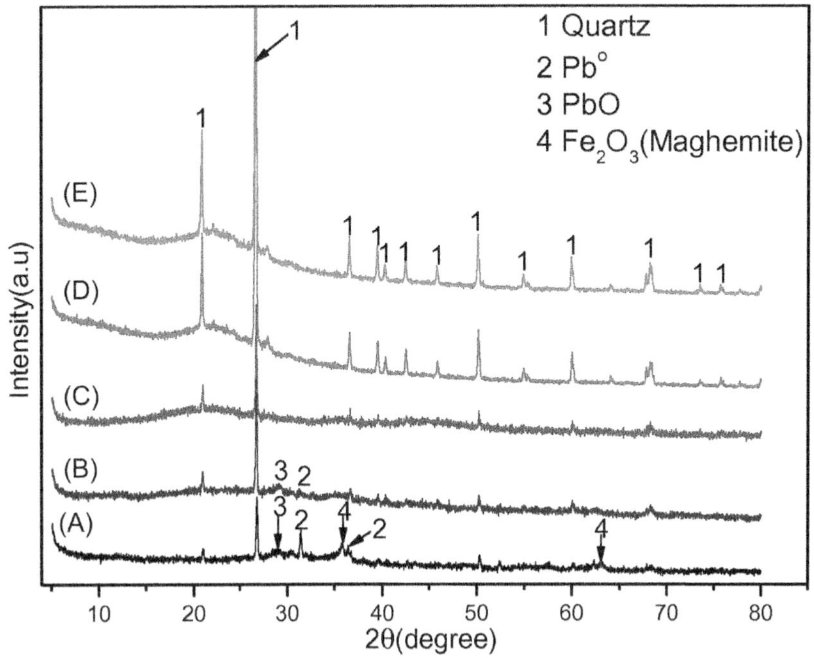

FIG. 6.19 XRD patterns of (a) spent Ni/Fe-HC with Pb(II) (500mg/L), (b) spent Ni/Fe-HC with Pb(II) (100mg/L) and initial pH = 3, (c) fresh Ni/Fe-HC, (d) spent single HC with Pb(II) (100mg/L) and initial pH = 3 and (e) spent single HC with Pb(II) (500mg/L). Conditions: t = 10h, dosage of Ni/Fe-HC = dosage of HC = 2g/L [211]. *(Data from Z. Tang, Y.F. Deng, T. Luo, Y.-S. Xu, N.M. Zhu, Enhanced removal of Pb(II) by supported nanoscale Ni/Fe on hydrochar derived from biogas residues, Chem. Eng. J. 292 (2016) 224–232. Copyright 2016, Elsevier.)*

Pb(0) was achieved by the catalytically generated atomic hydrogen rather than Ni/Fe nanoparticles as the electron donor. Ni/Fe-HC not only served as the effective adsorbent for Pb(II) removal but also functioned as the catalyst to produce atomic hydrogen. Wang et al. [161] reported that kaolinite-nZVI was synthesized by the liquid-phase reduction method with ambient atmosphere and used to adsorb Ni(II). The obtained kaolinite-nZVI particles exist on the clay surface or form the chain structure independently. The good dispersion of the swelling black sheer was noticed to fix to the kaolinite surface, and the nearly complete core-shell structure can prevent the iron oxide from added hydration. It was seen that the Ni(II) adsorption on kaolinite-nZVI was up to pH, following the pseudo-second-order kinetics model, and fitted the Langmuir isotherm model well. Kaolinite-nZVI showed the better adsorption capacity (9.24mg/g) than those of nZVI (8.69mg/g) and kaolinite (1.61mg/g). The walnut shell-supported nZVI (denoted as walnut-nZVI) was synthesized and served as the adsorbent for removing Ni(II) [212]. The equilibrium data was simulated by the Langmuir, Freundlich, and D-R models, which showed that the Langmuir isotherm was a better fit for describing the adsorption of Ni(II) than the other isotherm models. The above results suggested that the maximum adsorption capacity (327.9 mg/g for Ni(II)) is higher than other modified biomass waste adsorbents

6.5.5 Other heavy metal ions

Other metal ions of environmental importance such as Cu(II), Cd(II), Hg(II), Sb(V), Sb(III), and Ag(I) have been widely studied with nZVI technology [203, 213]. Cu(II) can also be reduced to Cu(I) with the use of nZVI, leading to Cu_2O formation. For instance, Karabelli et al. [213] prepared nZVI and used it to remove Cu(II) ions under all kinds of experimental conditions. According to the results of XPS and XRD analysis, Cu(II) ions were adsorbed mainly via the redox mechanism that led to the formation of Cu_2O and Cu(0). Liu et al. [214] reported that the D201-nZVI nanocomposite was applied for the removal of EDTA-chelated Cu(II) in water. The analysis of HPLC showed that the EDTA-chelated Cu(II) was gradually replaced by Fe(III) produced from the oxidation of Fe(0). Following that, the released Cu(II) was removed via the process of adsorption/precipitation and further reduced into the Cu(0) species, as formed by the analysis of XPS spectra.

Cd(II) could be removed onto nZVI and its composites via adsorption and coprecipitation. Boparai et al. [215] reported that nZVI particles were applied to study the Cd(II) removal in the concentration range from 25 to 450 mg/L. The highest removal capacity of nZVI toward Cd(II) was 769.2 mg/g at 297 K. Thermodynamic parameters (i.e., ΔG^0, ΔH^0, and ΔS^0) were also assessed. The whole adsorption process was endothermic and spontaneous in the natural environment. EDX analysis suggested the existence of Cd(II) ions on the surface of nZVI. Zhang et al. [216] used TEM, XRD, and XPS methods to characterize nZVI before and after removing Cd(II) (Fig. 6.20). Data further suggest that Cd(II) is sequestrated *via* adsorption or surface complex formation with no obvious Cd(II) reduction. In the investigation of Xue et al. [217], using rhamnolipid (RL)-coated nZVI (R-nZVI) in removing Cd(II) was studied. It verified that R-nZVI was highly effective in reducing Cd(II) to a stable fraction, along with reducing the mobility of metals. The increase of pH in nZVI and R-nZVI samples after treating the sediment suggested the related mechanism for Cd(II) immobilization. Enhancing the activities of urease and catalase, along with augmenting the incubation time, shows some degree of recovery in sediment metabolic function. The study offers new insights into the possible impacts of the application of R-nZVI on the Cd(II) immobilization in contaminated river sediments.

Ag(I) and Au(III) can also be immobilized by chemical reduction to their elemental forms. Zhang et al. [104] used nZVI to immobilize Ag(I) from water. Various characterization techniques such as TEM, XRD, and HR-XPS showed

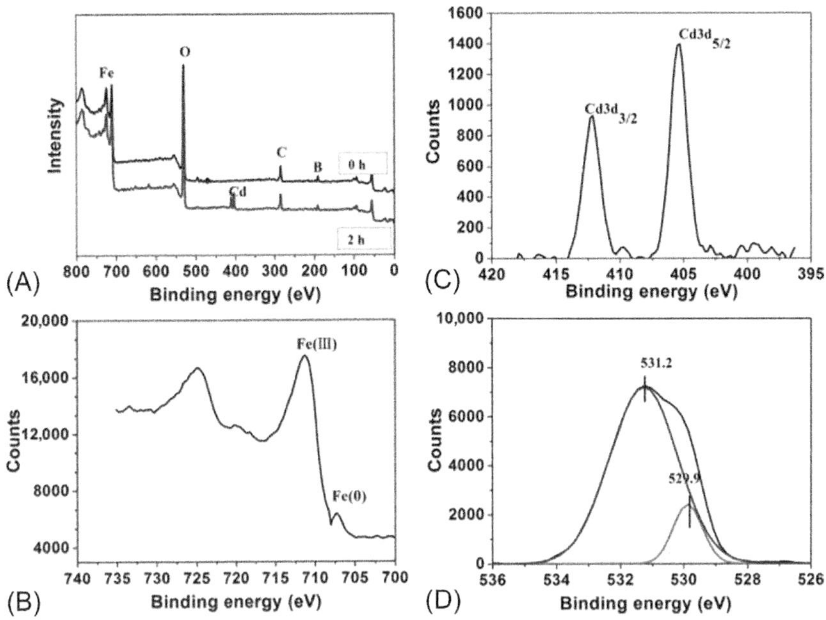

FIG. 6.20 (A) XPS broad survey of nZVI: freshly and reacted with 100mg/L Cd(II) for 2h, (B) XPS spectrum of Fe 2p of reacted nZVI, (C) Cd 3d XPS spectrum of nZVI reacted with 100mg/L cadmium for 2h at pH 6.5 and 25°C, (D) O 1s XPS spectrum on the surface of reacted nZVI [216]. *(Data from Y.L. Zhang, Y.T. Li, C.M. Dai, X.F. Zhou, W.X. Zhang, Sequestration of Cd(II) with nanoscale zero-valent iron (nZVI): characterization and test in a two-stage system, Chem. Eng. J. 244 (2014) 218–226. Copyright 2014, Elsevier.)*

that nZVI was successfully oxidized to the lepidocrocite/magnetite/maghemite and verified the formation of nanocrystalline silver. The formative Ag(0) nanostructures are fractal, spherical, and dendritic, respectively, in morphology analysis by FE-TEM images. Lin et al. [60] used the STEM-XEDS technique to analyze the enrichment behavior of Au(III) and Ag(I) ions from wastewater with nZVI. After reacting for just 10 s with Au(III), the used nZVI still holds the core-shell structure with the oxide shell decorated with brighter Au nanoparticles (Fig. 6.21A). The chemical reactivity of nZVI is superior at the initial stage with the movement of Au from aqueous solution into nZVI restricted by the mass transfer of Au(III) ions to the surface of nZVI. With rapid reduction of Au(III) ions and simultaneous Fe dissolution, ionization, and aging, pristine Au nanoparticles and nanoclusters are further dispersed on the immensely oxidized nZVIs (Fig. 6.21B and C). As a comparision with Au(III) ions, Ag(I) ions have correspondingly lower standard potentials and also require fewer electrons to be reduced into the homologous metallic phase. During the process of reacting with nZVI, Ag(I) ions are sorbed and reduced on the nZVI shell (Fig. 6.21E). The Ag seeds have been noticed to first deposit at the corners with sharp edges because of the space superiority of iron chains in competition for

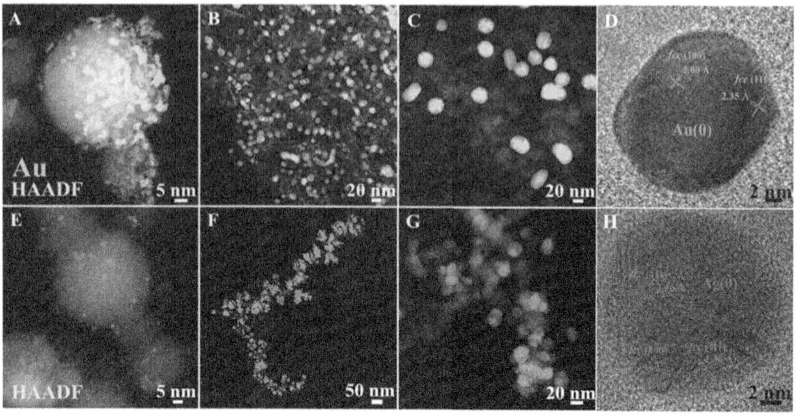

FIG. 6.21 Enrichment of Au and Ag from wastewater. STEM-XEDS mapping-HAADF image color overlays of the used nZVI after reactions with Au (A–C) and Ag (E–G) [60]. *(Data from L. Ling, X.Y. Huang, W.X. Zhang, Enrichment of precious metals from wastewater with core-shell nanoparticles of iron, Adv. Mater. 30 (2018) 1705703. Copyright 2017, Wiley.)*

the diffusing Ag(I) ions. The Ag seedlings sculpture the surface of the partly oxidized Fe(0) nanoparticles, and further accelerate the oxidation and evolution of nZVI (Fig. 6.21F and G). With high-resolution TEM images, it is confirmed that the brighter nanoparticles and clusters in the HAADF images are pristine metallic Au (Fig. 6.21D) and Ag (Fig. 6.21H). The decoration of Au and Ag nanoparticles on the surface of nZVI offer the vital evidence for the reaction mechanisms of precious-metal species enrichment, and also suggests a cost-effective method for the preparation, stabilization, and removal of Au and Ag nanoparticles.

The adsorption behavior of various rare earth element (REE) ions (i.e., Sc, Y, La-Lu) onto nZVI particles has also been studied in the report from Crane et al. [106]. The adsorption isotherms were carried out using REE-bearing acid mine drainage and a series of different concentrations of complex REE solutions. The highest adsorption capacity of Yb and La ions was calculated at 410 and 61 mg/g, respectively. Removing aqueous REE ions to ultratrace concentrations was also obtained after 30 min exposure in $\geq$0.5 g/L nZVI suspensions to 10 mg/L aqueous REE solutions. Rapid and near-complete removal ability was confirmed by exposing nZVI to the acid mine drainage. Nevertheless, the high concentration of nZVI was needed, along with removing the various REEs to $<$1 μg/L when exposed to nZVI at 4.0 g/L for 30 min. In all of systems, nZVI was selective for removing a high concentration of REE ions, combined with the HRTEM-EDS and XPS analysis, the mainly mechanism is surface mediated precipitation. All in all, the results verify that nZVI can be seen as having potential as an efficient and versatile agent for concurrent REE ion recovery and fractionation.

6.6 Conclusions

The application of nZVI and its composites for environmental remediation, especially for the adsorption of heavy metals and radionuclides, has attracted considerable attention. In this chapter, the advantages of using nZVI as the adsorbent or as the basis for new adsorbents lies primarily in the large surface area and the reduction property. The application of modified nZVI and supported-nZVI composites could lead to better removal performance of nZVI because they successfully restrict nZVI agglomeration, and supplied the nZVI nanoparticles with high reactivity directly into water contaminants. Many adsorption interactions between the nZVI-based materials and the metal ions have been characterized by the pseudo-second-order kinetic model. Adsorption isotherms usually are fitted by the Langmuir model and only the few number of cases are suitable to the Freundlich approach. Other kinetic models and isotherm models have been tested only sparingly with various degrees of success. Adsorption, reduction, precipitation, and coprecipitation phenomena are responsible for adsorbing radioactive and toxic metal ions by nZVI-based materials. Various analytical techniques involving XRD, XPS, FTIR, and EXAFS as well as Mössbauer spectroscopy and STEM-XEDS have been used and combined to explore the adsorption mechanisms.

Despite the efforts that have been made, there is still a need to find the evidence that holds the near-complete information upon which we can predict the mobility, reactivity, fate, and ecological impact of nZVI-based materials. Incorporation of this evidence and the expected reductions in production costs will be priceless for the successful application of nZVI and its composites in land remediation. Furthermore, from an economic point of view, more work should be done to recycle nZVI-based materials with better desorption and effective regeneration.

References

[1] J. Li, X.X. Wang, G.X. Zhao, C.L. Chen, Z.F. Chai, A. Alsaedi, T. Hayat, X.K. Wang, Metal-organic framework-based materials: superior adsorbents for the capture of toxic and radioactive metal ions, Chem. Soc. Rev. 47 (2018) 2322–2356.

[2] Y.D. Zou, X.X. Wang, A. Khan, P.Y. Wang, Y.H. Liu, A. Alsaedi, T. Hayat, X.K. Wang, Environmental remediation and application of nanoscale zero-valent iron and its composites for the removal of heavy metal ions: a review, Environ. Sci. Technol. 50 (2016) 7290–7304.

[3] G.X. Zhao, J.X. Li, X.M. Ren, C.L. Chen, X.K. Wang, Few-layered graphene oxide nanosheets as superior sorbents for heavy metal ion pollution management, Environ. Sci. Technol. 45 (2011) 10454–10462.

[4] J. Wen, Y. Fang, G.M. Zeng, Progress and prospect of adsorptive removal of heavy metal ions from aqueous solution using metal-organic frameworks: a review of studies from the last decade, Chemosphere 201 (2018) 627–643.

[5] S.E. Bailey, T.J. Olin, R.M. Bricka, D.D. Adrian, A review of potentially low-cost sorbents for heavy metals, Water Res. 33 (1999) 2469–2479.

[6] J.F. Liu, Z.S. Zhao, G.B. Jiang, Coating Fe3O4 magnetic nanoparticles with humic acid for high efficient removal of heavy metals in water, Environ. Sci. Technol. 42 (2008) 6949–6954.

[7] W. Yantasee, C.L. Warner, T. Sangvanich, R.S. Addleman, T.G. Carter, R.J. Wiacek, G.E. Fryxell, C. Timchalk, M.G. Warner, Removal of heavy metals from aqueous systems with thiol functionalized superparamagnetic nanoparticles, Environ. Sci. Technol. 41 (2007) 5114–5119.

[8] K.R. Zhu, C.L. Chen, H. Xu, Y. Gao, X.L. Tan, A. Alsaedi, T. Hayat, Cr(VI) reduction and immobilization by core-double-shell structured magnetic polydopamine@zeolitic idazolate frameworks-8 microspheres, ACS Sustain. Chem. Eng. 5 (2017) 6795–6802.

[9] D.L. Wu, S.H. Peng, K.L. Yan, B.B. Shao, Y. Feng, Y.L. Zhang, Enhanced As(III) sequestration using sulfide-modified nano-scale zero-valent iron with a characteristic core-shell structure: sulfidation and as distribution, ACS Sustain. Chem. Eng. 6 (2018) 3039–3048.

[10] K.C. Mondal, G. Jegadeesan, S.B. Lalvni, Removal of selenate by Fe and NiFe nanosized particles, Ind. Eng. Chem. Res. 43 (2004) 4922–4934.

[11] C.L. Chen, J. Hu, D.D. Shao, J.X. Li, X.K. Wang, Adsorption behavior of multiwall carbon nanotube/iron oxide magnetic composites for Ni(II) and Sr(II), J. Hazard. Mater. 164 (2009) 923–928.

[12] W.S.W. Ngah, M.A.K.M. Hanafiah, Removal of heavy metal ions from wastewater by chemically modified plant wastes as adsorbents: a review, Bioresour. Technol. 99 (2008) 3935–3948.

[13] J. Li, S.W. Zhang, C.L. Chen, G.X. Zhao, X. Yang, J.X. Li, X.K. Wang, Removal of Cu(II) and fulvic acid by graphene oxide nanosheets decorated with Fe_3O_4 nanoparticles, ACS Appl. Mater. Interfaces 4 (2012) 4991–5000.

[14] M. Ramezanzadeh, M. Asghari, B. Ramezanzadeh, G. Bahlakeh, Fabrication of an efficient system for Zn ions removal from industrial wastewater based on graphene oxide nanosheets decorated with highly crystalline polyaniline nanofibers (GO-PANI): experimental and ab initio quantum mechanics approaches, Chem. Eng. J. 337 (2018) 385–397.

[15] K.R. Zhu, C.L. Chen, M.W.C. Xu, K. Chen, X.L. Tan, M. Wakeel, N.S. Alharbi, In situ carbothermal reduction synthesis of Fe nanocrystals embedded into N-doped carbon nanospheres for highly efficient U(VI) adsorption and reduction, Chem. Eng. J. 331 (2018) 395–405.

[16] Y. Gao, K. Chen, X.L. Tan, X.K. Wang, A. Alsaedi, T. Hayat, C.L. Chen, Interaction mechanism of Re(VII) with zirconium dioxide nanoparticles archored onto reduced graphene oxides, ACS Sustain. Chem. Eng. 5 (2017) 2163–2171.

[17] X.L. Tan, M. Fang, L.Q. Tan, H.N. Liu, X.S. Ye, T. Hayat, X.K. Wang, Core-shell hierarchical $C@Na_2Ti_3O_7{\cdot}9H_2O$ nanostructures for the efficient removal of radionuclides, Environ. Sci. Nano 5 (2018) 1140–1149.

[18] N. Meunier, P. Drogui, C. Montane, R. Hausler, G. Mercier, J.F. Blais, Comparison between electrocoagulation and chemical precipitation for metals removal from acidic soil leachate, J. Hazard. Mater. 137 (2006) 581–590.

[19] D. Mohan, C.U. Pittman Jr., Arsenic removal from water/wastewater using adsorbents—a critical review, J. Hazard. Mater. 142 (2007) 1–53.

[20] M.J. McKie, L. Taylor-Edmonds, S.A. Andrews, R.C. Andrews, Engineered biofiltration for the removal of disinfection by-product precursors and genotoxicity, Water Res. 81 (2015) 196–207.

[21] Z.L. Li, H. Li, X.Y. Guan, J.J. Tang, Y.R. Yusran, Z. Li, M. Xue, Q.R. Fang, Y.S. Yan, V. Valtchev, S.L. Qiu, Three-dimensional ionic covalent organic frameworks for rapid, reversible, and selective ion exchange, J. Am. Chem. Soc. 139 (2017) 17771–17774.

[22] M.L. Feng, D. Sarma, X.H. Qi, K.Z. Du, X.Y. Huang, M.G. Kanatzidis, Efficient removal and recovery of uranium by a layered organic-inorganic hybrid thiostannate, J. Am. Chem. Soc. 138 (2016) 12578–12585.

[23] Z.J. Li, Z.W. Huang, W.L. Guo, L. Wang, L.R. Zheng, Z.F. Chai, W.Q. Shi, Enhanced photocatalytic removal of uranium(VI) from aqueous solution by magnetic TiO_2/Fe_3O_4 and its graphene composite, Environ. Sci. Technol. 51 (2017) 5666–5674.

[24] X.N. Liu, P.H. Du, W.Y. Pan, C.Y. Dang, T.W. Qian, H.F. Liu, W. Liu, D.Y. Zhao, Immobilization of uranium(VI) by niobate/titanate nanoflakes heterojunction through combined adsorption and solar-light-driven photocatalytic reduction, Appl. Catal. B Environ. 231 (2018) 11–22.

[25] C. Santhosh, V. Velmurugan, G. Jacob, S.K. Jeong, A.N. Grace, A. Bhatnagar, Role of nanomaterials in water treatment applications: a review, Chem. Eng. J. 306 (2016) 1116–1137.

[26] H.S. Zhu, X.L. Tan, L.Q. Tan, C.L. Chen, N.S. Alharbi, T. Hayat, M. Fang, X.K. Wang, Biochar derived from sawdust embedded with molybdenum disulfide for highly selective removal of Pb^{2+}, ACS Appl. Nano Mater. 1 (2018) 2689–2698.

[27] T.Z. Tong, S. Zhao, C.H. Boo, S.M. Hashmi, M.C. Elimelech, Relating silica scaling in reverse osmosis to membrane surface properties, Environ. Sci. Technol. 51 (2017) 4396–4406.

[28] S.L. Qiu, M. Xue, G.S. Zhu, Metal-organic framework membranes: from synthesis to separation application, Chem. Soc. Rev. 43 (2014) 6116–6140.

[29] L. Ke, P.F. Li, X. Wu, S.J. Jiang, M.B. Luo, Y.H. Liu, Z.G. Le, C.Z. Sun, S.Q. Song, Graphene-like sulfur-doped g-C_3N_4 for photocatalytic reduction elimination of UO_2^{2+} under visible light, Appl. Catal. B Environ. 205 (2017) 319–326.

[30] H.C. Gao, Y.M. Sun, J.J. Zhou, R. Xu, H.W. Duan, Mussel-inspired synthesis of polydopamine-functionalized graphene hydrogel as reusable adsorbents for water purification, ACS Appl. Mater. Interfaces 5 (2013) 425–432.

[31] D.R. Tonini, D.A. Gauvin, R.W. Soffel, W.P. Freeman, Achieving low mercury concentrations in chlor-alkali wastewaters, Environ. Prog. Sustain. 22 (2003) 1547–5921.

[32] P. Veerakumar, P. Thanasekaran, K.C. Lin, S.B. Liu, Biomass derived sheet-like carbon/palladium nanocomposite: an excellent opportunity for reduction of toxic hexavalent chromium, ACS Sustain. Chem. Eng. 5 (2017) 5302–5312.

[33] C.H. Lu, R.Y. Chen, X. Wu, M.F. Fan, Y.H. Liu, Z.G. Le, S.J. Jiang, S.Q. Song, Boron doped g-C_3N_4 with enhanced photocatalytic UO_2^{2+} reduction performance, Appl. Surf. Sci. 360 (2016) 1016–1022.

[34] S.B. Yang, J. Hu, C.L. Chen, D.D. Shao, X.K. Wang, Mutual effects of Pb(II) and humic acid adsorption on multiwalled carbon nanotubes/polyacrylamide composites from aqueous solutions, Environ. Sci. Technol. 45 (2011) 3621–3627.

[35] X.M. Ren, C.L. Chen, M. Nagatsu, X.K. Wang, Carbon nanotubes as adsorbents in environmental pollution management: a review, Chem. Eng. J. 170 (2011) 395–410.

[36] Y.B. Sun, J.X. Li, X.K. Wang, The retention of uranium and europium onto sepiolite investigated by macroscopic, spectroscopic and modeling techniques, Geochim. Cosmochim. Acta 140 (2014) 621–643.

[37] S. Babel, T.A. Kurniawan, Low-cost adsorbents for heavy metals uptake from contaminated water: a review, J. Hazard. Mater. 97 (2003) 219–243.

[38] D.L. Zhao, G.D. Sheng, J. Hu, C.L. Chen, X.K. Wang, The adsorption of Pb(II) on Mg_2Al layered double hydroxide, Chem. Eng. J. 171 (2011) 167–174.

[39] Y.Z. Hu, X.X. Wang, Y.D. Zou, T. Wen, X.L. Wang, A. Alsaedi, T. Hayat, X.K. Wang, Superior sorption capacities of Ca-Ti and Ca-Al bimetallic oxides for U(VI) from aqueous solutions, Chem. Eng. J. 316 (2017) 419–428.

[40] Z.M. Lv, S.M. Yang, H.S. Zhu, L. Chen, N.S. Alharbi, M. Wakeel, A. Wahid, C.L. Chen, Highly efficient removal of As(V) by using NiAl layered double oxide composites, Appl. Surf. Sci. 448 (2018) 599–608.
[41] H. Chen, X.X. Wang, J.X. Li, X.K. Wang, Cotton derived carbonaceous aerogels for the efficient removal of organic pollutants and heavy metal ions, J. Mater. Chem. A 3 (2015) 6073–6081.
[42] Y. Shen, X.Y. Zhu, B.L. Chen, Size effects of graphene oxide nanosheets on the construction of three-dimensional graphene-based macrostructures as adsorbents, J. Mater. Chem. A 4 (2016) 12106–12118.
[43] D.L. Zhao, X. Gao, C.N. Wu, R. Xie, S.J. Feng, C.L. Chen, Facile preparation of amino functionalized graphene oxide decorated with Fe_3O_4 nanoparticles for the adsorption of Cr(VI), Appl. Surf. Sci. 384 (2016) 1–9.
[44] Y. Mu, F. Jia, Z.H. Ai, L.Z. Zhang, Iron oxide shell mediated environmental remediation properties of nano zero-valent iron, Environ. Sci. Nano 4 (2017) 27–45.
[45] Y. Xie, D.M. Cwiertny, Influence of anionic cosolutes and pH on nanoscale zerovalent iron longevity: time scales and mechanisms of reactivity loss toward 1,1,1,2-tetrachloroethane and Cr(VI), Environ. Sci. Technol. 46 (2012) 8365–8373.
[46] D. Fan, R.P. Anitori, B.M. Tebo, P.G. Tratnyek, J.S. Lezama Pacheco, R.K. Kukkadapu, M.H. Engelhard, M.E. Bowden, L. Kovarik, B.W. Arey, Reductive sequestration of pertechnetate $^{99}TcC_4^-$ by nano zerovalent iron (nZVI) transformed by abiotic sulfide, Environ. Sci. Technol. 47 (2013) 5302–5310.
[47] D. Phillips, T.V. Nooten, L. Bastiaens, M. Russell, K. Dickson, S. Plant, J. Ahad, T. Newton, T. Elliot, R. Kalin, Ten year performance evaluation of a field-scale zero-valent iron permeable reactive barrier installed to remediate trichloroethene contaminated groundwater, Environ. Sci. Technol. 44 (2010) 3861–3869.
[48] M.H. Liu, Y.H. Wang, L.T. Chen, Y. Zhang, Z. Lin, $Mg(OH)_2$ supported nanoscale zero valent iron enhancing the removal of Pb(II) from aqueous solution, ACS Appl. Mater. Interfaces 7 (2015) 7961–7969.
[49] S. Tasharrofi, S.S. Hassani, H. Taghdisian, Z. Sobat, Environmentally friendly stabilized nZVI-composite for removal of heavy metals, in: New Polymer Nanocomposites for Environmental Remediation, Elsevier, 2018, pp. 623–642.
[50] F. Fu, D.D. Dionysiou, H. Liu, The use of zero-valent iron for groundwater remediation and wastewater treatment: a review, J. Hazard. Mater. 267 (2014) 194–205.
[51] R.W. Gillham, S.F. O'Hannesin, Enhanced degradation of halogenated aliphatics by zero-valent iron, Groundwater 32 (1994) 958–967.
[52] M.M. Scherer, S. Richter, R.L. Valentine, P.J.J. Alvarez, Chemistry and microbiology of permeable reactive barriers for in situ groundwater clean up, Crit. Rev. Environ. Sci. Technol. 30 (2000) 363–411.
[53] X.Q. Li, D.W. Elliott, W.X. Zhang, Zero-valent iron nanoparticles for abatement of environmental pollutants: materials and engineering aspects, Crit. Rev. Solid State Mater. Sci. 31 (2006) 111–122.
[54] J. Theron, J.A. Walker, T.E. Cloete, Nanotechnology and water treatment: applications and emerging opportunities, Crit. Rev. Microbiol. 34 (2008) 43–69.
[55] R.T. Wilkin, C. Su, R.G. Ford, C.J. Paul, Chromium-removal processes during groundwater remediation by a zerovalent iron permeable reactive barrier, Environ. Sci. Technol. 39 (2005) 4599–4605.
[56] B. Karn, T. Kuiken, M. Otto, Nanotechnology and in situ remediation: a review of the benefits and potential risks, Environ. Health Perspect. 117 (2009) 1813–1831.

[57] N.C. Mueller, J. Braun, J. Bruns, M. Cernik, P. Rissing, D. Rickerby, B. Nowack, Application of nanoscale zero valent iron (NZVI) for groundwater remediation in Europe, Environ. Sci. Pollut. Res. Int. 19 (2012) 550–558.

[58] A. Liu, J. Liu, J. Han, W.X. Zhang, Evolution of nanoscale zero-valent iron (nZVI) in water: microscopic and spectroscopic evidence on the formation of nano- and micro-structured iron oxides, J. Hazard. Mater. 322 (2017) 129–135.

[59] E.J. Reardon, R. Fagan, J.L. Vogan, A. Przepiora, Anaerobic corrosion reaction kinetics of nanosized iron, Environ. Sci. Technol. 42 (2008) 2420–2425.

[60] L. Ling, X.Y. Huang, W.X. Zhang, Enrichment of precious metals from wastewater with core-shell nanoparticles of iron, Adv. Mater. 30 (2018) 1705703.

[61] K. Sohn, S.W. Kang, S. Ahn, M. Woo, S.K. Yang, Fe(0) nanoparticles for nitrate reduction: stability, reactivity, and transformation, Environ. Sci. Technol. 40 (2006) 5514–5519.

[62] B.C. Reinsch, B. Forsberg, R.L. Penn, C.S. Kim, G.V. Lowry, Chemical transformations during aging of zerovalent iron nanoparticles in the presence of common groundwater dissolved constituents, Environ. Sci. Technol. 44 (2010) 3455–3461.

[63] R.A. Crane, T.B. Scott, Nanoscale zero-valent iron: future prospects for an emerging water treatment technology, J. Hazard. Mater. 211–212 (2012) 112–125.

[64] L. Ling, W.X. Zhang, Reactions of nanoscale zero-valent iron with Ni(II): three-dimensional tomography of the "Hollow Out" effect in a single nanoparticle, Environ. Sci. Technol. Lett. 1 (2014) 209–213.

[65] R.A. Crane, M. Dickinson, I.C. Popescu, T.B. Scott, Magnetite and zero-valent iron nanoparticles for the remediation of uranium contaminated environmental water, Water Res. 45 (2011) 2931–2942.

[66] M. Dickinson, T.B. Scott, The effect of vacuum annealing on the remediation abilities of iron and iron-nickel nanoparticles, J. Nanopart. Res. 13 (2011) 3699–3711.

[67] M. Krajewski, W.S. Lin, H.M. Lin, K. Brzozka, S. Lewinska, N. Nedelko, A. Slawska-Waniewska, J. Borysiuk, D. Wasik, Structural and magnetic properties of iron nanowires and iron nanoparticles fabricated through a reduction reaction, Beilstein J. Nanotechnol. 6 (2015) 1652–1660.

[68] P. Jiemvarangkul, W.X. Zhang, H.L. Lien, Enhanced transport of polyelectrolyte stabilized nanoscale zero-valent iron (nZVI) in porous media, Chem. Eng. J. 170 (2011) 482–491.

[69] D.E. Meyer, D. Bhattacharyya, Impact of membrane immobilization on particle formation and trichloroethylene dechlorination for bimetallic Fe/Ni nanoparticles in cellulose acetate membranes, J. Phys. Chem. B 111 (2007) 7142–7154.

[70] B. Schrick, B.W. Hydutsky, J.L. Blough, T.E. Mallouk, Delivery vehicles for zerovalent metal nanoparticles in soil and groundwater, Chem. Mater. 16 (2004) 2187–2193.

[71] A.R. Petosa, D.P. Jaisi, I.R. Quevedo, M. Elimelech, N. Tufenkji, Aggregation and deposition of engineered nanomaterials in aquatic environments: role of physicochemical interactions, Environ. Sci. Technol. 44 (2010) 6532–6549.

[72] S. Bhowmick, S. Chakraborty, P. Mondal, W.V. Renterghem, S. Van den Berghe, G. Roman-Ross, D. Chatterjee, M. Iglesias, Montmorillonite-supported nanoscale zero-valent iron for removal of arsenic from aqueous solution: kinetics and mechanism, Chem. Eng. J. 243 (2014) 14–23.

[73] Y.B. Zeng, H. Walker, Q.Z. Zhu, Reduction of nitrate by NaY zeolite supported Fe, Cu/Fe and Mn/Fe nanoparticles, J. Hazard. Mater. 324 (2017) 605–616.

[74] G.D. Sheng, P.J. Yang, Y.N. Tang, Q.Y. Hu, H. Li, X.M. Ren, B.W. Hu, X.K. Wang, Y.Y. Huang, New insights into the primary roles of diatomite in the enhanced sequestration

of UO_2^{2+} by zerovalent iron nanoparticles: an advanced approach utilizing XPS and EXAFS, Appl. Catal. B Environ. 193 (2016) 189–197.
[75] J.W. Chen, Z.M. Xiu, G.V. Lowry, P.J. Alvarez, Effect of natural organic matter on toxicity and reactivity of nano-scale zero-valent iron, Water Res. 45 (2011) 1995–2001.
[76] T. Phenrat, Y. Liu, R.D. Tilton, G.V. Lowry, Adsorbed polyelectrolyte coatings decrease Fe0 nanoparticle reactivity with TCE in water: conceptual model and mechanisms, Environ. Sci. Technol. 43 (2009) 1507–1514.
[77] T. Masciangioli, W.X. Zhang, Peer reviewed: environmental technologies at the nanoscale, Environ. Sci. Technol. 37 (2003) 102A–108A.
[78] T. Phenrat, N. Saleh, K. Sirk, R.D. Tilton, G.V. Lowry, Aggregation and sedimentation of aqueous nanoscale zerovalent iron dispersions, Environ. Sci. Technol. 41 (2007) 284–290.
[79] P.G. Tratnyek, R.L. Johnson, Nanotechnologies for environmental cleanup, Nano Today 1 (2006) 44–48.
[80] R.A. Crane, H. Pullin, J. Macfarlane, M. Silion, I.C. Popescu, M. Andersen, V. Calen, T.B. Scott, Field application of iron and iron-nickel nanoparticles for the ex situ remediation of a uranium-bearing mine water effluent, J. Environ. Eng. 40 (2015) 04015011.
[81] S.R. Kanel, D. Nepal, B. Manning, H. Choi, Transport of surface-modified iron nanoparticle in porous media and application to arsenic(III) remediation, J. Nanopart. Res. 9 (2007) 725–735.
[82] J. Biswal, S.P. Ramnani, S. Shirolikar, S. Sabharwal, Synthesis of guar-gum-stabilized nanosized silver clusters with γ radiation, J. Appl. Polym. Sci. 114 (2010) 2348–2355.
[83] T. Phenrat, N. Saleh, K. Sirk, H.J. Kim, R.D. Tilton, G.V. Lowry, Stabilization of aqueous nanoscale zerovalent iron dispersions by anionic polyelectrolytes: adsorbed anionic polyelectrolyte layer properties and their effect on aggregation and sedimentation, J. Nanopart. Res. 10 (2008) 795–814.
[84] R. Bardos, B. Bone, D. Elliott, N. Hartog, J. Henstock, P. Nathanail, A risk/benefit approach to the application of iron nanoparticles for the remediation of contaminated sites in the environment, in: Aiaa/asme/sae/asee Joint Propulsion Conference and Exhibit, 2011, pp. 425–426.
[85] Royal Society and Royal Academy of Engineering (UK), Nanoscience and Nanotechnologies: Opportunities and Uncertainties, Available from: http:/www.nanotec.org.uk/finalReport.htm, 2004.
[86] S.H. Xiang, W.C. Cheng, X.Q. Nie, C.C. Ding, F.C. Yi, A.M. Asiri, H.M. Marwani, Zero-valent iron-aluminum for the fast and effective U(VI) removal, J. Taiwan Inst. Chem. Eng. 85 (2018) 186–192.
[87] C. Jing, Y.L. Li, S. Landsberger, Review of soluble uranium removal by nanoscale zero valent iron, J. Environ. Radioact. 164 (2016) 65–72.
[88] Y.B. Hu, M. Zhang, R. Qiu, X.Y. Li, Encapsulating nanoscale zero-valent iron with a soluble $Mg(OH)_2$ shell for improved mobility and controlled reactivity release, J. Mater. Chem. A 6 (2018) 2517–2526.
[89] G. Vilardi, T. Mpouras, D. Dermatas, N. Verdone, A. Polydera, L.D. Palma, Nanomaterials application for heavy metals recovery from polluted water: the combination of nano zero-valent iron and carbon nanotubes. Competitive adsorption non-linear modeling, Chemosphere 201 (2018) 716–729.
[90] B.W. Hu, X. Mei, X. Li, J. Hu, D. Xu, J.Y. Ma, Y.Y. Huang, Decontamination of U(VI) from nZVI/CNF composites investigated by batch, spectroscopic and modeling techniques, J. Mol. Liq. 237 (2017) 1–9.

[91] T. Wen, X.X. Wang, J. Wang, Z.S. Chen, J.X. Li, J. Hu, T. Hayat, A. Alsaedi, B. Grambow, X.K. Wang, A strategically designed porous magnetic N-doped Fe/Fe_3C@C matrix and its highly efficient uranium(VI) remediation, Inorg. Chem. Front. 3 (2016) 1227–1235.

[92] H.B. Gu, H. Lou, D. Ling, B. Xiang, Z.H. Guo, Polystyrene controlled growth of zerovalent nanoiron/magnetite on a sponge-like carbon matrix towards effective Cr(VI) removal from polluted water, RSC Adv. 6 (2016) 110134–110145.

[93] X.S. Lv, X.Q. Xue, G.M. Jiang, D.L. Wu, T.T. Sheng, H.Y. Zhou, X.H. Xu, Nanoscale zero-valent iron (nZVI) assembled on magnetic Fe_3O_4/graphene for Chromium(VI) removal from aqueous solution, J. Colloid Interface Sci. 417 (2014) 51–59.

[94] K.S. Prasad, P. Gandhi, K. Selvaraj, Synthesis of green nano iron particles (GnIP) and their application in adsorptive removal of As(III) and As(V) from aqueous solution, Appl. Surf. Sci. 317 (2014) 1052–1059.

[95] C.M. Dai, Z. Zhou, X.F. Zhou, Y.L. Zhang, Removal of Sb(III) and Sb(V) from aqueous solutions using nZVI, Water Air Soil Pollut. 225 (2014) 1799.

[96] T. Liu, Y. Yang, Z.L. Wang, Y. Sun, Remediation of arsenic(III) from aqueous solutions using improved nanoscale zero-valent iron on pumice, Chem. Eng. J. 288 (2016) 739–744.

[97] W.B. Li, F.L. Fu, Z.C. Ding, B. Tang, Zero valent iron as an electron transfer agent in a reaction system based on zero valent iron/magnetite nanocomposites for adsorption and oxidation of Sb(III), J. Taiwan Inst. Chem. Eng. 85 (2018) 155–164.

[98] Y.D. Chen, S.H. Ho, D. Wang, Z.S. Wei, J.S. Chang, N.Q. Ren, Lead removal by a magnetic biochar derived from persulfate-ZVI treated sludge together with one-pot pyrolysis, Bioresour. Technol. 247 (2018) 463–470.

[99] J. Li, C.L. Chen, K.R. Zhu, X.K. Wang, Nanoscale zero-valent iron particles modified on reduced graphene oxides using a plasma technique for Cd(II) removal, J. Taiwan Inst. Chem. Eng. 59 (2016) 389–394.

[100] H.M. Deng, Z.J. Zhang, N. Xu, Y.H. Chen, H.H. Wu, T. Liu, H.P. Ye, Removal of Tl(I) ions from aqueous solution using Fe@Fe_2O_3 core-shell nanowires, Clean-Soil Air Water 44 (2016) 1214–1224.

[101] J. Li, C.L. Chen, R. Zhang, X.K. Wang, Nanoscale zero-valent iron particles supported on reduced graphene oxides by using a plasma technique and their application for removal of heavy-metal ions, Chem. Asian J. 10 (2015) 1410–1417.

[102] C.C. Ding, W.C. Cheng, Y.B. Sun, X.K. Wang, Effects of Bacillus subtilis on the reduction of U(VI) by nano Fe0, Geochim. Cosmochim. Acta 165 (2015) 86–107.

[103] J. Li, C.L. Chen, R. Zhang, X.K. Wang, Reductive immobilization of Re(VII) by graphene modified nanoscale zero-valent iron particles using a plasma technique, Sci. China Chem. 59 (2016) 150–158.

[104] Y.L. Zhang, J. Yan, C.M. Dai, Y.T. Li, Y. Zhu, X.F. Zhou, Sequestration of Ag(I) from aqueous solution as Ag(0) nanostructures by nanoscale zero valent iron (nZVI), J. Nanopart. Res. 17 (2015) 455.

[105] F.L. Fu, Z.H. Cheng, D.D. Dionysiou, B. Tang, Fe/Al bimetallic particles for the fast and highly efficient removal of Cr(VI) over a wide pH range: performance and mechanism, J. Hazard. Mater. 298 (2015) 261–269.

[106] R.A. Crane, D.J. Sapsford, Sorption and fractionation of rare earth element ions onto nanoscale zerovalent iron particles, Chem. Eng. J. 345 (2018) 126–137.

[107] M. Usman, J.M. Byrne, A. Chaudhary, S. Orsetti, K. Hanna, C. Ruby, A. Kappler, S.B. Haderlein, Magnetite and green rust: synthesis, properties, and environmental applications of mixed-valent iron minerals, Chem. Rev. 7 (2018) 3251–3304.

[108] G. Klingelhöfer, R.V. Morris, B. Bernhardt, D. Rodionov, P.A. de Souza Jr, S.W. Squyres, J. Foh, E. Kankeleit, U. Bonnes, R. Gellert, Athena MIMOS II Mössbauer spectrometer investigation, J. Geophys. Res. Planets 108 (2003) 8067.
[109] V.T.P. Vinod, J. Filip, S.K. Ittara, S. Wacławek, M. Cernik, Electrospun membrane composed of poly [acrylonitrile-co-(methyl acrylate)-co-(itaconic acid)] terpolymer and ZVI nanoparticles and its application for the removal of arsenic from water, RSC Adv. 6 (2016) 110288–110300.
[110] E. Petala, K. Dimos, A. Douvalis, T. Bakas, J. Tucek, R. Zboril, M.A. Karakassides, Nanoscale zero-valent iron supported on mesoporous silica: characterization and reactivity for Cr(VI) removal from aqueous solution, J. Hazard. Mater. 261 (2013) 295–306.
[111] G.D. Sheng, Y.N. Tang, W.S. Linghu, L.J. Wang, J.X. Li, H. Li, X.K. Wang, Y.Y. Huang, Enhanced immobilization of ReO_4^- by nanoscale zerovalent iron supported on layered double hydroxide via an advanced XAFS approach: implications for TcC_4^- sequestration, Appl. Catal. B Environ. 192 (2016) 268–276.
[112] S.R. Kanel, J.M. Greneche, H. Choi, Arsenic(V) removal kom groundwater using nano scale zero-valent iron as a colloidal reactive barrier material, Environ. Sci. Technol. 40 (2006) 2045–2050.
[113] D.X. Qian, Y.M. Su, Y.X. Huang, H.Q. Chu, X.F. Zhou, Y.L. Zhang, Simultaneous molybdate (Mo(VI)) recovery and hazardous ions immobilization via nanoscale zerovalent iron, J. Hazard. Mater. 344 (2018) 698–706.
[114] G.D. Sheng, A. Alsaedi, W. Shammakh, S. Monaquel, J. Sheng, X.K. Wang, H. Li, Y.Y. Huang, Enhanced sequestration of selenite in water by nanoscale zero valent iron immobilization on carbon nanotubes by a combined batch, XPS and XAFS investigation, Carbon 99 (2016) 123–130.
[115] L. Ling, B. Pan, W.X. Zhang, Removal of selenium from water with nanoscale zero-valent iron: mechanisms of intraparticle reduction of Se(IV), Water Res. 71 (2015) 274–281.
[116] L. Ling, X.Y. Huang, M. Li, W.X. Zhang, Mapping the reactions in a single zero-valent iron nanoparticle, Environ. Sci. Technol. 51 (2017) 14293–14300.
[117] M. Dickinson, T.B. Scott, The application of zero-valent iron nanoparticles for the remediation of a uranium-contaminated waste effluent, J. Hazard. Mater. 178 (2010) 171.
[118] R.A. Crane, M. Dickinson, T.B. Scott, Nanoscale zero-valent iron particles for the remediation of plutonium and uranium contaminated solutions, Chem. Eng. J. 262 (2015) 319–325.
[119] M. Arshadi, S. Foroughifard, J.E. Gholtash, A. Abbaspourrad, Preparation of iron nanoparticles-loaded Spondias purpurea seed waste as an excellent adsorbent for removal of phosphate from synthetic and natural waters, J. Colloid Interface Sci. 452 (2015) 69–77.
[120] F.L. Liu, J.H. Yang, J. Zuo, D. Ma, L.L. Gan, B.M. Xie, P. Wang, B. Yang, Graphene-supported nanoscale zero-valent iron: removal of phosphorus from aqueous solution and mechanistic study, J. Environ. Sci. 26 (2014) 1751–1762.
[121] Y.H. Cao, J.N. Huang, Y.H. Li, S. Qiu, J.R. Liu, A. Khasanov, M.A. Khan, D.P. Young, F. Peng, D.P. Cao, X.F. Peng, K.L. Hong, Z.H. Guo, One-pot melamine derived nitrogen doped magnetic carbon nanoadsorbents with enhanced chromium removal, Carbon 109 (2016) 640–649.
[122] L.L. Chen, S.J. Feng, D.L. Zhao, S.H. Chen, F.F. Li, C.L. Chen, Efficient sorption and reduction of U(VI) on zero-valent iron-polyaniline-graphene aerogel ternary composite, J. Colloid Interface Sci. 490 (2017) 197–206.
[123] Y.B. Sun, X.X. Wang, Y.J. Ai, Z.M. Yu, W. Huang, C.L. Chen, T. Hayat, A. Alsaedi, X.K. Wang, Interaction of sulfonated graphene oxide with U(VI) studied by spectroscopic analysis and theoretical calculations, Chem. Eng. J. 310 (2016).

[124] I. Pidchenko, K.O. Kvashnina, T. Yokosawa, N. Finck, S. Bahl, D. Schild, R. Polly, E. Bohnert, A. Rossberg, J. Göttlicher, Uranium redox transformations after U(VI) coprecipitation with magnetite nanoparticles, Environ. Sci. Technol. 51 (2017) 2217–2225.
[125] K.J. Cantrell, D.I. Kaplan, T.W. Wietsma, Zero-valent iron for the in situ remediation of selected metals in groundwater, J. Hazard. Mater. 42 (1995) 201–212.
[126] L.Q. Liang, W.J. Yang, X.H. Guan, J.L. Li, Z.J. Xu, J. Wu, Y.Y. Huang, X.Z. Zhang, Kinetics and mechanisms of pH-dependent selenite removal by zero valent iron, Water Res. 47 (2013) 5846–5855.
[127] J. Farrell, W.D. Bostick, R.J. Jarabek, J.N. Fiedor, Uranium removal from ground water using zero valent iron media, Groundwater 37 (1999) 618–624.
[128] C. Noubactep, A. Schöner, G. Meinrath, Mechanism of uranium removal from the aqueous solution by elemental iron, J. Hazard. Mater. 132 (2006) 202–212.
[129] S. Klimkova, M. Cernik, L. Lacinova, J. Filip, D. Jancik, R. Zboril, Zero-valent iron nanoparticles in treatment of acid mine water from in situ uranium leaching, Chemosphere 82 (2011) 1178–1184.
[130] Z.J. Li, L. Wang, L.Y. Yuan, C.L. Xiao, L. Mei, L.R. Zheng, J. Zhang, J.H. Yang, Y.L. Zhao, Z.T. Zhu, Efficient removal of uranium from aqueous solution by zero-valent iron nanoparticle and its graphene composite, J. Hazard. Mater. 290 (2015) 26–33.
[131] L. Ling, W.X. Zhang, Enrichment and encapsulation of uranium with iron nanoparticle, J. Am. Chem. Soc. 137 (2015) 2788–2791.
[132] C. Noubactep, Effect of selected ligands on the U(VI) immobilization by zerovalent iron, J. Radioanal. Nucl. Chem. 267 (2005) 13–19.
[133] Y.B. Sun, Q. Wang, S.T. Yang, G.D. Sheng, Z.Q. Guo, Characterization of nano-iron oxyhydroxides and their application in UO_2^{2+} removal from aqueous solutions, J. Radioanal. Nucl. Chem. 290 (2011) 643–648.
[134] T. Missana, M. Garcia-Gutierrez, C. Maffiotte, Experimental and modeling study of the uranium (VI) sorption on goethite, J. Colloid Interface Sci. 260 (2003) 291.
[135] A. Singh, J.G. Catalano, K.U. Ulrich, D.E. Giammar, Molecular-scale structure of uranium(VI) immobilized with goethite and phosphate, Environ. Sci. Technol. 46 (2012) 6594–6603.
[136] J.N. Fiedor, W.D. Bostick, R.J. Jarabek, J. Farrell, Understanding the mechanism of uranium removal from groundwater by zero-valent iron using X-ray photoelectron spectroscopy, Environ. Sci. Technol. 32 (2015) 1466–1473.
[137] F.S. Fateminia, C. Falamaki, Zero valent nano-sized iron/clinoptilolite modified with zero valent copper for reductive nitrate removal, Process Saf. Environ. Prot. 91 (2013) 304–310.
[138] Y. Su, A.S. Adeleye, A.A. Keller, Y. Huang, C. Dai, X. Zhou, Y. Zhang, Magnetic sulfide-modified nanoscale zerovalent iron (S-nZVI) for dissolved metal ion removal, Water Res. 74 (2015) 47–57.
[139] X.Y. Wang, P. Wang, J. Ma, H.L. Liu, P. Ning, Synthesis, characterization, and reactivity of cellulose modified nano zero-valent iron for dye discoloration, Appl. Surf. Sci. 345 (2015) 57–66.
[140] Y.B. Sun, C.C. Ding, W.C. Cheng, X.K. Wang, Simultaneous adsorption and reduction of U(VI) on reduced graphene oxide-supported nanoscale zerovalent iron, J. Hazard. Mater. 280 (2014) 399–408.
[141] C. Jing, Y.L. Li, R.P. Cui, J.L. Xu, Illite-supported nanoscale zero-valent iron for removal of 238U from aqueous solution: characterization, reactivity and mechanism, J. Radioanal. Nucl. Chem. 304 (2015) 859–865.

[142] H.B. Liu, M.X. Li, T.H. Chen, C.L. Chen, N.S. Alharb, T. Hayat, D. Chen, Q. Zhang, Y.B. Sun, New Synthesis of nZVI/C composites as an efficient adsorbent for the uptake of U(VI) from aqueous solutions, Environ. Sci. Technol. 51 (2017) 9227–9234.

[143] L. Zhu, D.P. Sheng, C. Xu, X. Dai, M.A. Silver, J. Li, P. Li, Y.X. Wang, Y.L. Wang, L.H. Chen, C.L. Xiao, J. Chen, R.H. Zhou, C. Zhang, O.K. Farha, Z.F. Chai, T.E. Albrecht-Schmitt, S.A. Wang, Identifying the recognition site for selective trapping of $^{99}TcC_4^-$ in a hydrolytically stable and radiation resistant cationic metal-organic framework, J. Am. Chem. Soc. 139 (2017) 14873–14876.

[144] E. Reinoso Maset, S.H. Sidhu, A. Fisher, A. Heydon, P.J. Worsfold, A.J. Cartwright, M.J. Keith-Roach, Effect of organic co-contaminants on technetium and rhenium speciation and solubility under reducing conditions, Environ. Sci. Technol. 40 (2006) 5472–5477.

[145] L.Y. Liang, B.H. Gu, X.P. Yin, Removal of technetium-99 from contaminated groundwater with sorbents and reductive materials, Sep. Technol. 6 (1996) 111–122.

[146] J.G. Darab, A.B. Amonette, D.S.D. Burke, R.D. Orr, S.M. Ponder, B. Schrick, T.E. Mallouk, W.W. Lukens, D.L. Caulder, D.K. Shuh, Removal of pertechnetate from simulated nuclear waste streams using supported zerovalent iron, Chem. Mater. 19 (2007) 5703–5713.

[147] D.M. Fan, R.P. Anitori, B.M. Tebo, P.G. Tratnyek, Oxidative remobilization of technetium sequestered by sulfide-transformed nano zerovalent iron, Environ. Sci. Technol. 48 (2014) 7409–7417.

[148] Y. Gao, C.L. Chen, H. Chen, R. Zhang, X.K. Wang, Synthesis of a novel organic-inorganic hybrid of polyaniline/titanium phosphate for Re(VII) removal, Dalton Trans. 44 (2015) 8917.

[149] Y.J. Sasaki, Y. Kitatsuji, T. Kimura, Highly selective extraction of TcC_4^-, ReO_4^-, and MoO_4^- by the new nigand, 2,2′-(Methylimino)bis(N,N-Dioctylacetamide) (MIDOA), Chem. Lett. 36 (2007) 1394–1395.

[150] S.Q. Yu, X.X. Wang, X.L. Tan, X.K. Wang, Sorption of radionuclides from aqueous systems onto graphene oxide-based materials: a review, Inorg. Chem. Front. 2 (2015) 593–612.

[151] D. Kim, A. Jo, H.M. Yang, B.K. Seo, K.W. Lee, T.S. Lee, Colorimetric detection and removal of radioactive Co ions using sodium alginate-based composite beads, J. Hazard. Mater. 326 (2017) 69.

[152] X.X. Wang, Y. Liu, H.W. Pang, S.J. Yu, Y.J. Ai, X.Y. Ma, G. Song, T. Hayat, A. Alsaedi, X.K. Wang, Effect of graphene oxide surface modification on the elimination of Co(II) from aqueous solutions, Chem. Eng. J. 344 (2018) 380–390.

[153] Y.H. Wang, B. Chi, C.L. Xiao, D.G. Chen, Y.J. Wang, M.H. Liu, Graphene oxide@-$Mg_3Si_4O_9(OH)_{10}$: a hierarchical layered silicate nanocomposite with superior adsorption capacity for enriching Eu(III), Chem. Eng. J. 338 (2018) 628–635.

[154] Q.L. Fang, S.X. Duan, J.F. Zhang, J.X. Li, K.C.F. Leung, Dual shelled Fe_3O_4/polydopamine hollow microspheres as an effective Eu(III) adsorbent, J. Mater. Chem. A 5 (2017) 2947–2958.

[155] C. Uzum, T. Shahwan, A.E. Eroglu, I. Lieberwirth, T.B. Scott, K.R. Hallam, Application of zero-valent iron nanoparticles for the removal of aqueous Co^{2+} ions under various experimental conditions, Chem. Eng. J. 144 (2008) 213–220.

[156] Y.L. Zhang, W. Chen, C.M. Dai, C.L. Zhou, X.F. Zhou, Structural evolution of nanoscale zero-valent iron (nZVI) in anoxic Co^{2+} solution: interactional performance and mechanism, Sci. Rep. 5 (2015) 13966.

[157] M. Xing, J.L. Wang, Nanoscaled zero valent iron/graphene composite as an efficient adsorbent for Co(II) removal from aqueous solution, J. Colloid Interface Sci. 474 (2016) 119–128.

[158] M. Xing, L.J. Xu, J.L. Wang, Mechanism of Co(II) adsorption by zero valent iron/graphene nanocomposite, J. Hazard. Mater. 301 (2016) 286–296.

[159] H.J. Zhu, H.Y. Gao, X.M. Huang, W.F. Kong, X.M. Yan, The uptake of europium by reduced graphene oxide-supported nanoscale zerovalent iron investigated by batch and modeling techniques, J. Environ. Chem. Eng. 3 (2015) 2974–2980.

[160] Y.L. Zhang, Y.M. Li, J.F. Li, G.D. Sheng, Y. Zhang, X.M. Zheng, Enhanced Cr (VI) removal by using the mixture of pillared bentonite and zero-valent iron, Chem. Eng. J. 185 (2012) 243–249.

[161] J. Wang, G.J. Liu, C.C. Zhou, T.F. Li, J.J. Liu, Synthesis, characterization and aging study of kaolinite-supported zero-valent iron nanoparticles and its application for Ni(II) adsorption, Mater. Res. Bull. 60 (2014) 421–432.

[162] L.N. Shi, Y. Zhou, Z.L. Chen, M. Megharaj, R. Naidu, Simultaneous adsorption and degradation of Zn^{2+} and Cu^{2+} from wastewaters using nanoscale zero-valent iron impregnated with clays, Environ. Sci. Pollut. Res. 20 (2013) 3639–3648.

[163] Z.H. Pang, Y. Liu, J. Luo, Y.T. Lei, Influence factors on removal of cadmium by montmorillonite supported nano zero-valent iron, in: Advanced Materials Research, Trans Tech Publ, 2013, pp. 539–542.

[164] Ç. Üzüm, T. Shahwan, A.E. Eroğlu, K.R. Hallam, T.B. Scott, I. Lieberwirth, Synthesis and characterization of kaolinite-supported zero-valent iron nanoparticles and their application for the removal of aqueous Cu^{2+} and Co^{2+} ions, Appl. Clay Sci. 43 (2009) 172–181.

[165] R.B. Fu, Y.P. Yang, Z. Xu, X. Zhang, X.P. Guo, D.S. Bi, The removal of chromium(VI) and lead(II) from groundwater using sepiolite-supported nanoscale zero-valent iron (S-NZVI), Chemosphere 138 (2015) 726–734.

[166] L.N. Shi, X. Zhang, Z.N. Chen, Removal of chromium(VI) from wastewater using bentonite-supported nanoscale zero-valent iron, Water Res. 45 (2011) 886–892.

[167] T. Liu, Z.L. Wang, X. Yan, B. Zhang, Removal of mercury(II) and chromium(VI) from wastewater using a new and effective composite: pumice-supported nanoscale zero-valent iron, Chem. Eng. J. 245 (2014) 34–40.

[168] N. Li, F.L. Fu, J.W. Lu, Z.C. Ding, B. Tang, J.B. Pang, Facile preparation of magnetic mesoporous $MnFe_2O_4@SiO_2$-CTAB composites for Cr(VI) adsorption and reduction, Environ. Pollut. 220 (2016) 1376–1385.

[169] E. Mahmaoodi, M.A. Behnajady, Synthesis of $Fe_3O_4@NiO$ core-shell nanocomposite by the precipitation method and investigation of Cr(VI) adsorption efficiency, Colloid. Surface A 538 (2017) 287–296.

[170] U.O. Aigbe, R. Das, H.H. Wei, V. Srinivasu, A. Maity, A novel method for removal of Cr(VI) using polypyrrole magnetic nanocomposite in the presence of unsteady magnetic fields, Sep. Purif. Technol. 194 (2018) 377–387.

[171] N.H. Kera, M. Bhaumik, K. Pillay, S.S. Ray, A. Maity, Selective removal of toxic Cr(VI) from aqueous solution by adsorption combined with reduction at a magnetic nanocomposite surface, J. Colloid Interface Sci. 503 (2017) 214–228.

[172] J. Wang, P.Y. Wang, H.H. Wang, J.F. Dong, W.Y. Chen, X.X. Wang, S.H. Wang, T. Hayat, A. Alsaedi, X.K. Wang, Preparation of molybdenum disulfide coated Mg/Al layered double hydroxide composites for efficient removal of chromium(VI), ACS Sustain. Chem. Eng. 5 (2017) 7165–7174.

[173] C.L. Chen, K.R. Zhu, K. Chen, A. Alsaedi, T. Hayat, Synthesis of Ag nanoparticles decoration on magnetic carbonized polydopamine nanospheres for effective catalytic reduction of Cr(VI), J. Colloid Interface Sci. 526 (2018) 1–8.

[174] I.M. Lo, C.S. Lam, K.C. Lai, Hardness and carbonate effects on the reactivity of zero-valent iron for Cr(VI) removal, Water Res. 40 (2006) 595–605.

[175] T. Liu, D.C. Tsang, I.M. Lo, Chromium(VI) reduction kinetics by zero-valent iron in moderately hard water with humic acid: iron dissolution and humic acid adsorption, Environ. Sci. Technol. 42 (2008) 2092–2098.

[176] X.Q. Li, J.S. Cao, W.X. Zhang, Stoichiometry of Cr(VI) immobilization using nanoscale zerovalent iron (nZVI): a study with high-resolution X-ray photoelectron spectroscopy (HR-XPS), Ind. Eng. Chem. Res. 47 (2008) 2131–2139.

[177] J.G. Shang, M.Z. Zong, Y. Yu, X.R. Kong, Q. Du, Q.J.H. Liao, Removal of chromium(VI) from water using nanoscale zerovalent iron particles supported on herb-residue biochar, J. Environ. Manag. 197 (2017) 331–337.

[178] Z.S. Chen, D.L. Wei, Q. Li, X.X. Wang, S.J. Yu, L. Liu, B. Liu, S.Y. Xie, J. Wang, D.Y. Chen, Macroscopic and microscopic investigation of Cr(VI) immobilization by nanoscaled zero-valent iron supported zeolite MCM-41 via batch, visual, XPS and EXAFS techniques, J. Clean. Prod. 181 (2018) 745–752.

[179] N. Ezzatahmadi, G.A. Ayoko, G.J. Millar, R. Speight, C. Yan, J.H. Li, S.Z. Li, J.X. Zhu, Y.F. Xi, Clay-supported nanoscale zero-valent iron composite materials for the remediation of contaminated aqueous solutions: a review, Chem. Eng. J. 312 (2017) 336–350.

[180] S.S. Zhu, S.H. Ho, X.C. Huang, D.W. Wang, F. Yang, L. Wang, C.Y. Wang, X.D. Cao, F. Ma, Magnetic nanoscale zerovalent iron assisted biochar: interfacial chemical behaviors and heavy metals remediation performance, ACS Sustain. Chem. Eng. 5 (2017) 9673–9682.

[181] S.R. Thakare, J. Thakare, P.T. Kosankar, Nanoscale zero valent iron supported on solid support as an advanced material for the environmental purification: a review, Res. J. Chem. Environ. 19 (2015) 77–84.

[182] Y.M. Li, J.F. Li, Y.L. Zhang, Mechanism insights into enhanced Cr(VI) removal using nanoscale zerovalent iron supported on the pillared bentonite by macroscopic and spectroscopic studies, J. Hazard. Mater. 227 (2012) 211–218.

[183] X.S. Lv, J. Xu, G.M. Jiang, X.H. Xu, Removal of chromium(VI) from wastewater by nanoscale zero-valent iron particles supported on multiwalled carbon nanotubes, Chemosphere 85 (2011) 1204–1209.

[184] Z. Wang, J. Yang, Y.S. Li, Q.X. Zhuang, J.L. Gu, In situ carbothermal synthesis of nanoscale zero-valent iron functionalized porous carbon from metal-organic frameworks for efficient detoxification of chromium(VI), Eur. J. Inorg. Chem. 2018 (2018) 23–30.

[185] J. Wu, H.S. Zhu, G. Liu, L.Q. Tan, X.Y. Hu, C.L. Chen, N.S. Alharbi, T. Hayat, X.L. Tan, Fabrication of core-shell CMNP@PmPD nanocomposite for efficient As(V) adsorption and reduction, ACS Sustain. Chem. Eng. 5 (2017) 4399–4407.

[186] W.C. Song, X.X. Wang, T. Wen, S.J. Yu, Y.D. Zou, Y.B. Sun, T. Hayat, X.K. Wang, Immobilization of As(V) in Rhizopus oryzae investigated by batch and XAFS techniques, ACS Omega 1 (2016) 899–906.

[187] T. Wen, J. Wang, S.J. Yu, Z.S. Chen, T. Hayat, X.K. Wang, Magnetic porous carbonaceous material produced from tea waste for efficient removal of As(V), Cr(VI), humic acid, and dyes, ACS Sustain. Chem. Eng. 5 (2017) 4371–4380.

[188] M.A.V. Ramos, W. Yan, X.Q. Li, B.E. Koel, W.X. Zhang, Simultaneous oxidation and reduction of arsenic by zero-valent iron nanoparticles: understanding the significance of the core-shell structure, J. Phys. Chem. C 113 (2009) 14591–14594.

[189] W.L. Zhang, I.M.C. Lo, L.M. Hu, C.P. Voon, B.L. Lim, W.K. Versaw, Environmental risks of nano zero-valent iron for arsenate remediation: impacts on cytosolic levels of inorganic phosphate and $MgATP^{2-}$ in arabidopsis thaliana, Environ. Sci. Technol. 52 (2018) 4385–4392.

[190] Z.T. Li, L. Wang, J. Meng, X.M. Liu, J.M. Xu, F. Wang, P. Brookes, Zeolite-supported nanoscale zero-valent iron: new findings on simultaneous adsorption of Cd(II), Pb(II), and As(III) in aqueous solution and soil, J. Hazard. Mater. 344 (2018) 1–11.

[191] K. Manquian-Cerda, E. Cruces, M.A. Rubio, C. Reyes, N. Arancibia-Miranda, Preparation of nanoscale iron (oxide, oxyhydroxides and zero-valent) particles derived from blueberries: reactivity, characterization and removal mechanism of arsenate, Ecotoxicol. Environ. Saf. 145 (2017) 69–77.

[192] T.H. Bui, C. Kim, S.P. Hong, J. Yoon, Effective adsorbent for arsenic removal: core/shell structural nano zero-valent iron/manganese oxide, Environ. Sci. Pollut. Res. Int. 24 (2017) 24235–24242.

[193] R. Yadav, A.K. Sharma, J.N. Babu, Sorptive removal of arsenite [As(III)] and arsenate [As(V)] by fuller's earth immobilized nanoscale zero-valent iron nanoparticles (F-nZVI): effect of Fe0 loading on adsorption activity, J. Environ. Chem. Eng. 4 (2016) 681–694.

[194] W.L. Yan, R. Vasic, A.I. Frenkel, B.E. Koel, Intraparticle reduction of arsenite (As(III)) by nanoscale zerovalent iron (nZVI) investigated with in situ X-ray absorption spectroscopy, Environ. Sci. Technol. 46 (2012) 7018–7026.

[195] I.A. Katsoyiannis, T. Ruettimann, S.J. Hug, pH dependence of fenton reagent generation and As(III) oxidation and removal by corrosion of zero valent iron in aerated water, Environ. Sci. Technol. 42 (2008) 7424–7430.

[196] C. Wu, W.Z. Liu, J. Zhang, S.Q. Chu, Z.Q. Shi, Z. Dang, Z. Lin, Mechanisms of synergistic removal of low concentration As(V) by nZVI@$Mg(OH)_2$ nanocomposite, J. Phys. Chem. C 121 (2017) 21411–21419.

[197] A.C. Scheinost, R. Kirsch, D. Banerjee, A. Fernandezmartinez, H. Zaenker, H. Funke, L. Charlet, X-ray absorption and photoelectron spectroscopy investigation of selenite reduction by FeII-bearing minerals, J. Contam. Hydrol. 102 (2008) 228–245.

[198] B.W. Hu, G.H. Chen, C.G. Jin, J. Hu, C.C. Huang, J. Sheng, G.D. Sheng, J.Y. Ma, Y.Y. Huang, Macroscopic and spectroscopic studies of the enhanced scavenging of Cr(VI) and Se(VI) from water by titanate nanotube anchored nanoscale zero-valent iron, J. Hazard. Mater. 336 (2017) 214–221.

[199] A.J. Bard, R. Parsons, J. Jordan, Standard Potentials in Aqueous Solution, Marcel Dekker, 1985.

[200] J.T. Olegario, N. Yee, M. Miller, J. Sczepaniak, B. Manning, Reduction of Se(VI) to Se(-II) by zerovalent iron nanoparticle suspensions, J. Nanopart. Res. 12 (2010) 2057–2068.

[201] Y.M. Li, W. Cheng, G.D. Sheng, J.F. Li, H.P. Dong, Y. Chen, L.Z. Zhu, Synergetic effect of a pillared bentonite support on SE(VI) removal by nanoscale zero valent iron, Appl. Catal. B Environ. 174-175 (2015) 329–335.

[202] C. Shan, X. Wang, X.H. Guan, F. Liu, W.M. Zhang, B.C. Pan, Efficient removal of trace Se(VI) by millimeter-sized nanocomposite of zerovalent iron confined in polymeric anion exchanger, Ind. Eng. Chem. Res. 56 (2017) 5309–5317.

[203] X.Q. Li, W.X. Zhang, Sequestration of metal cations with zerovalent iron nanoparticles—a study with high resolution X-ray photoelectron spectroscopy (HR-XPS), J. Phys. Chem. C 111 (2007) 6939–6946.

[204] S.M. Ponder, J.G. Darab, T.E. Mallouk, Remediation of Cr(VI) and Pb(II) aqueous solutions using supported, nanoscale zero-valent iron, Environ. Sci. Technol. 34 (2000) 2564–2569.

[205] U.A. Guler, M. Ersan, Cells modified with nZVI: a novel magnetic biosorbent for nickel removal from aqueous solutions, Desalin. Water Treat. 57 (2016) 7196–7208.

[206] W. Wang, Y.L. Hua, S.L. Li, W.L. Yan, W.X. Zhang, Removal of Pb(II) and Zn(II) using lime and nanoscale zero-valent iron (nZVI): a comparative study, Chem. Eng. J. 304 (2016) 79–88.

[207] N.M. Zhu, Y.S. Xu, L. Dai, Y.F. Zhang, G.Q. Hu, Application of sequential extraction analysis to Pb(II) recovery by zerovalent iron-based particles, J. Hazard. Mater. 351 (2018) 138–146.

[208] N. Chanthapon, S. Sarkar, P. Kidkhunthod, S. Padungthon, Lead removal by a reusable gel cation exchange resin containing nano-scale zero valent iron, Chem. Eng. J. 331 (2018) 545–555.

[209] X.Q. Li, W.X. Zhang, Iron nanoparticles: the core-shell structure and unique properties for Ni(II) sequestration, Langmuir 22 (2006) 4638–4642.

[210] M.Y. Fan, T.J. Li, J.W. Hu, R.S. Cao, Q. Wu, X.H. Wei, L.Y. Li, X.D. Shi, W.Q. Ruan, Synthesis and characterization of reduced graphene oxide-supported nanoscale zero-valent iron (nZVI/rGO) composites used for Pb(II) removal, Materials 9 (2016) 687.

[211] Z. Tang, Y.F. Deng, T. Luo, Y.S. Xu, N.M. Zhu, Enhanced removal of Pb(II) by supported nanoscale Ni/Fe on hydrochar derived from biogas residues, Chem. Eng. J. 292 (2016) 224–232.

[212] F.C. Yang, Y.Y. He, S.Q. Sun, Y. Chang, F. Zha, Z.Q. Lei, Walnut shell supported nanoscale Fe0 for the removal of Cu(II) and Ni(II) ions from water, J. Appl. Polym. Sci. 133 (2016) 43304.

[213] D. Karabelli, Ç. Üzüm, T. Shahwan, A.E. Eroğlu, T.B. Scott, K.R. Hallam, I. Lieberwirth, Batch removal of aqueous Cu^{2+} ions using nanoparticles of zero-valent iron: a study of the capacity and mechanism of uptake, Ind. Eng. Chem. Res. 47 (2008) 4758–4764.

[214] F. Liu, C. Shan, X.L. Zhang, Y.Y. Zhang, W.M. Zhang, B.C. Pan, Enhanced removal of EDTA-chelated Cu(II) by polymeric anion-exchanger supported nanoscale zero-valent iron, J. Hazard. Mater. 321 (2017) 290–298.

[215] H.K. Boparai, M. Joseph, D.M. O'Carroll, Kinetics and thermodynamics of cadmium ion removal by adsorption onto nano zerovalent iron particles, J. Hazard. Mater. 186 (2011) 458–465.

[216] Y.L. Zhang, Y.T. Li, C.M. Dai, X.F. Zhou, W.X. Zhang, Sequestration of Cd(II) with nanoscale zero-valent iron (nZVI): characterization and test in a two-stage system, Chem. Eng. J. 244 (2014) 218–226.

[217] W.J. Xue, D.L. Huang, G.M. Zeng, J. Wan, C. Zhang, R. Xu, M. Cheng, R. Deng, Nanoscale zero-valent iron coated with rhamnolipid as an effective stabilizer for immobilization of Cd and Pb in river sediments, J. Hazard. Mater. 341 (2017) 381–389.

Index

Note: Page numbers followed by "*f*" indicate figures, and "*t*" indicate tables.

A

B

N

O

P

Q

R

CPI Antony Rowe
Eastbourne, UK
June 18, 2019

About the Authors

Dr. Voyiadjis is a Member of the European Academy of Sciences, and Foreign Member of both the Polish Academy of Sciences, and the National Academy of Engineering of Korea. George Z. Voyiadjis is the Boyd Professor at the Louisiana State University, in the Department of Civil and Environmental Engineering. This is the highest professorial rank awarded by the Louisiana State University System. He is also the holder of the Freeport-MacMoRan Endowed Chair in Engineering. He joined the faculty of Louisiana State University in 1980. He is currently the Chair of the Department of Civil and Environmental Engineering. He holds this position since February of 2001. He also served from 1992 to 1994 as the Acting Associate Dean of the Graduate School. He currently also serves since 2012 as the Director of the Louisiana State University Center for GeoInformatics (LSU C4G; http://c4gnet.lsu.edu/c4g/).

Voyiadjis' primary research interest is in plasticity and damage mechanics of metals, metal matrix composites, polymers and ceramics with emphasis on the theoretical modeling, numerical simulation of material behavior, and experimental correlation. Research activities of particular interest encompass macro-mechanical and micro-mechanical constitutive modeling, experimental procedures for quantification of crack densities, inelastic behavior, thermal effects, interfaces, damage, failure, fracture, impact, and numerical modeling.

Dr. Voyiadjis' research has been performed on developing numerical models that aim at simulating the damage and dynamic failure response of advanced engineering materials and structures under high-speed impact loading conditions. This work will guide the development of design criteria and fabrication processes of high performance materials and structures under severe loading conditions. Emphasis is placed on survivability area that aims to develop and field a contingency armor that is thin and lightweight, but with a very high level of an overpressure protection system that provides low penetration depths. The formation of cracks and voids in the adiabatic shear bands, which are the precursors to fracture, are mainly investigated.

He has two patents, over 332 refereed journal articles and 19 books (11 as editor) to his credit. He gave over 400 presentations as plenary, keynote and invited speaker as well as other talks. Over sixty two graduate students (37 PhD) completed their degrees under his direction. He has also supervised numerous postdoctoral associates. Voyiadjis has been extremely successful in securing more than $30.0 million in research funds as a principal investigator/investigator from the National Science Foundation, the Department of Defense, the Air Force Office of Scientific Research, the Department of Transportation, National Oceanic and Atmospheric Administration (NOAA), and major companies such as IBM and Martin Marietta.

Affiliations and Expertise
Boyd Professor
Department of Civil and Environmental Engineering
Louisiana State University
Baton Rouge, LA, United States

Mohammadreza Yaghoobi is a research faculty in the Material Science and Engineering Department at University of Michigan, Ann Arbor. His primary research interest is in multiscale computational plasticity and damage mechanics of crystalline materials, composites, and ceramics with emphasis on the theoretical modeling, numerical simulation of material behavior, and experimental correlation. Research activities of particular interest include modeling at different length scales including atomistic simulation, crystal plasticity finite element method, and local and nonlocal continuum plasticity. Central to his research is serving as a lead developer of PRISMS-Plasticity software (http://www.prisms-center.org), which is an open-source parallel 3-D crystal plasticity and continuum plasticity finite element code (https://github.com/prisms-center/plasticity/).

Affiliations and Expertise
Research Faculty
Materials Science and Engineering Department
University of Michigan
Ann Arbor, MI, United States

Acknowledgments

The partial financial support provided by a grant from the National Science Foundation EPSCoR Consortium for Innovation in manufacturing and Materials, CIMM (Grant Number #OIA-1541079) at Louisiana State University is gratefully acknowledged by the authors.

The second author also wishes to acknowledge the partial support by the U.S. Department of Energy, Office of Basic Energy Sciences, Division of Materials Sciences and Engineering under Award #DE-SC0008637 as part of the Center for Predictive Integrated Structural Materials Science (PRISMS Center) at University of Michigan.

Finally, the authors want to thank Reem Abo Znemah and Yooseob Song for their help with proofreading the book.

Chapter 1

Introduction: Size effects in materials

In material science, size effects are described as the variation of material properties as the sample size changes. The size effects include various properties such as optical and Photocatalytic properties, thermal conductivity, Young's Modulus, material strength, diffusion processes, electrical conductivity, and magnetic properties. In this book, the size dependency of the material strength is addressed as size effects. The size effects underlying mechanisms depend on the nature of the considered material. The governing mechanisms of size effects in brittle materials, quasibrittle materials, and crystalline metals are different from each other. In the case of brittle materials, the size effects have a statistical nature. It may be considered to be originated from the random nature of the material strength that can be generally captured using Weibull statistical theory (Weibull, 1939). The basic assumptions of the theory, which was proposed by Weibull (1939), are the structure failure occurs by failing any element of the structure, and the strength of the structure elements is a random function that can be described by the Weibull distribution. The Weibull theory can successfully capture the size effects in brittle materials. In the case of quasibrittle materials, however, the nature of the size effects is deterministic and not statistical and it is due to the stress redistribution around the cracks. In the case of crystalline metallic structures, the sources of size effects are more versatile. The size effects in crystalline metallic samples can be related to some internal length scales, such as grain size, the average distance of second phase particles or precipitates, and dislocations mean free path. Finally, the external geometrical properties of the specimen such as the thin film thickness, pillar diameter, or indentation depth may lead to size effects.

1.1 Brittle materials

Studying the size effects in strength have been initiated during the Renaissance period by Leonardo da Vinci (1500s) who stated that the shorter the cord, the stronger it is for cords of equal thicknesses (Williams, 1957). Later on, Galileo (1638) rejected the rule proposed by da Vinci and stated that reducing the cord

Size Effects in Plasticity: From Macro to Nano. https://doi.org/10.1016/B978-0-12-812236-5.00001-3

length does not change the cord strength (Williams, 1957). Mariotte (1686) initiated the statistical theory of size effects in which he stated that the length should not affect the rope strength unless the longer rope has some defects which makes the longer rope weaker. Later on, Young (1807) rejected this theory using the deterministic point of view and related the wire strength to its cross section area. He stated that the wire strength does not depend on its length. Griffith (1921) observed the size effects during the mechanical testing of glass fibers where their nominal strength increases from 42,300 psi for the sample with diameter of 0.00542 in. to 491,000 psi for the diameter of 0.00013 in. He related the observed size effects to the defects and flaws in materials. The underlying mechanism for size effects proposed by Griffith (1921) was similar to the one originally stated by Mariotte (1686).

Weibull (1939) made a great contribution to size effects research topic by introducing the Weibull distribution to capture the sample strength based on its size. After the statistical framework of Weibull (1939), other researchers have applied, studied, and improved the method (see e.g. Bažant, 2005). The Weibull distribution can successfully capture the brittle materials that fail after the macroscopic crack initiation where their fracture process zone is very small. Fig. 1.1 illustrates the Weibull theory of size effects using a system of 1D structural elements. The Weibull theory describes the structure as a system in which the specimen fails by failure of the first element (Fig. 1.1). The failure probability of each structural element subjected to the uniaxial stress σ can be described as follows (Weibull, 1939):

$$f(\sigma) = \left(\frac{\sigma - \sigma_u}{\sigma_0}\right)^m, \tag{1.1}$$

where m is the Weibull modulus and σ_0 is a scale parameter. The element does not fail for $\sigma \le \sigma_u$. Accordingly, the failure probability of a structure composed of n small components at nominal strength σ_N can be described as follows:

$$1 - P_f(\sigma_N) = \prod_{i=1}^{n} [1 - f(\sigma_i)] \tag{1.2}$$

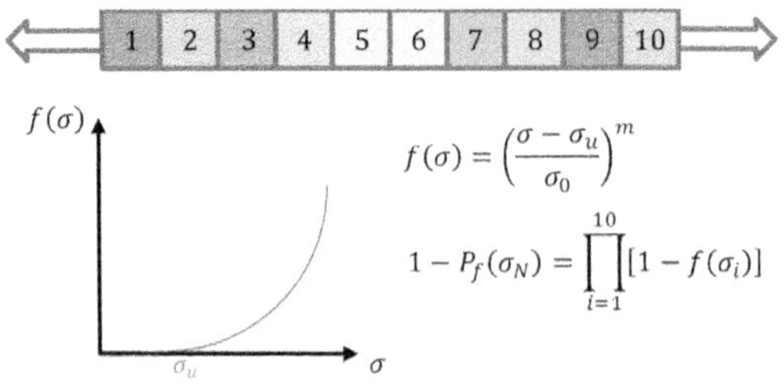

FIG. 1.1 A 1D bar with 10 elements each has a failure probability of $f(\sigma)$ defined according to the Weibull distribution. The failure probability of the structure $P_f(\sigma_N)$ is described using the Weibull theory that states the strength of the structure is equal to the strength of the weakest element.

In the case of a 3D continuous structure, the failure probability of the structure using the Weibull theory can be rewritten as follows (Bažant, 1999):

$$P_f(\sigma_N) = 1 - \exp\left\{-\int_V c[\boldsymbol{\sigma}(\mathbf{x})]dV(\mathbf{x})\right\}, \tag{1.3}$$

where $\boldsymbol{\sigma}$ is the stress tensor field at the failure, $\mathbf{x}$ is the position vector, V is the structure volume, and $c[\boldsymbol{\sigma}]$ is described as follows:

$$c[\boldsymbol{\sigma}] \approx \frac{\sum_i f(\sigma_i)}{V_0}, \tag{1.4}$$

where σ_i are the three eigenvalues of stress tensor $\boldsymbol{\sigma}$ and V_0 is the volume of each structural element. In the case of a 3D structure with uniform uniaxial stress, the average nominal strength $\overline{\sigma}_N$ and the coefficient of nominal strength variation ω can be obtained as follows:

$$\overline{\sigma}_N = \sigma_0 \Gamma\left(1 + m^{-1}\right)\left(\frac{V_0}{V}\right)^{1/m} \tag{1.5a}$$

$$\omega = \left[\frac{\Gamma\left(1 + 2m^{-1}\right)}{\Gamma^2\left(1 + m^{-1}\right)} - 1\right]^{1/2} \tag{1.5b}$$

where Γ is the gamma function. The average nominal strength captures size effects by incorporation of V which represents the effect of the structure size D. The parameter m can be calibrated against the experimental results using Eq. (1.5b). In the case of a general structure and stress, the Weibull theory introduces the size effects with a statistical nature as follows:

$$\overline{\sigma}_N \propto D^{-\frac{n}{m}} \tag{1.6}$$

where n is 1, 2, and 3 for 1D, 2D, and 3D spatial coordinates, respectively.

1.2 Quasibrittle materials

Unlike brittle materials, structures built from quasibrittle materials, such as concrete, rocks, cement mortars, ice, fiber-reinforced concrete, toughened ceramics, fiber composites, biological shells, cemented sands, and wood, may not fail after the initial crack nucleation, and the fracture process zone is not small. Accordingly, the size effects have a deterministic nature rather than a statistical one, and therefore, the Weibull distribution cannot capture the size effects. Since the fracture process zone is large in quasibrittle materials and the corresponding stress redistribution in the structure becomes important, it has been observed that the linear elastic fracture mechanics (LEFM) cannot capture the size effects in concrete (Leicester, 1969). In other words, LEFM predicts that the sample nominal strength σ_N is a function of D^{-n}, where $n = 1/2$;

however, Leicester (1969) observed that the value of n in the case of concrete is less than 1/2. Later on, Walsh (1972) showed that the size effects in concrete beams does not follow LEFM, and the size effects deviate from $\sigma_N \propto D^{-1/2}$. Hillerborg et al. (1976) developed the fictitious crack model to study the size effects in bending experiments on concrete specimens. The cohesive crack model is the most computationally efficient method to address the stress redistribution in solids due to the crack. They showed that the size effects in concrete have the deterministic nature that cannot be captured using the Weibull (1939). Bažant (1982) and Bažant and Oh (1983) developed the crack band model to investigate size effects in concrete samples. In this model, the fracture in quasibrittle materials is described using a single element that contains smeared cracking that its width is a material property. The crack band model ensures that the correct amount of energy is dissipated in the specimen using the material fracture energy G_f. The next generation of models to capture size effects in quasibrittle materials is the nonlocal models (Bažant et al., 1984). In the case of local models, the stress at a material point depends only on the strain at the same point. In the case of nonlocal models, however, the stress at a material point is a function of the strain at a region surrounding the material point. Accordingly, the model is able to capture the stress redistribution in the fracture process zone in quasibrittle materials.

Size effects in quasibrittle materials can be divided into two different types:

- Size effects in structures that have a deep notch or a large pre-existing stress-free crack.
- Size effects in structures that crack initiation results in failure.

There is a transition between these two size effect types that occur when the crack length at failure is neither small and negligible nor large enough to be categorized in one of two size effect types. Accordingly, some researchers have tried to develop a size effect model that can capture both types of size effects and their transition (Bažant, 1996; Bažant and Li, 1996; Bažant and Yu, 2009; Hoover and Bažant, 2014). In order to define the size effect governing mechanism in quasibrittle materials, two length scales are important, namely, the size of the structure and the size of the notch or crack at the structure failure. The relative size of the notch or crack with respect to the structure size dictates the governing mechanism of size effects, i.e. if it belongs to one of the two separate size effects types or if it is in the transition region. Next, the size of the structure defines the governing mechanism of the size effects. As an example, in the case of small specimens in which the sample size is comparable to the material length scale, a probability of a large crack nucleation decreases and there is no strength dependency on the specimen size. More details on the effects of sample size on the governing mechanism of size effects will be presented in Sections 1.2.1 and 1.2.2.

Concrete, as a quasibrittle material, is a widely used construction material that many experiments have been conducted to capture its size effects (Sabnis and Mirza, 1979; Bažant and Pfeiffer, 1987; Malvar and Warren, 1988; Rocco, 1995; Tang et al., 1996; Bažant and Planas, 1998; Karihaloo et al., 2003; Beygi et al., 2013). However, most of the experiments have targeted a limited range of sample properties. These experiments are not comparable to each other due to their different sample preparation procedures, such as sample age, curing conditions, incorporated materials, loading rates and experiment procedures. Hoover et al. (2013), however, conducted a comprehensive set of experiments on the concrete specimens, which have been prepared from a similar batch of concrete, with various beam dimensions and notch sizes. They captured both types of size effects and the transition between these two types. The related experiments conducted by Hoover et al. (2013) will be presented with more details in Sections 1.2.1 and 1.2.2.

1.2.1 Failure while the structure has a stable large crack or a deep notch

In the case of a structure with a deep notch or a stable large crack at its failure, strong deterministic size effects are observed. Accordingly, the size effects incorporate a characteristic length scale that controls its governing mechanisms of size effects (Fig. 1.2). In the case of specimens with the sample size comparable to the material length scale, the size effects vanish because the crack formation does not occur. However, in the case of very large structures in which the structure size is much larger than the material characteristic length, the material length scale is not a governing parameter anymore. Accordingly, the structure size effects can be captured using the LEFM and σ_N becomes a function of D^{-n}, where $n = 1/2$.

Bažant (1984) explained the deterministic size effects in a simple example using the concept of energy release. Fig. 1.3 illustrates two large and small structures containing a crack with the length of a during uniaxial tension experiment. The width of the fracture process zone is denoted by h. The energy release occurs incorporating two competitive mechanisms: the energy release in the crack band $E_{cb} \propto D\sigma_N^2/E$ and the strain energy release from the shaded area $E_{sa} \propto D^2\sigma_N^2/E$, where E is the beam elastic modulus. On the other hand,

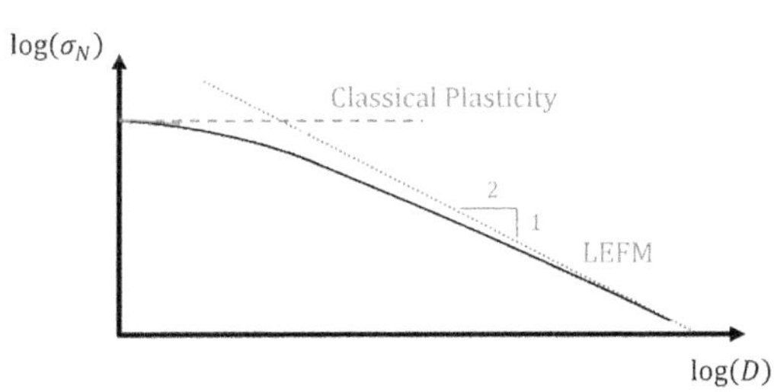

FIG. 1.2 Size effects in structures with a deep notch or a large stable crack.

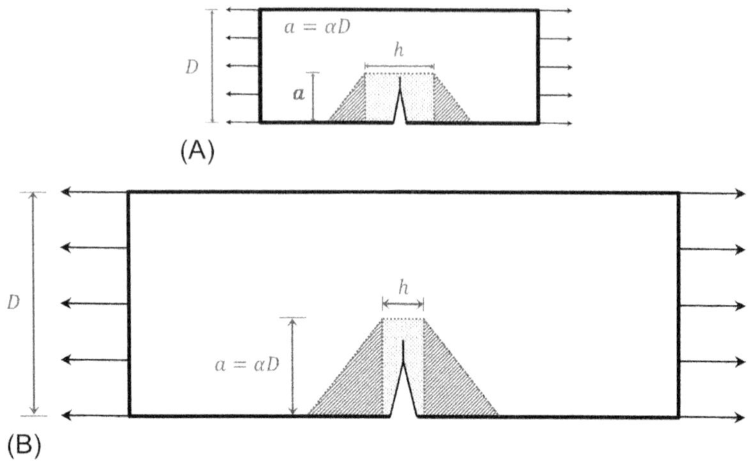

FIG. 1.3 Structures with a stable crack and regions of stress relief due to that crack: (A) small structure and (B) large structure.

the energy dissipated by the crack formation is $E_{crack} \propto G_f D$. In the case of small structure, the width of crack band h is comparable to the beam depth D. The area of the shaded regions and accordingly the energy released from the shaded regions become negligible compared to those of the crack band region. The size effects can be then written using the energy balance as follows:

$$E_{cb} = E_{crack} \to D\sigma_N^2/E \propto G_f D \to \sigma_N = \text{constant} \tag{1.7}$$

Eq. (1.1) shows that there is no size effects in the case of small beam containing a crack.

In the case of large structure, the energy released by the crack band E_{cb} becomes negligible compared to that of the shaded area E_{sa}. Accordingly, the energy balance defines the size effects as follows:

$$E_{sa} = E_{crack} \to D^2\sigma_N^2/E \propto G_f D \to \sigma_N \propto D^{-1/2} \tag{1.8}$$

Eq. (1.8) states that $\sigma_N \propto D^{-1/2}$, which is similar to the size effect description of LEFM.

Hoover and Bažant (2013) studied the size effects in concrete using the experiments conducted by Hoover et al. (2013). In the case of a structure which has a deep notch or a large stable crack at its failure, Fig. 1.4 shows the results presented by Hoover and Bažant (2013) for different notch size ratios of $\alpha_0 = 0.3$ and $\alpha_0 = 0.15$, where $a = \alpha_0 D$ is the notch size. They also showed that the size effects for the smaller structures follow the classical plasticity. In the case of large structures, however, the size effects follow the LEFM and $\sigma_N \propto D^{-1/2}$.

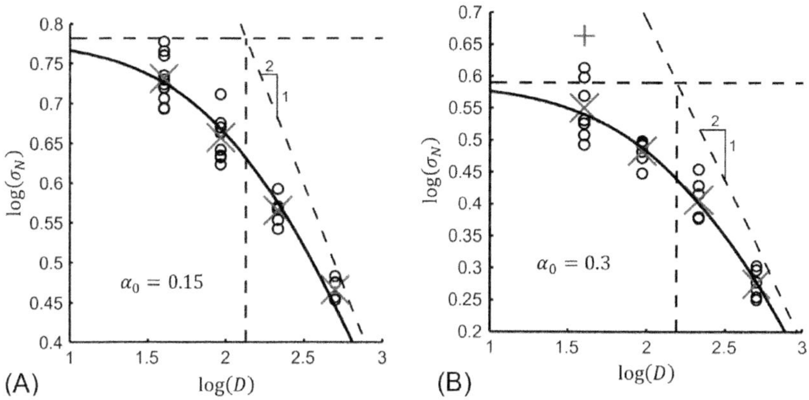

FIG. 1.4 The size effects in concrete beams with deep notches of (A) $\alpha_0 = 0.3$ and (B) $\alpha_0 = 0.15$. *(After Hoover, C.G., Bažant, Z.P., 2013. Comprehensive concrete fracture tests: size effects of types 1 and 2, crack length effect and postpeak. Eng. Fract. Mech. 110, 281–289.)*

Bažant (1984) derived an equation to capture the size effects in structures with a deep notch or a large stable crack using the concept of energy release as follows:

$$\sigma_N = \frac{Bf_t'}{\sqrt{1 + D/D_0}} \tag{1.9}$$

where f_t' is the material tensile strength and B and D_0 are the parameters that depend on the material fracture properties and structure shape. Eq. (1.9) can capture two limits of size effects. In other words, in the case of small structure size limit, i.e. $D \to 0$, $\sigma_N = Bf_t'$, and the size effects in structure vanish. On the other hand, in the case of large structure limit, i.e. $D/D_0 \gg 1$, $\sigma_N \propto D^{-1/2}$, and the size effects follow the LEFM. Bažant and Kazemi (1991) investigated the parameters of Eq. (1.9) using the LEFM approximation and related them to the material fracture properties as follows:

$$\sigma_N = \sqrt{\frac{E'G_f}{g'(\alpha_0)c_f + g(\alpha_0)D}} \tag{1.10}$$

where E' is the Young's modulus ($E' = E$) in the case of plane stress and $E' = E/(1 - \nu^2)$ for the case of plane stress, where ν is the Poisson's ratio. G_f is the initial fracture energy and c_f denotes the charactecteristic length which is around half of the fracture process zone. In addition, $g'(\alpha) = \partial g(\alpha)/\partial \alpha$ and $g(\alpha_0) = k^2(\alpha_0) = b^2DK_I^2/P$, where b is the thickness of the beam, K_I is the stress intensity factor, and P is the applied load.

1.2.2 Failure at crack initiation

The quasibrittle materials normally have a large stable crack at their failure. However, they sometimes fail at crack initiation before a stable large crack is formed. Although there is no deep notch or large stable crack in this type of failure, the Weibull theory still cannot capture the size effects in these structures. It is due to the fact that the fracture process zone is considerable. The nominal strength occurs when the microcracks coalesce and form a large crack in the boundary layer. Although the Weibull theory alone cannot handle the size effects in quasibrittle materials failure at crack initiation, it should be sometimes included in the size effects model. In other words, if the region of high stress is concentrated, such as a three-point bending experiment, the statistical part of size effects vanishes. However, if the experiment has a large region of high stress, such as a four-point bending experiment, the size effects will have both deterministic and statistical governing mechanisms. Depending on the size of the high stress region, the size effects for failure at crack initiation can be described as follows (Bažant and Novák, 2000):

$$\sigma_N = \sigma_{N,\infty}\left(1+\frac{rD_b}{D}\right)^{1/r} \quad \text{(no statistical size effects)} \tag{1.11}$$

$$\sigma_N = \sigma_{N,\infty}\left[\left(\frac{D_b}{D}\right)^{rn/m}+\frac{rD_b}{D}\right]^{1/r} \quad \text{(with statistical size effects)} \tag{1.12}$$

where $\sigma_{N,\infty}$ is the nominal strength for $D \rightarrow \infty$ when there is no statistical size effect, i.e. Eq. (1.11), D_b is the thickness of the boundary layer of cracking, r is a positive constant, n is 1, 2, and 3 for 1D, 2D, and 3D spatial coordinates, respectively, and m is the Weibull modulus. One should notice that in all practical cases, $rn/m \ll 1$ (Bažant and Novák, 2000). In the case of very small structures ($D \rightarrow 0$), however, Eqs. (1.11), (1.12) lead to an unlimited σ_N. To modify this mathematical issue, the material length scales l_s and l_p should be added as follows:

$$\sigma_N = \sigma_{N,\infty}\left(1+\frac{rD_b}{D+l_p}\right)^{1/r} \quad \text{(no statistical size effects)} \tag{1.13}$$

$$\sigma_N = \sigma_{N,\infty}\left[\left(\frac{D_b}{D+l_s}\right)^{rn/m}+\frac{rD_b}{D+l_p}\right]^{1/r} \quad \text{(with statistical size effects)} \tag{1.14}$$

It should be noted that both l_s and l_p are much smaller than D, i.e. $l_s, l_p \ll D$, in a way that the effects of l_s and l_p are negligible. The three asymptotic conditions of Eq. (1.14) are as follows:

- In the case of very small structure ($D \rightarrow 0$), Eq. (1.14) asymptotically approaches the deterministic size effects.
- In the case of very large structure ($D \rightarrow \infty$), Eq. (1.14) asymptotically approaches the statistical size effects.
- In the case of very large m, Eq. (1.14) reduces to the deterministic size effects relation.

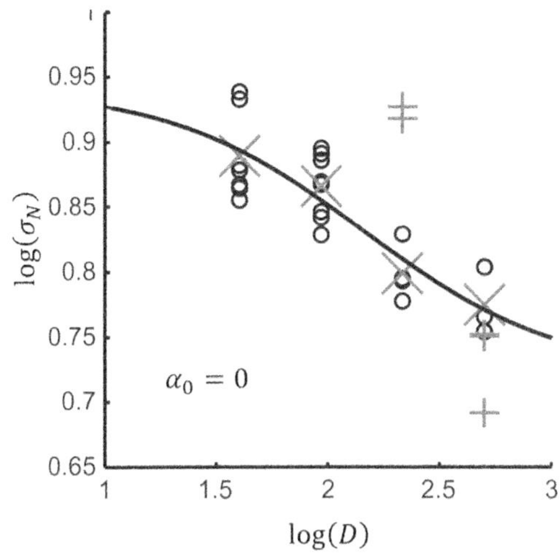

FIG. 1.5 The size effects in unnotched concrete beams ($\alpha_0 = 0$) fail at crack initiation. *(After Hoover, C.G., Bažant, Z.P., 2013. Comprehensive concrete fracture tests: size effects of types 1 and 2, crack length effect and postpeak. Eng. Fract. Mech. 110, 281–289.)*

Fig. 1.5 shows the size effects experiment conducted by Hoover and Bažant (2013) on unnotched beams ($\alpha_0 = 0$). They showed that the size effects in unnotched beams can be successfully captured using Eqs. (1.13), (1.14). Bažant and Novák (2000) compared their developed equation, i.e. Eq. (1.12), with $m = 12$ to capture both statistical and deterministic size effects against the available experimental results (Reagel and Willis, 1931; Wright, 1952; Nielsen, 1954; Linder and Sprague, 1956; Walker and Bloem, 1957; Sabnis and Mirza, 1979; Rocco, 1995; Rokugo et al., 1995), which is depicted in Fig. 1.6. Two asymptotic conditions of small and large structures are also included to study the developed size effects equation. The results show that the proposed equation can successfully capture the size effects observed in the previous experiments.

1.3 Crystalline metals

The strength enhancement of crystalline metallic structures is one of the primary target of material science and engineering. This can be reached through special fabrication procedures such as work hardening, grain size refinement, or precipitation hardening. Accordingly, the strength of metallic structures can be increased by two orders of magnitude. Size effects in crystalline metals are governed by the dislocations, as the primary deformation mechanism, and their interactions with one another, other defects such as grain boundaries, and precipitates or second phase particles. The size effects can be attributed to many different deformation and strengthening mechanisms. They can be originated

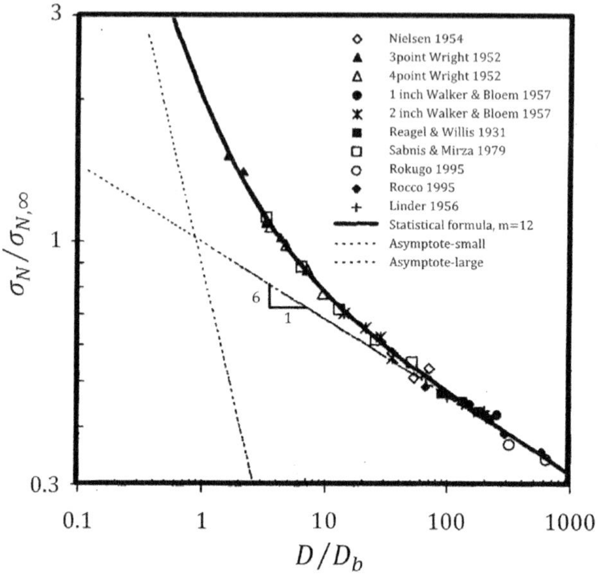

FIG. 1.6 Variation of normalized nominal strength $\sigma_N/\sigma_{N,\infty}$ versus the normalized structure size D/D_b. *(After Bažant, Z.P., Novák, D., 2000. Energetic–statistical size effect in quasibrittle failure at crack initiation. ACI Struct. J. 97, 381–392.)*

from some internal characteristic length scales, such as grain size or some external length scales, such as the film thickness, pillar diameter, and indentation depth. The size effects in crystalline metals are commonly categorized in two groups of intrinsic, related to the internal characteristic length scales such as grain size and dislocations mean free path, and extrinsic, originated from an external length scale such as thin film thickness or pillar diameter.

In more general cases, the crystalline metallic structures show size effects because of the interplay between these two size effects types. For example, thin films with different grain sizes, the nanocrystalline pillars, or nanoindentation of the polycrystalline samples. However, the interaction of the size effects mechanisms in these cases becomes complicated and more theoretical and experimental studies are required to shed light on this topic.

1.3.1 Intrinsic size effects

The intrinsic size effects is the dependency of material strength on the intrinsic microstructural properties such as grain size, the average distance of second phase particles or precipitates, and dislocations mean free path. Between all of the possible intrinsic size effects, two of them will be discussed here in more details that are the effects of the average distance of second phase particles or precipitates and the grain size on the crystalline metals strength.

1.3.1.1 Precipitates size

One of the first size effects which has been studied in material science and engineering is the dependency of material strength on the precipitate size for a given volume fraction of precipitates. Arzt (1998) reviewed the size effects due to the microstructural constraints in which the concept of obstacle spacing and dislocation curvature, i.e. the Orowan mechanism (Orowan, 1947), is incorporated to capture the effect of precipitate size on material strength. To elaborate this size effect, Fig. 1.7 depicts an array of hard objects with different diameters of D_1 and D_2 for a given volume fraction. The smaller hard objects ($D_1 < D_2$) leads to the smaller obstacle spacing ($L_1 < L_2$). The dislocation loop bypasses the hard obstacle if the equilibrium diameter of the curved dislocation d is smaller than the obstacle spacing L. In other words, plastic deformation occurs through the elongation and multiplication of dislocation loops that can fit between two obstacles. The equilibrium diameter of a dislocation for an elastic isotropic material can be described as follows:

$$d = \frac{2T}{b\tau} \approx \frac{Gb}{\tau} \tag{1.15}$$

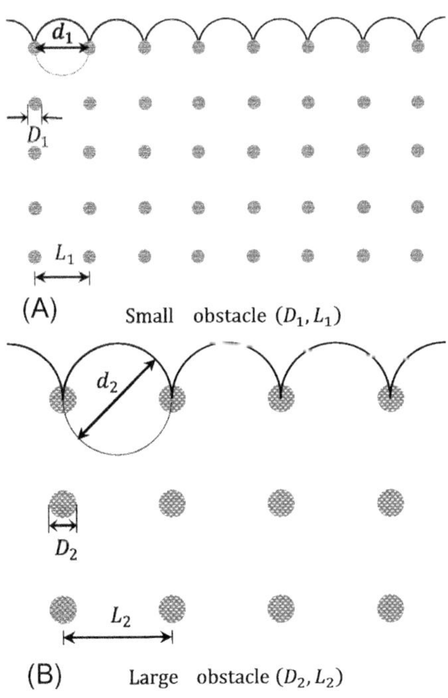

FIG. 1.7 Size effects due to the different obstacle size of: (A) small obstacle (D_1,L_1) and (B) large obstacle (D_2,L_2).

where τ is the shear stress, G is the shear modulus, b is the magnitude of Burgers vector, and T is the dislocation line tension, which can be simplified as $T \approx Gb^2/2$. Now, the minimum shear stress to activate the dislocation source, i.e. $d = L$, can be calculated as follows:

$$\tau \approx \frac{Gb}{L} \tag{1.16}$$

Accordingly, the sample with the smaller obstacle spacing L_1 (Fig. 1.7A) shows more strength compared to the one with the larger obstacle spacing L_2 (Fig. 1.7B), i.e. $\tau_2 < \tau_1$. This is just a simple first order description of size effects which states that higher strength can be reached for a finer dispersion.

Some other microstructural equations can be reproduced using the Orowan mechanism. For example, the Taylor hardening model relates the shear strength to the dislocation density ρ as follows:

$$\tau = \alpha G b \sqrt{\rho} \tag{1.17}$$

where α is a material parameter. In this case, the obstacles are forest dislocations and the obstacle spacing can be approximated by average dislocation source length L_{ave}, which can be obtained as follows:

$$L_{\text{ave}} = \frac{1}{\sqrt{\rho}} \tag{1.18}$$

Following Eq. (1.16), the material strength τ can be described as below:

$$\tau = \frac{Gb}{L_{\text{ave}}} = Gb\sqrt{\rho} \tag{1.19}$$

However, one should notice that the forest dislocations as obstacles are penetrable. Accordingly, Eq. (1.19) should be modified by using a reduction factor $\alpha < 1$ as follows:

$$\tau = \alpha G b \sqrt{\rho} \tag{1.20}$$

which is similar to the Taylor hardening model.

1.3.1.2 Grain size

1.3.1.2.1 Hall-Petch effect

One of the conventional processing methods to enhance the strength of metals is to modify the grain sizes. Hall (1951) initially addressed the effect of grain size on the strength of mild steel. He showed that the yield stress of mild steel is enhanced by decreasing its grain size. He also showed that the strength or material yield strength σ has a linear relation with the inverse root of grain size d_g, i.e., $\sigma - \sigma_0' \propto d_g^{-1/2}$, where σ_0' is the yield stress of single crystals (Hall, 1951). Hall (1951) justified the observed size effects using the pile-up theory proposed by Eshelby et al. (1951). Petch (1953) reached the same relation by studying the effect of grain size on the cleavage strength of iron and mild steel. He also attributed the observed size effect to the dislocation pile-up. Various experiments have been done on different metals to capture the variation of yield or flow stress versus the grain size, which is commonly known as the Hall-Petch effect. Fig. 1.8 presents the variation of yield or flow stress versus the inverse square root of the grain size, i.e., $d_g^{-1/2}$, for iron and steel (Hall, 1951; Petch, 1953; Armstrong et al., 1962; Kashyap and Tangri, 1997), copper (Feltham and Meakin, 1957; Hansen and Ralph, 1982), nickel (Thompson, 1977; Narutani and Takamura, 1991; Keller and Hug, 2008), and aluminum (Carreker and Hibbard, 1955; Hansen, 1977). Following the studies of Hall (1951) and Petch (1953), the general equation for the dependency of the material strength on grain size can be described as follows:

$$\sigma = \sigma_0 + K_{\mathrm{HP}} d_g^{-x} \tag{1.21}$$

where σ is the yield or flow stress, σ_0 is the corresponding stress for a very large grained bulk material or bulk single crystals, and K_{HP} is a material constant. The exponent x is a constant, where $0 < x < 1$. Many different experimental and theoretical studies have been performed to obtain the appropriate values of σ_0, K_{HP}, and x. Li et al. (2016) reviewed the conducted research to calibrate Eq. (1.21). They gathered the data bank of experimental data of flow or yield stress versus the grain size for different metals reported by many other researchers and conducted a statistical study to capture the best value of x. Besides the experimental studies, different theoretical models have been proposed to unravel the underlying mechanisms of the Hall-Petch effect. These studies include the dislocation pile-up model, dislocation generation from grain boundary ledges, dislocation density model, and non-homogenous plastic deformation model. These models are studied in more details in the following sections.

1.3.1.2.1.1 Dislocation pile-up model The dislocation pile-up model (Cottrell and Bilby, 1949; Eshelby et al., 1951) is the first theoretical model incorporated to support the Hall-Petch effect. The pile-up model states that the dislocations are moving inside the grain until they reach the grain boundary that stops their movement. Fig. 1.9 depicts the dislocation pile-up model. Since the dislocations cannot pass the grain boundary because of the mismatch between grains, the grain boundary acts as an obstacle. Accordingly, the dislocations form a pile-up behind the grain boundary with the length L_p ($L_\mathrm{p} = d_g/2$). The pile-up process continues until the stress concentration induced by the array of dislocations τ_p reaches a critical value of τ_cr. At this moment, the dislocations will pass the grain boundary and material yield occurs. To formulate this model, one can assume that the dislocation slip occurs as soon as the shear stress becomes larger than τ_0. The dislocations will be subjected to the slip stress τ_s which is:

$$\tau_\mathrm{s} = \tau - \tau_0 \tag{1.22}$$

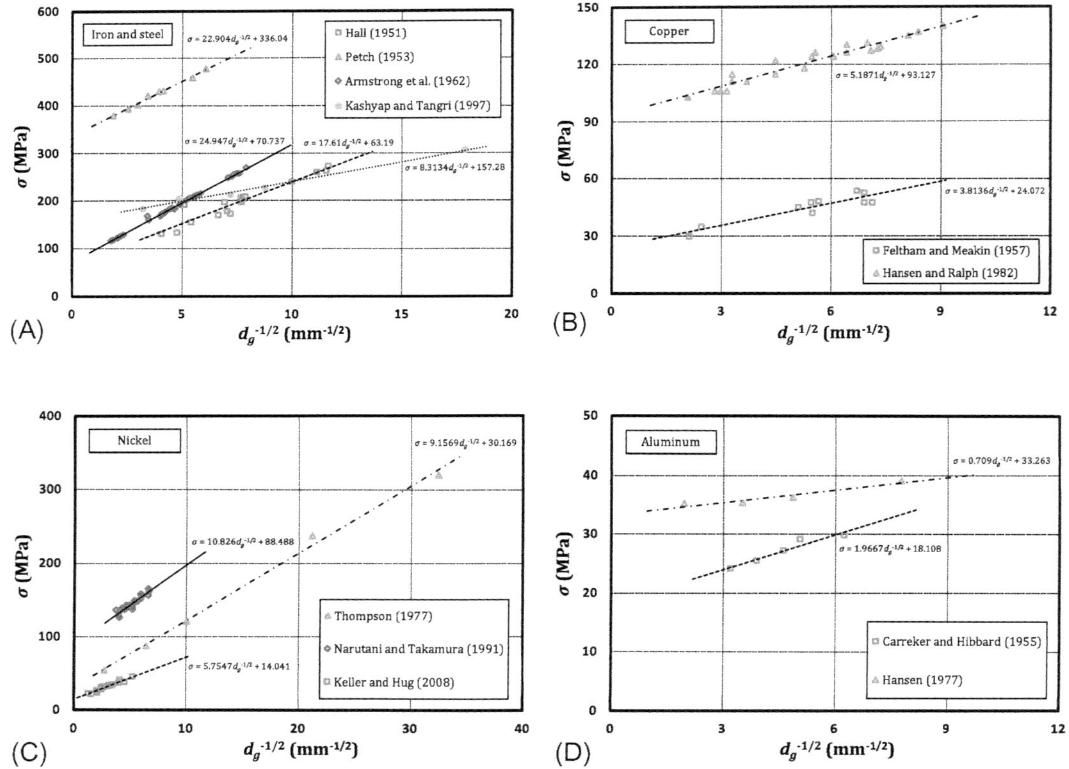

FIG. 1.8 Variation of yield or flow stress σ versus the inverse square root of grain size $d_g^{-1/2}$ for: (A) iron and steel, (B) copper, (C) nickel, and (D) aluminum.

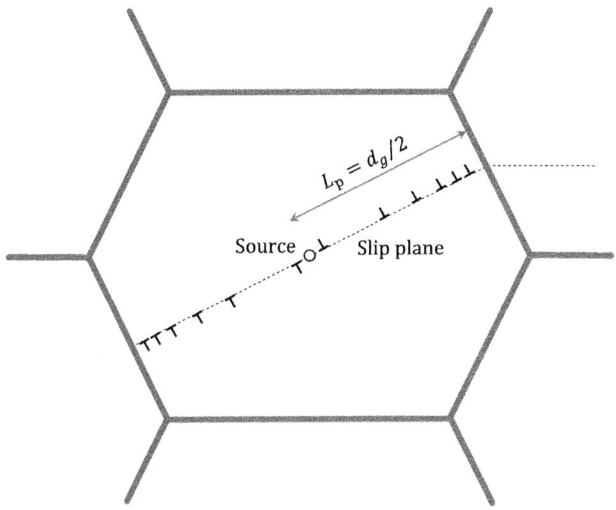

FIG. 1.9 Dislocation pile-up model to capture Hall-Petch effect.

The pileup stress induced by blocked dislocation can be calculated as follows:

$$\tau_p = n\tau_s \tag{1.23}$$

The number of dislocations n can be related to the pile-up length L_p as follows:

$$n = \frac{\tau_s L_p b}{T} \tag{1.24}$$

Incorporating Eq. (1.23) into Eq. (1.24), the pile-up stress can be obtained as follows:

$$\tau_p = \frac{\tau_s^2 L_p b}{T} \tag{1.25}$$

The material yields at $\tau_p = \tau_{cr}$. Accordingly, τ_s at yield can be calculated as follows:

$$\tau_s = \left(\frac{\tau_{cr} T}{L_p b}\right)^{1/2} \tag{1.26}$$

The shear stress at which the material yields τ_y can be obtained using Eqs. (1.22), (1.26) as follows:

$$\tau_y = \tau_0 + \left(\frac{\tau_{cr} T}{L_p b}\right)^{1/2} \xrightarrow{L_p = d_g/2} \tau_y = \tau_0 + \left(\frac{2\tau_{cr} T}{d_g b}\right)^{1/2} \tag{1.27}$$

For crystalline metals, the relation between the flow stress σ_y and shear stress τ_y can be described as follows (Kocks, 1970):

$$\sigma_y = A\tau_y \tag{1.28}$$

where A is a constant. Accordingly, using Eqs. (1.27), (1.28) and assuming $T = \overline{\alpha}Gb^2$, where $\overline{\alpha}$ is a constant of the order of unity, the relation between the flow stress σ_y and the grain size d_g can be stated as follows:

$$\sigma_y = \sigma_0 + \left(\frac{2\overline{\alpha}G\tau_{cr}b}{d_g}\right)^{1/2} \tag{1.29}$$

Comparing Eq. (1.29) with the Hall-Petch relation, i.e. Eq. (1.21), $x = 1/2$ and $K_{HP} = A(2\overline{\alpha}G\tau_{cr}b)^{1/2}$.

1.3.1.2.1.2 Dislocation generation from grain boundary ledges Another model that can justify the Hall-Petch effect is dislocation nucleation from grain boundary ledges. Li (1963) proposed that the grain boundary ledges may play as a source of dislocation nucleation (Fig. 1.10). The free energy of a ledge formation is a function of grain boundary misorientation, and it is formed in the cases of high angle grain boundaries. In this model, it is assumed that the yield stress is the one required to move the forest dislocations, which are nucleated from the grain boundary ledges, inside the grain. The dislocation density induced from the grain boundary ledges ρ_l can be approximated as follows (Li, 1963):

$$\rho_l = \frac{8m}{\pi d_g} \tag{1.30}$$

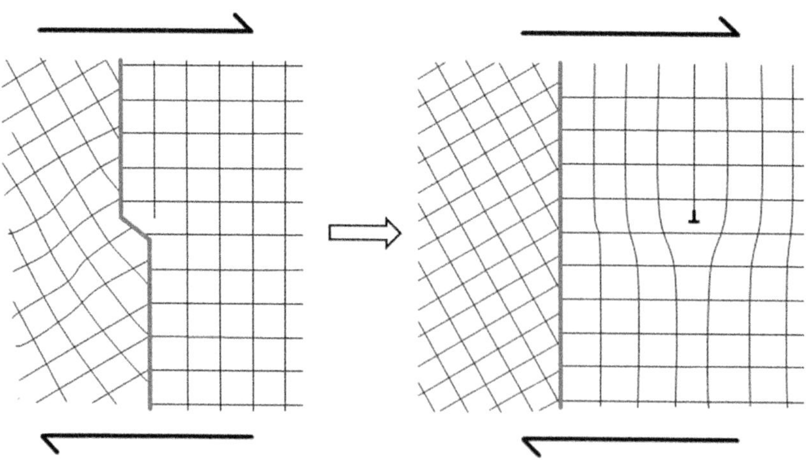

FIG. 1.10 Dislocation nucleation from the grain boundary ledge.

where m is the grain boundary ledge density, which is number per unit length or length per unit area. Considering the Taylor hardening model, i.e. $\tau \propto \sqrt{\rho}$, and assuming $\sigma = A\tau$, the flow stress can be written as follows:

$$\sigma - \sigma_0 = A\alpha Gb\left(\frac{8m}{\pi d_g}\right)^{1/2} \tag{1.31}$$

where α is the Taylor hardening model constant. Comparing Eq. (1.31) with Eq. (1.21), $x = 1/2$ and $K_{\mathrm{HP}} = A\alpha Gb(8m/\pi)^{1/2}$.

1.3.1.2.1.3 Dislocation density model Another model that has been presented to capture the Hall-Petch effect is to relate the grain size d_g to the dislocation mean free path $\bar{l}$ (Conrad, 1961). In this model, it has been assumed that the dislocation mean free path has a linear relation with the grain size as follows:

$$\bar{l} = \varsigma d_g \tag{1.32}$$

where ς is a constant. The plastic shear strain ε_{p} can be obtained from the mobile dislocation density ρ_{m} as follows (see, e.g., Hull and Bacon, 2011):

$$\varepsilon_{\mathrm{p}} = b\rho_{\mathrm{m}}\bar{l} = \varsigma b\rho_{\mathrm{m}} d_g \tag{1.33}$$

Assuming that the mobile dislocation density is a constant share of the total one, i.e. $\rho_{\mathrm{m}} = \phi\rho$, Eq. (1.33) can be rewritten as follows:

$$\varepsilon_{\mathrm{p}} = \varsigma b\phi\rho d_g \tag{1.34}$$

Assuming that the shear stress follows Taylor hardening model, i.e. $\tau \propto \sqrt{\rho}$, and $\sigma = A\tau$, one can obtain the relation between the flow stress and grain size as follows:

$$\sigma - \sigma_0 = A\alpha G\left(\frac{b\varepsilon_{\mathrm{p}}}{\varsigma\phi d_g}\right)^{1/2} \tag{1.35}$$

which is similar to the Hall-Petch relation. Comparing Eqs. (1.36), (1.21), $x = 1/2$ and $K_{\mathrm{HP}} = A\alpha G(b\varepsilon_{\mathrm{p}}/\varsigma\phi)^{1/2}$. Unlike the general Hall-Petch relation, K_{HP} is not a constant and varies with plastic strain ε_{p}. The dislocation density model predicts that there is no Hall-Petch effect at material yield, i.e. $K_{\mathrm{HP}}|_{(\varepsilon_{\mathrm{p}}=0)} = 0$. Accordingly, this model cannot capture the Hall-Petch effect at the initial phases of plasticity. However, for larger values of plastic strain, it may be incorporated to obtain an initial guess for K_{HP}.

1.3.1.2.1.4 Non-homogenous plastic deformation model Ashby (1970) incorporated the concept of geometrically necessary dislocations (GNDs) to capture the effect of grain size on material strength. He stated that the deformation of a polycrystalline sample, even a pure metal, is non-uniform. In this model, the deformation in the sample is divided into two parts; a uniform

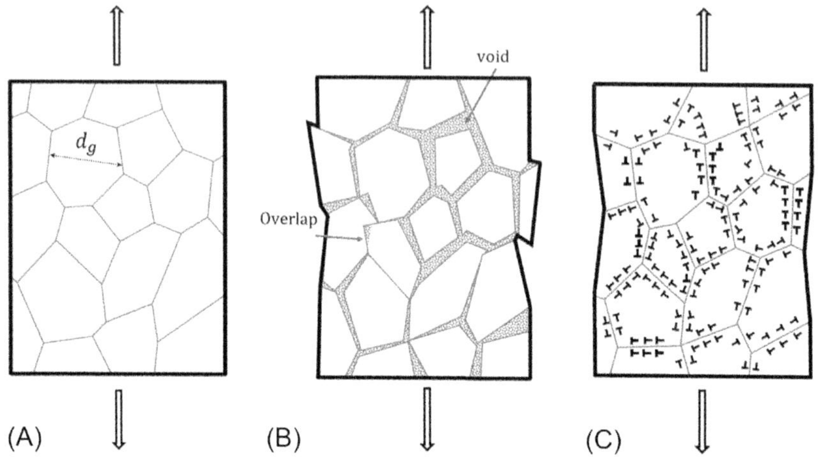

FIG. 1.11 Non-homogeneous plastic deformation of a polycrystal sample: (A) initial sample, (B) uniform deformation of grains leading to voids and overlaps formation, and (C) non-uniform and local deformation provided by GNDs to remove voids and overlaps.

deformation provided by the dislocation movement inside the grain, and a non-homogenous deformation which is the result of GNDs slip to produce the compatible deformation pattern between grains (Fig. 1.11). Ashby (1970) estimated the number of GNDs induced to accommodate the nun-uniform deformation as $d_g\bar{\epsilon}/4b$, where $\bar{\epsilon}$ is the applied uniaxial strain. In this estimation, he assumed that the voids and overlaps amount is always proportional to $d_g\bar{\epsilon}/2$. In the case of 2D sample, the grain area is approximately equal to d_g^2. Accordingly, the density of GNDs (ρ_G) can be obtained as follows:

$$\rho_G \cong \frac{\bar{\epsilon}}{4bd_g} \tag{1.36}$$

In the range of small strains, the total dislocation density can be replaced by the GND density, i.e. $\rho \cong \rho_G$ (Ashby, 1970). Accordingly, assuming that the shear stress follows Taylor hardening model, i.e. $\tau \propto \sqrt{\rho}$, and $\sigma = A\tau$, the flow stress can be described as a function of grain size as follows:

$$\sigma - \sigma_0 = A\alpha G\left(\frac{b\bar{\epsilon}}{4d_g}\right)^{1/2} \tag{1.37}$$

Comparing the general Hall-Petch effect, Eq. (1.21) with Eq. (1.37), one can obtain $x = 1/2$ and $K_{HP} = A\alpha G(b\bar{\epsilon}/4)^{1/2}$. Similar to the dislocation density model, K_{HP} is not a constant and varies with the applied uniaxial strain $\bar{\epsilon}$.

1.3.1.2.2 Inverse Hall-Petch effect

The general form of the Hall-Petch relation, i.e. Eq. (1.21), states that the material yield or flow stress increases as the grain size decreases. However, similar to all other material properties, the strength cannot become unlimited. Accordingly, Eq. (1.21) is not valid for polycrystalline samples with grain size smaller than some specific limit d_g^l. As the grain size decreases, two typical behaviors have been reported. One trend states that the material yield or flow stress reaches a plateau for smaller grain sizes. The other trend states that after some critical grain size, the material yield or flow stress decreases as the grain size decreases, which is commonly known as the inverse Hall-Petch effect. The inverse Hall-Petch was firstly reported by Chokshi et al. (1989) in which they studied the effects of grain size on the Vickers microhardness for nanocrystalline copper and palladium (Fig. 1.12). They attributed the inverse Hall-Petch effect to the diffusional creep occurring at room temperature for nanocrystalline materials. Later on, Weertman (1993) investigated the Vickers microhardness and tensile yield stress as the grain size varies for nanocrystalline palladium and copper. They did not observe the inverse Hall-Petch effect in any of their strength measurements (Fig. 1.12). In the case of nanocrystalline palladium, it was shown that the microhardness has a slight dependency on the grain size. Weertman (1993) stated that it is difficult to construct a Hall-Petch curve for the palladium tensile yield stress due to the uncertainty in the obtained results. However, it was observed that the curve is quite flat for nanocrystalline palladium. In the case of nanocrystalline copper, both microhardness and tensile yield stress follow the Hall-Petch relations. However, the slope of the curve is only one-seventh of the slope for copper with conventional grain sizes. They suggested that the inverse Hall-Petch effect observed

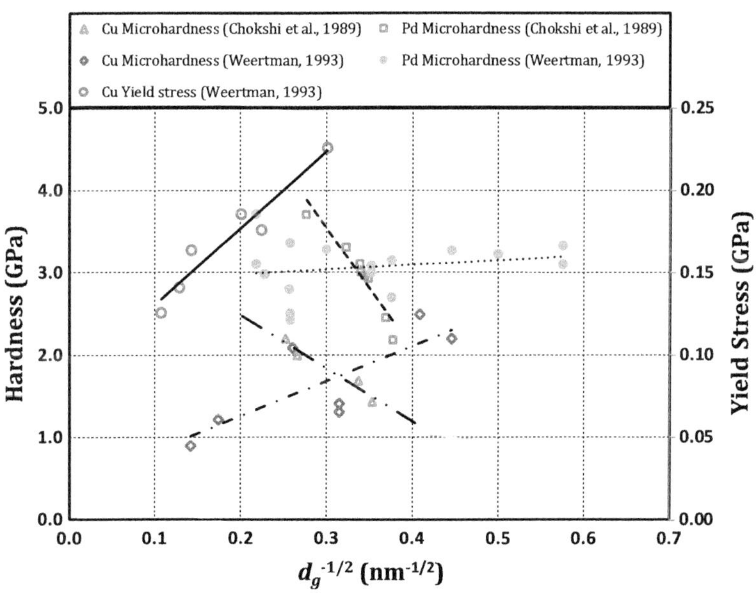

FIG. 1.12 Variation of hardness and tensile yield stress versus the inverse root of grain size $d_g^{-1/2}$ for nanocrystalline copper and palladium. The original experimental data has been reported by Chokshi et al. (1989) and Weertman (1993).

by Chokshi et al. (1989) is originated from the repeated annealing on one sample to obtain various grain size.

After Chokshi et al. (1989), several researchers have experimentally observed the inverse Hall-Petch effect. Fig. 1.13 shows the variation of flow stress σ versus the inverse root of grain size $d_g^{-1/2}$ for copper at 300 K subjected to strain rates in the range of $10^{-5}-10^{-3}$. The original data has been reprted by Henning et al. (1975), Merz and Dahlgren (1975), Hansen and Ralph (1982), Chokshi et al. (1989), Nieman et al. (1991), Embury and Lahaie (1993), Sanders et al. (1996, 1997), Cai et al. (1999), Huang and Saepen (2000), Conrad and Yang (2002). Fig. 1.13 includes the experimental data for the free-standing, vapor-deposited (VP) and electrodeposited (EP) copper films with $11 \leq d_g \leq 250$ μm which are scattered about the dashed line denoting an extrapolation of results reported by Hansen and Ralph (1982). Koch and Narayan (2001) investigated the validity of experiments demonstrating inverse Hall-Petch effect and concluded that only a few of them are reliable. The possible experimental errors that may alter the experimental results are the incomplete removal of porosity, poor powder-to-powder bonding, residual amorphous phase, and change in the composition and morphology. Koch and Narayan (2001) listed the experimental results with all details and possible artifacts (Table 1.1). They stated that the inverse Hall-Petch is real by considering a few promising experiments and atomistic simulation results.

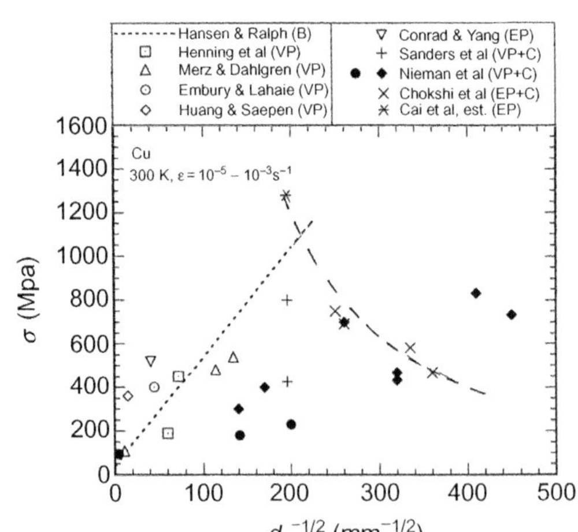

FIG. 1.13 Variation of flow stress versus σ the inverse root of grain size $d_g^{-1/2}$ for copper at 300 K subjected to the strain rates in the range of $10^{-5}-10^{-3}$. Symbols and $\times$ are for hardness with $\sigma = H/3$; the remaining symbols are for the flow stress at 0.01 tensile strain; EP denotes electrodeposits; VP denotes vapor-deposits; and +C denotes compacted powders. The original experimental data has been reported by Henning et al. (1975), Merz and Dahlgren (1975), Hansen and Ralph (1982), Chokshi et al. (1989), Nieman et al. (1991), Embury and Lahaie (1993), Sanders et al. (1996), Sanders et al. (1997), Cai et al. (1999), Huang and Saepen (2000), Conrad and Yang (2002). *(After Conrad, 2003.)*

TABLE 1.1 Description of experiments that show inverse Hall-Petch effect

Material	Preparation method	Grain size varied by annealing	Counter example in same material	Possible artifacts	Reference
Cu	IGC	Yes	Yes	Porosity	Chokshi et al. (1989)
Pd	IGC	Yes	Yes	Porosity	Chokshi et al. (1989)
Ni-P	Electrodeposition	No	No	Amorphous regions	Palumbo et al. (1990)
Ni-P	Crystallization of amorphous alloy	Yes	No	Changes in chemistry, morphology	Lu et al. (1990)
Ti-Al-Nb	Mechanical attrition	Yes	No	Changes in chemistry, porosity	Christman and Jain (1991)
$NbAl_3$	Mechanical attrition	Yes	No	Chemistry, particle bonding	Kim and Okazaki (1992)
TiAl	Sputtering, variation of IGC	Yes	No	Porosity	Chang et al. (1992)
$Fe_{80}Ni_{20}$	Electrodeposition	No	No		Cheung et al. (1994)
$(Fe, Co)_{33}Zr_{67}$	Crystallization of amorphous alloy	Yes	No	Amorphous regions	Alves et al. (1996)
Fe	Mechanical attrition	Yes	Yes	Porosity	Khan et al. (2000)
Ni	Electrodeposition	No	No		Erb (1995)
Zn	Laser ablation	No	No		Narayan (2000)

IGC denotes for inert gas-condensation method.
After Koch, C.C., Narayan, J., 2001. The inverse Hall-Petch effect-fact or artifact. Mater. Res. Soc. Symp. Proc. 634, B5 1.1–B5 1.11.

Chokshi et al. (1989) related inverse Hall-Petch effect to the diffusion creep rate for the nanocrystalline materials. They incorporated the Coble creep as the governing mechanism of deformation as follows:

$$\dot{\varepsilon} = \frac{150\Omega\delta D_{gb}\sigma}{\pi kTd_g^3} \tag{1.38}$$

where $\dot{\varepsilon}$ is the strain rate, Ω is the atomic volume, δ is the width of the grain boundary, D_{gb} is the diffusion coefficient of grain boundary, σ is the applied stress, k is the Boltzmann's constant, and T is the absolute temperature. Assuming $\Omega = 1.3 \times 10^{-29}$ m^3, $\delta = 1$ nm, $D_{gb} = 3 \times 10^{-9}e^{-62,000/RT}$ m^{2}s^{-1}, where $R = 8.31$ J mol^{-1} is the gas constant, $T = 300$ K, and $d_g = 5$ nm, Eq. (1.38) can be rewritten as follows (Chokshi et al., 1989):

$$\dot{\varepsilon} = 6 \times 10^{-11}\sigma \tag{1.39}$$

where $\dot{\varepsilon}$ and σ are considered in s^{-1} and Pa, respectively. For a specific grain size of $d_g = 5$ nm, if the stress changes from 100 to 1000 MPa, the strain rate at which the described diffusion process becomes the governing mechanism varies from 0.006 to 0.06 s^{-1}. Chokshi et al. (1989) did not consider the plastic deformation, which is not negligible, in their calculations. The Coble creep, however, has not been experimentally established and the predicted creep rate in nanocrystalline copper is two orders of magnitude smaller than the observed one (Nieman et al., 1991).

Besides the experimental results, atomistic simulation results have also supported the inverse Hall-Petch trend. Schiøtz et al. (1998) addressed the inverse Hall-Petch effect for nanocrystalline copper samples with the maximum simulated grain size of 6.56 nm during the uniaxial tension experiment using atomistic simulation. Schiøtz et al. (1998) reported that the large number of small sliding events in grain boundary is the dominant deformation mechanism. Fig. 1.14 shows the simulation results in which both yield stress and flow stress decrease as the grain size decreases. Schiøtz and Jacobsen (2003) conducted large atomistic simulations of copper during uniaxial deformation. They increased the size of largest grain to 48.6 nm. Accordingly, they were able to capture both Hall-Petch and inverse Hall-Petch effects as presented in Fig. 1.15. Swygenhoven and his co-workers (Van Swygenhoven and Caro, 1997; Van Swygenhoven et al., 2001a,b; Derlet et al., 2003; Hasnaoui et al., 2004; Derlet and Van Swygenhoven, 2002) have investigated the governing deformation mechanisms in nanocrystalline materials using atomistic simulation. Van Swygenhoven and Caro (1997) incorporated the atomistic simulation results to develop the following non-linear viscous model for plastic deformation of nanocrystalline materials at low temperature and high stresses:

$$\dot{\varepsilon} = \left(\frac{1}{d_g\beta}\right)\sigma \tag{1.40}$$

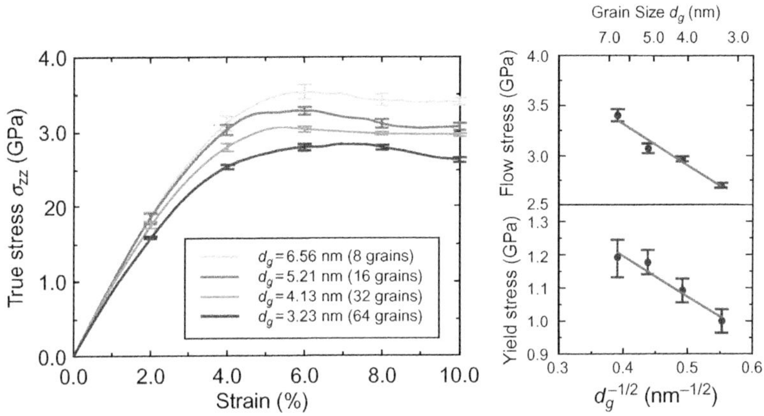

FIG. 1.14 The effect of grain size one the response of nanocrystalline copper with different grain size. *(After Schiøtz, J., Di Tolla F.D., Jacobsen, K.W., 1998. Softening of nanocrystalline metals at very small grain sizes. Nature 391, 561–563.)*

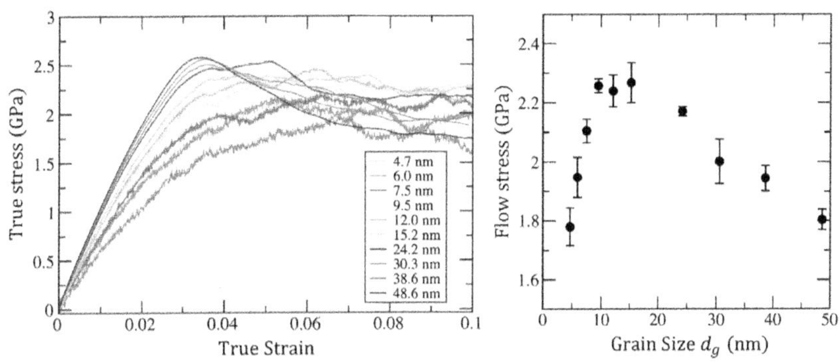

FIG. 1.15 The effect of grain size one the response of nanocrystalline copper with different grain size. *(After Schiøtz and Jacobsen, 2003.)*

where β characterizes the viscosity of a sample with $d_g = 1$ nm. Yamakov and coworkers (Yamakov et al., 2002, 2003a,b; Wolf et al., 2005) performed various atomistic simulations to investigate the effects of grain size on the controlling mechanisms of deformation in the case of nanocrystalline fcc metals. They have observed that by decreasing the grain size, the governing deformation mechanism changes from a dislocation-based to a grain boundary-based mechanism.

Different deformation mechanisms have been proposed to capture the deviation from the Hall-Petch relation including the breakdown in pile-up model (Pande et al., 1993; Hughes et al., 1986; Smith et al., 1994; Armstrong and Hughes, 1999), grain boundary sliding (Ball and Hutchinson, 1969; Hahn et al., 1997; Conrad and Narayan, 2000), phase mixture model (Bush, 1993;

Wang et al., 1995; Carsley et al., 1995; Kim et al., 2000; Fu et al., 2001; Voyiadjis and Deliktas, 2010), grain coalescence (Jiang and Jia, 1996; Jia et al., 2000, 2001; Wang et al., 2002), and shear band formation (Jia et al., 2003; Wei et al., 2004). Meyers et al. (2006) reviewed all these mechanisms in detail. Among these models, the breakdown in pile-up theory, grain boundary sliding, and phase mixture model will be discussed here in more details in the following sections.

1.3.1.2.2.1 Breakdown in dislocation pile-up model As explained in Section 1.3.1.2.1.1, the dislocation pile-up model was the first one introduced to justify the observed Hall-Petch effect. As Fig. 1.9 shows, dislocations are induced from the source and move toward the grain boundary. The stress concentration due to this pile-up against the grain boundary is the driving force of the pile-up model. As the grain size decreases, the number of dislocations inside the grain also drops. Accordingly, there is a grain size at which the number of dislocations are so few that the pile-up theory breaks down (Fig. 1.16).

For the first approximation, a grain size limit at which no more than one dislocation loop can be fit inside the grain is considered as the limit for the pile-up theory d_g^{cr}. Nieh and Wadsworth (1991) predicted the value of d_g^{cr} by introducing the critical equilibrium dislocation between two dislocations l_c as follows:

$$l_c = \frac{3Gb}{\pi(1-\nu)H} \tag{1.41}$$

where ν is Poisson's ratio and H is the hardness. In the case of $d_g < l_c$, the grain cannot contain more than one dislocation and no pile-up forms, i.e. $d_g^{cr} = l_c$.

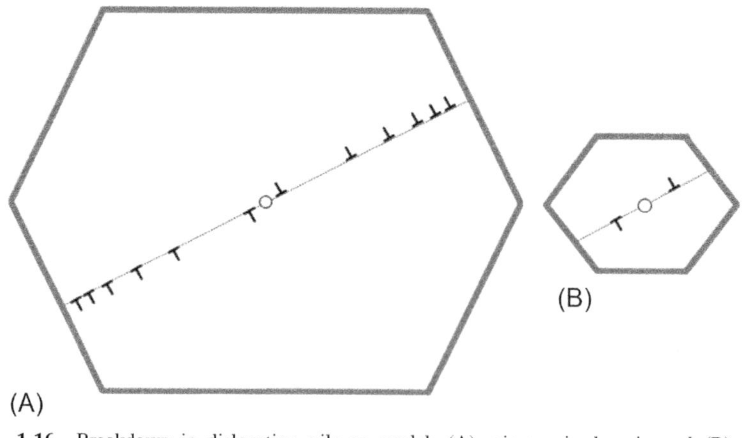

FIG. 1.16 Breakdown in dislocation pile-up model: (A) micron sized grain and (B) nano sized grain.

Pande et al. (1993) introduced a new model to predict the deviation from pile-up theory as the grain size decreases. The tip stress σ_{tip} for a dislocation pile-up can be obtained as follows (Fig. 1.17):

$$\sigma_{\text{tip}} = \frac{(n+m-1)\sigma}{m} \tag{1.42}$$

where σ is the applied stress, n is the number of dislocations, and mb is the Burgers' vector of the locked dislocation. The material yields when the tip stress σ_{tip} reaches the barrier stress σ^*, which is assumed to be a constant. Accordingly, the relation between the yield stress σ_y and the dislocation pile-up properties can be obtained as below:

$$\sigma_y = \frac{m\sigma^*}{(n+m-1)} \tag{1.43}$$

In order to obtain a relation between the yield stress σ_y and grain size d_g, the dislocation pile-up length L_p should be first related to the number of dislocations n. Pande et al. (1993) incorporated the following equation to relate the dislocation pile-up length L_p to the number of dislocations n:

$$L_p^{-1/2} = \sqrt{\frac{A}{2\sigma}}\left\{2(n+m-1)^{1/2} - \beta[4(n+m-1)]^{-1/6}\right\} \tag{1.44}$$

where $\beta = 1.85575$, and $A = Gb/(\pi\sigma^*)$ for a screw dislocation. Accordingly, the relation between the yield stress σ_y and dislocation pile-up length L_p can be reached by combinig Eqs. (1.43), (1.44) as follows:

$$\sigma_y = (2Am\sigma^*)L_p^{-\frac{1}{2}}f(L_p) \tag{1.45}$$

Fig. 1.18 compared the variation of scaled yield stress $\sigma_y/(G\sigma^*/\pi)^{1/2}$ versus the inverse square root of scaled dislocation pile-up length L_p/b predicted by the original Hall-Petch relation, the approximate size effect equation, i.e. Eq. (1.45), and the exact size effect relation predicted by the dislocation

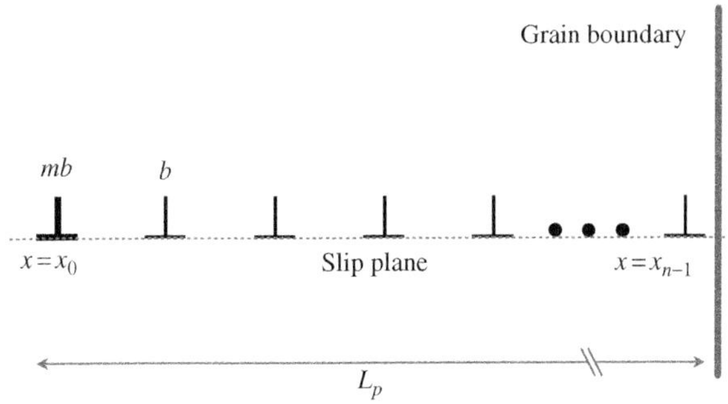

FIG. 1.17 An array of dislocations piled up against a grain boundary.

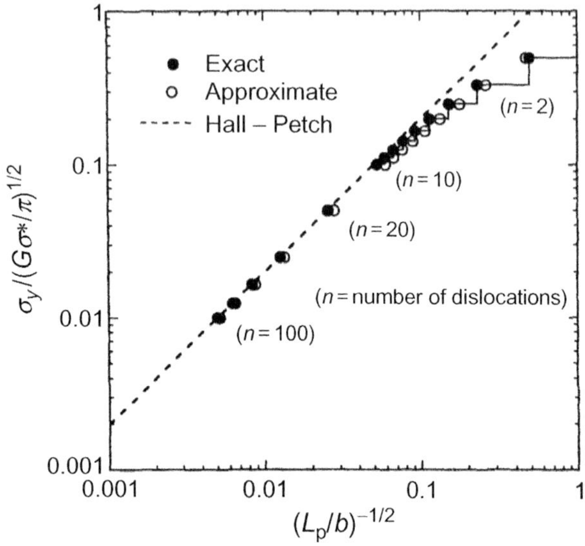

FIG. 1.18 Variation of scaled yield stress $\sigma_y/(G\sigma^*/\pi)^{1/2}$ versus the inverse square root of scaled dislocation pile-up length L_p/b predicted by original Hall-Petch relation, the approximate size effect equation, i.e. Eq. (1.42), and the exact size effect relation predicted by the dislocation pile-up equation developed by Mitchell et al. (1965). *(After Pande, C.S., Masumura, R.A., Armstrong, R.W., 1993. Pile-up based Hall-Petch relation for nanoscale materials. Nanostruct. Mater. 2, 323–331.)*

pile-up equation developed by Mitchell et al. (1965). The result shows that original Hall-Petch relation deviates from the exact and approximate predictions for $n < 20$. Also, in the cases of $n < 20$, both approximate and exact predictions exhibit discrete steps, and the difference between these predictions and Original Hall-Petch relation increases as the number of dislocations decreases.

1.3.1.2.2.2 Grain boundary sliding Grain boundary sliding (Fig. 1.19) has been introduced as one of the underlying mechanism of superplasticity, the phenomenon in which a polycrystalline material exhibits very high tensile elongations prior to failure (Nieh et al., 1997). The combination of grain boundary sliding with diffusion creep, grain boundary migration, or dislocation slip

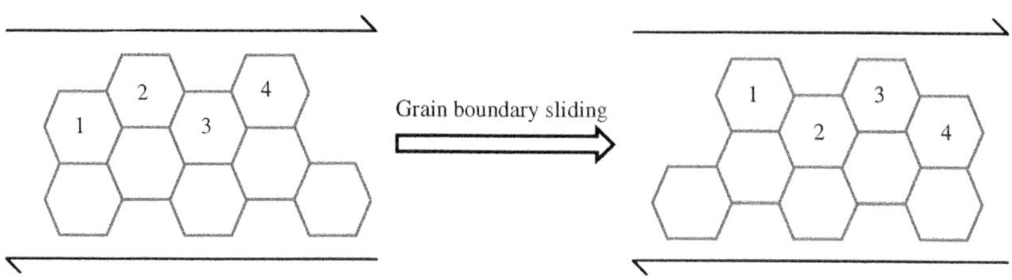

FIG. 1.19 Grain boundary sliding in polycrystalline materials.

and/or climb have been identified as superplasticity mechanisms (Davies et al., 1970). Cooperative grain boundary sliding, i.e., clusters of grains move cooperatively, has also been reported as a superplasticity deformation mechanism (Astanin et al., 1991; Yang et al., 1992), which is combined with the cooperative grain boundary migration and grain boundary rotation (Zelin and Mukherjee, 1995, 1996). As the grain size decreases, the number of dislocations fit in the grains decreases. Accordingly, the applied plastic flow cannot be sustained by the dislocations slip inside, and other mechanisms should govern the deformation process. In the case of nanocrystalline materials, the grain boundary sliding is the dominant deformation mechanism for $d_g^{cr} \approx 10-50$ nm (Conrad and Narayan, 2000).

Various theories have been proposed to incorporate the effect of grain boundary sliding into size effect models. Hahn and Padmanabhan (1997) and Hahn et al. (1997) developed the following relation to capture the inverse size effects due to the grain sliding:

$$H_v = H_{v0} - \frac{m_2}{d_g}\left(d_g - m_3\right)^{1/2} \text{for}\, d_g < d_g^{cr} \tag{1.46}$$

where H_v is the hardness, and m_2 and m_3 are the material properties. Eq. (1.46) is applicable for the grain size region where the grain boundary sliding is the dominant deformation mechanism. However, the model is largely phenomenological and contains many adjustable parameters (Conrad and Narayan, 2000).

Conrad and Narayan (2000) developed an atomistic size effect model for nanocrystalline materials with $d_g < d_g^{cr}$ in which the grain boundary sliding governs the deformation process. They developed the model based on the concept of thermally-activated shear and atomistic simulation results and compared the model with the available experiments. The macroscopic shear strain rate induced by atomic shear events at grain boundaries can be described as follows:

$$\dot{\gamma} = N_v A b \nu e^{\left[\frac{-\Delta G^*(\tau_e)}{KT}\right]} \tag{1.47}$$

where N_v is the number of places per unit volume where thermally-activated shear can occur, A is the area swept out per successful thermal fluctuation, ν is the vibration frequency, G^* is the free energy of activation, and $\tau_e = \tau - \tau_0$, where τ and τ_0 are the applied and threshold stresses, respectively. Eq. (1.44) can be further simplified by assuming $N_v = \delta/(d_g b^3)$, where $\delta \approx 3b$ is the width of the grain boundary, $A = b^2$, $\nu = \nu_D \approx 10^{13}\ s^{-1}$ is the Debye frequency, and $\Delta G^* = \Delta F^* - \tau_e V^*$, where F^* is the Helmholtz free energy and $V^* = b^3$ is the activation volume. Accordingly, Eq. (1.47) can be written as follows:

$$\dot{\gamma} = \frac{6b\nu_D}{d_g} \sinh\left(\frac{\tau_e V^*}{KT}\right) e^{\left[\frac{-\Delta F^*}{KT}\right]} \tag{1.48}$$

Fig. 1.20 compares Eq. (1.48) with the experiments conducted on Cu (Chokshi et al., 1989; Conrad and Narayan, 2000), Pd (Chokshi et al., 1989), Ni-P (McMahon and Erb, 1989; Lu et al., 1990); TiAl (Chang et al., 1991; Alstetter, 1993), and $Nb_{77}Al_{23}$ (Kim and Okazaki, 1992), in the region of $d_g < d_g^{cr}$. The activation volume V^* and Helmholtz free energy change ΔF^* were obtained using the presented experimental data. The results show that Eq. (1.48) can successfully capture the size effects in the region of $d_g < d_g^{cr}$.

Conrad (2003, 2004) defines three regions of grain size d_g for copper at temperature range of 77–373 K using the available data in the literature, where the governing deformation mechanisms in each region is distinct. All three regimes are depicted in Fig. 1.21. As the results show, the slope of the Hall-Petch curve in *regime I* is much larger than that of *regime II*. However, both slopes are still positive. In the case of *regime III*, on the other hand, the slope becomes negative which shows the inverse Hall-Petch effect. Conrad (2003, 2004) describes the three regimes as follows:

- *Regime I* ($d_g > \sim 10^{-6}$ m)

  The dislocation-pile up is the dominant strengthening mechanism. The total dislocation density ρ can be written as follows:

$$\rho = \rho_0 + \rho_{s,b} + \rho_{g,b} \tag{1.49}$$

where ρ_0 is the dislocation density stored for the sample with a very large grain size, $\rho_{s,\ b}$ is the additional dislocation density stored inside the grain due to the grain boundary, and $\rho_{g,\ b}$ is the geometrically necessary

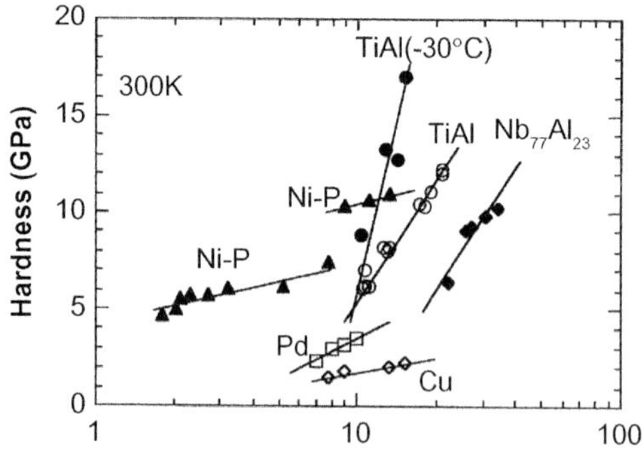

FIG. 1.20 Variation of hardness versus log(d_g) in the region of $d_g < d_g^{cr}$. The model developed by Conrad and Narayan (2000) is compared with the experimental results of Cu (Chokshi et al., 1989; Conrad and Narayan, 2000), Pd (Chokshi et al., 1989), Ni-P (McMahon and Erb, 1989; Lu et al., 1990); TiAl (Chang et al., 1991; Alstetter, 1993), and $Nb_{77}Al_{23}$ (Kim and Okazaki, 1992). *(After Conrad, H., Narayan, J., 2000. On the grain size softening in nanocrystalline materials. Scr. Mater. 42, 1025–1030.)*

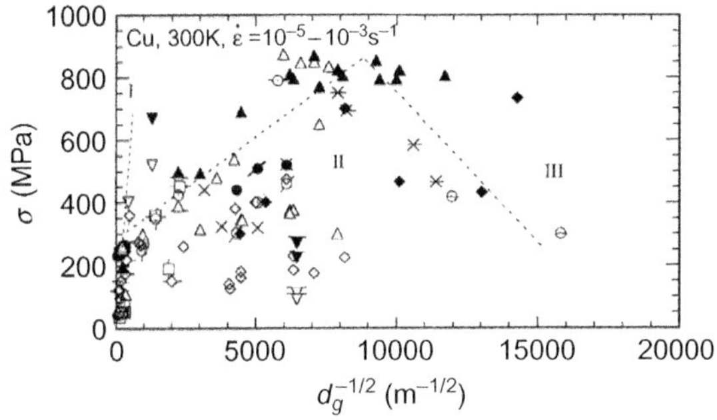

FIG. 1.21 Variation of flow stress σ versus the inverse root of grain size $d_g^{-1/2}$ for copper at 300 K subjected to strain rates in the range of $10^{-5}-10^{-3}$. *(After Conrad, H., 2004. Grain-size dependence of the flow stress of Cu from millimeters to nanometers. Metall. Mater. Trans. A Phys. Metall. Mater. Sci. 35A, 2681–2695.)*

dislocation induced because of non-homogenous deformation of polycrystalline sample. Conrad (2004) defined $\rho_{s,\ b}$ and $\rho_{g,\ b}$ as follows:

$$\rho_{s,b} = \frac{\beta_s \varepsilon}{bd_g}, \rho_{g,b} = \frac{\beta_g \varepsilon}{bd_g} \tag{1.50}$$

where ε is the tensile plastic strain, $\beta_s = 2-10$, and $\beta_g = 0.25$. Accordingly, the applied resolved shear stress τ can be defined using Taylor hardening model (Eq. 1.17) as follows:

$$\tau = \alpha Gb\sqrt{\rho} = \alpha Gb\sqrt{\rho_0 + \rho_b} \tag{1.51}$$

where $\rho_b = \rho_{s,\ b} + \rho_{g,\ b}$. Assuming $\sigma = A\tau$ and considering $\rho_b = \beta\varepsilon/bd_g$, where $\beta = \beta_s + \beta_g$, the flow stress σ can be obtained as follows:

$$\sigma = A\alpha Gb\left[\rho_0(T, \varepsilon, \dot{\varepsilon}) + \beta(T, \dot{\varepsilon})\varepsilon/bd_g\right]^{1/2} \tag{1.52}$$

Another way to consider grain boundary in *regime I* is to assume that the dislocations related to the grain boundary are located close to it. Accordingly, a composite model can be incorporated to capture the applied resolved shear stress τ as follows:

$$\tau = \alpha Gb\sqrt{\rho_0} + \alpha Gb\sqrt{\rho_b} \tag{1.53}$$

By considering $\sigma = A\tau$, the flow stress σ can be obtained as follows:

$$\sigma = A\alpha Gb\left\{\left[\rho_0(T, \varepsilon, \dot{\varepsilon})\right]^{1/2} + \left[\beta(T, \dot{\varepsilon})\varepsilon/bd_g\right]^{1/2}\right\} \tag{1.54}$$

Conrad (2004) stated that Eq. (1.52) is in better accord with copper experimental results. In other words, the excess dislocations induced by grain boundaries are mostly distributed throughout the grain.

- *Regime II* ($d_g \approx 10^{-8} - 10^{-6}$ m)
  The dislocations are still formed inside the grains. However, they cannot form a cell structure similar to *regime I*, and they are gliding on their slip planes. In this regime, the grain boundary shear is promoted by the pile-up dislocations. In addition, the total dislocation density is independent of grain size and can be described as follows:

$$\rho = \frac{\pi \tau_e \tau_\mu}{\alpha (Gb)^2} \tag{1.55}$$

where $\tau_e = \tau - \tau_\mu$ is the effective resolved shear stress acting on the pileup of n dislocations, and τ_μ is the long range internal stress due to the dislocations on the parallel slip planes. The plastic deformation process is assumed to be thermally activated shear of individual or small groups of atoms in the grain boundaries. Accordingly, the rate of plastic strain can be obtained as follows:

$$\dot{\gamma} = \dot{\gamma}_0 \exp\left[-\left(\frac{\Delta F^* - \frac{\pi V^* d_g \tau_e^2}{Gb}}{kT} \right) \right] \tag{1.56}$$

where ΔF^* is the difference in Helmholtz free energy for the process, V^* is the activation volume, k is the Boltzmann's constant, T is the absolute temperature, and $\dot{\gamma}_0$ is pre-exponential, which can be described as follows:

$$\dot{\gamma}_0 = \frac{\pi \tau_e \tau_\mu d_g \nu_D}{\alpha G^2 b} \tag{1.57}$$

where ν_D is the Debye frequency. The relation between the flow stress σ and grain size d_g can be obtained by rearranging Eq. (1.53) and considering $\sigma = A\tau$ as follows:

$$\sigma = A\alpha G b \rho^{1/2} + A d_g^{-1/2} \left\{ \left[\left(\Delta F^* - kT \ln\left(\frac{\dot{\gamma}_0}{\dot{\gamma}} \right) \right) \right]^{1/2} \left(\frac{Gb}{\pi V^*} \right)^{1/2} \right\} \tag{1.58}$$

- *Regime III* ($d_g \approx 10^{-8} - 10^{-6}$ m)
  In this regime, the dislocation slip inside the grain is not accountable for the material plastic deformation. Two mechanisms of Coble creep and grain boundary shear have been considered as the governing mechanisms of deformation in this regime. The tensile strain rate of Coble creep can be described as follows:

$$\dot{\varepsilon} = \frac{54 \Omega D_{gb} \sigma_e}{kT d_g^3} \tag{1.59}$$

where $\sigma_e = \sigma - \sigma_0$ is the effective tensile stress, σ_0 is threshold stress, Ω is the atomic volume, and D_{gb} is the diffusion coefficient of grain boundary.

In the case of grain boundary shear mechanism, the grain boundary shear strain rate $\dot{\gamma}$ can be defined as follows:

$$\dot{\gamma} = \frac{2\delta\nu_D}{d_g} \sinh\left(\frac{\tau_e^* V^*}{kT}\right) \exp\left(-\frac{\Delta F^*}{kT}\right) \tag{1.60}$$

where $\delta \approx 3b$ is the boundary width, $\tau_e^* = \tau - \tau_0$ is the effective thermal shear stress, and τ_0 is the threshold stress for shearing of grain boundary atoms. Considering $\tau_e^* V^* / kT \gg 1$, the sinh function can be approximated by an exponential and Eq. (1.57) can be rewritten as follows:

$$\tau_e^* = \left[\frac{kT}{V^*} \ln\left(\frac{\delta\nu_D}{\dot{\gamma}}\right) + \frac{\Delta F^*}{V^*}\right] + \frac{kT}{V^*} \ln\left(d_g\right) \tag{1.61}$$

Conrad (2004) compared two mechanisms of Coble creep and grain boundary shear, i.e., Eq. (1.59), (1.61), with the experimental results. He stated that the grain boundary shear mechanism is in better agreement with the experimental data compared to the Coble creep.

A major step in grain boundary sliding process is the compatibility of deformations. The grain boundary sliding cannot occur as simple as shown in Fig. 1.19. In other words, the plastic deformation should be incorporated to accommodate the grain boundary sliding. Adding the plastic term to the diffusion model can considerably change the material response (Fu et al., 2001). The schematic of grain boundary sliding in polycrystalline materials is illustrated in Fig. 1.22. Raj and Ashby (1971) developed a model in which the grain boundary sliding is accommodated by diffusion (Fig. 1.22C). Fu et al. (2001) demonstrated that the plastic deformation reduces the jaggedness of the interface and contributes to the grain boundary sliding accommodation (Fig. 1.22D). Fu et al. (2001) incorporated a plastic accommodation process for grain boundary sliding into their model as follows:

$$\tau_a = (\eta_i + \eta_D)\dot{\gamma} + \tau_p \tag{1.62}$$

where τ_a is the shear stress, $\dot{\gamma}$ is the shear strain rate, η_i is the intrinsic grain boundary viscosity, η_D is the diffusional component related to the accommodation process, and τ_p is the shear stress related to the plastic stress. As depicted in Fig. 1.22, since the grain boundary sliding is not fully compatible, an accommodation process is required. Raj and Ashby (1971) considered the diffusion as the governing deformation mechanism of the accommodation process. They defined the sliding rate $\dot{u}$ as follows:

$$\dot{u} = \frac{2\tau_a \Omega\, \lambda}{\pi\, kT\, h^2} D_V \left(1 + \frac{\pi \delta D_B}{\lambda\, D_V}\right) \tag{1.63}$$

where λ and h are the characteristics of the sinusoidal boundary (Fig. 1.22C), δ is the boundary thickness, and D_B and D_V are the volume and boundary diffusion

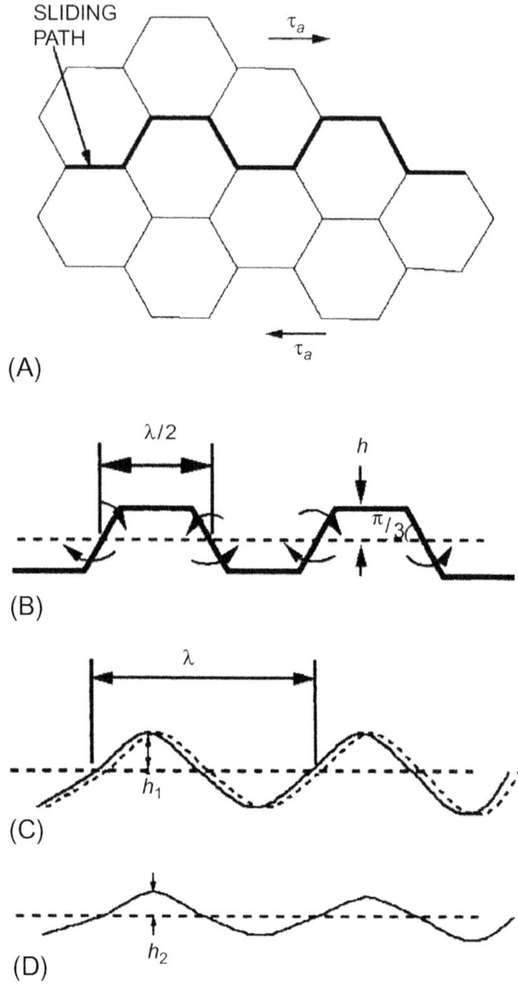

FIG. 1.22 (A) Grain boundary sliding in polycrystalline materials; (B) path of grain boundary sliding; (C) the migration of grain boundary by diffusion for the idealized sinusoidal grain boundary sliding path (Raj and Ashby, 1971); (D) The plastic strain which contributes to the grain boundary sliding accommodation by decreasing the wave amplitude, i.e., $h_2 < h_1$. *(After Fu, H.H., Benson, D. J., Meyers, M.A., 2001. Analytical and computational description of effect of grain size on yield stress of metals. Acta Mater. 49, 2567–2582.)*

coefficients, respectively. For temperatures significantly below half of the melting point, i.e., $T_m/2$, Eq. (1.63) can be simplified as follows:

$$\dot{u} \approx \frac{2\tau_a \Omega \delta D_B}{kT}\left(\frac{1}{h^2}\right) \tag{1.64}$$

Assuming that an identical sliding rate is induced by each grain boundary layer, $\dot{\gamma}$ can be obtained as follows:

$$\dot{\gamma} = \frac{\dot{u}}{2h} \approx \frac{\tau_a \Omega \delta D_B}{kT}\left(\frac{1}{h^3}\right) \tag{1.65}$$

Assuming $h = d_g/4$, The relation between $\dot{\gamma}$ and d_g can be described as follows:

$$\dot{\gamma} = \frac{64\tau_a \Omega \delta D_B}{kT}\left(\frac{1}{d_g^3}\right) \tag{1.66}$$

Fig. 1.23 depicts the variation of Log $(\dot{\gamma})$ versus Log(d_g) obtained using Eq. (1.66) and assuming values of $\delta = 0.75$ nm, $\Omega = 0.0087$ nm^3, $T = 298$ K, and $D_B = 10^{-21}$ m^2/s at $T/T_m = 0.225$. The procedure to calculate the stress levels for different grain sizes was elaborated by Fu et al. (2001). Fig. 1.23 shows that the diffusional grain boundary sliding can induce strain rates larger than 10^{-4} s^{-1} for $d_g \leq 10$ nm.

1.3.1.2.2.3 Phase mixture model A polycrystalline material can be considered as a composite material with various phases of grain interior, grain boundaries, triple junctions, and quadruple nodes (Fig. 1.24). Using the deformation mechanisms of each phase, the phase mixture model can then be incorporated to capture the deformation mechanism of a polycrystal. Several researchers have used phase mixture to address the grain size effect (Bush, 1993; Wang et al., 1995; Carsley et al., 1995; Kim et al., 2000; Fu et al., 2001; Voyiadjis and Deliktas, 2010). For the simplistic three-dimensional cubic composite model, the volume of fraction f for each component can be obtained as follows (Kim et al., 2000):

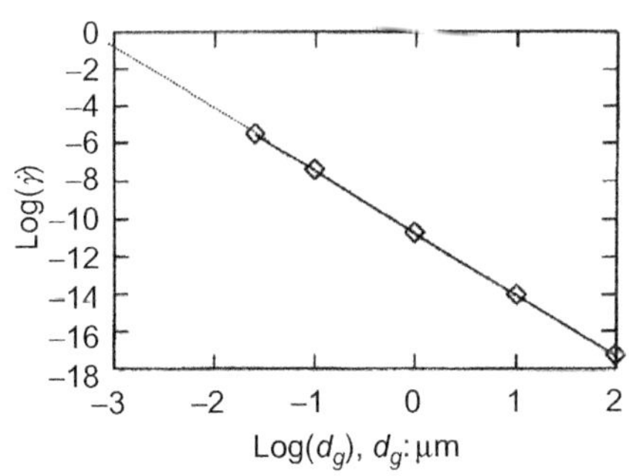

FIG. 1.23 Variation of Log $(\dot{\gamma})$ versus Log(d_g) induced by the diffusional grain boundary sliding; extrapolated results are illustrated by *dashed line. (After Fu, H.H., Benson, D.J., Meyers, M.A., 2001. Analytical and computational description of effect of grain size on yield stress of metals. Acta Mater. 49, 2567–2582.)*

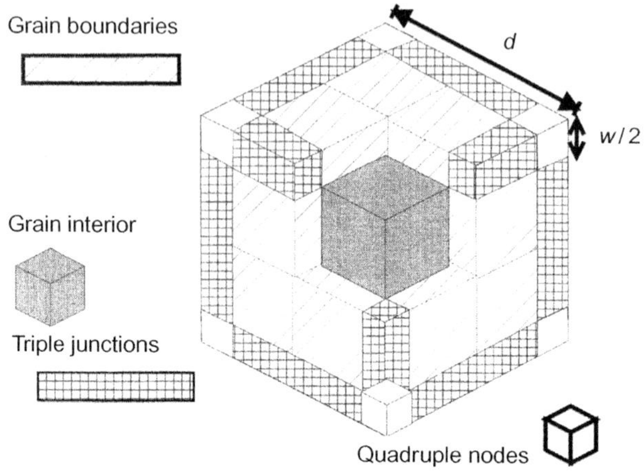

FIG. 1.24 Phase mixture model: a polycrystal can be considered as a composite material with various phases of grain interior, grain boundaries, triple junctions, and quadruple nodes. *(After Kim, H. S., Estrin, Y., Bush, M.B., 2000. Plastic deformation behaviour of fine-grained materials. Acta Mater. 48, 493–504.)*

$$f_{\text{gi}}=\frac{\left(d_g-w\right)^3}{d_g^3};\ f_{\text{gb}}=\frac{3w\left(d_g-w\right)^2}{d_g^3};\ f_{\text{tj}}=\frac{3\left(d_g-w\right)w^2}{d_g^3};\ f_{\text{qn}}=\frac{w^3}{d_g^3}; \tag{1.67}$$

where d_g is the grain size, $w/2$ is the grain boundary thickness, and $f_{\text{gi}}, f_{\text{gb}}, f_{\text{tj}}$, and f_{qn} are the volume fractions of grain interior, grain boundaries, triple junctions, and quadruple nodes, respectively. Fig. 1.25 depicts the variation of the volume fraction of each phase as the grain size changes in the case of $w=2$ nm. Fig. 1.25 shows how the grain interior volume fraction drops as the grain size decreases. On the other hand, the triple junctions and quadruple nodes volume fractions increases by decreasing the grain size. The grain boundaries volume fraction increases as the grain size decreases for grain sizes larger than ~6 nm. However, it starts dropping for grain size smaller than ~6 nm. Accordingly, the governing mechanisms of deformations varies as the grain size decreases due to the change in the phases of the volume fractions.

Kim et al. (2000) incorporated the simple mixture rule based on the components volume fractions to obtain the polycrystalline material strength σ as follows:

$$\sigma=f_{\text{gi}}\sigma_{\text{gi}}+f_{\text{gb}}\sigma_{\text{icgb}}+f_{\text{tj}}\sigma_{\text{tj}}+f_{\text{qn}}\sigma_{\text{qn}} \tag{1.68}$$

Although the atomic arrangement of grain boundaries, triple junctions, and quadruple nodes are different, Kim et al. (2000) modeled them using a similar diffusional creep mechanism. The total plastic strain rate can be defined as follows:

$$\dot{\varepsilon}^p=\dot{\varepsilon}^p_{c,u}+\dot{\varepsilon}^p_{c,b}+\dot{\varepsilon}^p_{c,l} \tag{1.69}$$

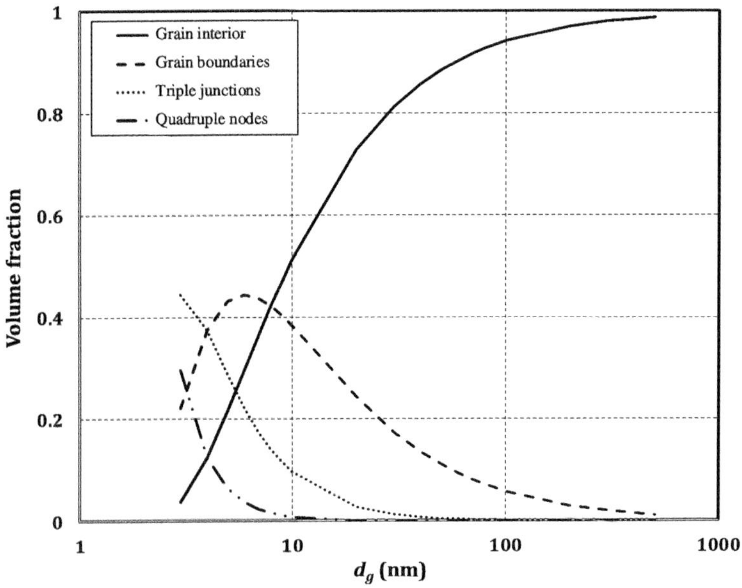

FIG. 1.25 Variation of the volume fraction of each phase in a polycrystalline material as the grain size (d_g) changes in the case of $w = 2$ nm.

where $\dot{\varepsilon}^p_{c,u}$, $\dot{\varepsilon}^p_{c,b}$, and $\dot{\varepsilon}^p_{c,l}$ are the plastic strain rate contributions from grain interior, boundary diffusion, and lattice diffusion, respectively. The plastic strain contribution from the grain interior is described using a dislocation density related model as follows:

$$\dot{\varepsilon}^p_{c,u} = \dot{\varepsilon}_* \left(\frac{\sigma}{\sigma_0}\right)^m Z^{-m/2} \tag{1.70}$$

where $\dot{\varepsilon}_*$ is a constant with the unit of strain rate, Z is the normalized dislocation density, σ_0 is the initial material strength, and m is a material parameter. The plastic strain contributions from the boundary diffusion $\dot{\varepsilon}^p_{c,b}$ and lattice diffusion $\dot{\varepsilon}^p_{c,l}$ can be obtained as follows:

$$\dot{\varepsilon}^p_{c,b} = 14\pi \frac{\Omega \sigma w D^{sd}_{bd}}{kT \ d_g^3} \tag{1.71}$$

$$\dot{\varepsilon}^p_{c,l} = 14\pi \frac{\Omega \sigma w D^{sd}_{ld}}{kT \ d_g^3} \tag{1.72}$$

where Q_{bd} and Q_{ld} are the activation energy for boundary and lattice self-diffusion, respectively. D^{sd}_{bd} and D^{sd}_{ld} are the grain boundary and lattice diffusivity, respectively. Fig. 1.26 shows the variation of the plastic strain rate contributions of grain interior, boundary diffusion, and lattice diffusion

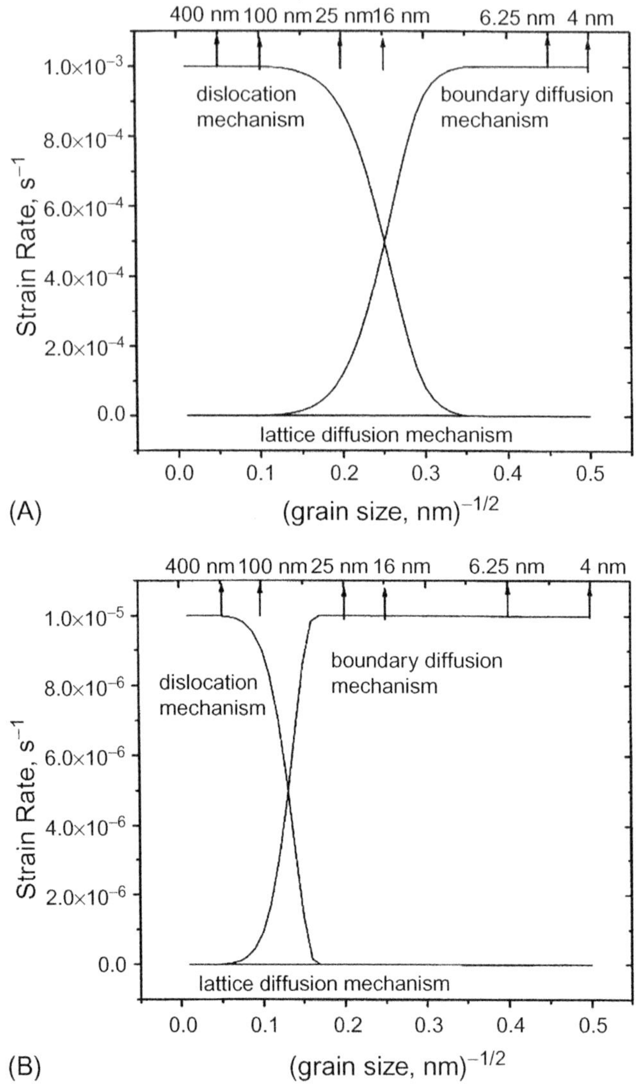

FIG. 1.26 Variation of the plastic strain rate contributions of grain interior, boundary diffusion, and lattice diffusion mechanisms as the grain size varies for two applied strain rates of: (A) $10^{-3}\ s^{-1}$ and (B) $10^{-5}\ s^{-1}$. *(After Kim, H.S., Estrin, Y., Bush, M.B., 2000. Plastic deformation behaviour of fine-grained materials. Acta Mater. 48, 493–504.)*

mechanisms as the grain size varies for two applied strain rates of $10^{-3}\ s^{-1}$ and $10^{-5}\ s^{-1}$. The results show that as the grain size decreases, the contributions of dislocation mechanism and boundary diffusion mechanism decreases and increases, respectively. The results also show that increasing the applied strain rate decreases the grain size at which the boundary diffusion becomes the dominant deformation mechanism. It is worth mentioning that the strain rate induced by the lattice diffusion mechanism is much lower than those induced by other mechanisms, and it can be neglected. Although the model proposed by Kim et al. (2000) was a great step to predict the effect of grain size for nanocrystalline materials, the compatibility of deformation was not considered in their model. Kim et al. (2000) compared their model to the experimental results of copper reported by Sanders et al. (1996) and Youngdahl et al. (1997) for applied strain rate of 10^{-3} s^{-1} and various grain sizes (Fig. 1.27).

Voyiadjis and Deliktas (2010) captured the effect of grain size on the material strength using the phase mixture model and considering three phases of grain core, grain boundary, and triple junctions. The thermodynamically consistent dislocation based inelastic deformation model and/or slip/separation model were incorporated to capture the behavior of each phase. The interfacial energy including the grain boundary sliding and separation was incorporated. The energetic and dissipative decomposition was used to accurately model the size effects. Two new internal variables are introduced to model the grain boundary sliding and separation. The representative volume element (RVE) used to model the idealized three phase nanocrystalline material is depicted in Fig. 1.28. The r symbol is incorporated as an indicator of the considered phases of grain core ($r = gc$), grain boundary ($r = gb$), and triple junctions ($r = tj$). Also, the symbol rs is used as an indicator of the interface surfaces including the interface between the grain core and boundary ($rs = cb$) and interface between the grain boundary and triple junctions ($rs = btj$).

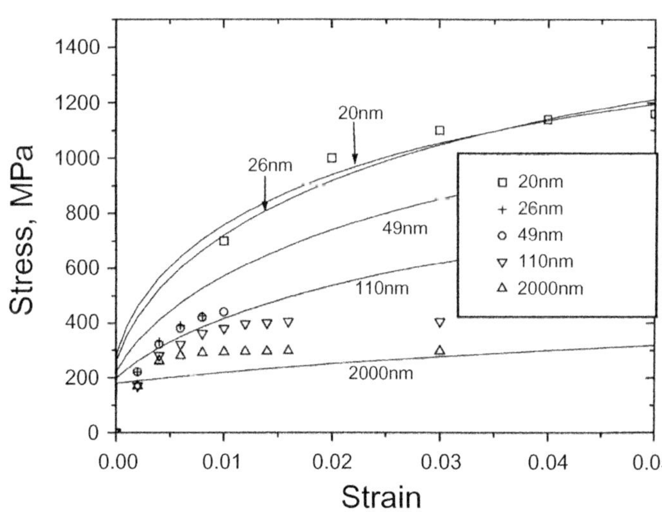

FIG. 1.27 Comparison of stress-strain curves obtained from the model developed by Kim et al. (2000) and experiments reported by Sanders et al. (1996) and Youngdahl et al. (1997) for applied strain rate of 10^{-3} s^{-1} and various grain sizes. *(After Kim, H.S., Estrin, Y., Bush, M.B., 2000. Plastic deformation behaviour of fine-grained materials. Acta Mater. 48, 493–504.)*

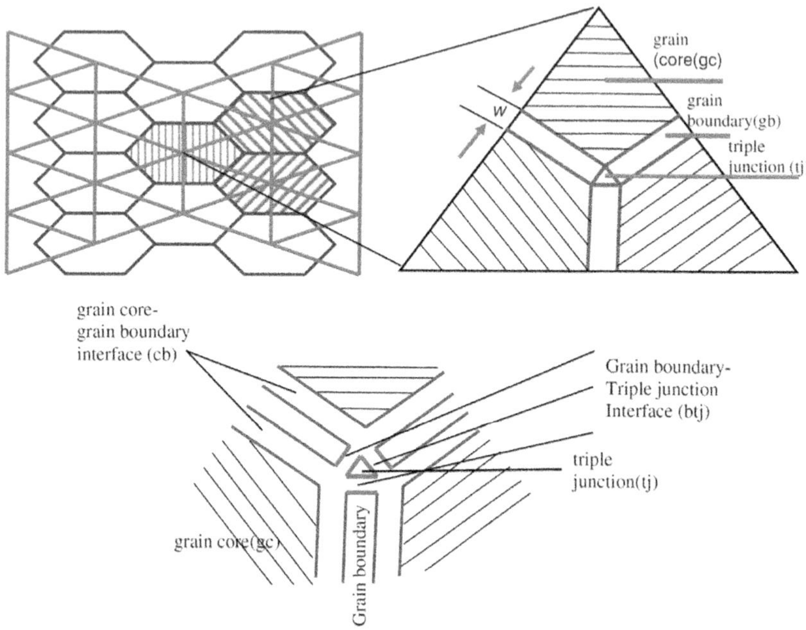

FIG. 1.28 Representative volume element (RVE) used to model the idealized three phase nanocrystalline material. *(After Voyiadjis, G.Z., Deliktas, B., 2010. Modeling of strengthening and softening in inelastic nanocrystalline materials with reference to the triple junction and grain boundaries using strain gradient plasticity. Acta Mech. 213, 3–26.)*

Voyiadjis and Deliktas (2010) assumed that each phase has the contributions from the elastic strain, plasticity internal variables and corresponding gradients, internal interface surface energy, and the two additional state variables at the interfaces. Accordingly, the principle of the virtual power $P_r^{\text{int}} = P_r^{\text{ext}}$ can be described as follows:

$$\begin{aligned}&\int_{V_r}\left(\sigma_{ij}\dot{\varepsilon}_{ij}^{e}+X_{ij}\dot{\varepsilon}_{ij}^{p}+S_{ijk}\dot{\varepsilon}_{ij,k}^{p}\right)dV_r+\sum_{rs}\int_{S_{rs}^{I}}\left[M_{ij}^{rs}\right]\left[\dot{\varepsilon}_{ij}^{p}\right]dS_{rs}^{I}\\&-\sum_{rs}\int_{S_{rs}^{c}}t_i v_i dS_{rs}^{c}+\sum_{rs}\int_{S_{rs}^{c}}\mathcal{W}^{rs}\dot{\theta}dS_{rs}^{c}+\sum_{rs}\int_{S_{rs}^{c}}\mathcal{J}_i^{rs}v_i dS_{rs}^{c}\\&=\sum_{rs}\int_{S_{rs}^{e}}\left(t_i^{rs}v_i+m_{ij}^{rs}\dot{\varepsilon}_{ij}^{p}\right)dS_{rs}^{e}\end{aligned} \tag{1.73}$$

where S_{rs}^{I}, S_{rs}^{c}, and S_{rs}^{e} are the internal subsurfaces, contact surfaces, and external surfaces, respectively. The contact surfaces are related to the separation and sliding. m_{ij} is the microtraction tensor which is power conjugate to $\dot{\varepsilon}_{ij}^{p}$. The displacement rate $\dot{u}$ is assumed to be continuous across the interface; however, there is a jump in plastic strain rate at the interfaces, i.e., $\left[\dot{\varepsilon}_{ij}^{p(rs)}\right]$, and $[M_{ij}^{rs}]$ is the corresponding power conjugate. In addition, t_i is the interface traction force vector. In Eq. (1.73), $\mathcal{W}^{rs}$ and $\mathcal{J}_i^{rs}$ are power conjugates to the rate of grain boundary separation $\dot{\theta}$ and sliding v_i, respectively. Through some manipulations and considering $\dot{\varepsilon}_{ij}^{p} = \dot{\varepsilon}_{ij} - \dot{\varepsilon}_{ij}^{e}$ and $\sigma_{ij}\dot{\varepsilon}_{ij}^{p} = \tau_{ij}\dot{\varepsilon}_{ij}^{p}$, one can obtain the governing equilibrium equations as follows:

$$\left.\begin{array}{l} \sigma_{ij,j} = 0 \\ \tau_{ij} = X_{ij} - S_{ijk,k} \end{array}\right\} \text{in } V_r \tag{1.74}$$

The standard and nonstandard boundary conditions on the S_{rs}^{e}, S_{rs}^{I}, and S_{rs}^{c} can be obtained as follows:

$$\left.\begin{array}{l} t_i^{rs} = \sigma_{ij} n_j^{rs} \\ m_{ij}^{rs} = S_{ijk} n_k^{rs} \end{array}\right\} \text{on } S_{rs}^{e} \tag{1.75}$$

$$\left.\begin{array}{l} [\sigma_{ij}]^{rs} n_j^{rs} = 0 \\ [M_{ij}]^{rs} + [S_{ijk}]^{rs} n_k^{rs} = 0 \end{array}\right\} \text{on } S_{rs}^{I} \tag{1.76}$$

$$\left.\begin{array}{l} -t_i^{rs} = \sigma_{ij} n_j^{rs} \\ \mathcal{W}^{rs} = 0 \\ \mathcal{J}_i^{rs} = 0 \end{array}\right\} \text{on } S_{rs}^{c} \tag{1.77}$$

Using the second law of thermodynamics and Clausius-Duhem inequality and decomposing the thermodynamic forces into the energetic and dissipative parts, i.e., $Y = Y^{(en)} + Y^{(dis)}$, the energetic thermodynamic forces can be obtained as follows:

$$\begin{array}{l} X_{ij}^{r(en)} = \rho \dfrac{\partial \psi_r}{\partial \varepsilon_{ij}^{p}}, S_{ijk}^{r(en)} = \dfrac{\partial \psi_r}{\partial \varepsilon_{ij,k}^{p}}, M_{ij}^{rs(en)} = \dfrac{\partial \psi_{rs}^{I}}{\partial \varepsilon_{ij}^{p(I)}}, \\ t_i^{rs(en)} = \dfrac{\partial \psi_{rs}^{c}}{\partial u_i}, \mathcal{W}^{rs(en)} = \dfrac{\partial \psi_{rs}^{c}}{\partial \theta}, \mathcal{J}_i^{rs(en)} = \dfrac{\partial \psi_{rs}^{c}}{\partial u_i} \end{array} \tag{1.78}$$

where ψ_r, ψ_{rs}^{I}, and ψ_{rs}^{c} are the Helmholtz free energies for the bulk material, internal subsurfaces, and contacting surfaces, respectively. The Lagrange multipliers can be used to find the stationary solution of the maximum principles problem for bulk material and interfaces. Accordingly, the internal state variables and constraint equations can be obtained as follows:

$$\begin{array}{l} \partial \varepsilon_{ij}^{p} = \dot{\lambda}^{r} \dfrac{\partial \phi^{r}}{\partial X_{ij}}, \partial \varepsilon_{ij,k}^{p} = \dot{\lambda}^{r} \dfrac{\partial \phi^{r}}{\partial S_{ijk}}, \varepsilon_{ij}^{p(I)} = \dot{\lambda}_{rs}^{I} \dfrac{\partial \phi_{rs}^{I}}{\partial M_{ij}^{rs(en)}}, \\ u_i = \dot{\lambda}_{rs}^{c} \dfrac{\partial \phi_{rs}^{c}}{\partial t_i^{rs(en)}}, \theta = \dot{\lambda}_{rs}^{c} \dfrac{\partial \phi_{rs}^{c}}{\partial \mathcal{W}^{rs(en)}}, u_i = \dot{\lambda}_{rs}^{c} \dfrac{\partial \phi_{rs}^{c}}{\partial \mathcal{J}_i^{rs(en)}}, \\ \dot{\lambda}^{r} \phi^{r} = 0, \dot{\lambda}_{rs}^{I} \phi_{rs}^{I} = 0, \dot{\lambda}_{rs}^{c} \phi_{rs}^{c} = 0 \end{array} \tag{1.79}$$

where $\dot{\lambda}^{r}$, $\dot{\lambda}_{rs}^{I}$, and $\dot{\lambda}_{rs}^{c}$ are the plastic multipliers for the bulk material, internal subsurfaces, and contacting surfaces, respectively. ϕ^{r}, ϕ_{rs}^{I}, and ϕ_{rs}^{c} are the yield functions for the bulk material, internal subsurfaces, and contacting surfaces, respectively. Finally, the dissipative thermodynamic forces can be obtained by introducing the dissipation potentials as follows:

$$\begin{aligned} X_{ij}^{r(dis)} &= \rho\frac{\partial \mathcal{D}_r}{\partial \varepsilon_{ij}^{p}}, S_{ijk}^{r(dis)} = \frac{\partial \mathcal{D}_r}{\partial \varepsilon_{ij,k}^{p}}, M_{ij}^{rs(dis)} = \frac{\partial \mathcal{D}_{rs}^{I}}{\partial \varepsilon_{ij}^{p(I)}}, \\ t_i^{rs(dis)} &= \frac{\partial \mathcal{D}_{rs}^{c}}{\partial u_i}, \mathcal{W}^{rs(dis)} = \frac{\partial \mathcal{D}_{rs}^{c}}{\partial \theta}, \mathcal{J}_i^{rs(dis)} = \frac{\partial \mathcal{D}_{rs}^{c}}{\partial u_i} \end{aligned} \tag{1.80}$$

where $\mathcal{D}_r$, $\mathcal{D}_{rs}^I$, and $\mathcal{D}_{rs}^c$ are the dissipation potentials for the bulk material, internal subsurfaces, and contacting surfaces, respectively. Voyiadjis and Deliktas (2010) implemented their model by discretizing the crystalline material using the Voronoi-tesselated grains.

After defining the governing equation of each phase, the homogenization scheme should be incorporated to capture the effective behavior. Voyiadjis and Deliktas (2010) incorporate a multilevel Mori-Tanaka averaging method to capture the homogenized material behavior. The localization and homogenization procedures are schematically illustrated in Fig. 1.29. The stress and strain concentration factors are used to define the localization relations as follows:

$$\begin{aligned} \langle \boldsymbol{\varepsilon} \rangle^{\mathrm{r}} &= \mathbf{A}^{\mathrm{r}} : \bar{\boldsymbol{\varepsilon}} \\ \langle \boldsymbol{\sigma} \rangle^{\mathrm{r}} &= \mathbf{A}^{\mathrm{r}} : \bar{\boldsymbol{\sigma}} \end{aligned} \tag{1.81}$$

The homogenization relations can be described as follows:

$$\begin{aligned} \langle \boldsymbol{\varepsilon} \rangle^2 &= \frac{1}{V_r}\int_{V_r} \boldsymbol{\varepsilon}(\mathbf{x})dV_r \text{with}\, \hat{\boldsymbol{\varepsilon}} = \sum_r c^r \langle \boldsymbol{\varepsilon} \rangle^r \\ \langle \boldsymbol{\sigma} \rangle^2 &= \frac{1}{V_r}\int_{V_r} \boldsymbol{\sigma}(\mathbf{x})dV_r \text{with}\, \hat{\boldsymbol{\sigma}} = \sum_r c^r \langle \boldsymbol{\sigma} \rangle^r \end{aligned} \tag{1.82}$$

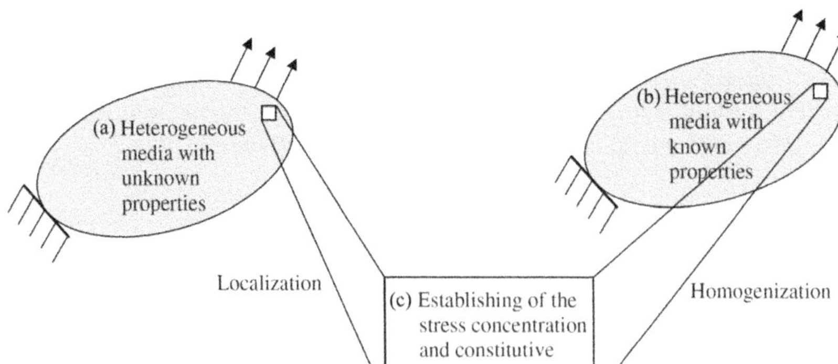

FIG. 1.29 Schematic illustration of the localization and homogenization procedures for heterogeneous media. *(After Voyiadjis, G.Z., Deliktas, B., 2010. Modeling of strengthening and softening in inelastic nanocrystalline materials with reference to the triple junction and grain boundaries using strain gradient plasticity. Acta Mech. 213, 3–26.)*

In order to homogenize the multi-phase system, a multilevel Mori-Tanaka averaging method is incorporated. First, the dilute approximation is incorporated which assumes that inhomogeneities are dilutely distributed in the matrix and do not interact with one another. Next, the Mori-Tanaka approximation is incorporated in which the interactions between inhomogeneities are approximated by embedding them in a medium with a homogenized response. Another dilute and Mori-Tanaka approximations are then applied (Fig. 1.30). To verify the capability of the model, Voyiadjis and Deliktas compared their model with the experimental data of Ni-P reported by Zhou et al. (2003) and the classical Hall-Petch prediction (Fig. 1.31). The results show that the model is capable of capturing both Hall-Petch and inverse Hall-Petch effects.

1.3.2 Extrinsic size effects

Material scientists and engineers have tried to strengthen the crystalline materials by altering the intrinsic properties of crystalline materials such as grain size, the average distance of second phase particles or precipitates, and dislocations mean free path. Accordingly, it was initially perceived that the ultimate strength of crystalline metals is governed by these internal characteristics that is independent of the sample size. However, further experiments have revealed that the ultimate strength of a crystalline material is also a function of the sample size. For example, it has been observed that the material ultimate strength during uniaxial experiments on small size pillars depends on the pillar diameter which can be described using a power law function. Accordingly, the extrinsic size effects, which is the dependency of material strength on the extrinsic properties such as thin film thickness, pillar diameter, and indentation depth are addressed in this section.

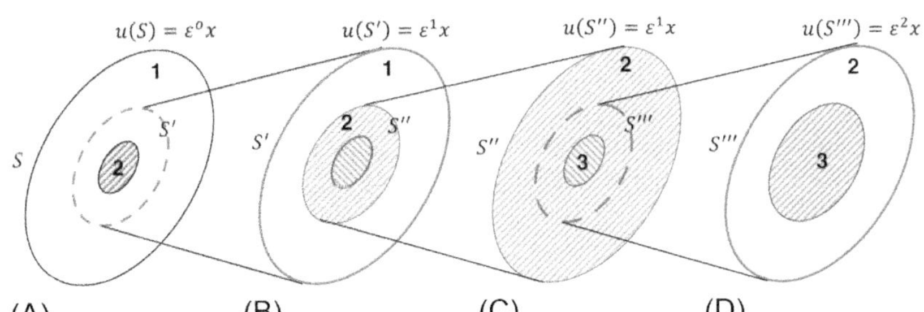

FIG. 1.30 Multilevel Mori-Tanaka averaging method: (A) dilute approximation, (B) Mori-Tanaka approximation, (C) dilute approximation, and (D) Mori-Tanaka approximation. *(After Voyiadjis, G.Z., Deliktas, B., 2010. Modeling of strengthening and softening in inelastic nanocrystalline materials with reference to the triple junction and grain boundaries using strain gradient plasticity. Acta Mech. 213, 3–26.)*

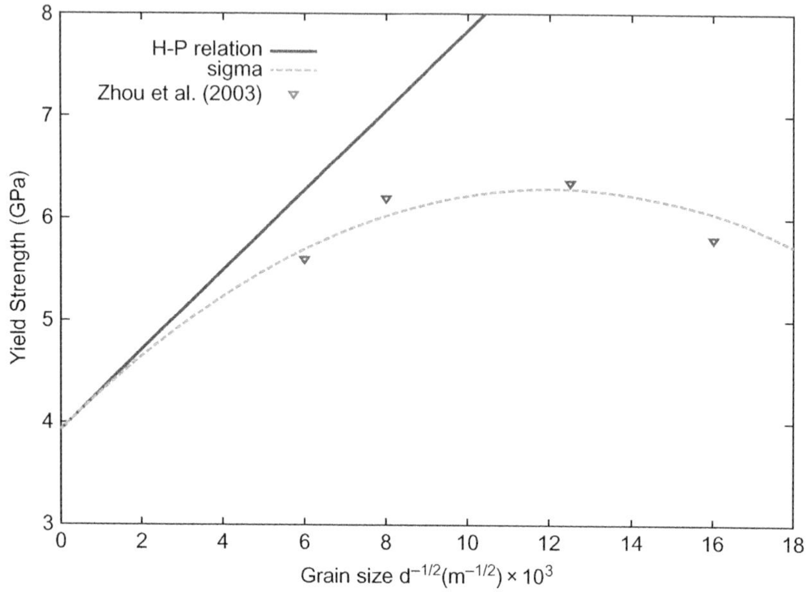

FIG. 1.31 Variation of yield strength versus the grain size for Ni-P. *(After Voyiadjis, G.Z., Deliktas, B., 2010. Modeling of strengthening and softening in inelastic nanocrystalline materials with reference to the triple junction and grain boundaries using strain gradient plasticity. Acta Mech. 213, 3–26.)*

1.3.2.1 Thin films

Thin films are structures with the thickness much smaller than the other two dimensions. Thickness, as an extrinsic characteristic length scale, may change the properties of thin films. As an example, the electrical and thermal conductivity in thin films depend on the film thickness in a way that for thicknesses smaller than the mean free path of electrons or phonons, they significantly drop in magnitude. In the case of material strength, the thickness is an important characteristic. The dependency of the thin film yield stress on thickness has been widely investigated.

The first model to relate the thin film yield shear stress τ and thickness H is the dislocation channeling mechanism. The thin film yield occurs by the motion of dislocations which are constrained to move inside the thin film. First, one can assume that the film surface is impenetrable to the dislocation (Fig. 1.32A). The yield occurs when a dislocation loop can fit inside the thin film:

$$d(\tau) = H' \tag{1.83}$$

where $d(\tau)$ is the equilibrium diameter of a dislocation loop for an elastic isotropic material. $H' = H/\sin(\phi)$ is the size variable, where ϕ is the angle

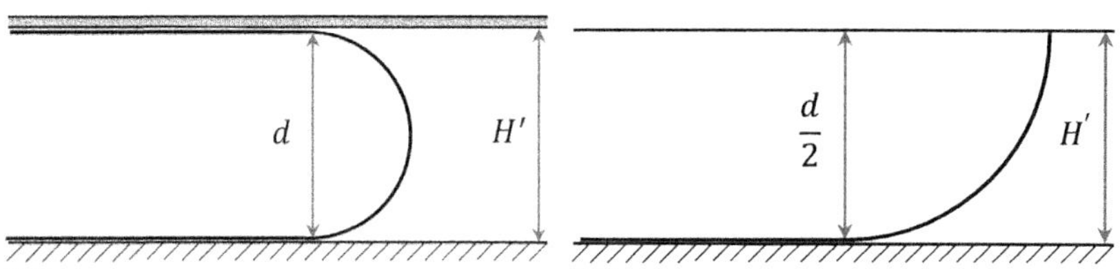

FIG. 1.32 Extrinsic size effects in thin films with: (A) impenetrable film surface and (B) free film surface.

between thin film normal and that of the dislocation loop plane. Accordingly, the yield stress can be described as follows:

$$\tau = \frac{G_f b}{H'} \tag{1.84}$$

where G_f is the thin film shear modulus.

In the case of free film surface (Fig. 1.32B), the surface induces image forces on dislocations. Accordingly, the yield stress occurs when half of the dislocation loop fits inside the thin film. The corresponding yield stress is half of the one described in Eq. (1.84). Nix (1989) stated that Eq. (1.84) underestimates the thin film yield stress. Freund (1987) and Nix (1989) developed a more accurate model to predict the thin film yield stress. The underlying mechanism of this modification is the additional dislocation line length in the interfaces induced by dislocation movement inside the thin film. A new effective shear modulus G_{eff} have been proposed using the energy balance between the work done by the dislocation and the energy stored in the additional dislocation line length as follows:

$$G_{eff} = \frac{G_f}{2\pi(1-\upsilon)}\left[\frac{G_s}{G_s + G_f}\ln\frac{\beta_1 H}{b} + \frac{G_O}{G_f + G_O}\ln\frac{\beta_2 H_O}{b}\right] \tag{1.85}$$

where o and s represent the oxide and substrate layers, respectively. υ denotes the thin film Poisson's ratio. β_1 and β_2 are constants relating to cut-off radii. H_O is the oxide layer thickness. In the case of stiff substrate, G_{eff} can be considerably larger than G_f which leads to higher yield stresses. Furthermore, the free surface case can be obtained by removing the second term in Eq. 1.85. Nix (1989) compared the modified model with the experimental results reported by Kuan and Murakami (1982) for Pb thin films (Fig. 1.33). The results show that the model can successfully capture the experimental results.

1.3.2.1.1 Interaction of size effects due to the thin film thickness and grain size

In order to accurately capture the size effects in polycrystalline thin films, both size effects mechanisms of thin film thickness and grain size should be

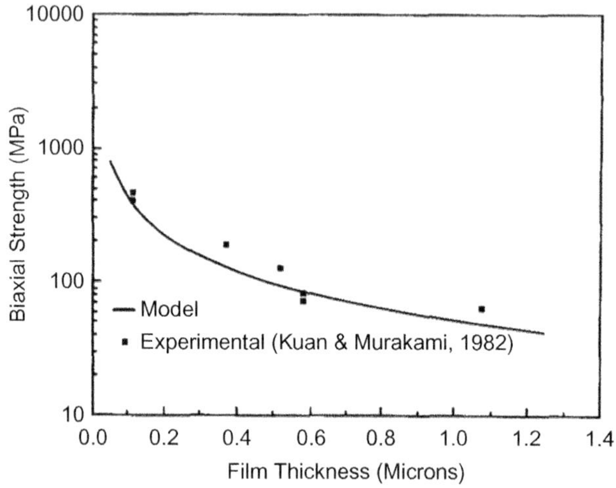

FIG. 1.33 Variation of biaxial strength versus the film thickness for Pb thin films. *(After Nix, W. D., 1989. Mechanical properties of thin films. Metall. Trans. A. 20, 2217–2245.)*

incorporated. Thompson (1993) proposed a simple model to capture the interaction of these size effects as follows:

$$\tau = \frac{G_f b}{d_g}\frac{\sin(\phi)}{4\pi(1-v)}\ln\frac{\beta_3 d_g}{b} \tag{1.86}$$

where the variation of yield shear stress versus the grain size does not follow the Hall-Petch effect. However, Keller et al. (1998) showed that the appropriate framework to address the interaction between the effects of grain size and film thickness is to consider each mechanism separately and then the total size effects can be described as the summation of the intrinsic and extrinsic ones. In other words, only part of the total size effects can be attributed to the extrinsic size effects due to the film thickness. The rest of the size effects can be addressed by using the Hall-Petch relation as follows:

$$\sigma - \sigma_{Nix} = K_{HP} d_g^{-1/2} \tag{1.87}$$

where σ_{Nix} is the stress predicted using the effective shear modulus G_{eff} described in Eq. (1.85). Fig. 1.34A shows the variation of $(\sigma - \sigma_{Nix})$ versus $d_g^{-1/2}$ for polycrystalline Cu thin films which shows that the Hall-Petch relation can successfully capture the difference between the Freund (1987) and Nix (1989) prediction of size effects due to the film thickness variation and the total size effects. Furthermore, the results for both passivated and unpassivated films can fit one straight line when the yield stress is normalized by G_{eff} (Fig. 1.34B). In other words, the passivated and unpassivated films show similar normalized stresses. It means that the effect of passivation was appropriately

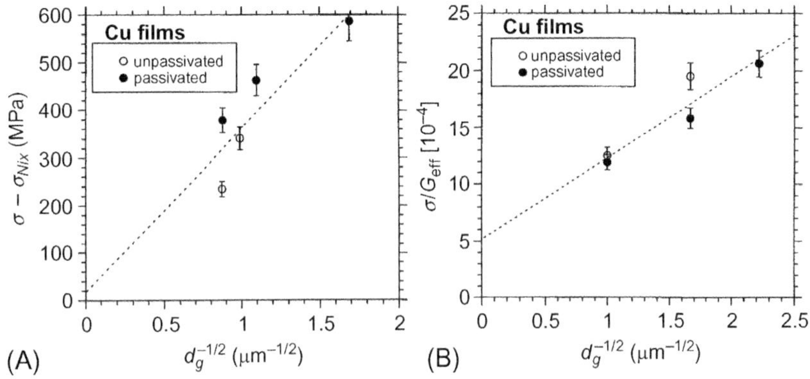

FIG. 1.34 Interaction between the effects of grain size and film thickness for copper polycrystalline thin films. (A) Variation of ($\sigma - \sigma_{Nix}$) versus the inverse root of grain size $d_g^{-1/2}$. (B) Variation of normalized stress versus the inverse of thin film thickness. *(After Keller, R.M., Baker, S.P., Arzt, E., 1998. Quantitative analysis of strengthening mechanisms in thin Cu films: effects of film thickness, grain size, and passivation. J. Mater. Res. 13, 1307–1317.)*

accounted by Freund (1987) and Nix (1989). However, which developed method is more accurate for polycrystalline thin films, and how size effect mechanisms interact are the questions which require further investigations in the future.

1.3.2.2 Pillars

Uchic et al. (2003, 2004) and Uchic and Dimiduk (2005) have introduced a micropillar compression test using the focused ion beam (FIB) machining to explore the effects of pillar size on the strength of crystalline materials. They observed that as the sample size decreases, the material strength considerably increases. For example, Uchic et al. (2004) reported the variation of stress-strain curves as the sample size changes for pure Ni at room temperature (Figs. 1.35 and 1.36). Fig. 1.35 shows that the material strength for a pillar with a diameter of 5 μm is around three times larger than that of the bulk sample. Fig. 1.36 shows that decreasing the sample size down to 0.5 μm can increase the strength up to15 times (at 5% strain). The reported experiments have shown that the extrinsic length scales should also be considered along with intrinsic ones to capture the size effects in crystalline materials.

Uchic et al. (2009) reviewed the pillar compression experiments with all required details about the experiment procedures and challenges, the possible artifacts, the reported experimental results, and possible governing mechanisms. After Uchic et al. (2009), two other groups (Kraft et al., 2010; Greer and De Hosson, 2011) have reviewed the size effects and deformation mechanisms in metallic samples of confined volumes that include micron and nano sized pillars. The pillar compression test is schematically represented in

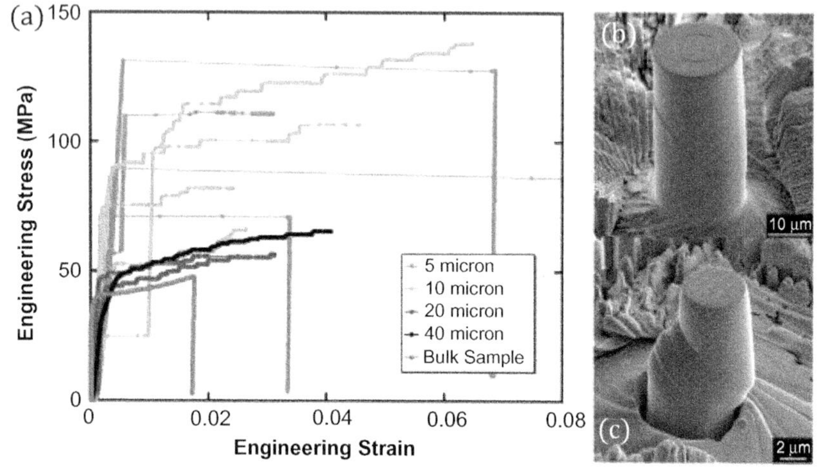

FIG. 1.35 Response of pure Ni pillars during uniaxial compression experiment at room temperature. (a) Stress-strain curves for pillars with diameters that vary from 40 to 5 μm and a bulk single crystal having approximate dimensions of 2.6 × 2.6 × 7.4 mm. (b) A scanning electron micrograph (SEM) image of a pillar with the diameter of 20 μm at 4% strain. (c) A SEM image of 5 μm diameter pillar at 19% strain during a rapid burst of deformation. *(After Uchic, M.D., Dimiduk, D.M., Florando, J.N., Nix, W.D., 2004. Sample dimensions influence strength and crystal plasticity. Science 305, 986–989.)*

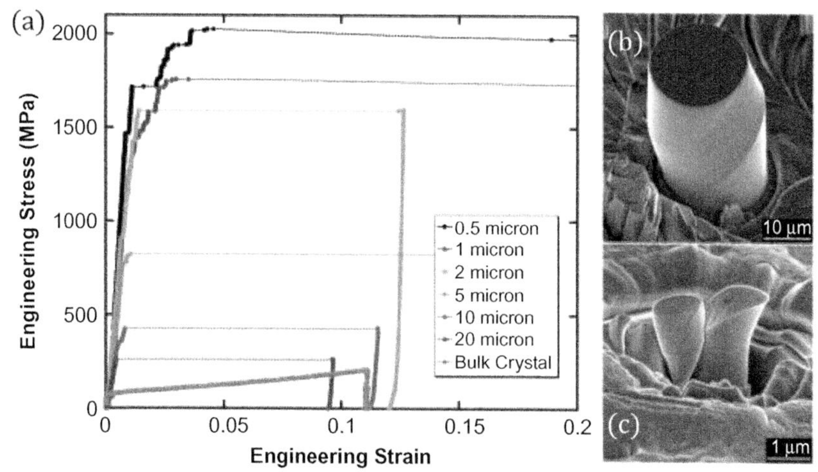

FIG. 1.36 Response of Ni_3Al-Ta pillars during uniaxial compression experiment at room temperature. (a) Stress-strain curves for pillars with diameters that vary from 20 to 0.5 μm and a bulk single crystal having approximate dimensions of 2.5 × 2.5 × 7.5 mm. (b) A SEM image of a pillar with the diameter of 20 μm at 10% strain during the rapid burst of deformation. (c) A SEM image of 1 μm diameter pillar after the experiment, where the top of the sample is completely sheared off during the rapid strain burst. *(After Uchic, M.D., Dimiduk, D.M., Florando, J.N., Nix, W.D., 2004. Sample dimensions influence strength and crystal plasticity. Science 305, 986–989.)*

Fig. 1.37. The experimental procedures mimic the conventional uniaxial compression test; however, it is modified to facilitate the experiment for micron sized samples. The primary difference is that pillars remain integrally attached to the bulk material, unlike the conventional experiments in which samples are freestanding. Nanoindenters with flat-punch tips are used to impose the compression. Besides the compression experiment, tension test has been also conducted on metallic samples of confined volumes. Fig. 1.38 depicts a SEM image of FIB-fabricated Cu pillar during the tension test at different strains.

Different cubic and non-cubic crystal structures including face centered cubic (FCC), body centered cubic (BCC), hexagonal close-packed (HCP),

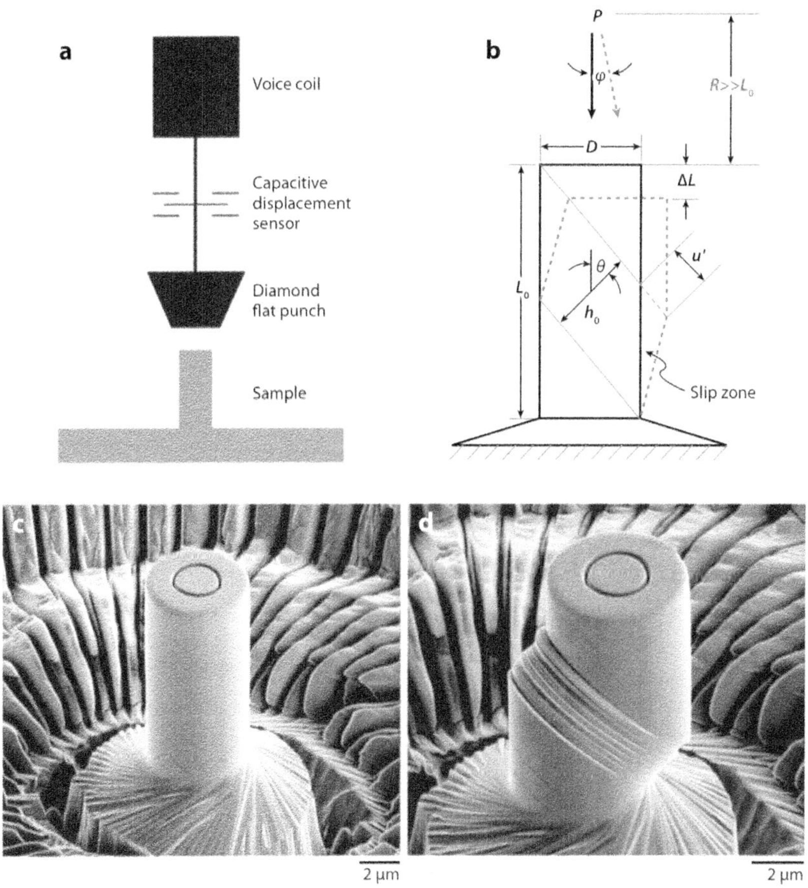

FIG. 1.37 (a) Schematic of the pillar compression experiment using a nanoindenter with a flat punch. (b) Schematic of the flow response of a microcrystal oriented for single slip. (c) SEM image of pure Ni pillar with the diameter of 5 μm. (d) SEM image of Ni pillar with the diameter of 5 μm after testing. *(After Uchic, M.D., Shade, P.A., Dimiduk, D.M., 2009. Plasticity of micrometer-scale single crystals in compression. Annu. Rev. Mater. Res. 39, 361–386.)*

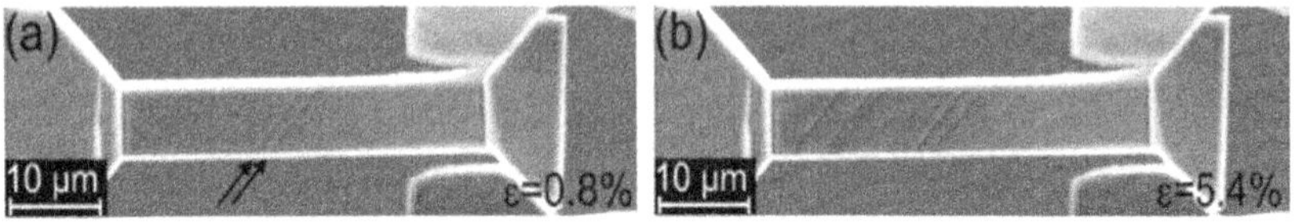

FIG. 1.38 A SEM image of FIB-fabricated Cu pillar with a diameter of 8 μm during the tension experiment at different strains of: (a) 0.8% and (b) 5.4%. *(After Kiener, D., Grosinger, W., Dehm, G., Pippan, R., 2008. A further step towards an understanding of size-dependent crystal plasticity: in situ tension experiments of miniaturized single-crystal copper samples. Acta Mater. 56, 580–592.)*

tetragonal metals, and shape memory alloys have been investigated using the pillar compression and tension tests. In this section, three crystal structures of FCC, BCC, and HCP are addressed in more details. The majorities of the initial investigations were focused on the FCC crystals. It was due to the fact that FCC metals have been extensively studied in the case of bulk samples and whiskers and the dislocation mechanisms have been greatly extracted from those studies. Furthermore, Au samples have been favorable since the oxide layers may alter the observed results during pillar compression experiments. It was observed that the relation between the sample yield or flow stress σ and the pillar diameter d can be captured using a power law as follows:

$$\sigma = Ad^{-n} \tag{1.88}$$

where A is a constant, and n is the power-law exponent, which is reported to be 0.61–0.97. In a different study, the strain of the sample strength is reported to vary from the proportional limit to 10% strain. In the case of pure FCC metals, the following equation can capture the size effects:

$$\frac{(\tau - \tau_0)b_{\text{ref}}}{K_s b_{metal}} = Bd^{-n} \tag{1.89}$$

where τ is the yield or flow shear stress, τ_0 is the reference stress, K_s is the anisotropic shear modulus, b_{ref} is the Burgers vector of a reference metal, b_{metal} is the Burgers vector of the considered metal, and B is a constant. Fig. 1.39 illustrates the results of pillar compression experiments for several FCC metals. Assuming that τ_0 is negligible and considering the Burgers vector of Ni as the reference one, a reasonable fit can be obtained for $n = 0.6$.

Although most of the pillar compression tests have been conducted on the FCC metals, some experimental studies have investigated the BCC metals (Fig. 1.40). BCC metals deformation mechanisms are more complex compared to those of the FCC metals. The response of BCC metals strongly depend on temperature, strain rate, and orientation. Furthermore, the Schmid's law breaks down in the case of BCC metals and slip plane can be different from the maximum resolved shear stress. Accordingly, they show tension-compression asymmetry. The cores of screw dislocations in BCC metals are non-planar, and both glide and non-glide shear stress components may lead to the

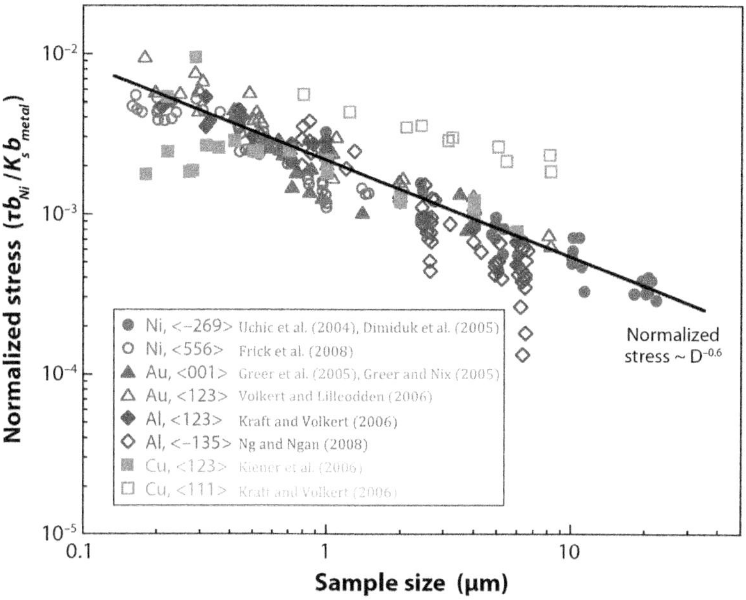

FIG. 1.39 Variation of normalized flow or yield shear stress $\tau b_{Ni}/K_s b_{metal}$ as a function of pillar diameter for different FCC metals. The original experimental data has been reported by Uchic et al. (2004), Dimiduk et al. (2005), Frick et al. (2008), Greer et al. (2005), Greer and Nix (2005), Volkert and Lilleodden (2006), Kraft and Volkert (2006), Ng and Ngan (2008), and Kiener et al. (2006). *(After Uchic, M.D., Shade, P.A., Dimiduk, D.M., 2009. Plasticity of micrometer-scale single crystals in compression. Annu. Rev. Mater. Res. 39, 361–386.)*

dislocation movements. Vitek (1968) showed that the threefold symmetry of screw dislocations leads to the dislocation core effect in BCC metals. Also, the nonplanar dislocation cores in BCC metals lead to the lower screw dislocation mobility (Vitek and Paidar, 2008).

Most of the pillar compression tests for BCC metals have been conducted on Mo single crystals. The sample preparation and experimental procedure is similar to those performed on FCC samples, and the observed results have shown great size effects as the sample size varies. Similar to FCC metals, the material strength increases as the sample size decreases (Fig. 1.40). Bei et al. (2007a,b, 2008a,b) have investigated the size effects in Mo pillars with diameters that vary from 360 to 1500 nm during compression test. They have studied different initial conditions from dislocation free (Bei et al., 2007b) to pre-strained pillars (Bei et al., 2008b). Fig. 1.41 depicts the effect of pre-straining on the extrinsic size effects for Mo pillars (Bei et al., 2008b). The results show that increasing the pre-straining, and consequently increasing the initial dislocation density, decreases the pillar yield strength.

Besides Mo pillars, size effects in three other BCC metals including Ta, W, Nb, and V have been investigated using pillar compression experiments

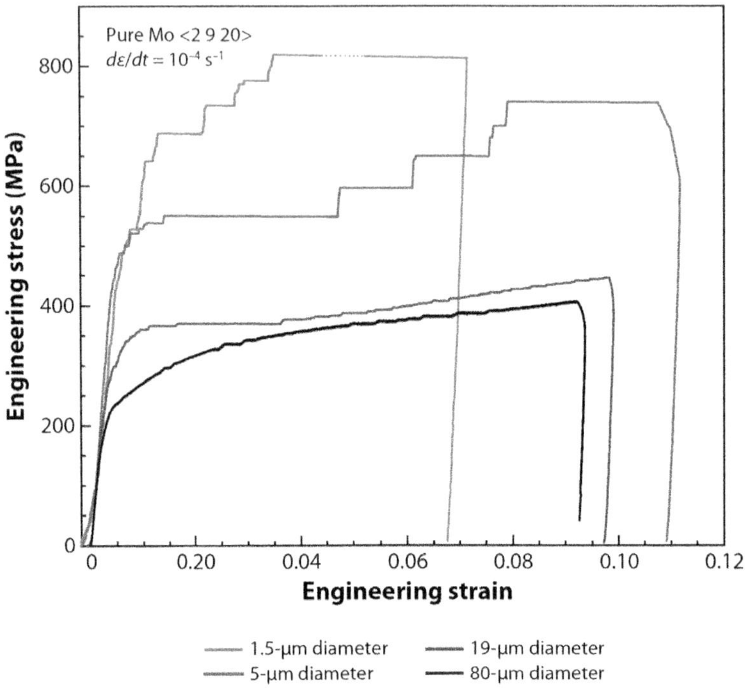

FIG. 1.40 Variation of the response for Mo ⟨2 9 20⟩-oriented pillar as its diameter changes. *(After Uchic, M.D., Shade, P.A., Dimiduk, D.M., 2009. Plasticity of micrometer-scale single crystals in compression. Annu. Rev. Mater. Res. 39, 361–386.)*

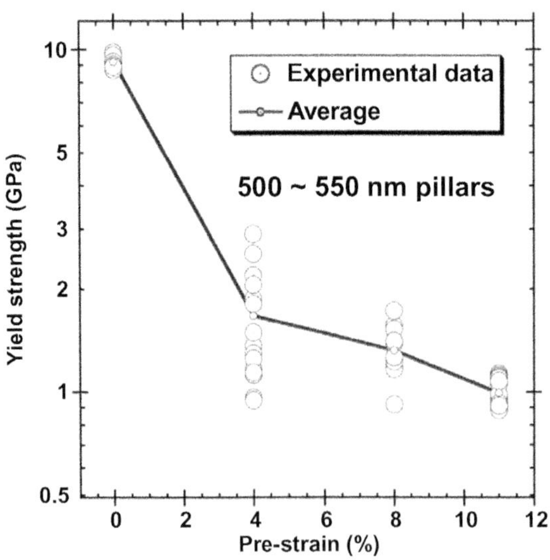

FIG. 1.41 Variation of Mo pillar yield strength versus the pre-strain for 500–550 nm pillar diameters. *(After Bei, H., Shim, S., Pharr, G.M., George, E.P., 2008b. Effects of pre-strain on the compressive stress–strain response of Mo-alloy single crystal micropillars. Acta Mater. 56, 4762–4770.)*

(Fig. 1.42). Schneider et al. (2009) investigated size effects in Mo, Ta, W, and Nb using the pillar compression test. They observed that for each material, the exponent of the power law equation for size effects correlate with the inverse of critical temperature. The underlying mechanism for this observation is the low mobility of screw dislocation at temperatures below the critical temperature (Schneider et al., 2009). They also observed that the studied BCC metals exhibit

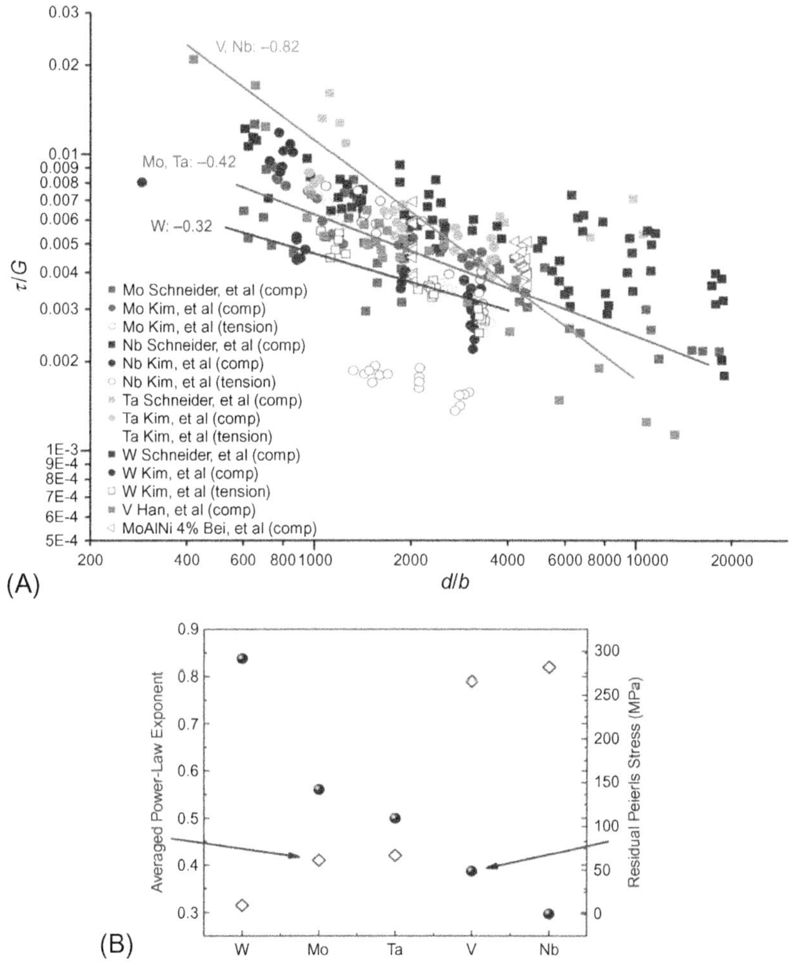

FIG. 1.42 Extrinsic size effects in BCC metals: (A) variation of normalized flow or yield shear stress τ/G as a function of normalized pillar diameter d/b for different BCC metals during uniaxial compression and tension experiments. The original experimental data has been reported by Schneider et al. (2009), Kim et al. (2010), Han et al. (2010), and Bei et al. (2007b). (B) Variation of average power law exponent and residual Peierls stress for different BCC metals. *(After Greer, J.R., De Hosson, J.T.M., 2011. Plasticity in small-sized metallic systems: intrinsic versus extrinsic size effect. Prog. Mater. Sci. 56, 654–724.)*

the least difference in strength for the smallest pillar size of 200 nm. Kim et al. (2010) investigated the effect of pillar size in W, Mo, Ta, and Nb. They observed that in the case of compression test, the strongest size effects belongs to Nb with the power law exponent of −0.93, and the rest of BCC metals exhibited nearly the same exponent around −0.44. In the case of tension test, however, Ta showed the strongest size effects with a power law exponent of −0.8. The rest of metals exhibited the exponents of −0.77, −0.58, and −0.43 for Nb, W, and Mo, respectively. They reported a strong tension-compression asymmetry in the size effects of the studied BCC metals. Han et al. (2010) investigated size effects in V using the pillar compression experiment. They reported strong size effects in V pillars with the power law exponent of −0.79.

The extrinsic size effects in metals with the hexagonal close-packed (HCP) crystal structure, as a non-cubic crystal structure, have been addressed using the pillar compression experiment (Fig. 1.43). Lilleodden (2010), Byer et al. (2010), and Ye et al. (2010) have been the first groups addressing the size effects in Mg, as a HCP metal. Lilleodden (2010) and Byer et al. (2010) observed that the governing deformation mechanism is a multiple slip on pyramidal planes, and twinning does not occur in pillars. Lilleodden (2010) observed that

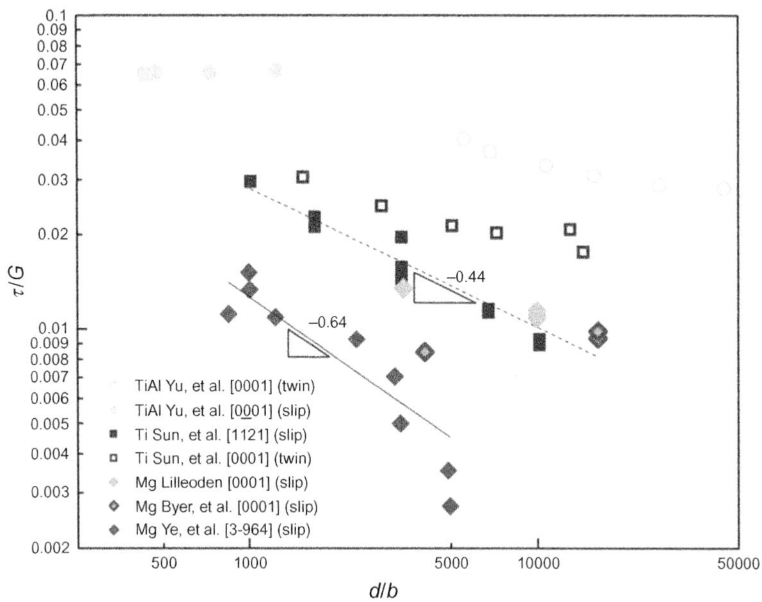

FIG. 1.43 Variation of normalized flow or yield shear stress τ/G as a function of normalized pillar diameter d/b for different HCP metals during uniaxial compression and tension experiments. The original experimental data has been reported by Lilleodden (2010), Byer et al. (2010), Ye et al. (2010), Yu et al. (2010), and Sun et al. (2011). *(After Greer, J.R., De Hosson, J.T.M., 2011. Plasticity in small-sized metallic systems: intrinsic versus extrinsic size effect. Prog. Mater. Sci. 56, 654–724.)*

independent of the sample size, a catastrophic shear instability occurs at stresses in the range of 440 – 450 MPa. As explained by Greer and De Hosson (2011), neither Lilleodden (2010) nor Byer et al. (2010) reported the initial densities of the tested pillars. The obtained experimental results, i.e. the variation of flow stress versus the pillar diameter, showed some discrepancies. Ye et al. (2010), however, considered the effect of pre-existing dislocations and studied the extrinsic size effects in pure Mg and Mg-0.2%Ce single crystals using in situ uniaxial compression experiments. They stated that two deformation mechanisms of sliding in basal plane and extension twining govern the size effects. It was observed that the stress required for the basal sliding to occur depends on the pillar diameter in a way that smaller pillars required higher stresses to initiate the basal sliding (Fig. 1.44).

Besides pure Mg and its alloy, size effects in single crystalline pure Ti and its alloy have been investigated using the pillar compression experiment. Yu et al. (2010) incorporated in situ uniaxial compression experiments to address size effects in Ti alloy single crystalline pillars. Depending on the pillar size, two distinct trends were observed. In the case of pillars with diameters larger than 1 μm, as the sample size decreases, the stress required to sustain the plastic flow using the twinning mechanisms increases. In the case of pillars with diameters smaller than 1 μm, however, the stresses are close to the theoretical strength and

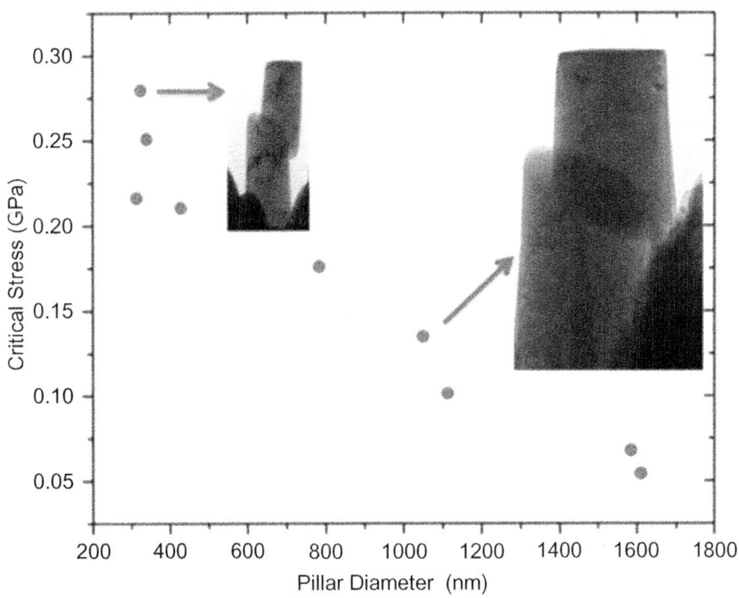

FIG. 1.44 Variation of the critical stress to initiate basal sliding versus the pillar diameter. *(After Ye, J., Mishra, R.K., Sachdev, A.K., Minor, A.M., 2010. In situ TEM compression testing of Mg and Mg–0.2 wt.% Ce single crystals. Scr. Mater. 64, 292–295.)*

deformation occurs on through crystallographic slip planes. In this range of sample sizes, no twinning was observed. Yu et al. (2010) capture the observed trends using a model in which two requirements should be satisfied for the twinning to occur as follows:

- The required stress for twinning dislocation to sweep the pillar cross section area in plane "n" should be provided. The applied stress for twinning should overcome the barriers and potential obstacles.
- A sufficient number of obstacles should be able to cross-slip from their original plane to the adjacent plane.

In the case of small pillars, i.e. smaller than 1 μm, the cross-slip of dislocations to the adjacent plane becomes impossible, and consequently, no twinning is observed for this range of pillar sizes. Sun et al. (2011) also investigated the size effects in pure Ti using pillar compression test. They observed that a pillar with a diameter of 300 nm can reach to the strength of 1040 MPa which is 16% of Ti theoretical strength. The obtained results were consistent with the previously observed size effects in HCP metals, i.e. the sample strength is inversely proportional to its size.

Different theoretical models have been proposed to capture the observed extrinsic size effects during the pillar compression and tension experiments. Source truncation, source exhaustion hardening, and weakest link theory have been proposed as the underlying mechanisms of extrinsic size effects (see, e.g., Uchic et al., 2009; Kraft et al., 2010). The mechanisms are described in more details in the following sections.

1.3.2.2.1 Source truncation

In the case of pillars with confined volumes, the double-ended dislocation sources are transformed into the single-ended ones due to the interaction of dislocations with the pillar free surfaces. The length of the new dislocation source is a function of pillar size in a way that as the sample size decreases, the length of dislocation sources also decreases. Accordingly, a smaller sample has a higher strength. This hardening mechanism is commonly termed as source truncation (Parthasarathy et al., 2007; Rao et al., 2007). The first mathematical modeling of source truncation was developed by Parthasarathy et al. (2007) by predicting the effective dislocation source length. In other words, as the pillar size varies, the effective dislocation source length changes which is the underlying mechanism of the observed size effects. The probability of a cylindrical pillar with n pins to have the maximum distance from the free surface equal to $\lambda_{\max}$ can be described as follows:

$$p(\lambda_{\max})d\lambda_{\max} = \left[1 - \frac{\pi(R-\lambda_{\max})(b-\lambda_{\max})}{\pi Rb}\right]^{n-1} \left(\frac{\pi[(R-\lambda_{\max})+(b-\lambda_{\max})]}{\pi Rb}\right) nd\lambda_{\max} \tag{1.90}$$

where R is the specimen radius, $b = R/\cos\beta$ is the major axis of the glide plane, and β is the angle between the primery slip plane and the loading axis. Using the probability function defined in Eq. (1.1), the mean effective source length $\bar{\lambda}_{max}$ can be obtained as below:

$$\bar{\lambda}_{max} = \int_0^R \lambda_{max} p(\lambda_{max}) d\lambda_{max}$$
$$= \int_0^R \left[1 - \frac{\pi(R-\lambda_{max})(b-\lambda_{max})}{\pi R b}\right]^{n-1} \left(\frac{\pi[(R-\lambda_{max})+(b-\lambda_{max})]}{\pi R b}\right) n\lambda_{max} d\lambda_{max} \tag{1.91}$$

In the next step, the critical resolved shear stress (CRSS) was related to $\bar{\lambda}_{max}$ as follows:

$$\text{CRSS} = \frac{\alpha G b}{\bar{\lambda}_{max}} + \tau_0 + 0.5 G b \sqrt{\rho} \tag{1.92}$$

where α is a constant, G is the shear modulus, b is the Burgers vector, τ_0 is the friction stress, and ρ is the dislocation density. The number of pins, n, is also predicted as follows:

$$n = \text{Integer}\left[\frac{L_{mobile}}{L_{ave}}\right] \tag{1.93}$$

where L_{ave} is the average length of dislocation segments, $L_{mobile} = \rho\pi R^2 h/s$ is the total length of mobile dislocations, in which, h is the height of the pillar, and s is the number of slip systems. Basically, Parthasarathy et al. (2007) and later Rao et al. (2007) related the size effects to the average largest source length $\bar{\lambda}_{max}$. Parthasarathy et al. (2007) also assumed that the CRSS is not nucleation limited and considered the required stress to move the partial dislocations as the upper bound of CRSS as follows:

$$\text{CRSS}(\max) = \frac{\gamma}{b} \tag{1.94}$$

where γ is the stacking fault energy. Fig. 1.45 compares the developed model with the experimental results for Ni (Uchic et al., 2004; Dimiduk et al., 2005) and Au (Greer et al., 2005; Volkert and Lilleodden, 2006) which shows that the model can successfully capture the extrinsic size effects.

1.3.2.2.2 Source exhaustion

The second mechanism of hardening occurs when the available mobile dislocations are not adequate to sustain the applied plastic flow. The loss of mobile dislocation density can occur due to the dislocation escape from the free surfaces, dislocation source shutdown, and mechanical annealing. Consequently, the applied stress should be increased to activate stronger sources and sustain the required plastic flow, which is termed as exhaustion hardening (see, e.g.,

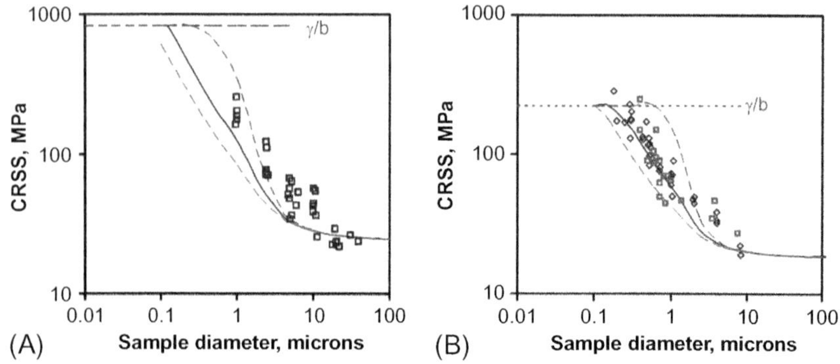

FIG. 1.45 Variation of critical resolved shear stress (CRSS) versus the pillar diameter for: (A) Ni and (B) Au. The *dotted lines* denote the lower and upper standard deviations from the mean as predicted by the model. The original experimental data has been reported by Uchic et al. (2004), Dimiduk et al. (2005), Greer et al. (2005), and Volkert and Lilleodden (2006). *(After Parthasarathy, T.A., Rao, S.I., Dimiduk, D.M., Uchic, M.D., Trinkle, D.R., 2007. Contribution to size effect of yield strength from the stochastics of dislocation source lengths in finite samples. Scr. Mater. 56, 313–316.)*

Rao et al., 2008). In the case of the pillar compression experiment, a dislocation-starved sample may be formed when all of the mobile dislocations leave the sample. Consequently, the deformation mechanism, which is commonly termed as the dislocation starvation hardening (Greer et al., 2005), is governed by source-limited activations. The dislocation starvation is a special case of source exhaustion. Both discrete dislocation dynamics (DDD) and molecular dynamics (MD) have been incorporated to address extrinsic size effects in crystalline pillars. Between all these simulations, some of them addresses source exhaustion mechanism including dislocation starvation. Here, the work conducted by Zhou et al. (2011) is addressed in more details.

Zhou et al. (2011) incorporated the 3-D DDD and investigated the governing mechanisms of deformation for metallic samples of different sizes. Various pillar diameters from 300 nm to 1000 nm have been incorporated. They modeled the Ni single crystal pillars during uniaxial compression experiment. The cross-slip was also incorporated in the model. Fig. 1.46 depicts the variation of engineering stress and dislocation density as the strain changes. In the case of pillar diameters of 300 nm and 500 nm, dislocation starvation occurs, which is not the case for a pillar diameter of 750 nm. After the dislocation starvation occurs, the response becomes elastic and hardening occurs due to the dislocation starvation. In the case of pillar diameter of 1000 nm, no dislocation starvation occurs. The observed hardening was related to the exhaustion hardening induced by dislocation interactions which shut off the dislocation sources.

Here, the framework presented by Zhou et al. (2011) to capture the dislocation starvation is elaborated. Zhou et al. (2011) followed the framework presented by Greer (2006) and developed a theoretical model to capture the

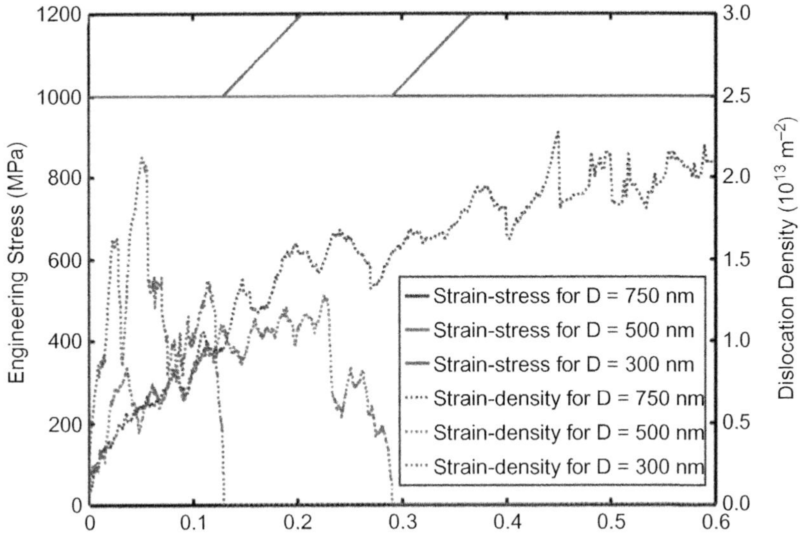

FIG. 1.46 Variation of engineering stress and dislocation density versus the strain. *(After Zhou, C., Beyerlein, I.J., LeSar, R., 2011. Plastic deformation mechanisms of fcc single crystals at small scales. Acta Mater. 59, 7673–7682.)*

dislocation starvation. In the case of pillars with confined volumes, as the sample size decreases, the rate of dislocations escaping from the pillar free surfaces increases while the rate of dislocation multiplication decreases. Accordingly, there is a critical sample size at which the dislocation starvation occurs for samples with smaller diameters. Fig. 1.47 depicts a pillar with a radius of r and primary slip plane oriented at an angle β. The elliptical glide plane has the major and minor radii of $a = r/\cos(\beta)$ and r, respectively. Accordingly, the dislocation loss rate can be described as an inverse function of the pillar radius r:

$$d\rho_{\text{loss}} = -\rho_{\text{mob}}\left(\frac{1}{2}\frac{\pi(a+r)\upsilon dt}{\pi ar}\right) = -\rho_{\text{mob}}\frac{\cos^2(\beta/2)\upsilon dt}{r} \tag{1.95}$$

where v is the velocity of dislocation and ρ_{mob} is the density of mobile dislocations. In Eq. (1.95), it is assumed that any dislocation within a distance υdt from free surfaces has a 50% escaping chance. The dislocation losses from annihilation or source shutdown are ignored in Eq. (1.95). The rate of dislocation multiplication can be defined as follows:

$$d\rho_{\text{mult}} = \rho_{\text{mob}}\frac{\upsilon dt}{L} \tag{1.96}$$

where L is the dislocation mean free path. Accordingly, the total dislocation content in the sample can be described as follows:

$$\rho = \rho_0 + \frac{[1/L - \cos^2(\beta/2)/r]}{b}\frac{\varepsilon_p}{M} \tag{1.97}$$

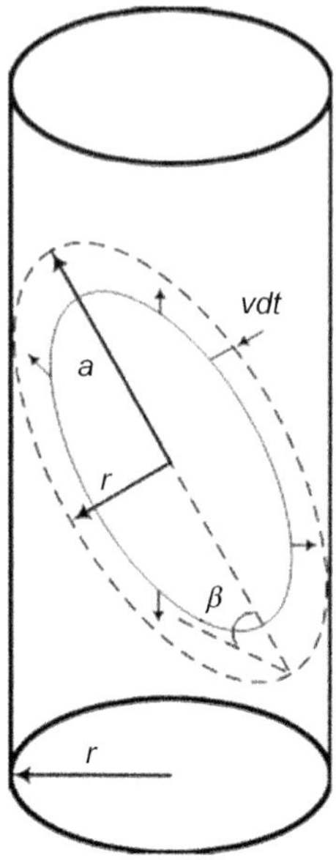

FIG. 1.47 A pillar contains a dislocation loop with the distance *vdt* from free surfaces. *(After Zhou, C., Beyerlein, I.J., LeSar, R., 2011. Plastic deformation mechanisms of fcc single crystals at small scales. Acta Mater. 59, 7673–7682.)*

where ρ_0 is the initial dislocation content, M is the Schmid factor, and $\varepsilon_p = \rho_{\text{mob}} bvdt$ is the plastic strain induced by the movement of mobile dislocations. In Eq. (1.97), if the term $[1/L - \cos^2(\beta/2)/r]$ becomes negative, the dislocation content decreases as the plastic strain increases, and eventually, the dislocation starvation occurs. As an approximation, let one assume $L \approx \rho^{-0.5}$, $\beta = 35^{\circ}$, and $\rho_0 = 2.0 \times 10^{13}\ \text{m}^{-2}$, the critical size is around 400 nm. Zhou et al. (2011) addressed two more size effects mechanisms including source truncation (section 1.3.2.2.1) and general source exhaustion in their work, which will be elaborated in more details in Chapter 4.

Based on the obtained results, Zhou et al. (2011) presented a physically based plastic deformation map for samples of confined volumes, which includes three sample size zones (Fig. 1.48). In Zone I, i.e., samples with very small length scales, the surface dislocations nucleation is the deformation mechanism, and dislocation starvation governs the hardening mechanism. In Zone III, i.e., large samples, the multiplication of internal dislocations governs the deformation mechanism. The exhaustion hardening caused by dislocation reactions can explain the observed intermittent plasticity. In Zone II, i.e., the transition zone, both surface dislocation nucleation and internal dislocations multiplication contribute to the plasticity. One should note that in this chapter, the dislocation starvation is considered as a special case of source exhaustion.

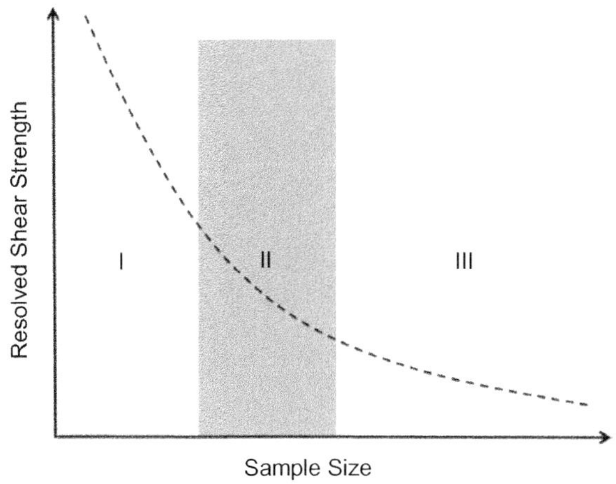

FIG. 1.48 Three zones of sample size. In Zone I, the surface dislocations nucleation is the governing deformation mechanism. In Zone II, both surface dislocations nucleation and multiplication of internal dislocations should be considered. In Zone III, the deformation is governed by multiplication of internal dislocations. *(After Zhou, C., Beyerlein, I.J., LeSar, R., 2011. Plastic deformation mechanisms of fcc single crystals at small scales. Acta Mater. 59, 7673–7682.)*

1.3.2.2.3 Weakest link theory

In the case of metallic samples of confined volumes such as pillars, the strength of the weakest slip plane present in the sample is increased by decreasing the sample size. Accordingly, the sample strength is increased, which is termed as the weakest link theory. Norfleet et al. (2008) investigated the size effects in Ni pillars with diameters varying from 1 – 20 μm during compression test (Fig. 1.49). They also measured the dislocation density in the undeformed and deformed bulk samples and micropillars. They proposed the following relation to capture the size effects in pillars considering both source truncation and forest hardening:

$$\tau = \tau_0 + k_s G \frac{\ln(\bar{\lambda}/b)}{(\bar{\lambda}/b)} + k_f G b \ \ln\left(\frac{1}{b\sqrt{\rho_f/2}}\right)\sqrt{\rho_f/2} \tag{1.98}$$

where τ is the flow shear stress, τ_0 is the lattice-friction stress, k_s is the source hardening constant, $\bar{\lambda}$ is the average dislocation source length, k_f denotes the average strength of the forest dislocations, and ρ_f is the scalar density of the forest dislocations, where $\rho_f = R\rho$. The value of R depend on the sample geometry and strain value. The accurate evaluation of R for each strain value and sample geometry is difficult. Accordingly, two values of $R = 1/2$ and $R = 3/4$ which are related to the strength lower and upper bounds, respectively, are considered. Fig. 1.50 compares the variation of the shear stress obtained from the presented model, both lower and upper bounds, with that of the experiments as the sample

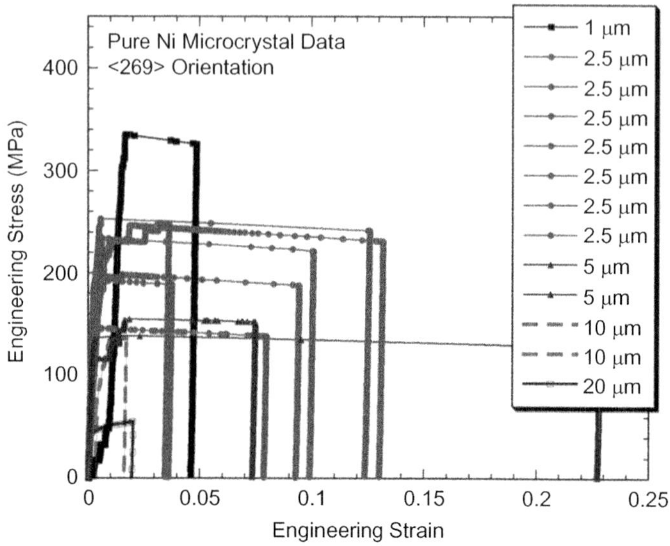

FIG. 1.49 Variation of stress versus strain for pure Ni ⟨2 6 9⟩-oriented pillars with different values of diameters. *(After Norfleet, D.M., Dimiduk, D.M., Polasik, S.J., Uchic, M.D., Mills, M.J., 2008. Dislocation structures and their relationship to strength in deformed nickel microcrystals. Acta Mater. 56, 2988–3001.)*

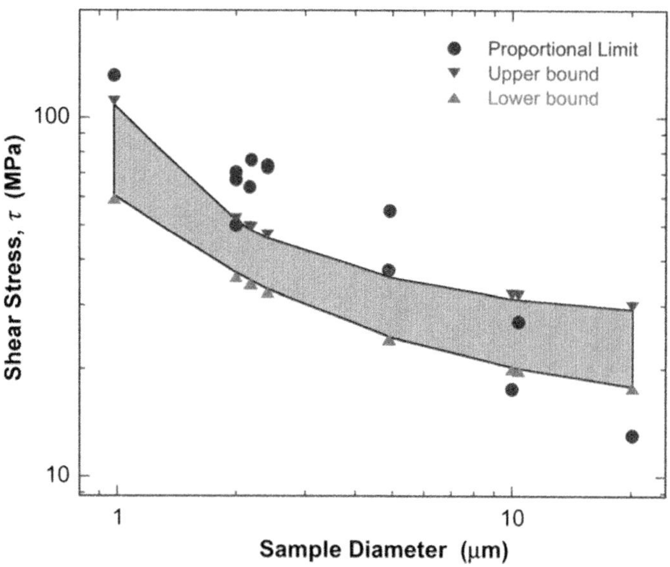

FIG. 1.50 Variation of proportional limit versus pillar diameter for experiment and lower and upper bounds of Eq. (1.98). *(After Norfleet, D.M., Dimiduk, D.M., Polasik, S.J., Uchic, M.D., Mills, M.J., 2008. Dislocation structures and their relationship to strength in deformed nickel microcrystals. Acta Mater. 56, 2988–3001.)*

size varies. The results show that the model can successfully capture size effects in samples with diameter larger than ~5 μm. In the cases of smaller samples, however, the model understimates the experimental results. Fig. 1.50 shows that there should be another hardening mechanism which is not considered in the presented model.

Norfleet et al. (2008) captured part of the difference between the model prediction and the experimental results by modifying the forest hardening mechanism to include the dislocation density correction measured for smaller samples. The results obtained from this modified method is compared with the experimental results in Fig. 1.51. Again, the modified model underestimates the strength of the smaller samples. The difference between the model and experimental results was attributed to the weakest link theory. The weakest link theory states that there is a mean-field limit ξ^* for any crystals. In the cases of samples larger than ξ^*, the dislocation ensemble behave according to the conventional forest hardening mechanism. In the cases of samples smaller than ξ^*, however, the mean-field forest hardening mechanism breaks down due to the new altered distribution of dislocations. In the samples smaller than ξ^*, the

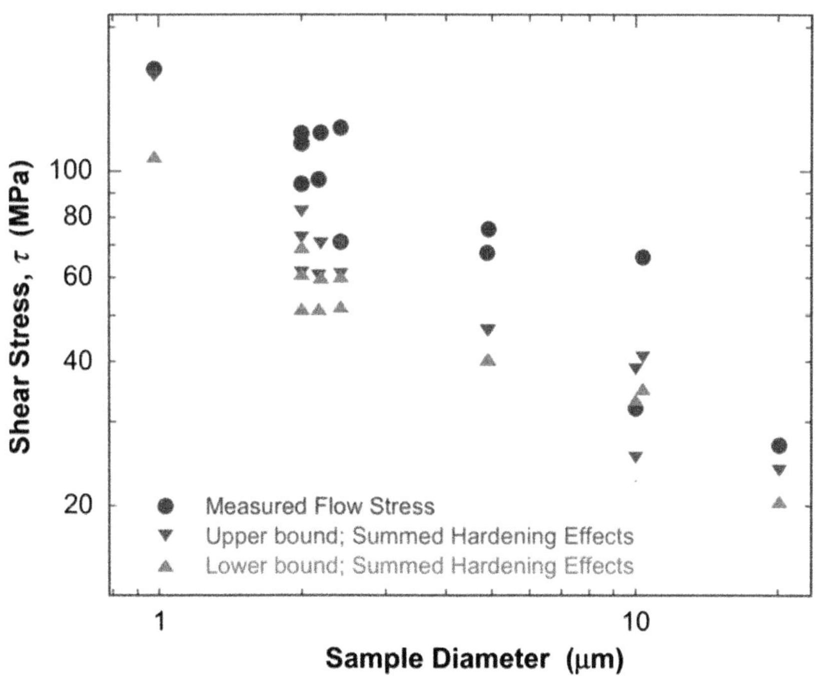

FIG. 1.51 Variation of proportional limit versus pillar diameter for experiment and lower and upper bounds of Eq. (1.98) corrected by increasing the dislocation density at smaller samples. *(After Norfleet, D.M., Dimiduk, D.M., Polasik, S.J., Uchic, M.D., Mills, M.J., 2008. Dislocation structures and their relationship to strength in deformed nickel microcrystals. Acta Mater. 56, 2988–3001.)*

longest dislocation sources, which are the weakest ones, are removed from the samples. Accordingly, the modified distribution of dislocation source length leads to a stronger forest-hardening contribution.

1.3.2.2.4 Interaction of size effects due to the pillar diameter and grain size

Few studies have addressed the interaction of extrinsic and intrinsic size effects for polycrystalline pillars with various diameters. Ng and Ngan (2009) investigated the effects of grain boundary in micron sized aluminum bi-crystal pillars. They reported a smoother response with shorter strain bursts and considerably higher work-hardening rates in the case of bi-crystal samples compared to those of single crystalline pillars. They incorporated TEM analysis and reported that the grain boundary increase the post-deformation dislocation density. The contribution of grain boundary to the dislocation storage inside the pillar leads to the higher work-hardening rate and sample strength.

Kunz et al. (2011) tested the nano sized aluminum bi-crystal pillars that contain high angle grain boundary oriented along the pillar axes. The observed trends were different from those reported by Ng and Ngan (2009) which shows that the pillar response depends on the pillar size and grain boundary properties. Jang and Greer (2011) investigated the effects of the grain boundary on the size effect in nanocrystalline Ni-4%W nanopillars with the grain size of 60 nm during uniaxial compression and tension experiments. It was observed that the smaller pillar has less strength, which is in contrast with the trend reported single crystalline pillars (Fig. 1.52). The observed trend was attributed the observed trend to the grain boundary-mediated deformation mechanisms including deformation mechanisms of grain rotation, grain boundary migration, and mechanical twinning.

1.3.2.3 Nanoindentation

Nanoindentation is one of the most popular experiments to investigate the behavior of materials at the micro and nano scales. During nanoindentation, a small hard indenter is pressed onto a sample, and the load versus the penetration depth is recorded. In the case of traditional indentation test, a hardness number is reported which represents the material strength. However, in the case of nanoindentation experiments, the results show that the hardness does not have a constant value. The variation of hardness versus the indentation depth depends on the indenter tip geometry and material properties. In the case of single crystal metal indented by conical or Berkovich tips, the hardness decreases as the indentation depth increases (Fig. 1.53).

Ashby (1970) introduced the concept of geometrically necessary and statistically stored dislocations, and captured the effect of grain size using this concept. He described that the dislocations are induced and stored for two reasons. One group of dislocations are required for the compatible deformation of

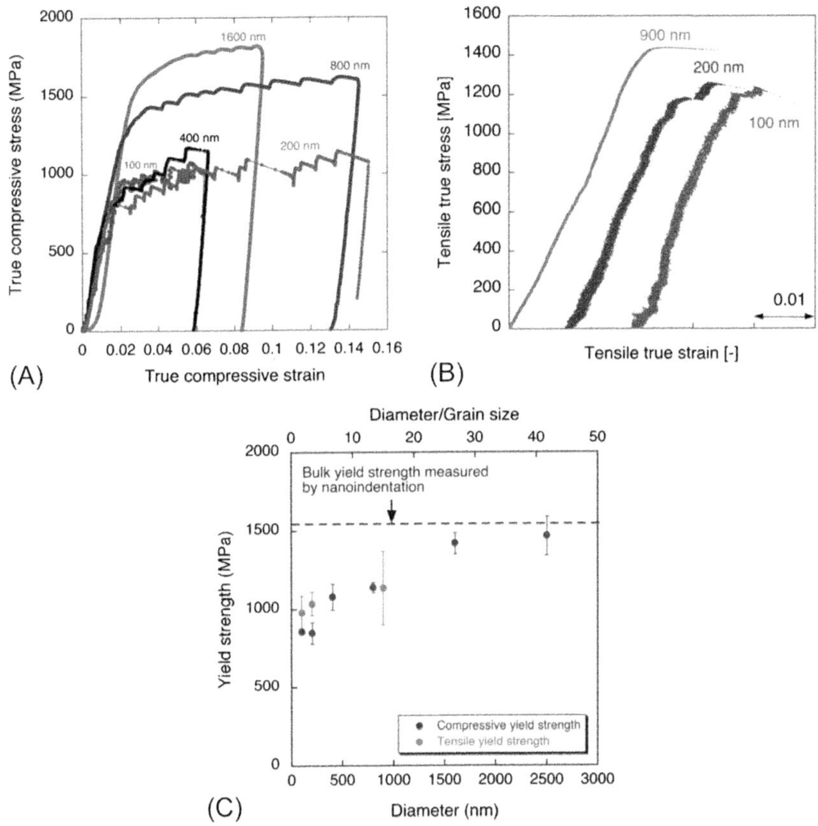

FIG. 1.52 Effects of grain boundary on the size effects in nanocrystalline Ni-4%W nanopillars with grain size of 60 nm. (A) Compressive response, (B) tensile response, and (C) variation of yield stress versus the pillar diameter during uniaxial compression and tension experiments. *(After Jang, D., Greer, J.R., 2011. Size induced weakening and grain boundary-assisted deformation in 60nm-grained Ni nano-pillar. Scr. Mater. 64, 77–80.)*

various parts that are geometrically necessary dislocations (GNDs). The other group of dislocations are accumulated by trapping one another in a random way that are statistically stored dislocations (SSDs). The size effects during nanoindentation is usually attributed to the role of the geometrically necessary dislocations that cause the enhanced hardening. Several researchers including Stelmashenko et al. (1993), De Guzman et al. (1993), Nix and Gao (1998), Swadener et al. (2002), and Pugno (2007) have developed a simple model based on the concept of GNDs to capture the size effects during nanoindentation (Fig. 1.54). One should note that the SSDs are excluded from Fig. 1.54; however, they contribute to the hardness. The imposed deformation is accommodated by GNDS nucleation, and the dislocation loops are assumed to be

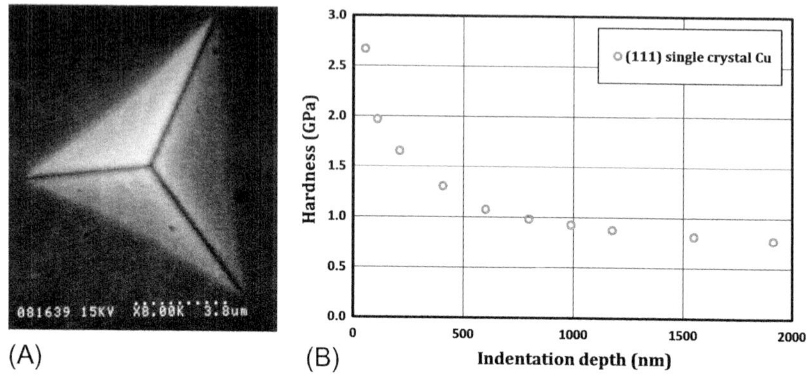

FIG. 1.53 Nanoindentation of a (111) single crystalline Cu sample indented with a Berkovich diamond indenter (A) a SEM micrographs of sample at indentation depth of 1500 nm and (B) variation of hardness versus the indentation depth. *(After McElhaney, K.W., Valssak, J.J., Nix, W.D., 1998. Determination of indenter tip geometry and indentation contact area for depth sensing indentation experiments. J. Mater. Res. 13, 1300–1306.)*

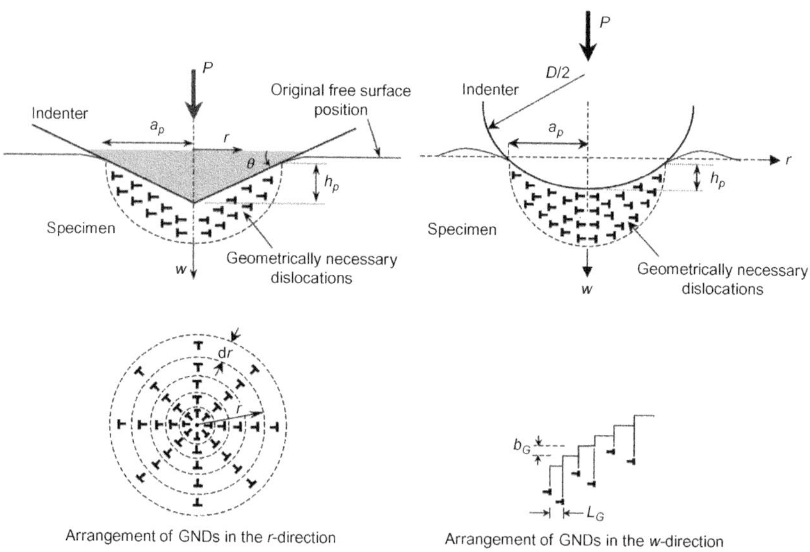

FIG. 1.54 Axisymmetric rigid conical and spherical indenters. Geometrically necessary dislocations created during the indentation process. The dislocation structure is idealized as circular dislocation loops. *(After Abu Al-Rub, R.K., Voyiadjis, G.Z., 2004. Analytical and experimental determination of the material intrinsic length scale of strain gradient plasticity theory from micro- and nano-indentation experiments. Int. J. Plast. 20, 1139–1182.)*

circular. The total GND length can be obtained as a function of the indentation depth h_p and the indenter shape as follows:

$$\lambda_G = \int_0^{a_p} \frac{2\pi r}{b_G} \left(\frac{dh}{dr} \right) dr \tag{1.99}$$

where a_p is the contact radius and b_G is the Burgers vector of GNDs. The effect of the indenter shape is incorporated in the term dh/dr. As an example, in the case of conical indenter, $dh/dr = \tan(\theta) = h/a_p$. Accordingly, the total GND length can be obtained as follows:

$$\lambda_G = \int_0^{a_p} \frac{2\pi rh}{b_G a_p} dr = \frac{\pi a_p h}{b_G} = \frac{\pi a_p^2 \tan(\theta)}{b_G} \tag{1.100}$$

In the case of spherical indenter, one can approximate the indenter by a paraboloid. Accordingly, the total GND length can be obtained as follows:

$$\lambda_G \approx \frac{4\pi}{3} \frac{a_p^3}{b_G D} \tag{1.101}$$

Yaghoobi and Voyiadjis (2016) obtained the precise total GND length for a spherical indenter as follows:

$$\lambda_G = \frac{2\pi}{b_G} \int_0^{a_p} \left(\frac{r^2}{\sqrt{R^2 - r^2}} \right) dr = \frac{2\pi}{b_G} \left[\frac{R^2}{2} \sin^{-1}\left(\frac{a_p}{R} \right) - \frac{1}{2} \left(a_p \sqrt{R^2 - a_p^2} \right) \right] \tag{1.102}$$

where $R = D/2$. The plastic zone can be defined as a hemisphere with the radius of $R_{pz} = fa_p$ in which f is a constant. Durst et al. (2005) stated that f is smaller than 3.5. They selected $f = 1.9$ in their simulations. The volume of the plastic zone can be defined as follows:

$$V = \frac{2}{3}\pi R_{pz}^3 = \frac{2}{3}\pi f^3 a_p^3 \tag{1.103}$$

The dislocation density can be stated as follows:

$$\rho_G = \lambda_G / V \tag{1.104}$$

The shear strength can be defined using the Taylor hardening model as follows:

$$\tau = \alpha G b \sqrt{\rho} \tag{1.105}$$

where G is the shear modulus, α is a material constant. The total dislocation density is function of ρ_G, ρ_S, and their interaction, where ρ_S is the density of SSDs. Abu Al-Rub and Voyiadjis (2004) proposed the following relation to capture the relation between the shear strength and ρ_G and ρ_S as follows:

$$\tau = \alpha_S G b_S \left[\rho_S^{\beta/2} + \left(\alpha_G^2 b_G^2 \rho_G / \alpha_S^2 b_S^2 \right)^{\beta/2} \right]^{1/\beta} \tag{1.106}$$

where the indices G and S designate GNDs and SSDs parameters, respectively, and β is a material parameter. Assuming the Tabor factor of 3, the indentation hardness can be obtained as follows:

$$H = 3\sigma = 3\sqrt{3}\tau = 3\sqrt{3}\alpha_S G b_S \left[\rho_S^{\beta/2} + \left(\alpha_G^2 b_G^2 \rho_G / \alpha_S^2 b_S^2 \right)^{\beta/2} \right]^{1/\beta} \tag{1.107}$$

As an example, in the case of single crystal sample indented by a conical indenter, $\rho_G = 3 \tan(\theta)/2f^3 a_p b_G$. As the indentation depth increases, the contact radius also increases which decreases the density of GNDs. Consequently, the hardness decreases as the indentation depth increases which is a typical size effect during nanoindentation (Fig. 1.53).

Voyiadjis and his coworkers (Abu Al-Rub and Voyiadjis, 2004; Voyiadjis et al., 2011; Voyiadjis and Zhang, 2015; Zhang and Voyiadjis, 2016) have captured the length scale required for strain gradient plasticity using a nanoindentation experiment. In the case of single crystal metals, Abu Al-Rub and Voyiadjis (2004) defined the non-local effective plastic strain $\hat{p}$ as follows:

$$\hat{p} = \left[p^\gamma + (l\eta)^\gamma \right]^{1/\gamma} \tag{1.108}$$

where p is the local effective plastic strain, l is the material length scale, η is the higher-order gradients of p, and γ is the material constant. The parameter η can be related to the density of GNDs ρ_G as follows:

$$\eta = \frac{\rho_G b_G}{\bar{r}} \tag{1.109}$$

where $\bar{r}$ is the Nye factor. The conventional flow stress σ can be defined as follows:

$$\sigma = \sigma_0 p^{1/m} \tag{1.110}$$

where σ_0 is the yield stress during uniaxial tension test. The non-local flow stress can be defined as follows:

$$\sigma = \sigma_0 \hat{p}^{1/m} = \sigma_0 \left[p^\gamma + (l\eta)^\gamma \right]^{1/m\gamma} \tag{1.111}$$

In the case of proportional loading, Abu Al-Rub and Voyiadjis (2004) defined the following relation between the plastic shear strain γ^p and the density of SSDs ρ_S as follows:

$$\gamma^p = b_S L_S \rho_S \tag{1.112}$$

where L_S is the mean free path of SSDs. Furthermore, they defined the plastic strain tensor ε_{ij}^p using the Schmidt's tensor M_{ij} as follows:

$$\varepsilon_{ij}^{p} = \gamma^{p} M_{ij} \tag{1.113}$$

The local effective plastic strain can be defined as follows:

$$p = \sqrt{\frac{2}{3}\varepsilon_{ij}^{p}\varepsilon_{ij}^{p}} \tag{1.114}$$

By incorporating Eqs. (1.112), (1.113), Eq. (1.114) can be rewritten as follows:

$$p = b_S L_S \rho_S \overline{M} \tag{1.115}$$

where $\overline{M} = \sqrt{\frac{2}{3}M_{ij}M_{ij}}$. Comparing Eqs. (1.107)–(1.115), one can obtain γ, m, and σ_0, as follows:

$$\gamma = \beta/2, m = 2, and\, \sigma_0 = \sqrt{3}\alpha_S G\sqrt{\frac{b_S}{L_S\overline{M}}} \tag{1.116}$$

and the material length scale l can be obtained as follows:

$$l = \hbar L_S \text{ where } \hbar = (\alpha_G/\alpha_S)^2 (b_G/b_S)\overline{M}\overline{r} \tag{1.117}$$

The interaction between the effects of the grain size, as an intrinsic length scale, with those of the indentation depth, as an external length scale, on the material strength have been addressed both theoretically and experimentally (Soer and De Hosson, 2005; Cao et al., 2006; Hou et al., 2008; Voyiadjis and Peters, 2010; Voyiadjis et al., 2010, 2011; Almasri and Voyiadjis, 2010; Voyiadjis and Zhang, 2015; Zhang and Voyiadjis, 2016). Here, the experimental and theoretical studies conducted by Voyiadjis and his coworkers (Voyiadjis and Peters, 2010; Voyiadjis et al., 2010, 2011; Almasri and Voyiadjis, 2010; Voyiadjis and Zhang, 2015; Zhang and Voyiadjis, 2016) are discussed in more details. The schematic of the interaction between the GNDs and grain boundaries for polycrystalline metals during nanoindentation is illustrated in Fig. 1.55. As the indenter penetrates, the GNDs are induced to accommodate the imposed displacement. However, the induced GNDs are piled up against the grain boundaries until the stress concentration induced by the array of dislocation pile-up reaches a critical value. At this moment, the dislocations will pass the grain boundary. Fig. 1.56 compares the nanoindentation response of a single crystalline Al sample with that of a polycrystalline one at strain rates of 0.1 s^{-1}. In the case of single crystalline sample, the results show that the hardness decreases as the indentation depth increases. In the of polycrystalline sample, however, there is an initial softening followed by a local hardening. The response becomes similar to that of the single crystalline sample by further increasing the indentation depth. Fig. 1.57 depicts the typical polycrystalline material response during nanoindentation. It was experimentally observed that the hardness initially

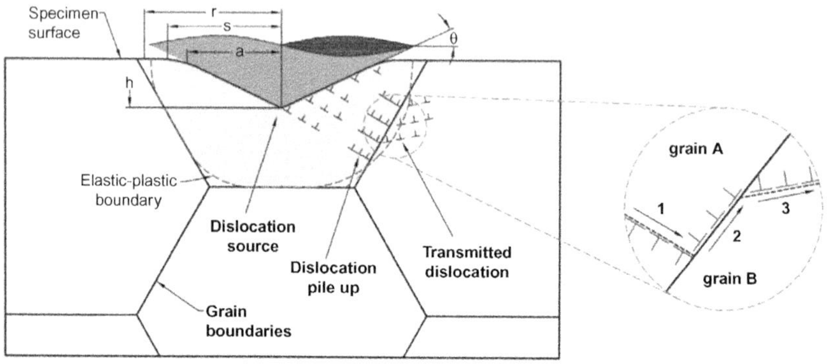

FIG. 1.55 The schematic of the interaction between the GNDs and grain boundaries for polycrystalline metals during nanoindentation. *(After Voyiadjis, G.Z., Zhang, C., 2015. The mechanical behavior during nanoindentation near the grain boundary in a bicrystal FCC metal. Mater. Sci. Eng. A 621, 218–228.)*

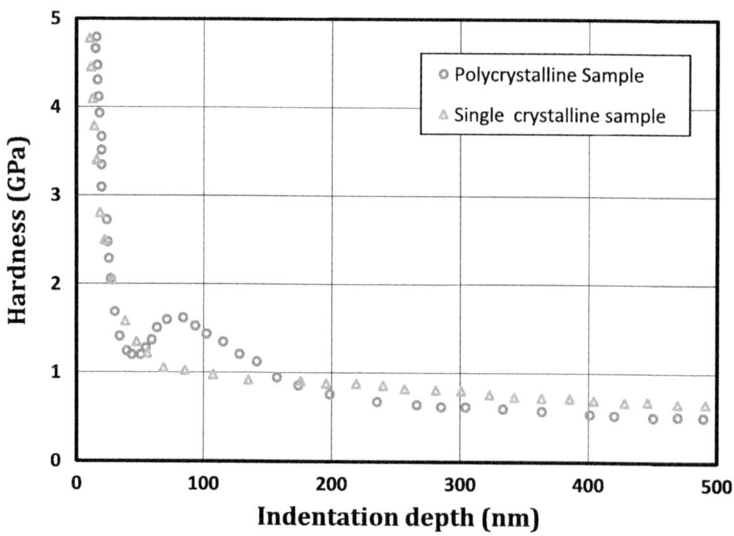

FIG. 1.56 Comparison of the nanoindentation response of a single crystalline Al sample with that of a polycrystalline one at strain rates of 0.1 s^{-1}. The original data was reported by Voyiadjis et al. (2011).

decreases which is related to the initial GNDs nucleation and movement (Region III). The GNDs start piling up against the grain boundaries which leads to the hardness enhancement (Region II). Eventually, the pile-up stress reaches the critical stress and GNDs will pass the grain boundaries and follow

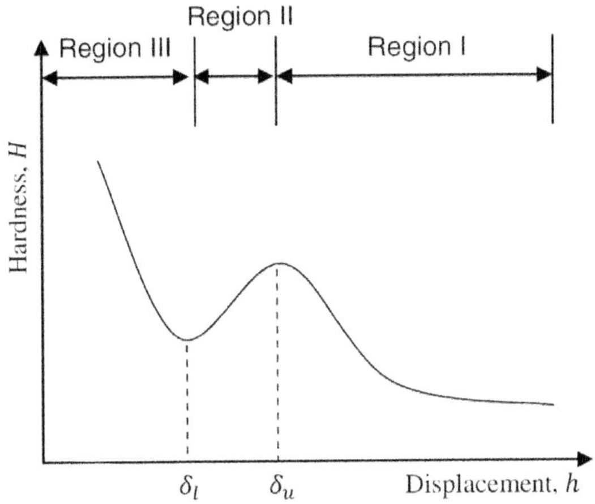

FIG. 1.57 The typical nanoindentation response of polycrystalline samples. *(After Almasri, A.H., Voyiadjis, G.Z., 2010. Nano-indentation in FCC metals: experimental study. Acta Mech. 209, 1–9.)*

the single crystalline samples in which the hardness decreases as the indentation depth increases.

Voyiadjis and Zhang (2015) and Zhang and Voyiadjis (2016) have studied the effects of grain boundary on the nanoindentation response for Al and Cu bi-crystal samples (Fig. 1.58). The results show that as the indentation distance to the grain boundary decreases, the hardening part of the response, i.e. Region II, occurs at lower indentation depth and exhibits larger maximum hardness at this region. Furthermore, as the indentation performed far from the grain boundary, the nanoindentation response becomes similar to that of single crystal samples. It is due to the fact that the GNDs do not feel the grain boundary. Accordingly, the grain boundary does not influence the nanoindentation response for indentations far from the grain boundary. Following the framework proposed by Abu Al-Rub and Voyiadjis (2004), Voyiadjis et al. (2011) presented a length scale based on the nanoindentation experiment which includes the effect of grain size as follows:

$$l = \left(\frac{\alpha_G}{\alpha_S}\right)^2 \left(\frac{b_G}{b_S}\right) \overline{M r} \left\{ \frac{\delta_1 e^{-(E_r/RT)}}{\left[1 + d_g \delta_2 p^{\left(1/\tilde{m}\right)}\right] \left[1 + \delta_3 (\dot{p})^q\right]} \right\} \tag{1.118}$$

where d_g is the average grain size, $1/\tilde{m}$ and q are the hardening exponents, δ_1, δ_2, and δ_3 are the material constants calibrated using the nanoindentation experiment, E_r is the activation energy, R is the gas constant, and T is the absolute temperature.

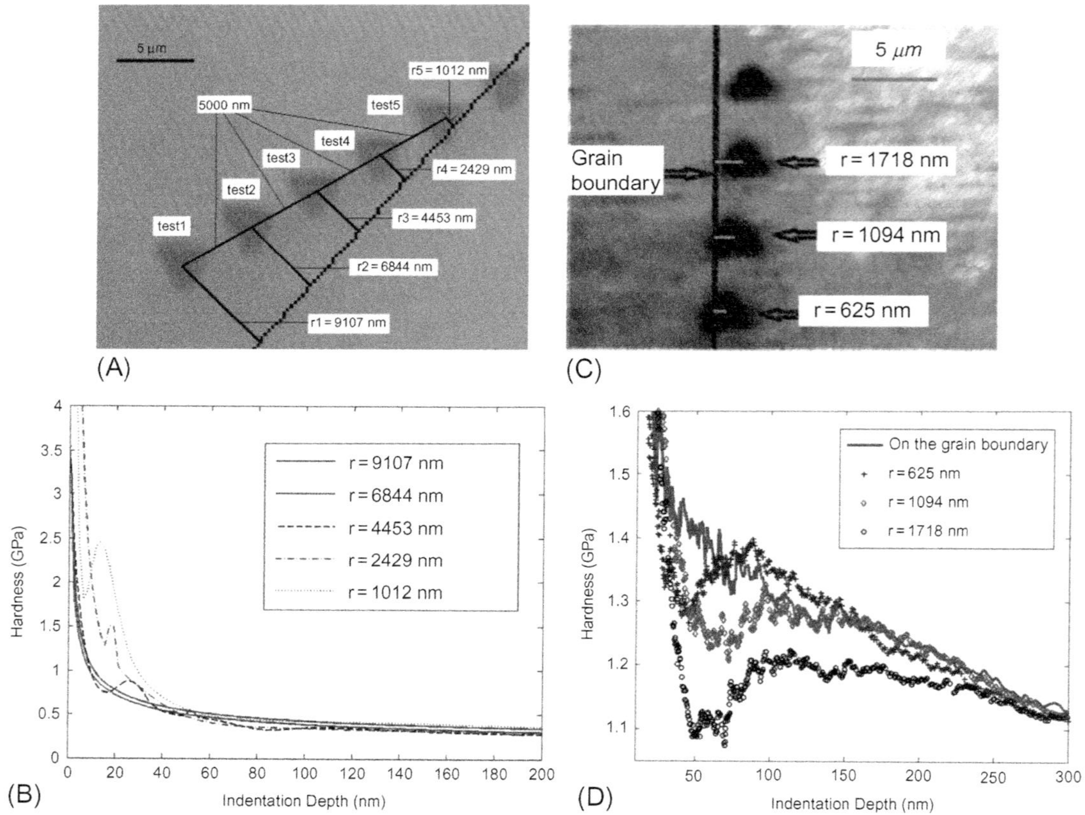

FIG. 1.58 Effect of grain boundary on the nanoindentation response of crystalline metals: (A) distances between the grain boundary and indentations for Al bi-crystal sample, (B) the effect of the distance from indentation to the grain boundary on the nanoindentation response of the Al bi-crystal sample, (C) distances between the grain boundary and indentations for Cu bi-crystal sample, and (D) the effect of the distance from indentation to the grain boundary on the nanoindentation response of the Cu bi-crystal sample. *(After Voyiadjis, G.Z., Zhang, C., 2015. The mechanical behavior during nanoindentation near the grain boundary in a bicrystal FCC metal. Mater. Sci. Eng. A 621, 218–228; Zhang, C., Voyiadjis, G.Z., 2016. Rate-dependent size effects and material length scales in nanoindentation near the grain boundary for a bicrystal FCC metal. Mater. Sci. Eng. A 659, 55–62.)*

References

Abu Al-Rub, R.K., Voyiadjis, G.Z., 2004. Analytical and experimental determination of the material intrinsic length scale of strain gradient plasticity theory from micro- and nano-indentation experiments. Int. J. Plast. 20, 1139–1182.

Almasri, A.H., Voyiadjis, G.Z., 2010. Nano-indentation in FCC metals: experimental study. Acta Mech. 209, 1–9.

Alstetter, C., 1993. Superplasticity of nanophase intermetallics. In: Nastasi, M., Parkin, D., Glaitar, H. (Eds.), Mechanical Properties and Deformation Behavior of Materials Having Ultra-Fine Microstructures. Kluver Academic Publishers, Dordrecht, pp. 381–399.

Alves, H., Ferreira, M., Köster, U., Müller, B., 1996. Mechanical properties of $(Fe,Co)_{33}Zr_{67}$ alloys. Mater. Sci. Forum 225–227, 769–774.

Armstrong, R.W., Hughes, G.D., 1999. The limiting strength properties of nanocrystalline materials. In: Chubng, Y.W. (Ed.), Advanced Materials for the Twenty-First Century: The Julia Weertman Symposium. TMS, Cincinnati, OH, pp. 409–420.

Armstrong, R., Codd, I., Douthwaite, R.M., Petch, N.J., 1962. The plastic deformation of polycrystalline aggregates. Philos. Mag. 7, 45–58.

Arzt, E., 1998. Size effects in materials due to microstructural and dimensional constraints: a comparative study. Acta Mater. 46, 5611–5626.

Ashby, M.F., 1970. The deformation of plastically non-homogeneous materials. Philos. Mag. 21, 399–424.

Astanin, V.V., Kaibyshev, O.A., Faizova, S.N., 1991. Cooperative grain boundary sliding under superplastic flow. Scr. Metall. Mater. 25, 2663–2668.

Ball, A., Hutchinson, M.M., 1969. Superplasticity in the aluminium–zinc eutectoid. J. Mater. Sci. 3, 1–7.

Bažant, Z.P., 1982. Crack band model for fracture of geomaterials. In: Eisenstein, Z. (Ed.), Proc. 4th Intern. Conf. on Num. Meth. in Geomechanics, Edmonton, AB. vol. 3, pp. 1137–1152.

Bažant, Z.P., 1984. Size effects in blunt fracture: concrete, rock, metal. J. Eng. Mech. 110, 518–535.

Bažant, Z.P., 1996. Size effect aspects of measurement of fracture characteristics of quasibrittle material. Adv. Cem. Based Mater. 4, 128–137.

Bažant, Z.P., 1999. Size effect on structural strength: a review. Arch. Appl. Mech. 69, 703–725.

Bažant, Z.P., 2005. Scaling of Structural Strength. Elsevier Butterworth-Heinemann, Burlington, MA.

Bažant, Z.P., Kazemi, M.T., 1991. Size dependence of concrete fracture energy determined by RILEM work-of-fracture method. Int. J. Fract. 51, 121–138.

Bažant, Z.P., Li, Z., 1996. Zero-brittleness size-effect method for one-size fracture test of concrete. J. Eng. Mech. 122, 458–468.

Bažant, Z.P., Novák, D., 2000. Energetic –statistical size effect in quasibrittle failure at crack initiation. ACI Struct. J. 97, 381–392.

Bažant, Z.P., Oh, B.H., 1983. Crack band theory for fracture of concrete. Mater. Struct. 16, 155–177.

Bažant, Z.P., Pfeiffer, P.A., 1987. Determination of fracture energy from size effect and brittleness number. ACI Mater. J. 84, 463–480.

Bažant, Z.P., Planas, J., 1998. Fracture and Size Effect in Concrete and Other Quasibrittle Materials. CRC Press.

Bažant, Z.P., Yu, Q., 2009. Universal size effect law and effect of crack depth on quasi-brittle structure strength. J. Eng. Mech. 135, 78–84.

Bažant, Z.P., Belytschko, T.B., Chang, T.P., 1984. Continuum model for strain softening. J. Eng. Mech. 110, 1666–1692.

Bei, H., Shim, S., Miller, M.K., Pharr, G.M., George, E.P., 2007a. Effects of focused ion beam milling on the nanomechanical behavior of a molybdenum-alloy single crystal. Appl. Phys. Lett. 2007, 91.

Bei, H., Shim, S., George, E.P., Miller, M.K., Herbert, E.G., Pharr, G.M., 2007b. Compressive strengths of molybdenum alloy micro-pillars prepared using a new technique. Scr. Mater. 57, 397–400.

Bei, H., George, E.P., Pharr, G.M., 2008a. Small-scale mechanical behavior of intermetallics and their composites. Mater. Sci. Eng. A 483–484, 218–222.

Bei, H., Shim, S., Pharr, G.M., George, E.P., 2008b. Effects of pre-strain on the compressive stress–strain response of Mo-alloy single crystal micropillars. Acta Mater. 56, 4762–4770.

Beygi, M.H.A., Kazemi, M.T., Nikbin, I.M., Javad, V.A., 2013. The effect of water to cement ratio on fracture parameters and brittleness of self-compacting concrete. Mater. Des. 50, 267–276.

Bush, M.B., 1993. Modelling of nanophase materials. Mater. Sci. Eng. A 161, 127–134.

Byer, C.M., Li, B., Cao, B., Ramesh, K.T., 2010. Microcompression of single-crystal magnesium. Scr. Mater. 62, 536–539.

Cai, B., Kong, Q., Lu, L., Lu, K., 1999. Interface controlled diffusional creep of nanocrystalline pure copper. Scr. Mater. 41, 755–759.

Cao, Y., Allameh, S., Nankivil, D., Sethiaraj, S., Otiti, T., Soboyejo, W., 2006. Nanoindentation measurements of the mechanical properties of polycrystalline Au and Ag thin films on silicon substrates: effects of grain size and film thickness. Mater. Sci. Eng. A 427, 232–240.

Carreker, R.P., Hibbard, W.R., 1955. Wright air development center technical report, 55–113. .

Carsley, J.E., Ning, J., Milligan, W.W., Hackney, S.A., Aifantis, E.C., 1995. A simple, mixtures-based model for the grain size dependence of strength in nanophase metals. Nanostruct. Mater. 5, 441–448.

Chang, H., Höfler, H.J., Altstetter, C.J., Averback, R.S., 1991. Synthesis, processing and properties of nanophase TiAl. Scr. Metall. Mater. 25, 1161–1166.

Chang, H., Altstetter, C.J., Averback, R.S., 1992. Characteristics of nanophase TiAl produced by inert gas condensation. J. Mater. Res. 7, 2962–2970.

Cheung, C., Palumbo, G., Erb, U., 1994. Synthesis of nanocrystalline permalloy by electrodeposition. Scr. Met. Mater. 31, 735–740.

Chokshi, H., Rosen, A., Karch, J., Gleiter, H., 1989. On the validity of the Hall-Petch relationship in nanocrystalline materials. Scr. Metall. 23, 1679–1684.

Christman, T., Jain, M., 1991. Processing and consolidation of bulk nanocrystalline titanium-aluminide. Scr. Met. Mater. 25, 767–772.

Conrad, H., 1961. In: Thames, G., Washburn, J. (Eds.), Electron Microscopy and Strength of Crystals. Interscience, New York, pp. 299–300.

Conrad, H., 2003. Grain size dependence of the plastic deformation kinetics in Cu. Mater. Sci. Eng. A 341, 216–228.

Conrad, H., 2004. Grain-size dependence of the flow stress of Cu from millimeters to nanometers. Metall. Mater. Trans. A Phys. Metall. Mater. Sci. 35A, 2681–2695.

Conrad, H., Narayan, J., 2000. On the grain size softening in nanocrystalline materials. Scr. Mater. 42, 1025–1030.

Conrad, H., Yang, D., 2002. Plastic deformation kinetics of electrodeposited Cu foil at low and intermediate homologous temperatures. J. Electron. Mater. 31, 304–312.

Cottrell, A.H., Bilby, B.A., 1949. Dislocation theory of yielding and strain ageing of iron. Proc. R. Soc. Lond. A Math. Phys. Sci. 62, 49–62.

da Vinci, L., 1500. See the Notebooks of Leonardo da Vinci (1945). vol. 1. Edward McCurdy, London, p. 546.

Davies, G.J., Edington, J.W., Cutler, C.P., Padmanabhan, K.A., 1970. Superplasticity: a review. J. Mater. Sci. 5, 1091–1102.
De Guzman, M.S., Neubauer, G., Flinn, P., Nix, W.D., 1993. The role of indentation depth on the measured hardness of materials. Mater. Res. Symp. Proc. 308, 613–618.
Derlet, P.M., Van Swygenhoven, H., 2002. Length scale effects in the simulation of deformation properties of nanocrystalline metals. Scr. Mater. 47, 719–724.
Derlet, P.M., Hasnaoui, A., Van Swygenhoven, H., 2003. Atomistic simulations as guidance to experiments. Scr. Mater. 49, 629–635.
Dimiduk, D.M., Uchic, M.D., Parthasarathy, T.A., 2005. Size-affected single-slip behavior of pure nickel microcrystals. Acta Mater. 53, 4065–4077.
Durst, K., Backes, B., Göken, M., 2005. Indentation size effect in metallic materials: correcting for the size of the plastic zone. Scr. Mater. 52, 1093–1097.
Embury, J.D., Lahaie, D.J., 1993. In: Natasi, M., et al., (Eds.), Mechanical Properties and Deformation Behavior of Materials Having Ultrafine Microstructure. Kluwer Academic Publisher, Dordrecht, pp. 287–301.
Erb, U., 1995. Electrodeposited nanocrystals: synthesis, properties and industrial applications. Nanostruct. Mater. 6, 533–538.
Eshelby, J.D., Frank, F.C., Nabarro, F.R.N., 1951. The equilibrium of linear arrays of dislocations. Philos. Mag. 42, 351–364.
Feltham, P., Meakin, J.D., 1957. On the mechanism of work hardening in face-centred cubic metals, with special reference to polycrystalline copper. Philos. Mag. 2, 105–112.
Freund, L.B., 1987. The stability of a dislocation threading a strained layer on a substrate. J. Appl. Mech. 54, 553–557.
Frick, C.P., Clark, B.G., Orso, S., Schneider, A.S., Arzt, E., 2008. Size effect on strength and strain hardening of small-scale [111] nickel compression pillars. Mater. Sci. Eng. A 489, 319–329.
Fu, H.H., Benson, D.J., Meyers, M.A., 2001. Analytical and computational description of effect of grain size on yield stress of metals. Acta Mater. 49, 2567–2582.
Galileo, G.L., 1638. Discorsi i Demostrazioni Matematiche Intorno à due Nuove Scienze. Elsevirii, Leiden, pp. 178–181 (English transl. by T. Weston, London 1730).
Greer, J.R., 2006. Bridging the gap between computational and experimental length scales: a review on nano-scale plasticity. Rev. Adv. Mater. Sci. 13, 59–70.
Greer, J.R., De Hosson, J.T.M., 2011. Plasticity in small-sized metallic systems: intrinsic versus extrinsic size effect. Prog. Mater. Sci. 56, 654–724.
Greer, J.R., Nix, W.D., 2005. Size dependence of mechanical properties of gold at the submicron scale. Appl. Phys. A Mater. Sci. Process. 80, 1625–1629.
Greer, J.R., Oliver, W.C., Nix, W.D., 2005. Size dependence of mechanical properties of gold at the micron scale in the absence of strain gradients. Acta Mater. 53, 1821–1830.
Griffith, A.A., 1921. The phenomena of rupture and flow in solids. Philos. Trans. 221A, 179–180.
Hahn, H., Padmanabhan, K.A., 1997. A model for the deformation of nanocrystalline materials. Philos. Mag. B 76, 559–571.
Hahn, H., Mondal, P., Padmanabhan, K.A., 1997. Plastic deformation of nanocrystalline materials. Nanostruct. Mater. 9, 603–606.
Hall, E.O., 1951. The deformation and ageing of mild steel: III. Discussion of results. Proc. Phys. Soc. B 64, 747–753.
Han, S.M., Bozorg-Grayeli, T., Groves, J., Nix, W.D., 2010. Size effects on strength and plasticity of vanadium nanopillars. Scr. Mater. 63, 1153–1156.
Hansen, N., 1977. The effect of grain size and strain on the tensile flow stress of aluminium at room temperature. Acta Metall. 25, 863–869.

Hansen, N., Ralph, B., 1982. The strain and grain size dependence of the flow stress of copper. Acta Metall. 30, 411–417.

Hasnaoui, A., Derlet, P.M., Van Swygenhoven, H., 2004. Interaction between dislocations and grain boundaries under an indenter—a molecular dynamics simulation. Acta Mater. 52, 2251–2258.

Henning, C., Boswell, F., Corbett, J., 1975. Mechanical properties of vacuum-deposited metal films—I. Copper films. Acta Metall. 23, 177–185.

Hillerborg, A., Modéer, M., Petersson, P.E., 1976. Analysis of crack formation and crack growth in concrete by means of fracture mechanics and finite elements. Cem. Concr. Res. 6, 773–782.

Hoover, C.G., Bažant, Z.P., 2013. Comprehensive concrete fracture tests: size effects of types 1 and 2, crack length effect and postpeak. Eng. Fract. Mech. 110, 281–289.

Hoover, C.G., Bažant, Z.P., 2014. Universal size-shape effect law based on comprehensive concrete fracture tests. J. Eng. Mech. 140, 473–479.

Hoover, C.G., Bažant, Z.P., Vorel, J., Wendner, R., Hubler, M.H., 2013. Comprehensive concrete fracture tests: description and results. Eng. Fract. Mech. 114, 92–103.

Hou, X.D., Bushby, A.J., Jennett, N.M., 2008. Study of the interaction between the indentation size effect and Hall–Petch effect with spherical indenters on annealed polycrystalline copper. J. Phys. D. Appl. Phys. 41.

Huang, H., Saepen, F., 2000. Tensile testing of free-standing Cu, Ag and Al thin films and Ag/Cu multilayers. Acta Mater. 48, 3261–3269.

Hughes, G.D., Smith, S.D., Pande, C.S., Johnson, H.R., Armstrong, R.W., 1986. Hall-Petch strengthening for the microhardness of twelve nanometer grain diameter electrodeposited nickel. Scr. Mater. 20, 93–97.

Hull, D., Bacon, D.J., 2011. Introduction to Dislocations, fifth ed. Elsevier Ltd.

Jang, D., Greer, J.R., 2011. Size-induced weakening and grain boundary-assisted deformation in 60nm-grained Ni nano-pillar. Scr. Mater. 64, 77–80.

Jia, D., Ramesh, K.T., Ma, E., 2000. Failure mode and dynamic behavior of nanophase iron under compression. Scr. Mater. 42, 73–78.

Jia, D., Wang, Y.M., Ramesh, K.T., Ma, E., Zhu, Y.T., Valiev, R.Z., 2001. Deformation behavior and plastic instabilities of ultrafine-grained titanium. Appl. Phys. Lett. 79, 611–613.

Jia, D., Ramesh, K.T., Ma, E., 2003. Effects of nanocrystalline and ultrafine grain sizes on constitutive behavior and shear bands in iron. Acta Mater. 51, 3495–3509.

Jiang, X., Jia, C.L., 1996. The coalescence of [001] diamond grains heteroepitaxially grown on (001) silicon. Appl. Phys. Lett. 69, 3902–3904.

Karihaloo, B.L., Abdalla, H.M., Xiao, Q.Z., 2003. Size effect in concrete beams. Eng. Fract. Mech. 70, 979–993.

Kashyap, B.P., Tangri, K., 1997. Hall–Petch relationship and substructural evolution in boron containing type 316 L stainless steel. Acta Mater. 45, 2383–2395.

Keller, C., Hug, E., 2008. Hall–Petch behaviour of Ni polycrystals with a few grains per thickness. Mater. Lett. 62, 1718–1720.

Keller, R.M., Baker, S.P., Arzt, E., 1998. Quantitative analysis of strengthening mechanisms in thin Cu films: effects of film thickness, grain size, and passivation. J. Mater. Res. 13, 1307–1317.

Khan, A.S., Zhang, H., Takacs, L., 2000. Mechanical response and modeling of fully compacted nanocrystalline iron and copper. Int. J. Plast. 16, 1459–1476.

Kiener, D., Motz, C., Schoberl, T., Jenko, M., Dehm, G., 2006. Determination of mechanical properties of copper at the micron scale. Adv. Eng. Mater. 8, 1119–1125.

Kim, D.K., Okazaki, K., 1992. Nano-crystalline consolidation of MA powders by EDC. Mater. Sci. Forum 88–90, 553–560.

Kim, H.S., Estrin, Y., Bush, M.B., 2000. Plastic deformation behaviour of fine-grained materials. Acta Mater. 48, 493–504.

Kim, J.Y., Jang, D., Greer, J.R., 2010. Tensile and compressive behavior of tungsten, molybdenum, tantalum and niobium at the nanoscale. Acta Mater. 58, 2355–2363.

Koch, C.C., Narayan, J., 2001. The inverse Hall-Petch effect-fact or artifact. Mater. Res. Soc. Symp. Proc. 634, B5 1.1–B5 1.11.

Kocks, U.F., 1970. The relation between polycrystal deformation and single-crystal deformation. Metall. Mater. Trans. B 1, 1121–1143.

Kraft, O., Volkert, C.A., 2006. Size effects on deformation and fatigue of thin films and small structures. In: Presented at CAMTEC, Cambridge, UK.

Kraft, O., Gruber, P., Mönig, R., Weygand, D., 2010. Plasticity in confined dimensions. Annu. Rev. Mater. Res. 40, 293–317.

Kuan, T.S., Murakami, M., 1982. Low temperature strain behavior of Pb thin films on a substrate. Metall. Trans. A 13, 383–391.

Kunz, A., Pathak, S., Greer, J.R., 2011. Size effects in Al nanopillars: single crystalline vs. bicrystalline. Acta Mater. 59, 4416–4424.

Leicester, R.H., 1969. The size effect of notches. In: Proc. 2nd Australasian Conf. Mech. Struct. Mater., Melbourne, pp. 4.1–4.20.

Li, J.M.C., 1963. Petch relation and grain boundary sources. Trans. Metall. Soc. AIME 227, 239–247.

Li, Y., Bushby, A.J., Dunstan, D.J., 2016. The Hall–Petch effect as a manifestation of the general size effect. Proc. Math. Phys. Eng. Sci. 472.

Lilleodden, E., 2010. Microcompression study of Mg (0 0 0 1) single crystal. Scr. Mater. 62, 532–535.

Linder, C.P., Sprague, I.C., 1956. Effect of depth of beams upon modulus of rupture of plain concrete. ASTM Proc. 55, 1062–1083.

Lu, K., Wei, W.D., Wang, J.T., 1990. Microhardness and fracture properties of nanocrystalline Ni-P alloy. Scr. Metall. Mater. 24, 2319–2323.

Malvar, J.L., Warren, G.E., 1988. Fracture energy for three-point-bend tests on single-edge-notched beams. Exp. Mech. 28, 266–272.

Mariotte, E., 1686. Traité du movement des eaux. Posthumously edited by M. de la Hire; English transl. by J. T. Desvaguliers, London 1718, p. 249; also Mariotte's collected works, 2nd ed., The Hague 1740.

McMahon, G., Erb, U., 1989. Bulk amorphous and nanocrystalline Ni-P alloys by electroplating. Microstruct. Sci. 17, 447–457.

Merz, M., Dahlgren, S., 1975. Tensile strength and work hardening of ultrafine-grained high-purity copper. J. Appl. Phys. 46, 3235–3237.

Meyers, M.A., Mishra, A., Benson, D.J., 2006. Mechanical properties of nanocrystalline materials. Prog. Mater. Sci. 51, 427–556.

Mitchell, T.E., Hecker, S.S., Smialek, R.L., 1965. Dislocation pile-ups in anisotropic crystals. Phys. Status Solidi 11, 585–594.

Narayan, J., 2000. Size and interface control of novel nanocrystalline materials using pulsed laser deposition. J. Nanopart. Res. 2, 91–96.

Narutani, T., Takamura, J., 1991. Grain-size strengthening in terms of dislocation density measured by resistivity. Acta Metall. Mater. 39, 2037–2049.

Ng, K.S., Ngan, A.H.W., 2008. Stochastic nature of plasticity of aluminum micropillars. Acta Mater. 56, 1712–1720.

Ng, K.S., Ngan, A.H.W., 2009. Deformation of micron-sized aluminum bi-crystal pillars. Philos. Mag. 89, 3013–3026.

Nieh, T.G., Wadsworth, J., 1991. Hall-Petch relation in nanocrystalline solids. Scr. Metall. Mater. 25, 955–958.

Nieh, T.G., Wadsworth, J., Sherby, O.D., 1997. Superplasticity in Metals and Ceramics. Cambridge University Press, Cambridge.

Nielsen, K.E.C., 1954. Effects of various factors on flexural strength of concrete test beams. Mag. Concr. Res. 6, 105–114.

Nieman, G.W., Weertman, J.R., Siegel, R.W., 1991. Mechanical behavior of nanocrystalline Cu and Pd. J. Mater. Res. 6, 1012–1027.

Nix, W.D., 1989. Mechanical properties of thin films. Metall. Trans. A. 20, 2217–2245.

Nix, W.D., Gao, H., 1998. Indentation size effects in crystalline materials: a law for strain gradient plasticity. J. Mech. Phys. Solids 46, 411–425.

Norfleet, D.M., Dimiduk, D.M., Polasik, S.J., Uchic, M.D., Mills, M.J., 2008. Dislocation structures and their relationship to strength in deformed nickel microcrystals. Acta Mater. 56, 2988–3001.

Orowan, E., 1947. Discussion, Symposium on Internal Stresses. Inst. Metals, London, p. 451.

Palumbo, G., Erb, U., Aust, K.T., 1990. Triple line disclination effects on the mechanical behaviour of materials. Scr. Met. Mater. 24, 2347–2350.

Pande, C.S., Masumura, R.A., Armstrong, R.W., 1993. Pile-up based Hall-Petch relation for nanoscale materials. Nanostruct. Mater. 2, 323–331.

Parthasarathy, T.A., Rao, S.I., Dimiduk, D.M., Uchic, M.D., Trinkle, D.R., 2007. Contribution to size effect of yield strength from the stochastics of dislocation source lengths in finite samples. Scr. Mater. 56, 313–316.

Petch, N.J., 1953. The cleavage strength of polycrystals. J. Iron Steel Inst. 174, 25–28.

Pugno, N.M., 2007. A general shape/size-effect law for nanoindentation. Acta Mater. 55, 1947–1953.

Raj, R., Ashby, M.F., 1971. On grain boundary sliding and diffusional creep. Metall. Trans. 2, 1113–1127.

Rao, S.I., Dimiduk, D.M., Tang, M., Parthasarathy, T.A., Uchic, M.D., Woodward, C., 2007. Estimating the strength of single-ended dislocation sources in micron-sized single crystals. Philos. Mag. 87, 4777–4794.

Rao, S.I., Dimiduk, D.M., Parthasarathy, T.A., Uchic, M.D., Tang, M., Woodward, C., 2008. Athermal mechanisms of size-dependent crystal flow gleaned from three-dimensional discrete dislocation simulations. Acta Mater. 56, 3245–3259.

Reagel, F.V., Willis, G.H., 1931. Effect of dimensions of test specimens on flexural strength of concrete. Public Roads 12, 37–46.

Rocco, C.G., 1995. Size Dependence and Fracture Mechanisms in the Diagonal Compression Splitting Test (Ph.D. thesis). Dept. Ciencia de Materiales, Universidad Politecnica de Madrid.

Rokugo, K., Uchida, Y., Katoh, H., Koyanagi, W., 1995. Fracture mechanics approach to evaluation of flexural strength of concrete. ACI Mater. J. 92, 561–566.

Sabnis, G.M., Mirza, S.M., 1979. Size effect in model concretes. J. Struct. Div. 105, 1007–1020.

Sanders, P.G., Eastman, J.A., Weertman, J.R., 1996. In: Suryanarayana, C., Singh, J., Froes, F.H. (Eds.), Processing and Properties of Nanocrystalline Materials. TMS, Warrendale, pp. 397–405.

Sanders, P.G., Youngdahl, C.J., Weertman, J.R., 1997. The strength of nanocrystalline metals with and without flaws. Mater. Sci. Eng. A 234–236, 77–82.

Schiøtz, J., Jacobsen, K.W., 2003. A maximum in the strength of nanocrystalline copper. Science 301, 1357–1359.

Schiøtz, J., Di Tolla, F.D., Jacobsen, K.W., 1998. Softening of nanocrystalline metals at very small grain sizes. Nature 391, 561–563.

Schneider, A.S., Kaufmann, D., Clark, B.G., Frick, C.P., Gruber, P.A., Monig, R., et al., 2009. Correlation between critical temperature and strength of small-scale bcc pillars. Phys. Rev. Lett. 103.

Smith, T.R., Armstrong, R.W., Hazzledine, P.M., Masamura, R.A., Pande, C.S., 1994. Pile-up based Hall-Petch considerations at ultra-fine grain sizes. Mater. Res. Soc. Proc. 362, 31–37.

Soer, W.A., De Hosson, J.T.M., 2005. Detection of grain-boundary resistance to slip transfer using nanoindentation. Mater. Lett. 59, 3192–3195.

Stelmashenko, N.A., Walls, M.G., Brown, L.M., Milman, Y.V., 1993. Microindentation on W and Mo oriented single crystals: an STM study. Acta Metall. Mater. 41, 2855–2865.

Sun, Q., Guo, Q., Yao, X., Xiao, L., Greer, J.R., Sun, J., 2011. Size effects in strength and plasticity of single crystalline titanium micropillars with prismatic slip orientation. Scr. Mater. 65, 473–476.

Swadener, J.G., George, E.P., Pharr, G.M., 2002. The correlation of the indentation size effect measured with indenters of various shapes. J. Mech. Phys. Solids 50, 681–694.

Tang, T., Bažant, Z.P., Yang, S., Zollinger, D., 1996. Variable-notch one-size test method for fracture energy and process zone length. Eng. Fract. Mech. 55, 383–404.

Thompson, A.W., 1977. Effect of grain size on work hardening in nickel. Meta 25, 83–86.

Thompson, C.V., 1993. The yield stress of polycrystalline thin films. J. Mater. Res. 8, 237–238.

Uchic, M.D., Dimiduk, D.M., 2005. A methodology to investigate size scale effects in crystalline plasticity using uniaxial compression testing. Mater. Sci. Eng. A 400–401, 268–278.

Uchic, M.D., Dimiduk, D.M., Florando, J.N., Nix, W.D., 2003. Exploring specimen size effects in plastic deformation of Ni_3(Al, Ta). Mater. Res. Soc. Symp. Proc. 753, BB1.4.1–BB1.4.6.

Uchic, M.D., Dimiduk, D.M., Florando, J.N., Nix, W.D., 2004. Sample dimensions influence strength and crystal plasticity. Science 305, 986–989.

Uchic, M.D., Shade, P.A., Dimiduk, D.M., 2009. Plasticity of micrometer-scale single crystals in compression. Annu. Rev. Mater. Res. 39, 361–386.

Van Swygenhoven, H., Caro, A., 1997. Plastic behavior of nanophase Ni: a molecular dynamics computer simulation. Appl. Phys. Lett. 71, 1652–1654.

Van Swygenhoven, H., Caro, A., Farkas, D., 2001a. Grain boundary structure and its influence on plastic deformation of polycrystalline FCC metals at the nanoscale: a molecular dynamics study. Scr. Mater. 44, 1513–1516.

Van Swygenhoven, H., Caro, A., Farkas, D., 2001b. A molecular dynamics study of polycrystalline fcc metals at the nanoscale: grain boundary structure and its influence on plastic deformation. Mater. Sci. Eng. A 309–310, 440–444.

Vitek, V., 1968. Intrinsic stacking faults in body-centred cubic crystals. Philos. Mag. 18, 773–786.

Vitek, V., Paidar, V., 2008. In: Hirth, J.P. (Ed.), Dislocation in Solids. In: vol. 14. Elsevier, p. 439.

Volkert, C.A., Lilleodden, E.T., 2006. Size effects in the deformation of submicron Au columns. Philos. Mag. 86, 5567–5579.

Voyiadjis, G.Z., Deliktas, B., 2010. Modeling of strengthening and softening in inelastic nanocrystalline materials with reference to the triple junction and grain boundaries using strain gradient plasticity. Acta Mech. 213, 3–26.

Voyiadjis, G.Z., Peters, R., 2010. Size effects in nanoindentation: an experimental and analytical study. Acta Mech. 211, 131–153.

Voyiadjis, G.Z., Zhang, C., 2015. The mechanical behavior during nanoindentation near the grain boundary in a bicrystal FCC metal. Mater. Sci. Eng. A 621, 218–228.

Voyiadjis, G.Z., Almasri, A.H., Park, T., 2010. Experimental nanoindentation of BCC metals. Mech. Res. Commun. 37, 307–314.

Voyiadjis, G.Z., Faghihi, D., Zhang, C., 2011. Analytical and experimental determination of rate- and temperature-dependent length scales using nanoindentation experiments. J Nanomech. Micromech. 1, 24–40.

Walker, S., Bloem, D.L., 1957. Studies of flexural strength of concrete-part 3: effects of variations in testing procedures. ASTM Proc. 57, 1122–1139.

Walsh, P.F., 1972. Fracture of plain concrete. Indian Concr. J. 46, 469–476.

Wang, N., Wang, Z., Aust, K.T., Erb, U., 1995. Effect of grain size on mechanical properties of nanocrystalline materials. Acta Metall. Mater. 43, 519–528.

Wang, Y.M., Ma, E., Chen, M.W., 2002. Enhanced tensile ductility and toughness in nanostructured Cu. Appl. Phys. Lett. 80, 2395–2397.

Weertman, J.R., 1993. Hall-Petch strengthening in nanocrystalline metals. Mater. Sci. Eng. A 166, 161–167.

Wei, Q., Kecskes, L., Jiao, T., Hartwig, K.T., Ramesh, K.T., Ma, E., 2004. Adiabatic shear banding in ultrafine-grained Fe processed by severe plastic deformation. Acta Mater. 52, 1859–1869.

Weibull, W., 1939. The phenomenon of rupture in solids. In: Proc. Royal Swedish Inst. Eng. Res. vol. 153, pp. 1–55.

Williams, E., 1957. Some observations of Leonardo, Galileo, Mariotte and others relative to size effect. Ann. Sci. 13, 23–29.

Wolf, D., Yamakov, V., Phillipot, S.R., Mukherjee, A., Gleiter, H., 2005. Deformation of nanocrystalline materials by molecular-dynamics simulation: relationship to experiments. Acta Mater. 53, 1–40.

Wright, P.J.F., 1952. Effect of method of test on flexural strength of concrete. Mag. Concr. Res. 11, 67–76.

Yaghoobi, M., Voyiadjis, G.Z., 2016. Atomistic simulation of size effects in single-crystalline metals of confined volumes during nanoindentation. Comput. Mater. Sci. 111, 64–73.

Yamakov, V., Wolf, D., Phillpot, S.R., Gleiter, H., 2002. Deformation twinning in nanocrystalline Al by molecular-dynamics simulation. Acta Mater. 50, 5005–5020.

Yamakov, V., Wolf, D., Phillpot, S.R., Gleiter, H., 2003a. Dislocation–dislocation and dislocation–twin reactions in nanocrystalline Al by molecular dynamics simulation. Acta Mater. 51, 4135–4147.

Yamakov, V., Wolf, D., Phillpot, S.R., Mukherjee, A.K., Gleiter, H., 2003b. Deformation mechanism crossover and mechanical behaviour in nanocrystalline materials. Philos. Mag. Lett. 83, 385–393.

Yang, H.S., Zelin, M.G., Valiev, R.Z., Mukherjee, A.K., 1992. Strain induced morphological changes of α_2 and β phases in Ti_3Al alloys during superplastic deformation. Scr. Metall. Mater. 26, 1707–1712.

Ye, J., Mishra, R.K., Sachdev, A.K., Minor, A.M., 2010. In situ TEM compression testing of Mg and Mg–0.2 wt.% Ce single crystals. Scr. Mater. 64, 292–295.

Young, T., 1807. A Course of Lectures on Natural Philosophy and the Mechanical Arts. vol. I. W. Savage, London, p. 144.

Youngdahl, C.J., Sanders, P.G., Eastman, J.A., Weertman, J.R., 1997. Scr. Mater. 37, 809.

Yu, Q., Shan, Z.W., Li, J., Huang, X., Xiao, L., Sun, J., et al., 2010. Strong crystal size effect on deformation twinning. Nature 463, 335–338.

Zelin, M.G., Mukherjee, A.K., 1995. Cooperative phenomena at grain boundaries during superplastic flow. Acta Metall. Mater. 43, 2359–2372.

Zelin, M.G., Mukherjee, A.K., 1996. Geometrical aspects of superplastic flow. Mater. Sci. Eng. A 208, 210–225.

Zhang, C., Voyiadjis, G.Z., 2016. Rate-dependent size effects and material length scales in nanoindentation near the grain boundary for a bicrystal FCC metal. Mater. Sci. Eng. A 659, 55–62.

Zhou, Y., Erb, U., Aust, K.T., Palumbo, G., 2003. The effects of triple junctions and grain boundaries on hardness and Young modulus in nanostructured Ni–P. Scr. Mater. 48, 825–830.

Zhou, C., Beyerlein, I.J., LeSar, R., 2011. Plastic deformation mechanisms of fcc single crystals at small scales. Acta Mater. 59, 7673–7682.

Further reading

El-Awady, J.A., Wen, M., Ghoniem, N.M., 2009. The role of the weakest-link mechanism in controlling the plasticity of micropillars. J. Mech. Phys. Solids 57, 32–50.

Kiener, D., Grosinger, W., Dehm, G., Pippan, R., 2008. A further step towards an understanding of size-dependent crystal plasticity: in situ tension experiments of miniaturized single-crystal copper samples. Acta Mater. 56, 580–592.

Li, J.C.M., 1961. High-angle tilt boundary—a dislocation core model. J. Appl. Phys. 32, 525–541.

McElhaney, K.W., Valssak, J.J., Nix, W.D., 1998. Determination of indenter tip geometry and indentation contact area for depth sensing indentation experiments. J. Mater. Res. 13, 1300–1306.

Meyers, M., Ashworth, E., 1982. A model for the effect of grain size on the yield stress of metals. Philos. Mag. A 46, 737–759.

Chapter 2

Nonlocal continuum plasticity

2.1 Introduction

In classical continuum theory, the constitutive model of a material point can be solely defined from certain characteristic properties at that point, such as strain, elastic properties, temperature, and defined internal variables. The coupling to the rest of continuum body is governed by the standard balanced equations and through the stress tensor. These families of continuum models are categorized as the local models. In the case of local continuum models, there is no length scale, and accordingly, scaling of the specimen size does not change the governing equations. In other words, the sample with very small length scale is described similarly as the large sample. In reality, however, due to the material heterogeneities, the material behavior is size dependent, and various examples for different material types were summarized in Chapter 1. In fact, as the sample size becomes comparable to the length of the nature of heterogeneities, such as dislocations length in metals and aggregate size in concrete, the governing mechanism of deformation becomes a function of sample length. Accordingly, at least one length scale should be incorporated into the material model to capture the size effects in different materials which leads to the nonlocal continuum models.

The local plasticity models, as the classical continuum model, cannot capture the phenomena related to the size effects. One of the well-documented issue with local plasticity models is the strain localization due to the softening behavior. Accordingly, the governing differential equations lose ellipticity and become ill-posed. As an example, the finite element solution of the problem becomes mesh sensitive and the deformation is localized in a narrow band. The width of deformation band, commonly termed the shear band, becomes a function of mesh size in a way that all the deformation is restricted to a narrow layer of elements. As the mesh size decreases, the width of the shear band decreases, which can go unrealistically small. One way to handle the problem is to define a length scale, based on the microstructural information of the material, which governs the width of shear band.

Various methods have been developed to incorporate the size effects in plasticity through different procedures. The nonlocal plasticity models can be

Size Effects in Plasticity: From Macro to Nano. https://doi.org/10.1016/B978-0-12-812236-5.00002-5

categorized as two families. First, the models which addresses size effects by incorporating additional kinematic variables in addition to strain including higher order gradient of displacement. Furthermore, the weighted spatial averages of strain can be also included in this category. Second, the nonlocal models which modify the constitutive equation by adding the gradients of the internal variables or their thermodynamic conjugates. Again, the weighted spatial averages of internal variables and their thermodynamic conjugates are accounted as the same group. Another way to categorize the nonlocal plasticity models is based on an operator used to capture size effects which includes the gradient and integral-type models. The Former type of nonlocal models uses differential operators to incorporate size effects, while the latter family incorporates the weighted spatial averages. In this chapter, local and nonlocal plasticity models will be discussed in both frameworks of small strain and finite strain continuum mechanics.

2.2 Small strain plasticity: Local models

Various local plasticity models have been developed to capture the constitutive behavior of materials. The basic concept is that at some point, the material behavior starts initiating plastic deformation in addition to the elastic deformation as expressed by the elasticity rule, i.e., Hooke's law. Two major families of plasticity models are defined with internal state variables and multisurface models. The nonlocal models are commonly the evolution of the local plasticity models with internal state variables, therefore, the local plasticity model is first addressed in more detail.

The plasticity models can be developed for both small and large strains depending on the load and consequently the strain applied on a material point. In this section, the small strain concept of plasticity is described in detail. The small strain plasticity can be described in two frameworks in terms of rate-independent and rate-dependent plasticity. This section describes the different elements of small strain plasticity including the additive decomposition of strain, yield criterion, plastic flow rule, hardening laws, and thermodynamically consistent plasticity models in the rate-independent framework. The difference between the rate-dependent and rate-independent plasticity will be then elaborated.

2.2.1 Strain additive decomposition

The idealized elastoplastic behavior of material during a uniaxial experiment is schematically illustrated in Fig. 2.1. The material is linear for the stress up to the initiation of yield at σ_y, which is a uniaxial yield stress, with the stress-strain slope equal to the elastic modulus E. The material behavior changes by further increasing the stress and the slope changes to the

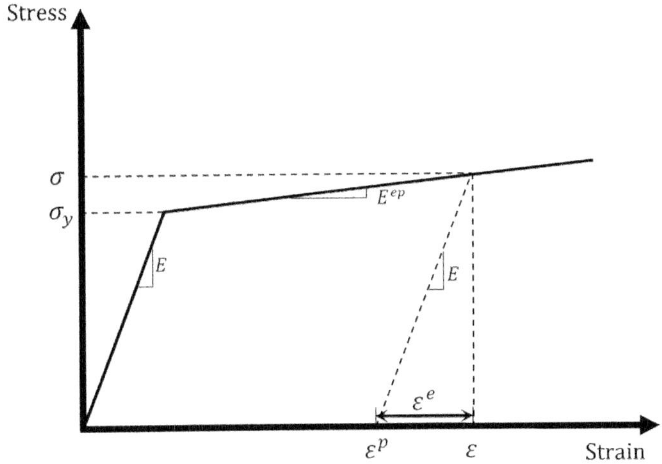

FIG. 2.1 The idealized elastoplastic behavior of material during a uniaxial experiment.

elastoplastic tangent modulus E^{ep}. During the unloading process, the material follows an elastic path with the slope of E. However, the material does not go back to the initial state and the residual strain of ε^p is induced. The total strain, as shown in Fig. 2.1, can be additively decomposed to two parts of elastic and plastic strains as follows:

$$\varepsilon = \varepsilon^e + \varepsilon^p \tag{2.1}$$

Fig. 2.1 shows that the relation between the stress and strain and can be described as follows:

$$\sigma = E\varepsilon^e = E(\varepsilon - \varepsilon^p) \tag{2.2}$$

The assumption of additive decomposition of strain can be extended to the general 3-D case for small strain plasticity as follows:

$$\varepsilon_{ij} = \varepsilon^e_{ij} + \varepsilon^p_{ij} \tag{2.3}$$

where ε_{ij}, ε^e_{ij}, and ε^p_{ij} are the total, elastic, and plastic strain tensors, respectively. The rate form of Eq. (2.3) can be described as follows:

$$\dot{\varepsilon}_{ij} = \dot{\varepsilon}^e_{ij} + \dot{\varepsilon}^p_{ij} \tag{2.4}$$

The uniaxial relation between the stress and strain described in Eq. (2.2) can also be generalized, which is known as Hooke's law, as follows:

$$\sigma_{ij} = C_{ijkl}\varepsilon^e_{kl} = C_{ijkl}\left(\varepsilon_{kl} - \varepsilon^p_{kl}\right) \tag{2.5}$$

where C_{ijkl} is the fourth order elasticity tensor.

2.2.2 Yield criterion

Yield criterion defines the stress or strain combinations in which a material yields and plasticity initiates. A yield criterion is commonly described as a function of stress state as follows:

$$f = f\left(\sigma_{ij}, l_1, l_2, \ldots, l_n\right) \tag{2.6}$$

where σ_{ij} is the Cauchy stress tensor and $l_1, l_2, \ldots, l_n$ are n material constants. For any stress tensor in which $f(\sigma_{ij}, l_1, l_2, \ldots, l_n) < 0$, the material remain elastic. In the case of isotropic materials, f is described as a function of stress and stress deviator tensor invariants I_1, J_2, and J_3, i.e., $f = f(I_1, J_2, J_3, l_1, l_2, \ldots, l_n)$. The stress and stress deviator tensors invariants I_1, J_2, and J_3 can be described as follows:

$$I_1 = \sigma_{kk} \tag{2.7}$$

$$J_2 = \frac{1}{2} s_{ij} s_{ji} \tag{2.8}$$

$$J_3 = \frac{1}{3} s_{ij} s_{jk} s_{ki} \tag{2.9}$$

where $s_{ij} = \sigma_{ij} - (I_1/3)\delta_{ij}$ is the stress deviator tensor, and δ_{ij} is the Kronecker delta. The invariants I_1, J_2, and J_3 are independent of the coordinate axis rotation.

It is also common that to define the yield criteria in the principal stresses space. One can find an orientation at a material point at which the stress tensor has no shear components, and the only available stresses are the ones normal to the plane. This plane is called a principal plane and its normal is called a principal direction. Accordingly, the stress tensor will have three normal components as follows:

$$\sigma_{ij} = \begin{bmatrix} \sigma_1 & 0 & 0 \\ 0 & \sigma_2 & 0 \\ 0 & 0 & \sigma_3 \end{bmatrix} \tag{2.10}$$

The definition of the yield criterion depends on the nature of plasticity for different materials. For example, in the case of pressure-independent materials such as metals, it has been observed that the yield surface is independent of the hydrostatic pressure. Accordingly, the yield surface can be defined as follows:

$$f = f(J_2, J_3, l_1, l_2, \ldots, l_n) \tag{2.11}$$

The most robust, famous, and simple yield surface of this category is the von Mises yield criterion, which can be described as follows:

$$f\left(J_2, \sigma_y\right) = \sqrt{3J_2} - \sigma_y \tag{2.12}$$

where σ_y is the uniaxial yield stress. The schematic of the von Mises yield surface in principal stress space is shown in Fig. 2.2. In the case of plane stress, i.e.,

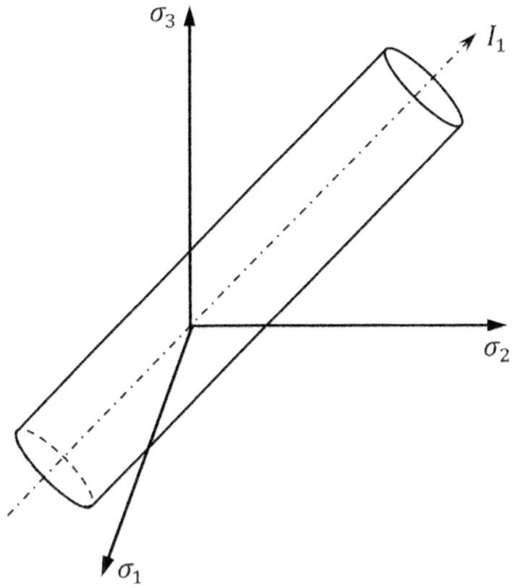

FIG. 2.2 Schematic of the von Mises yield surface in principal stress space.

$\sigma_3 = 0$, the von Mises yield surface becomes an ellipse (Fig. 2.3). The material may be ideal plastic or exhibit work hardening.

In the case of pressure-dependent materials such as concrete, rock, and soil, the yield criterion depends on the first stress tensor invariant I_1. Accordingly, the yield surface can be in the general form of $f = f(I_1, J_2, J_3, l_1, l_2, \ldots, l_n)$. The Drucker-Prager model extends the von Mises yield criterion by incorporating the effect of pressure as follows:

$$f(I_1, J_2, \sigma_y) = \alpha I_1 + \sqrt{J_2} - k \tag{2.13}$$

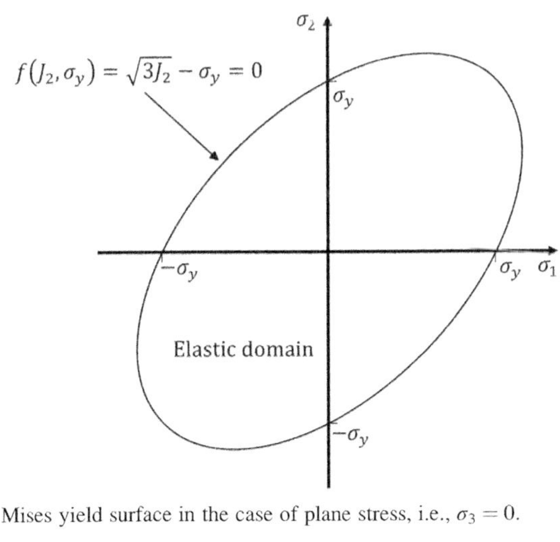

FIG. 2.3 Von Mises yield surface in the case of plane stress, i.e., $\sigma_3 = 0$.

where α and k are the material parameters, which can be calibrated using simple experiments. For example, considering the uniaxial tensile and compressive yield stress, σ_y^t and σ_y^c, respectively, one can obtain α and k by solving the following equations:

$$\alpha\sigma_y^t + \frac{1}{\sqrt{3}}\sigma_y^t - k = 0 \tag{2.14a}$$

$$-\alpha\sigma_y^c + \frac{1}{\sqrt{3}}\sigma_y^c - k = 0 \tag{2.14b}$$

Accordingly, α and k can be obtained as follows:

$$\alpha = \frac{\frac{1}{\sqrt{3}}\left(\sigma_y^c - \sigma_y^t\right)}{\sigma_y^t + \sigma_y^c} \tag{2.15a}$$

$$k = \sigma_y^t\left(\frac{2}{\sqrt{3}}\frac{\sigma_y^c}{\sigma_y^t + \sigma_y^c}\right) \tag{2.15b}$$

The schematic of the Drucker-Prager criterion in the principal stress space, which is a cone, is shown in Fig. 2.4.

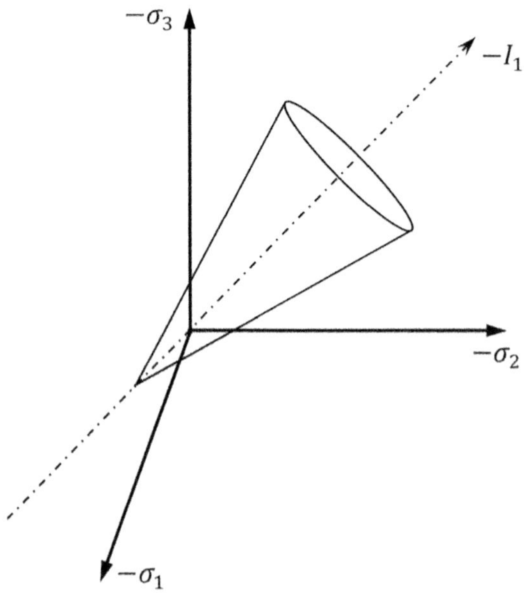

FIG. 2.4 Schematic of the Drucker-Prager criterion in the principal stress space.

2.2.3 Loading criteria

As the stress point is inside the yield surface, i.e., $f < 0$, the material is elastic. The plasticity initiates when the stress point reaches the yield surface, i.e., $f = 0$, and it continues as far as the point remains on the yield surface, which is defined as loading. However, if the stress point goes back inside the yield surface, plastic deformation stops, and the deformation becomes elastic, which is defined as unloading. The concept of loading and unloading can be mathematically defined as follows:

$$\begin{cases} f = 0 \text{ and } \dot{f} < 0 & \text{unloading} \\ f = 0 \text{ and } \dot{f} = 0 & \text{loading} \end{cases} \tag{2.16}$$

2.2.4 Plastic potential and flow rule

In the case of elastic behavior, it can be shown that the elastic strain can be obtained by differentiating the complementary energy density function Ω with respect to the stress tensor as follows (Chen and Han, 1988):

$$\varepsilon_{ij}^{e} = \frac{\partial \Omega}{\partial \sigma_{ij}} \tag{2.17}$$

Von Mises incorporated a similar framework and introduced the following plastic flow equation:

$$\dot{\varepsilon}_{ij}^{p} = \dot{\lambda}\frac{\partial F}{\partial \sigma_{ij}} \text{ or } d\varepsilon_{ij}^{p} = d\lambda\frac{\partial F}{\partial \sigma_{ij}} \tag{2.18}$$

The direction of the plastic strain increment (or plastic strain rate) is along the gradient to the yield surface, $\partial F/(\partial \sigma_{ij})$, in the stress space, $\dot{\lambda}$ (or $d\lambda$) is a positive scalar factor of proportionality, which is commonly called as the plastic or Lagrangian multiplier, and F is the plastic potential function that is convex, nonnegative, and zero at the origin. Eq. (2.18) states that the plastic strain increment tensor is normal to the hypersurface of $F(\sigma_{ij}) = 0$. The value of the plastic multiplier $\dot{\lambda}$ can be defined as follows:

$$\begin{cases} \dot{\lambda} = 0, & \text{if}(f < 0) \text{ or } (f = 0; \dot{f} < 0) \\ \dot{\lambda} > 0, & \text{if}(f = 0; \dot{f} = 0) \end{cases} \tag{2.19}$$

The plastic multiplier $\dot{\lambda}$ can be obtained using the consistency condition.

The plasticity models can be divided to two families of associated and non-associated flow rule models depending on the choice of the plastic potential function. If the yield surface f is chosen as the plastic potential function F, i.e., $f = F$, the model is associated. Otherwise, the plasticity model is non-associated. Associated plasticity models can successfully capture the plasticity in crystalline metals. However, soils and granular material demonstrate the non-linear volume change during hardening. For example, concrete materials shows both contraction and volume dilatation during the compressive loadings, which cannot be captured using the associated flow rule.

Mathematically, the associated flow rule can be defined as follows:

$$\dot{\varepsilon}_{ij}^{p} = \dot{\lambda}\frac{\partial f}{\partial \sigma_{ij}} \tag{2.20}$$

and the direction of the plastic strain rate tensor is normal to the yield surface, which is commonly termed as the normality hypothesis of plasticity. It will be shown in Section 2.2.7 that the normality hypothesis will maximize the thermodynamics dissipation potential. Furthermore, it can be easily shown that the Kuhn-Tucker conditions to maximize the plastic dissipation are similar to the normality hypothesis:

$$\dot{\lambda} f = 0, \dot{\lambda} \geq 0, f \leq 0 \tag{2.21}$$

It can be also shown that the normality hypothesis of plasticity leads to the uniqueness of the solution for the boundary-value problem for an elastoplastic material (Chen and Han, 1988).

2.2.5 Hardening rules

As the material plastically deforms, its mechanical properties varies, which can be exhibited as a change in the geometry and location of its yield surface. The resulting yield surface in a material with plastic deformation is commonly referred to the loading surface, which acts similar to the yield surface and it has all its properties. For example, in the case of a material point with a stress state located inside a loading surface, the response is elastic. The evolution of initial yield surface during the plastic deformation is governed by the hardening rule. Various hardening rules have been developed to capture the elastoplastic material behavior. Three common hardening rules of isotropic, kinematic, and mixed will be addressed here in more detail.

2.2.5.1 Loading criterion

The loading surface can be defined as a function of stress, plastic strain, and hardening parameter k as follows:

$$f = f\left(\sigma_{ij}, \varepsilon_{ij}^{p}, k\right) \tag{2.22}$$

where the hardening parameter k is a function of the plastic strain ε_{ij}^{p}. Hardening parameters will be further discussed below. In the case of material with hardening, the loading/unloading condition will be revisited. In order to define the loading/unloading condition, first the normal unit vector to the loading surface should be defined as follows:

$$N_{ij} = \frac{\partial f / \partial \sigma_{ij}}{\left(\frac{\partial f}{\partial \sigma_{kl}} \frac{\partial f}{\partial \sigma_{kl}} \right)^{1/2}} \tag{2.23}$$

The loading and consequently plastic deformation occurs if

$$\mathrm{f} = 0 \text{ and } N_{ij}\dot{\sigma}_{ij} > 0 \tag{2.24}$$

Eq. (2.24) states that the angle between the N_{ij} and σ_{ij} is acute. The unloading condition can be described as follows:

$$f = 0 \text{ and } N_{ij}\dot{\sigma}_{ij} < 0 \tag{2.25}$$

There is another loading condition which is commonly called as neutral loading, which is neither loading nor unloading, as follows:

$$f = 0 \text{ and } N_{ij}\dot{\sigma}_{ij} = 0 \tag{2.26}$$

No plastic deformation occurs in the case of neutral loading.

2.2.5.2 *Isotropic hardening*

The simplest hardening model can be demonstrated as the expansion of the yield surface without any distortion or translation, which is commonly termed isotropic hardening (Fig. 2.5). The loading surface of a material with isotropic hardening can be defined as follows:

$$f(\sigma_{ij}, k) = \bar{f}(\sigma_{ij}) - k(\varepsilon_p) \tag{2.27}$$

where ε_p is the effective plastic strain which controls the hardening, and it can be defined using two different methodologies. First, ε_p can be defined as the accumulated plastic strain as follows:

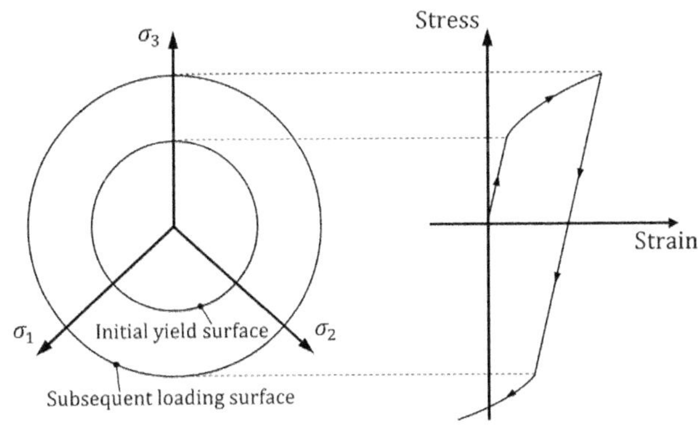

FIG. 2.5 Isotropic hardening: the stress-strain curve and π-plane representation during uniaxial cyclic loading.

$$\dot{\varepsilon}_p = C\sqrt{\dot{\varepsilon}^p_{ij}\dot{\varepsilon}^p_{ij}} \tag{2.28}$$

where $C = \sqrt{2/3}$ according to the uniaxial state of stress. Second, ε_p can be described using the plastic work per unit volume as follows in the absence of rate dependence:

$$\dot{W}_p = \sigma_e \dot{\varepsilon}_p \tag{2.29}$$

where σ_e is the effective stress defined based on the loading surface and $\dot{W}_p$ can be defined as shown below:

$$\dot{W}_p = \sigma_{ij}\dot{\varepsilon}^p_{ij} \tag{2.30}$$

In the case of von Mises loading surface, $\sigma_e = \sqrt{3J_2}$ and Eqs. (2.28), (2.29) lead to a similar $\dot{\varepsilon}_p$. Generally, however, Eqs. (2.28), (2.29) result in different definitions of effective plastic strain depending on the loading surface definition.

Different relations can be considered between the hardening parameter and effective plastic strain ε_p. For example, in the case of von Mises loading surface, the linear isotropic hardening can be selected as the simplest relation as follows:

$$\dot{k}(\varepsilon_p) = h\dot{\varepsilon}_p \rightarrow k(\varepsilon_p) = \sigma_{y0} + h\varepsilon_p \tag{2.31}$$

where h is a material constant. The most common isotropic hardening relation in the case of von Mises loading surface is as follows:

$$\dot{k}(\varepsilon_p) = b\{Q - [k(\varepsilon_p) - \sigma_{y0}]\}\dot{\varepsilon}_p \rightarrow k(\varepsilon_p) = \sigma_{y0} + Q(1 - e^{-b\varepsilon_p}) \tag{2.32}$$

where b and Q are the material constants defining the saturated value of hardening, i.e., $\max(k) = \sigma_{y0} + Q$, and the rate of hardening, respectively.

2.2.5.3 *Kinematic hardening*

Although the expansion of the loading surface can be captured using the isotropic hardening model, some phenomena including the rigid translation of loading surface cannot be modeled using this simple model. For example, the Bauschinger effect cannot be modeled using the simple expansion of the loading surface. In some materials, the increase in the yield stress in the direction of applied stress and plastic deformation leads to the decrease in the yield stress in the opposite direction (Fig. 2.6), which is commonly termed as Bauschinger effect. Accordingly, this effect can be considered as directional anisotropy. The isotropic hardening would predict the increase in yield stress in both directions and cannot capture this effect. To model these type of hardenings, the kinematic hardening can be introduced as a rigid body translation of loading surface, while the rest of geometrical properties of loading surface remain unchanged

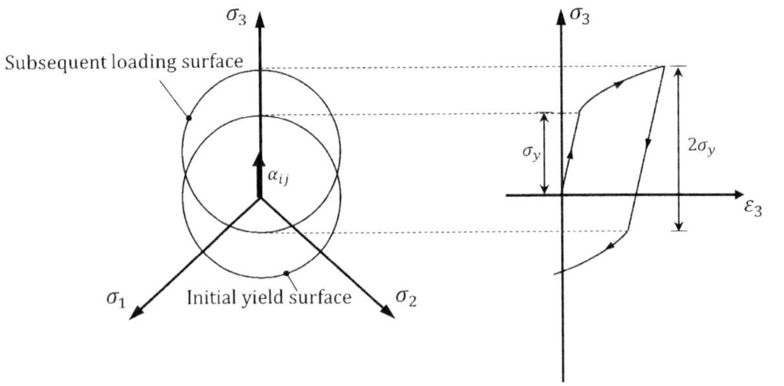

FIG. 2.6 Kinematic hardening: the stress-strain curve and π-plane representation during uniaxial cyclic loading. The increase in the yield stress in the direction of applied stress and plastic deformation leads to the decrease in the yield stress in the opposite direction, which is commonly termed as Bauschinger effect.

(Fig. 2.6). The loading surface with kinematic hardening can be described as follows:

$$f\left(\sigma_{ij}, \varepsilon_{ij}^{p}\right) = \bar{f}\left(\sigma_{ij} - \alpha_{ij}\right) - k \tag{2.33}$$

where k is a material constant and α_{ij} defines the center of the loading surface.

The next step is to define the relation between the center of the loading surface α_{ij} and plastic strain tensor ε_{ij}^{p}. The simplest kinematic hardening rule is assuming the linear relation between α_{ij} and ε_{ij}^{p}, which is commonly termed as Prager's hardening rule (Prager, 1955) as follows:

$$\dot{\alpha}_{ij} = c\dot{\varepsilon}_{ij}^{p} \tag{2.34}$$

where c is a material constant. However, this model has some inconsistencies in the case of subspace of stress, i.e., if some stress components are zero, and the rest are not. In this case, the model will dictate the distortion in the loading surface. Accordingly, Ziegler (1959) modified the Prager's kinematic hardening rule as follows:

$$\dot{\alpha}_{ij} = \dot{\mu}\left(\sigma_{ij} - \alpha_{ij}\right) \tag{2.35}$$

where $\dot{\mu}$ parameter depends on the plastic deformation history as follows:

$$\dot{\mu} = a\dot{\varepsilon}_{p} \tag{2.36}$$

where a is a material constant.

Besides the basic kinematic hardening models, several advanced kinematic hardening models have been developed. One of the most widely used kinematic hardening model was developed by Chaboche et al. (1979) and Chaboche (1986) which among other attributes is an assembly of several Armstrong and Frederick (1966). The idea is the foundation of many kinematic hardening models which have been later developed (see, e.g., Voyiadjis and Basuroychowdhury, 1998; Bari and Hassan, 2000; Chen and Jiao, 2004; Dafalias et al., 2008). Assuming the von Mises loading surface as follows:

$$f\left(\sigma_{ij}, \varepsilon_{ij}^{p}\right) = \sqrt{\frac{3}{2}\left(s_{ij} - \alpha_{ij}\right)\left(s_{ij} - \alpha_{ij}\right)} - \sigma_{y0} \tag{2.37}$$

The Chaboche kinematic hardening model can be described as follows:

$$\dot{\alpha}_{ij} = \sum_{k=1}^{M} \dot{\alpha}_{ij}^{k}, \; \dot{\alpha}_{ij}^{k} = \frac{2}{3} C^{k} \dot{\varepsilon}_{ij}^{p} - \gamma^{k} \alpha_{ij}^{k} \dot{\varepsilon}_{p} \tag{2.38}$$

where M is the number of decomposed Armstrong and Fredrick hardening rules, and C^k and γ^k are the material parameters.

2.2.5.4 *Mixed hardening*

The evolution of the loading surface may follow a combination of isotropic and kinematic hardening rules to capture both loading surface expansion and translation. Accordingly, a mixed hardening law can be described as follows:

$$f\left(\sigma_{ij}, \varepsilon_{ij}^{p}\right) = \bar{f}\left(\sigma_{ij} - \alpha_{ij}\right) - k\left(\varepsilon_{p}\right) \tag{2.39}$$

where the expansion of the loading surface is controlled by the term $k(\varepsilon_p)$ and the translation is governed by the term α_{ij}. One should note that the hardening rules are not always a simple combination of isotropic and kinematic hardenings. As an example, Voyiadjis and Foroozesh (1990) addressed a family of distortion yield models in which the geometry of loading surface evolves depending on the direction of the applied stresses and allows the yield surface to be distorted.

2.2.6 Incremental stress-strain relation for a material with mixed hardening

The incremental relation between the stress and strain can be described as follows:

$$\dot{\sigma}_{ij} = C_{ijkl}^{ep} \dot{\varepsilon}_{kl} \tag{2.40}$$

where C_{ijkl}^{ep} is the elastoplastic stiffness tensor. C_{ijkl}^{ep} depends on the stress state and loading history, and in the case of elastic deformation, it reduces to the elasticity tensor C_{ijkl}. In order to derive the elastoplastic stiffness tensor, one can start from the strain additive decomposition as follows:

$$\dot{\varepsilon}_{ij} = \dot{\varepsilon}_{ij}^{e} + \dot{\varepsilon}_{ij}^{p} \tag{2.41}$$

The relation between the stress and strain rates can be defined using Hooke's law as follows:

$$\dot{\sigma}_{ij}=C_{ijkl}\dot{\varepsilon}^{e}_{kl}=C_{ijkl}\left(\dot{\varepsilon}_{kl}-\dot{\varepsilon}^{p}_{kl}\right) \tag{2.42}$$

In the case of mixed hardening, the plastic strain rate is decomposed to two parts related to the isotropic and kinematic hardening, $\dot{\varepsilon}^{i}_{ij}$ and $\dot{\varepsilon}^{k}_{ij}$, respectively, as follows:

$$\dot{\varepsilon}^{p}_{ij}=\dot{\varepsilon}^{isotropic}_{ij}+\dot{\varepsilon}^{kinematic}_{ij} \tag{2.43}$$

where $\dot{\varepsilon}^{isotropic}_{ij}$ and $\dot{\varepsilon}^{kinematic}_{ij}$ can be defined as follows:

$$\dot{\varepsilon}^{isotropic}_{ij}=M\dot{\varepsilon}^{p}_{ij}=M\dot{\lambda}\frac{\partial F}{\partial\sigma_{ij}} \tag{2.44a}$$

$$\dot{\varepsilon}^{kinematic}_{ij}=a(1-M)\frac{\partial F}{\partial\sigma_{ij}} \tag{2.44b}$$

where $0\le M\le 1$ controls the mixture degree between the isotropic and kinematic hardenings. The consistency condition dictates that $\dot{f}=0$ during the plastic deformation. Considering Eq. (2.39) as the loading surface of a material with mixed hardening, the consistency condition can be written as below:

$$\dot{f}\equiv\frac{\partial f}{\partial\sigma_{ij}}\dot{\sigma}_{ij}+\frac{\partial f}{\partial\alpha_{ij}}\dot{\alpha}_{ij}+\frac{\partial f}{\partial k}\dot{k}=0 \tag{2.45}$$

The term related to the kinematic hardening can be generally written as follows:

$$\dot{\alpha}_{ij}=A_{ij}\dot{\lambda} \tag{2.46}$$

where A_{ij} depends on the kinematic hardening rule. For example, in the case of Ziegler's hardening rule,

$$\dot{\alpha}_{ij}=a(1-M)\left(\sigma_{ij}-\alpha_{ij}\right)\dot{\varepsilon}_{p}=a(1-M)\left(\sigma_{ij}-\alpha_{ij}\right)C\dot{\lambda}\sqrt{\frac{\partial F}{\partial\sigma_{kl}}\frac{\partial F}{\partial\sigma_{kl}}} \tag{2.47}$$

where the parameter a and C are described after Eqs. (2.28), (2.35), (2.36). Accordingly, A_{ij} for the Ziegler's hardening rule can be obtained as follows:

$$A_{ij}=a(1-M)\left(\sigma_{ij}-\alpha_{ij}\right)C\sqrt{\frac{\partial F}{\partial\sigma_{kl}}\frac{\partial F}{\partial\sigma_{kl}}} \tag{2.48}$$

The isotropic hardening parameter k is a function of the reduced effective plastic strain $\overline{\varepsilon}_{p}$, which is related to the effective plastic strain ε_{p} as follows:

$$\begin{aligned}\dot{\overline{\varepsilon}}_{p}&=C\sqrt{\dot{\varepsilon}^{isotropic}_{ij}\dot{\varepsilon}^{isotropic}_{ij}}=CM\sqrt{\dot{\varepsilon}^{p}_{ij}\dot{\varepsilon}^{p}_{ij}}=CM\dot{\varepsilon}_{p}=MC\dot{\lambda}\sqrt{\frac{\partial F}{\partial\sigma_{kl}}\frac{\partial F}{\partial\sigma_{kl}}}\overline{\varepsilon}_{p}=\int\dot{\overline{\varepsilon}}_{p}dt\\&=CM\int\sqrt{\dot{\varepsilon}^{p}_{ij}\dot{\varepsilon}^{p}_{ij}}dt=M\varepsilon_{p}\end{aligned} \tag{2.49}$$

Accordingly, the term $\dot{k}$ in Eq. (2.45) can be described as follows:

$$\dot{k}=\frac{dk}{d\overline{\varepsilon}_p}\dot{\overline{\varepsilon}}_p=\frac{dk}{d\overline{\varepsilon}_p}MC\dot{\lambda}\sqrt{\frac{\partial F}{\partial\sigma_{kl}}\frac{\partial F}{\partial\sigma_{kl}}} \tag{2.50}$$

The reduced stress tensor is introduced to simplify the calculations as follows:

$$\overline{\sigma}_{ij}=\sigma_{ij}-\alpha_{ij} \tag{2.51}$$

Accordingly,

$$\frac{\partial f}{\partial\sigma_{ij}}=\frac{\partial f}{\partial\overline{\sigma}_{kl}}\frac{\partial\overline{\sigma}_{kl}}{\partial\sigma_{ij}}=\frac{\partial f}{\partial\overline{\sigma}_{kl}}\delta_{ik}\delta_{jl}=\frac{\partial f}{\partial\overline{\sigma}_{ij}} \tag{2.52a}$$

$$\frac{\partial f}{\partial\alpha_{ij}}=\frac{\partial f}{\partial\overline{\sigma}_{kl}}\frac{\partial\overline{\sigma}_{kl}}{\partial\alpha_{ij}}=-\frac{\partial f}{\partial\overline{\sigma}_{kl}}\delta_{ik}\delta_{jl}=-\frac{\partial f}{\partial\overline{\sigma}_{ij}}=-\frac{\partial f}{\partial\sigma_{ij}} \tag{2.52b}$$

The consistency condition of Eq. (2.45) can be rewritten as follows:

$$\dot{f}=\frac{\partial f}{\partial\sigma_{ij}}C_{ijkl}\dot{\varepsilon}_{kl}-\overline{h}\dot{\lambda}=0 \tag{2.53}$$

where

$$\overline{h}=H_{kl}\frac{\partial F}{\partial\sigma_{kl}}+A_{kl}\frac{\partial f}{\partial\sigma_{kl}}-\frac{\partial f}{\partial k}\frac{dk}{d\overline{\varepsilon}_p}MC\sqrt{\frac{\partial F}{\partial\sigma_{kl}}\frac{\partial F}{\partial\sigma_{kl}}} \tag{2.54}$$

in which H_{kl} can be described as follows:

$$H_{kl}=\frac{\partial f}{\partial\sigma_{ij}}C_{ijkl} \tag{2.55}$$

The plastic multiplier $\dot{\lambda}$ can be obtained by solving the consistency relation of Eq. (2.53) as follows:

$$\dot{\lambda}=\frac{1}{\overline{h}}\frac{\partial f}{\partial\sigma_{ij}}C_{ijkl}\dot{\varepsilon}_{kl}=\frac{1}{\overline{h}}H_{kl}\dot{\varepsilon}_{kl} \tag{2.56}$$

The elastoplastic stiffness tensor C_{ijkl}^{ep} can be obtsined as follows:

$$C_{ijkl}^{ep}=C_{ijkl}-\frac{1}{\overline{h}}H_{kl}H_{ij}^{*} \tag{2.57}$$

where

$$H_{ij}^{*}=\frac{\partial F}{\partial\sigma_{nm}}C_{nmij} \tag{2.58}$$

The general stress-strain relation for a material with mixed hardening can be written as follows:

$$\begin{cases} \text{if}(f<0)\text{ or }\left(f=0;\ \dfrac{\partial f}{\partial \sigma_{ij}}C_{ijkl}\dot{\varepsilon}_{kl}\le 0\right) & \dot{\sigma}_{ij}=C_{ijkl}\dot{\varepsilon}_{kl} \\ \text{if}\left(f=0;\ \dfrac{\partial f}{\partial \sigma_{ij}}C_{ijkl}\dot{\varepsilon}_{kl}>0\right), & \dot{\sigma}_{ij}=C^{ep}_{ijkl}\dot{\varepsilon}_{kl} \end{cases} \tag{2.59}$$

2.2.7 Thermodynamically consistent plasticity models

Historically, the first generation of plasticity models are phenomenological and have not been developed according to the thermodynamics laws. These models may or may not satisfy the thermodynamics laws. The Clausius-Duhem inequality, which is a result of the second law of thermodynamics, can be incorporated to develop a thermodynamically consistent plasticity model. In the case of time-independent and elastoplastic materials, and assuming the additive decomposition, the Helmholtz free energy can be described as follows (Lemaitre and Chaboche, 1990):

$$\Psi=\Psi\left(\varepsilon_{ij}-\varepsilon^{p}_{ij},T,\xi_{ij}\right)=\Psi\left(\varepsilon^{e}_{ij},T,\xi_{ij}\right) \tag{2.60}$$

where the total strain tensor $\boldsymbol{\varepsilon}$ and temperature are the observable variables, and the plastic strain tensor ε^{p}_{ij}, elastic strain tensor ε^{e}_{ij}, and ξ_{ij} are the internal state variables. The internal state variables ξ_{ij} define the material state, and it can be related to the isotropic and kinematic hardening variables, p and $\varsigma^{(l)}_{ij}$ $(l=1,2,\ \dots\ ,L)$, respectively. The Helmholtz free energy can be rewritten as shown below:

$$\Psi=\Psi\left(\varepsilon^{e}_{ij},T,p,\varsigma^{(l)}_{ij}\right) \tag{2.61}$$

where p is the accumulated plastic strain, which can be defined as follows:

$$p=\int_0^t\sqrt{\frac{2}{3}\dot{\varepsilon}^{p}_{ij}\dot{\varepsilon}^{p}_{ij}}\,d\tau \tag{2.62}$$

The Clausius-Duhem inequality can be described as follows:

$$\sigma_{ij}\dot{\varepsilon}_{ij}-\rho\left(\dot{\Psi}+s\dot{T}\right)-q_i\frac{T_{,i}}{T}\ge 0 \tag{2.63}$$

where σ_{ij} is the Cauchy stress tensor, ρ is the material density, s is the specific entropy per unit mass, $\vec{q}$ is the vector of heat flux. The time derivative of the Helmholtz free energy $\dot{\Psi}$ can be described as follows:

$$\dot{\Psi}=\frac{\partial\Psi}{\partial\varepsilon^{e}_{ij}}\dot{\varepsilon}^{e}_{ij}+\frac{\partial\Psi}{\partial T}\dot{T}+\frac{\partial\Psi}{\partial p}\dot{p}+\sum_{l=1}^{L}\frac{\partial\Psi}{\partial\varsigma^{(l)}_{ij}}\dot{\varsigma}^{(l)}_{ij} \tag{2.64}$$

Accordingly, Eq. (2.64) can be substituted into Eq. (2.63) as follows:

$$\left(\sigma_{ij}-\rho\frac{\partial\Psi}{\partial\varepsilon_{ij}^{e}}\right)\dot{\varepsilon}_{ij}^{e}+\sigma_{ij}\dot{\varepsilon}_{ij}^{p}-\rho\left(\frac{\partial\Psi}{\partial T}+s\right)\dot{T}-\rho\frac{\partial\Psi}{\partial p}\dot{p}-\rho\sum_{l=1}^{L}\frac{\partial\Psi}{\partial\varsigma_{ij}^{(l)}}\dot{\varsigma}_{ij}^{(l)}-q_i\frac{T_{,i}}{T}\geq 0 \tag{2.65}$$

Eq. (2.65) defines the thermodynamic state laws as follows:

$$\sigma_{ij}=\rho\frac{\partial\Psi}{\partial\varepsilon_{ij}^{e}} \tag{2.66}$$

$$s=-\frac{\partial\Psi}{\partial T} \tag{2.67}$$

$$\alpha_{ij}^{(l)}=\rho\frac{\partial\Psi}{\partial\varsigma_{ij}^{(l)}} \tag{2.68}$$

$$R=\rho\frac{\partial\Psi}{\partial p} \tag{2.69}$$

Eqs. (2.66)–(2.69) define the relation between the observable and internal variables and their corresponding thermodynamic conjugate forces, which are summarized in Table 2.1. Accordingly, Eq. (2.65) can be rewritten as follows:

$$\sigma_{ij}\dot{\varepsilon}_{ij}^{p}-R\dot{p}-\sum_{l=1}^{L}\alpha_{ij}^{(l)}\dot{\varsigma}_{ij}^{(l)}-q_i\frac{T_{,i}}{T}\geq 0 \tag{2.70}$$

TABLE 2.1 Thermodynamic state variables and their corresponding conjugate forces

State variables		
Observable	**Internal**	**Associated conjugates**
ε_{ij}		σ_{ij}
T		s
	ε_{ij}^{e}	σ_{ij}
	ε_{ij}^{p}	$-\sigma_{ij}$
	$T_{,i}$	q_i
	p	R
	$\varsigma_{ij}^{(l)}$	$\alpha_{ij}^{(l)}$

Eq. (2.70) can be further simplified by assuming decoupling between plasticity and thermal dissipations as follows:

$$-q_i \frac{T_{,i}}{T} \geq 0 \tag{2.71}$$

$$\Pi = \sigma_{ij}\dot{\varepsilon}_{ij}^p - R\dot{p} - \sum_{l=1}^{L} \alpha_{ij}^{(l)} \dot{\varsigma}_{ij}^{(l)} \geq 0 \tag{2.72}$$

The plastic dissipation energy Π can be envisaged as the difference between the total energy due to plastic deformation, i.e., $\sigma_{ij}\dot{\varepsilon}_{ij}^p$, and the energy stored inside the volume element as isotropic and kinematic hardening. A complimentary formalism is required to describe the evolution of the internal state variables. Accordingly, a plastic potential, which is convex, nonnegative, and zero at the origin, should be defined as follows:

$$F = F\left(\sigma_{ij}, \alpha_{ij}^{(l)}, R\right) \tag{2.73}$$

The evolution of the internal state variables can be described as follows (Lemaitre and Chaboche, 1990):

$$\dot{\varepsilon}_{ij}^p = \dot{\lambda} \frac{\partial F}{\partial \sigma_{ij}} \tag{2.74}$$

$$\dot{\varsigma}_{ij}^{(l)} = -\dot{\lambda} \frac{\partial F}{\partial \alpha_{ij}^{(l)}} \tag{2.75}$$

$$\dot{p} = -\dot{\lambda} \frac{\partial F}{\partial R} \tag{2.76}$$

where $\dot{\lambda}$ is the plastic multiplier.

In the case of the associated flow rule, which is commonly termed the normality hypothesis, the loading surface becomes the plastic potential, i.e., $F = f$. During the plastic deformation, $f = 0$. Accordingly, one can maximize the plastic dissipation energy Π using the Lagrange multiplier method by building the objective function Ω as follows:

$$\Omega = \Pi - \dot{\lambda} f \tag{2.77}$$

In order to maximize Π, one should maximize the objective function Ω using the following equations:

$$\frac{\partial \Omega}{\partial \sigma_{ij}} = 0 \tag{2.78}$$

$$\frac{\partial \Omega}{\partial \alpha_{ij}^{(l)}} = 0 \tag{2.79}$$

$$\frac{\partial \Omega}{\partial R} = 0 \tag{2.80}$$

Considering Eqs. (2.77), (2.80), the evolution of the internal state variables can be obtained as follows:

$$\dot{\varepsilon}_{ij}^{p} = \dot{\lambda}\frac{\partial f}{\partial \sigma_{ij}} \tag{2.81}$$

$$\dot{\varsigma}_{ij}^{(l)} = -\dot{\lambda}\frac{\partial f}{\partial \alpha_{ij}^{(l)}} \tag{2.82}$$

$$\dot{p} = -\dot{\lambda}\frac{\partial f}{\partial R} \tag{2.83}$$

In other words, the associated flow rule maximizes the plastic dissipation energy Π.

2.2.8 Rate-dependent plasticity: Models with the von Mises yield surface

Until now, the rate-independent plasticity models have been considered in which the material response is independent of strain rate. In these models, the concept of pseudo-time is used merely to define the loading history and solving the nonlinear equation during that loading process. The rate-independent plasticity models are applicable if the material response does not depend on the applied loading rate. However, in general, the material response varies with the applied loading rates and is a function of actual time scale. Accordingly, the rate-dependent plasticity models, which are also termed as viscoplasticity models, should be incorporated to accurately capture the material properties. Different time-dependent phenomena have been reported including rate-dependence stress-strain curve, relaxation, and creep in which the material response is a function of actual time scale (Fig. 2.7).

The rate-dependent plasticity models follow the methodology similar to that of the rate-independent models discussed in Section 2.2. Accordingly, the total

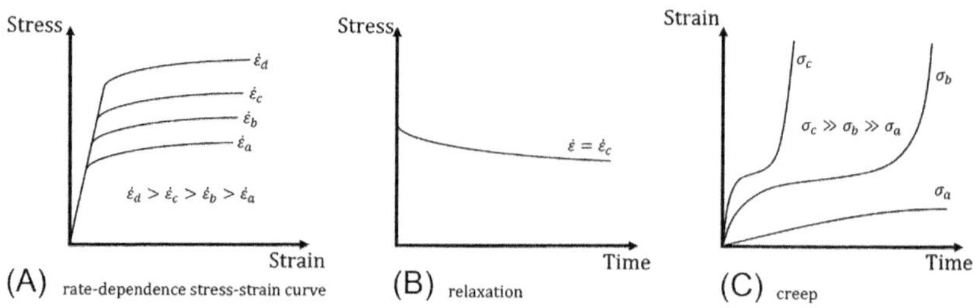

FIG. 2.7 Time-dependent phenomena in materials: (A) rate-dependence stress-strain curve, (B) relaxation, and (C) creep.

strain tensor ε_{ij} is decomposed to two tensors of elastic strain ε_{ij}^{e} and plastic strain ε_{ij}^{p} as follows:

$$\varepsilon_{ij} = \varepsilon_{ij}^{e} + \varepsilon_{ij}^{p} \tag{2.84}$$

Hooke's law defines the constitutive model of an elastoplastic material as follows:

$$\sigma_{ij} = C_{ijkl}\varepsilon_{kl}^{e} = C_{ijkl}\left(\varepsilon_{kl} - \varepsilon_{kl}^{p}\right) \tag{2.85}$$

The concept of the yield surface f defined in Section 2.2 can be incorporated to define the elastic domain for rate-dependent model, in a way that a material point with stress tensor located inside the yield surface is elastic. However, the yield surface may be greater than zero to describe the dynamic yield surface attributed to different strain rates. The von Mises yield surface can be defined as follows:

$$f\left(\sigma_{ij}, \varepsilon_{ij}^{p}\right) = \sqrt{\frac{3}{2}\left(s_{ij} - \alpha_{ij}\right)\left(s_{ij} - \alpha_{ij}\right)} - \sigma_{y}\left(\varepsilon_{p}\right) \tag{2.86}$$

where $\sigma_y(\varepsilon_p)$ and α_{ij} represent the expansion and the rigid translation of the loading surface, respectively. The plastic strain rate tensor can be defined by incorporating the associated flow rule as follows:

$$\dot{\varepsilon}_{ij}^{p} = \dot{\lambda}\frac{\partial f}{\partial \sigma_{ij}} = \dot{\lambda}\frac{\sqrt{\frac{3}{2}}\left(s_{ij} - \alpha_{ij}\right)}{\sqrt{\left(s_{ij} - \alpha_{ij}\right)\left(s_{ij} - \alpha_{ij}\right)}} \tag{2.87}$$

Unlike the rate-independent models, the plastic multiplier $\dot{\lambda}$ should be explicitly defined in the case of rate-dependent plasticity and the consistency condition should not be solved. Different models for plastic multiplier $\dot{\lambda}$ have been introduced. Here, three models of Bingham (1922), Perzyna (1966, 1971), and Peric (1993) will be discussed.

2.2.8.1 *Bingham model*

The simplest definition of plastic multiplier $\dot{\lambda}$ is described using the Bingham model (Bingham, 1922) as follows:

$$\dot{\lambda} = \begin{cases} \dfrac{\sqrt{\frac{3}{2}\left(s_{ij} - \alpha_{ij}\right)\left(s_{ij} - \alpha_{ij}\right)} - \sigma_{y}\left(\varepsilon_{p}\right)}{\xi} & \text{if } f \geq 0 \\ 0 & \text{if } f < 0 \end{cases} \tag{2.88}$$

where ξ is the viscosity parameter. In the case of a uniaxial loading with isotropic hardening, the plastic strain rate defined by the Bingham model can be described as follows:

$$\dot{\varepsilon}^{p} = \frac{1}{\xi}\left(|\sigma| - \sigma_{y}\right)\operatorname{sign}(\sigma) \tag{2.89}$$

Eq. (2.89) states that the plastic strain rate is a linear function of the uniaxial stress. In many real experiments, this relation is not linear. However, Eq. (2.89) can be applicable if a narrow range of stresses is considered.

2.2.8.2 Perzyna model

Perzyna (1966, 1971) proposed an explicit equation for the plastic multiplier $\dot{\lambda}$ as follows:

$$\dot{\lambda} = \begin{cases} \dfrac{1}{\mu}\left[\dfrac{\sqrt{\frac{3}{2}\left(s_{ij}-\alpha_{ij}\right)\left(s_{ij}-\alpha_{ij}\right)}}{\sigma_y\left(\varepsilon_p\right)} - 1\right]^{1/\epsilon} & \text{if} f \geq 0 \\ 0 & \text{if} f < 0 \end{cases} \tag{2.90}$$

where μ is the viscosity parameter for the Perzyna model and ϵ is the rate-sensitivity parameter. Eq. (2.90) is widely used in the commercial codes to model the rate-dependent plasticity. The Bingham model can be reproduced from Eq. (2.90) by substituting $\mu = \xi/\sigma_y$ and $\epsilon = 1$. The Perzyna model can reproduce the standard rate-independent plasticity model in the case of $\mu \to 0$ or vanishing strain rate. However, the Perzyna model cannot mimic the rate-independent model as the rate-sensitivity parameter ϵ approaches to zero. In this case, the model predicts the yield stress, which is twice that of the rate-independent model.

2.2.8.3 Peric model

Peric (1993) proposed a simple expression for the plastic multiplier $\dot{\lambda}$, which can recover the rate-independent plasticity in the cases of vanishing viscosity parameter and rate-sensitivity, as follows:

$$\dot{\lambda} = \begin{cases} \dfrac{1}{\mu}\left\{\left[\dfrac{\sqrt{\frac{3}{2}\left(s_{ij}-\alpha_{ij}\right)\left(s_{ij}-\alpha_{ij}\right)}}{\sigma_y\left(\varepsilon_p\right)}\right]^{1/\epsilon} - 1\right\} & \text{if} f \geq 0 \\ 0 & \text{if} f < 0 \end{cases} \tag{2.91}$$

Eq. (2.91) is a modification of the Perzyna model to recover the rate-independent models in the case of $\epsilon \to 0$. Again, the Bingham model can be reproduced from Eq. (2.90) by substituting $\mu = \xi/\sigma_y$ and $\epsilon = 1$.

2.2.9 Rate-dependent plasticity models without a yield surface

Unlike the rate-independent plasticity, yield surface is not an essential element of a rate-dependent plasticity model. In other words, in the case of rate-dependent models, the plasticity occurs as soon as the loading occurs, which can mimic many cases such as the material behavior at high temperatures. In this framework, the plastic strain rate can be obtained by defining an explicit expression for the plastic multiplier $\dot{\lambda}$.

The Norton's law is one of the famous secondary creep expressions in which the uniaxial plastic strain rate can be defined as follows:

$$\dot{\varepsilon}^p = \left(\frac{|\sigma|}{\vartheta}\right)^{\gamma} \text{sign}(\sigma) \tag{2.92}$$

where ϑ and γ are the temperature-dependent material parameters. Since the model does not define any yield surface, the plastic deformations initiates whenever the loading occurs. The generalization of Norton's law by assuming a von Mises plastic potential can be described as follows:

$$\dot{\lambda} = \left[\frac{\sqrt{\frac{3}{2}S_{ij}S_{ij}}}{\vartheta}\right]^{\gamma} \tag{2.93}$$

$$\dot{\varepsilon}_{ij}^p = \left[\frac{\sqrt{\frac{3}{2}S_{kl}S_{kl}}}{\vartheta}\right]^{\gamma} \frac{\sqrt{\frac{3}{2}}S_{ij}}{\sqrt{S_{mn}S_{mn}}} \tag{2.94}$$

One can recover the Bingham model with zero yield stress by substituting $\gamma = 1$ and $\xi = \vartheta$.

Lemaitre and Chaboche (1990) modifies the Norton's law as follows:

$$\dot{\lambda} = \left(\dot{\lambda}\right)_{Norton's\ Law} \exp\left\{\alpha\left[\sqrt{\frac{3}{2}S_{ij}S_{ij}}\right]^{\gamma+1}\right\} \tag{2.95}$$

$$\dot{\varepsilon}_{ij}^p = \left(\dot{\varepsilon}_{ij}^p\right)_{Norton's\ Law} \exp\left\{\alpha\left[\sqrt{\frac{3}{2}S_{ij}S_{ij}}\right]^{\gamma+1}\right\} \tag{2.96}$$

where α is another temperature-dependent parameter introduced by Lemaitre and Chaboche (1990) to improve the accuracy of the creep model.

A general rate-dependent plasticity model without yield surface can be described as follows:

$$\dot{\lambda}\left(\sigma_{ij}, t, T\right) = \dot{\lambda}_{\sigma_{ij}}\left(\sigma_{ij}\right)\dot{\lambda}_t(t)\dot{\lambda}_T(T) \tag{2.97}$$

Eq. 2.97 is a multiplicative decomposition of a plastic multiplier $\dot{\lambda}$ to three functions of stress $\dot{\lambda}_{\sigma_{ij}}\left(\sigma_{ij}\right)$, time $\dot{\lambda}_t(t)$, and $\dot{\lambda}_T(T)$.Various functions can be used for each of these functions (see Skrzypek, 1993). For example, the expression proposed by Lemaitre and Chaboche (1990), i.e., Eq. (2.95), can be incorporated as $\dot{\lambda}_{\sigma_{ij}}(\sigma_{ij})$.The Arrhenius type equation can be incorporated to define the temperature function $\dot{\lambda}_T(T)$ as follows:

$$\dot{\lambda}_T(T) = C\exp\left(\frac{-Q}{RT}\right) \tag{2.98}$$

where C is a material parameter, Q is the activation energy, and R is the gas constant. Boyle and Spence (1983) proposed an empirical expression for the plastic multiplier using the framework described in Eq. (2.97) as follows:

$$\dot{\lambda}(\sigma_{ij}, t, T) = C\exp\left(\frac{-Q}{RT}\right)t^A\left(\sqrt{\frac{3}{2}s_{ij}s_{ij}}\right)^B \tag{2.99}$$

where A and B are the material parameters.

2.3 Small strain plasticity: Nonlocal models

As discussed in Section 2.1, the local plasticity models cannot capture the size effects occurs in materials. Furthermore, problems containing the shear band become ill-posed using local plasticity models. Accordingly, nonlocal models should be incorporated to capture size effects and shear bands. The nonlocal plasticity models can be categorized based on the operator used to introduce nonlocality. In this section, two types of gradient and integral-type models will be elaborated. The gradient plasticity models incorporate differential operator, while integral-type plasticity models define nonlocality using the weighted spatial averages.

2.3.1 Gradient plasticity models

Gradient plasticity models are a family of nonlocal models which incorporate differential operators to capture the size effects and prevent localization. Depending on the methodology, the gradient plasticity models can be categorized into two separate groups. First, the gradient operator is applied on the strain, which can be first, second, or even higher order of strain gradients (Fleck and Hutchinson, 1993; Gurtin and Anand, 2005a; Voyiadjis and Deliktas, 2009a,b; Voyiadjis and Faghihi, 2012; Song and Voyiadjis, 2018). The second category of gradient plasticity models apply the differential operators on the internal variables of plasticity models (Aifantis, 1984a,b,c, 1987, 1989; Voyiadjis et al., 2001, 2004). In the case of the first family of gradient models, higher-order stresses should be introduced and incorporated in the equilibrium equations which are conjugate to the gradients of strains. In the case of second family of gradient models, no higher stresses are required to be included and new thermodynamic forces are conjugate to the gradients of internal variables which will not enter into the equilibrium equations. Accordingly, the first family are more demanding in a way that introducing gradients of strain and their conjugate, i.e., higher order stresses, modify the expression of internal and external work. However, gradients of internal variables and their conjugates solely modify the free energy expression. One should note that a model can use both frameworks by incorporating both the gradients of strain and internal variables (see, e.g., Zervos et al., 2001).

Another way to categorize the gradient models is the way the gradient terms are incorporated. In the case of explicit incorporation, the gradient terms are explicitly included in the free energy or the governing equations. Accordingly, that gradient term obtains information from the neighboring material points. However, the size of this coupling can be small, and accordingly, the coupling between a material point and the rest of the body is weak which makes the model weakly nonlocal. The second family of the gradient models implicitly include the gradient terms and couple the nonlocal term at a material point with the rest of body using additional governing equation.

Peerlings (1999) investigated the difference between the explicit and implicit gradient models in the case of nonlocal equivalent strain $\overline{\varepsilon}_{ij}$, which can be defined as follows:

$$\overline{\varepsilon}_{ij}(x_k) = \frac{1}{\Psi(x_k)} \int_{\Omega} \psi(y_k; x_k) \widetilde{\varepsilon}_{ij}(y_k) \mathrm{d}\Omega \tag{2.100}$$

where $\psi(y_i; x_i)$ is the weighting function, and $\Psi(x_i)$ can be described as follows:

$$\Psi(x_k) = \int_{\Omega} \psi(y_k; x_k) \mathrm{d}\Omega \tag{2.101}$$

The Gaussian function is usually selected as the weighting function, which can be described as follows:

$$\psi(r) = \frac{1}{(2\pi)^{\frac{3}{2}} l^3} \exp\left(\frac{-r^2}{2l^2}\right) \tag{2.102}$$

where l is an internal length scale which denotes the averaging volume. One should note that the weighting function is normalized in a way that:

$$\int_{\mathbb{R}^3} \psi(r) \mathrm{d}\Omega = 1 \tag{2.103}$$

The local strain tensor $\widetilde{\varepsilon}_{ij}(y_i)$ can be expanded into the following Taylor series:

$$\widetilde{\varepsilon}_{ij}(y_m) = \widetilde{\varepsilon}_{ij}(y_m) + \frac{\partial \widetilde{\varepsilon}_{ij}}{\partial x_k}(y_k - x_k) + \frac{1}{2!} \frac{\partial^2 \widetilde{\varepsilon}_{ij}}{\partial x_k \partial x_l}(y_k - x_k)(y_l - x_l) + \ldots \tag{2.104}$$

The Taylor expansion of nonlocal equivalent strain $\overline{\varepsilon}_{ij}$ can be obtained by substituting Eq. (2.104) into Eq. (2.100) as follows:

$$\overline{\varepsilon}_{ij}(x_m) = \widetilde{\varepsilon}_{ij}(x_m) + A_{kl} \frac{\partial^2 \widetilde{\varepsilon}_{ij}}{\partial x_k \partial x_l} + A_{klmn} \frac{\partial^4 \widetilde{\varepsilon}_{ij}}{\partial x_k \partial x_l \partial x_m \partial x_n} + \ldots \tag{2.105}$$

where the Taylor expansion coefficients can be obtained as follows:

$$A_{kl} = \frac{1}{2!\Psi(x_i)} \int_{\Omega} \psi(r)(y_k - x_k)(y_l - x_l) \mathrm{d}\Omega \tag{2.106}$$

$$A_{klmn} = \frac{1}{4!\Psi(x_i)} \int_{\Omega} \psi(r)(y_k - x_k)(y_l - x_l)(y_m - x_m)(y_n - x_n) \mathrm{d}\Omega \tag{2.107}$$

One should note that due to the isotropy of the weighting function, the odd derivatives of local strain tensor are vanished by substituting Eq. (2.104) into Eq. (2.100). Furthermore, the values of Taylor expansion coefficients are independent of the indices, and they can be dropped, i.e., A can be used instead of A_{kl}.

The nonlocal equivalent strain $\overline{\varepsilon}_{ij}$ can be defined using two methods of explicit and implicit. In the case of explicit framework, the fourth and higher order gradients of strain are neglected and Eq. (2.105) can be simplified as follows:

$$\frac{\overline{\varepsilon}_{ij}(x_m) = \widetilde{\varepsilon}_{ij}(x_m) + A_{kl}\partial^2 \widetilde{\varepsilon}_{ij}}{\partial x_k \partial x_l} \tag{2.108}$$

where $\frac{\partial^2}{\partial x_k \partial x_l}$ is the Laplacian operator. The Taylor expansion coefficient depends on the definition of weighting function. In the case of Gaussian function, $A_{kl} = (l^2/2)\delta_{kl}$, where δ_{kl} is the Kronecker delta.

Implicit framework of gradient model can be described by describing the Taylor expansion of nonlocal equivalent strain $\overline{\varepsilon}_{ij}$ as follows:

$$\overline{\varepsilon}_{ij}(x_m) - A_{kl}\frac{\partial^2 \overline{\varepsilon}_{ij}}{\partial x_k \partial x_l} = \widetilde{\varepsilon}_{ij}(x_m) + A_{klmn}\frac{\partial^4 \widetilde{\varepsilon}_{ij}}{\partial x_k \partial x_l \partial x_m \partial x_n} + \cdots \tag{2.109}$$

The final form can be obtained by neglecting the fourth and higher orders gradient as follows:

$$\overline{\varepsilon}_{ij}(x_m) - A_{kl}\frac{\partial^2 \overline{\varepsilon}_{ij}}{\partial x_k \partial x_l} = \widetilde{\varepsilon}_{ij}(x_m) \tag{2.110}$$

In the case of explicit methodology, i.e., Eq. (2.108), the nonlocal equivalent strain $\overline{\varepsilon}_{ij}$ can be explicitely obtained using the local counterpart $\widetilde{\varepsilon}_{ij}$. However, in the case of implicit framework, i.e., Eq. (2.110), the nonlocal equivalent strain $\overline{\varepsilon}_{ij}$ is the solution of a boundary value problem. Peerlings (1999) proved that the gradient model using Eq. (2.110), i.e., implicit framework, can replicate the integral form of nonlocality definition, i.e., Eq. (2.100), incorporating the weighting function of Green's function. He stated that the difference between the two frameworks is in the case of explicit scheme, the higher order gradients are totally neglected, while in the case of implicit method, they are included in the derivative of $\overline{\varepsilon}_{ij}$, i.e., $\frac{\partial^2 \overline{\varepsilon}_{ij}}{\partial x_k \partial x_l}$.

2.3.1.1 Gradient elasticity models

As explained in Section 2.3.1, the gradient operators can be applied on either strain or internal variables of plasticity models. In the latter case, the elastic part of the response is governed by the conventional elasticity and no gradient is included. In the case of former case, i.e., applying gradients on the strain, however, the gradient should be incorporated even in the elastic part. Accordingly, the strain gradient elasticity models should be defined to capture the elastic response.

Toupin (1962) and Mindlin (1964) pioneered the concept of strain gradient elasticity. One simple linear gradient elasticity model can be defined by assuming the free energy as follows:

$$\rho\Psi = \frac{1}{2}C_{ijkl}\varepsilon_{ij}\varepsilon_{kl} + \frac{l^2}{2}C_{ijkl}\eta_{ijm}\eta_{ikl} \tag{2.111}$$

where $\eta_{ijm} = \varepsilon_{ij,\,m}$ is the strain gradient and l is the material length scale. It is worth mentioning that the free energy of the classical elastic material is recovered in the case of $l = 0$. The Clausius-Duhem inequality for elastic materials can be reduced to the following equation:

$$\sigma_{ij}\dot{\varepsilon}_{ij} - \rho\dot{\Psi} = 0 \tag{2.112}$$

Accordingly, the constitutive equations of linear gradient elasticity model with the free energy of Eq. (2.111) can be described as follows:

$$\sigma_{ij} = \rho\frac{\partial\Psi}{\partial\varepsilon_{ij}} = C_{ijkl}\varepsilon_{kl} \tag{2.113}$$

$$\tau_{ijk} = \rho\frac{\partial\Psi}{\partial\eta_{ijk}} = l^2\left(C_{ijlm}\varepsilon_{lm}\right)_{,k} = l^2\sigma_{ij,k} \tag{2.114}$$

where τ_{kij} is the double stress. The total stress T_{ij} and surface stress Σ_{ij} can be described as follows:

$$T_{ij} = \sigma_{ij} - \tau_{ijk,k} \tag{2.115}$$

$$\Sigma_{ij} = \tau_{ijk}n_k \tag{2.116}$$

where n_k is the surface unit outward normal vector. The equilibrium equation can be written as follows:

$$T_{ij,j} + b_i = 0 \tag{2.117}$$

where b_i is the body force vector. The boundary conditions can be descried as follows:

$$T_{ij}n_j - \Sigma_{ij,j} + \left(\Sigma_{ij}n_l\right)_{,l}n_j = \bar{t}_i \text{ or } u_i = \bar{u}_i \tag{2.118}$$

$$\Sigma_{ij}n_j = \bar{q}_i \text{ or } u_{i,l}n_l = \overline{\frac{\partial u_i}{\partial n}} \tag{2.119}$$

where u_i is the displacement vector, t_i and q_i are the Cauchy and double stress traction vectors, respectively, and the overbar denotes the predefined values.

2.3.1.2 Gradient plasticity models: Fleck and Hutchinson

Fleck and Hutchinson (1993) presented a phenomenological strain gradient plasticity model using the concepts of couple stress theory (Toupin, 1962; Mindlin, 1964) with statistically and geometrically stored dislocations (Ashby, 1970). They introduced a single material length scale to capture the size effects. They enhanced the von Mises-type plasticity with the strain gradient. Consider the following von Mises plasticity model with the yield surface:

$$f\left(\sigma_{ij}, \varepsilon_{ij}^{p}\right) = \sqrt{\frac{3}{2} s_{ij} s_{ij}} - \sigma_y\left(\varepsilon_p\right) \tag{2.120}$$

The evolutions of plastic strain and stress are governed using the flow rule and consistency condition as defined in Sections 2.2.4 and 2.2.6.

Fleck and Hutchinson (1993) added the couple stress concept and defined the elastic strain state $(\varepsilon_{ij}^e, \eta_{ijk}^e)$ which is related to the stress state $(\sigma_{ij}, \tau_{ijk})$ as follows:

$$\sigma_{ij} = \rho \frac{\partial \Psi}{\partial \varepsilon_{ij}^{e}} = C_{ijkl} \varepsilon_{ij}^{e} \tag{2.121}$$

$$\tau_{ijk} = \rho \frac{\partial \Psi}{\partial \eta_{ijk}^{e}} = l^2 \left(C_{ijlm} \varepsilon_{lm}\right)_{,k} = l^2 \sigma_{ij,k} \tag{2.122}$$

They incorporated the assumption of additive decomposition of the strain gradient tensor as follows:

$$\eta_{ijk} = \eta_{ijk}^{e} + \eta_{ijk}^{p} \tag{2.123}$$

The rate form of Eq. (2.123) can be written accordingly as:

$$\dot{\eta}_{ijk} = \dot{\eta}_{ijk}^{e} + \dot{\eta}_{ijk}^{p} \tag{2.124}$$

The strain gradient tensor can be decomposed to the symmetric and asymmetric tensors as follows:

$$\eta_{ijk} = \eta_{ijk}^{S} + \eta_{ijk}^{A} \tag{2.125}$$

$$\eta_{ijk}^{S} = \frac{1}{3}\left(\eta_{ijk} + \eta_{jki} + \eta_{kij}\right) \tag{2.126}$$

where the superscripts S and A denote the symmetric and antisymmetric tensors, respectively. The symmetric and asymmetric double stress tensors, i.e., τ_{ijk}^{S} and τ_{ijk}^{A}, respectively, are similar to those of the strain gradient tensors as described in Eqs. (2.125), (2.126).

Fleck and Hutchinson (1997) substituted the conventional symmetric stress deviator tensor of s_{ij} in the yield surface, i.e., Eq. (2.120), with the vector Λ_i of 23 components including 5 local, i.e., independent components of s_{ij} and 18 components of $\mathbb{s}_{ijk}$, where $\mathbb{s}_{ijk}$ is the double stress deviator tensor. Similarly, they substituted the plastic strain rate tensor $\dot{\varepsilon}^p_{ij}$ by a 23 components vector of $\dot{\mathcal{E}}^p_i$ comprising the independent components of $\dot{\varepsilon}^p_{ij}$ and 18 components of $\dot{\eta}^p_{ijk}$. The nonlocal von Mises effective stress Γ_e can be defined as follows:

$$\Gamma_e = \sqrt{\frac{3}{2}s_{ij}s_{ij} + \sum_{I=1}^{3}\left[l_I^{-2}\mathbb{s}^{(I)}_{ijk}\mathbb{s}^{(I)}_{ijk}\right](\text{no sum on } I)} \tag{2.127}$$

where l_1, l_2, and l_3 are material constants related to the material length scale l, and the orthogonal decomposition of $\mathbb{s}_{ijk}$ into $\mathbb{s}^{(1)}_{ijk}$, $\mathbb{s}^{(2)}_{ijk}$, and $\mathbb{s}^{(3)}_{ijk}$ can be described as follows:

$$\mathbb{s}^{(1)}_{ijk} = \mathbb{s}^S_{ijk} - \frac{1}{5}\left(\delta_{ij}\mathbb{s}^S_{kpp} + \delta_{jk}\mathbb{s}^S_{ipp} + \delta_{ki}\mathbb{s}^S_{jpp}\right) \tag{2.128}$$

$$\mathbb{s}^{(2)}_{ijk} = \frac{1}{6}\left(e_{ikp}e_{jlm}\mathbb{s}_{lpm} + e_{jkp}e_{ilm}\mathbb{s}_{lpm} + 2\mathbb{s}_{ijk} - \mathbb{s}_{jki} - \mathbb{s}_{kij}\right) \tag{2.129}$$

$$\mathbb{s}^{(3)}_{ijk} = \frac{1}{6}\left(-e_{ikp}e_{jlm}\mathbb{s}_{lpm} - e_{jkp}e_{ilm}\mathbb{s}_{lpm} + 2\mathbb{s}_{ijk} - \mathbb{s}_{jki} - \mathbb{s}_{kij}\right) + \frac{1}{5}\left(\delta_{ij}\mathbb{s}^S_{kpp} + \delta_{jk}\mathbb{s}^S_{ipp} + \delta_{ki}\mathbb{s}^S_{jpp}\right) \tag{2.130}$$

The nonlocal yield surface can be defined as follows:

$$\bar{f}\left(\sigma_{ij}, \varepsilon^p_{ij}, \tau_{kji}, \eta^p_{ijk}\right) = \Gamma_e - \sigma_y\left(\mathcal{E}_p\right) \tag{2.131}$$

where $\mathcal{E}_p$ is the effective plastic strain including both local and nonlocal terms which can be defined as follows:

$$\mathcal{E}_p = \sqrt{\frac{3}{2}\dot{\varepsilon}^p_{ij}\dot{\varepsilon}^p_{ij} + \sum_{I=1}^{3}\left[l_I^2\dot{\eta}^{p(I)}_{ijk}\dot{\eta}^{p(I)}_{ijk}\right](\text{no sum on } I)} \tag{2.132}$$

where $\dot{\eta}^{p(1)}_{ijk}$, $\dot{\eta}^{p(2)}_{ijk}$, and $\dot{\eta}^{p(3)}_{ijk}$ are the orthogonal decomposition of $\dot{\eta}^p_{ijk}$ which can be defined similar to Eqs. (2.128)–(2.130).

Considering the associated flow rule, i.e., the plastic strain rate tensor is normal to the yield surface, and a linear relation between the plastic strain rate and the stress rate can be defined as follows:

$$\dot{\mathcal{E}}^p_i = \frac{1}{h(\Gamma_e)}\frac{\partial\bar{f}}{\partial\Lambda_i}\dot{\Gamma}_e \tag{2.133}$$

The function $h(\Gamma_e)$ describes the incorporated hardening law.

2.3.1.3 Gradient plasticity models: Aifantis and his co-workers

Aifantis (1984a,b,c, 1987, 1989) developed a family of strain gradient plasticity models in which the gradients of internal variables are incorporated instead of the strain gradients. Accordingly, the higher stresses are not introduced in these models, and the equilibrium equations are not changed. Thermodynamically, it means only the free energy potential is enhanced with the gradients of internal variables. Accordingly, it is assumed that the flow rule is the only part which should be enhanced and the elastic part and flow rule are similar to the local plasticity models. The flow rule can be modified by incorporating the second and fourth order gradients of the equivalent plastic strain. Aifantis originally incorporated only the second order gradient of equivalent plastic strain (Aifantis, 1984a,b,c, 1987). Zbib and Aifantis (1989) and Mühlhaus and Aifantis (1991) added the fourth order gradient later to address the salient features of the localized deformation pattern. To do this, the local von Mises yield surface, i.e., Eq. (2.120), is modified as follows:

$$f\left(\sigma_{ij}, \varepsilon_{ij}^{p}\right) = \sqrt{\frac{3}{2} s_{ij} s_{ij}} - \overline{\sigma}_y\left(\varepsilon_p\right) \tag{2.134}$$

where $\sigma_y\left(\overline{\varepsilon}_p\right)$ is the function of the equivalent plastic strain and its gradients as follows:

$$\overline{\sigma}_y\left(\varepsilon_p\right) = c_0 + c_1 \nabla^2 \varepsilon_p + c_2 \nabla^4 \varepsilon_p \tag{2.135}$$

where $c_i = c_i(\varepsilon_p)$. The simplest nonlocal model can be introduced by assuming $c_2 = 0$, $c_1 =$ constant, and $c_0 = \sigma_y(\varepsilon_p)$ as follows:

$$\overline{\sigma}_y\left(\varepsilon_p\right) = \sigma_y\left(\varepsilon_p\right) + c_1 \nabla^2 \varepsilon_p \tag{2.136}$$

2.3.1.4 Gradient ductile damage: Geers and coworkers

Geers and his coworkers (Geers et al., 2001; Engelen et al., 2003) followed the idea of nonlocal damage model developed by Peerlings (1999) to implicitly include the gradient terms. Accordingly, they developed an implicit gradient plasticity model using the concept of ductile damage. The yield surface is introduced as follows:

$$f\left(\sigma_{ij}, \varepsilon_{ij}^{p}\right) = \sqrt{\frac{3}{2} s_{ij} s_{ij}} - \left[1 - \omega_{\mathrm{p}}\left(\overline{k}\right)\right] \sigma_y\left(\varepsilon_p\right) \tag{2.137}$$

where $\omega_{\mathrm{p}}\left(\overline{k}\right)$ governs the ductile damage in the material as a function of nonlocal ductile damage parameter and $\sigma_y(\varepsilon_p)$ is a function of the equivalent plastic strain as follows:

$$\sigma_y\left(\varepsilon_p\right) = \sigma_{y0} + h\varepsilon_p \tag{2.138}$$

Geers et al. (2001) selected the nonlocal equivalent plastic strain $\overline{\varepsilon}_p$ as the governing variable of ductile damage which can be described using an additional governing equation as follows:

$$\overline{\varepsilon}_p - l^2 \nabla^2 \overline{\varepsilon}_p = \varepsilon_p \tag{2.139}$$

where l is the material length scale parameter. Eq. (2.139) is a Helmholtz type partial differential equation which has to be solved coupled to the equilibrium equation to obtain the nonlocal equivalent plastic strain $\overline{\varepsilon}_p$. The corresponding Neumann boundary condition of Eq. (2.139) can be defined as follows:

$$\overline{\varepsilon}_{p,i} n_i = 0 \; \partial\Omega \tag{2.140}$$

where n_i is the outward normal vector of the body surface $\partial\Omega$.

Geers et al. (2001) described the damage parameter $0 \le \omega_{\mathrm{p}}(\overline{\varepsilon}_p) \le 1$ as follows:

$$\omega_{\mathrm{p}}(\overline{\varepsilon}_p) = \begin{cases} 0 & \text{for}\, \overline{\varepsilon}_p \le (\varepsilon_p)_i \\ \dfrac{\overline{\varepsilon}_p - (\varepsilon_p)_i}{(\varepsilon_p)_c - (\varepsilon_p)_i} & \text{for } (\varepsilon_p)_i \le \overline{\varepsilon}_p \le (\varepsilon_p)_c \\ 1 & \text{for}\, \overline{\varepsilon}_p \ge (\varepsilon_p)_c \end{cases} \tag{2.141}$$

where $(\varepsilon_p)_i$ is the equivalent plastic strain at which the damage initiates and $(\varepsilon_p)_c$ is the equivalent plastic strain of the material failure. Eq. (2.141) states that the damage initiation is independent of incipient plasticity.

2.3.1.5 Gradient plasticity models: Gurtin and Anand

Gurtin and Anand (2005a) developed a strain gradient viscoplasticity model using the microstresses system which satisfies the microforce balance. The model was developed with the assumption of zero plastic rotation. They enhanced the conventional boundary conditions to incorporate the microscopic boundary conditions. Gurtin and Anand (2005a) incorporated the assumption of small deformation as follows:

$$u_{i,j} = H_{ij}^e + H_{ij}^p \tag{2.142}$$

where u_i is the displacement vector, $H_{ii}^p \equiv 0$, and $\varepsilon_{ij}^{e,p} = 1/2[H_{ij}^{e,p} + H_{ji}^{e,p}]$. One should notice that Eq. (2.142) is equivalent to the additive decomposition of the strain tensor presented in Eq. (2.3). Irrotational plastic deformation can be defined as follows:

$$W_{ij}^p = \frac{1}{2}\left[H_{ij}^p - \left(H_{ij}^p\right)^{\mathrm{T}}\right] \equiv 0 \rightarrow \varepsilon_{ij}^p \equiv H_{ij}^p \,\text{and}\, \varepsilon_{ii}^p \equiv 0 \tag{2.143}$$

Accordingly, Eq. (2.142) can be rewritten as follows:

$$u_{i,j} = H_{ij}^e + \varepsilon_{ij}^p \tag{2.144}$$

The rate form of Eq. (2.144) can be described as follows:

$$\dot{u}_{i,j} = \dot{H}^e_{ij} + \dot{\varepsilon}^p_{ij} \tag{2.145}$$

where $\dot{\varepsilon}^p_{ii} \equiv 0$. The gradient of plastic strain is incorporated in the model by introducing the Burgers tensor G_{ij} as follows:

$$G_{ij} = \left(\text{curl}\left[\varepsilon^p_{kl}\right]\right)_{ij} \tag{2.146}$$

which represents the role of macroscopic Burgers vector and G_{ij} is the ij component of the second order tensor curl $[\varepsilon^p_{kl}]$. Gurtin and Anand (2005a) include the effect of plastic strain tensor gradient $\varepsilon^p_{ij,\,k}$ into the Helmholtz free energy Ψ using G_{ij}.

Gurtin and Anand (2005a) incorporated the principle of virtual power to derive the material governing equations as follows:

$$P_{int} = P_{ext} \tag{2.147}$$

Eq. (2.147) should be satisfied for any arbitrary virtual velocity vector which satisfies Eqs. (2.142)–(2.145). The internal power P_{int} can be defined as follows:

$$P_{int} = \int_V \left(\sigma_{ij}\dot{\varepsilon}^e_{ij} + \mathcal{X}_{ij}\dot{\varepsilon}^p_{ij} + \mathcal{S}_{ijk}\dot{\varepsilon}^p_{ij,k}\right) dV \tag{2.148}$$

where σ_{ij} the Cauchy stress tensor, $\mathcal{X}_{ij}$ the plastic microstress tensor, and $\mathcal{S}_{ijk}$ the polar plastic microstress tensor are conjugate to $\dot{\varepsilon}^e_{ij}$, $\dot{\varepsilon}^p_{ij}$, and $\dot{\varepsilon}^p_{ij,k}$, respectively. The external power can be defined as follows:

$$P_{ext} = \int_V b_i v_i dV + \int_S \left(t_i v_i + m_{ij}\dot{\varepsilon}^p_{ij}\right) dS \tag{2.149}$$

where t_i and b_i are the traction and the external body force, respectively, which are conjugate to the arbitrary macroscopic velocity v_i. m_{ij} is the microtraction tensor conjugate to the plastic strain rate tensor due to the arbitrary macroscopic velocity. Considering Eqs. (2.147)–(2.149), the governing equations and corresponding boundary conditions can be obtained as follows:

$$\sigma_{ij,j} + b_i = 0 \text{ on } V \text{ (macroforce balance)} \tag{2.150}$$

$$s_{ij} = \mathcal{X}_{ij} - \mathcal{S}_{ijk,k} \text{ on } V \text{ (microforce balance)} \tag{2.151}$$

$$t_i = \sigma_{ij} n_j \text{ on } \partial S \text{ (macrotraction condition)} \tag{2.152}$$

$$m_{ij} = \mathcal{S}_{ijk} n_k \text{ on } \partial S \text{ (microtraction condition)} \tag{2.153}$$

where $s_{ij} = \sigma_{ij} - (I_1/3)\delta_{ij}$ is the stress deviator tensor.

Gurtin and Anand (2005a) incorporate the plasticity strain gradient into the Helmholtz free energy Ψ by including the Burgers tensor G_{ij}. They assumed that the free energy is a function of elastic strain tensor ε_{ij}^{e} and Burgers tensor G_{ij} as follows:

$$\Psi = \Psi\left(\varepsilon_{ij}^{e}, G_{ij}\right) \tag{2.154}$$

Assuming a purely mechanical theory, the increase in Ψ should be less than the power expended volume V, which can be described as follows:

$$\int_{V} \rho \dot{\Psi} dV \le P_{int} = P_{ext} \tag{2.155}$$

where ρ is the material density. One can define the local free energy inequality by substituting Eq. (2.148) into Eq. (2.155) as follows:

$$\rho \dot{\Psi} - \sigma_{ij}\dot{\varepsilon}_{ij}^{e} - \mathcal{X}_{ij}\dot{\varepsilon}_{ij}^{p} - \mathcal{S}_{ijk}\dot{\varepsilon}_{ij,k}^{p} \le 0 \tag{2.156}$$

One can write $\dot{\Psi}$ as follows:

$$\dot{\Psi} = \frac{\partial \Psi}{\partial \varepsilon_{ij}^{e}}\dot{\varepsilon}_{ij}^{e} + \frac{\partial \Psi}{\partial G_{ij}}\dot{G}_{ij} = \frac{\partial \Psi}{\partial \varepsilon_{ij}^{e}}\dot{\varepsilon}_{ij}^{e} + \frac{\partial \Psi}{\partial G_{ij}}\epsilon_{ipq}\dot{\varepsilon}_{jq,p}^{p} \tag{2.157}$$

where ϵ_{ipq} is the permutation tensor. Eq. (2.157) can be represented as follows:

$$\dot{\Psi} = \frac{\partial \Psi}{\partial \varepsilon_{ij}^{e}}\dot{\varepsilon}_{ij}^{e} + \frac{1}{\rho}P_{ijk}\dot{\varepsilon}_{ij,k}^{p} \tag{2.158}$$

where P_{jqp} is the thermodynamic defect stress tensor, which can be defined as follows:

$$P_{jqp} = \rho \frac{\partial \Psi}{\partial G_{ij}}\epsilon_{ipq} \tag{2.159}$$

Considering the fact that $\dot{\varepsilon}_{jq,p}^{p}$ is a symmetric-deviatoric in the first two subscripts, Eq. (2.158) can be rewritten as follows:

$$\dot{\Psi} = \frac{\partial \Psi}{\partial \varepsilon_{ij}^{e}}\dot{\varepsilon}_{ij}^{e} + \frac{1}{\rho}\mathcal{S}_{ijk}^{en}\dot{\varepsilon}_{ij,k}^{p} \tag{2.160}$$

where $\mathcal{S}_{ijk}^{en}$ is the energetic component of the polar plastic microstress tensor $\mathcal{S}_{ijk}$, which can be described as follows:

$$\mathcal{S}_{ijk}^{en} = \frac{1}{2}\left(P_{ijk} + P_{jik}\right) - \frac{1}{3}\delta_{ij}P_{rrk} \tag{2.161}$$

Accordingly, the polar plastic microstress tensor $\mathcal{S}_{ijk}$ can be additively decomposed to two corresponding energetic and dissipative parts, i.e., $\mathcal{S}_{ijk}^{en}$ and $\mathcal{S}_{ijk}^{dis}$, respectively, as follows:

$$\mathcal{S}_{ijk} = \mathcal{S}_{ijk}^{en} + \mathcal{S}_{ijk}^{dis} \tag{2.162}$$

The local free energy inequality can be rewritten by substituting Eq. (2.160) into Eq. (2.156) as follows:

$$\left(\rho\frac{\partial\Psi}{\partial\varepsilon_{ij}^{e}}-\sigma_{ij}\right)\dot{\varepsilon}_{ij}^{e}-\mathcal{X}_{ij}\dot{\varepsilon}_{ij}^{p}+\left(\mathcal{S}_{ijk}^{en}-\mathcal{S}_{ijk}\right)\dot{\varepsilon}_{ij,k}^{p}\leq 0 \tag{2.163}$$

The Cauchy stress tensor σ_{ij} can be obtained as follows:

$$\sigma_{ij}=\rho\frac{\partial\Psi}{\partial\varepsilon_{ij}^{e}} \tag{2.164}$$

Accordingly, the dissipation potential can be defined as follows:

$$\mathcal{D}=\mathcal{X}_{ij}\dot{\varepsilon}_{ij}^{p}+\mathcal{S}_{ijk}^{dis}\dot{\varepsilon}_{ij,k}^{p}\geq 0 \tag{2.165}$$

The constitutive models for $\mathcal{X}_{ij}$ and $\mathcal{S}_{ijk}^{dis}$ should satisfy Eq. (2.165). Gurtin and Anand (2005a) incorporated the following constitutive model:

$$\mathcal{X}_{ij}=S\left(\frac{\dot{\bar{\varepsilon}}_{p}}{\dot{\varepsilon}_{p}^{0}}\right)^{m}\frac{\dot{\varepsilon}_{ij}^{p}}{\dot{\bar{\varepsilon}}_{p}} \tag{2.166}$$

$$\mathcal{S}_{ijk}^{dis}=l_{dis}^{2}S_{Y}\left(\frac{\dot{\bar{\varepsilon}}_{p}}{\dot{\varepsilon}_{p}^{0}}\right)^{m}\frac{\dot{\varepsilon}_{ij,k}^{p}}{\dot{\bar{\varepsilon}}_{p}} \tag{2.167}$$

where l_{dis}, ε_{p}^{0}, m, and S_Y are the material constants which denote the dissipative material length scale, reference equivalent plastic strain rate, rate-sensitivity parameter, and initial plastic flow, i.e., yield stress, respectively. $\dot{\bar{\varepsilon}}_{p}$ and S are the nonlocal equivalent plastic strain rate and the resistance to the plastic flow, which is initially equal to S_Y, respectively, which can be defined as follows:

$$\dot{\bar{\varepsilon}}_{p}=\sqrt{\dot{\varepsilon}_{ij}^{p}\dot{\varepsilon}_{ij}^{p}+l_{dis}^{2}\dot{\varepsilon}_{ij,k}^{p}\dot{\varepsilon}_{ij,k}^{p}} \tag{2.168}$$

$$\dot{S}=H(S)\dot{\bar{\varepsilon}}_{p} \tag{2.169}$$

where $H(S)$ is the function that describes the material hardening.

Gurtin and Anand (2005a) incorporated a simple quadratic Helmholtz free energy Ψ as follows:

$$\Psi=\frac{1}{\rho}\mu\varepsilon_{ij}^{\prime e}\varepsilon_{ij}^{\prime e}+\frac{1}{2\rho}k\left(\varepsilon_{ii}^{e}\right)^{2}+\frac{1}{2\rho}\mu l_{en}^{2}G_{ij}G_{ij} \tag{2.170}$$

where μ, k, and l_{en} are the material constants denoting the elastic shear modulus, bulk modulus, and the energetic material length scale, respectively, and $\varepsilon_{ij}^{\prime e}=\varepsilon_{ij}^{e}-(\varepsilon_{ii}^{e}/3)\delta_{ij}$ is the strain deviator tensor. The first two terms of Eq. (2.170), i.e., $\frac{1}{\rho}\mu\varepsilon_{ij}^{\prime e}\varepsilon_{ij}^{\prime e}+\frac{1}{2\rho}k\left(\varepsilon_{ii}^{e}\right)^{2}$, are related to the elastic energy, while the last term, i.e., $\frac{1}{2\rho}\mu l_{en}^{2}G_{ij}G_{ij}$, is the representative of the defect energy.

The tensors σ_{ij}, P_{jqp}, and accordingly $\mathcal{S}^{en}_{jqp}$ can be obtained using Eqs. (2.146), (2.159), (2.161), (2.164), (2.170) as follows:

$$\sigma_{ij} = 2\mu\varepsilon'_{ij}e + k\left(\varepsilon^e_{ii}\right)\delta_{ij} \tag{2.171}$$

$$P_{jqp} = \mu l^2_{en} \epsilon_{ipq} G_{ij} = \mu l^2_{en}\left(\varepsilon^p_{jq,p} - \varepsilon^p_{jp,q}\right) \tag{2.172}$$

$$\mathcal{S}^{en}_{jqp} = \mu l^2_{en}\left[\varepsilon^p_{jq,p} - \frac{1}{2}\left(\varepsilon^p_{jp,q} + \varepsilon^p_{qp,j}\right) + \frac{1}{3}\delta_{jq}\varepsilon^p_{rp,r}\right] \tag{2.173}$$

Accordingly, the final form of the flow rule can be obtained using the Eqs. (2.151), (2.162), (2.166), (2.167), (2.173) as follows:

$$s_{ij} - \text{Energetic Backstress tensor} = \text{Dissipative Hardening tensor} \tag{2.174}$$

$$\text{Energetic Backstress} = -\mu l^2_{en}\left[\varepsilon^p_{ij,pp} - \frac{1}{2}\left(\varepsilon^p_{ik,jk} + \varepsilon^p_{jk,ik}\right) + \frac{1}{3}\delta_{ji}\varepsilon^p_{rk,rk}\right] \tag{2.175}$$

$$\text{Dissipative Hardening} = S\left(\frac{\dot{\bar{\varepsilon}}_p}{\dot{\varepsilon}^0_p}\right)^m \frac{\dot{\varepsilon}^p_{ij}}{\dot{\bar{\varepsilon}}_p} - l^2_{dis} S_Y \frac{\partial}{\partial x_k}\left[\left(\frac{\dot{\bar{\varepsilon}}_p}{\dot{\varepsilon}^0_p}\right)^m \frac{\dot{\varepsilon}^p_{ij,k}}{\dot{\bar{\varepsilon}}_p}\right] \tag{2.176}$$

The obtained flow rule is a second-order partial differential equation for the plastic strain ε^p_{ij} for a given stress deviator tensor s_{ij}.

2.3.1.6 *Gradient plasticity damage model: Voyiadjis and his co-workers*

Voyiadjis et al. (2001) developed a thermodynamically consistent gradient plasticity damage model in which the gradients of internal variables for both damage and plasticity are incorporated to capture the behavior of composite materials. Voyiadjis et al. (2004) extended the work to capture the strong coupling between nonlocal rate-dependent plasticity and nonlocal anisotropic rate-dependent damage for dynamic problems. First, the concept of effective stress space proposed by Kachanov (1958) is incorporated in the damage model. Fig. 2.8 elaborates the concept of effective stress space in a cylindrical bar. Fig. 2.8A shows a bar containing cracks and voids with the cross section A which is subjected to a uniaxial force F. The uniaxial stress σ can be obtained as $\sigma = F/A$. Fig. 2.8B shows a fictitious undamaged bar in which all the cracks and voids are excluded. Fig. 2.8B is commonly termed as the effective configuration. Considering a similar force F applied on the undamaged bar, the uniaxial stress can be obtained as $\hat{\sigma} = F/\hat{A}$. The scalar damage parameter ϕ can be defined as follows:

$$\phi = \frac{A - \hat{A}}{A} \tag{2.177}$$

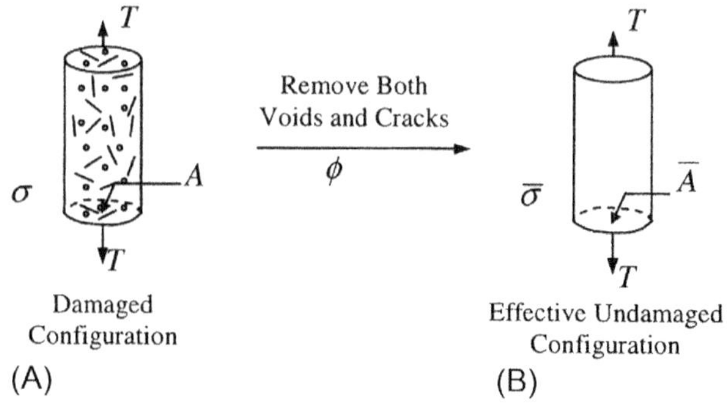

FIG. 2.8 A cylindrical bar subjected to uniaxial tension: (A) damaged configuration including voids and cracks. (B) Effective undamaged configuration by excluding both voids and cracks from the original sample. *(After Voyiadjis, G.Z, Al-Rub, R.K.A., Palazoto, A., 2004. Thermodynamic framework for coupling of non-local viscoplasticity and non-local anisotropic viscodamage for dynamic localization problems using gradient theory. Int. J. Plast. 20, 981–1038.)*

Accordingly, the stress in the effective configuration $\overline{\sigma}$ can be obtained as follows:

$$\hat{\sigma} = \frac{\sigma}{1-\phi} \tag{2.178}$$

The nonlocal damage variable $\overline{\phi}$ should be incorporated to handle different aspect of size effects and localization. Voyiadjis et al. (2004), however, incorporated an anisotropic damage model in which the effective stress tensor $\hat{\sigma}_{ij}$ and nominal stress tensor σ_{kl} are connected using a nonlocal fourth order tensor $\overline{M}_{ikjl}$ as follows:

$$\hat{\sigma}_{ij} = \overline{M}_{ikjl}\sigma_{kl} \tag{2.179}$$

where $\overline{M}_{ikjl}$ is described using the nonlocal second order damage tensor $\overline{\phi}_{ij}$ as follows:

$$\overline{M}_{ikjl} = 2\left[\left(\delta_{ik} - \overline{\phi}_{ik}\right)\delta_{jl} + \delta_{ik}\left(\delta_{jl} - \overline{\phi}_{jl}\right)\right]^{-1} \tag{2.180}$$

The evolution of $\overline{\phi}_{ij}$ will be defined later. The relation between the elasticity tensor in damaged and effective configurations can be defined as follows:

$$C_{ijkl} = \overline{M}^{-1}_{imjn}\hat{C}_{mnpq}\overline{M}^{-1}_{pkql} \tag{2.181}$$

where

$$\overline{M}^{-1}_{ikjl} = \frac{1}{2}\left[\left(\delta_{ik} - \overline{\phi}_{ik}\right)\delta_{jl} + \delta_{ik}\left(\delta_{jl} - \overline{\phi}_{jl}\right)\right] \tag{2.182}$$

The elastic energy equivalence hypothesis is incorporated which states that the elastic energy density is the same for both damaged and effective configurations, which leads to the following relation:

$$\hat{\varepsilon}^e_{ij} = \overline{M}^{-1}_{ikjl}\varepsilon^e_{kl} \tag{2.183}$$

Voyiadjis et al. (2001, 2004) incorporated a thermodynamically consistent methodology to develop a non-local model, which is the extension of the local plasticity model presented in Section 2.2.7. Accordingly, the Helmholtz free energy can be described as follows:

$$\Psi = \Psi\left(\varepsilon^e_{ij}, T, T_{,i}; \overline{\aleph}_k\right) \tag{2.184}$$

where $\overline{\aleph}_k$ can be described as follows:

$$\overline{\aleph}_k = \overline{\aleph}_k\left(\Xi_n, \nabla^2\Xi_n\right) \tag{2.185}$$

where Ξ_n is the set of internal variables defined to capture nonlocal rate-dependent plasticity damage model and $\nabla^2\Xi_n$ is their second-order gradients. The incorporated set of internal variables are as follows:

$$\Xi_n = \Xi_n\left(p, \varsigma_{ij}, r, \Gamma_{ij}, \phi_{ij}\right) \tag{2.186}$$

where p and α_{ij} describe the isotropic hardening and kinematic hardening of the rate-dependent plasticity model, respectively. Similarly, r and Γ_{ij} define the isotropic and kinematic evolution of damage surface, which can be defined similar to the plasticity yield surface. The second-order gradient set of internal state variables can be defined as follows:

$$\nabla^2\Xi_n = \nabla^2\Xi_n\left(\nabla^2 p, \nabla^2\varsigma_{ij}, \nabla^2 r, \nabla^2\Gamma_{ij}, \nabla^2\phi_{ij}\right) \tag{2.187}$$

One should note that including the second-order gradients of internal state variables requires to define the corresponding material length.

The Helmholtz free energy Ψ is additively decomposed to three potentials related to the thermo-elastic Ψ^{te}, rate-dependent thermo-plastic Ψ^{tvp}, and rate dependent thermo-damage Ψ^{tvd} models as follows:

$$\begin{aligned}\Psi = {} & \Psi^{te}\left(\varepsilon^e_{ij}, T, T_{,i}, \phi_{ij}, \nabla^2\phi_{ij}\right) + \Psi^{tvp}\left(T, T_{,i}, p, \varsigma_{ij}, \nabla^2 p, \nabla^2\varsigma_{ij}\right) \\ & + \Psi^{tvd}\left(T, T_{,i}, r, \Gamma_{ij}, \phi_{ij}, \nabla^2 r, \nabla^2\Gamma_{ij}, \nabla^2\phi_{ij}\right)\end{aligned} \tag{2.188}$$

The Clausius-Duhem inequality can be described as follows:

$$\sigma_{ij}\dot{\varepsilon}_{ij} - \rho\left(\dot{\Psi} + s\dot{T}\right) - q_i\frac{T_{,i}}{T} \geq 0 \tag{2.189}$$

where σ_{ij} is the Cauchy stress tensor, ρ is the material density, s is the specific entropy per unit mass, $\vec{q}$ is the vector of heat flux. Accordingly $\dot{\Psi}$ can be described as follows:

$$\dot{\Psi} = \frac{\partial \Psi}{\partial \varepsilon_{ij}^e}\dot{\varepsilon}_{ij}^e + \frac{\partial \Psi}{\partial T}\dot{T} + \frac{\partial \Psi}{\partial T_{,i}}\dot{T}_{,i} + \frac{\partial \Psi}{\partial \overline{\aleph}_k}\dot{\overline{\aleph}}_k \tag{2.190}$$

where the term $\frac{\partial \Psi}{\partial \overline{\aleph}_k}\dot{\overline{\aleph}}_k$ can be expressed as:

$$\frac{\partial \Psi}{\partial \overline{\aleph}_k}\dot{\overline{\aleph}}_k = \frac{\partial \Psi}{\partial \Xi_n}\dot{\Xi}_n + \frac{\partial \Psi}{\partial \nabla^2 \Xi_n}\nabla^2 \dot{\Xi}_n \tag{2.191}$$

and

$$\frac{\partial \Psi}{\partial \Xi_n}\dot{\Xi}_n = \frac{\partial \Psi}{\partial p}\dot{p} + \frac{\partial \Psi}{\partial \varsigma_{ij}}\dot{\varsigma}_{ij} + \frac{\partial \Psi}{\partial r}\dot{r} + \frac{\partial \Psi}{\partial \Gamma_{ij}}\dot{\Gamma}_{ij} + \frac{\partial \Psi}{\partial \phi_{ij}}\dot{\phi}_{ij} \tag{2.192}$$

$$\frac{\partial \Psi}{\partial \nabla^2 \Xi_n}\nabla^2 \dot{\Xi}_n = \frac{\partial \Psi}{\partial \nabla^2 p}\nabla^2 \dot{p} + \frac{\partial \Psi}{\partial \nabla^2 \varsigma_{ij}}\nabla^2 \dot{\varsigma}_{ij} + \frac{\partial \Psi}{\partial \nabla^2 r}\nabla^2 \dot{r} + \frac{\partial \Psi}{\partial \nabla^2 \Gamma_{ij}}\nabla^2 \dot{\Gamma}_{ij} + \frac{\partial \Psi}{\partial \nabla^2 \phi_{ij}}\nabla^2 \dot{\phi}_{ij} \tag{2.193}$$

Considering Eqs. (2.190)–(2.193), the Clausius-Duhem inequality, i.e., Eq. (2.189), can be rewritten as follows:

$$\left(\sigma_{ij} - \rho\frac{\partial \Psi}{\partial \varepsilon_{ij}^e}\right)\dot{\varepsilon}_{ij}^e + \sigma_{ij}\dot{\varepsilon}_{ij}^p - \rho\left(\frac{\partial \Psi}{\partial T} + s\right)\dot{T} - \rho\frac{\partial \Psi}{\partial T_{,i}}\dot{T}_{,i} - \rho\frac{\partial \Psi}{\partial \overline{\aleph}_k}\dot{\overline{\aleph}}_k - q_i\frac{T_{,i}}{T} \geq 0 \tag{2.194}$$

Eq. (2.194) defines the thermodynamic state laws as follows:

$$\sigma_{ij} = \rho\frac{\partial \Psi}{\partial \varepsilon_{ij}^e} \tag{2.195}$$

$$s = -\frac{\partial \Psi}{\partial T} \tag{2.196}$$

$$\frac{q_i}{\dot{T}} = \rho\frac{\partial \Psi}{\partial T_{,i}} \tag{2.197}$$

$$\Sigma_k = \rho\frac{\partial \Psi}{\partial \Xi_n}(k = 1, \ldots, 5) \tag{2.198}$$

$$\Sigma_k^g = \rho\frac{\partial \Psi}{\partial \nabla^2 \Xi_n}(k = 1, \ldots, 5) \tag{2.199}$$

Eqs. (2.195)–(2.199) define the relation between the observable and internal variables and their corresponding thermodynamic conjugate forces, which are summarized in Table 2.2. $\Sigma_k = \{Y_{ij}, R, \alpha_{ij}, K, H_{ij}\}$ and $\Sigma_k^g = \{Y_{ij}^g, R^g, \alpha_{ij}^g, K^g, H_{ij}^g\}$ are the conjugate forces of internal state variables $\Xi_n = \{\phi_{ij}, p, \varsigma_{ij}, r, \Gamma_{ij}\}$ and $\nabla^2\Xi_n = \{\nabla^2\phi_{ij}, \nabla^2 p, \nabla^2\varsigma_{ij}, \nabla^2 r, \nabla^2\Gamma_{ij}\}$, respectively. The superscript g denotes the thermodynamic force conjugate to the gradients of internal variables. The thermodynamic state laws of internal variables are summarized in Table 2.3.

TABLE 2.2 Thermodynamic state variables and their corresponding conjugate forces

State variables		
Observable	**Internal**	**Associated conjugates**
ε_{ij}		σ_{ij}
T		s
	ε_{ij}^{e}	σ_{ij}
	ε_{ij}^{p}	$-\sigma_{ij}$
	$T_{,i}$	q_i
	ϕ_{ij}, $\nabla^2\phi_{ij}$	$-Y_{ij}$, $-Y_{ij}^{g}$
	p	R, R^g
	ς_{ij}, $\nabla^2\varsigma_{ij}$	α_{ij}, α_{ij}^{g}
	r, $\nabla^2 r$	K, K^g
	Γ_{ij}, $\nabla^2\Gamma_{ij}$	H_{ij}, H_{ij}^{g}

TABLE 2.3 Thermodynamic state laws of internal variables

	Isotropic hardening	$R=\rho\frac{\partial\Psi}{\partial p}$	$R^g=\rho\frac{\partial\Psi}{\partial\nabla^2 p}$
Plasticity	Kinematic hardening	$\alpha_{ij}=\rho\frac{\partial\Psi}{\partial\varsigma_{ij}}$	$\alpha_{ij}^{g}=\rho\frac{\partial\Psi}{\partial\nabla^2\varsigma_{ij}}$
Damage	Isotropic hardening	$K=\rho\frac{\partial\Psi}{\partial r}$	$K^g=\rho\frac{\partial\Psi}{\partial\nabla^2 r}$
	Kinematic hardening	$H_{ij}=\rho\frac{\partial\Psi}{\partial\Gamma_{ij}}$	$H_{ij}^{g}=\rho\frac{\partial\Psi}{\partial\nabla^2\Gamma_{ij}}$
	Damage force	$Y_{ij}=-\rho\frac{\partial\Psi}{\partial\phi_{ij}}$	$Y_{ij}^{g}=-\rho\frac{\partial\Psi}{\partial\nabla^2\phi_{ij}}$

The final form of the Clausius-Duhem inequality can be stated as follows by substituting Eqs. (2.195)–(2.199) into Eq. (2.194) such that:

$$\Pi=\sigma_{ij}\dot{\varepsilon}_{ij}^{p}-R\dot{p}-R^g\nabla^2\dot{p}-\alpha_{ij}\dot{\varsigma}_{ij}-\alpha_{ij}^{g}\nabla^2\dot{\varsigma}_{ij}-K\dot{r}-K^g\nabla^2\dot{r}$$
$$-H_{ij}\dot{\Gamma}_{ij}-K_{ij}^{g}\nabla^2\dot{\Gamma}_{ij}+Y_{ij}\dot{\phi}_{ij}+Y_{ij}^{g}\nabla^2\dot{\phi}_{ij}-q_i\left(\frac{T_{,i}}{T}+\frac{\dot{T}_{,i}}{\dot{T}}\right)\geq 0 \qquad (2.200)$$

where Π is the dissipation energy which is non-negative. The dissipation energy can be additively decomposed as follows:

$$\Pi = \Pi^{vp} + \Pi^{vd} + \Pi^{th} \geq 0 \tag{2.201}$$

where Π^{vp}, Π^{vd}, and Π^{th} are the rate-dependent plastic, rate-dependent damage, and thermal dissipation energies, respectively, which can be described as follows:

$$\Pi^{vp} = \sigma_{ij}\dot{\varepsilon}_{ij}^{p} - R\dot{p} - R^{g}\nabla^{2}\dot{p} - \alpha_{ij}\dot{\varsigma}_{ij} - \alpha_{ij}^{g}\nabla^{2}\dot{\varsigma}_{ij} \geq 0 \tag{2.202}$$

$$\Pi^{vd} = -K\dot{r} - K^{g}\nabla^{2}\dot{r} - H_{ij}\dot{\Gamma}_{ij} - K_{ij}^{g}\nabla^{2}\dot{\Gamma}_{ij} + Y_{ij}\dot{\phi}_{ij} + Y_{ij}^{g}\nabla^{2}\dot{\phi}_{ij} \geq 0 \tag{2.203}$$

$$\Pi^{th} = -q_{i}\left(\frac{T_{,i}}{T} + \frac{\dot{T}_{,i}}{\dot{T}}\right) \geq 0 \tag{2.204}$$

Eq. (2.202) states that all the plastic works are dissipated and transformed to heat except the part which is stored in materials due to the hardening. Similarly, Eq. (2.203) states that all the work done by the damage process should be transformed to heat except the part stored in material as the hardening.

Based on the dissipation processes described in Eqs. (2.202)–(2.204), the complementary laws can be described as follows:

$$\Theta\left(\dot{\varepsilon}_{ij}^{p}, T, T_{,i}; \overline{\aleph}_{k}\right) = \Theta^{I}\left(\dot{\varepsilon}_{ij}^{p}; \overline{\aleph}_{k}\right) + \Theta^{th}(T, T_{,i}) \tag{2.205}$$

where Θ is the dissipation potential, and Θ^{I} and Θ^{th} are the dissipation potentials related to the inelastic deformation mechanisms including rate-dependent plasticity and rate-dependent damage, and thermal mechanisms, respectively. The Thermodynamic forces can be obtained from the dissipation potentials using the normality rule as follows:

$$\sigma_{ij} = \frac{\partial\Theta^{I}}{\partial\dot{\varepsilon}_{ij}^{p}} \tag{2.206}$$

$$\Sigma_{k} = -\frac{\partial\Theta^{I}}{\partial\dot{\Xi}_{n}}(k = 1, \ldots, 5) \tag{2.207}$$

$$\Sigma_{k}^{g} = -\frac{\partial\Theta^{I}}{\partial\nabla^{2}\dot{\Xi}_{n}}(k = 1, \ldots, 5) \tag{2.208}$$

$$\frac{q_{i}}{\dot{T}} = -\frac{\partial\Theta^{th}}{\partial T_{,i}} \tag{2.209}$$

The dual potentials Θ^{*} can be obtained using the Legendre-Fenchel transformation of the dissipation potential Θ as follows:

$$\Theta^{*}\left(\sigma_{ij}, \Sigma_{k}, \Sigma_{k}^{g}, \frac{q_{i}}{\dot{T}}\right) = \Pi - \Theta\left(\dot{\varepsilon}_{ij}^{p}, T, T_{,i}; \overline{\aleph}_{k}\right) = \Theta^{*I}\left(\sigma_{ij}, \Sigma_{k}, \Sigma_{k}^{g}\right) + \Theta^{*th}(T, T_{,i}) \tag{2.210}$$

where Θ^{*I} and Θ^{*th} are the dual potentials related to the inelastic deformation mechanisms including rate-dependent plasticity and rate-dependent damage, and thermal mechanisms, respectively. Accordingly, the governing equations of state variables can be described as follows:

$$\dot{\varepsilon}_{ij}^{p} = \frac{\partial \Theta^{*I}}{\partial \sigma_{ij}} \tag{2.211}$$

$$\dot{\Xi}_n = -\frac{\partial \Theta^{*I}}{\partial \Sigma_k}(k = 1, \ldots, 5) \tag{2.212}$$

$$\nabla^2 \dot{\Xi}_n = -\frac{\partial \Theta^{*I}}{\partial \Sigma_k^g}(k = 1, \ldots, 5) \tag{2.213}$$

$$\frac{T_{,i}}{\dot{T}} = -\frac{\partial \Theta^{*th}}{\partial q_i} \tag{2.214}$$

In order to define the evolution laws of state variables, the Helmholtz free energy Ψ and the dual potential Θ^* should be defined. As shown in Section 2.2.7, maximizing the dissipation potentials leads to the normality hypothesis, i.e., associated flow rules. The Lagrange multipliers method is incorporated to find the maximum of the dissipation potentials. Accordingly, two Lagrange multipliers of $\dot{\lambda}^{vp}$ and $\dot{\lambda}^{vd}$ are introduced, and the objective function Ω can be written as follows:

$$\Omega = \Pi^{vp} + \Pi^{vd} - \dot{\lambda}^{vp} f - \dot{\lambda}^{vd} g \tag{2.215}$$

where two constraints of $f = 0$ and $g = 0$ define the rate-dependent plasticity and damage loading surfaces, respectively. The criteria for maximizing the objective function Ω can be described as follows:

$$\frac{\partial \Omega}{\partial \sigma_{ij}} = 0 \text{ and } \frac{\partial \Omega}{\partial Y_{ij}} = 0 \tag{2.216}$$

Accordingly, the evolution of plastic strain and damage tensors can be obtained as follows:

$$\dot{\varepsilon}_{ij}^{p} = \dot{\lambda}^{vp}\frac{\partial f}{\partial \sigma_{ij}} + \dot{\lambda}^{vd}\frac{\partial g}{\partial \sigma_{ij}} = {}^{1}\dot{\varepsilon}_{ij}^{p} + {}^{2}\dot{\varepsilon}_{ij}^{p} \tag{2.217}$$

$$\dot{\phi}_{ij} = \dot{\lambda}^{vp}\frac{\partial f}{\partial Y_{ij}} + \dot{\lambda}^{vd}\frac{\partial g}{\partial Y_{ij}} = {}^{1}\dot{\phi}_{ij} + {}^{2}\dot{\phi}_{ij} \tag{2.218}$$

where ${}^{1}\dot{\varepsilon}_{ij}^{p} = \dot{\lambda}^{vp}\frac{\partial f}{\partial \sigma_{ij}}$, ${}^{2}\dot{\varepsilon}_{ij}^{p} = \dot{\lambda}^{vd}\frac{\partial g}{\partial \sigma_{ij}}$, ${}^{1}\dot{\phi}_{ij} = \dot{\lambda}^{vp}\frac{\partial f}{\partial Y_{ij}}$, and ${}^{2}\dot{\phi}_{ij} = \dot{\lambda}^{vd}\frac{\partial g}{\partial Y_{ij}}$.

The rate of isotropic hardening of plasticity and damage loading surfaces, i.e., $\dot{p}$ and $\dot{r}$, respectively, can be defined as follows;

$$\dot{p} = \sqrt{\frac{2}{3}\dot{\varepsilon}_{ij}^{p}\dot{\varepsilon}_{ij}^{p}} \tag{2.219}$$

$$\dot{r} = \sqrt{\dot{\phi}_{ij}\dot{\phi}_{ij}} \tag{2.220}$$

In order to capture nonlinear hardening of plasticity and damage observed in experiments, the non-associated flow rules are incorporated to describe the evolution of hardening variables of rate-dependent plasticity, i.e., $\dot{p}$ and $\dot{\alpha}_{ij}$, and rate-dependent damage, i.e., $\dot{r}$ and $\dot{\Gamma}_{ij}$, by introducing two potentials of rate-dependent plasticity F and rate-dependent damage G. The evolution of hardening variables, i.e., $\dot{p}$, $\dot{\alpha}_{ij}$, $\dot{r}$, and $\dot{\Gamma}_{ij}$, can be obtained using the normality rule, which is summarized in Table 2.4. Furthermore, the evolution laws of the second-order gradients of the hardening variables is presented in Table 2.4 by applying the corresponding operator on their local counterparts.

Voyiadjis et al. (2004) proposed the following Helmholtz free energy for the rate-dependent plasticity damage model:

$$\Psi^{te} = \frac{1}{2\rho}\varepsilon^e_{ij}C_{ijkl}\left(\phi_{pq}, \nabla^2\phi_{pq}\right)\varepsilon^e_{kl} - \frac{1}{\rho}\beta_{ij}\varepsilon^e_{ij}(T-T_r) - s_r(T-T_r) - \frac{1}{2}c(T-T_r)^2 - \frac{1}{2\rho T}k_{ij}T_{,i}T_{,j} \tag{2.221}$$

$$\rho\Psi^{tvp} = \frac{1}{2}a_1 p^2\vartheta + \frac{1}{2}b_1\left(\nabla^2 p\right)^2\vartheta + \frac{1}{2}a_2\varsigma_{ij}\varsigma_{ij}\vartheta + \frac{1}{2}b_2\nabla^2\varsigma_{ij}\nabla^2\varsigma_{ij}\vartheta \tag{2.222}$$

$$\rho\Psi^{tvp} = \frac{1}{2}a_3 r^2\vartheta + \frac{1}{2}b_3\left(\nabla^2 r\right)^2\vartheta + \frac{1}{2}a_4\Gamma_{ij}\Gamma_{ij}\vartheta + \frac{1}{2}b_4\nabla^2\Gamma_{ij}\nabla^2\Gamma_{ij}\vartheta \tag{2.223}$$

where $C_{ijkl}(\phi_{pq}, \nabla^2\phi_{pq})$ is the elasticity tensor in the damaged configuration, β_{ij} is the tangent conjugate tensor of thermal dilatation, c is the thermal expansion coefficient, s_r and T_r are the reference entropy and temperature, respectively, k_{ij} is the tensor of heat conductivity, a_k and $b_k (k=1, \ldots, 4)$ are the material parameters, and ϑ can be defined as follows:

$$\vartheta = 1 - \left(\frac{T}{T_m}\right)^n \tag{2.224}$$

TABLE 2.4 The evolution laws of internal state variables

Plasticity	Isotropic hardening	$\dot{p} = -\dot{\lambda}^{vp}\frac{\partial F}{\partial R}$	$\nabla^2\dot{p} = -\nabla^2\dot{\lambda}^{vp}\frac{\partial F}{\partial R} - \dot{\lambda}^{vp}\nabla^2\left(\frac{\partial F}{\partial R}\right)$
	Kinematic hardening	$\dot{\varsigma}_{ij} = -\dot{\lambda}^{vp}\frac{\partial F}{\partial \alpha_{ij}}$	$\nabla^2\dot{\varsigma}_{ij} = -\nabla^2\dot{\lambda}^{vp}\frac{\partial F}{\partial \alpha_{ij}} - \dot{\lambda}^{vp}\nabla^2\left(\frac{\partial F}{\partial \alpha_{ij}}\right)$
Damage	Isotropic hardening	$\dot{r} = -\dot{\lambda}^{vd}\frac{\partial G}{\partial K}$	$\nabla^2\dot{r} = -\nabla^2\dot{\lambda}^{vd}\frac{\partial G}{\partial K} - \dot{\lambda}^{vd}\nabla^2\left(\frac{\partial G}{\partial K}\right)$
	Kinematic hardening	$\dot{\Gamma}_{ij} = -\dot{\lambda}^{vd}\frac{\partial G}{\partial H_{ij}}$	$\nabla^2\dot{\Gamma}_{ij} = -\nabla^2\dot{\lambda}^{vd}\frac{\partial G}{\partial H_{ij}} - \dot{\lambda}^{vd}\nabla^2\left(\frac{\partial G}{\partial H_{ij}}\right)$
	Damage force	$\phi_{ij} = \nabla^2\dot{\lambda}^{vp}\frac{\partial f}{\partial Y_{ij}} + \dot{\lambda}^{vp}\nabla^2\left(\frac{\partial f}{\partial Y_{ij}}\right) + \nabla^2\dot{\lambda}^{vd}\frac{\partial g}{\partial Y_{ij}} + \dot{\lambda}^{vd}\nabla^2\left(\frac{\partial g}{\partial Y_{ij}}\right)$	

where T_m is the melting temperature and n is the material constant. The thermodynamic conjugate forces can be obtained using the defined Helmholtz free energy and following Eqs. (2.195)–(2.199) which are presented in Table 2.5.

Voyiadjis et al. (2004) incorporated an explicit version of gradients model to describe the nonlocal state variables as follows:

$$\dot{\overline{\Xi}}_n = \dot{\Xi}_n + \frac{1}{2} l_n^2 \nabla^2 \dot{\Xi}_n (n = 1, \ldots, 5) \tag{2.225}$$

where $\Xi_n = \{\phi_{ij}, p, \varsigma_{ij}, r, \Gamma_{ij}\}$, and $l_n (n = 1, \ldots, 5)$ are the length scales corresponding to the Ξ_n. Besides $\overline{Y}_{ij}$ which is described later, one can describe the evolution of other nonlocal thermodynamic conjugate forces, i.e., $\Sigma_n = \{\overline{R}, \overline{\alpha}_{ij}, \overline{K}, \overline{H}_{ij}\}$ as follows:

$$\dot{\overline{\Sigma}}_n = c_n \dot{\overline{\Xi}}_n \vartheta = c_n \dot{\Xi}_n \vartheta + \frac{1}{2} c_n l_n^2 \nabla^2 \dot{\Xi}_n \vartheta \; (n = 1, \ldots, 4) \tag{2.226}$$

where $c_n (n = 1, \ldots, 4)$ are the material parameters. On the other hand, the nonlocal thermodynamic conjugate forces can be described using their local counterpart and gradients as follows:

$$\dot{\overline{\Sigma}}_n = \dot{\Sigma}_n + \dot{\Sigma}_n^g = a_n \dot{\Xi}_n \vartheta + b_n \nabla^2 \dot{\Xi}_n \vartheta \; (n = 1, \ldots, 4) \tag{2.227}$$

Comparing Eqs. (2.226), (2.227), one can relate the material parameters as follows:

$$l_n = \sqrt{\frac{2b_n}{a_n}} \text{ and } c_n = a_n (n = 1, \ldots, 4) \tag{2.228}$$

Table 2.6 presents the evolution of local, gradient, and nonlocal thermodynamic conjugate forces of $\Sigma_n = \{R, \alpha_{ij}, K, H_{ij}\}$ with their corresponding length scales.

In order to define the evolution of state variables according to the evolution laws presented in Table 2.4, the rate-dependent plasticity and damage potentials, i.e., F and G, respectively, are defined as follows:

$$F = f + \frac{1}{2} k_1 \hat{\overline{R}}^2 + \frac{1}{2} k_2 \hat{\overline{\alpha}}_{ij} \hat{\overline{\alpha}}_{ij} \tag{2.229}$$

$$\mathrm{G} = g + \frac{1}{2} h_1 \overline{K}^2 + \frac{1}{2} h_2 \overline{H}_{ij} \overline{H}_{ij} \tag{2.230}$$

where k_1, k_2, h_1, and h_2 are material constants, and the effective tensors $\hat{\overline{R}}$ and $\hat{\overline{\alpha}}_{ij}$ are the rate-dependent plasticity hardening variables described in the effective configuration as follows:

$$\hat{\overline{R}} = \frac{\overline{R}}{1 - \overline{r}} \to \left(\hat{R} = \frac{R}{1 - \overline{r}} \text{ and } \hat{R}^g = \frac{R^g}{1 - \overline{r}} \right) \tag{2.231}$$

$$\hat{\overline{\alpha}}_{ij} = \overline{M}_{ikjl} \overline{\alpha}_{kl} \to \left(\hat{\alpha}_{ij} = \overline{M}_{ikjl} \alpha_{kl} \text{ and } \hat{\alpha}_{ij}^g = \overline{M}_{ikjl} \alpha_{kl}^g \right) \tag{2.232}$$

TABLE 2.5 Thermodynamic state laws of internal variables obtained using the Helmholtz free energy described in Eqs. (2.221)–(2.223)

Plasticity	Isotropic hardening	$R = a_1 p\vartheta$	$R^g = b_1(\nabla^2 p)\vartheta$
	Kinematic hardening	$\alpha_{ij} = a_2 \varsigma_{ij}\vartheta$	$\alpha_{ij}^g = b_2(\nabla^2 \varsigma_{ij})\vartheta$
Damage	Isotropic hardening	$K = a_3 r\vartheta$	$K^g = b_3(\nabla^2 r)\vartheta$
	Kinematic hardening	$H_{ij} = a_4 \Gamma_{ij}\vartheta$	$H_{ij}^g = b_4(\nabla^2 \Gamma_{ij})\vartheta$
	Damage force	$Y_{ij} = \frac{-\partial}{\partial \phi_{ij}}\left[\varepsilon_{mn}^e C_{mnpq}\varepsilon_{pq}^e - \beta_{mn}\varepsilon_{mn}^e(T - T_r)\right]$	$Y_{ij}^g = \frac{-\partial}{\partial \nabla^2 \phi_{ij}}\left[\varepsilon_{mn}^e C_{mnpq}\varepsilon_{pq}^e - \beta_{mn}\varepsilon_{mn}^e(T - T_r)\right]$

TABLE 2.6 Evolution of local, gradient, and nonlocal thermodynamic conjugate forces with their corresponding length scales

Plasticity	Isotropic hardening	$\dot{R} = a_1 \dot{p} \vartheta$	$\dot{R}^g = b_1 (\nabla^2 \dot{p}) \vartheta$	$\dot{\overline{R}} = \dot{R} + \frac{b_1}{a_1} \nabla^2 \dot{R}$	$l_1 = \sqrt{\frac{2b_1}{a_1}}$
	Kinematic hardening	$\dot{\alpha}_{ij} = a_2 \dot{\varsigma}_{ij} \vartheta$	$\dot{\alpha}^g_{ij} = b_2 \left(\nabla^2 \dot{\varsigma}_{ij} \right) \vartheta$	$\dot{\overline{\alpha}}_{ij} = \dot{\alpha}_{ij} + \frac{b_2}{a_2} \nabla^2 \dot{\alpha}_{ij}$	$l_2 = \sqrt{\frac{2b_2}{a_2}}$
Damage	Isotropic hardening	$\dot{K} = a_3 \dot{r} \vartheta$	$\dot{K}^g = b_3 (\nabla^2 \dot{r}) \vartheta$	$\dot{\overline{K}} = \dot{K} + \frac{b_3}{a_3} \nabla^2 \dot{K}$	$l_3 = \sqrt{\frac{2b_3}{a_3}}$
	Kinematic hardening	$\dot{H}_{ij} = a_4 \dot{\Gamma}_{ij} \vartheta$	$\dot{H}^g_{ij} = b_4 (\nabla^2 \dot{\Gamma}_{ij}) \vartheta$	$\dot{\overline{H}}_{ij} = \dot{H}_{ij} + \frac{b_2}{a_2} \nabla^2 \dot{H}_{ij}$	$l_4 = \sqrt{\frac{2b_4}{a_4}}$

where $\bar{r}$ is the nonlocal accumulative damage described as follows:

$$\bar{r} = \int_0^t \sqrt{\dot{\bar{\phi}}_{ij}\dot{\bar{\phi}}_{ij}} \tag{2.233}$$

The rate-dependent plasticity loading surface is defined in the effective configuration as follow:

$$f = \sqrt{\frac{3}{2}\left(\hat{s}_{ij} - \hat{\bar{\alpha}}_{ij}\right)\left(\hat{s}_{ij} - \hat{\bar{\alpha}}_{ij}\right)} - \hat{\sigma}_{yp}(T) - \hat{\bar{R}}\left(\hat{p}, \nabla^2\hat{p}, T\right) - \hat{\sigma}_v\left(\dot{\hat{p}}, \nabla^2\dot{\hat{p}}, \hat{p}, \nabla^2\hat{p}, T\right) \equiv 0 \tag{2.234}$$

where the effective initial yield stress $\bar{\sigma}_{yp}$, effective viscous overstress $\bar{\sigma}_v$, and effective accumulative plastic strain rate $\dot{\hat{p}}$ can be defined as follows:

$$\hat{\sigma}_{yp} = \vartheta \bar{Y}_0 \tag{2.235}$$

$$\hat{\sigma}_v = \left\langle \sqrt{\frac{3}{2}\left(\hat{s}_{ij} - \hat{\bar{\alpha}}_{ij}\right)\left(\hat{s}_{ij} - \hat{\bar{\alpha}}_{ij}\right)} - \hat{\sigma}_{yp}(T) - \hat{\bar{R}}\left(\hat{p}, \nabla^2\hat{p}, T\right) \right\rangle \tag{2.236}$$

$$\dot{\hat{p}} = \sqrt{\frac{2}{3}\dot{\hat{\varepsilon}}_{ij}^p \dot{\hat{\varepsilon}}_{ij}^p} \tag{2.237}$$

Where ϑ is defined in Eq. (2.224), $\bar{Y}_0$ is the effective initial yield stress at $T = 0$, $\dot{\hat{\varepsilon}}_{ij}^p = 0$, and zero rate, and $\dot{\hat{\varepsilon}}_{ij}^p$ is the rate of plastic strain tensor in the undamaged configuration. $\langle\rangle$ denotes the MacAuley brackets which can be described as follows:

$$\langle x \rangle = \frac{x + |x|}{2} \tag{2.238}$$

Eq. (2.234) states that the plasticity occurs for $\hat{\sigma}_v > 0$ which leads to $f = 0$, while $\hat{\sigma}_v = 0$ leads to $f < 0$ defining an elastic state.

The hypothesis of plastic strain energy equivalence can be used to relate the plastic strain rate in the damaged and undamaged configurations as follows:

$$\dot{\hat{\varepsilon}}_{ij}^p = \overline{M}_{ikjl}^{-1} \dot{\varepsilon}_{k1}^p \tag{2.239}$$

where $\overline{M}_{ikjl}^{-1}$ is defined in Eq. (2.182). Accordingly, the effective plastic strain rate tensor can be defined as follows:

$$\dot{\hat{\varepsilon}}_{ij}^p = \dot{\lambda}^{vp} \frac{\partial f}{\partial \hat{\sigma}_{ij}} \tag{2.240}$$

Considering Eqs. (2.234), (2.240), $\dot{\hat{p}} = \dot{\lambda}^{vp}$. The relation between the accumulative plastic strain rate in the damaged and undamaged configurations, i.e., $\dot{p}$ and $\dot{\hat{p}}$, repectively, can be obtained by applying the chain rule and considering Eq. (2.229) and Table 2.4 as follows:

$$\dot{p} = \frac{\dot{\hat{p}}}{1 - \bar{r}}\left(1 - k_1 \hat{\bar{R}}\right) \tag{2.241}$$

The Laplacian of $\dot{p}$, i.e., $\nabla^2\dot{p}$ can be then obtained as follows:

$$\nabla^2\dot{p} = \left[\frac{\nabla^2\dot{\hat{p}}}{1-\bar{r}} + \frac{\nabla^2 r}{(1-\bar{r})^2}\dot{\hat{p}}\right]\left(1-k_1\hat{\bar{R}}\right)\frac{-\dot{\hat{p}}}{1-\bar{r}}k_1\nabla^2\hat{R} \tag{2.242}$$

$\nabla^2\hat{R}$ can be obtained by considering the relation $\dot{R}_n^g = (b_1/a_1)\nabla^2\dot{R}_n$ (see Voyiadjis et al., 2004) along with Eq. (2.231) as follows:

$$\nabla^2\hat{R} = \frac{a_1}{b_1}\hat{R}^g + \frac{\nabla^2 r}{1-\bar{r}}\hat{R} \tag{2.243}$$

Voyiadjis et al. (2004) proposed a Perzyna-type viscoplastic model (Perzyna, 1963, 1966) as follows:

$$\dot{\bar{\lambda}}^{vp} = \frac{1}{\eta^{vp}}\left\langle\frac{\hat{\sigma}_v}{\hat{\sigma}_{yp}+\hat{\bar{R}}}\right\rangle^{m_1} \tag{2.244}$$

where $\dot{\bar{\lambda}}^{vp} = \dot{\hat{\bar{p}}}$ is the nonlocal viscoplastic multiplier, m_1 is the rate-sensitivity parameter, and η^{vp} is the material viscosity. The dynamic yield surface can be obtained using Eqs. (2.234), (2.244) as follows:

$$f = \sqrt{\frac{3}{2}\left(\hat{s}_{ij}-\hat{\bar{\alpha}}_{ij}\right)\left(\hat{s}_{ij}-\hat{\bar{\alpha}}_{ij}\right)} - \left[\hat{\sigma}_{yp}(T)+\hat{\bar{R}}\left(\hat{p},\nabla^2\hat{p},T\right)\right]\left[1+\left(\eta^{vp}\dot{\hat{\bar{p}}}\right)^{\frac{1}{m_1}}\right] \equiv 0 \tag{2.245}$$

Eq. (2.245) is the generalization of the Kuhn-Tucker loading/unloading conditions defined in Eq. (2.21).

The isotropic and kinematic hardening rules can be obtained using Table 2.6 and Eqs. (2.227), (2.231), (2.232), (2.241), (2.242), (2.243) as follows:

$$\dot{\hat{\bar{R}}} = \dot{\hat{R}} + \dot{\hat{R}}^g \tag{2.246}$$

$$\dot{\hat{R}} = \frac{a_1\vartheta}{(1-\bar{r})^2}\left(1-k_1\hat{\bar{R}}\right)\dot{\hat{p}} \tag{2.247}$$

$$\dot{\hat{R}}^g = \left\{\frac{\vartheta}{(1-\bar{r})^3}\left[b_1\left(1-2k_1\hat{R}-k_1\hat{R}^g\right)\nabla^2 r - k_1 a_1(1-\bar{r})\hat{R}^g\right]\right\}\dot{\hat{p}} + \left[\frac{b_1\vartheta}{(1-\bar{r})^2}\left(1-k_1\hat{\bar{R}}\right)\right]\nabla^2\dot{\hat{p}} \tag{2.248}$$

$$\dot{\hat{\bar{\alpha}}}_{ij} = \dot{\hat{\alpha}}_{ij} + \dot{\hat{\alpha}}_{ij}^g \tag{2.249}$$

$$\dot{\hat{\alpha}}_{ij} = \overline{M}_{ikjl}\overline{M}_{mknl}\left(a_2\dot{\hat{\varepsilon}}_{mn}^p - k_2 a_2\dot{\hat{p}}\hat{\bar{\alpha}}_{mn}\right)\vartheta \tag{2.250}$$

$$\begin{aligned}\dot{\hat{\alpha}}_{ij}^g = {} & \overline{M}_{ikjl}\overline{M}_{mknl}\left[b_2\nabla^2\dot{\hat{\varepsilon}}_{mn}^p - k_2 b_2\nabla^2\dot{\hat{p}}\hat{\bar{\alpha}}_{mn} - k_2\left(a_2\hat{\alpha}_{mn}^g - b_2\overline{M}_{pmqn}\nabla^2 M_{prqs}^{-1}\hat{\alpha}_{rs}\right)\dot{\hat{p}}\right]\vartheta \\ & + \overline{M}_{ikjl}\nabla^2\overline{M}_{mknl}\left(b_2\dot{\hat{\varepsilon}}_{mn}^p - k_2 b_2\dot{\hat{p}}\hat{\bar{\alpha}}_{mn}\right)\vartheta\end{aligned} \tag{2.251}$$

where

$$\nabla^2 \hat{\dot{\varepsilon}}_{ij}^{p} = \nabla^2 \hat{\dot{p}} \frac{\partial f}{\partial \hat{\sigma}_{ij}} \tag{2.252}$$

Using a similar methodology, Voyiadjis et al. (2004) presents the dynamic damage surface g as follows:

$$g = \sqrt{\left(\overline{Y}_{ij} - \overline{H}_{ij}\right)\left(\overline{Y}_{ij} - \overline{H}_{ij}\right)} - \left[l + \overline{K}\right]\left[1 + \left(\eta^{vd}\dot{\overline{\lambda}}^{vd}\right)^{\frac{1}{m_2}}\right] \equiv 0 \tag{2.253}$$

where m_2 is the viscodamage rate sensitivity parameter, and η^{vd} is the corresponding viscosity. The relation $l = l_0\vartheta$ is the initial damage threshold where l_0 is the initial damage threshold at $T = 0$ and zero damage strain and strain rate. The Kuhn-Tucker loading/unloading conditions then govern the viscodamage model as follows:

$$\dot{\lambda}^{vd} g = 0, \dot{\lambda}^{vd} \geq 0, g \leq 0 \tag{2.254}$$

The damage hardening rules can be obtained using the similar methodology incorporated to obtain the viscoplasticity hardening models as follows:

$$\dot{\overline{K}} = \dot{K} + \dot{K}^g \tag{2.255}$$

$$\dot{K} = a_3\vartheta\dot{\lambda}^{vd}\left(1 - h_1\overline{K}\right) \tag{2.256}$$

$$\dot{K}^g = \left[b_3\left(1 - h_1\overline{K}\right)\nabla^2\dot{\lambda}^{vd} - h_1 a_3 K^g \dot{\lambda}^{vd}\right]\vartheta \tag{2.257}$$

$$\dot{\overline{H}}_{ij} = \dot{H}_{ij} + \dot{H}_{ij}^g \tag{2.258}$$

$$\dot{H}_{ij} = \left[a_4\,{}^{2}\dot{\phi}_{ij} - h_2 a_4 \dot{\lambda}^{vd}\overline{H}_{ij}\right]\vartheta \tag{2.259}$$

$$\dot{H}_{ij}^g = \left\{b_4\nabla^2\left[{}^{2}\dot{\phi}_{ij}\right] - h_2 b_4 \overline{H}_{ij}\nabla^2\dot{\lambda}^{vd} - h_2 a_4 \dot{\lambda}^{vd} H_{ij}^g\right\}\vartheta \tag{2.260}$$

where ${}^{2}\dot{\phi}_{ij}$ is defined in Eq. (2.218), and $\nabla^2\left[{}^{2}\dot{\phi}_{ij}\right]$ can be defined as follows:

$$\nabla^2\left[{}^{2}\dot{\phi}_{ij}\right] = \nabla^2\dot{\lambda}^{vd}\frac{\partial g}{\partial Y_{ij}} \tag{2.261}$$

The final step is to obtain the nonlocal viscodamage force tensor $\overline{Y}_{ij}$ using Eq. (2.227) as follows:

$$\overline{Y}_{ij} = Y_{ij} + Y_{ij}^g \tag{2.262}$$

where Y_{ij} and Y_{ij}^g can be obtained using Table 2.5 as follows:

$$Y_{ij} = -\left[\frac{1}{2}\varepsilon_{rs}^{e} C_{mnkl}\varepsilon_{kl}^{e} - \varepsilon_{rs}^{e}\beta_{mn}\left(T - T_r\right)\right]\overline{M}_{manb} J_{arbsij} \tag{2.263}$$

$$Y_{ij}^g = -a\left[\frac{1}{2}\varepsilon_{rs}^{e} C_{mnkl}\varepsilon_{kl}^{e} - \varepsilon_{rs}^{e}\beta_{mn}\left(T - T_r\right)\right]\overline{M}_{manb} J_{arbsij} \tag{2.264}$$

where J_{arbsij} is a six-order constant tensor which can be defined as follows:

$$J_{arbsij} = -\frac{\partial \overline{M}_{arbs}^{-1}}{\partial \overline{\phi}_{ij}} = \frac{1}{2}\left(\delta_{ar}\delta_{bi}\delta_{sj} + \delta_{ai}\delta_{rj}\delta_{bs}\right) \tag{2.265}$$

The nonlocal viscodamage force tensor $\overline{Y}_{ij}$ can be obtained by substituting Eqs. (2.263), (2.264) into Eq. (2.262) as follows:

$$\overline{Y}_{ij} = -(1+a)\left[\frac{1}{2}\varepsilon_{rs}^{e} C_{mnkl}\varepsilon_{kl}^{e} - \varepsilon_{rs}^{e}\beta_{mn}(T - T_r)\right]\overline{M}_{manb}J_{arbsij} \tag{2.266}$$

Voyiadjis and his co-workers (Voyiadjis and Deliktas, 2009a,b; Voyiadjis and Faghihi, 2012; Song and Voyiadjis, 2018) evolved their gradient models and presented a family of thermodynamically consistent higher order SGP theories using the decomposition of the state variables into energetic and dissipative components. The model also includes the thermal effects to investigate the behavior of small-scale metallic materials. Furthermore, the effect of interface energy is incorporated into their strain gradient plasticity formulation to address various boundary conditions, strengthening and formation of the boundary layers.

2.3.2 Integral-type nonlocal plasticity models

The integral-type nonlocal plasticity models are elaborated in this section. Accordingly, the localization problem is addressed here using nonlocal models.

2.3.2.1 *Integral-type nonlocal softening models*

The simple local von Mises yield surface can be defined as follows:

$$f\left(\sigma_{ij}, \varepsilon_{ij}^{p}\right) = \sqrt{\frac{3}{2}\left(s_{ij} - \alpha_{ij}\right)\left(s_{ij} - \alpha_{ij}\right)} - \sigma_y(k) \tag{2.267}$$

In Eq. (2.267), $\sigma_y(k)$, i.e., the material yield stress, govern the softening in material as a linear function of nonlocal softening variable as follows:

$$\sigma_y = \sigma_0 + Hk \tag{2.268}$$

where H is commonly termed as the plastic modulus which is negative for material softening. The softening variable can be obtained as $k = \varepsilon_p$, where ε_p is the equivalent plastic strain, according to the consistency equation and Kuhn-Tucker conditions as described in Section 2.2.5.2. The nonlocal version of softening constitutive law described in Eqs. (2.267), (2.268) can be obtained by substituting the softening variable k by its nonlocal counterpart $\overline{k}$ in Eq. (2.268) as follows:

$$\sigma_y = \sigma_0 + H\overline{k} \tag{2.269}$$

Where $\overline{k} = \overline{\varepsilon}_p$, i.e., the nonlocal equivalent plastic strain, which can be described as follows:

$$\bar{k}(x_k) = \bar{\varepsilon}_p(x_k) = \frac{1}{\Psi(x_k)} \int_\Omega \psi(y_k; x_k) \varepsilon_p(y_k) \mathrm{d}\Omega \tag{2.270}$$

where $\psi(y_i; x_i)$ is the weighting function, and $\Psi(x_i)$ can be described as follows:

$$\Psi(x_k) = \int_\Omega \psi(y_k; x_k) \mathrm{d}\Omega \tag{2.271}$$

The Gaussian function is usually selected as the weighting function, which was described in Eq. (2.102).

Vermeer and Brinkgreve (1994) improved the simple linear nonlocal softening model by incorporating both local and nonlocal softening variable as follows:

$$\breve{k} = m\bar{k} + (1 - m)k \tag{2.272}$$

And the softening model can be defined as follows:

$$\sigma_y = \sigma_0 + H\breve{k} \tag{2.273}$$

Finally, the nonlinear softening model can be incorporated instead of its linear version as follows (Jirasek and Rolshoven, 2003):

$$\sigma_y = \sigma_0 + h\left(\breve{\mathrm{k}}\right) \tag{2.274}$$

where $h\left(\breve{k}\right)$ is the nonlinear softening law. Jirasek and Rolshoven (2003) investigated different types of nonlocal softening models described in Eqs. (2.269), (2.273), (2.274) and analyzed the localization in a one-dimensional case study.

2.3.2.2 *Integral-type nonlocal Gurson model*

Gurson (1977) presented a constitutive model to capture the damage induced by void growth in porous metals. The yield function can be defined as follows:

$$f\left(\sigma_{ij}, \sigma_y, p\right) = J_2 - \frac{1}{3}\left[1 + p^2 - 2p \cosh\left(\frac{\sigma_{kk}}{2\sigma_y}\right)\right]\sigma_y^2 \equiv 0 \tag{2.275}$$

where J_2 is the second stress deviator tensor invariant, p is the material porosity, and σ_y is the matrix yield strength in the absence of voids, which can be defined as follows:

$$\sigma_y = \sigma_0 + k \tag{2.276}$$

where σ_0 is the matrix initial yield surface and k defines the isotropic hardening. Gurson (1977) incorporated an associated flow rule, and plastic strain rate tensor can be defined using Eq. (2.20).

The Gurson model relates the softening to the porosity p which can be described as a function of plastic strain rate tensor $\dot{\varepsilon}_{ij}^p$ as follows:

$$\dot{p} = (1 - p)\dot{\varepsilon}_{kk}^p \tag{2.277}$$

Tvergaard (1981) incorporated and modified the Gurson yield surface for finite strain plasticity. The small strain version of the modified model can be described as follows:

$$f(\sigma_{ij}, \sigma_y, p) = \frac{3J_2}{\sigma_y^2} + 2pq_1 \cosh\left(\frac{q_2\sigma_{kk}}{2\sigma_y}\right) - 1 - p^2 q_3 \equiv 0 \tag{2.278}$$

where q_1, q_2, and q_3 are material constants. The modified Gurson model, i.e., Eq. (2.278), reproduces the original one, i.e., Eq. (2.275), for $q_1 = q_2 = q_3 = 1$.

Leblond et al. (1994) incorporated the modified Gurson yield surface by considering $q_1 = q$, $q_2 = 1$, and $q_3 = q^2$ and presented the integral-type nonlocal Gurson model (Gurson, 1977) by substituting the local porosity p with its nonlocal counterpart $\bar{p}$, which can be described as follows:

$$\dot{\bar{p}} = \frac{1}{\Psi(x_k)} \int_{\Omega} \psi(y_k; x_k)[1 - p(y_k)]\dot{\varepsilon}_{kk}^{p}(y_k)\mathrm{d}\Omega \tag{2.279}$$

where $\psi(y_i; x_i)$ is the weighting function, and $\Psi(x_i)$ is described in Eq. (2.271). Leblond et al. (1994) investigated some bifurcation phenomena using the presented models.

2.3.2.3 *Integral-type nonlocal plastic model: Bazant and Lin*

Bazant and Lin (1988) presented a nonlocal plasticity framework to capture the behavior of grouted soil. Instead of fully-nonlocal continuum model including higher order stresses and strains, Bazant and Lin (1988) incorporated the local equilibrium equations and applied the nonlocal concept just on the plastic strain tensor as follows:

$$\dot{\bar{\varepsilon}}_{ij}^{p}(x_k) = \frac{1}{\Psi(x_k)} \int_{\Omega} \psi(y_k; x_k)\dot{\varepsilon}_{ij}^{p}(y_k)\mathrm{d}\Omega \tag{2.280}$$

where $\dot{\bar{\varepsilon}}_{ij}^{p}$ and $\dot{\varepsilon}_{ij}^{p}$ are the nonlocal and local plastic strain rate tensors, respectively, $\psi(y_i; x_i)$ is the weighting function, and $\Psi(x_i)$ is described in Eq. (2.271). The local plastic strain rate tensor $\dot{\varepsilon}_{ij}^{p}$ can be described using the considered flow rule using Eq. (2.18). Next, the nonlocal plastic strain rate tensor $\dot{\bar{\varepsilon}}_{ij}^{p}$ can be obtained using Eq. (2.280). The relation between the stress and strain rates then can be defined using Hooke's law as follows:

$$\dot{\sigma}_{ij} = C_{ijkl}\left(\dot{\varepsilon}_{kl} - \dot{\bar{\varepsilon}}_{ij}^{p}\right) \tag{2.281}$$

Basically, the yield surface, flow rule, and consistency equations are still applicable. Bazant and Lin (1988) incorporated the Mohr-Coulomb yield surface along with associated flow rule to model the material behavior of grouted soil.

2.4 Finite strain plasticity: Local models

The kinematics and stress definition incorporated in small strain continuum mechanics are not applicable to the finite strain problems. Furthermore, many assumptions used in small strain problem are not valid in those of finite strain such as the additive decomposition of strains. Accordingly, the appropriate kinematics and stress definitions should be developed for finite strain problems. In this section, first, the kinematics of finite strain problem are elaborated. The concepts of material and spatial descriptions, deformation gradient, strain, polar decomposition, velocity and material time derivatives, and strain objectivity are described. Next, the concepts of equilibrium and stress measures are investigated for the finite strain analysis. This section includes the description of Cauchy stress tensor, equilibrium, principal of virtual work, Kirchhoff stress tensor, first and second Piola-Kirchhoff stress tensors, stress objectivity, and stress rate. The next part presents the finite strain hyperelasticity including the general aspects of the model, material objectivity, isotropic hyperelasticity models, and specific free energy function.

2.4.1 Kinematics

Kinematics is the investigation of the motion without considering the forces and stresses inducing that motion. In the case of small strain continuum mechanics, the difference between the undeformed and deformed bodies is negligible and one description is enough to capture the motion of the body. However, in the case of finite strain continuum mechanics, two descriptions related to the undeformed and deformed body should be introduced, which is commonly termed as material (Lagrangian) and spatial (Eulerian) descriptions, respectively. Accordingly, unlike the small strain theory in which one strain measure is defined, different strain measures can be defined in the case of finite strain theory. In this section, the concepts of the material and spatial descriptions, deformation, strain, their corresponding rates, and objectivity will be elaborated.

2.4.1.1 Material and spatial description

The major difference between the small and large strain problems is that in the case of the small strain problem, the difference between the deformed and undeformed configurations is negligible. Accordingly, one coordinate system is sufficient to describe the kinematics. In the case of finite strain problem, however, the difference is considerable. Accordingly, two coordinate systems of undeformed (material) and deformed (spatial) should be defined, which are generally termed as the Lagrangian and Eulerian descriptions, respectively. The material and spatial coordinate systems are shown in Fig. 2.9, which are denoted by X_I and x_i, respectively. The Cartesian basis of material and spatial coordinate systems are denoted by E_I and e_i. In the rest this chapter, the two coordinate systems are considered to be coincident. One should note that the spatial coordinate

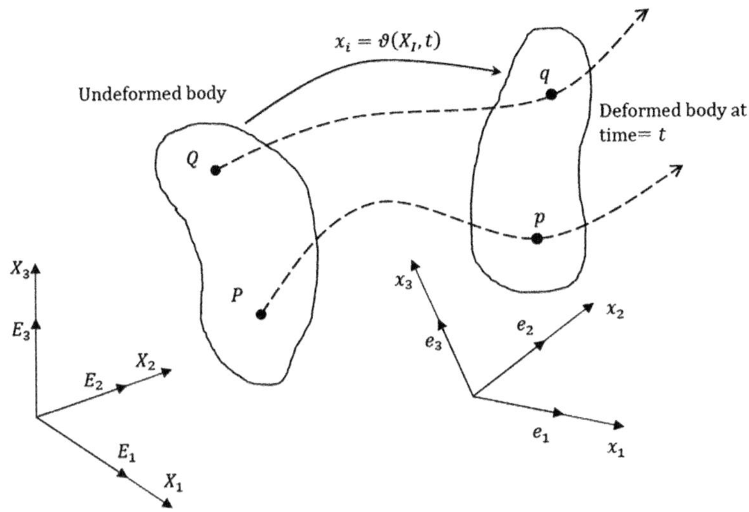

FIG. 2.9 Motion description of a deformable body in finite strain framework.

system becomes the material one at time $t = 0$. The relation between the spatial and material positions is described using the mapping ϑ as follows:

$$x_i = \vartheta(X_I, t), \tag{2.282}$$

Eq. (2.282) is commonly substituted using the following relation:

$$x_i = x_i(X_I, t), \tag{2.283}$$

Unlike the small strain framework that the displacement vector is infinitesimal compared to the size of the body, there is no constraint for finite strain continuum mechanics.

The variables can be defined based on either of Lagrangian or Eulerian descriptions. For example, a scalar variable Ξ can be defined as follows:

$$\text{Lagrangian (material) description}: \Xi = \Xi(X_I, t) \tag{2.284a}$$

$$\text{Eulerian (spatial) description}: \Xi = \Xi(x_i, t) \tag{2.284b}$$

Eq. (2.284a) controls the variation of Ξ for a particle initially located at X_I in time, while Eq. (2.284b) governs the variation of Ξ for a spatial position x_i in time.

2.4.1.2 *Deformation gradient*

The deformation gradient F_{iJ} is the most basic quantity of finite strain continuum mechanics, which describes the change in the relative position of two neighboring points during deformation. Consider material points P and Q in the undeformed body as shown in Fig. 2.9. The relative position of P and Q can be described as follows:

$$dX_I = X_I^Q - X_I^P \tag{2.285}$$

After deformation, the new positions of P and Q are p and q, respectively. Accordingly, the new relative position can be defined as follows:

$$dx_i = x_i^q - x_i^p = x_i\left(X_I^P + dX_I, t\right) - x_i\left(X_I^P, t\right) \tag{2.286}$$

The term $x_i(X_I^P + dX_I, t)$ can be approximated by considering the first two terms of the Taylor expansion as follows,

$$x_i\left(X_I^P + dX_I, t\right) \approx x_i\left(X_I^P, t\right) + \frac{\partial x_i}{\partial X_J} dX_J \tag{2.287}$$

The deformation gradient F_{iJ} is then defined as follows:

$$F_{iJ} = \frac{\partial x_i}{\partial X_J} \tag{2.288}$$

Accordingly, Eq. (2.286) can be rewritten as follows:

$$dx_i = F_{iJ} dX_J \tag{2.289}$$

The deformation gradient is a two-point tensor which relates a vector in the undeformed body to that of the deformed.

The operators which define the relation between Lagrangian and Eulerian quantities are commonly called the push forward and pull back operations. For example, in the case of Eq. (2.289), dx_i is the push forward of dX_j, and commonly stated as follows:

$$dx_i = \vartheta_*(dX_I)\delta_{Ii} = F_{iJ} dX_J \tag{2.290}$$

On the other hand, dX_j can be defined as the pull back of dx_i as follows:

$$dX_I = \vartheta_*^{-1}(dx_i)\delta_{Ii} = F_{Ij}^{-1} dx_j \tag{2.291}$$

where δ_{ij} is the Kronecker delta to make the indices consistent.

2.4.1.3 *Polar decomposition*

The deformation gradient F_{iJ} transforms a vector from undeformed body to that of deformed one. It can be decomposed to a pure rotation and stretch, which is commonly termed as polar decomposition. Depending on which one of transformation comes first, the polar decomposition can be applied in two ways as follows:

$$F_{iJ} = R_{iK} U_{KJ} \tag{2.292a}$$

$$F_{iJ} = V_{ik} R_{kJ} \tag{2.292b}$$

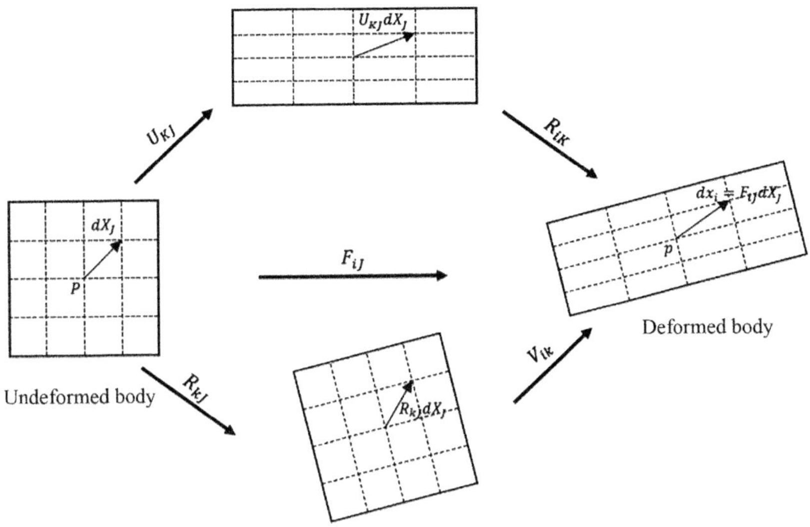

FIG. 2.10 Polar decomposition: describing the deformation gradient as a combination of rotation and stretch.

where R_{iJ} is a rotation tensor, and U_{KJ} and V_{ik} are the right and left (material and spatial) stretch tensors, respectively. Eq. (2.292a) describes the deformation gradient F_{iJ} by a stretch U_{KJ} followed by a rotation R_{iK}, while Eq. (2.292b) describes it by a rotation R_{kJ} followed by a stretch V_{ik} (Fig. 2.10). The rotation tensor R_{iJ} is a proper orthogonal transformation, i.e., $R_{iI}R_{iJ} = \delta_{IJ}$ and det $(R_{iJ}) = 1$.

The relation between the right and left stretch tensors can be defined as follows:

$$V_{ij} = R_{iK}U_{KL}R_{jL} \tag{2.293}$$

The right and left stretch tensors, i.e., U_{IJ} and V_{ij}, respectively, have the same set of eigenvalues $\{\lambda_1, \lambda_2, \lambda_3\}$. However, their eigenvectors are not the same, and can be related as follows:

$$n_i^\alpha = R_{iJ}N_J^\alpha \tag{2.294}$$

where $\boldsymbol{N}_I^\alpha$ and $\boldsymbol{n}_i^\alpha$ are the eigenvectors of U_{IJ} and V_{ij}, and $\alpha = 1, 2, 3$ for a 3-D problem. Accordingly, the material and spatial stretch tensors can be described using their eigenvalues and eigenvectors as follows:

$$U_{IJ} = \sum_{\alpha=1}^{3} \lambda_\alpha N_I^\alpha N_J^\alpha \tag{2.295a}$$

$$V_{ij} = \sum_{\alpha=1}^{3} \lambda_\alpha n_i^\alpha n_j^\alpha \tag{2.295b}$$

The eigenvalues $\{\lambda_1, \lambda_2, \lambda_3\}$ can be physically described. To do so, dx_i and dX_I can be described in the principal directions, n_i^1 and N_I^1, respectively, as follows:

$$dx_i = dl_1 n_i^1 \tag{2.296a}$$

$$dX_I = dL_1 N_I^1 \tag{2.296b}$$

The relation between dx_i and dX_I can be stated as follows:

$$dx_i = F_{iJ} dX_J = R_{iK} U_{KJ} dL_1 N_J^1 \xrightarrow{U_{KJ} N_J^1 = \lambda_1 N_K^1} dx_i = \lambda_1 R_{iK} N_K^1 dL_1 = \lambda_1 n_i^1 dL_1 \tag{2.297}$$

Comparing Eqs. (2.296a), (2.297), λ_1 can be defined as the ratio between the lengths in the deformed and undeformed bodies as follows:

$$\lambda_1 = \frac{dl_1}{dL_1} \tag{2.298}$$

The deformation gradient can also be stated in terms of eigenvalues and eigenvectors of U_{IJ} and V_{ij} as follows:

$$F_{iJ} = \sum_{\alpha=1}^{3} \lambda_\alpha n_i^\alpha N_J^\alpha \tag{2.299}$$

Accordingly, $\det(F_{iJ}) = \lambda_1 \lambda_2 \lambda_3 = J$, which is the Jacobian of the deformation gradient tensor. It can be shown that the ratio of the elemental volume in Eulerian description to that of the Lagrangian one is equal to the Jacobian of the deformation gradient tensor, i.e., $dv/dV = J$.

2.4.1.4 *Strain measures*

Rigid rotation of a body does not induce any stresses. Accordingly, kinematic quantities related to any strain measure should exclude the rigid rotation effects. Right Cauchy-Green tensor C_{IJ} is one of the deformation tensors which excludes rigid rotation as follows:

$$C_{IJ} = F_{kI} F_{kJ} = U_{KI} U_{KJ} \tag{2.300}$$

C_{IJ} can be physically described as the change in the square of relative position vector in an undeformed body to that of a deformed one as follows:

$$dx_i dx_i = dX_I C_{IJ} dX_J \tag{2.301}$$

One can also define the Right Cauchy-Green tensor C_{IJ} using the eigenvalues and eigenvectors of the material stretch tensor U_{IJ}, as described in Eq. (2.295a), as follows:

$$C_{IJ} = \sum_{\alpha=1}^{3} \lambda_\alpha^2 N_I^\alpha N_J^\alpha \tag{2.302}$$

Invariants of the right Cauchy-Green tensor C_{IJ} can be incorporated to define the constitutive material model of materials in finite strain continuum mechanics, which are:

$$I_1(C_{IJ}) = C_{KK} = \lambda_1^2 + \lambda_2^2 + \lambda_3^2 \tag{2.303}$$

$$I_2(C_{IJ}) = \lambda_1^2\lambda_2^2 + \lambda_2^2\lambda_3^2 + \lambda_3^2\lambda_1^2 \tag{2.304}$$

$$I_3(C_{IJ}) = \det(C_{IJ}) = \lambda_1^2\lambda_2^2\lambda_3^2 = J^2 \tag{2.305}$$

Left Cauchy-Green, also known as Finger, deformation tensor B_{ij} can be introduced using a similar framework as follows:

$$B_{ij} = F_{iK}F_{jK} = V_{ik}V_{jk} \tag{2.306}$$

The left Cauchy-Green deformation tensor B_{ij} can be defined using the eigenvalues and eigenvectors of spatial stretch tensor V_{ij}, as described in Eq. (2.295a), as follows:

$$B_{ij} = \sum_{\alpha=1}^{3} \lambda_\alpha^2 n_i^\alpha n_j^\alpha \tag{2.307}$$

One should note that the invariants of the left Cauchy-Green deformation tensor are the same as the right one described in Eqs. (2.303)–(2.305).

The scalar product of relative position vector can be considered as a measure of strain. However, similar to other quantities in finite strain continuum mechanics, the strain can be described in a material or spatial descriptions. In the case of material description, the strain E_{IJ} can be described as follows:

$$\frac{1}{2}(dx_i dx_i - dX_I dX_I) = dX_I E_{IJ} dX_J \tag{2.308}$$

Accordingly, the Lagrangian strain E_{IJ}, which is also termed as Green strain tensor, can be obtained as follows:

$$E_{IJ} = \frac{1}{2}(C_{IJ} - \delta_{IJ}) \tag{2.309}$$

In the case of the spatial description, the strain e_{ij} can be described as follows:

$$\frac{1}{2}(dx_i dx_i - dX_I dX_I) = dx_i e_{ij} dx_j \tag{2.310}$$

Rewriting Eq. (2.310), the Eulerian strain tensor e_{ij}, which is also termed as the Almansi strain tensor, can be obtained as follows:

$$e_{ij} = \frac{1}{2}\left(\delta_{ij} - B_{ji}^{-1}\right) \tag{2.311}$$

The Almansi strain tensor e_{ij} can be defined as the push forward of the Green strain tensor E_{KL} as follows:

$$e_{ij} = \vartheta_*(E_{IJ}) = F_{Ki}^{-1} E_{KL} F_{Lj}^{-1} \tag{2.312}$$

On the other hand, the Green strain tensor E_{IJ} can be described as the pull back of the Almansi strain tensor e_{kl} as shown below:

$$E_{IJ} = \vartheta_*^{-1}(e_{ij}) = F_{kI} e_{kl} F_{lJ} \tag{2.313}$$

It was later shown by Seth (1961, 1962) and Hill (1968) that a more general form of the material and spatial strain can be defined as follows:

$$E_{IJ}^{(m)} = \frac{1}{2m}\left(C_{IJ}^{m} - \delta_{IJ}\right); \tag{2.314a}$$

$$e_{ij}^{(n)} = \frac{1}{2n}\left(\delta_{ij} - B_{ij}^{n}\right); \tag{2.314b}$$

The Green strain tensor is a special case of Eq. (2.314a) for $m = 1$ and Eq. (2.314b) for $n = -1$, while the Almansi strain tensor is a special case of Eq. (2.314a) for $m = -1$ and Eq. (2.314b) for $n = 1$ respectively. The Biot strain tensor can be obtained for $m = 1/2$ as follows:

$$E_{IJ}^{(1/2)} = C_{IJ}^{1/2} - \delta_{IJ} \tag{2.315}$$

The material and spatial logarithmic strain can be obtained by substituting $m = 0$ and $n = 0$, respectively, as follows:

$$E_{IJ}^{(0)} = \frac{1}{2}(\ln C_{IJ}) = \sum_{\alpha=1}^{3} \ln \lambda_\alpha N_I^\alpha N_J^\alpha \tag{2.316a}$$

$$e_{ij}^{(0)} = \frac{1}{2}(\ln B_{IJ}) = \sum_{\alpha=1}^{3} \ln \lambda_\alpha n_i^\alpha n_j^\alpha \tag{2.316b}$$

One can relate the change in the elemental lengths squared using the Green and Almansi strain tensors. To do so, dx_i and dX_I can be defined as follows:

$$dx_i = dln_i \tag{2.317}$$

$$dX_I = dLN_I \tag{2.318}$$

where $N_I = dX_I/dL$ and $n_i = dx_i/dl$ are the unite vectors in the directions of dX_I and dx_i, respectively. Considering $dl^2 = dx_i dx_i$ and $dL^2 = dX_I dX_I$, Eq. (2.308) can be rewritten as follows:

$$\frac{(dl^2 - dL^2)}{dL^2} = N_I E_{IJ} N_J \tag{2.319}$$

Using the same approach, Eq. (2.310) can be rewritten as follows:

$$\frac{(dl^2 - dL^2)}{dl^2} = n_i e_{ij} n_j \tag{2.320}$$

2.4.1.5 Velocity

Similar to any other time-dependent phenomena, the quantities in finite strain continuum may be time dependent. Accordingly, they can be differentiated with respect to time. The first important time derivative quantity is the velocity v_i. The velocity can be described as follows:

$$v_i(X_I, t) = \frac{\partial x_i(X_I, t)}{\partial t} \tag{2.321}$$

One should note that the velocity v_i is a spatial vector. Accordingly, it is more appropriate to define it using the spatial position x_i as follows:

$$v_i(x_i, t) = \frac{\partial x_i\left[\vartheta^{-1}(x_I, t), t\right]}{\partial t} \tag{2.322}$$

where X_I is substituted by $\vartheta^{-1}(x_I, t)$ according to Eq. (2.282).

2.4.1.6 Material time derivative

Time derivative can be also applied to quantities described in the material description. Assume $G(X_I, t)$ is a quantity in the material description which follows the value of function for a specific material point with position X_i at $t = 0$ as the time varies. Accordingly, the material time derivative of $G(X_i, t)$ can be defined as follows:

$$\dot{G} = \frac{\partial G(X_I, t)}{\partial t} \tag{2.323}$$

The variation of an Eulerian quantity $g(x_i, t)$ with time is more complicated since the spatial position itself is a time-dependent vector. One can write the material derivative of $g(x_i, t)$ as follows:

$$\dot{g} = \frac{\partial g(x_i, t)}{\partial t} + \frac{\partial g(x_i, t)}{\partial x_i}\frac{\partial x_i(X_I, t)}{\partial t} = \frac{\partial g(x_i, t)}{\partial t} + \frac{\partial g(x_i, t)}{\partial x_i} v_i \tag{2.324}$$

The second term which incorporates the velocity vector v_i is commonly called the convective derivative.

2.4.1.7 Velocity gradient

The derivative of velocity with respect to the spatial coordinates is the velocity gradient tensor l_{ij}, which can be stated as follows:

$$l_{ij} = \frac{\partial v_i(x_i, t)}{\partial x_j} \tag{2.325}$$

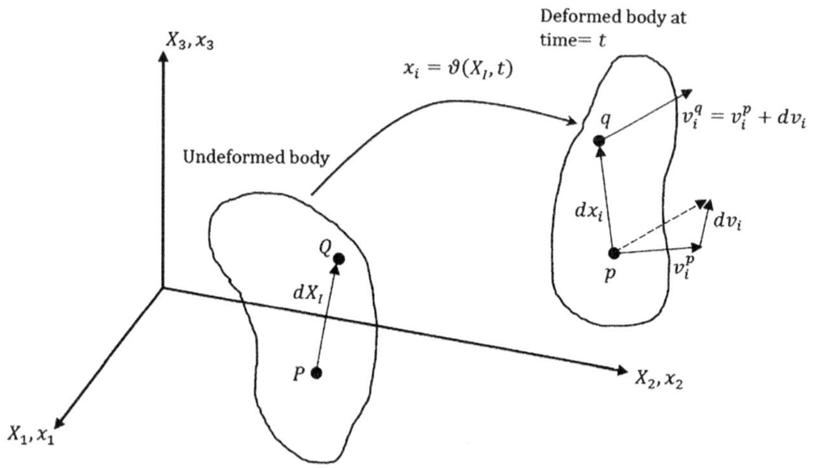

FIG. 2.11 The physical description of velocity gradient tensor.

The velocity gradient tensor l_{ij} is a spatial tensor. The physical meaning of velocity gradient tensor l_{ij} can be described using Fig. 2.11 as the relative velocity of a particle which is at point q of the deformed body with respect to that of a point p, i.e., dv_i, divided by their relative position vector dx_j. The time derivative of deformation gradient can be described using the velocity gradient tensor l_{ij} as follows:

$$\dot{F}_{iJ} = \frac{\partial v_i}{\partial X_J} = \frac{\partial v_i}{\partial x_k}\frac{\partial x_k}{\partial X_J} = l_{ik}F_{kJ} \tag{2.326}$$

Accordingly, the velocity gradient tensor l_{ij} can be described using the deformation gradient and its time derivative as follows:

$$l_{ij} = \dot{F}_{iJ}F_{Jj}^{-1} \tag{2.327}$$

2.4.1.8 *Rate of deformation*

Strain rate can be described as the rate of change in the square of relative position vector. To do so, first, one should obtain the rate of the right Cauchy-Green tensor $\dot{C}_{IJ}$ by applying the time derivative on Eq. (2.301) as follows:

$$\frac{d(dx_i dx_i)}{dt} = dX_I \dot{C}_{IJ} dX_J \tag{2.328}$$

The relation between the rate of the right Cauchy-Green tensor $\dot{C}_{IJ}$ and the Lagrangian strain rate tensor $\dot{E}_{IJ}$ can then be obtained as follows:

$$E_{IJ} = \frac{1}{2}(C_{IJ} - \delta_{IJ}) \rightarrow \dot{E}_{IJ} = \frac{1}{2}\dot{C}_{IJ} \tag{2.329}$$

The Lagrangian strain rate tensor can be rewritten by substituting Eq. (2.300) into Eq. (2.329) as follows:

$$\dot{E}_{IJ} = \frac{1}{2}\dot{C}_{IJ} = \frac{1}{2}\left(\dot{F}_{KI}F_{KJ} + F_{KI}\dot{F}_{KJ}\right) \tag{2.330}$$

It is common that instead of using the Lagrangian strain rate tensor $\dot{E}_{IJ}$, its spatial counterpart d_{ij} is incorporated as follows:

$$d_{ij} = \vartheta_*\left(\dot{E}_{IJ}\right) = F_{Ki}^{-1}\dot{E}_{KL}F_{Lj}^{-1} \tag{2.331}$$

Eq. (2.331) can be rewritten by incorporating Eqs. (2.326), (2.330) as follows:

$$d_{ij} = \frac{1}{2}\left(l_{ij} + l_{ji}\right) \tag{2.332}$$

Eq. (2.332) states that the rate of deformation tensor d_{ij} is the symmetric part of the velocity gradient tensor l_{ij}.

It should be noted that d_{ij} is the spatial counterpart of the Lagrangian strain rate tensor, which is not the same as the time derivative of the Eulerian strain tensor. The relation between d_{ij} and e_{ij} is commonly represented as follows:

$$d_{ij} = \mathcal{L}_v\left(e_{ij}\right) = \vartheta_*\left\{\frac{d\left[\vartheta_*^{-1}\left(e_{ij}\right)\right]}{dt}\right\} \tag{2.333}$$

where $\mathcal{L}_v(f)$ is the Lie time derivative of a spatial quantity f.

The rate of change in the elemental lengths squared in the deformed body can be related to d_{ij} as follows:

$$\frac{1}{2}\frac{d(dx_i dx_i)}{dt} = dx_i d_{ij} dx_j \tag{2.334}$$

Substituting Eq. (2.317) into Eq. (2.334), one can relate d_{ij} to the rate of extension of dl as follows:

$$\frac{1}{2}\frac{d(dl^2)}{dt} = dl^2 n_i d_{ij} n_j \rightarrow n_i d_{ij} n_j = \frac{1}{dl}\frac{d(dl)}{dt} = \frac{d\ \ln(dl)}{dt} \tag{2.335}$$

Eq. (2.335) states that d_{ij} defines the rate of the extension of dl. Accordingly, $d_{ij} = 0$ is the case of rigid body motion.

2.4.1.9 Spin tensor

The velocity gradient tensor l_{ij} can be additively decomposed to the symmetric and unsymmetric tensors as follows:

$$l_{ij} = d_{ij} + \omega_{ij} \tag{2.336}$$

where d_{ij} is the symmetric part of the velocity gradient tensor l_{ij} [Eq. (2.332)], while ω_{ij}, which is commonly termed as the spin tensor, is its asymmetric part, i.e., $\omega_{ij} = -\omega_{ji}$. The spin tensor ω_{ij} can be obtained by rewriting Eqs. (2.332), (2.336) as follows:

$$\omega_{ij} = l_{ij} - d_{ij} = \frac{1}{2}\left(l_{ij} - l_{ji}\right) \tag{2.337}$$

It can be proved that the spin tensor ω_{ij} can be related to the rotation tensor R_{ij} by substituting Eqs. (2.327), (2.292a) as follows (see Bonet and Wood, 2008):

$$\omega_{ij} = \dot{R}_{iK} R_{jK} \tag{2.338}$$

One can see that the spin tensor ω_{ij} is solely related to the rotation tensor R_{ij}. In other words, while the rate of deformation tensor d_{ij} is the representative of the contribution of stretching to the velocity gradient tensor l_{ij}, the spin tensor ω_{ij} is the contribution of rotary deformation.

2.4.1.10 *Strain objectivity*

The concept of objectivity in continuum mechanics states that the description of a physical phenomenon should be independent of the chosen coordinate system. Specifically, the effect of rigid body rotation is of interest. In other words, the rigid rotation should not change the quantities such as relative position vector, strain, and stress. To elaborate the concept of objectivity, a rigid rotation r_{ij} is applied to a deformed body as shown in Fig. 2.12. Accordingly, the relation

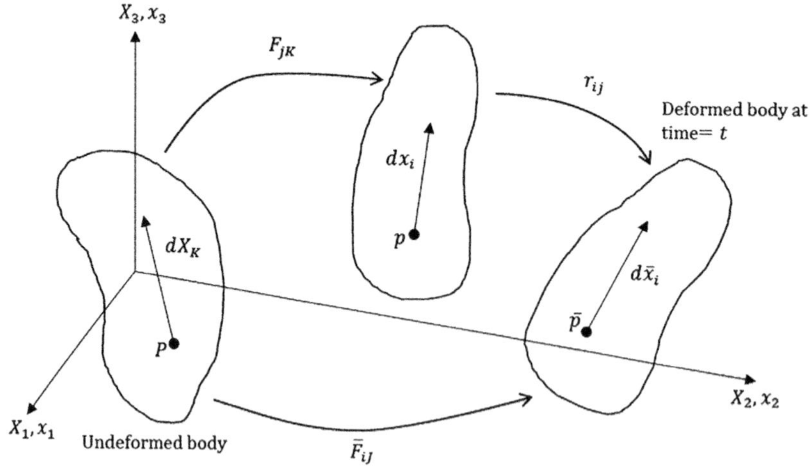

FIG. 2.12 Evaluating the concept of strain objectivity by applying a rigid rotation r_{ij} to a deformed body.

between the Eulerian and Lagrangian elemental vectors can be described as follows:

$$d\overline{x}_i = r_{ij} dx_j = r_{ij} F_{jK} dX_K \tag{2.339}$$

where dx_i and $d\overline{x}_i$ are the initial and rotated Eulerian elemental vectors, respectively, and dX_i is the Lagrangian elemental vectors. The objectivity of a vector means that its magnitude should be independent of rigid rotation. The Eulerian elemental vector dx_i is objective according to this definition. However, the velocity vector is not objective since the magnitude of v_i and $\overline{v}_i$ are not equal, which can be shown as follows:

$$\overline{v}_i = \frac{\partial \overline{x}_i}{\partial t} = r_{ij} v_i + \dot{r}_{ij} x_i \rightarrow |\overline{v}_i| \neq |v_i| \tag{2.340}$$

In the case of deformation and strain as second-order tensors, one can apply the rotation to the deformation gradient F_{ij} as follows:

$$\overline{F}_{iJ} = r_{ik} F_{kJ} \tag{2.341}$$

where $\overline{F}_{ij}$ is the deformation gradient for the rotated body. The rotated description of deformation and strain tensors can be defined using Eq. (2.341). In the case of Eulerian description, the deformation and strain tensors can be obtained by substituting Eq. (2.341) into Eqs. (2.306), (2.311) as follows:

$$\overline{B}_{ij} = r_{ik} B_{kl} r_{jl} \tag{2.342}$$

$$\overline{e}_{ij} = r_{ik} e_{kl} r_{jl} \tag{2.343}$$

The Almansi strain tensor e_{ij} is objective since it satisfies the following condition:

$$\frac{1}{2}\left(dl^2 - dL^2\right) = dx_i e_{ij} dx_j = d\overline{x}_i \overline{e}_{ij} d\overline{x}_j \tag{2.344}$$

In other words, the Almansi strain tensor e_{ij} is objective since it conserves the change in the square of elemental lengths during the rigid body rotation. Similar relation can be shown for the left Cauchy-Green deformation tensor B_{ij}. In the case of Lagrangian deformation and strain tensors, substituting Eq. (2.341) into Eqs. (2.300), (2.309) shows that $C_{IJ} = \overline{C}_{IJ}$ and $E_{IJ} = \overline{E}_{IJ}$. It can also be intuitively inferred since C_{IJ} and E_{IJ} are defined in the Lagrangian description, the rotation of the deformed body does not alter them.

The velocity gradient tensor l_{ij} is an example of non-objective second order tensor. To prove this, one can obtain the velocity gradient tensor of the rotated body $\overline{l}_{ij}$ by substituting Eq. (2.341) into the Eq. (2.327) as follows:

$$\overline{l}_{ij} = \dot{\overline{F}}_{iJ} \overline{F}_{Jj}^{-1} = r_{ik} l_{kl} r_{jl} + \dot{r}_{ik} r_{jk} \tag{2.345}$$

The velocity gradient tensor is non-objective due to the second term $\dot{r}_{ik} r_{jk}$. One can show, however, that the rate of deformation tensor d_{ij} is objective, i.e., $\overline{d}_{ij} = r_{ik} d_{kl} r_{jl}$.

2.4.2 Stress measures

Unlike small strain continuum mechanics with only one stress definition, different stress measures can be defined in finite strain theory including Cauchy stress, Kirchhoff stress, and first and second Piola-Kirchhoff stress tensors. The first stress definition is the Cauchy stress tensor which is defined using the spatial force and area. Using the material description for spatial quantities of force and area, Kirchhoff and first and second Piola-Kirchhoff stress tensors are defined using the concept of work conjugacy. In this section, first, different stress measures and their connections through the principal of virtual work will be addressed. Next, the definition of stress rates and the concept of stress objectivity will be elaborated.

2.4.2.1 Cauchy stress tensor

In continuum mechanics, stress at a point can be defined as the intensity of internal forces that neighboring particles of a continuous material apply on each other across imaginary internal surfaces. To elaborate the concept of Cauchy stress tensor, consider the resultant force vector $f_i^{(n_i)}$ in a deformed body at point p acting on the area Δa with the normal vector of n_i (Fig. 2.13). The stress vector, which is called traction vector t_i corresponding to the normal vector n_i can be described as follows:

$$t_i^{(n_i)} = \lim_{\Delta a \to 0} \frac{f_i^{(n_i)}}{\Delta a} \tag{2.346}$$

A stress tensor at a point defines the traction vectors at that point for normal vectors in all directions. However, it is not necessary to know the traction

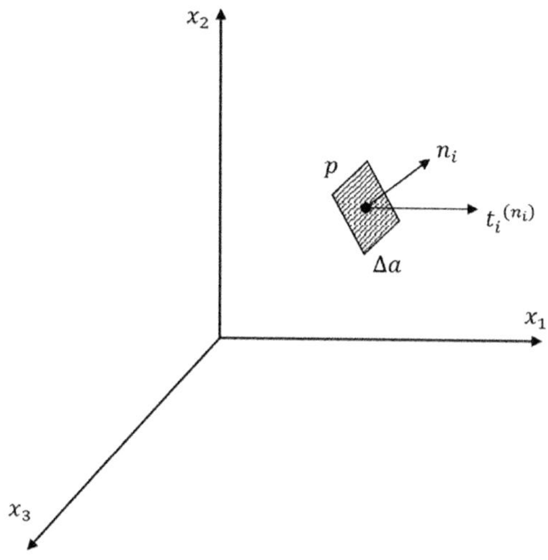

FIG. 2.13 The traction vector t_i corresponding to the normal vector n_i at point p.

vectors in all directions. One can define the Cauchy stress tensor using three tractions associated with three mutually perpendicular planes using the equilibrium conditions (Fig. 2.14). Fig. 2.15 shows an element *oabc* with three traction vectors of $t_i^{(1)}$, $t_i^{(2)}$, and $t_i^{(3)}$ acting on the surfaces *obc*, *oac*, and *oab* with the corresponding normal vectors of $e_i^{(1)}$, $e_i^{(2)}$, and $e_i^{(3)}$, respectively. Furthermore,

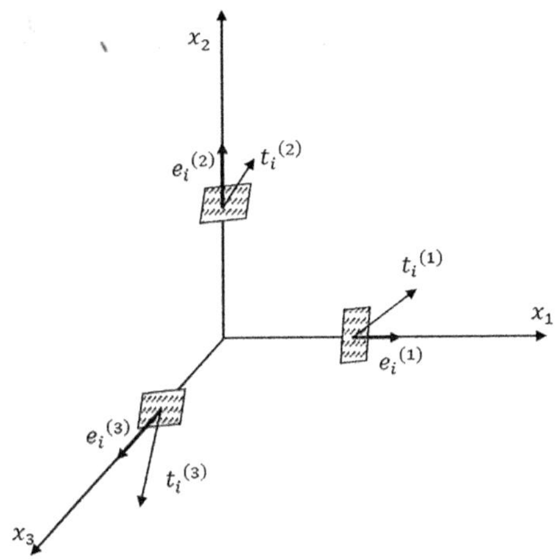

FIG. 2.14 Three traction vectors of $t_i^{(1)}$, $t_i^{(2)}$, and $t_i^{(3)}$ acting on three mutually perpendicular planes with the normal vectors of $e_i^{(1)}$, $e_i^{(2)}$, and $e_i^{(3)}$, respectively.

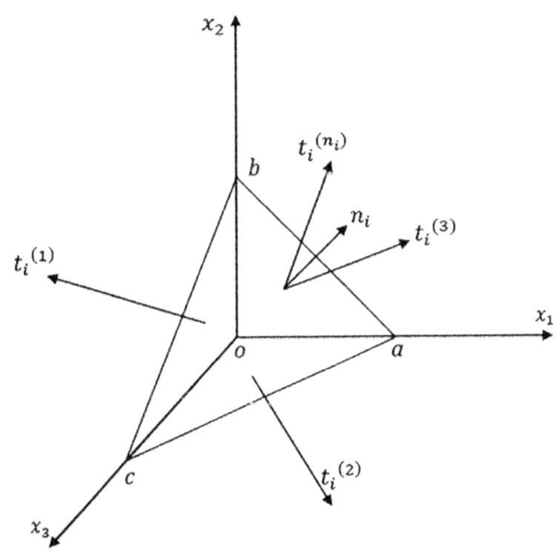

FIG. 2.15 The traction vector $t_i^{(n_i)}$ which is acting on the surface *abc* with the normal vector of n_i which can be described using the three traction vectors of $t_i^{(1)}$, $t_i^{(2)}$, and $t_i^{(3)}$ acting on the surfaces *obc*, *oac*, and *oab* with the normal vectors of $e_i^{(1)}$, $e_i^{(2)}$, and $e_i^{(3)}$, respectively.

the traction vector $t_i^{(n_i)}$ is acting on the surface *abc* with the normal vector of n_i which can be defined as follows:

$$n_i = \left[\cos\left(e_i^{(1)}, n_i\right), \cos\left(e_i^{(2)}, n_i\right), \cos\left(e_i^{(3)}, n_i\right)\right] \tag{2.347}$$

Considering Newton's third law, i.e., $t_i^{(-n_i)} = -t_i^{(n_i)}$, the equilibrium equation of the body *oabc* can be stated as follows:

$$t_i^{(n_i)}a - t_i^{(1)}a^{(1)} - t_i^{(2)}a^{(2)} - t_i^{(3)}a^{(3)} + f_i^{(v)}v = 0 \tag{2.348}$$

where a, $a^{(1)}$, $a^{(2)}$, and $a^{(3)}$ are the areas of the surfaces with the normal vectors of n_i, $e_i^{(1)}$, $e_i^{(2)}$, and $e_i^{(3)}$, respectively, $f_i^{(v)}$ is the force per unit volume acting on the body at point p and v is the volume of the body *oabc*. Eq. (2.348) can be rearranged as follows:

$$t_i^{(n_i)} = t_i^{(1)}\frac{a^{(1)}}{a} + t_i^{(2)}\frac{a^{(2)}}{a} + t_i^{(3)}\frac{a^{(3)}}{a} - f_i^{(v)}\frac{v}{a} \tag{2.349}$$

Substituting $a^{(l)}/a = \cos\left(e_i^{(l)}, n_i\right) = n_l$ and $v/a \to 0$, Eq. (2.349) can be rewritten as follows:

$$t_i^{(n_i)} = t_i^{(1)}n_1 + t_i^{(2)}n_2 + t_i^{(3)}n_3 \tag{2.350}$$

Accordingly, the traction at point p acting on an arbitrary plane with the normal vector of n_i can be always described using three tractions at the same point associated with three mutually perpendicular planes.

The Cauchy stress tensor components can be related to the traction vectors in each of the three mutually perpendicular planes. For example, in the case of the plane with the normal vector of $e_i^{(1)}$, Fig. 2.16 shows the components of the Cauchy stress tensor. One can see that the traction vector has two shear and one normal component. Accordingly, the traction vector $t_i^{(1)}$ can be defined as follows:

$$t_i^{(1)} = e_i^{(1)}\sigma_{11} + e_i^{(2)}\sigma_{12} + e_i^{(3)}\sigma_{13} = e_i^{(j)}\sigma_{1j} \tag{2.351}$$

Using the same methodology, one can define the traction vectors $t_i^{(2)}$ and $t_i^{(3)}$ as follows:

$$t_i^{(2)} = e_i^{(j)}\sigma_{2j} \tag{2.352}$$

$$t_i^{(3)} = e_i^{(j)}\sigma_{3j} \tag{2.353}$$

Substituting Eqs. (2.351)–(2.353) into Eq. (2.350), one can define the Cauchy stress tensor σ_{ij} as follows:

$$t_i^{(n_i)} = \sigma_{ij}n_j \tag{2.354}$$

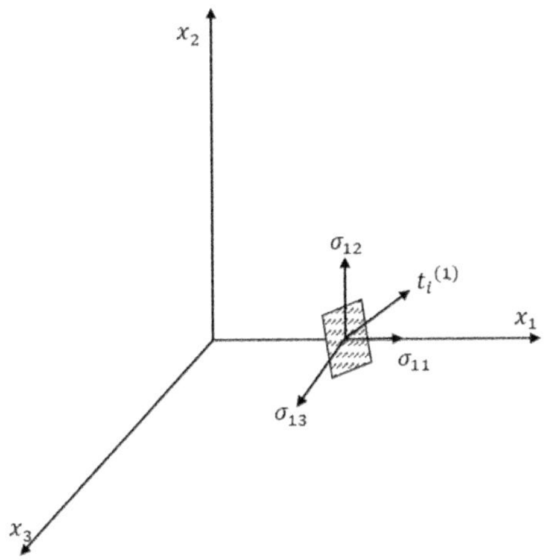

FIG. 2.16 The components of the Cauchy stress tensor corresponding to the traction vector $t_i^{(1)}$ acting on the surfaces with the normal vectors of $e_i^{(1)}$.

2.4.2.2 *Principle of virtual work*

In continuum mechanics, the principle of virtual work is incorporated to derive the weak form of the equilibrium equation. The static equilibrium equation in Eulerian description can be written as follows:

$$\sigma_{ij,j}+f_i=0 \tag{2.355}$$

where $\sigma_{ij,\,j}$ is the divergence of Cauchy stress tensor and f_i the body force vector per unit volume. Assuming the body to undergo an arbitrary but kinematically admissible virtual velocity δv_i, accordingly, the virtual work δW due to δv_i, which is a weak form of the static equilibrium equation in the Eulerian description, can be stated as follows:

$$\delta W=\int_v \left(\sigma_{ij,j}+f_i\right)\delta v_i dv=0 \tag{2.356}$$

In Eq. (2.356), the term $\sigma_{ij,\,j}\delta v_i$ can be rewritten as follows:

$$\sigma_{ij,j}\delta v_i=\left(\sigma_{ij}\delta v_i\right)_{,j}-\sigma_{ij}\left(\delta v_i\right)_{,j} \tag{2.357}$$

Eq. (2.356) can be rewritten using Eq. (2.357) as follows:

$$\int_v \sigma_{ij}\left(\delta v_i\right)_{,j}dv-\int_v \left(\sigma_{ij}\delta v_i\right)_{,j}dv-\int_v f_i\delta v_i dv=0 \tag{2.358}$$

Eq. (2.358) can be rewritten using the Gauss's theorem, as follows:

$$\int_v \sigma_{ij}\left(\delta v_i\right)_{,j}dv-\int_a n_j\sigma_{ij}\delta v_i da-\int_v f_i\delta v_i dv=0 \tag{2.359}$$

Considering $(\delta v_i)_{,j} = \delta l_{ij}$, $n_j\sigma_{ij} = t_i$, and $\sigma_{ij}\delta l_{ij} = \sigma_{ij}(\delta d_{ij} + \delta\omega_{ij}) = \sigma_{ij}\delta d_{ij}$ which is due to the symmetry of σ_{ij} and asymmetry of ω_{ij}, i.e., $\sigma_{ij}\delta\omega_{ij} = 0$, the Eulerian description of virtual work equation can be stated as follows:

$$\delta W = \int_v \sigma_{ij}\delta d_{ij} dv - \int_a t_i \delta v_i da - \int_v f_i \delta v_i dv = 0 \tag{2.360}$$

The virtual work done by internal stresses is represented by the first term of Eq. (2.360), i.e., $\int_v \sigma_{ij}\delta d_{ij} dv$, and σ_{ij} and d_{ij} are work conjugate in the Eulerian description. One can introduce other work conjugate pairs using material description.

2.4.2.3 *Kirchhoff stress tensor*

The Kirchhoff stress tensor can be introduced by describing the virtual work equation in terms of the undeformed volume and area as follows:

$$\delta W = \int_V J\sigma_{ij}\delta d_{ij} dV - \int_A T_i \delta v_i dA - \int_V F_i \delta v_i dV = 0 \tag{2.361}$$

where the relation between the volume of deformed and undeformed bodies are considered as $dv = JdV$, in which Jacobian $J = \det(F_{iJ})$, and $F_I = Jf_i\delta_{Ii}$ and $T_I = \delta_{Ii}t_i(da/dA)$ are the Lagrangian description of the body force per unit volume and traction vector per unit area, respectively. According to Eq. (2.361), one can introduce the Kirchhoff stress tensor τ_{ij} as the work conjugate to d_{ij} using the undeformed body volume and area as follows

$$\tau_{ij} = J\sigma_{ij} \tag{2.362}$$

2.4.2.4 *First Piola-Kirchhoff stress tensor*

Although the virtual work equation corresponding to the Kirchhoff stress tensor τ_{ij} definition is described with respect to the volume of the undeformed body, it still depends on τ_{ij} and d_{ij} with the Eulerian description. Accordingly, the virtual internal work term in Eq. (2.361) is rewritten by considering the fact that $\sigma_{ij}\delta d_{ij} = \sigma_{ij}\delta l_{ij}$ due to the symmetry of σ_{ij} as follows:

$$\delta W^{internal} = \int_V J\sigma_{ij}\delta l_{ij} dV = \int_V J\sigma_{ik}F_{Jk}^{-1}\delta\dot{F}_{iJ} dV \tag{2.363}$$

According to Eq. (2.363), one can introduce the first Piola-Kirchhoff stress tensor P_{iJ} as the work conjugate to $\dot{F}_{iJ}$ as follows:

$$P_{iJ} = J\sigma_{ik}F_{Jk}^{-1} \tag{2.364}$$

The virtual internal work term can be defined using the first Piola-Kirchhoff stress tensor P_{iJ} and its work conjugate $\dot{F}_{iJ}$ as $\delta W^{internal} = \int_V P_{iJ}\delta\dot{F}_{iJ} dV$.

One should note that the first Piola-Kirchhoff stress tensor P_{iJ} is an unsymmetric two-point tensor similar to the deformation gradient tensor F_{iJ}. To elaborate the concept of two-point tensor for the first Piola-Kirchhoff stress tensor P_{iJ}, Fig. 2.17 presents the sections of material in the undeformed and deformed body. Accordingly, the first Piola-Kirchhoff stress tensor P_{iJ} can be defined as follows:

$$dp_i = P_{iJ}dA_J \tag{2.365}$$

where $dA_J = N_J dA$. In other words, the first Piola-Kirchhoff stress tensor P_{iJ} is defined based on the force vector in the deformed body and area vector in the undeformed body. On the other hand, the Cauchy stress tensor σ_{ij} incorporates both the area and force in the deformed body as follows:

$$dp_i = \sigma_{ij}da_j \tag{2.366}$$

where $da_j = n_j da$.

2.4.2.5 Second Piola-Kirchhoff stress tensor

A true Lagrangian stress definition should consider both the area and force vectors in the undeformed body. Accordingly, instead of using the Eulerian force

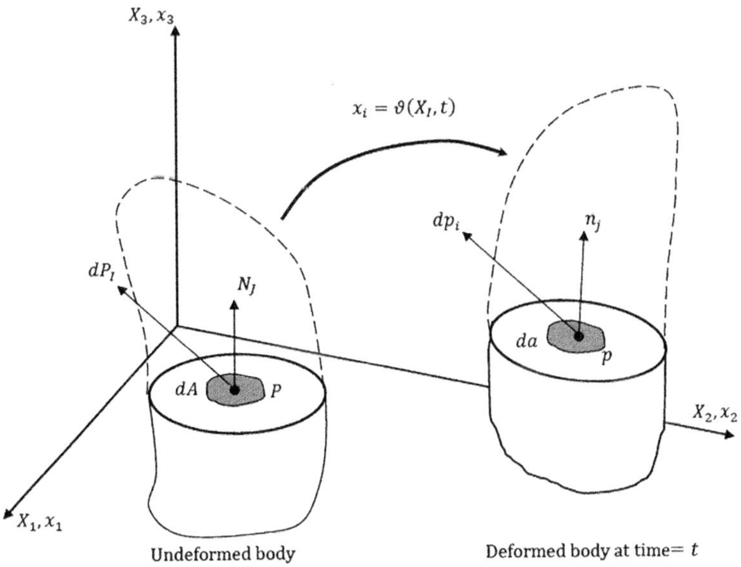

FIG. 2.17 Illustration of the concept of two-point tensor for the first Piola-Kirchhoff stress tensor P_{iJ}.

vector dp_i, the Lagrangian description $d\mathbb{P}_I$ should be used, which can be described as follows:

$$d\mathbb{P}_I = F_{Ij}^{-1} dp_j \tag{2.367}$$

The second Piola-Kirchhoff stress tensor S_{IJ} can be obtained using the Lagrangian description of force and area vectors as follows:

$$d\mathbb{P}_I = S_{IJ} dA_J \tag{2.368}$$

Comparing Eqs. (2.368), (2.366), the relation between the second Piola-Kirchhoff stress tensor S_{IJ} and the Cauchy stress tensor σ_{ij} can be obtained as follows:

$$S_{IJ} = JF_{Ik}^{-1}\sigma_{kl}F_{Jl}^{-1} \leftrightarrow \sigma_{ij} = J^{-1}F_{iK}S_{KL}F_{jL} \tag{2.369}$$

The pull back and push forward operators for second Piola-Kirchhoff and Cauchy stress tensors can be described as follows:

$$\sigma_{ij} = J^{-1}\vartheta_*(S_{IJ}) = J^{-1}F_{iK}S_{KL}F_{jL} \tag{2.370}$$

$$S_{IJ} = J\vartheta_*^{-1}(\sigma_{ij}) = JF_{Ik}^{-1}\sigma_{kl}F_{Jl}^{-1} \tag{2.371}$$

In order to find the work conjugate of the second Piola-Kirchhoff stress tensor S_{ij}, the virtual internal work term in Eq. (2.361) is rewritten by substituting σ_{ij} according to the Eq. (2.370) and dv with JdV as follows:

$$\delta W^{internal} = \int_v \sigma_{ij}\delta d_{ij} dv = \int_V F_{iK}S_{KL}F_{jL}\delta d_{ij} dV = \int_V S_{KL}\left(F_{iK}\delta d_{ij}F_{jL}\right) dV \tag{2.372}$$

Eq. (2.331) states that $F_{iK}\delta d_{ij}F_{jL} = \delta\dot{E}_{KL}$ and Eq. (2.372) can be written as follows:

$$\delta W^{internal} = \int_v \sigma_{ij}\delta d_{ij} dv = \int_V S_{IJ}\delta\dot{E}_{IJ} dV \tag{2.373}$$

Eq. (2.373) shows that the second Piola-Kirchhoff stress tensor S_{IJ} and the Lagrangian strain rate tensor $\dot{E}_{IJ}$ are work conjugate.

2.4.2.6 *Stress objectivity*

As discussed in Section 2.4.1.10, the objectivity of a second order tensor can be investigated by applying a rigid rotation r_{ij} to a deformed body. Applying the rotation tensor r_{ij} will lead to the rotated traction and surface normal vectors $\bar{t}_i$ and $\bar{n}_i$, respectively, which can be described as follows (Fig. 2.18):

$$\bar{t}_i = r_{ij} t_j \tag{2.374}$$

$$\bar{n}_i = r_{ij} n_j \tag{2.375}$$

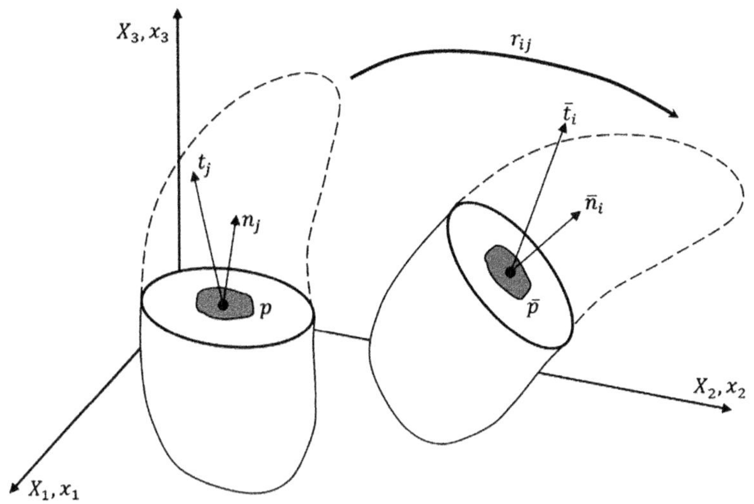

FIG. 2.18 Evaluating the concept of stress objectivity by applying a rigid rotation r_{ij} to a deformed body.

One can obtain the Cauchy stress tensor in the rotated body $\bar{\sigma}_{ij}$ by substituting Eq. (2.374) into the Eq. (2.354) as follows:

$$\bar{\sigma}_{ij} = r_{ik}\sigma_{kl}r_{jl} \tag{2.376}$$

Accordingly, the Cauchy stress tensor σ_{ij} is objective according to the definition of objectivity for second order tensors as presented in Section 2.4.1.10.

In the case of the second Piola-Kirchhoff stress tensor S_{IJ}, since it is described according to the undeformed body, one can intuitively consider it objective. In other words, the rotation of a deformed body will not change its definition and $\bar{S}_{IJ} = S_{IJ}$. The objectivity of the second Piola-Kirchhoff stress tensor S_{IJ} can be mathematically shown as follows:

$$\bar{S}_{IJ} = \overline{JF}^{-1}_{Ik}\bar{\sigma}_{kl}\bar{F}^{-1}_{Jl} = JF^{-1}_{Im}r_{km}r_{kp}\sigma_{pq}r_{lq}r_{lr}F^{-1}_{Jr} = JF^{-1}_{Im}\delta_{mp}\sigma_{pq}\delta_{qr}F^{-1}_{Jr} = S_{IJ} \tag{2.377}$$

2.4.2.7 *Stress rate*

It is stated that the Cauchy stress tensor σ_{ij} is objective. In the case of time derivative of σ_{ij}, i.e., $\dot{\sigma}_{ij}$, however, it can be shown that $\dot{\sigma}_{ij}$ is not objective and $\dot{\bar{\sigma}}_{ij} \neq r_{ik}\dot{\sigma}_{kl}r_{jl}$ as follows:

$$\dot{\bar{\sigma}}_{ij} = r_{ik}\dot{\sigma}_{kl}r_{jl} + \dot{r}_{ik}\sigma_{kl}r_{jl} + r_{ik}\sigma_{kl}\dot{r}_{jl} \tag{2.378}$$

Various objective stress rate tensors have been developed. The easiest way to obtain an objective stress rate tensor is to define it based on the rate of the second Piola-Kirchhoff stress tensor $\dot{S}_{IJ}$, which is independent of the applied rigid rotation. Accordingly, the Truesdell stress rate $\overset{\circ}{\sigma}_{ij}$ can be defined by transformation of $\dot{S}_{IJ}$ as follows:

$$\overset{\circ}{\sigma}_{ij} = J^{-1} F_{iK} \dot{S}_{KL} F_{jL} = J^{-1} F_{iK} \frac{d\left(J F_{Kp}^{-1} \sigma_{pq} F_{Lq}^{-1}\right)}{dt} F_{jL} \tag{2.379}$$

substituting $\dot{J} = J d_{kk}$ and $d(F_{Ji}^{-1})/dt = -F_{Ik}^{-1} l_{ki}$, the Truesdell stress rate $\overset{\circ}{\sigma}_{ij}$ can be obtained as follows:

$$\overset{\circ}{\sigma}_{ij} = \dot{\sigma}_{ij} - l_{ik}\sigma_{kj} - \sigma_{ik} l_{jk} + l_{ll}\sigma_{ij} \tag{2.380}$$

The Truesdell rate of the Kirchhoff stress tensor can be defined as follows:

$$\overset{\circ}{\tau}_{ij} = J \overset{\circ}{\sigma}_{ij} = \dot{\tau}_{ij} - l_{ik}\tau_{kj} - \tau_{ik} l_{jk} \tag{2.381}$$

Another objective stress rate is the Oldroyd stress rate which can be described as a Lie derivative of the Cauchy stress tensor as follows:

$$\sigma_{ij}^{\nabla} = \mathcal{L}_v\left(\sigma_{ij}\right) = F_{iK} \frac{d\left(F_{Kp}^{-1} \sigma_{pq} F_{Lq}^{-1}\right)}{dt} F_{jL} = \dot{\sigma}_{ij} - l_{ik}\sigma_{kj} - \sigma_{ik} l_{jk} \tag{2.382}$$

Performing the operations of pull back and push forward using F_{iJ} and F_{Ji}^{-1} leads to the convective stress rate $\sigma_{ij}^{\diamond}$ as follows:

$$\sigma_{ij}^{\diamond} = F_{Ki}^{-1} \frac{d\left(F_{pK} \sigma_{pq} F_{qL}\right)}{dt} F_{Lj}^{-1} = \dot{\sigma}_{ij} + l_{ki}\sigma_{kj} + \sigma_{ik} l_{kj} \tag{2.383}$$

In order to simply the stress rate tensors, one can neglect the stretch tensor and assume $F_{iJ} \approx R_{iJ}$. Accordingly, all three stress rate tensors of Truesdell, Oldroyd, and convective become the same, which is commonly termed as Green-Naghdi stress rate $\sigma_{ij}^{\square}$ and can be defined as follows:

$$\sigma_{ij}^{\square} = R_{iK} \frac{d\left(R_{pK} \sigma_{pq} R_{qL}\right)}{dt} R_{jL} = \dot{\sigma}_{ij} + \sigma_{ik} \dot{R}_{kL} R_{jL} - \dot{R}_{iK} R_{lK} \sigma_{lj} \tag{2.384}$$

The stress rate tensor can be further simplified by approximating the spin tensor as $\omega_{ij} \approx \dot{R}_{iK} R_{jK}$, which is called the Jaumann stress rate σ_{ij}^{Δ} and defined as follows:

$$\sigma_{ij}^{\Delta} = \dot{\sigma}_{ij} + \sigma_{ik}\omega_{kj} - \omega_{ik}\sigma_{kj} \tag{2.385}$$

2.4.3 Finite strain hyperelasticity

The deformation, strain and stress measures in finite strain continuum mechanics are discussed so far. The next step is to elaborate the constitutive equations which defines the mathematical relation between the stress and strain tensors.

In the case of small strain deformations, different material responses from the simplest one, i.e., nonlocal elastic material, to the complex ones such as nonlocal rate-dependent plasticity damage model are formulated in Sections 2.2 and 2.3. In this section, the constitutive relation is described in the context of finite strain continuum mechanics.

2.4.3.1 *Hyperelasticity*

In the special case that the material constitutive behavior depends solely on the undeformed and final states of the body, the material behavior becomes path independent. Accordingly, a specific free-energy function ψ, i.e. internal energy per unit mass, is only a function of F_{iJ} and X_J, which can be defined as follows:

$$\psi(F_{iJ}, X_J) = \frac{1}{\overline{\rho}} \int_t P_{iJ} \dot{F}_{iJ} dt \tag{2.386}$$

Eq. (2.386) is commonly termed as hyperelastic material model, where $\overline{\rho}$ is the Lagrangian description of the density. The time derivative of specific free-energy function $\dot{\psi}(F_{iJ}, X_J)$ can be formed using Eq. (2.386) as follows:

$$\dot{\psi}(F_{iJ}, X_J) = \frac{1}{\overline{\rho}} P_{iJ} \dot{F}_{iJ} \tag{2.387}$$

$\dot{\psi}(F_{iJ}, X_J)$ can be also defined using the time derivative of its functional form $\psi(F_{iJ}, X_J)$ as follows:

$$\dot{\psi}(F_{iJ}, X_J) = \frac{\partial \psi(F_{iJ}, X_J)}{\partial F_{kL}} \dot{F}_{kL} \tag{2.388}$$

Comparing Eqs. (2.387), (2.388), one can obtain the constitutive equation of a hyperelastic material as follows:

$$P_{iJ}(F_{iJ}, X_J) = \overline{\rho} \frac{\partial \psi(F_{iJ}, X_J)}{\partial F_{iJ}} \tag{2.389}$$

Eq. (2.391) states that the constitutive model of a hyperelastic material can solely be defined knowing F_{iJ} and X_J by defining the specific free-energy function $\psi(F_{iJ}, X_J)$, which solely depends on the undeformed and final states of the body. In the case of homogenous hyperelastic material, the dependency of the stress tensor on X_I can be excluded as follows:

$$P_{iJ}(F_{iJ}) = \overline{\rho} \frac{\partial \psi(F_{iJ})}{\partial F_{iJ}} \tag{2.390}$$

For the rest of the chapter, it is assumed that the material is homogenous.

2.4.3.2 *Material objectivity*

Material objectivity dictates that the constitutive equations should not depend on the rigid body rotations. Accordingly, the specific free-energy function ψ

should be a function of stretch tensor U_{IJ} and not the rotation tensor R_{iJ}. Considering the relation $C_{IJ} = U_{IK}U_{KJ}$ the specific free-energy function $\tilde{\psi}$ can be defined as a function of the right Cauchy-Green tensor C_{IJ} as follows:

$$\tilde{\psi} = \tilde{\psi}\,(C_{IJ}) \tag{2.391}$$

The Lagrangian internal energy per unit mass can be defined as follows:

$$\tilde{\psi}\,(C_{IJ}) = \frac{1}{\bar{\rho}}\int_t S_{IJ}\dot{E}_{IJ}dt = \frac{1}{2\bar{\rho}}\int_t S_{IJ}\dot{C}_{IJ}dt \tag{2.392}$$

where $\dot{E}_{IJ} = \dot{C}_{IJ}/2$. Accordingly, $\dot{\psi}(C_{IJ})$ can be defined as follows:

$$\dot{\tilde{\psi}}(C_{IJ}) = \frac{1}{2\bar{\rho}}S_{IJ}\dot{C}_{IJ} \tag{2.393}$$

$\dot{\tilde{\psi}}(C_{IJ})$ can be also defined using the time derivative of its functional form $\tilde{\psi}\,(C_{IJ})$ as follows:

$$\dot{\tilde{\psi}}(C_{IJ}) = \frac{\partial\tilde{\psi}}{\partial C_{IJ}}\dot{C}_{IJ} \tag{2.394}$$

One can obtain the Lagrangian constitutive equation by comparing Eqs. (2.393), (2.394) as follows:

$$S_{IJ} = 2\bar{\rho}\frac{\partial\tilde{\psi}}{\partial C_{IJ}} = \bar{\rho}\frac{\partial\tilde{\psi}}{\partial E_{IJ}} \tag{2.395}$$

One can obtain other stress measures of Cauchy stress σ_{ij}, Kirchhoff stress τ_{ij}, and first Piola-Kirchhoff tensors P_{iJ} using Eqs. (2.362), (2.364), (2.369), (2.395) as follows:

$$\sigma_{ij} = 2\bar{\rho}J^{-1}F_{iK}\frac{\partial\tilde{\psi}}{\partial C_{KL}}F_{jL} \tag{2.396}$$

$$\tau_{ij} = 2\bar{\rho}F_{iK}\frac{\partial\tilde{\psi}}{\partial C_{KL}}F_{jL} \tag{2.397}$$

$$P_{iJ} = 2\bar{\rho}F_{iK}\frac{\partial\tilde{\psi}}{\partial C_{KJ}} \tag{2.398}$$

2.4.3.3 *Isotropic hyperelasticity*

Until now, there is no constraint on the material constitutive model due to the material symmetry. The constitutive equation for isotropic materials is now discussed in which the specific free energy ψ is identical for any material direction. The isotropy can be mathematically represented as follows:

$$\psi(r_{ik}F_{kJ}) = \psi(F_{iJ}) \tag{2.399}$$

for all rotation tensors r_{ik}. Accordingly, one can simply show that the specific free energy of an isotropic hyperelastic material can be defined as follows:

$$\psi(F_{iJ}) = \psi(V_{ij}) = \psi(U_{IJ}) \tag{2.400}$$

Considering the relation $B_{ij} = V_{ik}V_{kj}$ and $C_{IJ} = U_{IK}U_{KJ}$, one can define the specific free energy of an isotropic hyperelastic material as a function of either left or right Cauchy-Green tensors, B_{ij} or C_{IJ}, respectively, as follows:

$$\tilde{\psi} = \tilde{\psi}\left(B_{ij}\right) = \tilde{\psi}\left(C_{IJ}\right) \tag{2.401}$$

Eq. (2.401) states the specific free energy of an isotropic hyperelastic material can be simply defined in either Lagrangian or Eulerian descriptions.

In the case of isotropic hyperelastic materials, the specific free energy can be described as a function of principal invariants of right or left Cauchy-Green tensors, i.e., $\{I_1(C_{IJ}), I_2(C_{IJ}), I_3(C_{IJ})\}$ or $\{I_1(B_{ij}), I_2(B_{ij}), I_3(B_{ij})\}$, respectively, as follows:

$$\tilde{\psi}\left(I_1(C_{IJ}), I_2(C_{IJ}), I_3(C_{IJ})\right) = \tilde{\psi}\left(I_1\left(B_{ij}\right), I_2\left(B_{ij}\right), I_3\left(B_{ij}\right)\right) \tag{2.402}$$

where the principal invariants of right and left Cauchy-Green tensors are the same which are presented in Eqs. (2.303)–(2.305).

The constitutive equation of an isotropic hyperelastic material in Lagrangian description can be defined using Eqs. (2.395), (2.402) as follows:

$$S_{IJ} = 2\overline{\rho}\frac{\partial \tilde{\psi}}{\partial C_{IJ}} = 2\overline{\rho}\left[\frac{\partial \tilde{\psi}}{\partial I_1(C_{KL})}\frac{\partial I_1(C_{KL})}{\partial C_{IJ}} + \frac{\partial \tilde{\psi}}{\partial I_2(C_{KL})}\frac{\partial I_2(C_{KL})}{\partial C_{IJ}} + \frac{\partial \tilde{\psi}}{\partial I_3(C_{KL})}\frac{\partial I_3(C_{KL})}{\partial C_{IJ}}\right] \tag{2.403}$$

where $\partial I_1(C_{IJ})/\partial C_{IJ}$, $\partial I_2(C_{IJ})/\partial C_{IJ}$, and $\partial I_3(C_{IJ})/\partial C_{IJ}$ can be obtained as follows:

$$\frac{\partial I_1(C_{KL})}{\partial C_{IJ}} = \delta_{IJ} \tag{2.404}$$

$$\frac{\partial I_2(C_{KL})}{\partial C_{IJ}} = 2C_{IJ} \tag{2.405}$$

$$\frac{\partial I_3(C_{KL})}{\partial C_{IJ}} = J^2 C_{IJ}^{-1} \tag{2.406}$$

The constitutive equation in the Lagrangian description can be then obtained by substituting Eqs. (2.404)–(2.406) in Eq. (2.403) as follows:

$$S_{IJ} = 2\overline{\rho}\left[\delta_{IJ}\frac{\partial \tilde{\psi}}{\partial I_1(C_{KL})} + 2C_{IJ}\frac{\partial \tilde{\psi}}{\partial I_2(C_{KL})} + J^2 C_{JI}^{-1}\frac{\partial \tilde{\psi}}{\partial I_3(C_{KL})}\right] \tag{2.407}$$

The constitutive equation in Eulerian description can be derived by substituting Eqs. (2.306), (2.369) into Eq. (2.407) as follows:

$$\sigma_{ij}=2\rho\left[B_{ij}\frac{\partial\widetilde{\psi}}{\partial I_1(B_{pq})}+2B_{ik}B_{kj}\frac{\partial\widetilde{\psi}}{\partial I_2(B_{pq})}+J^2\delta_{ij}\frac{\partial\widetilde{\psi}}{\partial I_3(B_{pq})}\right] \tag{2.408}$$

where $\rho=J^{-1}\overline{\rho}$ is the Eulerian description of the density, and $\frac{\partial\widetilde{\psi}}{\partial I_1(B_{pq})}$, $\frac{\partial\widetilde{\psi}}{\partial I_2(B_{pq})}$, and $\frac{\partial\widetilde{\psi}}{\partial I_3(B_{pq})}$ can be derived similar to Eqs. (2.404)–(2.406). The Kirchhoff stress tensor τ_{ij} can then be obtained using Eq. (2.362) as $\tau_{ij}=J\sigma_{ij}$.

2.4.3.4 *Specific free energy function*

Various specific free energy functions have been proposed to define the constitutive model of isotropic hyperelastic material. Two models of Mooney-Rivlin (see, Green and Zerna, 1954) and Ogden (1984) are presented here. In the case of rubbery materials, the specific free energy of the Mooney-Rivlin model (see, Green and Zerna, 1954) can be described as follows:

$$\overline{\rho}\widetilde{\psi}=A[I_1(C_{IJ})-3]+B[I_2(C_{IJ})-3] \tag{2.409}$$

where A and B are the material constants. The Mooney-Rivlin model becomes the neo-Hookean model in the case of $B=0$ (Green and Zerna, 1954).

The Mooney-Rivlin model cannot capture the behavior of rubbery materials for large strains. Ogden (1984) developed a model to overcome this issue as follows:

$$\overline{\rho}\widetilde{\psi}(\lambda_1,\lambda_2,\lambda_3)=\sum_{k=1}^{K}\frac{\mu_k}{\alpha_k}\left(\lambda_1^{\alpha_k}+\lambda_2^{\alpha_k}+\lambda_3^{\alpha_k}-3\right) \tag{2.410}$$

where the specific free energy is a summation of K terms with their corresponding material constants μ_k and α_k, and λ_1, λ_2 and λ_3 are the eigenvalues of stretch tensors, i.e., U_{IJ} and V_{ij}. The Mooney-Rivlin model can be retrieved from the Ogden model by substituting $K=2$, $\mu_1=2A$, $\mu_2=-2B$, $\alpha_1=2$, and $\alpha_2=-2$.

2.4.4 Finite strain local plasticity

In Section 2.2, the local plasticity models for small strain continuum mechanics are described. The core assumption for small strain local plasticity is that the strains are infinitesimal in a way that the total strain tensor can be additively decomposed in two tensors of elastic and plastic strain as described in Eq. (2.3). In the case of finite strain continuum mechanics, however, this assumption is no longer applicable. Accordingly, the whole plasticity concept should be revisited for finite strain problem.

As described in Section 2.2, the general framework of a local plasticity model in small strain continuum mechanics has five key elements of:

- A decomposition rule to define the share of elastic and plastic strains from the total one.
- A constitutive equation to define the relation between the elastic strain and stress.
- A yield criterion which defines the criteria for plasticity to initiate.
- A flow rule which defines the evolution of plastic strain.
- A hardening rule which governs the variation of mechanical properties as the material plastically deforms.

Similar elements can be defined for finite strain local plasticity models. In this section, the local plasticity framework for finite strain continuum mechanics is presented.

2.4.4.1 *Multiplicative decomposition*

In the case of finite strain plasticity, the additive decomposition of strains is not valid anymore and the elastic and plastic deformations are related through the multiplicative decomposition of the deformation gradients as follows:

$$F_{iJ} = F^e_{i\alpha} F^p_{\alpha J} \tag{2.411}$$

where $F^e_{i\alpha}$ and $F^p_{\alpha J}$ are the elastic and plastic deformation gradients, respectively. Fig. 2.19 illustrates the multiplicative decomposition. Accordingly, the plastic deformation gradient $F^p_{\alpha J}$ transforms the elemental vector dX_J in the undeformed body to $d\tilde{x}_\alpha$ in the intermediate state. Next, the elastic deformation gradient transforms the elemental vector $d\tilde{x}_\alpha$ to that of the deformed body dx_i. The mapping between the undeformed body and the intermediate state and between the intermediate state and the deformed body can be described as follows:

$$x_i = \vartheta\left(\tilde{x}_\alpha, t\right) \text{ or } x_i = x_i\left(\tilde{x}_\alpha, t\right) \tag{2.412}$$

$$\tilde{x}_\alpha = \tilde{\vartheta}\,(X_I, t)\; or\; \tilde{x}_\alpha = \tilde{x}_\alpha(X_I, t) \tag{2.413}$$

One would obtain the kinematics of the body using the mapping operators $\tilde{\vartheta}$ and ϑ. However, the forms of the operators $\tilde{\vartheta}$ and ϑ are generally not easy to be extracted.

The multiplicative decomposition of deformation gradient described in Eq. (2.411) can be micromechanically described as a pure sliding followed by the lattice rotation and distortion as shown in Fig. 2.20. The plastic deformation gradient $F^p_{\alpha J}$ governs the pure sliding while the elastic deformation gradient $F^e_{i\alpha}$ is responsible for lattice rotation and distortion.

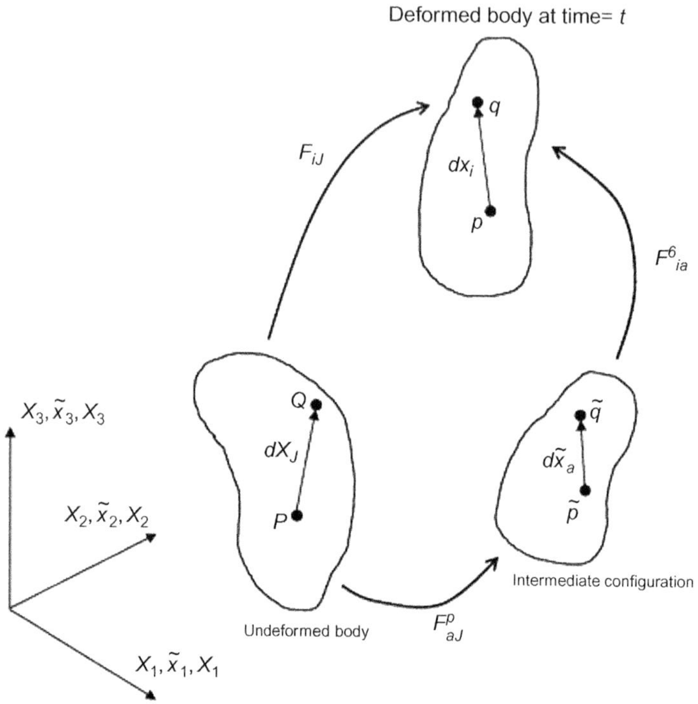

FIG. 2.19 Multiplicative decomposition of deformation gradient.

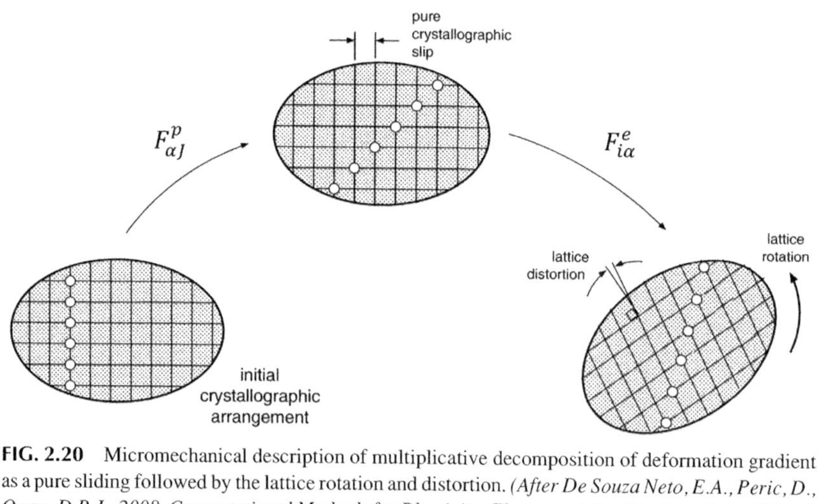

FIG. 2.20 Micromechanical description of multiplicative decomposition of deformation gradient as a pure sliding followed by the lattice rotation and distortion. *(After De Souza Neto, E.A., Peric, D., Owen, D.R.J., 2008. Computational Methods for Plasticity: Theory and Applications. John Wiley & Sons Ltd.)*

2.4.4.2 *Polar decomposition*

The polar decomposition of $F^p_{i\alpha}$ and $F^e_{\alpha J}$ can be written as follows:

$$F^p_{\alpha J} = R^p_{\alpha I} U^p_{IJ} = V^p_{\alpha\beta} R^p_{\beta J} \tag{2.414}$$

$$F^e_{i\alpha} = R^e_{i\beta} U^e_{\beta\alpha} = V^e_{ij} R^e_{j\alpha} \tag{2.415}$$

where the superscripts p and e designate plastic and elastic, respectively, $R^p_{\alpha I}$, U^p_{IJ}, and $V^p_{\alpha\beta}$ are the rotation, material stretch, and spatial stretch tensors, respectively, corresponding to the mapping operator $\tilde{\vartheta}$, and $R^e_{i\beta}$, $U^e_{\beta\alpha}$, and V^e_{ij} are the rotation, material stretch, and spatial stretch tensors, respectively, corresponding to the mapping operators ϑ. In Eq. (2.414), $F^p_{\alpha J} = R^p_{\alpha I} U^p_{IJ}$ states that the plastic deformation gradient, which is relating the intermediate body to the undeformed one, can be decomposed to a stretch U^p_{IJ} followed by a rotation $R^p_{\alpha I}$, while $F^p_{\alpha J} = V^p_{\alpha\beta} R^p_{\beta J}$ describes the plastic deformation as a rotation $R^p_{\beta J}$ followed by a stretch $V^p_{\alpha\beta}$. The same procedure can be considered for the elastic deformation gradient $F^e_{i\alpha}$ described in Eq. (2.415).

2.4.4.3 *Strain measures*

As described in Section 2.4.1.4, the strain definitions which exclude the rigid body rotations should be incorporated. Using the same framework described in Section 2.4.1.4, the total, elastic, and plastic right Cauchy-Green tensors, C_{IJ}, $C^e_{\alpha\beta}$, and C^p_{IJ}, respectively, can be described as follows:

$$C_{IJ} = F_{kI} F_{kJ} = U_{KI} U_{KJ} \tag{2.416a}$$

$$C^e_{\alpha\beta} = F^e_{i\alpha} F^e_{i\beta} = U^e_{\gamma\alpha} U^e_{\gamma\beta} \tag{2.416b}$$

$$C^p_{IJ} = F^p_{\alpha I} F^p_{\alpha J} = U^p_{KI} U^p_{KJ} \tag{2.416c}$$

Similarly, the total, elastic, and plastic left Cauchy-Green tensors, B_{ij}, B^e_{ij}, and $B^p_{\alpha\beta}$, respectively, can be described as follows:

$$B_{ij} = F_{iK} F_{jK} = V_{ik} V_{jk} \tag{2.417a}$$

$$B^e_{ij} = F^e_{i\alpha} F^e_{j\alpha} = V^e_{ik} V^e_{jk} \tag{2.417b}$$

$$B^p_{\alpha\beta} = F^p_{\alpha K} F^p_{\beta K} = V^p_{\alpha\gamma} V^p_{\beta\gamma} \tag{2.417c}$$

One useful equation which will be used later is the expansion of Eq. (2.417b) as follows:

$$B^e_{ij} = F^e_{i\alpha} F^e_{j\alpha} = F_{iI} \left(F^p_{I\alpha}\right)^{-1} \left(F^p_{J\alpha}\right)^{-1} F_{jJ} = F_{iI} \left(C^p_{IJ}\right)^{-1} F_{jJ} \tag{2.418}$$

Eq. (2.418) relates the elastic left Cauchy-Green tensor B^e_{ij} to the plastic right Cauchy-Green tensor C^p_{IJ}.

2.4.4.4 Velocity gradient, rate of deformation, and spin tensor

The velocity gradient is the derivative of velocity, as a spatial vector, with respect to the spatial coordinates, which can be stated as follows:

$$l_{ij} = \frac{\partial v_i}{\partial x_j} = \dot{F}_{iJ} F_{Jj}^{-1} \tag{2.419}$$

The elastic and plastic velocity gradient tensors, l_{ij}^e and l_{ij}^p, respectively, can be defined using two different frameworks. In both frameworks, the elastic velocity gradient is defined by mimicking the total velocity gradient defined in Eq. (2.419) as follows:

$$l_{ij}^e = \dot{F}_{i\alpha}^e \left(F_{\alpha j}^e\right)^{-1} \tag{2.420}$$

In the first framework, the work conjugacy is incorporated to define the plastic velocity gradient l_{ij}^p. Accordingly, the total internal work per unit volume $\dot{W}$ can be additively decomposed into the elastic and plastic parts as follows:

$$\dot{W} = \dot{W}^e + \dot{W}^p \tag{2.421}$$

where $\dot{W}$ and $\dot{W}^e$ can be defined as follows:

$$\dot{W} = \tau_{ij} l_{ij} \tag{2.422}$$

$$\dot{W}^e = \tau_{ij} l_{ij}^e \tag{2.423}$$

and $\dot{W}^p$, which is commonly called as the plastic dissipation rate, can be defined as follows:

$$\dot{W}^p = \dot{W} - \dot{W}^e = \tau_{ij} l_{ij}^p \tag{2.424}$$

$\dot{W}$, $\dot{W}^e$, and $\dot{W}^p$ can be rewritten as follows using Eqs. (2.411), (2.418)–(2.424) as follows:

$$\dot{W} = \tau_{ij} \left[\frac{1}{2} \frac{dB_{ik}^e}{dt} \bigg|_{C_{IJ}^p = \text{cons}} \left(B_{kj}^e\right)^{-1} \right] \tag{2.425}$$

$$\dot{W}^e = \tau_{ij} \left[\frac{1}{2} \frac{dB_{ik}^e}{dt} \left(B_{kj}^e\right)^{-1} \right] \tag{2.426}$$

$$\dot{W}^p = \tau_{ij} \left[-\frac{1}{2} \frac{dB_{ik}^e}{dt} \bigg|_{F_{iJ} = \text{cons}} \left(B_{kj}^e\right)^{-1} \right] \tag{2.427}$$

The plastic velocity gradient tensor l_{ij}^p can be obtained by comparing Eqs. (2.424), (2.427) as follows:

$$l_{ij}^p = -\frac{1}{2} \frac{dB_{ik}^e}{dt} \bigg|_{F_{iJ} = \text{cons}} \left(B_{kj}^e\right)^{-1} \tag{2.428}$$

The plastic velocity gradient tensor l^p_{ij} can be related to the plastic right Cauchy-Green tensor C^p_{IJ} and its rate $\dot{C}^p_{IJ}$ by incorporating Eqs. (2.418), (2.428) as follows:

$$l^p_{ij} = \frac{1}{2} F_{iJ} \left(C^p_{JK}\right)^{-1} \dot{C}^p_{KL} F^{-1}_{Lj} \tag{2.429}$$

The plastic velocity gradient tensor can be additively decomposed to symmetric and asymmetric tensors as follows:

$$l^p_{ij} = \frac{1}{2}\left(l^p_{ij} + l^p_{ji}\right) + \frac{1}{2}\left(l^p_{ij} - l^p_{ji}\right) = d^p_{ij} + \omega^p_{ij} \tag{2.430}$$

where d^p_{ij} is the symmetric plastic rate of deformation tensor and ω^p_{ij} is the asymmetric spin tensor, which can be described as follows:

$$d^p_{ij} = \frac{1}{2}\left(l^p_{ij} + l^p_{ji}\right) \tag{2.431}$$

$$\omega^p_{ij} = l^p_{ij} - d^p_{ij} = \frac{1}{2}\left(l^p_{ij} - l^p_{ji}\right) \tag{2.432}$$

In the case of isotropic material, it can be shown that the spin tensor $\omega^p_{ij} = 0$ and $d^p_{ij} = l^p_{ij}$. Furthermore, since the constitutive models are commonly defined in principal directions for isotropic materials, the elastic left Cauchy-Green deformation tensor B^e_{ij} and the plastic rate of deformation tensor d^p_{ij} can be defined as follows:

$$B^e_{ij} = \sum_{k=1}^{3} \left(\lambda^e_k\right)^2 \left(n^e_i\right)^k \left(n^e_j\right)^k \tag{2.433}$$

$$\begin{aligned} d^p_{ij} = l^p_{ij} &= -\frac{1}{2}\frac{dB^e_{ik}}{dt}\bigg|_{F_{iJ}=\text{cons}} \left(B^e_{kj}\right)^{-1} \\ &= -\frac{1}{2}\left.\frac{d\left[\sum_{k=1}^{3}\left(\lambda^e_k\right)^2\left(n^e_i\right)^k\left(n^e_j\right)^k\right]}{dt}\right|_{F_{iJ}=\text{cons}} \left[\sum_{p=1}^{3}\left(\lambda^e_p\right)^{-2}\left(n^e_i\right)^p\left(n^e_j\right)^p\right] \end{aligned} \tag{2.434}$$

where λ^e_k and $\left(n^e_i\right)^k$ are the eigenvalues and eigenvectors of the elastic spatial stretch tensor V^e_{ij}, respectively. Due to the material isotropy, i.e., $\omega^p_{ij} = 0$, Eq. (2.434) can be rewritten as follows:

$$\begin{aligned} d^p_{ij} = l^p_{ij} &= \sum_{k=1}^{3} -\frac{1}{2}\frac{d\left(\lambda^e_k\right)^2}{dt}\bigg|_{F_{iJ}=\text{cons}} \left(\lambda^e_k\right)^{-2}\left(n^e_i\right)^k\left(n^e_j\right)^k = \sum_{k=1}^{3} l^p_k \left(n^e_i\right)^k\left(n^e_j\right)^k \\ &= \sum_{k=1}^{3} d^p_k \left(n^e_i\right)^k\left(n^e_j\right)^k \end{aligned} \tag{2.435}$$

where $\{d_1^p, d_2^p, d_3^p\}$ and $\{l_1^p, l_2^p, l_3^p\}$ are the eigenvalues of d_{ij}^p and l_{ij}^p, respectively, which can be defined as follows:

$$d_k^p = l_k^p = -\frac{1}{2}\frac{d(\lambda_k^e)^2}{dt}\bigg|_{F_{iJ}=\text{cons}} (\lambda_k^e)^{-2} \tag{2.436}$$

The second framework to define the plastic velocity gradient l_{ij}^p is by following the general formulation of velocity gradient described in Eq. (2.419). Accordingly, the elastic and plastic velocity gradients, l_{ij}^e and $l_{\alpha\beta}^p$, respectively, can be defined as follows:

$$l_{ij}^e = \dot{F}_{i\alpha}^e \left(F_{\alpha j}^e\right)^{-1} \tag{2.437}$$

$$l_{\alpha\beta}^p = \dot{F}_{\alpha I}^p \left(F_{I\beta}^p\right)^{-1} \tag{2.438}$$

One can relate the velocity gradient tensor to those of the elastic and plastic ones as follows:

$$l_{ij} = l_{ij}^e + F_{i\alpha}^e l_{\alpha\beta}^p \left(F_{\beta j}^e\right)^{-1} \tag{2.439}$$

One should note that the velocity gradient l_{ij} is a spatial tensor, while the plastic velocity gradient $l_{\alpha\beta}^p$ is an intermediate tensor, and should be transformed to the spatial configuration using the elastic deformation gradient as $F_{i\alpha}^e l_{\alpha\beta}^p (F_{\beta j}^e)^{-1}$. The symmetric plastic rate of deformation tensor $d_{\alpha\beta}^p$ and asymmetric spin tensor $\omega_{\alpha\beta}^p$ can be described as follows:

$$d_{\alpha\beta}^p = \frac{1}{2}\left(l_{\alpha\beta}^p + l_{\beta\alpha}^p\right) \tag{2.440}$$

$$\omega_{\alpha\beta}^p = l_{\alpha\beta}^p - d_{\alpha\beta}^p = \frac{1}{2}\left(l_{\alpha\beta}^p - l_{\beta\alpha}^p\right) \tag{2.441}$$

Again, it can be shown that in the case of isotropic material, the spin tensor $\omega_{\alpha\beta}^p = 0$, i.e., $d_{\alpha\beta}^p = l_{\alpha\beta}^p$.

One should note that $d_{\alpha\beta}^p$ and $\omega_{\alpha\beta}^p$ are defined in the intermediate configuration. The plastic rate of deformation tensor $d_{\alpha\beta}^p$ can be described in principal directions as follows:

$$d_{\alpha\beta}^p = \sum_{k=1}^{3} d_k^p \left(n_\alpha^p\right)^k \otimes \left(n_\beta^p\right)^k \tag{2.442}$$

where $\{d_1^p, d_2^p, d_3^p\}$ are the eigenvalues of $d_{\alpha\beta}^p$ and $\left\{\left(n_\alpha^p\right)^1, \left(n_\alpha^p\right)^2, \left(n_\alpha^p\right)^3\right\}$ are the corresponding eigenvectors. One should note that d_k^p and n_α^p are defined in the intermediate configuration. It is common that the plasticity formulation is defined in the spatial configuration. Accordingly, the plastic rate of deformation tensor in the spatial configuration d_{ij}^p can be defined as follows:

$$d_{ij}^p = R_{i\alpha}^e d_{\alpha\beta}^p R_{j\beta}^e \tag{2.443}$$

The plastic rate of deformation tensor in the spatial configuration d^p_{ij} can be rewritten in the principal directions by substituting Eq. (2.442) into Eq. (2.443) as follows:

$$d^p_{ij} = \sum_{k=1}^{3} d^p_k \left(n^p_i\right)^k \left(n^p_j\right)^k \tag{2.444}$$

where $\left(n^p_i\right)^k = R^e_{i\alpha}\left(n^p_\alpha\right)^k$.

2.4.4.5 *Constitutive equation*

The hyperelasticity material model is commonly extended to obtain the constitutive model for an elastoplastic material. To do so, it is assumed that the specific free-energy function ψ is a function of the elastic right Cauchy-Green tensor $C^e_{\alpha\beta}$ as follows:

$$\psi = \psi\left(C^e_{\alpha\beta}\right) \tag{2.445}$$

Accordingly, different stress measures of the second Piola-Kirchhoff stress tensor S_{IJ}, Cauchy stress σ_{ij}, Kirchhoff stress τ_{ij}, and first Piola-Kirchhoff stress tensor P_{iJ}, respectively, can be obtained as follows:

$$S_{\alpha\beta} = 2\overline{\rho}\frac{\partial \widetilde{\psi}}{\partial C^e_{\alpha\beta}} \tag{2.446}$$

$$\sigma_{ij} = J^{e-1} F^e_{i\alpha} S_{\alpha\beta} F^e_{j\beta} \tag{2.447}$$

$$\tau_{ij} = F^e_{i\alpha} S_{\alpha\beta} F^e_{j\beta} \tag{2.448}$$

$$P_{i\beta} = F^e_{i\alpha} S_{\alpha\beta} \tag{2.449}$$

where $J^e = \det\left(F^e_{i\alpha}\right)$. One should note that the $S_{\alpha\beta}$ and $P_{i\beta}$ are defined in the intermediate state and not the undeformed state.

In the case of isotropic material, as described in Section 2.4.3.3, it is shown that the specific free energy can also be defined as a function of the left Cauchy-Green tensor B_{ij}. Accordingly, in the case of the isotropic plasticity model, one can define the specific free-energy function ψ as a function of the elastic left Cauchy-Green tensor B^e_{ij} as follows:

$$\psi = \psi\left(B^e_{ij}\right) \tag{2.450}$$

Similar to Eq. (2.408), the Cauchy stress tensor can be defined as follows:

$$\sigma_{ij} = 2\rho\left[B^e_{ij}\frac{\partial \widetilde{\psi}}{\partial I_1\left(B^e_{pq}\right)} + 2B^e_{ik}B^e_{kj}\frac{\partial \widetilde{\psi}}{\partial I_2\left(B^e_{pq}\right)} + J^2\delta_{ij}\frac{\partial \widetilde{\psi}}{\partial I_3\left(B^e_{pq}\right)}\right] \tag{2.451}$$

where the principal invariants of elastic left Cauchy-Green tensors, i.e., $I_1(B^e_{pq})$, $I_2(B^e_{pq})$, and $I_3(B^e_{pq})$, can be described similar to Eqs. (2.303)–(2.305) by substituting $\{\lambda^e_1, \lambda^e_2, \lambda^e_3\}$ instead of $\{\lambda_1, \lambda_2, \lambda_3\}$.

The Kirchhoff stress τ_{ij} can be defined using Eqs. (2.362), (2.451) as follows:

$$\tau_{ij} = J\sigma_{ij} = 2J\rho\left[B^e_{ij}\frac{\partial\tilde{\psi}}{\partial I_1\left(B^e_{pq}\right)} + 2B^e_{ik}B^e_{kj}\frac{\partial\tilde{\psi}}{\partial I_2\left(B^e_{pq}\right)} + J^2\delta_{ij}\frac{\partial\tilde{\psi}}{\partial I_3\left(B^e_{pq}\right)}\right] \tag{2.452}$$

The Kirchhoff stress tensor τ_{ij} is commonly used as the stress measure in the local plasticity formulation.

2.4.4.6 *Yield criterion*

As discussed in Section 2.2.2, the yield criterion defines the stress or strain conditions of material at which plasticity initiates. In the case of finite strain continuum mechanics, the yield surface can be defined as a function of the Kirchhoff stress tensor τ_{ij}. The yield surface f for isotropic materials can be described as $f = f(I_1, J_2, J_3, l_1, l_2, \ldots, l_n)$, where I_1 is the first invariant of Kirchhoff stress tensor and J_2 and J_3 are the second and third invariants of Kirchhoff stress deviator tensor, respectively, which can be defined as follows:

$$I_1 = \tau_{kk} \tag{2.453}$$

$$J_2 = \frac{1}{2}\tau'_{ij}\tau'_{ji} \tag{2.454}$$

$$J_3 = \frac{1}{3}\tau'_{ij}\tau'_{jk}\tau'_{ki} \tag{2.455}$$

where $\tau'_{ij} = \tau_{ij} - (\tau_{kk}/3)\delta_{ij}$ is the Kirchhoff stress deviator tensor, and δ_{ij} is the Kronecker delta.

The simplest model that can be incorporated is the von Mises yield surface which can be defined as follows:

$$f\left(J_2, \sigma_y\right) = \sqrt{3J_2} - \sigma_y \tag{2.456}$$

where σ_y is a material constant defining the uniaxial yield stress and J_2 is the second Kirchhoff stress deviator tensor invariant.

2.4.4.7 *Plastic potential and flow rule*

In the case of small strain theory plasticity, as shown in Section 2.2.4, the plastic strain tensor can be defined using a plastic potential function using Eq. (2.18). Furthermore, it is shown in Section 2.2.7 that the plastic dissipation energy can be maximized by selecting the yield surface as the plastic potential function.

Using the similar framework, one can define the flow rule in the finite strain plasticity theory as follows:

$$d_{ij}^p = l_{ij}^p = \dot{\lambda}\frac{\partial f}{\partial \tau_{ij}} \tag{2.457}$$

where, $\dot{\lambda}$ is a positive scalar factor of proportionality, which is commonly called as the plastic or Lagrangian multiplier. Eq. (2.457) dictates that the direction of the plastic rate of deformation tensor d_{ij}^p is normal to the yield surface, which is commonly termed as the normality hypothesis of plasticity. It can be easily shown that the Kuhn-Tucker conditions to maximize the plastic dissipation are similar to the normality hypothesis:

$$\dot{\lambda}f = 0, \dot{\lambda} \geq 0, f \leq 0 \tag{2.458}$$

In the case of von Mises yield surface defined in Eq. (2.456), the plastic rate of deformation tensor d_{ij}^p, which is equal to the plastic velocity gradient l_{ij}^p for isotropic materials, can be described as follows:

$$d_{ij}^p = l_{ij}^p = \dot{\lambda}\frac{\tau'_{ij}}{\sqrt{\frac{2}{3}\tau'_{ij}\tau'_{ji}}} \tag{2.459}$$

In the case of isotropic materials, as described in Section 2.4.4.4, it is common that the governing equations are described in the principal directions. Accordingly, the Kirchhoff stress tensor τ_{ij} can be described as follows:

$$\tau_{ij} = \sum_{k=1}^{3} \tau_k \left(n_i^e\right)^k \left(n_j^e\right)^k \tag{2.460}$$

where τ_k and $(n_i^e)^k$ are the eigenvalues and eigenvectors of the Kirchhoff stress tensor τ_{ij}. One should note that the eigenvectors of the Kirchhoff stress tensor τ_{ij} is similar to those of the elastic left Cauchy-Green deformation tensor B_{ij}^e, plastic rate of deformation tensor d_{ij}^p, and plastic velocity gradient l_{ij}^p [see, e.g. Eqs. (2.433), (2.435)]. Substituting Eqs. (2.435), (2.436), (2.460) into Eq. (2.457), one can define the evolution of the eigenvalues of d_{ij}^p and l_{ij}^p as follows:

$$d_k^p = l_k^p = -\frac{1}{2}\frac{d\left(\lambda_k^e\right)^2}{dt}\bigg|_{F_{iJ}=\text{cons}} \left(\lambda_k^e\right)^{-2} = -\frac{d\ \ln\lambda_k^e}{dt}\bigg|_{F_{iJ}=\text{cons}} = \dot{\lambda}\frac{\partial f}{\partial \tau_k} \tag{2.461}$$

For the case of the von Mises yield surface, Eq. (2.461) can be rewritten as follows:

$$d_k^p = l_k^p = -\frac{d\ \ln\lambda_k^e}{dt}\bigg|_{F_{iJ}=\text{cons}} = \dot{\lambda}\frac{\tau'_k}{\left[\frac{2}{3}\sum_{k=1}^{3}\left(\tau'_k\right)^2\right]^{1/2}} \tag{2.462}$$

where $\tau'_k = \tau_k - \sum_{k=1}^{3} \tau_k/3$ is the eigenvalues of the Kirchhoff stress deviator tensor τ_{ij}.

As an example, consider the constitutive material in the principal directions for nearly incompressible materials, which can be described as follows:

$$\psi = \psi\left(\lambda_1^e, \lambda_2^e, \lambda_3^e\right)$$
$$= \frac{\mu}{\overline{\rho}}\left[\left(\ln\frac{\lambda_1^e}{J^{e1/3}}\right)^2 + \left(\ln\frac{\lambda_2^e}{J^{e1/3}}\right)^2 + \left(\ln\frac{\lambda_3^e}{J^{e1/3}}\right)^2\right] + \frac{\mathscr{k}}{2\overline{\rho}}\left(\ln J^e\right)^2 \quad (2.463)$$

where $J^e = \det(F^e_{i\alpha})$, $\mathscr{k} = \lambda + \frac{2}{3}\mu$, and λ and μ are material constants. One can show that the eigenvalues of the Kirchhoff stress tensor τ_k can be obtained as follows (Bonet and Wood, 2008):

$$\tau_k = \overline{\rho}\frac{\partial \psi}{\partial \ln\lambda_k^e} \quad (2.464)$$

The eigenvalues of the Kirchhoff stress deviator tensor τ'_k can be obtained by substituting Eq. (2.462) into Eq. (2.464) as follows:

$$\tau'_k = \tau_k - \frac{\sum_{k=1}^{3} \tau_k}{3} = 2\mu \ln\lambda_k^e - \frac{2}{3}\mu \ln J^e \quad (2.465)$$

2.4.4.8 *Hardening rule*

A hardening rule governs the variation of the material strength, which is commonly represented by the material constants in the yield function, as the material plastically deforms. Change in the material strength can also be envisaged as a change in the geometry and/or the location of the yield surface. The evolved yield surface is commonly called the loading surface. The evolution of the loading surface may follow a combination of isotropic and kinematic hardening rules to capture both loading surface expansion and translation. Accordingly, a mixed hardening law can be described as follows:

$$f\left(\tau_{ij}, d_{ij}^p\right) = \overline{f}\left(\tau_{ij} - \alpha_{ij}\right) - k\left(\overline{\varepsilon}_p\right) \quad (2.466)$$

where $k\left(\overline{\varepsilon}_p\right)$ and α_{ij} govern the expansion and translation of the loading surface, respectively. $\overline{\varepsilon}_p$ is commonly called the equivalent plastic strain, which can be defined as follows:

$$\overline{\varepsilon}_p = \int \dot{\overline{\varepsilon}}_p dt \quad (2.467)$$

where the equivalent plastic strain rate $\dot{\overline{\varepsilon}}_p$ can be defined using a similar framework introduced in the case of the small strain plasticity, as described in

Section 2.2.5.2. For example, the von Mises type equivalent plastic strain rate $\dot{\bar{\varepsilon}}_p$ can be defined using the concept of plastic work per unit volume as follows:

$$\dot{W}^p = \tau_{ij} l_{ij}^p = \bar{\tau}\dot{\bar{\varepsilon}}_p \tag{2.468}$$

where $\bar{\tau} = \sqrt{3J_2}$ for the von Mises loading surface. Accordingly, the equivalent plastic strain rate $\dot{\bar{\varepsilon}}_p$ can be obtained by substituting Eq. (2.459) into Eq. (2.468) as follows:

$$\dot{\bar{\varepsilon}}_p = \sqrt{\frac{2}{3} l_{ij}^p l_{ij}^p} = \dot{\lambda} \tag{2.469}$$

The kinematic hardening parameter can also be defined by following a similar framework developed in the case of the small strain plasticity, as discussed in Section 2.2.5.3. For instance, the kinematic hardening rule developed by Ziegler (1959) for small strain plasticity can be extended for the finite strain framework as follows:

$$\mathcal{L}_v\left(\alpha_{ij}\right) = \dot{\mu}\left(\tau_{ij} - \alpha_{ij}\right) \tag{2.470}$$

where $\mathcal{L}_v(\square)$ is the Lie time derivative as defined in Eq. (2.333), and the parameter $\dot{\mu}$ depends on the plastic deformation history as follows:

$$\dot{\mu} = a\dot{\bar{\varepsilon}}_p \tag{2.471}$$

where a is a material constant. As described in Eq. (2.382), the Lie time derivative is incorporated to define the objective rate of α_{ij}.

2.4.4.9 *Thermodynamically consistent finite strain plasticity models*

As shown in Section 2.2.7, a class of plasticity models can be categorized as thermodynamically consistent models as they satisfy the basic thermodynamic laws. Similar to the small strain models, the same definition can be incorporated for finite strain plasticity. Consider the following Helmholtz free energy for an isotropic material:

$$\Psi = \Psi\left(e_{ij}^{e(0)}, T, \xi_{ij}\right) \tag{2.472}$$

where T is the temperature, ξ_{ij} are the internal state variables, and $e_{ij}^{e(0)}$ is the elastic spatial logarithmic strain tensor. The elastic spatial logarithmic strain tensor $e_{ij}^{e(0)}$ is commonly adopted to describe the Helmholtz free energy of isotropic materials and can be described similar to Eq. (2.316b) as follows:

$$e_{ij}^{e(0)} = \frac{1}{2}\left(\ln B_{ij}^e\right) = \sum_{\alpha=1}^{3} \ln \lambda_k^e \left(n_i^e\right)^k \left(n_j^e\right)^k \tag{2.473}$$

It is common to consider the isotropic and kinematic hardening variables p and $\varsigma_{ij}^{(l)}(l = 1, 2, \ldots, L)$, respectively, as the internal state variables, i.e.,

$\xi_{ij} = \{p, \varsigma_{ij}^{(l)}\}$. Accordingly, Helmholtz free energy described in Eq. (2.472) can be rewritten as follows:

$$\Psi = \Psi\left(e_{ij}^{e(0)}, T, p, \varsigma_{ij}^{(l)}\right) \tag{2.474}$$

The Clausius-Duhem inequality can be described as follows:

$$\tau_{ij} l_{ij} - \overline{\rho}\left(\dot{\Psi} + s\dot{T}\right) - q_i \frac{T_{,i}}{T} \geq 0 \tag{2.475}$$

where $\vec{q}$ is the vector of heat flux and s is the specific entropy per unit mass. One should note that $\tau_{ij} l_{ij} = \tau_{ij} d_{ij}$ due to the fact that τ_{ij} and d_{ij} are symmetric and $\omega_{ij} = l_{ij} - d_{ij}$ is antisymmetric. $\dot{\Psi}$ can be obtained by selecting the functional form presented in Eq. (2.474) as follows:

$$\dot{\Psi} = \frac{\partial \Psi}{\partial e_{ij}^{e(0)}} \dot{e}_{ij}^{e(0)} + \frac{\partial \Psi}{\partial T} \dot{T} + \frac{\partial \Psi}{\partial p} \dot{p} + \sum\nolimits_{l=1}^{L} \frac{\partial \Psi}{\partial \varsigma_{ij}^{(l)}} \mathcal{L}_v\left(\varsigma_{ij}^{(l)}\right) \tag{2.476}$$

where $\mathcal{L}_v(\square)$ is the Lie time derivative as defined in Eq. (2.333) which is used to obtain an objective rate of $\varsigma_{ij}^{(l)}$ [see Eq. (2.382)]. The first term in Eq. (2.476) can be further simplified by using Eqs. (2.418), (2.443), (2.473) as follows:

$$\begin{aligned} \frac{\partial \Psi}{\partial e_{ij}^{e(0)}} \dot{e}_{ij}^{e(0)} &= \frac{1}{2} \frac{\partial \Psi}{\partial e_{ij}^{e(0)}} \frac{\partial\left(\ln B_{ij}^{e}\right)}{\partial B_{kl}^{e}} \mathcal{L}_v\left(B_{kl}^{e}\right) = \frac{1}{2} \frac{\partial \Psi}{\partial e_{ij}^{e(0)}} \left(B_{im}^{e}\right)^{-1} \mathcal{L}_v\left(B_{mj}^{e}\right) \\ &= \frac{\partial \Psi}{\partial e_{ij}^{e(0)}} \left(d_{ij} - d_{ij}^{p}\right) \end{aligned} \tag{2.477}$$

One should note that the second framework of deriving the plastic velocity gradient l_{ij}^{p} introduced in Section 2.4.4.4 and Eqs. (2.437)–(2.444) is incorporated to define d_{ij}^{p}.

One can rewrite the Clausius-Duhem inequality by substituting Eq. (2.476) into Eq. (2.475) as follows:

$$\begin{aligned} &\left(\tau_{ij} - \overline{\rho} \frac{\partial \Psi}{\partial e_{ij}^{e(0)}}\right) d_{ij} + \overline{\rho} \frac{\partial \Psi}{\partial e_{ij}^{e(0)}} d_{ij}^{p} - \overline{\rho}\left(\frac{\partial \Psi}{\partial T} + s\right) \dot{T} - \overline{\rho} \frac{\partial \Psi}{\partial p} \dot{p} \\ &- \overline{\rho} \sum\nolimits_{l=1}^{L} \frac{\partial \Psi}{\partial \varsigma_{ij}^{(l)}} \mathcal{L}_v\left(\varsigma_{ij}^{(l)}\right) - q_i \frac{T_{,i}}{T} \geq 0 \end{aligned} \tag{2.478}$$

The following thermodynamic state laws can be derived by using Eq. (2.478) as follows:

$$\tau_{ij} = \overline{\rho} \frac{\partial \Psi}{\partial e_{ij}^{e(0)}} \tag{2.479}$$

$$s = -\frac{\partial \Psi}{\partial T} \tag{2.480}$$

$$\alpha_{ij}^{(l)} = \overline{\rho} \frac{\partial \Psi}{\partial \varsigma_{ij}^{(l)}} \tag{2.481}$$

$$R = \overline{\rho} \frac{\partial \Psi}{\partial p} \tag{2.482}$$

The relation between the observable and internal variables and their corresponding thermodynamic conjugate forces (Table 2.7) are described in Eqs. (2.479)–(2.482). Eq. (2.478) can be rewritten using Eqs. (2.479)–(2.482) as follows:

$$\tau_{ij} d_{ij}^{p} - R\dot{p} - \sum_{l=1}^{L} \alpha_{ij}^{(l)} \mathcal{L}_v \left(\varsigma_{ij}^{(l)} \right) - q_i \frac{T_{,i}}{T} \geq 0 \tag{2.483}$$

Eq. (2.483) can be further simplified by decoupling the plasticity and thermal dissipations as follows:

$$\Pi^{T} = -q_i \frac{T_{,i}}{T} \geq 0 \tag{2.484}$$

$$\Pi^{p} = \tau_{ij} d_{ij}^{p} - R\dot{p} - \sum_{l=1}^{L} \alpha_{ij}^{(l)} \dot{\varsigma}_{ij}^{(l)} \geq 0 \tag{2.485}$$

Eqs. (2.484), (2.485) define the dissipation due to the thermal and plasticity activations, Π^T and Π^p, respectively. The plastic dissipation energy Π^p can be described as the energy dissipated due to plasticity $\tau_{ij} d_{ij}^{p}$ excluding the plastic energy stored in the material as the isotropic and kinematic hardenings, i.e., $-R\dot{p} - \sum_{l=1}^{L} \alpha_{ij}^{(l)} \mathcal{L}_v \left(\varsigma_{ij}^{(l)} \right)$. Another equation is still required to describe the

TABLE 2.7 Finite strain thermodynamic state variables and their corresponding conjugate forces

State variables		Associated
Observable	Internal	conjugates
d_{ij}		τ_{ij}
T		s
	$e_{ij}^{e(0)}$	τ_{ij}
	d_{ij}^{p}	$-\tau_{ij}$
	$T_{,i}$	q_i
	p	R
	$\varsigma_{ij}^{(l)}$	$\alpha_{ij}^{(l)}$

evolution of internal state variables. To do so, a plastic potential, which is convex, non-negative, and zero at the origin, is defined as follows:

$$F = F\left(\tau_{ij}, \alpha_{ij}^{(l)}, R\right) \tag{2.486}$$

The evolution of the internal state variables can be described as follows (see Section 2.2.7):

$$d_{ij}^{p} = \dot{\lambda}\frac{\partial F}{\partial \tau_{ij}} \tag{2.487}$$

$$\mathcal{L}_v\left(\varsigma_{ij}^{(l)}\right) = -\dot{\lambda}\frac{\partial F}{\partial \alpha_{ij}^{(l)}} \tag{2.488}$$

$$\dot{p} = -\dot{\lambda}\frac{\partial F}{\partial R} \tag{2.489}$$

where $\dot{\lambda}$ is the plastic multiplier.

One can maximize the plastic dissipation energy Π^p using the Lagrange multiplier method by building the objective function Ω by assuming that the plastic potential and yield surface are the same, i.e., $F = f$, as follows:

$$\Omega = \Pi^p - \dot{\lambda} f \tag{2.490}$$

One should note that during the plastic deformation, $f = 0$. The plasticity dissipation Π^p can be maximized by maximizing the objective function Ω as follows:

$$\frac{\partial \Omega}{\partial \tau_{ij}} = 0 \tag{2.491}$$

$$\frac{\partial \Omega}{\partial \alpha_{ij}^{(l)}} = 0 \tag{2.492}$$

$$\frac{\partial \Omega}{\partial R} = 0 \tag{2.493}$$

Considering Eqs. (2.487)–(2.489), the evolution of the internal state variables can be obtained as follows:

$$d_{ij}^{p} = \dot{\lambda}\frac{\partial f}{\partial \tau_{ij}} \tag{2.494}$$

$$\mathcal{L}_v\left(\varsigma_{ij}^{(l)}\right) = -\dot{\lambda}\frac{\partial f}{\partial \alpha_{ij}^{(l)}} \tag{2.495}$$

$$\dot{p} = -\dot{\lambda}\frac{\partial f}{\partial R} \tag{2.496}$$

In other words, considering the yield surface as the plastic potential, i.e., $F = f$, maximizes the plastic dissipation energy Π^p.

2.4.4.10 Rate-dependent plasticity: Models with the von Mises yield surface

The finite strain rate-independent plasticity models have been presented so far. However, the finite strain rate-dependent plasticity models, which are also termed as viscoplasticity models, should be developed to accurately capture the rate dependent material properties. In this section, the finite strain rate-dependent plasticity models are developed for isotropic materials by following the general framework of the rate-independent finite strain plasticity model described in Sections 2.4.4.1–2.4.4.8. Accordingly, the multiplicative decomposition of the deformation gradient is incorporated as follows:

$$F_{iJ} = F^e_{i\alpha} F^p_{\alpha J} \tag{2.497}$$

The constitutive model of a finite strain elastoplastic isotropic material can be described as follows:

$$\tau_{ij} = \overline{\rho} \frac{\partial \Psi}{\partial e^{e(0)}_{ij}} \tag{2.498}$$

The yield surface f, as defined in Section 2.4.4.6, describes the elastic domain for the rate-dependent model. Unlike the rate-independent models where $f \leq 0$, the yield surface of a rate-dependent model may be greater than zero to describe the dynamic yield surface attributed to different strain rates. The von Mises yield surface can be defined as follows:

$$f\left(\tau_{ij}, d^p_{ij}\right) = \sqrt{\frac{3}{2}\left(\tau'_{ij} - \alpha_{ij}\right)\left(\tau'_{ij} - \alpha_{ij}\right)} - \sigma_y\left(\overline{\varepsilon}_p\right) \tag{2.499}$$

where $\sigma_y\left(\overline{\varepsilon}_p\right)$ and α_{ij} represent the expansion and the rigid translation of the loading surface, respectively. The plastic rate of deformation tensor d^p_{ij} can be defined as follows:

$$d^p_{ij} = l^p_{ij} = \dot{\lambda} \frac{\partial f}{\partial \tau_{ij}} = \dot{\lambda} \frac{\sqrt{\frac{3}{2}}\left(\tau'_{ij} - \alpha_{ij}\right)}{\sqrt{\left(\tau'_{kl} - \alpha_{kl}\right)\left(\tau'_{kl} - \alpha_{kl}\right)}} \tag{2.500}$$

Similar to the small strain rate-dependent plasticity models, the plastic multiplier $\dot{\lambda}$ should be explicitly defined in the case of rate-dependent plasticity and the consistency condition should not be solved. For example, three models of Bingham (1922), Perzyna (1966, 1971), Voyiadjis and Abed (2006), and Peric (1993) presented in Section 2.2.8 can be used to define the evolution of plastic multiplier $\dot{\lambda}$.

2.5 Finite strain plasticity: Nonlocal models

As discussed in Section 2.1, size effects in materials cannot be handled using local models. Furthermore, many aspects of nonlocal plasticity models are presented in Section 2.3 including the basics and advanced models. However, the presented models follow the small strain theory assumptions, which are not applicable to the finite strain theory. In this section, two nonlocal models of Gurtin and Anand (2005b), Voyiadjis and Abu Al-Rub (2006), and Abu Al-Rub and Voyiadjis (2006) are presented which incorporate the finite strain framework.

2.5.1 Gradient plasticity models: Gurtin and Anand

Gurtin and Anand (2005b) generalized their strain gradient viscoplastic nonlocal model which was developed using small strain assumptions (Gurtin and Anand, 2005a) and incorporated the framework of finite strain theory. They satisfied the microforce balance using the developed microstresses system. The model was developed for isotropic materials with the assumption of zero plastic rotation. They enhanced the conventional boundary conditions to incorporate the microscopic boundary conditions. Gurtin and Anand (2005b) incorporated the multiplicative decomposition of deformation gradient F_{iJ} as follows:

$$F_{iJ} = F^e_{i\alpha} F^p_{\alpha J} = x_{i,J} \tag{2.501}$$

where F_{iJ} is the deformation gradient which is multiplicatively decomposed into the elastic and plastic deformation gradients $F^e_{i\alpha}$, and $F^p_{\alpha J}$, respectively, and $x_i = x_i(X_I, t)$. The indices are chosen to show which coordinate system is considered, which are lower case alphabets for deformed configuration, upper case alphabets for undeformed configurations, and Greek alphabets for intermediate configuration (see Section 2.4.4.1). The velocity gradient can then be defined as follows:

$$l_{ij} = \dot{F}_{iJ} F^{-1}_{Jj} = l^e_{ij} + F^e_{i\alpha} l^p_{\alpha\beta} \left(F^e_{\beta j}\right)^{-1} \tag{2.502}$$

where l^e_{ij} and $l^p_{\alpha\beta}$ are the elastic and plastic velocity gradients, respectively, which can be defined as follows:

$$l^e_{ij} = \dot{F}^e_{i\alpha} \left(F^e_{\alpha j}\right)^{-1} \tag{2.503}$$

$$l^p_{\alpha\beta} = \dot{F}^p_{\alpha I} \left(F^p_{I\beta}\right)^{-1} \tag{2.504}$$

l^e_{ij} and $l^p_{\alpha\beta}$ can be additively decomposed into the symmetric and antisymmetric tensors as follows:

$$l^e_{ij} = d^e_{ij} + \omega^e_{ij} = \frac{1}{2}\left(l^e_{ij} + l^e_{ji}\right) + \frac{1}{2}\left(l^e_{ij} - l^e_{ji}\right) \tag{2.505}$$

$$l^p_{\alpha\beta} = d^p_{\alpha\beta} + \omega^p_{\alpha\beta} = \frac{1}{2}\left(l^p_{\alpha\beta} + l^p_{\beta\alpha}\right) + \frac{1}{2}\left(l^p_{\alpha\beta} - l^p_{\beta\alpha}\right) \tag{2.506}$$

where d^e_{ij} and $d^p_{\alpha\beta}$ are the elastic and plastic symmetric plastic rate of deformation tensors, respectively, and ω^e_{ij} and $\omega^p_{\alpha\beta}$ are the elastic and plastic asymmetric spin tensors, respectively. In the case of isotropic material, it can be shown that the spin tensor $\omega^p_{ij} = 0$ and $d^p_{\alpha\beta} = l^p_{\alpha\beta}$. Furthermore, Gurtin and Anand (2005b) focuses on materials which are plastically incompressible, i.e., det $F^p_{\alpha J} = 1$ and $l^p_{\gamma\gamma} = d^p_{\gamma\gamma} = 0$. In other words, $d^p_{\alpha\beta}$ is symmetric-deviatoric.

The polar decomposition of elastic deformation gradient $F^e_{i\alpha}$ can be described as follows:

$$F^e_{i\alpha} = R^e_{i\beta} U^e_{\beta\alpha} \tag{2.507}$$

where $R^e_{i\beta}$ and $U^e_{\beta\alpha}$ are the rotation and material stretch tensors, respectively. The elastic right Cauchy-Green tensors $C^e_{\alpha\beta}$ can then be described as follows:

$$C^e_{\alpha\beta} = F^e_{i\alpha} F^e_{i\beta} = U^e_{\gamma\alpha} U^e_{\gamma\beta} \tag{2.508}$$

Finally, the elastic material strain tensor $E^e_{\alpha\beta}$ can be obtained as follows:

$$E^e_{\alpha\beta} = \frac{1}{2}\left(C^e_{\alpha\beta} - \delta_{\alpha\beta}\right) \tag{2.509}$$

The plastic strain gradient model is incorporated in the model by introducing the Burgers tensor $G_{\alpha\beta}$ as follows:

$$G_{\alpha\beta} = F^p_{\alpha J}\left(\text{curl}\left[F^p_{\zeta K}\right]\right)_{J\beta} = F^p_{\alpha J} \epsilon_{IKJ} F^p_{\beta K,I} \tag{2.510}$$

which represents the role of the macroscopic Burgers vector. The term $(\text{curl}[F^p_{\zeta K}])_{J\beta}$ is the $J\beta$ component of the second order tensor curl$[F^p_{\zeta K}]$.

The internal power P_{int} can be defined by generalizing the small strain description of Gurtin and Anand (2005a) presented in Section 2.3.1.5 as follows:

$$P_{int} = \int_V \left(S^e_{i\alpha} \dot{F}^e_{i\alpha} + \mathcal{X}_{\alpha\beta} d^p_{\alpha\beta} + \mathcal{S}_{\alpha\beta\gamma} d^p_{\alpha\beta,\gamma}\right) dV \tag{2.511}$$

where $S^e_{i\alpha}$ is the macroscopic stress tensor conjugate to $\dot{F}^e_{i\alpha}$, $\mathcal{X}_{\alpha\beta}$ is the plastic microstress tensor conjugate to $d^p_{\alpha\beta}$, and $\mathcal{S}_{\alpha\beta\gamma}$ is the polar plastic microstress tensor conjugate to $d^p_{\alpha\beta,\gamma}$. In order to present the frame-indifferent description of internal power P_{int}, Gurtin and Anand (2005b) defined the elastic stress tensor $T^e_{\alpha\beta}$ which is conjugate to the elastic material strain tensor $E^e_{\alpha\beta}$ defined in Eq. (2.509) as follows:

$$S^e_{i\alpha} \dot{F}^e_{i\alpha} = T^e_{\alpha\beta} \dot{E}^e_{\alpha\beta} \tag{2.512}$$

Accordingly, the frame-indifferent internal power P_{int} can be defined by substituting Eq. (2.512) into Eq. (2.511) as follows:

$$P_{int} = \int_V \left(T^e_{\alpha\beta} \dot{E}^e_{\alpha\beta} + \mathcal{X}_{\alpha\beta} d^p_{\alpha\beta} + \mathcal{S}_{\alpha\beta J} d^p_{\alpha\beta,J}\right) dV \tag{2.513}$$

The external power P_{ext} can be defined as follows:

$$P_{ext} = \int_V b_i \dot{x}_i dV + \int_S \left(t_i \dot{x}_i + m_{\alpha\beta} d^p_{\alpha\beta} \right) dS \tag{2.514}$$

where t_i and b_i are the traction and xternal body force vectors conjugate to the arbitrary macroscopic velocity vector $\dot{x}_i$ respectively, and $m_{\alpha\beta}$ is the microtraction tensor conjugate to the plastic strain rate tensor due to the arbitrary macroscopic velocity. Since $d^p_{\alpha\beta}$ is symmetric-deviatoric, it is assumed that $m_{\alpha\beta}$ is also symmetric-deviatoric. One should note that the integrations incorporated to define P_{int} and P_{ext} in Eqs. (2.513), (2.514) are performed in the undeformed configuration.

The principle of virtual power is incorporated here assuming that the power related to each independent kinematics can be described using the associated force consistent with its own balance. However, the kinematic descriptions incorporated in internal and external power definition, i.e., $\dot{x}_i$, $\dot{F}^e_{i\alpha}$, and $d^p_{\alpha\beta}$ are not independent. In other words, they are constrained according to Eqs. (2.502)–(2.504) and zero spin tensor condition, i.e., $\omega^p_{ij} = 0$. These constraints can be reformulated as follows:

$$\dot{x}_{i,J} F^{-1}_{Jj} = \dot{F}^e_{i\alpha} \left(F^e_{\alpha j} \right)^{-1} + F^e_{i\alpha} d^p_{\alpha\beta} \left(F^e_{\beta j} \right)^{-1} \tag{2.515}$$

To address this issue, Gurtin and Anand (2005b) introduce the generalized virtual velocity $\mathcal{V}$ as follows:

$$\mathcal{V} = \left(\tilde{\dot{x}}_i, \tilde{\dot{F}}^e_{i\alpha}, \tilde{d}^p_{\alpha\beta} \right) \tag{2.516}$$

where $\tilde{\dot{x}}_i, \tilde{\dot{F}}^e_{i\alpha}$ and $\tilde{d}^p_{\alpha\beta}$ are the set of virtual velocities to be specified independently which satisfies Eq. (2.515) at some arbitrarily chosen but fixed time, in which x_i and $F^e_{i\alpha}$ are known. Accordingly, P_{int} and P_{ext} can be defined as follows:

$$P_{int} = \int_V \left(S^e_{i\alpha} \tilde{\dot{F}}^e_{i\alpha} + \mathcal{X}_{\alpha\beta} \tilde{d}^p_{\alpha\beta} + S_{\alpha\beta J} \tilde{d}^p_{\alpha\beta,J} \right) dV \tag{2.517}$$

$$P_{ext} = \int_V b_i \tilde{\dot{x}}_i dV + \int_S \left(t_i \tilde{\dot{x}}_i + m_{\alpha\beta} \tilde{d}^p_{\alpha\beta} \right) dS \tag{2.518}$$

Accordingly, the principle of virtual power can be incorporated to develop the material governing equations, which can be defined as follows:

$$P_{int} = P_{ext} \tag{2.519}$$

Eq. (2.519) should be satisfied for any set of generalized virtual velocity $\mathcal{V}$.

To obtain the macroscopic governing equations of the model, a generalized virtual velocity in which $\tilde{d}^p_{\alpha\beta} \equiv 0$ is considered. Accordingly, Eq. (2.515) can be rewritten as follows:

$$\tilde{\dot{F}}^e_{i\alpha} = \tilde{\dot{x}}_{i,J}\left(F^p_{J\alpha}\right)^{-1} \quad (2.520)$$

Considering Eqs. (2.517)–(2.520) and using the divergence theorem, the macroscopic power relation can be obtained as follows:

$$\int_V (S_{iJ,J} + b_i)\tilde{\dot{x}}_i dV + \int_S (t_i - S_{iJ}n_J)\tilde{\dot{x}}_i dS = 0 \quad (2.521)$$

where S_{iJ} is the first Piola-Kirchhoff stress tensor which can be defined as follows:

$$S_{iJ} = S^e_{i\alpha}\left(F^p_{J\alpha}\right)^{-1} = F^e_{i\alpha}T^e_{\alpha\beta}\left(F^p_{J\beta}\right)^{-1} \quad (2.522)$$

To obtain the microscopic governing equations of the model, a generalized virtual velocity in which $\tilde{\dot{x}}_i \equiv 0$ is considered. Accordingly, Eq. (2.515) can be rewritten as follows:

$$\tilde{\dot{F}}^e_{i\alpha} = -F^e_{i\beta}\tilde{d}^p_{\beta\alpha} \quad (2.523)$$

Considering Eqs. (2.517), (2.518), (2.519), (2.523) and using the divergence theorem, the microscopic power relation can be obtained as follows:

$$\int_S \left(m_{\alpha\beta} - \mathcal{S}_{\alpha\beta J}n_J\right)\tilde{d}^p_{\alpha\beta}dA - \int_V \left(\mathcal{X}_{\alpha\beta} - F^e_{i\alpha}S^e_{i\beta} - \mathcal{S}_{\alpha\beta J,J}\right)\tilde{d}^p_{\alpha\beta}dV = 0 \quad (2.524)$$

The macroscopic and microscopic governing equations and corresponding boundary conditions can be obtained using Eqs. (2.521), (2.524) and considering the fact that $d^p_{\alpha\beta}$ is symmetric-deviatoric as follows:

$$S_{iJ,J} + b_i = 0 \text{ on } V \text{ (macroforce balance)} \quad (2.525)$$

$$\text{symm}_0\left(F^e_{i\alpha}S^e_{i\beta}\right) = \mathcal{X}_{\alpha\beta} - \mathcal{S}_{\alpha\beta J,J} \text{ on } V \text{ (microforce balance)} \quad (2.526)$$

$$t_i = S_{iJ}n_J \text{ on } \partial S \text{ (macrotraction condition)} \quad (2.527)$$

$$m_{\alpha\beta} = \mathcal{S}_{\alpha\beta J}n_J \text{ on } \partial S \text{ (microtraction condition)} \quad (2.528)$$

where $\text{symm}_0(A_{ij}) = 1/2(A_{ij} + A_{ji}) - 1/3A_{kk}\delta_{ij}$. Eq. (2.526) can be rewritten by using Eqs. (2.508), (2.509), (2.512) as follows:

$$\text{symm}_0\left(C^e_{\alpha\gamma}T^e_{\gamma\beta}\right) = \mathcal{X}_{\alpha\beta} - \mathcal{S}_{\alpha\beta J,J} \text{ on } V \text{ (microforce balance)} \quad (2.529)$$

Eq. (2.529) can be rewritten by considering the fact that that $C^e_{\alpha\gamma}T^e_{\gamma\beta}$ is symmetric as follows:

$$\left(C^e_{\alpha\gamma}T^e_{\gamma\beta}\right)' = \mathcal{X}_{\alpha\beta} - \mathcal{S}_{\alpha\beta J,J} \text{ on } V \text{ (microforce balance)} \tag{2.530}$$

where $(C^e_{\alpha\gamma}T^e_{\gamma\beta})'$ is the deviatoric part of $C^e_{\alpha\gamma}T^e_{\gamma\beta}$.

Gurtin and Anand (2005b) incorporate the plasticity strain gradient into the Helmholtz free energy Ψ by including the Burgers tensor $G_{\alpha\beta}$. They assumed that the free energy is a function of elastic material strain tensor $E^e_{\alpha\beta}$ and Burgers tensor $G_{\alpha\beta}$ as follows:

$$\Psi = \Psi\left(E^e_{\alpha\beta}, G_{\alpha\beta}\right) \tag{2.531}$$

Assuming a purely mechanical theory, the second law of thermodynamics states that the increase in Ψ should be less than the power expended volume V, which can be described as follows:

$$\int_V \rho\dot{\Psi}\,dV \le P_{int} = P_{ext} \tag{2.532}$$

where ρ is the material density. One can define the local free energy inequality by substituting Eq. (2.513) into Eq. (2.532) as follows:

$$\rho\dot{\Psi} - T^e_{\alpha\beta}\dot{E}^e_{\alpha\beta} - \mathcal{X}_{\alpha\beta}d^p_{\alpha\beta} - \mathcal{S}_{\alpha\beta J}d^p_{\alpha\beta,J} \le 0 \tag{2.533}$$

Gurtin and Anand (2005b) considered the Helmholtz free energy Ψ as follows:

$$\Psi = \frac{1}{\rho}\mu E'^e_{\alpha\beta}E'^e_{\alpha\beta} + \frac{1}{2\rho}k\left(E^e_{\alpha\alpha}\right)^2 + \frac{1}{2\rho}\mu l^2_{en}G_{\alpha\beta}G_{\alpha\beta} \tag{2.534}$$

where μ, k, and l_{en} are the material constants denoting the elastic shear modulus, bulk modulus, and the energetic material length scale, respectively, and $E'^e_{\alpha\beta} = E^e_{\alpha\beta} - (E^e_{\gamma\gamma}/3)\delta_{\alpha\beta}$ is the elastic material strain deviator tensor. Accordingly, the elastic stress tensor $T^e_{\alpha\beta}$ can be obtained as follows:

$$T^e_{\alpha\beta} = 2\mu E'^e_{\alpha\beta} + kE^e_{\gamma\gamma}\delta_{\alpha\beta} \tag{2.535}$$

The thermodynamic defect stress $R^G_{\alpha\beta}$ which is conjugate to $\dot{G}_{\alpha\beta}$ can be described as follows:

$$R^G_{\alpha\beta} = \rho\frac{\partial\Psi}{\partial G_{\alpha\beta}} = \mu l^2_{en}G_{\alpha\beta} \tag{2.536}$$

Gurtin and Anand (2005b) showed that:

$$R^G_{\alpha\beta}\dot{G}_{\alpha\beta} = \mathcal{X}^{en}_{\alpha\beta}d^p_{\alpha\beta} + \mathcal{S}^{en}_{\alpha\beta J}d^p_{\alpha\beta,J} \tag{2.537}$$

where

$$\mathcal{X}_{\alpha\beta}^{en} = \text{symm}_0\left(R_{\alpha\gamma}^{G} G_{\beta\gamma} + R_{\gamma\alpha}^{G} G_{\gamma\beta}\right) = \mu l_{en}^2 \left(G_{\alpha\gamma} G_{\beta\gamma} + G_{\gamma\alpha} G_{\gamma\beta}\right)' \tag{2.538}$$

$$\mathcal{S}_{\alpha\beta J}^{en} = \text{symm}_0\left[\epsilon_{\alpha\varsigma\kappa} R_{\varsigma\beta}^{G} \left(F_{J\kappa}^{p}\right)^{-1}\right] = \mu l_{en}^2 \text{symm}_0\left[\epsilon_{\alpha\varsigma\kappa} G_{\varsigma\beta} \left(F_{J\kappa}^{p}\right)^{-1}\right] \tag{2.539}$$

where $(G_{\alpha\gamma}G_{\beta\gamma} + G_{\gamma\alpha}G_{\gamma\beta})'$ is the deviatoric part of $G_{\alpha\gamma}G_{\beta\gamma} + G_{\gamma\alpha}G_{\gamma\beta}$ and $\epsilon_{\alpha\varsigma\kappa}$ is the permutation tensor. The dissipative microstresses tensors can be defined as follows:

$$\mathcal{X}_{\alpha\beta}^{dis} = \mathcal{X}_{\alpha\beta} - \mathcal{X}_{\alpha\beta}^{en} \tag{2.540}$$

$$\mathcal{S}_{\alpha\beta J}^{dis} = \mathcal{S}_{\alpha\beta J} - \mathcal{S}_{\alpha\beta J}^{en} \tag{2.541}$$

Using the similar procedure presented in Section 2.3.1.5, the dissipation potential can be defined as follows:

$$\mathcal{D} = \mathcal{X}_{\alpha\beta}^{dis} d_{\alpha\beta}^{p} + \mathcal{S}_{\alpha\beta J}^{dis} d_{\alpha\beta,J}^{p} \geq 0 \tag{2.542}$$

The viscoplastic constitutive models for gradient plasticity model was presented by Gurtin and Anand (2005b) as follows:

$$\mathcal{X}_{\alpha\beta}^{dis} = S\left(\frac{d^p}{d_0}\right)^m \frac{d_{\alpha\beta}^{p}}{d^p} \tag{2.543}$$

$$\mathcal{S}_{\alpha\beta J}^{dis} = l_{dis}^2 S_Y \left(\frac{d^p}{d_0}\right)^m \frac{d_{\alpha\beta,J}^{p}}{d^p} \tag{2.544}$$

where l_{dis}, d_0, m, and S_Y are the material constants which denote the dissipative material length scale, reference equivalent plastic strain rate, rate-sensitivity parameter, and initial plastic flow, i.e., yield stress, respectively. d^p is the nonlocal equivalent plastic strain rate, and S is the resistance to the plastic flow initially equal to S_Y, which can be defined as follows:

$$d^p = \sqrt{d_{\alpha\beta}^{p} d_{\alpha\beta}^{p} + l_{dis}^2 d_{\alpha\beta,J}^{p} d_{\alpha\beta,J}^{p}} \tag{2.545}$$

$$\dot{S} = H(S) d^p \tag{2.546}$$

where $H(S)$ is a function that describes the material hardening.

The dissipation potential can be obtained by substituting Eqs. (2.543), (2.544) in Eq. (2.542) as follows:

$$\mathcal{D} = \left(\frac{d^p}{d_0}\right)^m \frac{S d_{\alpha\beta}^{p} d_{\alpha\beta}^{p} + l_{dis}^2 S_Y d_{\alpha\beta,J}^{p} d_{\alpha\beta,J}^{p}}{d^p} \geq 0 \tag{2.547}$$

The final form of the flow rule can be obtained using the Eqs. (2.530), (2.538), (2.539), (2.540), (2.541), (2.543), (2.544), (2.545), (2.546) as follows:

$$S_{\alpha\beta}^{FL} - \text{Energetic Backstress tensor} = \text{Dissipative Hardening tensor} \tag{2.548}$$

$$\text{Energetic Backstress} = \mu l_{en}^2 \left\{ \left(G_{\alpha\gamma}G_{\beta\gamma} + G_{\gamma\alpha}G_{\gamma\beta}\right)' - \frac{\partial}{\partial X_J}\left\{ \text{symm}_0 \left[\epsilon_{\alpha\varsigma\kappa} G_{\varsigma\beta} \left(F_{J\kappa}^p\right)^{-1} \right] \right\} \right\} \tag{2.549}$$

$$\text{Dissipative Hardening} = S\left(\frac{d^p}{d_0}\right)^m \frac{d_{\alpha\beta}^p}{d^p} - l_{dis}^2 S_Y \frac{\partial}{\partial X_J}\left[\left(\frac{d^p}{d_0}\right)^m \frac{d_{\alpha\beta,J}^p}{d^p}\right] \tag{2.550}$$

where $S_{\alpha\beta}^{FL}$ denotes the flow stress which can be defined as follows:

$$S_{\alpha\beta}^{FL} = S_{\beta\alpha}^{FL} \overset{\text{def}}{=} \left(C_{\alpha\gamma}^e T_{\gamma\beta}^e\right)' = \left(F_{i\alpha}^e S_{i\beta}^e\right)' = \left(F_{i\alpha}^e S_{iJ} F_{\beta J}^p\right)'$$

The obtained flow rule is a partial differential equation for the plastic deformation gradient tensor $F_{\alpha J}^p$.

2.5.2 Gradient plasticity damage model: Voyiadjis and his co-workers

Voyiadjis and Abu Al-Rub (2006) and Abu Al-Rub and Voyiadjis (2006) extended their model for small strain (Voyiadjis et al., 2004) to include the finite strain theory. The model captures the strong coupling between nonlocal rate-dependent plasticity and nonlocal anisotropic rate-dependent damage for finite strain dynamic problems. The fundamental of the finite strain model is similar to the small strain one presented in Section 2.3.1.6. Accordingly, here, the formulations which are different from the small strain theory are highlighted.

An anisotropic damage model is defined in which the Kirchhoff stress tensor τ_{ij} and its effective counterpart $\hat{\tau}_{ij}$ are connected using a nonlocal fourth order tensor $\overline{M}_{ikjl}$ as follows:

$$\hat{\tau}_{ij} = \overline{M}_{ikjl}\tau_{kl} \tag{2.551}$$

where $\overline{M}_{ikjl}$ is described using the nonlocal second order damage tensor $\overline{\phi}_{ij}$ as follows:

$$\overline{M}_{ikjl} = 2\left[\left(\delta_{ik} - \overline{\phi}_{ik}\right)\delta_{jl} + \delta_{ik}\left(\delta_{jl} - \overline{\phi}_{jl}\right)\right]^{-1} \tag{2.552}$$

Voyiadjis and Abu Al-Rub (2006) and Abu Al-Rub and Voyiadjis (2006) assumed that in the elastic strains are small. Accordingly, the additive decomposition of rate of deformation tensor d_{ij} into the elastic and viscoinelastic tensors, i.e., d_{ij}^e and d_{ij}^{vi}, respectively, is incorporated as follows:

$$d_{ij} = d_{ij}^e + d_{ij}^{vi} \tag{2.553}$$

Furthermore, the viscoinelastic rate of deformation tensor d_{ij}^{vi} is additively decomposed into the viscodamage and viscoplastic tensors, i.e., d_{ij}^{vp} and d_{ij}^{vd}, respectively, as follows:

$$d_{ij}^{vi} = d_{ij}^{vp} + d_{ij}^{vd} \tag{2.554}$$

The spatial strain tensor e_{ij} is also additively decomposed as follows:

$$e_{ij} = e_{ij}^{e} + e_{ij}^{vp} + e_{ij}^{vd} \tag{2.555}$$

Next, the Helmholtz free energy Ψ for finite strain problem is defined as follows:

$$\Psi = \Psi\left(e_{ij}^{e}, T, T_{,i}, J; \overline{\aleph}_{k}\right) \tag{2.556}$$

where $J = \det(F_{iJ})$ and $\overline{\aleph}_k$ are phenomenological nonlocal internal state variables, which can be described as follows:

$$\overline{\aleph}_k = \overline{\aleph}_k\left(\Xi_n, \nabla^2\Xi_n\right) \tag{2.557}$$

Similar to the small strain theory, the Helmholtz free energy Ψ can be additively decomposed into three potentials related to the thermo-elastic Ψ^{te}, rate-dependent thermo-plastic Ψ^{tvp}, and rate dependent thermo-damage Ψ^{tvd} models as follows:

$$\Psi = \Psi^{te}\left(e_{ij}^{e}, J^{e}, T, T_{,i}, \phi_{ij}, \nabla^2\phi_{ij}\right) + \Psi^{tvp}\left(T, T_{,i}, p, \varsigma_{ij}, \nabla^2 p, \nabla^2\varsigma_{ij}\right)$$

$$+ \Psi^{tvd}\left(T, T_{,i}, r, \Gamma_{ij}, \phi_{ij}, \nabla^2 r, \nabla^2\Gamma_{ij}, \nabla^2\phi_{ij}\right) \tag{2.558}$$

The Clausius-Duhem inequality can be described as follows:

$$\sigma_{ij}d_{ij} - \rho\left(\dot{\Psi} + s\dot{T}\right) - q_i\frac{T_{,i}}{T} \geq 0 \tag{2.559}$$

where σ_{ij} is the Cauchy stress tensor, ρ is the spatial density, s is the specific entropy per unit mass, and $\vec{q}$ is the vector of heat flux. The rest of the derivations are similar to those presented in Section 2.3.1.6.

2.5.2.1 Numerical applications

Several numerical examples were conducted by Voyiadjis and Abu Al-Rub (2006) to validate and test this computational framework and numerical algorithm using the coupled nonlocal viscoplastic damage model. The developed constitutive microdamage model was implemented in the explicit commercial finite element code ABAQUS/Explicit using the user material subroutine VUMAT.

2.5.2.1.1 Strip with a fixed edge

A two-dimensional initial boundary value problem for a specimen of length 100 mm and width 20 mm is considered as the first example. The bottom side of the specimen is fixed and the topside is movable. The loading is enforced by a velocity profile shown in Fig. 2.21 that acts at the free end of the specimen. Three mesh discretizations of 15 × 50, 25 × 70, and 30 × 100 meshes are used

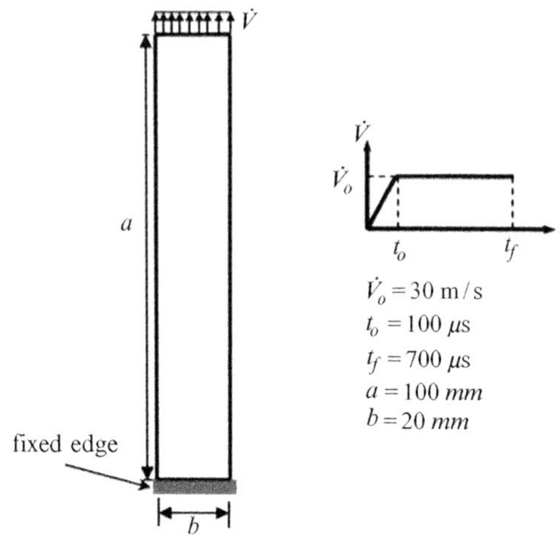

FIG. 2.21 Problem description of a strip with a fixed edge under impact tensile loading. *(After Voyiadjis, G.Z., Abu Al-Rub, R.K., 2006. A finite strain plastic-damage model for high velocity impacts using combined viscosity and gradient localization limiters: part II—numerical aspects and simulations. Int. J. Damage Mech. 15, 335–373.)*

with eight-noded rectangular elements. The model material constants of the target material of Weldox 460 E steel and for the projectile material of hardened Arne tool steel are listed in Table 2.8. The deformed configurations shown in Fig. 2.22 indicate the formation of a neck and a pronounced shear band of

TABLE 2.8 Material constants of the target material of Weldox 460 E steel and for the projectile material of hardened Arne tool steel.

Target material of Weldox 460 E steel			
$E = 200$ GPa	$a_1 = 400$ MPa	$h_1 = 0$	$r_0 = 0$
$v = 0.33$	$a_2 = 20$ GPa	$h_2 = 0$	$\gamma = 1.7$
$\overline{Y}_o = 490$ MPa	$a_3 = 0$	$\gamma = 0.9$	$m = 0.94$
$T_r = 295$ K	$a_4 = 0$	$\ell = 5$ μm	$n = 1$
$T_m = 1800$ K	$k_1 = 0.1$ MPa^{-1}	$\alpha_t = 1.1 \times 10^{-5}$/K	$\phi_{eq}^c = 0.30$
$\rho_o = 7850$ kg/m^3	$k_2 = 15$ GPa^{-1}	$c_p = 452$ J/kg. K	$P_{cutoff} = 160$ GPa
$\eta^v = 0.01$ s	$\eta_r = 0$	$c_v = 266$ J/kg. K	
Projectile material of hardened Arne tool steel			
$E = 204$ GPa	$v = 0.33$	$\rho_o = 7850$ kg/m^3	$Y_o = 1900$ MPa
$E_t = 15$ GPa	$\varepsilon_f = 2.15\%$		

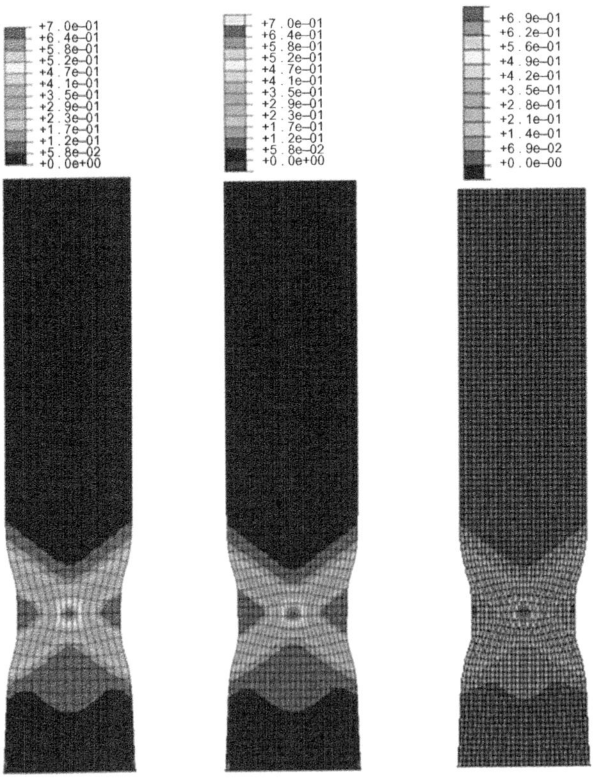

FIG. 2.22 Study of mesh sensitivity. Deformation patterns and contour plots of the effective viscoplastic strain for 15 × 50, 25 × 70, and 30 × 100 finite element discretizations. *(After Voyiadjis, G.Z., Abu Al-Rub, R.K., 2006. A finite strain plastic-damage model for high velocity impacts using combined viscosity and gradient localization limiters: part II—numerical aspects and simulations. Int. J. Damage Mech. 15, 335–373.)*

almost a mesh independent band width for the three finite element discretizations. Moreover, Fig. 2.22 shows that the magnitude and the distribution of the equivalent viscoplastic strain are almost independent of the mesh refinement.

2.5.2.1.2 Strip with a geometrical imperfection

Voyiadjis and Abu Al-Rub (2006) modeled the impact tensile loading of strip with geometrical imperfection. The geometric definition of the specimen is depicted in Fig. 2.23. The strip is constrained at the bottom, while a tensile velocity of 20 mm/s is imposed at the top. The constitutive parameters used in the computation are listed in Table 2.8 for the target material (Weldox 460 E steel). Two uniform meshes (mesh 1: 10 × 20, mesh 2: 20 × 40) are used with four-noded quadrilateral plane-strain elements. To avoid a homogeneous solution, we slightly increase the width of the specimen toward the top, so that

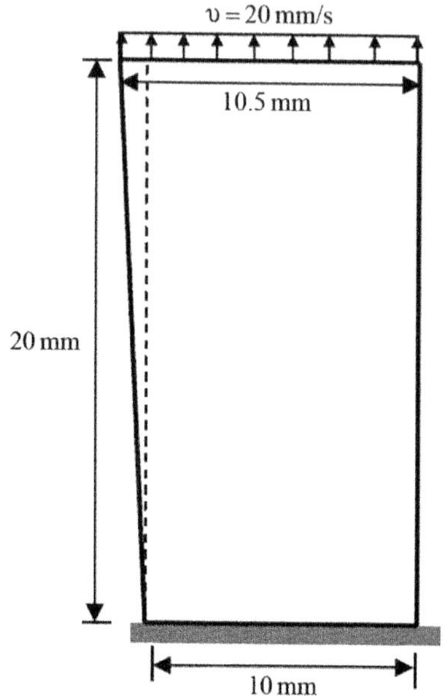

FIG. 2.23 Problem description of a strip with geometrical imperfection and subjected to low tensile impact velocity. *(After Voyiadjis, G.Z., Abu Al-Rub, R.K., 2006. A finite strain plastic-damage model for high velocity impacts using combined viscosity and gradient localization limiters: part II—numerical aspects and simulations. Int. J. Damage Mech. 15, 335–373.)*

the shear band will be initiated at the bottom left and develops with an inclination angle of 45°.

In Fig. 2.24A the deformation patterns for both meshes are plotted for a zero length-scale, $l = 0$. We observe that the width of the shear band is determined by the element size. Deformation is localized along a line of integration points. Mesh dependence is also obvious from the effective plastic strain plots in Fig. 2.25A. When the mesh is refined, the dissipated energy decreases. In Fig. 2.26A, the overall load-deflection responses are different for the two resolutions. Moreover, it can be shown from these results that the inclusion of the geometrical nonlinearity in the dissipation of the strain softening material cannot solve the discretization dependence.

To solve the mesh-sensitivity problem, a length-scale parameter of 2.5 mm ($l = 2.5$ mm) is assumed to keep the field equations well-posed. In Fig. 2.24B, the deformation patterns are plotted for the two considered meshes. By comparison with the results of Fig. 2.24A, we observe that the shear band has a finite width, which is almost independent of the finite element size. The inclination of the shear band is close to 45°. Also, a similar distribution of the effective plastic strain contours can be observed in Fig. 2.25B for the two meshes, which corroborates the mesh objectivity of results. In Fig. 2.26B, the overall

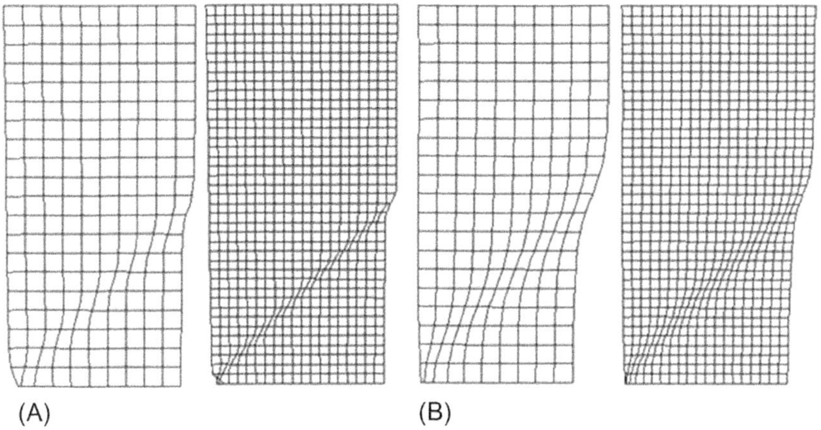

FIG. 2.24 The deformation patterns for 10 × 20 and 20 × 40 meshes: (A) results when $l = 0$ and (B) results when $l = 2.5$ mm. *(After Voyiadjis, G.Z., Abu Al-Rub, R.K., 2006. A finite strain plastic-damage model for high velocity impacts using combined viscosity and gradient localization limiters: part II—numerical aspects and simulations. Int. J. Damage Mech. 15, 335–373.)*

load-deflection responses is essentially identical for the two resolutions. Apparently, no visible discretization sensitivity is present in the results.

Since the formulation of the gradient dependent theory includes an internal length-scale as a material parameter, numerical solutions employing a strain-softening model are no longer dependent on numerical discretization. However, one limitation to the discretization remains. The width of the localization zone needs to be many times (roughly 6) larger than the element size for proper calculation of the plastic strain.

2.5.2.1.3 Hypervelocity impact induced damage in metals: Voyiadjis and his co-workers

The penetration and perforation mechanisms are interdisciplinary subjects and are based on the laws of conservation and compatibility. Penetration implies movement of a projectile into a target. Phenomenologically, as illustrated in Fig. 2.27, the penetration can be viewed as a process to generate a cone shaped macro crack in the material, in which, the kinetic energy of the penetrator is dissipated.

Penetration implies movement of a projectile into a target. Phenomenologically, the penetration can be viewed as a process to generate a cone shaped macro crack in the material, in which, the kinetic energy of the penetrator is dissipated. Perforation implies penetration all the way through a target. Projectile exit through a finite target is often accompanied by delamination and plugging. Delamination refers to a tensile failure parallel to the target rear surface

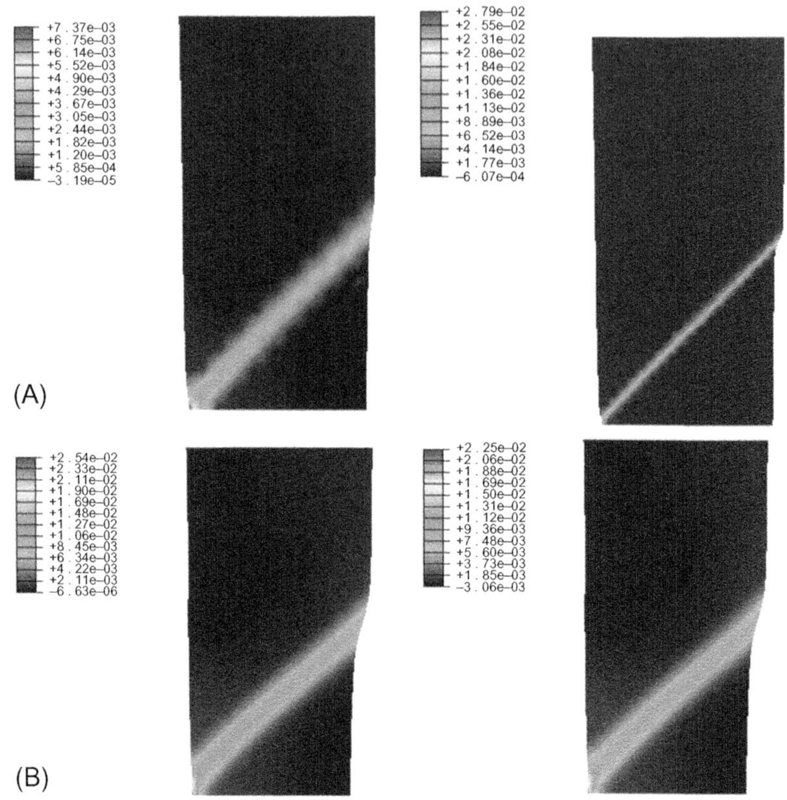

FIG. 2.25 The effective plastic strain contours for 10×20 and 20×40 meshes: (A) results when $l = 0$ and (B) results when $l = 2.5$ mm. *(After Voyiadjis, G.Z., Abu Al-Rub, R.K., 2006. A finite strain plastic-damage model for high velocity impacts using combined viscosity and gradient localization limiters: part II—numerical aspects and simulations. Int. J. Damage Mech. 15, 335–373.)*

and is often initiated by spall. However, it can also occur in quite thick targets, where it seems to be caused by shear bands around the projectile head that can propagate as cracks near the exit surface. In the event of impact there is an exchange of energy that takes place:

$$E_{\text{tans}} = E_{\text{plate}}^{\text{internal}} + E_{\text{plate}}^{\text{kinetic}} + E_{\text{eroded}}^{\text{internal}} + E_{\text{eroded}}^{\text{kinetic}} \tag{2.560}$$

Law of conservation is observed in any physical phenomena. Kinetic energy of the projectile is spent in raising the internal energy and kinetic energy of the plate and some part of the energy is lost in the form of eroded material. The amount of energy dissipated also differs with geometry. Blunt projectiles like cylinders are found to cause plugging because of pure shear failure, while the conical projectiles are found to cause petaling effects.

High velocity impact will localize compression of the composite and subsequently shearing the fiber and spalling of the resin during impact. Once

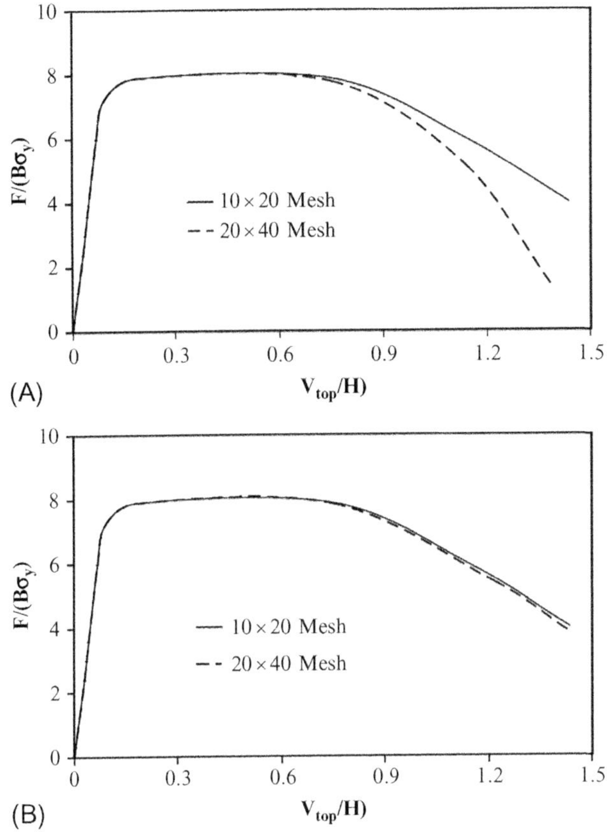

FIG. 2.26 The load-deflection diagrams for 10 × 20 and 20 × 40 meshes: (A) results when $l = 0$ and (B) results when $l = 2.5$ mm. *(After Voyiadjis, G.Z., Abu Al-Rub, R.K., 2006. A finite strain plastic-damage model for high velocity impacts using combined viscosity and gradient localization limiters: part II—numerical aspects and simulations. Int. J. Damage Mech. 15, 335–373.)*

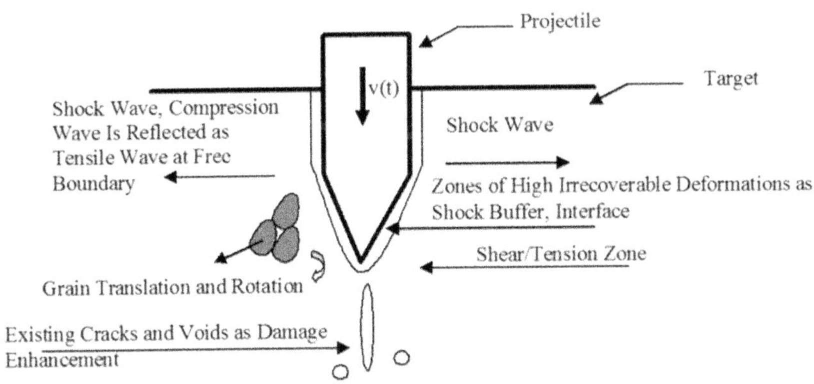

FIG. 2.27 Schematics illustration the penetration process. *(After Voyiadjis, G.Z., Abu Al-Rub, R. K., 2006. A finite strain plastic-damage model for high velocity impacts using combined viscosity and gradient localization limiters: part II—numerical aspects and simulations. Int. J. Damage Mech. 15, 335–373.)*

the projectile has slowed, the composite deforms causing fiber stretching, pull-out and delamination of the composite layers and thus lower the load carrying capacity.

The simulations were conducted using the gradient plasticity damage framework developed by Voyiadjis and Abu Al-Rub (2006) and Abu Al-Rub and Voyiadjis (2006) as presented in Section 2.5.2. Three sets of equations are needed to simulate the impact damage problems: a yield criterion that incorporates the strain rate and temperature effects on the material strength, an equation of state which accounts for material pressure sensitivity, and a damage model which is needed to simulate progressive failure and fracture. The model is based on the nonlocal gradient plasticity and gradient damage theories. It includes the nonlocal von Mises yield criterion, the non-associated flow rules, isotropic and anisotropic strain hardening, strain rate hardening, softening due to adiabatic heating and anisotropic damage evolution, and finally a path dependent equation of state.

Voyiadjis and Abu Al-Rub (2006) simulated a blunt projectile impacting a target to capture the hypervelocity impact induced damage in metals. The objective of this numerical example is to investigate if the proposed constitutive equations are able to describe the structural response to projectile impact damage when different failure modes are expected to occur. This is done by conducting numerical simulations of the experimental tests presented by Borvik et al. (2002) for a blunt projectile made of hardened Arne tool steel impacting a circular plate made of Weldox 460 E steel. The model material constants of the target material of Weldox 460 E steel and for the projectile material of hardened Arne tool steel are listed in Table 2.8. In these simulations a 4-node 2D axisymmetric element with one integration point and a stiffness based on hourglass control is used. Plot of the initial configuration, showing a part of the target plate and the blunt projectile just prior to impact, is shown in Fig. 2.28.

Borvik et al. (2001) simulations indicate the problem involving shear localization and plugging for blunt projectiles to be mesh size sensitive. A direct comparison between the numerical and experimental residual velocity curves for blunt projectile is shown in Fig. 2.29. As seen, there is good agreement with the experimental results. Moreover, Fig. 2.30 shows that the perforation times obtained from the different numerical simulations are similar and close to the experimental values estimated from high-speed camera images obtained by Borvik et al. (2002). Borvik et al. (2002) reported the high-speed camera images showing perforation of the target plate at impact velocities close to the ballistic limit as presented in Fig. 2.31. Voyiadjis and Abu Al-Rub (2006) reported the perforation of the target plate by a blunt projectile obtained from their simulations, as shown in Fig. 2.32, which shows that the model can capture the main features of the experiment conducted by Borvik et al. (2002). Furthermore, Fig. 2.32 depicts the contours of accumulated viscoinelastic strain on the deformed mesh. It can be seen that limited inelastic deformation occurs outside the localized shear zone. These plots clearly demonstrate that the numerical

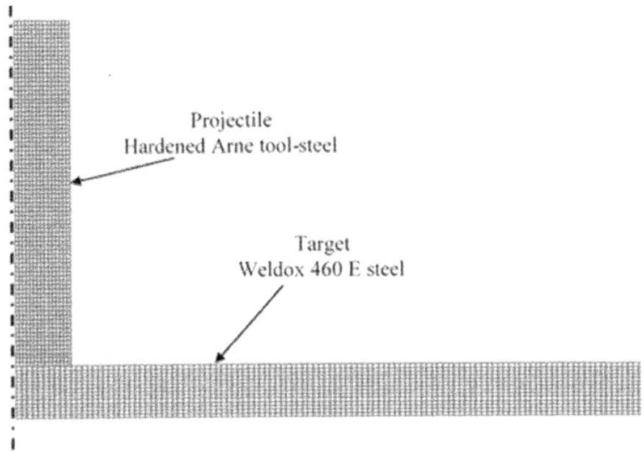

FIG. 2.28 Finite element mesh plot of the axisymmetric initial configuration just before impact. *(After Voyiadjis, G.Z., Abu Al-Rub, R.K., 2006. A finite strain plastic-damage model for high velocity impacts using combined viscosity and gradient localization limiters: part II—numerical aspects and simulations. Int. J. Damage Mech. 15, 335–373.)*

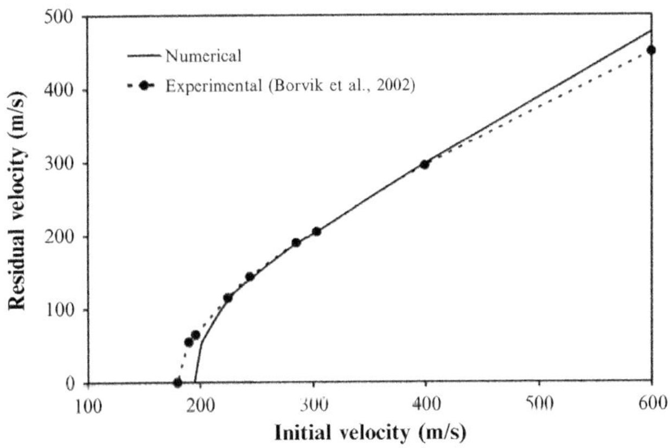

FIG. 2.29 A comparison between numerical and experimental results for the initial impact velocity vs residual projectile velocity by a blunt projectile. *(After Voyiadjis, G.Z., Abu Al-Rub, R.K., 2006. A finite strain plastic-damage model for high velocity impacts using combined viscosity and gradient localization limiters: part II—numerical aspects and simulations. Int. J. Damage Mech. 15, 335–373.)*

model qualitatively captures the overall physical behavior of the target during penetration and perforation. Notice also that in these plots, only a part of the complete target plate is shown.

The elements in the impacted area are significantly distorted but stable results are obtained. This delayed the damage evolution process and consequently the erosion of damaged elements. However, the upper nodes in a critical

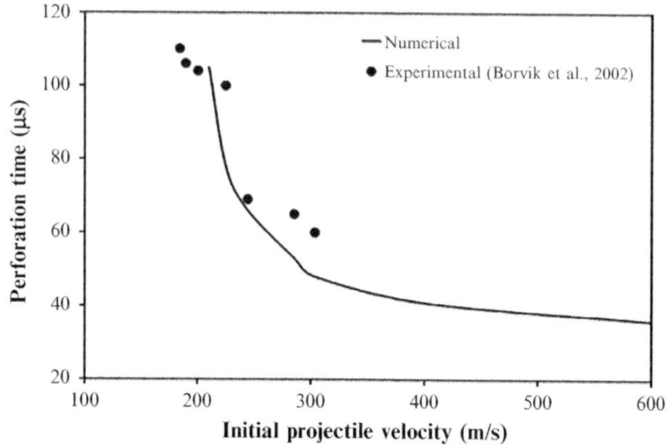

FIG. 2.30 Comparison between perforation times vs initial impact velocities by a blunt projectile. *(After Voyiadjis, G.Z., Abu Al-Rub, R.K., 2006. A finite strain plastic-damage model for high velocity impacts using combined viscosity and gradient localization limiters: part II—numerical aspects and simulations. Int. J. Damage Mech. 15, 335–373.)*

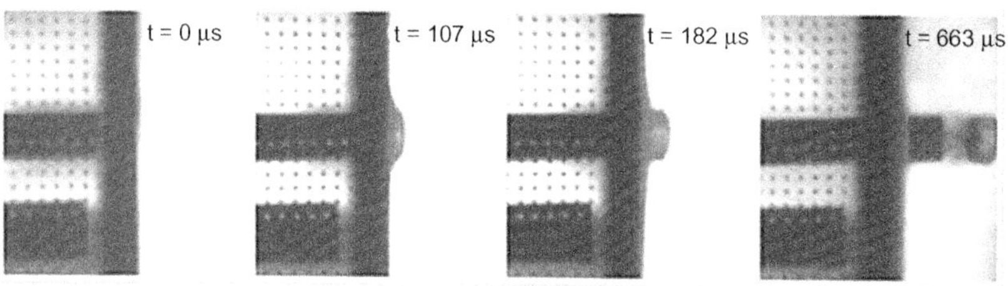

FIG. 2.31 High-speed camera images showing perforation of the target plate at impact velocities close to the ballistic limits with a blunt projectile. *(After Borvik, T., Langseth, M., Hopperstad, O. S., and Malo, K. A., 2002. Perforation of 12 mm thick steel plates by 20 mm diameter projectiles with flat, hemispherical and conical noses, part 1: experimental study. Int. J. Impact Eng. 27, 19–35.)*

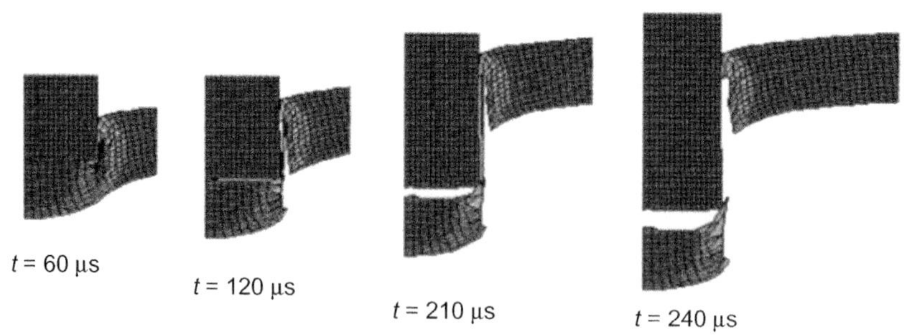

FIG. 2.32 Perforation of the target plate by a blunt projectile of initial impact velocity of 210 m/s using ALE meshing, plotted as contours of accumulated viscoinelastic strain at different times. The lighter region indicates an accumulated damage between 0.25 and 0.30. *(After Voyiadjis, G.Z., Abu Al-Rub, R.K., 2006. A finite strain plastic-damage model for high velocity impacts using combined viscosity and gradient localization limiters: part II—numerical aspects and simulations. Int. J. Damage Mech. 15, 335–373.)*

element do not penetrate the lower nodes, giving a stable solution and no error termination is encountered. Fig. 2.33 shows the element distortion in the target plate just after impact.

The final cross section of a target plate perforated by a blunt projectile at an impact velocity close to the ballistic limit is shown in Fig. 2.34, which is very similar to the experimental cross section obtained by Borvik et al. (2002).

FIG. 2.33 Penetration of the target plate by a blunt projectile of initial impact velocity of 300 m/s using ALE meshing, plotted as contours of accumulated viscoinelastic strain and showing details of element meshes just after adaptive remeshing. *(After Voyiadjis, G.Z., Abu Al-Rub, R.K., 2006. A finite strain plastic-damage model for high velocity impacts using combined viscosity and gradient localization limiters: part II—numerical aspects and simulations. Int. J. Damage Mech. 15, 335–373.)*

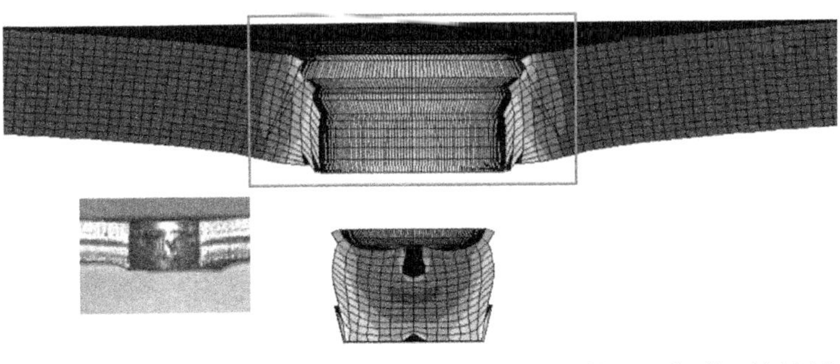

FIG. 2.34 Final cross section of the target plate perforated by a blunt projectile of initial impact velocity of 300 m/s using ALE meshing and plotted as contours of accumulated viscoinelastic strain. The lighter region part indicates an accumulated damage between 0.25 and 0.30. *(After Voyiadjis, G. Z., Abu Al-Rub, R.K., 2006. A finite strain plastic-damage model for high velocity impacts using combined viscosity and gradient localization limiters: part II—numerical aspects and simulations. Int. J. Damage Mech. 15, 335–373.)*

References

Abu Al-Rub, R.K., Voyiadjis, G.Z., 2006. A finite strain plastic-damage model for high velocity impacts using combined viscosity and gradient localization limiters, part I: theoretical formulation. Int. J. Damage Mech. 15, 293–334.

Aifantis, E.C., 1984a. Microscopic processes and macroscopic response. In: Desai, C.S., Gallagher, R.H. (Eds.), Mechanics of Engineering Materials. John Wiley, New York, pp. 1–22.

Aifantis, E.C., 1984b. On the microstructural origin of certain inelastic models. J. Eng. Mater. Technol. 106, 326.

Aifantis, E.C., 1984c. On the mechanics of modulated structures. In: Tsakalakos, T. (Ed.), Modulated Structure Materials. NATO ASI Series 83, Martinus Nijhoff, Dordrecht, pp. 357–385.

Aifantis, E.C., 1987. The physics of plastic deformation. Int. J. Plast. 3, 211–247.

Aifantis, E.C., 1989. Plasticity and self organization. In: Jinghong, F., Murakami, S. (Eds.), Advances in Constitutive Laws for Engineering Materials. vol. 2, pp. 893–905.

Armstrong, P.J., Frederick, C.O., 1966. A mathematical representation of the multiaxial Bauscinger effect. (Cegb Report No Rd/B/N 731, Berkeley Nuclear Laboratories).

Ashby, M.F., 1970. The deformation of plastically non-homogeneous materials. Philos. Mag. 21, 399–424.

Bari, S., Hassan, T., 2000. Anatomy of coupled constitutive models for ratcheting simulation. Int. J. Plast. 16, 381–409.

Bazant, Z.P., Lin, F.B., 1988. Nonlocal yield-limit degradation. Int. J. Numer. Methods Eng. 26, 1805–1823.

Bingham, E.C., 1922. Fluidity and Plasticity. McGraw-Hill, New York.

Bonet, J., Wood, R.D., 2008. Nonlinear Continuum Mechanics for Finite Element Analysis. Cambridge University Press.

Borvik, T., Hopperstad, O.S., Berstad, T., Langseth, M., 2001. Numerical simulation of plugging failure in ballistic penetration. Int. J. Solids Struct. 38, 6241–6264.

Borvik, T., Langseth, M., Hopperstad, O.S., Malo, K.A., 2002. Perforation of 12 mm thick steel plates by 20 mm diameter projectiles with flat, hemispherical and conical noses, part 1: experimental study. Int. J. Impact Eng. 27, 19–35.

Boyle, J.T., Spence, J., 1983. Stress Analysis for Creep. Butterworths, London.

Chaboche, J.L., 1986. Time-independent constitutive theories for cyclic plasticity. Int. J. Plast. 2, 149–188.

Chaboche, J.L., Dang Van, K., Cordier, G., 1979. Modelization of the strain memory effect on the cyclic hardening of 316 stainless steel. In: Proc. the 5th International Conf. on SMIRT, Div. L. Berlin, Germany.

Chen, W.F., Han, D.J., 1988. Plasticity for Structural Engineers. Springer, Berlin.

Chen, X., Jiao, R., 2004. Modified kinematic hardening rule for multiaxial ratcheting prediction. Int. J. Plast. 20, 871–898.

Dafalias, Y.F., Kourousis, K.I., Saridis, G.J., 2008. Multiplicative AF kinematic hardening in plasticity. Int. J. Solids Struct. 45, 2861–2880.

Engelen, R.A.B., Geers, M.G.D., Baaijens, F.P.T., 2003. Nonlocal implicit gradient-enhanced elastoplasticity for the modelling of softening behavior. Int. J. Plast. 19, 403–433.

Fleck, N.A., Hutchinson, J.W., 1993. A phenomenological theory for strain gradient effects in plasticity. J. Mech. Phys. Solids 41, 1825–1857.

Fleck, N.A., Hutchinson, J.W., 1997. Strain gradient plasticity. In: Hutchinson, J.W., Wu, T.Y. (Eds.), Advances in Applied Mechanics. In: vol. 33. Academic Press, New York, pp. 295–361.

Geers, M.G.D., Engelen, R.A.B., Ubachs, R.J.M., 2001. On the numerical modelling of ductile damage with an implicit gradient-enhanced formulation. Rev. Européenne Élém. Finis 10, 173–191.

Green, A.E., Zerna, W., 1954. Theoretical Elasticity. Oxford University Press.

Gurson, A.L., 1977. Continuum theory of ductile rupture by void nucleation and growth—part I: yield criteria and flow rule for porous media. J. Eng. Mater. Technol. 99, 2–15.

Gurtin, M.E., Anand, L., 2005a. A theory of strain-gradient plasticity for isotropic, plastically irrotational materials. Part I: small deformations. J. Mech. Phys. Solids 53, 1624–1649.

Gurtin, M.E., Anand, L., 2005b. A theory of strain-gradient plasticity for isotropic, plastically irrotational materials. Part II: finite deformations. Int. J. Plast. 21, 2297–2318.

Hill, R., 1968. On constitutive inequalities for simple materials—I. J. Mech. Phys. Solids 16, 229–242.

Jirasek, M., Rolshoven, S., 2003. Comparison of integral-type nonlocal plasticity models for strain-softening materials. Int. J. Eng. Sci. 41, 1553–1602.

Kachanov, L.M., 1958. Time of the rupture process under creep condition. Izv. Akad. Nauk. SSSR, Otd.Tekhn. Nauk. 8, 26–31.

Leblond, J.B., Perrin, G., Devaux, J., 1994. Bifurcation effects in ductile metals with no local damage. J. Appl. Mech. 61, 236–242.

Lemaitre, J., Chaboche, J.L., 1990. Mechanics of Solid Materials. Cambridge University Press, New York.

Mindlin, R.D., 1964. Micro-structure in linear elasticity. Arch. Ration. Mech. Anal. 16, 51–78.

Mühlhaus, H.B., Aifantis, E.C., 1991. A variational principle for gradient plasticity. Int. J. Solids Struct. 28, 845–857.

Ogden, R.W., 1984. Non-Linear Elastic Deformations. Ellis Horwood, Chichester.

Peerlings, R.H.J., 1999. Enhanced Damage Modelling for Fracture and Fatigue (Ph.D. thesis). Eindhoven University of Technology, Eindhoven.

Peric, D., 1993. On a class of constitutive equations in viscoplasticity: formulation and computational issues. Int. J. Numer. Methods Eng. 36, 1365–1393.

Perzyna, P., 1963. The constitutive equations for rate sensitive plastic materials. Quart. Appl. Math. 20, 321–332.

Perzyna, P., 1966. Fundamental problems in viscoplasticity. Adv. Appl. Mech. 9, 243–377.

Perzyna, P., 1971. Thermodynamic theory of viscoplasticity. Adv. Appl. Mech. 11, 313–354.

Prager, W., 1955. The theory of plasticity: a survey of recent achievements. Proc. Inst. Mech. Eng. 169, 41–57.

Seth, B.R., 1961. Generalized strain measure with applications to physical problems (MRC Technical Summary Report #248). Mathematics Research Center, United States Army, University of Wisconsin, pp. 1–18.

Seth, B.R., 1962. Generalized strain measure with applications to physical problems. In: IUTAM Symposium on Second Order Effects in Elasticity, Plasticity and Fluid Mechanics, Haifa.

Skrzypek, J.J., 1993. Plasticity and Creep. Theory, Examples and Problems. CRC Press, London.

Song, Y., Voyiadjis, G.Z., 2018. Small scale volume formulation based on coupled thermo-mechanical gradient enhanced plasticity theory. Int. J. Solids Struct. 134, 195–215.

Toupin, R.A., 1962. Elastic materials with couple-stresses. Arch. Ration. Mech. Anal. 11, 385–414.

Tvergaard, V., 1981. Influence of voids on shear band instabilities under plain strain conditions. Int. J. Fract. 17, 389–407.

Vermeer, P.A., Brinkgreve, R.B.J., 1994. A new effective non-local strain measure for softening plasticity. In: Chambon, R., Desrues, J., Vardoulakis, I. (Eds.), Localisation and Bifurcation Theory for Soils and Rocks. Balkema, Rotterdam, pp. 89–100.

Voyiadjis, G.Z., Abu Al-Rub, R.K., 2006. A finite strain plastic-damage model for high velocity impacts using combined viscosity and gradient localization limiters: part II—numerical aspects and simulations. Int. J. Damage Mech. 15, 335–373.

Voyiadjis, G.Z., Basuroychowdhury, I.N., 1998. A plasticity model for multiaxial cyclic loading and ratcheting. Acta Mech. 126, 19–35.

Voyiadjis, G.Z., Abed, F.H., 2006. A coupled temperature and strain rate dependent yield function for dynamic deformations of bcc metals. Int. J. Plast. 22, 1398–1431.

Voyiadjis, G.Z., Deliktas, B., 2009a. Mechanics of strain gradient plasticity with particular reference to decomposition of the state variables into energetic and dissipative components. Int. J. Eng. Sci. 47, 1405–1423.

Voyiadjis, G.Z., Deliktas, B., 2009b. Formulation of strain gradient plasticity with interface energy in a consistent thermodynamic framework. Int. J. Plast. 25, 1997–2024.

Voyiadjis, G.Z., Faghihi, D., 2012. Thermo-mechanical strain gradient plasticity with energetic and dissipative length scales. Int. J. Plast. 30–31, 218–247.

Voyiadjis, G.Z., Foroozesh, M., 1990. Anisotropic distortional yield model. J. Appl. Mech. 57, 537–547.

Voyiadjis, G.Z., Deliktas, B., Aifantis, E.C., 2001. Multiscale analysis of multiple damage mechanics coupled with inelastic behavior of composite materials. J. Eng. Mech. 127, 636–645.

Voyiadjis, G.Z., Al-Rub, R.K.A., Palazoto, A., 2004. Thermodynamic framework for coupling of non-local viscoplasticity and non-local anisotropic viscodamage for dynamic localization problems using gradient theory. Int. J. Plast. 20, 981–1038.

Zbib, H., Aifantis, E.C., 1989. On the localization and part-localization behaviour of plastic deformation. Res Mech. 23, 261–277.

Zervos, A., Papanastasiou, P., Vardoulakis, I., 2001. A finite element displacement formulation for gradient elastoplasticity. Int. J. Numer. Methods Eng. 50, 1369–1388.

Ziegler, H., 1959. A modification of Prager's hardening rule. Q. Appl. Math. 17, 55–65.

Further reading

De Souza Neto, E.A., Peric, D., Owen, D.R.J., 2008. Computational Methods for Plasticity: Theory and Applications. John Wiley & Sons Ltd.

Chapter 3

Nonlocal crystal plasticity

3.1 Introduction

Metals are generally available as polycrystalline materials consisting of many grains each has a specific crystallographic orientation. Accordingly, the simplest approximation of the polycrystalline metals can be described in two steps. First, the behavior of each grain should be investigated during loading. The constitutive model for each grain is commonly termed as the single crystal theory. Second, the response of the polycrystalline sample is the result of all constituent grains acting together. Accordingly, the response of each polycrystalline sample can be analyzed using a homogenization of the behavior of consisting grains, which have their own constitutive model.

Both small strain and finite strain theories have been developed to capture the single crystal response. In this chapter, for the conventional theories of crystal plasticity, i.e., local theories, the finite strain framework is incorporated to present a more general case. In the case of nonlocal crystal plasticity models, however, both finite strain and small strain models are presented.

To start the description of the single crystal constitutive model, some crystallographic concepts including slip, slip plane, slip direction, and slip system should be first elaborated upon. These concepts are described in the following sections.

3.1.1 Slip in metals

Deformation in metals can occur through two different mechanisms: glide and climb. In the former mechanism, dislocations move in the glide plane that includes both dislocation line and Burgers vector, while the later mechanism includes the movement of dislocations out of the glide plane. Slip is the result of the dislocation glide which is the conventional demonstration of plastic deformation in polycrystalline metals. Plastic deformation due to slip is the result of the successive sliding of one layer of atoms over another one, which occurs in specific crystallographic directions and planes commonly called slip direction and slip plane, respectively. Fig. 3.1 simply demonstrates the slip phenomenon in crystalline metals.

Size Effects in Plasticity: From Macro to Nano. https://doi.org/10.1016/B978-0-12-812236-5.00003-7

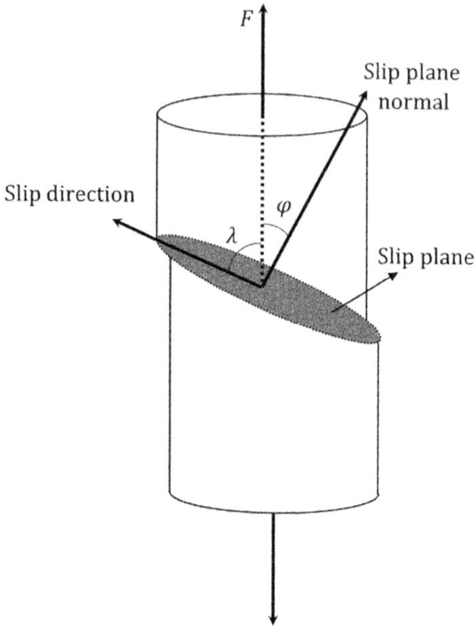

FIG. 3.1 Slip in crystalline metals.

The crystallographic properties of slip direction and plane depend on the material crystal structure. Generally, the slip planes are the planes with the highest atoms density. Accordingly, the space between those atomic layers and the next is maximum and the atomic layers can move easier relative to each other in those layers. In the case of slip direction, generally, it is the most packed atomic direction which leads to the shortest lattice translation vectors. Table 3.1 presents the slip planes and directions for three common crystal structures of face-centered cubic (FCC), body-centered cubic (BCC), and hexagonal

TABLE 3.1 Slip planes and directions for three common crystal structures of face-centered cubic (FCC), body-centered cubic (BCC), and hexagonal closed-packed (HCP)

Crystal structure	Slip direction	Slip plane
Face-centered cubic (FCC)	$\langle 110 \rangle$	$\{111\}$
Body-centered cubic (BCC)	$\langle 111 \rangle$	$\{110\}$, $\{112\}$, $\{123\}$
Hexagonal closed-packed (HCP)	$\langle 11\bar{2}0 \rangle$	$(0001), \{1\bar{1}00\}, \{1\bar{1}01\}$
	$\langle 11\bar{2}3 \rangle$	$\{1\bar{1}01\}$

closed-packed (HCP). A combination of a slip plane and direction is commonly called as a slip system. For example, according to Table 3.1, FCC metals have 12 slip systems of $\{111\}\langle 110\rangle$. A very important crystallographic property of a grain is the so-called Schmid factor which relates the resolved shear stress τ for a specific slip system to the applied force F as follows:

$$\tau = \frac{F}{A}\cos\varphi\cos\lambda \tag{3.1}$$

where A is the area of the cross-section, and φ and λ are the angles between the applied force F and the slip plane normal and the slip direction, respectively (Fig. 3.1). Accordingly, the Schmid factor dictates the value of shear stress on a specific slip system due to the applied loading.

3.2 Local crystal plasticity models

Taylor and Elam (1923, 1925) and Taylor (1938a,b) presented one of the pioneering works which relate the plastic flow to the crystallographic slip. The idea of relating the plasticity to the slip of different slip systems in the framework of small strain continuum model of single crystalline metals have been developed and pioneered by Mandel (1965) and Hill (1966). It further proceeds by incorporating the finite deformation framework in crystal plasticity developed by Rice (1971), Hill and Rice (1972), Asaro and Rice (1977), Asaro (1983a,b), Asaro and Needleman (1985), Anand and Kothari (1996), and Kalidindi et al. (1992). Many researchers, such as Havner (1992), Lubarda (2002), Dunne and Petrinic (2006), Bertram (2005), De Souza Neto et al. (2008), and Besson et al. (2009) have reviewed the local crystal plasticity models and presented the basic assumptions and formulations.

In the case of conventional crystal plasticity models, one of the basic assumptions is that there is no lattice distortion during the plastic deformation. Accordingly, similar to the local plasticity as discussed in Section 2.1, there is no length scale in these models, and accordingly, scaling of the specimen size does not change the governing equations. In other words, the sample with very small length scale is described similarly as the large sample. These families of crystal plasticity models are categorized as the local models in this book.

In this section, the general framework of the local crystal plasticity models is presented. First, the assumptions and formulations of the rate-independent crystal plasticity models are presented. Furthermore, the incremental relations of rate independent crystal plasticity models are elaborated upon. Next, the basic framework of rate-dependent crystal plasticity model is explained. Finally, the incremental relations of rate-dependent model are derived and presented.

3.2.1 Rate-independent crystal plasticity models

The rate-independent crystal plasticity model developed by Anand and Kothari (1996) is elaborated here. Similar to the rate-independent plasticity models described in Sections 2.2.2–2.2.7, the rate-independent crystal plasticity can be developed using the same framework. However, in the case of crystal plasticity, the plastic deformation is the result of the plastic activation in all active slip systems. The key idea of crystal plasticity is to link the plastic part of the velocity gradient, as a representative of macroscopic response, to the superposition of shear deformation induced by crystallographic slip on multiple slip, as a microscopic mechanisms of deformation, which can be described as follows:

$$l_{ij}^{p} = \sum_{\alpha} \dot{\gamma}^{\alpha} \operatorname{sign}(\tau^{\alpha}) S_{ij}^{\alpha} \tag{3.2}$$

where l_{ij}^{p} is the plastic velocity gradient and represents the macroscopic scale information. On the other hand, $\sum_{\alpha} \dot{\gamma}^{\alpha} \operatorname{sign}(\tau^{\alpha}) S_{ij}^{\alpha}$ governs the microscopic information in which $\dot{\gamma}^{\alpha}$ is the shearing rate on slip system α, τ^{α} is the shear stress acting on slip system α, and S_{ij}^{α} is the Schmid tensor for the slip system α, which can be defined as follows:

$$S_{ij}^{\alpha} = m_i^{\alpha} \otimes n_j^{\alpha} \tag{3.3}$$

where m_i^{α} and n_j^{α} are unit vectors of slip direction and slip plane normal in the deformed configuration, respectively. The acting shear stress on the slip system α can be obtained as follows:

$$\tau^{\alpha} = \sigma_{ij} S_{ij}^{\alpha} \tag{3.4}$$

where σ_{ij} is the Cauchy stress tensor.

The rate-independent plasticity rules are incorporated here for each slip system. Accordingly, the yield surface of slip system α can be defined as follows:

$$f^{\alpha} = |\tau^{\alpha}| - s^{\alpha} \tag{3.5}$$

where s^{α} is the slip resistance for slip system α.

The unit normal to the yield surface can be defined as (Anand and Kothari, 1996):

$$N_{ij}^{\alpha} = \operatorname{sign}(\tau^{\alpha}) \operatorname{sym}\left(S_{ij}^{\alpha}\right) \tag{3.6}$$

The elastic loading occurs, i.e., $\dot{\gamma}^{\alpha} = 0$, if $f^{\alpha} < 0$. The loading and consequently plastic deformation occurs in slip system α, i.e., $\dot{\gamma}^{\alpha} > 0$, if

$$f^{\alpha} = 0 \text{ and } N_{ij}^{\alpha} \ell_{ijkl} d_{kl} > 0 \tag{3.7}$$

where $\ell_{ijkl} d_{kl}$ is the trial stress rate, ℓ_{ijkl} is the forth order elasticity tensor, and d_{kl} is the rate of deformation tensor introduced in Section 2.4.1.8. Eq. (3.7) states that the angle between the N_{ij} and σ_{ij} is acute.

The unloading condition, i.e., $\dot{\gamma}^{\alpha} = 0$, can be described as follows:

$$f^{\alpha} = 0 \text{ and } N_{ij}^{\alpha} \dot{\sigma}_{ij} < 0 \tag{3.8}$$

There is another loading condition which is commonly called as neutral loading, which is neither loading nor unloading, as follows:

$$f^{\alpha} = 0 \text{ and } N^{\alpha}_{ij}\dot{\sigma}_{ij} = 0 \tag{3.9}$$

No plastic deformation occurs in the case of neutral loading, i.e., $\dot{\gamma}^{\alpha} = 0$ (Voyiadjis and Huang, 1996).

After determining the plastic shear slip for all slip systems, the evolution of slip resistance for slip system α, i.e., s^{α}, can be defined as follows:

$$\dot{s}^{\alpha} = \sum_{\beta} h^{\alpha\beta}\dot{\gamma}^{\beta} \tag{3.10}$$

where $h^{\alpha\beta}$ defines the variation of slip resistance for slip system α due to the plastic shear deformation in slip system β, which is commonly termed as the hardening moduli. Various formulations have been introduced to describe the hardening moduli. One of the simplest forms was developed by Asaro and Needleman (1985) as follows:

$$\begin{aligned} h^{\alpha\beta} &= qh^{\beta} \text{ for } \alpha \neq \beta \\ h^{\alpha\alpha} &= h^{\alpha} \text{ (no sum on } \alpha) \end{aligned} \tag{3.11}$$

where q is the latent hardening ratio, a value of $1 < q < 1.4$ is commonly selected, and h^{α} is the hardening rate for a single slip. Different functional forms have been developed to describe the hardening rate of a single slip. One the most conventional models to use has been developed by Brown et al. (1989), which is given as follows:

$$h^{\alpha} = h^{\alpha}_{0}\left\{1 - \frac{s^{\alpha}}{s^{\alpha}_{s}}\right\}^{a^{s}} \tag{3.12}$$

where h^{α}_{0}, s^{α}_{s}, and a^{s} are the material constants for slip system α.

3.2.1.1 *Incremental relation of rate-independent crystal plasticity models*

The incremental relation of rate-independent crystal plasticity models developed by Anand and Kothari (1996) is elaborated here. In the case of finite deformation rate-independent crystal plasticity model, the multiplicative decomposition is the first step as follows:

$$F_{iJ} = F^{e}_{i\xi}F^{p}_{\xi J} \tag{3.13}$$

where $F^{e}_{i\xi}$ and $F^{p}_{\xi J}$ are the elastic and plastic deformation gradients, respectively. One assumes that the deformed body at time t with the observable thermodynamics variables of $\{F_{iJ}(t), F^{e}_{i\xi}(t), F^{p}_{\xi J}(t), \sigma_{ij}(t), s^{\alpha}(t)\}$ undergoes an incremental deformation and the new deformed body at time $\tau = t + \Delta$ has the observable thermodynamics variables of $\{F_{iJ}(\tau), F^{e}_{i\xi}(\tau), F^{p}_{\xi J}(\tau), \sigma_{ij}(\tau), s^{\alpha}(\tau)\}$. In the case of metallic materials, the elastic stretch is generally negligible. Also, it is more common to define the stress state in terms of the second Piola-Kirchhoff stress tensor $S_{\eta\zeta}(\tau)$, which is defined in the intermediate state (see Section 2.4.4.1). Accordingly, the macroscopic constitutive model is described as follows:

$$S_{\eta\zeta}(\tau) = \ell_{\eta\zeta\xi\varrho} E^e_{\xi\varrho}(\tau) \tag{3.14}$$

where $\ell_{\eta\zeta\xi\varrho}$ is the fourth order elasticity tensor and $E^e_{\xi\varrho}(\tau)$ is the elastic Green strain tensor at time $= \tau$, which can be defined as follows:

$$E^e_{\xi\varrho}(\tau) = \frac{1}{2}\left[C^e_{\xi\varrho}(\tau) - \delta_{\xi\varrho}\right] \tag{3.15}$$

where $C^e_{\xi\varrho}(\tau)$ is the elastic right Cauchy-Green tensor at time τ, which can be described as follows:

$$C^e_{\xi\varrho}(\tau) = F^e_{i\xi}(\tau) F^e_{i\varrho}(\tau) \tag{3.16}$$

The relation between the second Piola-Kirchhoff stress tensor $S_{\eta\zeta}(\tau)$ and the Cauchy stress tensor $\sigma_{ij}(\tau)$ can be defined as follows:

$$S_{\eta\zeta}(\tau) = F^e_{\eta i}(\tau)^{-1} J^e \sigma_{ij}(\tau) F^e_{\zeta j}(\tau)^{-1} \tag{3.17}$$

where $J^e = \det[F^e_{i\xi}(\tau)]$. It is worth mentioning that the second Piola-Kirchhoff stress tensor $S_{\alpha\beta}(\tau)$ is conjugate to the elastic Green strain tensor $E^e_{\gamma\zeta}(\tau)$. Next, the resolved shear stress for slip system α is defined using the second Piola-Kirchhoff stress tensor $S_{\eta\zeta}(\tau)$ as follows:

$$\tau^\alpha = C^e_{\gamma\eta}(\tau) S_{\eta\zeta}(\tau) S^\alpha_{\gamma\zeta} \tag{3.18}$$

where $S^\alpha_{\gamma\zeta}$ is the Schmid tensor for slip system α in the intermediate state. In the case of negligible elastic stretches, Eq. (3.18) it is further simplified as follows:

$$\tau^\alpha = S_{\eta\zeta}(\tau) S^\alpha_{\eta\zeta} \tag{3.19}$$

The more applicable form of Eq. (3.19) can be obtained by assuming that the Schmid tensor in the intermediate state $S^\alpha_{\eta\zeta}$ is the same as the reference one, i.e., S^α_{IJ}, which can be defined as follows (Anand and Kothari, 1996):

$$S^\alpha_{IJ} = m^\alpha_I \otimes n^\alpha_J \tag{3.20}$$

where m^α_I and n^α_J are unit vectors of the slip direction and the slip plane normal in the reference configuration, respectively.

In order to numerically integrate the model, first, a trial state is defined by assuming that the incremental deformation that occurred form time $= t$ to time $= \tau$ is elastic. Accordingly, the trial elastic deformation gradient is obtained as follows:

$$F^e_{i\xi}(\tau)^{\mathrm{tr}} = F_{iJ}(\tau) F^p_{J\xi}(t)^{-1} \tag{3.21}$$

Accordingly, the trial elastic right Cauchy-Green tensor $C^e_{\xi\varrho}(\tau)^{\text{tr}}$ and trial elastic Green strain tensor $E^e_{\xi\varrho}(\tau)^{\text{tr}}$ are obtained as follows:

$$C^e_{\xi\varrho}(\tau)^{\text{tr}} = F^e_{i\xi}(\tau)^{\text{tr}} F^e_{i\varrho}(\tau)^{\text{tr}} \tag{3.22}$$

$$E^e_{\xi\varrho}(\tau)^{\text{tr}} = \frac{1}{2}\left[C^e_{\xi\varrho}(\tau)^{\text{tr}} - \delta_{\xi\varrho}\right] \tag{3.23}$$

The trial second Piola-Kirchhoff stress tensor $S_{\alpha\beta}(\tau)^{\text{tr}}$ is then obtained as follows:

$$S_{\eta\zeta}(\tau)^{\text{tr}} = \ell_{\eta\zeta\xi\varrho} E^e_{\xi\varrho}(\tau)^{\text{tr}} \tag{3.24}$$

Finally, the trial resolved shear stress is described as follows:

$$\tau^\alpha(\tau)^{\text{tr}} = S_{\gamma\beta}(\tau)^{\text{tr}} S^\alpha_{\gamma\beta} \tag{3.25}$$

where $S^\alpha_{\gamma\beta}$ is the same as $S^\alpha_{IJ} = m^\alpha_I \otimes n^\alpha_J$. Furthermore, it is common to assume that (Anand and Kothari, 1996):

$$\text{sign}\left[\tau^\alpha(\tau)^{\text{tr}}\right] = \text{sign}[\tau^\alpha(\tau)] \tag{3.26}$$

Accordingly, the incremental flow rule is defined as follows:

$$F^p_{\xi J}(\tau) = \left[\delta_{\xi\zeta} + \sum_\alpha \Delta\gamma^\alpha \,\text{sign}\left[\tau^\alpha(\tau)^{\text{tr}}\right] S^\alpha_{\xi\zeta}\right] F^p_{\zeta J}(t) \tag{3.27}$$

Anand and Kothari (1996) introduced the concept of inactive, potentially active, and active slip systems. If $\tau^\alpha(\tau)^{\text{tr}} - s^\alpha \leq 0$, slip system α is inactive and $\Delta\gamma^\alpha = 0$. However, if $\tau^\alpha(\tau)^{\text{tr}} - s^\alpha > 0$, slip system α is potentially active. The Active systems are the slip systems with $\Delta\gamma^\alpha > 0$, which should be defined among the potential active systems. After defining the potentially active slip systems, the slip resistances is updated as follows:

$$s^\alpha(\tau) = s^\alpha(t) + \sum_{\beta\in\mathcal{A}} h^{\alpha\beta}(t)\Delta\gamma^\beta \tag{3.28}$$

where $\mathcal{A}$ represents the set of active slip systems. Anand and Kothari (1996) neglected the higher order term of $\Delta\gamma^\beta$ and rewrote the consistency equation as follows:

$$|\tau^\alpha(\tau)| = \left|\tau^\alpha(\tau)^{\text{tr}}\right| - \sum_{\beta\in\mathcal{A}} \text{sign}\left[\tau^\alpha(\tau)^{\text{tr}}\right] \text{sign}\left[\tau^\beta(\tau)^{\text{tr}}\right] S^\alpha_{\eta\zeta} \ell_{\eta\zeta\xi\varrho} \,\text{sym}\left[C^e_{\xi\varsigma}(\tau)^{tr} S^\beta_{\varsigma\varrho}\right] \Delta\gamma^\beta \tag{3.29}$$

where $S^\alpha_{\eta\zeta}$ and $S^\beta_{\varsigma\varrho}$ are the Schmid tensors of slip systems α and β, respectively, in the intermediate state which are the same as those of the reference state [see, Eqs. (3.19), (3.20)]. Eq. (3.29) can be rewritten as follows:

$$\sum_{\beta\in\mathcal{A}} A^{\alpha\beta} x^\beta = b^\alpha \tag{3.30}$$

where:

$$A^{\alpha\beta} = h^{\alpha\beta}(t) + \text{sign}\left[\tau^{\alpha}(\tau)^{\text{tr}}\right]\text{sign}\left[\tau^{\beta}(\tau)^{\text{tr}}\right]S^{\alpha}_{\eta\varsigma}\ell_{\eta\varsigma\xi\varrho}\,\text{sym}\left[C^{e}_{\xi\varsigma}(\tau)^{tr}S^{\beta}_{\varsigma\varrho}\right] \tag{3.31}$$

$$b = \left|\tau^{\alpha}(\tau)^{\text{tr}}\right| - s^{\alpha}(t) > 0 \tag{3.32}$$

$$x^{\beta} = \Delta\gamma^{\beta} \tag{3.33}$$

After solving Eqs. (3.30)–(3.33), if $x^{\beta} = \Delta\gamma^{\beta} \le 0$, the slip system β should be removed from the active slip system set. Accordingly, Eqs. (3.30)–(3.33) should be updated and solved again until all active slip systems have $\Delta\gamma > 0$. After finalizing the values for $\Delta\gamma$ for each active slip system, the plastic deformation gradient is updated according to Eq. (3.27). Next, it should be checked if $\det[F^{p}_{\beta J}(\tau)] = 1$. Otherwise, one should normalize it as follows:

$$F^{p}_{\xi J}(\tau) = \left\{\det\left[F^{p}_{\zeta I}(\tau)\right]\right\}^{-\frac{1}{3}} F^{p}_{\xi J}(\tau) \tag{3.34}$$

Next, the elastic deformation gradient is updated using the multiplicative decomposition as follows:

$$F^{e}_{i\xi}(\tau) = F_{iJ}(\tau)F^{p}_{J\xi}(\tau)^{-1} \tag{3.35}$$

and the second Piola-Kirchhoff stress tensor $S_{\eta\varsigma}(\tau)$ can be obtained as follows:

$$S_{\eta\varsigma}(\tau) = S_{\eta\varsigma}(\tau)^{\text{tr}} - \sum_{\alpha\in\mathcal{A}}\Delta\gamma^{\alpha}\,\text{sign}\left[\tau^{\alpha}(\tau)^{\text{tr}}\right]\ell_{\eta\zeta\xi\varrho}\,\text{sym}\left[C^{e}_{\xi\varsigma}(\tau)^{tr}S^{\alpha}_{\varsigma\varrho}\right] \tag{3.36}$$

where $S^{\beta}_{\varsigma\varrho}$ is the Schmid tensor of the slip system α in the intermediate state which is the same as that of the reference state [see, Eqs. (3.19), (3.20)]. Accordingly, the slip resistances are updated using Eq. (3.28) and the updated Cauchy stress tensor $\sigma_{ij}(\tau)$ is obtained as below:

$$\sigma_{ij}(\tau) = F^{e}_{i\eta}(\tau)J^{e-1}S_{\eta\varsigma}(\tau)F^{e}_{j\varsigma}(\tau) \tag{3.37}$$

where $J^{e} = \det\,[F^{e}_{i\xi}(\tau)]$.

3.2.2 Rate-dependent crystal plasticity models

The rate-dependent crystal plasticity model developed by Kalidindi et al. (1992) is elaborated here. As discussed in Section 2.2.8, the rate-independent crystal plasticity models are applicable if the material response does not depend on the applied loading rate. However, in general, the material response depends on the applied loading rates. Accordingly, the rate-dependent crystal plasticity models are incorporated to accurately capture the material properties. Here, the general finite strain framework for rate-dependent crystal plasticity models is presented. First, the multiplicative decomposition is incorporated as follows:

$$F_{iJ} = F^{e}_{i\xi}F^{p}_{\xi J} \tag{3.38}$$

The flow rule for crystal plasticity model is described as follows:

$$\dot{F}^{p}_{\xi J} = l^{p}_{\xi\zeta} F^{p}_{\zeta J} \tag{3.39}$$

and the connection between the microscale and macroscale can be defined using the following equation:

$$l^{p}_{\xi\zeta} = \sum_{\alpha} \dot{\gamma}^{\alpha} S^{\alpha}_{\xi\zeta} \tag{3.40}$$

where $\dot{\gamma}^{\alpha}$ is the shearing rate on slip system α, and $S^{\alpha}_{\xi\zeta}$ is the Schmid tensor for slip system α in the intermediate state. The more applicable form of Eq. (3.40) can be obtained by assuming that the Schmid tensor in the intermediate state $S^{\alpha}_{\eta\zeta}$ is the same as in the reference one, i.e., S^{α}_{IJ}, which can be defined as follows (Anand and Kothari, 1996):

$$S^{\alpha}_{IJ} = m^{\alpha}_{I} \otimes n^{\alpha}_{J} \tag{3.41}$$

where m^{α}_{I} and n^{α}_{J} are unit vectors of slip direction and slip plane normal in the reference configuration, respectively.

The plastic shear rate for slip system α, $\dot{\gamma}^{\alpha}$, can be defined as a function of its active shear stress τ^{α} and slip resistance s^{α} as follows:

$$\dot{\gamma}^{\alpha} = f(\tau^{\alpha}, s^{\alpha}) \tag{3.42}$$

Various functional forms have been proposed to define the plastic shear rate. One of the simplest form one can select was developed by Hutchinson (1976) as follows:

$$\dot{\gamma}^{\alpha} = \dot{\gamma}_{0} \left| \frac{\tau^{\alpha}}{s^{\alpha}} \right|^{1/m} \text{sign}[\tau^{\alpha}] \tag{3.43}$$

where $\dot{\gamma}_{0}$ and m are the material parameters denoting the reference shearing rate and rate sensitivity of the material. The resolved shear stress for slip system α is defined using the second Piola-Kirchhoff stress tensor $S_{\eta\zeta}(\tau)$ as follows:

$$\tau^{\alpha} = C^{e}_{\gamma\eta} S_{\eta\zeta} S^{\alpha}_{\gamma\zeta} \tag{3.44}$$

The tensor, $S^{\alpha}_{\gamma\zeta}$ is the Schmid tensor for slip system α in the intermediate state and the second Piola-Kirchhoff stress tensor $S_{\eta\zeta}$ can be defined as follows:

$$S_{\eta\zeta} = \ell_{\eta\zeta\xi\varrho} E^{e}_{\xi\varrho} \tag{3.45}$$

where $\ell_{\eta\zeta\xi\varrho}$ is the fourth order elasticity tensor and $E^{e}_{\xi\varrho}$ is the elastic Green strain tensor, which can be defined as follows:

$$E^{e}_{\xi\varrho} = \frac{1}{2} \left[C^{e}_{\xi\varrho} - \delta_{\xi\varrho} \right] \tag{3.46}$$

The tensor, $C^{e}_{\xi\varrho}$ is elastic right Cauchy-Green tensor, which can be described as follows:

$$C^{e}_{\xi\varrho} = F^{e}_{i\xi} F^{e}_{i\varrho} \tag{3.47}$$

In the case of negligible elastic stretches, Eq. (3.18) is further simplified as follows:

$$\tau^{\alpha} = S_{\eta\zeta} S^{\alpha}_{\eta\zeta} \tag{3.48}$$

One should note that $S^{\alpha}_{\eta\zeta}$, i.e., the Schmid tensors of slip system α in the intermediate state, is similar to that of the reference state [see, Eqs. (3.19), (3.20)].

The last element of the rate-dependent crystal plasticity model is to define the evolution of the slip resistance for each slip system, which is expressed as follows:

$$\dot{s}^{\alpha} = \sum_{\beta} h^{\alpha\beta} \dot{\gamma}^{\beta} \tag{3.49}$$

where $h^{\alpha\beta}$ defines the variation of the slip resistance for slip system α due to the plastic shear slip in slip system β, which is commonly termed as the hardening modulus. Various formulations have been introduced to describe the hardening modulus. One of the simplest form is the one developed by Asaro and Needleman (1985) as follows:

$$\begin{aligned} h^{\alpha\beta} &= q h^{\beta} \text{ for } \alpha \neq \beta \\ h^{\alpha\alpha} &= h^{\alpha} \text{ (no sum on } \alpha) \end{aligned} \tag{3.50}$$

where q is the latent hardening ratio, which is commonly selected as $1 < q < 1.4$. Also, h^{α} is the hardening rate for a single slip. Different functional forms have been developed to describe the hardening rate of a single slip. One the most conventional model to use is developed by Brown et al. (1989), which is described as follows:

$$h^{\alpha} = h_0^{\alpha} \left\{ 1 - \frac{s^{\alpha}}{s_s^{\alpha}} \right\}^{a^s} \tag{3.51}$$

where h_0^{α}, s_s^{α}, and a^s are the material constants for the slip system α.

3.2.2.1 *Incremental relation of rate-dependent crystal plasticity models*

The incremental relation of rate-dependent crystal plasticity models developed by Kalidindi et al. (1992) is elaborated here. Similar to the incremental relation of rate-independent crystal plasticity models discussed in Section 3.1.1, one assumes that the deformed body at time t with the observable thermodynamics variables of $\{F_{iJ}(t), F^{e}_{i\xi}(t), F^{p}_{\xi J}(t), \sigma_{ij}(t), s^{\alpha}(t)\}$ undergoes an incremental deformation and new deformed body at time $\tau = t + \Delta t$ has the observable thermodynamics variables with the observable thermodynamics variables of $\{F_{iJ}(\tau), F^{e}_{i\xi}(\tau), F^{p}_{\xi J}(\tau), \sigma_{ij}(\tau), s^{\alpha}(\tau)\}$. In the case of finite deformation rate-dependent crystal plasticity model, the numerical integration starts from the flow rule described in Eq. (3.39) as follows:

$$F^{p}_{\xi J}(\tau) = \left[\delta_{\xi\zeta} + \Delta t l^{p}_{\xi\zeta}(\tau)\right] F^{p}_{\zeta J}(t) \tag{3.52}$$

Eq. (3.52) is rewritten by substituting Eq. (3.40) in Eq. (3.52) as follows:

$$F^{p}_{\xi J}(\tau) = \left[\delta_{\xi\zeta} + \sum_{\alpha} \Delta\gamma^{\alpha} S^{\alpha}_{\xi\zeta}\right] F^{p}_{\zeta J}(t) \tag{3.53}$$

where

$$\Delta\gamma^{\alpha} \equiv \Delta t \dot{\gamma}^{\alpha}(\tau^{\alpha}(\tau), s^{\alpha}(\tau)) \tag{3.54}$$

and $S^{\alpha}_{\xi\zeta}$, i.e., the Schmid tensors of slip system α in the intermediate configuration, is similar to that of the reference configuration [see, Eqs. (3.19), (3.20)], which is time-independent. Hence, one can say that the resolved shear stress τ^{α} on slip system α is solely a function of the second Piola-Kirchhoff stress tensor $S_{\eta\zeta}$ according to Eq. (3.48), and Eq. (3.54) can be rewritten as follows:

$$\Delta\gamma^{\alpha} \equiv \Delta t \dot{\gamma}^{\alpha}\left(\tau^{\alpha}\left[S_{\eta\zeta}(\tau)\right], s^{\alpha}(\tau)\right) \tag{3.55}$$

One can reformulate Eq. (3.52) as follows:

$$F^{p-1}_{J\xi}(\tau) = F^{p-1}_{J\zeta}(t)\left[\delta_{\zeta\xi} + \Delta t l^{p}_{\zeta\xi}(\tau)\right] \tag{3.56}$$

The second Piola-Kirchhoff stress tensor $S_{\eta\zeta}(\tau)$ is obtained by substituting Eqs. (3.38), (3.46), (3.47) into Eq. (3.45) as follows:

$$S_{\eta\zeta}(\tau) = \ell_{\eta\zeta\xi\varrho}\left\{\frac{1}{2}\left[C^{e}_{\xi\varrho}(\tau) - \delta_{\xi\varrho}\right]\right\} \tag{3.57}$$

where the elastic right Cauchy-Green tensor $C^{e}_{\xi\varrho}(\tau)$ is defined as follows:

$$C^{e}_{\xi\varrho}(\tau) = F^{p-1}_{J\xi}(\tau) F_{iJ}(\tau) F_{iK}(\tau) F^{p-1}_{K\varrho}(\tau) \tag{3.58}$$

One can substitute Eq. (3.56) into Eq. (3.58) as follows:

$$\begin{aligned} C^{e}_{\xi\varrho}(\tau) = {} & F^{p-1}_{J\xi}(t) F_{iJ}(\tau) F_{iK}(\tau) F^{p-1}_{K\varrho}(t) \\ & - \sum_{\alpha} \Delta\gamma^{\alpha} S^{\alpha}_{\zeta\xi} F^{p-1}_{J\zeta}(t) F_{iJ}(\tau) F_{iK}(\tau) F^{p-1}_{K\varrho}(t) \\ & - F^{p-1}_{J\xi}(t) F_{iJ}(\tau) F_{iK}(\tau) F^{p-1}_{K\zeta}(t) \sum_{\alpha} \Delta\gamma^{\alpha} S^{\alpha}_{\zeta\varrho} + O\left(\Delta\gamma^{\alpha 2}\right) \end{aligned} \tag{3.59}$$

Eq. (3.59) is further simplified by neglecting the $O(\Delta\gamma^{\alpha 2})$ term as follows:

$$\begin{aligned} C^{e}_{\xi\varrho}(\tau) = {} & F^{p-1}_{J\xi}(t) F_{iJ}(\tau) F_{iK}(\tau) F^{p-1}_{K\varrho}(t) \\ & - \sum_{\alpha} \Delta\gamma^{\alpha} S^{\alpha}_{\zeta\xi} F^{p-1}_{J\zeta}(t) F_{iJ}(\tau) F_{iK}(\tau) F^{p-1}_{K\varrho}(t) \\ & - F^{p-1}_{J\xi}(t) F_{iJ}(\tau) F_{iK}(\tau) F^{p-1}_{K\zeta}(t) \sum_{\alpha} \Delta\gamma^{\alpha} S^{\alpha}_{\zeta\varrho} \end{aligned} \tag{3.60}$$

Eq. (3.60) is reformulated as follows:

$$C^e_{\xi\varrho}(\tau) = A_{\xi\varrho} - \sum_\alpha \Delta\gamma^\alpha B^\alpha_{\xi\varrho} \tag{3.61}$$

where $A_{\xi\varrho}$ and $B^\alpha_{\xi\varrho}$ are defined as follows:

$$A_{\xi\varrho} = F^{p-1}_{J\xi}(t) F_{iJ}(\tau) F_{iK}(\tau) F^{p-1}_{K\varrho}(t) \tag{3.62}$$

$$B^\alpha_{\xi\varrho} = A_{\xi\zeta} S^\alpha_{\zeta\varrho} + S^\alpha_{\zeta\xi} A_{\zeta\varrho} \tag{3.63}$$

Again, it should be noted that the Schmid tensors of slip system α in the intermediate configuration is similar to that of the reference configuration [see, Eqs. (3.19), (3.20)]. Finally, the second Piola-Kirchhoff stress tensor $S_{\eta\zeta}(\tau)$ is obtained by substituting Eqs. (3.62) and (3.63) into Eq. (3.57) as follows:

$$S_{\eta\zeta}(\tau) = S_{\eta\zeta}(\tau)^{\text{tr}} - \sum_\alpha \Delta\gamma^\alpha \left(\tau^\alpha \left[S_{\eta\zeta}(\tau)\right], s^\alpha(\tau)\right) C^\alpha_{\eta\zeta} \tag{3.64}$$

where

$$S_{\eta\zeta}(\tau)^{\text{tr}} = \ell_{\eta\zeta\xi\varrho}\left\{\frac{1}{2}\left[A_{\xi\varrho} - \delta_{\xi\varrho}\right]\right\} \tag{3.65}$$

$$C^\alpha_{\eta\zeta} = \ell_{\eta\zeta\xi\varrho}\left(\frac{1}{2}\right) B^\alpha_{\xi\varrho} \tag{3.66}$$

One should note that $S_{\eta\zeta}(\tau)^{\text{tr}}$, $A_{\xi\varrho}$, $B^\alpha_{\xi\varrho}$, and $C^\alpha_{\eta\zeta}$ can be calculated using the observable thermodynamics variables at time $= t$ and the deformation gradient tensor at time $= \tau$, i.e., $F_{iJ}(\tau)$. However, $\Delta\gamma^\alpha$ is a function of the second Piola-Kirchhoff stress tensor and slip resistances at time $= \tau$, $S_{\eta\zeta}(\tau)$ and $s^\alpha(\tau)$, respectively, which should be determined. The updated slip resistances $s^\alpha(\tau)$ are obtained as follows:

$$s^\alpha(\tau) = s^\alpha(t) + \sum_\beta h^{\alpha\beta}\left(s^\beta(\tau)\right)\left|\Delta\gamma^\beta\left(\tau^\beta\left[S_{\eta\zeta}(\tau)\right], s^\beta(\tau)\right)\right| \tag{3.67}$$

One should solve Eqs. (3.64)–(3.67) which are a set of nonlinear equations in $S_{\eta\zeta}(\tau)$ and $s^\alpha(\tau)$. To do so, an n iterative framework with two different levels should be incorporated. First, $s^\alpha(\tau)$ is assumed to be fixed and equal to the best current estimate. Accordingly, the second Piola-Kirchhoff stress tensor at time $= \tau$, i.e., $S_{\eta\zeta}(\tau)$, is obtained by solving Eq. (3.64) using an iterative Newton-Raphson procedure as follows:

$$S_{\eta\zeta}(\tau)^{k+1} = S_{\eta\zeta}(\tau)^k - {\mathcal{K}^k_{\eta\zeta\xi\varrho}}^{-1} \mathcal{F}^k_{\xi\varrho} \tag{3.68}$$

where $S_{\eta\zeta}(\tau)^{k+1}$ and $S_{\eta\zeta}(\tau)^k$ are the second Piola-Kirchhoff stress tensor at time time $= \tau$ for $(k+1)^{\text{th}}$ and k^{th} iterations, respectively, and $\mathcal{K}^k_{\xi\varrho\eta\zeta}$ and $\mathcal{F}^k_{\xi\varrho}$ can be described as follows:

$$\mathcal{F}^k_{\xi\varrho} = S_{\eta\zeta}(\tau) - S_{\eta\zeta}(\tau)^{\text{tr}} + \sum_\alpha \Delta\gamma^\alpha\left(\tau^\alpha\left[S_{\eta\zeta}(\tau)\right], s^\alpha(\tau)\right) C^\alpha_{\eta\zeta} \tag{3.69}$$

$$\mathcal{K}^k_{\eta\zeta\xi\varrho} = \delta_{\eta\zeta\xi\varrho} + \sum_{\alpha} \frac{\partial\left(\Delta\gamma^\alpha\left(\tau^\alpha\left[S_{\eta\zeta}(\tau)\right], s^\alpha(\tau)\right)\right)}{\partial\tau^\alpha} C^\alpha_{\eta\zeta} \otimes S^\alpha_{\xi\varrho} \tag{3.70}$$

After obtaining the converged value of the second Piola-Kirchhoff stress tensor at time time $=\tau$, i.e., $S_{\eta\zeta}(\tau)$, the updated values of slip resistances are obtained using another iterative procedure as follows:

$$s^\alpha(\tau)^{k+1} = s^\alpha(t) + \sum_{\beta} h^{\alpha\beta}\left(s^\beta(\tau)^k\right)\left|\Delta\gamma^\beta\left(\tau^\beta\left[S_{\eta\zeta}(\tau)\right], s^\beta(\tau)^k\right)\right| \tag{3.71}$$

where $s^\alpha(\tau)^{k+1}$ and $s^\beta(\tau)^k$ are the slip resitances at time time $=\tau$ for $(k+1)^{\text{th}}$ and k^{th} iterations, respectively.

The elastic gradient tensor $F^e_{i\eta}(\tau)$ is calculated using the updated plastic gradient tensor $F^p_{\xi J}(\tau)$ and the multiplicative decomposition, i.e., Eq. (3.38). Finally, the updated Cauchy stress tensor $\sigma_{ij}(\tau)$ is obtained as below:

$$\sigma_{ij}(\tau) = F^e_{i\eta}(\tau) J^{e-1} S_{\eta\zeta}(\tau) F^e_{j\zeta}(\tau) \tag{3.72}$$

where $J^e = \det\left[F^e_{i\xi}(\tau)\right]$.

3.3 Homogenization models

As discussed in Section 3.1, the response of polycrystalline metals is obtained using a homogenization of the behavior of consisting grains, which have their own constitutive models. Various methods of homogenization have been developed to capture the response of polycrystalline metals including the original Taylor model (Taylor, 1938a), modified Taylor model including relaxation (Honeff and Mecking, 1981; Kocks and Chandra, 1982; Van Houtte, 1982), cluster models such as LAMEL (Van Houtte et al., 1999, 2002; Delannay et al., 2002), GIA (Crumbach et al., 2004, 2006a,b), and RGC (Tjahjanto et al., 2010), spectral finite elements schemes using fast Fourier transform (Moulinec and Suquet, 1998; Lebensohn, 2001), and crystal plasticity finite elements method (CPFEM) (see, e.g., Roters et al., 2010). Here, the first and last methods, i.e., the original Taylor model and crystal plasticity finite elements method (CPFEM), will be discussed in more detail.

3.3.1 Taylor model

Taylor model (Taylor, 1938a) is the simplest homogenization model for polycrystalline metals. The core kinematic assumption of the model is that all the grains undergo similar deformation gradient which is similar to the one imposed on the polycrystalline sample. One assumes a sample consisting of N grains, the Taylor model is then described as:

$$F_{iJ} = \mathrm{F}^1_{iJ} = \mathrm{F}^2_{iJ} = \mathrm{F}^3_{iJ} = \cdots = \mathrm{F}^N_{iJ} \tag{3.73}$$

While the deformation gradient of all grains is equal to the one which is imposed on the polycrystalline sample, i.e., F_{iJ}, the elastic and plastic deformation gradient tensors for each grain depend on the orientation on that grain, which is not essentially similar for all grains. One can rewrite Eq. (3.72) as follows:

$$F_{iJ} = F^{e\ 1}_{i\xi} F^{p\ 1}_{\xi J} = F^{e\ 2}_{i\xi} F^{p\ 2}_{\xi J} = F^{e\ 3}_{i\xi} F^{p\ 3}_{\xi J} = \cdots = F^{e\ N}_{i\xi} F^{p\ N}_{\xi J} \tag{3.74}$$

The Taylor model is also expressed in terms of the velocity gradients. In other words, the velocity gradient tensors of all grains are similar to the macroscopic velocity gradient tensor as follows:

$$l_{ij} = l^1_{ij} = l^2_{ij} = l^3_{ij} = \cdots = l^N_{ij} \tag{3.75}$$

Assuming that all the grains have the same volume, the homogenized Cauchy stress tensor σ_{ij} can be simply described as the average Cauchy stress tensor over all grains as follows:

$$\sigma_{ij} = \frac{1}{N} \sum_{k=1}^{N} {\sigma_{ij}}^k \tag{3.76}$$

3.3.2 Crystal plasticity finite element method

Among the different methods to capture the behavior of polycrystalline metallic samples, crystal plasticity finite element method (CPFEM) is the most accurate while demanding one. One now considers the polycrystalline sample presented in Fig. 3.2. The sample is discretized using the regular finite element (FE) using some FE meshes. Accordingly, the response of the sample to the specific imposed boundary conditions is obtained by solving the week form of the equilibrium equations. To do so, the material response at each Gauss integration point should be defined. In the case of CPFEM, first, the texture, i.e., the grain orientation, for each Gauss integration point should be determined. Finally, the material behavior at each Gauss integration point is dictated by the crystal plasticity model for each grain with the orientation specified for that integration point. One should note that this is the simplest version of CPFEM which only takes into account the effect of the texture while the effects of grain boundaries are neglected.

After obtaining the response of the sample using the CPFEM, one obtains the homogenized behavior of the sample using different homogenization schemes including the computational homogenization, mean-field theory, and cluster methods. Roters et al. (2010) elaborated different aspects of CPFEM. CPFEM has been incorporated to solve some very important problems in polycrystalline metallic samples including grain fragmentation, texture evolution, the evolution of dislocation density, recrystallization, damage initiation and propagation, twinning, and the grain size effect (Roters et al., 2010).

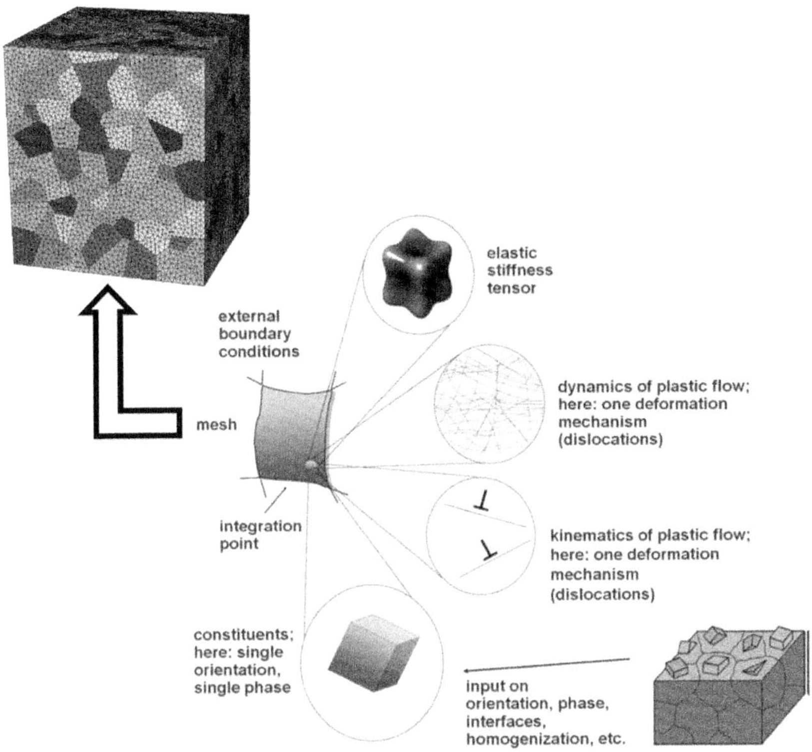

FIG. 3.2 Crystal plasticity finite elements simulation: different ingredients and finite element discretization of a polycrystalline sample. *(After Roters, F., Eisenlohr, P., Hantcherli, L., Tjahjanto, D. D., Bieler, T.R., Raabe, D., 2010. Overview of constitutive laws, kinematics, homogenization and multiscale methods in crystal plasticity finite-element modeling: theory, experiments, applications. Acta Mater. 58, 1152–1211.)*

3.4 Nonlocal crystal plasticity models

As discussed in Section 3.2, in the case of conventional crystal plasticity models, one of the basic assumptions is that there is no lattice distortion during the plastic deformation. Accordingly, there is no length scale in these models and these models cannot capture size effects, as discussed in Chapter 1. There are two general ideas of how to incorporate the size effects into the crystal plasticity models, which are very similar to the ones described for plasticity as discussed in Section 2.3.1. The first category of gradient plasticity models applies the differential operators on the internal variables of plasticity models (Acharya and Bassani, 2000; Han et al., 2005a,b; Ohashi, 2005; Dunne et al., 2007; Liang and Dunne, 2009). In the second category, the gradient operator is applied on the strain, which can be the first, second, or even higher order of strain gradients (Gurtin, 2002, 2006, 2008; Borg, 2007; Yefimov et al., 2004; Evers et al., 2004;

Arsenlis et al., 2004; Bayley et al., 2006; Levkovitch and Svendsen, 2006; Kuroda and Tvergaard, 2008a,b; Reddy, 2011; Yalçinkaya et al., 2012; Klusemann and Yalçinkaya, 2013).

In the case of the first family of gradient models, no higher stresses are required to be included and new thermodynamic forces are conjugate to the gradients of the internal variables which will not enter into the equilibrium equations. In the case of the second family of gradient models, higher-order stresses should be introduced and incorporated in the equilibrium equations which are conjugate to the gradients of strains. Accordingly, the second family is more demanding in a way that introducing gradients of strain and their conjugate, i.e., higher order stresses, they modify accordingly the expression of internal and external work. However, gradients of internal variables and their conjugates solely modify the free energy expression. In this section, the model of Han et al. (2005a,b) is presented for the first category of gradient crystal plasticity models and the model of Gurtin (2002) is elaborated for the second family of gradient models.

3.4.1 Gradient crystal plasticity models: Han, Gao, Huang, and Nix

Han et al. (2005a,b) developed a model to incorporate the effects of strain gradient in the rate-dependent crystal plasticity by modifying the internal variables. Accordingly, they modified the Taylor hardening model as a first step. In this method, they developed an equation to capture the density of geometrically necessary dislocations (GNDs) for each slip system. In the case of finite deformation model, the multiplicative decomposition is the first step as follows:

$$F_{iJ} = F^e_{i\xi} F^p_{\xi J} \tag{3.77}$$

where $F^e_{i\xi}$ and $F^p_{\xi J}$ are the elastic and plastic deformation gradients, respectively. The rest of the crystal plasticity constitutive model is similar to the ones presented in Eqs. (3.38)–(3.50). However, instead of Eq. (3.51), Han et al. (2005a,b) used a different hardening rule introduced by Zhou et al. (1993) as follows:

$$h^\alpha = h_0 \left\{ \frac{h_0 \gamma}{m \tau_0} + 1 \right\}^{m-1} \tag{3.78}$$

where h_0, τ_0, and m are the material constants, and γ is the total slip, which can be defined as follows:

$$\gamma = \int_t \sum_\alpha |\dot{\gamma}^\alpha| dt \tag{3.79}$$

In the case of classical local crystal plasticity, it is assumed that the plastic deformation gradient tensor $F^p_{\xi J}$ does not induce any lattice distortion. However, the strain gradient originated from the GNDs may lead to the crystal lattice distortion. To incorporate this lattice distortion, Han et al. (2005a,b) first introduced the concept of dislocation density tensor.

3.4.1.1 Dislocation density tensor

A concept of dislocation density tensor was developed by Han et al. (2005a,b) as a measure of GNDs. In this section, the curl operator of plastic deformation gradient is denoted as $\overline{G}_{J\xi}$, which can be defined as follows:

$$\left(\text{curl}\left[F^p_{\zeta K}\right]\right)_{J\xi} = \epsilon_{IKJ} F^p_{\xi K,I} = \overline{G}_{J\xi} \tag{3.80}$$

where ϵ_{IKJ} is the permutation tensor. The indices are defined in a way that E_J and $\widetilde{e}_\xi$ are the orthonormal unit vectors in the undeformed and intermediate configurations, respectively. $\overline{G}_{J\xi}$ is the $J\xi$ component of the second order tensor curl $F^p_{\zeta K}$. One should note that $\overline{G}_{J\xi}$ is different from the geometric dislocation tensor $G_{\alpha\beta}$ introduced by Gurtin (2002). The intermediate description of the net Burgers vector in a plane Π_0, which is defined in the undeformed configuration, can be described as follows:

$$b_\xi = \int_{\partial \Pi_0} F^p_{\xi J} dX_J \tag{3.81}$$

where $\partial \Pi_0$ is the boundary curve of the plane Π_0. One can rewrite Eq. (3.81) using the Stokes theorem as follows:

$$b_\xi = \int_{\Pi_0} \overline{G}_{J\xi} N_J d\Pi_0 \tag{3.82}$$

where N_J is the normal vector of the plane Π_0. In Eq. (3.82), the term $N_J d\Pi_0$ in the undeformed body can be related to that of the intermediate description as follows (Han et al., 2005a,b):

$$\widetilde{n}_\xi d\widetilde{\Pi} = J^p \left(F^p_{J\xi}\right)^{-1} N_J d\Pi_0 \tag{3.83}$$

where $J^p = \det[F^p_{\xi J}]$, $\widetilde{\Pi}$ is the plane in the intermediate state with $\widetilde{n}_\xi$ as its normal vector. Accordingly, Eq. (3.82) can be reformulated in the intermediate state as follows:

$$b_\xi = \int_{\Pi_0} (J^p)^{-1} \overline{G}_{J\xi} F^p_{\zeta J} \widetilde{n}_\zeta d\widetilde{\Pi} \tag{3.84}$$

One can rewrite Eq. (3.84) as follows:

$$b_\xi = \int_{\Pi_0} A^p_{\zeta\xi} \widetilde{n}_\zeta d\widetilde{\Pi} \tag{3.85}$$

where $A^p_{\zeta\xi}$ is the dislocation density tensor, which can be described as follows:

$$A^p_{\zeta\xi} = (J^p)^{-1} F^p_{\zeta J} \overline{G}_{J\xi} \tag{3.86}$$

Han et al. (2005a) showed that the definition of the dislocation density in the undeformed configuration, i.e., A^e_{IJ}, is equivalent to the one obtained in the intermediate configuration, i.e., $A^{p}_{\zeta\xi}$. Accordingly, one can drop the indices "e" and "p."

A physical interpretation of the dislocation density tensor $A_{\tilde{\zeta}\xi}$ can be presented by considering a dislocation with the length $d\tilde{l}$ within a body with the volume of $d\tilde{v}$, where both $d\tilde{l}$ and $d\tilde{v}$ are defined in the intermediate configuration, as follows:

$$A_{\zeta\xi} d\tilde{v} = d\tilde{l}\, t_{\zeta} b_{\xi} \tag{3.87}$$

where t_{ζ} represents the unit tangent vector of dislocation segment in the intermediate state. One can obtain the unit of the dislocation density from Eq. (3.87) which is m^{-1}. However, the conventional dislocation density has the dimension of m^{-2}. It is due to the fact that the dislocation density tensor $A_{\zeta\xi}$ is directly connected to the strain gradient using Eqs. (3.85), (3.86).

A linear combination of the Eq. (3.87) can be incorporated in the case of the multiple dislocation segments as follows (Han et al., 2005a):

$$A_{\zeta\xi} = \sum_{n} {\eta_A}^{n} t^{n}_{\zeta} b^{n}_{\xi} \tag{3.88}$$

where t^{n}_{ζ} and b^{n}_{ξ} are the nth dislocation unit vectors and ${\eta_A}^{n}$ is the corresponding dislocation density. One can generalize Eq. (3.88) using orthonormal base vectors of $\tilde{e}_{\xi}$ ($\xi = 1, 2, 3$) as follows:

$$A_{\zeta\xi} = \sum_{i} \sum_{j} {\eta_A}^{ij} d\tilde{l}\, \tilde{e}^{i}_{\zeta} \tilde{e}^{j}_{\xi} \; (\text{no summation over } i \text{ and } j) \tag{3.89}$$

The linear combination introduced in Eqs. (3.88), (3.89) is averaging out the statistically stored dislocations (SSDs). Accordingly, $A_{\zeta\xi}$ is a measure of GNDs in the material.

One way to link the dislocation density tensor to the conventional dislocation density interpretation is to use the norm of b^{α}_{ξ}, in which b^{α}_{ξ} is the net Burgers vector in a unit area of slip plane with the normal vector of $\tilde{n}^{\alpha}_{\zeta}$, as follows (Acharya et al., 2003):

$$\eta^{\alpha} = \left| b^{\alpha}_{\xi} \right| = \left| A_{\zeta\xi} \tilde{n}^{\alpha}_{\zeta} \right| \tag{3.90}$$

One can substitute Eq. (3.88) into Eq. (3.90) as follows:

$$\eta^{\alpha} = \left| \sum_{n} {\eta_A}^{n} t^{n}_{\zeta} b^{n}_{\xi} \tilde{n}^{\alpha}_{\zeta} \right| \tag{3.91}$$

The only nonzero terms of Eq. (3.91) is related to the dislocations with the tangential unit vectors, i.e., t^{n}_{ζ}, aligned with the normal unit vector of the slip system α, i.e., $\tilde{n}^{\alpha}_{\zeta}$.

3.4.1.2 Peach-Koehler force

One of the core relations of the dislocation mechanics is the Peach-Koehler force. The concept of a force acting on a dislocation is a method to capture its movement due to the applied stresses. The Peach-Koehler force in the deformed configuration can be defined as follows:

$$\xi_i = -\epsilon_{ijk} t_j \sigma_{kl} b_l \tag{3.92}$$

where ϵ_{ijk} is the permutation tensor, σ_{kl} is the Cauchy stress tensor, t_j is the unit tangent vector of the dislocation segment in the deformed state, and b_l is the Burgers vector for the deformed body. One can obtain the Peach-Koehler force in the intermediate state by assuming that the elastic stretch is negligible as follows:

$$\xi_\zeta = -\epsilon_{\zeta\eta\varsigma} t_\eta M_{\lambda\varsigma} b_\lambda \tag{3.93}$$

where t_η and b_λ are described in the intermediate state, and $M_{\varsigma\lambda}$ is the Mandel stress which can be described as follows:

$$M_{\lambda\varsigma} = -F^e_{\lambda k} \sigma_{kl} F^e_{l\varsigma} \tag{3.94}$$

One can rewrite Eq. (3.93) as follows:

$$\xi_\zeta = -\epsilon_{\zeta\eta\varsigma} A_{\eta\lambda} M_{\lambda\varsigma} \tag{3.95}$$

where $A_{\eta\lambda}$ can be defined as follows:

$$A_{\eta\lambda} = t_\eta b_\lambda \tag{3.96}$$

In the case of nth dislocation, the total Peach-Koehler force including the interaction with surrounding dislocations can be written as follows:

$$\xi^n_\zeta = -\epsilon_{\zeta\eta\varsigma} t^n_\eta \left(M_{\lambda\varsigma} + \sum_{j \neq n} M^j_{\lambda\varsigma} \right) b^n_\lambda \tag{3.97}$$

where the total stress is the superposition of the applied stress related to the nth dislocation $M_{\lambda\varsigma}$ and the stresses due to the interaction of the surrounding dislocations$\sum_{j \neq n} M^j_{\lambda\varsigma}$. The glide component of the force vector defined in Eq. (3.97) in the direction normal to the dislocation line and parallel to the slip plane can be described as follows:

$$\xi^n = \tilde{n}^\alpha_\varsigma \left(M_{\lambda\varsigma} + \sum_{j \neq n} M^j_{\lambda\varsigma} \right) b^n_\lambda \tag{3.98}$$

The net Peach-Koehler force of multiple dislocations can be obtained using Eq. (3.97) as follows:

$$\xi^{net}_\zeta = \sum_n \xi^n_\zeta = -\sum_n \epsilon_{\zeta\eta\varsigma} t^n_\eta \left(M_{\lambda\varsigma} + \sum_{j \neq n} M^j_{\lambda\varsigma} \right) b^n_\lambda \tag{3.99}$$

One can capture the effect of dislocation pile-up using Eq. (3.99) which shows that the total stress field, i.e., $M_{\varsigma\lambda}+\sum_{j\neq n}M^{j}_{\varsigma\lambda}$, can be much higher than the applied stress, i.e., $M_{\varsigma\lambda}$.

Crystal plasticity models commonly do not include the interaction of dislocations with each other, and accordingly, the general Peach-Koehler force can be described as follows:

$$\xi_{\zeta}=-\epsilon_{\zeta\eta\varsigma}A_{\eta\lambda}M_{\lambda\varsigma} \tag{3.100}$$

which is similar to the one described in Eq. (3.95) for an isolated dislocation. Eq. (3.100) can be interpreted as the Peach-Koehler force acting on GNDs due to the applied stress of $M_{\varsigma\lambda}$. On the other hand, the acting shear stress on the slip system α can be obtained as follows:

$$\tau^{\alpha}=M_{\varsigma\lambda}S^{\alpha}_{\varsigma\lambda} \tag{3.101}$$

where $S^{\alpha}_{\eta\zeta}$ is the Schmid tensor in the intermediate state, which can be defined as follows:

$$S^{\alpha}_{\varsigma\lambda}=m^{\alpha}_{\varsigma}n^{\alpha}_{\lambda} \tag{3.102}$$

where m^{α}_{ς} and n^{α}_{λ} are unit vectors of slip direction and slip plane normal to the slip system α in the intermediate configuration, respectively. One can obtain the Peach-Koehler force for the slip system α by substituting Eqs. (3.101), (3.102) into Eq. (3.100) as follows:

$$\xi^{\alpha}_{\zeta}=-\epsilon_{\zeta\eta\varsigma}\tau^{\alpha}A_{\eta\lambda}n^{\alpha}_{\lambda}m^{\alpha}_{\varsigma} \tag{3.103}$$

One can expand the dislocation density tensor $A_{\eta\lambda}$ over all slip systems as follows:

$$A_{\eta\lambda}=\sum_{\alpha}\left(\eta_{\odot}{}^{\alpha}m^{\alpha}_{\eta}m^{\alpha}_{\lambda}+\eta_{\vdash}{}^{\alpha}t^{\alpha}_{\eta}m^{\alpha}_{\lambda}\right) \tag{3.104}$$

where $\eta_{\odot}{}^{\alpha}$ is the density of screw dislocations, $\eta_{\vdash}{}^{\alpha}$ is the density of edge dislocations, and t^{α}_{η} can be described as follows:

$$t^{\alpha}_{\eta}=\epsilon_{\eta\zeta\varsigma}m^{\alpha}_{\zeta}n^{\alpha}_{\varsigma} \tag{3.105}$$

One should note that $\eta_{\odot}{}^{\alpha}$ and $\eta_{\vdash}{}^{\alpha}$ can have both positive and negative values. The Peach-Koehler force for the slip system α described in Eq. (3.103) can be rewritten by incorporating Eq. (3.104) as follows:

$$\xi^{\alpha}_{\zeta}=\epsilon_{\zeta\eta\varsigma}\tau^{\alpha}n^{\alpha}_{\eta}\sum_{\beta}m^{\alpha\beta}\left(\eta_{\odot}{}^{\beta}m^{\beta}_{\varsigma}+\eta_{\vdash}{}^{\beta}t^{\beta}_{\varsigma}\right) \tag{3.106}$$

where $m^{\alpha\beta}$ can be described as follows:

$$m^{\alpha\beta}=m^{\alpha}_{\varsigma}m^{\beta}_{\varsigma} \tag{3.107}$$

Eq. (3.106) can be simplified according to the fact that ξ_{ς}^{α} is parallel to the slip plane as follows:

$$\xi_{\varsigma}^{\alpha} = \tau^{\alpha}\left(\hat{\eta}_{\vdash}{}^{\alpha} m_{\varsigma}^{\alpha} - \hat{\eta}_{\odot}{}^{\alpha} t_{\varsigma}^{\alpha}\right) \tag{3.108}$$

where $\hat{\eta}_{\vdash}{}^{\alpha}$ is the effective density corresponding to the edge dislocations of the slip system α and $\hat{\eta}_{\odot}{}^{\alpha}$ is that of the screw dislocations. A measure of dislocation density can be obtained by taking the norm of the Peach-Koehler force for the slip system α as follows:

$$\left|\xi_{\varsigma}^{\alpha}\right| = |\tau^{\alpha}|\eta_{G}^{\alpha} \tag{3.109}$$

where η_{G}^{α} is the effective measure of GNDs density, which can be described as follows:

$$\eta_{G}^{\alpha} = \sqrt{\left(\hat{\eta}_{\vdash}{}^{\alpha}\right)^{2} + \left(\hat{\eta}_{\odot}{}^{\alpha}\right)^{2}} \tag{3.110}$$

One can also directly obtain η_{G}^{α} using Eqs. (3.103), (3.109) as follows:

$$\eta_{G}^{\alpha} = \left|\epsilon_{\zeta\eta\varsigma} A_{\eta\lambda} n_{\lambda}^{\alpha} m_{\varsigma}^{\alpha}\right| \tag{3.111}$$

Fleck et al. (1994) showed that in the case of a single slip system, the strain gradients normal to the slip plane does not induce any lattice distortion, and accordingly, does not change the hardening nor the dislocation density tensor $A_{\eta\lambda}$. The effective measure of GNDs density η_{G}^{α} described in Eq. (3.111) can satisfy the description of single slip system presented by Fleck et al. (1994). Han et al. (2005a,b) incorporated the definition of effective measure of GNDs density η_{G}^{α} described in Eq. (3.111) in their framework.

3.4.1.3 *Hardening description*

The Taylor hardening model relates the shear strength to the dislocation density as follows:

$$\tau = \overline{\alpha}\mu b\sqrt{\rho} \tag{3.112}$$

where μ is the shear modulus, b is the Burgers vector magnitude, ρ is the total dislocation density, and $\overline{\alpha}$ is a material constant. One can rewrite Eq. (3.112) by considering the total dislocation density as a summation of GNDs and SSDs densities, i.e., ρ_G and ρ_S, respectively, as follows:

$$\tau = \overline{\alpha}\mu b\sqrt{\rho_S + \rho_G} \tag{3.113}$$

According to dislocation mechanics, the velocity of dislocations can be related to the Peach-Koehler force as described in Eq. (3.98), e.g., $\xi^{n} \propto v^{n}$. Also, the velocity of the dislocations can be related to the plastic strain rate using the Orowan relation as follows:

$$\dot{\varepsilon}_p = \rho b v \tag{3.114}$$

In the case of crystal plasticity models, these relations should be defined for each slip system. In other words, the acting shear stress on the slip system α, i.e., τ^α, is related to the force acting on GNDs on that slip system, i.e., ξ_ζ^α. As explained in Eqs. (3.99), (3.100), crystal plasticity does not account for dislocation interactions in the calculation of dislocation forces. However, the effects of these interactions can be included in the average sense using the Taylor hardening model as described in Eq. (3.113). In other words, the interaction of dislocations with each other contributes to the slip resistance of each slip system. One can consider the following form for the plastic shear rate for slip system α:

$$\dot{\gamma}^\alpha = \dot{\gamma}_0 \left| \frac{\tau^\alpha}{s^\alpha} \right|^{1/m} \text{sign}[\tau^\alpha] \tag{3.115}$$

where s^α is the slip resistance for the slip system α, and $\dot{\gamma}_0$ and m are the material parameters denoting the reference shearing rate and rate sensitivity of the material, respectively. In the case of local crystal plasticity models, the slip resistance of the slip system can be related to the SSDs density using the Taylor hardening model as follows:

$$s^\alpha_{local} = \overline{\alpha}\mu b\sqrt{\rho^\alpha_S} \tag{3.116}$$

Accordingly, the SSDs density can be obtained using Eq. (3.116) as follows:

$$\rho^\alpha_S = \left(\frac{s^\alpha_{local}}{\overline{\alpha}\mu b} \right)^2 \tag{3.117}$$

Now, in the case of nonlocal crystal plasticity, one can incorporate the effect of GNDs density in the slip resistance as follows:

$$s^\alpha = s^0 \sqrt{\left(\frac{s^\alpha_{local}}{s^0} \right)^2 + l\eta^\alpha_G} \tag{3.118}$$

where s^0 is a material constant defining the reference slip resistance and l is the material length scale which can be defined as follows:

$$l = \frac{\overline{\alpha}^2 \mu^2 b}{(s^0)^2} \tag{3.119}$$

Considering $\mu/s^0 \approx 100$ and knowing the Burgers vector b is in the order of angstrom, the material length scale l is in the order of a micron. After introducing the nonlocal slip resistance in Eq. (3.118), the plastic shear rate for slip system α can be then obtained using Eq. (3.115).

3.4.1.4 *Small strain framework*

Han et al. (2005a) simplified the presented mode for the framework of small strains. Accordingly, instead of the multiplicative decomposition described in Eq. (3.77), the additive decomposition of the displacement gradient can be used as follows:

$$u_{i,j} = H^{e}_{ij} + H^{p}_{ij} \tag{3.120}$$

where u_i is the displacement vector, H^{e}_{ij} describes the lattice rotation, and H^{p}_{ij} represents the plastic distortion. One should note that $H^{p}_{ii} \equiv 0$. The elastic and plastic strain tensors, i.e., ε^{e}_{ij} and ε^{p}_{ij}, respectively, can be related to H^{e}_{ij} and H^{p}_{ij} as follows:

$$\varepsilon^{e,p}_{ij} = 1/2\left[H^{e,p}_{ij} + H^{e,p}_{ji}\right] \tag{3.121}$$

In the case of small strain crystal plasticity, the material properties from macro to the micro can be connected as follows:

$$H^{p}_{ij} = \sum_{\alpha} \gamma^{\alpha} S^{\alpha}_{ij} \tag{3.122}$$

where S^{α}_{ij} is the Schmid tensor for the slip system α, which can be defined as follows:

$$S^{\alpha}_{ij} = m^{\alpha}_{i} n^{\alpha}_{j} \tag{3.123}$$

One can substitute Eq. (3.122) into Eq. (3.121) and obtain the plastic strain tensor ε^{p}_{ij} as follows:

$$\varepsilon^{p}_{ij} = \sum_{\alpha} \gamma^{\alpha} \left(S^{\alpha}_{ij}\right)_{\text{sym}} \tag{3.124}$$

where $(S^{\alpha}_{ij})_{\text{sym}}$ can be defined as follows:

$$\left(S^{\alpha}_{ij}\right)_{\text{sym}} = 1/2\left[m^{\alpha}_{i} n^{\alpha}_{j} + m^{\alpha}_{j} n^{\alpha}_{i}\right] \tag{3.125}$$

The rate of slip resistance can be defined as follows:

$$\dot{s}^{\alpha} = \sum_{\beta} h^{\alpha\beta} \dot{\gamma}^{\beta} \tag{3.126}$$

In the case of small strain theory, the resolved shear stress for slip system α can be defined as follows:

$$\tau^{\alpha} = \sigma_{ij} S^{\alpha}_{ij} \tag{3.127}$$

where σ_{ij} is the Cauchy stress tensor. The dislocation density tensor for small strains can be simplified as follows:

$$A_{ij} = \left(\text{curl } \left[H^{p}_{kl}\right]\right)_{ij} = \sum_{\alpha} \epsilon_{ilk} \gamma^{\alpha}{}_{,l} n^{\alpha}_{k} m^{\alpha}_{j} \tag{3.128}$$

The dislocation density force for slip system α, i.e., ξ^{α}_{i}, can be written as follows:

$$\xi_i^\alpha = -\tau^\alpha \epsilon_{ijk} A_{jl} n_l^\alpha m_k^\alpha = \tau^\alpha \epsilon_{ijk} n_j^\alpha \sum_\beta \epsilon_{klm} m^{\alpha\beta} \gamma^\beta{}_{,l} n_m^\beta \tag{3.129}$$

where $m^{\alpha\beta}$ can be described as follows:

$$m^{\alpha\beta} = m_i^\alpha m_i^\beta \tag{3.130}$$

The effective measure of the GNDs density η_G^α can be obtained as follows:

$$\eta_G^\alpha = \left|\epsilon_{ijk} A_{jl} n_l^\alpha m_k^\alpha\right| = \left|\epsilon_{ijk} n_j^\alpha \sum_\beta \epsilon_{klm} m^{\alpha\beta} \gamma^\beta{}_{,l} n_m^\beta\right| \tag{3.131}$$

The GNDs density η_G^α can be incorporated to update the slip resistance using Eq. (3.118). Finally, the shear strain can be updated using Eq. (3.115).

Han et al. (2005a) started developing the model for a single slip system to capture the dislocation densities of the edge and screw dislocations for the slip system α, i.e., $\hat{\eta}_\vdash{}^\alpha$ and $\hat{\eta}_\odot{}^\alpha$, respectively. Accordingly, the relation between the material properties from macro scale to the micro scale can be captured as follows:

$$H_{ij}^p = \gamma^\alpha S_{ij}^\alpha \tag{3.132}$$

where S_{ij}^α is the Schmid tensor for the slip system α, which can be defined as follows:

$$S_{ij}^\alpha = m_i^\alpha n_j^\alpha \tag{3.133}$$

In the case of a single slip system and small strain framework, the dislocation density tensor A_{ij} can be defined as follows:

$$A_{ij} = \epsilon_{ilk} \gamma^\alpha{}_{,l} n_k^\alpha m_j^\alpha \tag{3.134}$$

where $\gamma^\alpha{}_{,l}$ is the gradient of the plastic shear deformation corresponding to the slip system α and ϵ_{ijk} is the permutation tensor. The dislocation density force can be written as follows:

$$\xi_i^\alpha = \tau^\alpha \epsilon_{ijk} \epsilon_{klm} n_j^\alpha \gamma^\alpha{}_{,l} n_m^\alpha \tag{3.135}$$

One can decompose $\gamma^\alpha{}_{,l}$ as follows:

$$\gamma^\alpha{}_{,l} = \gamma^\alpha{}_{,i} n_i^\alpha n_l^\alpha + \gamma^\alpha{}_{,i} m_i^\alpha m_l^\alpha + \gamma^\alpha{}_{,i} t_i^\alpha t_l^\alpha \tag{3.136}$$

where t_η^α can be described as follows:

$$t_\eta^\alpha = \epsilon_{\eta\zeta\varsigma} m_\zeta^\alpha n_\varsigma^\alpha \tag{3.137}$$

The final form of the dislocation density force can be obtained by substituting Eq. (3.137) into Eq. (3.135) as follows:

$$\xi_i^n = \tau^\alpha \left(-\gamma^\alpha{}_{,l} m_l^\alpha m_i^\alpha + \gamma^\alpha{}_{,l} t_l^\alpha t_i^\alpha\right) \tag{3.138}$$

One can obtain the dislocation densities of the edge and screw dislocations for the slip system α, i.e., $\hat{\eta}_{\vdash}{}^{\alpha}$ and $\hat{\eta}_{\odot}{}^{\alpha}$, respectively, by comparing Eq. (3.108) with Eq. (3.138) as follows: $\xi_{\varsigma}^{\alpha} = \tau^{\alpha}\left(\hat{\eta}_{\vdash}{}^{\alpha} m_{\varsigma}^{\alpha} - \hat{\eta}_{\odot}{}^{\alpha} t_{\varsigma}^{\alpha}\right)$

$$\hat{\eta}_{\vdash}{}^{\alpha} = -\gamma^{\alpha}{}_{,i} m_i^{\alpha} \tag{3.139}$$

$$\hat{\eta}_{\odot}{}^{\alpha} = -\gamma^{\alpha}{}_{,i} t_i^{\alpha} \tag{3.140}$$

The same procedure can be incorporated to calculate $\hat{\eta}_{\vdash}{}^{\alpha}$ and $\hat{\eta}_{\odot}{}^{\alpha}$ for multiple slip systems.

3.4.1.5 Plane strain bending

The dislocation network for a plain strain bending is shown in Fig. 3.3. The simplest case of crystal plasticity can be assumed by considering a single slip system with the following crystallographic properties:

$$m_i^{\alpha} = \begin{bmatrix} \cos\theta \\ \sin\theta \\ 0 \end{bmatrix} \tag{3.141}$$

$$n_i^{\alpha} = \begin{bmatrix} -\sin\theta \\ \cos\theta \\ 0 \end{bmatrix} \tag{3.142}$$

In the case of small strain framework, the displacement field of a plain strain bending problem with the curvature of κ can be obtained as follows:

$$u = \kappa x y \tag{3.143}$$

$$v = -\frac{1}{2}\kappa\left(x^2 + y^2\right) \tag{3.144}$$

$$w = 0 \tag{3.145}$$

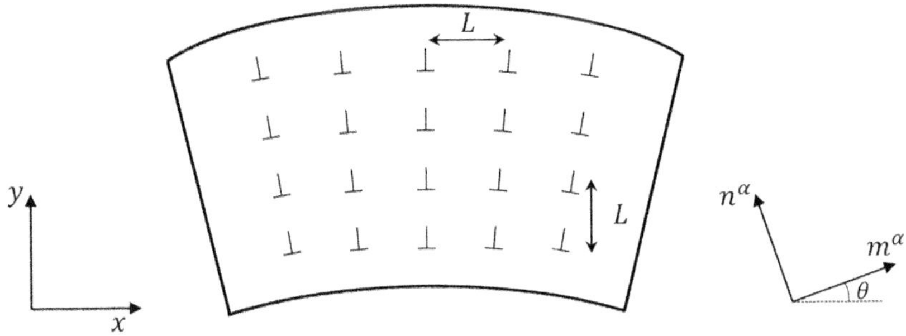

FIG. 3.3 Dislocation network for a plain strain bending.

where u, v, and w are the displacements in the x, y, and z directions, respectively. Accordingly, the nonzero strain tensor components can be obtained as follows:

$$\varepsilon_{11} = -\varepsilon_{22} = \kappa y \tag{3.146}$$

One can consider the following shear stress applied on the beam:

$$\tau_{ij} = \tau_0 \left(m_i^\alpha n_j^\alpha + n_i^\alpha m_j^\alpha \right) \tag{3.147}$$

The dislocation density force can then be obtained as follows:

$$\xi_i^n = \frac{b\tau_0}{L^2} \left(m_i^\alpha \cos\theta + n_i^\alpha \sin\theta \right) \tag{3.148}$$

The dislocation density force has two components, one normal to the slip plane, i.e., $b\tau_0 n_i^\alpha \sin\theta/L^2$, and one parallel to slip direction, i.e., $b\tau_0 m_i^\alpha \cos\theta/L^2$, where the former one does not contribute to the slip mechanism. Accordingly, the effective dislocation density force can be obtained as follows:

$$\xi_i^n = \frac{b\tau_0}{L^2} m_i^\alpha \cos\theta \tag{3.149}$$

which leads to the following effective GNDs density:

$$\eta_G^\alpha = \frac{b}{L^2} \cos\theta \tag{3.150}$$

In the case of $\theta = 90^\circ$, i.e., n_i^α is parallel with the beam axis, the effective GNDs density becomes zero, which is in line with the prediction of zero lattice distortion for $\theta = 90^\circ$ presented by Fleck et al. (1994).

3.4.2 Gradient crystal plasticity models: Gurtin

Gurtin (2002) developed a nonlocal rate-dependent crystal plasticity model by incorporating the GNDs density measure. In this model, higher-order stresses should be introduced and incorporated in the equilibrium equations which are conjugate to the gradients of strains. The basic kinematics for this model is similar to the one presented in Section 3.2.2 and 3.4.1 which is standard for the finite strain rate-dependent crystal plasticity models. The core equation is the multiplicative decomposition of deformation gradients as follows:

$$F_{iJ} = F_{i\xi}^e F_{\xi J}^p \tag{3.151}$$

where $F_{i\xi}^e$ and $F_{\xi J}^p$ are the elastic and plastic deformation gradients, respectively. The rest of the crystal plasticity constitutive model is similar to the one presented in Eqs. (3.38)–(3.50).

3.4.2.1 *The tensor of geometric dislocation*

Gurtin (2002) introduces the geometric dislocation tensor for an arbitrary plane with the unit normal vector of n_ξ^* as follows:

$$G_{\alpha\beta}=F^p_{\alpha J}\left(\text{curl}\left[F^p_{\zeta K}\right]\right)_{J\beta}=F^p_{\alpha J}\epsilon_{IKJ}F^p_{\beta K,I} \tag{3.152}$$

where the term $(\text{curl}[F^p_{\zeta K}])_{J\beta}$ is the $J\beta$ component of the second order tensor curl $[F^p_{\zeta K}]$. The next step is to define the rate of geometric dislocation $\dot{G}_{\alpha\beta}$. To do so, one can start from Eq. (3.152) as follows:

$$\dot{G}_{\alpha\beta}=\dot{F}^p_{\alpha J}\left(\text{curl}\left[F^p_{\zeta K}\right]\right)_{J\beta}+F^p_{\alpha J}\left(\text{curl}\left[\dot{F}^p_{\zeta K}\right]\right)_{J\beta} \tag{3.153}$$

In Eq. (3.153), first, one should derive the term $\left(\text{curl}\left[\dot{F}^p_{\zeta K}\right]\right)_{J\beta}$. To do so, one can define $\dot{F}^p_{\zeta K}$ as follows:

$$\dot{F}^p_{\zeta K}=l^p_{\zeta\xi}F^p_{\xi K} \tag{3.154}$$

Also, for the second order tensors of $l^p_{\zeta\xi}$ and $F^p_{\xi K}$ where $l^p_{\zeta\xi}$ is constant (Gurtin, 2002), one can write:

$$\left(\text{curl}\left[l^p_{\zeta\xi}F^p_{\xi K}\right]\right)_{J\beta}=\left(\text{curl}\left[F^p_{\zeta K}\right]\right)_{J\zeta}l^p_{\beta\zeta} \tag{3.155}$$

The term $\left(\text{curl}\left[\dot{F}^p_{\zeta K}\right]\right)_{J\beta}$ can be obtained by incorporating Eqs. (3.154), (3.155) as follows:

$$\left(\text{curl}\left[\dot{F}^p_{\zeta K}\right]\right)_{J\beta}=\left(\text{curl}\left[F^p_{\zeta K}\right]\right)_{J\zeta}l^p_{\beta\zeta}+\left(\text{curl}_{l^p}\left[l^p_{\zeta\xi}F^p_{\xi K}\right]\right)_{J\beta} \tag{3.156}$$

where $(\text{curl}_{l^p}[l^p_{\zeta\xi}F^p_{\xi K}])_{J\beta}$ is the $J\beta$ component of $\text{curl}[l^p_{\zeta\xi}F^p_{\xi K}]$ while $F^p_{\xi K}$ is considered as constant. In the case of crystal plasticity, Eq. (3.154) can be rewritten using Eq. (3.40) as follows:

$$\dot{F}^p_{\zeta K}=l^p_{\zeta\xi}F^p_{\xi K}=\sum_{\overline{\alpha}}\dot{\gamma}^{\overline{\alpha}}\left(S^{\overline{\alpha}}_{\zeta\xi}F^p_{\xi K}\right) \tag{3.157}$$

where $\sum_{\overline{\alpha}}$ is the summation over all slip systems. Accordingly, the term $(\text{curl}_{l^p}[l^p_{\zeta\xi}F^p_{\xi K}])_{J\beta}$ can be simplified as follows (Gurtin, 2002):

$$\left(\text{curl}_{l^p}\left[l^p_{\zeta\xi}F^p_{\xi K}\right]\right)_{J\beta}=\sum_{\overline{\alpha}}\epsilon_{JIK}\dot{\gamma}^{\overline{\alpha}}{}_{,I}F^p_{\zeta K}S^{\overline{\alpha}}_{\beta\zeta} \tag{3.158}$$

where $\dot{\gamma}^{\alpha}$ is a scalar field and $S^{\alpha}_{\xi\zeta}$ and $F^p_{\xi K}$ are constant. The rate of geometric dislocation $\dot{G}_{\alpha\beta}$ can be rewritten by substituting Eqs. (3.155), (3.156), (3.158) into Eq. (3.153) as follows:

$$\dot{G}_{\alpha\beta}=\dot{F}^p_{\alpha J}\left(\text{curl}\left[F^p_{\zeta K}\right]\right)_{J\beta}+F^p_{\alpha J}\left(\text{curl}\left[F^p_{\zeta K}\right]\right)_{J\zeta}l^p_{\beta\zeta}+F^p_{\alpha J}\sum_{\overline{\alpha}}\epsilon_{JIK}\dot{\gamma}^{\overline{\alpha}}{}_{,I}F^p_{\zeta K}S^{\overline{\alpha}}_{\beta\zeta} \tag{3.159}$$

One can further simplify Eq. (3.159) by incorporating Eq. (3.157) as follows:

$$\dot{G}_{\alpha\beta} = \sum_{\bar{\alpha}} \dot{\gamma}^{\bar{\alpha}} \left(S^{\bar{\alpha}}_{\alpha\xi} F^{p}_{\xi J} \right) \left(\text{curl} \left[F^{p}_{\zeta K} \right] \right)_{J\beta} + F^{p}_{\alpha J} \left(\text{curl} \left[F^{p}_{\zeta K} \right] \right)_{J\zeta} \sum_{\bar{\alpha}} \dot{\gamma}^{\bar{\alpha}} S^{\bar{\alpha}}_{\beta\zeta}$$
$$+ F^{p}_{\alpha J} \sum_{\bar{\alpha}} \epsilon_{JIK} \dot{\gamma}^{\bar{\alpha}}{}_{,I} F^{p}_{\zeta K} S^{\bar{\alpha}}_{\beta\zeta} \tag{3.160}$$

Eq. (3.160) can be rewritten by using Eq. (3.152) as follows:

$$\dot{G}_{\alpha\beta} = \sum_{\bar{\alpha}} \dot{\gamma}^{\bar{\alpha}} \left(S^{\bar{\alpha}}_{\alpha\xi} G_{\xi\beta} + G_{\alpha\zeta} S^{\bar{\alpha}}_{\beta\zeta} \right) + \sum_{\bar{\alpha}} F^{p}_{\alpha J} \epsilon_{JIK} \dot{\gamma}^{\bar{\alpha}}{}_{,I} F^{p}_{\zeta K} S^{\bar{\alpha}}_{\beta\zeta} \tag{3.161}$$

One can introduce $p^{\bar{\alpha}}_{\alpha\zeta}$ as follows:

$$p^{\bar{\alpha}}_{\xi} = F^{e}_{i\xi} \dot{\gamma}^{\bar{\alpha}}{}_{,i} = \left(F^{p}_{I\xi} \right)^{-1} \dot{\gamma}^{\bar{\alpha}}{}_{,I} \tag{3.162}$$

In the case of tensor notation (Gurtin, 2002):

$$\boldsymbol{p}^{\bar{\alpha}} = \boldsymbol{F}^{e\text{T}} \text{grad} \left[\dot{\gamma}^{\bar{\alpha}} \right] = \boldsymbol{F}^{p-\text{T}} \nabla \dot{\gamma}^{\bar{\alpha}} \tag{3.163}$$

where $\nabla \dot{\gamma}^{\bar{\alpha}} = \partial \dot{\gamma}^{\bar{\alpha}} / \partial X_L$. Also, one can show (Gurtin, 2002):

$$F^{p}_{\alpha J} \sum_{\bar{\alpha}} \epsilon_{JIK} \dot{\gamma}^{\bar{\alpha}}{}_{,I} F^{p}_{\zeta K} = \det \left[F^{p}_{\varsigma M} \right] \epsilon_{\alpha\beta\zeta} \left(F^{p}_{K\beta} \right)^{-1} \dot{\gamma}^{\bar{\alpha}}{}_{,K} \tag{3.164}$$

In the case of tensor notation (Gurtin, 2002):

$$\boldsymbol{F}^{p} \left(\nabla \dot{\gamma}^{\bar{\alpha}} \times \right) \boldsymbol{F}^{p\text{T}} = \det \left[\boldsymbol{F}^{p} \right] \left(\boldsymbol{F}^{p-\text{T}} \nabla \dot{\gamma}^{\bar{\alpha}} \right) \times = \left(\boldsymbol{F}^{p-\text{T}} \nabla \dot{\gamma}^{\bar{\alpha}} \right) \times = \boldsymbol{p}^{\bar{\alpha}} \times \tag{3.165}$$

where $\det[\boldsymbol{F}^{p}] = 1$. Accordingly, the third term of Eq. (3.161) can be rewritten as follows:

$$F^{p}_{\alpha J} \sum_{\bar{\alpha}} \epsilon_{JIK} \dot{\gamma}^{\bar{\alpha}}{}_{,I} F^{p}_{\zeta K} S^{\bar{\alpha}}_{\beta\zeta} = \sum_{\bar{\alpha}} \boldsymbol{p}^{\bar{\alpha}} \times \boldsymbol{S}^{\bar{\alpha}\text{T}} = \sum_{\bar{\alpha}} \epsilon_{\alpha\xi\zeta} p^{\bar{\alpha}}_{\xi} S^{\bar{\alpha}}_{\beta\zeta} \tag{3.166}$$

The final form of the rate of geometric dislocation $\dot{G}_{\alpha\beta}$ can be obtained by substituting Eq. (3.163) into Eq. (3.161) as follows:

$$\dot{G}_{\alpha\beta} = \sum_{\bar{\alpha}} \left[\dot{\gamma}^{\bar{\alpha}} \left(S^{\bar{\alpha}}_{\alpha\xi} G_{\xi\beta} + G_{\alpha\zeta} S^{\bar{\alpha}}_{\beta\zeta} \right) + \epsilon_{\alpha\xi\zeta} p^{\bar{\alpha}}_{\xi} S^{\bar{\alpha}}_{\beta\zeta} \right] \tag{3.167}$$

In the tensor notation:

$$\dot{\boldsymbol{G}} = \sum_{\bar{\alpha}} \left[\dot{\gamma}^{\bar{\alpha}} \left(\boldsymbol{S}^{\bar{\alpha}} \boldsymbol{G} + \boldsymbol{G} \boldsymbol{S}^{\bar{\alpha}\text{T}} \right) + \boldsymbol{p}^{\bar{\alpha}} \times \boldsymbol{S}^{\bar{\alpha}\text{T}} \right] \tag{3.168}$$

3.4.2.2 *The tensor of geometric dislocation: Pure dislocations*

The tensor of geometric dislocation $G_{\alpha\beta}$ can be represented using the linear combination of multiple dislocation segments as follows:

$$G_{\alpha\beta} = \sum_{n} \rho^{n} t^{n}_{\alpha} b^{n}_{\beta} \tag{3.169}$$

where ρ^n is the signed dislocation density of the nth dislocation segment, and t^n_α and b^n_β represent the unit tangent and Burgers vectors of the nth dislocation segment, respectively. In the case of screw dislocations, $t^n_\alpha = b^n_\alpha$, while in the case of edge dislocations t^n_α is normal to b^n_α. One can rewrite Eq. (3.169) by considering pure dislocations of screw and edge using orthonormal base vectors of $\tilde{e}_\alpha$ ($\xi = 1, 2, 3$) as follows:

$$G_{\alpha\beta} = \sum_{i=1}^{3}\sum_{j=1}^{3}\rho^{ij}\tilde{e}^{i}_{\alpha}\tilde{e}^{j}_{\beta} \text{ no summation over } i \text{ and } j \tag{3.170}$$

which is commonly called as Nye expansion (Gurtin, 2002).

The second expansion of tensor of geometric dislocations $G_{\alpha\beta}$ has been used by Kubin et al. (1992), Sun et al. (1998, 2000), and Arsenlis and Parks (1999) as follows:

$$G_{\alpha\beta} = \sum_{\overline{\alpha}}\left(\eta_{\odot}{}^{\overline{\alpha}} m^{\overline{\alpha}}_{\alpha} m^{\overline{\alpha}}_{\beta} + \eta_{\vdash}{}^{\overline{\alpha}} \boldsymbol{t}^{\overline{\alpha}}_{\alpha} m^{\overline{\alpha}}_{\beta}\right) \tag{3.171}$$

where $\eta_{\odot}{}^{\overline{\alpha}}$ is the density of screw dislocations, $\eta_{\vdash}\overline{\alpha}$ is the density of edge dislocations, and $\boldsymbol{t}^{\overline{\alpha}}_{\eta}$ can be described as follows:

$$\boldsymbol{t}^{\overline{\alpha}}_{\alpha} = \epsilon_{\alpha\zeta\varsigma} m^{\overline{\alpha}}_{\zeta} n^{\overline{\alpha}}_{\varsigma} \tag{3.172}$$

where $m^{\overline{\alpha}}_{\zeta}$ and $n^{\overline{\alpha}}_{\varsigma}$ are unit vectors of slip direction and slip plane normal of the slip system $\overline{\alpha}$ in the intermediate configuration, respectively. As Eq. (3.171) shows, the incorporated screw and edge dislocations are the only pure dislocations acting on the $\overline{\alpha}^{\text{th}}$ slip plane with the Burgers vector parallel to $m^{\overline{\alpha}}_{\beta}$.

One can decompose the Nye expansion of the geometric dislocation tensor $G_{\alpha\beta}$, as presented in Eq. (3.170), relative to a given arbitrary plane Π^* with the orthonormal base vectors of $\tilde{e}^{1}_{\alpha}$ and $\tilde{e}^{2}_{\alpha}$ and unit normal vector of $\tilde{e}^{3}_{\alpha} = n^*_\alpha$ as follows:

$$G_{\alpha\beta} = G^{\parallel}_{\alpha\beta} + G^{\perp}_{\alpha\beta} \tag{3.173}$$

$$G^{\parallel}_{\alpha\beta} = \sum_{i=1}^{2}\sum_{j=1}^{2}\rho^{ij}\tilde{e}^{i}_{\alpha}\tilde{e}^{j}_{\beta} + \sum_{i=1}^{2}\rho^{in}\tilde{e}^{i}_{\alpha}n^*_\beta \tag{3.174}$$

$$G^{\perp}_{\alpha\beta} = \rho^{nn}n^*_\alpha n^*_\beta + \sum_{i=1}^{2}\rho^{ni}n^*_\alpha\tilde{e}^{i}_{\beta} \tag{3.175}$$

where $G^{\parallel}_{\alpha\beta}$ and $G^{\perp}_{\alpha\beta}$ are corresponding to the edge and screw dislocations with the line direction parallel and perpendicular to Π^*, respectively. The decomposition presented in Eqs. (3.173)–(3.175) is commonly called as the expansion of geometric dislocation tensor $G_{\alpha\beta}$ relative to Π^*. One can define the relation between $G^{\parallel}_{\alpha\beta}$ and $G^{\perp}_{\alpha\beta}$ with the geometric dislocation tensor $G_{\alpha\beta}$ as follows:

$$G^{\parallel}_{\alpha\beta} = \mathbb{P}_{\alpha\zeta}G_{\zeta\beta} \tag{3.176}$$

$$G^{\perp}_{\alpha\beta} = n^*_\alpha \otimes g_\beta, \quad g_\beta = G_{\zeta\beta} n^*_\zeta \tag{3.177}$$

where g_β is the Burgers vector for surface elements with the normal vector of n^*_ζ and $\mathbb{P}_{\alpha\zeta}$ can be defined as follows:

$$\mathbb{P}_{\alpha\zeta} = \delta_{\alpha\zeta} - n^*_\alpha n^*_\zeta \tag{3.178}$$

where $\delta_{\alpha\zeta}$ is the Kronecker delta. Eqs. (3.176)–(3.178) show that the decomposition of $G_{\alpha\beta}$ to $G^{\parallel}_{\alpha\beta}$ and $G^{\perp}_{\alpha\beta}$ solely depends on Π^* and not the choice of $\tilde{e}^1_\alpha$ and $\tilde{e}^2_\alpha$. One should note that in general, g_β, as described in Eq. (3.177) is neither normal nor parallel to n^*_ζ, and it can be decomposed as follows:

$$g_\beta = \rho^{nn} n^*_\beta + \rho^{n1} \tilde{e}^1_\beta + \rho^{n2} \tilde{e}^2_\beta \tag{3.179}$$

3.4.2.3 Virtual power principle

In the case of a system with Π slip systems as $\overline{\alpha} = \{1, 2, \ldots, \Pi\}$, one can define Π rate of shear deformation as a vector $\dot{\gamma}_i = \left(\dot{\gamma}^1, \dot{\gamma}^2, \ldots, \dot{\gamma}^\Pi\right)$. The deformed body is denoted by $B(t) = x_i(\mathrm{B}_R, t)$ where B_R is corresponding to the undeformed body. Also, $D(t) = x_i(\mathrm{D}_R, t)$ is described as an arbitrary subregion of $B(t)$ which is connected to D_R as the corresponding undeformed subregion of B_R.

The principle of virtual power is incorporated here assuming that the power related to each independent kinematics can be described using the associated force consistent with its own balance.

In crystal plasticity, the flow rule is described according to Eqs. (3.39), (3.40). One can relate the velocity gradient tensor l_{ij} to those of the elastic and plastic ones as follows:

$$l_{ij} = l^e_{ij} + F^e_{i\alpha} l^p_{\alpha\beta} \left(F^e_{\beta j}\right)^{-1} \tag{3.180}$$

Eq. (3.180) can be rewritten by incorporating Eq. (3.40) as follows:

$$l_{ij} = F^e_{i\alpha} \sum_{\overline{\alpha}} \dot{\gamma}^{\overline{\alpha}} S^{\overline{\alpha}}_{\alpha\beta} \left(F^e_{\beta j}\right)^{-1} = l^e_{ij} + \sum_{\overline{\alpha}} \dot{\gamma}^{\overline{\alpha}} S^{\overline{\alpha}}_{ij} \tag{3.181}$$

where $S^{\overline{\alpha}}_{ij}$ is the Schmid tensor for the slip system $\overline{\alpha}$ which is pushed forward to the deformed state. Eq. (3.181) is the general constraint which all the velocity fields should satisfy.

In order to incorporate the principle of virtual power in crystal plasticity framework, Gurtin (2002) introduced the generalized virtual velocity $\mathcal{V}$ as follows:

$$\mathcal{V} = \left(\tilde{\dot{x}}_i, \tilde{\dot{\gamma}}_i, \tilde{l}^e_{ij}\right) \tag{3.182}$$

$\mathcal{V} = \left(\tilde{\dot{x}}_i, \tilde{\dot{\gamma}}_i, \tilde{l}^e_{ij}\right)$ is the set of virtual velocities to be specified independently which satisfies Eq. (3.181) at some arbitrarily chosen but fixed time, in which x_i,

$F^{e}_{i\alpha}$, and accordingly $F^{p}_{\xi J}$ are known. Gurtin defined a rigid set of generalized virtual velocity $\mathcal{V}_{\text{rigid}}$ as follows:

$$\tilde{\dot{x}}_i = a_i + \epsilon_{ijk}\omega_j x_k,\ \tilde{l}^{e}_{ij} = \epsilon_{ikj}\omega_k, \tilde{\dot{\gamma}}_i = 0 \tag{3.183}$$

where a_i and ω_j are vectors which are constant. One should note that the virtual velocity fields, i.e., $\left(\tilde{\dot{x}}_i, \tilde{\dot{\gamma}}_i, \tilde{l}^{e}_{ij}\right)$, are different from those of the actual body, i.e., $\left(\dot{x}_i, \dot{\gamma}_i, \tilde{l}^{e}_{ij}\right)$.

The principle of virtual power states that for an arbitrary subbody D undergoing a virtual velocity $\mathcal{V}$, which satisfies Eq. (3.181), the induced internal power due to the arbitrary virtual velocity is equal to that of the external power. First, one need to define the induced internal and external powers. The induced external power can be defined as follows:

$$P_{ext} = \int_V b_i \tilde{\dot{x}}_i dV + \int_S t_i \tilde{\dot{x}}_i dS + \sum_{\bar{\alpha}} \int_S m^{\bar{\alpha}} \tilde{\dot{\gamma}}^{\bar{\alpha}} dS \tag{3.184}$$

where the integration is performed over V and S, which are the volume and surface area of the subbody D. The first term of Eq. (3.184) represents the body force, the second term denotes the macroscopic surface traction t_i, and the third term denotes the microscopic surface traction for each slip system $m\bar{\alpha}$ which is conjugate to $\tilde{\dot{\gamma}}^{\bar{\alpha}}$.

The internal power due to the virtual velocity can be defined as follows:

$$P_{int} = \int_V T_{ij} \tilde{l}^{e}_{ij} dV + \sum_{\bar{\alpha}} \int_V \left(\mathcal{X}^{\bar{\alpha}} \tilde{\dot{\gamma}}^{\bar{\alpha}} + \mathcal{S}^{\bar{\alpha}}_i \tilde{\dot{\gamma}}^{\bar{\alpha}}_{,i}\right) dV \tag{3.185}$$

where the first term is the internal power induced by T_{ij} which is the lattice stress conjugate to the elastic velocity gradient l^{e}_{ij}, the second term is the contribution of the internal microforce for each slip system $\mathcal{X}^{\bar{\alpha}}$ which is conjugate to $\tilde{\dot{\gamma}}^{\bar{\alpha}}$, and the third term represents the contribution of the microstress for each slip system $\mathcal{S}^{\bar{\alpha}}_i$ which is conjugate to $\tilde{\dot{\gamma}}^{\bar{\alpha}}_i$.

The principle of virtual power can be incorporated to develop the material governing equations, which can be defined as follows:

$$P_{int} = P_{ext} \tag{3.186}$$

Eq. (3.186) should be satisfied for any set of generalized virtual velocity $\mathcal{V}$. Another constraint is that the governing equation should be objective, i.e., frame-indifference, which can be satisfied as follows:

$$P_{int} = 0 \text{ for any } \mathcal{V}_{\text{rigid}} \tag{3.187}$$

where the rigid generalized virtual velocity $\mathcal{V}_{\text{rigid}}$ is defined in Eq. (3.183).

First, one can assume no slip occurs, i.e., $\dot{\tilde{\gamma}}_i = 0$. Accordingly, $\tilde{l}^e_{ij} = \tilde{l}_{ij}$ and Eq. (3.186) can be rewritten by incorporating Eqs. (3.184), (3.185) as follows:

$$\int_V b_i \dot{\tilde{x}}_i dV + \int_S t_i \dot{\tilde{x}}_i dS = \int_V T_{ij} \tilde{l}_{ij} dV \tag{3.188}$$

Eq. (3.188) can be rewritten using the divergence theorem as follows:

$$\int_V \left(b_i + T_{ij,j}\right) \dot{\tilde{x}}_i dV + \int_S \left(t_i - T_{ij} n_j\right) \dot{\tilde{x}}_i dS = 0 \tag{3.189}$$

Eq. (3.189) should be valid for any arbitrary set of generalized virtual velocity $\mathcal{V}$ and subbody D. Accordingly, the following evolution laws can be derived:

$$t_i = T_{ij} n_j \text{ on } S \text{ (macrotraction condition)} \tag{3.190}$$

$$b_i + T_{ij,j} = 0 \text{ on } V \text{ (macroforce balance)} \tag{3.191}$$

Furthermore, Eq. (3.187) can be rewritten as follows:

$$\int_V \epsilon_{ikj} \omega_k T_{ij} dV = 0 \tag{3.192}$$

Eq. (3.192) should be valid for the arbitrary skew tensor $\epsilon_{ikj}\omega_k$ and subbody D. Accordingly, T_{ij} should be symmetric, i.e.,:

$$T_{ij} = T_{ji} \text{ on } V \text{ (macromoment balance)} \tag{3.193}$$

To obtain the microtraction condition and microforce balance, first, one can assume $\dot{\tilde{x}}_i = 0$ which leads to $\tilde{l}^e_{ij} = -\sum_{\overline{\alpha}} \dot{\tilde{\gamma}}^{\overline{\alpha}} S^{\overline{\alpha}}_{ij}$. Next, one needs to define the acting shear stress on the slip system $\overline{\alpha}$, i.e., $\tau^{\overline{\alpha}}$, can be obtained as follows:

$$\tau^{\overline{\alpha}} = T_{ij} S^{\overline{\alpha}}_{ij} \tag{3.194}$$

where $S^{\overline{\alpha}}_{ij}$ is the Schmid tensor for the slip system $\overline{\alpha}$ which is pushed forward to the deformed state. Accordingly, Eq. (3.186) can be rewritten by incorporating Eqs. (3.184), (3.185) as follows:

$$\sum_{\overline{\alpha}} \int_S m^{\overline{\alpha}} \dot{\tilde{\gamma}}^{\overline{\alpha}} dS = \sum_{\overline{\alpha}} \int_V \left[\left(\mathcal{X}^{\overline{\alpha}} - \tau^{\overline{\alpha}}\right) \dot{\tilde{\gamma}}^{\overline{\alpha}} + \mathcal{S}^{\overline{\alpha}}_i \dot{\tilde{\gamma}}^{\overline{\alpha}}_{,i} \right] dV \tag{3.195}$$

Eq. (3.195) can be rewritten using the divergence theorem as follows:

$$\sum_{\overline{\alpha}} \int_S \left(m^{\overline{\alpha}} - \mathcal{S}^{\overline{\alpha}}_i n_i\right) \dot{\tilde{\gamma}}^{\overline{\alpha}} dS + \sum_{\overline{\alpha}} \int_V \left[\left(\mathcal{S}^{\overline{\alpha}}_{i,i} + \tau^{\overline{\alpha}} - \mathcal{X}^{\overline{\alpha}}\right) \dot{\tilde{\gamma}}^{\overline{\alpha}} \right] dV = 0 \tag{3.196}$$

Eq. (3.196) should be valid for any arbitrary set of generalized virtual velocity $\mathcal{V}$ and subbody D. Accordingly, the following evolution laws can be derived:

$$m^{\overline{\alpha}} = \mathcal{S}^{\overline{\alpha}}_i n_i \text{ on } S \text{ (microtraction condition)} \tag{3.197}$$

$$\mathcal{S}^{\overline{\alpha}}_{i,i} + \tau^{\overline{\alpha}} - \mathcal{X}^{\overline{\alpha}} = 0 \text{ on } V \text{ (microforce balance)} \tag{3.198}$$

Eqs. (3.197), (3.198) should be satisfied for all slip systems.

The lattice stress T_{ij} contributes in both macro and micro evolution laws. In the case of the former one, it plays a similar role as Cauchy stress while in the latter one, it acts as the shear stress on the slip system $\overline{\alpha}$. $\mathcal{X}^{\overline{\alpha}}$ can be envisaged as the internal forces due to the creation, annihilation, and interaction of dislocations with each other. Finally, $\mathcal{S}^{\overline{\alpha}}_i$ can be interpreted as the traction vector induced due to the interaction of dislocations on the surface.

3.4.2.4 Second law of thermodynamics

Assuming a purely mechanical theory, the second law of thermodynamics states that the increase in Helmholtz free energy Ψ should be less than the power expended volume V, which can be described as follows:

$$\int_V \rho \dot{\Psi} J^{-1} dV - P_{ext} = -\int_V \wp dV \leq 0 \tag{3.199}$$

for an arbitrary subbody D, where ρ is the material density, $\wp \geq 0$ is the energy dissipation rate, and $J = F_{iJ}$. One can define the local free energy inequality by substituting Eq. (3.185) into Eq. (3.199) as follows:

$$\rho J^{-1} \dot{\Psi} - T_{ij} l^e_{ij} - \sum_{\overline{\alpha}} \left(\mathcal{X}^{\overline{\alpha}} \dot{\gamma}^{\overline{\alpha}} + \mathcal{S}^{\overline{\alpha}}_i \dot{\gamma}^{\overline{\alpha}}_{,i} \right) = -\wp \leq 0 \tag{3.200}$$

Instead of the term $T_{ij} l^e_{ij}$, one can use the elastic material strain tensor $E^e_{\alpha\beta}$ and the elastic stress tensor $T^e_{\alpha\beta}$, which can be defined as follows:

$$E^e_{\alpha\beta} = \frac{1}{2}\left(C^e_{\alpha\beta} - \delta_{\alpha\beta} \right) = \frac{1}{2}\left(F^e_{i\alpha} F^e_{i\beta} - \delta_{\alpha\beta} \right) \tag{3.201}$$

$$T^e_{\alpha\beta} = J \left(F^e_{\alpha i} \right)^{-1} T_{ij} \left(F^e_{\beta j} \right)^{-1} \tag{3.202}$$

where $C^e_{\alpha\beta}$ is the elastic right Cauchy-Green tensor. Accordingly,

$$T_{ij} l^e_{ij} = J^{-1} T^e_{\alpha\beta} \dot{E}^e_{\alpha\beta} \tag{3.203}$$

The final form of the local free energy inequality can be obtained by substituting Eq. (3.203) into Eq. (3.200) as follows:

$$\rho J^{-1} \dot{\Psi} - J^{-1} T^e_{\alpha\beta} \dot{E}^e_{\alpha\beta} - \sum\nolimits_{\overline{\alpha}} \left(\mathcal{X}^{\overline{\alpha}} \dot{\gamma}^{\overline{\alpha}} + \mathcal{S}^{\overline{\alpha}}_i \dot{\gamma}^{\overline{\alpha}}_{,i} \right) = -\wp \leq 0 \tag{3.204}$$

3.4.2.5 Constitutive theory

Gurtin (2002) developed the Helmholtz free energy Ψ for nonlocal crystal plasticity model by incorporating the tensor of geometric dislocation $G_{\alpha\beta}$.

He assumed that the free energy is a function of the elastic material strain tensor $E^e_{\alpha\beta}$ and geometric dislocation tensor $G_{\alpha\beta}$ as follows:

$$\Psi = \Psi\left(E^e_{\alpha\beta}, G_{\alpha\beta}\right) \tag{3.205}$$

Gurtin proposed the following Helmholtz free energy Ψ for nonlocal crystal plasticity model:

$$\Psi = \frac{1}{2\rho} E^e_{\alpha\beta} \mathbb{C}_{\alpha\beta\xi\zeta} E^e_{\xi\zeta} + \frac{1}{\rho}\psi\left(G_{\alpha\beta}\right) \tag{3.206}$$

where $\mathbb{C}_{\alpha\beta\xi\zeta}$ is the symmetric elasticity tensor and $\psi(G_{\alpha\beta})$ is the defect energy function.

The thermodynamic defect stress $R^G_{\alpha\beta}$ which is conjugate to $\dot{G}_{\alpha\beta}$ can be described as follows:

$$R^G_{\alpha\beta} = \rho \frac{\partial \Psi}{\partial G_{\alpha\beta}} \tag{3.207}$$

The defect-stress power can then be defined by incorporating Eq. (3.167) as follows:

$$R^G_{\alpha\beta}\dot{G}_{\alpha\beta} = \sum_{\overline{\alpha}} \left[\dot{\gamma}^{\overline{\alpha}} R^G_{\alpha\beta}\left(S^{\overline{\alpha}}_{\alpha\xi} G_{\xi\beta} + G_{\alpha\zeta} S^{\overline{\alpha}}_{\beta\zeta}\right) + \epsilon_{\alpha\xi\zeta} p^{\overline{\alpha}}_{\xi} S^{\overline{\alpha}}_{\beta\zeta} R^G_{\alpha\beta}\right] \tag{3.208}$$

Eq. (3.208) can be rewritten by incorporating Eqs. (3.162), (3.163), (3.209) as follows (Gurtin, 2002):

$$\epsilon_{\alpha\xi\zeta} p^{\overline{\alpha}}_{\xi} S^{\overline{\alpha}}_{\beta\zeta} R^G_{\alpha\beta} = p^{\overline{\alpha}}_{\xi} \epsilon_{\alpha\xi\zeta} n^{\overline{\alpha}}_{\zeta} R^G_{\alpha\beta} m^{\overline{\alpha}}_{\beta} \xrightarrow{\text{Tensor Notation}} \boldsymbol{p}^{\overline{\alpha}} \bullet \left(\boldsymbol{n}^{\overline{\alpha}} \times \boldsymbol{R}^G \boldsymbol{m}^{\overline{\alpha}}\right) \tag{3.209a}$$

$$R^G_{\alpha\beta}\dot{G}_{\alpha\beta} = \sum_{\overline{\alpha}} \left[\dot{\gamma}^{\overline{\alpha}} R^G_{\alpha\beta}\left(S^{\overline{\alpha}}_{\alpha\xi} G_{\xi\beta} + G_{\alpha\zeta} S^{\overline{\alpha}}_{\beta\zeta}\right) + F^e_{i\xi}\, \epsilon_{\alpha\xi\zeta} n^{\overline{\alpha}}_{\zeta} R^G_{\alpha\beta} m^{\overline{\alpha}}_{\beta} \dot{\gamma}^{\overline{\alpha}}{}_{,i}\right] \tag{3.209b}$$

The local free energy inequality can be obtained by substituting Eqs. (3.206), (3.209) into Eq. (3.204) as follows:

$$J^{-1}\left(\mathbb{C}_{\alpha\beta\xi\zeta} E^e_{\xi\zeta} - T^e_{\alpha\beta}\right)\dot{E}^e_{\alpha\beta} + \sum_{\overline{\alpha}} \left\{\left[J^{-1} R^G_{\alpha\beta}\left(S^{\overline{\alpha}}_{\alpha\xi} G_{\xi\beta} + G_{\alpha\zeta} S^{\overline{\alpha}}_{\beta\zeta}\right) - \mathcal{X}^{\overline{\alpha}}\right]\dot{\gamma}^{\overline{\alpha}}\right\}$$

$$+ \sum_{\overline{\alpha}} \left[\left(J^{-1} F^e_{i\xi}\, \epsilon_{\alpha\xi\zeta} n^{\overline{\alpha}}_{\zeta} R^G_{\alpha\beta} m^{\overline{\alpha}}_{\beta} - \mathcal{S}^{\overline{\alpha}}_{i}\right)\dot{\gamma}^{\overline{\alpha}}_{,i}\right] = -\wp \leq 0 \tag{3.210}$$

The constitutive equations for $T^e_{\alpha\beta}$, $\mathcal{X}^{\overline{\alpha}}$, and $\mathcal{S}^{\overline{\alpha}}_i$ can be obtained as functions of $F^e_{i\xi}$, $G_{\alpha\beta}$, $\dot{\gamma}^{\overline{\alpha}}$, and $\dot{\gamma}^{\overline{\alpha}}_{,i}$, where $\overline{\alpha} = \{1, 2, \ldots, A\}$ in the case of a system with A slip systems, by solving the inequality presented in Eq. (3.210). Eq. (3.210) should be valid for arbitrary values of $\dot{F}^e_{i\xi}$, $\dot{\gamma}^{\overline{\alpha}}$, and $\dot{\gamma}^{\overline{\alpha}}_{,i}$. Accordingly, the constitutive equations can be obtained as follows:

$$T^e_{\alpha\beta} = \mathbb{C}_{\alpha\beta\xi\zeta} E^e_{\xi\zeta} \tag{3.211}$$

$$S_i^{\bar{\alpha}} = J^{-1} F_{i\xi}^e \epsilon_{\alpha\xi\zeta} n_\zeta^{\bar{\alpha}} R_{\alpha\beta}^G m_\beta^{\bar{\alpha}} \tag{3.212}$$

The simplified form of the local free energy inequality can be obtained by substituting Eqs. (3.211), (3.212) into Eq. (3.210) as follows:

$$\sum_{\bar{\alpha}} \left\{ \left[\mathcal{X}^{\bar{\alpha}} - J^{-1} R_{\alpha\beta}^G \left(S_{\alpha\xi}^{\bar{\alpha}} G_{\xi\beta} + G_{\alpha\zeta} S_{\beta\zeta}^{\bar{\alpha}} \right) \right] \dot{\gamma}^{\bar{\alpha}} \right\} = \wp \geq 0 \tag{3.213}$$

In the case of the classical local theory, the energy dissipation rate $\wp$ can be defined as follows:

$$\wp = \sum_{\bar{\alpha}} H\left(\dot{\gamma}^{\bar{\alpha}}\right) s^{\bar{\alpha}} \dot{\gamma}^{\bar{\alpha}} \tag{3.214}$$

where $s^{\bar{\alpha}}$ is the slip resistance for the slip system $\bar{\alpha}$. In the case of local crystal plasticity theory, Gurtin (2002) assumed the same energy dissipation rate as described in Eq. (3.214). Accordingly, the constitutive equation for $\mathcal{X}^{\bar{\alpha}}$ can be defined as follows:

$$\mathcal{X}^{\bar{\alpha}} = \mathcal{X}_{\text{dissipative}}^{\bar{\alpha}} + \mathcal{X}_{\text{energetic}}^{\bar{\alpha}} = H\left(\dot{\gamma}^{\bar{\alpha}}\right) s^{\bar{\alpha}} + J^{-1} R_{\alpha\beta}^G \left(S_{\alpha\xi}^{\bar{\alpha}} G_{\xi\beta} + G_{\alpha\zeta} S_{\beta\zeta}^{\bar{\alpha}} \right) \tag{3.215}$$

where the first term, i.e., $\mathcal{X}_{\text{dissipative}}^{\bar{\alpha}} = H\left(\dot{\gamma}^{\bar{\alpha}}\right) s^{\bar{\alpha}}$, is the dissipative part of $\mathcal{X}^{\bar{\alpha}}$ and the second term, i.e., $\mathcal{X}_{\text{energetic}}^{\bar{\alpha}} = J^{-1} R_{\alpha\beta}^G \left(S_{\alpha\xi}^{\bar{\alpha}} G_{\xi\beta} + G_{\alpha\zeta} S_{\beta\zeta}^{\bar{\alpha}} \right)$, is its energetic part. One can define $H\left(\dot{\gamma}^{\bar{\alpha}}\right)$ using a viscosity function as follows:

$$H\left(\dot{\gamma}^{\bar{\alpha}}\right) = \left|\dot{\gamma}^{\bar{\alpha}}\right|^\delta \operatorname{sign}\left(\dot{\gamma}^{\bar{\alpha}}\right) \tag{3.216}$$

where $\delta > 0$ is a constant. The obtained constitutive equations presented in Eqs. (3.206), (3.212), (3.215) satisfy the second law of thermodynamics for any $H\left(\dot{\gamma}^{\bar{\alpha}}\right) \dot{\gamma}^{\bar{\alpha}} \geq 0$ including the one presented in Eq. (3.216).

The next step is to define the viscoplastic yield conditions. To do so, one can substitute Eqs. (3.212), (3.215) into the microforce balance presented in Eq. (3.198) as follows:

$$\tau^{\bar{\alpha}} = H\left(\dot{\gamma}^{\bar{\alpha}}\right) s^{\bar{\alpha}} + J^{-1} R_{\alpha\beta}^G \left(S_{\alpha\xi}^{\bar{\alpha}} G_{\xi\beta} + G_{\alpha\zeta} S_{\beta\zeta}^{\bar{\alpha}} \right) - \left(J^{-1} F_{i\xi}^e \epsilon_{\alpha\xi\zeta} n_\zeta^{\bar{\alpha}} R_{\alpha\beta}^G m_\beta^{\bar{\alpha}} \right)_{,i} \tag{3.217}$$

where the first term, i.e., $H\left(\dot{\gamma}^{\bar{\alpha}}\right) s^{\bar{\alpha}}$, describes the contribution of slip to hardening and the second and third terms define the contribution of energy stored by GNDs to hardening. The right hand side of the viscoplastic yield equation depends on $G_{\alpha\beta}$ and its gradient, and subsequently, on first and second gradients of the plastic gradient tensor $F_{\xi J}^p$ which leads to a nonlocal yield condition.

Gurtin (2002) proposed the simple defect energy function $\psi(G_{\alpha\beta})$ as follows:

$$\psi\left(G_{\alpha\beta}\right) = \frac{1}{2\rho} \mu G_{\alpha\beta} G_{\alpha\beta} \tag{3.218}$$

The thermodynamic defect stress $R^G_{\alpha\beta}$ can be obtained by substituting Eq. (3.218) into Eq. (3.207) as follows:

$$R^G_{\alpha\beta} = \rho \frac{\partial \Psi}{\partial G_{\alpha\beta}} = \mu G_{\alpha\beta} \tag{3.219}$$

Accordingly, $\mathcal{S}_i^{\bar{\alpha}}$ and $\mathcal{X}^{\bar{\alpha}}$ can be obtained as follows:

$$\mathcal{S}_i^{\bar{\alpha}} = \mu J^{-1} F^e_{i\xi} \epsilon_{\alpha\xi\zeta} n^{\bar{\alpha}}_{\zeta} G_{\alpha\beta} m^{\bar{\alpha}}_{\beta} \tag{3.220}$$

$$\mathcal{X}^{\bar{\alpha}} = H\left(\dot{\gamma}^{\bar{\alpha}}\right) s^{\bar{\alpha}} + \mu J^{-1} G_{\alpha\beta}\left(S^{\bar{\alpha}}_{\alpha\xi} G_{\xi\beta} + G_{\alpha\zeta} S^{\bar{\alpha}}_{\beta\zeta}\right) \tag{3.221}$$

The viscoplastic yield condition can be obtained by substituting Eq. (3.219) into Eq. (3.217) as follows:

$$\tau^{\bar{\alpha}} = H\left(\dot{\gamma}^{\bar{\alpha}}\right) s^{\bar{\alpha}} + \mu J^{-1} G_{\alpha\beta}\left(S^{\bar{\alpha}}_{\alpha\xi} G_{\xi\beta} + G_{\alpha\zeta} S^{\bar{\alpha}}_{\beta\zeta}\right) - \mu\left(J^{-1} F^e_{i\xi} \epsilon_{\alpha\xi\zeta} n^{\bar{\alpha}}_{\zeta} G_{\alpha\beta} m^{\bar{\alpha}}_{\beta}\right)_{,i} \tag{3.222}$$

where the second term, i.e., $\mu J^{-1} G_{\alpha\beta}\left(S^{\bar{\alpha}}_{\alpha\xi} G_{\xi\beta} + G_{\alpha\zeta} S^{\bar{\alpha}}_{\beta\zeta}\right)$, represents the energetic part of $\mathcal{X}^{\bar{\alpha}}$, i.e., $\mathcal{X}^{\bar{\alpha}}_{\text{energetic}}$, as defined in Eq. (3.215). One can define $\mathcal{X}^{\bar{\alpha}}_{\text{energetic}}$ using the expansion of $G_{\alpha\beta}$ based on the pure dislocations definition presented in Eq. (3.169) as follows:

$$\mathcal{X}^{\bar{\alpha}}_{\text{energetic}} = \mu J^{-1} S^{\bar{\alpha}}_{\alpha\beta} \sum_k \sum_n \rho^k \rho^n \left(b^n_{\xi} b^k_{\xi} t^k_{\alpha} t^n_{\beta} + t^n_{\xi} t^k_{\xi} b^k_{\alpha} b^n_{\beta}\right) \tag{3.223}$$

Eq. (3.222) states that the energetic part of $\mathcal{X}^{\bar{\alpha}}$, i.e., $\mathcal{X}^{\bar{\alpha}}_{\text{energetic}}$, is originated from the interaction of dislocations with each other.

The last part of the constitutive equation is to define the hardening model. In the case of local crystal plasticity, the hardening can be defined as follows:

$$\dot{s}^{\bar{\alpha}} = \sum_{\bar{\beta}} k^{\bar{\alpha}\bar{\beta}}\left(s^1, s^2, \ldots, s^A\right)\left|\dot{\gamma}\bar{\beta}\right| \tag{3.224}$$

where A is the number of slip systems and $\sum_{\bar{\beta}}$ is the summation over all slip systems. In the case of nonlocal crystal plasticity model, Gurtin (2002) modified Eq. (3.224) by assuming that $k^{\bar{\alpha}\bar{\beta}}$ is also a function of tensor of geometric dislocation $G_{\alpha\beta}$ as follows:

$$\dot{s}^{\bar{\alpha}} = \sum_{\bar{\beta}} k^{\bar{\alpha}\bar{\beta}}\left(s_i, G_{\alpha\beta}\right)\left|\dot{\gamma}^{\bar{\beta}}\right| \tag{3.225}$$

where A slip resistance as a vector of $s_i = (s^1, s^2, \ldots, s^A)$.

One can divide the hardening described in Eq. (3.217) into two parts of dissipative $\tau^{\bar{\alpha}}_{\text{dissipative}}$ and energetic $\tau^{\bar{\alpha}}_{\text{energetic}}$ as follows:

$$\tau^{\bar{\alpha}} = \tau^{\bar{\alpha}}_{\text{dissipative}} + \tau^{\bar{\alpha}}_{\text{energetic}} \tag{3.226}$$

$$\tau^{\overline{\alpha}}_{\text{dissipative}} = H\left(\dot{\gamma}^{\overline{\alpha}}\right)s^{\overline{\alpha}} \tag{3.227}$$

$$\tau^{\overline{\alpha}}_{\text{energetic}} = J^{-1}R^{G}_{\alpha\beta}\left(S^{\overline{\alpha}}_{\alpha\xi}G_{\xi\beta} + G_{\alpha\zeta}S^{\overline{\alpha}}_{\beta\zeta}\right) - \left(J^{-1}F^{e}_{i\xi}\,\epsilon_{\alpha\xi\zeta}n^{\overline{\alpha}}_{\zeta}R^{G}_{\alpha\beta}m^{\overline{\alpha}}_{\beta}\right)_{,i} \tag{3.228}$$

The hardening provided by Eq. (3.227), which is completed by Eq. (3.225), is solely dissipative, and it has a phenomenological nature. The only constraint for $\tau^{\overline{\alpha}}_{\text{dissipative}}$ is that the slip resistance $s^{\overline{\alpha}}$ should be nonnegative. Furthermore, the obtained hardening has no backstress.

Eq. (3.228) presents the energetic part of the hardening, i.e., $\tau^{\overline{\alpha}}_{\text{energetic}}$. The energetic term is originated from the microforce balance and thermodynamic constraints. Furthermore, the negative sign of $\tau^{\overline{\alpha}}_{\text{energetic}}$ provides the backstress on the slip system $\overline{\alpha}$.

3.4.2.6 Small strain framework

Gurtin (2002) simplified the developed model in the case of small strain framework. Accordingly, instead of the multiplicative decomposition as described in Eq. (3.77), the additive decomposition of displacement gradient can be used as follows:

$$u_{i,j} = H^{e}_{ij} + H^{p}_{ij} \tag{3.229}$$

where u_i is the displacement vector, H^{e}_{ij} describes the lattice rotation, and H^{p}_{ij} represents the plastic distortion. One should note that $H^{p}_{ii} \equiv 0$. The elastic and plastic strain tensors, i.e., ε^{e}_{ij} and ε^{p}_{ij}, respectively, can be related to H^{e}_{ij} and H^{p}_{ij} as follows:

$$\varepsilon^{e,p}_{ij} = 1/2\left[H^{e,p}_{ij} + H^{e,p}_{ji}\right] \tag{3.230}$$

In the case of small strain crystal plasticity, the material properties from macro to the micro can be connected as follows:

$$H^{p}_{ij} = \sum_{\overline{\alpha}} \gamma^{\overline{\alpha}} S^{\overline{\alpha}}_{ij} \tag{3.231}$$

where $S^{\overline{\alpha}}_{ij}$ is the Schmid tensor for the slip system $\overline{\alpha}$, which can be defined as follows:

$$S^{\overline{\alpha}}_{ij} = m^{\overline{\alpha}}_{i} n^{\overline{\alpha}}_{j} \tag{3.232}$$

The geometric dislocation tensor G_{ij} for small strains can be simplified as follows:

$$G_{ij} = \left(\text{curl }\left[H^{p}_{kl}\right]\right)_{ij} = \sum_{\alpha} \epsilon_{ilk}\gamma^{\overline{\alpha}}{}_{,l}n^{\overline{\alpha}}_{k}m^{\overline{\alpha}}_{j} \tag{3.233}$$

Accordingly, the constitutive equations can be described as follows:

$$\sigma_{ij} = \mathbb{C}_{ijkl}\varepsilon^{e}_{kl} \tag{3.234}$$

$$S_i^{\bar{\alpha}} = \epsilon_{ijk} n_j^{\bar{\alpha}} R_{kl}^G m_l^{\bar{\alpha}} \tag{3.235}$$

$$\mathcal{X}^{\bar{\alpha}} = H\left(\dot{\gamma}^{\bar{\alpha}}\right) s^{\bar{\alpha}} \tag{3.236}$$

where Eq. (3.234) is the famous Hooke's law, $\mathcal{X}^{\bar{\alpha}}$ does not depend on the geometric dislocation G_{ij}, and the thermodynamic defect stress R_{kl}^G can be described as follows:

$$R_{ij}^G = \rho \frac{\partial \Psi}{\partial G_{ij}} \tag{3.237}$$

One can consider the simple defect energy function $\psi(G_{ij})$ as follows:

$$\psi\left(G_{ij}\right) = \frac{1}{2\rho} \mu G_{ij} G_{ij} \tag{3.238}$$

The thermodynamic defect stress R_{kl}^G can be obtained by substituting Eq. (3.237) into Eq. (3.238) as follows:

$$R_{ij}^G = \rho \frac{\partial \Psi}{\partial G_{ij}} = \mu G_{ij} \tag{3.239}$$

Accordingly, Eq. (3.235) can be rewritten as follows:

$$S_i^{\bar{\alpha}} = \sum_{\bar{\beta}} M_{ij}^{\bar{\alpha}\bar{\beta}} \gamma_{,j}^{\bar{\beta}} \tag{3.240}$$

where

$$M_{ij}^{\bar{\alpha}\bar{\beta}} = \mu m_l^{\bar{\alpha}} m_l^{\bar{\beta}} \left(\delta_{ij} n_k^{\bar{\alpha}} n_k^{\bar{\beta}} - n_i^{\bar{\beta}} n_j^{\bar{\alpha}} \right) \tag{3.241}$$

in which δ_{ij} is the Kronecker delta. Finally, the yield function in the case of small strain framework can be defined as follows:

$$\tau^{\bar{\alpha}} + \sum_{\bar{\beta}} M_{ij}^{\bar{\alpha}\bar{\beta}} \left(\gamma^{\bar{\beta}}{}_{,j} \right)_{,i} = H\left(\dot{\gamma}^{\bar{\alpha}}\right) s^{\bar{\alpha}} \tag{3.242}$$

3.4.2.7 *Strict plane strain condition in small strain framework*

In the case of strict plain strain condition in the x-y plane, the plastic shear deformation in slip system $\bar{\alpha}$, i.e., $\gamma^{\bar{\alpha}}$, does not depend on the z direction. Accordingly, all the slip systems are planar as follows:

$$m_i^{\bar{\alpha}} e_i = 0, \quad n_i^{\bar{\alpha}} e_i = 0, \quad \epsilon_{ijk} m_j^{\bar{\alpha}} n_k^{\bar{\alpha}} = e_i \tag{3.243}$$

where e_i is the base vector in the z direction. Accordingly, the geometric dislocation tensor G_{ij} can be obtained as follows:

$$G_{ij} = \left(\text{curl } \left[H_{kl}^p\right]\right)_{ij} = e_i g_j \tag{3.244}$$

where g_j can be defined as follows:

$$g_j = \sum_{\overline{\alpha}} \gamma^{\overline{\alpha}}{}_{,l} m_l^{\overline{\alpha}} m_j^{\overline{\alpha}} \tag{3.245}$$

One can consider the simple defect energy function $\psi(g_i)$ as follows:

$$\psi(g_i) = \frac{1}{2\rho} \mu g_i g_i \tag{3.246}$$

Accordingly, one can describe the constitutive equation presented in Eqs. (3.240), (3.241) as follows:

$$S_i^{\overline{\alpha}} = \mu m_i^{\overline{\alpha}} \sum_{\overline{\beta}} m_l^{\overline{\alpha}} m_l^{\overline{\beta}} \gamma^{\overline{\beta}}_{,k} m_k^{\overline{\beta}} \tag{3.247}$$

Finally, the yield condition can be written as follows:

$$\tau^{\overline{\alpha}} + \mu \sum_{\overline{\beta}} m_l^{\overline{\alpha}} m_l^{\overline{\beta}} m_k^{\overline{\alpha}} \gamma^{\overline{\beta}}_{,km} m_m^{\overline{\beta}} = f\left(\dot{\gamma}^{\overline{\alpha}}\right) \tag{3.248}$$

In the case of strict plastic strain condition defined in Eq. (3.243), the elastic rotation can be defined as rotation about the z axis through an angle ϑ. One can neglect the gradients of lattice-strain and approximate g_j as follows:

$$g_j \approx \vartheta_{,j} \tag{3.249}$$

The constitutive equation of Eq. (3.247) and accordingly the yield condition presented in Eq. (3.248) can then be approximated as follows:

$$S_i^{\overline{\alpha}} \approx \mu \vartheta_{,l} m_l^{\overline{\alpha}} m_i^{\overline{\alpha}} \tag{3.250}$$

$$\tau^{\overline{\alpha}} + \mu m_k^{\overline{\alpha}} \vartheta_{,km} m_m^{\overline{\alpha}} \approx f\left(\dot{\gamma}^{\overline{\alpha}}\right) \tag{3.251}$$

Eq. (3.250) states that $S_i^{\overline{\alpha}}$ is a linear function of $\vartheta_{,l} m_l^{\alpha}$ which denotes the curvature of the slip line $\overline{\alpha}$, and the nonlocal term of the yield condition in Eq. (3.251) is $m_k^{\overline{\alpha}} \vartheta_{,km} m_m^{\overline{\alpha}}$ which denotes the change in this curvature in the slip direction of the slip system $\overline{\alpha}$. The nonlocal term, i.e., $m_k^{\overline{\alpha}} \vartheta_{,km} m_m^{\overline{\alpha}}$, plays the role of backstress in the yield function.

References

Acharya, A., Bassani, J.L., 2000. Lattice incompatibility and a gradient theory of crystal plasticity. J. Mech. Phys. Solids 48, 1565–1595.

Acharya, A., Bassani, J.L., Beaudoin, A., 2003. Geometrically necessary dislocations, hardening, and a simple gradient theory of crystal plasticity. Scripta Mater. 48, 167–172.

Anand, L., Kothari, M., 1996. A computational procedure for rate-independent crystal plasticity. J. Mech. Phys. Solids 44, 525–558.

Arsenlis, A., Parks, D.M., 1999. Crystallographic aspects of geometrically-necessary and statistically-stored dislocation density. Acta Mater. 47, 1597–1611.

Arsenlis, A., Parks, D.M., Becker, R., Bulatov, V.V., 2004. On the evolution of crystallographic dislocation density in non-homogeneously deforming crystals. J. Mech. Phys. Solids 52, 1213–1246.

Asaro, R.J., 1983a. Micromechanics of crystals and polycrystals. Adv. Appl. Mech. 23, 1–115.

Asaro, R.J., 1983b. Crystal plasticity. J. Appl. Mech. 50, 921–934.

Asaro, R.J., Needleman, A., 1985. Texture development and strain hardening in rate dependent polycrystals. Acta Metall. 33, 923–953.

Asaro, R.J., Rice, J.R., 1977. Strain localization in ductile single crystals. J. Mech. Phys. Solids 25, 309–338.

Bayley, C.J., Brekelmans, W.A.M., Geers, M.G.D., 2006. A comparison of dislocation induced back stress formulations in strain gradient crystal plasticity. Int. J. Solids Struct. 43, 7268–7286.

Bertram, A., 2005. Elasticity and Plasticity of Large Deformations. Springer.

Besson, J., Cailletaud, G., Chaboche, J.L., Forest, S., 2009. Non-linear Mechanics of Materials. vol. 167. Springer Science & Business Media.

Borg, U., 2007. A strain gradient crystal plasticity analysis of grain size effects in polycrystals. Eur. J. Mech. A Solids 26, 313–324.

Brown, S., Kim, K., Anand, L., 1989. An internal variable constitutive model for hot working of metals. Int. J. Plast. 5, 95–130.

Crumbach, M., Goerdeler, M., Gottstein, G., Neumann, L., Aretz, H., Kopp, R., 2004. Through-process texture modelling of aluminium alloys. Model. Simul. Mater. Sci. Eng. 12, S1–S18.

Crumbach, M., Goerdeler, M., Gottstein, G., 2006a. Modelling of recrystallisation textures in aluminium alloys: I. Model set-up and integration. Acta Mater. 54, 3275–3289.

Crumbach, M., Goerdeler, M., Gottstein, G., 2006b. Modelling of recrystallisation textures in aluminium alloys: II. Model performance and experimental validation. Acta Mater. 54, 3291–3306.

De Souza Neto, E.A., Peric, D., Owen, D.R.J., 2008. Computational Methods for Plasticity: Theory and Applications. John Wiley & Sons Ltd.

Delannay, L., Kalidindi, S.R., Van Houtte, P., 2002. Quantitative prediction of textures in aluminium cold rolled to moderate strains. Mater. Sci. Eng. A 336, 233–244.

Dunne, F., Petrinic, N., 2006. Introduction to Computational Plasticity, first ed. Oxford University Press, New York.

Dunne, F.P.E., Rugg, D., Walker, A., 2007. Lengthscale-dependent, elastically anisotropic, physically-based HCP crystal plasticity: application to cold-dwell fatigue in Ti alloys. Int. J. Plast. 23, 1061–1083.

Evers, L.P., Brekelmans, W.A.M., Geers, M.G.D., 2004. Non-local crystal plasticity model with intrinsic SSD and GND effects. J. Mech. Phys. Solids 52, 2379–2401.

Fleck, N.A., Muller, G.M., Ashby, M.F., Hutchinson, J.W., 1994. Strain gradient plasticity: theory and experiment. Acta Metall. Mater. 42 (2), 475–487.

Gurtin, M.E., 2002. A gradient theory of single-crystal viscoplasticity that accounts for geometrically necessary dislocations. J. Mech. Phys. Solids 50, 5–32.

Gurtin, M.E., 2006. The Burgers vector and the flow of screw and edge dislocations in finite-deformation single-crystal plasticity. J. Mech. Phys. Solids 54, 1882–1898.

Gurtin, M.E., 2008. A finite deformation, gradient theory of single-crystal plasticity with free energy dependent on densities of geometrically necessary dislocations. Int. J. Plast. 24, 702–725.

Han, C.S., Gao, H., Huang, Y., Nix, W.D., 2005a. Mechanism-based strain gradient crystal plasticity—I. Theory. J. Mech. Phys. Solids 53, 1188–1203.

Han, C.S., Gao, H., Huang, Y., Nix, W.D., 2005b. Mechanism-based strain gradient crystal plasticity—II. Analysis. J. Mech. Phys. Solids 53, 1204–1222.

Havner, K.S., 1992. Finite Plastic Deformation of Crystalline Solids. Cambridge University Press.
Hill, R., 1966. Generalized constitutive relations for incremental deformation of metal crystals by multislip. J. Mech. Phys. Solids 14, 95–102.
Hill, R., Rice, J.R., 1972. Constitutive analysis of elastic-plastic crystals at arbitrary strain. J. Mech. Phys. Solids 20, 401–413.
Honeff, H., Mecking, H., 1981. Analysis of the deformation texture at different rolling conditions. In: Nagashima, S. (Ed.), Proc. ICOTOM 6. In: vol. 1. The Iron and Steel Institute of Japan, pp. 347–355.
Hutchinson, J.W., 1976. Bounds and self-consistent estimates for creep of polycrystalline materials. Proc. R. Soc. Lond. A Math. Phys. Sci. 348, 101–127.
Kalidindi, S.R., Bronkhorst, C.A., Anand, L., 1992. Crystallographic texture evolution during bulk deformation processing of fcc metals. J. Mech. Phys. Solids 40, 537–569.
Klusemann, B., Yalçinkaya, T., 2013. Plastic deformation induced microstructure evolution through gradient enhanced crystal plasticity based on a non-convex helmholtz energy. Int. J. Plast. 48, 168–188.
Kocks, U.F., Chandra, H., 1982. Slip geometry in partially constrained deformation. Acta Metall. 30, 695–709.
Kubin, L.P., Canova, G., Condat, M., Devincre, B., Pontikis, V., Br Xechet, Y., 1992. Dislocation microstructures and plastic Fow: a 3D simulation. Solid State Phenom. 23–24, 455–472.
Kuroda, M., Tvergaard, V., 2008a. On the formulations of higher-order strain gradient crystal plasticity models. J. Mech. Phys. Solids 56, 1591–1608.
Kuroda, M., Tvergaard, V., 2008b. A finite deformation theory of higher-order gradient crystal plasticity. J. Mech. Phys. Solids 56, 2573–2584.
Lebensohn, R.A., 2001. N-site modeling of a 3D viscoplastic polycrystal using Fast Fourier Transform. Acta Mater. 49, 2723–2737.
Levkovitch, V., Svendsen, B., 2006. On the large-deformation and continuum-based formulation of models for extended crystal plasticity. Int. J. Solids Struct. 43, 7246–7267.
Liang, L., Dunne, F.P.E., 2009. GND accumulation in bi-crystal deformation: crystal plasticity analysis and comparison with experiments. Int. J. Mech. Sci. 51, 326–333.
Lubarda, V.A., 2002. Elastoplasticity Theory. CRC Press LLC, Boca Raton, FL.
Mandel, J., 1965. Generalisation de la theorie de la plasticite de W. T. Koiter. Int. J. Solids Struct. 1, 273–295.
Moulinec, H., Suquet, P., 1998. A numerical method for computing the overall response of nonlinear composites with complex microstructure. Comput. Methods Appl. Mech. Eng. 157, 69–94.
Ohashi, T., 2005. Crystal plasticity analysis of dislocation emission from micro voids. Int. J. Plast. 21, 2071–2088.
Reddy, B.D., 2011. The role of dissipation and defect energy in variational formulations of problems in strain-gradient plasticity. Part 2: single-crystal plasticity. Contin. Mech. Thermodyn. 23, 551–572.
Rice, J.R., 1971. Inelastic constitutive relations for solids: an internal variable theory and its application to metal plasticity. J. Mech. Phys. Solids 19, 433–455.
Roters, F., Eisenlohr, P., Hantcherli, L., Tjahjanto, D.D., Bieler, T.R., Raabe, D., 2010. Overview of constitutive laws, kinematics, homogenization and multiscale methods in crystal plasticity finite-element modeling: theory, experiments, applications. Acta Mater. 58, 1152–1211.
Sun, S., Adams, B.L., Shet, C.Q., Saigal, S., King, W., 1998. Mesoscale investigation of the deformation field of an aluminum bicrystal. Scr. Mater. 39, 501–508.
Sun, S., Adams, B.L., King, W.E., 2000. Observations of lattice curvature near the interface of a deformed aluminum bicrystal. Philos. Mag. A 80, 9–25.

Taylor, G.I., 1938a. Plastic strain in metals. J. Inst. Met. 62, 307–324.

Taylor, G.I., 1938b. Analysis of plastic strain in a cubic crystal. In: Stephen Timoshenko 60th Anniversary Volume. McMillan Co., New York, pp. 218–224.

Taylor, G.I., Elam, C.F., 1923. The distortion of an aluminum crystal during a tensile test. Proc. R. Soc. Lond. A Math. Phys. Sci. 102, 643–667.

Taylor, G.I., Elam, C.F., 1925. The plastic extension and fracture of aluminum single crystals. Proc. R. Soc. Lond. A Math. Phys. Sci. 108, 28–51.

Tjahjanto, D.D., Eisenlohr, P., Roters, F., 2010. A novel grain cluster-based homogenization scheme. Model. Simul. Mater. Sci. Eng. 18.

Van Houtte, P., 1982. On the equivalence of the relaxed Taylor theory and the Bishop-Hill theory for partially constrained plastic deformation of crystals. Mater. Sci. Eng. 55, 69–77.

Van Houtte, P., Delannay, L., Samajdar, I., 1999. Quantitative prediction of cold rolling textures in low-carbon steel by means of the LAMEL model. Texture Microstruct. 31, 109–149.

Van Houtte, P., Delannay, L., Kalidindi, S.R., 2002. Comparison of two grain interaction models for polycrystal plasticity and deformation texture prediction. Int. J. Plast. 18, 359–377.

Voyiadjis, G.Z., Huang, W., 1996. Modelling of single crystal plasticity with backstress evolution. Eur. J. Mech. A Solids 15, 553–573.

Yalçinkaya, T., Brekelmans, W.A.M., Geers, M.G.D., 2012. Non-convex rate dependent strain gradient crystal plasticity and deformation patterning. Int. J. Solids Struct. 49, 2625–2636.

Yefimov, S., Groma, I., van der Giessen, E., 2004. A comparison of a statistical-mechanics based plasticity model with discrete dislocation plasticity calculations. J. Mech. Phys. Solids 52, 279–300.

Zhou, Y., Neale, K.W., Toth, L.S., 1993. A modified model for simulating latent hardening during the plastic deformation of rate-dependent fcc polycrystals. Int. J. Plast. 9, 961–978.

Chapter 4

Discrete dislocation dynamics

4.1 Introduction

In crystalline metals, dislocations movement, nucleation, and interaction with one another and free surfaces govern the material behavior. One can study the size effects in crystalline metals using the discrete dislocation dynamics (DDD) simulation which directly includes dislocations as discrete line defects. Accordingly, the dislocations motion can be obtained by integration of equations of motion which include their interactions with other dislocations. Historically, similar to many other computational material science methods, DDD has been introduced in a 2-dimensional framework where dislocations are simplified as straight lines with infinite length (see, e.g. Lepinoux and Kubin, 1987; Ghoniem and Amodeo, 1988; Groma and Pawley, 1993; Van der Giessen and Needleman, 1995). The next generation of DDD model commonly known as 2.5-dimensional have been introduced in which some of the general 3-dimensional dislocation patterns including dynamic dislocation sources and obstacles and dislocation junction formation are incorporated in the 2-diemnsional model (Benzerga et al., 2004). Although simplified DDD simulation methods can provide valuable insight into many aspects of dislocations mechanisms, a true DDD simulation should be used to handle 3-dimensional dislocation mechanisms. Accordingly, 3-dimensional DDD models have been introduced in which dislocation loops can be discretized as a set of connected straight lines, which can be defined as pure edge or screw dislocations (Devincre and Kubin, 1994) or mixed edge and screw dislocations (Zbib et al., 1998), or curved splines (Ghoniem and Sun, 1999). Different 3-dimensional DDD codes have been developed to capture the 3D mechanisms of dislocations (Kubin et al., 1992; Verdier et al., 1998; Schwarz, 1999; Weygand et al., 2001; Zbib et al., 2002; Wang et al., 2006; Arsenlis et al., 2007). The theory of DDD simulation have been described in previous works (see, e.g., Zbib, 2012; Sills et al., 2016). In this chapter, the focus is on the application of DDD to capture size effect in crystalline metals. First, the investigation of size effects in crystalline metals using DDD simulation is extensively addressed here during micropillar compression test. In the last sections, the size effects during other experiments of microbending and nanoindentation are also briefly presented.

Size Effects in Plasticity: From Macro to Nano. https://doi.org/10.1016/B978-0-12-812236-5.00004-9

4.2 DDD simulation of size effects during micropillar compression experiment

Uchic et al. (2003, 2004) and Uchic and Dimiduk (2005) have developed the focused ion beam (FIB) machining technique to fabricate nano to micron sized metallic pillars. They tested different sizes of pillars during compression experiment, and studied the effects of sample size on them. They reported very strong size effects during compression for pure Ni in which the strength is almost three times larger than the bulk Ni (Fig. 1.35). The observed size effect was the smaller is stronger, i.e., a smaller sample shows a higher strength. In the literature, different mechanisms have been presented to capture the size effects during micropillar compression test including source truncation, source exhaustion, and weakest link theory (see, e.g., Uchic et al., 2009; Kraft et al., 2010), as described in Sections 1.3.2.2.1–1.3.2.2.3. These mechanisms are acting along the forest hardening mechanism, i.e., interaction of dislocations with each other, to control the change in material strength as the sample size changes. All of these mechanisms involve the interaction of dislocations with one another or other aspect of crystalline metals of confined volumes such as free surfaces. Accordingly, the DDD simulation, as a method which considers the dislocations as discrete line defects is very effective to study these size effect mechanisms during micropillar compression test. In the following sections, different mechanisms are addressed using the DDD studies available in the literature.

4.2.1 Source truncation

One major difference between micropillars and bulk samples is the role of free surfaces. In the case of pillars with confined volume, the double-ended dislocation sources interact with free surface and transform to the single-ended ones. Accordingly, the sample size controls the length of dislocation source through this transformation in a way that smaller sample leads to a smaller source. Smaller dislocation sources require higher stresses to be activated. This mechanism, which is commonly termed as source truncation (Parthasarathy et al., 2007; Rao et al., 2007), leads to size effects in which a smaller sample is stronger. Parthasarathy et al. (2007) addressed this mechanism by developing a formulation to capture the effective length of dislocation source. To do so, first, the probability of a cylindrical pillar with n pins to have the maximum distance from the free surface equal to $\lambda_{\max}$ can be described as follows:

$$p(\lambda_{\max})d\lambda_{\max} = \left[1 - \frac{\pi(R-\lambda_{\max})(b-\lambda_{\max})}{\pi Rb}\right]^{n-1}\left(\frac{\pi[(R-\lambda_{\max})+(b-\lambda_{\max})]}{\pi Rb}\right)nd\lambda_{\max} \tag{4.1}$$

where R is the specimen radius, $b = R/\cos\beta$ is the major axis of the glide plane, and β is the angle between the primery slip plane and the loading axis. The mean length of effective dislocation source $\overline{\lambda}_{\max}$ can then be obtained using Eq. (4.1) as follows:

$$\begin{aligned}\overline{\lambda}_{\max} &= \int_0^R \lambda_{\max} p(\lambda_{\max}) d\lambda_{\max} \\ &= \int_0^R \left[1 - \frac{\pi(R-\lambda_{\max})(b-\lambda_{\max})}{\pi Rb}\right]^{n-1} \left(\frac{\pi[(R-\lambda_{\max})+(b-\lambda_{\max})]}{\pi Rb}\right) n\lambda_{\max}\, d\lambda_{\max}\end{aligned} \tag{4.2}$$

Eq. (4.2) describes the relation between the sample size and the length of dislocation sources. Parthasarathy et al. (2007) related the critical resolved shear stress (CRSS) to $\overline{\lambda}_{\max}$ using the following equation:

$$\text{CRSS} = \frac{\alpha G b}{\overline{\lambda}_{\max}} + \tau_0 + 0.5 G b \sqrt{\rho} \tag{4.3}$$

where α is a constant, G is the shear modulus, b is the Burgers vector, τ_0 is the friction stress, and ρ is the dislocation density. They defined the number of pins, n, as follows:

$$n = \text{Integer}\left[\frac{L_{\text{mobile}}}{L_{\text{ave}}}\right] \tag{4.4}$$

where L_{ave} is the average length of dislocation segments, $L_{\text{mobile}} = \rho\pi R^2 h/s$ is the total length of mobile dislocations, h is the height of the pillar, and s is the number of slip systems. Parthasarathy et al. (2007) did not limit the CRSS due to nucleation and defined the upper bound of CRSS based on the required stress to move the partial dislocations as follows:

$$\text{CRSS}(\max) = \frac{\gamma}{b} \tag{4.5}$$

where γ is the stacking fault energy. Fig. 4.1 compares the model presented by Parthasarathy et al. (2007) with the experimental results of Ni (Uchic et al., 2004; Dimiduk et al., 2005) and Au (Greer et al., 2005; Volkert and Lilleodden, 2006). The results show that the presented formulation can successfully model the observed size effects.

In the analytical formulation presented by Parthasarathy et al. (2007), it was assumed that the dislocation line tension is isotropic. Also, it was assumed that the strength of single-ended dislocation sources of a length L can be described as $\alpha Gb/L$. Rao et al. (2007) incorporated the 3D DDD simulation to investigate the strength of single-ended and double-ended dislocation sources which have a key role in source truncation mechanism. The main purpose was to study the anisotropy of dislocation line tension and compare the strength of single-ended dislocation sources with double-ended ones. They used a DDD code ParaDis (Arsenlis et al., 2007) to capture the responses of single-ended and double-ended dislocation sources. Rao et al. (2007) modified the code to include the interaction of dislocations with free surface, which was termed SSA.

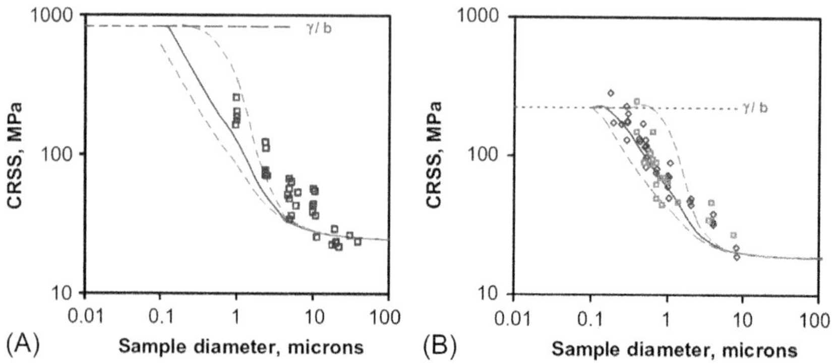

FIG. 4.1 Variation of critical resolved shear stress (CRSS) versus the pillar diameter for: (A) Ni and (B) Au. The *dotted lines* denote the lower and upper standard deviations from the mean as predicted by the model. The original experimental data has been reported by Uchic et al. (2004), Dimiduk et al. (2005), Greer et al. (2005), and Volkert and Lilleodden (2006). *(After Parthasarathy, T.A., Rao, S.I., Dimiduk, D.M., Uchic, M.D., Trinkle, D.R., 2007. Contribution to size effect of yield strength from the stochastics of dislocation source lengths in finite samples. Scr. Mater. 56, 313–316.)*

Furthermore, they addressed the effects of parallel simulation by using a serial version of ParaDis, which was termed the FFST method. They simulated four different types of samples:

- *Sample I:* Sample with double-ended dislocation source with isotropic and anisotropic line tension.
- *Sample II:* Sample with single-ended dislocation source with isotropic and anisotropic line tension.
- *Sample III:* Sample with single-ended dislocation source with different line orientations and anisotropic line tension.
- *Sample IV:* Sample with single-ended and double-ended dislocation sources of different length and anisotropic line tension.

The analytical data is available for the first simulation type within an infinite volume, which was compared to the simulation results. Also, assuming the isotropic line tension, one can consider the strength of a single-ended source to be half of a double-ended source with the same length. Rao et al. (2007) incorporated three dislocation source characters of edge, screw, and 30° mixed dislocation characters.

In the case of double-ended dislocation source in an infinite volume (*Sample I*), the critical resolved shear stress can be defined as follows:

$$\mathrm{CRSS}(\upsilon) = \frac{Gb}{L} C \left[\frac{3}{4(1-\upsilon)} + \frac{1}{4} \right] \tag{4.6}$$

where CRSS(υ) is a function of Poisson's ratio υ. In Eq. (4.6), C is a constant, G is the shear modulus, b is the Burgers vector, and L is length of the dislocation

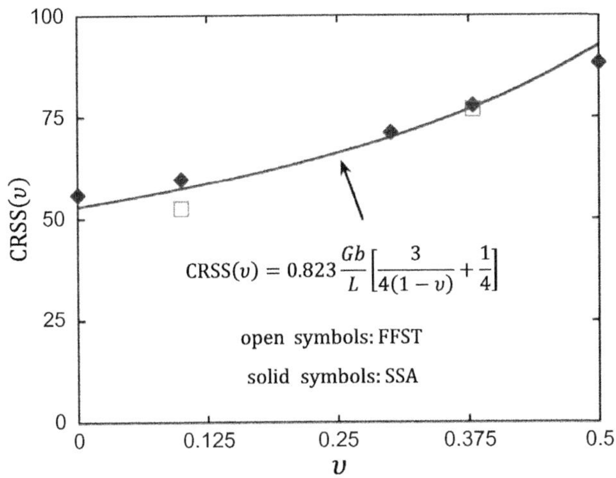

FIG. 4.2 Variation of critical resolved shear stress (CRSS) in MPa versus the Poisson's ratio υ: Comparison of the DDD simulation of a double-ended dislocation source of 30° mixed dislocation character with the analytical prediction of double-ended dislocation source in an infinite volume. *(After Rao, S.I., Dimiduk, D.M., Tang, M., Parthasarathy, T.A., Uchic, M.D., Woodward, C., 2007. Estimating the strength of single-ended dislocation sources in micron-sized single crystals. Philos. Mag. 87, 4777–4794.)*

source. Fig. 4.2 compares the results obtained from the DDD simulation of a double-ended source with a 30° mixed dislocation character with the analytical prediction of Eq. (4.6) by considering $C = 0.823$. The results show that the effect of confined volume of the sample is negligible for double-sided dislocation sources.

The next set of DDD simulations are samples with single-ended dislocation source with isotropic and anisotropic line tension (*Sample II*). Here, the critical resolved shear stress is obtained for the single-ended dislocation source with the length of $933b$. The single-ended dislocation source with a 30° mixed dislocation character is a straight line extending from a fixed node inside the cube to the surface. The strength of the single-ended source depends on the sign of applied stress which defines the way the source operates. It can be operated either "forward" or "reverse." As explained earlier, one can consider the strength of a single-ended source to be half of a double-ended source with the same length by assuming the isotropic line tension. Accordingly, a double-ended source with the same length are also simulated to compare the difference between the critical resolved shear stress of single-ended and double-ended dislocation sources. Fig. 4.3 compares the strength of the single-ended and double-ended dislocation sources as the Poisson's ratio υ varies. The strength of a double-ended source does not depend on the sign of applied stress. Fig. 4.3 shows that for the case of close to isotropic line energy, i.e., $\upsilon = 0 - 0.1$, the results follow the stated prediction and the strength of a single-ended source is close to the half of a

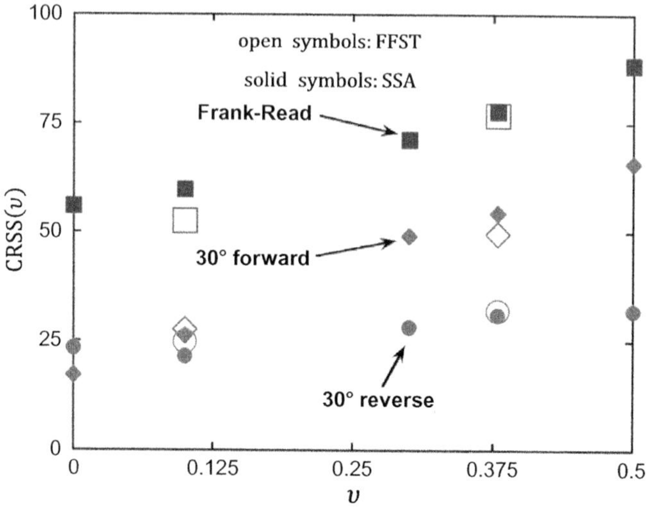

FIG. 4.3 Variation of critical resolved shear stress (CRSS) in MPa versus the Poisson's ratio υ: Comparison of the DDD simulation of a single-ended dislocation source of 30° mixed dislocation character with that of double-ended source. The length of the single-ended and double-ended source is similar. *(After Rao, S.I., Dimiduk, D.M., Tang, M., Parthasarathy, T.A., Uchic, M.D., Woodward, C., 2007. Estimating the strength of single-ended dislocation sources in micron-sized single crystals. Philos. Mag. 87, 4777–4794.)*

double-ended source strength with the same length. Furthermore, in this range of Poisson's ratio, the sign of the applied stress has a negligible effect on the strength of a single-ended dislocation source. For the larger values of υ, the critical resolved shear stress increases faster by increasing the Poisson's ratio in the case of the forward direction compared to the reverse one. Accordingly, for large values of Poisson's ratio, the critical resolved shear stress could be greater (forward direction) or smaller (reverse direction) than half of a double-ended source strength depending on the sign of the applied stress. Accordingly, the results show that the single-ended dislocation source with a fixed length could have a range of strengths.

In the third set of DDD simulations (*Sample III*), Rao et al. (2007) compares the responses of a single-ended source of a 30° mixed dislocation character with a single-ended one with pure screw dislocation character. The length of both sources is 933b, and the size of the simulation cube is 4000b with the shear modulus of $G = 59.9$ GPa and the Poisson's ratio of $\upsilon = 0.38$. The SSA scheme is incorporated to simulate the sample. Fig. 4.4 compares the responses of both single-ended sources. Again, the response of a source with a 30° mixed dislocation character depends on the sign of the applied stress. The forward operating single-ended source with a 30° mixed dislocation character demonstrates the initial elastic response, and after the stress reaches the value of 114 MPa, it shows a pure plastic behavior. The single-ended source with a pure screw

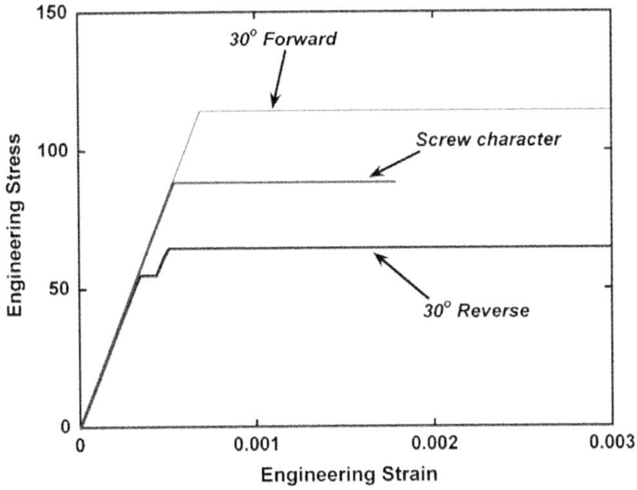

FIG. 4.4 Variation of the engineering stress in MPa versus the engineering strain: comparison of the DDD simulation of a single-ended dislocation source of 30° mixed dislocation character with the single-ended one with pure screw character. Again, the response of the single-ended source with a 30° mixed dislocation character depends on the sign of applied stress. *(After Rao, S.I., Dimiduk, D.M., Tang, M., Parthasarathy, T.A., Uchic, M.D., Woodward, C., 2007. Estimating the strength of single-ended dislocation sources in micron-sized single crystals. Philos. Mag. 87, 4777–4794.)*

character shows the similar pattern but with the smaller critical strength of 90 MPa. However, the reverse operating single-ended source with a 30° mixed dislocation character shows a small strain burst at the stress value of 57 MPa and then becomes perfectly plastic at the critical stress value of 64 MPa. Rao et al. (2007) also compared the initial stress free state and the final maximum stress one of the single-ended source of a 30° mixed dislocation character with the single-ended one of a pure screw character (Fig. 4.5).

The fourth set of simulations was the single-ended and double-ended dislocation sources of different lengths (*Sample IV*). The single-ended and double-ended dislocation sources with a 30° mixed dislocation character are generated with the lengths from 233b to 933b. The Poisson's ratio of $\upsilon = 0.38$ is selected. The results are fitted to the following equation:

$$\mathrm{CRSS}(L) = kG\frac{\ln(L/b)}{(L/b)} \tag{4.7}$$

where k is a constant. Fig. 4.6 compares the critical resolved shear stresses of the single-ended and double ended sources as their lengths change. One can see that Eq. (4.7) can successfully capture the variation of the critical resolved shear stresses versus the source length. With the values of $k = 0.06$ for reverse operating single-ended source, $k = 0.12$ for forward operating single-ended source, and $k = 0.18$ for the double-ended source.

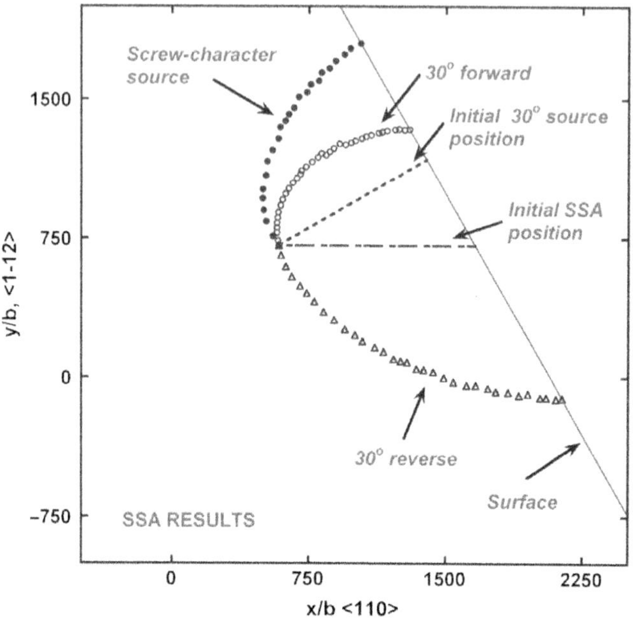

FIG. 4.5 Comparison of the initial stress free state and the final maximum stress one of the single-ended source of a 30° mixed dislocation character with the single-ended one of a pure screw character. *(After Rao, S.I., Dimiduk, D.M., Tang, M., Parthasarathy, T.A., Uchic, M.D., Woodward, C., 2007. Estimating the strength of single-ended dislocation sources in micron-sized single crystals. Philos. Mag. 87, 4777–4794.)*

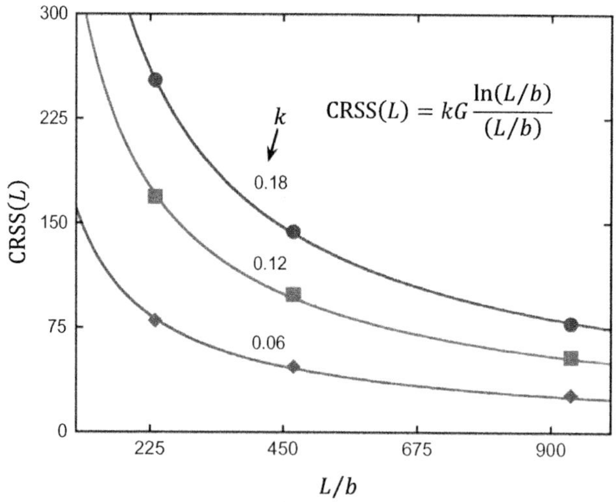

FIG. 4.6 Comparison of the critical resolved shear stresses of the single-ended and double ended sources as their lengths change. *(After Rao, S.I., Dimiduk, D.M., Tang, M., Parthasarathy, T.A., Uchic, M.D., Woodward, C., 2007. Estimating the strength of single-ended dislocation sources in micron-sized single crystals. Philos. Mag. 87, 4777–4794.)*

Another expression that is usually used to describe the variation of the critical shear stress versus the source length can be described as follows:

$$\mathrm{CRSS}(L) = \frac{\alpha G b}{L} \tag{4.8}$$

Eq. (4.8) is derived using the balance between the length of the dislocation line and its curvature. Rao et al. (2007) reported the constant α for the simulation of single-ended and double-ended dislocation sources with the length of $933b$ and different characters. The Poisson's ratio of $\upsilon = 0.38$ is selected. The SSA simulation scheme is incorporated for these samples. Table 4.1 presents the corresponding values of α. In the case of single-ended sources, the values of α vary from 0.4 to 0.84 in which the reverse operating source has the minimum value and forward operating source has the maximum value. In the case of double-ended sources, the values of α vary from 0.8 to 1.28 in which the source with the edge character has the minimum value and the source with the screw character has the maximum value. In the cases of sources with lengths different from $933b$, the constant α can be described using the values obtained for the length of $933b$, i.e., α_{933b}, as follows:

$$\alpha(L) = \alpha_{933b} \frac{\ln(L/b)}{6.84} \tag{4.9}$$

In the last step, Rao et al. (2007) conducted a statistical analysis and showed that value of $\alpha = 1$ which has been incorporated by Parthasarathy et al. (2007) slightly overestimates the role of source truncation mechanism in the modeled samples. The 3D DDD simulation results conducted by Rao et al. (2008) showed that there is another size effect mechanism, which is commonly termed

TABLE 4.1 **The value of α for the single-ended and double-ended sources with the length of $933b$, i.e., α_{933b}, which is obtained using the DDD simulation**

Source	α_{933b}
Forward operating single-ended (30° mixed dislocation character)	0.84
Reverse operating single-ended (30° mixed dislocation character)	0.4
Single-ended (screw character)	0.66
Double-ended (screw character)	1.28
Double-ended (30° mixed dislocation character)	1.19
Double-ended (edge character)	0.86

The original data was reported by Rao, S.I., Dimiduk, D.M., Tang, M., Parthasarathy, T.A., Uchic, M. D., Woodward, C., 2007. Estimating the strength of single-ended dislocation sources in micron-sized single crystals. Philos. Mag. 87, 4777–4794.

as source exhaustion. Parthasarathy et al. (2007) did not consider this mechanism and the slight difference between the model predictions and the experimental data can be captured using source exhaustion mechanism. The size effects mechanism of source exhaustion is described in the next section.

4.2.2 Source exhaustion

In the case of crystalline metals of confined volumes, if the mobile dislocation content is not enough to sustain the applied deformation, the applied stress should be increased to activate stronger dislocation sources and sustain the required plastic flow. This hardening mechanism is commonly termed as source exhaustion. The loss of mobile dislocation density can occur due to various mechanisms. In the case of sample with confined volumes, the dislocation escape from the free surfaces may result in the mobile dislocation content reduction (see, e.g., Rao et al., 2008). In the case of the pillar compression experiment, a dislocation-starved sample may be formed when all of the mobile dislocations leave the sample. Consequently, the deformation mechanism, which is commonly termed as dislocation starvation hardening (Greer et al., 2005), is governed by source-limited activations. Here, the dislocation starvation is considered as a special case of source exhaustion. Discrete dislocation simulation has been incorporated to address source exhaustion mechanism in crystalline pillars. Here, the work conducted by Rao et al. (2008) and Zhou et al. (2011) are addressed in more details.

Rao et al. (2008) incorporated the 3D discrete dislocations simulation (DDS) to capture size effects in Ni samples with the sizes in the range of 0.5 – 20 μm. Besides forest hardening mechanism, they reported two other size effects mechanisms of source truncation, which is addressed in the previous section, and source exhaustion. They related the source exhaustion mechanism to the break-down in the mean-field conditions of forest hardening mechanism in the samples with confined volumes. They used a DDD code ParaDis (Arsenlis et al., 2007) to capture the responses of single-ended and double-ended dislocation sources. Rao et al. (2008) followed their previous code in Rao et al. (2007) and modified the code to include the interaction of dislocations with free surface, which was termed SSA. The simulation samples are cube cells with the edge lengths of $a_0 = 10$ and 20 μm. Also, tetragonal cells are used to mimic the experiments with cylindrical samples. They set the initial dislocation density by populating a sample using Frank-Read (FR) sources. Table 4.2 presents the number of conducted simulations and number of initial FR sources for samples with different sizes and initial dislocation densities (ρ_0). In the case of smaller samples with 0.5, 1, and 4 μm cells, the scatter in critical stress is considerable from one random arrangement to another one. However, this scatter is insignificant, i.e., smaller than 5%, for larger samples.

In the first step, Rao et al. (2008) reported the stress-strain response of a tetragonal sample of 1 μm cell and aspect ratio of 2.8 with the initial dislocation

TABLE 4.2 Number of conducted simulations and number of initial Frank-Read (FR) sources for samples with different sizes and initial dislocation densities

Size (μm)	$\rho_0 = 7 \times 10^{11}\ m^{-2}$	$\rho_0 = 2 \times 10^{12}\ m^{-2}$	$\rho_0 = 10^{13}\ m^{-2}$
0.5	–	–	5 – 10, 12
1	5 – 10, 4	5 – 10, 12	5 – 10, 48
4	3 – 64	3, 192	3, 768
10 (tetragonal)	–	2, 1200	–
10 (cubic)	2, 200	2, 600	2, 2400
20 (cubic)	2, 800	2, 2400	–

The original data was reported by Rao, S.I., Dimiduk, D.M., Parthasarathy, T.A., Uchic, M.D., Tang, M., Woodward, C., 2008. Athermal mechanisms of size-dependent crystal flow gleaned from three-dimensional discrete dislocation simulations. Acta Mater. 56, 3245–3259.

density of $\rho_0 = 2 \times 10^{12}\ m^{-2}$. In order to compare the results of the discrete dislocations simulation with experimental data, the sample response obtained from the simulation are scaled from 59.9 GPa, which is used in simulations, to a shear modulus of 78.7 GPa. Accordingly, eight initial random microstructures are generated and simulated as shown in Fig. 4.7A. Also, the response of a cubic sample of 1 μm cell with a single FR source is also studied (Fig. 4.7B). The FR sources transformed into the single-ended dislocation sources at small strains, which is in line with source truncation mechanism. The simulation with the lower bound of stress equal to 140 MPa shows similar response as the case of single FR source with the same stress level including a linear elastic response followed by a perfectly plastic behavior. In the cases of the remaining seven samples, which has different initial microstructure but with the same initial dislocation density, the stress increases incrementally by increasing the strain until it reaches its critical value at some critical strains. Fig. 4.7A shows a considerable scatter in critical stress for each sample. The hardening mechanism which becomes active here is source exhaustion.

Fig. 4.8 compares the results of discrete dislocations simulation with those of the experiments for different sample sizes. In Fig. 4.8A, the simulation results include eight 1 μm cell samples, two 4 μm cell samples, a 10 μm cell sample, and a 20 μm cell sample. Again, the sample response obtained from the discrete dislocations simulation are scaled from 59.9 GPa, which is used in simulations, to a shear modulus of 78.7 GPa to compare the results of the simulation with the experimental data. Comparing the simulation and experimental results reported in Fig. 4.8A and B, respectively, one can see the size effects in which the smaller is stronger. Also, both simulation and experiment show the incremental increase

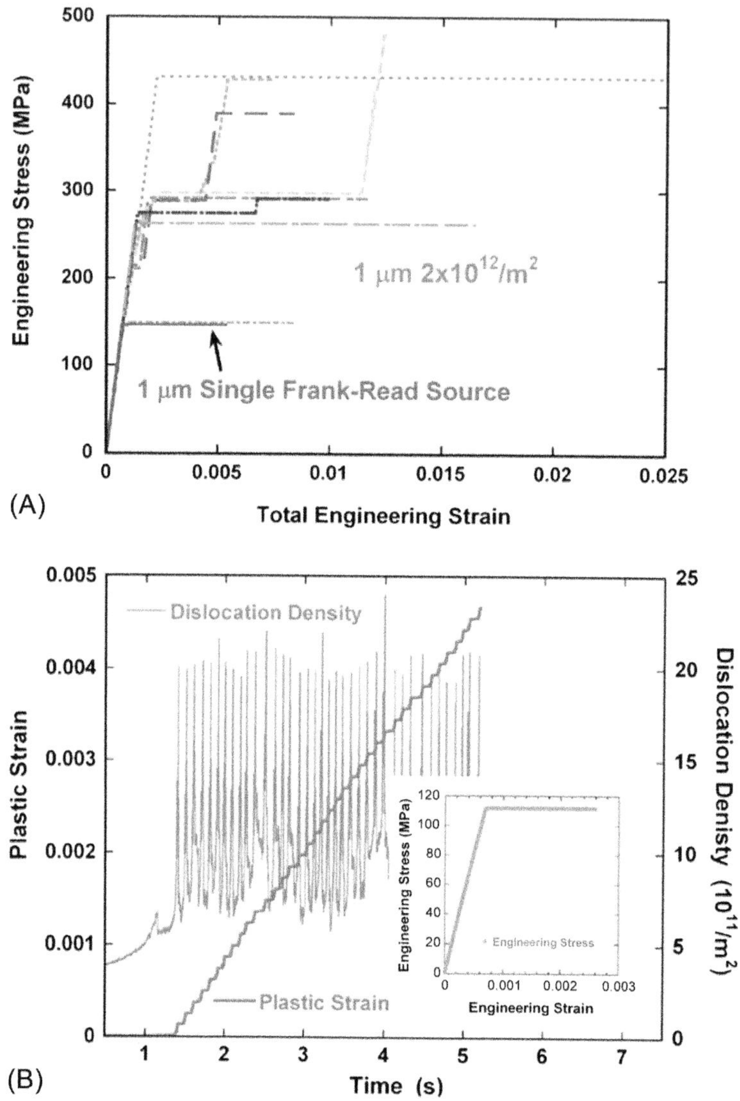

FIG. 4.7 (A) The responses of Ni tetragonal samples of 1 μm cell and aspect ratio of 2.8 with the initial dislocation density of $\rho_0 = 2 \times 10^{12}\ \mathrm{m}^{-2}$ and (B) the variations of plastic strain and dislocation density versus time and stress versus strain for a cubic sample of 1 μm cell with a single FR source. *(After Rao, S.I., Dimiduk, D.M., Parthasarathy, T.A., Uchic, M.D., Tang, M., Woodward, C., 2008. Athermal mechanisms of size-dependent crystal flow gleaned from three-dimensional discrete dislocation simulations. Acta Mater. 56, 3245–3259.)*

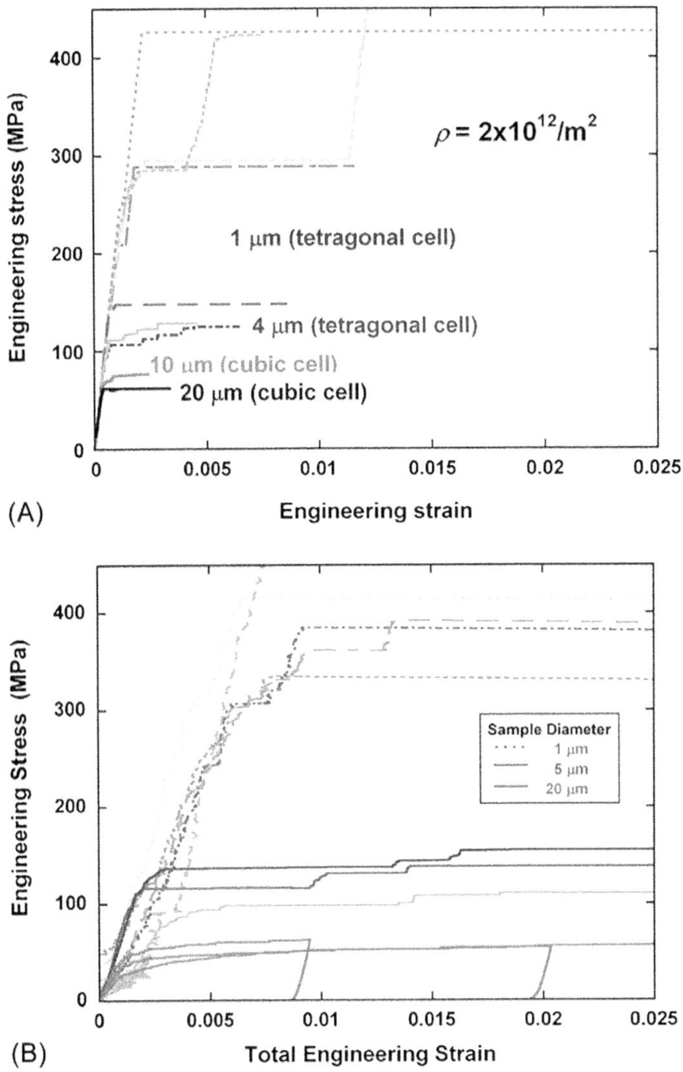

FIG. 4.8 Comparison of the responses of the Ni samples with sizes from 1 to 20 μm: (A) the discrete dislocations simulation of Ni samples with the initial dislocation density of $\rho_0 = 2 \times 10^{12}\ m^{-2}$ and (B) experimental results of Ni microcrystals. *(After Rao, S.I., Dimiduk, D.M., Parthasarathy, T.A., Uchic, M.D., Tang, M., Woodward, C., 2008. Athermal mechanisms of size-dependent crystal flow gleaned from three-dimensional discrete dislocation simulations. Acta Mater. 56, 3245–3259.)*

in flow stress which can be attributed to the source exhaustion mechanism (Rao et al., 2008). Also, the overall hardening rate of the larger sample with 10 and 20 μm cell samples is significantly smaller than the 1 μm cell samples.

In conclusion, Rao et al. (2008) reported two hardening mechanisms of source truncation and source exhaustion. The first one is due to the transformation of double-ended sources to the single-ended ones and their activation. The later mechanism, i.e., source exhaustion, is the origin of stochastic variation in flow stress. This happens when the dislocation density of the sample is as large as the real crystals and dislocations interact with each other.

The second work which is elaborated here was presented by Zhou et al. (2011). They used the 3D DDD simulation to study the size effects in Ni single crystal pillars with diameters from 300 to 1000 nm during uniaxial compression experiment. To increase the accuracy of the simulation, they included the cross-slip into the DDD simulation. Fig. 4.9 depicts the variation of engineering stress and dislocation density as the strain changes. Dislocation starvation is the governing mechanism of size effects for pillars with diameters of 300 and 500 nm. The response becomes elastic again following the dislocation starvation. Increasing the sample diameter to 750 and 1000 nm, however, no more dislocation starvation was observed. In these range of pillar diameters, Zhou et al. (2011) attributed the observed hardening to the exhaustion hardening induced by dislocation interactions which shut off the dislocation sources.

Following the formulation introduced by Greer (2006), Zhou et al. (2011) presented a theoretical framework to address dislocation starvation. They stated that by decreasing the pillar diameter, the rate of dislocation multiplication

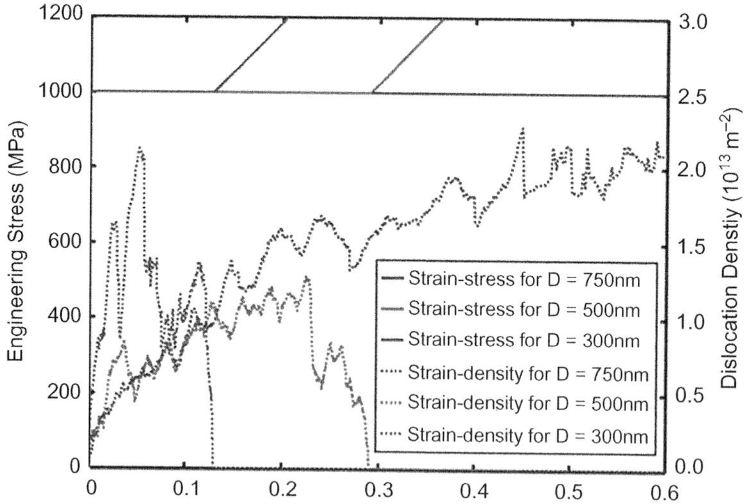

FIG. 4.9 Variation of engineering stress and dislocation density versus the strain. *(After Zhou, C., Beyerlein, I.J., LeSar, R., 2011. Plastic deformation mechanisms of fcc single crystals at small scales. Acta Mater. 59, 7673–7682.)*

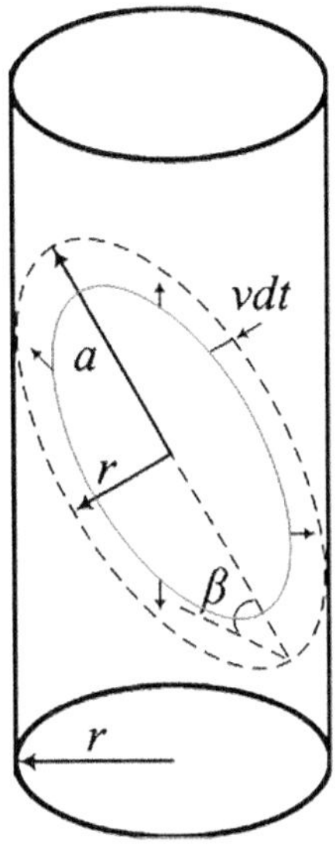

FIG. 4.10 A pillar contains a dislocation loop with the distance vdt from free surfaces. *(After Zhou, C., Beyerlein, I.J., LeSar, R., 2011. Plastic deformation mechanisms of fcc single crystals at small scales. Acta Mater. 59, 7673–7682.)*

decreases. On the other hand, it becomes easier for dislocation to leave the sample for smaller diameters. There is a critical diameter at which the rate of dislocation multiplication becomes the same as the rate of dislocations leaving the pillar. In the case of a sample with the size smaller than the described critical diameter, dislocation starvation becomes the governing mechanism of size effects. As an example, a pillar of radius r and primary slip plane oriented at an angle β is shown in Fig. 4.10. The glide plane becomes an ellipse with the major and minor radii of $a = r/\cos(\beta)$ and r, respectively. Zhou et al. (2011) defines the rate of dislocation loss $d\rho_{\text{loss}}$ equal to the rate of dislocations leaving the sample as follows:

$$d\rho_{\text{loss}} = -\rho_{\text{mob}}\left(\frac{1}{2}\frac{\pi(a+r)\upsilon dt}{\pi ar}\right) = -\rho_{\text{mob}}\frac{\cos^2(\beta/2)\upsilon dt}{r} \tag{4.10}$$

where v is the velocity of dislocation and ρ_{mob} is the density of mobile dislocations. As shown in Eq. (4.10), Zhou et al. (2011) assumed that $d\rho_{\text{loss}}$ is proportional to r^{-1}. Furthermore, they assumed that any dislocation within a distance υdt from free surfaces has a 50% escaping chance. Other sources of dislocation loss including annihilation or source shutdown are not considered in Eq. (4.10). Zhou et al. (2011) defines the dislocation multiplication rate as below:

$$d\rho_{\text{mult}} = \rho_{\text{mob}}\frac{\upsilon dt}{L} \tag{4.11}$$

where L is the dislocation mean free path. The dislocation content can be obtained by considering two operations of dislocation loss and multiplication as below:

$$\rho = \rho_0 + \frac{[1/L - \cos^2(\beta/2)/r]}{b}\frac{\varepsilon_p}{M} \tag{4.12}$$

where ρ_0 is the initial dislocation content, M is the Schmid factor, and $\varepsilon_p = \rho_{mob} bvdt$ is the plastic strain induced by the movement of mobile dislocations. According to Eq. (4.12), the values of the dislocation mean free path L, pillar radius r, and the angle β govern the size effects mechanism. In the case of negative $[1/L - \cos^2(\beta/2)/r]$, the dislocation density decreases as the plastic strain increases, and the dislocation starvation becomes the dominant mechanism of size effects. The critical pillar diameter is around $D = 400$ nm in the case of $L \approx \rho^{-0.5}$, $\beta = 35^\circ$, and $\rho_0 = 2.0 \times 10^{13}\ \text{m}^{-2}$. After the dislocation starvation model, Zhou et al. (2011) addressed the Source truncation mechanism using single arm dislocation (SAD) model developed by Parthasarathy et al. (2007) (see, Section 4.2.1). However, they stated the SAD model focuses on the incipient plasticity and cannot capture the intermittency in plastic flow, strain bursts, and discrete load increases after the initial yield.

In order to investigate the source exhaustion mechanism, Zhou et al. (2011) conducted a DDD simulation of a pillar with diameter of $D = 1$ μm which has a realistic initial dislocation pattern. To do so, they cut the sample from a deformed bulk sample and relaxed the sample. It leads to a sample with initial dislocation density of $1.7 \times 10^{13}\ \text{m}^{-2}$, which is close to the experimental observations. Fig. 4.11 shows the variation of stress and dislocation density as the compressive strain increases. The results show that there is some plasticity activation even for small loadings, which can be related to some surface or weakly entangled dislocations leaving the sample leading to the dislocation density reduction, as shown in Fig. 4.11. The results also show some strain bursts followed by elastic response, and the observed hardening is induced by exhaustion hardening. This can be attributed to the dislocation sources trapped by dislocation reaction. In order to further elaborate the observed phenomenon, Zhou et al. (2011) reports the dislocation structure before and after load increase, i.e., exhaustion hardening as shown in Fig. 4.12. The strain at which the structure is reported is marked by an arrow in Fig. 4.11. Fig. 4.12A shows that the gliding of two sources of s1 and s2 provide the plasticity in the sample. Eventually, a dislocation junction is formed as two sources getting closer to each other, which is shown in Fig. 4.12B and C. To sustain the applied plastic flow, the applied stress should be increased to activate more dislocation sources. Accordingly, the applied stress is increased until the source s3 is activated as shown in Fig. 4.12D. After the activation of source s3, the required plasticity is achieved by glide of the source s3 under the constant applied stress.

Zhou et al. (2011) addressed three mechanisms of dislocation starvation, source truncation (SAD model), and source exhaustion for metallic samples

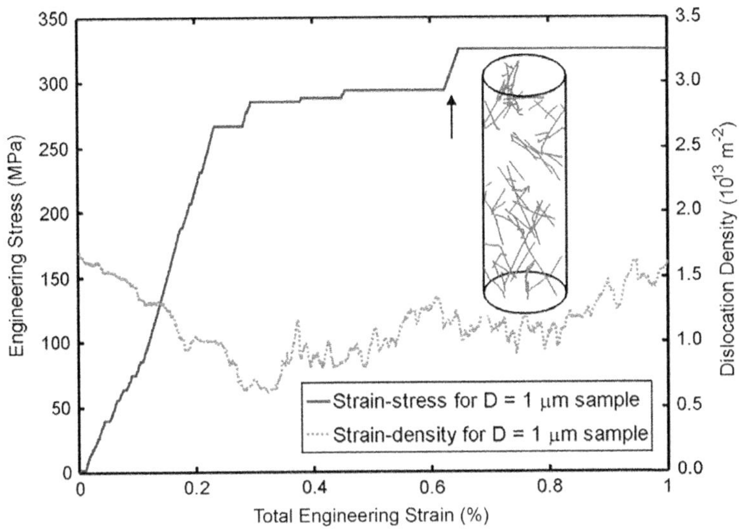

FIG. 4.11 Variation of stress and dislocation density versus the strain for the pillar with diameter of $D = 1$ μm. *(After Zhou, C., Beyerlein, I.J., LeSar, R., 2011. Plastic deformation mechanisms of fcc single crystals at small scales. Acta Mater. 59, 7673–7682.)*

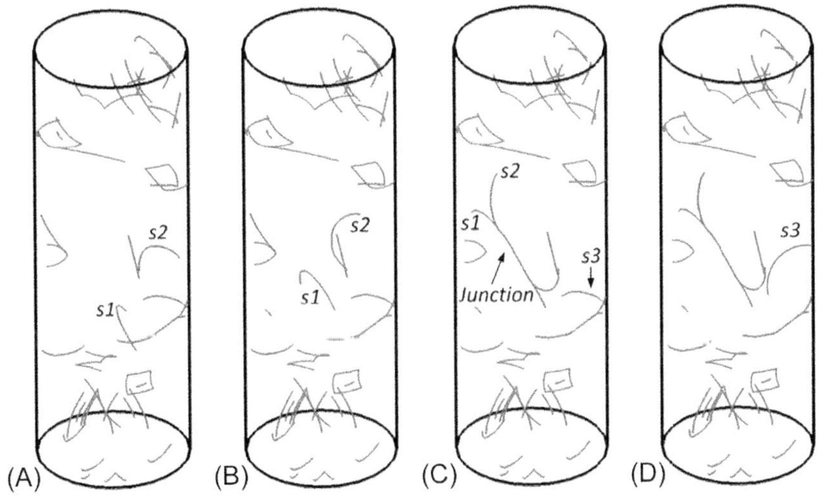

FIG. 4.12 The dislocation structure before and after exhaustion hardening for the pillar with diameter of $D = 1$ μm. *(After Zhou, C., Beyerlein, I.J., LeSar, R., 2011. Plastic deformation mechanisms of fcc single crystals at small scales. Acta Mater. 59, 7673–7682.)*

of confined volumes. Based on the obtained results, Zhou et al. (2011) presented a physically based plastic deformation map for samples of confined volumes, which includes three sample size zones (Fig. 4.13). They stated that the SAD model is applicable to all three regions to capture the incipient plasticity. After

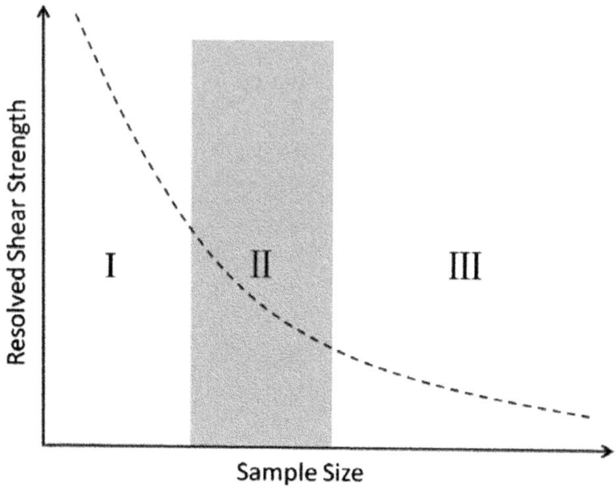

FIG. 4.13 Three zones of sample size. In Zone I, the surface dislocations nucleation is the governing deformation mechanism. In Zone II, both surface dislocations nucleation and multiplication of internal dislocations should be considered. In Zone III, the deformation is governed by multiplication of internal dislocations. *(After Zhou, C., Beyerlein, I.J., LeSar, R., 2011. Plastic deformation mechanisms of fcc single crystals at small scales. Acta Mater. 59, 7673–7682.)*

the yield onset, the plasticity is governed by nucleation of surface dislocations and/or internal source multiplication. In Zone I, i.e., samples with very small length scales, the surface dislocations nucleation is the deformation mechanism, and dislocation starvation governs the hardening mechanism. In Zone III, i.e., large samples, the multiplication of internal dislocations governs the deformation mechanism. The exhaustion hardening caused by dislocation reactions can explain the observed intermittent plasticity. In Zone II, i.e., the transition zone, both surface dislocation nucleation and internal dislocations multiplication contribute to the plasticity. One should note that in this chapter, the dislocation starvation is considered as a special case of source exhaustion.

4.2.3 Weakest link theory

Another mechanism of size effects in pillars of confined volumes is commonly referred as weakest link theory. This mechanism attributes the size effects in pillars during the compression to the strength of the weakest slip plane present in the sample. It states that the strength of this weakest link increases as the pillar size decreases. Accordingly, the strength of the sample increases as the pillar diameter decreases. In this section, first, the work conducted by Norfleet et al. (2008) is elaborated in which the weakest link theory is introduced. Next, a set of DDD simulation performed by El-Awady et al. (2009) to capture the weakest link mechanism is presented.

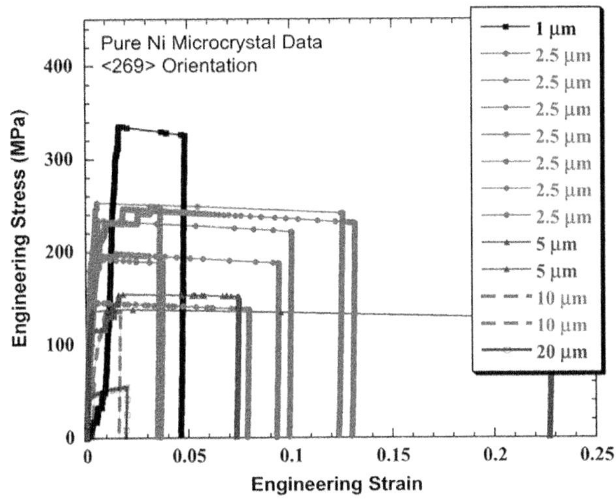

FIG. 4.14 Variation of stress versus strain for pure Ni ⟨2 6 9⟩-oriented pillars with different values of diameters. *(After Norfleet, D.M., Dimiduk, D.M., Polasik, S.J., Uchic, M.D., Mills, M.J., 2008. Dislocation structures and their relationship to strength in deformed nickel microcrystals. Acta Mater. 56, 2988–3001.)*

Norfleet et al. (2008) introduced the weakest link theory to capture the difference between the observed experimental responses with those predicted by their theoretical model. They studied size effects in Ni pillars with the diameters varying from 1 to 20 μm during compression test (Fig. 4.14). In order to unravel the underlying mechanisms of size effects, they also measured the dislocation density in the undeformed and deformed bulk samples and micropillars. First, they tried to capture the observed size effects using the following relation:

$$\tau = \tau_0 + k_s G \frac{\ln(\bar{\lambda}/b)}{(\bar{\lambda}/b)} + k_f G b \ \ln\left(\frac{1}{b\sqrt{\rho_f/2}}\right)\sqrt{\rho_f/2} \tag{4.13}$$

where τ is the flow shear stress, τ_0 is the lattice-friction stress, k_s is the source hardening constant, $\bar{\lambda}$ is the average dislocation source length, k_f denotes the average strength of the forest dislocations, and ρ_f is the scalar density of the forest dislocations, where $\rho_f = R\rho$. The value of R depend on the sample geometry and strain value. The accurate evaluation of R for each strain value and sample geometry is difficult. Accordingly, two values of $R = 1/2$ and $R = 3/4$ which are related to the strength lower and upper bounds, respectively, were considered. Eq. (4.13) include both source truncation and forest hardening mechanisms. Fig. 4.15 compares the experimental results reported by Norfleet et al. (2008) compared to their theoretical model described in Eq. (4.13). In the case of pillars with diameters larger than ~5 μm, Eq. (4.13) can predict the sample response. However, for samples with smaller diameters, the experiment shows higher strength compared to Eq. (4.13).

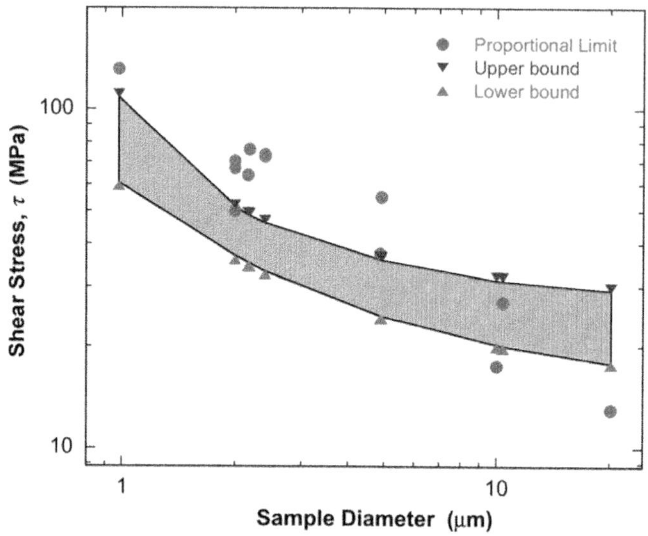

FIG. 4.15 Variation of proportional limit versus pillar diameter for experiment and lower and upper bounds of Eq. (4.13). *(After Norfleet, D.M., Dimiduk, D.M., Polasik, S.J., Uchic, M.D., Mills, M.J., 2008. Dislocation structures and their relationship to strength in deformed nickel microcrystals. Acta Mater. 56, 2988–3001.)*

In order to explain the difference between the experimental observation and the theoretical model, Norfleet et al. (2008) modified the forest hardening mechanism to include the dislocation density correction measured for smaller samples. The results obtained from this modified method are compared with the experimental results in Fig. 4.16. The modified model still underestimates the strength of the smaller samples. Fig. 4.16 shows that there should be another hardening mechanism which is not considered in the presented model.

The differences between the model and experimental results was attributed to the weakest link theory. The weakest link theory states that there is mean-field limit ξ^* for any crystals. In the cases of samples larger than ξ^*, the dislocation ensemble behave according to the conventional forest hardening mechanism. In the cases of samples smaller than ξ^*, however, the mean-field forest hardening mechanism breaks down due to the new altered distribution of dislocations. In the samples smaller than ξ^*, the longest dislocation sources, which are the weakest ones, are removed from the samples. Accordingly, the modified distribution of dislocation source length leads to a stronger forest-hardening contribution.

The second work which is presented here was conducted by El-Awady et al. (2009). They performed a set of 3D DDD to capture the size effects in Ni pillars during compression test. The pillar diameters vary from $D = 0.25 - 5$ μm. The initial dislocation densities vary from $\rho_0 = 1 \times 10^{12}$ m^{-2} to 50×10^{12} m^{-2}. They assumed the initial dislocation network as a randomly distributed

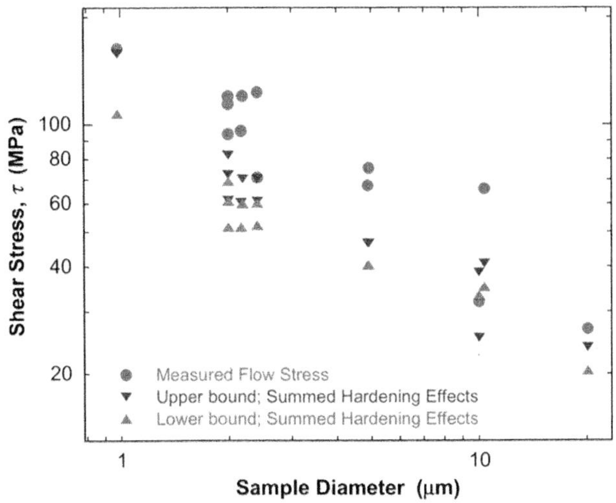

FIG. 4.16 Variation of proportional limit versus pillar diameter for experiment and lower and upper bounds of Eq. (4.13) corrected by increasing the dislocation density at smaller samples. *(After Norfleet, D.M., Dimiduk, D.M., Polasik, S.J., Uchic, M.D., Mills, M.J., 2008. Dislocation structures and their relationship to strength in deformed nickel microcrystals. Acta Mater. 56, 2988–3001.)*

Frank-Read sources and single-ended sources with different slip planes. They incorporated a two-parameter Weibull distribution to define the initial length of dislocation sources as follows:

$$\mathrm{f}(\lambda) = \frac{\beta}{\theta}\left(\frac{\lambda}{\theta}\right)^{\beta-1} e^{-(\lambda/\theta)} \text{ for } \lambda \geq 0 \tag{4.14}$$

where $\beta > 0$ defines the shape of the distribution and $0 < \theta < D$ is the mean of the distribution length. First, El-Awady et al. (2009) studied the effect of the incorporated length distribution for an initial dislocation density of $\rho_0 = 3 \times 10^{12}\ \mathrm{m}^{-2}$. Different pillar diameters of $D = 0.25$, 0.5, 0.75, and 1 μm are selected. Fig. 4.17 shows the response of each pillar using different distribution parameters. The results show that the distribution parameters have a clear effect on the sample response. El-Awady et al. (2009) stated that the mean of the distribution length θ has a key role in material response in a way that $\theta \sim D/2$ leads to a lower bound of material strength. Accordingly, as θ decreases, the pillar strength increases.

In the next step, El-Awady et al. (2009) developed a set of scaling laws. To do so, they conducted a statistical investigation on samples with different radii of $D = 0.25 - 5$ μm, initial dislocation densities of $\rho_0 = 1 \times 10^{12}\ \mathrm{m}^{-2} - 50 \times 10^{12}\ \mathrm{m}^{-2}$, and different Weibull distribution parameters. Fig. 4.18 shows the variation of statistical average of flow strength at 0.5% strain, i.e., σ_f, versus the pillar diameter D, square root of the dislocation density at incipient plasticity $\sqrt{\rho}$, multiplication of pillar diameter and average activated source length $D\langle\lambda\rangle$,

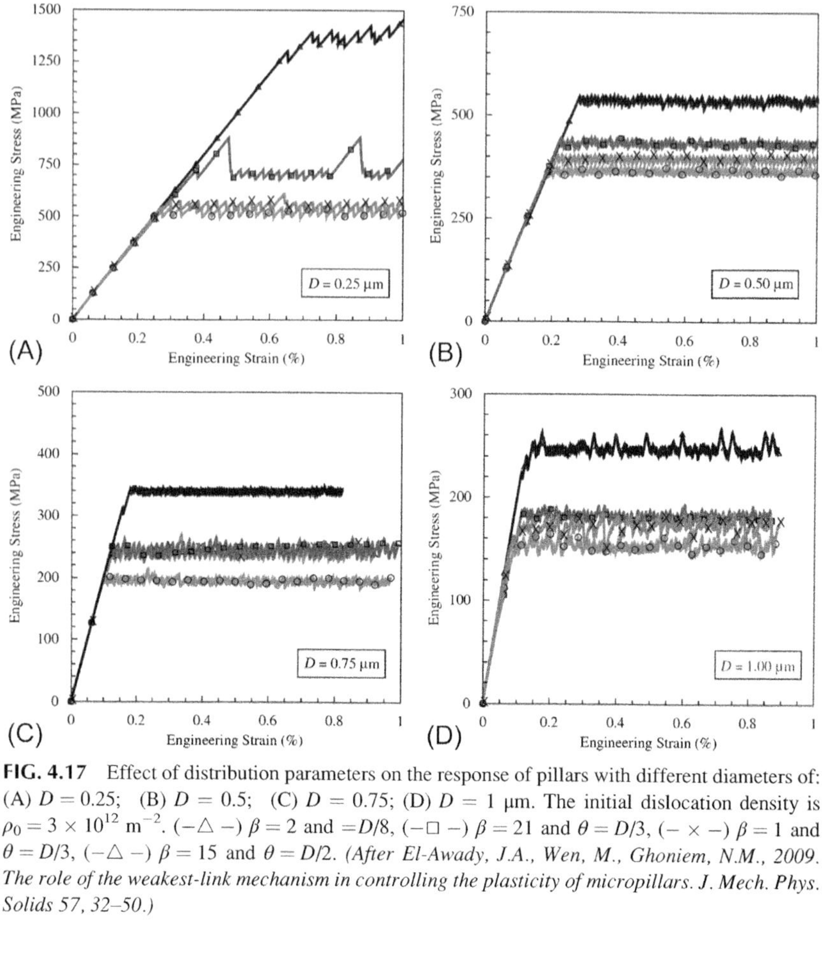

FIG. 4.17 Effect of distribution parameters on the response of pillars with different diameters of: (A) $D=0.25$; (B) $D=0.5$; (C) $D=0.75$; (D) $D=1$ μm. The initial dislocation density is $\rho_0=3\times10^{12}$ m^{-2}. (−△−) $\beta=2$ and $=D/8$, (−□−) $\beta=21$ and $\theta=D/3$, (− × −) $\beta=1$ and $\theta=D/3$, (−△−) $\beta=15$ and $\theta=D/2$. *(After El-Awady, J.A., Wen, M., Ghoniem, N.M., 2009. The role of the weakest-link mechanism in controlling the plasticity of micropillars. J. Mech. Phys. Solids 57, 32–50.)*

and average activated source length $\langle\lambda\rangle$. In the case of variation of σ_f versus D which is shown in Fig. 4.18A, the best fit with the correlation factor of $r=0.84$ can be described as follows (El-Awady et al., 2009):

$$\sigma_f = 222.65D^{-0.69} \tag{4.15}$$

where σ_f and D have the units of MPa and μm, respectively. The correlation factor of $r=0.84$ shows that in addition to pillar diameter, there should be another important factor which governs the size effects. In the case of variation of σ_f versus $\sqrt{\rho}$ which is shown in Fig. 4.18B, no reasonable correlation can be defined. El-Awady et al. (2009) stated that based on the obtained results, one can say that the Taylor hardening model cannot capture the size effects in the samples investigated in their work. In the case of variation of σ_f versus $D\langle\lambda\rangle$

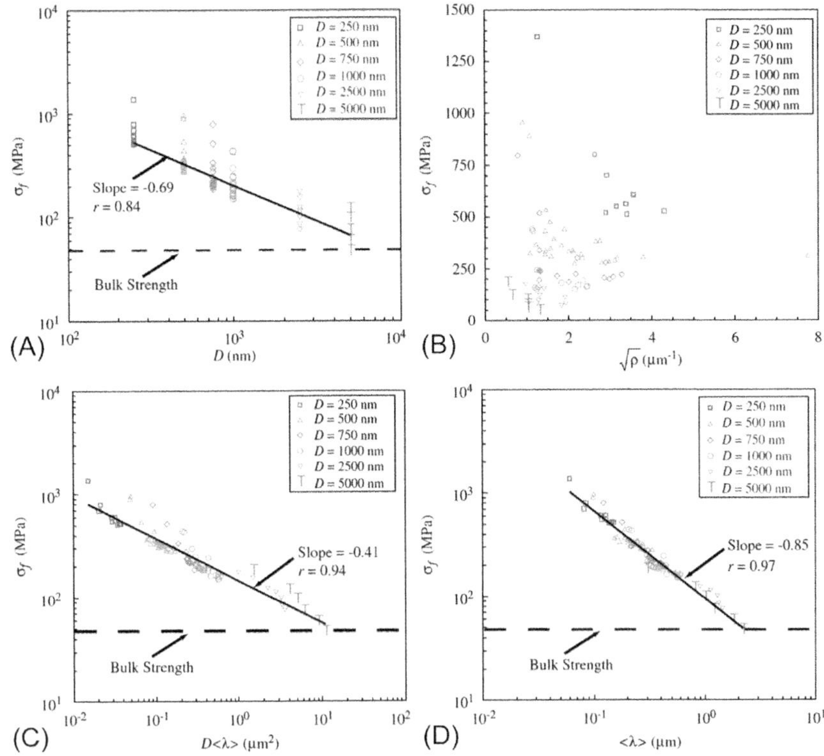

FIG. 4.18 The variation of statistical average of flow strength at 0.5% strain, i.e., σ_f, versus: (A) pillar diameter D; (B) square root of the dislocation density at incipient plasticity $\sqrt{\rho}$; (C) multiplication of pillar diameter and average activated source length $D\langle\lambda\rangle$; (D) average activated source length $\langle\lambda\rangle$. *(After El-Awady, J.A., Wen, M., Ghoniem, N.M., 2009. The role of the weakest-link mechanism in controlling the plasticity of micropillars. J. Mech. Phys. Solids 57, 32–50.)*

which is shown in Fig. 4.18C, the best fit with the correlation factor of $r = 0.94$ can be described as follows:

$$\sigma_f = 146.23(D\langle\lambda\rangle)^{-0.41} \tag{4.16}$$

El-Awady et al. (2009) showed that one can enhance the correlation by excluding the pillar diameter. Accordingly, in the case of variation of σ_f versus $\langle\lambda\rangle$ which is shown in Fig. 4.18D, the best fit with the correlation factor of $r = 0.97$ can be described as follows:

$$\sigma_f = 94.93(\langle\lambda\rangle)^{-0.85} \tag{4.17}$$

Accordingly, the variation of σ_f versus $\langle\lambda\rangle$ has the smallest scatter among all investigated factors.

El-Awady et al. (2009) analyzed the experimental data reported by Dimiduk et al. (2005) and Frick et al. (2008) and extracted the Weibull distribution parameters of $\beta = 21$ and $\theta = D/25$ for the dislocation network. The flow strength obtained from these parameters are located in the middle of experimental values. Also, the value of $\theta = D/25$ shows that the dislocation sources with the length comparable to the radius of pillar are rare.

El-Awady et al. (2009) attributed the observed size effects to the weakest link theory. To elaborate their theory, they incorporated the probability distribution function (PDF) of pinned dislocation length reported by Mughrabi (1976) in the case of bulk FCC Cu, as shown in Fig. 4.19. One can incorporate the same PDF for dislocation network of a pillar. El-Awady et al. (2009) stated that for pillars with diameters of $D > 2$ μm, the source length distribution is similar to the bulk material one, which is described using a solid blue line. For a pillar with diameters of $D < 2$ μm, the distribution should be modified. For example, in the case of a pillar with the diameter of $D = 1$ μm of the dislocation networks for single-ended and double-ended dislocation sources can be renormalized according to the green and red dashed lines, respectively.

The first type of sources which become activated are surface-pinned ones since they are not pinned to any interior points. They provide the plastic flow by leaving the sample. Next, the stress increases until the double-ended sources with the length larger than the pillar radius becomes activated. They glide until they transform to single-ended sources with the length smaller than the pillar radius. Then the stress increases elastically until this single-ended sources with the length smaller than the pillar radius becomes activated. The sample with smaller diameter, the source length distribution of the bulk material should

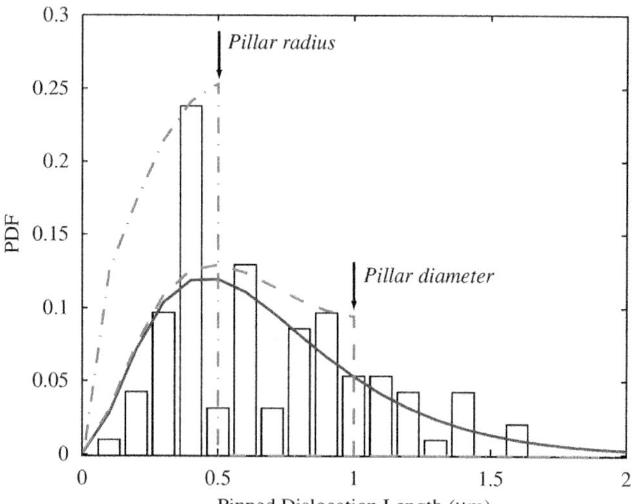

FIG. 4.19 Probability distribution function of dislocation link-length. *(After El-Awady, J.A., Wen, M., Ghoniem, N.M., 2009. The role of the weakest-link mechanism in controlling the plasticity of micropillars. J. Mech. Phys. Solids 57, 32–50.)*

be modified to incorporate the effect of free surfaces in the pillars. Accordingly, as the pillar diameter decreases, the length of the dislocation with largest length, which is the weakest link, decreases. Accordingly, the material strength increases. Within the framework of weakest link theory, El-Awady et al. (2009) studied three factors to study the size effects which are the cross-slip activation, effect of crystal rotation, and change in the crystal geometry. One can read their work to understand different aspects of these factors.

4.3 DDD simulation of size effects during microbending and nanoindentation experiments

4.3.1 Microbending experiment

Microbending is an important experiment to study the size effects in metallic samples of confined volumes. Stolken and Evans (1998) conducted the microbending experiment of nickel foils for different thicknesses of $t = 12.5 - 50\ \mu m$. They showed that as the sample length scale decreases, the sample strength increases (Fig. 4.20). Haque and Saif (2003) incorporated the MEMS-based testing techniques to study the size effects of 100, 150, 200, and 485 nm thick freestanding Aluminum samples during microbending (Fig. 4.21). They observed the same size effects and the sample strength increases by decreasing its thickness. They incorporated the mechanism-based strain gradient (MSG) plasticity model to capture the observed size effects. Fig. 4.22 compares the size effects observed during the experiment versus the MSG model predictions. The results show that the MSG model can capture the experimental size effects in which the smaller sample is stronger. Uchic et al. (2003, 2004) and Uchic and Dimiduk (2005) have developed the focused ion beam (FIB) machining technique to fabricate nano to micron sized metallic pillars. Motz et al. (2005) was the first group that used the FIB technique to fabricate the metallic samples of confined volumes for microbending experiment. Copper single crystal samples with thicknesses of $t = 1.0 - 7.5\ \mu m$ with an $\{111\}\langle 011\rangle$ orientation were fabricated using the FIB technique (Fig. 4.23). They used a nanoindenter to apply the bending load as shown in Fig. 4.23B. Motz et al. (2005) studied the dependency of sample strength on its thickness. Fig. 4.24 shows the variation of flow stress as the sample thickness varies. The obtained results show a slight increase of flow stress as the sample size decreases for beams with thicknesses larger than $t = 5.0\ \mu m$. In the cases of smaller samples, however, the increase in flow stress is dramatic as the sample thickness decreases. Ultimately, the strength reaches almost 1 GPa for the smallest sample. A line is also fit to the obtained results which shows almost an inverse relation between the flow stress and sample thickness, i.e., flow stress $\propto \sim t^{-1}$. Motz et al. (2008) again addressed the size effects of copper single crystal samples of confined volumes during microbending experiment and observed the scaling behavior of *flow stress* $\propto t^{-1}$. Later on, Kiener et al.

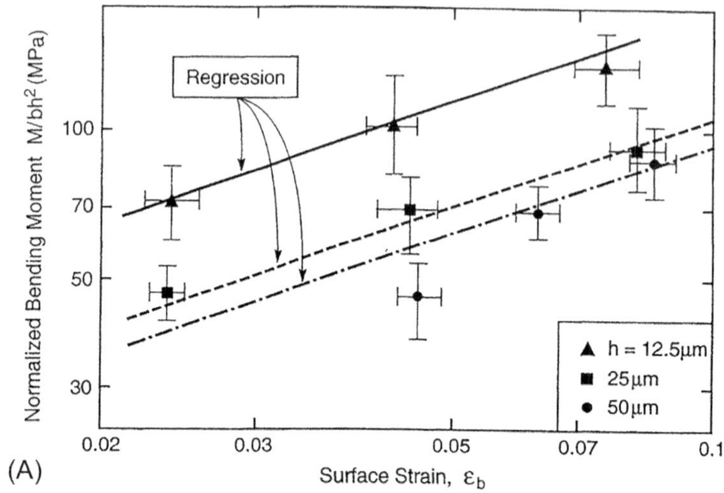

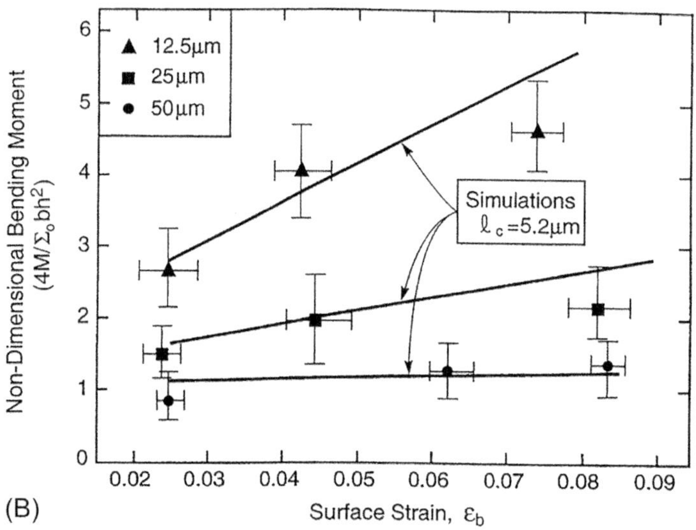

FIG. 4.20 Variation of bending moment versus the surface strain for three foil thicknesses of $t = 12.5$, 25, and 50 μm: (A) The normalized moment M/bh^2; (B) The non-dimensional moment $M/\Sigma_0 bh^2$. *(After Stolken, J.S., Evans, A.G., 1998. A microbend test method for measuring the plasticity length scale. Acta Mater. 46, 5109–5115.)*

(2010) used the same technique to fabricate single crystal copper samples of confined volume to study the size effects during cyclic microbending experiment.

Discrete dislocation dynamics (DDD) simulation is a powerful method to address the size effect during the microbending experiment. It directly includes dislocations as discrete line defects. Accordingly, one can unravel the

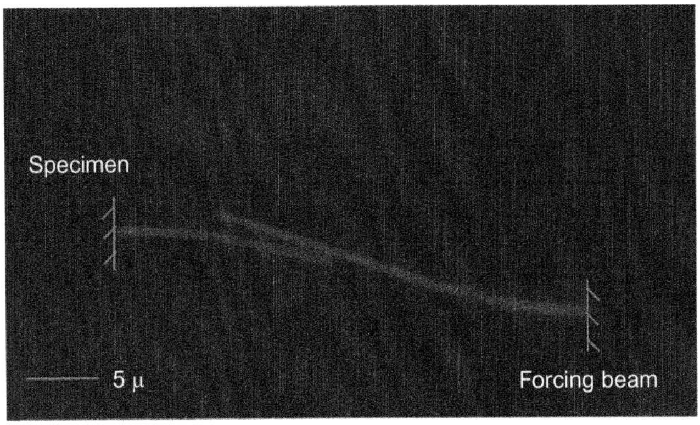

FIG. 4.21 Optical micrograph of microbending experiment of a 200 nm thick Aluminum specimen. The load is applied using a forcing beam, which is made of thermally grown silicon oxide. *(After Haque, M.A., Saif, M.T.A., 2003. Strain gradient effect in nanoscale thin films. Acta Mater. 51, 3053–3061.)*

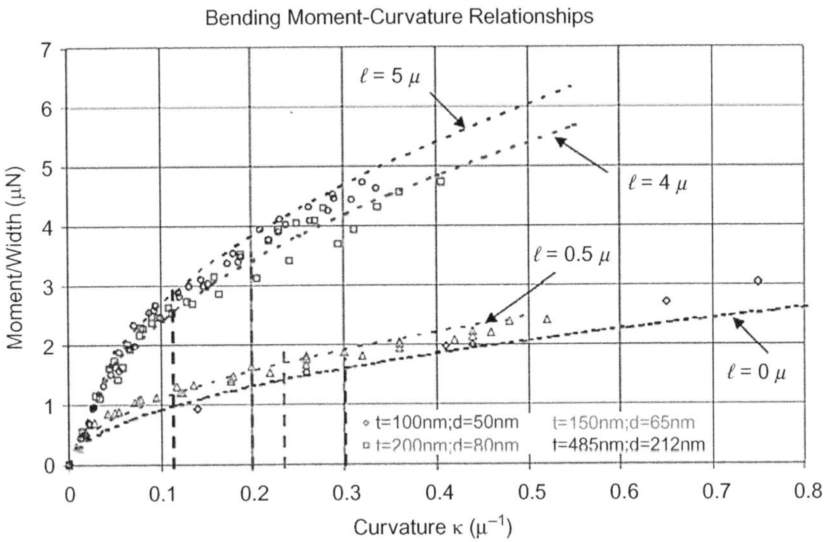

FIG. 4.22 Size effects during microbending experiment for 100, 150, 200, and 485 nm thick free-standing Aluminum samples: comparing the experimental results *(data points)* versus the MSG *(dashed lines)*. *Vertical dashed vertical lines* are the tensile fracture strains for samples with the thicknesses of 485, 200, 150, and 100 nm thick specimens. *(After Haque, M.A., Saif, M.T.A., 2003. Strain gradient effect in nanoscale thin films. Acta Mater. 51, 3053–3061.)*

underlying mechanisms of size effects using DDD simulation. Similar to any other DDD simulation, the study of microbending test has started using 2D DDD simulations (Yefimov et al., 2004; Hou et al., 2008; Fan et al., 2013). Fig. 4.25 shows the dislocation network obtained using a 2D DDD simulation

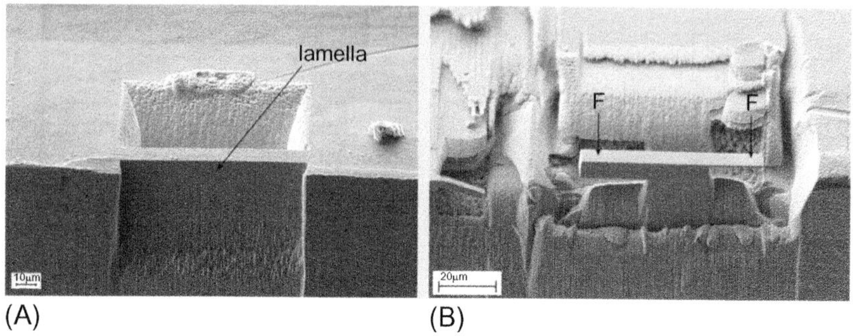

FIG. 4.23 Using the FIB-technique to fabricate single crystal sample of confined volumes for microbending experiment: (A) preparing a thick lamella close to the sample's edge and (B) the final sample is cutting out from the thick lamella. *(After Motz, C., Schoberl, T., Pippan, R., 2005. Mechanical properties of micro-sized copper bending beams machined by the focused ion beam technique. Acta Mater. 53, 4269–4279.)*

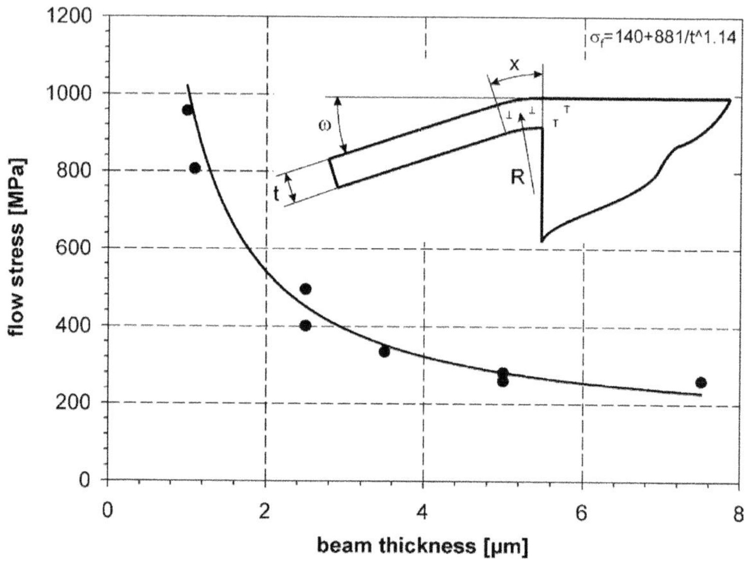

FIG. 4.24 Size effects during microbending experiment: variation of flow stress as the sample thickness varies. The *fitted line* shows almost an inverse relation between the flow stress and sample thickness. *(After Motz, C., Schoberl, T., Pippan, R., 2005. Mechanical properties of micro-sized copper bending beams machined by the focused ion beam technique. Acta Mater. 53, 4269–4279.)*

of microbending experiment compared to the total dislocation density ρ obtained using the nonlocal continuum crystal plasticity theory (Yefimov et al., 2004). Although simplified 2D DDD simulation methods can provide valuable insight into many aspects of size effects during microbending experiment, a true DDD simulation should be used to handle 3D dislocation mechanisms. Several 3D DDD simulations have addressed the size effects during

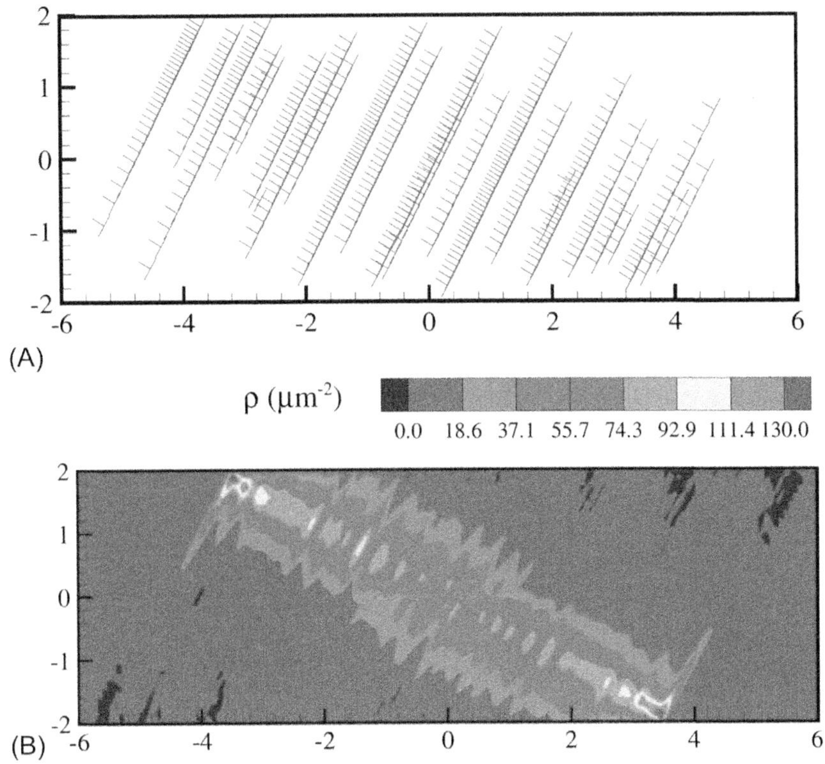

FIG. 4.25 The simulation results of microbending experiment at the applied rotation angle $\theta = 0.015$: (A) distribution of dislocations obtained using the 2D DDD simulation and (B) total dislocation density ρ obtained using the nonlocal continuum crystal plasticity theory. *(After Yefimov, S., van der Giessen, E., Groma, I., 2004. Bending of a single crystal: discrete dislocation and nonlocal crystal plasticity simulations. Model. Simul. Mater. Sci. Eng. 12, 1069.)*

microbending test (Akasheh et al., 2007; Motz et al., 2008; Kiener et al., 2010; Motz and Dunstan, 2012; Aifantis et al., 2012; Crone et al., 2014). In this section, the work conducted by Motz et al. (2008) will be elaborated.

Motz et al. (2008) addressed the size effects of copper single crystal samples of confined volumes during microbending test using both experiment and 3D DDD simulation. In the case of microbending experiment, single crystal copper samples with various thicknesses of $t = 1.0 - 7.5$ μm have been tested (Fig. 4.26). They reported the variation of flow stress at a bending angle of ~20° versus the beam thickness to show the size effects during microbending experiment (Fig. 4.27). They fit two curves to the obtained results, one without offset and one with offset, which leads to the exponents of −0.75 and −1.1, respectively. In the next step, they incorporated the 3D DDD simulation to capture the size effects during microbending experiment. Four different beam thicknesses of $t = 0.50$, 0.75, 1.0, and 1.5 μm are simulated. Furthermore, the simulation incorporates the cross-slip

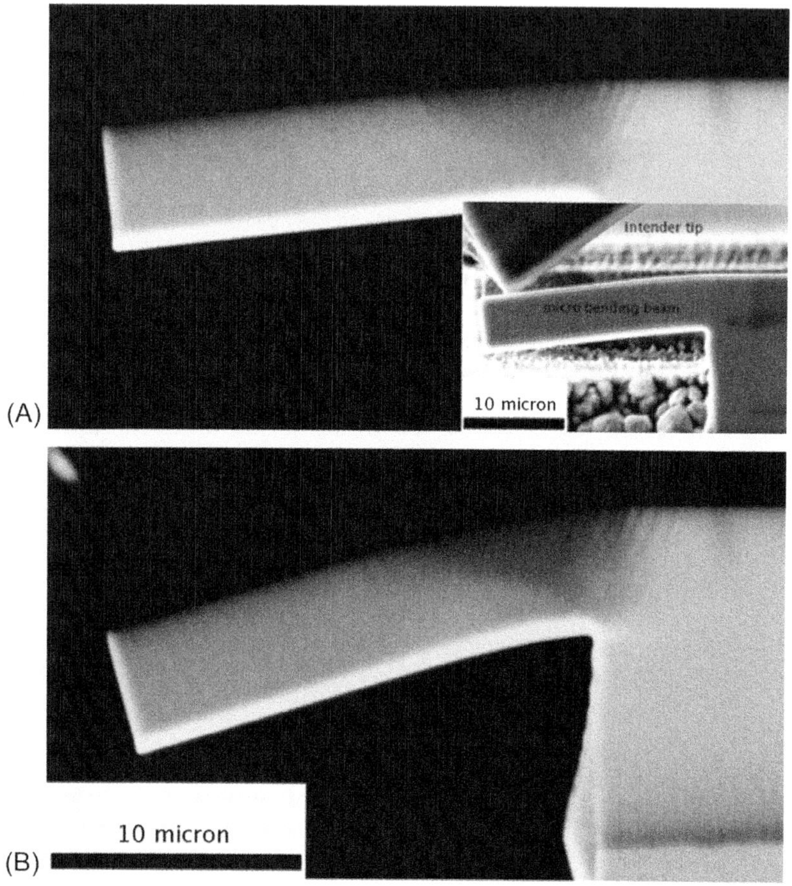

FIG. 4.26 SEM micrographs of a bending beam at a bending angle of: (A) 4° and (B) 16.7°. *(After Motz, C., Weygand, D., Senger, J., Gumbsch, P., 2008. Micro-bending tests: a comparison between three-dimensional discrete dislocation dynamics simulations and experiments. Acta Mater. 56 (9), 1942–1955.)*

mechanism. The initial dislocation density of $\rho_0 = 2.0 \times 10^{13}\ \mathrm{m}^{-2}$ is used which includes the Frank-Read sources with the length of $500a$. Motz et al. (2008) used the finite element method (FEM) discretization of $8 \times 8 \times 24$ brick elements with 20-node. Two different sets of boundary conditions are used to model the micro-bending experiment as follows:

- *Boundary condition (i):* A cantilever with one end fixed and the other end loaded. This boundary conditions lead to a linear bending moment. This set of boundary conditions mimics the incorporated experimental set-up.
- *Boundary condition (ii)*: Applying opposite bending moments on two sides of a free beam. This set of boundary conditions leads to a constant bending moment.

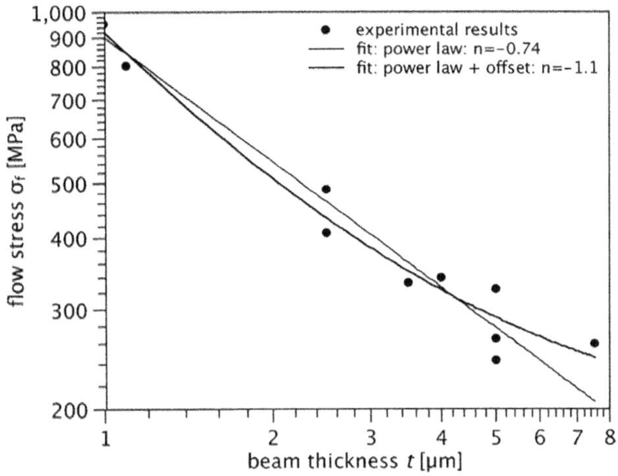

FIG. 4.27 Size effects during microbending experiment: variation of flow stress at a bending angle of ~20° versus the sample thickness. *(After Motz, C., Weygand, D., Senger, J., Gumbsch, P., 2008. Micro-bending tests: a comparison between three-dimensional discrete dislocation dynamics simulations and experiments. Acta Mater. 56 (9), 1942–1955.)*

Fig. 4.28A depicts the variation of normalized bending moment versus the normalized displacement for different beam sizes by selecting the second set of boundary conditions which leads to a constant bending moment. The observed size effects are in line with those of the experiments, and a smaller sample is stronger. Motz et al. (2008) investigated the different aspects of DDD simulation in the case of the sample with the thickness of $t = 0.50$ μm including the effect of cross-slip, FEM discretization, and using a random initial dislocation structure, as shown in Fig. 4.28B. However, Motz et al. (2008) observed no major change in the bending response of the sample due to these changes.

Fig. 4.29 shows the variation of dislocation density ρ versus both the normalized total and plastic displacements for different beam thicknesses of $t = 0.50, 0.75, 1.0$, and 1.5 μm. The dislocations density increases from the initial value of $\rho_0 = 2.0 \times 10^{13}\ \mathrm{m}^{-2}$ to the final value of $\sim 1.0 \times 10^{14}\ \mathrm{m}^{-2}$ as the bending increases. As the total displacement increases, the dislocation density increases for all samples. Variation of dislocation density versus plastic displacement, however, gives a better insight for different sample sizes in a way that the slope of the change in dislocation density versus the change in plastic displacement is proportional to the inverse of the beam thickness. As described by Motz et al. (2008), this is in line with the results of the strain gradient plasticity model where the density of the geometrically necessary dislocations (GNDs) should be proportional to the inverse of the beam thickness, i.e., $\rho \propto t^{-1}$.

In the last step, Motz et al. (2008) investigated the size effects during microbending test for different thicknesses of $t = 0.50, 0.75, 1.0$, and 1.5 μm

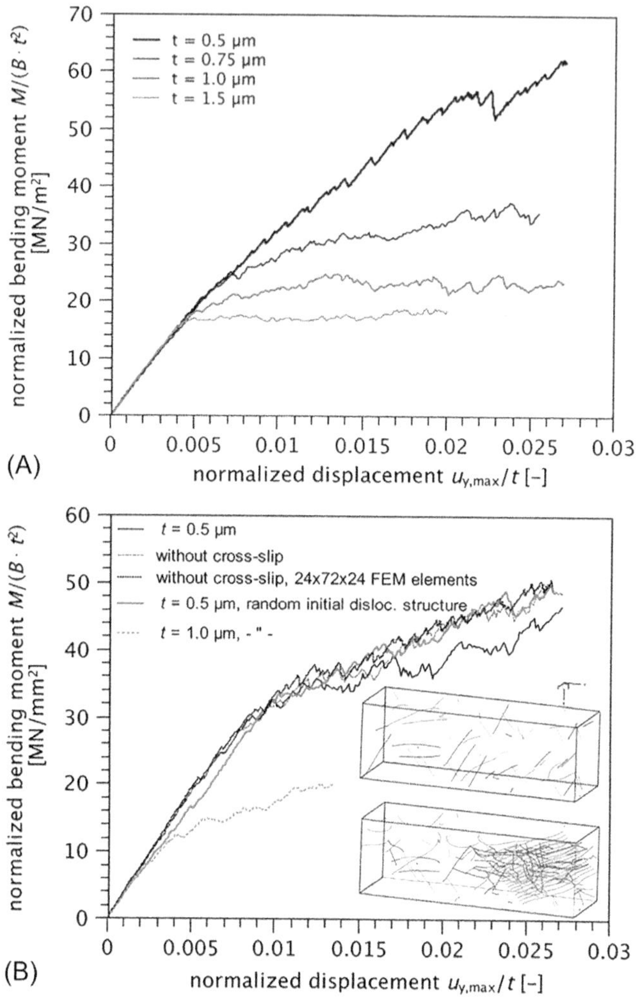

FIG. 4.28 3D DDD simulation of size effects during microbending experiment: (A) variation of normalized bending moment versus the normalized displacement for samples with different thicknesses of $t = 0.50, 0.75, 1.0$, and 1.5 μm and (B) effects of cross-slip, FEM discretization, and initial dislocation structure on the microbending response of the sample with the thickness of $t = 0.50$ μm. The initial and final dislocation structures of sample with initial random structure and thickness of $t = 0.50$ μm are also depicted. *(After Motz, C., Weygand, D., Senger, J., Gumbsch, P., 2008. Micro-bending tests: a comparison between three-dimensional discrete dislocation dynamics simulations and experiments. Acta Mater. 56 (9), 1942–1955.)*

at plastic displacement equal to 0.01 of the sample thickness by selecting both *Boundary condition (i)* and *Boundary condition (ii)* (Fig. 4.30). The scaling exponents of −1.1 and −1.2 are obtained for *Boundary conditions (i)* and *Boundary conditions (ii)*, respectively, which are close to the experimental

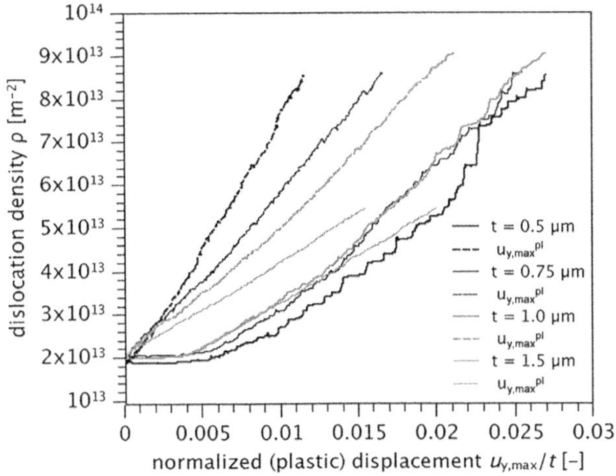

FIG. 4.29 Variation of dislocation density ρ versus both the normalized total *(solid lines)* and plastic *(dashed lines)* displacements for different beam thicknesses of $t = 0.50$, 0.75, 1.0, and 1.5 μm. *(After Motz, C., Weygand, D., Senger, J., Gumbsch, P., 2008. Micro-bending tests: a comparison between three-dimensional discrete dislocation dynamics simulations and experiments. Acta Mater. 56 (9), 1942–1955.)*

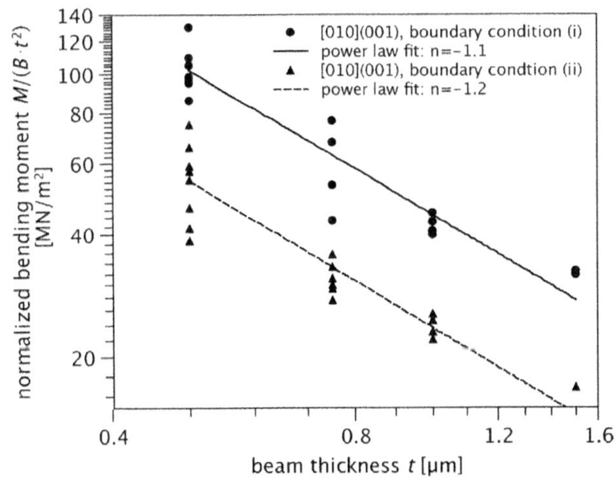

FIG. 4.30 Size effects during microbending experiment: variation of normalized bending moment at plastic displacement equal to 0.01 of the sample thickness versus the sample thickness for the [010](001) beam orientation by selecting both *Boundary condition (i)* and *Boundary condition (ii)*. *(After Motz, C., Weygand, D., Senger, J., Gumbsch, P., 2008. Micro-bending tests: a comparison between three-dimensional discrete dislocation dynamics simulations and experiments. Acta Mater. 56 (9), 1942–1955.)*

scaling exponents of −0.75 and −1.1. However, one should note that a direct comparison between the DDD simulation and experiment is not possible due to different samples size ranges. Furthermore, the bending angle at which the experimental scaling law is obtained is different from that of the simulation.

4.3.2 Nanoindentation experiment

Nanoindentation is one of the most popular experiments to study the behavior of materials at the micro and nano scales. During nanoindentation, a small hard indenter is pressed onto a sample, and the load versus the penetration depth is recorded. Traditionally, a hardness number was reported as a result of indentation test. However, the hardness value varies during the nanoindentation test, which is commonly called as size effects. For example, In the case of single crystal metal indented by conical or Berkovich tips, the hardness decreases as the indentation depth increases (Fig. 1.53). Size effects during nanoindentation is briefly addressed in Section 1.3.2.3, and it is elaborately presented in Chapter 5. Here, the focus is on the discrete dislocation dynamics (DDD) simulation of nanoindentation.

Similar to microbending experiment, the DDD studies of indentation experiment have started using 2D simulations (Widjaja et al., 2005, 2007a,b; Balint et al., 2006; Kreuzer and Pippan, 2004, 2007; Ouyang et al., 2008, 2010a,b). Fig. 4.31 shows the dislocation network obtained using a 2D DDD simulation of indentation experiment (Ouyang et al., 2010b). 2D DDD simulation addressed various aspects of indentation experiments such as indenter size,

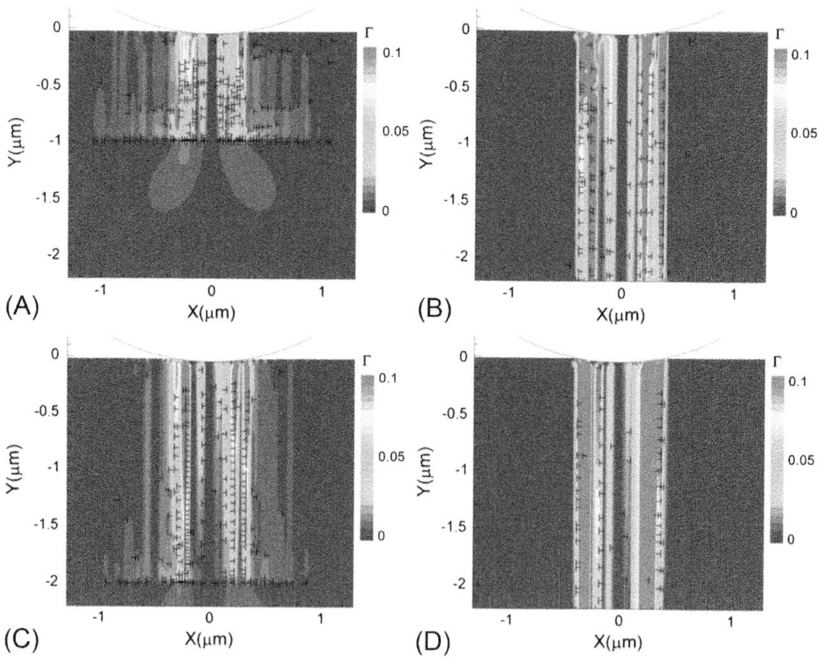

FIG. 4.31 2D DDD simulation of indentation: The distributions of dislocation and slip quantity Γ at indentation depth of $h = 0.05$ μm in thin films with different thicknesses of: (A) 1 μm, (B) 2 μm, (C) 4 μm, and (D) 8 μm. *(After Ouyang, C., Li, Z., Huang, M., Fan, H., 2010b. Cylindrical nano-indentation on metal film/elastic substrate system with discrete dislocation plasticity analysis: a simple model for nano-indentation size effect. Int. J. Solids Struct. 47, 3103–3114.)*

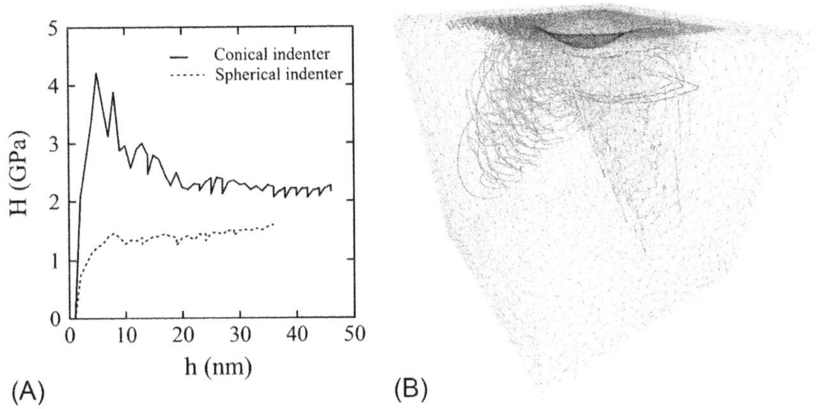

FIG. 4.32 The 3D DDD simulation of nanoindentation experiment on copper single crystal samples: (A) size effects during nanoindentation using both conical and spherical indenters and (B) dislocation microstructure during nanoindentation using a spherical indenter, where the mesh displacement is magnified. *(After Po, G., Mohamed, M.S., Crosby, T., Erel, C., El-Azab, A., Ghoniem, N., 2014. Recent progress in discrete dislocation dynamics and its applications to micro plasticity. J. Mater. 66, 2108–2120.)*

grain size, depth of indentation, and the effect of grain boundary. The 3D DDD simulation is also incorporated to investigate the nanoindentation experiment (Fivel et al., 1997, 1998; Po et al., 2014; Gagel et al., 2016; Crone et al., 2018). Fig. 4.32 shows the size effects during nanoindentation of copper single crystal samples using both conical and spherical indenters along with the dislocation microstructure obtained using 3D DDD simulation (Po et al., 2014). Bertin et al. (2018) developed a spectral scheme to investigate the nanoindentation experiment using the 3D DDD simulation. Recently, Lu et al. (2019) incorporated 3D DDD to investigate the effect of grain boundary on the indentation response of an aluminum bicrystal sample. In this section, the work conducted by Fivel et al. (1998) will be elaborated.

Fivel et al. (1998) combined the 3D discrete dislocation simulation and the FEM to investigate the nanoindentation test. The simulations were validated by direct comparison with experiment. Accordingly, Fivel et al. (1998) performed a set of 3D DDD simulation of nanoindentation test on a single crystalline copper cube with the length of 2 μm indented by a spherical indenter with the radius of $R = 420$ nm. Fig. 4.33 compares the experimental response, pure elastic finite element (FE) solution, and the discrete dislocation simulation. Fivel et al. (1998) reported the dislocation microstructure at the indentation depth of 50 nm obtained using the simulation as shown in Fig. 4.34. In the final step, Fivel et al. (1998) presented the dislocation microstructure obtained from simulation at the indentation depth of 50 nm according to several diffracting vectors as shown in Fig. 4.35. These results can be directly be compared to the experimental results obtained by Fivel et al. (1998) using the transmission electron microscopy (TEM) as shown in Fig. 4.36. Considering the plastic zone as a hemisphere, Fivel et al. (1998) reported its radius as 800 nm at the indentation

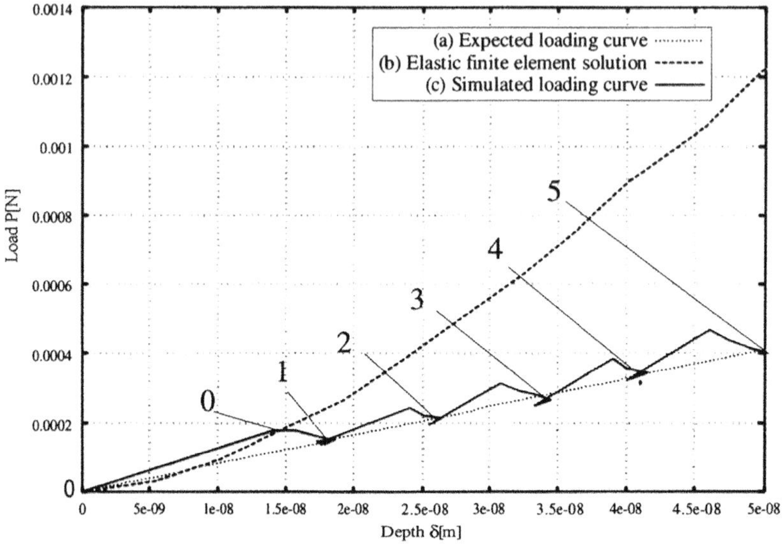

FIG. 4.33 Simulation of the nanoindentation experiment of a single crystal copper sample: comparison of the experimental response, pure elastic finite element (FE) solution, and the discrete dislocation simulation. *(After Fivel, M.C., Robertson, C.F., Canova, G.R., Boulanger, L., 1998. Three-dimensional modeling of indent-induced plastic zone at a mesoscale. Acta Mater. 46, 6183–6194.)*

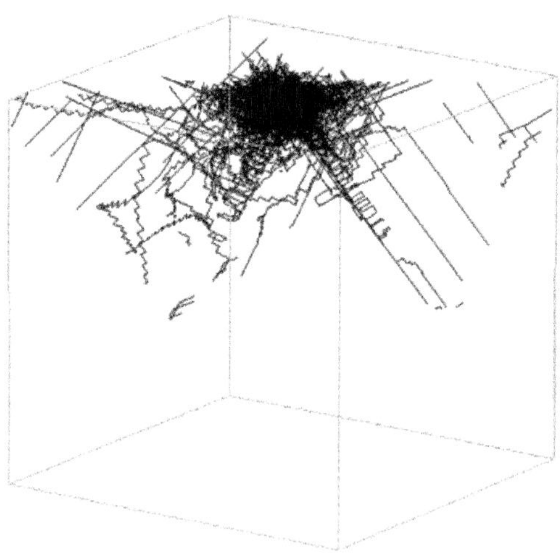

FIG. 4.34 Simulation of the nanoindentation experiment of a single crystal copper sample: dislocation microstructure at the indentation depth of 50 nm. *(After Fivel, M.C., Robertson, C.F., Canova, G.R., Boulanger, L., 1998. Three-dimensional modeling of indent-induced plastic zone at a mesoscale. Acta Mater. 46, 6183–6194.)*

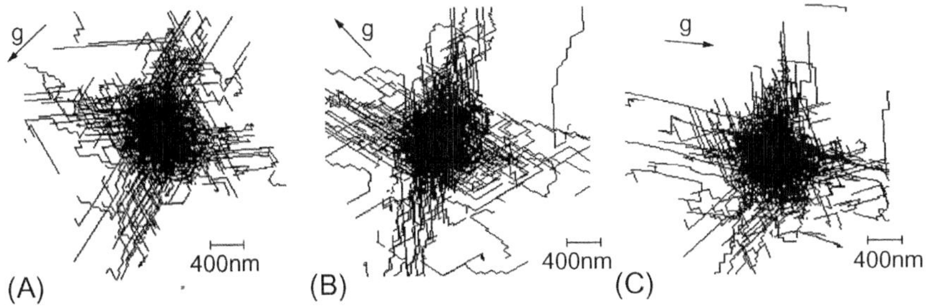

FIG. 4.35 Dislocation microstructure obtained from simulation at the indentation depth of 50 nm according to several diffracting vectors of: (A) **g** = (020), (B) **g** = (200), and (C) **g** = $(1\bar{1}1)$. *(After Fivel, M.C., Robertson, C.F., Canova, G.R., Boulanger, L., 1998. Three-dimensional modeling of indent-induced plastic zone at a mesoscale. Acta Mater. 46, 6183–6194.)*

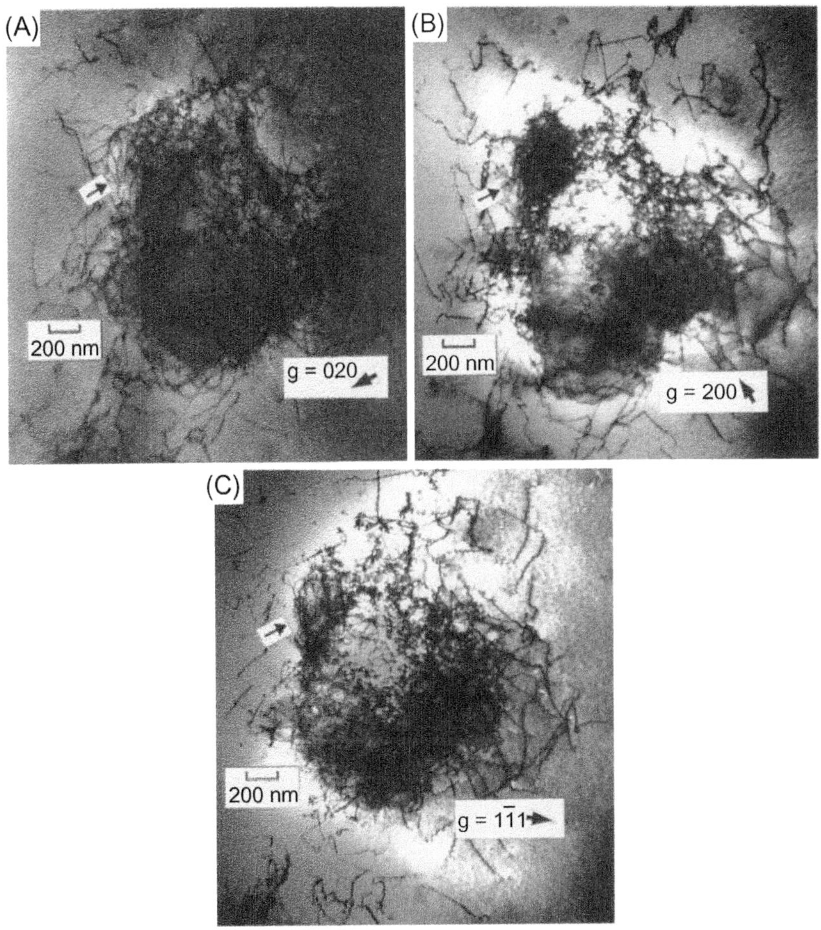

FIG. 4.36 TEM micrographs of the plastic zone at the indentation depth of 50 nm according to several diffracting vectors of: (A) **g** = (020), (B) **g** = (200), and (C) **g** = $(1\bar{1}1)$. *(After Fivel, M.C., Robertson, C.F., Canova, G.R., Boulanger, L., 1998. Three-dimensional modeling of indent-induced plastic zone at a mesoscale. Acta Mater. 46, 6183–6194.)*

depth of 50 nm using the simulation visualization results. This value compares quite well with the theoretical plastic zone radius of 800 nm, which is close to the TEM observed dislocation microstructure at the indentation depth of 50 nm (Fivel et al., 1998). Fivel et al. (1998) also reported that the dislocation microstructure obtained from the simulations is in good agreement with that of the experiments obtained using TEM.
(After El-Awady, J.A., Wen, M., Ghoniem, N.M., 2009. The role of the weakest-link mechanism in controlling the plasticity of micropillars. J. Mech. Phys. Solids 57, 32–50.)

References

Aifantis, K.E., Weygand, D., Motz, C., Nikitas, N., Zaiser, M., 2012. Modeling microbending of thin films through discrete dislocation dynamics, continuum dislocation theory, and gradient plasticity. J. Mater. Res. 27, 612–618.

Akasheh, F., Zbib, H.M., Ohashi, T., 2007. Multiscale modelling of size effect in FCC crystals: discrete dislocation dynamics and dislocation-based gradient plasticity. Philos. Mag. 87, 1307–1326.

Arsenlis, A., Cai, W., Tang, M., Rhee, M., Oppelstrup, T., Hommes, G., Pierce, T.G., Bulatov, V.V., 2007. Enabling strain hardening simulations with dislocation dynamics. Model. Simul. Mater. Sci. Eng. 15 (6), 553–595.

Balint, D.S., Deshpande, V.S., Needleman, A., Van der Giessen, E., 2006. Discrete dislocation plasticity analysis of the wedge indentation of films. J. Mech. Phys. Solids 54 (11), 2281–2303.

Benzerga, A.A., Brechet, Y., Needleman, A., Van der Giessen, E., 2004. Incorporating three dimensional mechanisms into two-dimensional dislocation dynamics. Model. Simul. Mater. Sci. Eng. 12 (1), 159–196.

Bertin, N., Glavas, V., Datta, D., Cai, W., 2018. A spectral approach for discrete dislocation dynamics simulations of nanoindentation. Model. Simul. Mater. Sci. Eng. 26.

Crone, J.C., Chung, P.W., Leiter, K.W., Knap, J., Aubry, S., Hommes, G., Arsenlis, A., 2014. A multiply parallel implementation of finite element-based discrete dislocation dynamics for arbitrary geometries. Model. Simul. Mater. Sci. Eng. 22(3).

Crone, J.C., Munday, L.B., Ramsey, J.J., Knap, J., 2018. Modeling the effect of dislocation density on the strength statistics in nanoindentation. Model. Simul. Mater. Sci. Eng. 26.

Devincre, B., Kubin, L.P., 1994. Simulations of forest interactions and strain hardening in FCC crystals. Model. Simul. Mater. Sci. Eng. 2, 559.

Dimiduk, D., Koslowski, M., LeSar, R., 2005. Size-affected single-slip behavior of pure nickel microcrystals. Scr. Mater. 53, 4065–4077.

El-Awady, J.A., Wen, M., Ghoniem, N.M., 2009. The role of the weakest-link mechanism in controlling the plasticity of micropillars. J. Mech. Phys. Solids 57, 32–50.

Fan, H., Wang, Q., Khan, M.K., 2013. Cyclic bending response of single- and polycrystalline thin films: two dimensional discrete dislocation dynamics. Appl. Mech. Mater. 275–277, 132–137.

Fivel, M., Verdier, M., Canova, G., 1997. 3D simulation of a nanoindentation test at a mesoscopic scale. Mater. Sci. Eng. A 234, 923–926.

Fivel, M.C., Robertson, C.F., Canova, G.R., Boulanger, L., 1998. Three-dimensional modeling of indent-induced plastic zone at a mesoscale. Acta Mater. 46, 6183–6194.

Frick, C., Clark, B., Orso, S., Schneider, A., Arzt, E., 2008. Size effect on strength and strain hardening of small-scale [111] nickel compression pillars. Mater. Sci. Eng. A 489, 319–329.

Gagel, J., Weygand, D., Gumbsch, P., 2016. Formation of extended prismatic dislocation structures under indentation. Acta Mater. 111, 399–406.

Ghoniem, N.M., Amodeo, R., 1988. Computer simulation of dislocation pattern formation. Solid State Phenom. 3–4, 377–388.

Ghoniem, N.M., Sun, L.Z., 1999. Fast sum method for the elastic field of 3-D dislocation ensembles. Phys. Rev. B 60 (1), 128–140.

Greer, J.R., 2006. Bridging the gap between computational and experimental length scales: a review on nano-scale plasticity. Rev. Adv. Mater. Sci. 13, 59–70.

Greer, J.R., Oliver, W.C., Nix, W.D., 2005. Size dependence of mechanical properties of gold at the micron scale in the absence of strain gradients. Acta Mater. 53, 1821–1830.

Groma, I., Pawley, G.S., 1993. Role of the secondary slip system in a computer simulation model of the plastic behavior of single crystals. Mater. Sci. Eng. A 164, 306–311.

Haque, M.A., Saif, M.T.A., 2003. Strain gradient effect in nanoscale thin films. Acta Mater. 51, 3053–3061.

Hou, C., Li, Z., Huang, M., Ouyang, C., 2008. Discrete dislocation plasticity analysis of single crystalline thin beam under combined cyclic tension and bending. Acta Mater. 56, 1435–1446.

Kiener, D., Motz, C., Grosinger, W., Weygand, D., Pippan, R., 2010. Cyclic response of copper single crystal micro-beams. Scr. Mater. 63, 500–503.

Kraft, O., Gruber, P., Mönig, R., Weygand, D., 2010. Plasticity in confined dimensions. Annu. Rev. Mater. Res. 40, 293–317.

Kreuzer, H.G.M., Pippan, R., 2004. Discrete dislocation simulation of nanoindentation: the effect of moving conditions and indenter shape. Mater. Sci. Eng. A 387–389, 254–256.

Kreuzer, H.G.M., Pippan, R., 2007. Discrete dislocation simulation of nanoindentation: indentation size effect and the influence of slip band orientation. Acta Mater. 55, 3229–3235.

Kubin, L.P., Canova, G., Condat, M., Devincre, B., Pontikis, V., Bréechet, Y., 1992. Dislocation microstructures and plastic flow: a 3-D simulation. Solid State Phenom. 23–24, 455–472.

Lepinoux, J., Kubin, L.P., 1987. The dynamic organization of dislocation structures: a simulation. Scr. Metall. 21 (6), 833–838.

Lu, S., Zhang, B., Li, X., Zhao, J., Zaiser, M., Fan, H., Zhang, X., 2019. Grain boundary effect on nanoindentation: a multiscale discrete dislocation dynamics model. J. Mech. Phys. Solids 126, 117–135.

Motz, C., Dunstan, D.J., 2012. Observation of the critical thickness phenomenon in dislocation dynamics simulation of microbeam bending. Acta Mater. 60, 1603–1609.

Motz, C., Schoberl, T., Pippan, R., 2005. Mechanical properties of micro-sized copper bending beams machined by the focused ion beam technique. Acta Mater. 53, 4269–4279.

Motz, C., Weygand, D., Senger, J., Gumbsch, P., 2008. Micro-bending tests: a comparison between three-dimensional discrete dislocation dynamics simulations and experiments. Acta Mater. 56 (9), 1942–1955.

Mughrabi, H., 1976. Observation of pinned dislocation arrangements by transmission electron microscopy (TEM). J. Microsc. Spectrosc. Electron 1, 571–584.

Norfleet, D.M., Dimiduk, D.M., Polasik, S.J., Uchic, M.D., Mills, M.J., 2008. Dislocation structures and their relationship to strength in deformed nickel microcrystals. Acta Mater. 56, 2988–3001.

Ouyang, C., Li, Z., Huang, M., Hou, C., 2008. Discrete dislocation analyses of circular nanoindentation and its size dependence in polycrystals. Acta Mater. 56, 2706–2717.

Ouyang, C., Huang, M., Li, Z., Hu, L., 2010a. Circular nano-indentation in particle-reinforced metal matrix composites: simply uniformly distributed particles lead to complex nano-indentation response. Comput. Mater. Sci. 47, 940–950.

Ouyang, C., Li, Z., Huang, M., Fan, H., 2010b. Cylindrical nano-indentation on metal film/elastic substrate system with discrete dislocation plasticity analysis: a simple model for nano-indentation size effect. Int. J. Solids Struct. 47, 3103–3114.

Parthasarathy, T.A., Rao, S.I., Dimiduk, D.M., Uchic, M.D., Trinkle, D.R., 2007. Contribution to size effect of yield strength from the stochastics of dislocation source lengths in finite samples. Scr. Mater. 56, 313–316.

Po, G., Mohamed, M.S., Crosby, T., Erel, C., El-Azab, A., Ghoniem, N., 2014. Recent progress in discrete dislocation dynamics and its applications to micro plasticity. J. Mater. 66, 2108–2120.

Rao, S.I., Dimiduk, D.M., Tang, M., Parthasarathy, T.A., Uchic, M.D., Woodward, C., 2007. Estimating the strength of single-ended dislocation sources in micron-sized single crystals. Philos. Mag. 87, 4777–4794.

Rao, S.I., Dimiduk, D.M., Parthasarathy, T.A., Uchic, M.D., Tang, M., Woodward, C., 2008. Athermal mechanisms of size-dependent crystal flow gleaned from three-dimensional discrete dislocation simulations. Acta Mater. 56, 3245–3259.

Schwarz, K.W., 1999. Simulation of dislocations on the mesoscopic scale. J. Appl. Phys. 85 (1), 108–129.

Sills, R.B., Kuykendall, W.P., Aghaei, A., Cai, W., 2016. Weinberger, C.R., Tucker, G.J. (Eds.), Multiscale Materials Modeling for Nanomechanics. Springer.

Stolken, J.S., Evans, A.G., 1998. A microbend test method for measuring the plasticity length scale. Acta Mater. 46, 5109–5115.

Uchic, M.D., Dimiduk, D.M., 2005. A methodology to investigate size scale effects in crystalline plasticity using uniaxial compression testing. Mater. Sci. Eng. A 400–401, 268–278.

Uchic, M.D., Dimiduk, D.M., Florando, J.N., Nix, W.D., 2003. Exploring specimen size effects in plastic deformation of $Ni_3(Al, Ta)$. Mater. Res. Soc. Symp. Proc. 753, BB1.4.1–BB1.4.6.

Uchic, M.D., Dimiduk, D.M., Florando, J.N., Nix, W.D., 2004. Sample dimensions influence strength and crystal plasticity. Science 305, 986–989.

Uchic, M.D., Shade, P.A., Dimiduk, D.M., 2009. Plasticity of micrometer-scale single crystals in compression. Annu. Rev. Mater. Res. 39, 361–386.

Van der Giessen, E., Needleman, A., 1995. Discrete dislocation plasticity: a simple planar model. Model. Simul. Mater. Sci. Eng. 3 (5), 689–735.

Verdier, M., Fivel, M., Groma, I., 1998. Mesoscopic scale simulation of dislocation dynamics in FCC metals: principles and applications. Model. Simul. Mater. Sci. Eng. 6 (6), 755–770.

Volkert, C.A., Lilleodden, E.T., 2006. Size effects in the deformation of submicron Au columns. Philos. Mag. 86, 5567–5579.

Wang, Z., Ghoniem, N.M., Swaminarayan, S., LeSar, R., 2006. A parallel algorithm for 3D dislocation dynamics. J. Comput. Phys. 219 (2), 608–621.

Weygand, D., Friedman, L.H., van der Giessen, E., Needleman, A., 2001. Discrete dislocation modeling in three-dimensional confined volumes. Mater. Sci. Eng. A 309–310, 420–424.

Widjaja, A., Van der Giessen, E., Needleman, A., 2005. Discrete dislocation modelling of submicron indentation. Mater. Sci. Eng. A 400–401, 456–459.

Widjaja, A., Van der Giessen, E., Needleman, A., 2007a. Discrete dislocation analysis of the wedge indentation of polycrystals. Acta Mater. 55, 6408–6415.

Widjaja, A., Needleman, A., Van der Giessen, E., 2007b. The effect of indenter shape on sub-micron indentation according to discrete dislocation plasticity. Model. Simul. Mater. Sci. Eng. 15, S121.

Yefimov, S., van der Giessen, E., Groma, I., 2004. Bending of a single crystal: discrete dislocation and nonlocal crystal plasticity simulations. Model. Simul. Mater. Sci. Eng. 12, 1069.

Zbib, H.M., 2012. Introduction to discrete dislocation dynamics. In: Sansour, C., Skatulla, S. (Eds.), Generalized Continua and Dislocation Theory. In: CISM Cources and Lectures, vol. 537. Springer, New York, pp. 289–317.

Zbib, H.M., Rhee, M., Hirth, J.P., 1998. On plastic deformation and the dynamics of 3-D dislocations. Int. J. Mech. Sci. 40 (2–3), 113–127.

Zbib, H.M., Rhee, M., Hirth, J.P., 2002. A multiscale model of plasticity. Int. J. Plast. 18, 1133–1163.

Zhou, C., Beyerlein, I.J., LeSar, R., 2011. Plastic deformation mechanisms of fcc single crystals at small scales. Acta Mater. 59, 7673–7682.

Chapter 5

Molecular dynamics

5.1 Introduction

The size effects in materials can be addressed by modeling them with full atomistic details using molecular dynamics (MD) (Lidorikis et al., 2001; Nakano et al., 2001). Accordingly, the size effects governing mechanisms can be captured with the resolution of atomistic mechanisms. In this case, the simulation precision is controlled by the accuracy of the interatomic potentials. Although MD can capture the atomistic mechanisms, the simulation time scale and length scale is limited. However, increasing the efficiencies of the parallel MD codes along with rising new powerful Supercomputers expand the limits of time and size this method can address. In this chapter, due to the huge success of using MD to model the behavior of crystalline metals, the focus is on the investigation of size effects in crystalline metals. Various size effects studies of crystalline metals has been conducted. As an example, the atomistic simulation of grain size effects has been addressed in Section 1.3.1.2. However, in this chapter, the focus is on the MD simulation of size effects during nanoindentation and micropillar compression experiments. In the case of nanoindentation, the topic has just been briefly presented in Section 1.3.2.3. Accordingly, the elaborate description of the conventional and advanced experiments and theoretical models of the indentation size effects will be presented to clarify the necessity and value of MD simulation. Next, the atomistic simulation of the nanoindentation experiment will be presented. In the case of size effects during micropillar compression test, a comprehensive experimental review and theoretical models is presented in Section 1.3.2.2. Accordingly, the focus in this chapter is solely on the MD simulation.

5.2 Molecular dynamics simulation of size effects during nanoindentation experiment

5.2.1 Nanoindentation size effects: Conventional experimental observations and theoretical models

Size effects during nanoindentation can be investigated using the geometrically self-similar indenter tips. According to the local plasticity models, the hardness

Size Effects in Plasticity: From Macro to Nano. https://doi.org/10.1016/B978-0-12-812236-5.00005-0

should be independent of indentation depth. However, the experimental observations depict that the hardness varies as the indentation depth changes, which is commonly termed as size effects during nanoindentation (Pharr et al., 2010; Mott, 1957; Bückle, 1959; Gane, 1970; Upit and Varchenya, 1973; Chen and Hendrickson, 1973; Tabor, 1986; Stelmashenko et al., 1993; De Guzman et al., 1993; Nix and Gao, 1998; Swadener et al., 2002; Lim and Chaudhri, 1999; Bushby and Dunstan, 2004; Spary et al., 2006; Durst et al., 2008; Gerberich et al., 2002; Bull, 2003; Zhu et al., 2008; Kiener et al., 2009; Sangwal, 2000; King, 1987; Voyiadjis and Yaghoobi, 2017b,c). The characteristic lengths of size effects during the nanoindentation is the indentation depth. In other words, the indentation hardness varies as the indentation depth changes. The majority of size effects studies during nanoindentation experiments provided a size effect trend in which the hardness decreases as the indentation depth increases. Fig. 5.1 shows the size effects in several nanoindentation experiments available in the literature, in which, the hardness related to each set of experiments is normalized using the microhardness H_0, which is the indentation hardness at large indentation depths at which there is no more size effects.

The size effects during nanoindentation is assigned to the variation of hardness versus the indentation depth. However, there is another size effect phenomenon related to the nanoindentation experiment in the case of spherical indenters, which is the variation of hardness versus the indenter tip radius. Accordingly, the size effects characteristic length of this phenomenon is the radius of the indenter tip. The experimental observations have shown that as the indenter tip radius increases, the hardness decreases (Swadener et al., 2002; Lim and Chaudhri, 1999; Bushby and Dunstan, 2004; Spary et al., 2006; Durst et al., 2008).

5.2.2 Theoretical models of nanoindentation size effects

The concept of geometrically necessary dislocations (GNDs) developed by Ashby (1970) is the most common method to address the size effects during nanoindentation (Stelmashenko et al., 1993; De Guzman et al., 1993; Nix and Gao, 1998; Swadener et al., 2002; Pugno, 2007). Ashby (1970) categorized dislocations into two groups of geometrically necessary dislocations (GNDs) and statistically stored dislocations (SSDs). GNDs accommodate the imposed displacement to satisfy the compatibility, while the SSDs are located with respect to each other in a random way. The fundamental model of Nix and Gao (1998) incorporated the concept of GNDs to capture size effects during nanoindentation (Fig. 1.54). As described in Section 1.3.2.3, the total length of GNDs during nanoindentation in the case of a conical indenter with the angle of θ, see Fig. 1.54, can be described as follows:

$$\lambda_G = \int_0^{a_p} \frac{2\pi r h}{b_G a_p} \mathrm{d}r = \frac{\pi a_p h}{b_G} = \frac{\pi a_p^2 \tan(\theta)}{b_G} \tag{5.1}$$

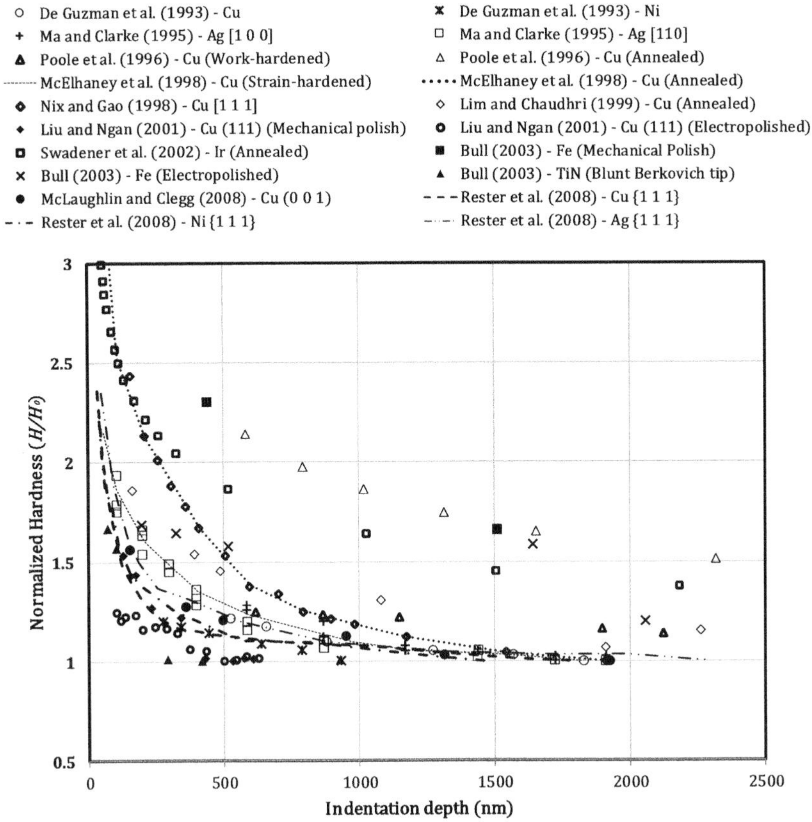

FIG. 5.1 Size effect during nanoindentation. The original experimental data has been reported by De Guzman et al. (1993), Ma and Clarke (1995), Poole et al. (1996), McElhaney et al. (1998), Nix and Gao (1998), Lim and Chaudhri (1999), Liu and Ngan (2001), Swadener et al. (2002), Bull (2003), McLaughlin and Clegg (2008), and Rester et al. (2008). *(After Voyiadjis, G.Z., Yaghoobi, M., 2017b. Review of nanoindentation size effect: experiments and atomistic simulation. Crystals 7, 321.)*

where h is the indentation depth, a_p is the contact radius, and b_G is the Burgers vector of GNDs. The GNDs density ρ_G can be defined as the total GNDs length λ_G divided by the plastic zone volume V, which is a hemisphere with the radius of $R_{pz} = a_p$, as follows:

$$\rho_G = \frac{\lambda_G}{V} = \frac{3h}{2b_G a_p^2} = \frac{3}{2b_G h}\tan^2(\theta) \tag{5.2}$$

Nix and Gao (1998) considered similar Burgers vector for GNDs and SSDs, i.e., $b_G = b_S = b$, and simplified the total dislocation density as a summation of GNDs and SSDs densities as follows:

$$\rho = \rho_S + \rho_G \tag{5.3}$$

The interactions between SSDs and GNDs are neglected in Eq. (5.3). Accordingly, one can obtain the hardness using the Taylor hardening model for a material with the shear modulus of G as follows:

$$H = 3\sqrt{3}\alpha G b\sqrt{\rho} = 3\sqrt{3}\alpha G b\sqrt{\rho_G + \rho_S} \tag{5.4}$$

where α is a material constant. One can presents the variation of hardness versus the indentation depth in the case of conical indenter by substituting Eq. (5.2) into Eq. (5.4) as follows:

$$\frac{H}{H_0} = \sqrt{1 + \frac{h^*}{h}} \tag{5.5}$$

where H_0 and h^* are the hardness originated from the interactions of SSDs and the material characteristic length, respectively, which can be defined as follows:

$$H_0 = 3\sqrt{3}\alpha G b\sqrt{\rho_S} \tag{5.6}$$

$$h^* = \frac{81}{2} b\alpha^2 \tan^2(\theta)\left(\frac{G}{H_0}\right)^2 \tag{5.7}$$

Nix and Gao (1998) linked the strain gradient plasticity to the nanoindentation experiment using the following definition of strain gradient:

$$\chi \equiv \frac{\tan(\theta)}{a_p} \tag{5.8}$$

The length scale related to the strain gradient model can then be defined as follows:

$$l \equiv b\left(\frac{3G}{H_0}\right)^2 \tag{5.9}$$

Swadener et al. (2002) followed the work of Nix and Gao (1998) and considered a more general indenter tip geometry of:

$$h = Ar^n \tag{5.10}$$

where A is a constant and $n > 1$. Accordingly, the total GNDs length during nanoindentation experiment can be obtained as follows:

$$\lambda_G = \int_0^{a_p} \frac{2\pi r}{b}\left(\frac{\mathrm{d}h}{\mathrm{d}r}\right)\mathrm{d}r = \frac{2\pi n A}{b(n+1)} a_p^{n+1} \tag{5.11}$$

Again, it is assumed in Eq. (5.11) that $b_G = b_S = b$. The GNDs density ρ_G can then be defined as the total GNDs length λ_G divided by the plastic zone volume V as follows:

$$\rho_G = \frac{\lambda_G}{V} = \frac{3nA}{b(n+1)} a_p^{n-2} = \frac{3nA^{(2/n)}}{b(n+1)} h^{(1-2/n)} \tag{5.12}$$

In the case of an indenter tip with general tip, Pugno (2007) presented an equation for the total GNDs length λ_G formation during the nanoindentation experiment as follows:

$$\lambda_G = \frac{\Omega - A}{b} = \frac{S}{b} \tag{5.13}$$

in which the indentation surface is modeled as a combination of discrete steps formed as a result of dislocation loop beneath the indenter as presented in Fig. 5.2. In Eq. (5.13), S and A are the total lengths of vertical and horizontal surfaces, respectively, and $\Omega = A + S$. The GNDs density ρ_G can then be defined as the total GNDs length λ_G divided by the plastic zone volume V as follows:

$$\rho_G = \frac{\lambda_G}{V} = \frac{S}{bV} \tag{5.14}$$

Eq. (5.14) states that the GNDs dislocation density can be related to the ratio of the indentation surface to the plastic zone volume.

Swadener et al. (2002) investigated the size effects during nanoindentation experiment with spherical indenter tips. They approximated the spherical indenter using a parabola as follows:

$$h = \frac{r^2}{2R_p} \tag{5.15}$$

Accordingly, one can obtain the total GNDs density in the case of spherical indenter tip by substituting Eq. (5.13) into Eq. (5.12) as follows:

$$\rho_G = \frac{\lambda_G}{V} = \frac{1}{bR_p} \tag{5.16}$$

Eq. (5.14) states that the hardness does not depend on the indentation depth and there should be no size effects with the characteristic length of indentation depth. However, Swadener et al. (2002) studied the size effects with the characteristic length of indenter tip radius by following the same framework incorporated to obtain Eq. (5.5). Accordingly, they obtained the dependency of the hardness on the indenter tip radius as follows:

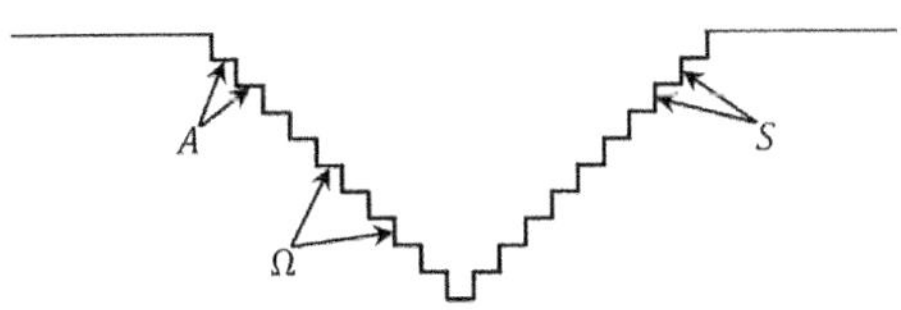

FIG. 5.2 Approximating the indentation surface as a summation of discrete steps: A is the projected contact area, Ω is the contact surface, and $S = \Omega - A$. *(After Voyiadjis, G.Z., Yaghoobi, M., 2017b. Review of nanoindentation size effect: experiments and atomistic simulation. Crystals 7, 321.)*

$$\frac{H}{H_0} = \sqrt{1 + \frac{R^*}{R_p}} \tag{5.17}$$

where R^* is defined as follows:

$$R^* = \frac{\bar{r}}{b\rho_S} \tag{5.18}$$

Fig. 5.3 compares the experimental results obtained by Swadener et al. (2002) with their theoretical prediction, which is obtained using the framework presented by Nix and Gao (1998). In the case of large indentation depths and indenter tip radii, the theoretical model can successfully predict the size effects. However, the difference between the theoretical predictions and experimental results are considerable in the case of small indentation depths and tip radii.

Feng and Nix (2004) observed the same trend in the case of nanoindentation of MgO. Fig. 5.4 compares the experimentally observed nanoindentation size effects with the theoretical model of Nix and Gao (1998). Again, the results show that the model cannot capture the size effects for small indentation depths. They proposed three factors for the breakdown of Nix and Gao (1998) size

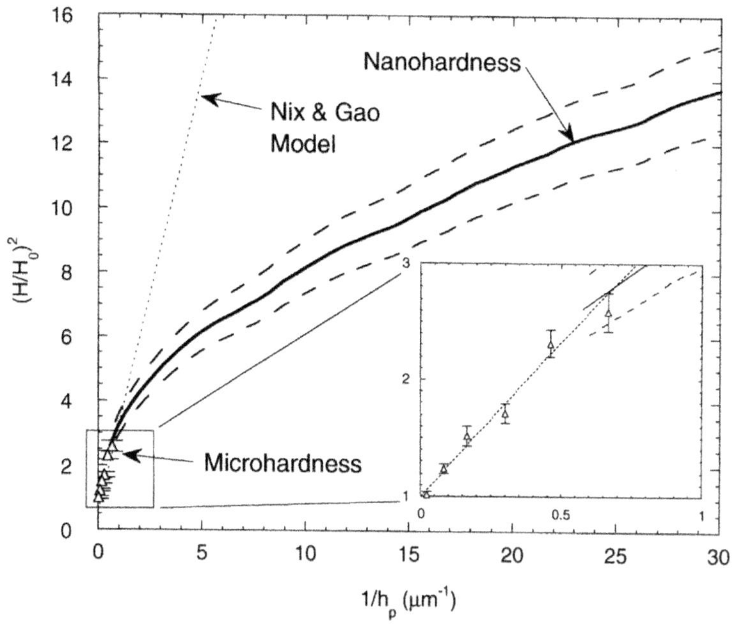

FIG. 5.3 Indentation size effect in annealed iridium measured with a Berkovich indenter (Δ and solid line) and comparison of experiments with the Nix and Gao (1998) model ($H_0 = 2.5$ GPa; $h^* = 2.6$ μm). The *dashed lines* represent +/− one standard deviation of the nanohardness data. *(After Swadener, J.G., George, E.P., Pharr, G.M., 2002. The correlation of the indentation size effect measured with indenters of various shapes. J. Mech. Phys. Solids 50, 681–694.)*

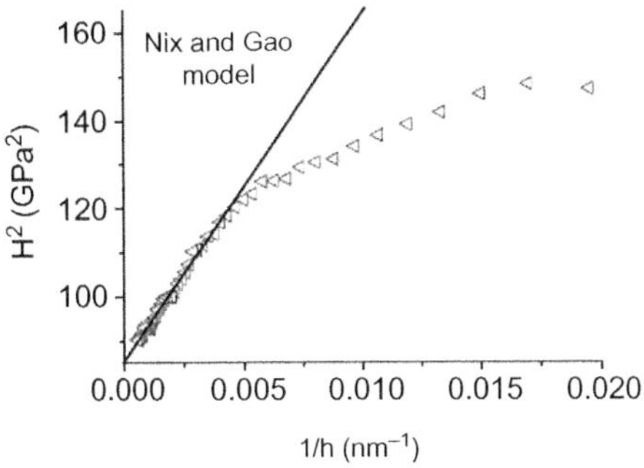

FIG. 5.4 Indentation size effect in polished MgO: experimental results versus the Nix and Gao (1998) model ($H_0 = 9.24$ GPa; $h^* = 92.5$ nm). *(After Feng, G., Nix, W.D., 2004. Indentation size effect in MgO. Scr. Mater. 2004, 51, 599–603.)*

effects theory at small indentation depths which are bluntness of the indenter tip, Peierls stress, and plastic zone size. Feng and Nix (2004) stated that the first two factors are not responsible of the discrepancies between the Nix and Gao (1998) theoretical prediction and the indentation size effects for small depths. They attributed the discrepancies to the last factor. Accordingly, Nix and Gao (1998) presented the following plastic zone radius as follows:

$$R_{pz} = fa_p \tag{5.19}$$

where a_p is the contact radius and f is a function of indentation depth. Unlike the original Nix and Gao theory (1998) which considers a constant plastic zone radius of $R_{pz} = a_p$, Eq. (5.19) states that the plastic zone radius is not constant and varies with the indentation depth. One should note that $f \to 1$ for the large indentation depths and Eq. (5.19) reproduces the original Nix and Gao theory.

The new plastic zone volume can be obtained using Eq. (5.19) as follows:

$$V = \frac{2}{3}\pi R_{pz}^3 = \frac{2}{3}\pi f^3 a_p^3 \tag{5.20}$$

Feng and Nix (2004) modified the Nix and Gao model (1998) using the plastic zone volume described in Eq. (5.20) and compared the obtained results versus their experiments as shown in Fig. 5.5. The results show that the modified model can capture the indentation size effects for all range of indentation depths. Durst et al. (2005) modified the Nix and Gao model (1998) by using the same procedure. Instead of using a fitted curve for the radius of plastic zone, however (Feng and Nix, 2004), they incorporated a constant $f = 1.9$ in their work and showed that the nanoindentation size effects in single crystal and ultrafine-grained copper can be captured using this model (Durst et al., 2005).

Huang et al. (2006) modified the Nix and Gao model (1998) to capture the size effects for small indentation depths. First, they introduced the concept of local density of GNDs $(\rho_G)_{local}$. Next, they defined a cap for the local density of

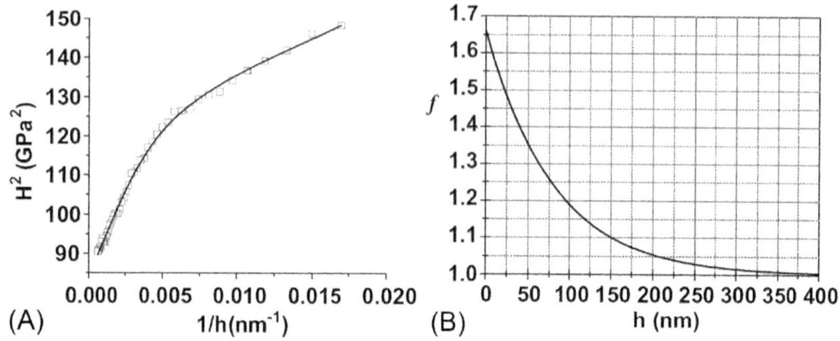

FIG. 5.5 Indentation size effect in polished MgO: (A) experimental results versus the modified Nix and Gao model presented by Feng and Nix (2004) ($H_0 = 9.19$ GPa; $h^* = 102$ nm) and (B) variation of f versus h. *(After Feng, G., Nix, W.D., 2004. Indentation size effect in MgO. Scr. Mater. 2004, 51, 599–603.)*

GNDs $\rho_G^{\max}$, which is defined as a material constant. Finally, the average density of GNDs ρ_G can be obtained by averaging the local GND density $(\rho_G)_{\text{local}}$ over the plastic zone volume. Huang et al. (2006) incorporated the same plastic zone radius as Nix and Gao (1998) which is $R_{pz} = a_p$. In the case of the conical indenter, the local density of GNDs $(\rho_G)_{\text{local}}$ can be defined as follows:

$$(\rho_G)_{\text{local}} = \rho_G^{\max} \text{ if } h < h_{\text{nano}} \tag{5.21a}$$

$$(\rho_G)_{\text{local}} = \begin{cases} \rho_G^{\max} & \text{for} \quad r < \dfrac{h_{\text{nano}}}{\tan(\theta)} \\ \dfrac{\tan(\theta)}{br} & \text{for} \quad \dfrac{h_{\text{nano}}}{\tan(\theta)} \le r \le a_p = \dfrac{h}{\tan(\theta)} \end{cases} \tag{5.21b}$$

where h_{nano} is the size effects of the characteristic length for nanoindentation, which is obtained as follows:

$$h_{\text{nano}} = \frac{\tan^2(\theta)}{b\rho_G^{\max}} \tag{5.22}$$

The average density of GNDs ρ_G can be calculated as follows:

$$\rho_G = \rho_G^{\max} \begin{cases} 1 & \text{if } h < h_{\text{nano}} \\ \dfrac{3h_{\text{nano}}}{2h} - \dfrac{h_{nano}^3}{2h^3} & \text{if } h < h_{\text{nano}} \end{cases} \tag{5.23}$$

Huang et al. (2006) incorporated the strain gradient plasticity model along with the modified GNDs density to capture the indentation size effects in MgO and Ir. Fig. 5.6 compares the model predictions versus the experiments of Swadener et al. (2002) and Feng and Nix (2004). The maximum local GNDs densities of $\rho_G^{\max} = 1.28 \times 10^{16}\,\text{m}^{-2}$ and $\rho_G^{\max} = 9.68 \times 10^{14}\,\text{m}^{-2}$ were considered for MgO and Ir, respectively. Fig. 5.6 shows that the model presented by Huang et al. (2006) can successfully capture the size effects for all range of indentation depths.

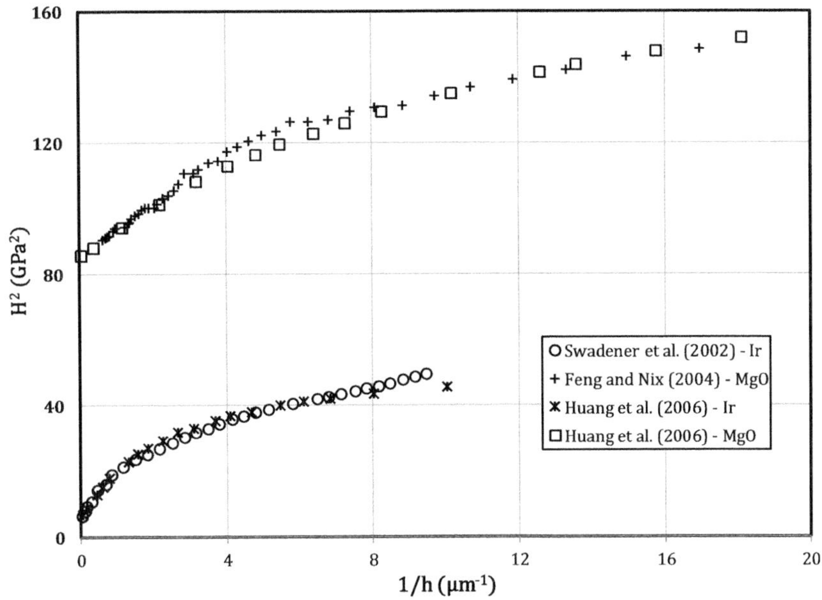

FIG. 5.6 Comparison between the indentation size effect captured using the strain gradient plasticity model developed by Huang et al. (2006) using a cap for GND density versus the experimental results in Ir and MgO ($\rho_G^{max}{}_{Ir} = 9.68 \times 10^{14}$ m^{-2}, $\rho_G^{max}{}_{MgO} = 1.28 \times 10^{16}$ m^{-2}). *(After Voyiadjis, G.Z., Yaghoobi, M., 2017b. Review of nanoindentation size effect: experiments and atomistic simulation. Crystals 7, 321.)*

5.2.3 Nanoindentation size effects: Recent experimental observations and theoretical models

The indentation size effects theories presented in Section 5.2.2, such as Nix and Gao model (1998), assume that the interaction of dislocations with each other, which is also known as forest hardening mechanism, govern the size effects. Accordingly, these models relate the variation of indentation hardness to the change in the GNDs content induced by the strain gradient during nanoindentation experiment. Some nanoindentation experiments, such as the ones conducted by Swadener et al. (2002) and Feng and Nix (2004), showed that the Nix and Gao model (1998) cannot capture the size effects during nanoindentation for small indentation depths (See Figs. 5.3 and 5.4). Some efforts have been done to modify the model by modifying the plastic zone size (see e.g., Feng and Nix, 2004; Durst et al., 2005) and applying a cap for GNDs content during nanoindentation (Huang et al., 2006). In these models, the density of GNDs have been modified using phenomenological frameworks including curve fitting to capture size effects for small indentation depths. However, the modified models still attributed the nanoindentation size effects for small depths to the interaction of dislocations with one another, i.e., forest hardening, and does not propose a new size effects mechanism.

In the case of nanoindentation experiment of crystalline metals, in order to investigate the size effects mechanisms for small depths, the dislocation network pattern should be studied. To do so, three methods of backscattered electron diffraction (EBSD) (Kiener et al., 2009, 2006; Zaafarani et al., 2006, 2008; Dahlberg et al., 2014, 2017; Rester et al., 2008, 2007; Demir et al., 2009), convergent beam electron diffraction (CBED) (McLaughlin and Clegg, 2008), and X-ray microdiffraction (μXRD) (Yang et al., 2004; Larson et al., 2007, 2008; Feng et al., 2008) have been developed and incorporated to capture the dislocation structure pattern. These experiments obtain the Nye dislocation tensor using the local lattice orientations map with high spatial resolutions. Next, they calculated the GND density using the obtained Nye dislocation tensor. Only a limited number of experimental studies have incorporated these methods to capture the size effects during nanoindentation due to the complexities of these methods. Kiener et al. (2006), McLaughlin and Clegg (2008), and Demir et al. (2009) incorporated these methods and observed that the forest hardening mechanism along with the strain gradient theory cannot capture the indentation size effects for small indentation depths. Kiener et al. (2006) conducted nanoindentation tests on tungsten and copper samples using a Vickers indenter. Next, they incorporated EBSD to capture the lattice misorientation. Finally, they selected the misorientation angle as a measure of GNDs density. Fig. 5.7 plots the variation of maximum misorientation angle ω_f versus the indentation size in tungsten and copper. The results show that the maximum misorientation angle ω_f increases as the indentation depth increases. However, based on the strain gradient model and forest hardening mechanism, the GNDs density and accordingly the GNDs density should be decreased as the indentation depth increases. Kiener et al. (2006) showed that the indentation size effects for small depths cannot be captured using the strain gradient model and forest hardening mechanism. In other words, the conventional size effects model including Nix and Gao model (1998) cannot capture the size effects for small indentation depths. McLaughlin and Clegg (2008) incorporated the CBED to investigate the indentation size effects in single crystal copper using a Berkovich tip. The total misorientation was measured for two indentation forces of 5 and 15 mN. McLaughlin and Clegg (2008) reported that both values of overall and neighbor to neighbor misorientations are significantly larger in the case of the indentation load of 15 mN compared to that of the 5 mN. On the other hand, the hardness corresponds to the indentation force of 5 mN is larger than hardness measured for indentation force of 15 mN. Similar to Kiener et al. (2006) and McLaughlin and Clegg (2008) observed that the models based on the forest hardening mechanism along with the strain gradient model, including Nix and Gao model (1998), cannot capture size effects for small indentation depths. McLaughlin and Clegg (2008) stated that the governing mechanism of size effects for shallow indentation depths is the lack of dislocation sources. The introduced mechanism is similar to the source exhaustion mechanism presented in Section 1.3.2.2.2.

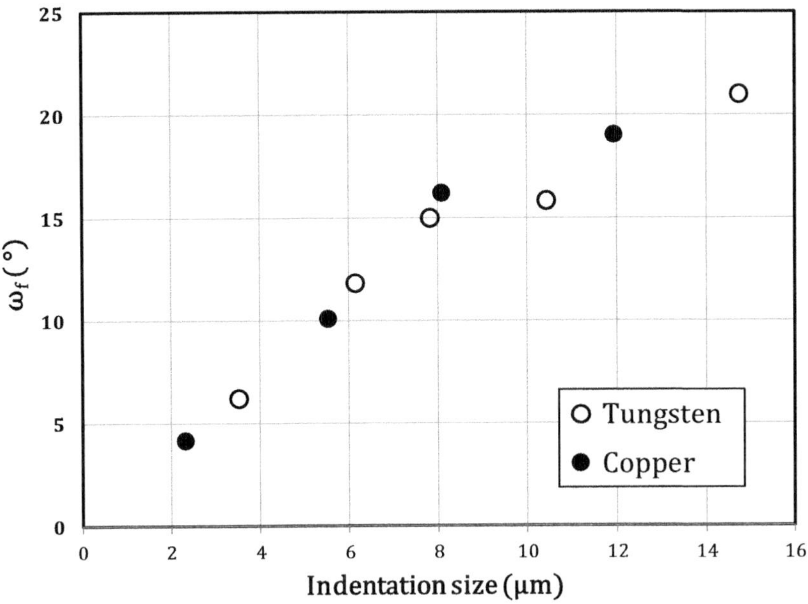

FIG. 5.7 Variation of the maximum misorientation angle ω_f versus the indentation size in tungsten and copper. The original data was reported by Kiener et al. (2006). *(After Voyiadjis, G.Z., Yaghoobi, M., 2017b. Review of nanoindentation size effect: experiments and atomistic simulation. Crystals 7, 321.)*

Demir et al. (2009) incorporated the EBSD tomography to study the indentation size effects in a single crystal copper for small indentation depths. The indenter geometry is conical with a spherical tip. The GND density contours of five cross sections, which are equally spaced, corresponding to four indentation depths is presented in Fig. 5.8. The GNDs density contours show that unlike the conventional assumption of homogenous dislocation network beneath the indenter, the dislocation structure is heterogeneous. In the next step, the GNDs densities of four different indentation depths was reported by Demir et al. (2009), which is obtained using 50 separate 2-D EBSD sections. Fig. 5.9 presents the variation of GNDs density along with the variation of hardness versus the indentation depth. Fig. 5.9 indicates that the GNDs density decreases as the indentation depth decreases. The hardness, on the other hand, increases by decreasing the depth. One can see that as the GNDs density increases, the indentation hardness decreases. The observed trend cannot be captured using the model developed based on the forest hardening mechanism, such as Nix and Gao model (1998). In other words, the forest hardening mechanism would lead to an increase in hardness for increase in GNDs density, which is the opposite of the observed trend. The experimental results of Demir et al. (2009) showed the breakdown of forest hardening mechanism for small indentation depths. Demir

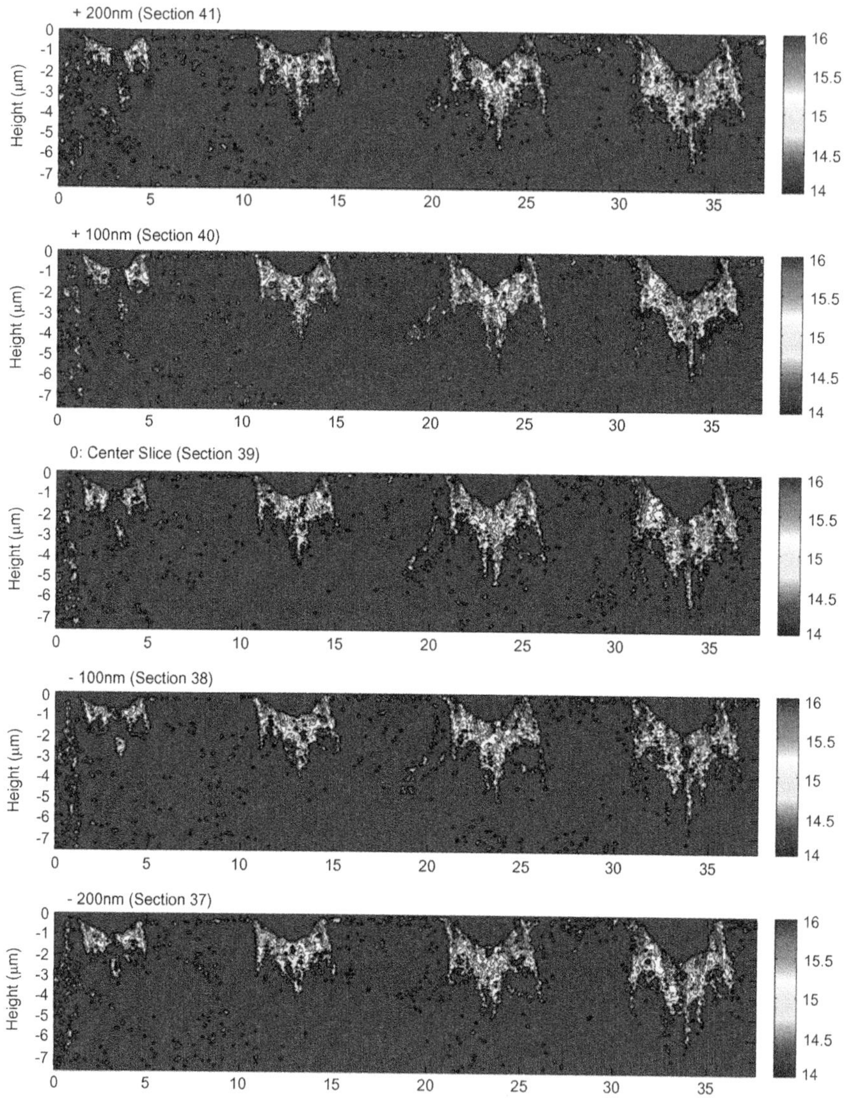

FIG. 5.8 Five equally spaced cross-sections (center slice, ±100 nm, ±200 nm) through the four indents. Color code: GND density in decadic logarithmic scale (m-2). *(After Demir, E., Raabe, D., Zaafarani, N., Zaefferer, S., 2009. Investigation of the indentation size effect through the measurement of the geometrically necessary dislocations beneath small indents of different depths using EBSD tomography. Acta Mater. 57, 559–569.)*

et al. (2009) introduced the reduction of dislocation segment lengths as the governing mechanism of the observed size effects. This size effects mechanism has the same nature as the source truncation mechanism as introduced in Section 1.3.2.2.2. One should note that the SSDs density is not accounted in the obtained conclusion. However, Demir et al. (2009) stated that the imposed

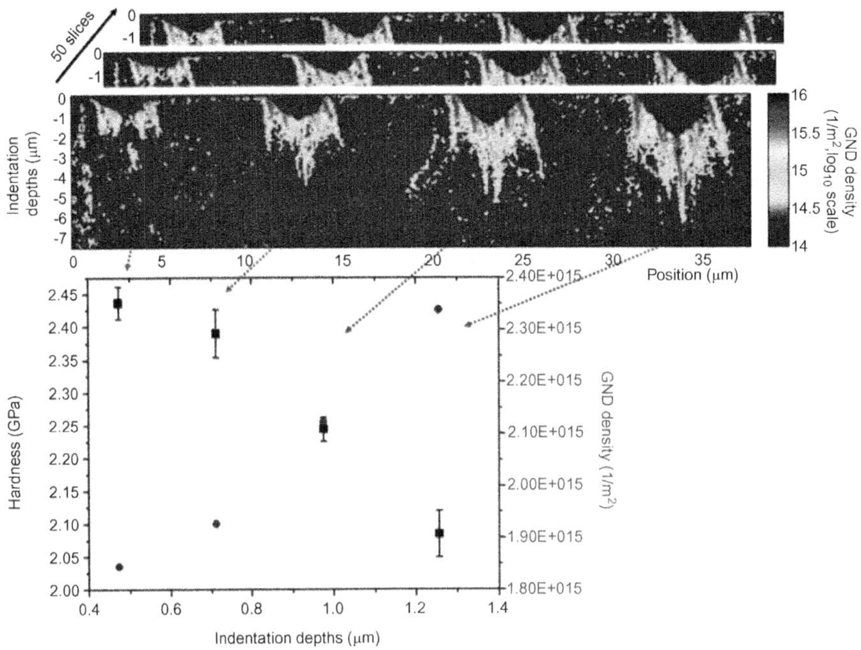

FIG. 5.9 The variations of hardness and GND density versus the indentation depth. *(After Demir, E., Raabe, D., Zaafarani, N., Zaefferer, S., 2009. Investigation of the indentation size effect through the measurement of the geometrically necessary dislocations beneath small indents of different depths using EBSD tomography. Acta Mater. 57, 559–569.)*

displacement decreases as the indentation depth decreases. Consequently, the enhanced hardening at small indentation cannot be attributed to the role of SSDs.

Feng et al. (2008) investigated the defects evolution during the indentation experiment of single crystal copper using the μXRD method. The selected indenter tip was Berkovich. Unlike the experimental observations of Kiener et al. (2006), McLaughlin and Clegg (2008), and Demir et al. (2009), Feng et al. (2008) observed that the experimental results can be captured using the forest hardening mechanism along with the strain gradient theory. Feng et al. (2008) reported that both GNDs density and hardness decreases as the indentation depth increases. However, unlike Demir et al. (2009) which obtained the GND density pattern below the indenter, Feng et al. (2008) approximated the strain gradient. The GNDs density is then a function of the plastic zone radius R_{pz}, which was considered as $R_{pz} = fa_p$, where a_p is the contact radius and f is a constant. Demir et al. (2009) showed that the dislocation network beneath the indenter is heterogeneous. Consequently, the conclusion made by Feng et al. (2008) may not be accurate.

Two trends of experimental observations have been reported. One trend of experimental results presented by Kiener et al. (2006), McLaughlin and

Clegg (2008), and Demir et al. (2009) show that the conventional forest hardening mechanisms cannot be incorporated for small indentation depths. The other trend presented by Feng et al. (2008) stated that the forest hardening mechanism can capture the indentation size effects even for small indentation depths. Accordingly, in order to clarify the governing mechanisms of size effects during nanoindentation, the meso-scale simulations, such as the atomistic simulation, can be incorporated.

5.2.4 Molecular dynamics simulation of nanoindentation

Recent experimental results reported controversial observations regarding to the indentation size effects. One group of researcher including Kiener et al. (2006), McLaughlin and Clegg (2008), and Demir et al. (2009) showed that the conventional size effects model including the forest hardening mechanism along with the strain gradient model cannot capture the indentation size effects for small indentation depths. On the other hand, Feng et al. (2008) showed that conventional indentation size effects can be applied even for small indentation depths. In order to shed light to these discrepancies and unravel the governing mechanism of size effects during nanoindentation, the sample can be modeled as a cluster of atoms. The molecular dynamics (MD) is a powerful tool that can be incorporated as a pseudo-experimental tool to address the size effect during the nanoindentation. Many deformation mechanisms during nanoindentation of metallic thin films have been captured using MD. Incorporating the MD simulation, Kelchner et al. (1998) investigated the defect nucleation and evolution of Au during nanoindentation. The surface step effects on the response of Au during nanoindentation were investigated by Zimmerman et al. (2001) using atomistic simulation. Lee et al. (2005) conducted a comprehensive study on the defect nucleation and evolution patterns during nanoindentation of Al and tried to explain the nanoindentation response using those patterns. Hasnaoui et al. (2004), Jang and Farkas (2007), and Kulkarni et al. (2009) has studied the interaction between the dislocations and GB during nanoindentation experiment using molecular dynamics. Yaghoobi and Voyiadjis (2014) investigated the effects of the MD boundary conditions on the sample response and defect nucleation and evolution patterns during nanoindentation. Voyiadjis and Yaghoobi (2015) investigated the theoretical models developed to capture the size effects during nanoindentation using MD. Yaghoobi and Voyiadjis (2016a) investigated the governing mechanisms of size effects during nanoindentation using MD. Voyiadjis and Yaghoobi (2016) incorporated large scale MD to study the GB effects on the material strength as the grain size varies.

In this section, first, the simulation methodology of the atomistic simulation of nanoindentation is elaborated. Next, the effects of different incorporated boundary conditions for the molecular dynamics simulation of the nanoindentation is investigated (Yaghoobi and Voyiadjis, 2014). In the next step, different aspects of size effects during nanoindentation including the variation of plastic zone size and hardness versus the indentation depth for various indenter tip geometries are captured using atomistic simulation and compared against the theoretical models (Voyiadjis and Yaghoobi, 2015). The governing mechanisms of indentation size effects in thin films of confined volumes is then addressed using a spherical indenter tip (Yaghoobi and Voyiadjis, 2016a). Finally, the effects of grain boundary and grain size is addressed using the atomistic simulation (Voyiadjis and Yaghoobi, 2016).

5.2.4.1 Molecular simulation methodology

In the case of N interacting monoatomic molecules, the Newton's equation of motion can be defined as follows:

$$m_i\ddot{\mathbf{r}}_i = -\nabla_i U + \mathbf{f}_i, i = 1,2,\ldots,N \tag{5.24}$$

where $\ddot{\mathbf{r}}_i$ is the second time derivative of ith particle trajectory $\mathbf{r}_i$, m_i is the mass of ith particle, $\mathbf{f}_i$ is an external force on the ith particle, and

$$\nabla_i U = \left(\frac{\partial}{\partial x_i} + \frac{\partial}{\partial y_i} + \frac{\partial}{\partial z_i}\right) U, i = 1,2,\ldots,N \tag{5.25}$$

where $U(\mathbf{r}_1, \mathbf{r}_2, \ldots, \mathbf{r}_N)$ is the potential energy. A metallic sample can be modeled according to Eq. (5.24) by considering each atom as a mass point. Eq. (5.24) governs the position and velocity of each atom. Accordingly, during MD simulation, Eq. (5.24) should be numerically solved. To model a nanometer sized sample, the number of atoms is in the order of million. If one tries to model a micron sized sample, the number of atoms order grows to billion. Accordingly, the MD code should be efficiently parallelized and along with very powerful supercomputers, one should be able to model a sample close to 0.5 μm (see, e.g., Yaghoobi and Voyiadjis, 2017, 2018. However, in the case of strain rate, MD simulation still has considerable limitations and it cannot model the quasi-static simulations.

Several interatomic potentials have been used to model the crystalline metals including Lennard-Jones (LJ), Morse, embedded-atom method (EAM), and modified embedded-atom method (MEAM).

The Lennard-Jones potential E^{LJ} is defined as follows:

$$E^{\mathrm{LJ}}(r_{ij}) = 4\varepsilon\left[\left(\frac{\sigma}{r_{ij}}\right)^{12} - \left(\frac{\sigma}{r_{ij}}\right)^{6}\right] \tag{5.26}$$

where σ is the distance from the atom at which $E^{\mathrm{LJ}} = 0$ and ε is the potential well depth. A cutoff distance should be selected for L-J potential.

Morse potential E^{Morse} is defined as below:

$$E^{\mathrm{Morse}}(r_{ij}) = D\left\{e^{\left[-2\alpha\left(r_{ij}-r_0\right)\right]} - 2e^{\left[-\alpha\left(r_{ij}-r_0\right)\right]}\right\} \tag{5.27}$$

where D is the cohesive energy, α is the elastic parameter, and r_0 is the equilibrium distance.

The embedded-atom method (EAM) is the most common function to capture the interatomic potential of crystalline metals (Daw and Baskes, 1984). The EAM potential E^{EAM} can be described as follows:

$$E^{\text{EAM}}(r_{ij}) = \frac{1}{2}\sum_{i,j} V(r_{ij}) + \sum_{i} F(\rho_i), \rho_i = \sum_{i \neq j} \varphi(r_{ij}) \tag{5.28}$$

where $V(r_{ij})$ is the pair interaction potential, $F(\rho_i)$ denotes the embedding potentials, and $\varphi(r_{ij})$ is a function which is defined using the electron charge density.

The modified embedded-atom method (MEAM) potential is the modification of EAM potential which is defined as below (Baskes, 1992):

$$E^{\text{MEAM}}(r_{ij}) = \frac{1}{2}\sum_{i,j} V(r_{ij}) + \sum_{i} F(\rho'_i) \tag{5.29}$$

where

$$\rho'_i = \sum_{i \neq j} \varphi(r_{ij}) + \frac{1}{2}\sum_{k, j \neq i} f_{ij}(r_{ij}) f_{ik}(r_{ik}) g_i(\cos\theta_{ijk}) \tag{5.30}$$

θ_{ijk} is the angle between the ith, j^{th}, and k^{th} atoms. The explicit 3-body term is incorporated by introducing the functions f_{ij}, f_{ik}, and g_i.

During the indentation, after defining the sample as a cluster of atoms, the indenter itself can also be captured with full atomistic details. However, the indenter itself is commonly made from a very hard material such as diamond, which has negligible deformation compared to the metallic sample itself. Accordingly, the indenter can be modeled as repulsive potential. This can save a computational cost since Eq. (5.24) should not be solved for indenter anymore. Two common repulsive forces to mimic the interaction between the indenter and sample are as follows:

- The repulsive force between the spherical indenter and the sample can be defined as follows. (see, e.g., Yaghoobi and Voyiadjis, 2014):

$$\begin{aligned} F^{\text{ind}}(r) &= -K^{\text{ind}}(r - R)^2 \quad \text{for } r < R \\ F^{\text{ind}}(r) &= 0 \quad \text{for } r \geq R \end{aligned} \tag{5.31}$$

where K^{ind} is the force constant, r is the atomic distance to the indenter surface, and R is the radius of indenter.

- The interaction of the indenter with general geometry and the sample can be modeled using the following repulsive potential (see, e.g., Voyiadjis and Yaghoobi, 2015):

$$E^{\text{ind}}(r) = \varepsilon^{\text{ind}}(r - r_c)^2 \; r < r_c \tag{5.32}$$

where ε^{ind} is the force constant, r is the distance between the particle and indenter surface, and r_c is the cutoff distance.

One of the challenging aspect of nanoindentation experiment is to evaluate the correct contact area. This measurement directly influences the hardness value obtained from the experiment. An important advantageous of MD simulation of nanoindentation is that the contact area can be precisely obtained. To do so, one identifies the atom which are in direct contact with the indenter. The atoms should be then projected on the 2D plane. The contact area can be obtained using triangulation from the projected atoms.

Another challenge of nanoindentation experiment is to define the precise value of indentation depth, which may be different from the indenter displacement. In the case of MD simulation, however, this issue can be addressed using the precise contact area and geometrical relations, which depend on the indenter geometry. In the case of conical indenter, the indentation depth h is calculated as follows:

$$h = \frac{(a_c - a_0)}{\tan\theta} \tag{5.33}$$

where θ is the cone semi-angle, $a_c = \sqrt{A/\pi}$ is the contact radius, and $a_0 = r_2 + r_c(1/\cos\theta - \tan\theta)$.

In the case of the spherical tip, the indentation can be calculated as follows:

$$h = R - \sqrt{R^2 - a_c^2} \tag{5.34}$$

One common case of experimental tips are the conical indenter with the spherical tip. In this case, the indentation depth is divided to two parts. In the region where the sample is solely in contact with the spherical tip, Eq. (5.34) can be used to calculate the indentation depth. In the region where both the spherical and conical parts are in contact with the sample, the indentation depth can be defined as follows:

$$h = \frac{(a_c - a_0)}{\tan(\theta/2)} + h_0 \tag{5.35}$$

where h_0 is the depth at which the indenter geometry changes from spherical to conical, and a_0 is the contact radius at h_0 (Fig. 5.10). In the cases of cylindrical and right square prismatic indenters, one can approximate the indentation depth h with tip displacement d, i.e., $h \approx d$.

MD simulation only provides the atomic trajectories and velocities, which is the raw data of atomistic simulation. However, in order to capture the size effects mechanisms, one should be able to extract the defects structure from this raw data. In order to visualize the defects, several methods have been introduced such as energy filtering, bond order, centrosymmetry parameter, adaptive common neighbor analysis, Voronoi analysis, and neighbor distance analysis which

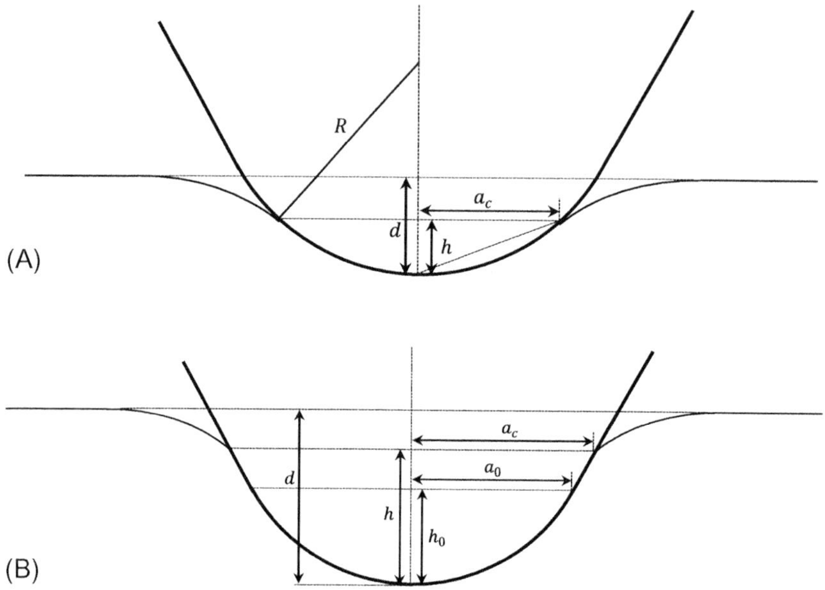

FIG. 5.10 The true indentation depth h for: (A) spherical part of the indenter and (B) conical part of the indenter. *(After Yaghoobi, M., Voyiadjis, G.Z., 2016a. Atomistic simulation of size effects in single-crystalline metals of confined volumes during nanoindentation. Comput. Mater. Sci. 111, 64–73.)*

have been compared with each other by Stukowski (2012). Furthermore, the Crystal Analysis Tool developed by Stukowski and his coworkers (Stukowski and Albe, 2010; Stukowski et al., 2012; Stukowski, 2012, 2014) have been incorporated to extract the dislocations from the atomistic data. Here, centrosymmetry parameter (*CSP*) and Crystal Analysis Tool are explained in more detail. *CSP* is defined as follows (Kelchner et al., 1998):

$$CSP = \sum_{i=1}^{N_p} \left| \mathbf{R}_i + \mathbf{R}_{i+N_p} \right|^2 \tag{5.36}$$

where $\mathbf{R}_i$ and $\mathbf{R}_{i+N_p}$ are vectors from the considered atom to the ith pair of neighbors, and N_p depends on the crystal structure. For example, $N_p = 6$ for FCC materials. *CSP* is equal to zero for perfect crystal structure. However, the atomic vibration introduces a small *CSP* for atoms which are not defects. Accordingly, a cutoff should be introduced in a way that if $CSP_i < CSP_{\text{cutoff}}$, the ith atom is not considered as a defect (Yaghoobi and Voyiadjis, 2014). Also, point defect is removed to clearly illustrate stacking faults. Second, the MD outputs can be post-processed using the Crystal Analysis Tool (Stukowski and Albe, 2010; Stukowski, 2012, 2014; Stukowski et al., 2012). The common-neighbor analysis method (Faken and Jonsson, 1994) is the basic idea of this code. The code is able to calculate the dislocations information such as the Burgers vector and total dislocation length. To extract the required information, the Crystal Analysis Tool constructs a Delaunay mesh which connects all atoms. Next, using the constructed mesh, the elastic deformation gradient tensor is obtained. The code defines the dislocations using the fact that the elastic deformation gradient does not have a unique value when a tessellation element intersects a dislocation.

5.2.4.2 Boundary conditions effects

In order to simulate the nanoindentation experiment using MD, a very important step is to appropriately assign a set of boundary conditions to the sample which appropriately captures the test condition. The selected BCs may influence the observed behavior during the nanoindentation experiment. In the literature, four different sets of boundary conditions have been incorporated to simulate the nanoindentation experiments as follows (Fig. 5.11):

- *BC1*: Fixing some atomic layers at the sample bottom to act as a substrate, using free surface for the top and periodic boundary conditions for the remaining surfaces (see e.g. Nair et al., 2008; Kelchner et al., 1998; Zimmerman et al., 2001).
- *BC2:* Fixing some atomic layer at the surrounding surfaces and using free surfaces for the sample top and bottom (see e.g. Medyanik and Shao, 2009; Shao and Medyanik, 2010).

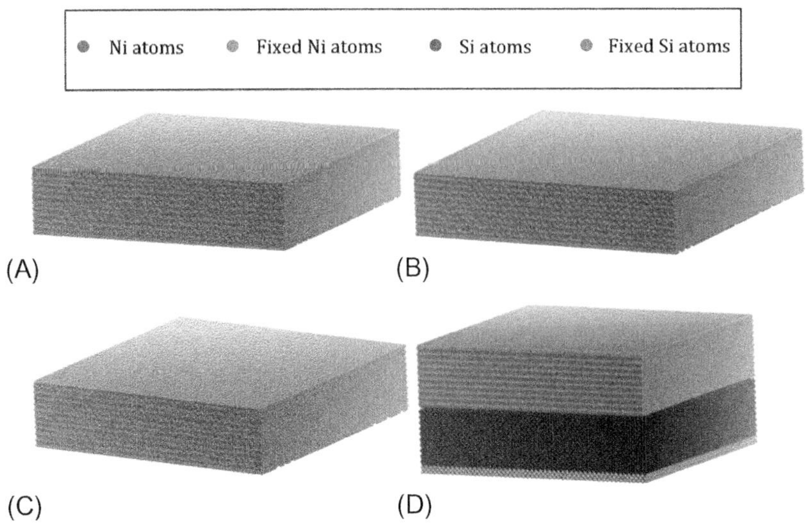

FIG. 5.11 Boundary conditions of thin films (A) BC1, (B) BC2, (C) BC3, and (D) BC4. *(After Yaghoobi, M., Voyiadjis, G.Z., 2014. Effect of boundary conditions on the MD simulation of nanoindentation. Comput. Mater. Sci. 95, 626–636.)*

- *BC3:* Using free surface for the sample top and bottom, incorporating the periodic boundary conditions for the remaining surfaces, and putting a substrate under the thin film (see e.g. Peng et al., 2010).
- *BC4:* Incorporating the free surfaces for the sample top and bottom, using periodic boundary conditions for the remaining surfaces, and equilibrating the sample by adding some forces (see e.g. Li et al., 2002; Lee et al., 2005).

Yaghoobi and Voyiadjis (2014) elaborately investigated the described choices of boundary conditions for MD simulation of nanoindentation using a spherical indenter. In their study, different sizes of sample thickness (t_f) and indenter radius (R). Accordingly, they addresses the effects of boundary conditions on the dislocation nucleation and evolution along with the indentation hardness and elastic modulus. The molecular dynamics code LAMMPS (Plimpton, 1995) is selected to conduct the simulation, which is well-stablished and highly parallel MD code developed at Sandia National Laboratories. The velocity Verlet scheme is selected to numerically integrate the Newton's equation of motion described in Eq. (5.24). The interatomic potentials are used to capture intermolecular forces, which can be described as follows:

- The interaction of Nickel atoms with each other (Ni-Ni).
- The interaction of Silicon atoms with each other (Si-Si).
- The interaction of Nickel atoms with Silicon ones (Ni-Si).

The three interactions were modeled using the embedded-atom method (EAM) potential for Ni-Ni interaction, Tersoff potential for Si-Si interaction, and Lennard-Jones (LJ) potential for Ni-Si interaction. The Ni-Ni interaction is modeled using the EAM potential parameterized by Mishin et al. (1999). To capture the Si-Si interaction, a three-body Tersoff potential (Tersoff, 1988) were chosen. Table 5.1 presents the potential parameters of Si. The LJ potential E^{LJ} was used to model the Ni-Si interaction and required parameters ($\varepsilon_{\mathrm{Ni-Si}}$ and $\sigma_{\mathrm{Ni-Si}}$) are presented in Table 5.2. The cutoff distance of 2.5σ was selected for L-J potential. The indenter was spherical modeled using the repulsive force F^{ind}

TABLE 5.1 Tersoff potential parameters of Si-Si (Yaghoobi and Voyiadjis, 2014)

$A = 3264.7$ eV	$B = 95.373$ eV	
$\lambda_1 = 3.2394\ \mathring{A}^{-1}$	$\lambda_2 = 1.3258\ \mathring{A}^{-1}$	
$\alpha = 0$	$\beta = 0.33675$	$n = 22.956$
$c = 4.8381$	$d = 2.0417$	$h = 0.0000$
$\lambda_3 = \lambda_2$	$R = 3.0\ \mathring{A}$	$D = 0.2\ \mathring{A}$

TABLE 5.2 LJ potential parameters of Ni-Ni, Si-Si, and Ni-Si (Yaghoobi and Voyiadjis, 2014)

	ε (J)	$\sigma(\mathring{A})$
Ni-Ni	8.3134e−20	2.2808
Si-Si	2.7904e−21	3.8260
Ni-Si	1.5231e−20	3.0534

presented in Eq. (5.8). The centrosymmetry parameter (*CSP*) was incorporated to visualize the defects with the cutoff equal to 1.5 ($CSP_{\text{cutoff}} = 1.5$).

Yaghoobi and Voyiadjis (2014) observed that two mechanisms of indentation and bending govern the primary stages of dislocation nucleation and evolution, which also depend on the sample thickness. Three defect patterns were reported as illustrated in Fig. 5.12. The characteristics of these three patterns can be identified as follows:

- *Type I*: The location of initial defect nucleation is beneath the indenter. Two faces of embryonic dislocation loops evolve on $\left(1\,\bar{1}\,1\right)$ and $\left(1\,1\,\bar{1}\right)$ planes. Eventually, a tetrahedral sessile lock is formed. This dislocation pattern is controlled by the indentation mechanism.
- *Type II*: Again, the location of initial nucleation is beneath the indenter. However, the defect evolution occurs on the plane which is parallel to (1 1 1), which is the indentation plane. A dislocation pattern similar to *Type I* starts evolving as the indentation depth increases. Both bending and indentation mechanisms are important in this pattern.
- *Type III*: The initial dislocation is nucleated at the sample bottom. The dislocations are evolved on {1 1 1} planes while they are moving toward the sample top. Bending is the dominant mechanism of deformation.

The nanoindentation response of the samples with defect pattern presented in Fig. 5.12 is illustrated in Fig. 5.13. The response of the sample with the defect pattern of *Type I* is presented in Fig. 5.13A. Two faces of embryonic dislocation loops evolve on $\left(1\,\bar{1}\,1\right)$ and $\left(1\,1\,\bar{1}\right)$ planes. Another load relation happens following the formation of the tetrahedral sessile lock. Fig. 5.13B shows the nanoindentation response of the sample corresponding to the defect pattern of *Type II*. Again, the first load relaxation occurs following the initial defect nucleation beneath the indenter. The defect evolution occurs on the plane which is parallel to (1 1 1), which is the indentation plane. In this mechanism, both mechanisms of bending and indentation are significant. Accordingly, the defect evolution pattern becomes complex which leads to a complicated nanoindentation response. Fig. 5.13C depicts the response of the sample with the defect

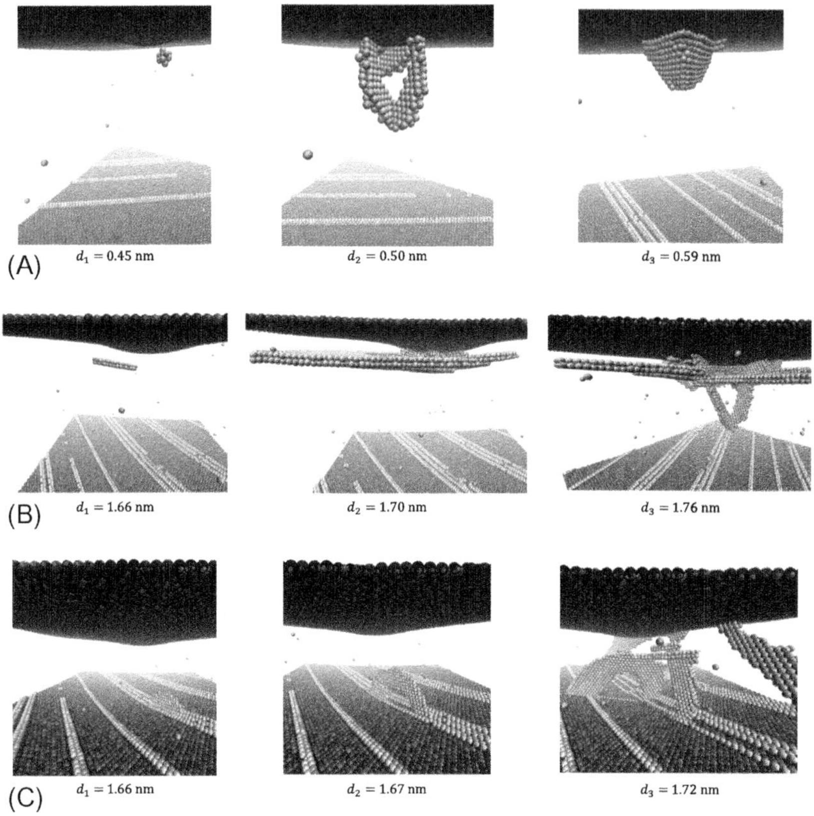

FIG. 5.12 Defect nucleation and evolution of (A) Type I, (B) Type II, and (C) Type III. *(After Yaghoobi, M., Voyiadjis, G.Z., 2014. Effect of boundary conditions on the MD simulation of nanoindentation. Comput. Mater. Sci. 95, 626–636.)*

pattern of *Type III* during nanoindentation. The initial dislocation nucleation leads to the first load relaxation, which is located at the bottom of the sample. Afterwards, the oscillatory response was reported which is due to the complexity of dislocation pattern. The general trend, however, still dictates the increase in indentation force as the indentation proceeds.

Yaghoobi and Voyiadjis (2014) investigates the effects of film thickness t_f and indenter radius R on the microstructural mechanisms of deformation. In the case of sample with BC1, irrespective of the selected sample thickness or indenter radius, the film does not experience bending and the defects nucleation and evolution pattern is always *Type I*. However, in the case of sample with BC2 and BC3, all three patterns of *Type I*, *Type II*, and *Type III* are possible depending on the sample thickness and indenter radius. Yaghoobi and Voyiadjis (2014) showed that the controlling geometrical parameter is the value of the indenter radius divided by the film thickness, i.e., R/t_f. In other words, the indentation is

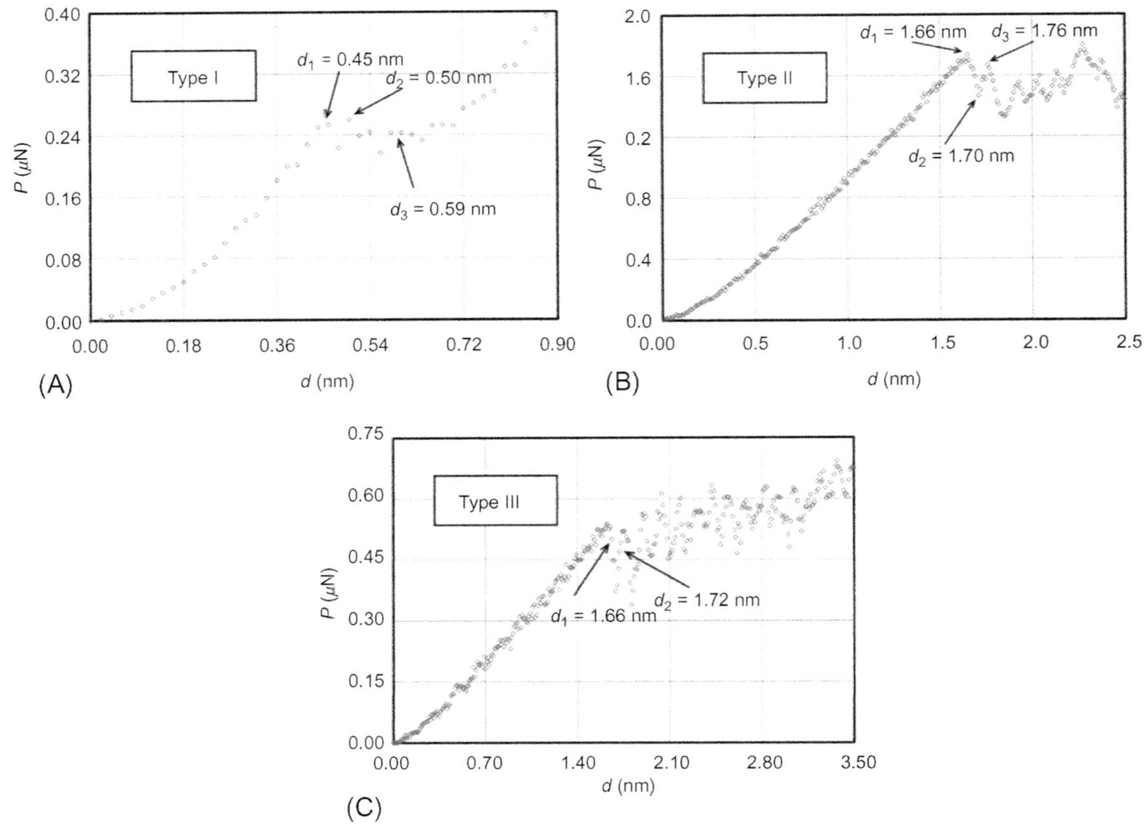

FIG. 5.13 Nanoindentation responses of thin films with different dislocation nucleation patterns of (A) Type I, (B) Type II, and (C) Type III. *(Reprinted with permission from Yaghoobi, M., Voyiadjis, G.Z., 2014. Effect of boundary conditions on the MD simulation of nanoindentation. Comput. Mater. Sci. 95, 626–636.)*

the dominant mode for samples with very small R/t_f, accordingly which leads to the defect pattern of *Type I*. As the value of R/t_f increases, the role of bending mode cannot be neglected anymore. Consequently, the defect pattern of *Type I* occurs. The bending mode becomes dominant by further increasing the value of R/t_f which leads to the defect pattern of *Type III*. In the case of sample with BC4 and for very small values of R/t_f, indentation is only deformation mode and defect pattern of *Type I* is observed. Increasing the value of R/t_f, the bending mode cannot be neglected and defect pattern of *Type II* is reported.

Yaghoobi and Voyiadjis (2014) also studied the contact pressure of the plasticity initiation p_m^y during the nanoindentation. The MD simulation results showed that the p_m^y depends on the choice of the selected boundary conditions, film thickness, and indenter radius. Yaghoobi and Voyiadjis (2014) capture this effect by incorporating the three extracted defect patterns. They showed that the contact pressure of the plasticity initiation p_m^y is independent of film thickness for the defect pattern of *Type I*. In the cases of defect patterns of *Type II* and *Type III*, however, the contact pressure of the plasticity initiation p_m^y decreases by decreasing the sample thickness. Finally, Yaghoobi and Voyiadjis (2014) reported that the samples with the defect pattern of *Type I* has the largest and sample with the defect pattern of *Type III* has the smallest contact pressure at the time of plasticity initiation.

5.2.4.3 Comparing MD results with theoretical models

Different theoretical models have been proposed to capture the dislocation content during nanoindentation test. As described in Section 5.2.2, the concept of geometrically necessary dislocations (GNDs) developed by Ashby (1970) is incorporated to capture the dislocation length during indentation. The model developed by Nix and Gao (1998) is a milestone in these category of models which predicts the variation of GNDs length versus the indentation depth in the cases of conical indenter, which is presented in Eq. (5.1). Pugno (2007) extends the model to predict the general geometry, which is presented in Eq. (5.13). Accordingly, in the cases of cylindrical and right square prismatic indenters, the variation of GNDs length versus the indentation depth can be obtained as follows:

$$\lambda_{cy} = \frac{2\pi a_c h}{b} \tag{5.37}$$

$$\lambda_{pr} = \frac{4ch}{b} \tag{5.38}$$

where b is the magnitude of the Burgers vector and $c = \sqrt{A}$.

Voyiadjis and Yaghoobi (2015) investigated the theoretical models of dislocation length prediction during nanoindentation with different tip geometries of cylindrical, right square prismatic, and conical. To do so, they incorporated MD as a pseudo-experimental tool and they modeled the nanoindentation of a

Ni thin film. They simulated a relatively large Ni sample to remove the size effects of the boundary conditions on the obtained results with the dimensions of 1200 nm, 1200 nm, and 600 nm along $[1\,\bar{1}\,0]$, $[1\,1\,\bar{2}]$, and [1 1 1] directions, respectively. The sample consists of 80 million atoms. The indenter geometries are as follows:

- $r_1 = 4.8$ nm is the indentation surface radius of the cylindrical indenter.
- A 7.5 nm × 7.5 nm square is the indentation surface of the right square prismatic indenter.
- $r_2 = 0.3$ nm is the smaller radius of blunt conical indenter with the cone semi-angle of $\theta = 56.31^{\circ}$.

The MD simulation is performed using the code of LAMMPS (Plimpton, 1995). The velocity Verlet scheme is selected to numerically integrate the Newton's equation of motion described in Eq. (5.24). Among the defined set of boundary conditions investigated in Section 5.2.4.2, BC4 is selected to mimic the nanoindentation test. The Ni-Ni interatomic interaction is captured using the EAM potential developed by Mishin et al. (1999). Eq. (5.32) is used to model the repulsive interaction between the Ni thin film and the indenter. The Crystal Analysis Tool (Stukowski and Albe, 2010; Stukowski, 2012, 2014; Stukowski et al., 2012) is finally used to post-process the MD outputs and extract the dislocation network.

The nucleation and evolution of defect was the first aspect of nanoindentation experiment that Voyiadjis and Yaghoobi (2015) investigated. Fig. 5.14 illustrates the dislocation nucleation and evolution which leads to the dislocation loops emission in the case of cylindrical indenter. OVITO (Stukowski, 2010) was selected to visualize the extracted dislocation network. The atoms without any defects, i.e., perfect crystal, was excluded in order to obtain a clear representation of dislocations and stacking faults. Different dislocation types of Shockley, Hirth, and stair-rod partial dislocations and perfect dislocations were formed during nanoindentation, which are illustrated with the colors of green, yellow, blue, and red, respectively. One of the significant aspects of dislocation structure evolution is the formation of prismatic dislocations loops and their movement. Fig. 5.15 depicts the prismatic dislocation loops movements along three directions of $[\bar{1}\,0\,\bar{1}]$, $[\bar{1}\,\bar{1}\,0]$, and $[0\,\bar{1}\,\bar{1}]$, which is dictated by the crystal structure.

The next issue Voyiadjis and Yaghoobi (2015) addressed was the variation of plastic zone size during nanoindentation. As discussed in Section 5.2.2, Feng and Nix (2004) and Durst et al. (2005) stated that the original assumption for plastic zone radius to be equal to the contact radius, i.e., $R_{\mathrm{pz}} = a_p$, which was introduced by Nix and Gao (1998), cannot capture the indentation size effects. Accordingly, Voyiadjis and Yaghoobi (2015) incorporated the atomistic simulation data to investigate the variation of plastic zone size during nanoindentation. In order to quantitatively measure the plastic zone size, Voyiadjis and Yaghoobi (2016) proposed that the plastic zone solely includes

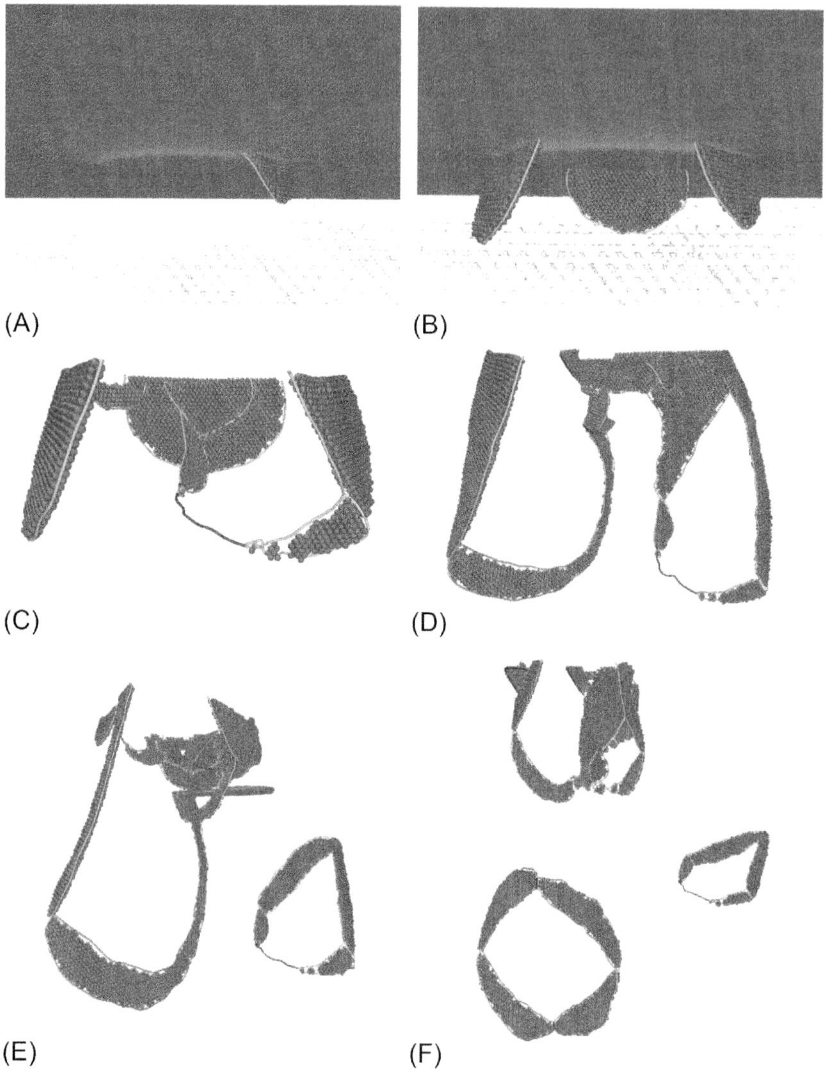

FIG. 5.14 Defect nucleation and evolution of Ni thin film indented by the cylindrical indenter at (A) $d \approx 0.70$ nm, (B) $d \approx 0.86$ nm, (C) $d \approx 0.96$ nm, (D) $d \approx 1.02$ nm, (E) $d \approx 1.05$ nm, and (F) $d \approx 1.12$ nm. *(After Voyiadjis, G.Z., Yaghoobi, M., 2015. Large scale atomistic simulation of size effects during nanoindentation: dislocation length and hardness. Mater. Sci. Eng. A 634, 20–31.)*

the main dislocation network beneath the indenter, and the prismatic dislocation loops detached from the main network should not be considered inside the plastic zone. Fig. 5.16 depicts the variation of plastic zone radius as the indentation proceeds. In order to investigate the theoretical models, Voyiadjis and Yaghoobi (2015) used the plastic zone radius proposed by Feng and Nix (2004) as described in Eq. (5.19), i.e., $R_{pz} = fa_p$. Accordingly, they studied

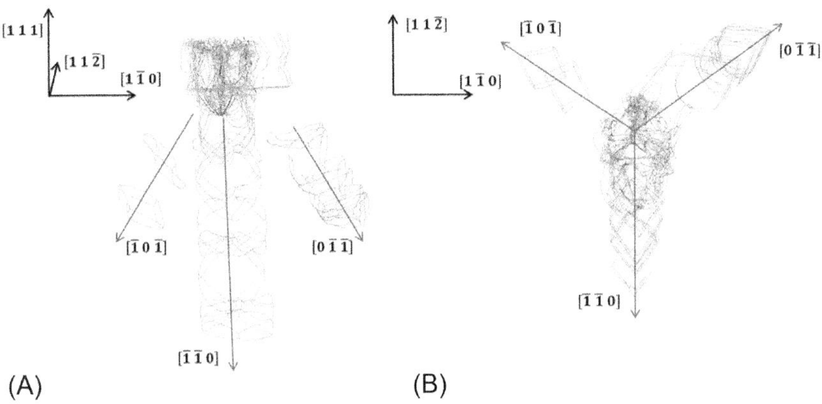

FIG. 5.15 Prismatic loops forming and movement in Ni thin film indented by the cylindrical indenter during nanoindentation (A) side view and (B) top view. *(After Voyiadjis, G.Z., Yaghoobi, M., 2015. Large scale atomistic simulation of size effects during nanoindentation: dislocation length and hardness. Mater. Sci. Eng. A 634, 20–31.)*

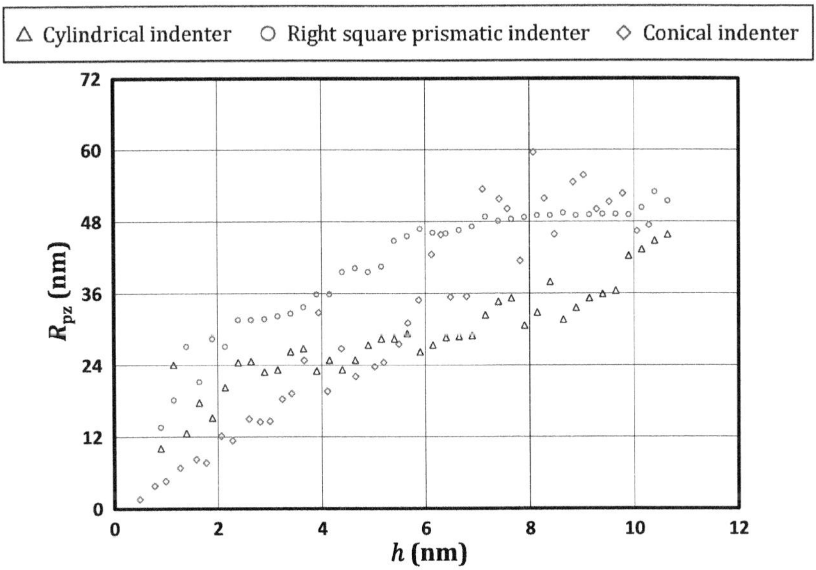

FIG. 5.16 Radius of plastic zone (R_{pz}) versus the indentation depth of thin films indented by the indenters with different geometries of cylindrical, right square prismatic, and conical. *(After Voyiadjis, G.Z., Yaghoobi, M., 2015. Large scale atomistic simulation of size effects during nanoindentation: dislocation length and hardness. Mater. Sci. Eng. A 634, 20–31.)*

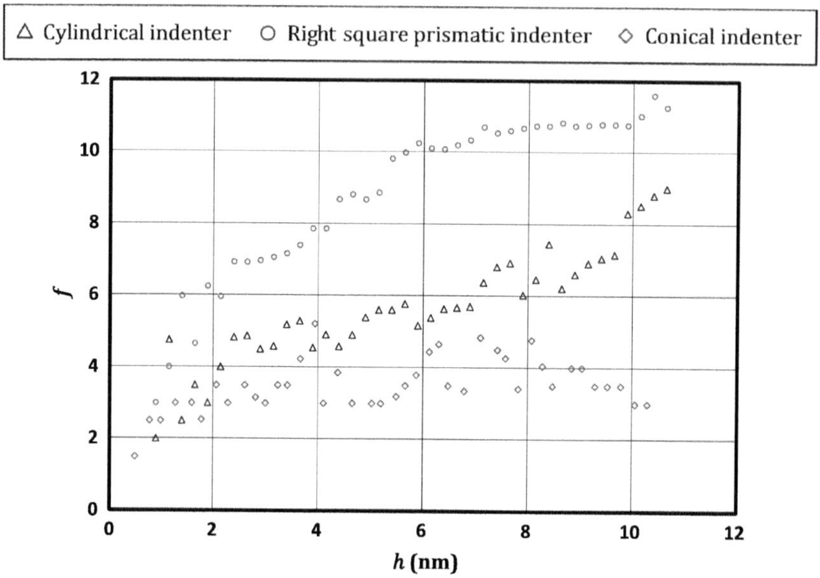

FIG. 5.17 The parameter *f* versus the indentation depth of thin films indented by the indenters with different geometries of cylindrical, right square prismatic, and conical. *(After Voyiadjis, G.Z., Yaghoobi, M., 2015. Large scale atomistic simulation of size effects during nanoindentation: dislocation length and hardness. Mater. Sci. Eng. A 634, 20–31.)*

the variation of f during the nanoindentation test for different tip geometries (Fig. 5.17). They used the equivalent contact radius in the case of right square prismatic indenters as $a_p = \sqrt{A/\pi}$, where A is the contact radius. In the case of conical indenter, the results show that the value of f oscillates around a constant value, and $1.5 \leq f \leq 5.2$. However, in the cases of cylindrical and right square prismatic indenters, value of f increases as the indentation depth increases, and $2 \leq f \leq 9$ and $3 \leq f \leq 11.6$ for cylindrical and right square prismatic indenters, respectively.

Quite novel and important results reported by Voyiadjis and Yaghoobi (2015) was the comparison between the dislocation length predicted by theoretical models and the one captured using the MD simulation. Fig. 5.18 compares the variation of dislocation length as the indentation depth increases between the theoretical models and MD simulation for three indenter geometries of cylindrical, right square prismatic, and conical. Fig. 5.18 shows that the theoretical models can accurately capture the dislocation length variation during nanoindentation and agree well with that of MD simulation. However, Voyiadjis and Yaghoobi (2015) observed some slight differences between the dislocation length obtained using MD and the one obtained using the theoretical models. They attributed these discrepancies to the following arguments:

- The theoretical models developed to capture the indentation size effects predict the GNDs length and not the total dislocation length, which also

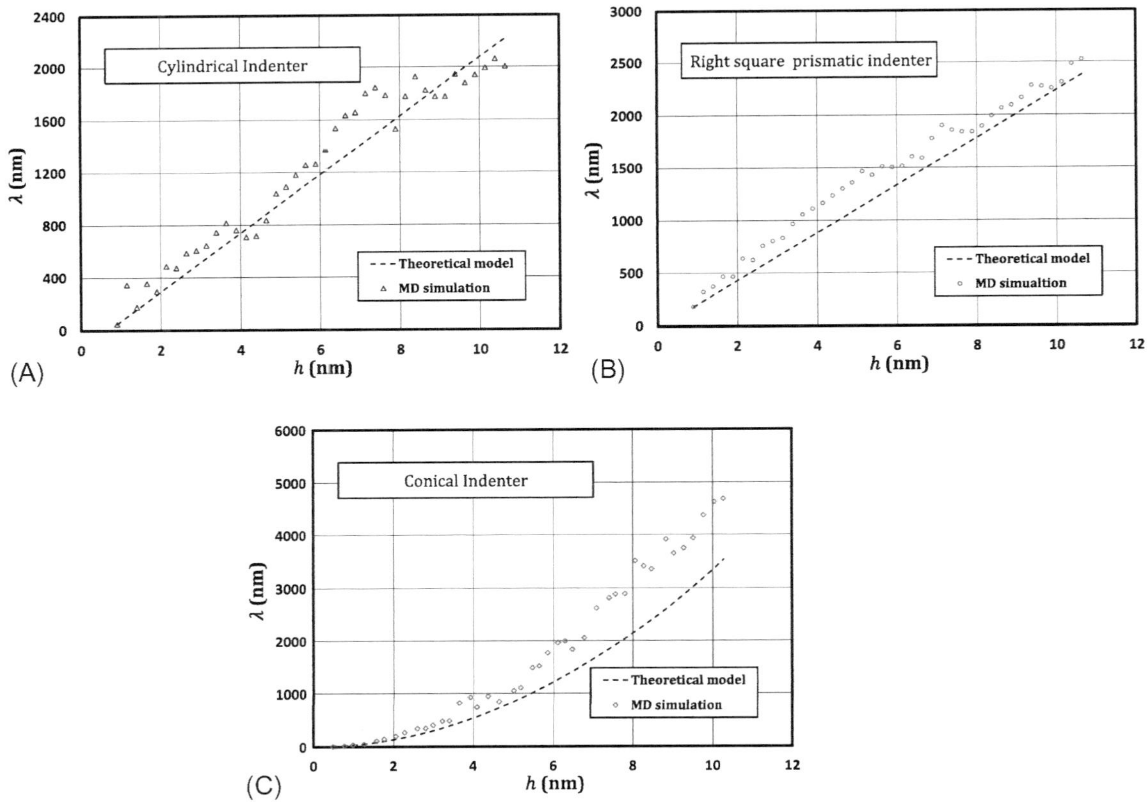

FIG. 5.18 Total dislocation length obtained from simulation and theoretical models in samples indented by the (A) cylindrical indenter, (B) right square prismatic indenter, and (C) conical indenter. *(After Voyiadjis, G.Z., Yaghoobi, M., 2015. Large scale atomistic simulation of size effects during nanoindentation: dislocation length and hardness. Mater. Sci. Eng. A 634, 20–31.)*

includes SSDs. MD simulation, however, extracts the total dislocation length. Accordingly, the dislocation length obtained by MD is generally larger than the one calculated using the theoretical models.
- During the calculation of GNDs dislocation length using theoretical models, one should incorporate a Burgers vector. Voyiadjis and Yaghoobi (2016) used the value for Shockley partial dislocations which was the dominant dislocation type. Accordingly, they neglected the difference between the Burgers vector of other dislocation types including stair-rod and Hirth partial and perfect dislocations with that of the Shockley partial dislocations.
- In the case of MD simulation, Voyiadjis and Yaghoobi (2015) reported the total dislocation length as those which are located inside the plastic zone and excluded the prismatic loops separating from the main body of dislocations.

5.2.4.4 Size effects in small length scales during nanoindentation

As discussed in Section 5.2.2, the conventional models developed to capture the indentation size effects, such as Nix and Gao model (1998), incorporates the forest hardening model along with the strain gradient model. The Nix and Gao model cannot capture the indentation size effects for small depths. Feng and Nix (2004) and Huang et al. (2006) tried to enhance the model; however, the proposed modifications are highly phenomenological. Also, they still attributed the size effects for small indentation depth to the interaction of dislocations with each other, i.e., forest hardening mechanism. As discussed in Section 5.2.3, recent advancements in experimental schemes allow researchers to quantitatively capture the dislocation content. Accordingly, a few studies have been conducted to investigate the variation of dislocation density during nanoindentation. The results presented in the literature can be categorized into two categories. One branch of studies, including Kiener et al. (2006), McLaughlin and Clegg (2008), and Demir et al. (2009), showed that the conventional size effects model, i.e., forest hardening mechanism along with the strain gradient model, cannot capture size effects for small indentation depths. On the other hand, Feng et al. (2008) showed that the conventional theories can capture the indentation size effects even for small indentation depths.

Yaghoobi and Voyiadjis (2016a) used MD simulation of nanoindentation as a pseudo-experiment to unravel the real indentation size effects for small indentation depths. To do so, they chose a single crystal Ni sample with the dimensions of 120 nm, 120 nm, and 60 nm along $[1\bar{1}0]$, $[11\bar{2}]$, and [1 1 1] directions, respectively. Yaghoobi and Voyiadjis (2016a) selected a conical indenter with a spherical tip which is similar to the one used in the experiment of Demir et al. (2009). The rest of MD simulation scope and methodology is similar to the one presented in Section 5.2.4.3.

In the first step, the dislocation length variation during nanoindentation obtained from the MD simulation was compared with the theoretical models. While the sample is in contact with the spherical part of the indenter, two expressions of Swadener et al. (2002) and Yaghoobi and Voyiadjis (2016a) can be used. Swadener et al. (2002) approximated the spherical indenter with a parabola and obtained the total length of GNDs during nanoindentation as follows:

$$\lambda_{sp} \approx \frac{2\pi}{3}\frac{a_p^3}{bR} \tag{5.39}$$

The approximation, however, is only applicable for small indentation depths. Yaghoobi and Voyiadjis (2016a) introduced a theoretical equation to predict the dislocation length of sample indented by a spherical tip using the precise geometry of the indenter. The total dislocation length can be described as below (Yaghoobi and Voyiadjis, 2016a):

$$\begin{aligned}\lambda_{sp} &= \int_0^{a_p} \frac{2\pi r}{b}\left(\frac{\mathrm{d}h}{\mathrm{d}r}\right)\mathrm{d}r = \frac{2\pi}{b}\int_0^{a_p}\left(\frac{r^2}{\sqrt{R^2-r^2}}\right)\mathrm{d}r \\ &= \frac{2\pi}{b}\left[\frac{R^2}{2}\sin^{-1}\left(\frac{a_p}{R}\right) - \frac{1}{2}\left(a_p\sqrt{R^2-a_p{}^2}\right)\right]\end{aligned} \tag{5.40}$$

In the case of a sample being in contact with both spherical and conical parts of the indenter, the total GNDs length can be modified as follows:

$$\lambda_{co} = \lambda_{sp}\big|_{h=h_0} + \int_{a_0}^{a_p} \frac{2\pi r}{b\tan\left(\dfrac{\theta}{2}\right)}\mathrm{d}r = \lambda_{sp}\big|_{h=h_0} + \frac{\pi\left(a_p{}^2 - a_0{}^2\right)}{b\tan\left(\dfrac{\theta}{2}\right)} \tag{5.41}$$

where $\lambda_{sp}|_{h=h_0}$ is the GNDs dislocation length described by Eqs. (5.39), (5.40) at indentation depth of $h = h_0$, at which the transition occurs in the indenter geometry from the spherical to a conical, and a_0 is the contact radius at indentation depth of $h = h_0$. Fig. 5.19 compares the dislocation length obtained from atomistic simulation with those calculated from the approximate and precise theoretical models during nanoindentation. The GNDs length calculated from the theoretical model is a lower bound for the total dislocation length obtained from MD, which includes both GNDs and SSDs.

Yaghoobi and Voyiadjis (2016a) studied the variation of dislocation density during the nanoindentation using the theoretical models presented in Eqs. (5.39), (5.40). To do so, they defined the plastic zone according to the scheme introduced by Durst et al. (2005), in which the plastic zone is a hemisphere with a radius of $R_{pz} = fa_p$, where $f = 1.9$. First, they considered the same plastic zone and calculated the dislocation density as follows:

$$\rho = \lambda / V \tag{5.42}$$

where V is the plastic zone volume. They compared the dislocation density obtained from the approximated and precise dislocation length obtained from Eqs. (5.39), (5.40), respectively (Fig. 5.20). The results show that approximating the spherical indenter as a parabola (Swadener et al., 2002) results in a

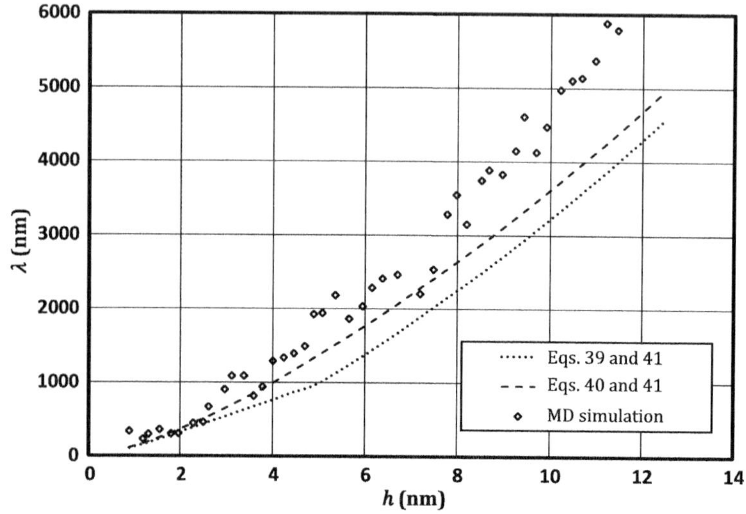

FIG. 5.19 Comparison between the dislocation lengths obtained from theoretical models and MD simulation during nanoindentation. *(After Yaghoobi, M., Voyiadjis, G.Z., 2016a. Atomistic simulation of size effects in single-crystalline metals of confined volumes during nanoindentation. Comput. Mater. Sci. 111, 64–73.)*

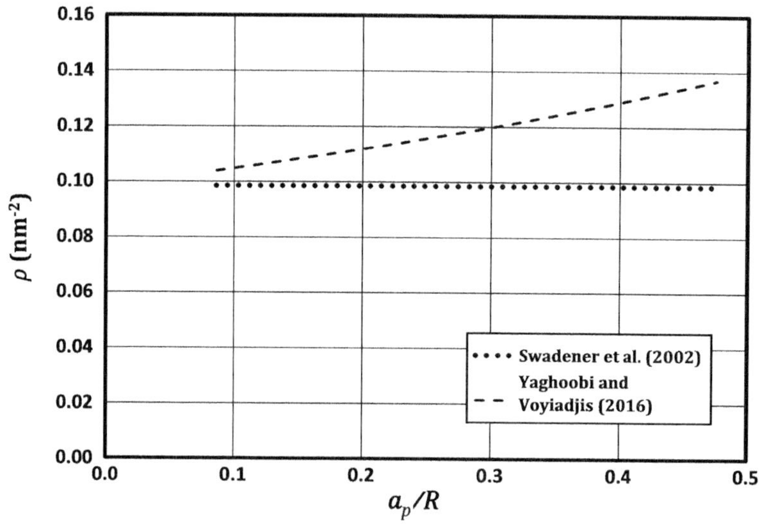

FIG. 5.20 Theoretical GNDs density obtained from the approximate (Swadener et al., 2002) and exact (Yaghoobi and Voyiadjis, 2016a) geometries of the indenter versus the normalized contact radius a_p/R. *(After Yaghoobi, M., Voyiadjis, G.Z., 2016a. Atomistic simulation of size effects in single-crystalline metals of confined volumes during nanoindentation. Comput. Mater. Sci. 111, 64–73.)*

constant dislocation density. However, the dislocation density obtained from the precise geometry (Yaghoobi and Voyiadjis, 2016a) increases as the indentation proceeds. Also, the results show that by choosing a parabola gives a lower bound for the dislocation density obtained by using the precise geometry.

In the next step, Yaghoobi and Voyiadjis (2016a) investigated the variation of dislocation density by using the MD simulation result. They considered the plastic zone as a hemisphere with a radius of $R_{pz} = fa_p$, where various values of $f = 1.5, 2.0, 2.5, 3.0$, and 3.5 were selected. The volume of plastic zone then can be precisely obtained by excluding the indenter volume (V_{indenter}) as follows:

$$V = (2/3)\pi(fa_c)^3 - V_{\text{indenter}} \tag{5.43}$$

The dislocation density can then be calculated using Eqs. (5.42), (5.43). Fig. 5.21 shows the dislocation density evolution as the indentation proceeds for different values of plastic zone sizes. The results show that the dislocation density increases during nanoindentation which is similar to the trend obtained from the theoretical model considering the precise indenter geometry as presented in Eq. (5.40) (see Fig. 5.20).

Fig. 5.22 depicts the curve of $p_m - h$, i.e., mean contact pressure ($p_m = P/A$) versus the indentation depth. The mean contact pressure p_m is equivalent to the hardness H in the plastic region. In the elastic region, the mean contact

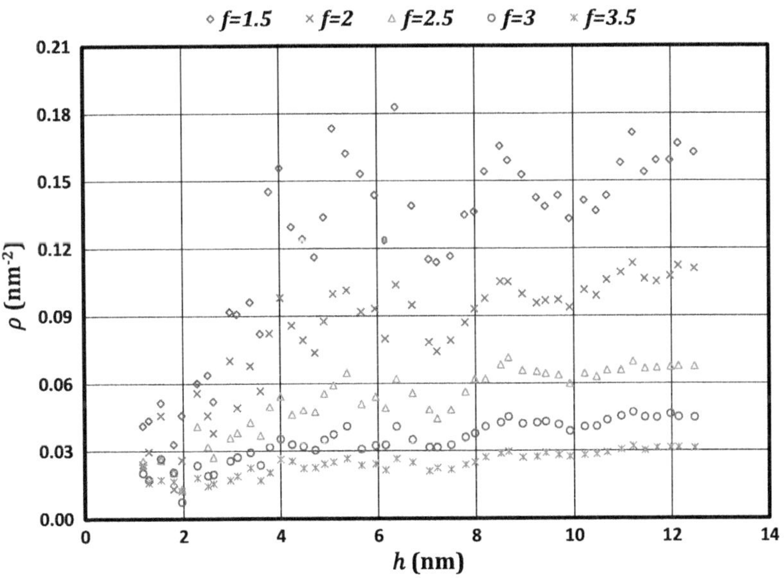

FIG. 5.21 Dislocation density obtained from MD simulation for different values of f during nanoindentation. *(After Yaghoobi, M., Voyiadjis, G.Z., 2016a. Atomistic simulation of size effects in single-crystalline metals of confined volumes during nanoindentation. Comput. Mater. Sci. 111, 64–73.)*

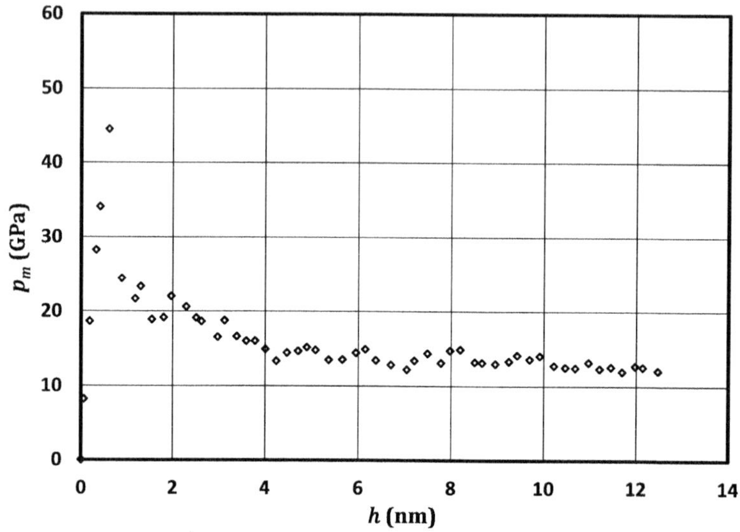

FIG. 5.22 Variation of mean contact pressure p_m as a function of indentation depth h. *(After Yaghoobi, M., Voyiadjis, G.Z., 2016a. Atomistic simulation of size effects in single-crystalline metals of confined volumes during nanoindentation. Comput. Mater. Sci. 111, 64–73.)*

pressure p_m increases as the indentation depth h increases. This trend can be captured using the Hertzian theory, which is valid until the first dislocation is nucleated. Afterwards, the results show that the mean contact pressure p_m and equivalently the hardness decreases as the indentation depth increases.

In the next step, Yaghoobi and Voyiadjis (2016a) investigated the governing mechanism of indentation size effects using the atomistic data. First, they studied the forest hardening mechanism, as it has been conventionally considered to govern the indentation size effects. Forest hardening mechanism is responsible for size effects in bulk-sized materials which attributed the size effects to the dislocations interactions with one another. The Taylor model and its derivations is commonly incorporated to capture the forest hardening mechanism. Voyiadjis and Abu Al-Rub (2005) modified the original Taylor model to separate the contributions of SSDs and GNDs to the material shear strength τ as follows:

$$\tau = \alpha_S \mu b_S \sqrt{\rho}$$
$$\rho = \left[\rho_S^{\beta/2} + \left(\alpha_G^2 b_G^2 \rho_G / \alpha_S^2 b_S^2\right)^{\beta/2}\right]^{2/\beta} \tag{5.44}$$

where α is a constant, μ is the shear modulus, and the indices G and S designate GNDs and SSDs parameters, respectively. Eq. (5.44) represents the forest hardening mechanism in which the material strength increases as the dislocation density increases due to the interaction of dislocations with one another. According to the MD simulation, dislocation density increases as the indentation depth increases (see Fig. 5.21). On the other hand, the hardness obtained from the atomistic simulation decreases by proceeding the nanoindentation (see Fig. 5.22). One can see that hardness decreases as the dislocation density increases. The results show that the forest hardening mechanism cannot model the observed indentation size effects. Consequently, the atomistic simulation results reported by Yaghoobi and Voyiadjis (2016a) supported the experiments conducted by Kiener et al. (2006), McLaughlin and Clegg (2008), and Demir et al. (2009) and showed that the conventional size effects model, i.e., forest hardening mechanism, cannot capture size effects for small indentation depths.

In order to address the sources of size effects at small indentation depths, Yaghoobi and Voyiadjis (2016a) studied the pattern of defect formation during nanoindentation. Fig. 5.23 illustrates the initial stages of defects formation. The results show that the dislocation sources are mainly created using the cross-slip mechanism. Elongation of dislocations pinned at their ends provides the required dislocation length to sustain the imposed deformation. Dislocation loops are then released by pinching off of screw dislocations and glide along the three directions of $\left[\bar{1}\,\bar{1}\,0\right]$, $\left[0\,\bar{1}\,\bar{1}\right]$, and $\left[\bar{1}\,0\,\bar{1}\right]$. After the initial dislocation nucleation, the available dislocation length is insufficient to sustain the imposed deformation, and the source exhaustion controls the size effects. Consequently, the required stress reduces as the dislocation length and density increase during nanoindentation. The dislocation density and length are eventually reaching the value required to sustain the imposed plastic flow, and the hardness tends to a constant value. Also, the interaction of dislocation with each other becomes important by increasing the dislocation length. However, the dislocation density reaches a constant value and forest hardening mechanism does not lead to any size effects.

5.2.4.5 *Effects of grain boundary on the nanoindentation response of thin films*

One of the key characteristics of the polycrystalline metals is the grain boundary (GB). GB governs many aspect of the polycrystalline material response. Various researchers investigated the role of GB on the material behavior, which has been summarized in some noticeable review works, such as Meyers et al. (2006), Koch et al. (2007), and Zhu et al. (2008). Effects of GB on the material response depends on the grain size. In the case of coarse-grained polycrystalline samples, the grains are large enough to accommodate enough dislocations. In other words, the interactions of dislocations with each other and GB govern the size effects. In the case of coarse-grained samples, the grain size effect is commonly described using the Hall-Petch effects which relates the size effects to the dislocation pile-up. Accordingly, as the grain size decreases, the material strength increases. In the region of very fine grains, however, the Hall-Petch relation breaks down. Other GB mechanisms such as rotation and sliding will

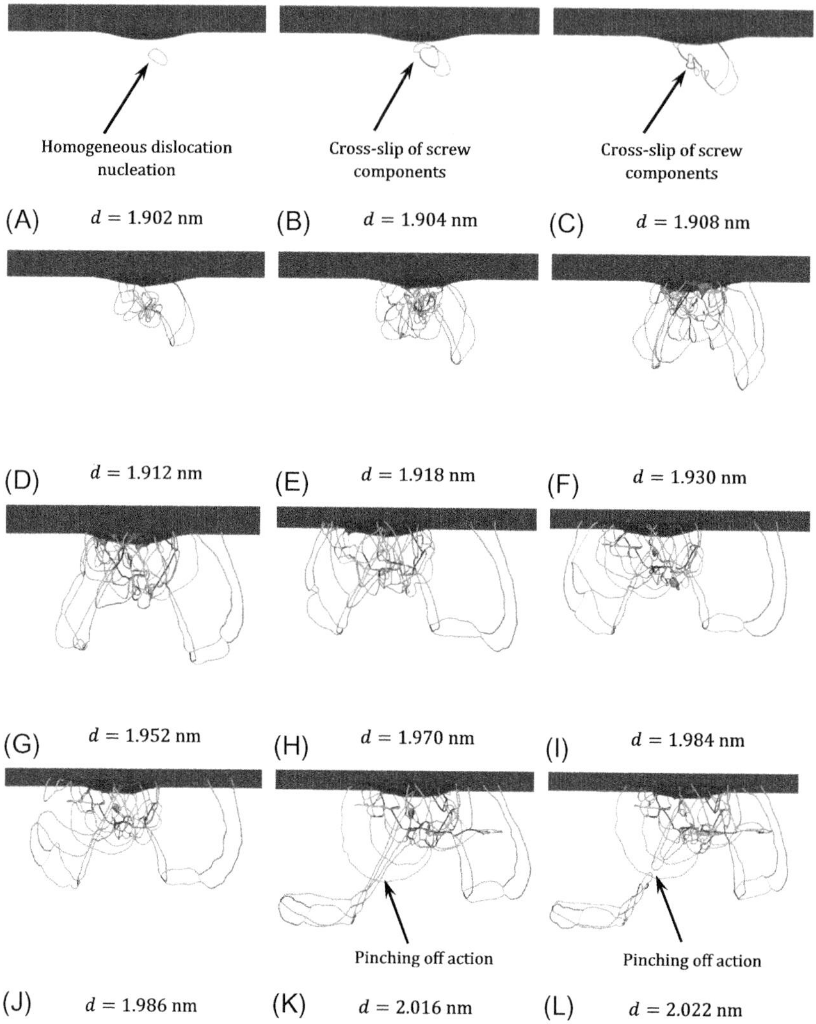

FIG. 5.23 Dislocation nucleation and evolution at small tip displacements: (A) initial homogeneous dislocation nucleation beneath the indenter which has a Burgers vector of $1/6[\bar{2}\,1\,\bar{1}]$ (Shockley partial dislocation); (B–J) cross slip of screw components which produces new pinning points; (K, L) first loop is released by pinching off action. *(After Yaghoobi, M., Voyiadjis, G.Z., 2016a. Atomistic simulation of size effects in single-crystalline metals of confined volumes during nanoindentation. Comput. Mater. Sci. 111, 64–73.)*

be responsible for size effects. In this region, as the grain size decreases, the material strength also decreases (Meyers et al., 2006; Koch et al., 2007; Zhu et al., 2008).

The effects of GB on the nanoindentation size effects is one of the challenging research topics. Some experiments which addresses the effect of GB on the nanoindentation response is presented in Section 1.3.2.3. Similar to the case of the size effects during nanoindentation for a single crystalline material, atomistic simulation can be used to capture the GB effects on the nanoindentation response of the metallic samples. Several mechanisms of dislocation reflection, transmission, and absorption were investigated by De Koning et al. (2003) by incorporating the atomistic simulation. Hasnaoui et al. (2004) studied the interaction between the dislocations and GB during nanoindentation experiment using molecular dynamics. Jang and Farkas (2007) conducted the atomistic simulation of bi-crystal nickel thin film nanoindentation and observed that the GBs can contribute to the nanoindentation hardness. Kulkarni et al. (2009), however, observed that the GBs mainly reduce the hardness of the metallic samples. They showed that the CTB has the least hardness reduction compared to the other types of GBs (Kulkarni et al., 2009). Tsuru et al. (2010) investigated the effect of indenter distance from the GB using MD. Stukowski et al. (2010) conducted MD simulation of nanoindentation for metallic samples with twin boundaries and observed that the effects of twin GBs on the material response depends on the unstable stacking fault and twin boundary migration energies. Sangid et al. (2011) proposed an inverse relation between the GB energy barrier and GB energy based on the MD simulation results.

Effects of GB on the response of thin film during nanoindentation has been studied by many researchers (Hasnaoui et al., 2004; Jang and Farkas, 2007; Kulkarni et al., 2009; Tsuru et al., 2010). However, a study which addresses a wide range of grain sizes is not a trivial task due to the MD simulation limitations. Voyiadjis and Yaghoobi (2016) incorporated the large scale MD to study the effects of grain size and grain boundary geometry on the nanoindentation response. They incorporated Ni thin films with two sizes of 24 nm × 24 nm × 12 nm (S1) and 120 nm × 120 nm × 60 nm (S2). Four symmetric tilt boundaries of $\sum 3\,(1\,1\,1)\,[1\,\bar{1}\,0]\,(\theta = 109.5^{\circ})$, $\sum 11\,(1\,1\,3)\,[1\,\bar{1}\,0]\,(\theta = 50.5^{\circ})$, $\sum 3\,(1\,1\,2)\,[1\,\bar{1}\,0]\,(\theta = 70.5^{\circ})$, and $\sum 11\,(3\,3\,2)\,[1\,\bar{1}\,0]\,(\theta = 129.5^{\circ})$, and three assymetric tilt boundaries of $\sum 11$ (2 2 5)/(4 4 1) $(\varphi = 54.74^{\circ})$, $\sum 3\,(1\,1\,2)/(\bar{5}\,\bar{5}\,2)\,(\varphi - 19.47^{\circ})$, and $\sum 3$ (1 1 4)/(1 1 0) $(\varphi = 35.26^{\circ})$ were generated at the two third of the sample from bottom to compare the governing mechanisms of size effects with those of the single crystal thin films. θ and φ are the interface misorientation and inclination angles, respectively. The spherical indenters with two different radii of $R_1 = 10$ nm and $R_2 = 15$ nm were modeled using the repulsive potential E^{ind} presented in Eq. (5.32). The procedure to generated and equilibrate the GBs were elaborated by Voyiadjis and Yaghoobi (2016). The equilibrium structures of grain boundaries are illustrated in Fig. 5.24 using the *CSP*. The remaining simulation methodology is similar to the Section 5.2.4.3.

Fig. 5.25 depicts the variation of mean contact pressure p_m during nanoindentation for S1 thin films, i.e. the smaller samples. It can be observed that the GB generally reduces the material strength for S1 thin films. However, in the case of coherent twin boundary (CTB), i.e. $\sum 3\,(1\,1\,1)\,[1\,\bar{1}\,0]$, the hardness is slightly enhanced for some indentation depths. Generally, in the cases of smaller thin films, i.e. S1 samples, the CTB has the best performance. Further

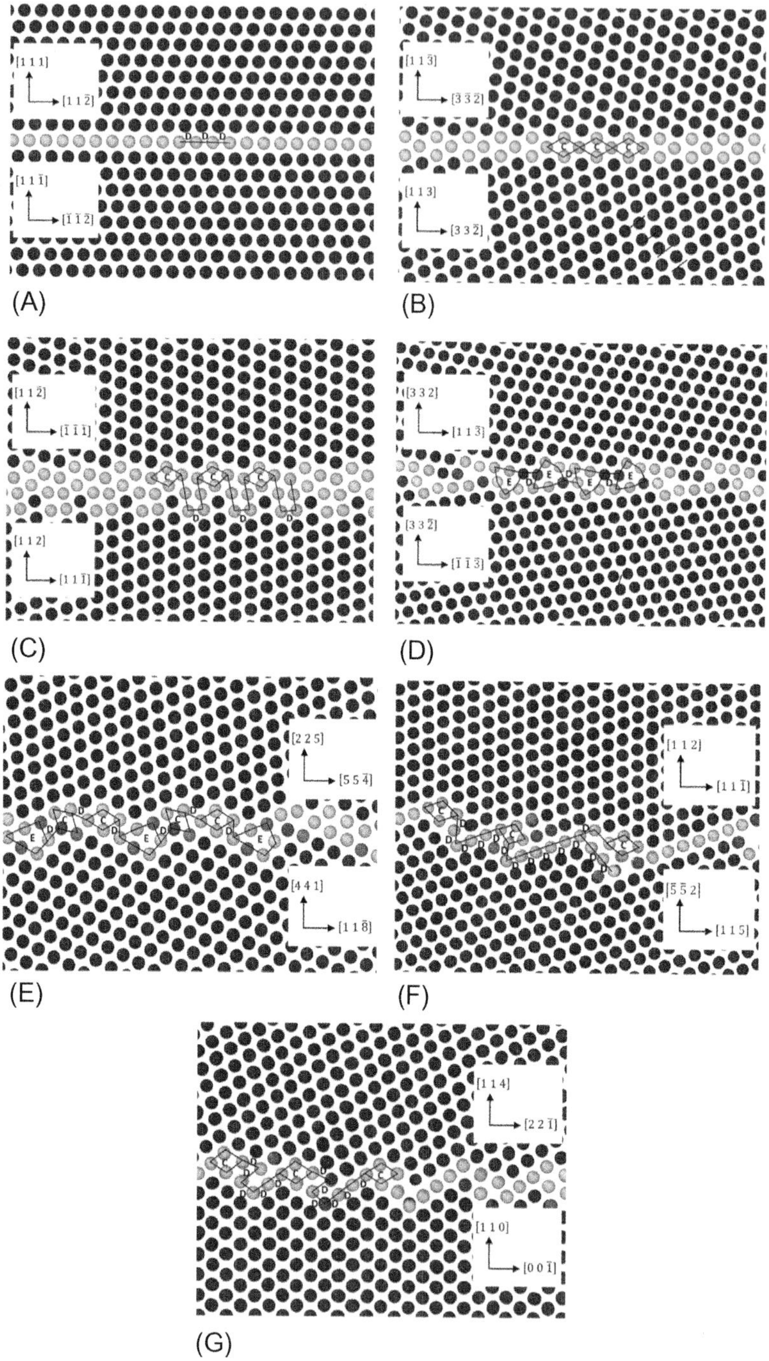

FIG. 5.24 The equilibrium structure of the symmetric and asymmetric tilt grain boundaries of (A) $\sum 3\,(1\,1\,1)\,[1\,\bar{1}\,0]\,(\theta = 109.5^\circ)$, (B) $\sum 11\,(1\,1\,3)\,[1\,\bar{1}\,0]\,(\theta = 50.5^\circ)$, (C) $\sum 3\,(1\,1\,2)\,[1\,\bar{1}\,0]\,(\theta = 70.5^\circ)$, (D) $\sum 11\,(3\,3\,2)\,[1\,\bar{1}\,0]\,(\theta = 129.5^\circ)$, (E) $\sum 11$ (2 2 5)/(4 4 1) ($\varphi = 54.74^\circ$), (F) $\sum 3\,(1\,1\,2)/(\bar{5}\,\bar{5}\,2)\,(\varphi = 19.47^\circ)$, and (G) $\sum 3$ (1 1 4)/(1 1 0) ($\varphi = 35.26^\circ$), along $[1\,\bar{1}\,0]$ axis. *(After Voyiadjis, G.Z., Yaghoobi, M., 2016. Role of grain boundary on the sources of size effects. Comput. Mater. Sci. 117, 315–329.)*

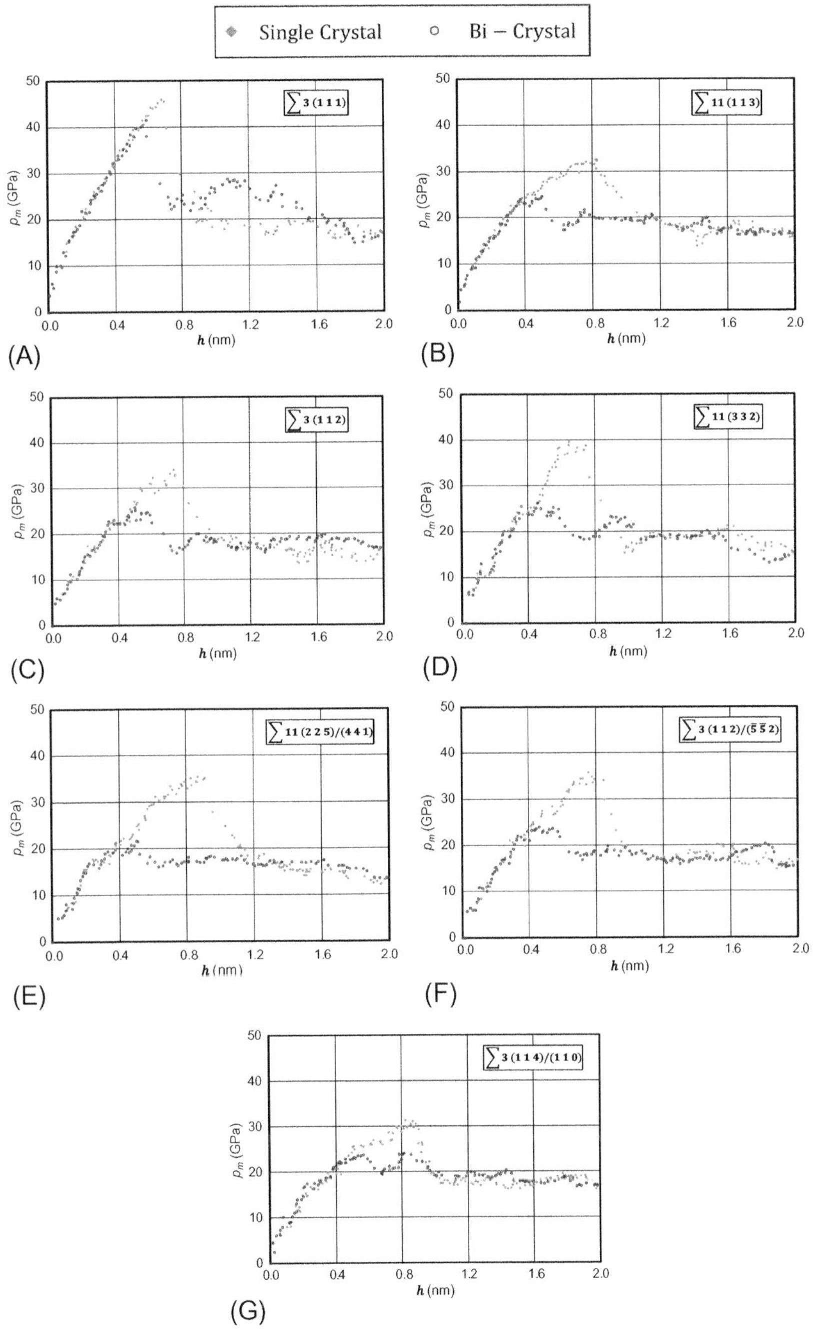

FIG. 5.25 Variation of mean contact pressure p_m as a function of indentation depth h for S1 single crystal and their related bi-crystal samples with the grain boundaries of (A) $\sum 3$ (1 1 1), (B) $\sum 11$ (1 1 3), (C) $\sum 3$ (1 1 2), (D) $\sum 11$ (3 3 2), (E) $\sum 11$ (2 2 5)/(4 4 1), (F) $\sum 3(1 1 2)/(\bar{5}\,\bar{5}\,2)$, and (G) $\sum 3$ (1 1 4)/(1 1 0). *(After Voyiadjis, G.Z., Yaghoobi, M., 2016. Role of grain boundary on the sources of size effects. Comput. Mater. Sci. 117, 315–329.)*

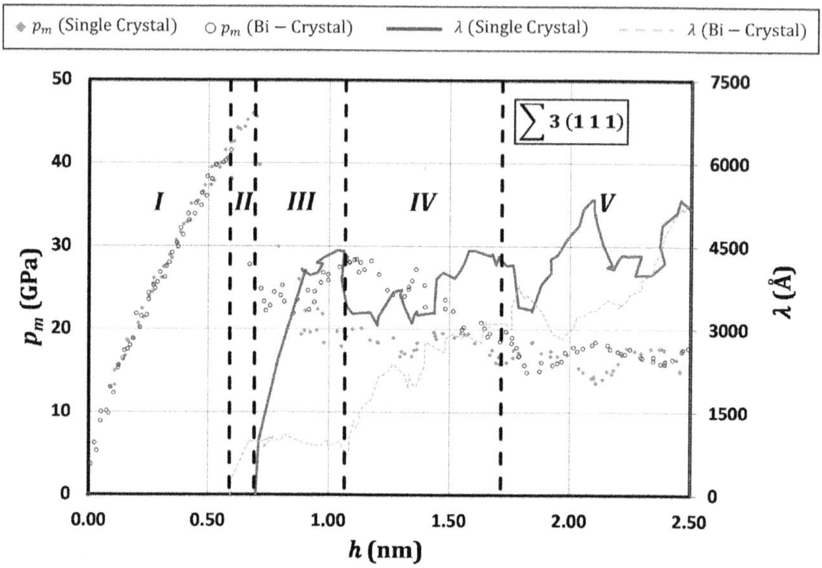

FIG. 5.26 Variation of mean contact pressure p_m and dislocation length λ as a function of indentation depth h for S1 bicrystal sample with $\sum 3\ (1\ 1\ 1)$ GB and its related single crystal sample. *(After Voyiadjis, G.Z., Yaghoobi, M., 2016. Role of grain boundary on the sources of size effects. Comput. Mater. Sci. 117, 315–329.)*

investigation was conducted by depicting the variation of dislocation length λ during nanoindentation. Fig. 5.26 illustrates the variation of p_m and λ during nanoindentation for CTB. Voyiadjis and Yaghoobi (2016) divided the nanoindentation response to five different regions:

- *Region I*: The bi-crystal and single crystal thin films show similar responses during the initial indentation phase which are elastic, and CTB is the only defect that exists in the bi-crystal thin film.
- *Region II*: In this region, the dislocation nucleation occurs for the bi-crystal thin film beneath the indenter following by a stress relaxation while the single crystal sample remains elastic. In the case of bi-crystal thin film, the size effects is initially governed by the dislocation nucleation and source exhaustion. The dislocation density increases during nanoindentation which decreases the required stress to sustain the imposed plastic flow. Consequently, the hardness decreases by nucleation and evolution of new dislocations.
- *Region III*: The plasticity is initiated in single crystal thin film beneath the indenter following by a stress relaxation. The thin film strength reduces according to the dislocation nucleation and source exhaustion mechanisms. The dislocation content does not change for bi-crystal thin film. Accordingly, the stress should be increased to sustain the imposed deformation

based on the source exhaustion mechanism. The dislocations eventually reach the GB which blocks the dislocations. However, the blocked dislocations do not contribute to the strength.

- *Region IV*: The strength in both single and bicrystal thin films is decreased by increasing the dislocation length which follows the source exhaustion mechanism. However, the influence of the source exhaustion mechanism decreases as the dislocation length increases which decreases the slope of the hardness reduction. Also, the dislocations which are blocked by the GB start dissociating into the next grain.
- *Region V:* In this region, the available dislocation content is sufficient to sustain the imposed plastic flow and no further stress reduction occurs. Also, the single and bi-crystal thin films reaches a similar hardness which shows that the dislocation blockage by GB does not have any contribution to the size effects.

The structures of dislocations in different regions are illustrated in Fig. 5.27 for bi-crystal thin film with CTB and related single crystal sample. Fig. 5.27A and B show the dislocation structure in *Region II* at which the single crystal sample is defect free and the nucleation occurs beneath the indenter for bi-crystal thin film. The results show that the dominant mechanism of dislocation multiplication is cross-slip. Cross-slip introduces the new pinning points and provides the required dislocation length to sustain the plastic flow. Dislocations are elongated while they are pinned at their ends. Fig. 5.27C and D illustrate the dislocation structure in *Region III* while the cross-slip is still the governing mechanism of deformation for both single and bi-crystal thin films. Fig. 5.27D depicts the dislocation blockage by CTB. Fig. 5.27E and F illustrate the dislocation structure in *Region IV*. Many dislocation multiplications are observed in both single and bi-crystal thin films which are induced according to the cross-slip mechanism. Fig. 5.27F shows the initial dislocation dissociation into the next grain in the case of bi-crystal sample which is a Shockley partial dislocation with the Burgers vector of $\frac{1}{6}\left[\bar{1}\bar{2}\bar{1}\right]$. In the case of *Region V*, Fig. 5.27G and H depict the dislocation structure which shows enough dislocation length is provided to sustain the imposed deformation. Also, the interaction of dislocations with each other cannot be neglected anymore.

Fig. 5.28 shows the variation of p_m and λ during nanoindentation for the S1 samples with different GBs and their related single crystal thin films. In contrast to CTB, Fig. 5.28 shows that the first stress relaxation does not occur immediately after the first dislocation nucleation. The first large stress relaxation occurs with the first jump in dislocation density for single crystal thin films. The nature of first apparent strength drop in thin films with GB is more complicated due to the interaction of dislocations with GB. Fig. 5.28 shows that the GB decreases the depth at which the first large stress relaxation occurs, and the bi-crystal thin films have larger dislocation length at that depth compared to the single crystal samples. The GB itself can be a source of dislocation nucleation which can be

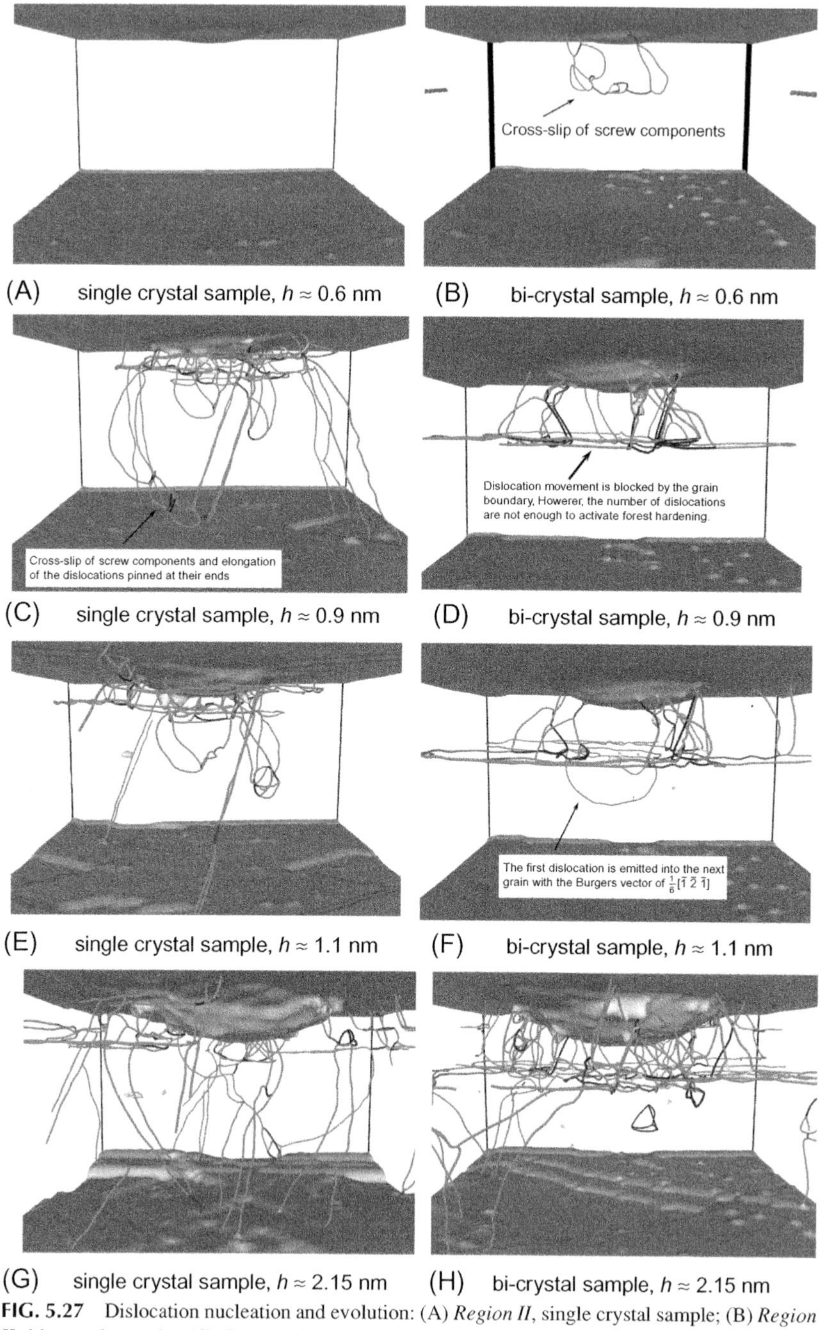

FIG. 5.27 Dislocation nucleation and evolution: (A) *Region II*, single crystal sample; (B) *Region II*, bi-crystal sample; (C) *Region III*, single crystal sample; (D) *Region III*, bi-crystal sample; (E) *Region IV*, single crystal sample; (F) *Region IV*, bi-crystal sample; (G) *Region V*, single crystal sample; (H) *Region V*, bi-crystal sample. *(After Voyiadjis, G.Z., Yaghoobi, M., 2016. Role of grain boundary on the sources of size effects. Comput. Mater. Sci. 117, 315–329.)*

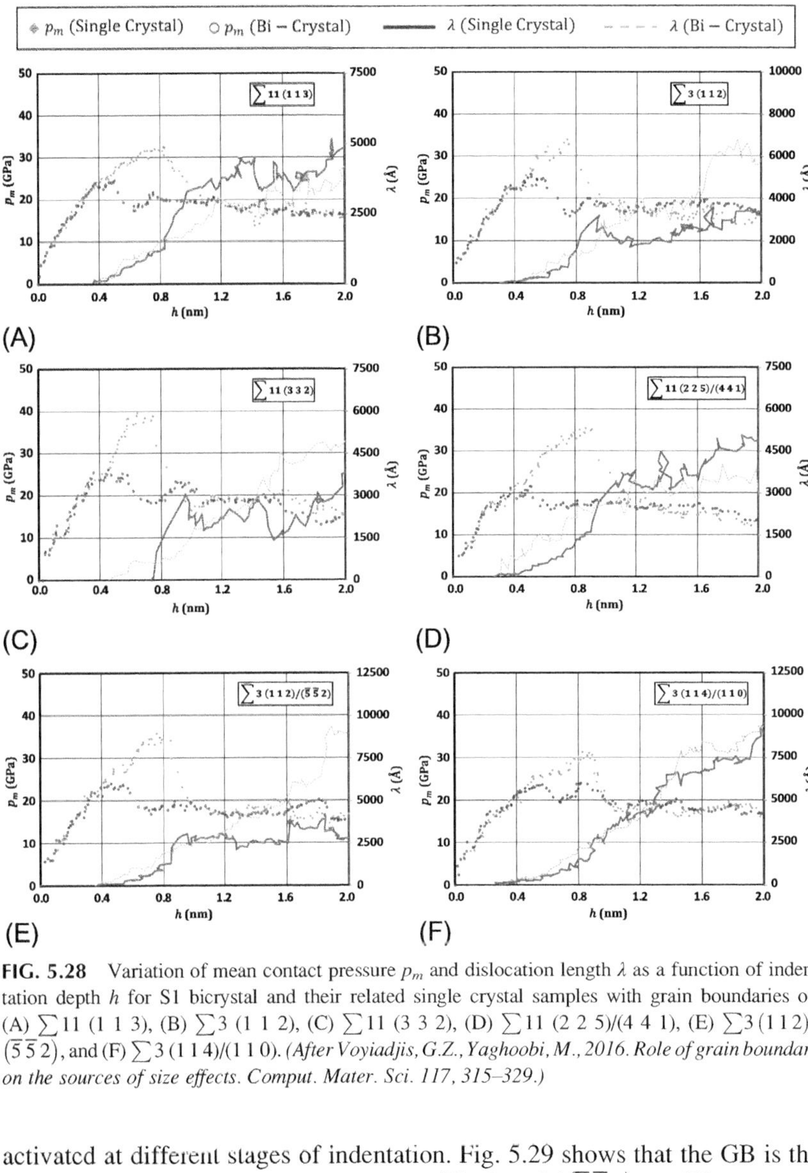

FIG. 5.28 Variation of mean contact pressure p_m and dislocation length λ as a function of indentation depth h for S1 bicrystal and their related single crystal samples with grain boundaries of: (A) $\sum 11$ (1 1 3), (B) $\sum 3$ (1 1 2), (C) $\sum 11$ (3 3 2), (D) $\sum 11$ (2 2 5)/(4 4 1), (E) $\sum 3\,(1\,1\,2)/(\bar{5}\,\bar{5}\,2)$, and (F) $\sum 3$ (1 1 4)/(1 1 0). *(After Voyiadjis, G.Z., Yaghoobi, M., 2016. Role of grain boundary on the sources of size effects. Comput. Mater. Sci. 117, 315–329.)*

activated at different stages of indentation. Fig. 5.29 shows that the GB is the initial source of dislocation nucleation for $\sum 3\,(1\,1\,2)/(\bar{5}\,\bar{5}\,2)$ and $\sum 11$ (2 2 5)/(4 4 1) GBs, i.e. the initial dislocation nucleation occurs from the GB and not beneath the indenter. The nucleated dislocations are Shockley partial dislocations with the Burgers vectors of $\frac{1}{6}[1\,1\,\bar{2}]$ and $\frac{1}{6}[1\,1\,2]$ for the GBs of $\sum 3\,(1\,1\,2)/(\bar{5}\,\bar{5}\,2)$ and $\sum 11$ (2 2 5)/(4 4 1), respectively. If the dislocation nucleation from GB occurs at the initial steps of dislocation nucleation and evolution, it will severely decrease the thin film strength which can be noted for $\sum 3\,(1\,1\,2)/(\bar{5}\,\bar{5}\,2)$ and $\sum 11$ (2 2 5)/(4 4 1) GBs in Fig. 5.28D and E. The size

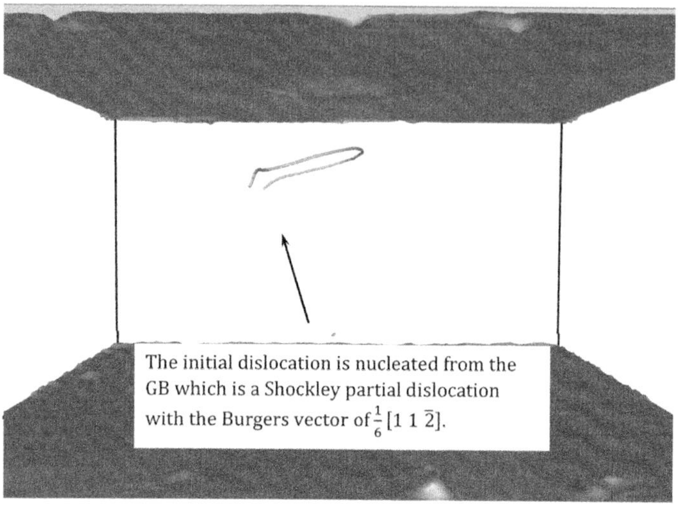

(A) bi-crystal sample with $\Sigma 3\ (1\ 1\ 2)/(\bar{5}\ \bar{5}\ 2)$ GB, $h \approx 0.37$ nm

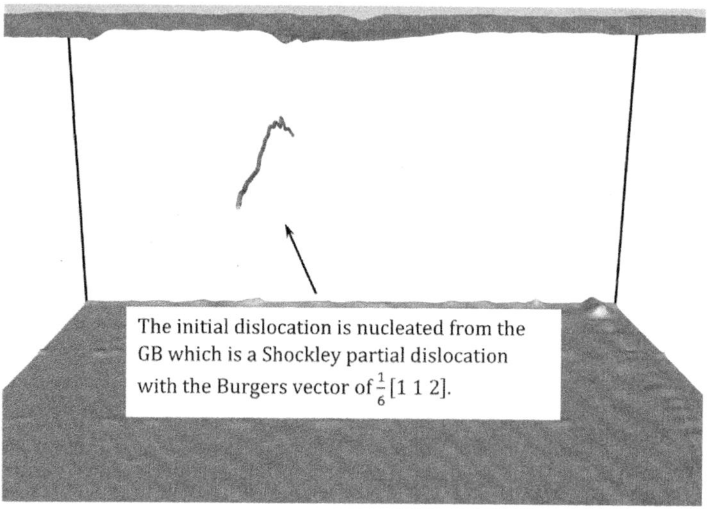

(B) bi-crystal sample with $\Sigma 11(2\ 2\ 5)/(4\ 4\ 1)$ GB, $h \approx 0.43$ nm

FIG. 5.29 Dislocation nucleation from the GB: (A) bi-crystal sample with $\sum 3\,(1\,1\,2)/(\bar{5}\,\bar{5}\,2)$ GB, $h \approx 0.37$ nm; (B) bi-crystal sample with $\sum 11\,(2\,2\,5)/(4\,4\,1)$ GB, $h \approx 0.43$ nm. *(After Voyiadjis, G.Z., Yaghoobi, M., 2016. Role of grain boundary on the sources of size effects. Comput. Mater. Sci. 117, 315–329.)*

effects during nanoindentation can be described for all GBs incorporating the variation of total dislocation length and dislocation visualization during nanoindentation. The results show that the source exhaustion is the controlling mechanism of size effects for the initial stages of dislocation nucleation and evolution. Increasing the total dislocation length, however, the required dislocation length for sustaining the imposed deformation is provided and the source exhaustion mechanism becomes less dominant. Also, the dislocation interactions with each other becomes non-negligible by increasing the dislocation content. Eventually, both bi-crystal and their related single crystal thin films reaches a similar hardness which indicates that the dislocation pile-up does not enhance the hardening in the cases of studied S1 samples.

Fig. 5.30 shows mean contact pressure versus indentation depth in the cases of S2 samples, i.e. larger samples. The initial responses of both single crystal and bi-crystal thin films are similar. However, GB enhances the hardness for higher indentation depths. In order to unravel the underlying mechanisms of size effects for S2 samples, the variations of mean contact pressure and total dislocation length should be studied. Fig. 5.31 compares the mean contact pressure and total dislocation density of the single crystal thin film with those of the bi-crystal sample with CTB. The nanoindentation response can be divided to three different regions:

- *Region I*: There is no plasticity at this region. GB does not change the nanoindentation response of thin film.
- *Region II*: The plasticity is initiated beneath the indenter for both single and bi-crystal thin films following by a sharp stress relaxation. After the initial nucleation, the source exhaustion governs the size effects and the required stress to maintain the plastic flow decreases by increasing the total dislocation length. In this region, the GB does not significantly change the total dislocation length and consequently the hardness. The dominancy of source exhaustion decreases during nanoindentation as more dislocations are provided to sustain the imposed deformation. Accordingly, the hardness reduction slope decreases during nanoindentation. Eventually, the dislocations reach the GB, and the GB blocks the dislocation movements.
- *Region III*: Enough dislocation length is provided to sustain the imposed deformation, and the source exhaustion hardening is not active anymore. The interactions of dislocations with each other and GB become important by increasing the dislocation content. Also, the number of dislocations blocked by GB becomes considerable and the produced pile-up enhances the sample strength. Consequently, the GB enhances the nanoindentation response of thin film for S2 sample.

The dislocation visualization of the S2 thin film with and without CTB is illustrated in Fig. 5.32 during nanoindentation. Fig. 5.32A and B illustrate that the initial dislocation is homogeneously nucleated beneath the indenter which is a Shockley partial dislocation with the Burgers vector of $\frac{1}{6}[2\,\bar{1}\,1]$. The results show that the GB does not change the nucleation pattern. After the initial nucleation, Fig. 5.32C and D show that the cross-slip is the controlling mechanism of deformation which increases the number of dislocation sources and provides the required dislocation content. The effects of GB is still negligible on the dislocation pattern. The dislocation loops are induced by cross-slipping and pinching off of screw dislocations as the indentation depth increases. The induced loops

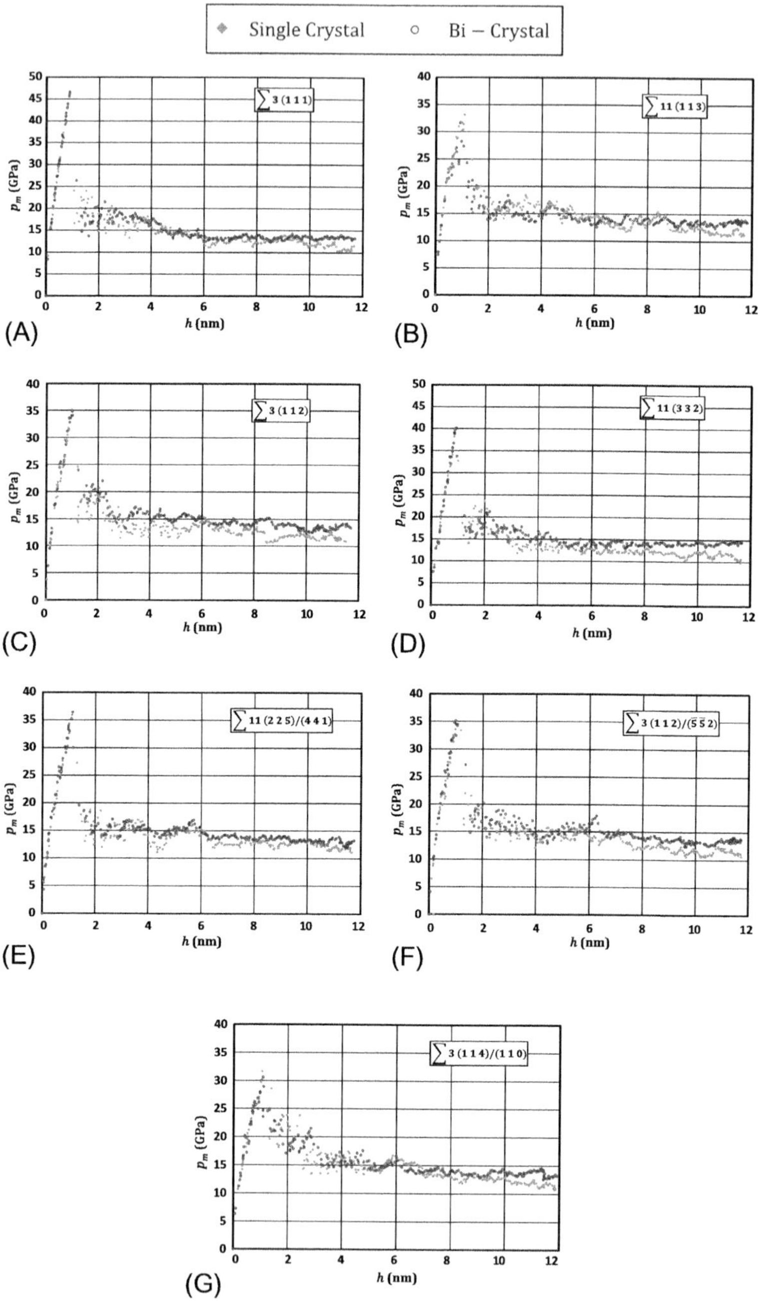

FIG. 5.30 Variation of mean contact pressure p_m as a function of indentation depth h for S2 single crystal and their related bi-crystal samples with the grain boundaries of (A) $\sum 3$ (1 1 1), (B) $\sum 11$ (1 1 3), (C) $\sum 3$ (1 1 2), (D) $\sum 11$ (3 3 2), (E) $\sum 11$ (2 2 5)/(4 4 1), (F) $\sum 3\,(1\,1\,2)/(\bar{5}\,\bar{5}\,2)$, and (G) $\sum 3$ (1 1 4)/(1 1 0). *(After Voyiadjis, G.Z., Yaghoobi, M., 2016. Role of grain boundary on the sources of size effects. Comput. Mater. Sci. 117, 315–329.)*

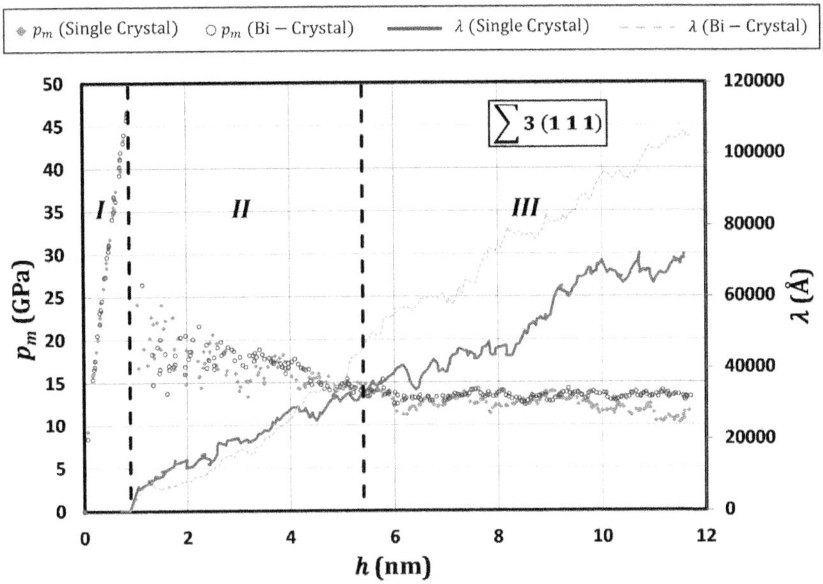

FIG. 5.31 Variation of mean contact pressure p_m and dislocation length λ as a function of indentation depth h for S2 bicrystal sample with $\sum 3\ (1\ 1\ 1)$ GB and its related single crystal sample. *(After Voyiadjis, G.Z., Yaghoobi, M., 2016. Role of grain boundary on the sources of size effects. Comput. Mater. Sci. 117, 315–329.)*

are moving downward which are blocked by GB. Consequently, the GB starts to change the pattern of dislocation evolution. Fig. 5.32E and F illustrate the dislocation loops movement in thin films with and without GB, respectively. The dislocation blocked by GB are eventually emitting into the next grain by increasing the indentation depth. Fig. 5.32H illustrates the first dislocation emitting into the next grain which is a Shockley partial dislocation with the Burgers vector of $\frac{1}{6}\left[1\,\overline{2}\,1\right]$. Fig. 5.32I and J show the dislocation visualization of the sample at the higher indentation depths for thin film with and without GB, respectively. Although some dislocations are emitted into the next grain, the visualization results show a considerable pile-up behind the GB, while the dislocations are moving downward freely for single crystal thin film.

Fig. 5.33 shows the variation of p_m and λ during nanoindentation for the S2 samples with different GBs and their related single crystal thin films. The observed microstructural behavior for CTB can be incorporated for all other GBs except $\sum 11\ (3\ 3\ 2)$ and $\sum 11\ (2\ 2\ 5)/(4\ 4\ 1)$ GBs. In the cases of two latter GBs, the GB enhances the hardness while the total dislocation length of thin film with GB is very close to the one without GB. The observed discrepancy is due to the fact the total dislocation length is not an appropriate factor to study the forest hardening mechanism. In the case of source exhaustion hardening, the total dislocation length dictates the amount of stress required to sustain the

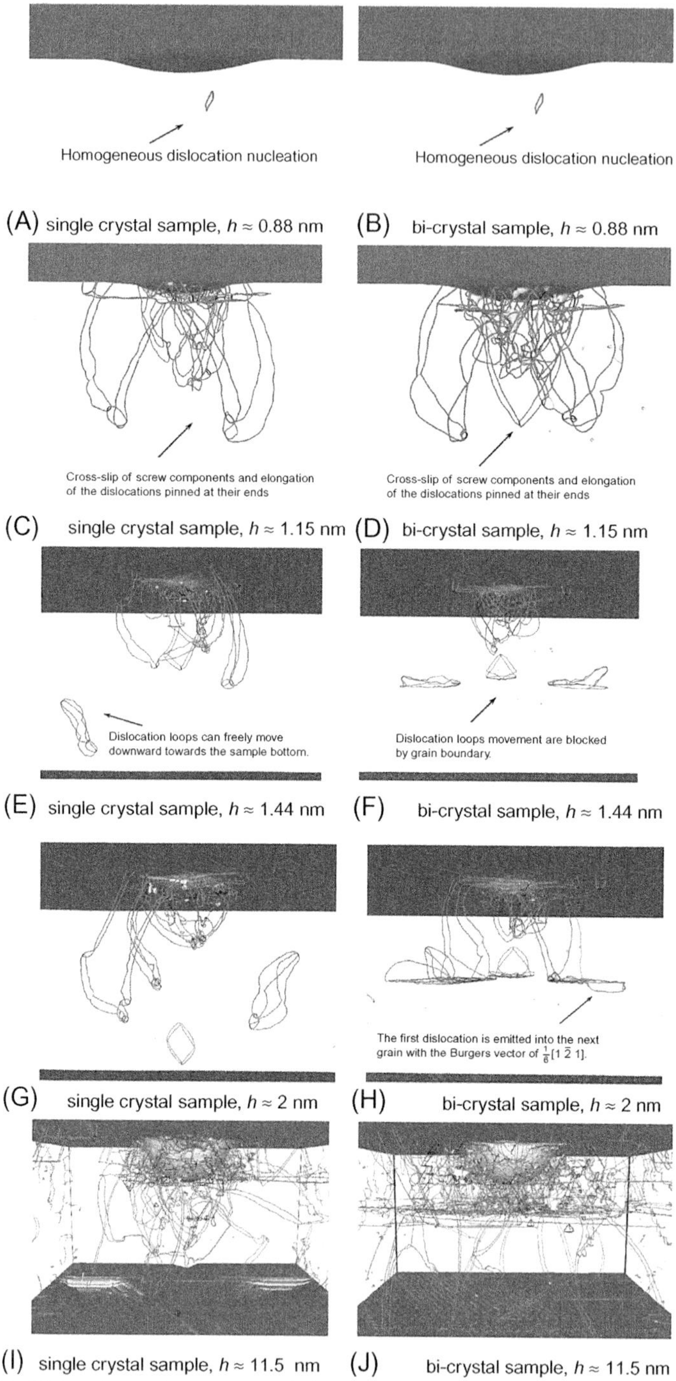

FIG. 5.32 Dislocation nucleation and evolution: (A) single crystal sample, $h \approx 0.88$ nm; (B) bi-crystal sample, $h \approx 0.88$ nm; (C) single crystal sample, $h \approx 1.15$ nm; (D) bi-crystal sample, $h \approx 1.15$ nm; (E) single crystal sample, $h \approx 1.44$ nm; (F) bi-crystal sample, $h \approx 1.44$ nm; (G) single crystal sample, $h \approx 2.03$ nm; (H) bi-crystal sample, $h \approx 2.03$ nm; (I) single crystal sample, $h \approx 11.5$ nm; (J) bi-crystal sample, $h \approx 11.5$ nm. *(After Voyiadjis, G.Z., Yaghoobi, M., 2016. Role of grain boundary on the sources of size effects. Comput. Mater. Sci. 117, 315–329.)*

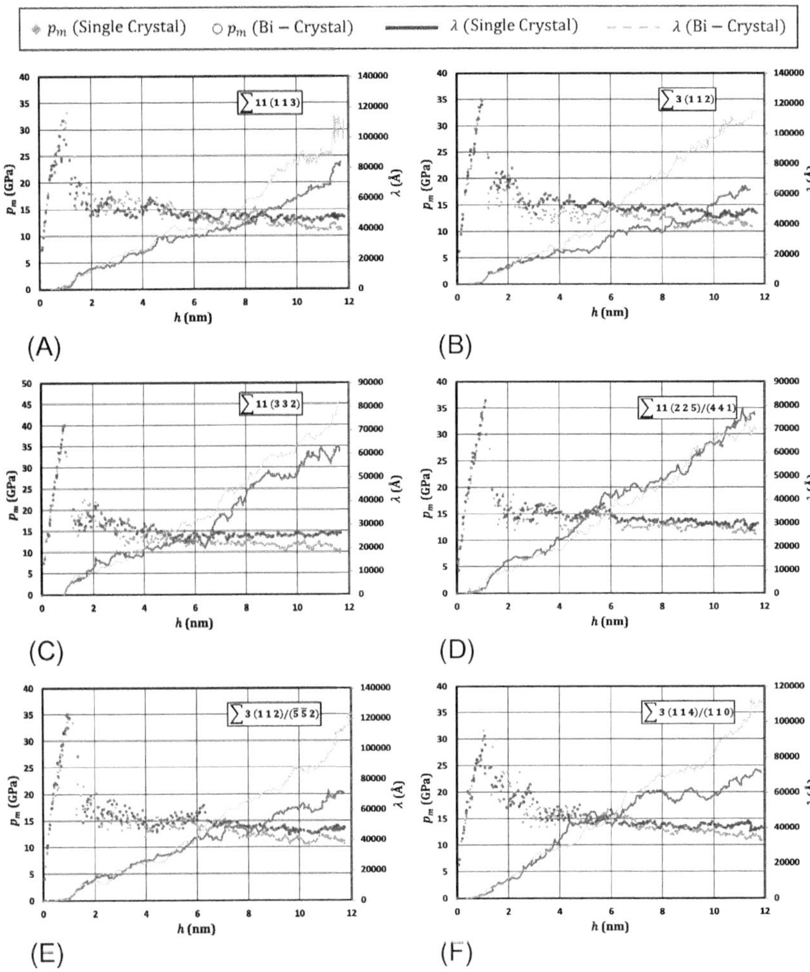

FIG. 5.33 Variation of mean contact pressure p_m and dislocation length λ as a function of indentation depth h for S2 bicrystal and their related single crystal samples with grain boundaries of: (A) $\sum 11$ (1 1 3), (B) $\sum 3$ (1 1 2), (C) $\sum 11$ (3 3 2), (D) $\sum 11$ (2 2 5)/(4 4 1), (E) $\sum 3$(1 1 2)/($\bar{5}\,\bar{5}$ 2), and (F) $\sum 3$ (1 1 4)/(1 1 0). *(After Voyiadjis, G.Z., Yaghoobi, M., 2016. Role of grain boundary on the sources of size effects. Comput. Mater. Sci. 117, 315–329.)*

plastic flow. On the other hand, the density of dislocation in the plastic zone should be taken as the representative factor for the forest hardening mechanism. Voyiadjis and Yaghoobi (2016) assumed that the plastic zone is located in the upper grain. Accordingly, the total dislocation length in the upper grain λ_{upper} should be investigated during nanoindentation. The dislocations located in the upper one third is considered for single crystal thin film, and the obtained results are compared with those of bi-crystal thin films. Fig. 5.34 compares the

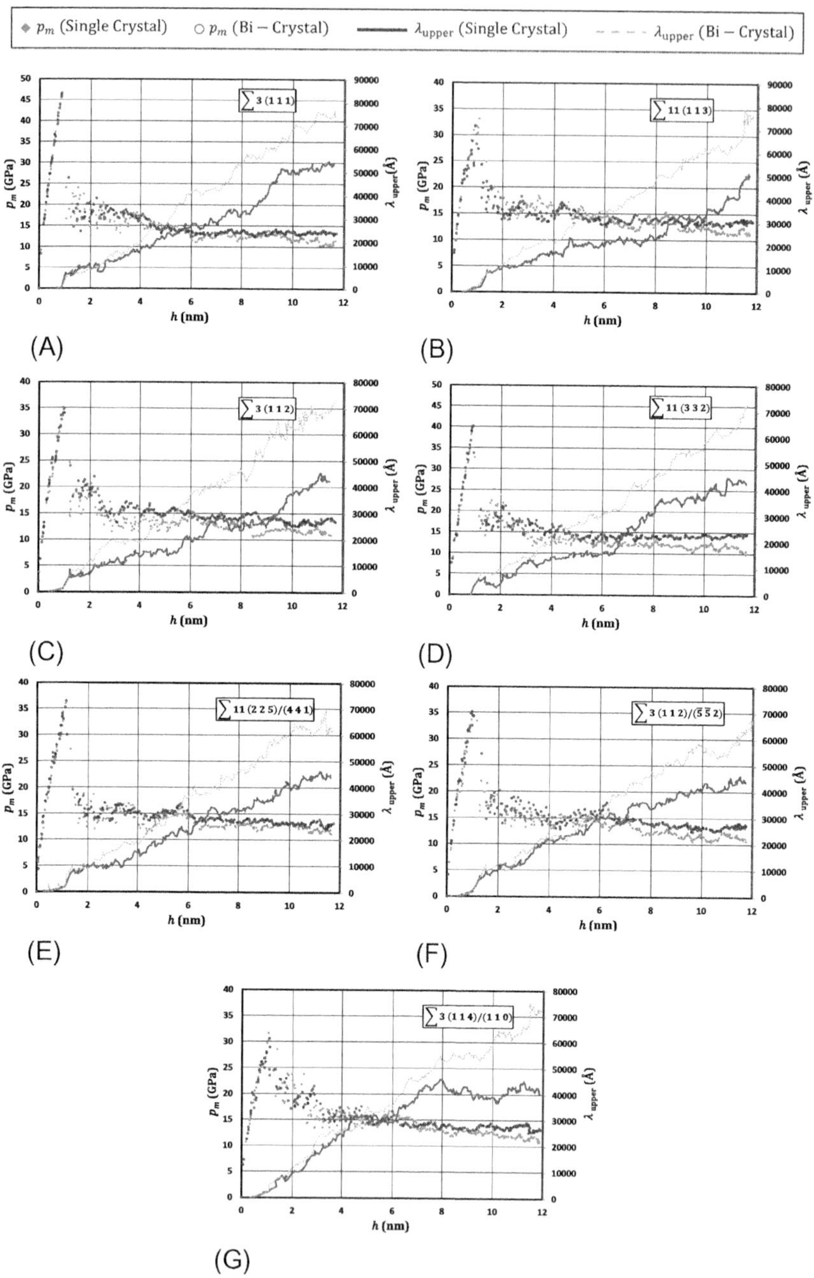

FIG. 5.34 Variation of mean contact pressure p_m and total dislocation length located in the upper grain λ_{upper} as a function of indentation depth h for S2 bicrystal and their related single crystal samples with grain boundaries of (A) $\sum 3$ (1 1 1), (B) $\sum 11$ (1 1 3), (C) $\sum 3$ (1 1 2), (D) $\sum 11$ (3 3 2), (E) $\sum 11$ (2 2 5)/(4 4 1), (F) $\sum 3\,(1\,1\,2)/(\bar{5}\,\bar{5}\,2)$, and (G) $\sum 3$ (1 1 4)/(1 1 0). *(After Voyiadjis, G.Z., Yaghoobi, M., 2016. Role of grain boundary on the sources of size effects. Comput. Mater. Sci. 117, 315–329.)*

variations of mean contact pressure and the total dislocation length located in the plastic zone during nanoindentation for thin films with and without GB. The results show that the GB increases the total dislocation length located in the plastic zone and consequently enhances the hardness according to the forest hardening mechanism. The results show that the main role of GB in the cases of large thin films, i.e. S2 samples, is to modify the pattern of dislocation in a way that increases the dislocation density located in the plastic zone and accordingly strengthens the thin films. One should note that the strain rates incorporated in the atomistic simulation are much higher than those selected for experiments. Accordingly, the interpretation of the obtained results should be carefully handled. The applied strain rate can influence both hardening mechanisms and dislocation network properties (see, e.g., Yaghoobi and Voyiadjis, 2016b, 2017; Voyiadjis and Yaghoobi, 2017a). In other words, one should ensure that the observed mechanisms are not artifacts of the high strain rates used in the atomistic simulation.

5.3 Molecular dynamics simulation of size effects during micropillar compression experiment

5.3.1 Size effects during micropillar compression experiment

Recent experimental advancements have allowed researchers to investigate the size effects mechanism for very small length scales. One of these experiments is micropillar compression test, as presented in Section 1.3.2.2. Micropillar compression test was introduced by Uchic et al. (2003, 2004). The focused ion beam (FIB) machining is used to fabricate nano to micron sized metallic pillars. Accordingly, the effects of pillar size can be investigated using micropillar compression test. They observed that the material response varies as the pillar size changes. Uchic et al. (2004) investigated the strain-stress response and reported the variation of stress-strain curves as the sample size changes for pure Ni at room temperature (Fig. 1.35). The results show that the pillar strength is almost three times larger than the bulk Ni. Accordingly, besides forest hardening, other mechanisms of size effects should be responsible for the observed results. As described in Sections 1.3.2.2.1–1.3.2.2.3, three size effects mechanisms of source truncation, source exhaustion, and weakest link theory (see, e.g., Uchic et al., 2009; Kraft et al., 2010) have been developed to capture the size effects during micropillar compression experiments. Accordingly, besides forest hardening mechanism, any of these size effects mechanisms could potentially be responsible for size effects in metallic samples of small length scales. In order to further investigate the governing mechanisms of size effects during micropillar compression test, the meso-scale simulations, such as the atomistic simulation, can be incorporated.

5.3.2 Molecular dynamics simulation of micropillar compression experiment

As described in Section 5.2.4, molecular dynamics (MD) is a very powerful tool to capture the size effects mechanisms in metallic sample. MD can capture the atomistic mechanism governing size effects which cannot be captured using simulation with larger length scales. Molecular dynamics is able to accurately capture free surfaces, which is a very important feature in micropillars. Additionally, MD simulation is able to accurately capture the cross-slip mechanism. Using very powerful supercomputers and highly efficient MD codes, samples with the size of 0.5 μm can be modeled (see, e.g., Yaghoobi and Voyiadjis, 2017, 2018). However, in the case of strain rate, MD simulation still has considerable limitations and it cannot model the quasi-static simulations. Accordingly, MD has been widely used to study the behavior of metallic pillar during high rate deformations. Bringa et al. (2006) studied the high rate deformation mechanisms of copper pillars during compression experiment. They incorporated a very large sample consisting of 352 million atoms and were able to obtain the experimentally reported mechanisms. Diao et al. (2006) incorporated MD simulation and studied the size dependency of the nanowire yield strength using various sizes of nanowires. They reported that in the cases of nanowires with small length scales, the size-dependency of nanowire yield strength is governed by the surface stress. Weinberger and Cai (2008) captured a dislocation multiplication mechanism in BCC pillars using MD simulation. They stated that as the dislocation leaves the sample, it multiplies itself as a result of surface stress. Sansoz (2011) conducted a series of atomistic simulations on pillars with the diameter range of 10.8–72.3 nm. Sansoz (2011) used a new scheme to generate a sample with physical initial density of defects instead of using a defect free sample. In the range of simulated samples, Sansoz (2011) reported that independent of the pillar diameter, both mechanisms of dislocation exhaustion and source-limited activation should be considered. Sansoz (2011) showed that in the cases of pillars with pre-exististing defects, the dislocation exhaustion mechanism is active for small strain. At some strains, which depend on the pillar size, the source-limited mechanism becomes the dominant mechanism of deformation. Weinberger et al. (2012) excluded the effects of high strain rates from the atomistic simulation using the transition state theory. They investigated the dislocation nucleation stress and failure mechanisms of pillars with different cross-section geometries. Weinberger and Tucker (2012) used the MD simulation to study the stability of a single arm source in a nanopillar. They concluded that the dislocation arm is not stable enough to create static pinning points. Accordingly, in order to study the size effects in nanopillars in the case of stable single arm source, they created artificial pining points. They observed that decreasing the pillar diameter increases its strength. Tucker et al. (2013) incorporated the atomistic simulation of nanopillars to investigate the effects of grain boundary. They showed that the grain boundary itself can become a dislocation nucleation source. Xu et al. (2013) investigated the effects of crystallographic orientation in Al nanopillars with the samples consisting of 10^4–10^7 atoms during the compression test.

In this section, first, the size effects in FCC pillars during the high rate compression test will be investigated (Yaghoobi and Voyiadjis, 2016b). Accordingly, different size effects mechanisms of dislocation starvation, dislocation exhaustion, and forest hardening will be addressed. The effects of average dislocation source length and pre-straining will be studied. In the next step, the coupling effects of size and strain rate will be addressed during the micropillar compression test (Voyiadjis and Yaghoobi, 2017a). In addition to the stress-strain and dislocation density-strain curves, the dislocation length distribution will be studied to extract the underlying mechanisms of size effects.

5.3.2.1 Size effects in FCC pillars during the high rate compression test

5.3.2.1.1 Molecular simulation methodology

In this section, the size effects in FCC pillars during the high rate compression test will be investigated (Yaghoobi and Voyiadjis, 2016b). The large scale atomistic simulation is used to investigate the deformation mechanisms. The MD simulation and methodology is mostly similar to the one used for nanoindentation, as presented in Section 5.2.4.1. The parallel code LAMMPS (Plimpton, 1995) is incorporated to conduct the MD simulation of Ni pillars. The embedded-atom method (EAM) potential obtained by Mishin et al. (1999) is incorporated to capture the Ni-Ni atomic interaction. The wide range of sample sizes should be generated to study the controlling mechanism of size effects. Here, the pillars sizes are changed in two different ways. First, the pillar diameter is changed while the height is similar for all pillars. Samples with height of $H = 45$ nm and different diameters of $D = 22.5$, 45, 90, and 135 nm are generated. Next, the pillar height is varied while the diameter is kept fixed for all pillars. The pillars with the diameters of 22.5 and 135 nm and heights of 30, 45, and 75 nm are generated to study the effects of pillar height on the controlling mechanisms of size effects. Finally, the response of a very large sample with the height of 0.3 μm and diameter of 0.15 μm is generated. The largest sample generated contains around 487 million atoms. The pillars are circular cross sections with the axis along [1 1 1] direction.

The pillar boundary conditions should be carefully selected to accurately capture the dislocation density during the simulation. The boundary conditions at the surroundings, along $[1\,\bar{1}\,0]$ and $[1\,1\,\bar{2}]$ directions, are set free. Along the loading direction, however, four different types of boundary conditions can be used:

- The periodic boundary conditions are used along the loading direction. The pillar loading procedure is conducted by shrinking the simulation box along the loading direction. Using these boundary conditions, a dislocation which leaves the bottom surface will enter from the upper one.

- The free surface is incorporated for the top, and some atomic layers at the bottom are fixed to simulate the substrate. The pillar can be uniaxially compressed using a large flat indenter. The dislocations cannot pass the bottom surface due to the fixed atomic layers which leads to a slight overestimation of dislocation density.
- The third type of boundary conditions is to model the pillar with its substrate, which is from the same material. Again, the displacement is imposed on the pillar using a flat indenter. This boundary conditions type is able to precisely capture the dislocation content. However, the substrate itself may have some plastic deformation, which is not easy to predict or exclude from the total deformation of the pillar. The substrate deformation is not uniform, and it is a function of pillar dimensions and applied strain. Hence, the pillar strain cannot be easily interpreted.
- The boundary conditions for top and bottom surfaces are set free. The substrate is simulated using a prescribed potential wall. The compressive displacement is then applied using a large flat indenter.

Yaghoobi and Voyiadjis (2016b) tested all four types of boundary conditions and observed that the last one is the most appropriate choice to study the size effects of pillar in which the dislocation density should be precisely captured during the simulation. The interaction between the indenter and Ni atoms is modeled using the repulsive potential as presented in Eq. (5.32). The indenter applies the displacement-controlled compressive load in a way to induce the strain rate of 3.33e8 s^{-1}. The Lennard-Jones (LJ) potential is incorporated to model the interaction between the substrate and Ni atoms which is defined as presented in Eq. (5.26). The LJ parameters are selected to model the Si substrate, which are $\bar{\varepsilon}_{Ni-Si} = 1.5231e - 20$ J and $\bar{\sigma}_{Ni-Si} = 3.0534$ ˉ. The LJ potential cutoff distance is chosen as $2.5\bar{\sigma}$. The velocity Verlet algorithm with the time step of 5 fs is used to numerically integrate the equations of motion. The NVT ensemble is selected to simulate the compression test (Plimpton, 1995). The precise cross section should be calculated during the simulation to obtain the true stress. Here, the contact area is obtained using the triangulation method described in Section 5.2.4.1.

In order to study the effects of pillar initial structure on the governing mechanisms of size effects, two different types of samples are generated as follows:

- First, the pillars are initially defect free and have perfect lattice. The samples are relaxed for 100 ps with the temperature increasing from 1 to 300 K. The samples are then relaxed for 100 ps at 300 K.
- Second, the pillars are pre-strained. The initial relaxation procedure is similar to the defect free samples. Next, the pillars are uniaxially loaded up to the strain equal to 0.12. Then the samples are unloaded and relaxed for 100 ps at 300 K.

The dislocation structures are extracted at each step using the Crystal Analysis Tool developed by Stukowski and his co-workers (Stukowski and Albe, 2010; Stukowski, 2012, 2014; Stukowski et al., 2012). The dislocation visualization and their information processing, such as the total dislocation length and density calculations, are conducted using the software OVITO (Stukowski, 2010) and Paraview (Henderson, 2007).

5.3.2.1.2 Size effects in FCC pillars

Fig. 5.35 presents the variation of true stress (σ) versus the strain (ε) in the cases of pillars with the height of 45 nm and several diameters of 22.5, 45, 90, and 135 nm. In order to study the size effects, the initial dislocation nucleation stress is not considered as the strength of the sample. It is due to the fact that this stress is not realistic and only available in perfect MD samples, which does not have any kind of defect. Instead, the stress after the initial nucleation is incorporated to study the size effects. The pillars compressive responses are initially elastic and independent of pillar diameter. After the elastic region, however, the stress shows strong size effects. Fig. 5.35 shows that in the case of D from 22.5 to 90 nm, as the diameter decreases the strength increases. However, there is

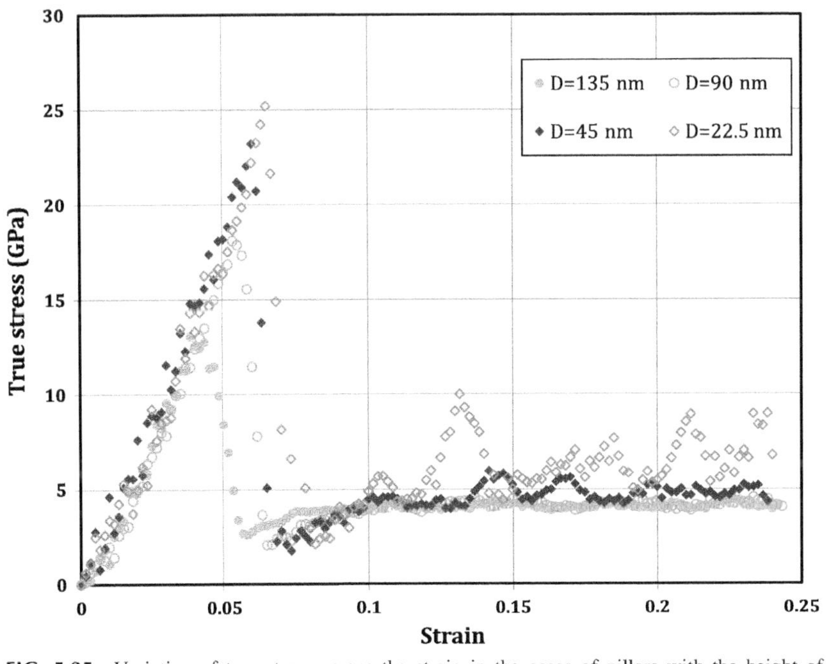

FIG. 5.35 Variation of true stress versus the strain in the cases of pillars with the height of $H = 45$ nm and different diameters of $D = 22.5, 45, 90,$ and 135 nm. *(After Yaghoobi, M., Voyiadjis, G.Z., 2016b. Size effects in FCC crystals during the high rate compression test. Acta Mater. 121, 190–201.)*

nearly no size effects by increasing the pillar diameter from 90 to 135 nm. The variation of dislocation density should be investigated to unravel the controlling mechanisms of observed size effects. Fig. 5.36 shows the variations of dislocation density and true stress versus the strain. Fig. 5.36A shows that the size effects in the pillar with the diameter of 22.5 nm are controlled by the source-limited activation, i.e. dislocation starvation. In other words, each time, the mobile dislocations are driven out of the sample, which locally increases the stress until another dislocation nucleation and evolution occurs leading to a stress relaxation. In the case of the pillar with the diameter of $D=45$ nm, Fig. 5.36B shows that the exhaustion hardening mechanism controls the size effects in a way that as the dislocation density decreases, the stress increases. Some small local jumps are observed in true stress which occur because some dislocations leave the pillar from the free surfaces. However, the peak of these local changes are smaller than those observed in the pillar with the diameter of $D=22.5$ nm. After the elastic response and following stress relaxation, Fig. 5.36C and D show that there is no jump in the stress-strain curves in the

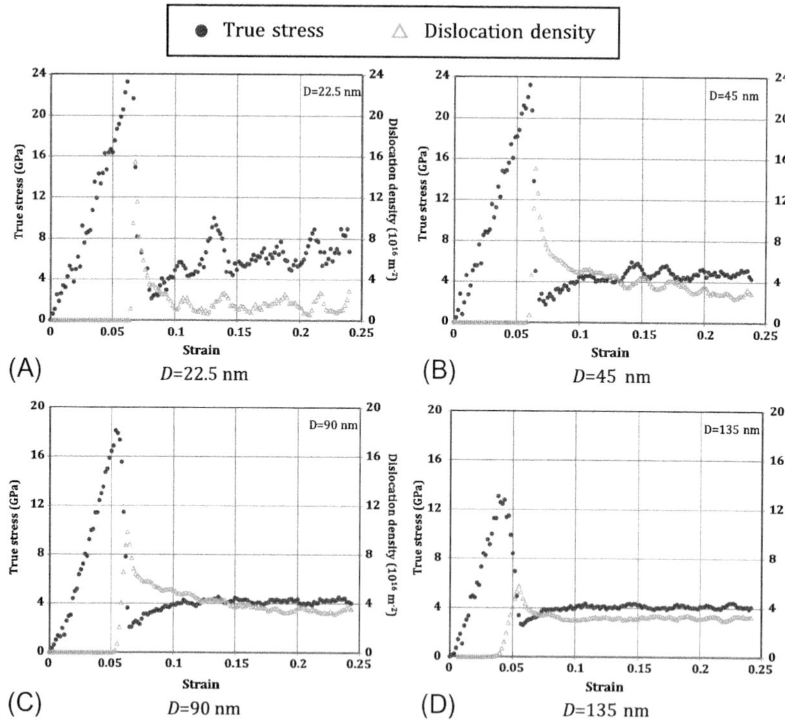

FIG. 5.36 The compressive responses of pillars with the height of 45 nm and different diameters of: (A) $D=22.5$ nm, (B) $D=45$ nm, (C) $D=90$ nm, and (D) $D=135$ nm. *(After Yaghoobi, M., Voyiadjis, G.Z., 2016b. Size effects in FCC crystals during the high rate compression test. Acta Mater. 121, 190–201.)*

case of the pillars with the diameters of 90 and 135 nm, which is due to the smooth variation of dislocation density. The observed responses are different from experimental results and discrete dislocation dynamics (DDD) simulations, which are conducted at much slower rates. The smooth variation of dislocation density is due to the high dislocation density induced to sustain the imposed high rate deformations. The results show that both pillars eventually reach a steady response which is not size dependent. However, they reach the steady states response throughout the different procedures in a way that the response of the pillar with larger diameter, i.e. $D = 135$ nm, becomes steady at smaller strains. Two pillars with the diameters of 90 and 135 nm eventually tend to the same stress. It means that if the sample height is kept fixed, there is a limit that increasing the sample diameter will not further change the response. Fig. 5.36C and D show that increasing the pillar diameter from 90 to 135 nm does not change the dislocation density after the dislocation density reaches its steady state.

The responses of pillars with three heights of 30, 45, and 75 nm and two diameters of 22.5 and 135 nm are compared during the compression test to study the effects of pillar height on the controlling mechanisms of size effects. Fig. 5.37A and B present the variations of true stress and dislocation density versus the strain, respectively, in the cases of samples with the diameter of 22.5 nm and different heights of 30, 45, and 75 nm. The results show that the size effects governing mechanism is dislocation starvation for all samples, and the general trend is nearly independent of pillar height. In other words, when the dislocation starvation is the dominant mechanism of size effects, changing the sample height does not considerably alter the pillar response. However, the values of local stress peak are slightly different based on the density of dislocations remains in the sample during the starvation step. For example, the maximum stress peak happens in the case of the pillar with the height of 75 nm at $\varepsilon \approx 0.18$ which is related to the minimum dislocation density occurring slightly prior to the same strain. Fig. 5.38 depicts the response of pillars with the diameter of 135 nm and different heights of 30, 45, and 75 nm during the compression test. Fig. 5.38A shows that as the pillar height decreases, the strength of the pillar increases. As shown in Fig. 5.38B, the dislocation starvation does not occur, and the source-limited activation is not the governing mechanism of size effects. In the cases of samples with heights of 30 and 45 nm, Fig. 5.38B shows that the pillars have nearly the similar dislocation density as the strain varies. However, the required stress to maintain the observed dislocation density increases as the pillar height decreases. It is due to the fact that as the pillar height decreases, dislocations reach the sample top and bottom surfaces faster and they are not able to sustain more deformation. Accordingly, more dislocations should be nucleated to maintain the plastic flow which increases the required stress. Lee et al. (2009) studied the pillars with low height to diameter ratios during the compression experiment and observed that the slip on $\{1\ 1\ 1\}$ planes could be inhibited at the top and bottom surfaces. They also observed that

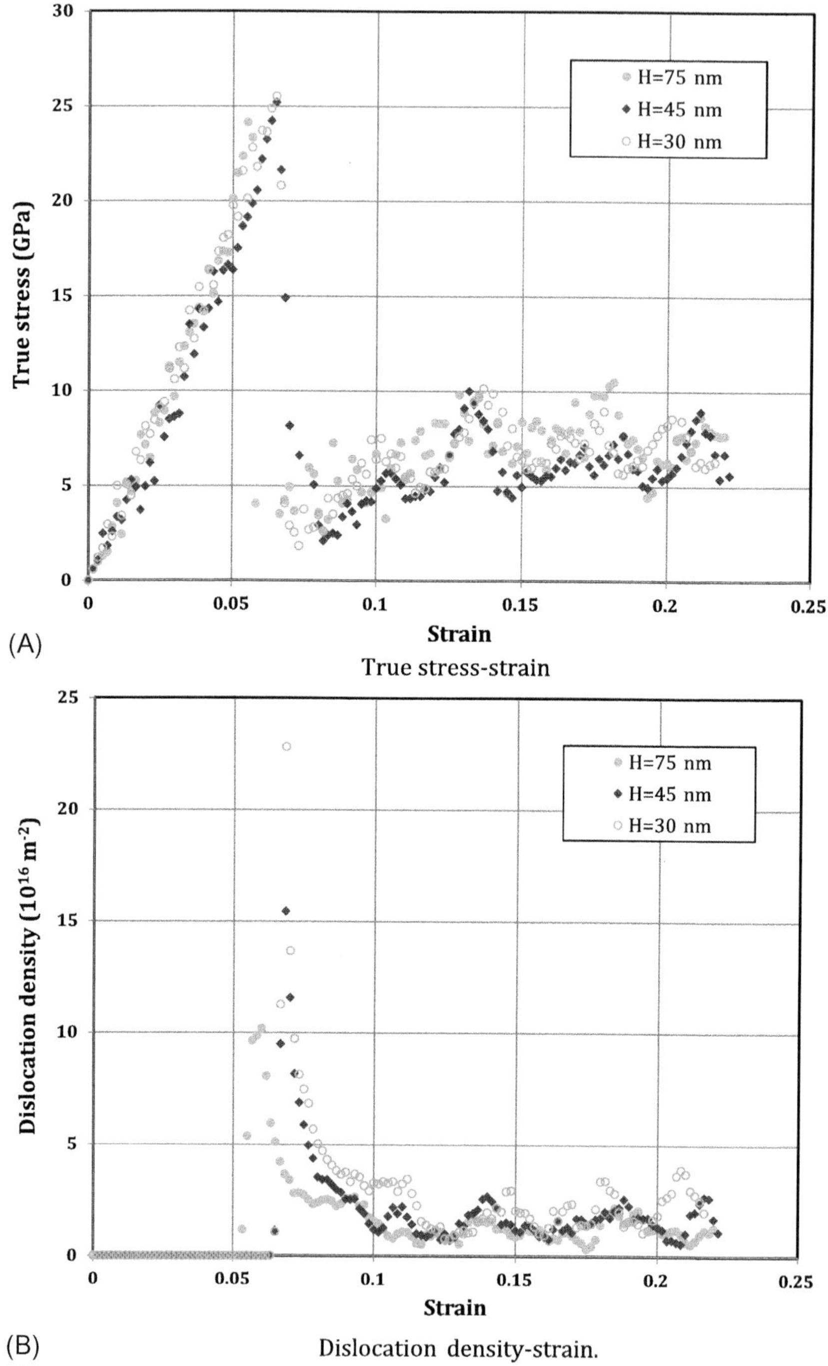

FIG. 5.37 Compressive response of pillars with the diameter of 22.5 and different heights of 30, 45, and 75 nm: (A) true stress-strain and (B) dislocation density-strain. *(After Yaghoobi, M., Voyiadjis, G.Z., 2016b. Size effects in FCC crystals during the high rate compression test. Acta Mater. 121, 190–201.)*

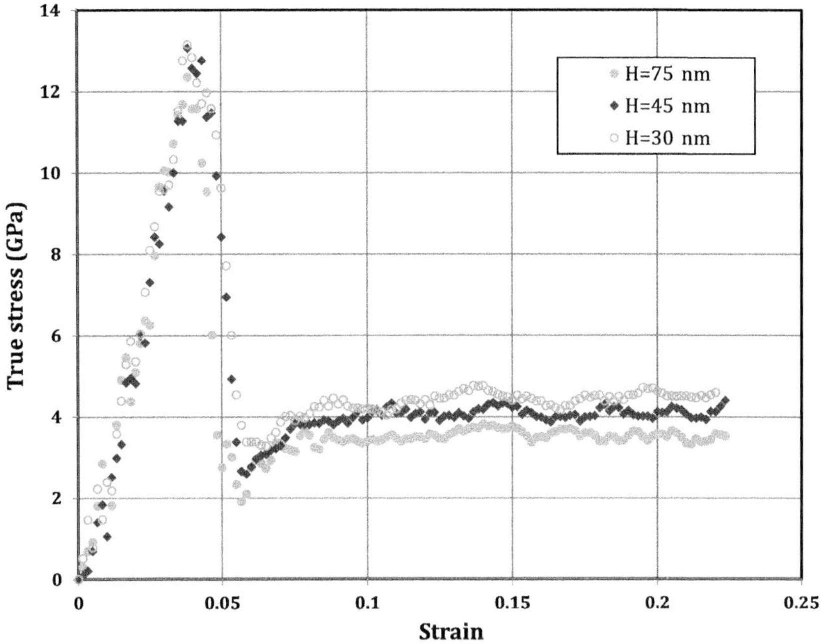

(A) True stress-strain

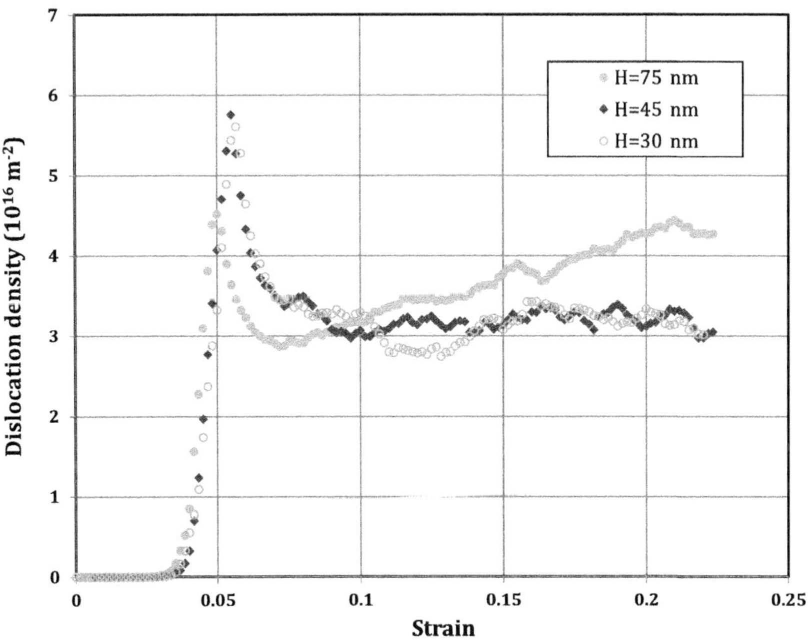

(B) Dislocation density-strain.

FIG. 5.38 Compressive response of pillars with the diameter of 135 and different heights of 30, 45, and 75 nm: (A) true stress-strain and (B) Dislocation density-strain. *(After Yaghoobi, M., Voyiadjis, G.Z., 2016b. Size effects in FCC crystals during the high rate compression test. Acta Mater. 121, 190–201.)*

the strong adhesion between pillar and substrate induces a multi-axial stress state (Lee et al., 2009). In the case of the current simulation, however, the potential selected to model the substrate does not induce adhesion. In the case of the pillar with the height of 75 nm, due to the similar procedure, the sample strength is less than those of the pillars with the heights of 30 and 40 nm. Also, the dislocation density of sample with the height of 75 nm is higher than those of the shorter pillars which provides more dislocations to sustain the plastic flow and decreases the required stress.

In the case of the pillar with the height of 75 nm and after the initial dislocation nucleation phase, the results show that the dislocation density increases as the strain increases. If all the dislocations are mobile, the required stress to sustain the plastic flow should decrease as the dislocation density increases. However, the stress reaches a steady state with a constant value. It is due to the fact that the dislocation density, which is presented in Fig. 5.38, is the total dislocation density including both the mobile and immobile dislocations. In other words, in the case of sample with the height of 75 nm, the increase in total dislocation density as the strain increases is mainly due to the increases in the immobile dislocation density. Hence, the stress does not decreases because what releases the stress is an increase in mobile dislocation density and not the total one. In order to investigate the proposed explanation, the variation of immobile dislocation density verses the strain should be studied. In the case of fcc metals, the mobile dislocations include the majority of Shockley partial dislocations with the Burgers vector of $1/6\langle 1\ 1\ 2\rangle$ and perfect dislocations with the Burgers vector of $1/2\langle 1\ 1\ 0\rangle$. On the other hand, the immobile dislocations include the Hirth, stair-rod, and Frank partial dislocations with the Burgers vectors of $1/3\langle 0\ 0\ 1\rangle$, $1/6\langle 0\ 1\ 1\rangle$, and $1/3\langle 1\ 1\ 1\rangle$, respectively, and some of the Shockley partial dislocations. Here, the density of immobile dislocations ρ_{immobile} is approximated by the total dislocation density excluding the perfect and Shockley partial dislocations, which is denoted by $\rho_{\text{immobile}}{}'$. Fig. 5.39 presents the variation of $\rho_{\text{immobile}}{}'$ versus the strain in the cases of pillars with the diameter of 135 nm and heights of 30, 45, and 75 nm. In order to have the exact value of immobile dislocation density, the immobile Shockley partial dislocations should also be considered. However, there is no applicable way to distinguish between the mobile and immobile Shockley partial dislocations. In the case of the pillar with the height of 75 nm and after the initial dislocation nucleation phase, Fig. 5.39 shows that the $\rho_{\text{immobile}}{}'$ increases as the strain increases. The results show that, however, the $\rho_{\text{immobile}}{}'$ does not vary in the cases of smaller pillars. It is worth mentioning that the immobile dislocations trap other mobile dislocations and locally immobilized them. Hence, larger $\rho_{\text{immobile}}{}'$ leads to the larger fraction of immobile Shockley partial dislocations.

In the final step, the size effects mechanisms of a pillar with the height of 0.3 μm and diameter of 0.15 μm (*S1*) are studied. Fig. 5.40 compares the compressive response of the *S1* sample to the pillar with $D = 135$ nm and $H = 75$ nm (*S2*). The results show that although the dislocation density of *S1*

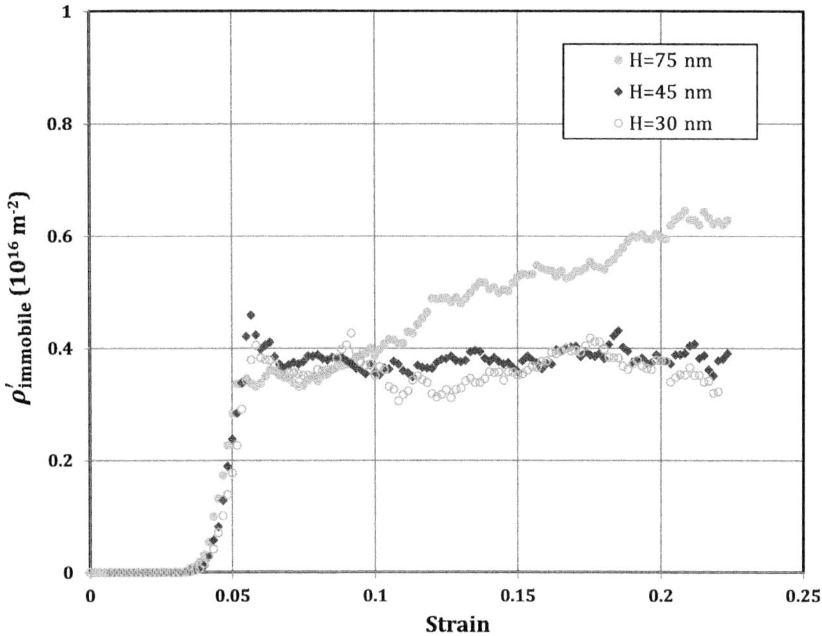

FIG. 5.39 Variation of $\rho'_{immobile}$ versus the strain in the cases of pillars with the diameter of 135 nm and different heights of 30, 45, and 75 nm. *(After Yaghoobi, M., Voyiadjis, G.Z., 2016b. Size effects in FCC crystals during the high rate compression test. Acta Mater. 121, 190–201.)*

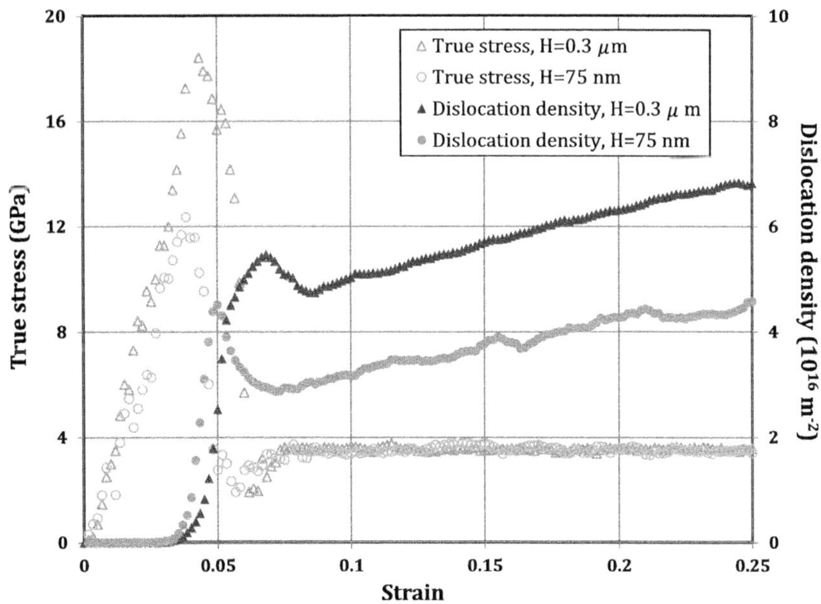

FIG. 5.40 Compressive response of pillars with the heights of 75 nm and 0.3 μm and diameters of 135 nm and 0.15 μm, respectively. *(After Yaghoobi, M., Voyiadjis, G.Z., 2016b. Size effects in FCC crystals during the high rate compression test. Acta Mater. 121, 190–201.)*

is larger than that of the *S2*, besides the initial dislocation nucleation stress, there is no size effects in strength. In other words, in the absence of strain gradient and in the high rate deformations, there is no more size effect in the case of pristine pillars larger than the one with D = 135 nm and H = 75 nm. It is worth mentioning that part of the differences in dislocation density of the two samples is due to their different aspect ratios. However, as the results indicate, it does not lead to any size effects. Compared to the experimental results obtained at quasi-static rates, such as Greer et al. (2005) and Espinosa et al. (2005), the results show that increasing the strain-rate decreases the size of pillars at which there is no more size effects.

In order to study the microstructure of different size effects governing mechanisms, the dislocation visualization is incorporated here for the largest and smallest simulated pillar heights, which are $H = 30$ nm and $H = 0.3$ μm, with the height to diameter ratio of 2:1. Fig. 5.41 presents the compressive response of the pillar with the heights of 30 nm and diameter of 15 nm. The dislocation nucleation and evolution patterns at various strains are shown in Fig. 5.42. The color of Shockley, Hirth, and stair-rod partial dislocations and perfect dislocations are green, yellow, blue, and red, respectively. The initial dislocation nucleation occurs at the free surfaces. The results show that the dominant dislocation multiplication mechanism is the dislocation nucleation from the free surfaces

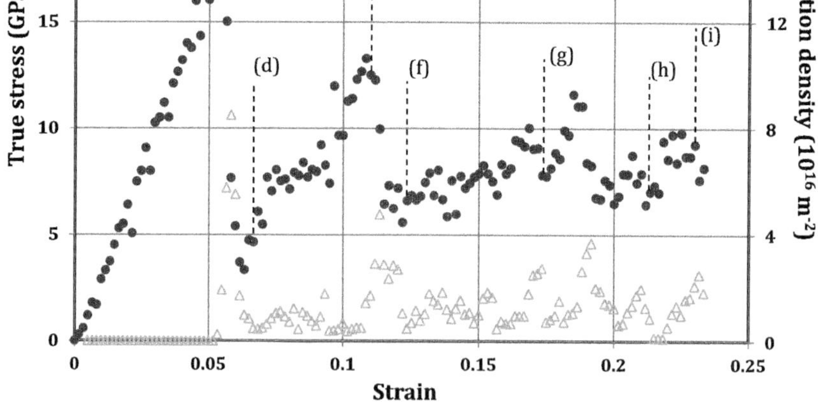

FIG. 5.41 Compressive response of pillar with the diameter of 15 nm and height of 30 nm. *(After Yaghoobi, M., Voyiadjis, G.Z., 2016b. Size effects in FCC crystals during the high rate compression test. Acta Mater. 121, 190–201.)*

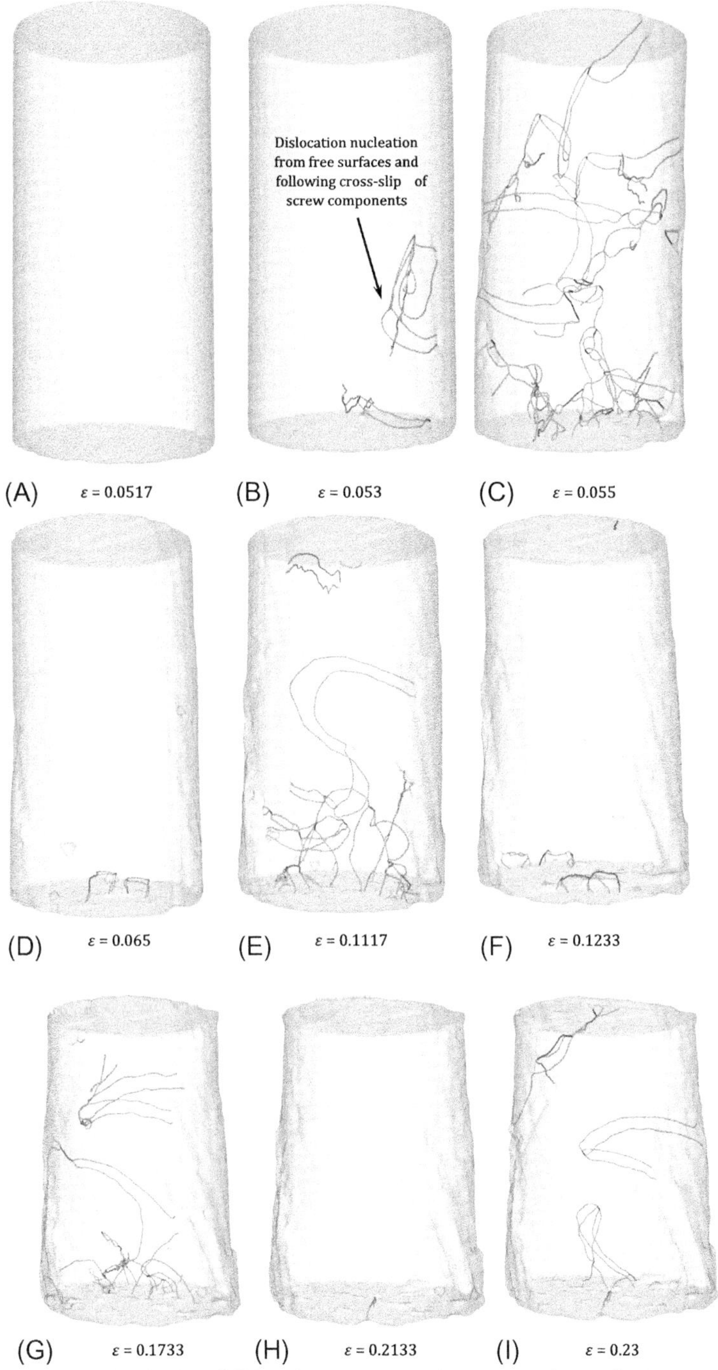

FIG. 5.42 The visualization of dislocations structure during compression test in the case of pillar with the diameter of 15 nm and height of 30 nm at different strains of: (A) $\varepsilon = 0.0517$, (B) $\varepsilon = 0.053$, (C) $\varepsilon = 0.055$, (D) $\varepsilon = 0.065$, (E) $\varepsilon = 0.1117$, (F) $\varepsilon = 0.1233$, (G) $\varepsilon = 0.1733$, (H) $\varepsilon = 0.2133$, and (I) $\varepsilon = 0.23$. *(After Yaghoobi, M., Voyiadjis, G.Z., 2016b. Size effects in FCC crystals during the high rate compression test. Acta Mater. 121, 190–201.)*

which follows by cross-slip, which is termed as surface cross-slip (Rao et al., 2013). Hussein et al. (2015) observed a similar deformation mechanism during 3-D DDD simulation of fcc crystals. However, since the pillar diameter is so small, cross-slip cannot create many pining points before the dislocations leave the sample from another free surface. The results show that each time, the dislocation starvation occurs, the strength required to sustain the plastic flow starts increasing. New dislocations are then nucleated and elongated which releases the stress as shown in Fig. 5.42B, C, E, G, and I. Dislocations eventually leave the pillar which leads to another starvation, as shown in Fig. 5.42D, F, and H. Fig. 5.43 illustrates the dislocation nucleation and evolution of the pillar with the height of 0.3 μm and diameter of 0.15 μm. Fig. 5.43B shows that the initial dislocation nucleation occurs from the free surfaces which follows by the cross-slip. However, as shown in Fig. 5.43C, the dislocation starvation does not occur in the pillar. In order to investigate the dislocation multiplication mechanism, Fig. 5.44 shows the dislocations in a small block of pillar at $\varepsilon = 0.1416$, which is presented in Fig. 5.43C. Fig. 5.44 shows several cross-slips which increase the number of pinning points. Next, the elongation of the dislocations which are pinned at their ends provides the dislocation length required to sustain the imposed plastic flow. The results show that if the dislocations do not immediately leave the sample, the dominant dislocation multiplication mechanism is

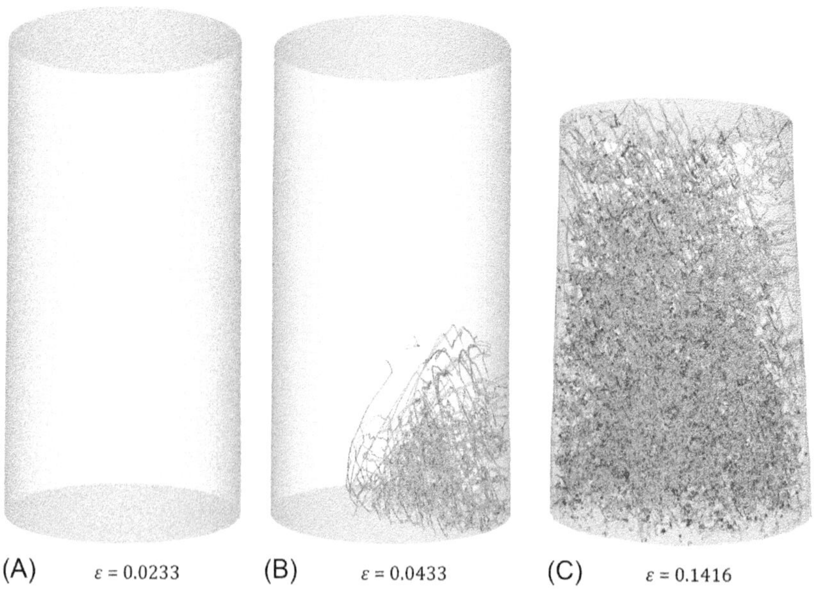

FIG. 5.43 The visualization of dislocations structure during compression test in the case of pillar with the diameter of 0.15 μm and height of 0.3 μm at different strains of: (A) $\varepsilon = 0.0233$, (B) $\varepsilon = 0.0433$, and (C) $\varepsilon = 0.1416$. *(After Yaghoobi, M., Voyiadjis, G.Z., 2016b. Size effects in FCC crystals during the high rate compression test. Acta Mater. 121, 190–201.)*

FIG. 5.44 A selected block of dislocations in the case of pillar with the diameter of 0.15 μm and height of 0.3 μm at $\varepsilon = 0.1416$. *(After Yaghoobi, M., Voyiadjis, G.Z., 2016b. Size effects in FCC crystals during the high rate compression test. Acta Mater. 121, 190–201.)*

cross-slip. It is worth mentioning that in the case of quasi-static experiment, if the sample is large enough that the dislocation starvation does not occur, the plastic deformation mechanism is governed by truncated dislocation sources operations (Zhou et al., 2011; Ryu et al., 2015).

Sansoz (2011) presented a model which stated that all nanopillars experience both deformation mechanisms of source-limited activation, i.e. dislocation starvation, and mobile dislocation exhaustion, i.e. exhaustion hardening. It was demonstrated that as the strain increases, the governing mechanism of deformation will change from the dislocation exhaustion to the source-limited activation (Sansoz, 2011). It was also stated that all samples will experience the phase which the dislocation exhaustion is the dominant mechanism of deformation (Sansoz, 2011). However, Figs. 5.35 and 5.36 show that in the case of sample with very small diameter, such as D = 22.5 nm, the governing mechanism of size effects is solely dislocation starvation. The results also show that the dislocation exhaustion is the only governing mechanism for pillars with a large diameter, such as $D = 135$ nm. Sansoz (2011) incorporated the pillars which contain the initial dislocations. In order to study the effect of pillar initial structure on the governing mechanisms of size effects, the responses of pre-strained pillars with the height of 45 nm and diameters of 22.5 and 135 nm, which have been initially loaded and unloaded, are studied during the compression test. Fig. 5.45A and B present the variations of stress and dislocation density versus

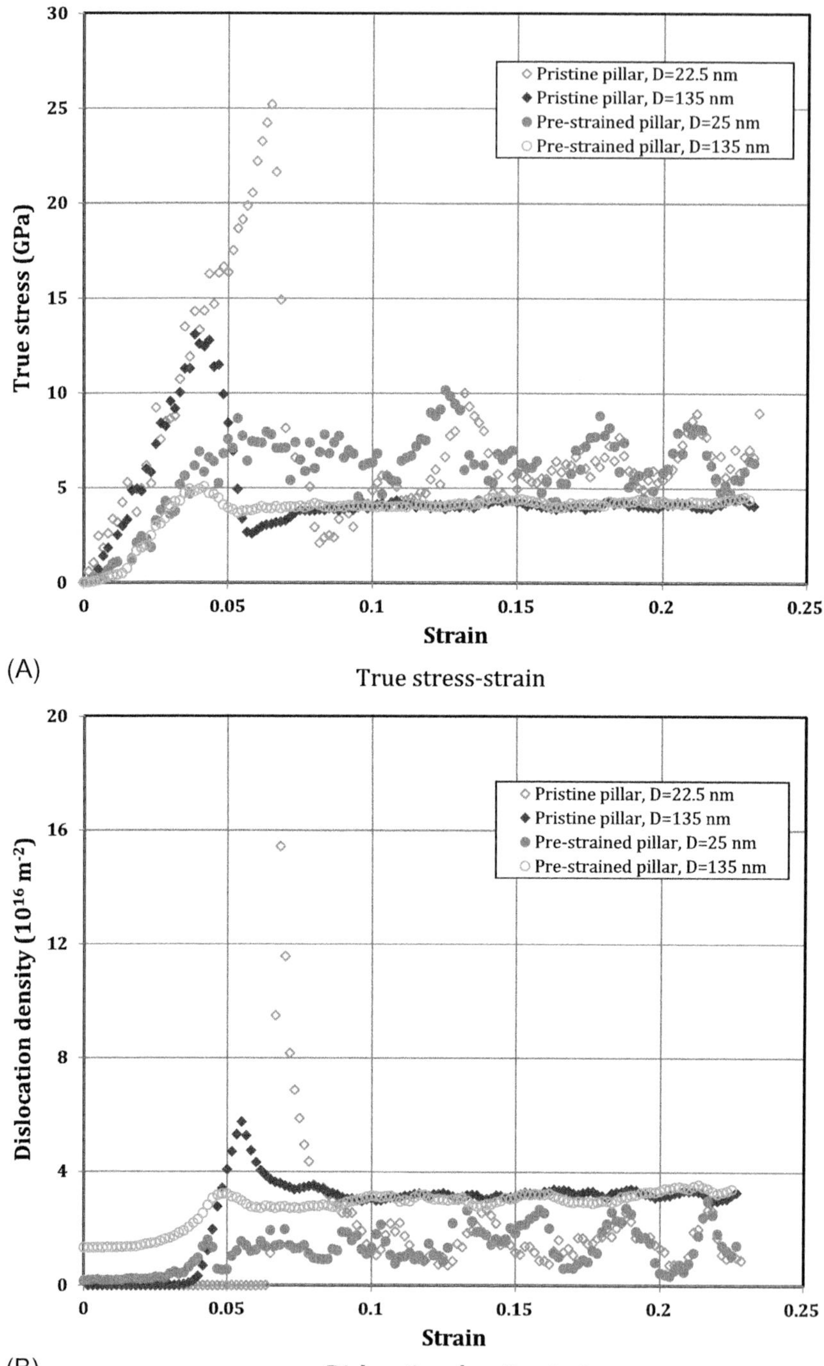

FIG. 5.45 Compressive response of pristine and pre-strained pillars with the height of 45 nm and different diameters of 22.5 and 135 nm: (A) true stress-strain and (B) dislocation density-strain. *(After Yaghoobi, M., Voyiadjis, G.Z., 2016b. Size effects in FCC crystals during the high rate compression test. Acta Mater. 121, 190–201.)*

the strain, respectively, during the compression test for both pristine and pre-strained pillars. In the case of larger sample, i.e. pillar with D = 135 nm and H = 45 nm, the pre-straining only alters the initial part of the response, and both stress and dislocation density become similar for the pristine and pre-strained pillars after the initial phase of dislocation nucleation. In the cases of smaller sample, i.e. pillars with D = 22.5 nm and H = 45 nm, besides the initial phase of dislocation nucleation, the pre-straining slightly shift the location of stress peaks. However, the size effects is still governed by the source-limited activation. The results show that the controlling mechanisms of size effects are independent of sample initial structure in the cases of pillars with diameters of 22.5 and 135 nm and height of 45 nm. Fig. 5.46 compares the response of the pre-strained sample with height of 0.3 μm and diameter of 0.15 μm to that of the pristine pillar to investigate the effect of initial structure in the case of larger pillars. The results show that the pre-straining increases the dislocation density which triggers the forest hardening mechanism. This shows that the strength increases as the strain increases due to the increase in dislocation density which activates the mechanism of dislocation interaction with each other.

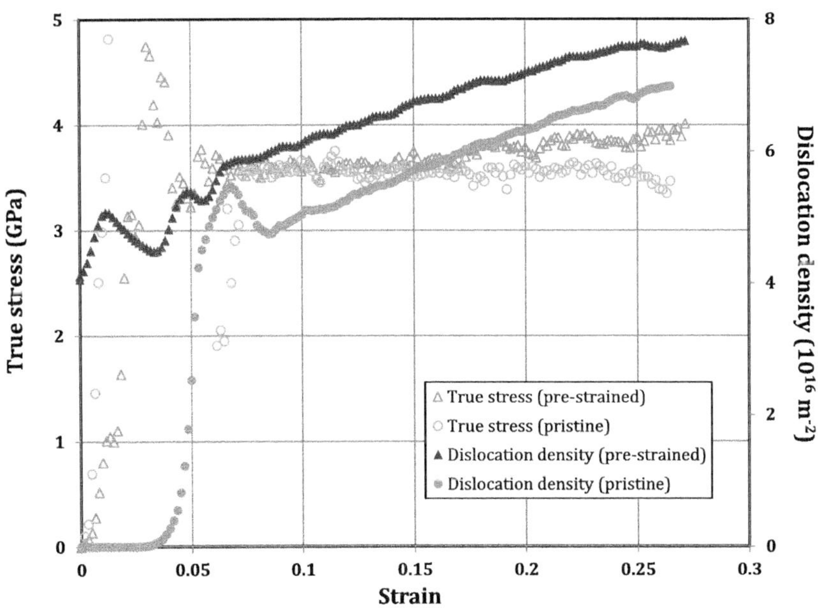

FIG. 5.46 Compressive response of pre-strained and pristine pillars with the height of 0.3 μm and diameter of 0.15 μm. *(After Yaghoobi, M., Voyiadjis, G.Z., 2016b. Size effects in FCC crystals during the high rate compression test. Acta Mater. 121, 190–201.)*

5.3.2.2 *Coupling effects of size and strain rate in FCC pillars*

Voyiadjis and Yaghoobi (2017a) incorporated large MD simulation to investigate the coupling effects of size and strain rates in FCC pillars during compression test. They investigated the distribution of dislocation length formed during the compression to unravel the underlying mechanisms of size effects. Various researchers have related the size effects to some characteristics of dislocation length. The first mathematical prediction of effective dislocation source length was proposed by Parthasarathy et al. (2007). They introduced the probability of a cylindrical sample with n pins to have the maximum distance from the free surface equal to λ_{max} as follows:

$$p(\lambda_{max})d\lambda_{max} = \left[1-\frac{\pi(R-\lambda_{max})(b-\lambda_{max})}{\pi Rb}\right]^{n-1}\left(\frac{\pi[(R-\lambda_{max})+(b-\lambda_{max})]}{\pi Rb}\right)nd\lambda_{max} \quad (5.45)$$

where R is the specimen radius, $b = R/\cos\beta$ is the major axis of the glide plane, and β is the angle between the primery slip plane and the loading axis. Using the probability function defined in Eq. (5.45), the mean effective source length $\bar{\lambda}_{max}$ can be obtained as below (Parthasarathy et al., 2007):

$$\bar{\lambda}_{max} = \int_0^R \lambda_{max}p(\lambda_{max})d\lambda_{max} = \int_0^R \left[1-\frac{\pi(R-\lambda_{max})(b-\lambda_{max})}{\pi Rb}\right]^{n-1}\left(\frac{\pi[(R-\lambda_{max})+(b-\lambda_{max})]}{\pi Rb}\right)n\lambda_{max}d\lambda_{max} \quad (5.46)$$

In the next step, the CRSS was related to $\bar{\lambda}_{max}$ as follows:

$$\text{CRSS} = \frac{\alpha Gb}{\bar{\lambda}_{max}} + \tau_0 + 0.5Gb\sqrt{\rho} \quad (5.47)$$

where α is a constant, G is the shear modulus, b is the Burgers vector, τ_0 is the friction stress, and ρ is the dislocation density. The number of pins, n, is also predicted as follows (Parthasarathy et al., 2007):

$$n = \text{Integer}\left[\frac{L_{mobile}}{L_{ave}}\right] \quad (5.48)$$

where L_{ave} is the average length of dislocation segments, $L_{mobile} = \rho\pi R^2 h/s$ is the total length of mobile dislocations, h is the height of the pillar, and s is the number of slip systems. Basically, Parthasarathy et al. (2007) and later Rao et al. (2008) related the size effects to the average largest source length $\bar{\lambda}_{max}$. El-Awady et al. (2008, 2009)El-Awady (2014) related the size effects to the mean dislocation length L_{ave}. El-Awady (2014) proposed an equation to obtain L_{ave} using the pillar diameter and initial dislocation density ρ_0 as follows:

$$L_{ave} = \frac{bD\sqrt{\rho_0}}{\overline{\beta}} for\ \rho_0 < \rho_0^{cr} \tag{5.49a}$$

$$L_{ave} = \frac{1}{\overline{\alpha}\sqrt{\rho_0}} for\ \rho_0 \geq \rho_0^{cr} \tag{5.49b}$$

where ρ_0^{cr} is the critical initial dislocation density at which L_{ave} becomes independent of sample size, and $\overline{\alpha}$ and $\overline{\beta}$ are dimensionless constants.

Voyiadjis and Yaghoobi (2017a) addressed the coupling effects of size and strain rate by investigating the dislocation length distribution. To do so, they conducted large atomistic simulation using the parallel code LAMMPS (Plimpton, 1995) to study the size and strain rate effects. The embedded-atom method (EAM) potential developed by Mishin et al. (1999) is used to model the Ni-Ni atomic interaction. Two pillar with circular cross section, heights of 90 nm and 300 nm, and aspect ratio of *Length: Diameter* = 2: 1 are modeled, which consist of 13 million and 487 million atoms, respectively. The axis of pillars is along the [1 1 1] direction. The boundary conditions are selected similar to the one described in Section 5.3.2.1.1 in which the top and bottom surfaces are set free. A prescribed potential wall is incorporated to simulate the substrate. A large flat indenter is used to impose the compressive displacement with three strain rates of $\dot{\varepsilon}_1 = 6.66e8\,s^{-1}$, $\dot{\varepsilon}_2 = 3.33e8\,s^{-1}$, and $\dot{\varepsilon}_3 = 1.665e8\,s^{-1}$. The repulsive potential of Eq. (5.32) is incorporated to model the interaction between the indenter and Ni atoms. The Si substrate is modeled using the Lennard-Jones (LJ) potential as presented in Eq. (5.26). The LJ parameters are $\overline{\varepsilon}_{Ni-Si} = 1.5231e-20\,J$ and $\overline{\sigma}_{Ni-Si} = 3.0534^{-}$ with the cutoff distance equal to $2.5\overline{\sigma}$. The velocity Verlet algorithm with the time step of 5 fs is used to numerically integrate the equations of Motion The simulation is conducted using the NVT ensemble (Plimpton, 1995). The triangulation method described in Section 5.2.4.1 is incorporated to capture the precise cross section area during the simulation. The crystal Analysis tool (Stukowski and Albe, 2010; Stukowski, 2012, 2014; Stukowski et al., 2012) is used to extract the dislocation structure from the atomic trajectories. The dislocation network is visualized and analyzed using the software Paraview (Henderson, 2007) and OVITO (Stukowski, 2010).

Since the samples are initially defect free at the start of MD simulation, which is not true in the real experiments, the stress after the initial nucleation is incorporated to study the size effects. Fig. 5.47 presents the variations of true stress (σ) and dislocation density (ρ) versus the strain (ε) in the case of the smaller pillar at three different strain rates of $\dot{\varepsilon}_1 = 6.66e8\,s^{-1}$, $\dot{\varepsilon}_2 = 3.33e8\,s^{-1}$, and $\dot{\varepsilon}_3 = 1.665e8\,s^{-1}$. Fig. 5.47 shows that in the case of the smaller pillar, the sample strength is nearly independent of the strain rate. However, the dislocation density increases as the strain rate increases. The results indicate that the dislocation density is not an appropriate measure to study the size effects in the case of metallic samples of confined volumes. Fig. 5.48 presents the $\sigma - \varepsilon$

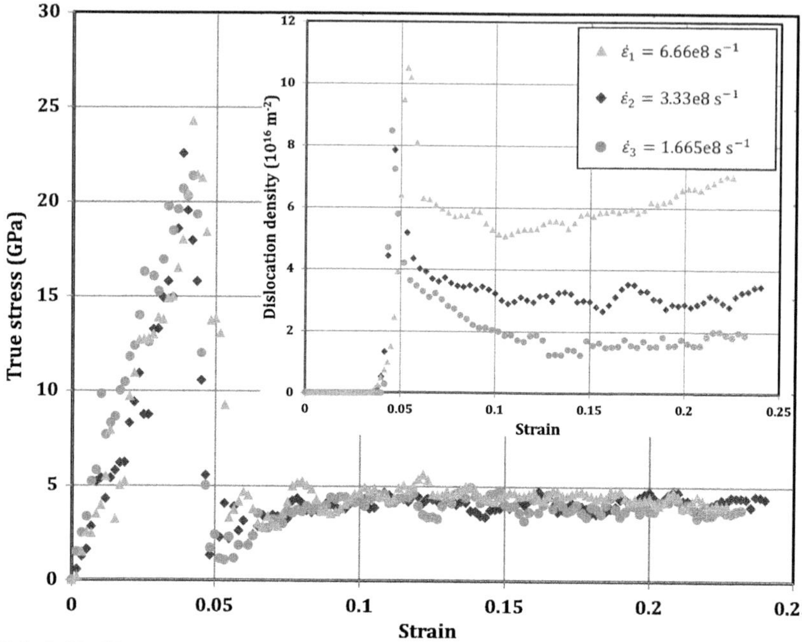

FIG. 5.47 The compressive responses of the pillar with the diameter of 45 nm at different strain rates of $\dot{\varepsilon}_1 = 6.66\text{e}8\,\text{s}^{-1}$, $\dot{\varepsilon}_2 = 3.33\text{e}8\,\text{s}^{-1}$, and $\dot{\varepsilon}_3 = 1.665\text{e}8\,\text{s}^{-1}$. *(After Voyiadjis, G.Z., Yaghoobi, M., 2017a. Size and strain rate effects in metallic samples of confined volumes: dislocation length distribution. Scr. Mater. 130, 182–186.)*

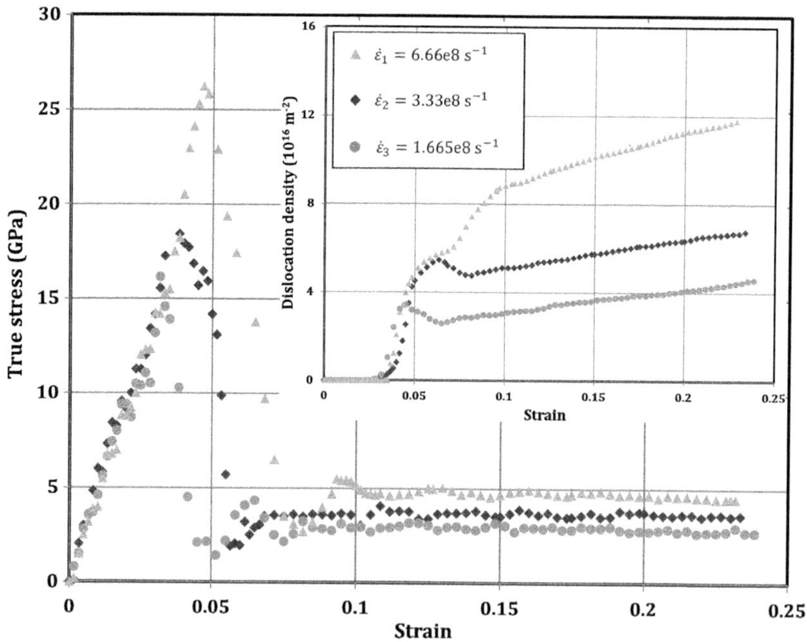

FIG. 5.48 The compressive responses of the pillar with the diameter of 150 nm at different strain rates of $\dot{\varepsilon}_1 = 6.66\text{e}8\,\text{s}^{-1}$, $\dot{\varepsilon}_2 = 3.33\text{e}8\,\text{s}^{-1}$, and $\dot{\varepsilon}_3 = 1.665\text{e}8\,\text{s}^{-1}$. *(After Voyiadjis, G.Z., Yaghoobi, M., 2017a. Size and strain rate effects in metallic samples of confined volumes: dislocation length distribution. Scr. Mater. 130, 182–186.)*

and $\rho - \varepsilon$ in the case of the larger pillar at three different strain rates of $\dot{\varepsilon}_1 = 6.66e8\,s^{-1}$, $\dot{\varepsilon}_2 = 3.33e8\,s^{-1}$, and $\dot{\varepsilon}_3 = 1.665e8\,s^{-1}$. The dislocation density shows a similar trend compared to that of the smaller sample as the strain rate varies. However, the results show that the strength of the sample decreases by decreasing the strain rate. Fig. 5.49 compares the responses of pillars with different sizes during the compression test for different strain rates. The results show that as the strain rate increases, less size effects are observed in the simulated samples. This implies that increasing the strain rate decreases the size effects.

In order to capture the observed results, the dislocation length distribution is studied for different sample sizes subjected to various strain rates. The distribution of each sample is obtained by averaging the distributions at five strain values of 0.1, 0.125, 0.15, 0.175, and 0.2. Fig. 5.50 presents the variation of probability density function (PDF) versus the dislocation link length in both pillar sizes at different strain rates. The vertical axis of Fig. 5.50, i.e. PDF, is presented in logarithmic form. In the case of dislocation length distribution function of bulk materials, the maximum PDF occurs for some dislocation length in the middle of the distribution, and the dislocation length distribution can be approximated by a Weibull distribution function, see e.g. El-Awady et al. (2009). In the case of samples of confined volumes, however, the results show that most of the dislocation lengths are populated in the first length bean, which has the smallest length. In other words, the maximum PDF occurred at the smallest dislocation length bean. It is due to the activity of cross-slip as the major deformation mechanism in the samples with confined volumes. To verify the proposed description, the dislocation visualization of both pillar sizes at $\varepsilon = 0.2$ for strain rate of $\dot{\varepsilon}_2$ is presented as an example in Fig. 5.51. The visualization results show that the main mechanism of deformation is cross-slip which produces many small dislocations. A similar trend can be observed in the results obtained by Hussein et al. (2015) in which many small dislocations are produced by incorporating the cross-slip in the DDD formulation.

In the case of the smaller pillar, the values of average dislocation length L_{ave} are 25.29 $\mathring{A}$, 27.03 $\mathring{A}$, and 27.2 $\mathring{A}$ at strain rates of $\dot{\varepsilon}_1$, $\dot{\varepsilon}_2$, and $\dot{\varepsilon}_3$, respectively. In the case of the larger sample, the average dislocation length values are 21.09 $\mathring{A}$, 23.2 $\mathring{A}$, and 24.75 $\mathring{A}$ at strain rates of $\dot{\varepsilon}_1$, $\dot{\varepsilon}_2$, and $\dot{\varepsilon}_3$, respectively. As described by El-Awady (2014), L_{ave} can be a function of dislocation density and sample size depending on the dislocation density of the sample. Considering the order of dislocation density which is $10^{16}\,m^{-2}$, the L_{ave} should follow Eq. (5.49b) in which the L_{ave} is independent of the sample size and has an inverse relation with the dislocation density. The results show that L_{ave} independent of the sample size, i.e. L_{ave} of the larger pillar are close to that of the smaller one and even the smaller pillar has slightly larger L_{ave}. Also, as the strain rate increases, L_{ave} decreases which is due to the fact that increasing the strain rate increases the dislocation density. Eqs. (5.49a) and (5.49b), which was proposed by El-Awady (2014), can be micromechanically justified based on the dislocation

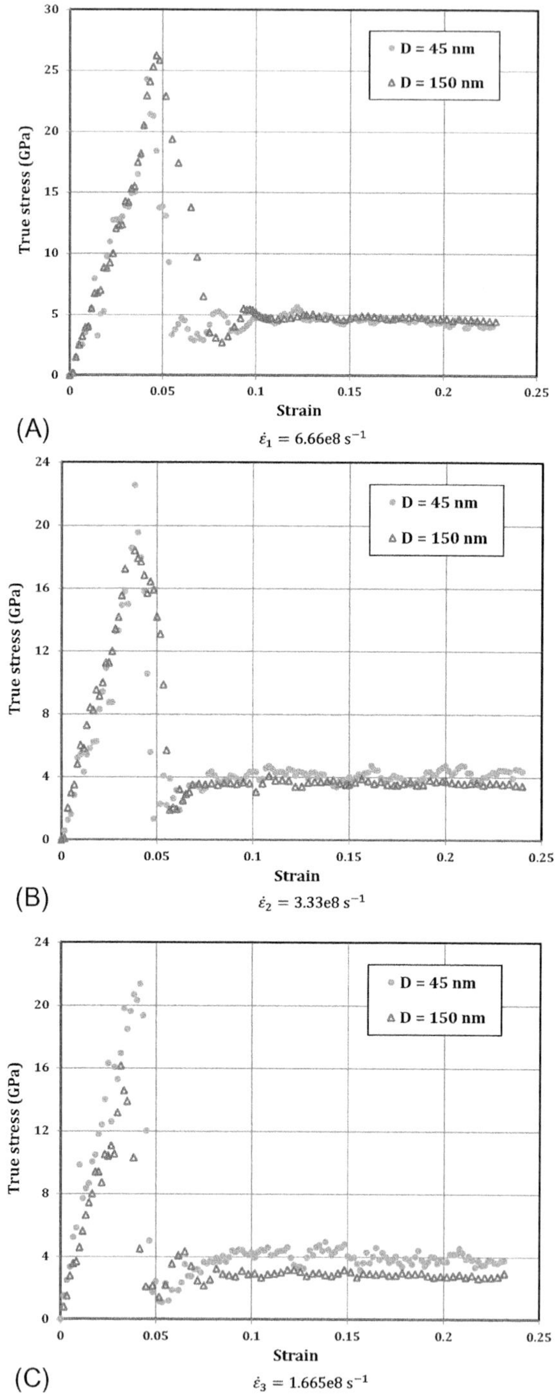

FIG. 5.49 Variation of true stress versus the strain in the cases of the pillars with the diameters of 45 and 150 nm at different strain rates of: (A) $\dot{\varepsilon}_1 = 6.66e8\,s^{-1}$, (B) $\dot{\varepsilon}_2 = 3.33e8\,s^{-1}$, and (C) $\dot{\varepsilon}_3 = 1.665e8\,s^{-1}$. *(After Voyiadjis, G.Z., Yaghoobi, M., 2017a. Size and strain rate effects in metallic samples of confined volumes: dislocation length distribution. Scr. Mater. 130, 182–186.)*

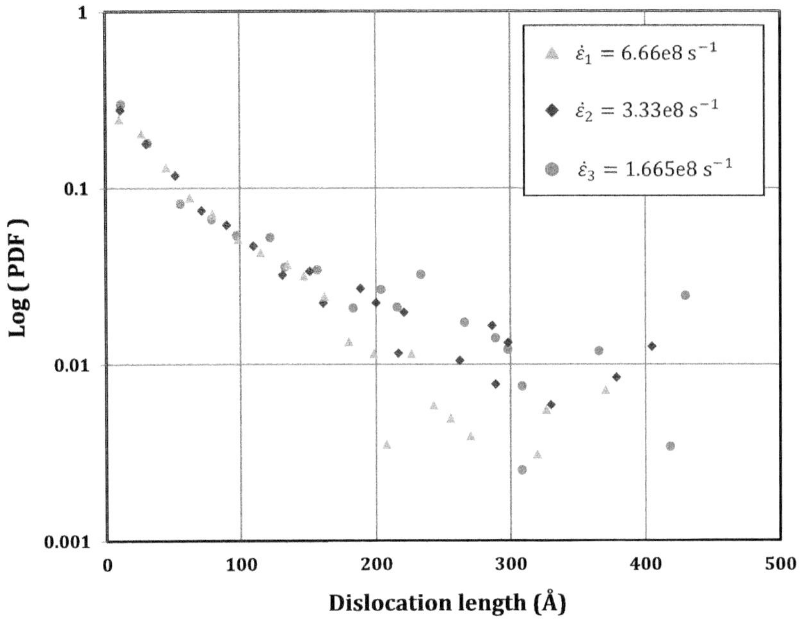

(A) D = 45 nm

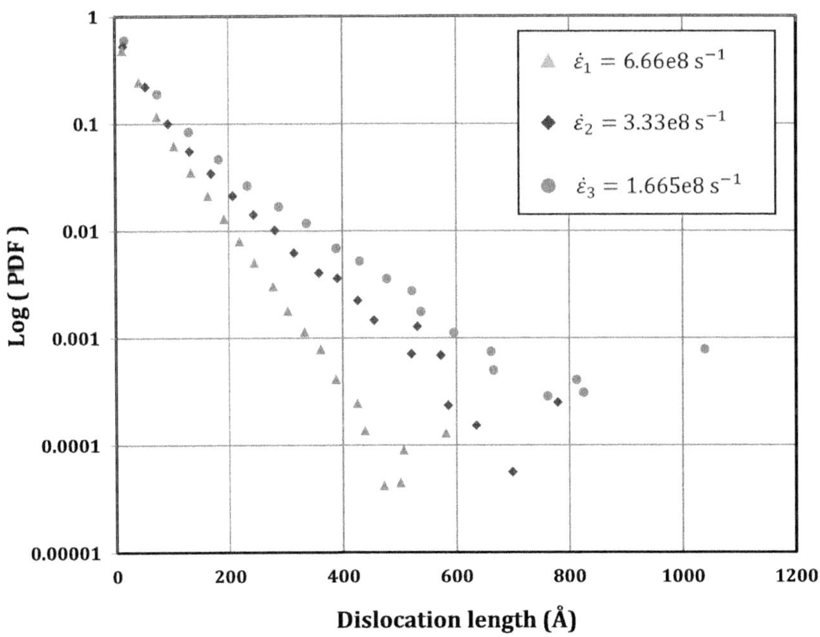

(B) D=150 nm

FIG. 5.50 Probability distribution function of dislocation lengths at different strain rates and pillars with the diameters of: (A) D = 45 nm and (B) D = 150 nm. *(After Voyiadjis, G.Z., Yaghoobi, M., 2017a. Size and strain rate effects in metallic samples of confined volumes: dislocation length distribution. Scr. Mater. 130, 182–186.)*

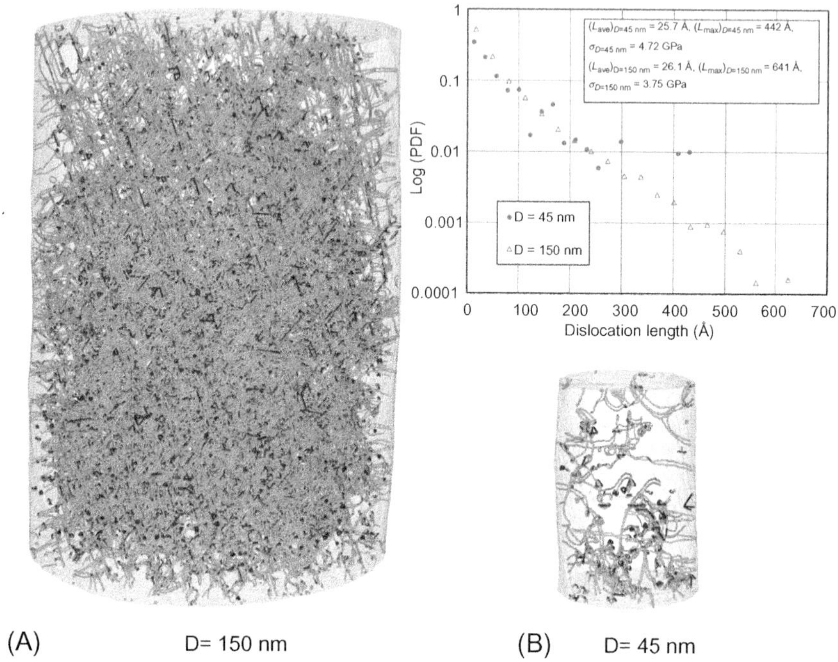

FIG. 5.51 The dislocation network at $\varepsilon = 0.2$ for the strain rate of $\dot{\varepsilon}_2 = 3.33e8\ s^{-1}$ and pillar diameters of: (A) $D = 150$ nm and (B) $D = 45$ nm. *(After Voyiadjis, G.Z., Yaghoobi, M., 2017a. Size and strain rate effects in metallic samples of confined volumes: dislocation length distribution. Scr. Mater. 130, 182–186.)*

network characteristics. In the region of small dislocation densities, i.e., Eq. (5.49a), increasing the sample size and dislocation density increases the chance of larger dislocation formation. In the region of high dislocation densities, i.e., Eq. (5.49b), however, increasing the dislocation density increases the chance of dislocations colliding with each other and dislocation refinement which decrease the dislocation length and consequently L_{ave}.

The values of the $\frac{(L_{ave})_{\dot{\varepsilon}_3}}{(L_{ave})_{\dot{\varepsilon}_1}}$ for smaller and larger pillars are 1.08 and 1.17 which are very close to each other. However, the strength of the smaller sample does not change as the strain rate varies while the larger sample exhibits significant strain rate effects. The results show that the strain rate effects cannot be captured using L_{ave}. Another way to reach the same conclusion is since L_{ave} is a function of dislocation density, and dislocation density variation is not capable of capturing size effects as shown in Fig. 5.47. Accordingly, L_{ave} is also not an appropriate dislocation network property to study the size effects.

Another dislocation network property which can be incorporated to study the size effects is the largest dislocation length L_{max}. Here, the L_{max} is averaged at five strain values of 0.1, 0.125, 0.15, 0.175, and 0.2 for each pillar size and strain rate. The values of L_{max} for the smaller sample and different rates of $\dot{\varepsilon}_1$,

$\dot{\varepsilon}_2$, and $\dot{\varepsilon}_3$ are 377.9 $\mathring{A}$, 415 $\mathring{A}$, and 432.9 $\mathring{A}$, respectively. In the case of the larger sample, L_{max} values are 605.8 $\mathring{A}$, 783.4 $\mathring{A}$, and 1095.6 $\mathring{A}$ for strain rates of $\dot{\varepsilon}_1$, $\dot{\varepsilon}_2$, and $\dot{\varepsilon}_3$, respectively. The values of the $\frac{(L_{max})_{\dot{\varepsilon}_3}}{(L_{max})_{\dot{\varepsilon}_1}}$ for smaller and larger pillars are 1.14 and 1.81 which shows a great difference compared to the values of $\frac{(L_{ave})_{\dot{\varepsilon}_3}}{(L_{ave})_{\dot{\varepsilon}_1}}$. As an example, in the case of Fig. 5.50, the true stresses for smaller and larger pillars at $\varepsilon = 0.2$ and strain rate of $\dot{\varepsilon}_2$ are $\sigma_{D=45\ nm} = 4.72$ GPa and $\sigma_{D=150\ nm} = 3.75$ GPa, respectively, where $\sigma_1/\sigma_2 = 1.26$. The average dislocation length for the smaller and larger pillars are $(L_{ave})_{D=45\ nm} = 25.7\ \mathring{A}$ and $(L_{ave})_{D=150\ nm} = 26.1\ \mathring{A}$, respectively, which are nearly similar. The maximum dislocation length, on the other hand, for the smaller and larger samples are $(L_{max})_{D=45\ nm} = 442\ \mathring{A}$ and $(L_{max})_{D=150\ nm} = 641\ \mathring{A}$, respectively, which shows that the sample with smaller L_{max} has the larger strength. The results show that the maximum dislocation length L_{max} is the appropriate dislocation network property to study the size effects and not the L_{ave}. It is due to the fact that there are tremendous small dislocations induced by the cross-slip in the samples of confined volumes, which was also observed by Hussein et al. (2015). Accordingly, the average dislocation length is highly influenced by the small dislocations, and the effect of maximum source length on L_{ave} diminishes. The maximum dislocation length L_{max}, on the other hand, is fully capable of capturing size and strain rate effects. For example, the results observed in Fig. 5.48 can be fully explained by L_{max}. The L_{max} in the larger pillar divided by that of the smaller pillar is 1.6, 1.89, and 2.53 at different strain rates of $\dot{\varepsilon}_1$, $\dot{\varepsilon}_2$, and $\dot{\varepsilon}_3$ which shows that increasing the strain rate decreases the size effects by decreasing the difference between the L_{max} for samples of different sizes.

References

Ashby, M.F., 1970. The deformation of plastically non-homogeneous materials. Philos. Mag. 21, 399–424.

Baskes, M.I., 1992. Modified embedded-atom potentials for cubic materials and impurities. Phys. Rev. B 46, 2727.

Bringa, M., Rosolankova, K., Rudd, R.E., Remington, B.A., Wark, J.S., Duchaineau, M., Kalantar, H., Hawreliak, J., Belak, J., 2006. Shock deformation of face-centred-cubic metals on subnanosecond timescales. Nat. Mater. 5, 805.

Bückle, H., 1959. Progress in microindentation hardness testing. Metall. Rev. 4, 49–100.

Bull, S.J., 2003. On the origins of the indentation size effect. Z. Metallkd. 94, 787–792.

Bushby, A.J., Dunstan, D.J., 2004. Plasticity size effects in nanoindentation. J. Mater. Res. 19, 137–142.

Chen, C.C., Hendrickson, A.A., 1973. Microhardness phenomena in silver. In: Westbrook, J.H., Conrad, H. (Eds.), The Science of Hardness Testing and Its Research Applications. vol. 21, pp. 274–290.

Dahlberg, C.F.O., Saito, Y., Öztop, M.S., Kysar, J.W., 2014. Geometrically necessary dislocation density measurements associated with different angles of indentations. Int. J. Plast. 54, 81–95.

Dahlberg, C.F.O., Saito, Y., Öztop, M.S., Kysar, J.W., 2017. Geometrically necessary dislocation density measurements at a grain boundary due to wedge indentation into an aluminum bicrystal. J. Mech. Phys. Solids 105, 131–149.

Daw, M.S., Baskes, M.I., 1984. Embedded-atom method: derivation and application to impurities, surfaces, and other defects in metals. Phys. Rev. B 29, 6443–6453.

De Guzman, M.S., Neubauer, G., Flinn, P., Nix, W.D., 1993. The role of indentation depth on the measured hardness of materials. Mater. Res. Soc. Symp. Proc. 308, 613–618.

De Koning, M., Kurtz, R.J., Bulatov, V.V., Henager, C.H., Hoagland, R.G., Cai, W., Nomura, M., 2003. Modeling of dislocation–grain boundary interactions in FCC metals. J. Nucl. Mater. 323, 281–289.

Demir, E., Raabe, D., Zaafarani, N., Zaefferer, S., 2009. Investigation of the indentation size effect through the measurement of the geometrically necessary dislocations beneath small indents of different depths using EBSD tomography. Acta Mater. 57, 559–569.

Diao, J., Gall, K., Dunn, M.L., Zimmerman, J.A., 2006. Atomistic simulations of the yielding of gold nanowires. Acta Mater. 54, 643–653.

Durst, K., Backes, B., Göken, M., 2005. Indentation size effect in metallic materials: correcting for the size of the plastic zone. Scr. Mater. 52, 1093–1097.

Durst, K., Göken, M., Pharr, G.M., 2008. Indentation size effect in spherical and pyramidal indentations. J. Phys. D. Appl. Phys. 41.

El-Awady, J.A., 2014. Unravelling the physics of size-dependent dislocation-mediated plasticity. Nat. Commun. 6, 5926.

El-Awady, J.A., Biner, S.B., Ghoniem, N.M., 2008. A self-consistent boundary element, parametric dislocation dynamics formulation of plastic flow in finite volumes. J. Mech. Phys. Solids 56 (2008), 2019–2035.

El-Awady, J.A., Wen, M., Ghoniem, N.M., 2009. The role of the weakest-link mechanism in controlling the plasticity of micropillars. J. Mech. Phys. Solids 57, 32–50.

Espinosa, H., Berbenni, S., Panico, M., Schwarz, K.W., 2005. An interpretation of size-scale plasticity in geometrically confined systems. Proc. Natl. Acad. Sci. 102, 16933–16938.

Faken, D., Jonsson, H., 1994. Systematic analysis of local atomic structure combined with 3D computer graphics. Comput. Mater. Sci. 2, 279–286.

Feng, G., Nix, W.D., 2004. Indentation size effect in MgO. Scr. Mater. 51, 599–603.

Feng, G., Budiman, A.S., Nix, W.D., Tamura, N., Patel, J.R., 2008. Indentation size effects in single crystal copper as revealed by synchrotron X-ray microdiffraction. J. Appl. Phys. 104.

Gane, N., 1970. The direct measurement of the strength of metals on a submicrometer scale. Proc. R. Soc. Lond. Ser. A 317, 367–391.

Gerberich, W.W., Tymiak, N.I., Grunlan, J.C., Horstemeyer, M.F., Baskes, M.I., 2002. Interpretations of indentation size effect. J. Appl. Mech. 69, 433–442.

Greer, J.R., Oliver, W.C., Nix, W.D., 2005. Size dependence of mechanical properties of gold at the micron scale in the absence of strain gradients. Acta Mater. 53, 1821–1830.

Hasnaoui, A., Derlet, P.M., Van Swygenhoven, H., 2004. Interaction between dislocations and grain boundaries under an indenter–a molecular dynamics simulation. Acta Mater. 52, 2251–2258.

Henderson, A., 2007. Paraview Guide, A Parallel Visualization Application. Kitware Inc. http://www.paraview.org.

Huang, Y., Zhang, F., Hwang, K.C., Nix, W.D., Pharr, G.M., Feng, G., 2006. A model for size effects in nanoindentation. J. Mech. Phys. Solids 54, 1668–1686.

Hussein, A., Rao, S., Uchic, M.D., Dimiduk, D.M., El-Awady, J., 2015. Microstructurally based cross-slip mechanisms and their effects on dislocation microstructure evolution in fcc crystals. Acta Mater. 85, 180–190.

Jang, H., Farkas, D., 2007. Interaction of lattice dislocations with a grain boundary during nanoindentation simulation. Mater. Lett. 61, 868–871.

Kelchner, C.L., Plimpton, S.J., Hamilton, J.C., 1998. Dislocation nucleation and defect structure during surface indentation. Phys. Rev. B 58, 11085–11088.

Kiener, D., Pippan, R., Motz, C., Kreuzer, H., 2006. Microstructural evolution of the deformed volume beneath microindents in tungsten and copper. Acta Mater. 54, 2801–2811.

Kiener, D., Durst, K., Rester, M., Minor, A.M., 2009. Revealing deformation mechanisms with nanoindentation. JOM 61, 14–23.

King, R.B., 1987. Elastic analysis of some punch problems for a layered medium. Int. J. Solids Struct. 23 (12), 1657–1664.

Koch, C.C., Ovid'ko, I.A., Seal, S., Veprek, S., 2007. Structural Nanocrystalline Materials: Fundamentals and Applications. Cambridge University Press, Cambridge.

Kraft, O., Gruber, P., Mönig, R., Weygand, D., 2010. Plasticity in confined dimensions. Annu. Rev. Mater. Res. 40, 293–317.

Kulkarni, Y., Asaroa, R.J., Farkas, D., 2009. Are nanotwinned structures in FCC metals optimal for strength, ductility and grain stability? Scr. Mater. 60, 532–535.

Larson, B.C., El-Azab, A., Yang, W., Tischler, J.Z., Liu, W., Ice, G.E., 2007. Experimental characterization of the mesoscale dislocation density tensor. Philos. Mag. 87, 1327–1347.

Larson, B.C., Tischler, J.Z., El-Azab, A., Liu, W., 2008. Dislocation density tensor characterization of deformation using 3D X-ray microscopy. J. Eng. Mater. Technol. 130.

Lee, Y., Park, J.Y., Kim, S.Y., Jun, S., 2005. Atomistic simulations of incipient plasticity under Al (111) nanoindentation. Mech. Mater. 37, 1035–1048.

Lee, S.W., Han, S.M., Nix, W.D., 2009. Uniaxial compression of fcc Au nanopillars on an MgO substrate: the effects of prestraining and annealing. Acta Mater. 57, 4404–4415.

Li, J., Van Vliet, K.J., Zhu, T., Yip, S., Suresh, S., 2002. Atomistic mechanisms governing elastic limit and incipient plasticity in crystals. Nature 418, 307–310.

Lidorikis, E., Bachlechner, M.E., Kalia, R.K., Nakano, A., Vashishta, P., Voyiadjis, G.Z., 2001. Coupling-of-length-scales approach for multiscale atomistics-continuum simulations: atomistically-induced stress distributions in Si/Si_3N_4 nanopixels. Phys. Rev. Lett. 87 (8) (paper 086104, six pages).

Lim, Y.Y., Chaudhri, M.M., 1999. The effect of the indenter load on the nanohardness of ductile metals: an experimental study on polycrystalline work-hardened and annealed oxygen-free copper. Philos. Mag. A 79, 2979–3000.

Liu, Y., Ngan, A.H.W., 2001. Depth dependence of hardness in copper single crystals measured by nanoindentation. Scr. Mater. 44, 237–241.

Ma, Q., Clarke, D.R., 1995. Size dependent hardness of silver single crystals. J. Mater. Res. 10, 853–863.

McElhaney, K.W., Valssak, J.J., Nix, W.D., 1998. Determination of indenter tip geometry and indentation contact area for depth sensing indentation experiments. J. Mater. Res. 13, 1300–1306.

McLaughlin, K.K., Clegg, W.J., 2008. Deformation underneath low-load indentations in copper. J. Phys. D. 41.

Medyanik, S.N., Shao, S., 2009. Strengthening effects of coherent interfaces in nanoscale metallic bilayers. Comput. Mater. Sci. 45, 1129–1133.

Meyers, M.A., Mishra, A., Benson, D.J., 2006. Mechanical properties of nanocrystalline materials. Prog. Mater. Sci. 51, 427–556.

Mishin, Y., Farkas, D., Mehl, M.J., Papaconstantopoulos, D.A., 1999. Interatomic potentials for monoatomic metals from experimental data and ab initio calculations. Phys. Rev. B 59, 3393–3407.

Mott, B.W., 1957. Microindentation Hardness Testing. Butterworths, London.

Nair, A.K., Parker, E., Gaudreau, P., Farkas, D., Kriz, R.D., 2008. Size effects in indentation response of thin films at the nanoscale: a molecular dynamics study. Int. J. Plast. 24, 2016–2031.

Nakano, A., Bachlechner, E., Kalia, R.K., Lidorikis, E., Vashishta, P., Voyiadjis, G.Z., Cambell, T.J., Ogata, S., Shimojo, F., 2001. Multiscale simulation of nanosystems. Comput. Sci. Eng. 3 (4), 56–66 (IEEE Computer Society and the American Institute of Physics).

Nix, W.D., Gao, H.J., 1998. Indentation size effects in crystalline materials: a law for strain gradient plasticity. J. Mech. Phys. Solids 46, 411–425.

Parthasarathy, T.A., Rao, S.I., Dimiduk, D.M., Uchic, M.D., Trinkle, D.R., 2007. Contribution to size effect of yield strength from the stochastics of dislocation source lengths in finite samples. Scr. Mater. 56, 313–316.

Peng, P., Liao, G., Shi, T., Tang, Z., Gao, Y., 2010. Molecular dynamic simulations of nanoindentation in aluminum thin film on silicon substrate. Appl. Surf. Sci. 256, 6284–6290.

Pharr, G.M., Herbert, E.G., Gao, Y., 2010. The indentation size effect: a critical examination of experimental observations and mechanistic interpretations. Annu. Rev. Mater. Res. 40, 271–292.

Plimpton, S., 1995. Fast parallel algorithms for short-range molecular dynamics. J. Comput. Phys. 117, 1–19.

Poole, W.J., Ashby, M.F., Fleck, N.A., 1996. Micro-hardness of annealed and work-hardened copper polycrystals. Scr. Mater. 34, 559–564.

Pugno, N.M., 2007. A general shape/size-effect law for nanoindentation. Acta Mater. 55, 1947–1953.

Rao, S.I., Dimiduk, D.M., Parthasarathy, T.A., Uchic, M.D., Tang, M., Woodward, C., 2008. Athermal mechanisms of size-dependent crystal flow gleaned from three-dimensional discrete dislocation simulations. Acta Mater. 56, 3245–3259.

Rao, S.I., Dimiduk, D.M., Parthasarathy, T.A., Uchic, M.D., Woodward, C., 2013. Atomistic simulations of surface cross-slip nucleation in face-centered cubic nickel and copper. Acta Mater. 61, 2500–2508.

Rester, M., Motz, C., Pippan, R., 2007. Microstructural investigation of the volume beneath nanoindentations in copper. Acta Mater. 55, 6427–6435.

Rester, M., Motz, C., Pippan, R., 2008. Indentation across size scales: a survey of indentation-induced plastic zones in copper {111} single crystals. Scr. Mater. 59, 742–745.

Ryu, I., Cai, W., Nix, W.D., Gao, H., 2015. Stochastic behaviors in plastic deformation of face-centered-cubic micropillars governed by surface nucleation and truncated source operation. Acta Mater. 95, 176–183.

Sangid, M.D., Ezaz, T., Sehitoglu, H., Robertson, I.M., 2011. Energy of slip transmission and nucleation at grain boundaries. Acta Mater. 59, 283–296.

Sangwal, K., 2000. On the reverse indentation size effect and microhardness measurement of solids. Mater. Chem. Phys. 63, 145–152.

Sansoz, F., 2011. Atomistic processes controlling flow stress scaling during compression of nanoscale face-centered-cubic crystals. Acta Mater. 59, 3364–3372.

Shao, S., Medyanik, S.N., 2010. Dislocation–interface interaction in nanoscale FCC metallic bilayers. Mech. Res. Commun. 37, 315–319.

Spary, I.J., Bushby, A.J., Jennett, N.M., 2006. On the indentation size effect in spherical indentation. Philos. Mag. 86, 5581–5593.

Stelmashenko, N.A., Walls, M.G., Brown, L.M., Milman, Y.V., 1993. Microindentation on W and Mo oriented single crystals: an STM study. Acta Metall. Mater. 41, 2855–2865.

Stukowski, A., 2010. Visualization and analysis of atomistic simulation data with OVITO–the Open Visualization Tool. Model. Simul. Mater. Sci. Eng. 18.

Stukowski, A., 2012. Structure identification methods for atomistic simulations of crystalline materials. Model. Simul. Mater. Sci. Eng. 20.

Stukowski, A., 2014. Computational analysis methods in atomistic modeling of crystals. JOM 66, 399–407.

Stukowski, A., Albe, K., 2010. Extracting dislocations and non-dislocation crystal defects from atomistic simulation data. Model. Simul. Mater. Sci. Eng. 18.

Stukowski, A., Albe, K., Farkas, D., 2010. Nanotwinned FCC metals: strengthening versus softening mechanisms. Phys. Rev. B. 82.

Stukowski, A., Bulatov, V.V., Arsenlis, A., 2012. Automated identification and indexing of dislocations in crystal interfaces. Model. Simul. Mater. Sci. Eng. 20.

Swadener, J.G., George, E.P., Pharr, G.M., 2002. The correlation of the indentation size effect measured with indenters of various shapes. J. Mech. Phys. Solids 50, 681–694.

Tabor, D., 1986. Indentation hardness and its measurement: some cautionary comments. In: Blau, P.J., Lawn, B.R. (Eds.), Microindentation Techniques in Materials Science and Engineering. In: ASTM STP 889, pp. 129–159.

Tersoff, J., 1988. New empirical approach for the structure and energy of covalent systems. Phys. Rev. B 37, 6991–7000.

Tsuru, T., Kaji, Y., Matsunaka, D., Shibutani, Y., 2010. Incipient plasticity of twin and stable/unstable grain boundaries during nanoindentation in copper. Phys. Rev. B 82.

Tucker, G.J., Aitken, Z.H., Greer, J.R., Weinberger, C.R., 2013. The mechanical behavior and deformation of bicrystalline nanowires. Model. Simul. Mater. Sci. Eng. 21.

Uchic, M.D., Dimiduk, D.M., Florando, J.N., Nix, W.D., 2003. Exploring specimen size effects in plastic deformation of Ni3(Al, Ta). Mater. Res. Soc. Symp. Proc. 753, BB1.4.1–BB1.4.6.

Uchic, M.D., Dimiduk, D.M., Florando, J.N., Nix, W.D., 2004. Sample dimensions influence strength and crystal plasticity. Science 305, 986–989.

Uchic, M.D., Shade, P.A., Dimiduk, D.M., 2009. Plasticity of micrometer-scale single crystals in compression. Annu. Rev. Mater. Res. 39, 361–386.

Upit, G.P., Varchenya, S.A., 1973. The size effect in the hardness of single crystals. In: Westbrook, J.H., Conrad, H. (Eds.), The Science of Hardness Testing and Its Research Applications. vol. 10, pp. 135–146.

Voyiadjis, G.Z., Abu Al-Rub, R.K.A., 2005. Gradient plasticity theory with a variable length scale parameter. Int. J. Solids Struct. 42, 3998–4029.

Voyiadjis, G.Z., Yaghoobi, M., 2015. Large scale atomistic simulation of size effects during nanoindentation: dislocation length and hardness. Mater. Sci. Eng. A 634, 20–31.

Voyiadjis, G.Z., Yaghoobi, M., 2016. Role of grain boundary on the sources of size effects. Comput. Mater. Sci. 117, 315–329.

Voyiadjis, G.Z., Yaghoobi, M., 2017a. Size and strain rate effects in metallic samples of confined volumes: dislocation length distribution. Scr. Mater. 130, 182–186.

Voyiadjis, G.Z., Yaghoobi, M., 2017b. Review of nanoindentation size effect: experiments and atomistic simulation. Crystals 7, 321.

Voyiadjis, G.Z., Yaghoobi, M., 2017c. Size effects during nanoindentation: molecular dynamics simulation. In: Voyiadjis, G.Z. (Ed.), Handbook of Nonlocal Continuum Mechanics for Materials and Structures. Springer International Publishing, pp. 1–38.

Weinberger, C.R., Cai, W., 2008. Surface-controlled dislocation multiplication in metal micropillars. Proc. Natl. Acad. Sci. U. S. A. 105, 14304–14307.

Weinberger, C.R., Tucker, G.J., 2012. Atomistic simulations of dislocation pinning points in pure face-centered-cubic nanopillars. Model. Simul. Mater. Sci. Eng. 20.

Weinberger, C.R., Jennings, A.T., Kang, K., Greer, J.R., 2012. Atomistic simulations and continuum modeling of dislocation nucleation and strength in gold nanowires. J. Mech. Phys. Solids 60, 84–103.

Xu, S., Guo, Y.F., Ngan, A.H.W., 2013. A molecular dynamics study on the orientation, size, and dislocation confinement effects on the plastic deformation of Al nanopillars. Int. J. Plast. 43, 116–127.

Yaghoobi, M., Voyiadjis, G.Z., 2014. Effect of boundary conditions on the MD simulation of nanoindentation. Comput. Mater. Sci. 95, 626–636.

Yaghoobi, M., Voyiadjis, G.Z., 2016a. Atomistic simulation of size effects in single-crystalline metals of confined volumes during nanoindentation. Comput. Mater. Sci. 111, 64–73.

Yaghoobi, M., Voyiadjis, G.Z., 2016b. Size effects in FCC crystals during the high rate compression test. Acta Mater. 121, 190–201.

Yaghoobi, M., Voyiadjis, G.Z., 2017. Microstructural investigation of the hardening mechanism in FCC crystals during high rate deformations. Comput. Mater. Sci. 138, 10–15.

Yaghoobi, M., Voyiadjis, G.Z., 2018. The effects of temperature and strain rate in FCC and bcc metals during extreme deformation rates. Acta Mater. 151, 1–10.

Yang, W., Larson, B.C., Pharr, G.M., Ice, G.E., Budai, J.D., Tischler, J.Z., Liu, W., 2004. Deformation microstructure under microindents in single-crystal Cu using three-dimensional X-ray structural microscopy. J. Mater. Res. 19, 66–72.

Zaafarani, N., Raabe, D., Singh, R.N., Roters, F., Zaefferer, S., 2006. Three dimensional investigation of the texture and microstructure below a nanoindent in a Cu single crystal using 3D EBSD and crystal plasticity finite element simulations. Acta Mater. 54, 1863–1876.

Zaafarani, N., Raabe, D., Roters, F., Zaefferer, S., 2008. On the origin of deformation-induced rotation patterns below nanoindents. Acta Mater. 56, 31–42.

Zhou, C., Beyerlein, I.J., LeSar, R., 2011. Plastic deformation mechanisms of fcc single crystals at small scales. Acta Mater. 59, 7673–7682.

Zhu, T.T., Bushby, A.J., Dunstan, D.J., 2008. Materials mechanical size effects: a review. Mater. Technol. 23, 193–209.

Zimmerman, J.A., Kelchner, C.L., Klein, P.A., Hamilton, J.C., Foiles, S.M., 2001. Surface step effects on nanoindentation. Phys. Rev. Lett. 87.

Further Reading

Almasri, A.H., Voyiadjis, G.Z., 2010. Nano-indentation in FCC metals: experimental study. Acta Mech. 209, 1–9.

Corcoran, S.G., Colton, R.J., Lilleodden, E.T., Gerberich, W.W., 1997. Anomalous plastic deformation at surfaces: nanoindentation of gold single crystals. Phys. Rev. B 55, 16057–16060.

Demir, E., Raabe, D., Roters, F., 2010. The mechanical size effect as a mean-field breakdown phenomenon: example of microscale single crystal beam bending. Acta Mater. 58, 1876–1886.

Humphrey, W., Dalke, A., Schulten, K., 1996. VMD: visual molecular dynamics. J. Mol. Graph. 14, 33–38.

Kysar, J.W., Briant, C.L., 2002. Crack tip deformation fields in ductile single crystals. Acta Mater. 50, 2367–2380.

Kysar, J.W., Gan, Y.X., Morse, T.L., Chen, X., Jones, M.E., 2007. High strain gradient plasticity associated with wedge indentation into face-centered cubic single crystals: geometrically necessary dislocation densities. J. Mech. Phys. Solids 55, 1554–1573.

Soer, W.A., De Hosson, J.T.M., 2005. Detection of grain-boundary resistance to slip transfer using nanoindentation. Mater. Lett. 59, 3192–3195.

Suresh, S., Nieh, T.G., Choi, B.W., 1999. Nanoindentation of copper thin films on silicon substrates. Scr. Mater. 41, 951–957.

Chapter 6

Future evolution: Multiscale modeling framework to develop a physically based nonlocal plasticity model for crystalline materials

6.1 Introduction

In the previous chapters, the study of size effects have been elaborated at different length scales from the nonlocal continuum models, which can be applied to the real-life applications, to the atomistic simulation, which can be incorporated to extract the atomistic mechanisms of size effects. Each of these methods can handle certain ranges of time scale and length scale. Accordingly, each method can capture specific characteristics of the problem. The idea of multiscale modeling is to bridge the gap between different length scales to benefit the advantages of various length scales without paying the price of substantial simulations. In this chapter, a novel idea of multiscale framework is presented to develop a physically based nonlocal plasticity model to capture size and rate effects in micron-sized crystalline samples assisted by experiments and atomistic simulation. The results of the current framework can shed light on the role of dislocations at the smaller length scales for crystalline metals. Ultimately, this chapter presents a new framework to develop a new physically based nonlocal continuum plasticity model for crystalline metals, which is able to capture different mechanisms of strain rate and size effects using both experiments and molecular dynamics (MD) simulation.

The primary result of the current framework is to develop a new physically based nonlocal continuum plasticity model for micron-sized crystalline metals to capture the different mechanism of strain rate and size effects, which is a function of length scale, strain rate, grain size, and dislocation density. The results obtained from the conducted indentation and microbending experiments

Size Effects in Plasticity: From Macro to Nano. https://doi.org/10.1016/B978-0-12-812236-5.00006-2

and large scale MD simulations of nanoindentation and micropillar compression can be incorporated to develop the model. Besides the primary goal, the presented framework can stretch the current experimental knowledge of plasticity in metallic samples of confined volumes by conducting two important experiments of indentation and microbending and monitoring the evolution of dislocations. Finally, valuable information on the strain rate and size effects in micron-sized crystalline metals can be provided from a systematic parametric study using large scale MD for samples with various sizes, materials, grain sizes, crystal structures, and crystallographic orientations subjected to different strain rates.

The results of this framework can greatly help the industry by providing a very efficient scientific based design scheme. Last but not least, the new strain rate and size effects mechanisms can greatly contribute to the fundamentals of mechanics of materials at smaller length scales while it incorporates the results from both experiments and MD simulations. In the current chapter, first, the overviews and objective of the proposed framework is presented. Next, the technical background and preliminary analysis corresponding o the proposed framework is elaborated. The proposed framework is then presented in the next step. Finally, the proposed multiscale framework is divided into three separate research tasks.

6.2 Overview and objectives of the multiscale modeling framework

In Chapters 1–5 and this chapter, size effects in materials have been addressed using methods with different length scales from nonlocal continuum mechanics, which can simulate real size problems, to atomistic simulation, which can capture the atomistic mechanisms of size effects. As the length and time scales of the simulation method decreases, more details can be captured. However, the size of the sample which can be simulated decreases. Developing a multi-scale framework that can bridge the gap between different length and time scale and its corresponding simulation method is of great interest in the material science and engineering community. Accordingly, a simulation can benefit the information provided by the small length and time scale simulations while model a real size sample using the nonlocal continuum mechanics.

In recent years, micron-sized metallic devices have significantly impacted many different industries including information technology, biomedicine, bridge and highway monitoring assessment, optical coatings, biological implants, data storage media, photovoltaic cells, and intelligent control and automation. As an example, thin films and microelectromechanical systems have resulted in a significant expansion of the capabilities of several industries including information technology, intelligent control and automation, biomedicine, and optical coatings. The mechanical behavior of metallic samples at microscale is different from that of the bulk samples which remains a great challenge that attracts many researchers. The governing mechanisms of strain rate and size effects on strength, i.e. varying the material strength as the strain rate and sample size change, for bulk materials are different from those of the micron-sized samples. Presenting a model that can capture the strain rate and size effects at various length scales can greatly help the industries by providing scientific based design schemes.

In the case of bulk samples, the interaction of dislocations with one another and other defects such as grain boundaries is responsible for size effects in strength that is commonly termed as the forest hardening. Accordingly, the strain gradient triggers the size effects mechanism, which can be captured using the forest hardening mechanism. The Taylor-like hardening models are usually incorporated to capture the forest hardening which states that the strength increases as the dislocation density increases (Nix and Gao, 1998; Abu Al-Rub and Voyiadjis, 2004; Voyiadjis and Abu Al-Rub, 2007; Voyiadjis et al., 2011). In recent years, new experimental techniques have been developed to measure the density of geometrically necessary dislocations (GNDs) (Sun et al., 2000; Kysar and Briant, 2002; El-Dasher et al., 2003; Kiener et al., 2006; Kysar et al., 2007, 2010; McLaughlin and Clegg, 2008; Pantleon, 2008; Rester et al., 2008; Zaafarani et al., 2008; Demir et al., 2009; Wheeler et al., 2009; Dahlberg et al., 2014, 2017; Ruggles et al., 2016). The experimental methods are developed based on the relation between GND density and spatial gradients of plastic slip (Nye, 1953; Kondo, 1964; Fox, 1966). These experimental methods incorporate the spatially-resolved methods to measure lattice distortion via diffraction-based methods. Experimental results from the recent studies demonstrate new phenomena in metallic samples which cannot be explained using the available theories. Demir et al. (2009, 2010) showed that the Taylor hardening model fails to justify the results of the nanoindentation and microbending of microscale metallic samples. Demir et al. (2009) conducted nanoindentation of Cu single crystal and measured the GND densities at various indentation depths. Fig. 6.1 shows both hardness and GND densities at different indentation depths. The results show that as the indentation depth decreases the GND density decreases. According to the Taylor hardening model, as the GND density decreases the hardness should also decrease (Nix and Gao, 1998; Abu Al-Rub and Voyiadjis, 2004; Voyiadjis et al., 2011). However, the results show that the hardness increases as the indentation depth decreases (Demir et al., 2009). In other words, as the GND density decreases the hardness increases. Demir et al. (2010) also investigated the Taylor hardening model in the microbending experiment on the Cu single crystal. Again, the results showed the failure of the Taylor hardening model. Voyiadjis and his coworkers (Voyiadjis and Almasri, 2009; Voyiadjis et al., 2011; Voyiadjis and Zhang, 2015; Zhang and Voyiadjis, 2016) have also conducted several nanoindentation experiments to investigate the effects of indentation depth, grain boundary, and strain rate on the mechanical response of micron-sized metallic samples.

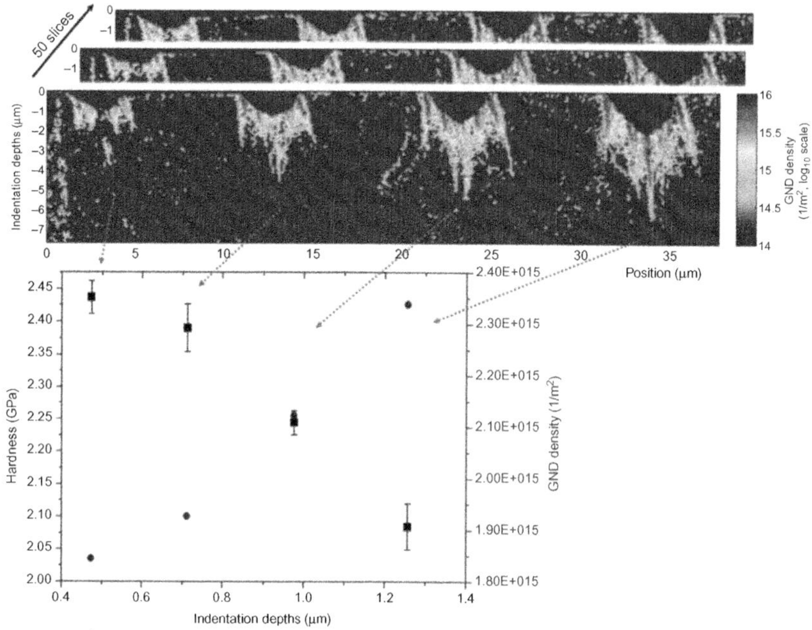

FIG. 6.1 The total GND densities below each indent obtained by summation over all 50 individual 2-D EBSD sections together with the measured hardness. *(After Demir, E., Raabe, D., Zaafarani, N., Zaefferer, S., 2009. Investigation of the indentation size effect through the measurement of the geometrically necessary dislocations beneath small indents of different depths using EBSD tomography. Acta Mater. 57 (2), 559–569.)*

The size effects have been originally attributed to the presence of the strain gradient. However, Uchic et al. (2003, 2004) introduced a micropillar compression test without any strain gradients using the focused ion beam (FIB) machining and observed that the specimens still show strong size effects. As an example, Fig. 6.2 illustrates the typical size effects during the micropillar compression test (Greer et al., 2005). These results show that the controlling mechanism of size effects in micropillars is not the forest hardening. Uchic et al. (2009), Kraft et al. (2010), Greer and De Hosson (2011), and Greer (2013) have reviewed different types of size effects models in metallic samples of confined volumes. Three models of source exhaustion hardening (Rao et al., 2008; El-Awady, 2015), source truncation (Parthasarathy et al., 2007; Rao et al., 2007), and weakest link theory (Norfleet et al., 2008; El-Awady et al., 2009) are usually incorporated to describe these size effects (Greer, 2013). Some phenomenological models have also been proposed to capture different mechanisms of size effects (see, for example, Norfleet et al., 2008). However, there is still no physically based model developed using the experimental observations or simulation results which is able to capture all the governing mechanisms of size effects.

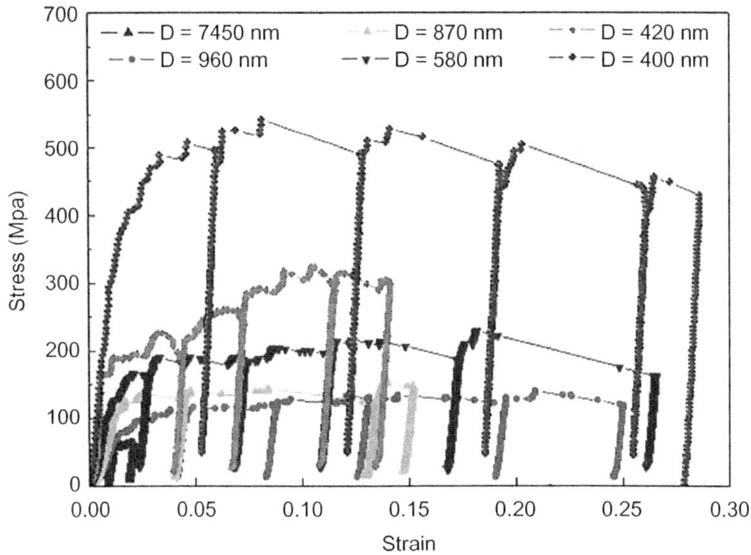

FIG. 6.2 Stress-strain behavior of ⟨001⟩-oriented single crystal gold pillars: flow stresses increase significantly for pillars with a diameter of 500 nm and less. *(After Greer, J.R., Oliver, W.C., Nix, W.D., 2005. Size dependence of mechanical properties of gold at the micron scale in the absence of strain gradients. Acta Mater. 53 (6), 1821–1830.)*

The strain rate effects on the response of materials are crucial for many applications including armor design, automobile collisions, and projectile impact. The dislocation mechanics in high rate deformation is different from those with slower rates. In the case of quasi-static deformations, the shear strain rate can be described using the usual Orowan equation (Meyers et al., 2009; Armstrong and Li, 2015). In the case of high strain rate deformations, however, the shear strain rate should be calculated from the modified Orowan equation considering the rate of dislocation nucleation (Meyers et al., 2009; Armstrong and Li, 2015). In other words, the dislocation mechanics in high strain rate deformations is governed by dislocation generation at the propagating shock front. The material response to high-rate deformations has been investigated using various experiments including the split-Hopkinson pressure bar (SHPB), pulsed laser loading, and high intensity laser facilities coupled with X-ray diffraction techniques (Meyers et al., 2009; Armstrong and Li, 2015). Various models have been proposed to capture the governing mechanisms of deformation observed at different strain rate experiments (Meyers, 1994; Meyers et al., 2003, 2009; Armstrong et al., 2007; Armstrong and Walley, 2008; Armstrong and Li, 2015; Gurrutxaga-Lerma et al., 2015). However, the focus has been more on the deformation mechanisms, and the coupling of strain rate effects and size effects have not been thoroughly studied.

In the case of nano-sized metallic samples, the simulations and experiments showed that the dislocation starvation and surface dislocation nucleation is responsible for strain rate and size effects (Shan et al., 2008; Zhou et al., 2010; Sansoz, 2011; Jennings et al., 2011; Yaghoobi and Voyiadjis, 2016b). Since the individual dislocation movements are important in dislocation starvation, conventional continuum models cannot consider this mechanism. Also, the continuum model is incapable of modeling the surface dislocation sources. In the case of micron-size metallic samples, however, dislocation starvation does not occur according to the experimental and numerical investigations (Norfleet et al., 2008; Zhou et al., 2010; Sansoz, 2011; Yaghoobi and Voyiadjis, 2016b). Furthermore, the importance of surface dislocation nucleation and surface stresses decreases as the sample size increases (Jennings et al., 2011; Diao et al., 2006), and the collective dislocation behavior becomes the major deformation mechanism for micron-sized samples (Fig. 6.3). Although continuum models cannot capture the strain rate and size effects in nano-sized metallic samples due to the surface effects and discrete nature of dislocations, the micron-sized sample can be modeled using a modified nonlocal continuum plasticity model by introducing new material length scales.

Numerical simulations can play a complementary role to the experimental results. The nature and physical properties of the dislocations should be fully investigated to model the size effect in micron-sized metallic samples. Molecular dynamics (MD) is a powerful tool to simulate various tests with the atomistic details. Nowadays, using very powerful supercomputers and efficient and massively parallel codes, the samples with the dimensions of up to 0.3 μm can be simulated using MD. In the case of nanoindentation, Kelchner et al. (1998) conducted atomistic simulation of nanoindentation to study the dislocation nucleation of Au. Using MD simulations, Zimmerman et al. (2001) investigated the effects of surface step on the nanoindentation test of Au, and Li et al. (2002) studied the dislocation nucleation and evolution of Cu and Al during the nanoindentation. Lee et al. (2005) simulated the defects nucleation and evolution of Al during the nanoindentation. Yaghoobi and Voyiadjis (2014) studied the effect of boundary conditions on the MD simulation of nanoindentation by incorporating various boundary conditions and thicknesses. Voyiadjis and Yaghoobi (2015) studied the relation between the dislocation density and hardness during nanoindentation of metallic samples. Yaghoobi and Voyiadjis (2016a) also studied the sources of size effects in nanosize single crystal Ni thin films during nanoindentation using large scale atomistic simulation. They showed that the dislocation nucleation and source exhaustion are mainly responsible for size effects during the nanoindentation of Ni samples of confined volumes at lower indentation depths (Yaghoobi and Voyiadjis, 2016a). Voyiadjis and Yaghoobi (2016) incorporated large scale MD to model the nanoindentation of bicrystalline thin films and studied the effects of various grain boundaries on the controlling mechanisms of size effects as the sample length scale increases.

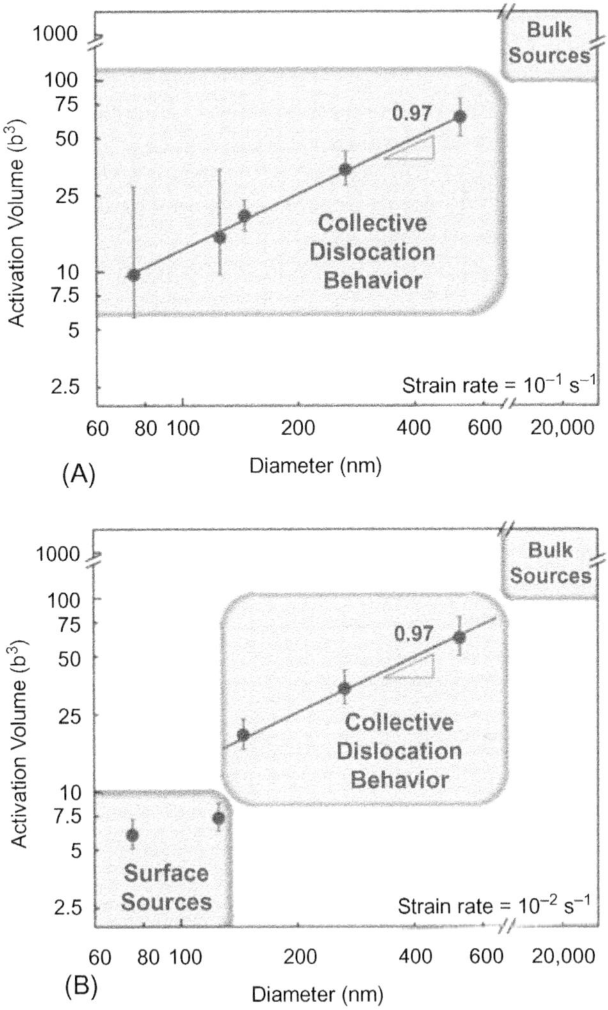

FIG. 6.3 The extracted activation volumes and plasticity mechanisms for each pillar diameter at strain rates of (A) $10^{-1}\,s^{-1}$ and (B) $10^{-2}\,s^{-1}$. *(After Jennings, A.T., Li, J., Greer, J.R., 2011. Emergence of strain-rate sensitivity in Cu nanopillars: transition from dislocation multiplication to dislocation nucleation. Acta Mater. 59, 5627–5637.)*

The MD simulation of micropillar deformation mechanisms has also been investigated by many researchers. Bringa et al. (2006) incorporated the large scale atomistic simulation to study the deformation mechanisms of FCC metallic samples at high strain rate compression tests. They were able to capture the experimentally observed deformation mechanisms of copper during the high strain rate experiments by incorporating a very large sample which contains 352 million atoms. It was concluded that the yield strength of the sample with very small length scales is mainly governed by the surface stress. Sansoz (2011) comprehensively simulated the size effects of pillars with diameters in the range of 10.8–72.3 nm with the periodic height of 30 nm. Yaghoobi and Voyiadjis (2016b) investigated the different mechanisms of size effects in FCC metallic pillars during high rate compression tests using large scale atomistic simulation. Voyiadjis and Yaghoobi (2017) studied the dislocation length distribution in pillars with different sizes during compression tests with different strain rates using large scale atomistic simulation.

The aim of this framework is to develop a new physically based nonlocal continuum plasticity model for FCC crystalline metals by introducing new length scales to capture the strain rate and size effects observed in the experiments and atomistic simulations. This requires extensive and advanced numerical and experimental studies. The new model can be implemented in ABAQUS (Hibbitt et al., 2009), which is a well-known FE software, as UMAT, UEL, VUMAT, and VUEL subroutines, and it can be validated against the results obtained from experiments and atomistic simulations. Large scale atomistic simulations of nanoindentation and micropillar compression tests can be conducted on single crystalline and polycrystalline metallic samples, and the dislocation density and pattern can be extracted for different stages of loading. Based on the obtained numerical and experimental results, the effects of sample length scale, strain rate, dislocation density and pattern, grain size, and crystallographic orientation on the material strength can be studied, and the governing mechanisms of strain rate and size effects can be investigated. Experiments of indentation and microbending should be conducted on single crystal and bicrystal FCC metallic samples. In the case of indentation experiment, different strain rates can be incorporated to investigate both size and strain rate effects. In the case of microbending test, samples with different length scales can be tested to address size effect during bending. The pattern of total GND dislocation density can then be extracted during the experiments using Electron Backscatter Diffraction (EBSD) analysis. Besides the primary target which is to develop a new physically based nonlocal continuum plasticity, the new experimental procedures and atomistic simulation results can shed light to some fundamental and vital unanswered questions in mechanics of materials regarding the coupling between size and strain rate effects in metallic samples of confined volumes.

6.3 Multiscale framework

The designed multiscale framework requires both numerical modeling and experiment. In the case of the numerical modeling, the atomistic mechanisms are extracted using the lower length scale simulation of molecular dynamics. Next, these mechanisms are incorporated inside the nonlocal continuum models to simulate real life problem. Along numerical simulations, experimental tools should be incorporated to validate and explore the numerical simulation at both lengths scales of micro and macro, i.e., atomistic simulation and nonlocal continuum modeling. In this section, more details about each of these step are presented. Finally, the proposed framework is presented here including the precise description of goals and targets this framework can achieve.

6.3.1 Molecular dynamics simulation

One approach to investigate the governing atomistic process of size effects in crystalline metals is to simulate the sample with the full atomistic details using MD. Many deformation mechanisms of metallic thin films can be captured using MD. During MD simulation, Newton's equations of motion are solved for all the atoms in the metallic sample. The interaction between the atoms is described using a predefined potential. MD only calculates the atomic trajectories and velocities. Accordingly, the dislocation properties should be obtained from the atomic trajectories using some post-processing methods. As an example, part of the simulations are reported by Yaghoobi and Voyiadjis (2016a,b) is presented here. First, the governing mechanisms of size effects are studied during nanoindentation. A Ni thin film is simulated using the classical molecular dynamics. The sample dimensions used are 120, 120, and 60 nm. The simulation details and methodology can be found in Yaghoobi and Voyiadjis (2016a). Fig. 6.4 presents the variation of the mean contact pressure ($p_m = P/A$), which is equivalent to the hardness H in the plastic region, as a function of indentation depth h. In the elastic region, Fig. 6.4 shows that p_m increases as the indentation depth increases. However, after the initiation of plasticity, Fig. 6.4 shows that the mean contact pressure, i.e. hardness, decreases as the indentation depth increases.

The plastic zone is defined as a hemisphere with the radius of $R_{pz} = fa_c$ in which f is a constant. In the case of MD simulation, the dislocation density is

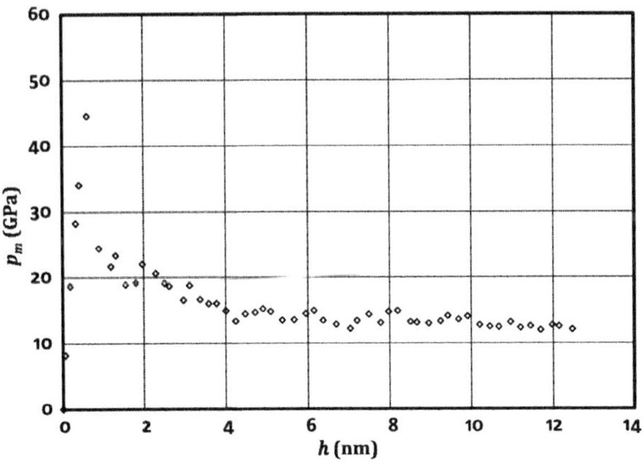

FIG. 6.4 Variation of mean contact pressure p_m as a function of indentation depth h. *(After Yaghoobi, M., Voyiadjis, G.Z., 2016a. Atomistic simulation of size effects in single-crystalline metals of confined volumes during nanoindentation. Comput. Mater. Sci. 111, 64–73.)*

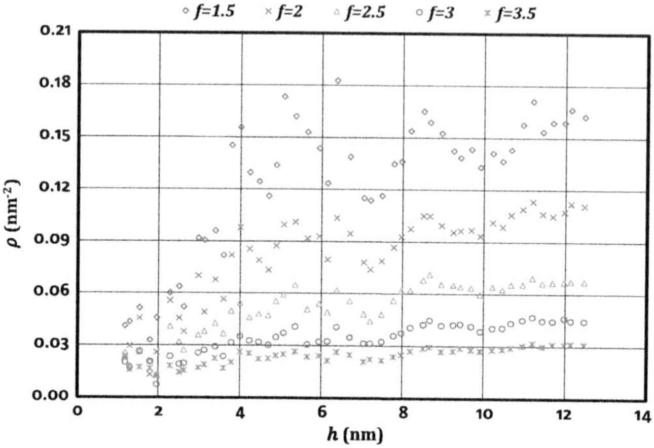

FIG. 6.5 Dislocation density obtained from MD simulation for different values of f during nanoindentation. *(After Yaghoobi, M., Voyiadjis, G.Z., 2016a. Atomistic simulation of size effects in single-crystalline metals of confined volumes during nanoindentation. Comput. Mater. Sci. 111, 64–73.)*

measured using the dislocation length located in the plastic zone divided by the volume of the plastic zone. Five different values of $f = 1.5, 2.0, 2.5, 3.0,$and 3.5 are chosen to investigate the effect of plastic zone size on the dislocation density during nanoindentation. The dislocation density ρ versus the indentation depth h is plotted in Fig. 6.5. The results indicate that as the indentation depth increases, the dislocation density also increases for different values of f. Since the dislocation density increases as the indentation depth increases (Fig. 6.5), the hardness should also increase according to the Taylor hardening model. However, Fig. 6.4 shows that as the indentation depth increases, the hardness decreases. In other words, the results show that the forest hardening model cannot capture the size effect in the case of the simulated sample.

The dislocation visualization is shown in Fig. 6.6 for three indentation depths. Figs. 6.4–6.6 show that the forest hardening mechanism does not govern

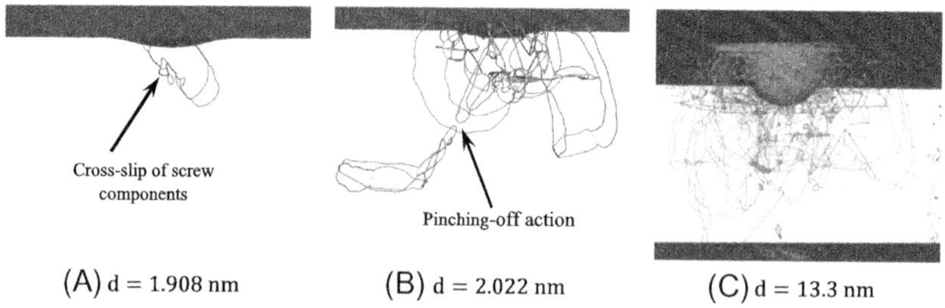

FIG. 6.6 Dislocation nucleation and evolution at tip displacements of (A) d = 1.908 nm, (B) d = 2.022 nm, and (C) d = 13.3 nm. *(After Yaghoobi, M., Voyiadjis, G.Z., 2016a. Atomistic simulation of size effects in single-crystalline metals of confined volumes during nanoindentation. Comput. Mater. Sci. 111, 64–73.)*

size effects in the nanoscale samples during nanoindentation. The resulting mobile dislocation density is insufficient and the applied stress should be increased to sustain the plastic deformation. Hence, the source exhaustion hardening is the governing mechanism of size effects (Rao et al., 2008; El-Awady, 2015). By increasing the indentation depth, the dislocation density increases as well, which provides more dislocation sources. Consequently, the applied stress required to sustain flow during nanoindentation decreases, i.e. hardness decreases as indentation depth increases. At higher indentation depths, the dislocation density is large enough to activate the forest hardening (Fig. 6.6C). However, since the dislocation density tends to a constant value at high indentation depths, the hardness becomes nearly constant.

The size effect during micropillar compression experiment on the Ni pillar with the height of 0.3 μm and diameter of 0.15 μm, which contains around 487 million atoms, is also presented here as the second example. The simulation details and methodology are similar to the first example, i.e., nanoindentation experiment, except for the selected boundary conditions. In the case of Ni pillar with pre-straining, it is observed that the pre-straining activates the forest hardening mechanism in which the interaction of dislocations with each other controls the size effects (Fig. 6.7).

6.3.2 Experiments

6.3.2.1 Indentation and microbending experiments

Indentation is a widely used technique to probe the mechanical properties, such as hardness and elastic stiffness of solid state materials, via measuring their surface response to penetration of a probe with known geometry and imposed load. The high-resolution capacitive gauges and actuators enable the instrument to

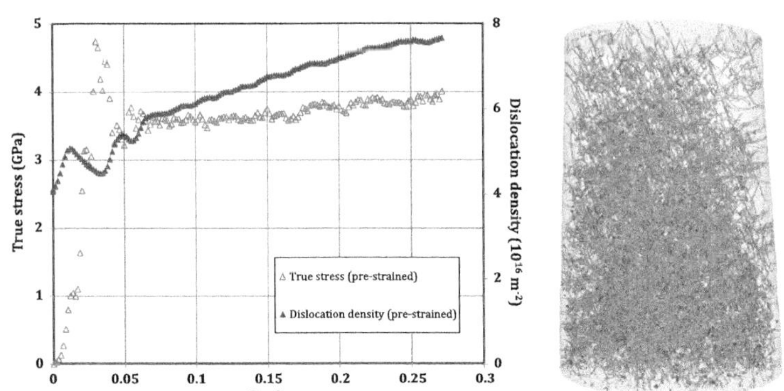

(A) (B)

FIG. 6.7 (A) Compressive response of pre-strained pillar with the height of 0.3 μm and diameter of 0.15 μm and (B) dislocation visualization of the same pillar at $\varepsilon = 0.12$. *(After Yaghoobi, M., Voyiadjis, G.Z., 2016b. Size effects in fcc crystals during the high rate compression test. Acta Mater. 121, 190–201.)*

continuously control and monitor the load and displacement of the indenter as it is driven into and withdrawn from a surface material. Both size effects and strain rate effects can be investigated using the indentation experiment. In addition to the indentation experiment, the microbending experiment can be conducted to capture the size effects. A well-controlled micro cantilever bending experiment can be conducted to achieve a more homogeneous deformation state. Different sample sizes can be used during microbending to capture the size effects. The details of the proposed experiments are presented in the Section 6.4.2.

6.3.2.2 *Electron backscatter diffraction analysis*

To investigate the underlying hardening mechanisms which govern the size and strain rate effects in micron-sized metallic samples, an experimental scheme should be incorporated that can satisfactorily capture the dislocation evolution during the indentation and microbending experiments. Electron Backscatter Diffraction (EBSD) analysis, as a well-established accessory for the Scanning Electron Microscope (SEM), has been recently utilized to shed light on the local crystallographic texture misorientations that can be manifested as the nucleation of GNDs during the plastic deformation (Kysar et al., 2007, 2010; Dahlberg et al., 2014, 2017; Ruggles et al., 2016). Kysar et al. (2007, 2010), Dahlberg et al. (2014, 2017), Ruggles et al. (2016), and Sarac et al. (2016) have developed a method to capture the GNDs content in crystalline metals in the case of plane strain deformation state. In this method, the in-plane lattice rotations of the considered region are calculated from the crystallographic orientation maps obtained from EBSD measurements. Next, the Nye dislocation density tensor and the associated GND densities introduced by the plastic deformation are calculated.

The lattice rotation about the x_1, x_2, and x_3 coordinates is defined as ω_1, ω_2, and ω_3, respectively. The x_1, x_2, and x_3 are the global coordinates and x_1', x_2', and x_3' are the local ones (Kysar et al., 2007). The crystal lattice curvature tensor, κ_{ij}, defined by Nye (1953) is:

$$\kappa_{ij} = \frac{\partial \omega_i}{\partial x_j} \tag{6.1}$$

Due to plane strain conditions, one has $\omega_1 = \omega_2 = 0$ and $\omega_3 = \omega_z$, which is the in-plane rotation angle measured with EBSD. Accordingly, the non-zero components of the Nye tensor are $\kappa_{31} = \partial\omega_3/\partial x_1$ and $\kappa_{32} = \partial\omega_3/\partial x_2$, which can be determined by numerical differentiation of the crystal lattice rotation angles in the global coordinate system. The Nye tensor is in turn directly related to the weighted sum of GND densities on all slip systems as (Dahlberg et al., 2017):

$$\alpha_{ij} = \sum_{m=1}^{N_e} \rho_{G(e)}^{(m)} b^{(m)} s_i^{(m)} t_j^{(m)} + \sum_{m=1}^{N_s} \rho_{G(s)}^{(m)} b^{(m)} s_i^{(m)} s_j^{(m)} \tag{6.2}$$

where N_e and N_s are the number of unique edge and screw dislocation components, respectively, in the crystal, $\rho_{G(e)}^{(m)}$ and $\rho_{G(s)}^{(m)}$ are edge and screw components, respectively, of the GND density on slip system m, and $b^{(m)}$ is the magnitude of Burgers vector. Furthemore, $\boldsymbol{n}^{(m)}$ and $\boldsymbol{s}^{(m)}$ are the unit slip plane normal vector and unit slip direction vector, respectively, on slip system m, and $\boldsymbol{t}^{(m)} = \boldsymbol{s}^{(m)} \times \boldsymbol{n}^{(m)}$. Kysar et al. (2010) defined the total GND density as an L^1-norm of the GND densities as follows:

$$\rho_G^{\text{tot}} = \sum_{i=1}^{I} \left|\rho_G^i\right| \tag{6.3}$$

where I is the total number of edge and screw slip systems. Kysar et al. (2007, 2010), Dahlberg et al. (2014, 2017), Ruggles et al. (2016), and Sarac et al. (2016) have incorporated EBSD to capture the total GND density in metallic samples. As an example, Fig. 6.8 illustrates the spatially-resolved measurements of the total GND density for single crystal Cu (Kysar et al., 2007) and bicrystal Al (Dahlberg et al., 2017) samples during the wedge indentation.

6.3.3 Continuum modeling of strain rate and size effects

6.3.3.1 Forest hardening mechanism

The interaction of dislocations with one another and with GBs governs the size effects in bulk metals which is termed as the forest hardening mechanism. A Taylor hardening model is usually incorporated to capture the forest hardening mechanism, which is described as follows (Nix and Gao, 1998):

$$\tau = \alpha\mu b\sqrt{\rho} = \alpha\mu b\sqrt{\rho_G + \rho_S} \tag{6.4}$$

where ρ is the dislocation density, μ is the shear modulus, b is the magnitude of the dislocation Burgers vector, and α is a constant. There are few modifications for the Taylor hardening model such as using different Burgers vector for geometrically necessary dislocations (GNDs) and statistically stored dislocations (SSDs) and modifications in exponents as follows (Voyiadjis and Abu Al-Rub, 2005):

$$\tau = \alpha_S\mu b_S\sqrt{\rho} = \alpha_S\mu b_S\left[\rho_S^{\beta/2} + \left(\alpha_G^2 b_G^2 \rho_G / \alpha_S^2 b_S^2\right)^{\beta/2}\right]^{1/\beta} \tag{6.5}$$

where the indices G and S subscripts designate GNDs and SSDs parameters, respectively. However, all equations derived from the Taylor hardening model have the same trend in which the strength increases as the dislocation density increases.

6.3.3.2 Effects of dislocation source length

Parthasarathy et al. (2007) proposed a model to describe the size effects due to the stochastic variations in dislocation source lengths, sample volume, and truncation of sources at the nearby free surfaces. Source activation is governed by

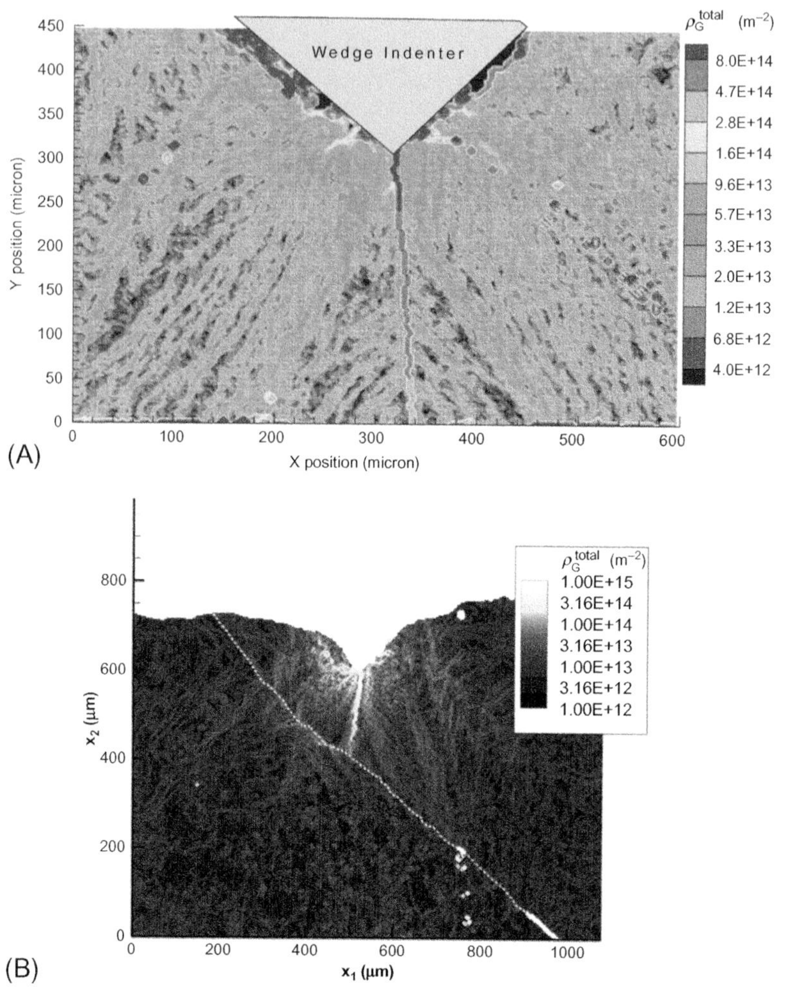

FIG. 6.8 The total geometrically necessary dislocation density during wedge indentation of: (A) single crystal Cu and (B) bicrystal Al. *((A) After Kysar, J.W., Gan, Y.X., Morse, T.L., Chen, X., Jones, M.E., 2007. High strain gradient plasticity associated with wedge indentatio into face-centered cubic single crystals: geometrically necessary dislocations densities. J. Mech. Phys. Solids 55 (7), 1554–1573. (B) After Dahlberg, C.F.O., Saito, Y., Öztop, M.S., Kysar, J.W., 2017. Geometrically necessary dislocation density measurements at a grain boundary due to wedge indentation into an aluminum bicrystal. J. Mech. Phys. Solids 105, 131–149.)*

the easiest source operation that is the source with the largest length. Therefore, given a random distribution of sources generated from the initial dislocation density, an average effective source length $\overline{\lambda}$ is related to an effective source stress $\overline{\tau}_s$ as follows (Parthasarathy et al., 2007):

$$\overline{\tau}_s = k_s \mu \frac{\ln\left(\overline{\lambda}/b\right)}{\left(\overline{\lambda}/b\right)} \tag{6.6}$$

where k_s is a source hardening constant. Yaghoobi and Voyiadjis (2016b) showed that the dislocation source length controls the strain rate and size effects even if the dislocation truncation does not occur in the sample.

6.3.3.3 *Strain-rate sensitivity and activation volume*

An empirical fit of $\sigma = \sigma_0 \dot{\varepsilon}^m$ is usually incorporated to describe the strain-rate sensitivity, where m is a material parameter. A more precise way of defining the strain-rate sensitivity can be described using the Arrhenius form equation as follows (Jennings et al., 2011):

$$\dot{\gamma} = \dot{\gamma}_0 \exp\left(-\frac{Q - \tau\Omega(\tau, T)}{k_B T}\right) \tag{6.7}$$

where $\dot{\gamma}$ is the shear strain rate, τ is the applied shear stress, $\dot{\gamma}_0$ is a source's attempt frequency constant, Q is the activation energy, k_B is Boltzmann's constant, Ω is the activation volume, and T is the temperature. The concept of activation volume which describes how the activation energy changes with shear stress can be described as follows:

$$\Omega \equiv -\left.\frac{\partial Q}{\partial \tau}\right|_T \equiv k_B T \frac{\partial \ln(\dot{\gamma})}{\partial \tau} \tag{6.8}$$

Activation volume is an important variable and the deformation mechanism and strain rate sensitivity can be determined using the sample size and activation volume. For example, Fig. 6.3 shows how the activation volume controls three different regions of plasticity mechanisms including surface sources, collective dislocation behavior, and bulk sources for two different strain rates. Since the micron-sized metallic samples can be studied in this research, the mechanism of surface nucleation can be neglected. Jennings et al. (2011) showed that as the sample length scale changes, not only does the material strength increase, but the strain rate dependence of FCC materials changes as well.

6.3.3.4 *Nonlocal continuum plasticity model*

A nonlocal continuum plasticity model based on the higher-order strain gradient plasticity (SGP) theory has been developed by Voyiadjis and his coworkers (Voyiadjis and Faghihi, 2012, 2013; Voyiadjis and Song, 2017; Voyiadjis et al., 2017) to investigate the behavior of small-scale metallic volumes. In a physically based nonlocal continuum plasticity model, a material length scale should be introduced based on the material microstructural characteristics. Figs. 6.9 and 6.10 show an example in which the developed model captures the size effects during plane strain bulge test, which was conducted by Xiang and Vlassak (2006). Voyiadjis and his co-workers (Voyiadjis and Abu Al-Rub, 2005; Voyiadjis and Almasri, 2009; Faghihi and Voyiadjis, 2010; Voyiadjis et al., 2011; Voyiadjis and Zhang, 2015) developed a methodology to obtain a physically based length scale to use in their higher-order SGP theory.

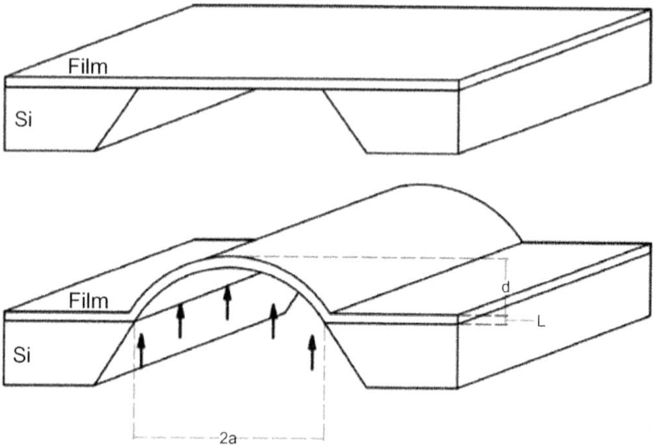

FIG. 6.9 Perspective views of a typical bulge test sample before and after a uniform pressure is applied to one side of the membrane. *(After Voyiadjis, G.Z., Faghihi, D., 2012. Thermo-mechanical strain gradient plasticity with energetic and dissipative length scales. Int. J. Plast. 30–31, 218–247.)*

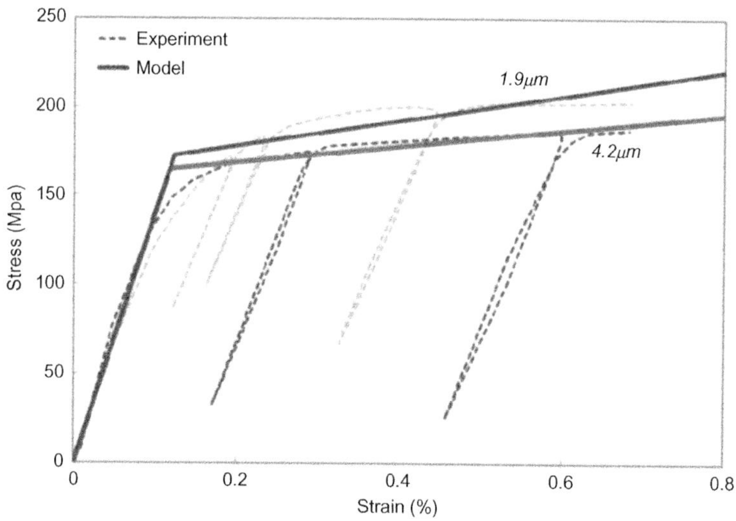

FIG. 6.10 The film thickness effect in electroplated Cu films: the comparison of model predictions presented by Voyiadjis and Faghihi (2012) with the experimental measurements reported by Xiang and Vlassak (2006). *(After Voyiadjis, G.Z., Faghihi, D., 2012. Thermo-mechanical strain gradient plasticity with energetic and dissipative length scales. Int. J. Plast. 30–31, 218–247.)*

Accordingly, Voyiadjis and Zhang (2015) introduced a material length scale parameter ℓ for gradient isotropic hardening plasticity based on the results of the nanoindentation experiments, which can be described as follows:

$$\ell = \left(\frac{\alpha_G}{\alpha_S}\right)^2 \left(\frac{b_G}{b_S}\right) M\bar{r} \left(\frac{\delta_1 d e^{-(E_r/R_g T)}}{\left(1+\delta_2 d p^{(1/m)}\right)\left(1+\delta_3 (\dot{p})^q\right)}\right) \tag{6.9}$$

where $\bar{r}$ is the Nye factor, M is the Schmid factor which is usually taken to be 0.5, d is the average grain size, D is the macroscopic characteristic size of the specimen, p is the equivalent plastic strain, $\dot{p}$ is the equivalent plastic strain rate, m is the hardening exponent and δ_1, δ_2 and δ_3 are material parameters that need to be determined through experimental results (Voyiadjis and Zhang, 2015).

The principle of virtual power has been commonly used to derive the equations for the local equation of motion and the nonlocal microforce balance for volume V as well as the equations for local traction force and nonlocal microtraction condition for the external surface S respectively (see e.g., Voyiadjis and Song, 2017). The internal parts of the principle of virtual power for the bulk P_{int} and for the grain boundary P^I_{int} are expressed in terms of the energy contributions in the arbitrary subregion of the volume V and the arbitrary subsurface of the grain boundary S^I respectively as follows (Voyiadjis and Song, 2017):

$$P_{int} = \int_V \left(\sigma_{ij}\dot{\varepsilon}^e_{ij} + \mathcal{X}_{ij}\dot{\varepsilon}^p_{ij} + \mathcal{S}_{ijk}\dot{\varepsilon}^p_{ij,k} + \mathcal{A}\dot{T} + \mathcal{B}_i\dot{T}_{,i} \right) dV \tag{6.10}$$

$$P^I_{int} = \int_{S^I} \left(\mathbb{M}^{I\mathcal{G}_1}_{ij}\,\dot{\varepsilon}^{pI\mathcal{G}_1}_{ij} + \mathbb{M}^{I\mathcal{G}_2}_{ij}\,\dot{\varepsilon}^{pI\mathcal{G}_2}_{ij} \right) dS^I \tag{6.11}$$

where the superscripts e, p, and I are used to express the elastic state, the plastic state, and the grain boundary respectively. The internal power for the bulk in the form of Eq. (6.10) is defined using the Cauchy stress tensor σ_{ij}, the backstress $\mathcal{X}_{ij}$ conjugate to the plastic strain rate $\dot{\varepsilon}^p_{ij}$, the higher order microstress $\mathcal{S}_{ijk}$ conjugate to the gradients of the plastic strain rate $\dot{\varepsilon}^p_{ij,k}$, and the generalized stresses $\mathcal{A}$ and $\mathcal{B}_i$ conjugate to the temperature rate $\dot{T}$ and the gradient of the temperature rate $\dot{T}_{,i}$ respectively.

The external parts of the principle of virtual power for the bulk and grain boundary are given by (Voyiadjis and Song, 2017):

$$P_{ext} = \int_V b_i v_i dV + \int_S \left(t_i v_i + m_{ij}\dot{\varepsilon}^p_{ij} + a\dot{T} \right) dS \tag{6.12}$$

$$P^I_{ext} = \int_{S^I} \left\{ \left(\sigma^{\mathcal{G}_2}_{ij} n^I_j - \sigma^{\mathcal{G}_1}_{ij} n^I_j \right) v_i + \mathbb{S}^{I\mathcal{G}_2}_{ijk} n^I_k \dot{\varepsilon}^{pI\mathcal{G}_2}_{ij} - \mathbb{S}^{I\mathcal{G}_1}_{ijk} n^I_k \dot{\varepsilon}^{pI\mathcal{G}_1}_{ij} \right\} dS^I \tag{6.13}$$

where t_i and b_i are traction and the external body force conjugate to the macroscopic velocity v_i respectively. By using the principle of virtual power that the external power is equal to the internal power ($P_{int} = P_{ext}$, $P^I_{int} = P^I_{ext}$), the equations for balance of linear momentum and nonlocal microforce balance for V, the equations for local surface traction conditions and nonlocal microtraction conditions on S, and the interfacial macro- and microforce balances at the grain boundary can be obtained respectively.

It is further assumed in the proposed work that the thermodynamic microstress quantities $\mathcal{X}_{ij}$, $\mathcal{S}_{ijk}$, $\mathcal{A}$ and $\mathbb{M}^I_{ij}$ are decomposed into the energetic and the dissipative components such that:

$$\mathcal{X}_{ij} = \mathcal{X}^{en}_{ij} + \mathcal{X}^{dis}_{ij};\ \mathcal{S}_{ijk} = \mathcal{S}^{en}_{ijk} + \mathcal{S}^{dis}_{ijk};\ \mathcal{A} = \mathcal{A}^{en} + \mathcal{A}^{dis};\ \mathbb{M}^I_{ij} = \mathbb{M}^{I,en}_{ij} + \mathbb{M}^{I,dis}_{ij} \tag{6.14}$$

The energetic and dissipative parts of the thermodynamic microstresses can be obtained respectively by using the following formulations (Voyiadjis and Song, 2017):

$$
\begin{aligned}
&\text{Energetic } \sigma_{ij} = \rho\frac{\partial \Psi}{\partial \varepsilon^{e}_{ij}};\ \mathcal{X}^{en}_{ij} = \rho\frac{\partial \Psi}{\partial \varepsilon^{p}_{ij}};\ \mathcal{S}^{en}_{ijk} = \rho\frac{\partial \Psi}{\partial \varepsilon^{p}_{ij,k}};\ \mathcal{A}^{en} = \rho\left(\mathcal{s} + \frac{\partial \Psi}{\partial T}\right);\\
&\mathcal{B}_{i} = \rho\frac{\partial \Psi}{\partial T_{,i}};\ \mathbb{M}^{I,en}_{ij} = \rho\frac{\partial \Psi^{I}}{\partial \varepsilon^{pI}_{ij}}\\
&\text{Dissipative } \mathcal{X}^{dis}_{ij} = \frac{\partial \mathcal{D}}{\partial \dot{\varepsilon}^{p}_{ij}};\ \mathcal{S}^{dis}_{ijk} = \frac{\partial \mathcal{D}}{\partial \dot{\varepsilon}^{p}_{ij,k}};\ \mathcal{A}^{dis} = \frac{\partial \mathcal{D}}{\partial \dot{T}};\ -\frac{q_i}{T} = \frac{\partial \mathcal{D}}{\partial T_{,i}};\ \mathbb{M}^{I,dis}_{ij} = \frac{\partial \mathcal{D}^{I}}{\partial \dot{\varepsilon}^{pI}_{ij}}
\end{aligned}
\tag{6.15}
$$

where Ψ and $\mathcal{D}$ are the Helmholtz free energy function and the dissipation potential respectively. In the proposed work, one or more energetic and dissipative length scales can be involved in the functional form of Ψ and $\mathcal{D}$ to account for the rate effects and size effects. The user-element subroutines UMAT and VUMAT in the commercial finite element packages ABAQUS/standard and ABAQUS/explicit, respectively, can be used to define the mechanical constitutive behavior.

6.3.4 Proposed framework

As outlined in the preceding, the field of material modeling of metallic samples with confined volumes is an essential component of any strategy for design of micron-sized metallic devices. The lack of fundamental understanding and quantification of the atomic scale deformation mechanisms and processes is certainly disconcerting, given the major investments of the scientific communities in material design and multiscale modeling and simulation. This is largely due to the lack of systematic experimental procedures and meso-scale simulations to provide the required information to develop physically based nonlocal continuum models which can be utilized to analyze the mechanical behavior of micron-sized metallic devices. Although some nonlocal plasticity models have been developed by now (see, e.g., Voyiadjis and Faghihi, 2012, 2013; Voyiadjis and Song, 2017; Voyiadjis et al., 2017), the incorporated hardening models and length scales are partly phenomenological due to the lack of a systematic set of experiments and simulations which addresses the evolution of dislocation network in the cases of micron-sized metallic samples.

The primary goal of this framework is to develop a physically based nonlocal continuum plasticity model based on the inputs obtained from the conducted experiments and atomistic simulations. A second goal is to unravel the underlying mechanisms of size and strain rate effects in micron-sized metallic samples using both experiments and atomistic simulations. Besides the two primary goals, the results of this work can stretch the current experimental knowledge of plasticity in metallic samples of confined volumes by conducting two important experiments of indentation and microbending and monitoring the dislocation pattern using EBSD analysis. The effect of strain rate on dislocation evolution and pattern can be experimentally investigated using EBSD analysis. Furthermore, very few atomistic simulations have been performed on metallic samples with length scales close to the micron to address the size and strain rate effects, including the ones performed by the Yaghoobi and Voyiadjis (2016a,b), Voyiadjis and Yaghoobi (2015, 2017). It is due to the fact that the MD simulation of large metallic samples is computationally very demanding. The post-processing of the micron-sized sample is also not a trivial task. Accordingly, a systematic parametric study for samples with different sizes in the order 0.3 μm, with various materials, grain sizes, crystal structure, and crystallographic orientation subjected to different strain rates can be conducted which can provide valuable information for strain rate and size effects in micron-sized metallic samples. The proposed concept is divided into three major complimentary tasks, which are described in the next section.

6.4 Required research steps to develop the multiscale framework

As explained in Section 6.3, three aspects of the proposed multiscale framework is atomistic simulation, indentation and microbending experiments, and nonlocal continuum models. In this section, the description of required simulation and experiments to achieve this framework is presented.

6.4.1 Large scale MD simulation and post processing

Step 1: Atomistic simulation of micron-sized samples and the post-processing require tremendous computational power. Although few preliminary studies have been conducted on the samples with length scale in the order of 0.3 μm during the nanoindentation and micropillar compression experiments, the conducted simulations were solely conducted to show the capability of large scale atomistic simulation to capture some features of size and strain rate effects in the case of micron-sized metallic samples. Furthermore, until now, a systematic atomistic study which investigates the important characteristics of micron-sized metallic samples including the selected material, crystal structure, grain size, crystallographic orientation, and strain rate during nanoindentation and micropillar compression tests have not been conducted. In order to incorporate the atomistic results into a physically based nonlocal continuum plasticity model, however, one needs a systematic and comprehensive set of results from large scale MD simulation.

In this step, a systematic study of single crystalline and polycrystalline Ni, Al, and Cu samples with different crystallographic orientations can be conducted during nanoindentation and micropillar compression experiments.

The sample length scale can be in the order of 0.3 μm. The effect of grain size can be also addressed by changing the grain size from 5 to 50 nm. The velocity Verlet algorithm can be incorporated to integrate Newton's equations of motion. The parallel code LAMMPS can be used to perform the MD simulations (Plimpton, 1995). All the simulations can be conducted at room temperature, i.e., 300 K. The embedded-atom method (EAM) (Daw and Baskes, 1984) and modified embedded-atom method (MEAM) (Baskes, 1992) potentials can be incorporated to model the interaction of metallic samples atoms with each other. The Si substrate can be modeled using the Lennard-Jones (LJ) potential, and the indenter can be modeled using a repulsive potential for both nanoindentation and micropillar compression experiments (Voyiadjis and Yaghoobi, 2017). In the case of nanoindentation experiment, different indenter geometries of conical, spherical, and flat punch can be incorporated to study the effect of indenter geometry on the nanoindentation response. Furthermore, in some cases, the indenter itself can be modeled as a cluster of atoms to study the effect of repulsive potential used for indenter on the obtained results. The precise contact area, which is an essential ingredient of both nanoindentation and micropillar compression experiments, can be obtained using the triangulation method (Yaghoobi and Voyiadjis, 2014).

The Crystal Analysis Tool developed by Stukowski and Albe (2010), Stukowski et al. (2012), and Stukowski (2014) can be incorporated to visualize the dislocations and provide additional information such as dislocation length and Burgers vector. The software OVITO (Stukowski, 2010) and Paraview (Henderson, 2007) can be used to visualize the defects and analyze the dislocation information. Using the obtained data, the dislocation density and pattern can be obtained for nanoindentation and micropillar compression experiments. Furthermore, the interaction between dislocations and governing mechanisms of size and strain rate effects can be investigated. Finally, physically based hardening mechanisms which can capture both size and strain rate effects in micron-sized samples can be formulated to be incorporated in a nonlocal plasticity model.

6.4.2 Indentation and microbending experiments

Step 2: The indentation and microbending experiments can be conducted to investigate the size and strain rate effects in micron-sized metallic samples. Both experiments can be conducted in the plane strain deformation state according to the formulation of GND density calculation described in Section 6.3.2.2. The indentation can be conducted on the single crystal and bicrystal Al and Cu. The bicrystal sample contains a symmetric tilt Coincident Site Lattice (CSL) Σ43 (335)[110] grain boundary. The sample preparation details for both single and bicrystal samples can be found in Kysar et al. (2007) and Dahlberg et al. (2017). Two cylindrical indenters with radii of 150 and 300 μm and two wedge indenters with 90° and 120° included angles can be made of tungsten carbide

(WC) bonded by a ferrous alloy and cut to shape via EDM. To study the effect of strain rates, one set of experiments can be under quasistatic deformation rates (e.g., ~5 μm s^{-1}) can be performed in an Instron. Another set of experiments can take place in a home-built drop test instrumented with an accelerometer from which the force as a function of position can be calculated which induces the strain rate of order 10^4 s^{-1}. Experiments by Dahlberg et al. (2017) of wedge indentation at quasistatic displacement rates in pure aluminum crystals dissipate approximate 0.1 J of energy to induce an indentation to the depth of 200 μm. A 1 kg mass falling 1 cm generates kinetic energy of this order. The sample dimension can be approximately 1 cm cube, similar to Dahlberg et al. (2017), and the maximum indentation depths can be of the order of 200 μm. The total GND dislocation pattern can be obtained using EBSD analysis at different indentation depths to capture the dislocation content evolution during the indentation. The as-deformed orientation of the crystal lattice on the newly exposed surface is measured with EBSD on a JEOL 5600 SEM. A Si single crystal is used for detector orientation and projection parameter calibration. Following calibration, the Kikuchi diffraction patterns are obtained at a 20 kV accelerating voltage at working distance of 12 mm. The measurements are made on a 3 μm square raster over an area of about 1 × 1 mm. The Kikuchi diffraction patterns are processed with HKL Channel 5 software to determine the as-deformed crystallographic orientation of the specimen in terms of Euler angles at each measurement position. The overall experimental uncertainty of the angular orientation is about ±0.5° as discussed in Gardner et al. (2011). The Euler angles do not represent a unique lattice configuration because of the symmetries in the FCC crystal lattice. Thus an algorithm based on quaternion algebra, similar in spirit to the work by Gupta and Agnew (2010), is implemented in Matlab to post-process the Euler angle data and determine the crystallographic orientation in the deformed configuration. The lattice rotation is determined at each measurement point by comparing the measured as-deformed crystallographic orientation with the known crystallographic orientation of the undeformed crystal lattice. By using quaternions to describe the rotation, it is straightforward to decompose the lattice rotation into out-of-plane and in-plane components.

The microbending experiments can be performed on the single crystal Al and Cu. Several different beam thicknesses ranging from 10 to 100 μm can be tested to study the sample size effects. The cantilever is deflected by positioning a micromanipulator tip perpendicularly against its free end (end loaded cantilever beam). The force from the micromanipulator acts perpendicular only at the start of the experiment. With increasing beam deflection, the force angle decreases inevitably but has no noticeable influence on the deformation conditions of the cantilever. Limitations to the range of the micromanipulator tip prevents a full deflection of the cantilever. Approximately 25% of the free end do not impact the anvil and remains virtually undeformed. End loaded cantilever beams of this size deform differently compared to their macroscopic counterparts. Most of the imposed strain tends to accumulate around the fixed end, leading rather to buckling than to bending. A well-controlled micro cantilever bending experiment can be conducted in a mildly constrained manner by milling a quarter-circle-shaped anvil below the actual cantilever to achieve a more homogeneous deformation state. This particular arrangement forces the cantilever to adopt a constant bending radius of curvature. Due to this constraint, a tangential point between the beam and the anvil exists, which gradually shifts from the fixed toward the free end with increasing deflection. This prevents already deformed volumes between the fixed end and the tangential point to take up further deformation. The GND density can be calculated in each sample using EBSD analysis at different beam deflections. The EBSD analysis is similar to the one incorporated in the indentation test. In addition to the GND density calculations, the radius of curvature of the specimen in the unloaded configuration can be measured.

6.4.3 Development, implementation, and validation of a new nonlocal continuum plasticity model

Step 3: Based on the information obtained from the conducted MD simulations (*Step 1*) and experiments (*Step 2*), a set of new length scales can be introduced which are related to the sample size, applied strain rates, grain size and distribution, and dislocation structures. Next, the new length scales are incorporated in the nonlocal continuum plasticity model. In addition to the new length scales, the modified hardening mechanisms developed based on the results of experiments and atomistic simulations can be included in the nonlocal model. The new nonlocal continuum model can be able to capture the hardening mechanisms observed at both bulk and micron-sized samples. Finally, the size and strain rate effects can be studied using the developed nonlocal continuum plasticity model, and the model can be validated against the conducted experiments and MD simulations.

First, in order to study the size effects, two experiments of microbending and indentation can be simulated using the FEM. The user-element subroutine UMAT in the commercial finite element package ABAQUS/standard can be used to define the mechanical constitutive behavior. The user-element subroutine UEL in the commercial finite element package ABAQUS/standard can be developed with the displacement field u and the plastic strain field ε^p as independently discretized nodal degrees of freedom in order to numerically solve the proposed nonlocal continuum plasticity model. The increments in nodal displacements and plastic strains can be computed by solving global system of linear equations as follows (Voyiadjis and Song, 2017):

$$\underbrace{\begin{bmatrix} K^{el}_{uu} & K^{el}_{u\varepsilon^p} \\ K^{el}_{\varepsilon^p u} & K^{el}_{\varepsilon^p \varepsilon^p} \end{bmatrix}}_{\boldsymbol{K}^{el}} \begin{Bmatrix} \Delta\mathcal{U}^{\xi}_{u} \\ \Delta\mathcal{E}^{\xi}_{\varepsilon^p} \end{Bmatrix} = \begin{Bmatrix} (r_u)_{\xi} \\ (r_{\varepsilon^p})_{\xi} \end{Bmatrix} \tag{6.16}$$

where $\mathcal{U}_u^{\xi}$, $\mathcal{E}_{\varepsilon^p}^{\xi}$ and $(r_u)_{\xi}$, $(r_{\varepsilon^p})_{\xi}$ are the nodal values and the nodal residuals of the displacement and the plastic strain at node ξ respectively, and $\boldsymbol{K}^{el}$ is the Jacobian matrix.

The material constants of the model related to the size effects can be calibrated using the results of the conducted atomistic simulations (*Step 1*) and experiments (*Step 2*). In the case of microbending, due to the horizontal displacement of the loading point during the bending test, horizontal forces can arise, which depend on the lateral stiffness of the indentation device and the friction between indenter and beam surface. Two extreme conditions can be considered: lateral free indenter (no lateral forces) and lateral fixed indenter (highest lateral forces) and the obtained results can be compared. In the case of indentation, beside the wedge indenter, two indenter geometries of spherical and Berkovich indenter can be modeled on top of the cube. A finer mesh is used in the region which can be in contact with the indenter. The interaction between the indenter and sample can be modeled using the contact module implemented in ABAQUS. The precise contact area can be obtained from the deformed meshes using the triangulation method.

In the next step, in order to study the effect of strain rate, the indentation experiments can be modeled using the explicit FEM. The user-element subroutine VUMAT in the commercial finite element package ABAQUS/explicit can be used to define the mechanical constitutive behavior. The user-element subroutine VUEL in the commercial finite element package ABAQUS/explicit can be developed with the displacement field u and the plastic strain field ε^p as independently discretized nodal degrees of freedom in order to numerically solve the proposed nonlocal continuum plasticity model. The material constants of the model related to the strain rate effects can be calibrated using the results of the conducted experiments and atomistic simulations. In the next step, different strain rates can be applied in the cases of indentation experiments and the results are then compared with those of the experiments and MD simulation. The indentation problem set-up for FE simulation can be similar to the ones conducted by the MD simulations and experiments.

References

Abu Al-Rub, R.K., Voyiadjis, G.Z., 2004. Analytical and experimental determination of the material intrinsic length scale of strain gradient plasticity theory from micro- and nano-indentation experiments. Int. J. Plast. 20 (6), 1139–1182.

Armstrong, R.W., Li, Q., 2015. Dislocation mechanics of high-rate deformations. Metall. Mater. Trans. A. 46, 4438–4453.

Armstrong, R.W., Walley, S.M., 2008. High strain rate properties of metals and alloys. Int. Mater. Rev. 53, 105–128.

Armstrong, R.W., Arnold, W., Zerilli, F.J., 2007. Dislocation mechanics of shock-induced plasticity. Metall. Mater. Trans. A. 38, 2605–2610.

Baskes, M.I., 1992. Modified embedded-atom potentials for cubic materials and impurities. Phys. Rev. B. 46, 2727–2742.

Bringa, M., Rosolankova, K., Rudd, R.E., Remington, B.A., Wark, J.S., Duchaineau, M., Kalantar, H., Hawreliak, J., Belak, J., 2006. Shock deformation of face-centred-cubic metals on subnanosecond timescales. Nat. Mater. 5, 805–809.

Dahlberg, C.F.O., Saito, Y., Öztop, M.S., Kysar, J.W., 2014. Geometrically necessary dislocation density measurements associated with different angles of indentations. Int. J. Plast. 54, 81–95.

Dahlberg, C.F.O., Saito, Y., Öztop, M.S., Kysar, J.W., 2017. Geometrically necessary dislocation density measurements at a grain boundary due to wedge indentation into an aluminum bicrystal. J. Mech. Phys. Solids. 105, 131–149.

Daw, M.S., Baskes, M.I., 1984. Embedded-atom method: derivation and application to impurities, surfaces, and other defects in metals. Phys. Rev. B. 29, 6443–6453.

Demir, E., Raabe, D., Zaafarani, N., Zaefferer, S., 2009. Investigation of the indentation size effect through the measurement of the geometrically necessary dislocations beneath small indents of different depths using EBSD tomography. Acta Mater. 57 (2), 559–569.

Demir, E., Raabe, D., Roters, F., 2010. The mechanical size effect as a mean-field breakdown phenomenon: example of microscale single crystal beam bending. Acta Mater. 58, 1876–1886.

Diao, J., Gall, K., Dunn, M.L., Zimmerman, J.A., 2006. Atomistic simulations of the yielding of gold nanowires. Acta Mater. 54, 643–653.

El-Awady, J.A., 2015. Unravelling the physics of size-dependent dislocation-mediated plasticity. Nat. Commun. 6.

El-Awady, J.A., Wen, M., Ghoniem, N.M., 2009. The role of the weakest-link mechanism in controlling the plasticity of micropillars. J. Mech. Phys. Solids. 57, 32–50.

El-Dasher, B.S., Adams, B.L., Rollett, A.D., 2003. Viewpoint: experimental recovery of geometrically necessary dislocation density in polycrystals. Scr. Mater. 48 (2), 141–145.

Faghihi, D., Voyiadjis, G.Z., 2010. Size effects and length scales in nanoindentation for body-centred cubic materials with application to iron. Proc. Inst. Mech. Eng. N J. Nanoeng. Nanosyst. 224, 5–18.

Fox, N., 1966. A continuum theory of dislocations for single crystals. IMA J. Appl. Math. 2 (4), 285–298.

Gardner, C.J., Kacher, J., Basinger, J., Adams, B.L., Öztop, M.S., Kysar, J.W., 2011. Techniques and applications of the simulated pattern adaptation of Wilkinson's method for advanced microstructure analysis. Exp. Mech. 51 (8), 1379–1393.

Greer, J.R., 2013. Size-dependent strength in single-crystalline metallic nanostructures. In: Espinosa, H.D., Bao, G. (Eds.), Nano and Cell Mechanics: Fundamentals and Frontiers. John Wiley & Sons, Ltd, Chichester.

Greer, J.R., De Hosson, J.T.M., 2011. Plasticity in small-sized metallic systems: intrinsic versus extrinsic size effect. Prog. Mater. Sci. 56, 654–724.

Greer, J.R., Oliver, W.C., Nix, W.D., 2005. Size dependence of mechanical properties of gold at the micron scale in the absence of strain gradients. Acta Mater. 53 (6), 1821–1830.

Gupta, V., Agnew, S., 2010. A simple algorithm to eliminate ambiguities in EBSD orientation map visualization and analyses: application to fatigue crack-tips/wakes in aluminum alloys. Microsc. Microanal. 16, 831–841.

Gurrutxaga-Lerma, B., Balint, D.S., Dini, D., Eakins, D.E., Sutton, A.P., 2015. Attenuation of the dynamic yield point of shocked aluminum using elastodynamic simulations of dislocation dynamics. Phys. Rev. Lett. 114.

Henderson, A., 2007. Paraview Guide, A Parallel Visualization Application. Kitware Inc. http://www.paraview.org.

Hibbitt, D., Karlsson, B., Sorensen, P., 2009. ABAQUS Analysis User's Manual Version 6.9, USA.

Jennings, A.T., Li, J., Greer, J.R., 2011. Emergence of strain-rate sensitivity in Cu nanopillars: transition from dislocation multiplication to dislocation nucleation. Acta Mater. 59, 5627–5637.

Kelchner, C.L., Plimpton, S.J., Hamilton, J.C., 1998. Dislocation nucleation and defect structure during surface indentation. Phys. Rev. B. 58, 11085–11088.

Kiener, D., Pippan, R., Motz, C., Kreuzer, H., 2006. Microstructural evolution of the deformed volume beneath microindents in tungsten and copper. Acta Mater. 54 (10), 2801–2811.

Kondo, K., 1964. On the analytical and physical foundations of the theory of dislocations and yielding by the differential geometry of continua. Int. J. Eng. Sci. 2 (3), 219–251.

Kraft, O., Gruber, P., Mönig, R., Weygand, D., 2010. Plasticity in confined dimensions. Annu. Rev. Mater. Res. 40, 293–317.

Kysar, J.W., Briant, C.L., 2002. Crack tip deformation fields in ductile single crystals. Acta Mater. 50 (9), 2367–2380.

Kysar, J.W., Gan, Y.X., Morse, T.L., Chen, X., Jones, M.E., 2007. High strain gradient plasticity associated with wedge indentatio into face-centered cubic single crystals: geometrically necessary dislocations densities. J. Mech. Phys. Solids. 55 (7), 1554–1573.

Kysar, J.W., Saito, Y., Öztop, M., Lee, D., Huh, W., 2010. Experimental lower bounds on geometrically necessary dislocation density. Int. J. Plast. 26 (8), 1097–1123.

Lee, Y., Park, J.Y., Kim, S.Y., Jun, S., 2005. Atomistic simulations of incipient plasticity under Al (111) nanoindentation. Mech. Mater. 37, 1035–1048.

Li, J., Van Vliet, K.J., Zhu, T., Yip, S., Suresh, S., 2002. Atomistic mechanisms governing elastic limit and incipient plasticity in crystals. Nature. 418, 307–310.

McLaughlin, K.K., Clegg, W.J., 2008. Deformation underneath low-load indentations in copper. J. Phys. D. 41(7).

Meyers, M.A., 1994. Dynamic Behavior of Materials. John Wiley, Hoboken, NJ.

Meyers, M.A., Gregori, F., Kad, B.K., Schneider, M.S., Kalantar, D.H., Remington, B.A., Ravichandran, G., Boehly, T., Wark, J.S., 2003. Laser-induced shock compression of monocrystalline copper: characterization and analysis. Acta Mater. 51, 1211–1228.

Meyers, M.A., Jarmakani, H., Bringa, E.M., Remington, B.A., 2009. Dislocations in shock compression and release. In: Hirth, J.P., Kubin, L. (Eds.), Dislocations in Solids. Elsevier, pp. 91–197.

Nix, W.D., Gao, H.J., 1998. Indentation size effects in crystalline materials: a law for strain gradient plasticity. J. Mech. Phys. Solids. 46 (3), 411–425.

Norfleet, D.M., Dimiduk, D.M., Polasik, S.J., Uchic, M.D., Mills, M.J., 2008. Dislocation structures and their relationship to strength in deformed nickel microcrystals. Acta Mater. 56, 2988–3001.

Nye, J.F., 1953. Some geometrical relations in dislocated crystals. Acta Metall. 1 (2), 153–162.

Pantleon, W., 2008. Resolving the geometrically necessary dislocation content by conventional electron backscattering diffraction. Scr. Mater. 58 (11), 994–997.

Parthasarathy, T.A., Rao, S.I., Dimiduk, D.M., Uchic, M.D., Trinkle, D.R., 2007. Contribution to size effect of yield strength from the stochastics of dislocation source lengths in finite samples. Scr. Mater. 56, 313–316.

Plimpton, S., 1995. Fast parallel algorithms for short-range molecular dynamics. J. Comput. Phys. 117, 1–19.

Rao, S.I., Dimiduk, D.M., Tang, M., Parthasarathy, T.A., Uchic, M.D., Woodward, C., 2007. Estimating the strength of single-ended dislocation sources in micron-sized single crystals. Philos. Mag. 87, 4777–4794.

Rao, S.I., Dimiduk, D.M., Parthasarathy, T.A., Uchic, M.D., Tang, M., Woodward, C., 2008. Athermal mechanisms of size-dependent crystal flow gleaned from three-dimensional discrete dislocation simulations. Acta Mater. 56, 3245–3259.

Rester, M., Motz, C., Pippan, R., 2008. The deformation-induced zone below large and shallow nanoindentations: a comparative study using EBSD and TEM. Philos. Mag. Lett. 88 (12), 879–887.

Ruggles, T.J., Fullwood, D.T., Kysar, J.W., 2016. Resolving geometrically necessary dislocation density onto individual dislocation types using EBSD-based continuum dislocation microscopy. Int. J. Plast. 76, 231–243.

Sansoz, F., 2011. Atomistic processes controlling flow stress scaling during compression of nanoscale face-centered-cubic crystals. Acta Mater. 59, 3364–3372.

Sarac, A., Oztop, M.S., Dahlberg, C.F.O., Kysar, J.W., 2016. Spatial distribution of the net burgers vector density in a deformed single crystal. Int. J. Plast. 85, 110–129.

Shan, Z.W., Mishra, R.K., Syed Asif, S.A., Warren, O.L., Minor, A.M., 2008. Mechanical annealing and source-limited deformation in submicrometre-diameter Ni crystals. Nat. Mater. 77, 115–119.

Stukowski, A., 2010. Visualization and analysis of atomistic simulation data with OVITO–the Open Visualization Tool. Model. Simul. Mater. Sci. Eng. 18. http://www.ovito.org/.

Stukowski, A., 2014. Computational analysis methods in atomistic modeling of crystals. JOM. 66, 399–407.

Stukowski, A., Albe, K., 2010. Extracting dislocations and non-dislocation crystal defects from atomistic simulation data. Model. Simul. Mater. Sci. Eng. 18.

Stukowski, A., Bulatov, V.V., Arsenlis, A., 2012. Automated identification and indexing of dislocations in crystal interfaces. Model. Simul. Mater. Sci. Eng. 20.

Sun, S., Adams, B.L., King, W.E., 2000. Observations of lattice curvature near the interface of a deformed aluminium bicrystal. Philos. Mag. A. 80 (1), 9–25.

Uchic, M.D., Dimiduk, D.M., Florando, J.N., Nix, W.D., 2003. Exploring specimen size effects in plastic deformation of $Ni_3(Al,Ta)$. Mater. Res. Soc. Symp. Proc. 753, 27–32.

Uchic, M.D., Dimiduk, D.M., Florando, J.N., Nix, W.D., 2004. Sample dimensions influence strength and crystal-plasticity. Science. 305, 986–989.

Uchic, M.D., Shade, P.A., Dimiduk, D.M., 2009. Plasticity of micrometer-scale single crystals in compression. Annu. Rev. Mater. Res. 39, 361–386.

Voyiadjis, G.Z., Abu Al-Rub, R.K., 2005. Gradient plasticity theory with a variable length scale parameter. Int. J. Solids Struct. 42 (14), 3998–4029.

Voyiadjis, G.Z., Abu Al-Rub, R.K., 2007. Nonlocal gradient-dependent thermodynamics for modeling scale-dependent plasticity. Int. J. Multiscale Comput. Eng. 5 (3–4), 295–323.

Voyiadjis, G.Z., Almasri, A.H., 2009. Variable material length scale associated with nanoindentation experiments. J. Eng. Mech. 135 (3), 139–148.

Voyiadjis, G.Z., Faghihi, D., 2012. Thermo-mechanical strain gradient plasticity with energetic and dissipative length scales. Int. J. Plast. 30–31, 218–247.

Voyiadjis, G.Z., Faghihi, D., 2013. Gradient plasticity for thermo-mechanical processes in metals with length and time scales. Philos. Mag. 93 (9), 1013–1053.

Voyiadjis, G.Z., Song, Y., 2017. Effect of passivation on higher order gradient plasticity models for non-proportional loading: energetic and dissipative gradient components. Philos. Mag. 97, 318–345.

Voyiadjis, G.Z., Yaghoobi, M., 2015. Large scale atomistic simulation of size effects during nanoindentation: dislocation length and hardness. Mater. Sci. Eng. A. 634, 20–31.

Voyiadjis, G.Z., Yaghoobi, M., 2016. Role of grain boundary on the sources of size effects. Comput. Mater. Sci. 117, 315–329.

Voyiadjis, G.Z., Yaghoobi, M., 2017. Size and strain rate effects in metallic samples of confined volumes: dislocation length distribution. Scr. Mater. 130, 182–186.

Voyiadjis, G.Z., Zhang, C., 2015. The mechanical behavior during nanoindentation near the grain boundary in a bicrystal FCC metal. Mater. Sci. Eng. A. 621, 218–228.

Voyiadjis, G.Z., Faghihi, D., Zhang, C., 2011. Analytical and experimental detemination of rate, and temperature dependent length scales using nanoindentation experiments. J. Nanomech. Micromech. 1 (1), 24–40.

Voyiadjis, G.Z., Song, Y., Park, T., 2017. Higher-order thermomechanical gradient plasticity model with energetic and dissipative components. J. Eng. Mater. Technol. 139(2).

Wheeler, J., Mariani, E., Piazolo, S., Prior, D.J., Trimby, P., Drury, M.R., 2009. The weighted Burgers vector: a new quantity for constraining dislocation densities and types using electron backscatter diffraction on 2D sections through crystalline materials. J. Microsc. (Oxford). 233 (3), 482–494.

Xiang, Y., Vlassak, J.J., 2006. Bauschinger and size effects in thin-film plasticity. Acta Mater. 54 (20), 5449–5460.

Yaghoobi, M., Voyiadjis, G.Z., 2014. Effect of boundary conditions on the MD simulation of nanoindentation. Comput. Mater. Sci. 95, 626–636.

Yaghoobi, M., Voyiadjis, G.Z., 2016a. Atomistic simulation of size effects in single-crystalline metals of confined volumes during nanoindentation. Comput. Mater. Sci. 111, 64–73.

Yaghoobi, M., Voyiadjis, G.Z., 2016b. Size effects in fcc crystals during the high rate compression test. Acta Mater. 121, 190–201.

Zaafarani, N., Raabe, D., Roters, F., Zaefferer, S., 2008. On the origin of deformation-induced rotation patterns below nanoindents. Acta Mater. 56 (1), 31–42.

Zhang, C., Voyiadjis, G.Z., 2016. Rate-dependent size effects and material length scales in nanoindentation near the grain boundary for a bicrystal FCC metal. Mater. Sci. Eng. A. 659, 55–62.

Zhou, C., Biner, S.B., LeSar, R., 2010. Discrete dislocation dynamics simulations of plasticity at small scales. Acta Mater. 58, 1565–1577.

Zimmerman, J.A., Kelchner, C.L., Klein, P.A., Hamilton, J.C., Foiles, S.M., 2001. Surface step effects on nanoindentation. Phys. Rev. Lett. 87(16).

Further reading

Cui, Y.N., Liu, Z.L., Zhuang, Z., 2013. Dislocation multiplication by single cross slip for FCC at submicron scales. Chin. Phys. Lett. 30.

Espinosa, H., Berbenni, S., Panico, M., Schwarz, K.W., 2005. An interpretation of size-scale plasticity in geometrically confined systems. Proc. Natl. Acad. Sci. U. S. A. 102, 16933–16938.

Gao, Y., Ruestes, C.J., Urbassek, H.M., 2014. Nanoindentation and nanoscratching of iron: atomistic simulation of dislocation generation and reactions. Comput. Mater. Sci. 90, 232–240.

Ghazi, N., Kysar, J.W., 2016. Experimental investigation of plastic strain recovery and creep in nanocrystalline copper thin films. Exp. Mech. 56 (8), 1351–1362.

Hay, J.C., Pharr, G.M., 2000. ASM Handbook for Mechanical Testing and Evaluation. vol. 8. ASM International, Materials Park, OH, pp. 202–232.

Hutchinson, J.W., 2012. Generalizing J(2) flow theory: fundamental issues in strain gradient plasticity. Acta Mech. Sinica. 28, 1078–1086.

Jang, H., Farkas, D., 2007. Interaction of lattice dislocations with a grain boundary during nanoindentation simulation. Mater. Lett. 61, 868–871.

Kiener, D., Minor, A.M., 2011. Source truncation and exhaustion: insights from quantitative in situ TEM tensile testing. Nano Lett. 11, 3816–3820.

Messerschmidt, U., Bartsch, M., 2003. Generation of dislocations during plastic deformation. Mater. Chem. Phys. 81, 518–523.

Mishin, Y., Farkas, D., Mehl, M.J., Papaconstantopoulos, D.A., 1999. Interatomic potentials for monoatomic metals from experimental data and ab initio calculations. Phys. Rev. B. 59, 3393–3407.

Oliver, W.C., Pharr, G.M., 1992. An improved technique for determining hardness and elastic modulus using load and displacement sensing indentation experiments. J. Mater. Res. 7, 1564–1583.

Tucker, G.J., Aitken, Z.H., Greer, J.R., Weinberger, C.R., 2013. The mechanical behavior and deformation of bicrystalline nanowires. Model. Simul. Mater. Sci. Eng. 21.

Weinberger, C.R., Cai, W., 2008. Surface-controlled dislocation multiplication in metal micropillars. Proc. Natl. Acad. Sci. 105, 14304–14307.

Xu, S., Guo, Y.F., Ngan, A.H.W., 2013. A molecular dynamics study on the orientation, size, and dislocation confinement effects on the plastic deformation of Al nanopillars. Int. J. Plast. 43, 116–127.

Yaghoobi, M., Voyiadjis, G.Z., 2017. Microstructural investigation of the hardening mechanism in fcc crystals during high rate deformations. Comput. Mater. Sci. 138, 10–15.

Index

Note: Page numbers followed by *f* indicate figures and *t* indicate tables.

G

K

L

M

N

Q

R

S

T

U

V

CPI Antony Rowe
Eastbourne, UK
November 20, 2019

CHAPTER 1

Mechanism design for legged locomotion systems

Giuseppe Carbone[a], Marco Ceccarelli[b]
[a]DIMEG, University of Calabria, Rende, Italy
[b]LARM2: Laboratory of Robot Mechatronics, Department of Industrial Engineering, University of Rome Tor Vergata, Rome, Italy

1 Introduction

Locomotion is defined as the "Movement or the ability to move from one place to another" [1]. More in particular in IFToMM terminology [2], it is defined as "autonomous, internally driven change of location of human being, animals, or machines during which base of support and center of mass of the body are displayed" with more details in the several types of locomotion as per the environment in which it is performed.

Locomotion is fundamental to the survival of many animal species including humans. The mechanics and performance of locomotion varies significantly as function of the environment in which locomotor behaviors are executed, which can be divided into terrestrial, aquatic, aerial, as outlined in refs. [3–6]. Terrestrial locomotion can be achieved with legs, wheels, and crawlers. Legged locomotion is the most widely used solution for terrestrial locomotion in nature as it is the most effective speedy and versatile when it operates in a rough terrain or in presence of obstacles. The energy efficiency of legged locomotion might significantly vary among animals and machines. Wheeled/crawler locomotion is instead preferred for vehicles on flat surfaces as it can be more easily controlled, and it can be more energy efficient.

A large literature reports a wide range of legged locomotion systems, which have been designed for a wide variety of applications as indicated for example in refs. [7, 8]. For example, there are legged locomotion systems, which are used for entertainment purposes; other solutions for carrying heavy loads on hills or rough terrains, or even for carrying humans while overcoming stairs or other architectonic barriers. The common limits of legged locomotion systems are high costs and complex design and operation, which often prevent a widespread in the market, even if they have been

https://doi.org/10.1016/B978-0-12-815659-9.00001-9

inspired by very successful examples in nature. Accordingly, efforts should be made to improve user-friendliness, user-printed design and operation, costs of the solutions for legged locomotion systems, with activities since the very early design stage.

Generally, legged systems can be slow and more difficult to design and operate with respect to mobile machines that are equipped with crawlers or wheels. But, legged robots are more suitable for rough terrain, where obstacles of any size can appear. In fact, the use of wheels or crawlers limits the size of the obstacle that can be climbed, to half the diameter of the wheels. On the contrary, legged machines can overcome obstacles that are comparable with the size of the machine leg.

This chapter provides useful considerations for the design of legged locomotion systems by focusing at their mechanism synthesis for specific applications with suitable low-cost user-friendly features. After a general overview on design requirements and design process, several examples are reported as based on over 20 years of experiences by the authors.

2 Characteristics of legged locomotion

Legged locomotion is the basis for several different types of movement such as walking, running, and jumping. Walking and running, in which the body is carried well off the surface on which a body is moving (substrate), occur only in arthropods and vertebrates. Running (cursorial) vertebrates are characterized by elongated lower legs and feet and by reduction and fusion of toes. Saltatory locomotion, movement by leaping, hopping, or jumping, is found in a number of insects and vertebrates.

Only arthropods (e.g., insects, spiders, and crustaceans) and vertebrates have developed a means of rapid surface locomotion. In both groups, the body is raised above the ground and moved forward by means of a series of jointed appendages, the legs. Because the legs provide support as well as propulsion, the sequences of their movements must be adjusted to maintain the body's center of gravity within a zone of support; if the center of gravity is outside this zone, the animal loses its balance and falls. It is the necessity to maintain stability that determines the functional sequences of limb movements, which are similar in vertebrates and arthropods. The apparent differences in the walking and slow running gaits of these two groups are caused by differences in the tetrapodal (four-legged) sequences of vertebrates and in the hexapodal (six-legged) or more sequences of arthropods. Although many legs increase stability during locomotion, they also appear to reduce the maximum speed of locomotion. Whereas the fastest vertebrate gaits are asymmetrical, arthropods cannot have asymmetrical gaits, because the movements of the legs would interfere with each other.

The cycle of limb movements is the same in both arthropods and vertebrates. During the propulsive, or retractive, stage, which begins with footfall and ends with foot liftoff, the foot and leg remain essentially stationary as the body pivots forward over the leg. During the recovery, or protractive, stage, which begins with foot liftoff and ends with footfall, the body remains essentially stationary as the leg moves forward. The advance of one leg is a step; a stride is composed of as many steps as there are legs. During a stride, each leg passes through one complete cycle of retraction and protraction, and the distance that the body travels is equal to the longest step in the stride. The speed of locomotion is the product of stride length and duration of stride. Stride duration is directly related to retraction: the longer the propulsive stage, the more time is required to complete a stride and the slower is the gait. A gait is the sequence of leg movements for a single stride. For walking and slow running, gaits are generally symmetrical—i.e., the footfalls are regularly spaced in time. The gaits of fast-running vertebrates, however, tend to be asymmetrical—i.e., the footfalls are irregularly spaced in time.

For example, the different gaits of insects are based on the synchrony of leg movements on the left (*L*) and right (*R*) sides of the animal. The wave of limb movement for each side passes anteriorly; the posterior leg protracts first, then the middle leg, and finally the anterior leg, producing the sequence *R*3 *R*2 *R*1 or *L*3 *L*2 *L*1. There is no limb interference, because the legs of one side do not have footfalls along the same longitudinal axis. The slowest walking gait of insects is the sequence *R*3 *R*2 *R*1 followed by the sequence *L*3 *L*2 *L*1. As the rate of protraction increases, the protractive waves of the right and left sides begin to overlap. Eventually, the top speed is reached when the posterior and anterior legs of one side move synchronously. This gait occurs because the protraction times for all legs are constant, the intervals between posterior and middle legs and between middle and anterior legs are constant, and the interval between posterior and anterior legs decreases with faster movements. Other gaits are possible in addition to those indicated above by altering the synchrony between left and right sides.

The limb movements of centipedes and millipedes follow the same general rules as those of insects. The protraction waves usually pass from posterior to anterior. Because each leg is slightly ahead of its anteriorly adjacent leg during the locomotory cycle, one leg touches down or lifts off slightly before its anteriorly adjacent one. This coordination of limb movement produces metachronal waves, the frequency of which equals the duration of the complete protractive and retractive cycle. The length of the wave is directly proportional to the phase lag between adjacent legs.

Cursorial (running) vertebrates are characterized by short, muscular upper legs and thin, elongated lower legs. This adaptation decreases the duration of the retractive–protractive cycle, thereby increasing the animal's speed. Because the leg's cycle is analogous to the swing of a pendulum, reduction of weight at the end of the leg increases its speed of oscillation. Cursorial mammals commonly use either the pace or the trot for steady, slow running. The highest running speeds, such as the gallop, are obtained with asymmetrical gaits. When galloping, the animal is never supported by more than two legs and occasionally is supported by none. The fastest runners, such as cheetahs or greyhounds, have an additional no-contact phase following hind foot contact.

In cursorial birds and lizards, both of which are bipedal, the feet are enlarged to increase support and the body axis is held perpendicular to the ground, so that the center of gravity falls between the feet or within the foot-support zone. The running gait is, of course, a simple alternation of left and right legs. In lizards, however, bipedal running must begin with quadrupedal (four-footed) locomotion. As the lizard runs on all four legs, it gradually builds up sufficient speed so that its head end tilts up and back, after which it then runs on only its two hind legs.

The structure of the legs can be very different among different types of animals in term of anatomy but the kinematic design and functioning can be recognized with common characters that can be summarized in having a foot with a space mobility of at least three d.o.f.s and an ovoid-like trajectory of foot reference point with the step size S and step height H, as shown in Fig. 1. In particular, such a kinematic design of a leg can be characterized by a spherical joint S at the hip articulation and revolute joints R at knee and ankle articulations. The step size gives the capability of the locomotion motion, while the step height H gives indication of overpassing obstacles. Both parameters can be increased on purpose during operation, but H is the main characteristic of a leg that makes legged systems very flexible, mainly in environments crowded with obstacles.

The foot of a leg has the function of contact and interaction with the ground in order to provide a proper action and reaction in locomotion, both for force exchange and stable positioning. The peculiarities of foot anatomy and operation are also significant in differentiating and specializing animals for the specific environments in which they usually operate. The anatomy of

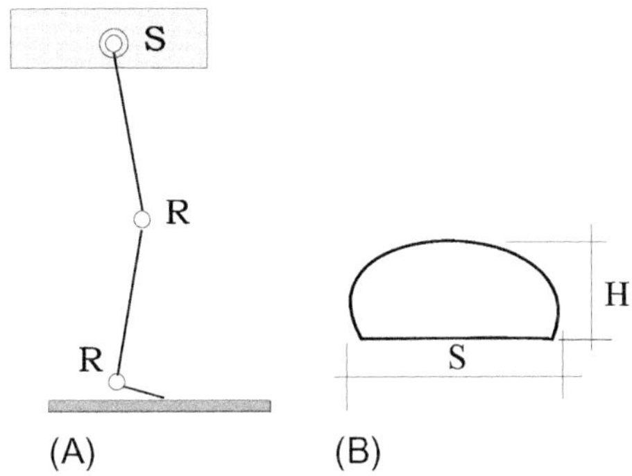

Fig. 1 Main characteristics of a anthropomorphic leg: (A) Kinematic design; (B) foot point path.

legs in nature can be summarized with a bone structure with joint articulations, a complex system of muscles, and a complex neurological system.

In replicating leg anatomy and functionality in walking robots, bone structure is often used as a reference structure for kinematic design with motion properties for smooth operation, payload capability, and nature-like actions.

Design problems for leg mechanisms can be outlined by tackling the mechanics of robots, as in ref. [9]. Mechanical design aspects can be formulated by using models like the one in Fig. 2, in attaching the following problems:

- design compactness and light weight
- motion synchronization for step size and lift height
- balancing actions and dynamics response
- ground-foot contact and impact
- actuation and forward velocity

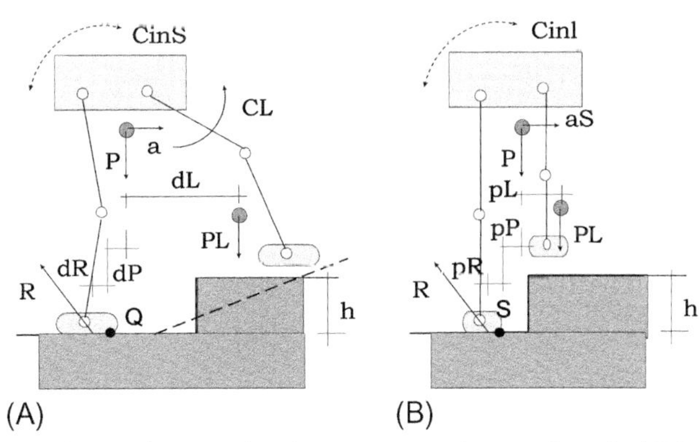

Fig. 2 A scheme for conditions in leg design overpassing an obstacle: (A) Sagittal plane; (B) front view.

- motion planning (also for obstacle avoidance)
- sensored and controlled interactions during walking
- payload capacity and environment interaction.

The problems for overcoming obstacle can be formulated with conditions referring to the schemes in Fig. 2, when referring to climbing a step as or for obstacle overpass.

The static equilibrium gives the necessary conditions for walking and overcoming obstacles of the maximum allowed step height. In a sagittal plane the equilibrium can be expressed referring to Fig. 2A as

$$\begin{aligned} & R_h - P\frac{a}{g} - PL\frac{a_L}{g} \geq 0 \\ & R_v - P - PL \geq 0 \\ & R_v dR - PdP - PL(dP + dL) - C_{inS} \geq 0 \end{aligned} \tag{1}$$

where R_h and R_v are the horizontal and vertical components of R contact reaction; P is robot weight; PL is leg weight; a is motion acceleration and g is the gravity; dL, dR, and dP are the indicated distances; CL is the torque actuating a leg; C_{inS} is the sagittal component of the inertial torque due to waist balancing movement. Point Q is considered the foot contact point about which the system will rotate in a possible fall. In the front plane, the equilibrium condition can be expressed referring to Fig. 2B as

$$\begin{aligned} & R_l - P\frac{aS}{g} \geq 0 \\ & R_v - P\frac{a_S}{g} \geq 0 \\ & R_v pR - PpP - PL(pP + pL) - C_{inl} \geq 0 \end{aligned} \tag{2}$$

where R_l is the lateral component of R; a_S is the lateral acceleration of robot body; C_{inl} is the lateral component of the inertial torque of waist balancing movement. Point S is the foot contact point about which the system will rotate in a possible fall. The components R_h and R_1 refer to friction actions at the foot contact area. By using Eqs. (1), (2) conditions can be formulated for design and operation features that are useful to overcome obstacles of height h. From geometric viewpoints the obstacle/step of height h can be overpassed when the leg moves with the condition

$$l_1(1 - \cos\phi_1) + l_2(1 - \cos\phi_2) > h \tag{3}$$

in which l_1 and l_2 are the lengths of leg body links, whose angles ϕ_1 and ϕ_2 are given with respect to a vertical line. The design problem for overpassing obstacles can be formulated by using conditions like those in Eqs. (1)–(3) to properly size the leg links, and to give proper mobility ranges and actions of the leg operation.

3 Existing legged locomotion systems

The first ideas to design and implement legged locomotion vehicles date back to very old times. Since ancient times devices have been designed for transportation of passengers and materials, often with the use of animals. In addition to load transportation (including carrying passengers), there have been challenges for locomotion purposes, with attempts to replicate and/or mimic solutions in nature, as seen with a large variety of animals. Transportation systems were developed mainly as wheeled systems that are today very successful in vehicle technology. Readers can refer to corresponding literature for the history of vehicles, like cars, trucks, motorcycles and so on.

With the development of Robotics, locomotion systems and walking machines have received stronger and stronger attention, with a large number of inventions and designs mainly within the last three decades. But most of the concepts and even structural designs have evolved, even unconsciously, from past solutions. Historical studies have been published on the evolution of walking machines in general, also within the history of robotics, whose literature is reported with basic references, like for example in refs. [10–12].

In modern times, two leading researchers Ichiro Kato and Miomir Vukobratovic achieved pioneering works in the field of legged robots in early 70s. Namely, Kato and his team at Waseda University, Tokyo, in Japan achieved the first anthropomorphic walking robot, WABOT 1, which was demonstrated in 1973, as reported in ref. [13]. In parallel, Vukobratovic and his team at the Mihailo Puppin Institute, Belgrade, Yugoslavia, designed the first active legged exoskeletons together with the related Zero Moment Point theory, as celebrated in ref. [14].

In the recent past several walking machines have been developed for several different purposes mainly in research laboratories. Significant examples of walking machines are shown in Figs. 3–13:

- Honda robot ASIMO, Fig. 3A;
- Sony robot SDR-4X, Fig. 3B;
- Waseda robot WABIAN-RV, Fig. 3C;
- Waseda biped locomotor WL-16R, Fig. 4;
- CSIC robot RIMHO2, Fig. 5A;
- Ambulatory Robotic Lab. robot Scout II, Fig. 5B;
- Hirose & Yoneda Robotic Lab. robot TITAN VIII, Fig. 6A;
- Intelligent Machines and Special Robotics Institute robot WorkPartner, Fig. 6B;
- Chiba University COMET II, Fig. 7A;

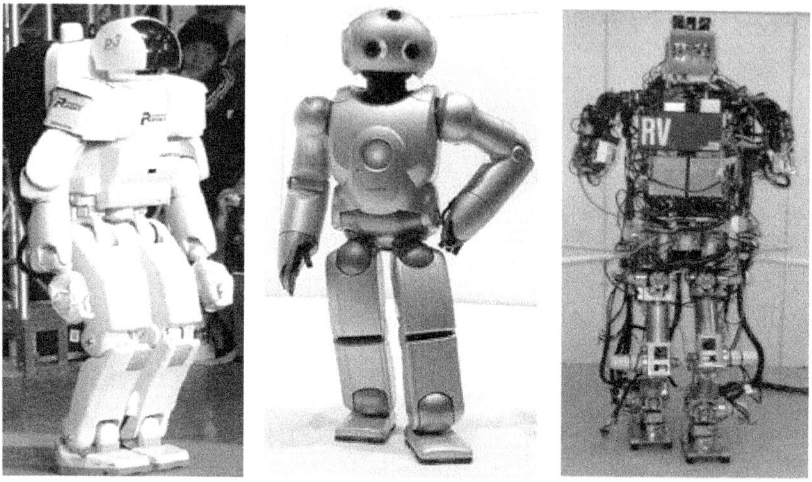

(A) (B) (C)

Fig. 3 Examples of current biped walking machines: (A) Honda robot ASIMO; (B) Sony Robot SDR-4X; (C) Waseda robot WABIAN-RV.

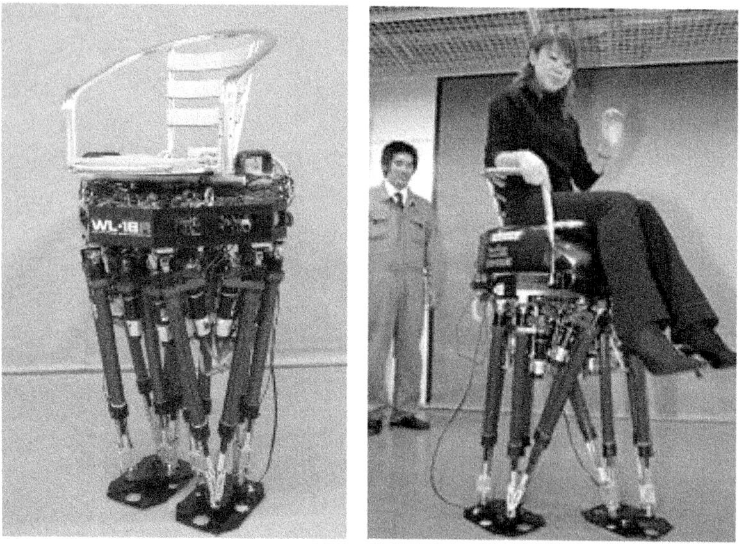

(A) (B)

Fig. 4 Waseda biped locomotor WL-16R with non anthropomorphic legs: (A) a side view; (B) carrying a human.

(A) (B)

Fig. 5 Examples of four-leg walking machines: (A) RIMHO2; (B) Scout II.

(A) (B)

Fig. 6 Examples of four-leg walking machines with wheels: (A) TITAN VIII; (B) WorkPartner.

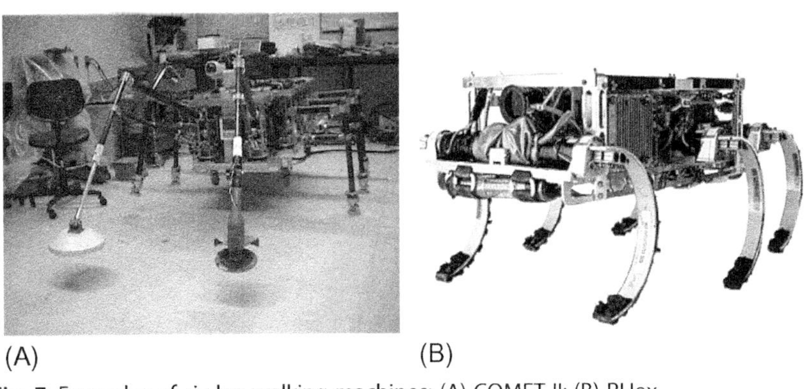

(A) (B)

Fig. 7 Examples of six-leg walking machines: (A) COMET II; (B) RHex.

(A) (B)

Fig. 8 Examples of six-leg walking machines for outdoor applications: (A) Walking forest machine; (B) Adaptive Suspension Vehicle.

Fig. 9 The low-cost four-leg robot AIBO.

- Ambulatory Robotic Lab. robot Rhex, Fig. 7B;
- Plustech Ltd. Walking Forest Machine, Fig. 8A;
- Ohio State University Adaptive Suspension Vehicle, Fig. 8B;
- Sony robot AIBO, Fig. 9.

ASIMO (Advanced Step in Innovative MObility), Fig. 3A, has been built by Honda company in Japan in the year 2000. It is a biped humanoid robot with 26 actuate degrees of freedom. Its size, weight and ranges of mobility have been designed to mimic as much as possible a human child and to move

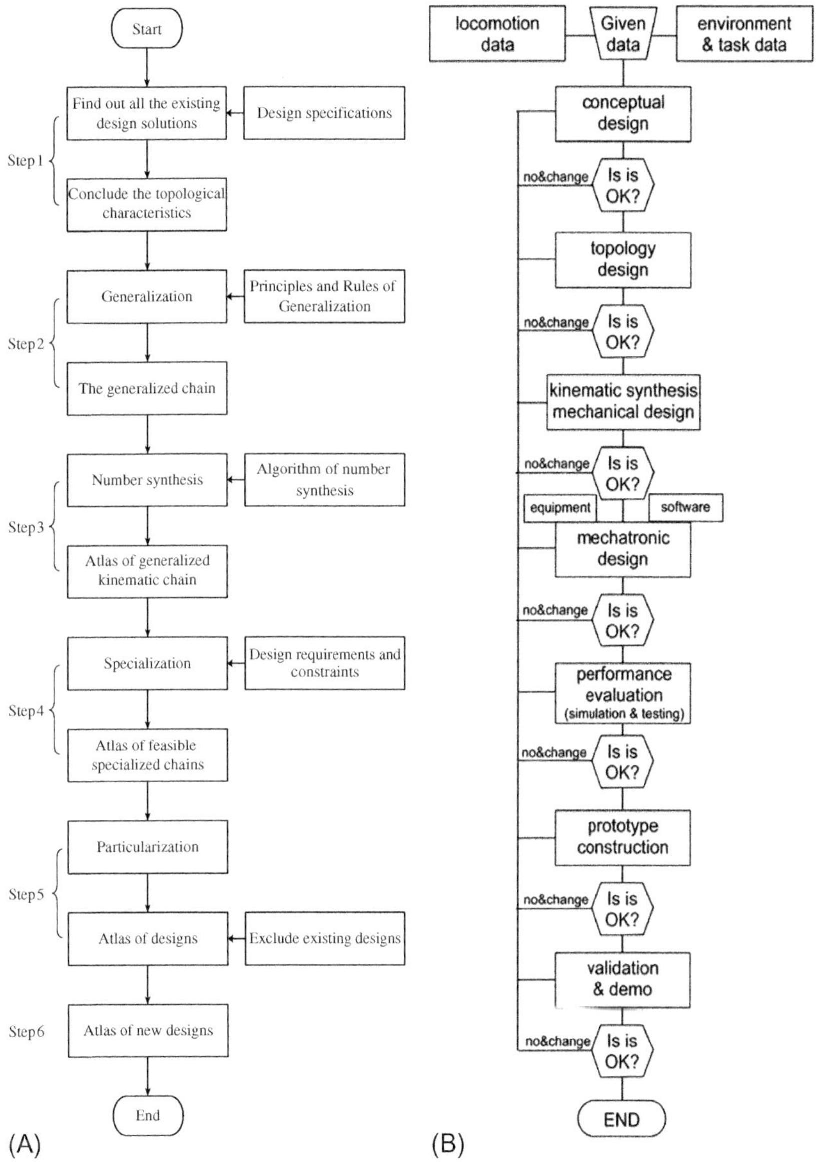

Fig. 10 Flowcharts for design procedure of locomotion legged systems: (A) a topology search design; (B) a general scheme.

freely within human living environments. ASIMO is able to ascend and descend stairs, to walk by following different patterns, to avoid obstacles, to grasp objects, to interact with humans by means of sound and image recognition. It is equipped with on board batteries for a continuous operating time of about 30 min.

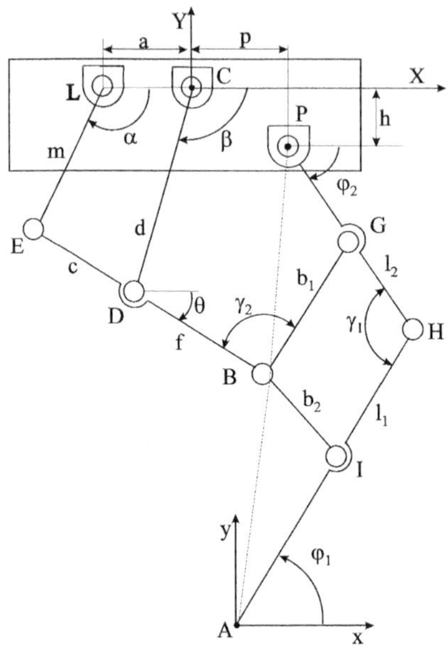

(A)

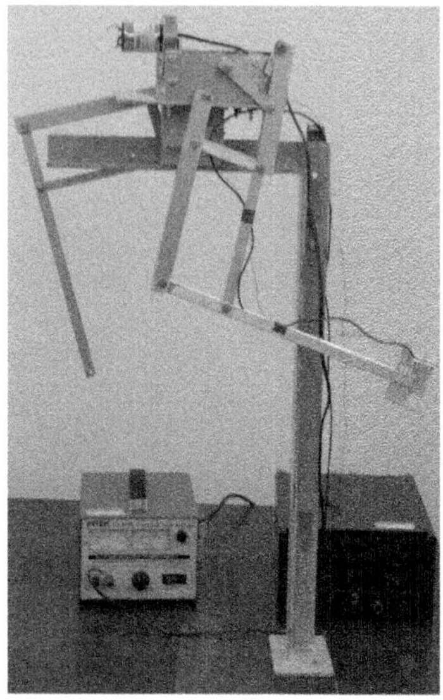

(B)

Fig. 11 The so-called Chebyshev-linkage leg developed at LARM: (A) a first prototype; (B) a kinematic diagram.

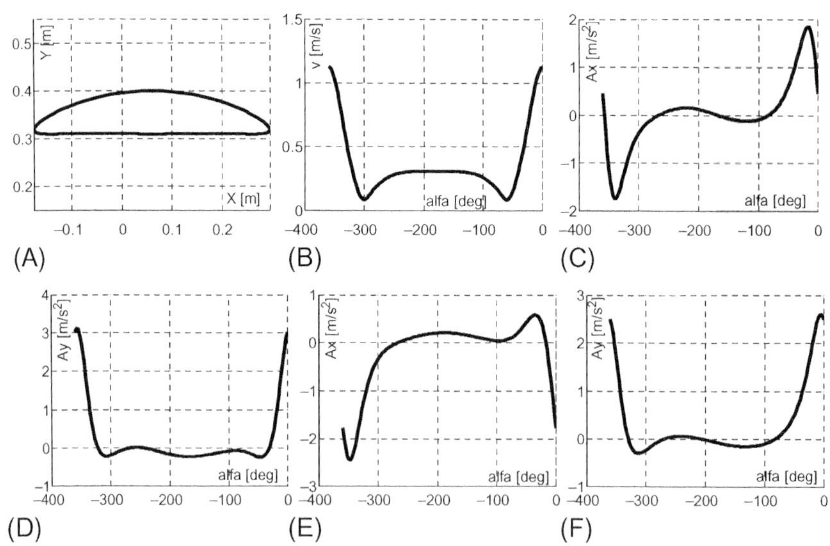

Fig. 12 Simulation results for the walking characteristics of the 1-DOF leg in Fig. 11 with $p = 20$ mm and $h = -30$ mm: (A) Point C trajectory; (B) Point A velocity; (C) acceleration a_{AX}; (D) acceleration a_{AY}; (E) acceleration a_{Ax}; (F) acceleration a_{Ay}.

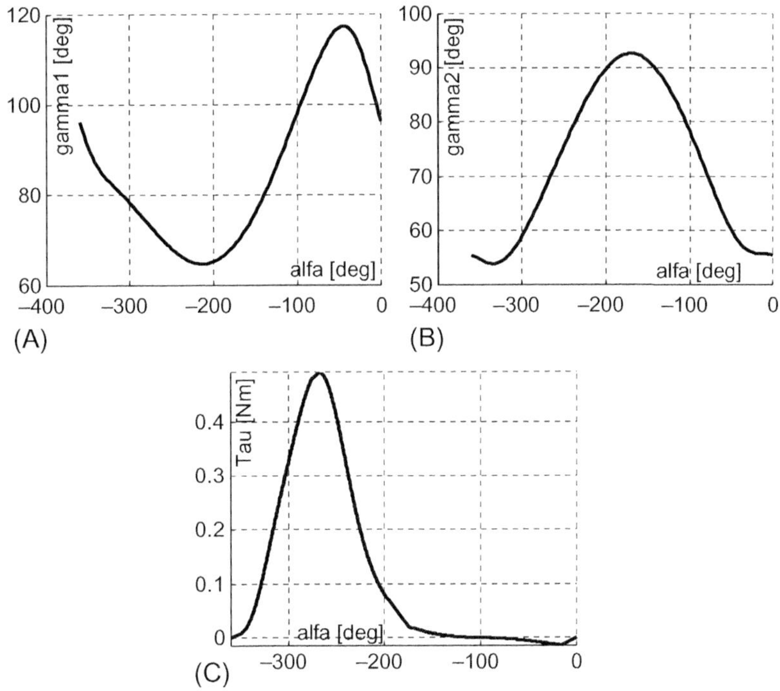

Fig. 13 Simulation results for the walking characteristics of the 1-DOF leg in Fig. 11 with $p = 20$ mm and $h = -30$ mm: (A) transmission angle γ_1; (B) transmission angle γ_2; (C) actuating torque τ.

SDR-4X (Sony Dream Robot version 4X), Fig. 3B, has been built at Sony in the year 2002. It is a small biped humanoid robot that has been designed for entertainment purposes. It has a total of 38 degrees of freedoms. It can walk on irregular pavement (up to 10 mm) and tilted surface (up to 10°) and can prevent falling when an external pressure is applied. Image and sound recognition features have been also included in its capacities. It is equipped with on board batteries for a continuous operating time of 2 h approximately.

WABIAN-RV (Waseda Biped humANoid Refined V), Fig. 3C, is a biped humanoid robot that has been developed for human-robot cooperation work at Waseda University, in Tokyo in the year 2002. It is the last version of WABIAN series started since 1972 [15]. It has a total of 43 dofs. The size and motion range of each link has been designed to be as human like as possible. A variety of walking modes are operated as dynamic forward and backward walking, marching in place, dancing, carrying a load, and emotional walking. Its software includes an on-line pattern generator, image and sound recognition features. It requires an external power supply.

WL-16R (Waseda-Leg No.16 Refined), Fig. 4, has been designed at Waseda University in Tokyo in the year 2004. It is composed of two 6-dof legs with parallel architecture design. It is a bipedal robot with only lower limbs that can dynamically walk independently. Its upper body can be added by users according to their use purposes. In particular, it would be applicable to the welfare field as a walking wheelchair or as a walking support machine that is able to walk up and down stairs carrying or assisting a human. It is equipped with an on-board Nickel Metal Hydride battery for a continuous operating time of 1 h approximately.

The RIMHO II walking robot, Fig. 5A, has been developed from the Industrial Automation Institute-CSIC and the CIEMAT in Madrid since 1993. It is a quadruped-walking machine of the insect type. Its four legs are based on a three-dimensional Cartesian pantograph mechanism. The RIMHO walking robot can perform both discontinuous and wave gaits over irregular terrain including slopes and stairs and has been tested also over natural terrZain.

SCOUT II, Fig. 5B, has been developed at Ambulatory Robotic Laboratory in Montreal since 1998. It is composed of four legs. Each leg has one active degree of freedom only. A spring and a passive knee are added in order to provide two additional passive degrees of freedom for each leg. These passive degrees of freedom make the Scout II capable of achieving dynamic running similar to gallop and trot. SCOUT II is fully autonomous having on board power, computing and sensing. Other features include an on-board pan-tilt camera system and laser sensors.

TITAN VIII, Fig. 6A has been built at Hirose & Yoneda Robotic Lab. in Tokyo in the year 1996. TITAN VIII is a walking machine having four legs. The leg mechanism is composed of a planar two degrees of freedom mechanism and a rotating mechanism which rotates this planar mechanism. Wires and pulleys are used for the power transmission within the leg. The feet of TITAN VIII can be used also as wheels in order to achieve faster motion on flat surfaces.

WorkPartner, Fig. 6B, is a four leg mobile robot that has been built at the Intelligent Machines and Special Robotics Institute in Helsinki in 2000. The locomotion system of WorkPartner is hybrid. In fact, it is possible to move by means of legs only, with legs and wheels powered at the same time or with wheels only. WorkPartner is equipped with two arms having three degrees of freedom arms and a two-degree of freedom camera head. Several sensors have been installed on board such as potentiometers, force sensors, inclinometers, gyro, accelerometers, ultrasonic sensors, laser scanner. A combustion engine and four batteries are also installed on board for a continuous operating time of 30 min approximately.

COMET II in Fig. 7A has been developed at Chiba University in Tokyo in 2002. It can be used as fully autonomous system or teleoperated by a human for demining tasks. It is equipped with two manipulators that are used for mine detection and grass cutting. Several sensors are installed on board such as metal detector, radar, cameras, force sensors, potentiometers. The implemented software includes obstacle avoidance features. Power supply is provided by a gasoline power generator for outdoor operation or from an external power supply for indoor laboratory tests.

RHEx robot, Fig. 7B, has been developed at Ambulatory Robotic Lab. in Montreal since 1999. It has six legs with only one degree of freedom. The leg has a very simple design. It is made by heat shaped Delrin rods and it has a soft foot at its free end. RHEx robot is equipped with gyros accelerometers, and optical encoders. It is capable of achieving a wide variety of dynamically dextrous tasks, such as walking, running, leaping over obstacles, climbing stairs. Two batteries are on board installed for a continuous operating time of about 10 min.

The Walking Forest Machine, Fig. 8A, has been developed at Plustech Ltd. since 1995 for outdoor forest harvesting tasks. It is composed of six articulated legs. It can move forward, backward, sideways and diagonally. It can also turn in place and step over obstacles. Depending on the irregularity of the terrain, the operator can adjust both the ground clearance of the machine and height of each step. The operator-friendly controls are incorporated in a joystick that controls direction of movement, traveling speed, step height and gait, and the ground clearance.

Adaptive Suspension Vehicle, Fig. 8B, has been developed at Ohio State University since early 80s. It is composed of six articulated legs. Each leg has three active degrees of freedom. It has been designed for walking on rough terrains by carrying a maximum load of over 2000 N. It is equipped with gyros, laser sensors and a computer vision system that is used for adapting the gait to the environment. It can work either in teleoperated or operator-on-board mode by using active compliance control algorithms.

AIBO robot in Fig. 9 is a four-leg robot for entertainment purposes. It has been developed by Sony since early 90s and is available on the market since 2003 at a price of about 2000 Euros. Its design has four legs with three degrees of freedom each. It is equipped with two microphones, a speaker, touch sensors, infrared sensors, accelerometers that provide AIBO with very high human-robot interaction capabilities. Rechargeable batteries are on board installed allowing for a continuous operating time of 1.5h, approximately.

The reported examples in Figs. 3–9 give an overview of the variety of walking systems that have been developed all around the world with different solutions for different applications. It is worth noting that most of them (with the exception of AIBO) are not yet available in the market but they are under further development in Research Labs.

Generally, legged systems can be slow and more difficult to design and operate with respect to machines that are equipped with crawlers or wheels. But legged robots are more suitable for rough terrain, where obstacles of any size can appear. In fact, the use of wheels or crawlers limits the size of the obstacle that can be climbed, to half the diameter of the wheels. On the contrary, legged machines can overcome obstacles that are comparable with the size of the machine leg. Therefore, hybrid solutions that have legs and wheels at the same time have been also developed as shown for example in Fig. 6A and B. This type of walking machines may range from wheeled devices to true walking machines with a set of wheels. In the first case, the suspensions are arms working like legs to overcome particularly difficult obstacles. In the second case wheels are used to enhance the speed when moving on flat terrain.

4 Design considerations for legged locomotion systems

Challenging design problems concerning traditional subjects for overall robot design can be specifically indicated for locomotion legged systems in

- mechanical efficiency, for mechanical transmission and actuating power
- light design, with large payload and limited size
- static accuracy, for better grasping configurations
- dynamic response, for controlling impulsive actions and balanced dynamics

In addition, new problems can be identified for new solutions in

- topological mechanism structures, for new enhanced designs
- materials, for better mechanical design and environment interaction
- tribology issues, for reduction of wear and longer accurate functioning with limited friction
- energy sustainable solutions, for better attention to energy saving and recycling of wasted components

The short above lists are aimed to indicate that new issues and reconsiderations of past experiences, even with more subjects, will be the challenges in the future mechanism design for legged mobile robots within increased mechatronic components.

Future developments and more widespread use of walking machines may depend on the enhancement of leg design in the many mechatronic aspects that can be approached with new perspectives and technologies. Challenges and trends for designing leg mechanisms will be focused on improvements mainly in compact design, efficiency, payload capability, flexibility and environment impact and they can be attached mainly in the following topics:

- topology of mechanism structures
- formulation for high-speed computation
- light compact mechanical design
- interactions with environments and users or operators
- advanced sensorization
- adaptable control systems
- smart materials and components
- anthropomorphic and nature-inspired solutions for friendly user-oriented solutions

The above considerations emphasize expected advances in traditional performance, but perhaps new applications will be invented for walking machines as a function of the new technology that will be made available in the future. In addition, future developments and applications will generate the rise of new problems and new emerging topics from other fields. A final aspect to be considered can be considered not only nature-like designs but developments beyond standard solutions.

As per kinematic design of leg mechanisms, a design procedure may include not only the traditional dimensional synthesis for given data referring to motion requirements, as not only for foot point and body. Traditional techniques of mechanism design (synthesis) can be used when formulating the design problem properly as motion guiding problem, as reported for example in refs. [16, 17].

A specific topology search procedure can be proposed for the development of a conceptual design by looking to all the feasible mechanism structures as summarized in Fig. 10A, consisting of six steps as in ref. [18]. Step 1: Finding out all the existing design solutions that can fully satisfy required design specifications for locomotion and tasks and identifying the topological characteristics of these existing designs. Step 2: Selecting one of these existing solutions and transforming it into its corresponding generalized chain according to predetermined rules of generalization. Step 3: By using an algorithm for number synthesis, synthesizing an atlas of generalized chains that have the same number of links and joints as the generalized chain obtained in step 2. Step 4: By using a suitable algorithm of specialization, assigning links and joints to each generalized chain generated in step 3 in order to obtain an atlas of all the feasible specialized chains that satisfy the design specifications and constraints. Step 5: Particularizing each feasible specialized chain obtained in step 4 into its corresponding mechanical design to get an atlas of mechanical devices. Step 6: Identifying and excluding existing designs from the atlas of mechanical devices obtained in step 5, to get an atlas of new feasible designs satisfying required design specifications.

The above design considerations and challenges can be summarized in a general design procedure like the one in Fig. 10B, in which the mechanical design of a locomotion legged system is a key aspect since the mechanical nature of the locomotion, although the design solution is an integrated solution of several other components with a mechatronic structure and operation.

In particular, a design procedure can be started with a proper deep analysis and identification of the locomotions aims and the characteristics of the environment and tasks for which the system is designed. The conceive of a conceptual design may be outlined form different activity, including previous experiences and looking at the existing solutions. This wìcan be considered a starting point of a new design with challenges in matching the prescribed requirements and expected results, even beyond the given data. For this purpose, it is convenient to look for an optimized solution that can be obtained from several specific approaches. Form the authors' experience the topology search can be considered useful to investigate all the possible solution that can be derived from the conceptual design also with the possibility to find new design ideas. This is part of the creative design activity that can give challenging results in identifying solutions that are beyond just the given data. Once the topology structure is chosen for the design developments the other activities can be carried out sequentially, as indicated in the flowchart in Fig. 10B with traditional or innovative procedures. Thus, the dimensional design can be worked out using the algorithms of kinematic synthesis and then the mechanical structure can be shaped in all its components with CAD designs and simulations. After having the mechanical structure defined the legged system can be completed with the equipment and corresponding software making it with the necessary abilities and flexibility. Thus, in this phase the sensors, control devices, computer facilities and other equipment as required for the task and environment can be chosen or even designed on purpose for the locomotion system under development. The complete design of a legged locomotion system as a robot system can be developed with a mechatronic design in which all the components and aspects are integrated in giving the necessary features of motion, sensing, reactintin and so on. After the design is defined a check of the performance can be worked out through simulation with numerical results and also through testing activities to characterize the built prototype. The construction of the prototype may require adjustments and manufacturing problems up to finalize the design process with a validation giving the characteristic data of the solution and usually a demo is used both to confirm practically the obtained results and to exhibit the prototype for possible applications and exploitations.

In the flowchart of Fig. 10B it is indicated that after each phase a check is planned to verify the design progress and in case changes are required in the previous phases as in an iterative process towards an optimal solution that will consider all the possibilities for a final improved solution. In each of these phases, peculiarities of the attached problems and used approaches can lead to different solutions by looking both to the mechanics of locomotion and mechatronic design of a solution.

5 Illustrative examples

Illustrative examples are reported from the direct experience of the authors with systems that they developed in the last two decades. At LARM locomotion legged systems were conceived, designed, and tested by looking at mechanical robotic design with features of low-cost solutions and easy operation. In the following main solutions are discussed to show those experiences with the aim to indicate the experience in developing legged locomotion systems.

5.1 Biped robot with Chebyshev-linkage legs

Basic considerations for a low-cost leg design can be outlined as referring to the facts that the leg should generate an approximately straight-line trajectory for the foot point with respect to the body; the leg should have an easy robust mechanical design; and it should have the minimum number of DOFs to ensure the motion capability. At LARM the so-called Chebyshev-linkage leg has been developed with the above features as in ref. [19].

The designed leg mechanism is shown in Fig. 11 with its mechanical design that is based on a Chebyshev four-bar-linkage, a five-bar linkage, and a pantograph mechanism. For such a mechanism, the leg motion can be performed by using one actuator only. The leg has been designed by considering compactness, modularity, light weight, reduced number of DOF as basic objectives to achieve the walking operation.

Numerical and experimental results show that proper walking features can be obtained when points C and P in Fig. 11 are not coincident. The main characteristic of the proposed leg design consists in a fully-rotative actuation at point L to obtain the suitable trajectory of point B with one motor only that run continuously without any speed regulation. Furthermore, the trajectory of point B, and consequently, point A can be suitably modified by changing the design parameters. In particular, better features can be obtained if the transmission angles γ_1 and γ_2 have suitable values.

Dimension of the leg prototype in Fig. 1 are 400 mm high, 40 mm × 250 mm so that the leg has a maximum lift of 80 mm and the step is of 470 mm.

In Figs. 12 and 13 the basic operation features of the Chebyshev-pantograph leg mechanism are reported in terms of simulation results to show the feasibility of the low-cost easy operation design that have been experienced successfully by using a commercial DC motor without motion control equipment, since the capability of full rotatability of the crank link m.

The computed walking motion sequences of the single DOF biped robot in Fig. 11B with the leg mechanism in Fig. 11A in a biped walking gait are shown in Figs. 14 and 15. Fig. 14 shows the computed motion sequence of the biped robot in sagittal and horizontal planes where the dashed line represents the left leg and the solid line represents the right leg. The right foot grasps the ground and the left leg swings in the air. Two configurations are shown in the figure with actuation angle $\alpha 1 = 270°$ and $\alpha 1 = 90°$. In Fig. 14B the triangles represent the right foot and the rectangles represent the left foot in the horizontal plane. A black circle represents that the foot grasps the ground and the corresponding leg is in the propelling phase, otherwise it represents the foot in the air and the leg is in a non-propelling phase. A logic flowchart of the walking gaits for the biped robot is listed in Fig. 15 as per motion programming purposes with the two leg mechanisms operating sequentially in propelling and non-propelling phases to obtain a proper forward motion of the biped robot.

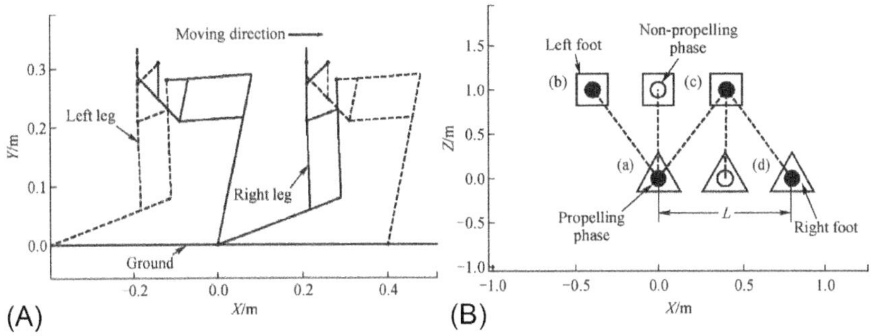

Fig. 14 Motion sequences of walking for biped in Fig. 11B: (A) in sagittal plane; (B) in horizontal plane.

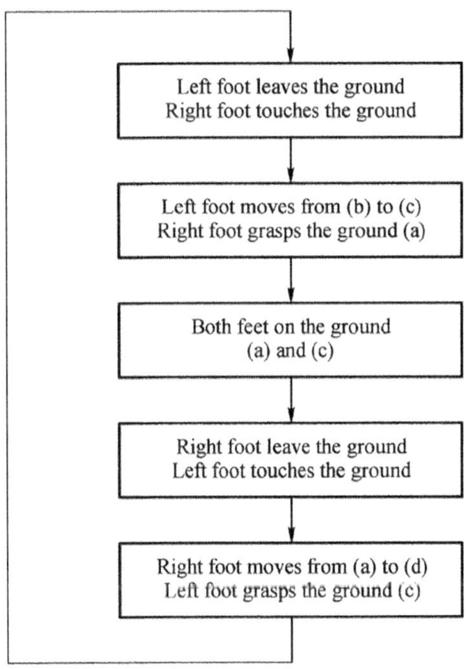

Fig. 15 Flowchart for the walking gait of the biped robot in Fig. 11B as referring to Fig. 14.

Fig. 16 shows the snapshots of the walking sequence of the biped robot prototype during a lab test. Experiment results show that the biped robot walks just like a "drunk-man." But with a step length L that is almost equal to the dimension of the leg mechanism.

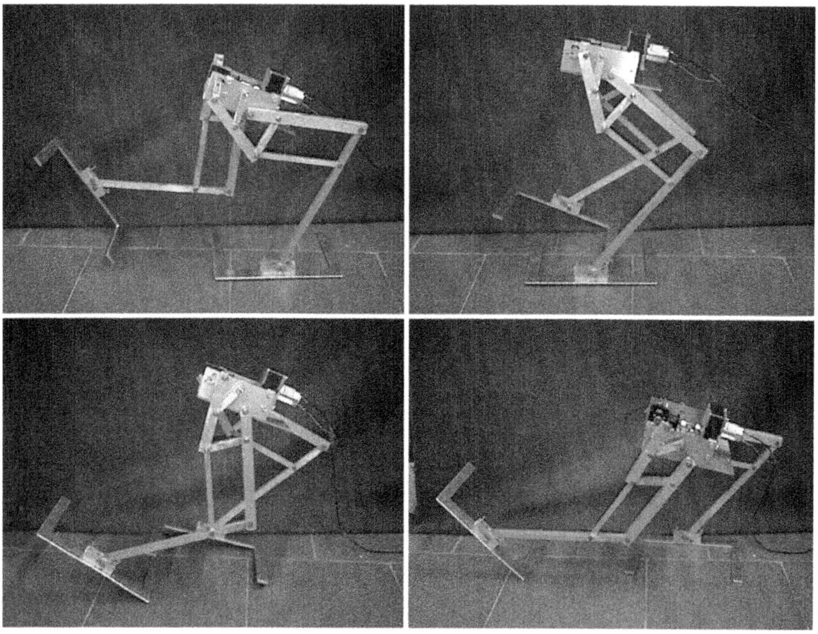
Fig. 16 Snapshots of a walking test by the LARM biped prototype of Fig. 11B.

5.2 LARM hexapod series

At LARM a modular anthropomorphic leg has been designed by defining a single link module that can be easily connected with other modules and can have inside all the needed actuators, transmissions and sensors as in ref. [20]. Fig. 17 shows the proposed design for a single link module by using conic gears and timing belt transmissions. The main components of a single link module are:
- the body of the module;
- a dc motor with reduction gear train;
- two conic gears or a timing belt transmission;
- two mechanical switches.

The link modules can be also properly oriented with respect to the others in order to achieve the required pitch, jaw or roll motions. A link module can be also easily modified in order to drive a wheel in the foot.

Dimension of a built prototype leg that is composed by three modules and one wheel in the foot, is high 500mm and has a cross-section of 60mm × 60mm, Fig. 17 with a capability of a maximum lift of 155mm and a step of 310mm when each joint can rotate +/− 90°. The leg module is the basis for the design of hexapode with low-cost user-oriented features that has been developed since 2000, [20].

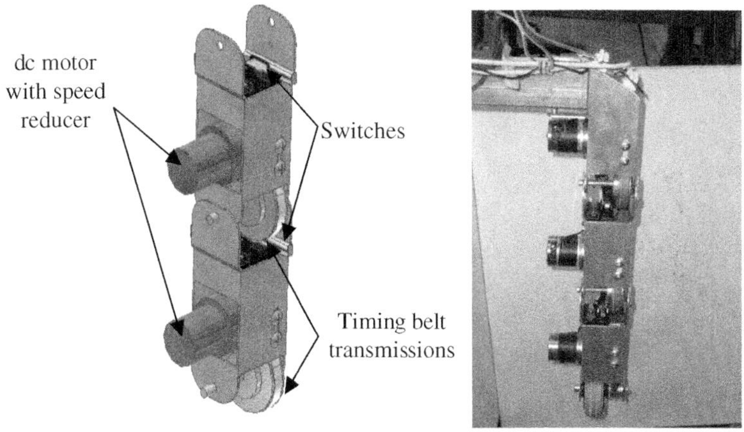

(A) (B)

Fig. 17 The LARM modular anthropomorphic leg: (A) the mechanical design; (B) a built prototype.

LARM Hexapod family is a series of hybrid legged-wheeled mobile robots that has been designed and built with three version as reported in Fig. 18, [21]. Main characteristics of the LARM Hexapod series are based on the combination of legs and wheels as well as the use of low-cost control architectures to achieve user-friendly solutions. LARM Hexapods have been applied for different inspection operations in non-accessible places. The programming of walking is designed from the analysis of elementary actions to control the operation of the actuators by using signals by suitable switches for the leg mobility [22]. An example of the versatility of the hexapode and its motion planning is shown in Fig. 18 as referring on the obstacle overpassing through a suitable leg motion (Fig. 19).

(A) (B) (C)

Fig. 18 The LARM hexapod robots: (A) version I, (B) version II, (C) version III.

Fig. 19 Snapshots of locomotion of LARM Hexapod versIIIbis in an obstacle overcoming.

5.3 HeritageBot platform with parallel mechanism legs

HeritageBot Platform is a design solution, which integrates a legged mobile robot and a drone module in order to get a robotic platform that has mobility capability both on terrain and aerial areas and the possibility to be equipped with sensors and instrumentation for applications in Cultural Heritage frames, Fig. 20A, as reported in ref. [23].

The HeritageBot Platform has been designed with three modules Fig. 20B: the first module is designed for the control and operation equipment (including batteries and communication hardware) including specific sensors and instrumentation that are needed by the users for specific tasks; the second module is a quadcopter drone system for small flight capability to help avoiding obstacles and to increase payload/stability capacity; the third module is a tripod parallel architecture as locomotion walking system with main features for a very high payload and a wide step range. With such a design, HeritageBot Platform is able to operate in narrow spaces, in presence of obstacles comparable with the HeritageBot platform size while avoiding high pressures or damages on the operation surface.

A final prototype of HeritageBot Platform, as version III, has been built at LARM as a proof-of-concept device, Fig. 21, whose overall cost has been limited to less than 10,000 Euros. Commercial components were purchased for the control and operation hardware, batteries, propellers, actuators, and cables and connectors. Main frame of first module and all the remaining structural components have been manufactured via 3D printing The overall size is contained in a box of 50 × 50 × 50 cm with a weight of 5.0 kg when equipped, including batteries of 2.0 kg for an operation duration of 2 h while walking at about 250 m/h and flying at less than 1 m height with a tilting capability of 55°. The prototype is equipped with a sonar for obstacle detection, thermal and barometer sensors for environment monitoring, and a telecommunication of 80 m.

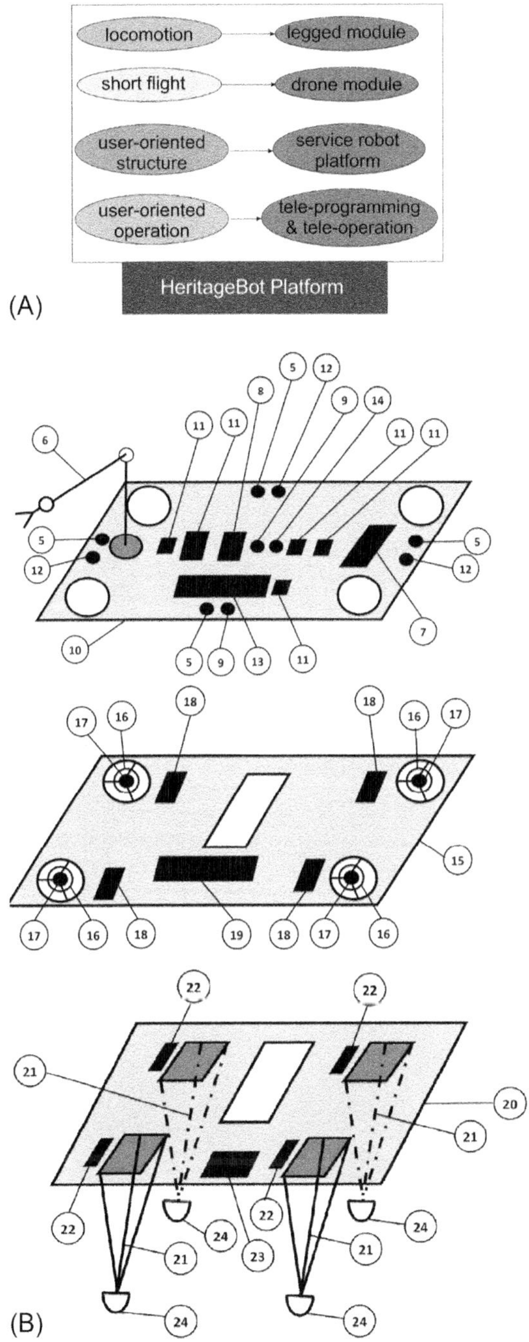

Fig. 20 The conceptual design of the HeritageBot Platform: (A) a scheme design; (B) a modular design.

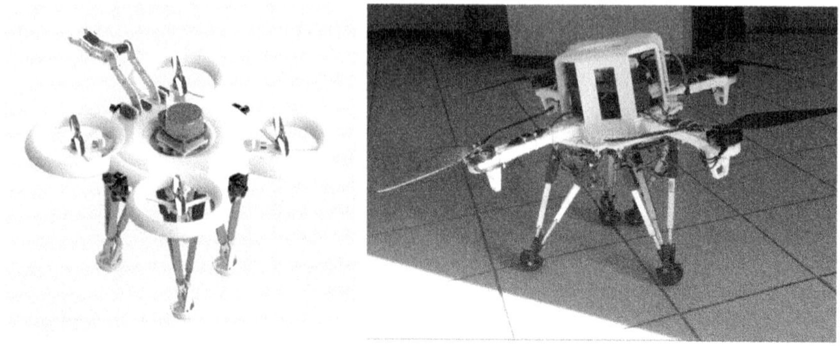

(A) (B)

Fig. 21 A prototype design of HeritageBot Platform: (A) a CAD design; (B) a built unit.

The leg structure is capable of a large walking step as depending of the stroke of the linear actuators that ensure also a high payload with stiff behavior. As indicated in the modular design, the legged locomotor is fully equipped on board with the necessary control units and sensors by leaving specie for additional equipment as necessary for the application. The flight module is installed on top of the central plate with independent control units and sensors. It is designed for small flight that can increase the locomotion capability in over passing obstacles and can increase the payload capacity of the platform when used as floating unit.

Fig. 22 shows an outdoor test of the HeritageBot Platform in a typical motion operation combining walking with a small flight [24]. The combination of walking and small flight makes the system useful in many applications where a single capability cannot be sufficient, beside giving more possibility of motion.

(A) (B)

Fig. 22 An outdoor test of the prototype of HeritageBot platform III: (A) outdoor walking operation; (B) small flight operation.

5.4 Exoskeleton leg mechanism

Particular legged systems are the exoskeleton devices that are recently under development for several applications like motion assistance, training exercising, and rehabilitation therapies.

In order to develop a low-cost exoskeleton with a fairly simple mechanism structure it can be convenient to simplify and adapt existing solutions with linkage architectures. At LARM linkage solutions have been designed as based on a pantograph and Chebyshev linkage for efficient reproduction of human walking with fairly easily manufactured robust systems and simple operation features [25]. A pantograph is useful for human walking reproduction since it can amplify properly an input motion from the body frame to the foot point. A Chebyshev linkage can be a proper choice for producing an input "human-like" path of ovoid shape with high mechanical efficiency and fairly simple motion control. In Fig. 23 examples of experienced mechanism designs for leg exoskeleton are shown as based on the above combination of a pantograph with a Chebyshev four-far linkage. The solution in Fig. 23A has a high number of links and joints and the ankle joint is missing. Nevertheless, this exoskeleton mechanism shows a good stability during walking. The mechanism design in Fig. 23B has the actuation link (1) on the front and no actuation mechanism on ankle joint, but it cannot permit to place a human subject between exoskeleton legs and the actuator, although the driving link can be placed in the back side. A feasible solution can be identified as in Fig. 23C, with a knee F-joint.

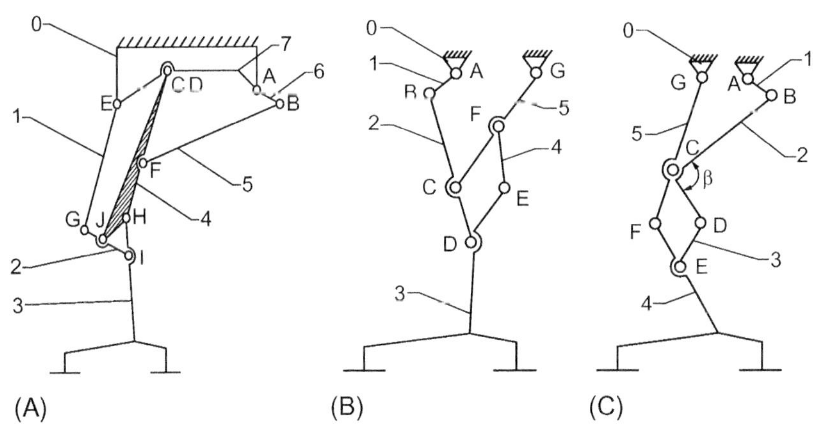

Fig. 23 Solutions of leg mechanisms with pantograph: (A), (B) a new solution; (C) the LARM design.

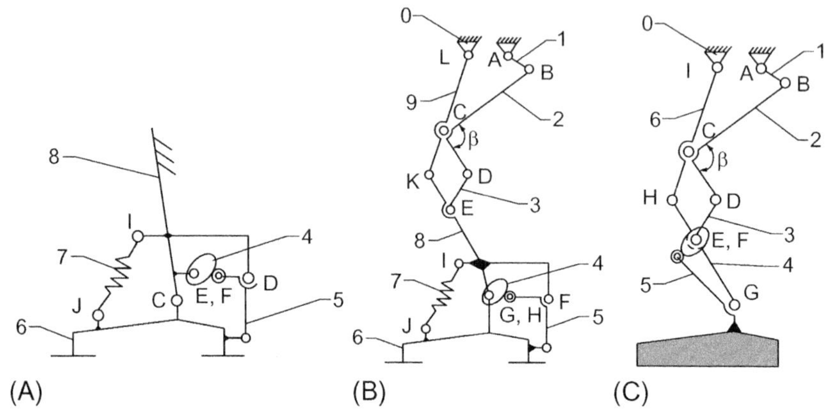

Fig. 24 A design of a leg exoskeleton mechanism with ankle joint: (A) the ankle cam mechanism; (B) the overall solution; (C) the LARM design.

For the leg exoskeleton solutions in Fig. 23 it is necessary only one actuator that can be placed in joint A on body 0 (the frame). The links can be made adjustable in order to adapt a solution to any human body. It can be observed that the ankle joint for these solutions is neglected due to the impossibility to place a mechanism near the ankle region, as in most of the leg exoskeletons. However, for the ankle joint an equivalent mechanism can be proposed as a cam mechanism, as in the design scheme in Fig. 24 with four links, seven revolute joints, and one cam mechanism.

Previous leg mechanisms in Figs. 23 and 24B are based on pantograph and Chebyshev mechanisms, and they have been considered a starting point for developing a solution for a human locomotion assisting system, even in a topology search design procedure. The LARM leg exoskeleton mechanism is presented in Fig. 23C with the additional solution for articulating the motion of ankle joint in coordination with knee and hip motions. The cam mechanism has been placed with a joint at the point E with the possibility to choose any point on the shank link by using a proper cam profile to perform an angular motion as much as similar to the human one.

Another LARM leg exoskeleton mechanism has been developed for a specific lower limb design [26]. Most of the currently developed lower extremity exoskeletons are bulky and have limited torque and power making the exoskeletons not fully portable, especially for paralyzed human subjects and their rehabilitation. Therefore, a mechanism solution can be a convenient solution for a lower limb exoskeleton with three degrees-of-freedom that can be actuated by electric motors, one rotational servomotor

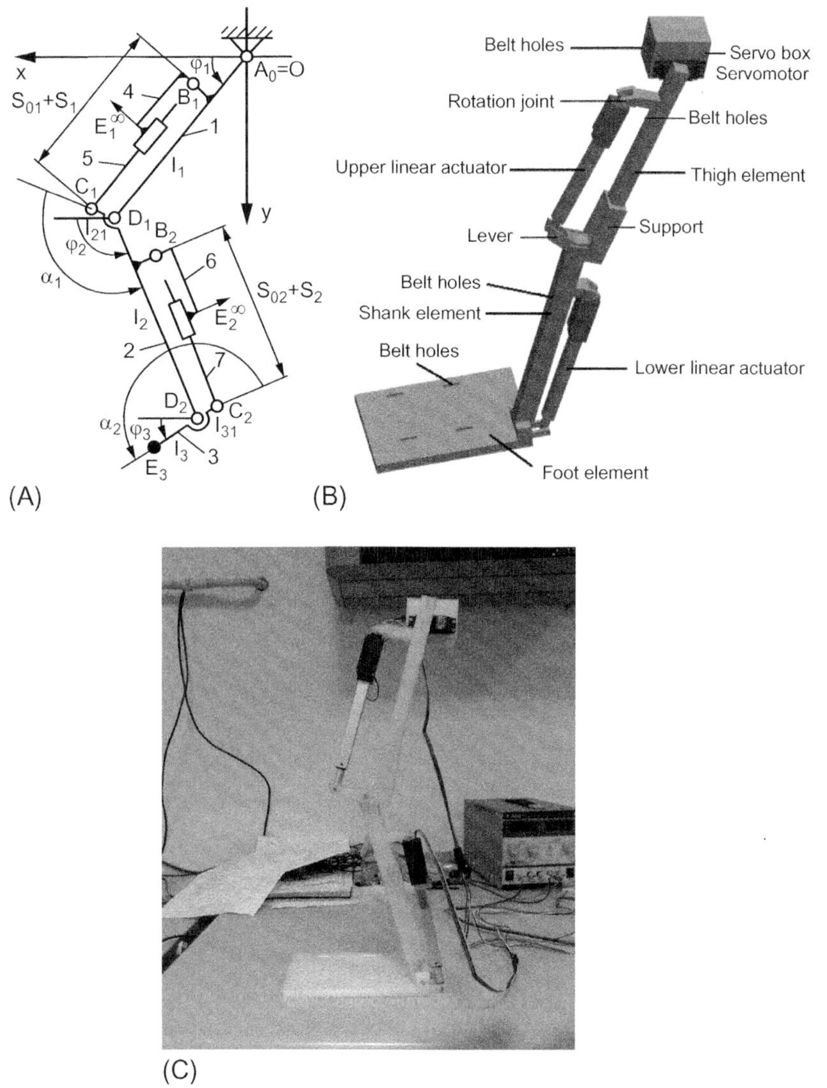

Fig. 25 The LARM lower limb exoskeleton: (A) a kinematic design; (B) a CAD design; (C) a prototype.

for the hip joint and two linear actuators for the leg links. The kinematic design is shown in Fig. 25 where the whole exoskeleton weight is sustained by a support belt that is located at the waist of the user.

The mechanism design is actuated by two linear actuators that are installed in parallel inverted slider-crank chains between thigh and shank,

shank and foot respectively as in the schemes in Fig. 25A and B. The translational motion of the linear actuators provides the rotation of the knee and foot joint. The built prototype in Fig. 25C is a scaled model of a leg exoskeleton, which was assembled by using market components and 3D printing manufacturing. For the actuation of the knee and ankle joints two Firgelli L16 linear actuators were used, with a maximum stroke of 100 mm.

6 Conclusions

This chapter presents legged locomotion systems as based on mechanism design by surveying existing solutions and indicating design issues as from the experiences of the authors with a variety of solutions and applications. The large variety of mechanism solutions is illustrated also by examples that refer to a design procedure that has been outlined by considering the characteristics of locomotion and features for user/task-oriented design and operation.

References

[1] Oxford Dictionary, Locomotion, Oxford University Press, Oxford, UK, 2019.
[2] IFToMM, special issue. Standardization and terminology, Mech. Mach. Theory 38 (2003) 7–10.
[3] M. Ceccarelli, E.M. Kececi (Eds.), Designs and Prototypes of Mobile Robots, In: ASME Press Robotics Engineering Book SeriesASME, 2015.
[4] J. Liu, M. Tan, X.G. Zhao, Legged robots-an overview, Trans. Inst. Meas. Control. 29 (2) (2007) 185–202.
[5] A. Morecki, K.J. Waldron, Human and Machine Locomotion, Springer, New York, 1997.
[6] B. Siciliano, O. Khatib (Eds.), Springer Handbook of Robotics, Springer, Cham, 2016.
[7] G. Carbone, M. Ceccarelli, Legged robotic systems, in: Cutting Edge Robotics, Intech, Wien, 2005, pp. 553–576.
[8] E.M. Kececi, M. Ceccarelli (Eds.), Mobile Robots for Dynamic Environments, In: ASME Press Robotics Engineering Book SeriesASME, 2015 ISBN: 9780791860526.
[9] M. Ceccarelli, Fundamentals of Mechanics of Robotic Manipulation, Kluwer/Springer, Dordrecht, 2004.
[10] E. Bautista Paz, M. Ceccarelli, J. Echavarri Otero, J.J. Munoz Sanz, A brief illustrated history of machines and mechanisms, in: Science and Engineering, Book Series on History of Machines and Machine Science, vol. 10, Springer, Dordrecht, 2010.
[11] M. Ceccarelli, A historical perspective of robotics toward the future, Fuji Int. J. Robot. Mechatronics 13 (3) (2001) 299–313.
[12] M.E. Rosheim, Robot Evolution, Wiley, New York, 1994.
[13] I. Kobrinski et al., (Ed.), First CISM-IFToMM ROMANSY (Udine 5-8 September 1973), Springer-Verlag, Wien, 1974.
[14] M. Vukobratović, B. Borovac, Zero-moment point—thirty five years of its life, Int. J. Humanoid Rob. 1 (1) (2004) 157–173.

[15] S. Hashimoto, A. Takanishi (Eds.), Humanoid Robotics Institute at Waseda University, Waseda University Publishing Office, Tokyo, 2008.
[16] C. Lopez-Cajùn, M. Ceccarelli, Mecanismos, Trillas, Mexico City, 2010.
[17] J.E. Shigley, G.R. Pennock, J.J. Uicker, Theory of Machines and Mechanisms, McGraw-Hill, New York, 2015.
[18] T. Li, M. Ceccarelli, A topology search for a new LARM leg mechanism, in: Proceedings of IFToMM-FeIbIM International Symposium on Mechatronics and Multibody Systems MUSME 2011, Editorial Polytechnic University of Valencia, Valencia, 2011, pp. 77–94.
[19] C. Liang, M. Ceccarelli, Y. Takeda, Operation analysis of a Chebyshev-Pantograph leg mechanism for a single DOF biped robot. Front. Mech. Eng. 7 (4) (2012) 357–370, https://doi.org/10.1007/s11465-012-0340.
[20] G. Carbone, M. Ceccarelli, A low-cost easy-operation hexapod walking machine, Int. J. Adv. Robot. Syst. 5 (2) (2008) 161–166.
[21] F. Tedeschi, G. Carbone, Design issues for hexapod walking robots, Int. J. Robot. 3 (2) (2014) 181–206.
[22] G. Carbone, A. Shrot, M. Ceccarelli, Operation strategy for a low-cost easy operation Cassino hexapod, Appl. Bionics Biomech. 4 (4) (2007) 149–156.
[23] M. Ceccarelli, D. Cafolla, M. Russo, G. Carbone, HeritageBot platform for service in cultural heritage frames. Int. J. Adv. Robot. Syst. 15 (4) July 1.(2018). https://doi.org/10.1177/1729881418790692.
[24] M. Ceccarelli, D. Cafolla, M. Russo, G. Carbone, Prototype and testing of HeritageBot platform for service in cultural heritage, in: M. Ceccarelli et al., (Ed.), New Activities for Cultural Heritage, Springer International Publishing AG, 2017, pp. 104–112.
[25] C. Copilusi, M. Ceccarelli, G. Carbone, Design and numerical characterization of a new leg exoskeleton for motion assistance, Robotica 33 (2015) 1147–1162.
[26] C. Iancu, M. Ceccarelli, E.-C. Lovasz, Design and lab tests of a scaled leg exoskeleton with electric actuators. in: Advances in Service and Industrial Robotics—Proceedings of the 26th International Conference on Robotics in Alpe-Adria-Danube Region RAAD 2017, Springer, 2017, pp. 719–726, https://doi.org/10.1007/978-3-319-61276-8_76.

Further reading

[27] M. Ceccarelli, Mechanism design for robots, in: Proceedings of the 11th IFToMM International Symposium on Science of Mechanisms and Machines (SYROM'13 Brasov), Springer, Dordrecht, 2013, pp. 1–8.
[28] M. Ceccarelli, Leg mechanisms, (Chapter 1).in: M. Ceccarelli, E.F. Kececı (Eds.), Designs and Prototypes of Mobile Robots, ASME Press Robotics Engineering Book SeriesASME, 2015, pp. 1–21.

CHAPTER 2

Gait analysis and regeneration by means of principal component analysis and its application to kinematic design of wearable walking assist device for hemiplegics

Daisuke Matsuura, Yuta Chounan, Masaki Omata, Yusuke Sugahara, Yukio Takeda
Laboratory of Mechanical Systems Design, Department of Mechanical Engineering, School of Engineering, Tokyo Institute of Technology, Tokyo, Japan

1 Introduction

Approximately 1.18 million patients suffer from stroke in Japan in the year 2014. Stroke is one of the major reasons for elderly to need nursing care. After stroke, patients may have possibility to become hemiplegia. In order to recover such situation, rehabilitation is quite important. In the development of practical and wearable assist device for walking rehabilitation, mechanism design of such device which can perform cooperative movement of limbs of human (patient) with allowable costs should be done. To this end, a systematic methodology to reflect the characteristics of human locomotion on the design of such mechanism is demanded.

In our previous works, walking assist devices using crutches for paraplegics and hemiplegics have been proposed, and structural design and motion synthesis for several gaits for step ascending and descending gaits as well as straight walking gaits on a horizontal floor, have been carried out [1–5].

As a design tool of such assistive devices, systematic procedure for structural design and motion synthesis using reference motions obtained from healthy subjects would be helpful. In this article, a method for analysis of motion obtained by motion capture experiments and regeneration based on the principal component analysis are introduced. As an example of the application of the systematic method, a design and motion synthesis of a

Design and Operation of Human Locomotion Systems
https://doi.org/10.1016/B978-0-12-815659-9.00002-0

wearable walking assist device for hemiplegics which has functions of changing step length and walking direction on the flat floor will be described.

2 Gait analysis and generation based on principal component analysis

2.1 Gait measurement using motion capture system

As the first step of human gait analysis, motion of the gait should be digitized and corrected into a computer. For this purpose, optical motion capture systems are widely used these days because of their ease of use and scalability. Fig. 1 illustrates a typical setup of 3D-motion capture system. A number of cameras are surrounding a measurement volume in which human subjects' or any other moving objects' motion can move. The human subjects are wearing a number of spherical target markers that are covered with retroreflective material. Each camera has a ring-shape infra-red (IR) LED illumination around its lens so that target markers in its field of view can be easily found since projected IR illumination is reflected toward the light source, and other in-coming lights can be easily filtered out. Recorded position of target markers are projected onto each camera's screen coordinate then transmitted to a host computer. The host computer is in charge of correcting positions of target markers from all cameras, and reconstructing them into a 3D-coordinate frame. In order to carry out this 3D reconstruction, all cameras' positions against a specific world coordinate should be calibrated. This process can be done by using specific calibration procedures according to each motion capture system, typically using L-shaped- or cubic-calibrators that have a couple of target markers of which distances are pre-calibrated. In addition, T-shaped wand is also used to compensate for distortions of each camera's optical system within the working volume.

In addition to the target marker's position, other measurement information such as ground reaction forces (GRF) measured by force plates can be simultaneously corrected. In order to synchronize a lot of frame grabbing by cameras and additional measurement signals, trigger signal of control box is commonly used.

2.2 Gait analysis based on principal component analysis

Corrected information by above-mentioned motion capture system such as time trajectory of target markers and ground reaction force can be analyzed by using mathematical analysis methods. Key point of such methods is how effective and easy they are in finding proper explanatory variables. Since

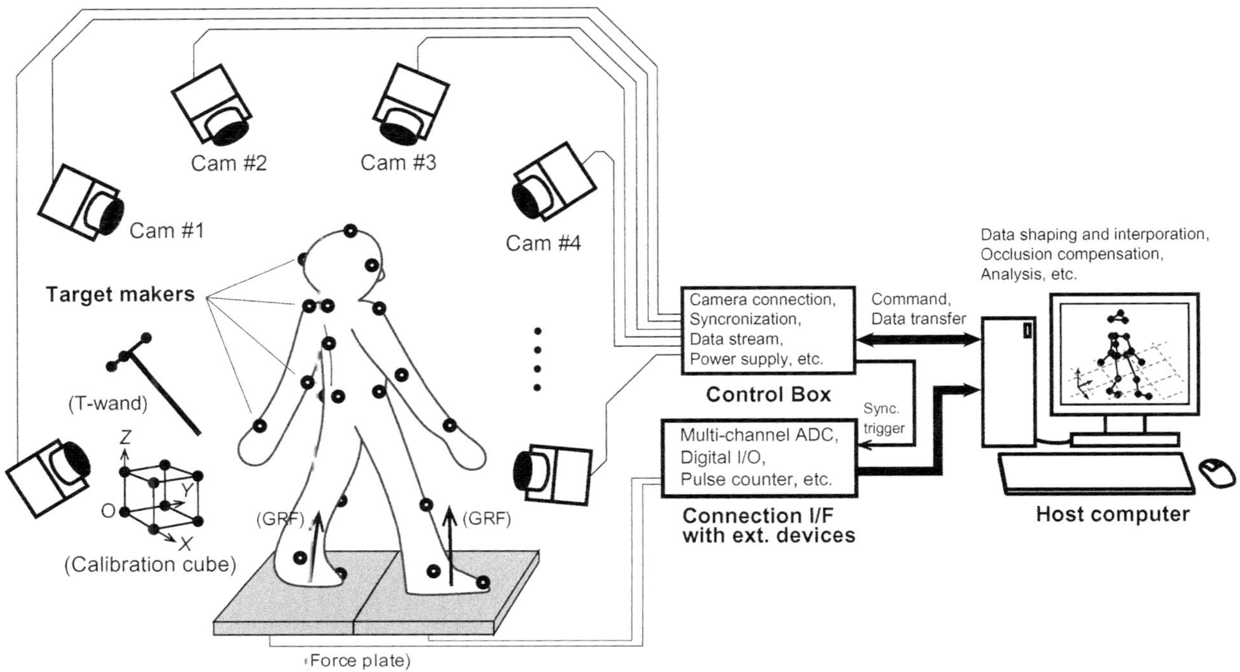

Fig. 1 Composition of optical motion capture system.

human's gait is a complex phenomenon, purpose of the measurements is valid such like extracting personal characteristics (according to one's age, sex, body height or weight, disease, etc.,) motion pattern characteristics (walking, turning, running, etc.,) change of motion due to the effects of wearing robotic devices, etc., In order to achieve this, effectiveness in reducing the number of explanatory variables while taking various measured information into account is also important. When one is aiming to utilize the analysis method also for the motion generation of robotic assistive devices and prosthesis, the method should work both in direct direction and inverse direction.

Principal Component Analysis (PCA) is one of the latest and most effective methods satisfying above requirements. Mathematically, it is a kind of orthogonal linear transformation to transform the database matrix Z into a new coordinate frame of which 1st axis is aligned to the most significant explanatory component. This can be achieved by obtaining a set of eigenvectors and eigenvalues. The eigenvectors are called *principal components*, and the eigenvalues are called *factor loadings* (or *factor scores*). Fig. 2 shows a basic flow of data analysis and synthesis using PCA. First of all, a database matrix Z is constructed by arranging a number of N component vectors (say, N is a number of trials), $\boldsymbol{z}_{n\ (n=1,\ \ldots,N)}$, which has p particular arguments. Since each of the p columns of Z can be different kind of information such as position, force, and so on, each of the n rows should be normalized (z-scored). After constructing the database matrix, order reduction, and extraction of explanatory vectors and variables are performed.

As the result, Z can be written as a dot product of principal component scores $\boldsymbol{t}_n$ and factor loadings $\boldsymbol{w}_m$ as $\boldsymbol{t}_n = \boldsymbol{z}_n\ \boldsymbol{w}_m$, for each $n = 1, \ldots, N$, $m = 1, \ldots, M$, where M is the number of principal components.

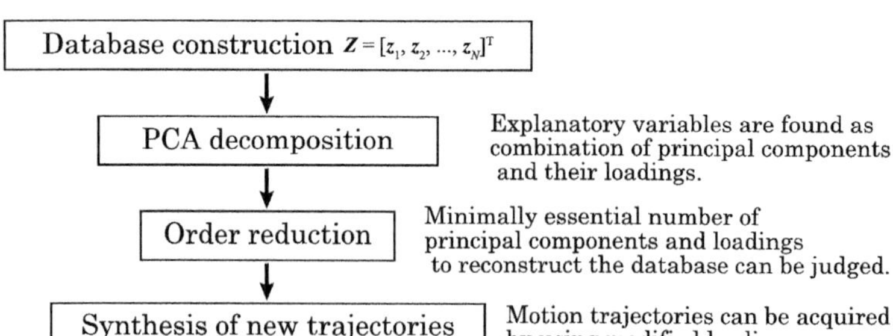

Fig. 2 Flowchart of principal component analysis.

This first component can be found by solving the following maximizing problem:

$$\boldsymbol{w}_1 = \arg\max\left\{\sum_{n=1}^{N}(t_1)_n^{\ 2}\right\} = \arg\max\left\{\boldsymbol{w}^{\mathrm{T}} Z^{\mathrm{T}} Z\boldsymbol{w}\right\},\ \text{while}\ \|\boldsymbol{w}\| = 1. \quad (1)$$

Since the constraint $\|\boldsymbol{w}\| = 1$, Eq. (1) can be written as

$$\boldsymbol{w}_1 = \arg\max\left\{\frac{\boldsymbol{w}^{\mathrm{T}} Z^{\mathrm{T}} Z\boldsymbol{w}}{\boldsymbol{w}^{\mathrm{T}}\boldsymbol{w}}\right\} \quad (2)$$

A standard result for a positive semidefinite matrix, $Z^{\mathrm{T}}Z$, is that the quotient's maximum possible value is the largest eigenvalue of the matrix, which occurs when $\boldsymbol{w}$ is the corresponding eigenvector. In this way, first component $\boldsymbol{w}_1$ can be found. After that, further principal components can be found recursively. In this time, database $\boldsymbol{Z}$ should be rewritten as

$$\hat{Z}_m = Z - \sum_{s=1}^{m-1} Z\boldsymbol{w}_s\boldsymbol{w}_s^{\ \mathrm{T}}, \quad (3)$$

which represents the remaining part of the database independent from the first principal component. Further principal components can be obtained by using rewritten Eq. (1) as

$$\boldsymbol{w}_m = \arg\max\left\{\frac{\boldsymbol{w}^{\mathrm{T}}\hat{Z}_m^{\ \mathrm{T}}\hat{Z}_m\boldsymbol{w}}{\boldsymbol{w}^{\mathrm{T}}\boldsymbol{w}}\right\} \quad (4)$$

As the result, full principal components decomposition can be given as

$$T = ZW, \quad (5)$$

where W is a p by p matrix whose columns are the eigenvectors of $\mathrm{Z}^{\mathrm{T}}\mathrm{Z}$. This decomposition can also reversely work as

$$Z = TW^{\mathrm{T}}\left(\text{since } W \text{ is an orthogonal matrix, } W^{-1} = W^{T}\right). \quad (6)$$

By modifing each component of W^{-1} from original ones, one can synthesize new (modified) data set. When a meaningful relationships between demanded features and principal components can be found, enhancement or decline of that feature can be done by modifying relating factor loadings.

2.3 Numerical example (design of assistive force trajectory for squatting motion)

As a numerical example of the motion analysis based on PCA and its application to design of motion assistance strategy, multiple human subjects' lower-limb torques while squatting motion was analyzed, and modified motion to calculate suitable assistive body-lifting torque was synthesized. First of all, multiple subjects' squatting motion was recorded by using an optical motion capture system (MAC3D system with 10 of EAGLE IR-cams, Motion Analysis Corp.). Ground reaction force was also recorded simultaneously by using force plates. During each of the iteration of trial motions, body-lifting force of constant (but different in each trial) strength was applied through a harness fixed to the subject's trunk. Experimental setup and subject's schematic model are shown in Fig. 3A and B. From

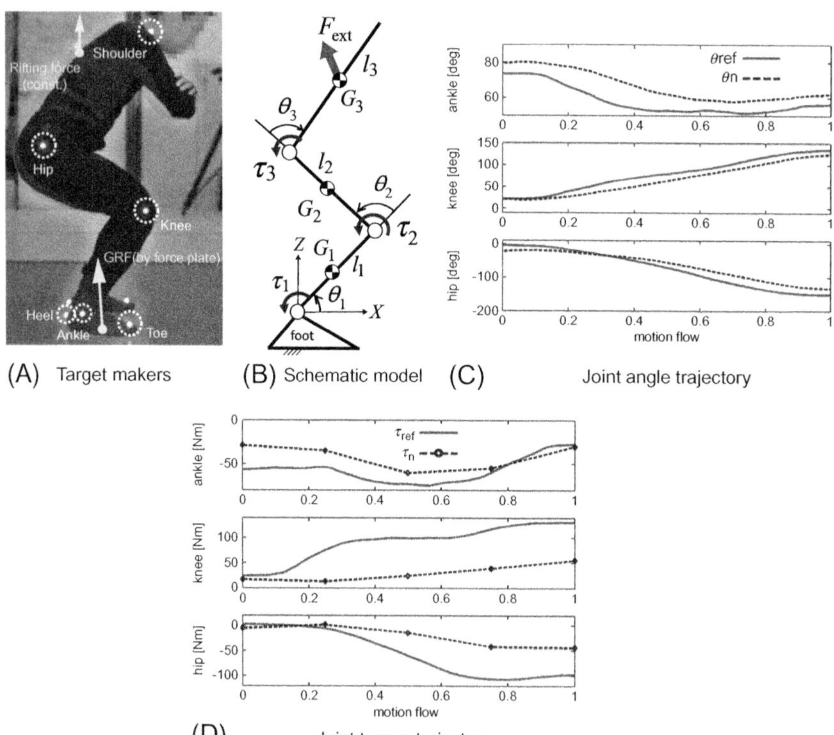

Fig. 3 Setup and acquired results of a squatting motion capture experiment. (A) Target makers, (B) schematic model, (C) joint angle trajectory, (D) joint torque trajectory.

acquired results of N-times trials, a database matrix Z was constructed as followings:

$$Z = \begin{bmatrix} z_1 \\ \vdots \\ z_N \end{bmatrix}, \text{in which } z_i = \left[\theta_{1,1}\ \theta_{2,1}\ \theta_{3,1}\ \cdots\ \theta_{1,100}\ \theta_{2,100}\ \theta_{3,100}\ \tilde{\tau}_1\ \tilde{\tau}_2\ \tilde{\tau}_3\ \tilde{F}_X\ \tilde{F}_Y\right], \qquad (7)$$

where $\theta_{j,\tilde{t}}$, $\tilde{\tau}_j$, and $\tilde{F}_{X,Y}$ are the joint angles of the j-th joint in normalized time, $1 \leq \tilde{t} \leq 100$, mean joint torques of the j-th joint, and lifting force's horizontal and vertical components, respectively (one of those variables' trajectories are plotted in Fig. 3C and D with a subscript "ref"). The joint torques and lifting force were normalized based on the length of each subject's leg length and body weight. After carrying out the PCA decomposition, factor loadings corresponding to the three joint torques, $\tilde{\tau}_{1,2,3}$, at five representative time frame, $\tilde{t} = 0, 0.25, 0.5, 0.75, 1.0$, are obtained as shown in Fig. 4A. From this figure, it can be seen that 1st principal component is the most dominant factor for the knee joint torque. Namely, synthesis of a suitable lifting force trajectory to reduce knee joint torque can be done by carrying out inverse PCA decomposition with a modified factor loading matrix, in which the factor loading for 1st principal component is increased. For example, modified joint displacement and joint torque trajectories obtained by using a 20% increased 1st factor loading are shown as dotted lines in Fig. 3C and D (with the subscript "n"). By comparing the two results in these figures, the knee joint torque in modified trajectories (dotted line) was remarkably made smaller than the original one (thick line). Following the same calculation procedure, lifting force trajectory to achieve the modified trajectory was also obtained as shown in Fig. 4B. The trajectory is determined by connecting force amplitude at each of five representative time frames which is calculated by static analysis using the modified joint angles at each time frame obtained by reverse PCA decomposition.

3 Kinematic synthesis of a wearable walking assist device for hemiplegics

In this section, a systematic design process of an exoskeleton-type walking assist device is presented. The design was aiming to achieve minimum number of joints and actuators in order to guarantee an intrinsic safety, simple and inexpensive structure. The device is divided into two pieces so that a part of

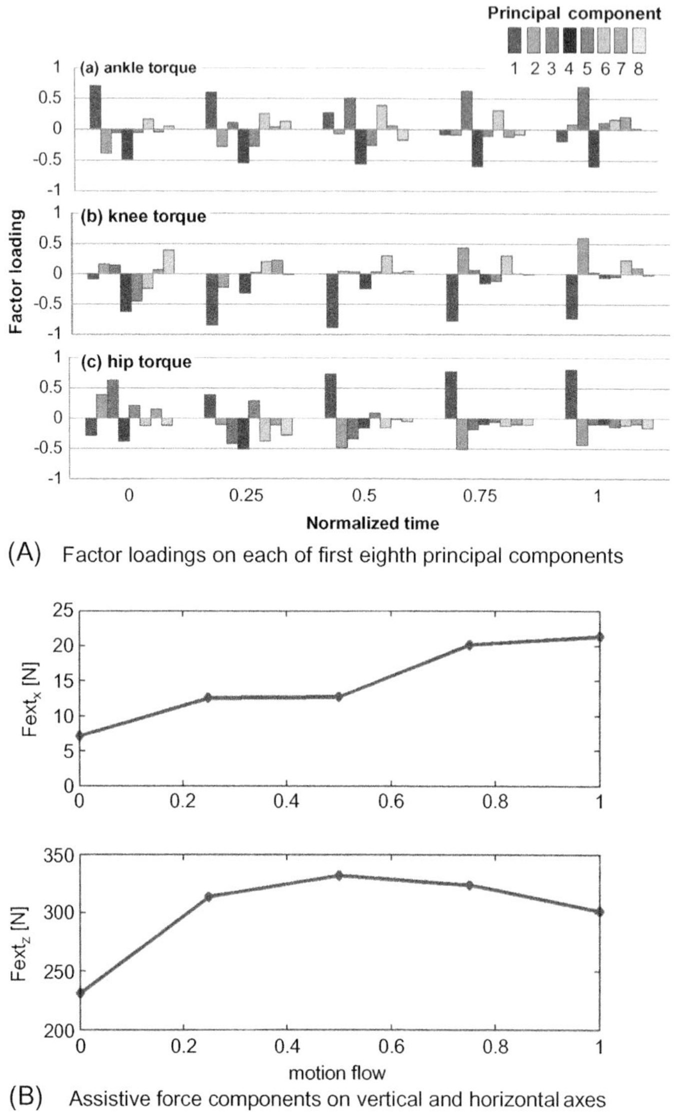

Fig. 4 Obtained factor loadings and synthesized assistive force trajectory to reduce generative knee torque reduction. (A) Factor loadings on each of first eighth principal components, (B) assistive force components on vertical and horizontal axes.

the device can be removed when user's body function has been recovered. Since the obtained mechanism was designed to achieve natural human gaits including changings of step lengths and walking directions. After deciding mechanism composition, decision of the number and location of minimally essential active joints and generation of input trajectories for those joints were performed by using PCA decomposition and its inverse calculation with modified factor loadings. 3D-motion capture experiment with healthy subjects were carried out to acquire essential datasets. Detail of the above process is discussed.

3.1 Mechanism design of wearable walking assist device

Since the walking assist device for hemiplegics is aiming to assist whole-lower limb motions of a user's disabled side, it should be able to drive one's hip, knee and ankle joints. Those three joints can be represented by a spherical, revolute and again spherical joint, respectively, as shown in the leftmost of Fig. 5. In order to design an assist device that can generate a whole-lower limb motion, and is allowed to be separated into several components, one can find out four different patterns as shown in Figs. 5(i)–(iv) (in each of the patterns, numbers in the ellipses identify the components). Among those candidates, pattern (ii) could be picked up as a suitable one, since it is generally said that among the disabled joints of hemiplegics, ones closer to trunk, which is hip joint in this case, tend to be recovered earlier than the others. Accordingly, hip joint assist unit tends to become unnecessary at first although it is the heaviest part among the device, and therefore it should be removable from the remaining components of the device.

Based on the above consideration, mechanisms to assist hip joint, and knee and ankle joints were figured out as shown in Fig. 6. For ease of designing, those mechanisms were composed of revolute joints only. In addition, in the upper mechanism shown in Fig. 6A, the first rotation axis, A, is parallelized to a user's hip joint's extension/flection axis, and third, fourth axes, C and D, are in parallel. At last, the fifth axis, E, is parallelized to the inner/outer rotation axis. As the result, rotational motion around the axis

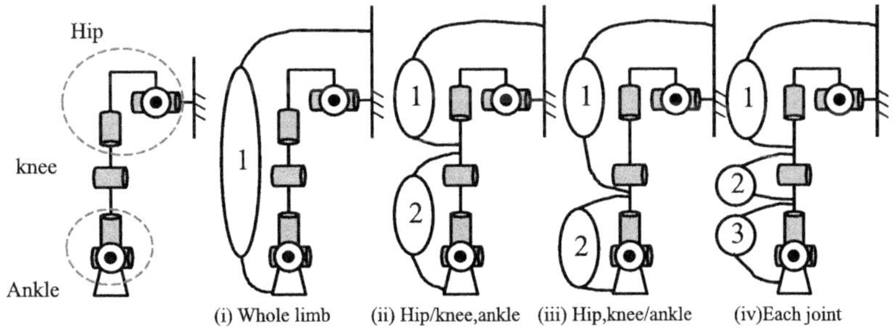

Fig. 5 A kinematic model of human lower limb (leftmost) and four possible separation patterns of the walking assist device (i)–(iv).

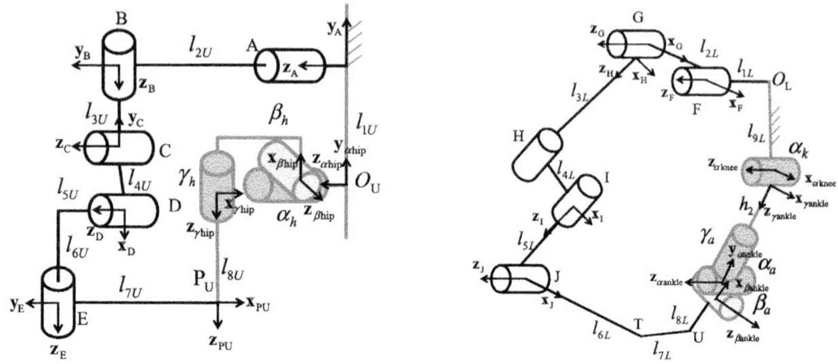

(A) Upper mechanism (for hip joint)

(B) Lower mechanism (for knee and ankle joints)

Fig. 6 Actual joint composition of the two spatial mechanisms for walking assist mechanism for hemiplegics. (A) Upper mechanism (for hip joint), (B) lower mechanism (for knee and ankle joints).

perpendicular to both axes of D and E is always constrained. In the lower mechanism shown in Fig. 6B, axes F and G, and H and I are parallelized to each other. As the result, rotational motion around the axis perpendicular to both axes of G and H is always constrained. As the result, aduction/abduction of user's hip and ankle joints is naturally constrained by the two mechanisms, and remaining axes of the whole-lower limb can be driven. In order to make a decision where to attach minimally essential actuators and to synthesize input trajectories for them, PCA were applied.

3.2 Joint input trajectory synthesis based on PCA

In this section, joint input trajectories that can achieve straight gait with variable step length and turning gait on a horizontal ground are synthesized. First of all, database matrix was constructed by following steps.

(a) Motion capture experiments using a marker sets shown in Fig. 7B were carried out with seven subjects. For each of subjects, five-different walking patterns that were [i] Straight walk without specifying step length, [ii] Straight walk with as large a step length as possible, [iii] Straight walk with small a step length as possible, [iv] Walk while turning right, [v] Walk while turning left, were recorded (each of the patterns was performed five times).

(b) From acquired data, each joint angle, and translation and angular velocity of the trunk, v_{cx} and ω_{cz}, were calculated. Fig. 7B and C show one of

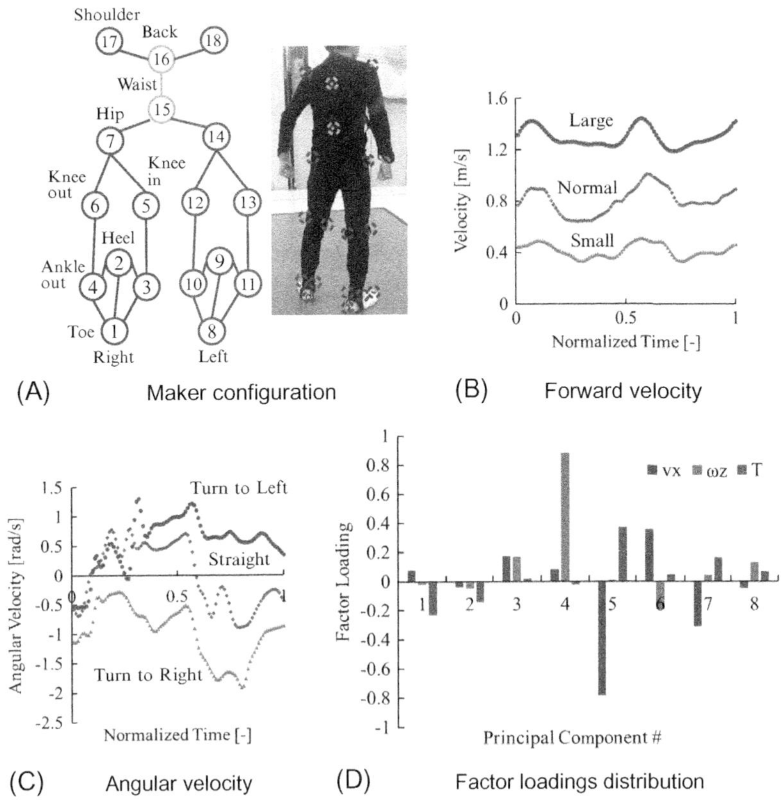

Fig. 7 Motion capture setup and acquired results. (A) Maker configuration. (B) Forward velocity. (C) Angular velocity. (D) Factor loadings distribution.

the obtained trajectories of translation and angular velocities. From those trajectories, walking pattern evaluation measures were calculated as

$$v_{cx} = \sum_{\tilde{t}=1}^{100} v_{x,\tilde{t}}, \ \omega_{cz} = \sum_{\tilde{t}=1}^{100} \omega_{z,\tilde{t}} \tag{8}$$

that approximately correspond to translational and rotational displacement among entire trials.

(c) By using the acquired joint angles and evaluation measures in Eq. (8), database matrix Z was constructed with z-scored element vectors:

$$Z = \begin{bmatrix} \widetilde{\mathbf{z}}_1 \\ \vdots \\ \widetilde{\mathbf{z}}_N \end{bmatrix}, \text{ where } \mathbf{z}_n = \begin{bmatrix} \boldsymbol{\alpha}_{1,n} & \boldsymbol{\beta}_{1,n} & \boldsymbol{\gamma}_{1,n} & \cdots & \boldsymbol{\gamma}_{6,n} & v_{cx,n} & \omega_{cz,n} & T_n \end{bmatrix} \quad (9)$$

where $\boldsymbol{\alpha}$, $\boldsymbol{\beta}$, $\boldsymbol{\gamma}$ are angular displacement vectors on the axes of flexion and extension, adduction and abduction, and inner and outer rotation, and T is the length of gait cycle. Since each of angular displacement vectors includes 100 scalars, one z_n vector includes 1803 scalars in total.

After carrying out the PCA decomposition, factor loadings with respect to the three evaluation measures, v, ω, and T are obtained as shown in Fig. 7D. As shown in Table 1, cumulative contribution ratio up to eighth principal component passed over 80%. It could thus be said that up to eighth principal components can sufficiently explain the experimental results. In addition, the fourth principal component has a strong positive relationship between the angular velocity, the fifth component has a strong negative relationship between the translational velocity, and also a mid-strong positive relationship between the gait cycles. These relationships can also be observed in the distribution of factor loadings plotted in Fig. 8A and B. Circled numbers in each figure represent one of the above-mentioned five gait patterns. Factor loading against fourth component has significant peak on the pattern [iv] and [v], in which turning right and left had performed. On the other hand, that against fifth component has a peak at the pattern [iii] in which subjects had tried to walk with a smaller step length than the other patterns. From these results, it was predictable that input joint trajectories to achieve step length changing and turning can be synthesized by using reverse PCA decomposition with modified factor loadings in which fourth and fifth loadings were promptly changed. Figs. 9 and 10 illustrate numerical examples. By changing the fourth factor loading in positive and negative direction, heading direction was changed to left and right, respectively, as shown in Fig. 9. On the other hand, by changing the fifth factor loading in positive and negative direction, step length became small and large step lengths, respectively, as shown in Fig. 10.

Table 1 Contribution ratio and cumulative contribution ratio of each principal component

ID number of PC	#1	#2	#3	#4	#5	#6	#7	#8
Contribution ratio	20.6	19.3	15.4	7.61	6.24	5.50	3.79	2.77
Cumulative contribution ratio	20.6	39.9	55.3	62.9	69.1	74.6	78.4	81.2

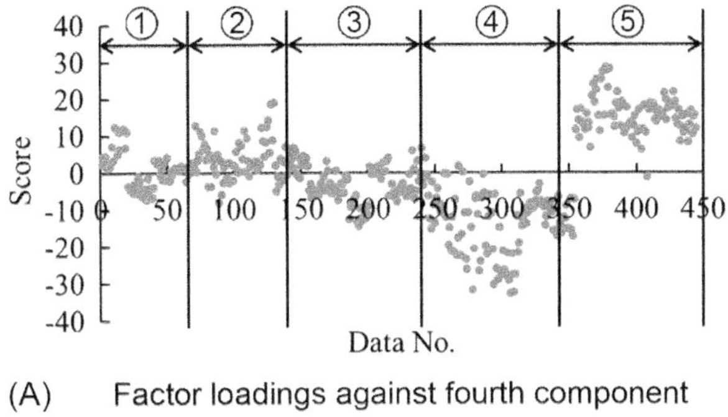

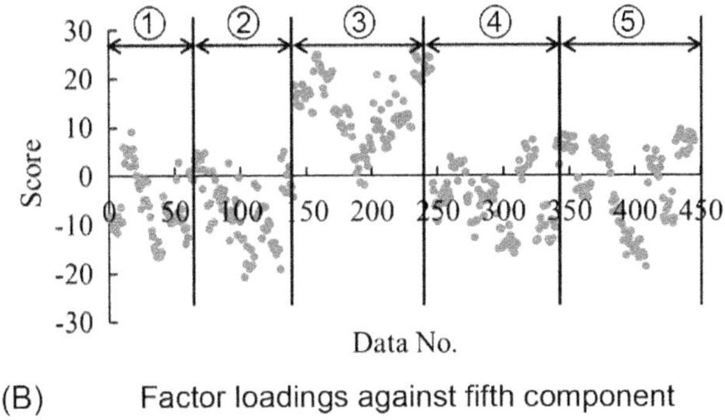

Fig. 8 Distribution of factor loadings. (A) Factor loadings against fourth component, (B) factor loadings against fifth component.

3.3 Selection of joints to be active ones

In above section, synthesis of angular input trajectory for each joint was discussed. Remaining problem is which joint should become active to establish a fine driving capability. This problem can be solved by a consequence of PCA-based motion synthesis. Tables 2 and 3 show the ranges of motion (ROM) of each lower limb axes, when the fourth and fifth factor loadings had modified that $w_4 = [-30, -20, -10, 0, 10, 20, 30]$ and $w_5 = [-20, -10, 0, 10, 20]$. It can be said that axes having a larger ROM than others are more sensitive to the key factor loadings, and it is thus meaningful to actuate those axes to achieve the target motion. By picking up axes having larger standard deviations of ROM, flexion/extension and inner/outer rotation of hip joint, adduction/abduction of knee joint, and flexion/extension and inner/outer rotation of ankle joint can be selected as significant ones. Joint trajectory generation using PCA can be again performed by using angular displacements around only above five axes. In the obtained result, new factor

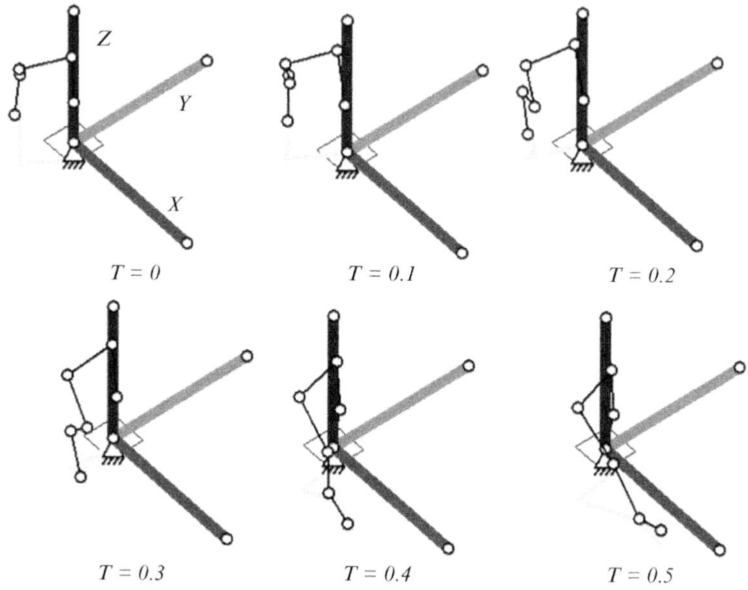

Fig. 9 Result of motion generation ($w_4 = 20$).

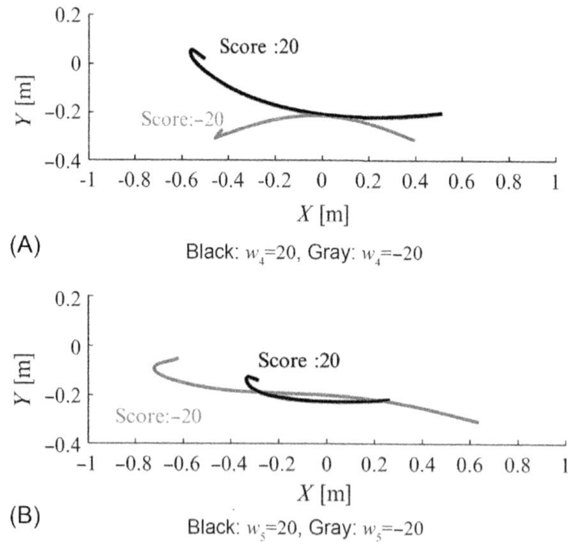

Fig. 10 Ankle trajectories with modified loadings. (A) Black: $w_4 = 20$, Gray: $w_4 = -20$ (B) Black: $w_5 = 20$, Gray: $w_5 = -20$.

Table 2 Range of motion with modified fourth factor loading w_4 [deg]

	Hip joint			Knee joint			Ankle joint		
	α	β	γ	α	β	γ	α	β	γ
Mean value	35.1	5.16	18.2	49.1	3.93	6.09	16.4	4.46	18.5
SD	0.937	2.16	10.6	0.464	1.01	1.68	1.89	0.649	5.50

Table 3 Range of motion with modified fifth factor loading w_5 [deg]

	Hip joint			Knee joint			Ankle joint		
	α	β	γ	α	β	γ	α	β	γ
Mean value	34.9	4.65	11.8	49.7	4.90	5.95	20.2	3.87	17.1
SD	9.06	2.56	2.12	11.7	1.71	1.20	7.32	0.539	6.05

loading distribution is obtained as Fig. 11, and cumulative contribution ratio up to seventh principal components passed over 80%. Fourth PC has a negative relationship to the translational velocity, and sixth PC has a positive relationship to the angular velocity. Just like the above case, turning left and right can be achieved by changing the sixth factor loading, as shown in Fig. 12, and step length change can be achieved by changing the fourth factor loading, as shown in Fig. 13. The selected five axes are therefore sufficiently significant to achieve the desired gaits, and joints of the wearable assist device parallel to those axes, namely A, B, F, G, and H should be actuated.

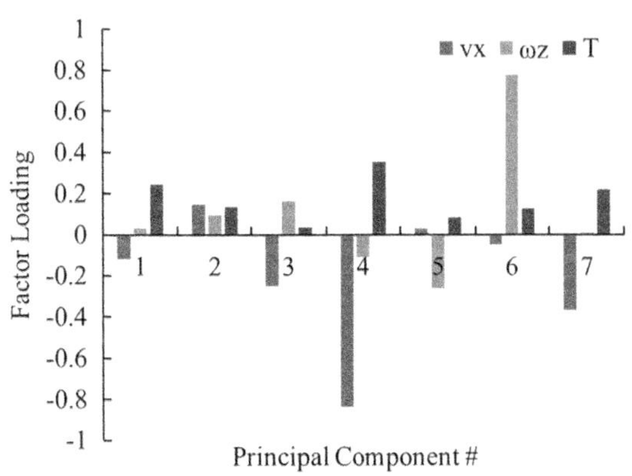

Fig. 11 Factor loadings only with significant PCs.

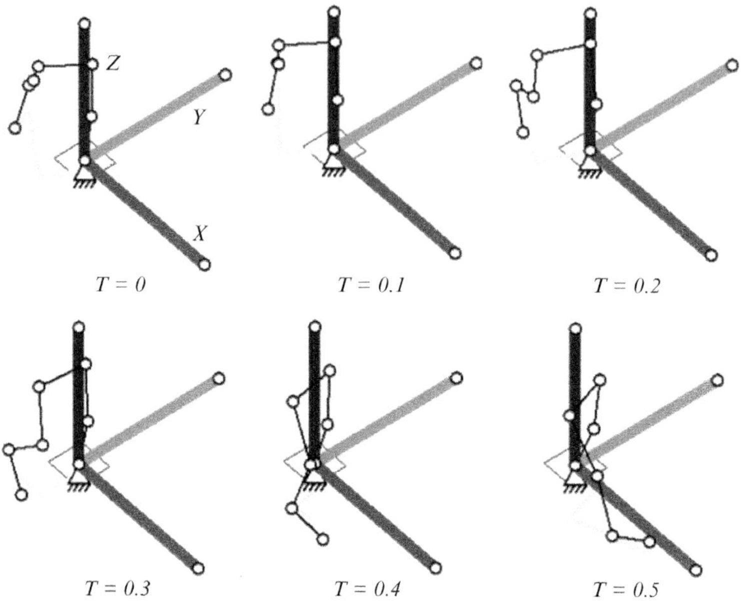

Fig. 12 Result of motion generation ($w_4 = 20$).

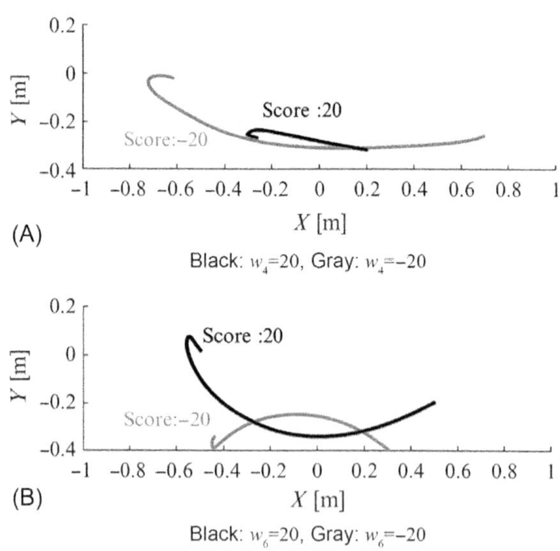

Fig. 13 Ankle trajectories with modified loadings. (A) Black: $w_4 = 20$, Gray: $w_4 = -20$ (B) Black: $w_6 = 20$, Gray: $w_6 = -20$.

4 Conclusions

This article presents a systematic method useful in the structural design and motion synthesis of motion assist exoskeleton. Analysis procedure of complex human motions with multiple degrees of freedom for different functions and regeneration method of arbitrary motions from the analyzed data by adjusting parameters based on the principal component analysis were formulated and explained using an example. Through an illustrative example on the design of walking assist device for hemiplegics, the practical application scheme was shown.

References

[1] H. Masaru, O. Makoto, S. Shuta, T. Yukio, Development of a walking assist machine using crutches (composition and basic experiments), J. Mech. Sci. Technol. 24 (2010) 245–248.
[2] T. Iwaya, Y. Takeda, M. Ogata, M. Higuchi, Development of a walking assist machine using crutches—motion for ascending and descending steps, in: New Trends in Mechanism Science—Analysis and Design, Mechanism and Machine Science, vol. 5, Springer, 2010, pp. 319–326.
[3] D. Matsuura, R. Funato, M. Ogata, M. Higuchi, Y. Takeda, Efficiency improvement of walking assist machine using crutches based on gait-feasible region analysis, Mech. Mach. Theory 84 (2015) 126–133.
[4] D. Matsuura, R. Funato, Y. Chounan, K. Tharasrisuthi, Y. Takeda, Synthesis and analysis of the mechanism for walking assist machine for persons with hemiplegia, in: L.J. De Vin, J. Solis (Eds.), Proceedings of the 14th Mechatronics Forum International Conference, Mechatronics June 13–19. 2014, Karlstad, Sweden, 2014, pp. 303–309.
[5] Y. Takeda, D. Matsuura, M. Higuchi, S. Sato, T. Iwaya, M. Ogata, R. Funato, Mechanism design and control of a simple and low-cost walking assist machine, in: L.J. De Vin, J. Solis (Eds.), Proceedings of the 14th Mechatronics Forum International Conference, Mechatronics June 13–19. 2014, Karlstad, Sweden, 2014, pp. 555–562.

CHAPTER 3

Multibody dynamics for human-like locomotion

Mario Acevedo[a], Hiram Ponce[b]
[a]Universidad Panamericana, Facultad de Ingeniería, Zapopan, Jalisco, Mexico
[b]Universidad Panamericana, Facultad de Ingeniería, Ciudad de México, Mexico

1 Introduction

The human body can be considered as multibody system. The general methods from multibody dynamics have been applied to study its motion just as has been done for machines and mechanisms. The study of machines that walk had probably its origin at the end of the 19th century. An early walking model appeared in about 1870. It used a linkage designed by Chevyshev to move the body along a straight horizontal path while the feet moved up and down to exchange support during stepping [1]. But the performance of such machines was limited by their fixed patterns of motion, and by the late 1950s, it had become clear that linkages providing fixed motion would not suffice, control would be needed.

Digitally controlled legged robots started to appear in the late 1960s. But by the end of the 1970s were still limited to quasistatic gaits, that is, slow walking motions with the center of mass (CoM) of the robot always kept above its feet. Digitally controlled legged robots started to appear in the late 1960s. Dynamic legged locomotion started with the first dynamically walking bipedal robot in Tokyo University, and the hopping and running monopedal, bipedal and quadrupedal robots developed at the MIT [2]. Meanwhile it was demonstrated that stable dynamic walking motions could be obtained by pure mechanical means, developing the concept of *passive dynamic walking*.

Analysis and design of legged robots remained as matter of research in universities and laboratories until the P2 humanoid robot from Honda appeared, followed in 2000 by the Asimo humanoid robot. The progress over the last decades has been remarkable and how to make legged robots walk and run dynamically is understood, but how best to make them walk and run efficient is still an open issue. And here is where multibody dynamics for human-like locomotion come into relevance.

Multibody dynamics for human-like locomotion is a topic of great relevance in current research for a number of fields. It could help in illness

Design and Operation of Human Locomotion Systems
https://doi.org/10.1016/B978-0-12-815659-9.00003-2

diagnosis and to restore or improve lost of mobility, rehabilitation. In robotics can be very helpful to develop human-like robots. There is an area where these both field meet, the prosthetic and orthotic sector [3]. Different efforts to develop software tools to use multibody in this areas has been done, see for example, Sherman et al. [4].

Robotics for rehabilitation treatment is an emerging field which is expected to grow. In this area efforts are done for replacing the physical training effort of a therapist, allowing more intensive repetitive motions and delivering therapy at a reasonable cost, and to assess quantitatively the level of motor recovery by measuring force and movement patterns. Here although passive robotic rehabilitation devices are less complex, are limited compared to active devices where a multibody model is essential. See Díaz et al. [5] for a complete count on applications.

One of the most popular use of multibody dynamics in the analysis of real motions, where the inverse dynamics problem is solved for a model animated with some acquired motion, see Gong et al. [6]. As results driving torques at joints are obtained. Inverse and forward dynamics is used for true motion prediction, a problem representing a great challenge currently being a topic of intensive research. In addition compliance at joints is a topic of special attention somehow unattended in research, but having great influence [7].

In the case of humanoid robots, methods from multibody dynamics can be applied to simulate their motion. Practical difficulties can appear in general due to the unstable nature of gait, see, as example, Khusainov et al. [8], and the use of foot-ground contact models [9].

Dynamic balancing concepts have also contributed to human-like locomotion, see, as examples, Takayuki et al. [10] and Stojicevic et al. [11]. In this chapter the application of a force-balanced mechanism is proposed as a leg to be part of a biped robot. Stability is analyzed through the application of learning approaches based on an artificial intelligence, namely artificial hydrocarbon networks (AHN). Modeling and results from multibody dynamics simulation are presented.

2 Stability in human-like locomotion

The problem of equilibrium is critical for planning, control, and analysis of legged robots [2]. Control algorithms for legged robots use the equilibrium criteria to avoid falls [12, 13]. Regardless of the application, the computational efficiency of the equilibrium tests is critical, as computation time is often the bottleneck of control algorithms. To achieve static equilibrium (i.e., a dynamic wrench equal to zero), the total wrench of gravity and contact forces must therefore be equal to zero. This means that the CoM of the mechanism projects vertically inside the convex hull of the contact points when these points are located in the same horizontal plane. To comply with the previous condition, it is necessary to calculate the horizontal momentum rotation with respect to the CoM whose location, in general, has also to be calculated for every moment. Although for arbitrary contact geometries, more complex and computationally expensive techniques are required to check equilibrium, some requiring to do it 100–1000 times per second [14].

Designing equilibrium controllers for legged robots is a challenging problem [15]. Inverted pendulums, ballbots, hoppers, and cart-pole systems have been revised in the literature [15–17]. It has been shown that nonlinear or more complex control systems have to be designed to balance those, which it also complicates the computational cost and demands robust actuators [17, 18].

In this work, we propose a force-balanced mechanism as a building element for the synthesis of legged robots that can be easily balance controlled. The mechanism has two degrees of freedom (DOFs), in opposition to the more traditional one DOF linkages generally used as legs in robotics [19]. As the mechanical system is balanced, the CoM of the whole system is located in a specific location within the mechanism. This facilitates the efficient use of the "projection of the center of mass" criterion with the aid of a counter rotating inertia, and reduces the number of calculations required by the control algorithm to check equilibrium and correct the CoM position. We conducted different experiments to balance the mechanism and to track unstable set-point positions. To do so, we implemented proportional error controllers with different strategies as well as learning approaches, based on an artificial intelligence method namely AHN.

3 One-leg mechanism model

The proposed mechanism is presented in Fig. 1. It is a closed-loop mechanism formed by two inverted double-pendula balanced by force [20]. The right side pendulum is formed by bars OA and CF, while the left one is formed by bars OB and DG. Both double pendula have a counterweight (green disc) at points F and G, respectively.

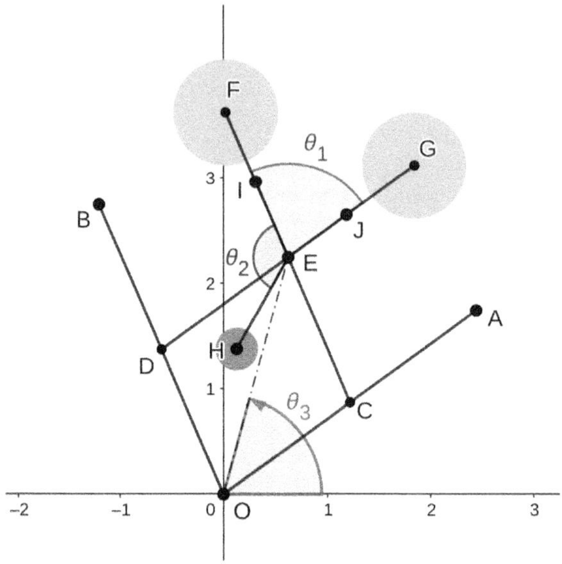

Fig. 1 Proposed one-leg mechanism.

This configuration of the mechanism makes its CoM located exactly at point E. Thus, it is very easy to check the equilibrium condition: in this case to maintain the CoM of the system in a vertical position.

The mechanism has five DOFs, if it is not in contact to the ground, otherwise it has two DOFs when point O is fixed (the case of lock up) or three DOFs when point O is sliding. The driven DOFs are the angle θ_1, which allows to stretch and shrink the leg, and the angle θ_2, moving a counter inertia defined by the bar EH and a mass that helps the system to equilibrate.

In this way, the CoM of the whole leg system can be positioned in such a way that the static equilibrium is reached in a tilted pose, where the angle θ_3 is different from 90 degrees.

3.1 Kinematics

The dynamic equations of motion of the mechanism have been obtained following a multibody approach. In this case the multibody system is constructed by a group of rigid bodies, which depend on the kinematic constraints and on the applied forces.

3.1.1 Kinematics

If the proposed planar mechanism is made up of b moving rigid bodies, the number of Cartesian generalized coordinates is $n = 3 \times b$. Thus the vector of generalized coordinates for the system can be written as Eq. (1) where $\mathbf{q}_b = [x_b, y_b, \phi_b]^T$, in this case for all $b = 1, \ldots, 5$:

$$\mathbf{q} = [\mathbf{q}_1, \mathbf{q}_2, \mathbf{q}_3, \mathbf{q}_4, \mathbf{q}_5]^T \tag{1}$$

On the other hand, a revolute kinematic joint between bodies i and j introduces a pair of constraints that in general can be described by Eq. (2); where $\mathbf{r}_i$ is the position vector of the CoM of body i, $\mathbf{A}_i$ is its rotation matrix and $\mathbf{s}_i'$ is the local coordinates vector that positions the kinematic pair with respect to its local reference frame. Similar description can be done for the terms corresponding to body j, thus we can write:

$$\Phi_m = (\mathbf{r}_i + \mathbf{A}_i\mathbf{s}_i') - (\mathbf{r}_j + \mathbf{A}_j\mathbf{s}_j') \tag{2}$$

The complete set of m kinematic constraints, dependent on the generalized Cartesian coordinates, can be expressed as:

$$\boldsymbol{\Phi}(\mathbf{q}, t) = 0 \tag{3}$$

The first derivative of Eq. (3) with respect to time is used to obtain the velocities, Eq. (4), while the second derivative of Eq. (3) with respect to time yields the accelerations equations as Eq. (5); where $\boldsymbol{\Phi}_q$ is the Jacobian matrix, $\mathbf{b}$ and $\mathbf{c}$ are the velocities and accelerations respectively, containing time-dependent expressions.

$$\boldsymbol{\Phi}_q\dot{\mathbf{q}} = \mathbf{b} \tag{4}$$

$$\boldsymbol{\Phi}_q\ddot{\mathbf{q}} = \mathbf{c} \tag{5}$$

3.2 Dynamics

The equations of motion for the constrained multibody system are obtained applying the virtual power principle [21], and can be expressed by:

$$\mathbf{M}\ddot{\mathbf{q}} + \boldsymbol{\Phi}_q^T\boldsymbol{\lambda} = \mathbf{f} \tag{6}$$

where $\mathbf{M}$ is the diagonal mass matrix, $\boldsymbol{\lambda}$ is the vector of Lagrange multipliers directly associated with the reactions at the joints, and $\mathbf{f}$ is the vector of applied external forces.

For dynamics analysis, the kinematic constraint equations determine the algebraic configuration, while the dynamical behavior can be defined by the second-order differential equations. Therefore, Eqs. (5), (6) are arranged to form of differential-algebraic equations as (Eq. 7):

$$\begin{bmatrix} \mathbf{M} & \boldsymbol{\Phi}_q^T \\ \boldsymbol{\Phi}_q & \mathbf{0} \end{bmatrix} \begin{bmatrix} \ddot{\mathbf{q}} \\ \boldsymbol{\lambda} \end{bmatrix} = \begin{bmatrix} \mathbf{f} \\ \boldsymbol{\gamma} \end{bmatrix} \tag{7}$$

having to solve the positions problem (3) and the velocities problem (4) every specific number of time steps to reduce the accumulation error in numerical simulations to obtain an accurate solution.

For the case of the one-leg mechanism proposed three different cases can be analyzed: luck up or stiction (when point O is fixed to the ground at some location), sliding (when point O is in contact with the ground and moving in the horizontal direction), and contact (when point O loss contact with the ground because gait simulation is in process). These alternatives will be described in the following sections.

3.2.1 Lock up or stiction

In this case Eq. (7) can be modified to include the stiction constraint equation:

$$\boldsymbol{\Phi}^r = 0 \tag{8}$$

thus Eq. (7) is modified as:

$$\begin{bmatrix} \mathbf{M} & \boldsymbol{\Phi}_q^T & \boldsymbol{\Phi}_q^{rT} \\ \boldsymbol{\Phi}_q & \mathbf{0} & \mathbf{0} \\ \boldsymbol{\Phi}_q^r & \mathbf{0} & \mathbf{0} \end{bmatrix} \begin{bmatrix} \ddot{\mathbf{q}} \\ \boldsymbol{\lambda} \\ \boldsymbol{\lambda}^r \end{bmatrix} = \begin{bmatrix} \mathbf{f} \\ \boldsymbol{\gamma} \\ \boldsymbol{\gamma}^r \end{bmatrix} \tag{9}$$

Taking as reference to Fig. 1, all terms in Eq. (9) can be obtained. Body 1 is the bar OA, body 2 the bar OB, body 3 the bar CF, body 4 the bar DG, and body 5 bar EH. For this case the complete set of m kinematic constraints, $m = 12$, dependent on the $n = 15$ generalized Cartesian coordinates can be expressed as:

$$\boldsymbol{\Phi}(\mathbf{q}) = \begin{bmatrix} l_{OD} \cdot \cos(\phi_2) - l_{OC} \cdot \cos(\phi_1) - x_2 + x_1 \\ l_{OD} \cdot \sin(\phi_2) - l_{OC} \cdot \sin(\phi_1) - y_2 + y_1 \\ y_2 - l_{OD} \cdot \sin(\phi_2) \\ l_{CI} \cdot \cos(\phi_3) - x_3 + x_1 \\ l_{CI} \cdot \sin(\phi_3) - y_3 + y_1 \\ l_{DJ} \cdot \cos(\phi_4) - x_4 + x_2 \\ l_{DJ} \cdot \sin(\phi_4) - y_4 + y_2 \\ l_{EJ} \cdot \cos(\phi_4) - l_{EI} \cdot \cos(\phi_3) - x_4 + x_3 \\ l_{EJ} \cdot \sin(\phi_4) - l_{EI} \cdot \sin(\phi_3) - y_4 + y_3 \\ l_{HE} \cdot \cos(\phi_5) - l_{EI} \cdot \cos(\phi_3) - x_5 + x_3 \\ -l_{HE} \cdot \sin(\phi_5) - l_{EI} \cdot \sin(\phi_3) - y_5 + y_3 \end{bmatrix} = 0 \tag{10}$$

where l_{MN} is the distance between points M and N. The stiction constraint is:

$$\boldsymbol{\Phi}^{r}(\mathbf{q}) = [x_2 - l_{OD} \cdot \cos(\phi_2)] = 0 \tag{11}$$

Eq. (10) can be reorganized in a single set of constraints: $\boldsymbol{\Phi}^{s} = \mathbf{0}$, to facilitate calculation. So the Jacobian matrix of the reorganized system, $\boldsymbol{\Phi}^{s}_{\mathbf{q}}$ can be expressed as:

$$\boldsymbol{\Phi}^{s}_{\mathbf{q}} = \begin{bmatrix} \mathbf{I}_2 & \mathbf{r}_{OC} & -\mathbf{I}_2 & -\mathbf{r}_{OD} & \mathbf{0}_2 & \mathbf{0}_1 & \mathbf{0}_2 & \mathbf{0}_1 & \mathbf{0}_2 & \mathbf{0}_1 \\ \mathbf{0}_2 & \mathbf{0}_1 & \mathbf{I}_2 & \mathbf{r}_{OD} & \mathbf{0}_2 & \mathbf{0}_1 & \mathbf{0}_2 & \mathbf{0}_1 & \mathbf{0}_2 & \mathbf{0}_1 \\ \mathbf{I}_2 & \mathbf{0}_1 & \mathbf{0}_2 & \mathbf{0}_1 & -\mathbf{I}_2 & \mathbf{r}_{CI} & \mathbf{0}_2 & \mathbf{0}_1 & \mathbf{0}_2 & \mathbf{0}_1 \\ \mathbf{0}_2 & \mathbf{0}_1 & \mathbf{I}_2 & \mathbf{0}_1 & \mathbf{0}_2 & \mathbf{0}_1 & -\mathbf{I}_2 & \mathbf{r}_{DJ} & \mathbf{0}_2 & \mathbf{0}_1 \\ \mathbf{0}_2 & \mathbf{0}_1 & \mathbf{0}_2 & \mathbf{0}_1 & \mathbf{I}_2 & -\mathbf{r}_{EI} & -\mathbf{I}_2 & \mathbf{r}_{EJ} & \mathbf{0}_2 & \mathbf{0}_1 \\ \mathbf{0}_2 & \mathbf{0}_1 & \mathbf{0}_2 & \mathbf{0}_1 & \mathbf{I}_2 & -\mathbf{r}_{EI} & \mathbf{0}_2 & \mathbf{0}_1 & -\mathbf{I}_2 & \mathbf{r}_{HE} \end{bmatrix} \tag{12}$$

where $\mathbf{I}_2$ is the 2 × 2 identity matrix, $\mathbf{0}_2$ is the 2 × 2 zero matrix, $\mathbf{0}_1$ is the 2 × 1 zero vector, and:

$$\begin{aligned} \mathbf{r}_{OC} &= l_{OC}\begin{bmatrix} \sin(\phi_1) \\ -\cos(\phi_1) \end{bmatrix}; \mathbf{r}_{OD} = l_{OD}\begin{bmatrix} \sin(\phi_2) \\ -\cos(\phi_2) \end{bmatrix}; \mathbf{r}_{CI} = l_{CI}\begin{bmatrix} \sin(\phi_3) \\ -\cos(\phi_3) \end{bmatrix} \\ \mathbf{r}_{EI} &= l_{EI}\begin{bmatrix} -\sin(\phi_3) \\ \cos(\phi_3) \end{bmatrix}; \mathbf{r}_{DJ} = l_{DJ}\begin{bmatrix} \sin(\phi_4) \\ -\cos(\phi_4) \end{bmatrix}; \mathbf{r}_{EJ} = l_{EJ}\begin{bmatrix} \sin(\phi_4) \\ -\cos(\phi_4) \end{bmatrix} \\ \mathbf{r}_{HE} &= l_{HE}\begin{bmatrix} \sin(\phi_5) \\ \cos(\phi_5) \end{bmatrix} \end{aligned}$$

In this case $\mathbf{b} = 0$ because the constraints do not depend explicitly on time, and $\mathbf{c}$ can be expressed as:

$$\mathbf{c} = \begin{bmatrix} l_{OD}\cos(\phi_2)\omega_2^2 - l_{OC}\cos(\phi_1)\omega_1^2 \\ l_{OD}\sin(\phi_2)\omega_2^2 - l_{OC}\sin(\phi_1)\omega_1^2 \\ -l_{OD}\cos(\phi_2)\omega_2^2 \\ -l_{OD}\sin(\phi_2)\omega_2^2 \\ -l_{CI}\cos(\phi_3)\omega_3^2 \\ -l_{DJ}\sin(\phi_3)\omega_3^2 \\ -l_{DJ}\cos(\phi_4)\omega_4^2 \\ -l_{CI}\sin(\phi_4)\omega_4^2 \\ l_{EI}\cos(\phi_3)\omega_3^2 - l_{EJ}\cos(\phi_4)\omega_4^2 \\ l_{EI}\sin(\phi_3)\omega_3^2 - l_{EJ}\sin(\phi_4)\omega_4^2 \\ l_{HE}\sin(\phi_5)\omega_5^2 + l_{EI}\cos(\phi_3)\omega_3^2 \\ -l_{HE}\cos(\phi_5)\omega_5^2 + l_{EI}\sin(\phi_3)\omega_3^2 \end{bmatrix} \tag{13}$$

The diagonal mass matrix is:

$$\mathbf{M}=\begin{bmatrix} m_1\mathbf{I}_2 & \mathbf{0}_1 & \mathbf{0}_2 & \mathbf{0}_1 & \mathbf{0}_2 & \mathbf{0}_1 & \mathbf{0}_2 & \mathbf{0}_1 & \mathbf{0}_2 & \mathbf{0}_1 \\ \mathbf{0}_1^T & J_1 & \mathbf{0}_1^T & 0 & \mathbf{0}_1^T & 0 & \mathbf{0}_1^T & 0 & \mathbf{0}_1^T & 0 \\ \mathbf{0}_2 & \mathbf{0}_1 & m_2\mathbf{I}_2 & \mathbf{0}_1 & \mathbf{0}_2 & \mathbf{0}_1 & \mathbf{0}_2 & \mathbf{0}_1 & \mathbf{0}_2 & \mathbf{0}_1 \\ \mathbf{0}_1^T & 0 & \mathbf{0}_1^T & J_2 & \mathbf{0}_1^T & 0 & \mathbf{0}_1^T & 0 & \mathbf{0}_1^T & 0 \\ \mathbf{0}_2 & \mathbf{0}_1 & \mathbf{0}_2 & \mathbf{0}_1 & m_3\mathbf{I}_2 & \mathbf{0}_1 & \mathbf{0}_2 & \mathbf{0}_1 & \mathbf{0}_2 & \mathbf{0}_1 \\ \mathbf{0}_1^T & 0 & \mathbf{0}_1^T & 0 & \mathbf{0}_1^T & J_3 & \mathbf{0}_1^T & 0 & \mathbf{0}_1^T & 0 \\ \mathbf{0}_2 & \mathbf{0}_1 & \mathbf{0}_2 & \mathbf{0}_1 & \mathbf{0}_2 & \mathbf{0}_1 & m_4\mathbf{I}_2 & \mathbf{0}_1 & \mathbf{0}_2 & \mathbf{0}_1 \\ \mathbf{0}_1^T & 0 & \mathbf{0}_1^T & 0 & \mathbf{0}_1^T & 0 & \mathbf{0}_1^T & J_4 & \mathbf{0}_1^T & 0 \\ \mathbf{0}_2 & \mathbf{0}_1 & \mathbf{0}_2 & \mathbf{0}_1 & \mathbf{0}_2 & \mathbf{0}_1 & \mathbf{0}_2 & \mathbf{0}_1 & m_5\mathbf{I}_2 & \mathbf{0}_1 \\ \mathbf{0}_1^T & 0 & \mathbf{0}_1^T & 0 & \mathbf{0}_1^T & 0 & \mathbf{0}_1^T & 0 & \mathbf{0}_5^T & J_5 \end{bmatrix} \tag{14}$$

where m_i and J_i are the mass and moment of inertia of body i, respectively. Finally,

$$\mathbf{f}=\begin{bmatrix} 0 \\ -g\cdot m_1 \\ 0 \\ 0 \\ -g\cdot m_2 \\ 0 \\ 0 \\ -g\cdot m_3 \\ \tau_1+\tau_2 \\ 0 \\ -g\cdot m_4 \\ -\tau_1 \\ 0 \\ -g\cdot m_5 \\ -\tau_2 \end{bmatrix} \tag{15}$$

where g is the acceleration of gravity, τ_1 is the applied torque associated to angle θ_1 between bodies 3 and 4, and τ_2 is the applied torque associated to angle θ_2 between bodies 3 and 5.

3.2.2 Sliding

In the case of sliding constraints change as in Eq. (10), because point O can move in the x direction. The number of DOFs is increased in one.

Unlike lock up, in sliding the external forces vector depend on some elements from vector $\boldsymbol{\lambda}$, so Eq. (6) change to:

$$\mathbf{M}\ddot{\mathbf{q}}+\boldsymbol{\Phi}_q^T\boldsymbol{\lambda}=\mathbf{f}+\mathbf{f}(\boldsymbol{\lambda}) \tag{16}$$

where $\mathbf{f}(\boldsymbol{\lambda})$ are the forces dependent on some joint reactions. In our case the friction force at point O depend on the reaction between contact point O and the ground, the reaction associated with the third constraint in Eq. (10).

To solve this multibody dynamics equations it is necessary to use an iterative process to calculate the correct reactions at the sliding joints, as described in García de Jalón and Bayo [21]. In Eq. (16), $\mathbf{f}(\boldsymbol{\lambda})$ can depend on joint geometry and position. Thus the external forces vector is expressed as:

$$\mathbf{f} = \begin{bmatrix} -\mu \cdot sing(\dot{x}_O)\lambda_1 \\ -g \cdot m_1 \\ \mu \cdot sing(\dot{x}_O)\lambda_1(y_1 - OC\sin(\phi_1)) \\ 0 \\ -g \cdot m_2 \\ 0 \\ 0 \\ -g \cdot m_3 \\ \tau_1 + \tau_2 \\ 0 \\ -g \cdot m_4 \\ -\tau_1 \\ 0 \\ -g \cdot m_5 \\ -\tau_2 \end{bmatrix} \tag{17}$$

where μ is the friction coefficient, $sign(\dot{x}_O)$ is the sign of the velocity of point O in the x direction, and λ_1 is the ground reaction (the normal force). The reactions at the joints can be calculated following an algorithm based on the fixed point iteration:

Algorithm 1 Iterative process to calculate joint reactions in sliding condition

Estimate $\boldsymbol{\lambda}^0$ and compute $\mathbf{f}(\boldsymbol{\lambda})$

repeat

  Calculate $\boldsymbol{\lambda}$ from Eq. (6) adjusted with $\mathbf{f}(\boldsymbol{\lambda})$

  If $abs(\boldsymbol{\lambda} - \boldsymbol{\lambda}^0) > tolerance$, $\boldsymbol{\lambda} = \boldsymbol{\lambda}^0$

until $abs(\boldsymbol{\lambda} - \boldsymbol{\lambda}^0) < tolerance$

3.3 Biped robot, gait design

The proposed leg mechanism can be part of a new concept for a biped robot, presented at Fig. 2, the counterweights are omitted for clarity. This robot has a trunk contributing with three DOFs: the displacement of point A, coordinates (x_1, y_1) and its orientation with respect to the horizontal line, ϕ_1. The robot also has two of the proposed one-leg mechanism connected to the trunk and contributing with four additional DOFs: the angle between the two upper bars of red leg θ_1, the angle between the two upper bars of blue leg θ_3, the angle between the right upper bar of the red leg and the trunk θ_2, and the angle between the right upper bar of the red leg and the trunk θ_4, a balancing pendulum connected to the trunk also contributing with one additional DOF, θ_5. In total this system has eight DOFs.

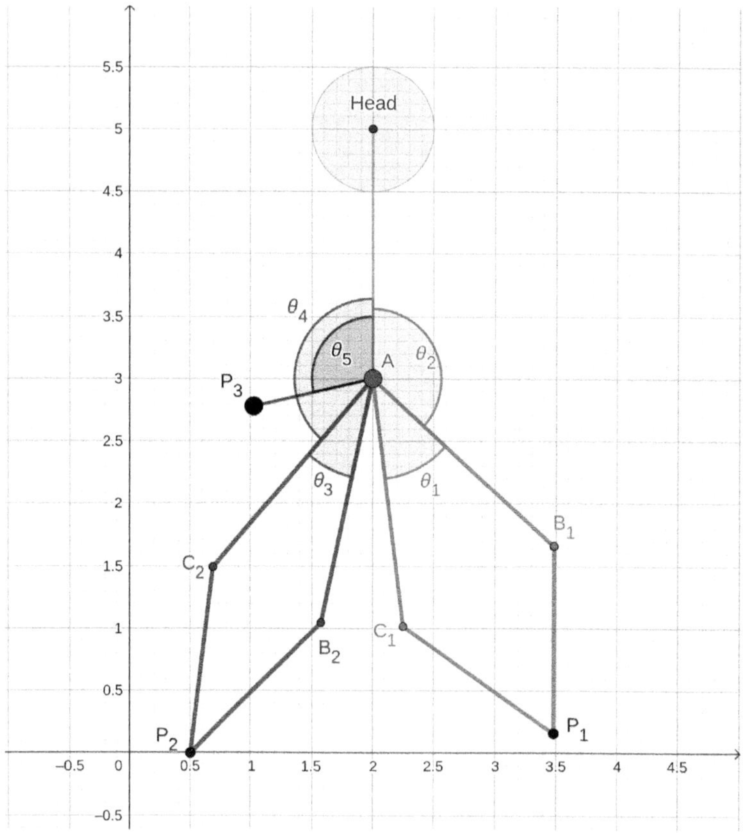

Fig. 2 Biped robot in the plane, using two of the proposed leg-mechanism.

For this system the balancing principles can also help in designing the gate. Lets considering that point P_2 is in contact to the ground, the robot really has six DOFs. Then we can define a specific trajectory to point P_1, being careful to impose starting and ending zero velocity and acceleration. In this way we avoid impact between P_1 and the ground.

Then it is possible to define the motion of point A by reducing the shaking force of the robot by the optimal control of the acceleration of the total mass center of moving links. This means minimizing the norm the CoM acceleration along the trajectory. This can be done using the "bang-bang" profile, as is described in Briot et al. [22], by defining a specific gate time. The shaking force can be calculated straightforward following the procedure described in Acevedo [23].

To define the complete motion of the robot it is possible to define a constant value to angle θ_5 and minimize the shaking moment imposing counter rotation of both legs. This is very easy as the location of the CoM of the system is located at point A.

Once the complete motion of the system is defined, it is possible to use Eq. (6) to solve the inverse dynamics problem as:

$$\boldsymbol{\Phi}_q^T \boldsymbol{\lambda} = \mathbf{f} - \mathbf{M}\ddot{\mathbf{q}} \tag{18}$$

4 Control of and learning the balancing task

At first glance, the designing process of the balancing control system for the one-leg mechanism might be seen as a challenging task [15]. Nevertheless, the force-balanced mechanical design provides a simple way to take it over. Thus, in this section, we propose two approaches to solve the balancing problem in the mechanism: (i) designing an intelligent controller and (ii) learning control actions from scratch using reinforcement learning (RL). We introduce not only these strategies, but also we describe AHN that suit the balancing task under disturbances.

4.1 Overview of artificial hydrocarbon networks

In machine learning, AHN algorithm is a supervised learning method inspired in the inner mechanisms and interactions of chemical hydrocarbon compounds [24]. This method aims to model data points like packages of information, called molecules. The interaction among these units allows capturing the nonlinearity of data correlation. From this point o view, an artificial hydrocarbon compound is built and it can be seen as a net of molecules. If required, more than one artificial hydrocarbon compound can be added up to finally get a mixture of compounds.

More precisely in AHN, the molecule is the basic unit of processing information. It performs an output response $\varphi(x)$ due to an input $x \in \mathbb{R}^k$, as expressed in Eq. (19) where $v_C \in \mathbb{R}$ represents a carbon value, $h_{i,r} \in \mathbb{C}$ are the hydrogen values attached to this carbon atom, and d represents the number of hydrogen atoms in the molecule.

$$\varphi(x) = v_C \sum_{r=1}^{k} \prod_{i=1}^{d \leq 4} (x_r - h_{i,r}) \tag{19}$$

Consider then that molecules are unsaturated (i.e., $d < 4$), then they are able to join with other unsaturated molecules forming chains. In AHN, these chains are namely hydrocarbon compounds or simply compounds. Throughout this chapter, compounds are made of n molecules: a linear chain of $(n-2)$ CH_2 molecules with two CH_3 molecules, one at each side of the CH_2-chain, as shown in Eq. (20), where CH_d-symbol represents a molecule with d hydrogen [25].

$$CH_3 - CH_2 - \cdots - CH_2 - CH_3 \tag{20}$$

A piecewise function ψ denoted as Eq. (21) is associated to the compound; where $L_t = \{L_{t,1}, \ldots, L_{t,k}\}$ for all $t = 0, \ldots, n$ are bounds where molecules can act over the input space. This function represents the behavior of the compound due to an input x. Other compound functions have been developed as presented in Ponce et al. [25], but the piecewise function has been the most adopted one [26–29].

$$\psi(x) = \begin{cases} \varphi_1(x) & L_{0,r} \leq x_r < L_{1,r} \\ \cdots & \cdots \\ \varphi_n(x) & L_{n-1,r} \leq x_r \leq L_{n,r} \end{cases} \tag{21}$$

At last, different compounds can be selected and added up to form complex structures called mixtures. In AHN, a mixture is a linear combination of behavior compounds ψ_j in finite ratios α_j, namely stoichiometric coefficients or simply weights of compounds, as expressed in Eq. (22).

$$S(x) = \sum_{j=1}^{c} \alpha_j \psi_j(x) \tag{22}$$

For training purposes, the least squares error is used for obtaining carbon and hydrogen values, while bounds are computed using a gradient descent method with learning rate $0 < \eta < 1$ based on the energy of adjacent molecules [25]. To this end, AHN is trained using the so-called AHN algorithm. Details can be found in Refs. [24, 25, 27, 28].

4.2 Artificial organic controller for balancing the one-leg mechanism

Intelligent control systems able to work well in nonlinear and dynamic systems, such as those related to changes in the operating point, environmental noise, disturbances, uncertainty in sensor measurements, miscalibration, among other factors [30]. Intelligent control systems are primary built on artificial intelligence approaches in order to tackle the above-mentioned issues. Particularly, literature reports several intelligent control systems to handle uncertainties and disturbances, for example, fuzzy controllers [31], neural controllers [30], and neuro-fuzzy controllers [32]. Recently, artificial organic controllers (AOCs) have also been proposed [29, 33]. In this regard, we propose to use an AOC to tackle the balancing problem in the one-leg mechanism, since AOC has been proved to be very effective in uncertain domains [29, 33–35].

4.2.1 Artificial organic controllers

An AOC is an intelligent control system that performs the control law using an ensemble method namely fuzzy-molecular inference (FMI) system [34], as shown in Fig. 3. FMI consists of three steps: fuzzification, fuzzy inference engine, and defuzzification based on AHN.

Fuzzification and fuzzy inference engine steps are quite similar to fuzzy logic. An input x is mapped to a set of fuzzy sets, using membership functions. Then, an inference operation, represented as a fuzzy rule, is applied to obtain a consequent value y_p. Considering, the pth fuzzy rule R_p denoted as

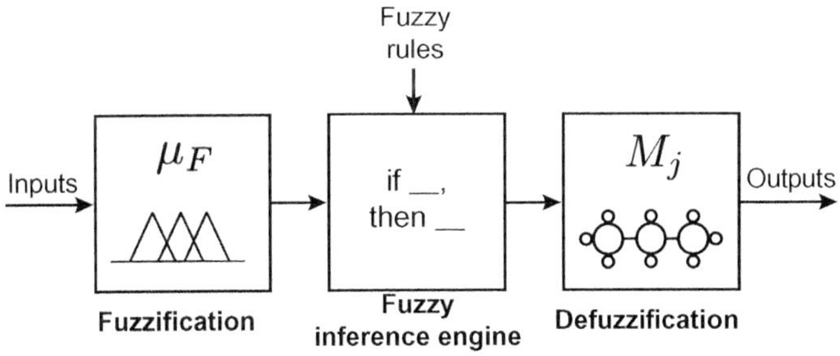

Fig. 3 Block diagram of the fuzzy-molecular inference system.

Eq. (23), inference computes y_p in terms of an artificial hydrocarbon compound with n molecules, M_j, each one with function compound $\varphi_j(x)$ for all $j = 1, \ldots, n$. In this work, the membership value of y_p is calculated using the min function, expressed as $\mu_\Delta(x_1, \ldots, x_k)$, over the fuzzy inputs.

$$\begin{aligned} R_p: \quad & \text{if } x_1 \in A_1 \wedge \cdots \wedge x_k \in A_k, \\ & \text{then } y_p = \varphi_j(\mu_\Delta(x_1, \ldots, x_k)) \end{aligned} \tag{23}$$

In the defuzzification step, it computes the crisp output value y, using the center of gravity approach [34], as expressed in Eq. (24).

$$y = \frac{\sum_p \mu_\Delta(x_1, \ldots, x_k) \cdot y_p}{\sum_p \mu_\Delta(x_1, \ldots, x_k)} \tag{24}$$

4.2.2 Design of the artificial organic controller

Two torque inputs are expected for the one-leg mechanism: τ_1 and τ_2. In order to design a controller for the balancing task in this system, we decided to decoupled the two inputs such that τ_1 will regulate the distance R between point E and the origin O, and τ_2 will regulate the angle θ_3 (see Fig. 1).

To design the AOC for balancing the one-leg mechanism, we propose to use a proportional (P) controller for each torque input. Thus, a P-based AOC, or P-AOC, is developed as follows. First, the error signal $e(t)$ is considered as input to the P-AOC with three fuzzy partitions: "negative" (N), "zero" (Z), and "positive" (P). Fig. 4 shows the input membership functions used for the both controlled variables.

Then, the fuzzy rules for the P-AOC are proposed to be as the set summarized in Table 1. These rules consider the proportional action of the controller. Finally, a hydrocarbon compound of three molecules is proposed for

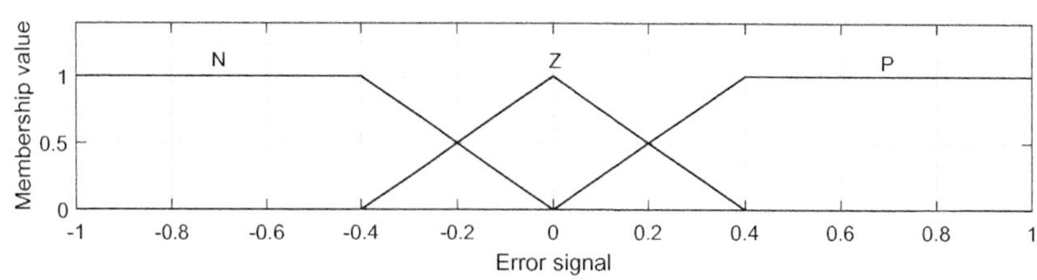

Fig. 4 Input membership functions of the error signal $e(t)$, for both distance R and angle θ_3.

Table 1 Fuzzy rules in the P-AOC

Error signal: $e(t)$	Control signal: $u(t)$
N	M_N
Z	M_Z
P	M_P

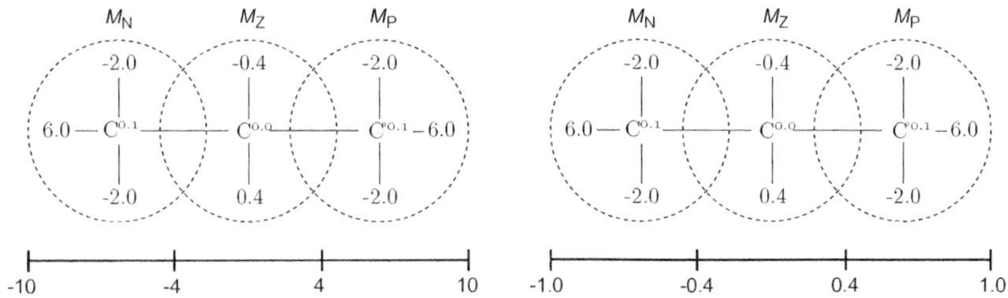

Fig. 5 Molecular partitions for the output signal $u(t)$: (*left*) of the distance value R and (*right*) of the angle θ_3.

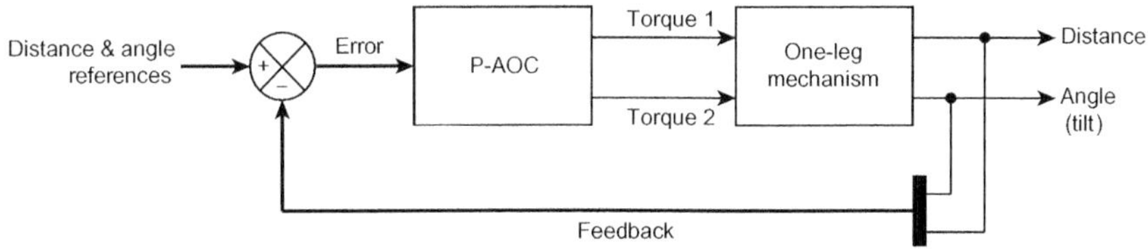

Fig. 6 Block diagram of the balancing control system using the P-AOC strategy.

the defuzzification step. This compound represents a molecular partition of the output signal of the control law $u(t)$, considering: "negative" (M_N), "zero" (M_Z) and "positive" (M_P). Fig. 5 shows the two output hydrocarbon compounds, one for each controlled variable.

To this end, Fig. 6 shows the overall balancing control system for the one-leg mechanism using the P-AOC strategy.

4.3 AHN-based reinforcement learning for balancing the one-leg mechanism

Alternatively, a learning approach is proposed for tackling the balancing task in the one-leg mechanism using RL. The goal of RL is to produce/learn a policy π (i.e., a sequence of actions) that is performed by an agent or robot to reach a goal. Furthermore, RL in continuous states and actions has been studied in limited occasions [36]. Moreover, for robotics, continuous RL is required for learning complex tasks [37], especially when control expertise is limited or the system is not known completely.

Thus, we propose the usage of AHN as a continuous RL method, namely AHN-RL, to find a strategy for balancing the one-leg mechanism from scratch. It consists in four main steps: (i) initial data collection, (ii) learning the dynamics model, (iii) policy search, and (iv) updating the training dataset. Algorithm 2 depicts the overall functionality of the AHN-RL.

In this work, we consider the dynamics function model $\hat{f}_{\omega}$ of a continuous system as Eq. (25) where $\mathbf{s}_t \in \mathbb{R}^D$ represents the state and $\mathbf{a}_t \in \mathbb{R}^F$ the action applied in time t.

$$\hat{\mathbf{s}}_{t+1} = \mathbf{s}_t + \hat{f}_{\omega}(\mathbf{s}_t, \mathbf{a}_t) \tag{25}$$

The dynamics function $\hat{f}_{\omega}$ is proposed to be parameterized with AHN, where the parameter vector ω represents the hydrogen, carbon, and bound values over the hydrocarbon compound (i.e., $\{h, v_C, L\}$ for all molecules). Then, $\hat{\mathbf{s}}_{t+1}$ represents the predicted next state that occurs after the predicted change in state $\mathbf{s}_t$ over the time step Δt.

4.3.1 Initial data collection

For training the AHN in the AHN-RL, the first step is to collect data from the system. Initially, a set of roll-outs of random action signals is applied to retrieve data. The samples considers tuples of the form $(\mathbf{s}_{t-1}, \mathbf{a}_{t-1}) \in \mathbb{R}^{D+F}$ as training inputs and differences $\Delta\mathbf{s}_t = \mathbf{s}_t - \mathbf{s}_{t-1} \in \mathbb{R}^D$ as training outputs, collected in a dataset $\mathcal{D}$.

4.3.2 Learning of the dynamics model

Then, the dynamics function model $\hat{f}_{\omega}$ is trained by minimizing the error (26) using the gradient descent method over the training dataset $\mathcal{D}$. As described below, the dataset is updated several times to increase the accuracy of the learning process.

$$E(\omega) = \frac{1}{2} \sum_{(\mathbf{s}_{t-1}, \mathbf{a}_{t-1}, \mathbf{s}_t) \in \mathcal{D}} \| \Delta\mathbf{s}_t - \hat{f}_{\omega}(\mathbf{s}_{t-1}, \mathbf{a}_{t-1}) \|^2 \tag{26}$$

4.3.3 Policy search

Once the dynamics model is learned, a policy search method is applied to the learned model $\hat{f}_{\omega}$. For simplicity, we use the random-sampling shooting method described in Rao [38] and employed in Nagabandi et al. [36]. This policy evaluation considers that at each time step t, it generates K candidate action sequences of H actions each. Then, the learned dynamics model is applied to predict the resulting future states from running each action sequence and evaluating it in a prior cost function $c(\mathbf{s}_t, \mathbf{a}_t)$ that encodes the task. The optimal action sequence is then selected, but only the first action $\mathbf{a}_t$ is executed. Then, a replan at the next step is done, as suggested in the model predictive control (MPC) [36].

4.3.4 Updating of the dataset

Lastly, tuples $(\mathbf{s}_{t-1}, \mathbf{a}_{t-1})$ and $\Delta\mathbf{s}_t$ resulting from this policy evaluation are collected in $\mathcal{D}_{\hat{f}}$ and added to the current training dataset $\mathcal{D}$ in order to update the learned model with this new information.

At last, the proposed method iterates over the dynamics model training, policy search, and collection of new training data, until the task is learned.

Algorithm 2 Proposed AHN-based RL method

Apply random control signals and collect initial training dataset $\mathcal{D}$.
repeat
  Learn the dynamics model $\hat{f}_\omega$ using $\mathcal{D}$.
  $\pi \leftarrow$ Policy search using $\hat{f}_\omega$ and MPC.
  $\mathcal{D}_{\hat{f}} \leftarrow$ Collect new data $(\mathbf{s}_{t-1}, \mathbf{a}_{t-1})$ and $\Delta\mathbf{s}_t$ from π.
  Add new data to the training dataset: $\mathcal{D} \leftarrow \mathcal{D} \cup \mathcal{D}_{\hat{f}}$.
until task learned

5 Experimental results

We numerically simulated the one-leg mechanism (Fig. 1). The distances $OA = OB = CF = DG = 30$ cm, $OC = CA = 15$ cm, $OD = DB = 15$ cm, $CE = EF = 15$ cm, $DE = EG = 15$ cm, and $EH = 12$ cm. The physical parameters of the bodies are presented in Table 2.

To analyze the performance of the above-mentioned control and learning approaches, we applied the two AHN-based strategies for tackling the balancing problem in the one-leg mechanism. To compare the performance of the AHN-based strategies, we developed: (i) a conventional P-controller as expressed in Eq. (27) with proportional gain $K_P = \text{diag}\{1, 10\}$ and (ii) a

Table 2 Physical parameters of the mechanism bodies

Body	Mass (kg)	Moment of inertia (kg m^2)
OA	0.0915	0.0007621
OB	0.0915	0.0007621
CF	0.0915	0.0007621
DG	0.0915	0.0007621
Disc at F	0.0915	0.000097
Disc at G	0.0915	0.000097
Counter inertia EH	0.271	0.0000736

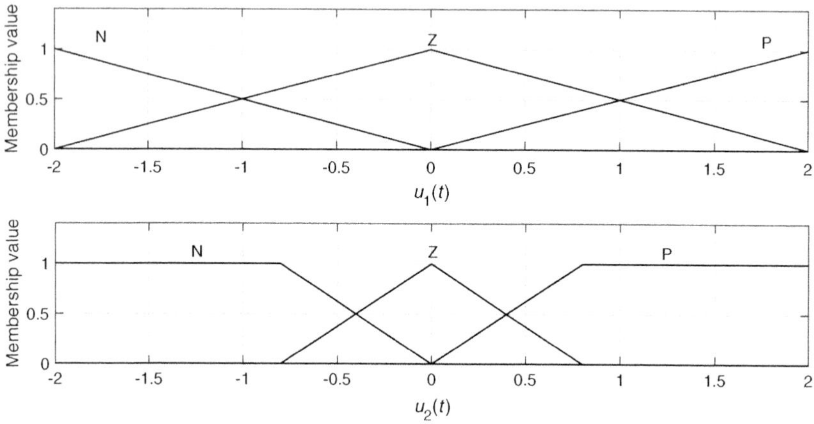

Fig. 7 Output membership functions of the control signal *u*(*t*): (*top*) for the distance value *R* and (*bottom*) for the angle θ_3.

fuzzy P-controller using almost the same input membership functions reported in Fig. 4, with fuzzy rules of Table 1, and the proposed output membership functions like the ones shown in Fig. 7. Notice that the output control signal $u(t)$ represents the applied torques $u(t) = [\tau_1(t), \tau_2(t)]$.

$$u(t) = K_P e(t) + u(t-1) \tag{27}$$

For comparison purposes, four common performance indices for control systems were applied [39]: the integral of squared error (28), the integral of absolute error (29), the integral of time multiply squared error (30), and the integral of time multiply absolute error (31), where e is the error signal at time t.

$$ISE = \int_0^{\infty} e^2(t)dt \tag{28}$$

$$IAE = \int_0^{\infty} |e(t)| dt \tag{29}$$

$$ITSE = \int_0^{\infty} te^2(t) dt \tag{30}$$

$$ITAE = \int_0^{\infty} t|e(t)| dt \tag{31}$$

5.1 Control of balancing task under disturbances

This experiment aims to measure the output response of the P-AOC under instant disturbances of 20N horizontal force, one per 500 ms, during 10 s. The one-leg mechanism has initial conditions at distance $R = 0.21$ m and angle $\theta_3 = 90$ degrees.

Fig. 8 shows a comparison between the output response of the P-AOC and the other controllers, and Table 3 summarizes the performance indices for this experiment. As notice, all the controllers have almost the same output response. Particularly to P-AOC, it performs better than the fuzzy-P controller in terms of the indices, and it performs the fastest response (see Fig. 8). In addition, Fig. 9 shows the torques τ_1 and τ_2 applied to the one-leg mechanism using the different controllers. It can be seen that P-AOC has advantage in targeting the distance value R more than the other controllers since it applies less torque to maintain the reference. Also, in terms of balancing associated to the angle θ_3, Fig. 9 shows that P-AOC generates the fastest torque τ_2. It is important to note that the actions performed assured that the one-leg mechanism did not slip.

5.2 Control of balancing task under noisy conditions

The goal of this experiment is to measure the output response of the P-AOC under noisy conditions (10% NSR) in addition to the disturbances of 20N horizontal force applied in the same way as the previous experiment. Again, the one-leg mechanism has initial conditions at distance $R = 0.21$ m and angle $\theta_3 = 90$ degrees.

Fig. 10 shows a comparison between the output response of the P-AOC and the other controllers under noisy conditions, and Table 4 summarizes the performance indices for this experiment. In terms of the P-AOC, it can be observed that performs slightly better for controlling the angle θ_3 than the others. But, notice that all the controllers can deal with the noisy conditions. Moreover, Fig. 11 depicts the applied torques when using in noisy

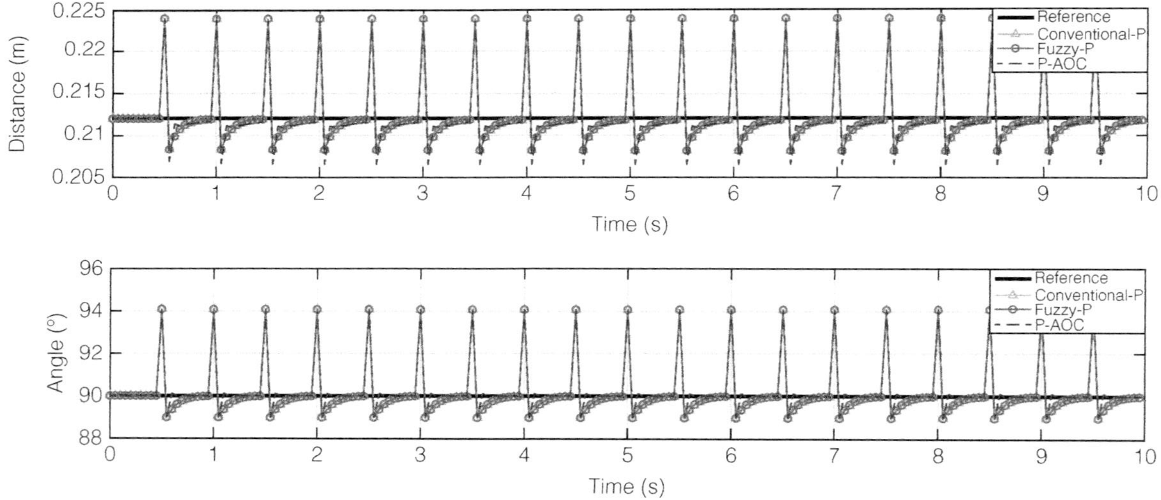

Fig. 8 Comparative output response for the balancing task under disturbances.

Table 3 Performance indices of the controllers under disturbances

Controller type	Distance R				Angle θ_3			
	ISE	IAE	ITSE	ITAE	ISE	IAE	ITSE	ITAE
Conventional-P	0.0002	0.0216	0.1104	0.0008	0.0053	0.1141	0.5767	0.0266
Fuzzy-P	0.0009	0.0278	0.0032	0.0900	0.0678	0.2828	0.2248	0.9598
P-AOC	0.0008	0.0182	0.0025	0.0536	0.0306	0.1322	0.1060	0.4395

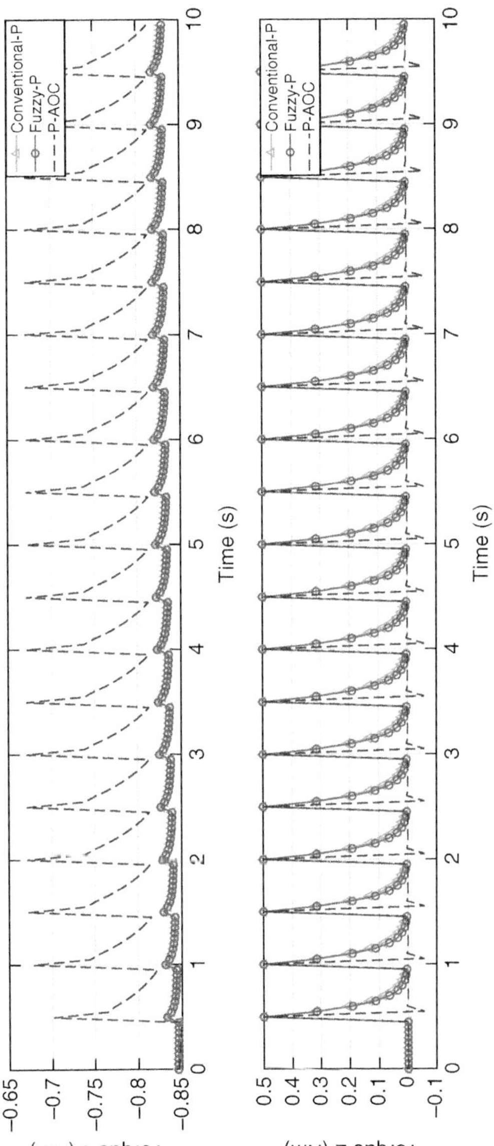

Fig. 9 Applied torques to the one-leg mechanism using the controllers.

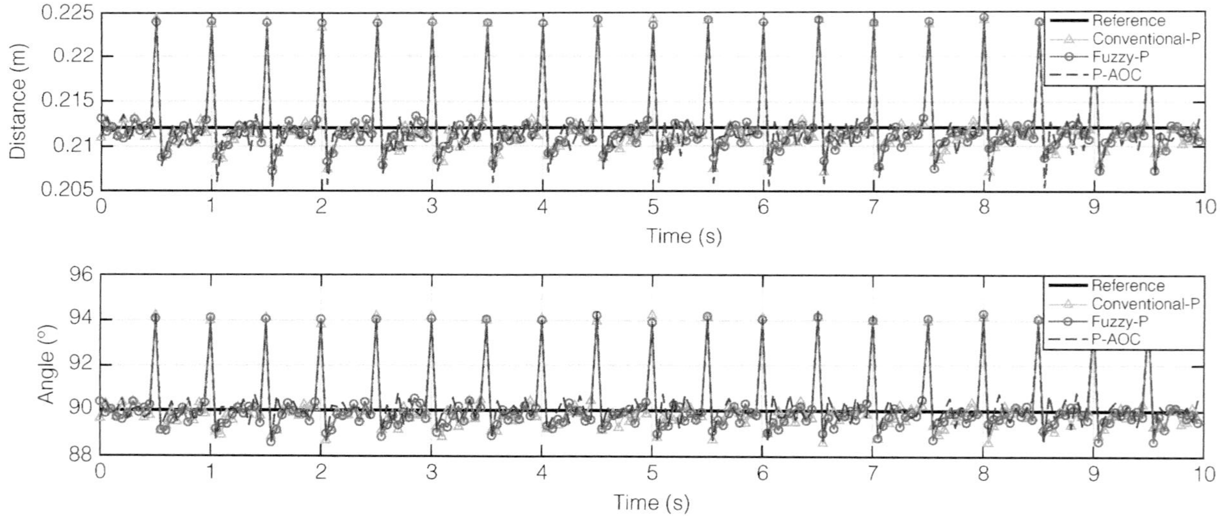

Fig. 10 Comparative output response for the balancing task under noisy conditions.

Table 4 Performance indices of the controllers under noisy conditions

Controller type	Distance R				Angle θ_3			
	ISE	IAE	ITSE	ITAE	ISE	IAE	ITSE	ITAE
Conventional-P	0.0002	0.0232	0.1169	0.0008	0.0055	0.1279	0.6396	0.0275
Fuzzy-P	0.0002	0.0230	0.1149	0.00080	0.0055	0.1277	0.6322	0.0275
P-AOC	0.0002	0.0252	0.1270	0.0009	0.0054	0.1291	0.6487	0.0271

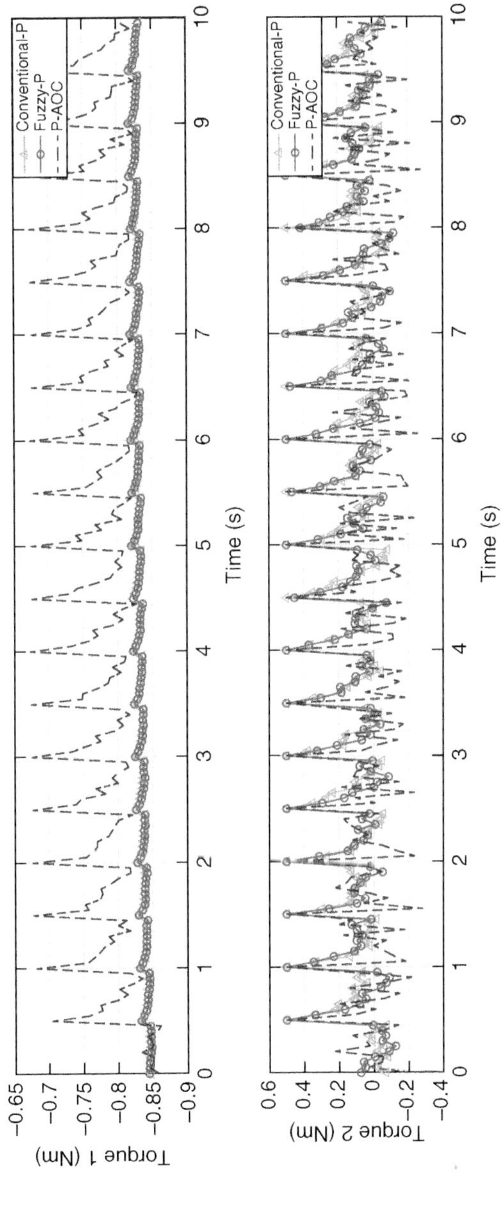

Fig. 11 Applied torques to the one-leg mechanism using the controllers.

conditions. Again, the P-AOC performs better in terms of τ_1 and reacts faster in τ_2. It is remarkable to say that the actions performed assured that the one-leg mechanism did not slip.

5.3 Disturbance rejection

Another concern about balancing the one-leg mechanism is to determine the maximum disturbance load rejection when applying a controller. In this regard, this experiment considers to measure the maximum horizontal force that can be applied to the system before sliding and falling when using the P-AOC. Thus, the one-leg mechanism is subjected to instant disturbances of horizontal forces starting at time $t=0$ s with $F=0$N and incrementing the force by +20N at a rate of 500 ms through a maximum period of 10 s. Again, the one-leg mechanism has initial conditions at distance $R=0.21$ m and angle $\theta_3=90$ degrees.

Assuming that slipping occurs when the magnitude of the resultant horizontal force R_x in the contact point is greater than the friction force f, the condition expressed in Eq. (32) is true, where μ is the static constant friction and R_y is the normal force. In that sense, Fig. 12 shows the resultant horizontal force R_x in comparison with the friction force f aiming to determine when the mechanism is slipping or not, considering $\mu=1$ for rubber on dry-concrete contact. As notice, all the controllers can reject disturbances above 200N, and particularly P-AOC can reduce significantly the resultant horizontal force R_x that is applied to the contact point for disturbances in the range from 0 to 160N, in contrast with the other controllers.

$$|R_x| > |f = \mu R_y| \tag{32}$$

To this end, it is remarkable to say that P-AOC can deal with disturbances and noisy conditions, specifically for the balancing task in the one-leg mechanism. It can also observe that P-AOC rejects larger disturbances than the other control strategies. Furthermore, the mechanical balance of the one-leg mechanism provides an easy way to control it, even with the conventional-P control strategy.

5.4 Learning of balancing task

Alternatively, we ran an experiment for learning the balancing task in the one-leg mechanism from scratch and without any prior knowledge. This experiment aims to determine the performance of the output response of the policy π obtained through the AHN-RL approach.

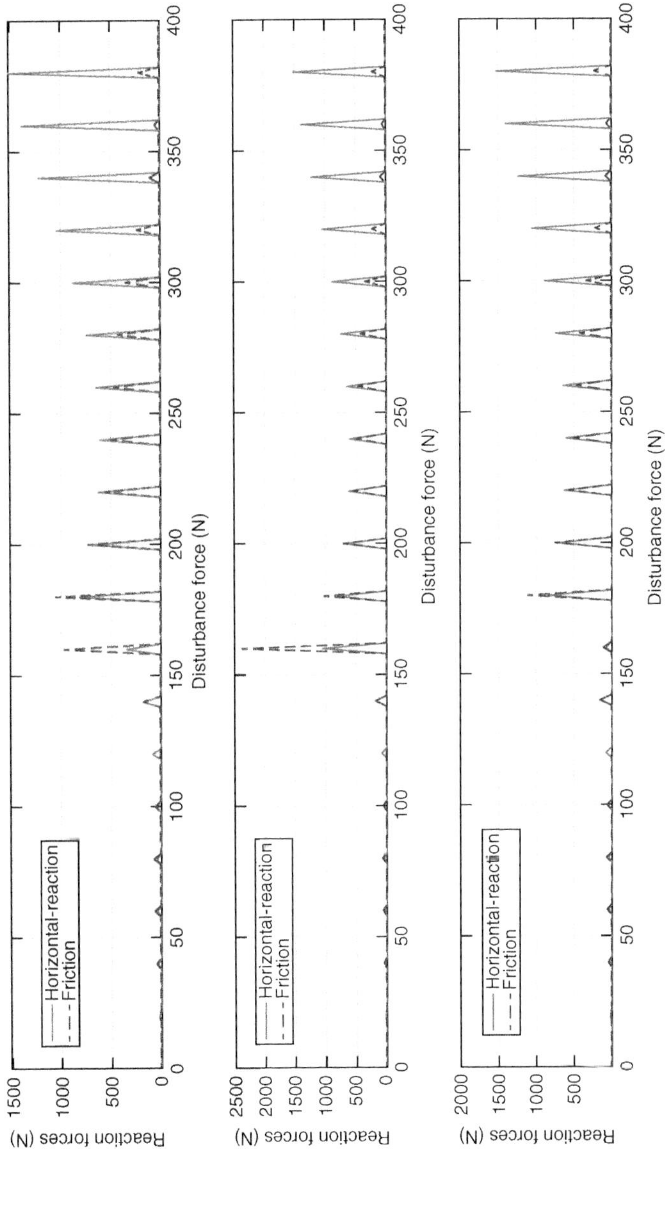

Fig. 12 Graph showing the slipping condition when applying the controllers: (*top*) conventional-P, (*middle*) fuzzy-P, and (*bottom*) P-AOC.

Thus, the balancing task was set up as follows. A state of the mechanism considers the distance R and the angle θ_3 such that $\mathbf{s}_t = (R_t, \theta_{3_t})$. The goal state is $\mathbf{s}_{goal} = (R, \theta_3)$ and the cost function $c(\mathbf{s}_t, \mathbf{a}_t)$ is expressed as Eq. (33) where *penalty* is a positive value greater than zero that punishes the slipping behavior in the mechanism.

$$c(\mathbf{s}_t, \mathbf{a}_t) = \begin{cases} \sqrt{\sum_i (\mathbf{s}_{goal,i} - \mathbf{s}_{t,i})^2} & |R_x| \leq |f = \mu R_y| \\ penalty & |R_x| > |f = \mu R_y| \end{cases} \tag{33}$$

Initial dataset $\mathcal{D}$ collected 10 roll-outs with random initial states $\mathbf{s}_{initial} = (0.21, 90 \pm 15)$ for coverage. Five iterations were fixed for training AHN. We set the AHN with 10 molecules and learning rate $\eta = 0.01$. We preprocessed the training data by normalizing it to be mean 0 and standard deviation 1 ensuring that the loss function weights all parts of the state equally. For the policy search, we set $K = 100$ candidate sequences and $H = 3$ horizon steps. After training, we performed an online policy search π via MPC with $K = 3$ and $H = 3$ and the usage of the trained AHN-RL.

For the training step, Fig. 13 shows the estimation of the AHN-model in contrast with the target values from $\mathcal{D}$. As depicted, it can be observed that estimates highly approximate to the target values, obtaining an accuracy of 99.42% which can be considered well trained.

After that, the policy π was computed. Fig. 14 shows the output response of the performed π found by AHN-RL over the two control experiments explained earlier: output response under disturbances and under noisy conditions. Table 5 summarizes the performance indices for the entire experiment. As shown, the output response of the AHN-RL controller under disturbances and noisy conditions is almost the same. In terms of the distance value, it is not reached but the indices confirm a slight variation less than 0.0008 ISE and 0.0039 ITAE, considered good metrics. In terms of the angle, it performs well in both disturbances and noisy conditions proving that the AHN-RL controller can learn to balance the one-leg mechanism from scratch.

Moreover, Fig. 15 shows the torques applied during the two experiments. As shown, the torques are similar in both conditions, but actually the applied torques are the smallest ones over the entire experimentation (i.e., $\tau_1 \in [-0.1, 0.1]$ and $\tau_2 \in [-0.05, 0.05]$). This particular behavior gives insights about the powerful of learning a task using the RL approach.

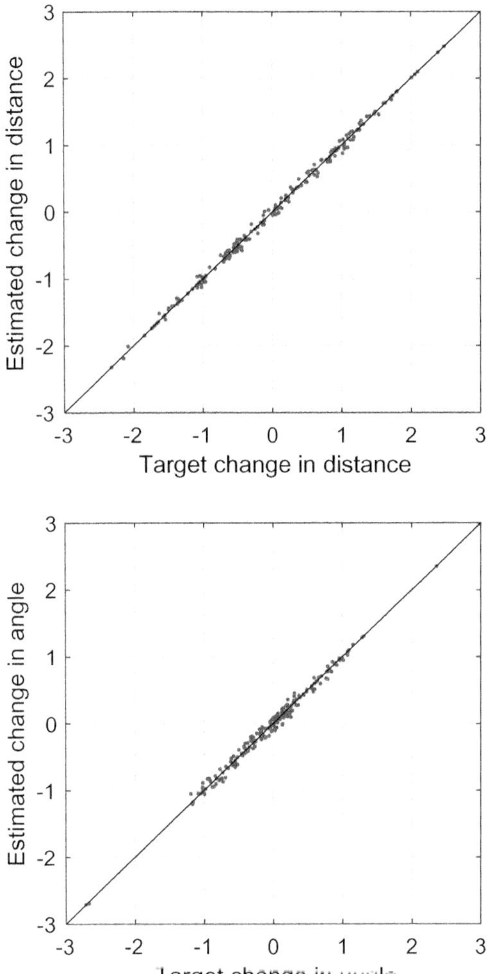

Fig. 13 Comparison between estimates and targets of the AHN-based dynamics model $\hat{f}_\theta$: (*top*) changes in distance R, and (*bottom*) changes in angle θ_3.

6 Conclusions

This chapter presented a force-balanced one-leg mechanism for easy controlling. The proposed design was modeled and simulated. In addition, we proposed to use AHN for controlling and learning the balancing task in the mechanism. Experimental results provided information to conclude that the proposed mechanism is easily controlled by simple P-control

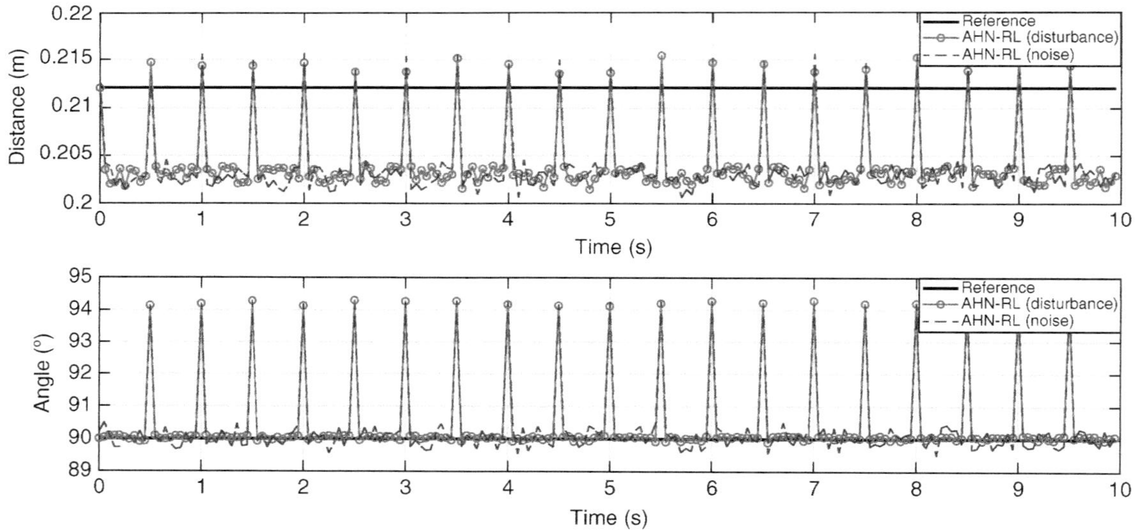

Fig. 14 Comparative output response for the balancing task using AHN-RL.

Table 5 Performance indices of the AHN-RL controller

AHN-RL controller	Distance R				Angle θ_3			
	ISE	IAE	ITSE	ITAE	ISE	IAE	ITSE	ITAE
Disturbance	0.0008	0.0847	0.4255	0.0039	0.0052	0.0788	0.3944	0.0258
Noisy conditions	0.0008	0.0865	0.4304	0.0039	0.0053	0.1026	0.5133	0.0264

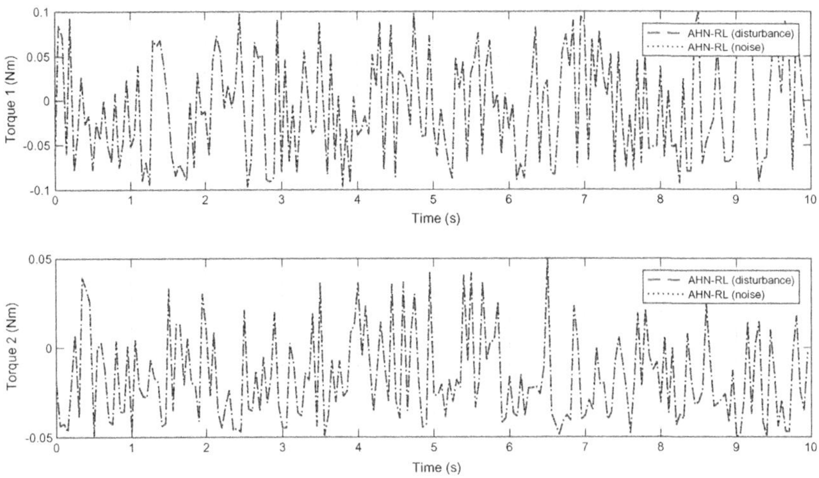

Fig. 15 Applied torques to the one-leg mechanism using AHN-RL.

strategies. In addition, both P-AOC control and AHN-RL learning approaches performed well when the mechanism was subjected to disturbances and noisy conditions. Also, a disturbance rejection analysis was done to validate the performance of the controllers.

For future work, we will investigate controllability and learnability of more complex robotic tasks such as walking, jumping and climbing stairs using this mechanism in legged robots. Also, we will characterize the dynamics of the one-leg mechanism in detail.

Acknowledgments

This research has been funded by the Universidad Panamericana through the grant "Fomento a la Investigación UP 2017", under project code UP-CI-2017-ING-MX-03, and it has been partially supported by the Google Research Awards for Latin America 2017.

References

[1] M.H. Raibert, Legged robots, Commun. ACM 29 (6) (1986) 499–514.

[2] P.B. Weiber, R. Tredake, S. Kuindersma, Modeling and control of legged robots, in: B. Siciliano, O. Khatib (Eds.), Springer Handbook of Robotics, Springer International Publishing, second ed, Switzerland, 2015 (Chapter 48).

[3] J. Cuadrado, U. Lugris, F. Michaud, F. Mouzo, Role of multibody dynamics based simulation in human, robotic and hybrid locomotion benchmarking, in: Workshop on Benchmarking Bipedal Locomotion, 2014 IEEE-RAS Int. Conference on Humanoid Robots, Poster, Madrid, Spain, 2014.

[4] M.A. Sherman, A. Seth, S.L. Delp, Simbody: multibody dynamics for biomedical research, Procedia Iutam 2 (2011) 241–261.
[5] I. Díaz, J.J. Gil, E.E. Sánchez, Lower-limb robotic rehabilitation: literature review and challenges, J. Robot. 2011 (1) (2011) 1–11.
[6] D. Gong, J. Shao, Y. Li, G. Zuo, Study of human-like locomotion for humanoid robot based on human motion capture data, in: 2016 IEEE International Conference on Robotics and Biomimetics (ROBIO), 2016, pp. 933–938.
[7] D. Torricelli, J. Gonzalez, W. Maarten, R. Jiménez-Fabián, B. Vanderborght, M. Sartori, S. Dosen, D. Farina, D. Lefeber, J.L. Pons, Human-like compliant locomotion: state of the art of robotic implementations, Bioinspiration Biomim. 11 (5) (2016) 051002.
[8] R. Khusainov, I. Shimchik, I. Afanasyev, E. Magid, Toward a human-like locomotion: modelling dynamically stable locomotion of an anthropomorphic robot in Simulink environment, in: 2015 12th International Conference on Informatics in Control, Automation and Robotics (ICINCO), Colmar, France, 2015.
[9] J. Jackson, C. Hass, B. Fregly, Development of a subject-specific foot-ground contact model for walking, J. Biomech. Eng. 138 (9) (2016) 091002 (12 pages).
[10] N. Takayuki, Y. Araki, H. Omohi, A new gravity compensation mechanism for lower limb rehabilitation, in: 2009 International Conference on Mechatronics and Automation, 2009, pp. 943–948.
[11] M. Stojicevic, M. Stoimenov, Z. Jeli, A bipedal mechanical walker with balancing mechanism, Tech. Gazette 25 (1) (2018) 118–124.
[12] J.R. Rebula, P.D. Neuhaus, B.V. Bonnlander, M.J. Johnson, J.E. Pratt, A controller for the LittleDog quadruped walking on rough terrain, in: Proceedings—IEEE International Conference on Robotics and Automation, 2007, pp. 1467–1473.
[13] E. Yoshida, O. Kanoun, C. Esteves, J.P. Laumond, Task-driven support polygon reshaping for humanoids, in: Proceedings of the 2006 6th IEEE-RAS International Conference on Humanoid Robots, HUMANOIDS, 2006, pp. 208–213.
[14] A. Del Prete, S. Tonneau, N. Mansard, Fast algorithms to test static equilibrium for legged robots, in: IEEE International Conference on Robotics and Automation (ICRA), Stockholm, Sweden, 2016, pp. 1601–1607.
[15] E. Najafi, G.A.D. Lopes, R. Babuska, Balancing a legged robot using state-dependent Riccati equation control, IFAC Proc. 47 (3) (2014) 2177–2182.
[16] F. Grasser, A. D'Arrigo, S. Colombi, A.C. Rufer, JOE: a mobile, inverted pendulum, IEEE Trans. Ind. Electron. 49 (1) (2002) 107–114.
[17] M. Kashki, J. Zoghzoghy, Y. Hurmuzlu, Adaptive control of inertially actuated bouncing robot, IEEE Trans. Robot. 33 (3) (2017) 509–522.
[18] P. Wensing, A. Wang, S. Seok, D. Otten, J. Lang, S. Kim, Proprioceptive actuator design in the MIT Cheetah: impact mitigation and high-bandwidth physical interaction for dynamic legged robots, IEEE/ASME Trans. Mechatron. 22 (5) (2017) 2196–2207.
[19] K. Komoda, H. Wagatsuma, Energy-efficacy comparisons and multibody dynamics analyses of legged robots with different closed-loop mechanisms, Multibody Syst. Dyn. 40 (2017) 123–153.
[20] V. van der Wijk, J. Herder, Synthesis of dynamically balanced mechanisms by using counter-rotary countermass balanced double pendula, ASME J. Mech. Des. 131 (11) (2009) 111003 (8 pages).
[21] J. García de Jalón, E. Bayo, Kinematic and Dynamic Simulation of Multibody Systems: The Real-Time Challenge, Springer-Verlag, New York, NY, 1994.
[22] S. Briot, V. Arakelian, J.-P. Le Baron, Shaking force minimization of high-speed robots via centre of mass acceleration control, Mech. Mach. Theory 57 (2012) 1–12.
[23] M. Acevedo, An efficient method to find the dynamic balancing conditions of mechanisms: planar systems, in: ASME International Design Engineering Technical

Conferences and Computers and Information in Engineering Conference, vol. 5B: 39th Mechanisms and Robotics Conference, 2015.
[24] H. Ponce, P. Ponce, Artificial organic networks, in: 2011 IEEE Conference on Electronics, Robotics and Automotive Mechanics, IEEE, Cuernavaca, Morelos, Mexico, 2011, pp. 29–34.
[25] H. Ponce, P. Ponce, A. Molina, Artificial organic networks: artificial intelligence based on carbon networks, in: Studies in Computational Intelligence, vol. 521, Springer, New York, NY, 2014.
[26] H. Ponce, P. Ponce, A. Molina, Adaptive noise filtering based on artificial hydrocarbon networks: an application to audio signals, Expert Syst. Appl. 41 (14) (2014) 6512–6523.
[27] H. Ponce, P. Ponce, A. Molina, The development of an artificial organic networks toolkit for LabVIEW, J. Comput. Chem. 36 (7) (2015) 478–492.
[28] H. Ponce, A novel artificial hydrocarbon networks based value function approximation in hierarchical reinforcement learning, in: 15th Mexican International Conference on Artificial Intelligence, Springer, 2016, pp. 211–225.
[29] H. Ponce, P. Ponce, A. Molina, A novel robust liquid level controller for coupled-tanks systems using artificial hydrocarbon networks, Expert Syst. Appl. 42 (22) (2015) 8858–8867.
[30] D. Qian, S. Tong, H. Lium, X. Liu, Load frequency control by neural-network-based integral sliding mode for nonlinear power systems with wind turbines, Neurocomputing 173 (2016) 875–885.
[31] A. Roose, S. Yahyam, H. Al-Rizzo, Fuzzy-logic control of an inverted pendulum on a cart, Comput. Electr. Eng. 61 (2017) 31–47.
[32] M. Stogiannos, A. Alexandridis, H. Sarimveis, Model predictive control for systems with fast dynamics using inverse neural models, ISA Trans. 72 (2018) 161–177.
[33] P. Ponce, H. Ponce, A. Molina, Doubly fed induction generator (DFIG) wind turbine controlled by artificial organic networks, Soft Comput. 22 (9) (2018) 2867–2879.
[34] H. Ponce, P. Ponce, A. Molina, Artificial hydrocarbon networks fuzzy inference system, Math. Probl. Eng. 2013 (2013) 1–13.
[35] A. Molina, H. Ponce, P. Ponce, G. Tello, M. Ramirez, Artificial hydrocarbon networks fuzzy inference systems for CNC machines position controller, Int. J. Adv. Manuf. Technol. 72 (9–12) (2014) 1465–1479.
[36] A. Nagabandi, G. Yang, T. Asmar, G. Kahn, S. Levine, R.S. Fearing, Neural network dynamics models for control of under-actuated legged millirobots (2017) arXiv:1711.05253.
[37] E.M. de Cote, E.O. Garcia, E.F. Morales, Transfer learning by prototype generation in continuous spaces, Adapt. Behav. 24 (6) (2016) 464–478.
[38] A. Rao, A survey of numerical method for optimal control, Adv. Astronaut. Sci. 135 (2009) 497–528.
[39] S.-K. Oh, W. Pedrycz, S.-B. Rho, T.-C. Ahn, Parameter estimation of fuzzy controller and its application to inverted pendulum, Eng. Appl. Artif. Intell. 17 (1) (2004) 37–60.

CHAPTER 4

Human lower limb operation tracking via motion capture systems

Med Amine Laribi, Said Zeghloul
Department of GMSC, Pprime Institute CNRS, ENSMA, University of Poitiers, UPR 3346, Poitiers, France

1 Introduction

The motion capture systems have become an essential tool and widely used by researchers in their works. Basically, the motion analysis was used by doctors and physical therapists for kinematics analysis and biomechanics, sports medicine and physical therapy, including human gait analysis and injury rehabilitation. This chapter describes content from fundamentals up to recent advances on human lower limb analysis that our team made in the past few years based on motion capture system.

Numerous studies and design of biped humanoid robots have been inspired through the observation of human locomotion. As the problem of biped locomotion is solved by nature during a long time of human being's evolution.

Human walking is quite smooth and efficient which could be explained by the optimal use of the gravity by humans [1]. In general, the requested energy during normal walking is rather weak. In addition, human locomotion like presents an optimal energy consumption [2]. In steady state, the basic characteristics of the walking gestures are linked to features of somewhere symmetric and periodic walking.

Many research works have focused on the study of the kinematics and dynamics of human locomotion by applying several approaches.

For example, a wire tracking system called CaTraSys (Cassino Tracking System), that has been designed and built at LARM [3–5], has been used to measure large displacements and orientation variations of complex movable multibody systems. This measuring system is capable to determine the trajectory of the human limb extremity during walking operation [6]. An optical based wearable goniometer for human joint measurement has been

Design and Operation of Human Locomotion Systems
https://doi.org/10.1016/B978-0-12-815659-9.00004-4

proposed by Lim et al. [7]. The motivation behind this device was to have a compact, low cost, wearable goniometer that can be worn for extended monitoring of human joint movement.

Motion capture systems have been used to study and record the human gestures and convert them to a specific digital model. This method, increasingly used, was adapted to military, entertainment, sports, and medical applications, and for validation of computer vision [8] and robotics.

Laribi et al. used a motion capture method to study the expert gestures aiming at identifying a maximum orientation of a medical probe. The purpose was to perform an optimal synthesis of a new spherical parallel mechanism used as a structure for a probe holder for application to a tele-echography chain [9].

The same approach of motion capture is applied for the case of minimally invasive surgery to identify the workspace [10]. The recording of the movement of surgeon is performed with equipment dedicated to laparoscopy learning techniques. The synthesis of movement and the biomechanics were introduced into the design methodology for the development of a minimally invasive surgery robot manipulator.

Boutin et al. have developed an auto-adaptable method to imitate a human gait with a humanoid robot. The human walking pattern is recorded by a motion capture system and two force platforms. The proposed method can be used for different humanoid robots of different sizes [11].

The chapter is organized as follows: the Section 2 focuses on the analysis of human walking. A human walk cycle as well as his two main phases the stance phase and the swing phase, have been described, respectively. Gait parameters and lower extremity angles are defined at the end of this section. The quantitative gait analysis is presented in Section 3, which is an important tool for studying and quantifying patterns of locomotion. In this section, the experimental layout and a general markers configuration fitted to lower body of an adult human are presented. In this section, we present a simple markers configuration that can be easily implemented for gait evaluations.

Section 4 presents a human walking analysis which is carried out to get information on kinematics and dynamics of the lower limbs during a walking cycle that will would be used in the design of the training machine. Section 5 addresses an experimental analysis of overcoming obstacle in human walking performed by means of a motion capture system. Finally, Section 6 provides the conclusions drawn from the analysis of study of human walking via motion capture.

2 Analysis of human walking

2.1 Overview of human walking

A gait cycle is defined as the interval of time between any of the repetitive events of walking. The two legs have the same series of events, with a phase shift of one half cycle [12]. Such an event could include the point when the foot initially contacts the ground, to when the same foot contacts the ground again.

A human walk cycle has two main phases: Stance phase and Swing phase. During stance time the foot is on the ground, over the swing time the foot is no longer in contact with the ground [13]. The sub phases of stance and swing is presented in Fig. 1.

2.1.1 Stance phase

The stance phase represents about 60% of the gait cycle. Describes the entire time the foot is in contact with the ground and the limb is bearing weight. This phase is begins with the initial contact of the foot on the ground, and concludes when the ipsilateral foot leaves the ground. The stance period takes about 0.6s during an average walking speed [14].

Within the stance period, two tasks and four intervals are recognized. The two tasks include weight acceptance and single limb support. The weight acceptance task occurs during the first 10% of the stance period. The single limb support, also called double limb stance, when the current limb is on the ground and the other limb is in the air. The four intervals include loading response, mid stance, terminal stance and pre-swing.

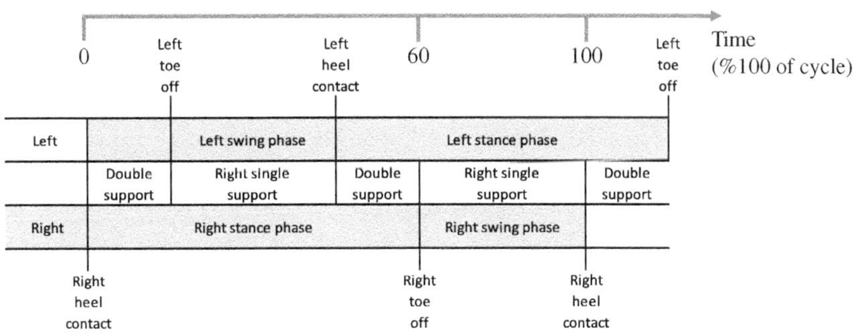

Fig. 1 Sub phases of stance and swing.

2.1.2 Swing period

The swing period constitutes approximately 40% of the gait cycle. Describes the period when the foot is not in contact with the ground. This phase begins as the foot is lifted from the ground and ends with initial contact with the ipsilateral foot.

Within the swing period, one task and four intervals are recognized. The task involves limb advancement. The four intervals include pre-swing, initial swing, mid-swing, and terminal swing.

2.1.3 Gait parameters

During a gait cycle, feet placements provide some useful distance parameters. Step length is measured as the distance between the same points of 1 ft on successive footprints. Stride length, on the other hand, is the distance between successive points of foot-to-floor contact of the same foot.

The gait can be characterized by its cadence. It is defined as the number of separate steps taken in a certain time. Normal cadence is between 90 and 120 steps per minute (Fig. 2).

2.1.4 Lower extremity angles in gait cycle

Two types of angles are used in the measurement of joint angles by the biomechanics community: absolute angles and relative angles. The first are the angles between a body segment and a fixed coordinate system. The relative angles or intersegmental angles are taken between longitudinal axes of two adjacent body segments.

The definition of angles at the beginning of the examining joint motions represent a major importance.

3 Quantitative gait analysis

Quantitative gait analysis is an important tool for studying and quantifying patterns of locomotion. Several practical computer-aided methods

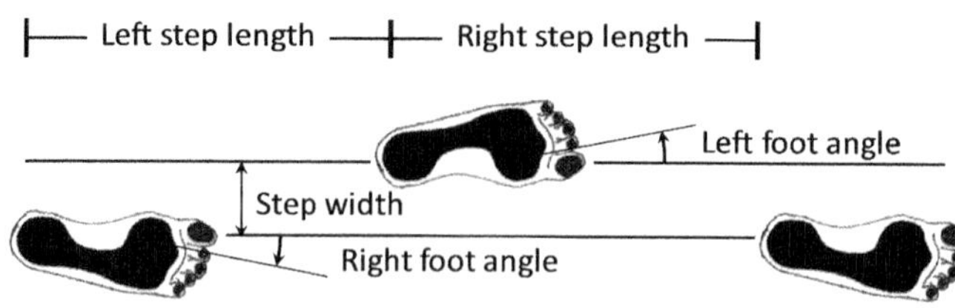

Fig. 2 Distance parameters of the gait cycle with reference to the feet footprints.

in current use provide relative orientation of segments either directly or as a derived parameter from measurements of relative position of segments.

These methods are based on the use of several cameras provide accurate three-dimensional spatial positions of reflective skin markers placed on key anatomical sites on the lower extremities. From these positional data, the relative angular rotation of the individual body segments is derived using analytical techniques based on a biomechanical model of the lower extremity.

The gait analysis presented in this work has been performed in the biomechanics laboratory of PPRIME institute (Poitiers University, France), using an optical motion capture system and force platforms.

Our motion capture method is based on the experience of biomechanics community [3] especially for the choice of marker sets and segment reference definitions. Computer graphics [4] focus on the use of motion capture and synthesis movement to generate three-dimensional realistic movements for virtual models including non anthropomorphic models as well.

3.1 An experimental layout

The motion measurement device used in this work is the Vicon MX infrared optical system. Typical configuration of the system is reported in Fig. 3. Two information are implied during the gait capture. The position tracking of a set of reflective markers placed on a moving subject on the one hand, and the recording of the ground reaction force on the other.

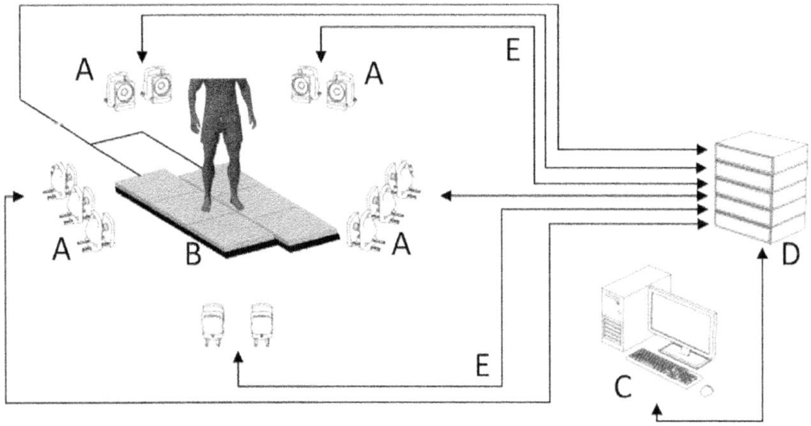

Fig. 3 Typical configuration of the motion capture system VICON: A-MX cameras, B-Force plates, C-Host PC with Nexus Vicon software, Vicon Ethernet card and network Ethernet card, D-MX connectivity units, E-MX cables.

All cameras are connected to acquisition platform units, which sends information to a Host PC with Nexus software for processing. The system supports up to 20 MX cameras (4 megapixel @ 370 fps) and in the present work only 12 has been used. Each camera is positioned in a way that its field of view covers as much as possible the captured volume.

Each marker must be seen at least by two cameras during the capture. In fact, the position of a marker is computed by the Vicon Nexus software based on triangulation process [15]. All markers are described by x, y and z components of coordinates at each moment in a reference coordinates frame.

In the middle part of the walking zone, six force plates, which can provide 6D interaction wrench measurements, are placed on the floor in order to record ground reaction force at a sampling frequency of 1000 Hz (see Fig. 3). Both position and force information are synchronized by the MX system.

3.2 Markers configuration

The markers configuration described by the number as well as the position of markers is usually linked to the application to study. The marker positions here was designed with a minimum of markers to simplify the identification of marker trajectories.

The position of markers (4 cm in diameter, weighing 4.4 g, developed in this study) was selected to satisfy the rigid body assumption as well as other practical requirements based on the experience of the biomechanics community. The requirement is useful especially for the choice of marker sets and segment reference definition [16]. Computer graphics focus on the use of motion capture and synthesis movement to generate three-dimensional realistic movements for virtual models [17].

Markers are used in such a way as the kinematic data of human body are obtained. Used markers configuration is shown in Fig. 4. Here it is just focused on the lower body of human. In order to obtain the orientation between different segments, at least three markers are needed to be attached to each segment. The lower body has been divided into three parts and six segments (see Fig. 4). The three parts are defined by the pelvis and the two limbs left and right. The six segments are left thigh, left shank, left foot, right thigh, right shank, and right foot.

Thus, total 21 markers are used and 9 markers are placed on each limb. Note that R and L prefix are used to distinguish, respectively, the markers of the right and left limbs.

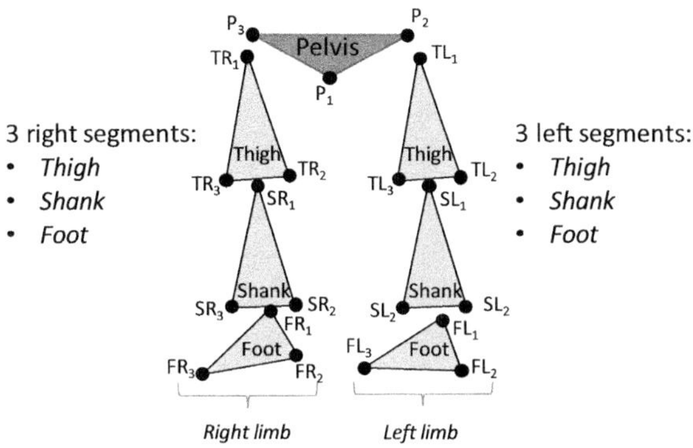

Fig. 4 Used markers configuration for the lower body.

In particular, markers P_1, P_2 and P_3 are placed at the pelvis, TR_1, TR_2 and TR_3 are placed at the thigh, SR_1, SR_2 and SR_3 are placed at the shank and finally FR_1, FR_2 and FR_3 are placed at the foot. Markers on the left limb are attached similarly as the right limb.

Note that these reflective markers are placed in anatomical locations that have less important soft tissue artifact, i.e., the movement of the skin relative to the bones.

In the next, two examples of quantitative analysis will be presented and aiming each one for a specific purpose. The first quantitative gait analysis is performed to be used as input to the inverse dynamic model of a gait training machine based on a cable robot. The second quantitative gait analysis is carried out to study experimentally and characterize the overcoming obstacle in human walking.

4 Human walking analysis

The knowledge of the kinematics and dynamics of the lower limbs during a walk cycle is a valuable information required to the design of the training machine [18]. A quantitative gait experiment is performed in this aim with a subject equipped with a set of reflective markers achieving a walking movement. All reflective marker trajectories are used to calculate the orientation of the lower limb segments.

At the beginning, a static trial is recorded and the subject is asked to do not move. This motionless record is required to identify and label each marker. Hereafter, the candidate was asked to perform 4 min of free walking along the captured walkway.

The first step of the post-processing is to locate the center of rotation (CoR) of the hip, knee and ankle joints. In this purpose, the candidate achieved a number of cycles of flexion/extension of each joint of the lower limbs resulting in six motion trials. These motions are in close angular ranges to those occurring during a normal walking.

4.1 Joint centers location

By examining the recorded flexion/extension motion trial, the first step of the post-processing is to locate the center of rotation (CoR) of the hip, knee and ankle joints. Various approaches have been developed in the literature to identify the CoR linked to the hip joint known as the hip joint center (HJC). These approaches are divided into two groups: predictive approaches [19, 20] and functional methods [12, 21, 22].

The predictive approaches compute the hip joint center by relying on the position of palpated anatomical landmarks [23], i.e., the joint center is an empirical relation between bone landmarks.

However, functional methods use kinematic and geometric proprieties to locate the CoR between the pelvis and the thigh. Such methods are of two types: sphere fitting methods and transformation techniques [21].

The first method fits the trajectory of a marker on the thigh to a sphere, consequently the sphere's center is the CoR.

In order to write the motion of a thigh marker in relation to the pelvis, a transformation is applied from a global frame to a pelvic local frame. Consequently, the trajectory of such marker will sweep the surface of a sphere. An optimization method based on geometric formulation is used to fit a cloud of point to a sphere. The sphere fit method is selected for the identification of the hip joint center. The sphere fitting functional methods is recommended by the International Society of Biomechanics [24] and its accuracy is reported in several works [21, 22, 25].

The formulation of the sphere fit method is given for the case of the right hip joint center. The local coordinate system is computed using the coordinates of three markers P_1, P_2 and P_3 located on the pelvis. The marker P_3 is considered as the center of the local frame $\mathcal{F}_p$. The coordinates of each marker located on the hip is computed using the homogenous transformation matrix $\mathbf{T}^{pg}$ given by Eq. (1), in relation to the local reference frame $\mathcal{F}_p$.

$$\mathbf{T}^{pg} = \begin{bmatrix} \mathbf{R}^{pg} & \mathbf{L}_{pg} \\ 000 & 1 \end{bmatrix} = \begin{bmatrix} (\mathbf{R}^{gp})^{-1} & -(\mathbf{R}^{gp})^{-1}\mathbf{P}_3 \\ 000 & 1 \end{bmatrix} \tag{1}$$

Where, $\mathbf{R^{pg}} = [\mathbf{x_p}\ \ \mathbf{y_p}\ \ \mathbf{z_p}]$ is the rotation matrix defined by $\mathbf{x_p}$ $\mathbf{y_p}$ and $\mathbf{z_p}$ the coordinates of frame $\mathcal{F}_p$ axes and P_3 is the coordinates of its origin.

The rigid transformation, defined by the matrix $\mathbf{T^{pg}}$, applied to markers TR_2 and TR_3 leads to a sphere surface centered at the hip joint center (Fig. 5).

The geometric sphere fitting algorithm is applied to compute the right hip joint center, the point C_h. This algorithm consists on the minimization of an objective function defined as follows:

$$\text{minimize}: \sum_{i=1}^{n} (\|\mathrm{TR}_i - C_h\| - R)^2 \tag{2}$$

With $R = \frac{1}{n} \bullet \sum_{i=1}^{n} \| TR_i - C_h \|$

The final hip joint center is obtained by computing the mean values between the two centers obtained from each marker TR_2 and TR_3. Thus, the global position of the hip joint center, $\mathbf{H}$, was computed using the point C_h from the local frame $\mathcal{F}_p$ to the global frame $\mathcal{F}_g$: $\mathbf{H} = \mathbf{T^{gp}} \cdot C_h$.

The CoR for the knee and ankle joints can be estimated by computing a mean position between the two markers (TR_2, TR_3) and (SR_2, SR_3), respectively. Since, these joints are described by a revolute joint each

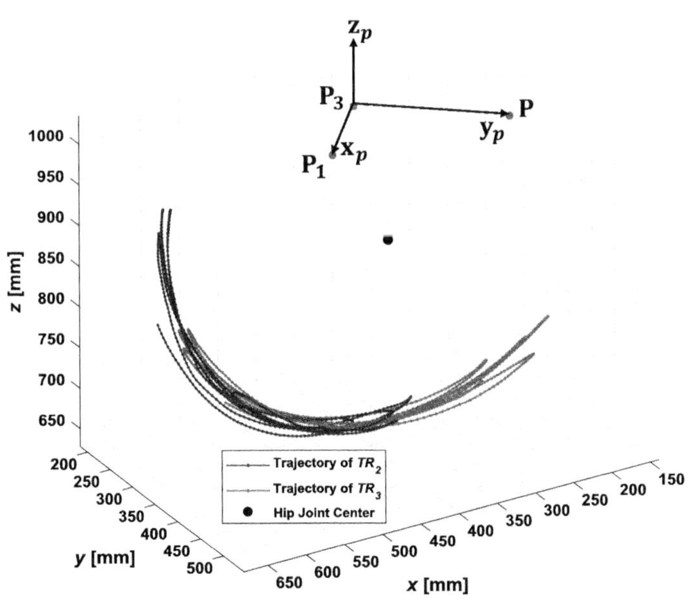

Fig. 5 Hip joint center using sphere fitting functional method.

one, we chose to use the sphere fitting method to define the CoR. The center for the knee joint as well as the center of the ankle joint are computed in a similar way as the hip joint center.

Within the walking motion record, once all the center of rotations are computed using the positions of the reflective markers as a function of time, the angular positions are defined with respect to a reference position as shown on Fig. 6 and given by Eqs. (3)–(5).

$$\theta_p = (\widehat{z, C_h C_k}) \tag{3}$$

$$\theta_p = (\widehat{C_h C_k, C_k C_a}) \tag{4}$$

$$\theta_p = \left(\widehat{C_k C_a, FR_2 FR_3}\right) - \frac{\pi}{2} \tag{5}$$

The computed angular positions for one gait cycle are depicted in Fig. 7. Thus, if we normalize the duration to 100% of one cycle gait, the position in function of time is converted to the one given in Fig. 8.

In sagittal plane, the displacement of the pelvis is obtained based on the vertical position of any marker that belongs to the pelvis. The vertical position of the pelvis normalized to 100% of one cycle gait is depicted in Fig. 9.

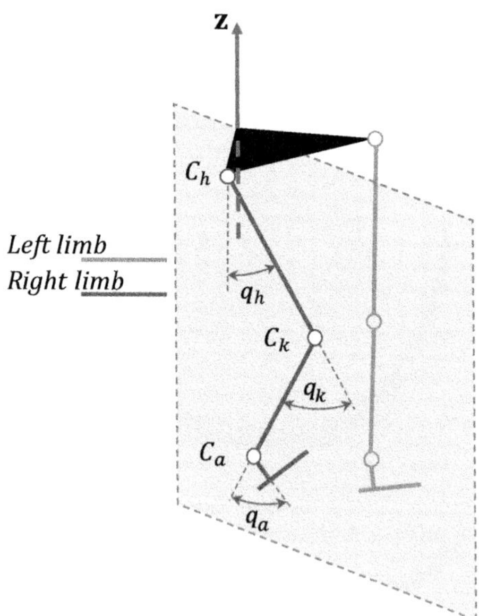

Fig. 6 The angular joints for lower limb.

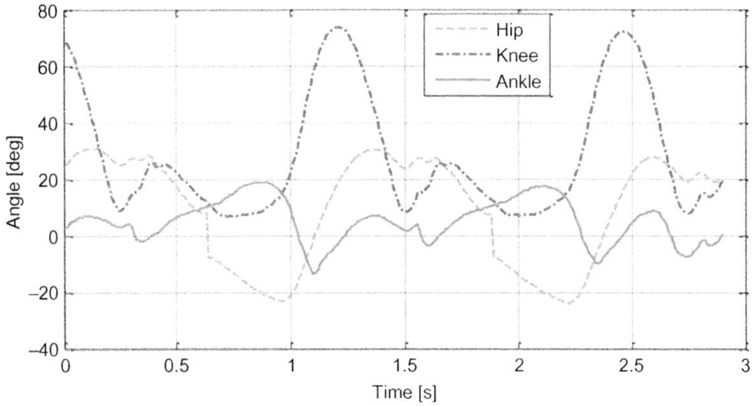

Fig. 7 Angular positions of the lower limb for one gait cycle in function of time.

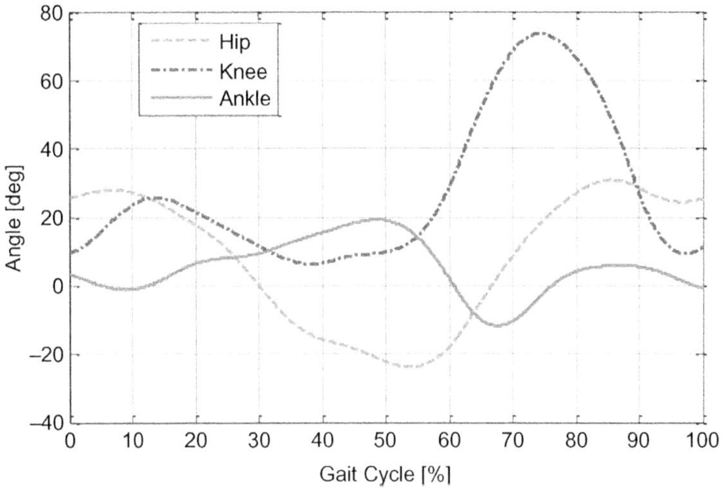

Fig. 8 Angular positions of the lower limb in percentage during one gait cycle.

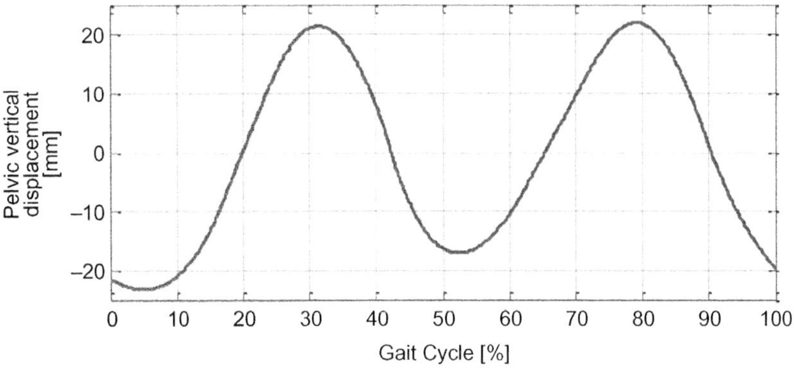

Fig. 9 Vertical position of the pelvis in percentage during one gait cycle.

4.2 Ground reaction force

The ground reaction force is recorded using the force platform running at a frequency of 1000 Hz. The reaction force given by the platform in the form of a three-dimensional vector.

The components included in the sagittal plane of both feet are the vertical reaction (R_v) and the anterior-posterior reaction (R_{ap}). These reactions are positive when directions are upward and forward, respectively (Fig. 10).

The vertical reaction R_v is positive during the stance phase and linked to the body weight. The curve of R_v presents two peaks which describe each one upward acceleration and deceleration of the center of motion of the body. The anterior-posterior reaction is negative during the first half of the stance phase resulting due to braking and positive for the second half describing propulsion. The two components (R_v and R_{ap}) vanishes when the foot is swinging.

Similarly as the angular positions, if we normalize the duration to 100% of one cycle gait, the ground reaction in function of time is converted to the one given in Fig. 11. Here, the forces were normalized in relation to the body weight.

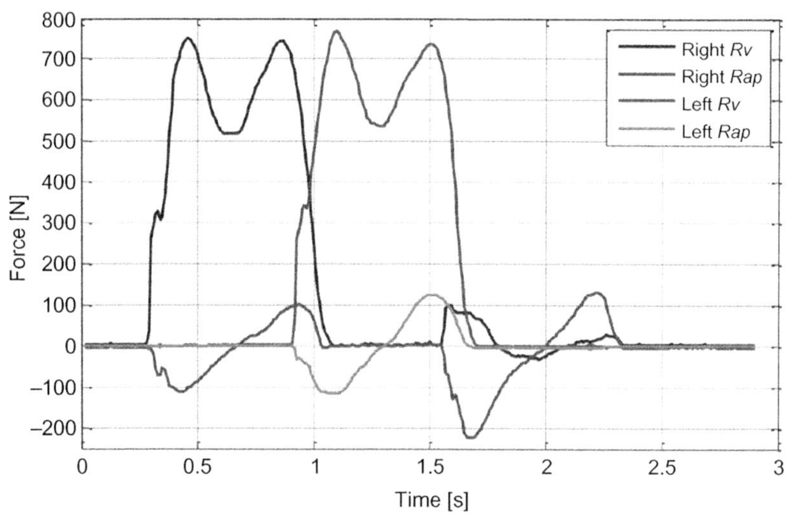

Fig. 10 Ground reaction force in function of time for the 2 ft.

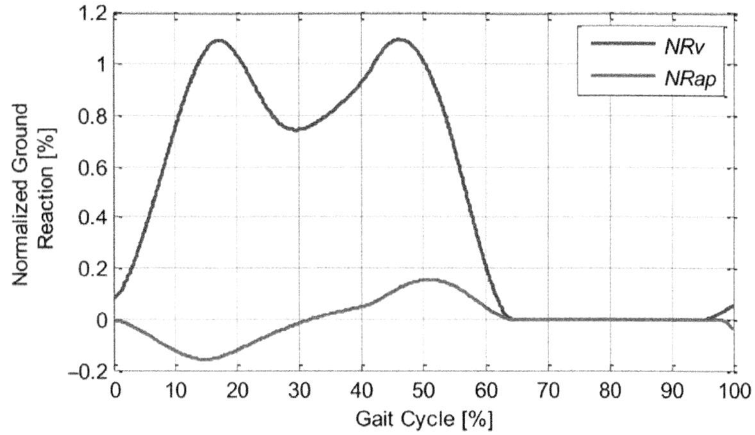

Fig. 11 Normalized ground reaction during a gait cycle.

5 Analysis of obstacle overcoming

In this section, an experimental characterization of human biped walking in obstacle overcoming is carried out. A motion capture system has been used for analyzing this typical walking pattern of humans [26].

A normal subject is asked to walk in an overcoming scenario. A cuboid was placed in front of the subject on the walking zone presented in Fig. 3. The obstacle is described by its length, width and height as follows 0.1 m, 0.5 m, and 0.25 m, respectively. A screenshot sequence of the lower limb in actually environment as well as its reconstruction in Vicon Nexus environment, is depicted in Figs. 12 and 13.

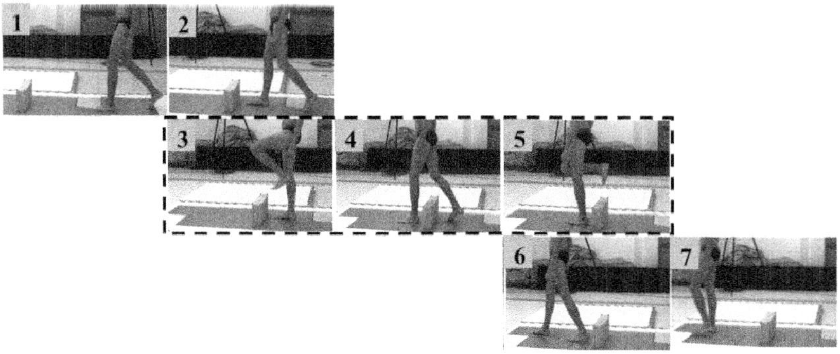

Fig. 12 Screenshot sequence of obstacle overcoming in the biomechanics laboratory of PPRIME institute.

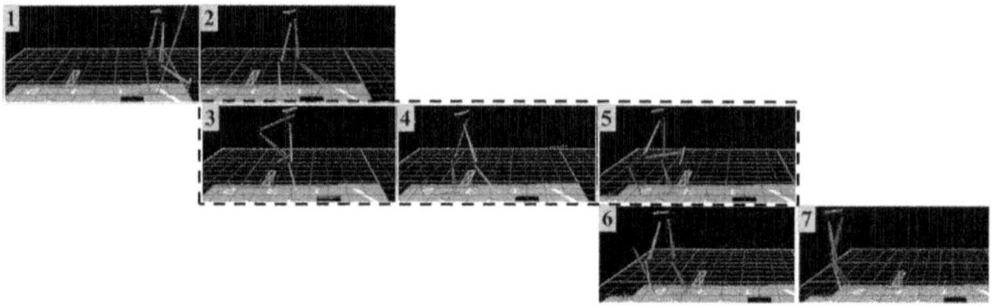

Fig. 13 Screenshot sequence of obstacle overcoming in Vicon Nexus environment after reconstruction.

Through these acquired data of the walking test, the characterization of human walking gait in overcoming obstacle will be performed. The kinematic data are in terms of displacement, velocity and acceleration. Thanks to the use of the force sensors, the ground reaction forces are measured as well [26]. Ground reaction force is expected to be equal on the 2 ft in case of normal feature. In case of overcoming obstacle, the left foot will involve much bigger ground reaction force than the right foot.

An overview of the kinematics of human biped locomotion has been studied in Section 5.1. An analysis of the gait of obstacle overcoming is proposed in Section 5.2.

5.1 Overview of biped walking

In presence of obstacle, human walking is adapted in order to overcome possible collision which is an obvious consequence. In this section, experimental data is presented to show the walking change linked to the cuboid overcoming.

In normal walk, the stride length is approximately around 1 m and it will change in case of presence of obstacle. In the present case, a cuboid is placed at $x = 1.5$ m. The step length has been changed to reach a suitable touchdown point, point M on Fig. 14, before reaching the obstacle. The left leg overcomes the cuboid first and the right leg is used as supporting leg. The step width is equal to 0.85 m along x-axis. Overcoming the cuboid leads to decreasing the right step. According to the size of the cuboid, the both step length and height have been adjusted.

To illustrate the obstacle overcoming, sequence of the both legs positions in Cartesian space is reconstructed under Matlab. The coordinates referring to different markers placed at the hip, knee, ankle, heel and foot joint, as shown in Fig. 4, are used. The legs movement, left leg and right leg, as well

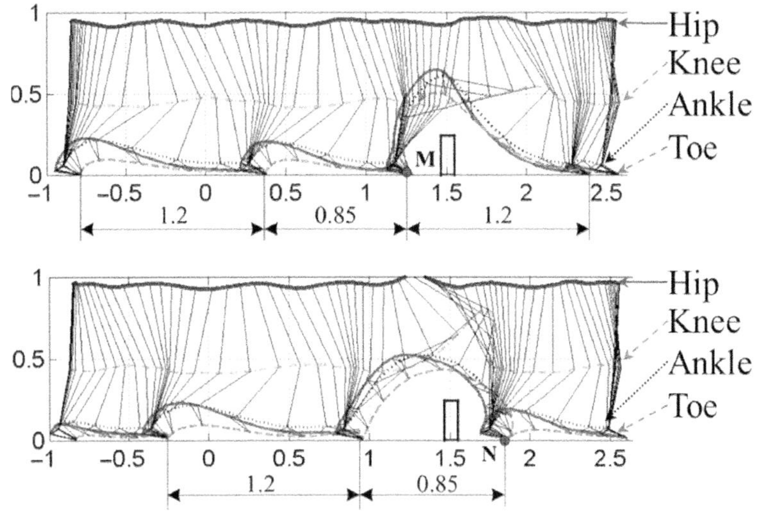

Fig. 14 The legs movement in sagittal plane (*x,z*) in [m]: left right (*up*) and left (*down*).

as the joint centers during the gait cycle are plotted separately in sagittal plane and given by Fig. 14 (up and down respectively).

As shown in Fig. 14, a suitable choice of an adequate touchdown point, noted by M, by the supporting leg is mandatory to overcome successfully the obstacle. Therefore, stride length has been decreased from 1.2 to 0.85 m. Immediately after the left foot overpassing the top of the cuboid, stride length was decreased for the purpose of avoiding to make the right foot an oversize stride length, which is important for keeping balance.

According to the recorded motion within the test, all markers placed on the two legs present variations in the position through the three planes (frontal, transversal and sagittal) but with different amplitude degree. Through the graphics on Fig. 14, one can observe a uniform motion in sagittal plane for the hip marker. The knee marker presents slightly the same behavior as the hip marker with the exception during the overcoming phase. These remarks are available for the two legs. A displacement around 0.1 m is recorded for the lower body in transverse plane. The curves of the y-coordinates of markers are given in Fig. 15 for the left and right legs.

The duration of the obstacle overcoming phase is about 2.5 s. It started at about 5 s and finished at 7.5 s. These remarkable moment is deduced from the curves of ankle and heel markers in the sagittal plane. The z-coordinates of different markers are given on Fig. 16.

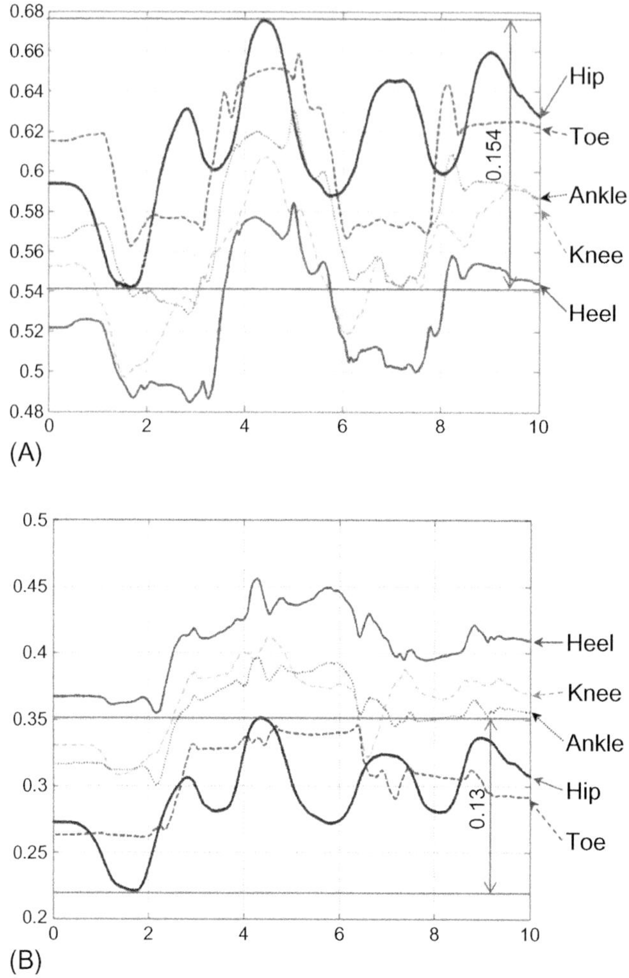

Fig. 15 Y-coordinates of the five representative markers in [m] function of time in [s]. (A) Left leg and (B) Right leg.

The velocities of the hip, the knee and the ankle present periodic feature during the walking cycle. In addition, the amplitudes of the three velocities along x-axis, sagittal plane, are approximately equal to 1.1 m/s, 2 m/s, and 3 m/s, respectively. The ankle displacement is more faster than the hip displacement having the lowest velocity. The plots in Fig. 17 show that the velocity for each joint is more important along the walking direction, x axis, than in the other two directions, y and z axes. One observes that the velocity curve of the ankle presents a symmetry behavior along the walking direction.

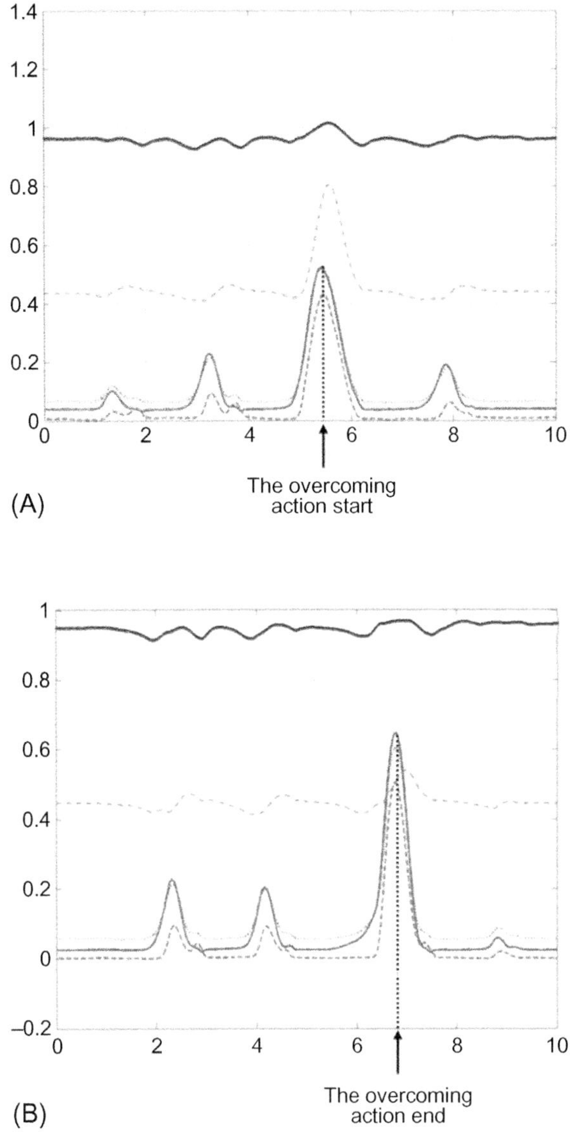

Fig. 16 Z-coordinates of the five representative markers in [m] function of time, with starting and ending overcoming action. (A) Left leg and (B) Right leg.

Similarly to velocity, all the accelerations of the three markers linked to the hip, the knee and the ankle and present a periodic behavior. During the overcoming obstacle and as shown by the curves on Fig. 18, the lowest acceleration is supported by the hip and the highest acceleration is presented but the ankle, 3.5 m/s^2 and 22 m/s, respectively.

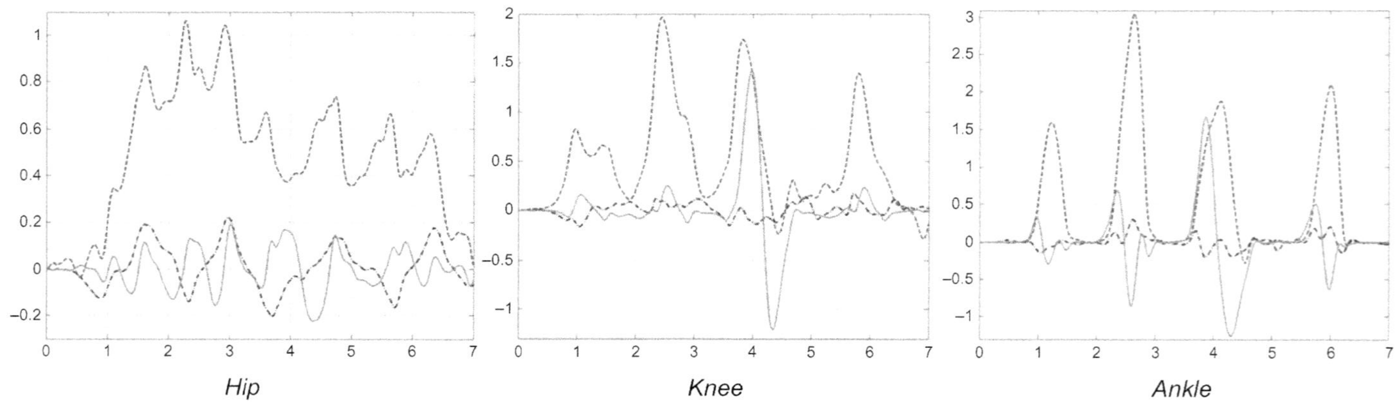

Fig. 17 Velocity curves of the hip, the knee and the ankle during the walking cycle of the left leg [m/s] (x-axis: *red (dotted line)*, y-axis: *blue (dashed line)*, z-axis: *green (solid line)*).

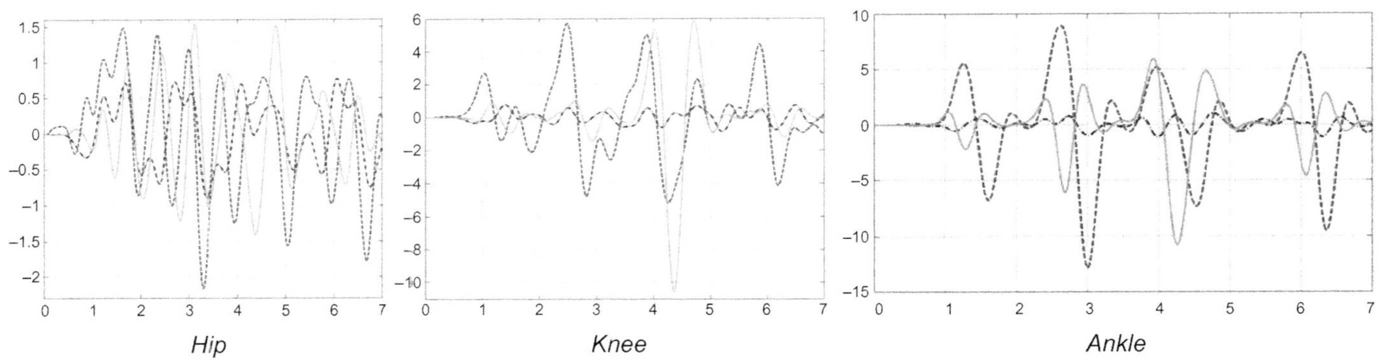

Fig. 18 Acceleration curves of the hip, the knee and the ankle during the walking cycle of the left leg [m/s^2] (x-axis: *red (dotted line)*, y-axis: *blue (dashed line)*, z-axis: *green (solid line)*).

5.2 Obstacle overcoming analysis

The trajectories of the markers placed on the foots are analyzed in this section to study the overcoming phase. One notes that two kinds of walking gaits are occurred during overcoming scenario. The first kind concerns the phase before and after overcoming obstacle phase which will be called normal gait. The second is the fitted gait interposed between two normal gaits and used over overcoming phase.

The normal gait is shown on Fig. 19 through the trajectories of the three markers placed on the ankle, the heel and the toe. These trajectories are characterized each one by a height measured using the horizontal position along the z axis. The heel trajectory as well as the ankle trajectory present a similar height variation and a similar behavior. Furthermore, the toe moves during the swing phase close to the ground leading to a small impact force in the contact with the ground at the beginning and the end of the step. The maximal heights of all trajectories for one step normal gait are depicted on Fig. 19.

The recorded trajectories of the three markers placed on each foot, left and right, during obstacle overcoming phase are shown on the Fig. 20. Similarly to the normal gait, the markers are placed at on the ankle, the heel and the toe, respectively. The fitted gait is characterized by the step length and height. The toe moves during the swing phase with a height around 0.6 m for the 2 ft. A more important impact force is expected in the contact with the ground comparing to the normal gait. This height should be upper than the obstacle height, defined in Section 5.1, to avoid collision. Here, we can define the minimal size gait to overcome the obstacle successfully, of which parameters L and H are the step length and height. These parameters are fixed for a given type of obstacle.

Fig. 21 depicts the fixed parameters for the case of prismatic obstacle as considered in following study. Therefore, a security domain delimited by the trajectories of three markers could be defined, right toe, left hell and left toe, respectively. This security domain is obtained by combining the trajectories of the 2 ft and keeping the small area covering the obstacle. If the security domain decrease, which means that the feet are close the obstacle, then the subject should be more careful to overcome the obstacle and should reduce his walking velocity. The quality of gait stability as well as the walking velocity is linked to the security domain. The suitable choice of one, security domain, should enhance the others, gait stability and walking velocity.

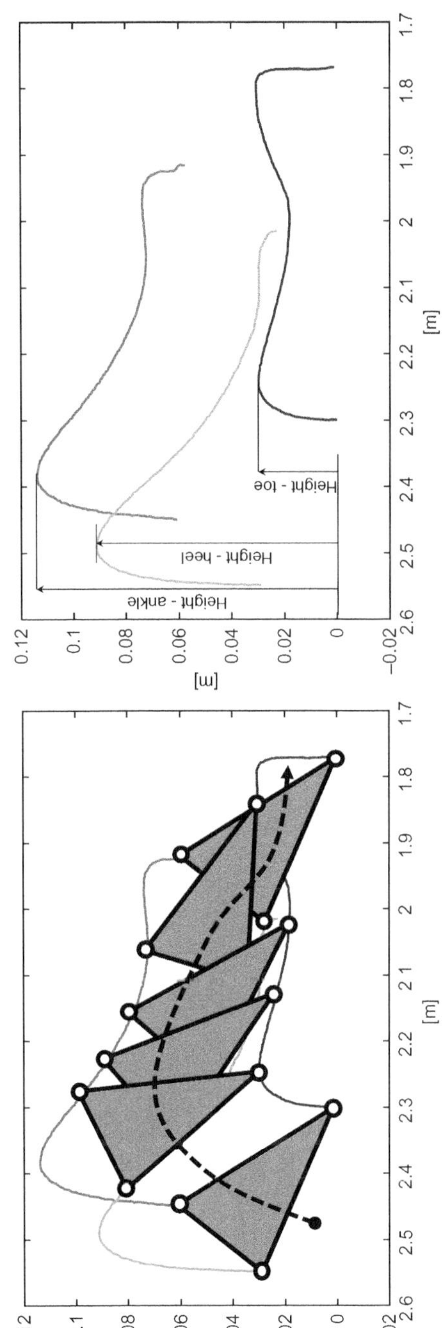

Fig. 19 Heights and the markers trajectories of 1 ft for one step of normal gait.

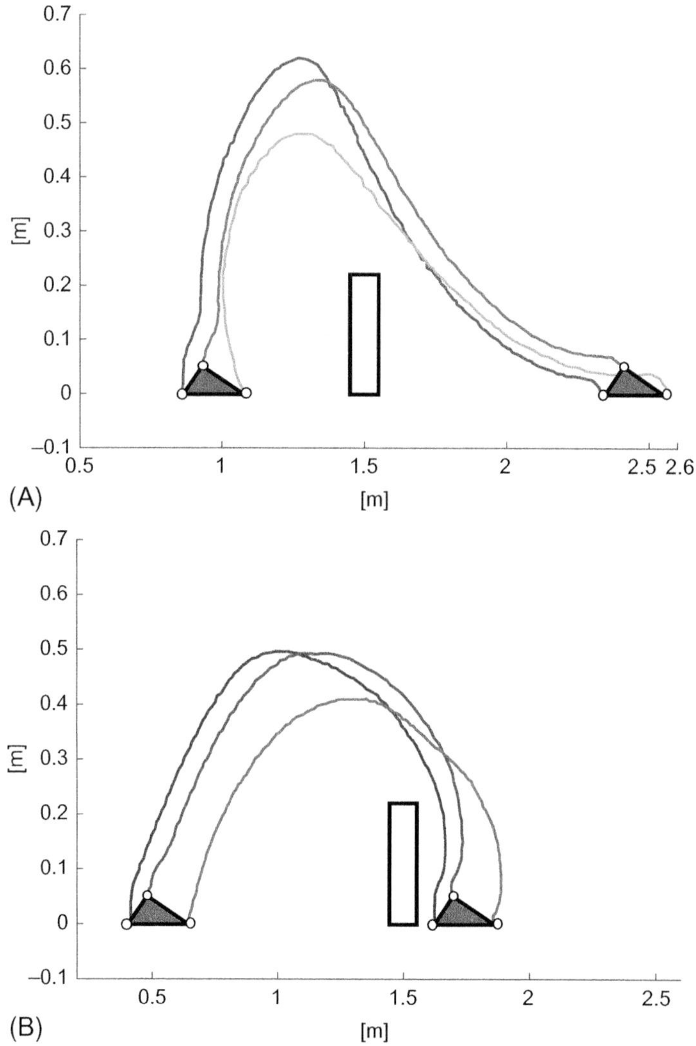

Fig. 20 Trajectories of the markers during the obstacle overcoming phase (A) Right foot, (B) Left foot.

6 Conclusion

This chapter reports the study of the human lower limb tracking via motion capture system from fundamentals up to recent advances. A human walk cycle has been introduced through the description of the stance phase and swing phase. In addition, the parameters and lower extremity angles

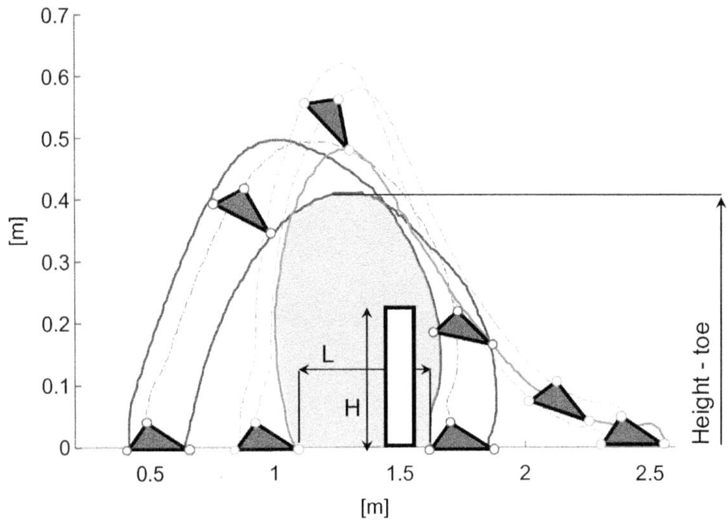

Fig. 21 Security domain description (left foot in *red* (*dark gray* in print version) and right foot in *green (light gray in print version)*).

linked to the gait cycle have been defined. A general markers configuration, based on the experience of biomechanics community, fitted to lower body of an adult human has been presented and applied to establish the quantitative gait analysis. Which represents a valuable tool for studying and quantifying patterns of locomotion. The proposed markers configuration has been applied and exposed through two examples addressed by our team in the past few years. The two applications have been chosen to illustrate how can the gait analysis can be useful.

The first application concerns the design of a training machine. The conducted analysis of human walking has led to the joint trajectories of the lower limbs, and to the reaction forces between the feet and the ground. These crucial data will then be used in a dynamic analysis of the cable robot actuating the lower limb to generate the required motion for a normal gait.

The second application focuses on the experimental analysis of overcoming obstacle in human walking. The normal walking gait, as well as the obstacle overcoming gait, have been typified by means of kinematic performances which will provide useful databases for the implementation of motion planning in case of overcoming obstacles for humanoid robots.

Acknowledgments

This work has been sponsored by the French government research program Investissements d'Avenir through the Robotex Equipment of Excellence (ANR-10-EQPX-44). It is also supported by the Nouvelle Aquitaine Region 2014-2020, in partnership with the European Union (FEDER/ERDF, European Regional Development Fund).

References

[1] F. Faure, G. Debunne, M.P. Cani-Gascuel, F. Multon, Dynamic analysis of human walking, in: Eurographics Workshop on Computer Animation and Simulation, Budapest (Hungary), 2–3 September, 1997, pp. 53–65.
[2] M. Adrian, J. Cooper, Biomechanics of Human Movement, Benchmark Press, Indianapolis, IN, 1995.
[3] M. Ceccarelli, M.E. Toti, E. Ottaviano, CATRASYS (Cassino tracking system): a new measuring system for workspace evaluation of robots, in: Proc. Eighth International Workshop on Robotics in AlpeAdria-Danube Region RAAD'99, Munich, Germany, June 17–19, 1999, pp. 19–24.
[4] M. Ceccarelli, E. Ottaviano, M. Toti, Experimental determination of robot workspace by means of CATRASYS (Cassino tracking system), in: Thirteenth CISM-IFToMM Symposium on Theory and Practice of Robots and Manipulators, Wien, Austria, 2000, pp. 85–92.
[5] E. Ottaviano, M. Ceccarelli, M. Toti, C. Avila Carrasco, CaTraSys (Cassino tracking system): a wire system for experimental evaluation of robot workspace, Fuji Int. J. Robot. Mechatronics 14 (1) (2002) 78–87.
[6] E. Ottaviano, M. Ceccarelli, F. Palmucci, An application of CaTraSys, a cable-based parallel measuring system for an experimental characterization of human walking, Robotica Cambridge University Press. 28 (2010) 119–133.
[7] C.K. Lim, Z.Q. Luo, I.-M. Chen, S.H. Yeo, A low cost wearable optical-based goniometer for human joint monitoring, Front. Mech. Eng. 6 (2011) 13–22.
[8] D. Noonan, P. Mountney, D. Elson, A. Darzi, G.Z. Yang, A stereoscopic fibroscope for camera motion and 3D depth recovery during minimally invasive surgery, in: Proceeding of ICRA, 2009, pp. 4463–4468.
[9] M.A. Laribi, T. Essomba, S. Zeghloul, G. Poisson, Optimal synthesis of new spherical parallel mechanism for application to tele-echography chain, in: ASME, International Design Engineering Technical Conferences (IDETC), Washington, DC, 28–31 August, 2011.
[10] M.A. Laribi, T. Rivière, M. Arsicault, S. Zeghloul, A design of slave surgical robot based on motion capture. in: The 2012 IEEE International Conference on Robotics and Biomimetics (ROBIO2012), December 11–14, 2012, Guangzhou, China, 2012, pp. 600–605, https://doi.org/10.1109/ROBIO.2012.6491032.
[11] L. Boutin, A. Eon, S. Zeghloul, P. Lacouture, An auto-adaptable algorithm to generate human-like locomotion for different humanoid robots based on motion capture data, in: International Conference on Intelligent Robots and Systems (IROS), Taipei, Taiwan, 18–22 October, 2010, pp. 1256–1261.
[12] S. Upadhyaya, W.-S. Lee, Survey of formal methods of hip joint center calculation in human studies, APCBEE Proc. 7 (2013) 27–31.
[13] C.L. Vaughan, B.L. Davis, L. Christopher, J.C.O. Connor, Dynamics of Human Gait, Kiboho Publishers, 1999.
[14] M.W. Whittle, Gait Analysis: An Introduction, Butterworth-Heinemann, 2014.

[15] R. Bartlett, Introduction to Sports Biomechanics Analysing Human Movement Patterns, Routledge, 2007.
[16] G.W. Cavanagh, ISB recommendations for standardization in the reporting of kinematic data, J. Biomech. 28 (10) (1995) 1257–1261.
[17] M. Gleicher, Retargetting motion to new characters, in: Proceedings of the ACM SIGGRAPH Conference on Computer Graphics, 1998, pp. 33–42.
[18] H. Lamine, M.A. Laribi, S. Bennour, L. Romdhane, S. Zeghloul, Design study of a cable-based gait training machine, J. Bionic Eng. 14 (2) (2017) 232–244.
[19] A.L. Bell, R.A. Brand, D.R. Pedersen, Prediction of hip joint centre location from external landmarks, Hum. Mov. Sci. 8 (1) (1989) 3–16.
[20] A.L. Bell, A comparison of the accuracy of several hip center location prediction methods, J. Biomech. 23 (6) (1990) 617–621.
[21] R.M. Ehrig, W.R. Taylor, G.N. Duda, M.O. Heller, A survey of formal methods for determining the centre of rotation of ball joints, J. Biomech. 39 (15) (2006) 2798–2809.
[22] S.J. Piazza, N. Okita, P.R. Cavanagh, Accuracy of the functional method of hip joint center location: effects of limited motion and varied implementation, J. Biomech. 34 (2001) 967–973.
[23] H. Kainz, C.P. Carty, L. Modenese, R.N. Boyd, D.G. Lloyd, Estimation of the hip joint centre in human motion analysis: a systematic review, Clin. Biomech. 30 (4) (2015) 319–329.
[24] G. Wu, S. Siegler, P. Allard, C. Kirtley, A. Leardini, D. Rosenbaum, M. Whittle, D.D. D'Lima, L. Cristofolini, H. Witte, ISB recommendation on definitions of joint coordinate system of various joints for the reporting of human joint motion—part I: ankle, hip, and spine, J. Biomech. 35 (4) (2002) 543–548.
[25] J.L. Hicks, J.G. Richards, Clinical applicability of using spherical fitting to find hip joint centers, Gait Posture 22 (2) (2005) 138–145.
[26] T. Li, M. Ceccarelli, M. Luo, M.A. Laribi, S. Zeghloul, An experimental analysis of overcoming obstacle in human walking, J. Bionic Eng. 11 (4) (2014) 497–505.

CHAPTER 5

Design and operation of exoskeletons for limb replacement or performance enhancement

Aleksandar Rodić, Svemir Popić, Miloš Jovanović
Robotics Department, Mihajlo Pupin Institute, University of Belgrade, Belgrade, Serbia

1 Introduction

The movement and activity significantly impact the quality of life of individuals and life expectancy because they promote and maintain many physiological functions in humans, such as circulatory, respiratory, cardio-vascular system, neuro-muscular system, etc., Pathologic appearances and injuries of the skeletal neuro-musculature system significantly reduce psycho-physical abilities of human that also indirectly affects the socio-economic status of the person. Care about people with physical disorders or with physical limitations, in terms of loss or partial loss of ability to stand and walk, is of great importance both for individuals as well for community to which individual belongs. There is a great interest of researchers and industry engaged in development and selling of medical aids, for continuous improvement of know-how and technology in order to design robotic assistive systems for replacing human physical handicap. There are not only the science and engineering that make crucial factors in development of medical devices. Significant impacts to progress in this field have the proper legal and ethical regulations that are involved indirectly in distribution and use of this type of medical aids due to the very strict ethical and legislative requirements and rather restrictive standards. For this reason, the progress in this field is neither fast nor stepwise.

Paraplegia is a motor and sensory impaired function of the lower extremities. One of the major impairments caused by paraplegia is a muscular weakness, i.e., neuro-muscular atrophy. The consequence of that is loss of stability and mobility resulting in inability to stand and walk. Spinal cord injury is the most common cause of paraplegia which may be either

Design and Operation of Human Locomotion Systems
https://doi.org/10.1016/B978-0-12-815659-9.00005-6

complete or incomplete. In the case of a complete spinal cord injury all sensory-motor functions below the injured region are lost. Patient at the current level of medicine development has unfortunately no hope of functional recovery. In the case of an incomplete spinal cord injury some sensory and motor functions are partially preserved in the area below the region of the injury. Patients with incomplete spinal cord injuries are able to significantly recover their functions after successfully conducted physical therapy. The use of robots and medical aids in the rehabilitation process is very effective to patients with respect to the physical treatment consisting of long-term repetition of motion which ultimately leads to the recovery and re-establish nerve pathways and strengthening musculature system. Robots, apart to carrying out physical exercises, are also used for acquisition indices of patient's physical condition (power, strength, speed, stability, etc.). By using the wearable robotic devices, such as exoskeletons in medical therapy, it additionally provides necessary motivation of patients by offering them ability to perform more realistic movements and more freedom of mobility which positively affects the human mental condition.

In the chapter we deal with mechanical design and control of wearable robotic prosthesis for limb replacements and body performance enhancement. First, it will be discussed some biomechanical aspects of natural human walking of healthy persons in Section 2. These findings serve to better understand the phenomenon of human gait and to better comprehend the possibilities of generating artificial human gait, i.e., assisted gait enabled by use of the active medical leg prosthesis—exoskeleton. In Section 3, it is presented the archival material regarding to the early development of the orthotic devices and exoskeletons from the early 1970s of the last century. These solutions represent fundamentals for the contemporary research in this field of wearable robotic science. In the same section it is also given a brief survey of the recent developments of wearable robotic devices. In Section 4, the control aspects will be considered. Discussion section is about context in which the results obtained in early 1970s of the last century were inspired the recent design and engineering solutions. So this is a review contribution, explaining how first active prosthetic devices were born and how they initiated further developments in this filed and generally field of humanoid robotics happened 50 years ago.

2 Biomechanical aspects of biped locomotion

Human body represents in mechanical sense a multi-body large-scale dynamic system of variable geometry. It consists of muscular-skeletal and nervous system which includes bones, ligaments, tendons, cartilage and attached muscles. It gives to the body a basic structure and ability for movements. Human body has in total 244 degrees of freedom [1]. There are about 230 joints in the body and most of them have one degree of freedom (DOF). Some joints have multiple degrees of freedom, for example, the hip and the shoulder joints have at least 3 DOF. These 244 DOF are controlled by 630 muscles [1]. It is assumed that the muscles have no freedom of their own and are constrained by the joints.

The observations carried out in this section are focused to the biomechanical aspects of anthropomorphic bipedal walk being the topics of this chapter are dedicated to design and operation of exoskeletons for limb replacement or performance enhancement. Aiming to this goal, our consideration begins from the experimental measurements of human subjects conducted in a dedicated biomechanical, motion caption laboratory (Fig. 1) [2–5].

For the purpose of experimental measurements, the healthy examinees of different age and gender and different physical constitution have been engaged. Subjects were instructed to demonstrate practically various

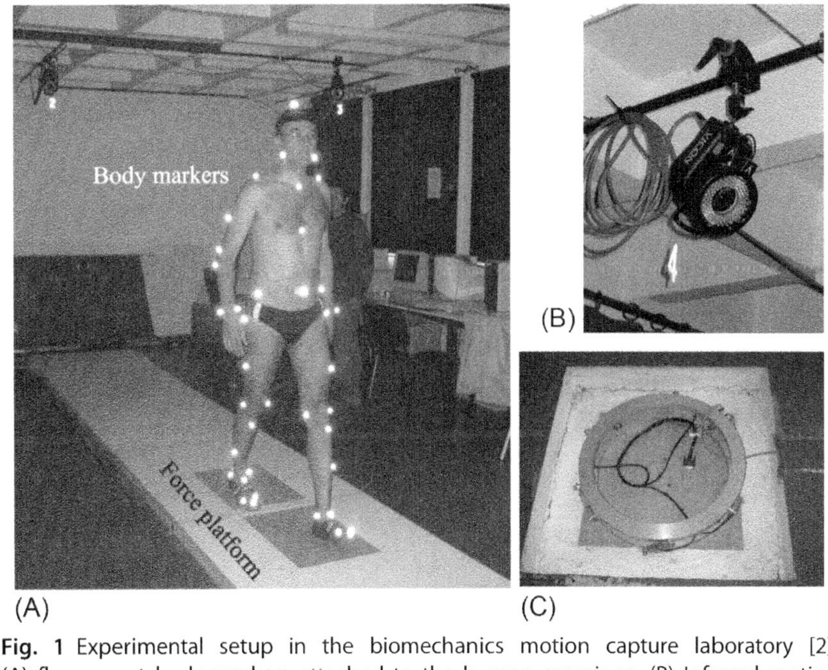

Fig. 1 Experimental setup in the biomechanics motion capture laboratory [2]: (A) fluorescent body markers attached to the human examinee, (B) Infrared motion capture camera, and (C) 3-axis force platform.

performances of human walking by changing the gait length, width, speed, foot lifting height, etc. The results of the experimental measurements were stored in the database that was used later for studying walking kinematics and dynamics with aim to acquire a specific knowledge about the biomechanics of human bipedal locomotion. Some characteristic results of human motion capture experiments are presented in Fig. 2A–D. These results, as well as complementary ones, were used for generation of modeling and simulation software entitled "Humanoid Robot Simulation Platform (HRSP)" [6] aimed to the research of human locomotion or locomotion of biped robot mechanisms such as humanoids and wearable robotic devices.

Experimental results shown in Fig. 2A–D illustrate that even human body has a natural symmetry of the right and the left part, locomotion is not always consequently symmetric. That indicates certain pathological changes due to some early signs of the disease or due to the physical impairments of joints and limbs as consequence of previous injuries or health problems. The motion capture experiments are useful tool that can help scientists and engineers to acquire some "hidden problems" in human anatomy and to identify corresponding indicators of pathologic states connected to the joints or muscles. In spite of that, some important indicators of body physical conditions cannot be acquired experimentally at all. For example, the payload torques of body joints cannot be measured nor experimentally acquired in spite of that these values are of great importance for the design of wearable prosthetic devices. Due to these facts, the modeling and simulation of human biped locomotion are of great benefits for research of human locomotion and manipulative functions. The HRSP software toolbox [6] was used for study and analysis of the leg joint torques generated for different types of human gait. The software also enables some early performance validation of leg prosthesis or exoskeleton kinematics and dynamics in the design phase before experimental verification with end-user. Simulation tests help designer to better fit the device to patient needs and body size.

Due to the aforementioned reasons we used the HRSP software for evaluation of different gait performances by using different body parameters (e.g., body size, mass, distance between joints, etc.). The goal is to identify which joints are subjected to the excessive load and how loads, i.e., torques are distributed per particular joints. This analysis helps designers to make an appropriate choice of the prosthetic mechanism actuators and

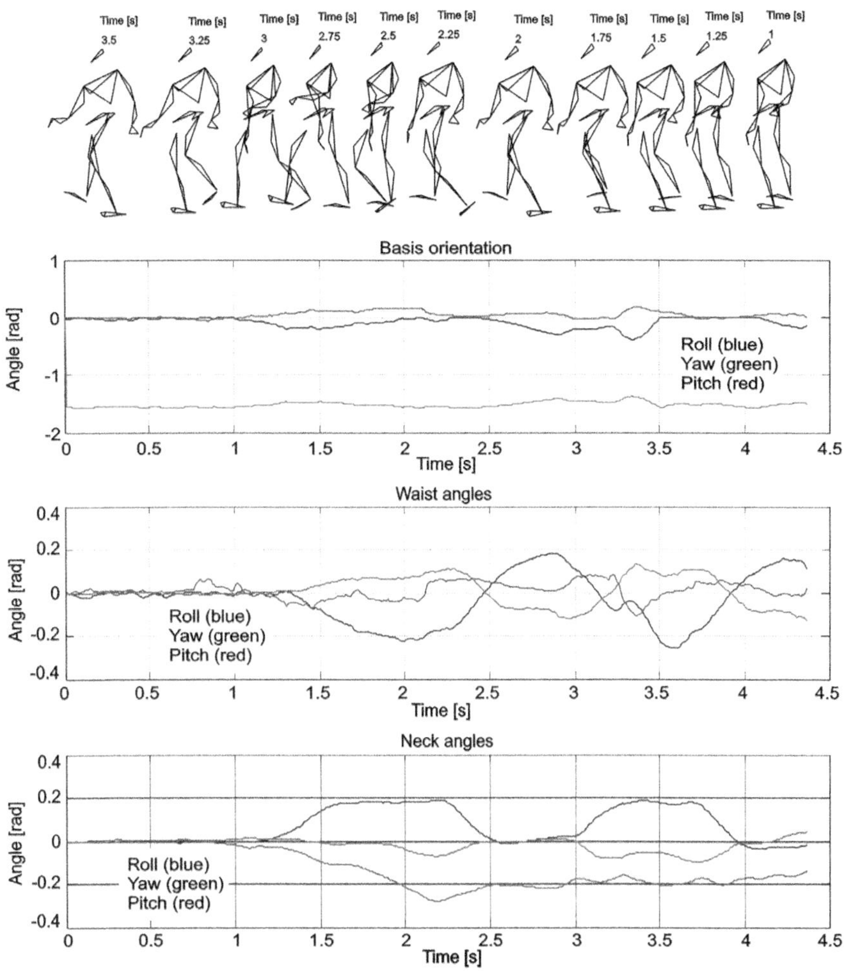

Fig. 2 (A) Experimental results of human motion capture; Snapshots of human body contours defined by the body markers (top figure) in the human locomotion experiments; Attitude deflection of the basic link (upper graph); Attitude deflection of the trunk (rotation in the waist joint) (middle graph); Attitude deflection of the head link, i.e., rotation in the neck joint (bottom graph); (B) Experimental results of human motion capture; Acquired foot cycloids (upper plot); Acquired landing speed of the foot-attached markers (middle plot); Acquired landing acceleration of the foot-attached markers; (C) Experimental results of human motion capture. Heave motion of the hip link in the sagittal plane (upper plot). Acquired right and left foot cycloids—natural cycle (middle plots). Ground reaction forces—vertical components upon the feet soles (bottom plot). (D) Experimental results of human motion capture; Attitude deflection of the hip link of human body (top plot); Right leg joint angles captured in the roll, pitch and yaw direction.

(Continued)

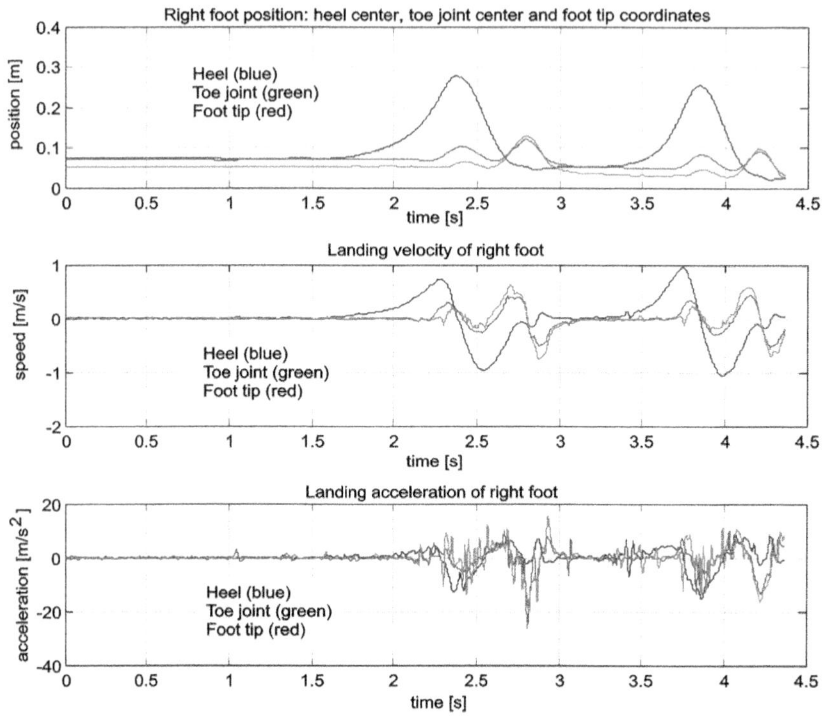

Fig. 2, Cont'd

(Continued)

complementary mechanical structure accessories such as springs, dampers, leverage mechanisms, bumpers, etc., Simulation experiments enable to:

- Calculate internal and external forces that act in the particular leg joints (hip, knee and ankle) for the given patient motion. The gait is imposed by defining the feet cycloids (frequency, length and foot lifting height) and gait speed;
- Determine leg trajectories in the joint space when the external forces and torques act on the body limbs are known (e.g., payload in hands, ground reaction forces when foot strikes ground support, etc.,);
- Determine forces in the leg joints when the body is in the rest, i.e., in standing position (gravity loads);
- Determine the stability indices such as Zero Moment Point (ZMP) or Center of Pressure (CoP) at the different walking rates moving on the flat pad or climbing the stairs. This chapter does not specially consider the problems of gait stability being the patients using prosthetic devices are compelled to utilize auxiliary medical aids to ensure the gait stability (e.g., crutches or passive walker, etc.).

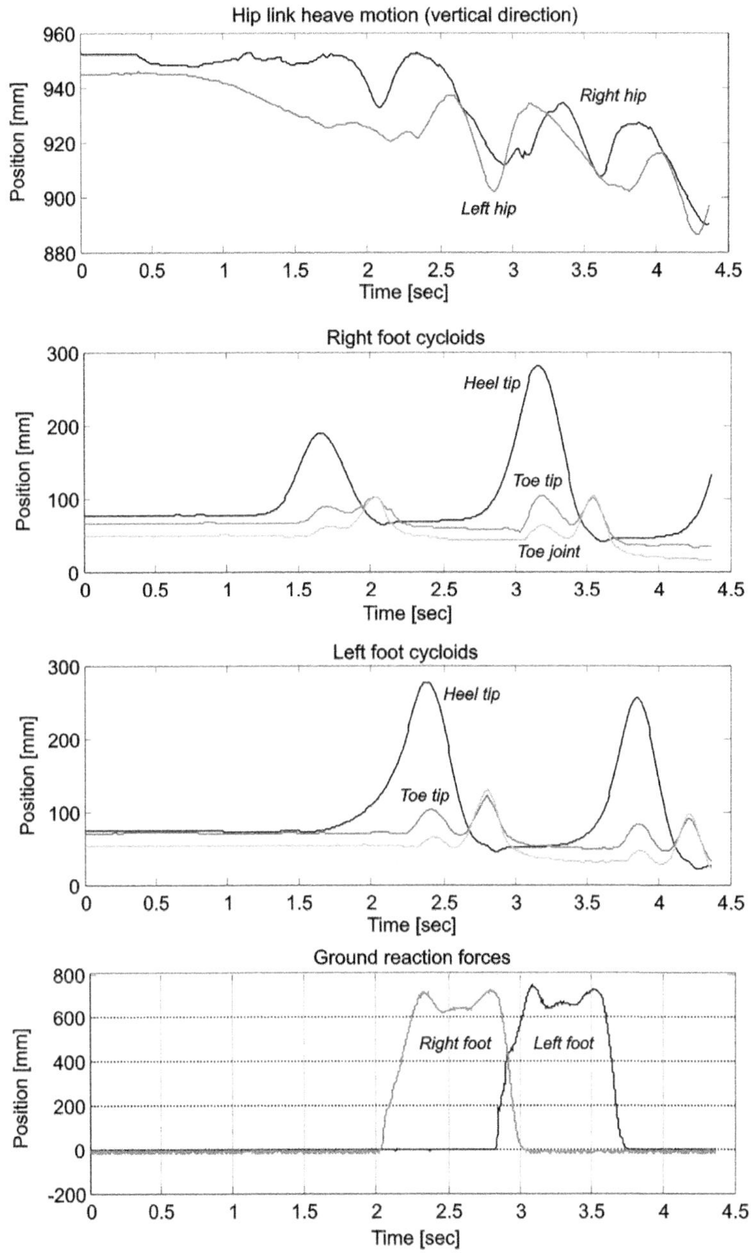

Fig. 2, Cont'd

(Continued)

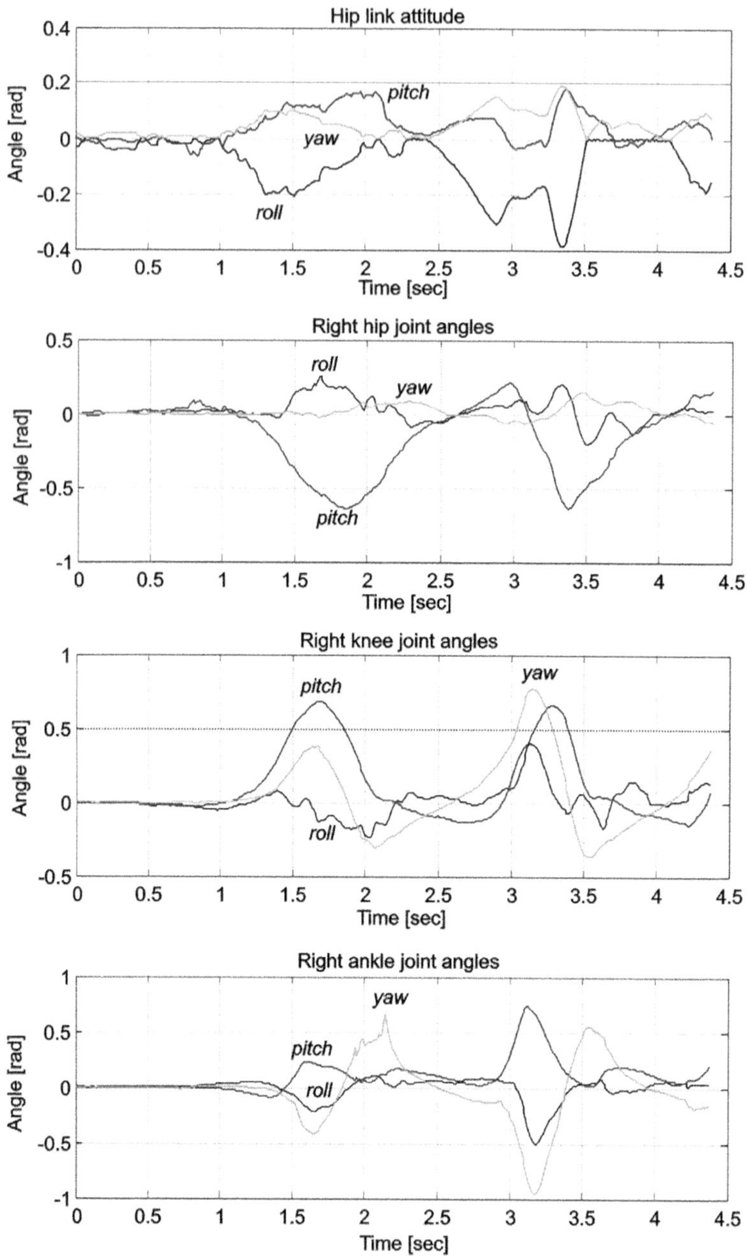
Hip link attitude
pitch
yaw
roll
Angle [rad]
Time [sec]
Right hip joint angles
roll
yaw
pitch
Angle [rad]
Time [sec]
Right knee joint angles
pitch
yaw
roll
Angle [rad]
Time [sec]
Right ankle joint angles
yaw
pitch
roll
Angle [rad]
Time [sec]

Fig. 2, Cont'd

Simulation experiments considered in this chapter are aimed to explore how much the body constitution of individuals influence the loads of particular leg joints, generating joint forces and torques, and thus, how much the joint payloads influence design parameters of the leg prosthetic devices. Manufacturers of the wearable prosthetic devices offer various customized solutions of exoskeletons and lower limb prostheses that are anatomically fitted to the patient size (body weight and height). However, there are no universal design solutions for these devices. For these reasons, it is of great importance for design of exoskeletons and lower limbs prosthetic devices to check (before production of the device) for the person who need to use a prosthetic device, if the actuators and complementary passive elements (e.g., springs, dampers, leverage mechanisms, bumpers, etc.) are enough robust and powerful to ensure regular supported walk or climbing stairs.

Bearing in mind previous consideration, the analysis of several typical use cases of planar walk was done using the HRSP software toolbox for modeling and simulation of bipedal locomotion. Two individuals of different body constitution are selected to accomplish the human gait experiments. Since, in general, it is not possible to measure payload in the particular leg joints during a walk in a non-invasive way, it was done by simulation experiment where the leg joint torques are determined by using of the HRSP. The physical body feature characteristics of the individuals subjected to the experiments are adjusted empirically as explained in [7,8]. Two individuals with fine healthy conditions, whose body parameters are known, were selected for the purpose of simulation experiments: (A) an adult male of age 21, of height 173 cm and weight 71 kg, and (B) an adult, male of age 44 years, height 190 cm and body weight of 85 kg. The chosen cases (persons (A) and (B)) are simulated in several walking experiments (by introducing different gait speeds and gait cycloids of various lengths and lifting heights) moving on a flat surface or climbing stairs. The imposed gait parameters are specified in Table 1 and the considered simulation examples are presented in Table 2.

For purpose of simulation it was used a simplified kinematic model of human body (biped locomotion mechanism) with 38 DOF and with a branched kinematical structure with five kinematic chains (two legs, two arms and torso chain consisting of pelvis, trunk and head with the neck). Each leg is represented by a kinematic mechanism of 7 DOF: (i) 3 rotations in the hip joint, about x, y and z axis (for roll, pitch and yaw, i.e., twist motion), (ii) 1 rotation of the knee joint in the pitch direction (about the y-axis), (iii) 2 rotations in the ankle joint for rolling and pitching, and

Table 1 Human gait parameters used in the simulation experiments for case study

Case	Person	Type of gait	Gait parameters
I	B	Planar gait	Length = 0.60 m; Speed = 4 steps/s; Height = 0.08 m
II	B	Climbing stairs	Length = 0.30 m; Speed = 2 steps/s; Height = 0.13 m
III	A	Planar gait	Length = 0.55 m; Speed = 4 steps/s; Height = 0.08 m
IV	A	Climbing stairs	Length = 0.30 m; Speed = 2 steps/s; Height = 0.13 m

Table 2 Overview of the simulation examples considered in the chapter that take into account the use cases defined in Table 1 and different body masses

Example	Case	Body mass (kg)
1	I	85
2	III	71
3	III	71 + 25%
4	IV	71

(iv) 1 rotation in the toe joint for flexion about the y-axis in the pitch direction. Simulation examples, defined in Table 2, were accomplished using the HRSP software toolbox. The joint torques in the particular leg joints were calculated in three particular directions—the roll, pitch and yaw. The joint angles (movements) of a healthy person are symmetric for the right and left leg and the legs' joint torques are similar in the corresponding joints of both limbs. The simulation results obtained by using model of biped locomotion mechanism with 38 DOF for the examples specified in Tables 1 and 2 are displayed in Table 3. Complementary graphical presentation of the calculated leg joint torques is given in Fig. 3.

By analysis of the obtained simulation results shown in Fig. 3 the following findings can be made. It is obvious that the hip joint of a leg in the pitch-direction is the most loaded joint being it carries on the majority of the entire body mass in both, lighter person (A) and heavier person (B) for the case of a planar walking (Examples 1 and 2, Table 3). The person "B" of larger body mass is in this example taller than the person "A" and also with longer legs, suffers significantly due to the higher payloads in the hips. The corresponding comparative analysis between the leg joint payloads of these two observed persons is presented graphically by the charts given in Fig. 4.

Table 3 Simulation results: The leg joint torques obtained by simulation of biped locomotion experiments specified in Tables 1 and 2

		Torques (Nm)		
Example	**Joint**	**Roll**	**Pitch**	**Yaw**
1	Hip	23	957	8
	Knee	0	443	0
	Ankle	41	57	0
2	Hip	21	660	6
	Knee	0	320	0
	Ankle	37	40	0
3	Hip	27	825	5
	Knee	0	402	0
	Ankle	46	51	0
4	Hip	18	509	3
	Knee	0	1131	0
	Ankle	29	267	0

In further consideration it was analyzed how the body weight and the type of gait (planar walk or climbing stairs) influence joint payloads (forces and torques) in the lower limbs of human body. By analysis of the results given in Table 3 (Examples 1–3), it is evident that the person with 25% increasing in body mass has significantly higher joint torques than person without the extra mass. The comparison of the joint torques is presented by the corresponding charts given in Fig. 4. The following conclusion can be derived from the charts presented in Fig. 4. This means that in determining the size and performance of medical aid the patient height will not only play a role, but also the patient's body weight. Therefore, for each patient it must be accurately examined the all anatomical parameters before choose the right type of the wearable robotic prosthetic device and determine which type of power-driven actuators should meet the dynamic requirements of biped locomotion and gait stability.

The latest analysis derived in the section refers to a comparison of the loads acting in the particular leg joints of the human walking on a flat surface as well as climbing stairs. Unlike the previous case, when it was found that the hip joint was the most loaded joint in the case of planar walking, during climbing stairs human extremely burdens the knee joint. The largest torque in the knee joint is generated in the pitch direction as documented in Table 3 (Example 4). The hip joint in this case is less loaded than it was in the case of a planar walk considered in the Example 2. In other words, the knee joint is

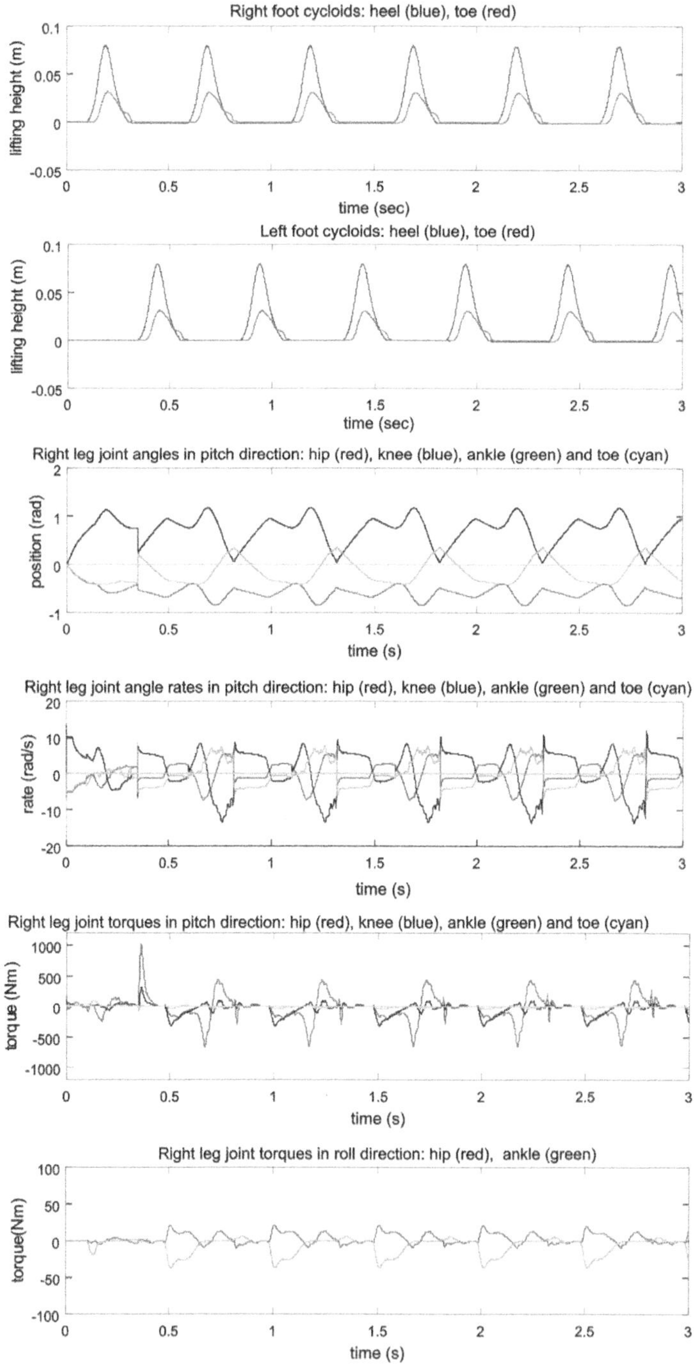

Fig. 3 Plots of simulation results obtained for the person specified as "A" with body mass 71 kg and accomplishing a planar gait whose parameters are defined in Table 1, Case III.

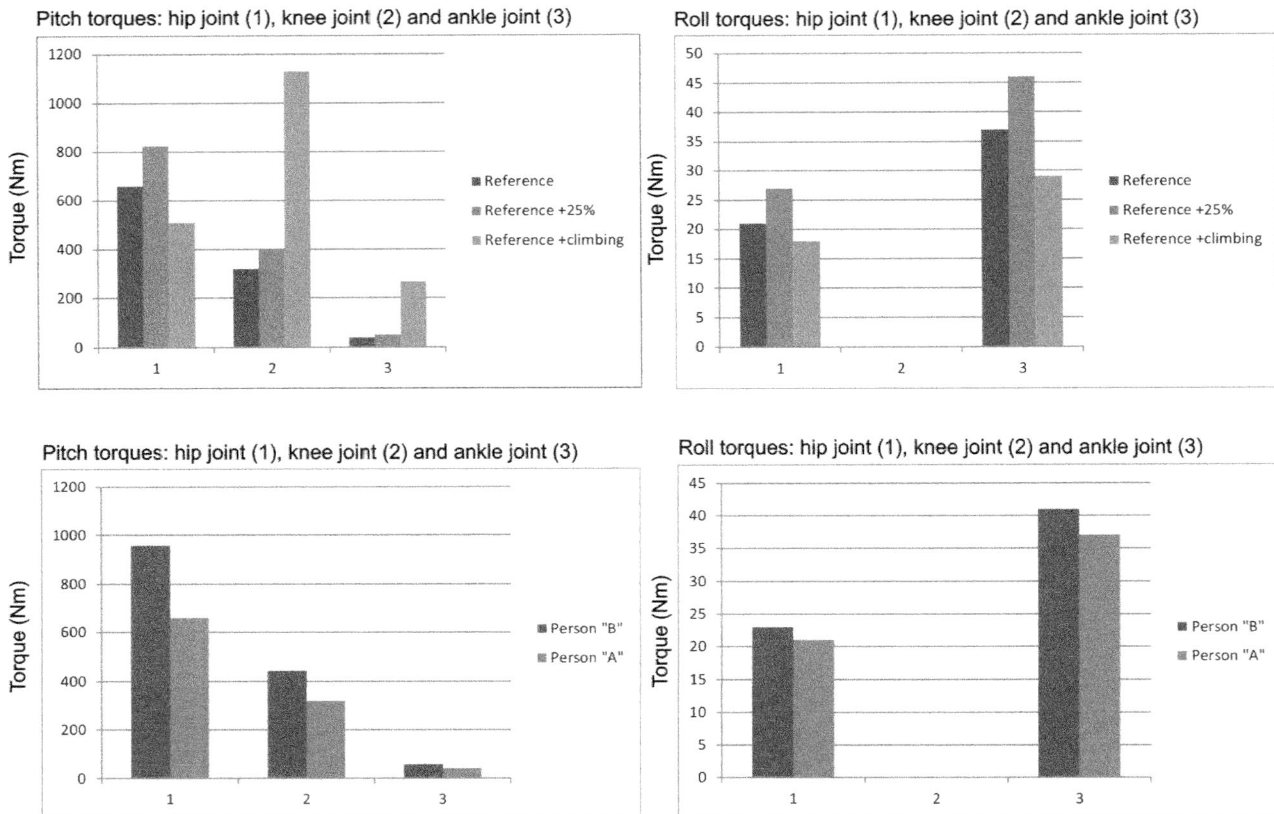

Fig. 4 Simulation results: comparison of joint torques in the hip joint (1), knee joint (2), and ankle joint (3). The flexion in the toe joints is considered in simulation experiments as immobile.

four times more burdened. This proves that the climbing stairs is more demanding task to be accomplished by a wearable robotic prosthetic device than performing a planar walk. The matter can be additionally complicated if the person is heavy. Due to these facts, designer might be thinking that the knee joint has to be supplementary strengthen by some passive or semi-active supporting elements such as for example pneumatic springs or shock absorbers.

The experimental and corresponding simulation results presented in the section, together with a short biomechanical analysis derived, serve to better understand issues that will be discussed in the following sections of the chapter. More useful information about natural human gait and synthesis of artificial anthropomorphic biped locomotion and gait stability can be found in [9–23].

The power-driven actuators used in active prosthesis and exoskeleton applications are required to provide high torques while operating at high speeds. This requires use of larger and heavier actuators. However, the available area around the joint is very limited for integrating actuators. Because, it is necessary to select the compact forms of actuators that can provide required torque-speed values for the targeted movements. The speed-torque characteristics of the joints for targeted movements have to be known for proper selection of actuators. The speed-torque values of the lower limbs joints can be determined from simulation by use of the HRSP toolbox as described earlier.

3 Mechanical design and operation of exoskeletons

Originally, the exoskeleton represents an external skeleton that fully or complementary reinforces the internal structure of the organism (body limbs). If the internal skeleton structure, i.e., endoskeleton, is not enough powerful or if it is additionally weakened due to the disease or trauma occurrence, it has to be additionally strengthened by some supporting device. Unfortunately, it is only possible from the outside. Immobilization of injuries or fractures is a way of implementing the external stiffeners as a passive support meaning. But if one needs to enable or intensify the movement of one or more locomotor organs (especially the hands and feet) or the whole body, then the use of exoskeletons is the only way to do it. In that sense, there are partial exoskeletons (orthotics arms, legs, etc.,) or complete exoskeletons to strengthen whole body. Historically, exoskeletons have occurred primarily with the intention of helping people with neuromuscular diseases. However, in the recent years increasingly emphasizes is on development of exoskeletons as an aid in carrying out severe physical work [24].

Also, exoskeletons have found application in the military industry such as the BLEEX exoskeleton whose application significantly increases the performance of soldiers at the battlefield, primarily in long-term relief of carrying on the heavy loads [25]. Also, the Panasonic, utilizing robot technology, has developed the assist robot to support everyone, from the elderly to workers in performing heavy tasks [26].

Basic operation principle of a lower extremity exoskeleton is transferring the body weight to the ground by means of the exoskeleton mechanism. In such a way exoskeleton eliminates the gravity effects upon the user. So, the exoskeleton mechanism has to be strong enough to carry both its own weight and the body weight of the device user. The exoskeletons should satisfy the conflicting demands to be as light as possible due to ergonomics, to use small, compact actuators and have low power consumption requirements. In addition, a critical point for exoskeleton application regards to easy adjusting of mechanism adapting to different sized users. The second critical point in the design of exoskeletons is appropriate choice of actuators. The selected actuators must be capable of meeting the speed and torque requirements for the targeted motions, as well as being enough compact and lightweight to be ergonomic and acceptable to user with limited physical abilities. In addition, an exoskeleton should be able to accomplish the targeted movements in a manner that will provide minimal inconvenience to the user. To achieve that, the degrees of freedom and the range of joint motion have to be selected properly. It is known that by increasing the degrees of freedom and range of motion of the exoskeleton mechanism provides more anthropomorphic and consequently more comfortable use of the prosthetic device. However, by releasing all degrees of freedom of the exoskeleton mechanism is not safe for users with very limited muscular capabilities in their legs since it causes involuntary movements. For devices in contact with the user, safety is a very important criterion. It has to be ultimately satisfied being persons with physical limitations has not sufficient mobility and power to keep stable by themselves. Therefore, all electrical and mechanical safety precautions have to be considered during the design process for preventing user from injury.

In the following considerations it will be taken into account all prior notices and analyses for the purpose of a general overview of state-of-the-art in this area from the very beginning of development in the 1970s until today. It will be interesting to compare the state of technology today and in the time when the first research was done many years ago, how much progress has been made and which are still outstanding issues that inhibit the rapid progress in the field. What are the problems in development of exoskeletons and other wearable robotic prosthetic devices that were not still solved and waiting for new ideas, technology and materials?

The beginning of development of the humanoid robotics coincided with the beginning of development of the world's first active exoskeletons at the Mihajlo Pupin Institute in 1969 [27, 28], under the guidance of Prof. Vukobratović [29]. Legged locomotion systems were developed first. Also, the first theory on bipedal locomotion systems was developed in this institute in the frame of active exoskeletons. Hence, it can be said that active exoskeletons were the predecessors of the modern high-performance humanoid robots. The present-day active exoskeletons are developed as the systems for enhancing capabilities of the natural human skeletal system. The most successful version of an active exoskeleton for rehabilitation of paraplegics and similar disabled persons, pneumatically powered and electronically programmed, was realized and tested at Belgrade Orthopedic Clinic in 1972. One specimen was delivered to the Central Institute for Traumatology and Orthopedics, Moscow, in the frame of the USSR-Yugoslav inter-state scientific cooperation. From 1991 the exoskeleton belongs to the basic fund of Polytechnic Museum (Moscow) and State Museum Fund of Russian Federation. It is exhibited in the frame of the museum's exposition dedicated to the development of automation and cybernetics [28]. During the research which was lasted more than two decades there were developed several types of the exoskeleton devices. It was analyzed different constructive solutions of the devices, primarily concern with to the power source required to drive the device actuators, then, number of mechanical degrees of freedom that were directly related to the patient power-driven joints with respect to kinematic and dynamic constraints to the achievement of dynamic motion. The original exoskeletons and active suit developed in 1970s in Mihajlo Pupin Institute can be seen in the Museum of Science (Belgrade), Republic of Serbia.

Prof. Miomir Vukobratović with his associates in the Robotics Laboratory, Mihajlo Pupin Institute—Belgrade noticed in the late 1960s that a number of patients with diseases of the type of paraplegia or some forms of muscular dystrophy can be helped if their feebler bodies are pulled on the device from the outside whose function is to mechanically enable accomplishing of movements as similar as possible to the movement of

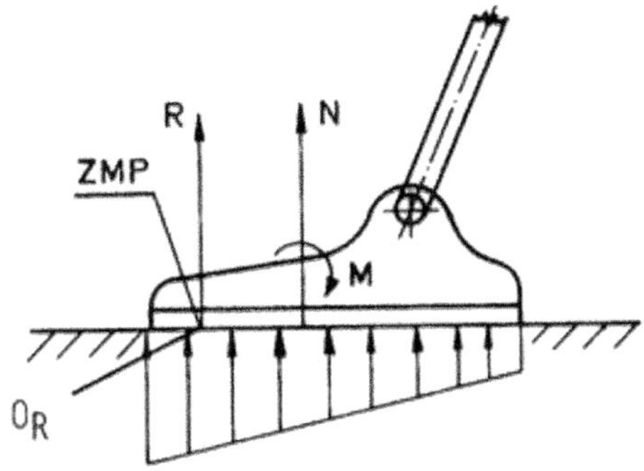

Fig. 5 Load distribution along the foot.

the healthy people [28]. In addition, such a mechanical device does not harm or further undermine in any way the integrity of the patient.

In parallel with the states feedback including loads feedback at powered joints of legged locomotion robots and particularly of biped mechanisms, it is essential for dynamic stability of the overall system to control ground reaction forces at the contacts of the feet and the ground. For instance, with the biped robot in the single support phase, shown in Fig. 5, it is possible to replace all elementary vertical forces by their resultant. Let the point O_R (Fig. 5) represent the point at which the sum of moments is equal zero, so that this point where only force is acting is called Zero-Moment Point (ZMP) [30–34].

The equations of dynamic equilibrium for the biped mechanism can be derived for ZMP. Therefore, it has been made possible to solve this very specific problem of applied mechanics. Namely, for any other point except for ZMP, equations of dynamic equilibrium would contain unknown dynamic reaction forces, and thus it would be impossible to solve the problem of dynamics modeling in the class of legged, particularly biped locomotion robots. But, if one integrates the equations written for the ZMP, then it becomes possible to calculate the reaction forces, as they depend on all internal coordinates, velocities, and accelerations of the whole mechanism.

3.1 What is the essence of the semi-inverse method?

A next decisive step in modeling and control of legged, particularly biped locomotion robots was the introduction of the semi-inverse method [31–34]. The conditions of dynamic equilibrium with respect to the coordinate frame located at the Zero Moment Point give three relations between the generalized coordinates and their derivatives. As the whole system has n degrees of freedom ($n > 3$), the trajectories of the ($n - 3$) coordinates can be prescribed so as to ensure the dynamic equilibrium of the overall system (the trunk motion including the arms if the biped robot is under consideration). If there is some supplementary ZMPs (like passive joints of the biped arms), then for every additional ZMP another three equilibrium conditions are available. Thus, when applied to the problem of investigating the dynamics of biped systems, the motion of the links is partly known, while the unknown moments are equal zero. Vanishing of the given moment results from the equilibrium conditions about the supporting point (ZMP) and about the joints of passive links.

Active (powered) humanoid mechanisms have been under development and scientific and professional concern of the Robotics Laboratory of the Mihajlo Pupin Institute in Belgrade, since the late 1960s, when the first in the world active experimental, pneumatically-driven exoskeleton for producing artificial "sliding" gait was developed and tested (Fig. 6).

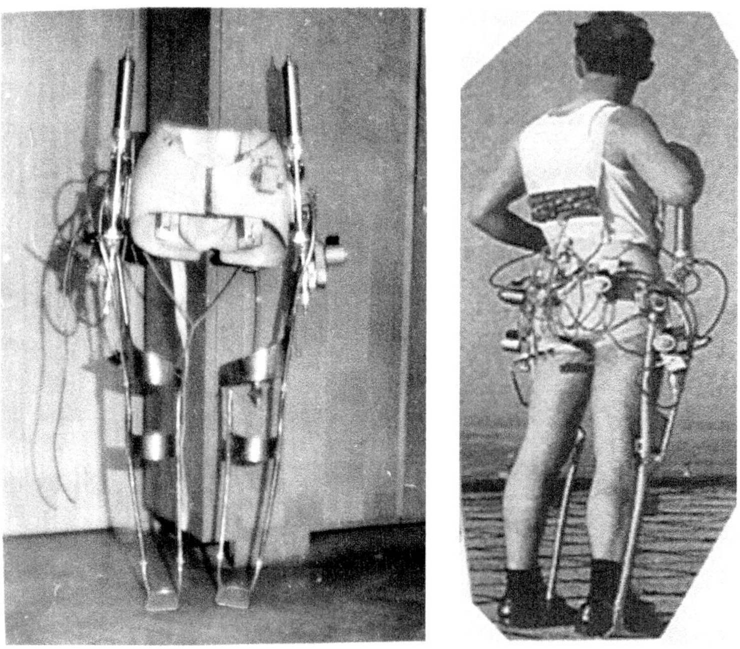

Fig. 6 The first active exoskeleton made in the world, pneumatically powered and partly kinematically programmed for generating quasi anthropomorphic gait. The active exoskeleton realized in 1969 was predecessor of several the more complex exoskeletons for severely handicapped persons.

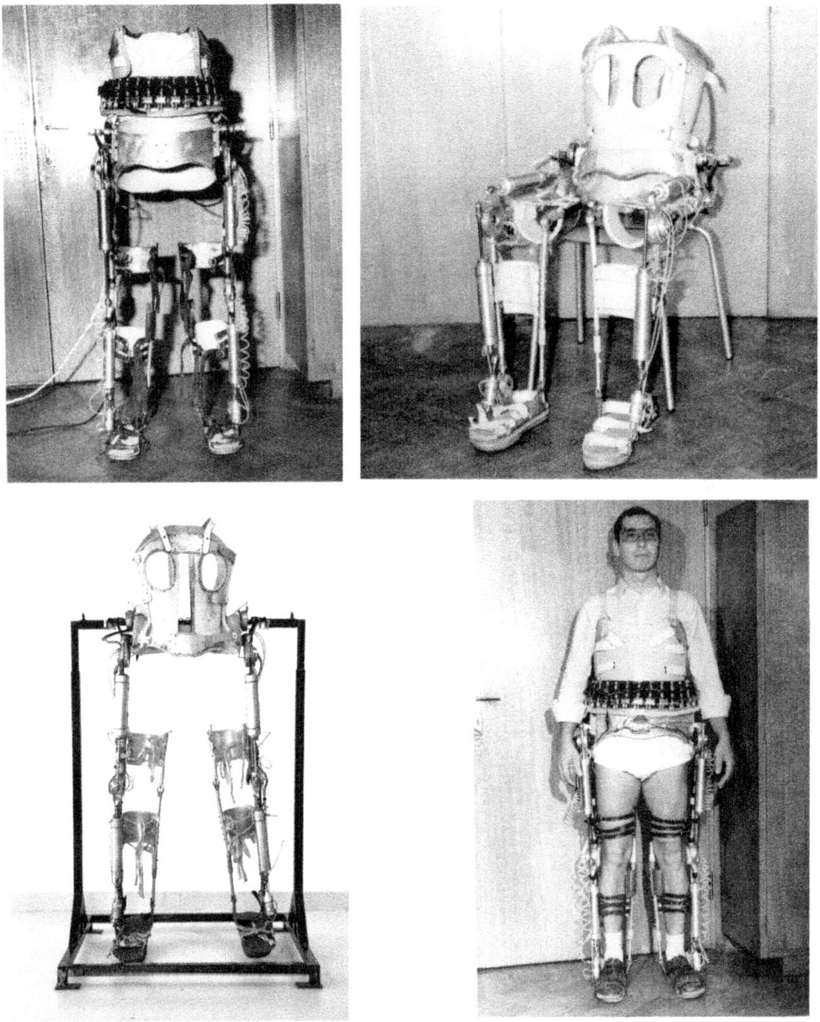

Fig. 7 The first pneumatically driven and clinically tested active exoskeleton in the world made for rehabilitation of paraplegics and other disabled persons, pneumatically powered and electronically programmed, realized and tested at Belgrade Orthopedic Clinic in 1972.

A most successful version of active exoskeleton for rehabilitation of paraplegics and similar disabled persons, pneumatically powered and electronically programmed, designed and realized [35–38] in Belgrade (see Fig. 7), was tested at the Belgrade Orthopedic Clinic in 1972. A specimen has been delivered to the Central Institute for Traumatology and Orthopedics,

Moscow in the frame of USSR – Yugoslavia inter-state scientific collaboration. From 1991 this exoskeleton belongs to the basic fund of the Polytechnic Museum (Moscow) and the State Museum Fund of the Russian Federation. This exoskeleton is included in the list of the General Museum Fund of the Russian Federation as a "relic of science and technology of this century that must be preserved for next generations" (conclusion from the statement of Professor G.G. Grigorian, General Director of the Polytechnical Museum, Moscow).

In order to underline the direct relationships between the theoretical results such as the semi-inverse method and the method of zero-moment-point (ZMP) and practical results in the domain of rehabilitation robotics, it is presented in this section the active exoskeleton described in more details with focus to some important practical results.

This first pneumatically driven exoskeleton (Fig. 6) had only one degree of freedom of motion which was located in the hip joint. Displacement of the knee joint is enabled by a kinematic mechanism, i.e., by coupled movements, since it was, based on the previously obtained measurements for certain type of movements, calculated depending on the knee joint displacement from the corresponding hip joint motion. In the ankle joint, there was only a passive cylindrical hinge. These are certainly the first exoskeletons in the world where human movement was tested. To make it possible, and to avoid friction of the swinging foot after another supporting leg, the movement was very slow with a slight tilting of the body to the side of the supporting leg as to provide sufficient space for transfer of the swinging leg. Swinging leg accomplishes the so-called "half-step." In order to facilitate the structure of the pneumatic exoskeleton as much as possible, the manufacturing of the dual action pneumatic cylinders as well as the shock absorbers were done at own production in laboratory (as the non-standard components) made of thin steel pipes. Cylinder heads are made of duralumin. Control of pneumatic cylinders, and thus locomotion with the help of the exoskeleton, is ensured by solenoid valves.

Initial results achieved during the examination have indicated the important facts and suggested development of next, more complex model of the exoskeleton in which the leg prosthetic mechanism had three degrees of freedom, plus one degree of freedom of motion for lateral movement compensation (3 + 1 DOF per leg). This model (Fig. 7), appeared in 1972 and it was tested at the Orthopedic Clinic in Belgrade. To facilitate the structure of the device and there are also all the pneumatic cylinders are made in our own laboratory. Thereby, the knowledge applied has been acquired during testing of the first prototype model.

As an interesting fact it can be mentioned that these specially designed cylinders weighed only 0.5 kg but they were powerful and generate force of 1500 N at high air-pressure of 10 bars. Although the main task of the project was to develop exoskeleton, during the system development the other elements were advanced as well, necessary for reliable and safe use of the exoskeleton. Thus, for example, it was demonstrated that the best results of air sealing are achieved by means of the Teflon sleeve, while by application of the O-rings, shortly after the start of use, leaking of the fluid appeared. Also, it was experimented with different types of lubricants to find out which lubricant provides the best results in sense of relation lubricating—sealing. At the same time, there were analyzed also various gases: air, carbon dioxide, nitrogen and nitrous oxide. Although the best results were achieved with the carbon dioxide, because of problems with a pressure regulation, in the further researches it was used the air. During the tests, the structure has also adjusted and further facilitated and by application of contemporary control methods and novel electronic components, the pressure in the installation was lowered to 6–7 bar.

As a successor of the electro-pneumatic exoskeletons, developed in the Mihajlo Pupin Institute in the period of 1969–1972, it was designed and engineered in 1974 the first electro-mechanical robotic prosthetic device of this type (Fig. 8). This new-type exoskeleton served mainly to evaluate and develop electromechanical drives to be implemented with active orthotic devices, as the "active suit" or active arm orthosis for example.

R&D work in the field of rehabilitation robotics continued in the following years (1978–1982), but less intensely. The "Active Suit" was created in 1978. It was a modular semi-soft active orthotic device for dystrophic patients, electromechanically powered and programmed and controlled by using a micro-computer (Fig. 9). It was successfully used in testing rehabilitation and in research. It was realized as part of a project funded by American organizations SRS (Social Rehabilitation Service) and NSF (National Scientific Foundation), as part of the intensive scientific cooperation between USA and Yugoslavia. This device was sent to Texas Rehabilitation Center in Huston for testing. After that, in 1982, a newer model was created. It was the first in the world active orthotic modular device for lower extremities of the dystrophy patients and the first in the world exoskeletal electronic arm for dystrophic patients in the advanced stages of the disease, controlled using a joystick. Both devices were bought by the Texas Rehabilitation Center in Huston. In this way, the scientific community has been informed about the initial results in the field of the so-called active rehabilitation using an exoskeleton-type robots.

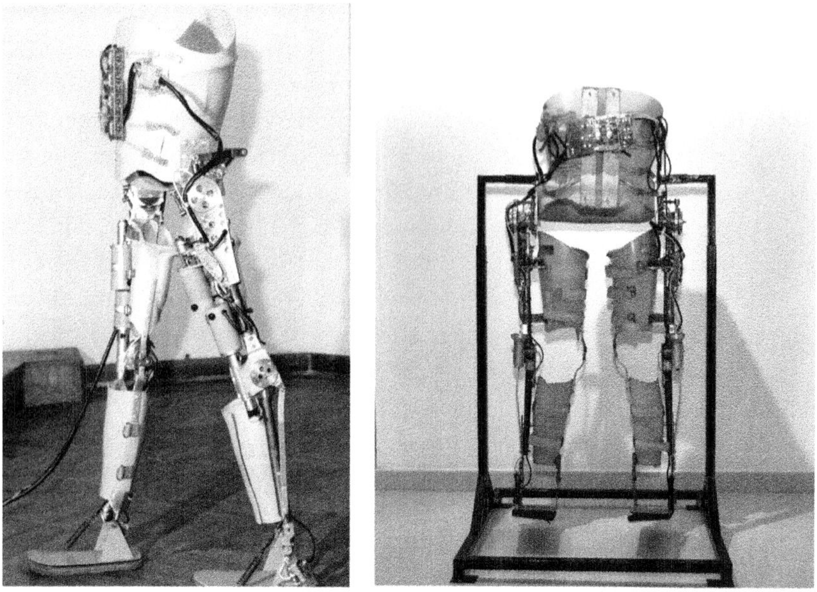

Fig. 8 The active exoskeleton with electromechanical drives, electronically programmed, built and tested in 1974.

Twenty years ago, in the Institute Mihajlo Pupin has started up the initiative to re-design and to develop new electro-mechanic exoskeleton where were planned to be applied contemporary control techniques and to be used a more efficient power-actuation system, sensors and orthopedic materials. Such a re-designed exoskeleton of modular type is shown in Fig. 10.

Basis of the modular solution represented a reinforced corset that pulled in an appropriate manner and fastened to the patient's body and after that, other necessary elements were added on it. The movements of the hip and knee joints are constructively solved by using compact, plate motors in combination with a Harmonic Drive gearing. The measurement of position was provided by use of optical encoder. Such a construction was proposed in order to, as much as possible reduce the lateral width of exoskeleton. As far as the ankle joint, there are suggested three possible solutions: (i) with a plate motor with harmonic-drive reducer, (ii) with a linear actuator of ball-screw drive type, and (iii) a passive solution. Compared to some other exoskeletons known at that time, the IMP exoskeleton had additional active lateral compensation (deviation from the vertical axis to the left and to the

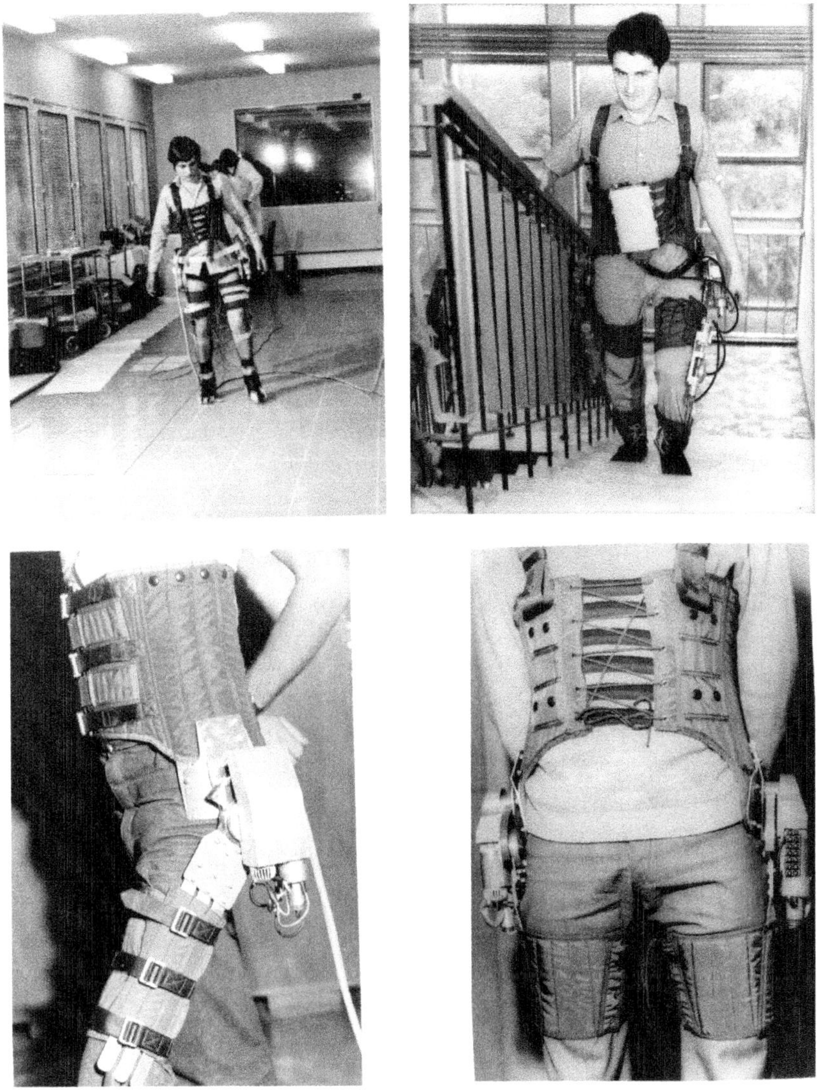

Fig. 9 The "Active Suit," modular, semi-soft active orthotic device for dystrophic. It was made in 1978. Electromechanically driven and micro-computer programmed and controlled. It was delivered to the Texas Rehabilitation Center, Houston, for evaluation purposes.

right with respect to the direction of movement), which would contribute to a better anthropomorphic locomotion. This, more complex but more complete exoskeleton solution has existed with the first Pupin exoskeletons that were presented in previous presentations.

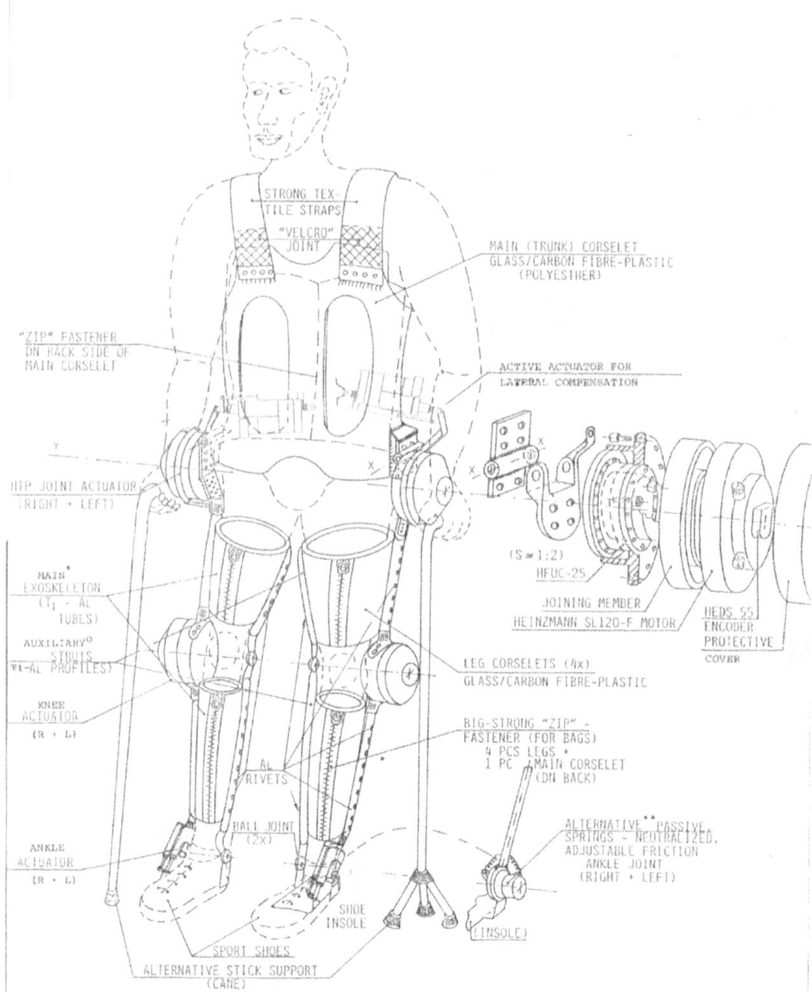

Fig. 10 The conceptual drawing of the IMP re-designed electrically powered modular exoskeleton (Archival material).

In Fig. 11 it is represented the preliminary drawing of the novel type of body corset for exoskeleton. According to the requirements the corset, as the main supporting element of the whole structure, had to be easy to fit to the patient. The idea was that the patient, in extreme cases may oneself to use it, of course depend on other, personal psycho-physical capabilities. At the time, when development was doing, it was envisaged to develop an exoskeleton on the basis of modern, easily processed materials, reinforced by glass or carbon fibers particularly in the sensitive areas of the increased stress. The pelvis belt was further strengthened because of its physical connection

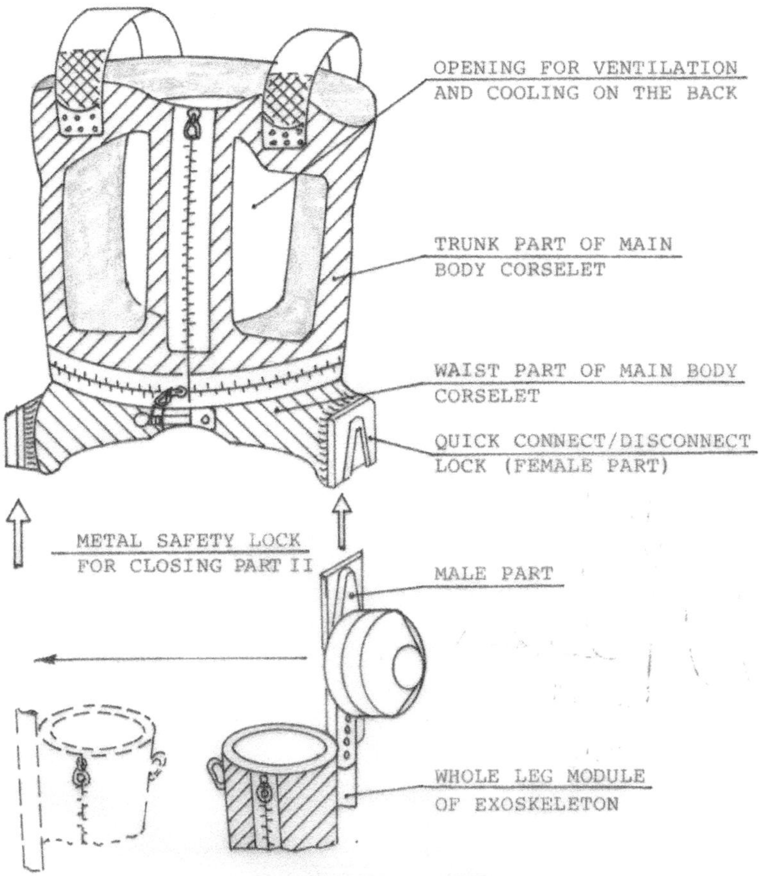

Fig. 11 Preliminary drawing of the new corset (Archival material)

to the power-drives as well as for installation of reliable quick detachable coupling. As indicated above, the construction was designed as a modular structure, easily adaptable to any potential user. The corset of exoskeleton and inserts, physical reinforcements of the thigh and lower leg, were intended to be made in the orthopedic clinic in order to be fully adapted to the user. The size of the driving motors and gear reducer were also selected on the basis of dynamic analysis to be accomplished according to the anatomic requirements of the individual patient. Some interesting details regarding to a new conceptually set exoskeleton are presented in Figs. 12–15.

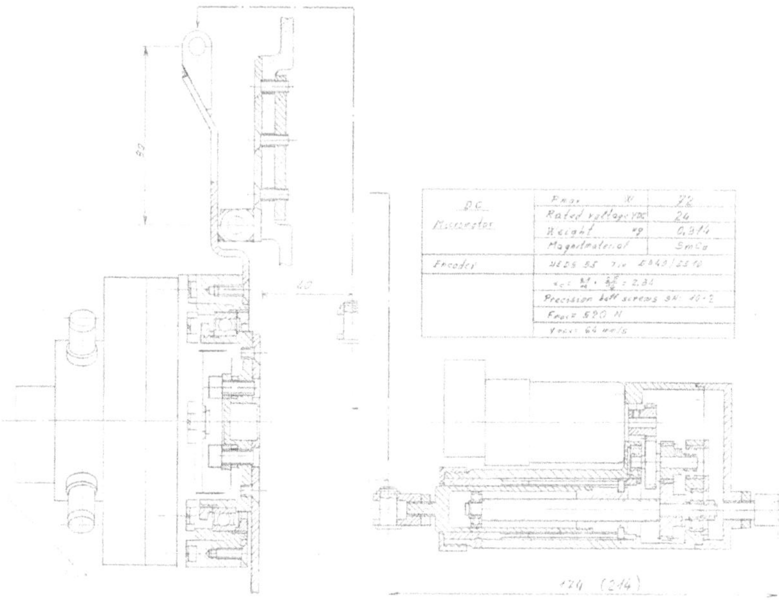

Fig. 12 The drawings of actuators (left side: a disc motor with reducer and an HD encoder and the connection with orthopedic parts; on the right side: a linear actuator for lateral compensation of the ankle joint and a DC motor, ball screw spindle and optical encoder) (Archival material).

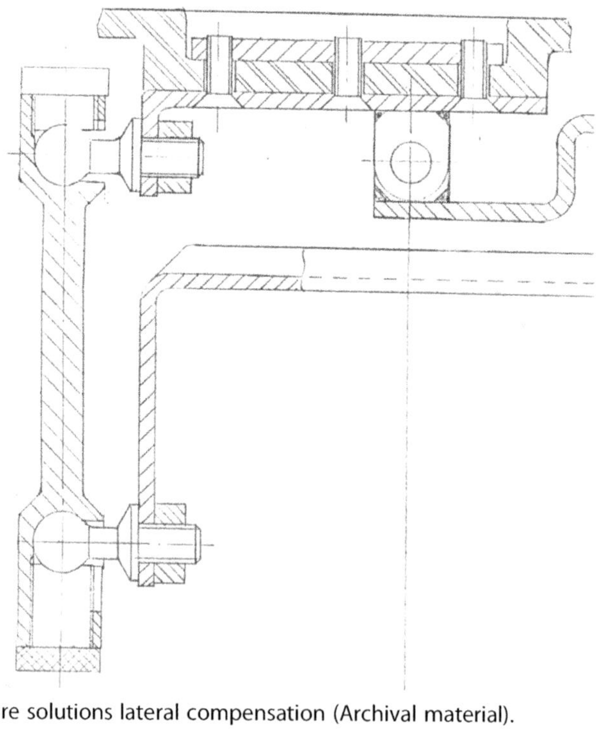

Fig. 13 Figure solutions lateral compensation (Archival material).

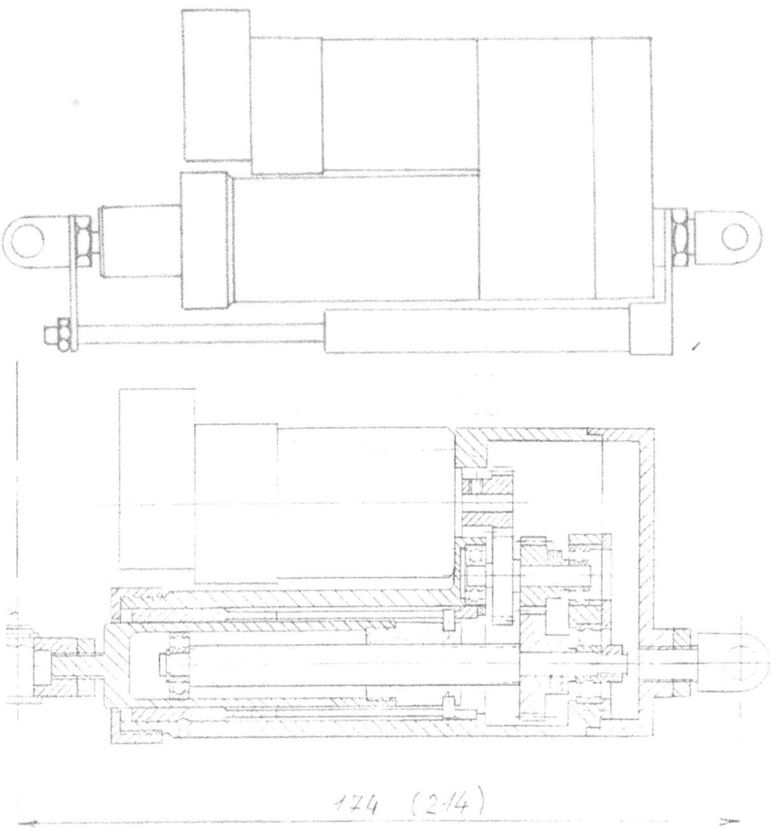

Fig. 14 Drawings of linear actuators (left side: with a linear potentiometer, right side: with optical encoder) (Archival material).

Parallel to the research and development of the lower limbs prosthetic devices at the Institute Mihajlo Pupin, it was also working there on development some other orthotic devices for upper body limbs. The arm orthosis was designed as a special type of exoskeleton intended for patients with progressive muscular dystrophy whose medical conditions (absence of a muscle spasm, etc.,) allowed the use of such kind of robotic device. Orthosis had 3 degrees of freedom, one in the shoulder and two at the elbow. They allowed to the patient (Fig. 16) to perform the following basic operations, i.e., personal needs (e.g., eating, grooming, etc.,). The small, high-speed DC motors, with specially-made miniature cylindrical and worm gearboxes, were controlled by using the small bars, so-called joysticks. Joystick is moved up by the fingers of the other, free arm. Specifically, it was shown that in the course of the disease it retains a residual strength in the fingers, which is

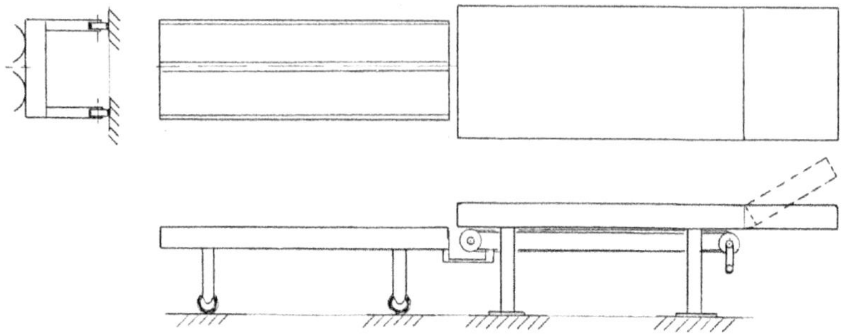

Phase 1: Working bed with extension and mechanism for pulling leg parts of exoskeleton.

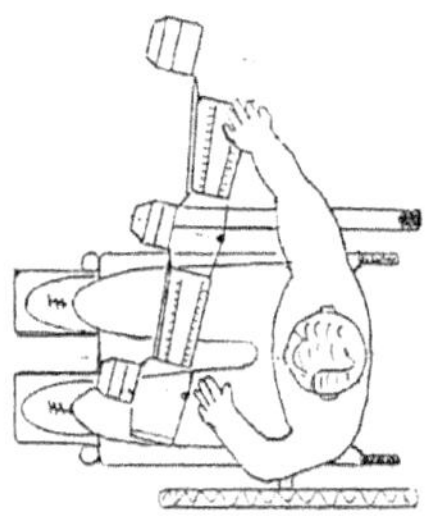

Phase 1: User sitting in the wheelchair brings exoskeleton parts to a working bed.

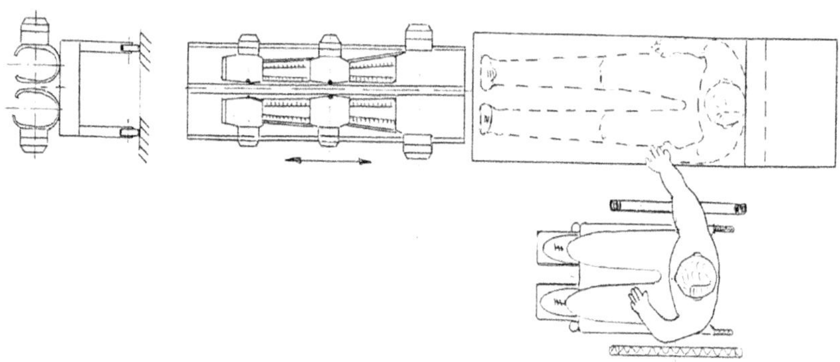

Phase 2: User shifting his body to sitting position by power of his arms.
Phase 12: User shifting his body from working bed to wheelchair by power of his arms.
Phase 13: same as Phase 12.

Fig. 15 (A) Drafts principle of use the exoskeleton presented in Figs. 10–14 with phase descriptions of manipulation (Archival material). (B) Drafts principle of use the exoskeleton presented in Figs. 10–14 with phase descriptions of manipulation (Archival material). (C) Drafts principle of use the exoskeleton presented in Figs. 10–14 with phase descriptions of manipulation (Archival material).

(Continued)

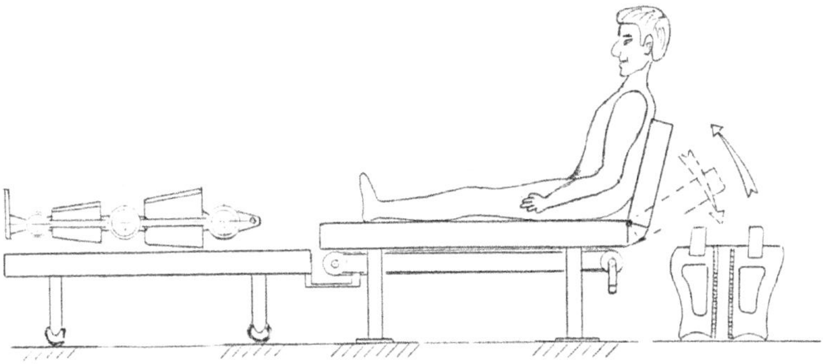

Phase 3: User aligning his legs with the leg corselets after lifting the back rest into the upright position.
Phase 12: User, after having pulled off the leg parts (modules III, IV from Fig05.11), takes off the body corselet and turns into lateral sitting position.

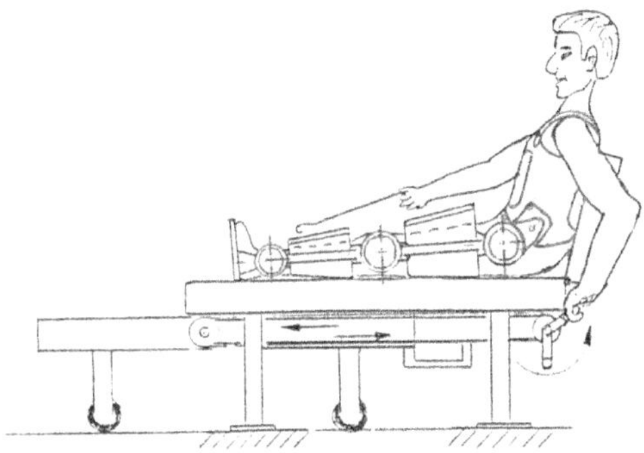

Phase 4: The user, in a sitting position, pulls on the leg parts of exoskeleton onto his legs.
Phase 5: The user takes the body corselet beside the working bed and puts it on.
Phase 10: The user puts by the power of his arms his legs into the exoskeleton corselets sitting onto the working bed.
Phase 11: The user pulls off the leg parts of the corsolet by turning the crank handle of the mechanism.

Fig. 15, Cont'd

(Continued)

enough powerful to run the joystick lever and thus moving the arm orthosis. Otherwise, the orthosis itself was mounted on the back (frame) of wheelchair. The orthosis received the power from the battery and it is easily mounted to the impaired arm by help of the Velcro.

As a specific form of orthopedic devices in the Institute Pupin are designed and developed active leg prosthetic devices of the PROMONA series. Prototypes of the PROMONA-1 leg above knee prosthesis (Fig. 17) appeared in the mid 1990s. It was aimed at investigating the possibility of making active upper knee prosthesis in which the rotation of the

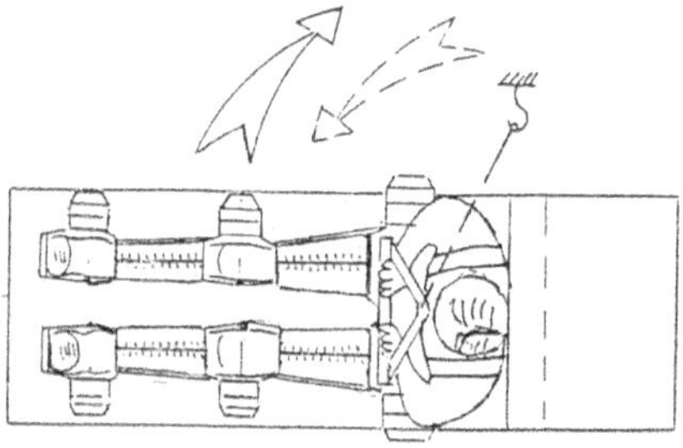

Phase 6: The user shifts his legs into the sitting position from the bed.
Phase 9: The user lifts his leg into the sitting position on the bed.

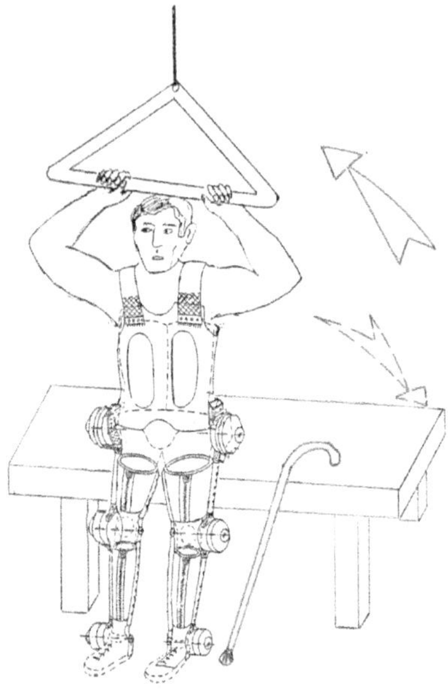

Phase 7: The user lifts his body to standing position.
Phase 8: The user lowers his body to sitting position on the working bed.

Fig. 15, Cont'd

knee achieved using servo motors. DC motor is powered directly from the AC adapter and the first tests have shown that it is possible to improve the solution with appropriate changes in the structure and with the improvement of the control system. Specifically, this first solution had limited bending angle because of the short trapezoidal thread which pretended rotational movement of the motor into reciprocating movement of the spindle. The patient is thereby adapted to the speed of bending the knee without being able to influence a change of pace. Due to a small angle of bending and to

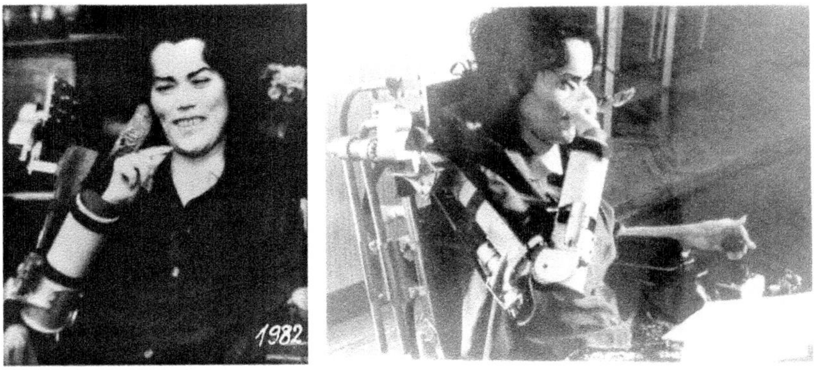

Fig. 16 Electrical arm orthosis developed in Mihajlo Pupin Institute in 1972 (Archival material).

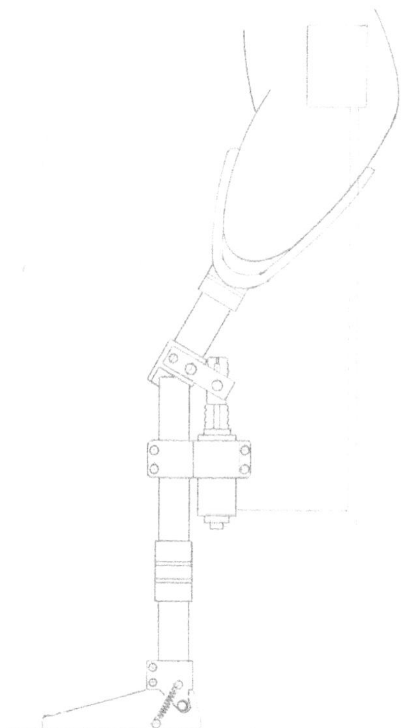

Fig. 17 The PROMONA-1 active leg above knee prosthesis.

avoid getting stuck fingers it was made passive ankle that would spring loaded foot bent upward and prevented the occurrence. The observed defects are significantly improved in the following, highly successful embodiment PROMONA-2 (Figs. 18 and 19).

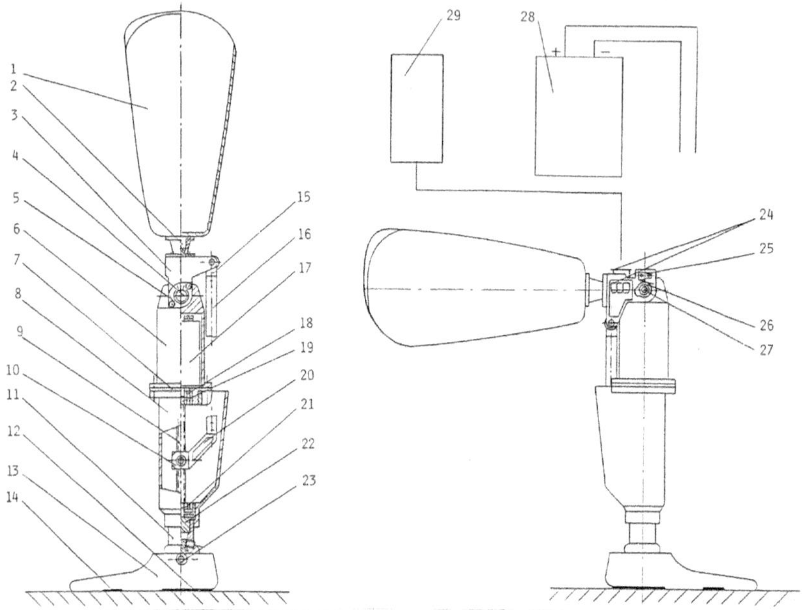

Fig. 18 PROMONA-2 active above knee leg prosthesis—technical drawings.

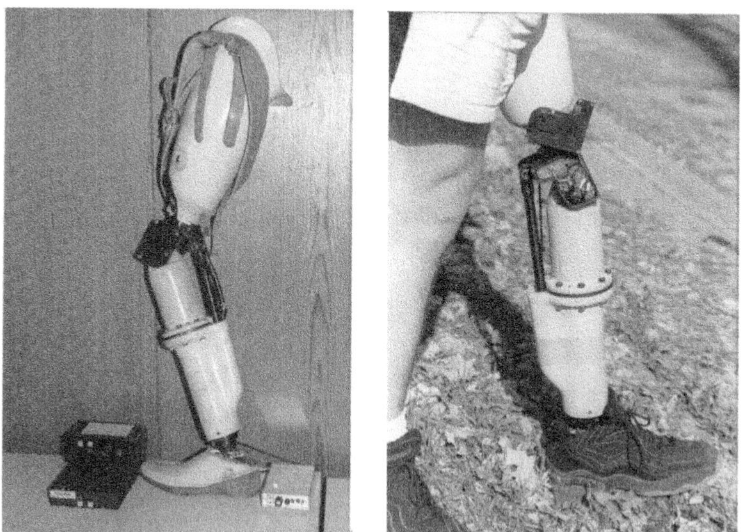

Fig. 19 PROMONA-2 active above knee leg prosthesis—prototype of the device.

The first identified shortcomings were that the external positioning of the DC motor with a short lead screw prevents taking of the sitting position of the user. Therefore, the designers decided to implement a certain anthropomorphic construction. The motor and gearbox, realized in terms of linear ball-threaded spindle of high degree of efficiency, was situated insides of the construction. This was achieved primarily to protect the system from external influences and unwanted damage. Even, the armor of the prosthesis takes form of a human leg. In the Figs. 18 and 19 there are given technical positions, i.e., parts of the above knee prosthesis. They are (Fig. 18): Tray (1), which approaches the amputated leg—the stump via standard prosthetic joint (2) that is coupled to the fork (3) which is a cylindrical joint (4) attached to the upper casing (6) in which is housed a DC motor (17). The output shaft of the motor, via a coupling (19) is connected to the ball recirculation (9) which is mounted on the underside of the sleeve (8) by means of radial and axial ball bearings (18, 21). With spindle is moving a nut (10) which is fixed to a lever (20) which via a lever (16) and (15) transmits a linear movement of the nut to the fork (3) that does a rotation of the cylindrical joint (4).

Length of the ball-spindle and the fork arm determine the bending angle of the knee and it is, in this case $\geq 90°$. Of the other specified parts, there are also (Fig. 18) the incremental encoder (5) which measures the rotation of the knee joint, adaptive tube (11) the length of which adjusts to the height of the patient as well as the inner spacer (22) which are functionally connected each-other, the passive cylindrical ankle joint (23), an artificial foot (13) and the micro-switches set in the feet (13, 14). By activation of micro-switches in the heel and under the big finger, the operation of the motor is triggered. In addition to being one of the main elements in the transmission of movement, the fork (3) is at the same time, an enclosure in which electronics is located (24), micro switches of the end positions (25), the cam mechanisms as triggers (26), and optical switches (opt couplers) for defining the zero position of the prosthesis (27). Power supply of the system is provided through a set of NiCd battery of 24VDC, while the control box (29) was used for the selection of gait programs and monitoring of the battery charge status and signaling errors i.e., warnings. In use there were several programs for different gait speeds as well as program to climb up and down the stairs.

PROMONA 2, active prosthesis, was aimed for disabled persons with above knee amputation. It enables a planar walking, upstairs and downstairs and walking on inclined and irregular terrains. It consists of the mechanical part of the system, the control system and a power supply. Several types of motion of the leg prosthesis have been programmed and the disabled patient could select one of the available (desired one) programs. The program selection was accomplished through a small control box, which was carried on the patient belt. The system was powered by a DC servo motor, which performed a knee joint rotation through the ball screw reducer and a lever system. An optical shaft encoder for position and velocity control, and two optical switches as limiters are incorporated into the microprocessor-based control system. The main start-stop switch was incorporated into the heel of the artificial foot. The height of the prosthesis could be adopted according to the user demands. Specific orthopedic parts (the socket and the foot) were linked to the powered part of the prosthesis by the standard connections. The main advantage of this active prosthesis, with respect to the passive one, is that it provides capability of regular upstairs and downstairs walking and the patient the patient was less bothered during walking. Average walking speed on a flat ground was ~0.9 m/s. Weight of the batteries for 2 h of gait autonomy was 1.7 kg (NiCd) and 0.6 kg corresponding lithium, respectively. Also, the additional performances were: rotation of the ankle joint was up to 30°, the prosthesis total weight was ~5.5 kg (that is close to the mass of the disabled leg).

3.2 Recent design solutions—State-of-the-art

To date, it was developed a number of different variants of exoskeletons for rehabilitation of patients affected by paraplegia. Generally speaking, the exoskeletal devices can be systematized in two categories. The first type is based on the exoskeleton leg orthosis and the system to support body weight combined with a conveyor belt for exercise. Leg's ligaments apply only the force necessary to make movement of the patient in the regular walking requested form. These systems are used only in clinical conditions because they require more space for implementation and supervision by an expert. The leg orthosis keep the pelvis of the patient in a fixed position which causes deviation of the walk kinematics from the natural model. These systems do not care about the locomotion stability and serve only to strengthen the impaired limb motor function. Known exoskeletons of this type are LOCOMAT [39–42], LOPES [43–45] and ALEX [46–48]. Another type of exoskeletons is ambulatory devices. These systems are used to simultaneously carry on both, the body weight as well mass of the prosthetic device. By use of such a type of exoskeleton users have more freedom of movement in the open space. These exoskeletons are used as assistive devices to support an impaired locomotion of patients in their daily activities. By use of these devices the patients are able to improve the gait stability, maintain overall body balance and strengthen extremities by exercising. The examples of the second type of exoskeletons are discussed in the text to follow.

Hybrid Assistive Limb (HAL) [49–51] was developed for the purpose of rehabilitation or living support of persons who have disorders in the lower limbs and whose leg muscles are weak. Full body versions of the HAL exoskeleton were also developed to assist humans in accomplishing heavy labor or for the purpose of the outdoor rescue tasks. The lower body model of the device weighs about 12kg and the full body model weighs about 23kg. The batteries of the HAL exoskeleton intended for rehabilitation provide the power for 60–90min in normal operation regime. The hip and the knee joints of the HAL are powered in sagittal plane. Joint actuators are composed of DC servomotors and harmonic drives. A hybrid control system of the HAL exoskeleton consists of an autonomous posture controller and a complementary power-assisted controller. The intended motion of the user is identified by using electromyography (EMG) sensors and ground reaction force sensors. The drawback of this control system is that it requires adaptation and adjustment to a specific user. ReWalk [52] has two different models developed for rehabilitation and life support of spinal cord impaired patients. The hip and knee joints of the ReWalk exoskeleton are powered in the sagittal plane by using DC motors. It comprises a wearable brace support suit, which integrates DC motors, rechargeable batteries, sensors and a computer control system. The device is about 20kg weight. Pre-programmed motion control strategy is used to control of ReWalk. Changes in the user's center of gravity are used to initiate and maintain walking processes. The user also has a remote control placed in the arm for selecting different tasks, such as: sit-to-stand or climbing stairs. The VANDERBILT exoskeleton [53, 54] was developed to provide gait assistance to patients with spinal cord injuries. The device weights about 12kg. Hip and knee joints are actuated by brushless DC motors. Control is based on postural information measured on the device. A lithium-polymer battery of 29.6V and 3.9Ah brings 60min of autonomy for continuous walk at a speed of 0.8km/h. The EKSO is a gait training exoskeleton intended for medically supervised use by individuals with various levels of paralysis or hemiparesis. EKSO weights approximately 20kg and has a maximum speed of 3.2km/h with a battery life of 6h. It can execute sit-to-stand and stand-to-sit operations and walk in a straight line. EKSO uses a gesture-based human-machine interface to determine the user's gestural intentions and then acts accordingly. The WSE [55] was developed to support walking of partially or entirely disabled individuals. The total weight of WSE is about 18.5 kg. A 24 V DC motors powered by Li–Po battery pack are used for actuation of WSE. The pre-programmed motion control strategy is used to control of WSE. Adaptive network-based fuzzy logic controllers are used to control of joint motions of WSE. Changes in the user's center of gravity are used to initiate and maintain walking processes. The Li–Po battery pack of WSE can provide the necessary power for about 3 h in normal operation. The second generation prototype of WSE (WSE-2) has improved technical performances [56]. The exoskeleton has two degrees of freedom per leg. Hip and knee joints are actuated in the sagittal plane by using DC servomotors. Additional effort supplied by user's arms through crutches is defined as user support rate. Experimentally determined user support rate values were considered in actuator torque computations for achieving a realistic actuator selection. A custom-embedded system was used to control exoskeleton. Reference joint trajectories were determined by using clinical gait analysis. Three-loop cascade controllers with current, velocity and position feedback are designed for controlling the joint motions of the exoskeleton. A non-linear ARX model is used to determine controller parameters. Overall performance and an assistive effect of WSE-2 are experimentally investigated by conducting tests with a paraplegic patient.

In the recent years there have been many attempts in the world to create a device that will not only help to the sick people in performing their daily life activities, but will help healthy people to facilitate their physical efforts, especially in lifting heavy objects. The Panasonic's exoskeleton AWN-03 [57, 58] is a device that helps lift and move loads up to 15 kg. Two robust AC motors are built in at the hip joints of the mechanism. They are driven by electricity from the Li-Ion battery pack of 48.1 V. Batteries are carried by man on the back. The weight of the device is about 6 kg without battery and harness and the autonomy of the device is about 8 h of continuous operation. No matter what the propaganda film [57] shows some workers that easily manipulate loads by relaxing their body structure and muscles of the spine, based on our previous practical experiences achieved during testing the above-knee prosthesis PROMONA-2 [59], the ballast that user constantly wears on the back does not give us a great hope about effectiveness of this type of device in spite of implementing contemporary technology. From our experience, patients who tested the above knee prosthesis PROMONA-2 also carried on the NiCd batteries weighing about 2 kg on the back. These batteries have provided to one driving motor the autonomy of just 1 h and the user who wearied PROMONA-2 already felt tired after an hour of operation. For this reason, it is unlikely that, even by use of the modern technology, it can get some significant results in terms of operational autonomy.

Exoskeleton "Phoenix" [60, 61] is also an interesting attempt to solve the problem of walking by deal with two driving motors in the hip joints of the mechanism. Weighing a mere 25.12 kg this exoskeleton allows movement at a speed of 1.7 km/h with the autonomy of operation of about 4 h with the recommended rest of up to 8 h. However, the physical effort of the patient when using this device refers to the physical load that one has to spend to stabilize the weight of his/her own body and the additional weight of the mechanism. The additional supports in crutches, in whose arms are built in certain commands that support the control of the exoskeleton, are necessary for gait stabilization and safety (in a physical and psychological sense), nevertheless this motion was slow. The considered device has certain rehabilitation characteristics, but moving patient from one place to another requires a considerable effort of the individual so that the exoskeleton is more acceptable for the persons with more powerful arms and torso [60].

Berkeley Lower Extremity Exoskeleton (BLEEX) [26, 62] intended to carry on larger payloads was developed for military purposes under the DARPA development program. Its purpose is to enable a soldier carrying on large loads in the long run, without additional effort and fatigue because payload is transferred to the frame structure—the corset. The corset of exoskeleton has its own drive, it is monitored and ensured the stability of locomotion by help of a computer system, and user, who is mainly a well-trained and healthy person, represents a "system for guidance" of the motorized exoskeleton mechanism. More details about the device can be found in the literature [62]. What the designers of this system have observed was that use of the electric drives requires high power consumption and thus the large batteries as additional ballast. The batteries nevertheless were modern technology and every day less in size and lighter, still do not provide sufficient long autonomy of motion. If during this motion it is done in a free space without the possibility of energy amendments, the constraints are high. Therefore, it was designed exoskeleton system with the hydraulic actuators that provide significantly higher power while decreasing the construction mass. The driving forces for the hydraulic pumps and distributors were obtained from a small inner combustion engine that is part of the structure. A few liters of gasoline do not pose a large additional weight for a healthy and trained soldier and provides significant additional autonomy. The only problem in this case represents the noise generated by the inner combustion engine and that is impossible to be completely dimmed. Also, the problem of the exhaust gases from the engine discharged directly next to the human exists in spite of the fact they were reduced with new type of motors. Therefore, it is questionable the application of this type of exoskeleton in real operating conditions due to the aforementioned drawbacks.

The TWIICE exoskeleton [63], a product of research of scientists and engineers from EPFL is currently one of the most advanced technical solutions of these devices that have recently been realized. According to the basic idea of the device it is applicable to a variety of neuro-physical pathologies, morphology and physical needs of patients. The supporting structure is made of carbon fiber and weighs about 15 kg. It enables complete movement of the legs with the possibility of climbing up and down stairs. The autonomy of operation of the device is about 3 h at a speed of 1.3 km/h [63, 64]. TWIICE exoskeleton has only 2 DOF, one in the hip and another at the knee joint. This motion is substantially dominant when stepping (as previously elaborated in the Section 2 of this chapter) but a problem with climbing stairs is difficult to be successfully overcome without adequate active drive in the ankle joint or at least some additional kinematic (functional) connection between ankle joint and knee joint. This kinematic deficiency was attempted to be solved by introducing an auxiliary rounded feet as a kind of constructive solution [64]. More details about the controller architecture and crutch remote are available at [64] and the video showing exoskeleton during exercising is presented in [65].

AUTONOYO [66, 67] represents an innovative solution designed to support independent living of people with muscular-motor deficiencies through improved mobility. The AUTONOYO is an active orthotic suit, motorized with intelligent interface, intended for unmatched freedom of motion. This wearable robotic device is accessible being it solves the problem of necessity of additional mechanical support such as passive walker (roller) or crutches. It allows to be applied with people who suffer from weakness of upper extremities. The device is modular and adaptable as it allows adaptation to different sizes of patients with different levels of impairment. The device allows more freedom to patient in sense that his/her hands are uncommitted and can be engaged for other activities (e.g., use of cellular phone, transfer of personal belongings, opening doors, etc.). By mounting the motors about the trunk it was achieved reduction of the device inertia, mass decreasing of the lower segments and better dynamic properties of the entire system. By providing corresponding physical support, device allows walking, transition from sitting to standing as well as climbing up and down the stairs. This significantly enhances the mobility of patients and their living autonomy. An additional degree of freedom of movement in the lateral direction allows user more freedom of motion and more natural walk. The device has ability to maintain the balance of locomotion based on permanent estimating of stability conditions and additional corrective actions before its losing. The back drivability of the transmission enables the user to interact in a smooth and intuitive way with the exoskeleton.

4 Control of exoskeletons

Control techniques used in control of exoskeletons are generally characterized by the methods used in determination of motion intended by the user. Most of the control techniques used in exoskeletons need interaction signals between user and device for identification of the user movement intention. However, majority of paraplegic patients are not able to generate an effective user-exoskeleton interaction. Another method utilized in detection of user's intentions regards to exploitation of the EMG signals. However, there are technical difficulties in implementing EMG-based control techniques since the EMG signals are subjected to extreme noise and require extensive signal processing. A pre-defined motion control (PMC) technique is an alternative and frequently used method for controlling exoskeletons. In PMC control, the user selects the intended motion by him/herself. The PMC technique is characterized by a reduced computational complexity, hardware complexity and sensor requirements. For implementation of PMC, the reference motion database (which includes information about sitting, standing up and walking kinematic parameters) is generated by use of corresponding software interface (anthropomorphic biped gait generator, e.g., HRSP [6]). Commonly used, cascade PID controllers are designed for motion control and tracking error compensation of the hip, knee and optionally ankle joints. The cascade controller comprises three feedback loops: (i) current loop, (ii) velocity loop, and (iii) position loop (Fig. 20). Both, the current loop and the velocity loop are inner sub-control loops while the position

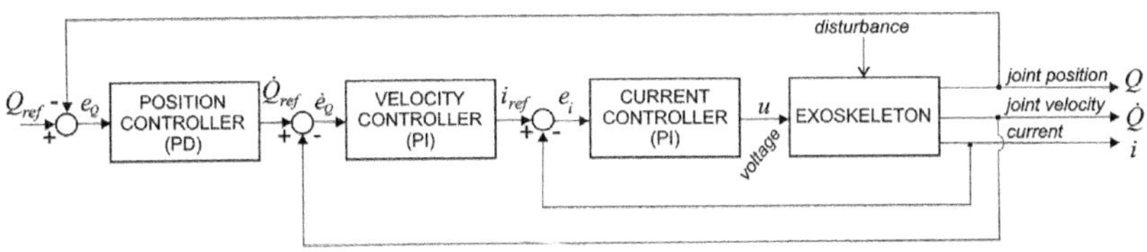

Fig. 20 The cascade PID controller comprising three feedback loops.

loop is a primary control loop. The current loop is used to limit the current (e.g., shaft torque) of actuator by keeping it constant under the maximum allowed value at the start and stop, and it optimizes the variation of current. The velocity loop is used to enhance the ability to resist disturbances in load and to suppress fluctuations in velocity.

PI type controllers are used in both the current and the velocity loop. The position loop is used to ensure good dynamic tracking performance and static position accuracy. The PD-type controller is used in the position loop. The choice of controller parameters for a non-linear system is quite complicated. Linearization of a system is a way of determining controller parameters. If it is difficult to linearize a highly non-linear system then use of a linearizable non-linear autoregressive exogenous (NARX) model was implemented instead in determination of the controller parameters as presented in [56].

In the text to follow it is given an overview of the various control methods that have been applied since development of the first exoskeleton until to the present days. The first exoskeletons, developed in the late 1960s and early 1970s (Figs. 6–8) had limited control capabilities due to the modest processor technology available in this time. These first wearable robotic systems are remotely controlled by wire. They have been connected by wired cable to the global (high-level) controller of the system. The entire control system is presented in details in Fig. 21. As it is shown, the human is involved in the control loop to regulate speed, direction of motion and attitude (body orientation) in the sagittal and frontal plane. These first exoskeletons were aimed to help persons who couldn't move their legs due to illness or injuries. The local control system, based on the ZMP feedback loop [34], was

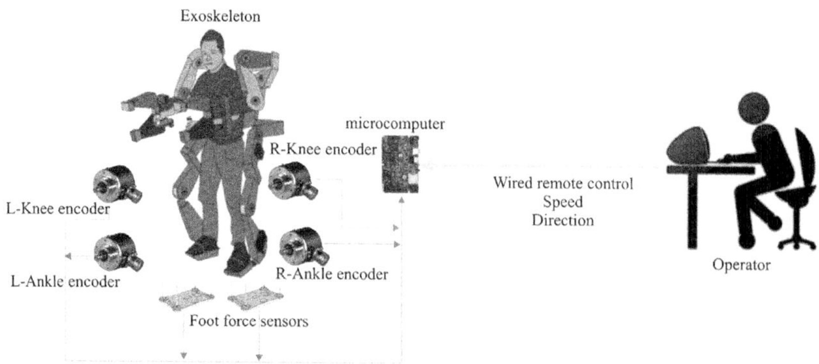

Fig. 21 High-level control scheme of the simplest exoskeleton control system implemented with early exoskeleton from the late 1960s.

responsible to preserve dynamical balance of the coupled system patient-exoskeleton and to predict the conditions of overturning. The sensors implemented in the feet and exoskeleton joints are: force sensors and position sensors. Their role was to acquire the system position (attitude) and ZMP location within the supporting polygon.

Nowadays, modern exoskeletons have various and very complex control architecture depending on exoskeleton implementation. If the exoskeleton is intended for body power enhancement, the control system has to be very robust and dynamic, because the healthy people have no body motion limitations and exoskeletons are used to enlarge the body strength (to pull load more powerful or to carry on extra weight, etc.,) for special purposes as it has been already discussed in the chapter. In this case, the control system has to be capable to establish dynamic balance of the overall system due to external force/load and to minimize the posture movement due to dynamical coupling between human body and wearable robotic device. In other words, person with the attached exoskeleton has to have satisfactory mobility and also to have enhanced power capability. These two opposing demands need very sophisticated control system with numerous sensors implemented into the system. One such control system is presented by its high-level scheme in Fig. 22. The exoskeleton system presented in Fig. 22 is equipped with large number of sensors. There is 3-axis force sensor attached to the foot to be used for detection of the ZMP location. Each link into the exoskeletal system has 3-axis torque sensor as well as 3-axis gyro sensor together with 3-axis inertial sensor (accelerometer). Also, every joint of the mechanism has 3-axis local accelerometer together with absolute encoder (Fig. 23). Such control-sensing system is capable and enough sensitive to determine external perturbation which can occur to the whole human-exoskeleton system in order to preserve dynamic balance and protect human from falling down and injuries. Human wearing the exoskeleton is an inner operator who freely moves his body. By measuring corresponding dynamic variables as consequence of motion, control system of the exoskeleton is capable to determine actual position and velocity of the entire system. In this case, dynamic (model-based) control algorithm combined with adaptive or knowledge-based control algorithm has to be implemented to compensate own system dynamics as well as system uncertainties such as external perturbations and other uncertainties existing in the system. The ultimate goal of control system is to prevent overturning, i.e., to maintain stability.

Exoskeletal systems are also developed for persons who have limited mobility or serious neuro-muscular impairments that cause partial or

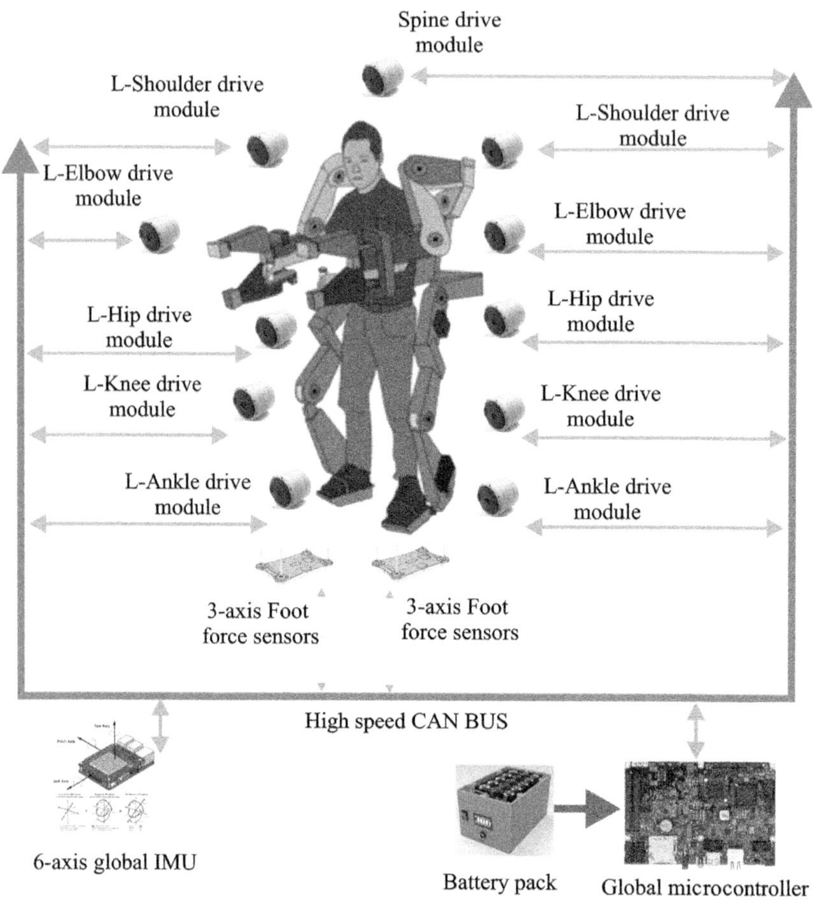

Fig. 22 High-level control scheme of the exoskeleton for body power enhancement.

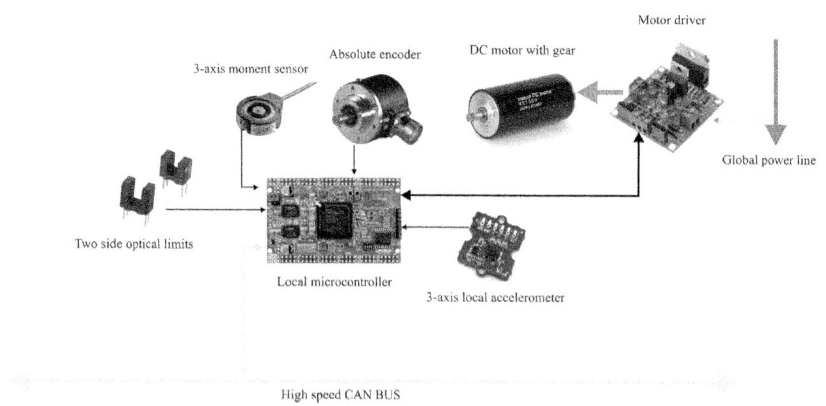

Fig. 23 Control structure of a single power driving module of the exoskeleton dedicated for body power enhancement.

complete inability of body movements. Such persons need medical medical aid, i.e., treatment due to rehabilitation. For that purpose, the implemented control system need to be more sophisticated and requires some special safety actions embedded in the controller to prevent possible injuries. For persons who have limited mobility capabilities, control system has to establish dynamical balance as well as path control. These persons commonly have very poor leg and/or arm control and coordination and accordingly the exo-skeleton control scheme differs from the control schemes presented in Figs. 22 and 23. If the person can move hand and fingers, a simple joystick control can be implemented to manage desired motion. Such a control system is presented in Fig. 24. The person coupled with the exoskeleton is equipped with a standard joystick. Joystick can be mounted on the frame of the pelvis link of exoskeleton or more recently it is built in the handles of the supporting crutches. Exoskeleton is equipped with numerous sensors as already described in the previous consideration and shown in Fig. 22.

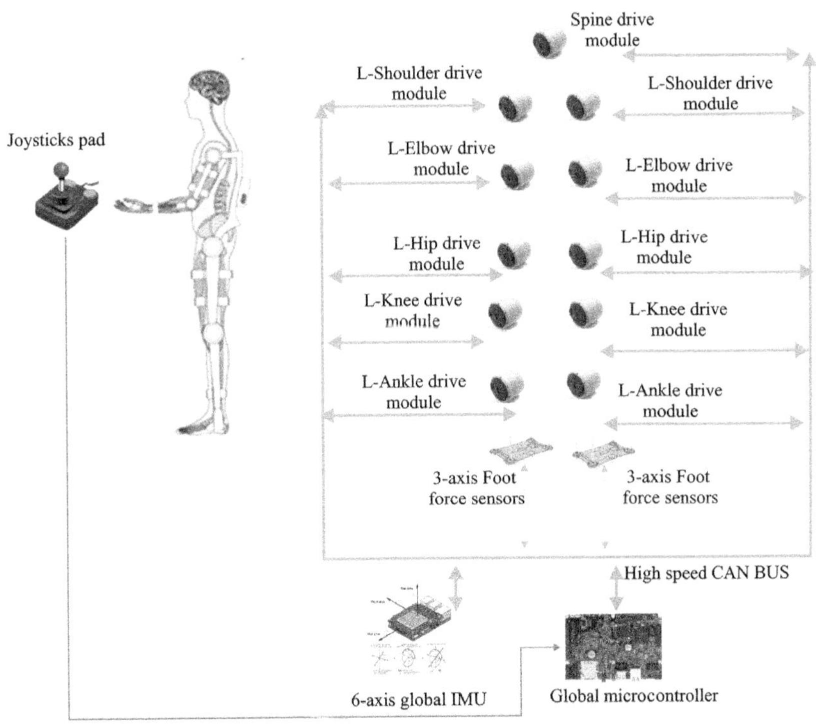

Fig. 24 Exoskeleton controlled by a joystick.

This type of controller has certain level of intelligence in sense that it is capable to detect weather the movement of exoskeleton segments is in accordance with a natural, biological leg limbs movement. The exoskeleton controller has to check, by measuring load in the powered exoskeleton joints (electric current of the motor or driving torque), if there is no accidentally occurrence of the muscle spasm or irritation in the leg muscles while at the same time motors in the exoskeleton joints continue forcefully to carry out the planned motion. This could cause serious injuries of the leg tissues.

Brain-computer interface (BCI) is a system that uses signals emanating from a human or animal brain in order to control objects outside the physiology of same human or animal body. The basic idea is to generate a brain state which will be manifested as brain signal to control appearances on a computer screen, or sound objects, or physical objects with a mass (such as robots, wheelchairs, and prostheses). Electroencephalogram (EEG) is a classical modality of obtaining a brain signal, but other ways of recording brain signals (e.g., magneto encephalogram) are also used in researches. If the person is totally disabled (general muscular insufficiency), then BCI controller can be implemented. A possible solution is brain control combined with head movement control. This type of control system is presented in Fig. 25. A specially equipped sensorial cap is capable to record the brain signals and thus to determine brain functionality. A 3-axis gyro sensor and 3-axis acceleration sensor are also implemented into the cap. Person can control the exoskeleton using his minds and head movements. Exoskeleton system is also equipped with numerous previously explained sensors to detect ZMP, acceleration of exoskeleton links and acceleration in the joints of the whole mechanism. Control system consists of special neural network and complementary AI algorithms implemented into the control system. Some initial training of the person is needed preliminary. Training requires mind training and neural network training. Mind training is procedure to determine which brain signals are responsible for corresponding extremity movements. This type of brain signals are unique for each person and have to be recorded and implemented into the neural network module of the control system. The neural network interpreter has to be separately trained for simple primitives of movement as well as for complex movements for the particular links of the exoskeletons as well as for the exoskeletons as a whole. As it is presented in Fig. 25, the control system is quite complex. It is capable to prevent accidental overturning and patient injuries thanks to the dynamic measurement of forces and torques in the joints of the exoskeleton during the motion.

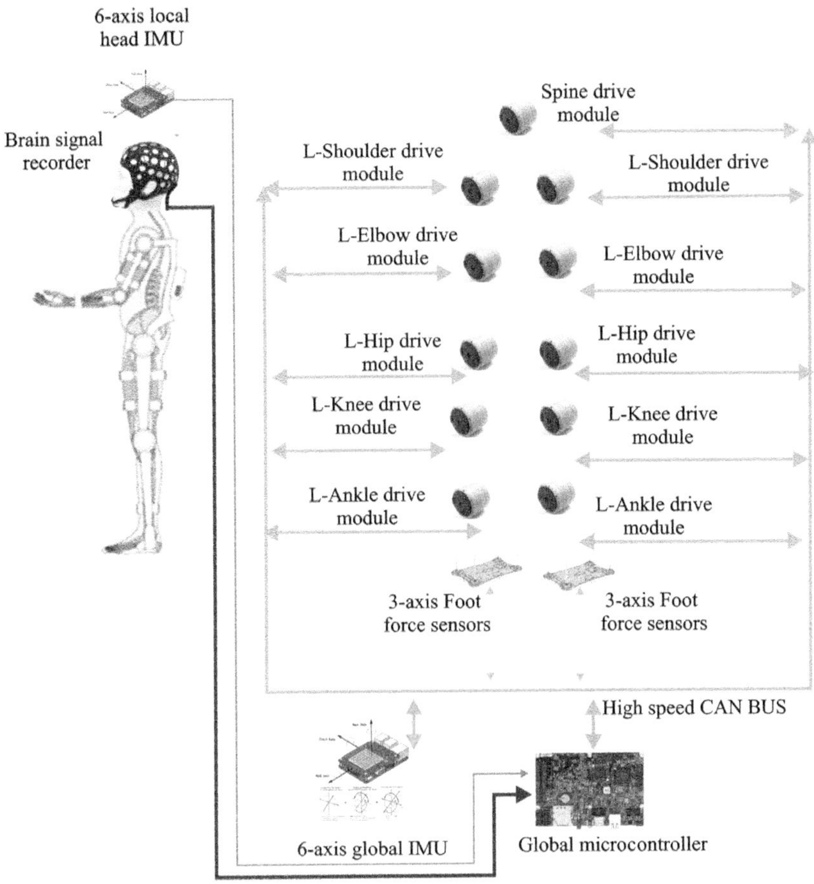

Fig. 25 Exoskeleton controlled by the Brain Computer Interface.

5 Conclusion

We are in the fifth decade since the first active exoskeletons were appeared. Lot of research and development efforts in the meantime has been done in this field. If one would like to make an objective analysis and assessment of how much really we advanced in this area then we need to define the appropriate criteria against which to do an objective analysis. At the end of this chapter it is given the list of the most significant criteria for objective assessment of technical performances and quality of this type of wearable robotic devices. These are: (i) mass (weight), (ii) number of mechanical degrees of freedom for motion, (iii) power and energy efficiency, (iv) maximal speed,

(v) stability, (vi) sensory diversity, (vii) control capabilities and complexity, (viii) operation range (indoor and/or outdoor), (ix) autonomy of operation, (x) ergonomics, esthetics and social acceptability of prosthetic device, etc., The fore mention criteria can serve for objective assessment different prosthetic devices and comparison of their price-performance ratio.

Bearing in mind the previous set of criteria, the following conclusions can be derived. Although it has passed such a long time nothing crucial has changed compared to the pioneering solutions made 50 years ago. The advancements in this field were incremental and not stepwise as probably it was expected. The novelties that can be mentioned are in application of new, lightweight, composite and ergonomic materials, in use of more compact size and efficient driving actuators, in implementation of more efficient and smaller size battery storages, in fusion of heterogeneous sensors, in developing new intelligent control algorithms, implementation of wireless communication between technical modules, etc., In spite of that, all main problems that have been identified as problems from the pioneer period of development still remained as unsolved more or less. Problem of heavy mechanism, problem of low power actuators if wish to implement a small size ones, problem of limited autonomy of operation, problem of stability, etc., All these problems are looking for new radical technology solutions for the next decades. Because of that, the wheelchair manufacturers still have not lost the battle in the transport of patients.

References

[1] V.M. Zatsiorsky, B.I. Prilutsky, Biomechanics of Skeletal Muscles, Human Kinetics, ISBN-13: 978-0-7360-8020-0. 2012.

[2] A. Rodić, K. Addi, G. Dalleau, Modeling of impact phenomena with biped locomotion mechanisms—theory and experiments, Sci. Tech. Rev. YU ISSN 1820-0206, UDC: 355/35962. LVIX (1) (2009).

[3] A. Rodić, K. Addi, G. Dalleau, Adaptive bio-inspired control of humanoid robots—from human locomotion to artificial biped gait of high perforomences (Chapter in the book), in: Contemporary Robotics—Challenges and Solutions, In-Tech, 2009, pp. 275–300. www.intechweb.org, ISBN-978-953-307-038-4.

[4] A. Rodić, M. Vukobratović, K. Addi, G. Dalleau, Contribution to the modeling of non-smooth, multi-point contact dynamics of biped locomotion—theory and experiments, Robotica Cambridge University Press, ISSN: 0263–5747. 26 (02) (2008) 157–175.

[5] A. Rodić, K. Addi, G. Dalleau, Contribution to the mathematical modeling of multi-point, non-smooth impact/contact dynamics of human gait, in: Programe Resume of the 6th International Conference AIMS 2006 "Dynamic Systems, Differential equations and Applications", University Poitiers, Poitiers, France, 2006, pp. 101–102.

[6] A. Rodić, Humanoid Robot Simulation Platform (HRSP)—Customized Software for Modeling and Simulation of Biped Locomotion Mechanisms, Software Toolbox

for Matlab/Simulink, http://www.pupin.rs/RnDProfile/hrsp.html, 2007. Retrieved 2018-07-22.
[7] P. de Leva, Adjustements to Zatsiorsky-Seluyanov's segment inertia parameter, J. Biomech. 29 (9) (1996) 1223–1230.
[8] C-Motion, WIKI Documentation, Adjusted Zatsiorsky-Seluyanov's segment inertia parameters, https://www.c-motion.com/v3dwiki/index.php?title=Adjusted_Zatsiorsky-Seluyanov%27s_segment_inertia_parameters (Retrieved 2019-07-11)
[9] G. Mester, A. Rodić, Autonomous locomotion of humanoid robots in presence of static and mobile obstacles—trajectory prediction, path planning, control and simulation, (Chapter in the book) in: J. Kacprzyk (Ed.), Studies in Computational Intelligence, Towards Intelligent Engineering and Information Technology, Part III Robotics, vol. 243/2009, Springer, 2009, pp. 279–293. ISBN 978–3–642-03736-8, Library of Congress: 2009933683, https://doi.org/10.1007/978-3-642-03737-5.
[10] M. Vukobratović, A. Rodić, Contribution to the integrated control of biped locomotion mechanisms, Int. J. Humanoid Rob. World Scientific Publishing Company, New Jersey, London, Singapore, ISSN: 0219–8436. 4 (1) (2007) 49–95.
[11] D. Katić, A. Rodić, M. Vukobratović, Hybrid dynamic control algorithm for humanoid robots based on reinforcement learning, J. Intell. Robot. Syst. Springer, ISSN: 0921-0296. 51 (1) (2008) 3–30.
[12] M. Vukobratović, V. Potkonjak, A. Rodić, Contribution to the dynamic study of humanoid robots interacting with dynamic environment, Robotica Cambridge University Press, ISSN: 0263-5747, United Kingdom. 22 (2004) 439–447.
[13] A. Rodić, M. Vukobratović, Control of dynamic balance of biped locomotion mechanisms in service tasks requiring appropriate trunk postures, Problemi Masinostroenia i Avtomatizacii, Int. J. Eng. Autom. Probl. ISSN: 0234-6206, ISSN: 0234-6206. 5 (1) (2006) 4–22.
[14] D. Katić, A. Rodić, M. Vukobratović, Reinforcement learning control algorithm for humanoid robot walking, Int. J. Inf. Syst. Sci. ISSN: 1708-296X. 4 (2) (2007) 256–267.
[15] K. Addi, A. Rodić, Impact dynamics in biped locomotion analysis: two modelling and implementation approaches, AIMS J. Math. Biosci. Eng. 7 (3) (2010) 479–504.
[16] M. Vukobratović, B. Borovac, A. Rodić, D. Katić, V. Potkonjak, A bio-inspired approach to the realization of sustained humanoid motion, Int. J. Adv. Robot. Syst. InTech, 9 (2012) 201–212. 1729-8806, https://doi.org/10.5772/52419.
[17] M. Petrović, A. Rodić, Synthesis of adaptive impedance control for biped robot mechanisms, J. Autom. Control Published by Faculty of Electrical Engineering, University of Belgrade, ISSN: 0354-124X. 18 (1) (2008) 15–18.
[18] D. Katić, A. Rodić, Dynamic control algorithm for biped walking based on policy gradient fuzzy reinforcement learning, in: Proceedings of the 17th IFAC World Congress Seoul, Korea. 2008, pp. 1717–1722.
[19] D. Katić, A. Rodić, Policy gradient fuzzy reinforcement learning control of humanoid walking, in: Proceedings of 2nd IFAC International Conference on Intelligent Control Systems and Signal Processing Istanbul, Turkey, September. 2009.
[20] D. Katić, A. Rodić, E.B. Korochano, Hybrid control algorithm for humanoid robots walking based on episodic reinforcement learning, in: World Automation Congress WAC 2012, ISSN: 2154-4824, Print ISBN: 978-1-4673-4497-5, Puerto Vallarta, Mexico, June 24–28. 2012.
[21] A. Rodić, K. Addi, E.B. Korochano, Bio-inspired control of biped robot locomotion by anthropomorphic leg impedance modulation, in: World Automation Congress WAC 2012, Puerto Vallarta, ISSN: 2154-4824, Print ISBN: 978-1-4673-4497-5, Mexico, June 24–28. 2012.
[22] V. Potkonjak, M. Tomic, A. Rodic, V. Antoska, Human and humanoid motion—distinguish Between safe and risky mode of locomotion, in: 10th International IFAC

Symposium on Robot Control, Dubrovnik, Croatia, 2012, pp. 524–529. https://doi.org/10.3182/20120905-3-HR-2030.00098.
[23] M. Vukobratović, A. Rodić, Contribution to the integrated control of artificial human gait, in: Proceedings of 2nd Serbian-Hungarian Joint Symposium on Intelligent Systems, Subotica, Serbia & Montenegro, 2004.
[24] Panasonic Assist Robot, Youtube, Panasonic Assist Robot, https://www.youtube.com/watch?v=zGmymin7d0o, (Retrieved 2019-07-11).
[25] K. Siciliano, Springer Handbook of Robotics, Springer-Verlag Berlin Heidelberg, ISBN: 978-3-540-23957-4. 2008.
[26] Panasonic power-lifting exoskeletons, WIRED portal, https://www.wired.co.uk/article/panasonic-robot-suit-exoskeleton, (Retrieved 2019-07-11).
[27] Powered exoskeleton, Wikipedia, Powered exoskeleton, https://en.wikipedia.org/wiki/Powered_exoskeleton, (Retrieved 2019-07-11).
[28] M. Vukobratović, When Were Active Exoskeletons Actually Born? Robotics Laboratory, 2007, http://www.pupin.rs/RnDProfile/pdf/exoskeletons.pdf (PDF) Retrieved 2018-07-22.
[29] World pioneers of Robotics—Miomir Vukobratović, https://en.wikipedia.org/wiki/Miomir_Vukobratovi%C4%87. Retrieved 2019-07-11.
[30] M. Vukobratović, D. Juričić, A Contribution to the Synthesis of Biped Gait, IFAC Symp, Technical and Biological Problem of Control, Yerevan, USSR, 1968.
[31] M. Vukobratović, D. Juričić, Contribution to the synthesis of biped gait, IEEE Trans. Biomed. Eng. BME-16 (1) (1969) 1–6.
[32] M. Vukobratović, Y. Stepanenko, On the stability of anthropomorphic systems, Math. Biosci. 15 (1972) 1–37.
[33] D. Juričić, M. Vukobratović, Mathematical modeling of a bipedal walking system, in: ASME Publication 72-WA/BHF-13, Winter Annual Meeting, New York, 1972, pp. 26–30.
[34] M. Vukobratović, How to control artificial anthropomorphic systems, IEEE Trans. Syst. Man Cybern. SMC-3 (5) (1973) 497–507.
[35] M. Vukobratović, D. Hristić, Z. Stojiljković, Development of active anthropomorphic exoskeletons, Med. Biol. Eng. 12 (1) (1974) 66–80.
[36] M. Vukobratović, D. Hristić, V. Cirić, M. Zečević, Analysis of energy demand distribution with in anthropomorphic systems, Trans. of the ASME, Series G, J. Dyn. Syst. Meas. Control. 95 (4) (1973) 402–413.
[37] M. Vukobratović, D. Hristić, Z. Stojiljković, Development of active anthropomorphic exoskeletons, Med. Biol. Eng. 12 (1) (1974) 66–80.
[38] M. Vukobratović, D. Hristić, Active orthoses of lower extremities, Orthop. Technic. 4 (1985).
[39] J. Hidler, D. Nichols, M. Pelliccio, K. Brady, D.D. Campbell, J.H. Kahn, T.G. Hornby, Multicentre randomized clinical trial evaluating the effectiveness of the Lokomat in subacute stroke, Neurorehabil. Neural Repair 23 (1) (2009) 5–13.
[40] J. Hidler, W. Wisman, N. Neckel, Kinematic trajectories while walking within the Lokomat robotic gait-orthosis, Clin. Biomech. 23 (10) (2008) 1251–1259, https://doi.org/10.1016/j. clinbiomech.
[41] G. Colombo, M. Joerg, R. Schreier, V. Diet, Treadmill training of paraplegic patients using a robotic orthosis, J. Rehabil. Res. Dev. 37 (6) (2000) 693–700.
[42] G. Colombo, M. Joerg, V. Diet, Driven gait orthosis to do locomotor training of paraplegic patients, in: Design, Control and Applications of Mechatronic Systems in Engineering, vol. 4, Engineering in Medicine and Biology Society, 2000, pp. 3159–3163.
[43] R. Ekkelenkamp, J. Veneman, H.V.D. Kooij, LOPES: a lower extremity powered exoskeleton. in: IEEE International Conference on Robotics and Automation

ICRA2007, 2007, pp. 3132–3133. https://doi.org/10.1109/ROBOT.2007.363952 Roma, Italy10–14 April 2007.
[44] J. Veneman, R. Kruidhof, E.E.G. Hekman, R. Ekkelenkamp, E.V. Asseldonk, H.V. D. Kooij, Design and evaluation of the LOPES exoskeleton robot for interactive gait rehabilitation, IEEE Trans. Neural Syst. Rehabil. Eng. 15 (3) (2007) 379–386.
[45] J. Veneman, E.V. Asseldonk, R. Ekkelenkamp, F.V.D. Helm, H.V.D. Kooij, Evaluation of the effect on walking of balance-related degrees-of-freedom in a robotic gait training device, in: IEEE 10th International Conference on Rehabilitation Robotics, pp. 868–875, https://doi.org/10.1109/ICORR.2007.4428526 Noordwijk, Netherlands, 2007.
[46] S.K. Banala, S.K. Agrawal, S.H. Kim, J.P. Scholz, Novel gait adaptation and neuromotor training results using an active leg exoskeleton, IEEE/ASME Trans. Mechatron. 15 (2) (2010) 216–225.
[47] K.N. Winfree, P. Stegall, S.K. Agrawal, Design of a minimally constraining, passively supported gait training exoskeleton: Alex II, in: IEEE 12th International Conference on Rehabilitation Robotics, Zurich, Switerland, 29 June–1 July, 2011, pp. 1–6.
[48] D. Zanoto, G. Rosati, S. Spagnol, P. Stegall, S. Agrawal, Effects of complementary auditory feedback in robot-assisted lower extremity motor adaptation, IEEE Trans. Neural Syst. Rehabil. Eng. 21 (5) (2013) 775–786.
[49] H. Kawamoto, Y. Sankai, Power assist method based on phase sequence and muscle force condition for HAL, Adv. Robot. 19 (7) (2005) 717–734.
[50] S. Lee, Y. Sankai, Power assist control for walking aid with HAL-3 based on EMG and impedance adjustment around knee join, in: IEEE/RSJ International Conference on Intelligent Robots and Systems, Lausanne, Switerland, 30 September–4 October, 2002, pp. 1499–1504.
[51] A. Tsukahara, Y. Hasegawa, Y. Sankai, Standing-up motion support for paraplegic patient with robot suit HAL, in: IEEE 11th International Conference on Rehabilitation Robotics, Kyoto, Japan, 23–26 June, 2009, pp. 211–217.
[52] A. Esquenazi, M. Talaty, A. Packel, M. Saulino, The ReWalk powered exoskeleton to restore ambulatory function to individuals with thoracic-level motor-complete spinal cord injury, IEEE Trans. Neural Syst. Rehabil. Eng. 91 (11) (2012) 911–921.
[53] R.J. Farris, H.A. Quintero, M. Goldfarb, Preliminary evaluation of a powered lower limb orthosis to aid walking in paraplegic individuals, IEEE Trans. Neural Syst. Rehabil. Eng. 19 (6) (2011) 652–659.
[54] R.J. Farris, H.A. Quintero, S.A. Murray, K.H. Ha, C. Hartigan, M. Goldfarb, Preliminary assessment of legged mobility provided by a lower limb exoskeleton for persons with paraplegia, IEEE Trans. Neural Syst. Rehabil. Eng. 22 (3) (2014) 482–490.
[55] U. Onen, F.M. Botsali, M. Kalyoncu, M. Tinkir, N. Yilmaz, Y. Sahin, Design and actuator selection of a lower extremity exoskeleton, IEEE/ASME Trans. Mechatron. 19 (2) (2013) 623–632.
[56] Ü. Önen, F.M. Botsalı, M. Kalyoncu, Y. Şahin, M. Tınkır, Design and motion control of a lower limb robotic exoskeleton, in: Design, Control and Applications of Mechatronic Systems in Engineering, INTEH, In the book (Chapter 7), 2016, pp. 135–152.
[57] Panasonic Assist Robot, https://www.youtube.com/watch?v=zGmymin7d0o. n.d. Retrieved 2018-07-16.
[58] AWN-03 Panasonic Exoskelet, n.d. https://news.panasonic.com/global/stories/2016/44969.html.
[59] Above-Knee Prosthesis, http://www.pupin.rs/RnDProfile/promona-2.html, n.d. Retrieved 2018-07-16.
[60] Suit X, Exoskeleton, https://www.youtube.com/watch?v=qTxxwLWsMoA. Retrieved 2018-07-16.

[61] Suit-X, Fenix Exoskelet, http://www.suitx.com/phoenix-medical-exoskeleton-europe. Retrieved 2018-07-16.
[62] BLEEX, Exoskeleton, http://bleex.me.berkeley.edu/research/exoskeleton/bleex/. Retrieved 2018-07-16.
[63] TWIICE, Exoskeleton, http://twiice.ch/. Retrieved 2018-07-16.
[64] T. Vouga, R. Baud, J. Fasola, M. Bouri, H. Bleuler, TWIICE—A lightweight lower-limb exoskeleton for complete paraplegics, in: 2017 International Conference on Rehabilitation Robotics (ICORR), 2017, pp. 1639–1645.
[65] TWIICE Exoskeleton Video, n.d. https://www.youtube.com/watch?v=PphYGkNENGw. Retrieved 2018-07-16.
[66] A. Ortlieb, M. Bouri, H. Bleuler, AUTONOMYO: design challenges of lower limb assistive device for elderly people, multiple sclerosis and neuromuscular diseasess, in: Wearable Robotics: Challenges and Trends, Springer, Cham, 2017, pp. 439–443.
[67] AUTONOMYO Video Presentation, n.d. https://www.autonomyo.ch/econcept. Retrieved 2018-07-16.

Index

Note: Page numbers followed by *f* indicate figures and *t* indicate tables.

W

Z

CPI Antony Rowe
Eastbourne, UK
November 20, 2019

Contents

About the Authors

Alessandro Santoni worked at the International Monetary Fund (IMF) in the Financial Crisis Division and European Central Bank (ECB) manager responsible for Onsite Inspections and beforehand manager for Financial Crisis Management Interventions. Beforehand, he worked in the Banking industry at Goldman Sachs, BMPS and ABN. He has a PhD from the University of Siena, a Master in International Economics & Management from SDA Bocconi and an EMBA from Columbia University, LBS and Hong Kong University. He has three degrees in Economics, History and Political Science from the University of Siena. He teaches Risk Management, Financial Crimes at Bayes University of London, Frankfurt School of Management and University of St. Gallen. Certified Financial Crime Specialist (CFCS) and AML Specialists (CAMS). He is a co-author of the book "Corporate Governance in the Banking Sector".

Federico Salerno worked as a banking supervisor and beforehand for Financial Crisis Management Interventions in the ECB's Single Supervisory Mechanism (SSM) in Frankfurt. He previously worked as equity research analyst, covering financial institutions ranging from small to large caps at both bulge brackets and boutiques, mainly out of Milan, London, and Paris. He held a Master in International Economics & Management from SDA Bocconi in Milan and degrees in Law and Political Science from the University of Siena. He was a CFA charter holder, CFA Institute.

List of Tables

Banks M&A: Strategy and Valuation

Valuation in Resolution

Valuation in Liquidation

Loan Valuation

Financial Assets Valuation

Solvent Wind-Down

Introduction

Abstract

Bank valuation analysis is very different from corporates, with greater emphasis to balance sheet examination. Complexity emerges also from the public role of banks and the potential impact from their default. This led to an enormous quantity of rules designed to manage the risk arising from this business. Difficulties emerge also from the different approach needed during the life cycles determined by diverse macroeconomic contexts and idiosyncratic events. In this book, we provide practical guidance of banks' valuation methodologies juxtaposing practical examples to each valuation model to enrich the theoretical discussion and help practitioners build their own spreadsheets.

Overview

> If you owe the bank $100 that's your problem. If you owe the bank $100 million, that's the bank's problem. JP Getty

According to the Basel-based Bank for International Settlements (BIS),[1] banks are institutions whose business is to receive deposits and/or close substitutes for deposits and grant credits or invest in securities on their own account. This immediately characterizes the business model and the financial statement of banks as very different from other industries. Concepts such as working capital and operating income take on a wholly different meaning. The approach to financial analysis of a Bank is also very different from corporates, with greater emphasis on balance sheet examination. A profitable bank usually can generate profits from the spread between interest income obtained from the credit versus the cost of funding (deposits) net of

[1] Guide to the International Banking Statistics, Bank for International Settlements, Basel, Switzerland, 2000, Part III—Glossary of Terms.

A. Santoni, F. Salerno, *How to Value a Bank*, Springer Texts in Business and Economics, https://doi.org/10.1007/978-3-031-43872-1_1

other costs (including operating costs, cost of losses from non-performing credit and taxes). This apparently easy business model is, indeed, a complex one. The public role of banks and the potential impact from their default in society led to an enormous quantity of rules designed to manage the risk arising from this business.

Given their peculiarities, banks and other financial services firms (such as insurance or fintech companies) can be particularly challenging to value and may require a wholly different approach relative to the methods used for industrial corporations. Investors must be aware of several potential pitfalls before embracing in banks valuation. In this book, we provide practical guidance of banks' valuation methodologies. Chapter "Main Features of the Banking Business Model" describes the peculiarities of banks' business model and the regulatory framework governing their risk management.

In chapter "Valuing a Bank in Going Concern", we delve into some more peculiarities of the banking business model and then describe the most popular valuation techniques for banks as a going concern. These are the time-honoured DDM and DCF, peer group multiples (price–earnings ratio in our example), excess return models or economic value-added valuation method (EVA) and their variant, the warranted equity value (WEV). We have tried to juxtapose practical examples to each valuation model to enrich the theoretical discussion and help practitioners build their own spreadsheets.

The chapter also contains a description of how to carry out a sum-of-the-parts valuation (SOTP) including the allocation of available capital by business area as well as the description of a value map and the statistical valuation model based on the regression it provides. Given the capital-driven nature of banks' business model, we find EVA and SOTP models particularly fitting. We also include a description of our bottom-up cost of equity model, which can be used, for instance, in a SOTP.

Valuing a Bank in an merger and acquisition (M&A) scenario can be challenging, and several factors and points of views must be considered. The starting point of an M&A valuation can be the traditional methodologies used for valuing a bank in a normal scenario. In chapter "Banks M&A: Strategy and Valuation", we investigate some of the critical aspects of a Banking M&A and the valuation implications from two points of view: the shareholders' point of view and the supervisors' point of view. From the shareholders' point of view, the key aspects to be considered are the calculation of the synergies arising from the M&A and the potential earning accretion/dilution. On the other side, Banks M&A are always assessed by supervisors on a case-by-case basis. The main objectives from the supervisor's point of view will be to ascertain, to the extent possible, the sustainability of the business model of the combined entity.

In April 2014, following the early 2000 financial crisis, European Authorities published the Bank Recovery and Resolution Directive (BRRD). The Bank Recovery and Resolution Directive has been adopted to provide authorities with comprehensive and effective tools to deal with failing banks at national level and with cooperation arrangements to tackle cross-border banking failures. The overarching objective of the BRRD resolution regime is to make sure a bank can be resolved swiftly with minimal risk to financial stability. Banks' valuation methodologies, utilized by Supervisors and Resolution Authority to trigger the failing of a bank and the best resolution tool to utilize, constitute the key aspects of the BRRD. In chapter "Valuation in Resolution", we revise the key principles of these valuation methodologies providing a real-case simulation study.

While resolution is applied for systemic or banks with critical functions, normal insolvency or liquidation proceedings are the default procedures when a bank is failing. Liquidation value is the value of the assets that remain if the company goes from a going concern to a gone concern status. In the liquidation process, creditors, lenders, and shareholders based on seniority of their claims will receive the proceeds from the liquidator of the company's assets. Banking assets included in liquidation value usually include loans, financial assets, tangible assets like real estate, equipment, investment. In chapter "Valuation in Liquidation", we provide a practical example of liquidation valuation.

In chapter "Loan Valuation", we deep dive on key aspects of loan valuation. We analyse the performing and non-performing loan valuation methodologies. On the latter, we evaluate the different strategies to address NPL disposals as fire sales and the methodologies used by third parties when acquiring them and the methodologies used by Supervisors (as ECB) in their NPL manual.

The most complex aspect of banks valuation is however the financial asset assessment (chapter "Financial Assets Valuation"). Accounting valuation is the process of valuing a company's assets and liabilities for financial reporting purposes. Several accounting-valuation methods may be used while preparing financial statements to value financial instruments (and non-financial). Accounting International Financial Reporting Standards (IFRS) require market transactions being recorded at fair value. To obtain a prudent value of the assets, regulation requires the fair value to account for additional valuation adjustments (AVA). In chapter "Financial Assets Valuation", we analyse prudent valuation methodologies for financial assets.

In chapter "Solvent Wind Down", we focus on some of the valuation and operational aspects of winding down a bank's trading book portfolio. The topic follows SRB SWD guidance and FSB on the "solvent wind-down of derivatives and trading portfolios" and provides a deep dive on valuation principles and exit strategies currently considered by industry practitioners when designing a solvent wind-down plan. It also provides the reader with an overview of the most relevant concepts underpinning the valuation or pricing concepts (i.e. "Fair value," "realisable value", and "Solvent Wind Down value"). Ultimately it shows that the cost to wind-down a trading portfolio beyond the usual accounting carrying value might be mostly driven by wind-down operating costs (incl. Liquidity and funding costs) and two main pricing components, namely the Capital Valuation Adjustment and Margining Valuation Adjustment. We advocate that recovery plans and resolution plans should, contrary to the current practice, properly factor in these additional costs.

In summary, this book touched upon the key valuation methodologies of a Bank starting from its licensing process, the equity valuation, and the M&A approach up to the possible wind-down. We investigated resolution and liquidation valuation methodologies focusing on performing and non-performing loan, fair value, and AVA for financial assets.

Main Features of the Banking Business Model

Abstract

Banks are very different from other industries. The business model is mainly based on collecting deposits from customers and granting credits or invest in securities on their own account. This apparently easy business model is, in fact, a complex one. The public role of banks and the potential impact from their default in society has led regulators to introduce a significant quantity of rules designed to mitigate the risks arising from this business. Given their peculiar business model, banks can be particularly challenging to value and require a wholly different approach relative to industrial entities.

Key Differences Between Banks and OFIs

According to the Basel-based Bank for International Settlements (BIS),[1] banks are defined as institutions whose business is to receive deposits and/or close substitutes for deposits and grant credits or invest in securities on their own account. Financial institutions, other than banks, according to the BIS[2] engage primarily in the provision of financial services and activities auxiliary to financial intermediation, such as fund management. Non-bank financial corporations, sometimes referred to as "shadow banks", include special purpose vehicles, hedge funds, securities brokers, money market funds, pension funds, insurance companies, financial leasing corporations, central counterparties (CCPs), unit trusts, other financial auxiliaries, and other captive financial institutions.

Thus, it can be argued that, according to the BIS, the key difference between a Bank and other financial institutions is that a Bank collects deposits. The protection

[1]Guide to the International Banking Statistics, Bank for International Settlements, Basel, Switzerland, 2000, Part III—Glossary of Terms.

[2]https://www.bis.org/statistics/glossary.

A. Santoni, F. Salerno, *How to Value a Bank*, Springer Texts in Business and Economics, https://doi.org/10.1007/978-3-031-43872-1_2

of the customer deposits and ensuring the financial stability in the country are the main tasks of supervisors and regulators. In most markets, banks represent the largest kind of financial institution by asset size, followed by insurance companies and asset managers. Incidentally, we point out that the difference between banks and insurance companies is threefold:

1. Insurance liabilities are long term and substantially less liquid, which renders insurers less prone to "runs" than banks
2. On the other hand, insurers' liabilities are estimated with probabilistic assumptions, which can lead to the need to revise them upwards and hence to P&L losses even many years after the contract is entered into. Asbestos claims in US insurance are a case in point
3. The assets of insurance companies are largely quoted financial products, which allows markets to form a good idea about the evolution of net asset value. In times of high volatility, this can, arguably, reinforce contagion effects
4. As regards their P&L account, insurers are characterized by the inversion of the revenues-costs cycle. They collect revenues (premiums) long before they sustain the costs (claims) associated with said revenues. It can even be the case that the cost is never incurred (e.g. a life insurance policyholder dies before the contractual age to withdraw his/her savings and counter-insurance is not present; a motor insurance contract expires with no accident and no claims). As a matter of fact, this is a defining feature of the insurance business and not just a difference to banks.

Opening a Bank

The process of setting up a banking business starts with a written authorization from the local authority (usually a Central Bank). In Europe, this kind of authorization is granted by the European Central Bank (ECB) in consultation with the National Supervisory Authority under Article 4(1) of the SSM Regulation. There are several criteria that need fulfilling. The main one is constituted by the minimum or initial capital to be invested. The entrepreneur must demonstrate that he/she can commit a minimum or initial capital, which depends on the nature of the intended business, but in general, at least EUR 5 m in the Eurozone. Additionally, the authorization application must contain a business plan showing that the Bank can be sustainably profitable, along with a description of its internal control systems and risk management processes. In addition, as regards personnel, the application should contain a description of the organizational structure and of qualified resources. In the management board, the Bank should appoint people who have acquired sufficient knowledge and practical experience during their professional careers.

Valuation of Financial Projections for Licensing Approval

Supervisors have a detailed assessment process to validate the licensing request.[3] This valuation process should allow supervisors to assess the viability and sustainability of the business model. The applicant request should contain financials covering at least three years' projections of balance sheet and profit and loss account statements. The projections should contain the base case scenario and an adverse one to evaluate the potential impact on capital and liquidity ratios. The projections should contain the macroeconomic expectations, as for example GDP, loan growth, interest rates, unemployment rates. The assumptions used should be based on data provided by Institutions such as IMF, World Bank, ECB. There are two main points of attention on the assessment of the valuation: (1) the capacity of the applicant to have adequate funding sources to face possible negative liquidity scenarios; (2) adequate capital level to sustain absorb losses from loans and/or financial assets depending on the business model of the applicant.

Example of Initial Capital Requirement Valuation

The examples below illustrate the capital expected for the authorization when the authority establishes a higher threshold for the initial capital requirement. The initial minimum capital of EUR 5 m is used as the minimum threshold. The authority may have, given the business profile of the applicant, a higher yearly minimum threshold, in our example starting at EUR 7 mln. The highest amount capital requirement in the third year (EUR 14 mln in our example) is added to the projected cumulative losses of the first three years (EUR 7 mln in our example), for a total of EUR 21 mln, which is the amount of capital expected at the time of authorization for the applicant (Table 1).

Banks and Capital Requirements

The main reason why Banks are required to hold capital is represented by the need to protect depositor–holders from potential losses stemming from unexpected losses (typically in a bank's assets). Capital, therefore, constitutes a buffer against losses that could originate from loans defaulted upon, a decline in the market value of financial assets, losses from fraud, rogue trading, and others. An additional reason for capital requirements is, arguably, represented by the need to reduce contagion and negative feedback loops to other financial institutions as well as to shield taxpayers from losses.

[3] https://www.bankingsupervision.europa.eu/ecb/pub/pdf/ssm.201901_guide_assessment_credit_inst_licensing_appl.en.pdf.

Table 1 Valuation of minimum capital requirement for licensing

EUR mln	Years 1	2	3
Initial Capital Requirement	5	5	5
Supervisor Minimum Capital required	7	10	14
Plus		+	
Losses accumulated for the start up	3	5	7
Capital Required for Authorization	**10**	**15**	**21**

Source: Authors own's elaboration

Regulatory capital that a Bank needs to hold is predominantly in the form of common stock, paid-in capital as well as retained earnings. The amount of capital that a bank needs to hold depends mainly on its dimension as well as on the amount of risk that it is taken on and off its balance sheet. In this sense, the amount of assets a bank owns is the starting point to calculate how much capital a bank needs to hold. Assets have different risks and hence risk weights. Internal models or standardized approaches are used to calculate risk-weighted assets, starting from (mainly) on-balance sheet exposures. Off-balance sheet credit exposures are converted to on-balance sheet equivalents using credit conversion factors.

How are Capital Requirements Determined?

Following turmoil in international currency and banking markets caused by a host of well-known banking crises (notably the failure of Bankhaus Herstatt in Germany in mid-1974) the Central Bank Governors of the Group of Ten countries, at the end of 1974, created the Basel Committee on Banking Supervision. This aims to develop global regulatory standards for banks and seeks to strengthen micro- and macro-prudential supervision. The capital of the Bank for International Settlements (BIS) is currently underwritten by 63 central banks and monetary authorities from around the world. In June 2004, the Committee released a detailed capital framework aimed at setting minimum capital requirements for Banks, which sought to develop and expand the standardized rules set out of the 1988 Accord. This agreement is known as "Basel II", the revised framework comprised three pillars of which the minimum capital requirement was the first one. The other two comprise:

- Supervisory review of an institution's capital adequacy and internal assessment process

- Effective use of disclosure as a lever to strengthen market discipline and encourage sound banking practices

The capital requirements under which the minimum is set consist of three main elements for a Bank:

- Minimum capital requirements, known as Pillar 1 capital requirements
- An additional capital requirement, known as the Pillar 2 capital requirement
- Buffer requirements

The minimum total capital requirement (called Pillar 1 requirement) or total capital ratio is set by BIS at 8% of banks' risk-weighted assets. The ratio is composed of two elements: the capital as numerator and risk-weighted assets (RWA) as denominator.[4] RWA are the results of the total assets a bank has, multiplied by their respective risk factors (risk weights). Risk factors reflect how risky a certain asset type is and is perceived to be according to historical average and statistical models (we will see later how these are modelled). For example, a mortgage loan that is secured with collateral (typically a dwelling) is less risky—has a lower risk factor—than an uncollateralized loan. As a result, a bank needs to hold less capital to cover for such a mortgage loan than it does to cover for the uncollateralized loan.

The second element to calculate capital requirements (above the required minimum) is constituted by Pillar 2 requirements. This is where Bank Supervisors play a key role. Indeed, supervisors analyse the idiosyncratic risk of each Bank and may demand it holds additional capital above and beyond regulatory minimum if they conclude that its risks are not sufficiently covered by minimum capital requirements.

Since minimum and additional capital requirements are binding there would be legal consequences if they are not adhered to, the magnitude of which will depend on the amount of the breach and the capacity of the Bank to close this gap. The maximum consequence could be the withdrawal of the license (Table 2).

Risks Associated with Banking Activity

Banks have an important social and economic function impact as they are supposed to protect savings while utilizing them to provide the necessary resources for the economy to grow. In a simplified way, Banks collect resources from depositors which have generally short maturity (most are sight deposits), limited in amount and

[4] Assets weighted according to their riskiness. Generally, the higher the risk, the higher the weight. It serves as the denominator of capital ratios (e.g. CET1 ratio and Total Capital Ratio). US regulators continue to rely on non-risk-weighted measures as well, mainly the leverage ratio. Since an institution is required to hold capital sufficient to cover the higher of the two sets of indicators (both risk weighted and not), the importance of risk-weighted capital indicators ends up being somewhat lower in some jurisdictions such as the US.

Table 2 Banks capital composition

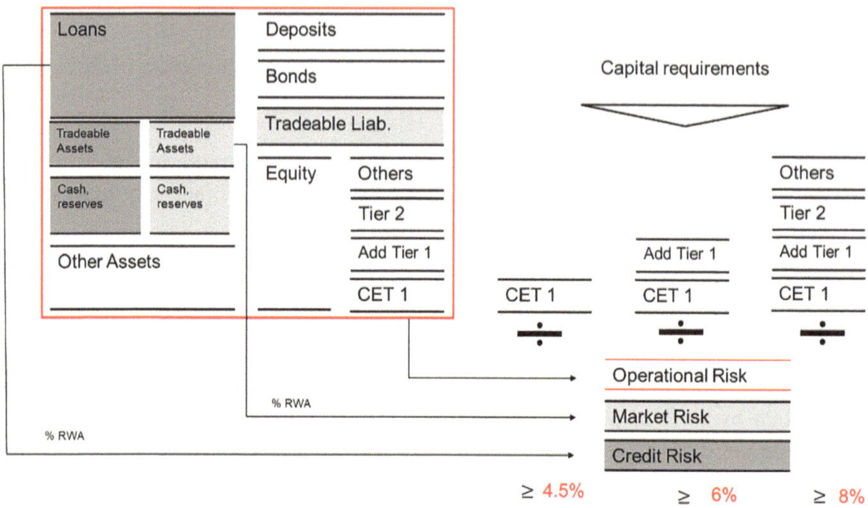

Source: Authors

with, up to a certain level, guarantee to be reimbursed (the deposit guarantee amount depends on local regulations). Once collected these resources, for which an interest is recognized to depositors, Banks' job is to invest these liabilities into assets. Banks have a vast array of choices to make. They can invest by lending money to borrowers (via loans), by buying financial assets, or by investing in real estate. Many of these investments have characteristic that could create risk management issues for Banks. These investments often have long maturities (e.g.: 20 years mortgages, 10 years government bonds), can be large and the repayment is, at times, uncertain (Table 3).

The risks associated with banking activity are defined by BIS as:

Credit Risk is defined as the risk of default on a debt derived from the incapacity of a borrower to fulfil the required payments. Borrowers usually pay monthly instalments and part of the capital initially provided by the Bank. If the borrower stops making these payments, then the Bank would need first to provide provisions on the number of payments missed and then eventually, when found out that for the borrower it would be impossible to repay the loan, writes off the loan.

Liquidity Risk: Liquidity is defined as the ability of a bank to fund increases in assets and meet obligations as they come due, without incurring unacceptable losses. The so-called Bank run materialized when a few depositors decided to withdraw the sum deposited in the Bank at the same time. This could make impossible for the Bank to meet the obligation to repay the deposit since the Bank may not have enough cash or liquid assets to sell.

Market risk is the risk of losses resulting from changes in the prices of instruments such as bonds, shares, and currencies. Market risk arises also from interest rate risk

Table 3 Risks associated with the banking activity

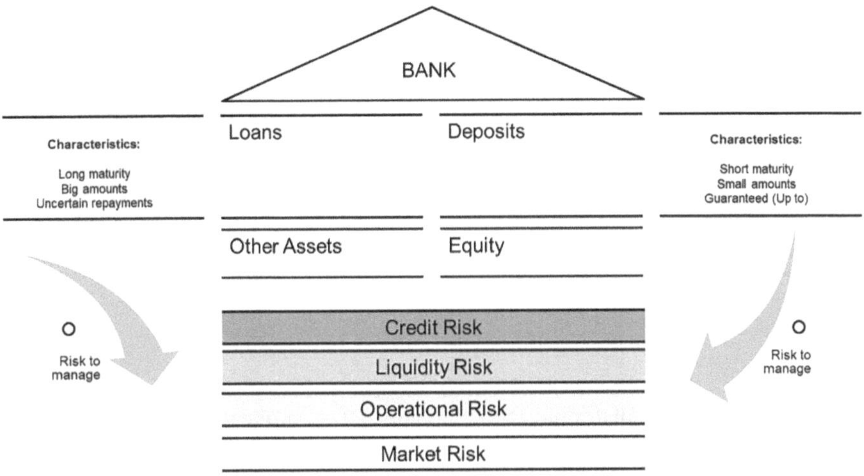

Source: Authors

associated with a mismatch between fixed and floating interest rates at which assets and liabilities are priced.

Operational risk: The Basel Committee defines the operational risk as the "risk of loss resulting from inadequate or failed internal processes, people and systems or from external events". The losses from operational risk could arise from internal fraud, external fraud, physical damages to Banks infrastructures, IT incidents, legal issues.

Business Models

Banks' business model can differentiate a lot ranging from pure retail bank dedicated to selling via local branches basic financial services, such as checking and savings accounts to investment banking specialized in more sophisticated products as underwriting new debt and equity securities for all types of customers ranging from government to corporates. An alternative way to differentiate business models is to distinguish between an originate-to-hold model typical of European banks and an originate-to-distribute model typical of US banks.

In the first case, the business model is quite simple with banks collecting deposits and use them for providing loans monetizing their competitive advantage by applying a spread between interest paid and received.

In this model, banks accept the credit default risk of the counterpart by holding the asset on their balance sheet. The risk management activity will be focused on monitoring borrowing customers' behaviour while protecting the depositors by holding appropriate levels of capital to cover unexpected risk.

Table 4 Originate-to-hold-EU style

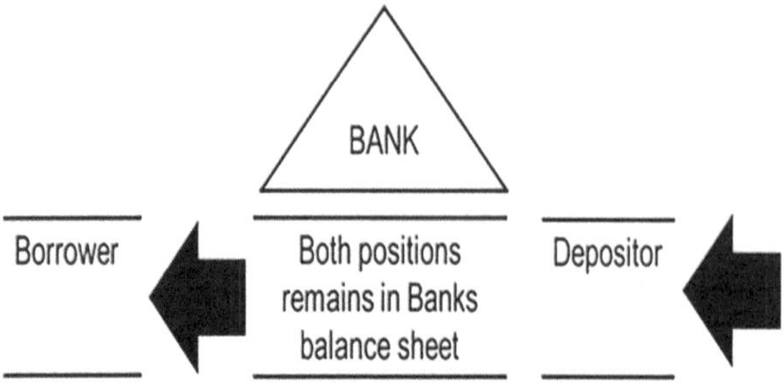

Source: Authors

In the originate-to-distribute model, Banks act as brokers in credit risk. Banks collects deposits and use them to provide loans. However, Banks do not keep the loans on their balance sheet but rather sell the loans. Loans can be bundled into a mortgage pool as MBS (mortgage-backed securities) or asset pool as ABS (asset-backed securities) to SPV (special purpose vehicle). These financial instruments as MBS, ABS can be issued by third-party financial company such as Investment Banks or, in the case of USA, Government sponsored enterprises (GSE) known as Fannie Mae or Freddie Mac. SPV shares are distributed to investors that receive cash on monthly basis from the monthly instalments paid by the initial borrowers. This model allows to significantly increase the volumes of loans provided to the economy since initial deposits are leveraged several times. Moreover, by offloading the assets out of balance sheet, Banks reduce the associated risk. The scheme allows the banks to (1) increase leverage and liquidity; (2) hold less capital since the asset side of the banks' balance sheets usually incorporated high-rated (AAA) securities which required no significant capital reserves (low RWA); (3) increase the profitability since upfront fees on sale of the loans (a significant portion of the loans overall profitability) are multiplied several times (Table 4 and 5).

There are several reasons why EU Banks are not replicating the US model. Among the most important ones is that to operate such a business model a very developed secondary capital market is needed, with instruments as SPVs and investors willing to purchase and hold certain kinds of securities. Moreover, the presence in US of Government agency as Fannie Mae and Freddie Mac, created by the US Congress to provide liquidity, stability, and affordability to the mortgage market significantly facilitate this mechanism. Fannie Mae and Freddie Mac buy mortgages from Banks and/or package the loans into mortgage-backed securities (MBS) that may be sold or retained. Thanks to this mechanism, banks can raise cash to engage in further lending. This process has the additional benefit of attracting to the secondary mortgage market investors who might not otherwise invest in this kind of assets, providing further liquidity for the housing market and decreasing the cost of borrowing for the consumers.

Table 5 Originate-to-distribute-US Style

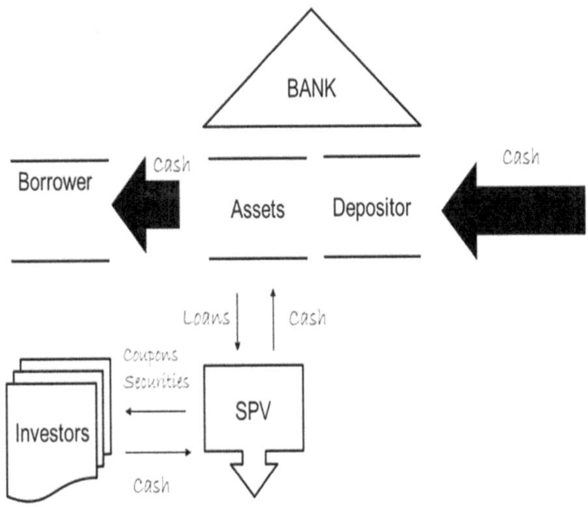

Source: Authors

The difference of the two models is so relevant that the level of US securitization in the period 2007–2017 has been ranging between 7 and 10 times the EU ones.[5]

[5]Lucia Quaglia. It Takes Two to Tango: The European Union and the International Governance of Securitization in Finance. JCMS 2021.

Valuing a Bank in Going Concern

Abstract

In common with other financial institutions, banks exhibit some specific features relative to industrial companies. This has led the analyst community to opt for certain valuation methods rather than others, over the years. In this chapter, we investigate the most common methodologies used by analysts in valuing Banks in going concern status. The warranted equity value (WEV) method, an adaptation of the EVA approach, is used to value both the entity as a whole or individual business areas in a sum-of-the-parts. It responds to the need to take a bank's capital absorption and capital position into account. The DDM is a universal method which suits banks as well as industrials. And so is the free cash flow to equity (FCFE), in a sense an expansion of the DDM in that it adds potential dividends to actual dividends.

Valuation Methodologies for Banks

We touch upon the key differences between a bank and industrial entities, including the implications of such differences for the valuation of banks and then delve into the most widely used valuation methodologies for banks. These are:

1. Dividend discount model (DDM)
2. Discounted free cash flow to equity (DFCFE)
3. Valuation with market multiples
4. Statistical approach with a value map
5. Warranted equity value method (WEV)
6. Excess return (EVA)

We present examples of various valuation approaches for a hypothetical bank, which we call "XYZ" but point out that the numbers shown in the valuation

A. Santoni, F. Salerno, *How to Value a Bank*, Springer Texts in Business and Economics, https://doi.org/10.1007/978-3-031-43872-1_3

examples are there purely for illustration purposes and do not refer to any actual entity.

We discuss our preferred approach to the allocation of capital, an essential step for a sum-of-the-parts valuation. Having recapped the result of the various valuation approaches, we show an example of a sum-of-the-parts. Hence, after briefly discussing the effects of inflation on banks, we close the chapter with a description of our bottom-up cost of equity model.

How Banks Differ from Corporates: Valuation Implications

We highlight three key differences between banks and industrial companies:

1. Capital
2. Balance sheet, namely debt and property, plant & equipment
3. Cash flow

Capital as a Scarce Resource for Financial Institutions

As mentioned in the previous chapter, one of the key differences between banks and industrial companies, in fact, arguably, the most significant, is the fact that banks face regulatory constraints as for their use of capital in the business. Beginning with the first Basel agreement, banks were required to hold 8% of RWA as Tier One capital (mainly equity). Regulation aside, financial institutions must keep their capital ratios at levels consistent with a solid rating agencies' rating to prevent funding costs from rising to levels that might endanger an institution's profitability.

Debt

Secondly, debt can be considered as part of ordinary working capital for a financial institution: an industrial entity typically issues debt to fund investment in plants & equipment which will generate cash flow. This is not the case for a bank or insurance company: issuing debt is part and parcel of a financial institution's day-to-day activity aimed at attracting the public's savings and thus perform its core business of financial intermediation.[1]

[1] A. Damodaran, Investment valuation, Chap. 21, where the author states that some banks see debt as "raw material" to be transformed into high value-added financial products to be sold to clients. Indeed, we notice the statement made by a UK bank's head of mortgages in the wake of the turmoil in the mortgage market of September 2022, as reported by the Financial Times: "we were taken aback by how quickly some people withdrew their products". "if you are a supplier of any product and relying on raw materials, you have a period of time in which you should be planning. You shouldn't need to act in a knee-jerk way".

Table 1 DCF formula

$$\text{Discounted Cash Flow} = \frac{CF^1}{(1+dr)^1} + \frac{CF^2}{(1+dr)^2} + \cdots + \frac{CF^n}{(1+dr)^n}$$

CF = Cash Flow

dr = discount rate

Source: Authors own elaboration

Fixed Assets

Conversely, a bank is unlikely to have significant investment in Plant & Equipment. In this respect, leverage for a financial institution should be assessed with respect to the presence/absence of excess capital.

Cash Flow

Cash flow differs from industrials to banks: fixed assets are likely to account for a small share of the total in a financial institution. Therefore, depreciation is likely to be minor. Bar the presence of intangibles, e.g. goodwill from acquisitions, stated earnings are usually a good proxy for cash flow for a bank.

These peculiarities impact the way banks can be valued. Looking at free cash flow valuation methods, we notice, for instance, that only the free cash flow to equity method is applicable to banks (see below).

Valuation Implications: Free Cash Flow Valuation and DDM Models for Banks

The value of any asset is represented by the sum of the cash flows generated by the asset, discounted to the present at an interest rate consistent with the riskiness of said cash flow. For a listed equity, the cash flow (or dividends if a more conservative dividend discount model (DDM) is used) should be discounted at the bank's cost of equity (see Table 1).

We distinguish among two methods, in line with the doctrine:[2]

1. Free cash flow to equity
2. Free cash flow to assets (FCFA). Or, alternatively, free cash flow to the firm (FCFF)

[2] A. Damodaran, Investment Valuation: Tools and Techniques for Determining the Value of any Asset, University Edition, Wiley, Chap. 14 (free cash flow to equity discount models) and Chap. 15 (free cash flow to the firm), Wiley Finance Series, April 2012. And, by the same author: https://pages.stern.nyu.edu/~adamodar/pdfiles/eqnotes/fcff.pdf

Free Cash Flow to Equity (FCFE)

Banks can be valued using a free cash flow to equity method. As mentioned, a bank is unlikely to have significant investments in Plant & Equipment, requiring depreciation. The latter will normally be a small item on a bank's P&L, which implies that the difference between FCFE and stated earnings will be limited. This is a key difference relative to an industrial entity, for which depreciation can represent a substantial item. Other factors include:

1. Capital strain. Incremental capital required to run the business should be subtracted from free cash flow in that it cannot be distributed as dividends but must be retained to satisfy regulatory capital requirements.
2. Add excess capital (or subtract a capital shortfall to the value of the business determined with a DCF). To avoid double-counting, the analyst should ideally identify the portion of the cash flow generated by excess capital and remove it from the valuation.
3. No differently to what is the case with industrial entities, intangibles' amortization and write-downs should be added back to earnings to calculate the FCFE.

It is easy to infer from the formula for the free cash flow valuation model, that the terminal value is going to represent the bulk of the overall valuation, in most instances. The analyst should, therefore, make sure that the inputs for the terminal value are carefully chosen:

Estimated earnings and hence the cash flow should not be overly optimistic but, rather, sustainable.

The terminal growth in earnings should not exceed growth in nominal GDP in the bank's reference markets.

As said, it must also be borne in mind that part of the institution's earnings must be reinvested in the business to satisfy regulatory requirements and, consequently, cannot be distributed. Therefore, the cash flow used for the terminal value should be adjusted to reflect this.

Free Cash Flow to Assets (Firm)

The free cash flow to assets approach is hardly applicable to banks given that, for a bank, it is conceptually impossible to identify financial debt separate from the core business.[3]

[3]Damodaran: Investment valuation, Wiley finance, University edition, Chap. 21 (Valuing financial services firms). According to the author, many banks see financial debt as a sort of "raw material", to be moulded into higher value financial products.

Dividend Discount Models (DDMs)

The DDM is one of the most established valuation methods for listed shares. It is more conservative than discount free cash flows to equity (DCFE) in that it considers only the portion of earnings that is distributed. To the contrary, DCFE considers potential dividends as well as actual ones. DDM models can be used for listed banks just as with any other listed entity.

To use a DDM, the analyst should project stated earnings into the future (3–5 years) along with a sustainable dividend payout ratio. Then, he/she should discount the stream of dividends back to the present at the institution's cost of equity.

It must be borne in mind that not all earnings may be distributable: a portion of these must be retained to support growth in assets, e.g. because of new loans being extended, hence growth in the loan book or because the bank's asset management arm is enjoying net inflows, etc.

As regards the terminal value in the DDM formula, the same considerations we expressed for DCF still apply (see above).

Even in the DDM, the analyst should, therefore, make sure that the inputs for the terminal value are carefully chosen:

- Estimated earnings and hence the cash flow should not be overly optimistic but, rather, sustainable
- The terminal growth in earnings should not exceed growth in nominal GDP in the bank's reference markets. A growth rate higher than nominal GDP would, indeed, imply that the bank's market share is continuously increasing, arguably an aggressive assumption
- It must also be borne in mind that part of an institution's earnings must be reinvested in the business to satisfy regulatory requirements and, consequently, cannot be distributed. The dividend payout used for the terminal value should be adjusted to reflect this[4]

The terminal value should hence be calculated according to the following formula:

$$\text{Terminal Value} = \frac{\text{Last Explicit Dividend} \times (1 + g)}{(K_e - g)}$$

Where *k* is the cost of equity and g is the long term (LT) growth rate in dividends. where:

[4]A. Damodaran: Investment valuation, Wiley finance, University edition, Chap. 21 (Valuing financial services firms), Chap. 12 (estimating terminal value), and Chap. 13 (Dividend discount models). Also, See Koller, Tim, Goedhart, Marc, Wessels, David (McKinsey & co.): Valuation: measuring and managing the value of companies, Chap. 14 (estimating continuing value).

g indicates the nominal growth rate sustainable over the long term ("Sustainable Growth Rate") and was calculated based on estimated growth in real terms plus inflation.

Ke indicates the discount rate, equal to the Cost of Equity.

Dividend Discount Model with Excess Capital

As an alternative to our preferred method of treating excess capital (see below), the analyst can assume that excess capital will be returned to shareholders via ordinary dividends, hence a higher payout than would otherwise have been the case.

A FCFE method will be preferrable when the bank's dividend is unsustainably high but estimates about future payout ratios are unreliable.[5]

We, show, below, an example of a DDM valuation:

1. The cost of equity is obtained with a capital asset pricing model (CAPM) and is equal to 10%
2. Perpetual g (2%) is equal to the LT growth in nominal GDP. Hence, we implicitly assume that the bank's market share will remain unchanged
3. Free cash flow is obtained by subtracting the increase in required capital (CET1) from the accounting free cash flow, the latter being constituted by earnings plus depreciation, amortization, and goodwill write-downs
4. There is no excess capital nor shortfall and the pension fund is adequately funded

DDM Valuation of Hypothetical Bank XYZ: An Example

The terminal value is obtained with the formula shown above. The analyst can run a sensitivity analysis of the terminal value price-earnings (PE)[6] for various levels of k and g (Tables 2 and 3).

DDM Valuation of a Bank: Cost of Equity and Fair Value Per Share

We show, below, the derivation of the CAPM cost of equity and fair value per share. This is compared to the current share price to detect an over/undervaluation, which can inform a trading strategy (Table 4).

As a "sanity check", we calculate PE and P/TE implied in our valuation, i.e. the multiples at which the bank would trade if it was valued exactly in line with our

[5]A Damodaran: Investment valuation, Chap. 14 (Free cash flow to equity discount models)

[6]Market price divided by earnings per share. These can be estimates of future years' earnings, in which case we will have a forward PE, or data from the past, in which case we have a trailing PE. PE ratios along with P/TE are arguably the most popular ratios to compare banks' market valuation.

Table 2 DDM valuation of Bank XYZ—example

Year		1	2	3	4	5	6	7
(eur mn)	2022A	2023E	2024E	2025E	2026E	2027E	2028E	2029E
Earnings	7500	7690	7885	7964	8044	8125	8206	8289
Amortisation & depreciation	200	200	200	200	200	200	200	200
Goodwill write-downs	400	401	402	402	402	402	402	402
Cash flow	*8100*	*8291*	*8487*	*8566*	*8646*	*8727*	*8808*	*8891*
Risk weighted assets	4,12,000	4,19,828	4,27,805	4,35,933	4,44,216	4,49,546	4,54,941	4,60,400
CET1 capital (@ 12% CET1 ratio)	49,440	50,379	51,337	52,312	53,306	53,946	54,593	55,248
Increase in CET1 capital	0	939	957	975	994	640	647	655
Free cash flow to equity (FCFE)		*7352*	*7529*	*7590*	*7652*	*8087*	*8161*	*8236*
Dividends paid out		*4411*	*4668*	*4858*	*5050*	*5337*	*5223*	*5106*
Payout ratio		57%	59%	61%	63%	66%	64%	62%
Dividends as % of FCFE		60%	62%	64%	66%	66%	64%	62%
Discount factor		0.910	0.829	0.754	0.687	0.625	0.569	0.518
NPV of dividends		4015	3869	3665	3468	3337	2972	2645

Source: Authors

Table 3 Sensitivity analysis of Terminal Value PE for various levels of COE and G

COE/G	6.85%	8.85%	*9.85%*	10.85%	11.85%
3.0%	26.0	17.1	14.6	12.7	11.3
2.5%	23.0	15.7	13.6	12.0	10.7
2.0%	*20.6*	*14.6*	***12.7***	*11.3*	*10.2*
1.5%	18.7	13.6	12.0	10.7	9.7
1.0%	17.1	12.7	11.3	10.2	9.2
0.5%	15.7	12.0	10.7	9.7	8.8

Source: Authors

Table 4 DDM valuation of Bank XYZ—example (II)

Risk-free rate	4%
Beta	1.3
Risk premium	4.5%
Cost of equity	10%
Perpetual growth rate	2.0%
Terminal PE	12.7
Terminal value	33,699
Total value	*57,671*
Terminal value as % of total value	58%
Shares outstanding (mn)	3000
Value per share	19.2
Current share price	6.0
Upside potential	220%
Implied PE (2024)	7.3
Implied P/TE (2023)	1.14

Source: Authors

valuation for it. Such multiples can be compared with those of peers to glean whether we have been too harsh or too lenient in valuing the bank. Implied multiples that are much lower than the peer group average, for instance, might suggest that we have erred on the side of caution in our valuation. Vice versa, implied multiples that are above peers might warn us that we have been too optimistic in our estimates and/or valuation assumptions.

FCFE Valuation

The DDM valuation shown above can easily be converted to a DCFE one by replacing the projected stream of dividends with the free cash flows to equity. A DCFE will generally value the firm more generously than a DDM, given that the cash flow stream includes potential dividends along with actual ones. According to A. Damodaran, the difference between the two represents the value of the control premium for the firm: a new management could theoretically increase dividends to keep in line with cash flows. If a market is characterized by openness in corporate control, then it is correct to use a DCFE to value a firm (Table 5).[7]

Market Multiples

The analyst risks running into a circularity problem when using market multiples. Therefore, we do not recommend using these as a primary valuation tool. Nevertheless, market multiples can usefully be employed as a control method in the valuation of the whole entity or individual business areas, especially in markets where a significant number of listed peers is present for a division/business area. PEs[8] and P/BV are the most used market multiples. Unlike what is the case for an industrial entity, it makes limited sense to refer to earnings before interest and taxes/earnings before interest, taxes, depreciation, and amortization (EBIT/EBITDA) for banks: interest income and expense are an integral part of the core business. On the other hand, depreciation is usually a minor item given the lack of sizeable, fixed assets.

- We also use PE methodology as a "sanity check" for our divisional values in sum-of-the-parts approaches. This makes sense when a significant number of listed peers exist for a division or business area. Similarly, the analyst can check whether multiples implied in his valuation are consistent with those paid in recent M&A transactions in the same market. Some analysts use price to pre-provision profits to neutralize the volatility of loan loss provisions (LLP). The assumption is that loan loss rates for banks operating in the same market will converge towards similar levels over the medium term. For example, some equity research houses estimate LLP at around 40 bps of loans in the period 2023–24, for European banks. This compares with LLP of >100 bps for the years immediately following the 2006–08 financial crisis.
- A variant of the popular PE ratio is represented by the price/earnings-to-growth (PEG). This is none other than a PE divided by a growth rate. The PEG allows the analyst to rank stocks based not only on their relative price like a PE but incorporate different growth prospects too.
- Last, the analyst should extend the comparison to other multiples such as those based on asset under management (AuM) for asset management or embedded value for insurance operations.

Research houses regularly calculate valuation sheets with key market data and multiples. The sheet in the example below shows:

[7] A. Damodaran, Investment valuation, Chap. 14. FCFE Valuation versus Dividend Discount Model Valuation.

[8] Forward and trailing PE—see Glossary for definitions.

Table 5 DCFE valuation of Bank XYZ—example

Year		1	2	3	4	5	6	7
(eur mn)	2022A	2023E	2024E	2025E	2026E	2027E	2028E	2029E
Earnings	7000	7177	7359	7433	7508	7583	7659	7736
Amortisation & depreciation	200	200	200	200	200	200	200	200
Goodwill write-downs	400	401	402	402	402	402	402	402
Cash flow	*7600*	*7778*	*7961*	*8035*	*8110*	*8185*	*8261*	*8338*
Risk weighted assets	4,12,000	4,19,828	4,27,805	4,35,933	4,44,216	4,49,546	4,54,941	4,60,400
CET1 capital(@ 12% CET1 ratio)	49,440	50,379	51,337	52,312	53,306	53,946	54,593	55,248
Increase in CET1 capital	0	939	957	975	994	640	647	655
Free cash flow to equity (FCFE)		*6839*	*7004*	*7059*	*7116*	*7545*	*7614*	*7683*
DIvidends paid out		4103	4342	4518	4696	4980	4873	4764
Payout ratio		57%	59%	61%	63%	66%	64%	62%
Dividends as % of FCFE		60%	62%	64%	66%	66%	64%	62%
Discount factor		0.910	0.829	0.754	0.687	0.625	0.569	0.518
NPV of FCFE		6226	5804	5326	4887	4717	4333	3980
Risk-free rate	4%							
Beta	1.3							
Risk premium	4.5%							
Cost of equity	10%							
Perpetual growth rate	2.0%							
Terminal PE	12.7							
Terminal value	50,707							
Total value	*85,979*							
Terminal value as % of total value	59%							
Shares outstanding (mn)	3000							
Value per share	28.7							

Current share price	6.0								
Upside potential	378%								
Implied PE (2024)	11.7								
Implied P/TE (2023)	1.71								

Source: Authors

- Market capitalization
- PE stated and adjusted
- Book value (BV)
- Adjusted book value, corresponding to tangible equity (TE), or adjusted net asset value
- Price to BV and price to TE(P/TE)
- Return on stated equity (ROE) and return on tangible equity (ROTE) (Table 6)

As implied by finance theory, there is normally a strong correlation between the return on capital generated by a (listed) business and the price to capital multiple that investors are willing to pay for its shares. Based on such a correlation, it is possible to build a "value map" for a peer group of comparable banks belonging to the same market. Incidentally, we point out that the "correct" correlation is between the P/Capital and value added, i.e. the spread between the return on capital and the institution's cost of equity. We can, however, make the simplifying assumption that that the cost of equity for banks belonging to the same market is very similar, which allows us to focus on returns only.

The value map shown in the example below is a linear regression of P/TE against return on tangible equity (Table 7).

PE Valuation: An Example

PE valuation is the ratio between the Price of a stock versus the earning per share (EPS). EPS is calculated on 2 or 3 years forward-looking perspective. To estimate the forward-looking EPS, the analyst should start from latest announced balance sheet and P&L figures.

The key metrics of the relative valuation are listed in Table 8.

To obtain the forward-looking estimates, we suggest the following steps:

Step 1 Forecast balance sheet growth. The support materials for the analysis are: (1) macro estimates (as GDP, industrial production, consumer confidence, and other macroeconomic indicators). In many cases, it is possible to find a statistical correlation between macro indicator growth and the potential loans growth. The multiplication factor is also very much correlated with the loan's penetration in the country (lower the penetration of loan over GDP and higher is the multiplication factor). Remaining asset growth can be forecasted based on the bank's management business plan, bank's expectation on possible market trends and marketing strategy.

Step 2 Determine required equity by forecasting the required CET1 equity it is necessary to meet the capital requirements.

Step 3 Forecast required funding mix by possibly matching estimated asset growth and liability plus equity. The financing of asset expansion, partially done via capital requirement, would come from a mix of liabilities that varies between banks and may be linked to factors such as the strength of the depositor base,

Table 6 Banking sector valuation sheet—example

	Price	Shares outst (m)	Mkt cap		EpS adj.	BVpS						ROTE
	Eur	m	EUR m	EpS	Eur	Eur	adj.BVpS	PE	PE adj.	P/TE	ROE	Eur
Bank 1	25	1234	30,850	2.08	1.9	12.5	8.9	12.0	13.3	2.80	16.7%	8.0%
Bank 2	49	2015	98,735	4.45	4.0	24.5	16.9	11.0	12.2	2.90	18.2%	10.0%
Bank 3	27	260	7020	3.38	3.0	13.5	12.9	8.0	8.9	2.10	25.0%	25.0%
Bank 4	20	2126	42,520	1.11	1.0	10.0	8.0	18.0	20.0	2.50	11.1%	5.0%
Bank 5	22	1982	43,604	3.14	2.8	11.0	9.2	7.0	7.8	2.40	28.6%	7.0%
Bank 6	23	871	20,033	7.67	6.9	11.5	19.0	3.0	3.3	1.21	66.7%	12.0%
Bank 7	34	1914	65,076	1.89	1.7	17.0	37.8	18.0	20.0	0.90	11.1%	18.0%
Bank 8	18	1907	34,326	4.50	4.1	9.0	7.2	4.0	4.4	2.50	50.0%	24.0%
Bank 9	6	900	5400	0.67	0.6	3.0	2.4	9.0	10.0	2.50	22.2%	4.0%
Bank 10	46	1512	69,552	2.09	2.3	23.0	18.4	22.0	20.0	2.50	9.1%	11.0%
Bank 11	25	333	8325	1.32	1.2	12.5	10.0	19.0	21.1	2.50	10.5%	20.0%
Bank 12	20	1924	38,480	1.00	0.9	10.0	8.0	20.0	22.2	2.50	10.0%	15.0%
Bank 13	28	1583	44,324	9.33	8.4	14.0	11.2	3.0	3.3	2.50	66.7%	8.0%
Bank 14	19	955	18,145	2.11	1.9	9.5	7.6	9.0	10.0	2.50	22.2%	7.0%
Bank 15	29	1526	44,254	2.42	2.2	14.5	11.6	12.0	13.3	2.50	16.7%	2.0%

Source: Authors

Table 7 Value map

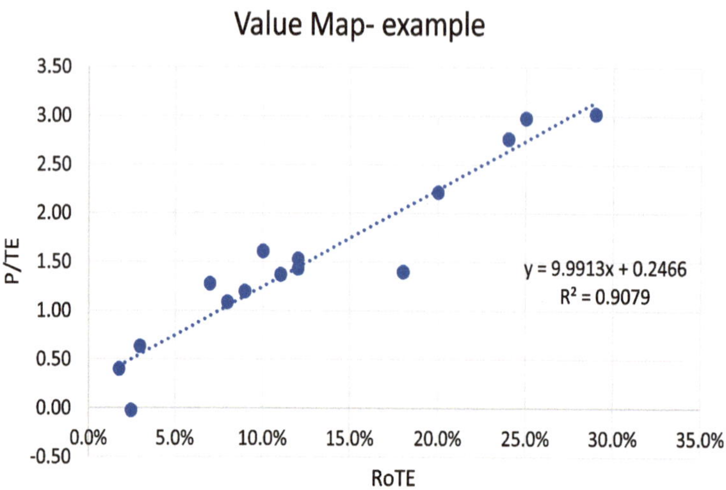

Source: Authors

Table 8 Valuation metrics

Valuation metric	Formula
Net income	Profits
EPS	Earnings per share = Net income/# shares
Book value	Shareholder equity
Tangible book	Shareholder equity-intangibles (e.g. goodwill)
Book value per share (BVPS)	Book value/# shares
ROE	Net income/book
ROTE	Net income/tangible book
Return on assets (ROA)	Net income/assets
Dividend per share (DPS)	Dividend/# shares
Valuation multiple	*Formula*
P/E	Current price/EPS
P/BV	Current price/BVPS
P/TBV	Current price/tangible BVPS

Source: Authors

access to wholesale markets and access to bond markets. Banks would need to comply with certain requirement as Liquidity Coverage Ratios (LCR),[9] Net Stable Funding ratio (NSFR)[10] or less technical requirement as loan to deposit ratios. Using this metrics can help to calculate the required funding mix.

Step 4 Forecast future profit and loss. Once calculated the new balance sheet it would be time to calculate the P&L impact derived from the additional assets and the new liability mix. In addition to the new balance sheet mix, it would be necessary to factor the expected interest rate trend and the factors influencing margin. Commission trend would be influenced by banks' commercial policies, competition, and asset growth.

When a bancassurance agreement is in place, the bank normally receives distribution fees based on agreed sales targets. Additionally, it is not uncommon for a bank to manage the funds backing insurance reserves. This is especially common for unit-linked reserves, which are normally invested in funds run by the bank's asset management arm. Fees from this activity might be reported both under the umbrella of insurance commissions and asset management. The analyst should identify if overlapping exist and remove the double-counting from the valuation.

Trading income would be impacted by the market trends and volatility. Operating expenses can be forecast using, ceteris paribus, the inflation rate or investments decision (e.g.: branches opening, hiring, bonuses decisions, depreciation, and amortization expected).

Losses on loans are linked to PDs, LGDs, etc., like loan portfolio valuation methodology (see chapter "Loan Valuation"). Macro trends are crucial to calculate losses on loans. In many cases, it is possible to find a statistical correlation between macro indicator (e.g.: unemployment, GDP) and the potential losses.

- The starting level of credit quality might be low after years in which a bank has posted above-average growth rates in its loan book and market share gains.
- The expected evolution of default rates. In turn, this will largely be driven by GDP growth and changes in the level of interest rates, with the latter being positively correlated with default rates (higher interest rates increase the probability of default on mortgages and corporate loans as well as consumer credit) and the former negatively (better GDP growth should lead to lower default rates thanks to lower unemployment and higher average corporate profits).

A decline in property prices can exacerbate issues of credit quality in that: (a) privates are more inclined to default on their mortgage obligations if they fall

[9]BIS: LCR is designed to ensure that banks hold a sufficient reserve of high-quality liquid assets (HQLA) to allow them to survive a period of significant liquidity stress lasting 30 calendar days. HQLA are cash or assets that can be converted into cash quickly through sales (or by being pledged as collateral) with no significant loss of value.

[10]BIS: NSFR is defined as the amount of available stable funding relative to the amount of required stable funding. This ratio should be equal to at least 100% on an ongoing basis.

into negative equity; (b) a decline in house prices could lead to lower recovery rates for banks.

Bank-specific issues also matter too, of course. For example, a strengthening of the risk management function may be beneficial for credit quality and so could the disposal of a portfolio of bad loans.

For a European portfolio composed mainly of mortgages and loans to SMEs, we would argue that the average loan loss rate (i.e. loan loss provisions/net loans) should be, for EU Banks, on the base of the 1990–2020 average, circa 60 bps with peaks as high as 180 in the worst years of the credit cycle and lows at 20–30 bps in the best ones. Assuming a loss given default (LGD) of 50%, this implies a default rate just below 2%. Observing historical data on default rates as provided by rating agency S&P, such a number corresponds to the probability of default of a speculative-grade corporate (See graph below on historical default rates).

The analyst should attempt to isolate the positive effect of recoveries on loan loss provisions since recoveries tend to appear near the peak of the credit cycle and usually do not last very many quarters. Lastly, we point out that LLPs should not be seen in isolation: an institution might exhibit higher than average LLPs, but this could be offset by lower fixed costs and/or better revenue margins. What matters at the end of the day is return on capital, or, better, its spread over the cost of equity (Table 9).

Business plans can help the analyst form an idea about normalized earnings and so can interactions with the bank's management. The results of the model should provide the forward EPS. The 2–3 years forward EPS would be the denominator of our PE valuation methodology. Differently from DDM, DCF, and WEV, the price earning valuation is a relative valuation methodology. We show, below, an example of a valuation using with peers' multiples, in this case PE ratios. Five peers are selected, and an average PE calculated. This is applied to forward earnings, year 3 in our example, to obtain a valuation (Table 10).

Table 9 Global default rates—S&P

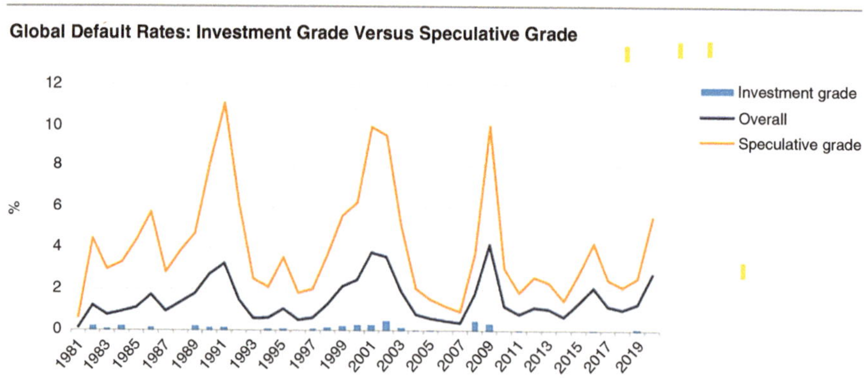

Source: S&P

Table 10 PE valuation

Multiples PE	
Bank 1	9.8
Bank 2	8.5
Bank 3	15.7
Bank 4	12.8
Bank 5	8.9
Average	*11.14*
Earnings (FY3)	7433
Valuation	*82,802*

Source: Authors

PE analysis is a relative analysis. Analyst should be careful in comparing Banks in different countries using the same PE multiple.

Various factors explain the generally higher multiples of US banks relative to their European counterparts. The key factor is arguably represented by differences in the business model. In particular.:

1. The different structure of the mortgage market. In US it is controlled by two GSEs (Government Sponsored Enterprises) Fannie Mae and Freddie Mac consolidating c. half of outstanding mortgages.[11] On the contrary, most mortgages are retained on balance sheet by European banks, which, therefore leads to a higher capital absorption and lowers free cash flow.
2. A higher reliance by US banks on fee business in the capital markets business lines. Again, this implies a lower capital absorption.

Both factors contribute to US banks operating with lower levels of required capital and, hence, higher cash flow and better returns, which, in turn, justify higher multiples, both on earnings (PE) and capital (P/BV P/TE).

Additional factors are represented by the fact that the US banking market is ampler and more concentrated and by macro factors such as the higher GDP growth as well as a less depressed level of interest rates over the past several years.

[11] J. A. Soler Martin, S. Carbo Valverde: market value of banks, an international comparison, Madrid, April 2020. https://biblioteca.cunef.edu/files/documentos/TFG_Jose_Antonio_Soler_Martin.pdf

Banks and Inflation

As concerns a bank's balance sheet, inflation should largely be a "wash" in that neither loans on the asset side, nor deposits or the value of issued debt on the liabilities side, stand to be affected, as such.[12]

As regards the P&L, things may, on the contrary, look rather different in that several second-order effects kick in: a context of high inflation might compel the Central Bank to act, rising interest rates. A context of higher rates should benefit the net interest margin if, as it typically happens, banks can re-price their asset side faster and more decidedly than competitive pressures force them to pass higher rates to their depositors. Additionally, if the increase in rates leads to a steepening yield curve, banks should profit by engaging in more aggressive maturity transformation. However, a context of high inflation coupled with a depreciating domestic currency might create a difficult environment for certain products. We point out, for example, that volatile and rapidly rising domestic interest rates, particularly after the September 2022 minibudget, forced UK lenders to pull many existing fixed-rate mortgages from the market as these products had suddenly become very difficult to hedge. Fixed-rate mortgages are typically hedged with interest-rate swaps.[13]

Building a Value Map

A "value map" is created by regressing a price over capital multiple (price to tangible equity—P/TE in our example) against a profitability indicator (ROTE in our example) having the same denominator. The regression allows the analyst to understand, briefly, which listed shares are undervalued: those sitting below the regression line, ceteris paribus, these should trade at a higher price to tangible equity, given their returns, i.e., sit on the regression line.

Conversely, shares of banks depicted above the regression line are, ceteris paribus, overvalued. We point out, however, that:

1. The "ceteris paribus" qualification is arguably a strong one: various company-specific factors can justify a discount to fair value. This is something that requires fundamental analysis.
2. As mentioned above, P/capital multiples should be regressed against risk-adjusted returns rather than simple returns.

The value map is a purely statistical tool and, no matter the strength of the correlation, (high, in our example, as exemplified by 90% R squared), should not be relied upon to make investment decisions, in our opinion. Nevertheless, the value

[12] See T. Koller, M. Goodhart, D. Wessels (Mc Kinsey & co.): Valuation: measuring and managing the value of companies, Chap. 26 (inflation).

[13] ECB working paper series (September 2018): Who bears interest rate risk?

Table 11 Value map

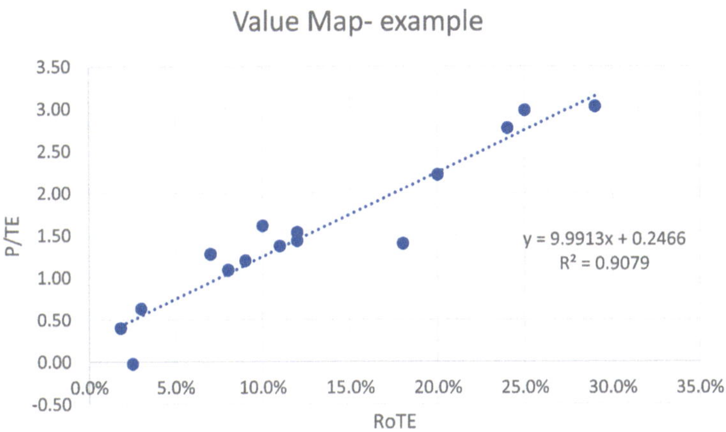

Source: Authors

Table 12 Valuation of Bank XYZ with regression

EUR 0.000	
Alpha	0.25
Beta	9.91
ROTE 3YE	0.15
Fair P/TE	1.76
TE 2025E	52,312
Valuation	91,845

Source: Authors

map can support the analyst as an intuitive guide to identify stocks that may be undervalued or overvalued and hence help the analyst identify a smaller group of banks on which it may be worth carrying out additional and time-consuming fundamental research (Table 11).

Inserting our bank's ROTE (independent variable) in the regression formula, we obtain a fair P/TE (the dependent variable), hence a valuation for the equity of our bank. Since the key driver for value creation is the spread of returns over the cost of equity, a theoretically more correct version of the value map would use the spread (ROTE-COE) as the explanatory variable. Alternatively, the analyst could risk-adjust returns by subtracting the value of credit default swaps from the ROTE. This method, however, fails to consider a properly calculated cost of equity (Table 12).[14]

[14] HSBC Global Research, Greek banks, July 2020

The Warranted Equity Value (WEV) Method

Ensuing from the same principle of high correlation between returns and price to capital returns, we introduce the warranted equity value (WEV) as a commonly used valuation tool for banks. Indeed, the warranted equity value approach makes perfect sense if we want to value a capital-driven business model such as that of a bank. The WEV can be applied to the institution as a whole or to the entity's individual business areas in a sum-of-the-parts approach (see below). The WEV is an extension of the economic value-added valuation method (EVA) in that:

- It uses the normalized (sustainable) return on capital as its main input
- Compare this to the hurdle rate (cost of equity). For our approach to derive the cost of equity, see below

The WEV applies a "fair" P/capital multiple to divisional capital or to the bank's overall equity. The fair P/capital is, driven by value added, i.e., the spread of the (sustainable) returns on capital over and above the cost of equity (COE). The higher the sustainable return and the lower the volatility of returns (and hence the COE), the higher the valuation, consistently with finance theory. To estimate a sustainable return on capital (ROE), the analyst should use a process like the EPS-PE methodology. He/she should look for normalized earnings, i.e. remove the effect of one-off factors from earnings and project earnings some years into the future. To obtain the Return on Tangible Equity (ROTE) analyst should divide the forward-looking estimated EPS with the Capital (less goodwill).

As for the estimation of the cost of equity, factor models such as the CAPM or Fama–French can be used (See chapter below).

The formula for the WEV is represented by a perpetuity of the value (V) created by the business, i.e. the return on capital (tangible equity for a bank) over and above the cost of equity (k). where g represents the sustainable LT growth in economic value, (i.e. the time during which the current spread of ROTE-COE can be sustained), while TE is constituted by capital (tangible equity). The size of the spread of ROTE-G over COE-g creates a "fair" P/TE multiple, which is applied to tangible equity.

$$V = \frac{\text{ROTE} - g}{K - g} \times \text{TE}.$$

Alternatively, V can be derived as the perpetuity of sustainable value added, i.e., by first calculating the difference between return on TE and COE in numerical terms (e.g. in EUR m) and then dividing by (k-g).[15]

[15]For Economic Value added, see: A. Damodaran, Investment valuation, Chap. 32; And T. Koller, M. Goedhart, D. Wessels (McKinsey): Valuation (1990), Chap. 38. The authors define value added in the banking sector in an interesting way: value added on loans, for instance, is defined as (a) net interest margin less (b) MOR (matched opportunity rate), i.e. the return that the bank could have

Table 13 WEV valuation of Bank XYZ—an example

ROTE	15%
Cost of equity	10%
Perpetual g	2%
Fair P/TE	1.7
TE	52,312
Valuation	88,121

Source: Authors

The following is an example of a valuation carried out with a warranted equity value. The key inputs are represented by:

- The bank's sustainable return on tangible equity (14%)
- A CAPM-derived cost of equity (10%)
- A perpetual growth of 2%. This should refer to the growth in capital which can be deployed at the assumed ROTE, though the growth rate in earnings is sometimes used as a proxy

The WEV formula returns a fair P/tangible equity of 1.5×, well above 1×. Since the spread between returns and cost of equity is largely positive, this makes sense (Table 13).

Excess Return Models

Excess return models constitute a generalization of the WEV: it sums the PV of excess returns (economic value added) to equity holders to existing capital, to obtain an equity valuation. Unlike in the WEV, which is based on a perpetuity, implicitly if the inputs to the valuation will stay the same, in the excess return models, EVAs estimated year-by-year analytically.[16]

The valuation in the example below has the beginning of 2023 as a reference date. We have used the same inputs as in our DDM and DCFE examples above, wherever possible, e.g. our CAPM-derived cost of equity of 10% (Table 14).

Capital—How to Treat It

Firstly, the analyst should allocate capital by division, generally using 8–12% of divisional RWA[17] as a driver. This is in line with somewhat above regulatory minimums. It is important not to penalize riskier business areas such as investment

earned on investments with a similar duration and risk profile less (c) a tax penalty on required equity capital and on the portion of loans funded by equity.

[16] see: A. Damodaran, Investment valuation, Chap. 21; excess return models.

[17] RWA: https://www.investopedia.com/terms/r/riskweightedassets.asp.

Table 14 Excess return valuation of Bank XYZ—example

Year		1	2	3	4	5	6	7
(eur mn)	2022A	2023E	2024E	2025E	2026E	2027E	2028E	2029E
Earnings	7500	7690	7885	7964	8044	8125	8206	8289
Tangible equity	49,440	50,379	51,337	52,312	53,306	53,946	54,593	55,248
ROTE	15.2%	15.3%	15.4%	15.2%	15.1%	15.1%	15.0%	15.0%
COE	10%	10%	10%	10%	10%	10%	10%	10%
Excess return		5.4%	5.5%	5.4%	5.2%	5.2%	5.2%	5.2%
EVA		2727	2828	2811	2793	2811	2829	2847
Discount factor		0.91	0.83	0.75	0.69	0.63	0.57	0.52
NPV		2483	2343	2121	1918	1757	1610	1475
Invested capital	49,440							
Sum of discounted EVAS 2023–2028	12,233							
Terminal value	18,789							
Total value	*80,461*							
Shares outstanding (mn)	3000							
Value per share	26.8							

Source: Authors

banking by allocating too much capital to them. This might happen, for example, if the analyst chooses to allocate more capital than average (e.g. 12% of RWA versus 8% for other divisions to a business area which is already more RWA-intensive than the rest of the group).

Secondly, analysts can derive the value of each division on the basis of different valuation methods: e.g. a WEV for capital-intensive business areas, a DCF for cash-generating businesses such as asset management. Market multiples when a significant number of listed peers exists for a division or business area.

Third, we can derive a figure for excess capital/capital shortfall. This is given by the difference between available shareholders' capital, and the sum of divisional allocated capital. Other sources of capital such as minorities, subordinated debt, super-subordinated notes included in shareholders' equity and preferred shares, which are usually included within minorities, should all be removed from available capital for valuation purposes in that they are not supplied by common equity holders. Regarding the definition of available capital, to be prudent, we recommend taking the lower of tangible net asset value and common equity Tier One capital.

Lastly, we discount the face value of excess capital by 25%, to consider agency costs. The discount is, however, subjective: the analyst could very well use a zero discount if he/she believes that management is going to deploy excess capital at a rate that is in line with the institution's cost of equity or is likely to return excess capital to shareholders in full and reasonably quickly. Some banks might show a capital shortfall under this definition. We discount this at 15%, to reflect the fact that part of the shortfall could be covered by subordinated debt, whose face value is reduced by the tax shield. On the contrary, the analyst might choose to apply a higher discount if management has a poor track record as per value creation via M&A and/or the distribution of excess capital is penalized from a tax standpoint.

Capital and a Word of Caution on the Danish Compromise: A Numerical Example

Special caution should be used when assessing the existence of a capital surplus for financial conglomerates using the Danish compromise. We cite (Intesa Sanpaolo Pillar III 2019, pg 13 re: own funds) for the definition of the "Danish compromise" (DC):

Article 49 of Regulation (EU) No 575/2013 (CRR) grants financial conglomerates the option to activate (with the prior permission of the competent authorities) the so-called Danish Compromise, which permits banks that hold own funds instruments in insurance companies not to deduct those significant investments from Common Equity Tier 1 Capital (CET 1[18]), weighing them at 370% for the purposes of RWA, reducing capital absorption. As a "financial conglomerate" with a Parent Company of a banking group—Intesa Sanpaolo Spa—which controls the Intesa Sanpaolo Vita Insurance Group, on 9 May 2019, the Intesa Sanpaolo Group submitted the request for authorization to the competent supervisory authorities and received permission from the ECB on 9 September 2019 to calculate the Group's consolidated capital ratios by risk weighing the insurance investment instead of deducting it, with effect from the report as at 30 September 2019. The amount of T2 subordinated instruments issued by the Group's insurance companies and held by (Intesa Sanpaolo Pillar III 2019, pg 13 re: own funds). Ceteris paribus, the DC allows a conglomerate to free up a substantial amount of equity capital. If the analyst expects that, for whatever reason, that supervisors might restrict the use of the DC, therefore, he/she should apply a discount to the capital freed-up by the DC, in our view. Let us assume that a bank with EUR 82 m in common equity Tier One (CET1) capital and EUR 1000 m RWA owns 95% in an insurance company having an equity capital of EUR 40 m. The bank's CET1 ratio would be EUR 82/1000 = 12%.

[18]CET1 capital: https://www.investopedia.com/terms/c/common-equity-tier-1-cet1.asp

Table 15 Capital relief from the Danish compromise—a numerical example

Bank	Base	Ordinary regime	Danish C.		
CET1 capital	120	82	120		
RWA	1000	1000	1140.6	370%	14%
CET1 ratio	*12%*	*8%*	*11%*		
Insurance capital	40				
Stake owned	95%				

Source: Authors

Table 16 Summary of valuation methods

	Value	Per share			
Method	EUR mn	Eur/share	Upside	Implied PE	Implied P/TE
DDM	57,671	19.2	−20%	7.2	1.10
DCFE	85,979	28.7	19%	10.8	1.64
PE	82,802	27.6	15%	10.4	1.58
EVA	80,461	26.8	12%	10.1	1.54
WEV	88,121	29.4	22%	11.1	1.68
Statistical	91,845	30.6	28%	11.5	1.76
Average	*81,147*	*27.0*	*13%*	*10.2*	*1.55*

Source: Authors

Under the ordinary regime, the bank would have to deduct from its CET1, its stake (95%) in the insurance company's capital, or 95% of EUR 40 m. If RWA remains unchanged, this would lead to the CET1 ratio declining a hefty 4PP to 8%.

Under the DC, on the contrary, the amount of CET1 capital is almost unchanged. The stake held in the insurance company is risk weighted at 370%, which increases total RWA by 14%. The new CET1 ratio is 11%, down only 1 pp versus 4 pp under the CRR's general regime. The application of the DC allows the institution to reap a sizeable 3 pp higher CET One (Table 15).

Assuming a target CET1 is, e.g., 10%, the conglomerate will have excess capital in the base case and a shortfall in the other cases, especially with no DC.

We show, below, a recap of the valuation methods used in our examples, along with a theoretical upside potential, implied PE, and implied P/TE (Table 16).

The statistical method values bank XYZ at EUR 30.6/share, with an implied PE of 11.5x and a P/TE of 1.76x. It is the most generous valuation methodology. Of the four fundamental "absolute" valuation methodologies (DDM, DCFE, EVA, and WEV), the latter appears to be somewhat more generous than the others. We point out, however, that the WEV valuation of EUR 29.4/share is not too distant from the others and, in fact, rather close to the EUR 28.7/share of the DCFE. Given the theoretical formulaic identity of EVA and DCF, we find this reassuring.[19] Unsurprisingly, the DDM valuation is the most conservative, valuing Bank X at a modest

[19] See T. Koller, M. Goedhart, D. Wessels (McKinsey & co.): Valuation: measuring and managing the value of companies, appendix A (discounted economic profit equals discounted free cash flow)

7x PE and only marginally above its tangible equity. In general, we take comfort from the fact that values are rather concentrated around the mean value of EUR 27/share: the standard deviation of the distribution is less than EUR 3.5 and the coefficient of variation (standard deviation divided by the mean) is a modest 16%. Our relative valuation methodology (peers' PE) returns a value of EUR 27.6/share, not far from the average.

A Sum-of-the-Part Valuation: A Practical Example

In the previous examples, we have valued our hypothetical bank's equity in its entirety. We will now discuss a sum-of-the-parts approach, where each business area is allocated its own slice of tangible equity and valued independently. A WEV can thus be used for most banking businesses (retail, investment banking), a DCF for highly cash-generating businesses such as asset management, embedded value for insurance operations.

As mentioned, we then discount the face value of excess capital by 25%, to consider agency costs. The discount is, however, subjective: the discount could very well be zero if the analyst believes that management is going to deploy excess capital at a rate which is in line with the institution's cost of equity or is likely to return excess capital to shareholders in full and reasonably soon. On the contrary, the analyst might choose to apply a higher discount if management has a poor track record as per value creation via M&A and/or the distribution of excess capital is penalized from a tax standpoint. In the opposite case, a bank might show a capital shortfall. We discount this at 15%, to consider the fact that part or all of the shortfalls could be covered by subordinated debt, whose negative value for shareholders is reduced by the tax shield (Table 17).

The first two columns on the left-hand side contain essential economic information pertaining to each of the bank's division or business area, i.e. earnings and risk-weighted assets (RWA). The analyst should use normalized earnings, removing the effect of one-off factors and projecting earnings 2–3 years into the future, if possible.

Business plans can help the analyst form an idea about normalized earnings and so can interactions with the bank's management. The loan loss rate (LLP/net loans) should be normalized to reflect the risk of the loan portfolio.

It is a warranted equity value approach (WEV) that we apply to most banking divisions/business areas. However, as mentioned above, this need not necessarily be the case: a DCF can be used to value low capital-intensity, high cash flow businesses such as asset management, embedded value for insurance operations.

Market Multiples as a "Control" Method

We also use PE as a "sanity check" for our divisional values. This helps the analyst ensure that the values of the institution's business areas are consistent:

Table 17 SOTP valuation of Bank XYZ

Division EUR m	Retail banking	Investment banking	Asset management	Total-operations	Excess capital	Total-bank
(a) Earnings	150	130	190	470	20	*490*
RWA	15,000	8667	0	23,667	8000	*55,333*
Capital/RWA	9%	12%			11%	Nm
(b) Capital	1350	1040	175	2565	5130	*10,260*
(c)(a/(b) ROC	11.1%	12.5%	Nm	18.3%	Nm	4.8%
	11.1%	12.5%	108.6%	18.3%	0.4%	4.8%
Rf	5%	5%	5%	5%	5%	5%
Beta	1.1	1.3	3		0.8	
Risk premium	4%	4%	4%		4%	*4%*
(d) CoE	9%	10%	17%	5%	5%	*5%*
(e)c/(d) fair P/BV	1.2	1.2	Na		0.75	
(b)x (d) = (f) valuation	1596	1275	2850		3848	***9568***
PE (X)	10.6	9.8	15.0	0.0	192.4	
Notes			DCF for AM		25%discount	

Source: Authors

- With those of listed peers in the same or similar markets, if available
- With the multiples paid in recent M&A transactions. To ensure this, the analyst should extend the comparison to other suitable multiples such as those based on AuM for asset management or embedded value for insurance operations

For banks having a defined-benefit pension fund, a shortfall (usually disclosed in the notes to the annual Report) should be subtracted from the valuation in that it is shareholders that are responsible for plugging in the shortfall.

ROE Dupont Decomposition for Banks

The classic Dupont ROE decomposition can be extended to banks, with only minor variations (Table 18).

Relative to an analysis valid for an industrial entity, we have added only a few, mainly P&L-related, ratios:

Table 18 Bank-ROE decomposition

P&L	(EUR mn)	
Net revenues	1000	
Expenses	−561	
Gross operating profit	*439*	
Llp	−129	
EBT	*310*	
Taxes	−30	
Minority interests	*−6*	
Net income	*274*	
Balance sheet	(EUR mn)	
Net loans	21,500	
Total assets	46,500	
Deposits	36,500	
Shareholders' equity	10,000	
Ratio		*Comment*
Net income/EBT	88%	Share of income not absorbed by taxes & minorities
EBT/GOP	71%	Loss absorption capacity
GOP/net revenues	44%	1—cost-income ratio
Asset turnover	2%	Not much sense for a bank
Leverage	4.65	
Loan loss rate (basis points)	60	RE: LLP/net loans
ROE	2.7%	
	2.7%	

Source: Authors

- Net income/EBT, which represents the share of pre-tax profits not lost to taxes and minority interests.
- Gross operation profit (EBT/GOP), a measure of loss absorption capability. The complement to 1 of these ratios is indeed the share of profits lost to loan loss provisions (including write-offs, known in Europe as the main component of the cost of risk).

Cost of Equity

While its power to predict equity returns is too low according to part of the doctrine, the single-factor capital asset pricing model continues to be widely used by practitioners, including bank analysts. Its main advantage is its simplicity: betas can be obtained from several data providers such as Bloomberg and DataStream while the market risk premium is derived from historical data series and conventionally set at 3–6%.

Given the fact that debt is part and parcel of a financial institution's business, levered betas are of little significance and little used by practitioners.

For non-listed institutions, the analyst may resort to using the cost of equity of peers listed in the same or similar markets.

Factor Models for COE

ECB researchers have used the CAPM as well as other factor models (e.g.: Fama–French) to estimate the cost of equity of Euro area banks (See: ECB Occasional paper series, January 2021. Measuring the cost of equity of Euro area banks).

The mean cost of equity is estimated at 7.33% with the CAPM and 12.45% with the Fama–French factor model. Results are consistent with Coco (contingent convertible debt) yields.

Bottom-Up Approach to the COE for the SOTP

Cost of equity based on division/geography/leverage:

Our WEV model compares normalized returns on (divisional) capital with divisional costs of equity. These are obtained by summing up country risk-free rates and division-specific betas times a market risk premium[20]

[20] It should be borne in mind that the analyst should use perspective betas for valuation. Historical betas should be seen as a useful starting point but not necessarily accepted and used as God's immutable will, as some analysts seem want to do.

1. Country risk-free rates correspond to rates on long-term (10Y) sovereign debt. We recommend using the risk-free rate of the country where the business is based.
2. Division betas can be derived by listed pure play businesses. While the analyst is allowed a degree of discretion in imputing the divisional betas, these should be consistent: divisions with more volatile returns, e.g., investment banking, should receive a higher beta than businesses with more defensive profiles, e.g. asset management or retail banking.
3. We use a 4% risk premium across the board.[21]

The bank's cost of equity is then derived "bottom up" as the weighted average of divisional cost of equity, with the weighting driven by the amount of capital allocated to each division. Excess capital can be assumed to have a cost of equity equal to the risk-free rate.

This bottom-up approach leads to different estimates for the cost of equity, depending on each bank's business composition, geographic diversification, and leverage, given by the size of the excess capital or capital shortfall. This is entirely consistent with finance theory, in our view.

Consistency Between Capital, Liquidity, and Cost of Equity

An important sanity check consists in making sure that institutions with riskier business models are, ceteris paribus, valued using a higher cost of equity than less risky ones.

The analyst should consider, in particular:

1. A bank's capitalization (common equity tier one, total capital ratios), leverage (leverage ratio)
2. Liquidity and funding (LCR, NSFR, loan to deposit ratio)

Consistency Between COCO Yields and Cost of Equity

If a bank has issued COCO bonds (contingent convertible debt), then the yield on Cocos should be considered as a floor for the cost of equity: Cocos combine elements of fixed-income and equity investments and are, therefore, less risky than pure common equity. Hence, the required return on Cocos must logically be lower

[21] See T. Koller, M. Goedhart, D. Wessels (McKinsey & co.): Valuation: measuring and managing the value of companies, Appendix G, where the (historical)global risk premium is calculated to be 4.5% for the years 1967–2016 and 4.2% for 1900–2016. T bills are used as the risk-free asset.

On the cost of equity, see also: A Damodaran, Investment valuation, Chap. 4 (estimating risk parameters and cost of financing), Chap. 7 (riskless rates and risk premiums), & Chap. 8 (the cost of equity and capital).

than that of equity instruments, consistently with the principle of investors' risk aversion.

Reference

Koller, T., Goedhart, M., & Wessels, D. (2010). *Valuation: Measuring and managing the value of companies*. McKinsey & Co..

Banks M&A: Strategy and Valuation

Abstract

Valuing a Bank in an M&A scenario can be challenging, and several factors and points of views would need to be considered. The starting point of an M&A valuation can be the traditional methodologies used for valuing a Bank in a normal scenario. In this chapter, we investigate some of the critical aspects of a Banking M&A and the valuation implications from two points of view: the shareholders' point of view and the supervisors' point of view. From the shareholders' point of view, the key aspects to be considered are the calculation of the synergies arising from the M&A and the potential earning accretion/dilution. On the other side, Banks M&A are always assessed by supervisors on a case-by-case basis. The main objectives from their point of view will be to ascertain, to the extent possible, the sustainability of the business model of the combined entity. In this sense, the analysis of the potential goodwill impairment represents a key component of the consideration.

Rationale for Banking M&A

In the past 20 years, the two main drivers for Banking M&A were cost reduction in mature markets and expansion into higher growth markets. Acquisition of smaller companies, usually in the same line of business, that presents strategic value was often regarded as the best way to get scale and critical mass in niche products and selected geographies. At the same time, where critical mass is seen as unobtainable or where the underlying profitability of the operation is seen as inferior companies proceeded to selected disposals. There are however other several factors that could incite a company's management to launch a takeover bid. This could involve:

1. Deploying a capital surplus
2. Expand product and service offering

A. Santoni, F. Salerno, *How to Value a Bank*, Springer Texts in Business and Economics, https://doi.org/10.1007/978-3-031-43872-1_4

3. Expand customer base
4. Desire to find new sources of top line growth by entering new regions
5. Access the target's complementary products or regional coverage
6. A need to reach economies of scale
7. Exploit a valuation premium (i.e. a Bank is trading at premium multiples relative to the sector average providing it with a strong currency for a payment in shares)
8. Funding synergies: it may be beneficial for a bank whose funding is mainly institutional and, perhaps, concentrated towards a small number of providers, to purchase a bank with more diversified funding sources and a lower loan to deposit ratio
9. Tax efficiency

Valuation considerations are very important to allow management to justify an offer to acquire a competitor. In the event of a hostile bid, the management of the buyer might argue that the two companies are complementary and that the merger, thanks to the synergies, would create shareholder value. The price paid for the target would be crucial for the determination of the positive shareholder value for the buyers as we will see in the example.

A key metric to judge a merger or an acquisition are the synergies that the deal would be able to produce. Synergies can be defined as the value that the combination of two entities is able to produce following their merger. Synergies can be defined revenue synergies when they are the results of top line value creation (eg: they could come from cross selling, share of wallet increases thanks to new products, new market access). On the other side, synergies deriving from cost cutting are defined as cost synergies. They usually are generated from economies of scale and scope. There is also a third source of synergy that can be defined as pure financial synergies that could be generated by RWA optimization, tax benefit, reduction in cost of funding, reduction in cost of capital among others. Deutsche Bank analysts analysed[1] the merger and acquisition track record of European Banks in the period 1999–2015. They found that cost synergies represented 33% of the target cost base on average in the in-market M&A of the European Banks. For the cross-border M&A, they found that the cost synergies represented 17% of the target cost base. The difference derives from lower complementarities reducing the cost cutting opportunities (Table 1).

For US Banks, expected in-market cost savings for transactions reached 35% (of the seller's non-interest expenses) versus 25% in 2010.[2]

There could be some easy hanging fruits synergies in case the target presents features as:

1. Governance issues that create incoherent or unconvincing strategies
2. Distrust by the market as results of bad investor communication strategy

[1]Deutsche Bank (2015).

[2]Olsen and Palmer (2019).

Table 1 Costs synergies from Banks M&A (1999–2015)

In-market M&A	Synergies % target costs
Italy	25%
Iberia	44%
Greece	8%
Nordic	27%
Germany	57%
France	22%
UK and Irish	29%
Average	*30%*
EU cross-border transaction	*17%*

Source: Author, DB, Reuters, Bloomberg

3. Weak revenue generation as a result of lack of product cross selling or ownership of a limited clients' share of wallet. This means to be easier for the buyer to rectify if the weakness is the result of managerial or marketing mistakes rather than issues affecting the base, e.g. product overpenetration
4. High-cost base as a result of inefficiencies

From a pure valuation point of view, the most common characteristics observed for the search of the ideal target for an M&A deal would include:

- Attractive valuation—a discount valuation versus the sector characterized by a 12 m or 24 m forward P/E below sector median.
- Low expected growth—with analyst estimating EPS growth below sector median.
- Low dividend yield expectation characterized by low earning generation or by lack of capital.

Once the deal has been agreed in the case of a friendly M&A or the deal has been decided in the case of a hostile take-over is decided or agreed there are however two main points of view to consider: the supervisor point of view and the market point of view.

M&A Transaction from the Supervisor's Point of View

Once a deal has been decided by the buyer, one of the most important next steps would be to obtain the green light from the Supervisors (Central Bank). Supervisors' assess M&A deals on a case-by-case basis. The main objectives from its point of view will be to ascertain, to the extent possible, the sustainability of the business model of the combined entity. Key criteria for the assessment would be from one side the managerial capacity to execute the deal and from the financial point of view, capital, liquidity, business model, and risk profile (including quality of assets and operational risk) of the combined entity.

For this purpose, Supervisors[3] will investigate the business plan presented by the management mainly focusing on:

- Business plan assumptions. These are expected to be conservative and appropriately justified and documented by solid microeconomic and macroeconomic assumptions that form the baseline scenario
- Company is expected to provide tangible signs of capacity of turnaround capacity based on experience and truck record
- The targets are expected to comply with applicable regulatory requirements and supervisory measures at each year of the plan
- Targets should avoid any reliance to optimistic revenue targets based on unproved sale or marketing capacity or excessive risk taking
- Cost synergies should not rely on layoffs plan that is difficult to obtain unions green light thus creating frictions or higher than expected layoffs costs
- A liquidity and funding structure, consistent with macro and micro business developments assumptions and the company risk appetite
- A detailed IT integration plan
- Provide a detailed and credible roadmap and timeline to achieve the targets announced

Among the criteria particularly important for the Supervisor, two of them have relevance: governance and risk management. The governance and organizational structure of the business combination is expected[4] at the level of the management body to have a clear decision-making capacity and an adequate composition with a clear allocation of responsibilities and decision-making processes as governance arrangements of the business combination. The merger should be managed by a strong leadership team with a proven track record while the consolidation plan includes the timely integration of the risk management and internal control framework, in particular the mitigation of execution risk.

M&A Transaction from the Market's Point of View: Key Criteria on Top of Regulators

From a shareholder's point of view, the criteria that a manager would need to consider when decides to launch a deal would include the following:

[3] https://www.bankingsupervision.europa.eu/legalframework/publiccons/pdf/consolidation/ssm.guideconsolidation_draft.en.pdf.

[4] Section 2 of ECB Banking Supervision: SSM Supervisory Priorities 2020, the SSM supervisory statement on governance and risk appetite, and Section 5 Element 2: Internal governance and risk management of the SSM SREP Methodology.

1. Will the deal be EPS accretive or dilutive (these aspects need to be explained to shareholders after allowing for a possible a) capital increase and/or share swap b) synergies)?
2. How are the synergies calculated?

   Management would need to consider the following potential sources of synergies:
   Costs synergies

- Branches overlap and the cost to close
- Headquarters (HQs) to reduce
- Product companies to merge
- Number of employees per branch to optimize the distribution
- IT to integrate or not
- One-off costs (usually ca 5% of combined cost base)

Revenue synergies

- Market share pre and post
- Product company's diversification
- Funding gap (e.g. cost of funding difference between the two Banks)
- Liquidity mismatch (e.g. Mismatch in maturity to cover)
- Customer spread gap in general
- Products penetration
- AUM penetration per branch
- Quality of the revenue line of the target (% of trading income)

Management should verify if due diligence has been performed on asset quality and other legacy. On asset quality, the buyer might want to pay special attention to the NPL coverage gap, under various standpoints:

- Split NPL secured/unsecured portfolio
- NPL vintage gap. The longer an NPL (with similar collateral) stays in the balance sheet of a Bank, the lower the probability to be recovered
- Other risks such as legal and money laundering risks

Capital impacts from the deal could concern:

- RWA density. The buyer could have acquired expertise in calculating RWA with internal models, which the target may be lacking, thus having to rely on standardized models for the calculation of RWA. The acquiror might, therefore, be able to re-calibrate the target's RWA based on its more sophisticated and efficient models. This could lead to lower capital requirements for the combined entity. The benefit might as well stem from the acquiror's getting hold and leveraging better expertise in the target entity.

Table 2 M&A valuation, value created

Market value (day announcement)	US $ bn	
Bank buyer	80	A
Bank target	50	B
Combined pre deal value	130	C = A + B
Pre-tax synergies	3	D
Post tax synergies	2	E
Value of synergies based on		
10× 2Y FW PE	20	E*10
After-tax one-off costs	−5	G
Combined post deal value	145	H = C + E + G
% Market value created	12%	H/C-1

Source: Authors

- Goodwill/badwill: A business combination transaction may result in the generation of goodwill or badwill. If the market valuation of a bank is below book value, badwill will be generated. In principle, Banking Supervisors recognize duly verified accounting badwill as capital from a prudential perspective. However, Supervisor expects it to be used to increase the sustainability of the business model of the combined entity. This happens when the bank uses badwill to increase provisions for non-performing loans, to cover transaction, integration costs, or other investments. Or simply retains it to boost capital ratios. On the contrary, supervisors' frown upon badwill being distributed to shareholders as dividends or buybacks.

Once the deal has been approved by the Supervisors and the synergies properly calculated by the management, then the most important step would be to convince equity analysts and the shareholders on the value creation capacity of the deal.

One of the most common and easy techniques to evaluate the value added from an M&A relies on a PE valuation methodology.

The steps are the following:

Step 1: Consider the market cap pre-announcement of the two company (a)

Step 2: Consider the synergies announced by the banks ($bn) net of tax and net of one-off costs and apply the average PE multiple of the banks under consideration to obtain the total net synergies created by the deal (b)

Step 3: Sum (a) and (b) to get the expected new market cap of the new bank created from the deal

Below is an example of a friendly full shares' merger between two Banks (Table 2).

Several caveat to this calculation should be made:

- Synergies should be discounted for their expected feasibility and credibility. Usually cost synergies are considered credible because more tangible. On the contrary, the credibility of revenue synergies is lower due to the higher execution risks as such are usually heavily discounted.
- Time of execution. Synergies should be discounted for the expected timing of execution. If the PE valuation is 2 years forward, then the synergies that can be added to the valuation are the 2 years announced while the remaining should be subject to a special valuation assumption.

Accretive Versus Dilutive Deal

Acquisitions sometimes involve a capital increase. This is because the buyer must finance the deal while maintaining the regulatory capital required by the Supervisor and other market stakeholders as rating agencies and equity analysts. To evaluate the possible accretive or dilutive impact on the buyer's earnings per share it is important to compare the buyer EPS pre-acquisition with the EPS post-acquisition. The EPS post-acquisition can be calculated assuming the earnings estimated for the target bank are summed up to the synergies excepted. The new future earnings will be divided by the new number of shares (post capital increase). In the example below, the acquisition represented an accretive deal for the buyer with a positive impact (EPS accretion) of 22% already from year 3. This matrix is usually used by analysts and traders to evaluate in the short term an M&A deal. For the value creation of a deal in the long term, a valuation methodology as DCF analysis should be used (Table 3).

Table 3 Bank acquisition simulation

	EUR millions	Year 1	Year 2	Year 3
A	Net profits (NP) Bank acquirer	1000	1050	1100
	NP bank target	500	540	550
	Post tax one-off charges	−400		
	Post tax synergies	60	80	140
B	Pro forma Bank acquirer NP	1160	1670	1790
C	# Shares Bank acquirer	1500	1500	1500
D	Pro forma # shares	2000	2000	2000
E = A/C	Bank acquirer EPS	0.667	0.700	0.733
F=B/D	New Bank acquirer EPS	0.580	0.835	0.895
F/E	Accretion/dilution	−13%	19%	22%

Source: Authors

NPV Valuation in M&A

Below we provide a simplification of a Net Present Value valuation methodology for an M&A transaction mainly focusing on capital deficit versus the target CET 1. The process foresees the following steps (Table 4):

1. Identification of the cost of equity for the calculation of the discount rate for the terminal value, long-term growth (in line with long-term GDP growth of the country where the Bank operates), and the common equity target in line with the Supervisors' expectations.
2. Estimates of the RWA growth and the common equity excess/deficit of capital versus the Supervisors' expectations. In case of capital deficit, the model should include the potential impact.
3. Estimates of net profits and dividend for the full period.
4. Calculation of the TV on the base of adjusted terminal value net profits and final RWA versus target CET1.

$$\text{Terminal value} = (\text{Adjusted Net Profit TV})^{*}\ (1+g)/(\text{coe}-g) - [\text{RWA final } 2028^{*}\text{CET target}^{*}g^{*}0.33/(\text{coe}-g)]$$

5. Calculation of the Core Equity Value as:

$$\text{Core Equity value} = \text{Sum FCF } 2024\text{–}2028 + \text{Excess/deficit initial capital} + \text{NPV TV}$$

Premium over Deposits

According to the S&P Global Market Intelligence data[5] the median M&A tangible book premium to core deposits during the period 2017–2022 in US ranged from over 9% in the period 2017–2018 to 6.9% in 2022 and an average 1.5× PBV. Range varies depending on how Banks are looking to M&A to solve funding pressures, making targets with low-cost core deposit bases more valuable. Premium to core deposits is a valuable "sanity check" valuation method (Table 5).

An additional step to a more granular valuation metric would be to add to core deposits valuation a clean book analysis. This analysis would involve a valuation of the loan book portfolio, with a special focus on NPL, and compare the current coverage versus peers' comparison. This would allow us to discount the possible

[5] https://www.spglobal.com/marketintelligence/en/news-insights/latest-news-headlines/funding-arms-race-drives-up-m-a-pricing-for-banks-with-low-cost-deposits-73457534.

Table 4 NPV valuation (example with fictitious numbers)

	NPV valuation	2023	2024	2025	2026	2027	2028
i	*Risk-free rate TV*	4.02%					
ii	*Beta TV*	1.17					
iii	*Equity risk premium*	4.00%					
	*Cost of equity (I + ii*iii)*	8.70%					
	Long-term growth net profits	1.00%					
	Long-term growth RWA	1.00%					
iv	*Target CET 1 2025*	9.00%					
v	*Capital buffer*	0.12%					
	CET 1 2025 (iv + v)	9.12%					
a)	RWA initial		120,829	121,762	125,041	128,449	144,706
b)	RWA final		121,762	125,041	128,449	144,706	148,709
c)	CET 1 initial		8.18%	8.66%	9.12%	9.60%	9.10%
d)	CET 1 final		8.66%	9.12%	9.60%	9.10%	9.53%
	ROE		10.87%	11.42%	11.90%	11.13%	11.52%
e)	Excess/deficit capital initial (c-target CET1%) * a		−987	−408	147	772	151
f)	Remuneration excess/deficit e*(risk-free rate) * (1-tax rate)		−26.5	−11.0	4.0	20.8	4.0
g)	Excess/deficit capital final (d-target CET1%) * b		−408	147	772	151	791
h)	Net profits		1172	1316	1467	1452	1603
i)	Dividends		586	658	733	726	802
j)	Capital increases		0				
k)	Net profits adjusted (net of f). i-f		1199	1327	1463	1431	1599
l)	Delta BV due to distribution excess cap/rec = g-e-j		578	556	625	−622	641
n)	Free cash flow distributable = k + l		1777	1882	2088	809	2240
o)	Discount factor = 1/(1 + i t)t−0.5		0.96	0.88	0.81	0.75	0.69
p)	NPV free cash flow = n*o		1705	1661	1695	604	1539
q)	Sum FCF 2024–2028		7203				
r)	Excess/deficit initial capital		−987				
s)	Net profits TV		1619				

(continued)

Table 4 (continued)

	NPV valuation	2023	2024	2025	2026	2027	2028
t)	Remuneration excess/deficit capital in TV		6.05				
u)	Terminal value (s-t) (1 + g)/(coe-g)-[RWA final 2028*CET target*g*0.33/(coe-g)]		20,587				
v)	Discount factor TV		0.69				
w)	NPV TV = u*v		14,144				
x)	Core equity value = q + r + w		20,360				
y)	Tangible book 2023		10,550				
z)	P/TBV implicit x/y		1.93				
	Current PBV		1.16				
	Upside		66.8%				

Source: Authors

Table 5 Example M&A tangible book premium to core deposits

EUR mln	2022	Value
Book value	1000	
Intangibles	200	
Tangible book value	800	800
Core deposits	20,000	
% Premium	6.90%	1380
Total value		2180

Source: Authors

hidden losses derived from under-collateralization of the portfolio or overvaluation of the collateral.

The analysis would involve the haircut of the book value for the hidden NPL losses.

Capital Increase and TERP

An important practical consideration in the analysis of an M&A deal would derive from the calculation of the Theoretical Ex-Rights price or the so-called TERP. Investors , in order to decide to acquire the new shares issued, would likely compare the TERP to the spot share price share and, as a consequence, the current valuation with the implied valuation of the new shares.

The steps to calculate the TERP are:

Table 6 Simulation TERP

	EUR	
A	Old price	10
B	Market cap	15,000
C	# Shares Bank acquirer	1500
	P/E old Bank	15.00
D	Capital increase	3000
E	Assumed discount to current price	30%
F = A*(1-E)	Right price issue	7
G = D/F	New shares issued	429
H = G + C	New outstanding shares	1929
(B + D)/H	TERP	9.33

Source: Authors

A. Market capitalization of the Bank buyer before the capital increase
B. Total number of new shares issued multiplied by the price of issuance (cash raised)
C. New number of shares as the sum of the old shares issued plus the new shares issued

The formula for the calculation of the TERP is:

$$(A + B)/C$$

Theoretical Ex-Rights Price (TERP) can so be defined as the "theoretical" price of a single share of a company immediately after a rights issue.

Usually, TERP would be set lower than the market value of a share prior to the rights issue. This is because, to attract new investors, new shares issued will be sold at a price below the prevailing market price. Keeping the same example of Table 1, we can investigate the TERP (Table 6).

Goodwill Valuation

One of the most important items to be valued post M&A is goodwill, that is the intangible asset that is created when one company acquires another company for a price greater than its net asset. The International Accounting Standards (IAS 36) establishes that the recoverable amount of goodwill acquired in a business combination must be assessed each year.

IAS 36 states that an asset must not be carried in the financial statements at more than the highest amount to be recovered through its use or sale. If the carrying amount exceeds the recoverable amount, the asset is described as impaired. The entity must reduce the carrying amount of the asset to its recoverable amount and recognize an impairment loss. IAS 36 establishes that the recoverable value is the higher between the market value/transfer net of costs to sell and the value in use:

- The market value/transfer net of costs to sell represents the price at which the asset could be sold net of directly attributable charges.
- The value in use, on the other hand, represents the present value of the expected future income flows attributable to the asset being valued.

From a methodological point of view, the different valuation approaches can be traced back to the following fundamental categories:

- Analytical criteria (to determine the value in use)
- Market criteria (to determine the fair value)

As regards the analytical methods, the valuation of the value in use as a criterion for determining the recoverable value implies the application of methodologies based on the estimate of the dividends that the cash-generating units (CGUs) can produce in the future (discounted dividend method or Dividend Discount Model—DDM, in the meaning of "Excess Capital").

The Dividend Discount Model is a variant of the cash flow method. This method, in the "Excess Capital" variant, establishes that the economic value of a financial company is given by the discounting of a flow of dividends determined on the basis of the minimum capital constraints imposed by the Supervisory Authority.

Among the market criteria, the two main methods are:

- The method of stock prices/market multiples
- The method of comparable transactions

The method of stock prices/market multiples is based on the analysis of listed companies as a company being valued. Due to the high volatility of the multiples themselves and the fact that they do not reflect control, they have not been used in recent years. The method of comparable transactions is based on an analysis like that of market multiples and on the relationship between the prices expressed in the context of company/company branch transactions and the assets and income figures relating to these companies.

The assessments carried out with the analytical criterion (value in use) of the DDM foresee the identification of the key parameters to use in the valuation of the value in use:

1) The Cost of Equity (Ke), 2) the long-term growth rate (g), 3) the last published CET 1 ratio, and 4) the Supervisory target of CET 1. The differences between the CET 1 estimated and the Supervisor target will determine the maximum dividend that can be distributed. The analysis of the distributable dividend can be based on a three-stage model:

1. On the last historical financial account published by the Group
2. On the 4 year forward-looking projections of the individual banks/companies of the Group based on company budget
3. On the further extrapolation for the fifth year which was constructed according to the following assumptions:
   - Annual growth rate of risk-weighted assets (RWA) in line with the long-term growth rate
   - Maintenance of the Intermediation Margin/RWA ratio constant at the levels of the last projected year
   - Value adjustments on loans estimated assuming constant (versus latest estimated the cost of risk)
   - Operating costs: growth rate equal to the annual growth rate expected for the latest estimated year

The Terminal Value is therefore constructed assuming as a reference the final year net profit (last year of explicit projection) increased by long-term growth.

In summary, the DDM method to estimate the value in use of the Banking Assets makes it possible to estimate the equity value (EV) by discounting:

- Future cash flows represented by dividends (Divi) in the period covered by the explicit projections and
- A closing cash flow (Terminal Value, TV) which represents the discounting of perpetual cash flows (with a view to the companys going concern) after the explicit projections which are calculated by multiplying the net profit in the last year of the projections by a long-term coefficient (coefficient g).

The Excess Capital variant provides that future cash flows are determined considering the minimum capital constraints imposed by the Supervisory Authority.

$$EV = \sum\nolimits_{i=1}^{n} \frac{Divi}{(1 + Ke)(i)} + TV$$

References

Deutsche Bank. (2015). M&A: Poor track record, rich opportunities.
Olsen and Palmer. (2019). *Bank Director*. MA Monitor.

Valuation in Resolution

Abstract

In April 2014, following the early 2000 financial crisis, European Authorities published the Bank Recovery and Resolution Directive (BRRD). The Bank Recovery and Resolution Directive was adopted to provide authorities with comprehensive and effective tools to deal with failing banks at national level and provide cooperation arrangements to tackle cross-border banking failures. The overarching objective of the BRRD resolution regime is to make sure a bank can be resolved swiftly with minimal risk to financial stability. Banks' valuation methodologies, which are used by Supervisors and Resolution Authority to trigger the failure of a bank and the best resolution tool to utilize, constitute the key aspects of the BRRD. In this chapter, we will revise the key principles of these valuation methodologies.

Resolution and Valuation Methodologies

Resolution Authorities should ensure that a fair, prudent, and realistic valuation of the Bank's assets and liabilities is carried out before deciding to write down or convert relevant capital instruments. The accuracy of the valuation is essential to limit litigation risk from creditors.

If the valuation is inaccurate, valuing assets above a fair value might require additional capital injections in the future. Alternatively, if assets are valued below their fair value, it might require the Resolution Authority to compensate creditors.

Valuation has been formalized by different Resolution Authorities, as for example the SRB and the PRA (Bank of England), into three types of valuations (Table 1).

Valuation 1 is based on the standard financial reporting obligations using normal accounting practice for going concern banks. The Valuation is intended to determine whether the bank is in a position that justifies entering resolution. As such it introduces a higher degree of prudence in the value of the assets.

A. Santoni, F. Salerno, *How to Value a Bank*, Springer Texts in Business and Economics, https://doi.org/10.1007/978-3-031-43872-1_5

Table 1 Valuation pre- and post-resolution

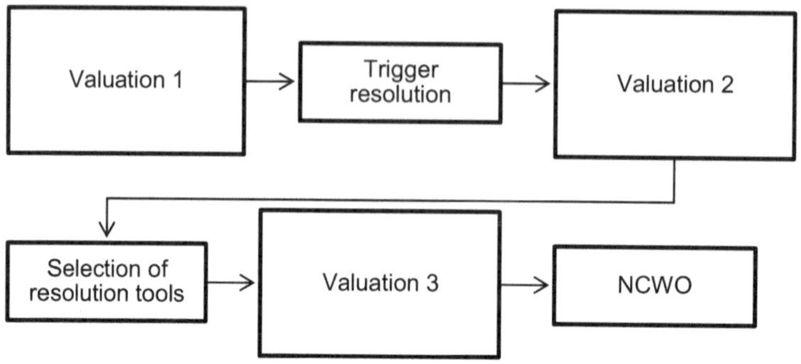

Source: Authors

Valuation 2 is a forward-looking exercise. It is intended to provide an accurate economic value of the assets to confirm that all losses are properly and fully recognized. This kind of valuation goes beyond the accounting recognition of the losses since it aims to identify all the losses which might materialize in the resolution process. This process envisages a range of potential resolution actions with the valuer that may submit several valuations for each resolution action.

In the case of capital injections, Valuation 2 shall also provide an estimate of the post-conversion equity value of new shares issued. In the case of bail-in tool, Valuation 2 should quantify the extent of the eligible liabilities written down. If the chosen resolution tool is the Bridge Bank, then Valuation 2 should provide information on the value of assets, liabilities, or other instruments to be transferred. Very similarly in the case of sale of business tool, Valuation 2 should provide an estimate of the value of the assets on sale.

Valuation 3 comes after resolution, and it provides an estimate of the potential valuation gaps between the resolution tool adopted and liquidation scenario. If the creditors had been better off under normal insolvency proceedings, then the "no creditor worse off" rule would not have been respected as such they would reserve a compensation by the Resolution Authority (Articles 20(16)–(18) SRMR). To avoid any conflicts of interest, this kind of valuation would need to be carried out by an independent person assessor in a timely manner after the resolution actions have taken effect.

A Practical Example

A practical example will facilitate the understanding of the different steps of the valuation methodologies.

Assuming financial institutions reporting a loss that triggers resolution actions by the local Resolution Authority. Being a significant institution for the country and

Table 2 Valuation 1

	Opening value	Valuation 1-FOLTF	Adjustments
Assets (EUR m)			
Cash	100	100	0
Securities	50	50	0
Loans	650	600	−50
Investments	90	80	−10
Real estate & equipment	110	100	−10
Intangibles	50	50	0
Other assets	25	20	−5
Total assets	1075	1000	−75
Liabilities			
Deposits	700	700	0
Subordinated debt	280	280	0
Pension liabilities	5	5	0
Accrued expenses	11	10	−1
Other liabilities	10	10	0
Total liabilities	1006	1005	−1
P&L			
Equity	69	−5	−74
Total liabilities and equity	1075	1000	−75

Source: Authors

having critical services the Authority decided to apply a resolution instead of liquidation. The first step, Valuation 1, was based on the assessment of the failing or likely failing of the Bank. The Authority assesses the capital position, the liquidity position, and other requirements for continuing authorization.

EBA Guidelines (EBA-CP-2014-22 (CP on GL on failing or likely to fail)) provide guidance on the objective elements that should guide competent authorities and resolution authorities in determining that:

- An institution infringes, or is likely to infringe in the near future, requirements for continuing authorization in a way that would justify the withdrawal of its banking license by the competent authority, including but not limited to because it has incurred or is likely to incur losses that will deplete all or a significant amount of its own funds.
- An institution's assets are, or there are objective elements to support the determination that its assets will be, in the near future, less than its liabilities.
- An institution is, or is likely to be in the near future, unable to pay its debts or other liabilities as they fall due.

In our example, the institution has incurred losses that depleted all or a significant amount of its own funds. Since no private solution was available then the Bank was judged failing or likely to fail (FOLTF), resolution was triggered (Table 2).

Table 3 Valuation 2

	Valuation 1-FOLTF	Economic value	Valuation adjustment
Assets (EUR m)			
Cash	100	100	0
Securities	50	40	−10
Loans	600	500	−100
Investments	80	70	−10
Real estate & equipment	100	90	−10
Intangibles	50	45	−5
Other assets	20	18	−2
Total assets	1000	863	−137
Liabilities			
Deposits	700	700	0
Subordinated debt	280	280	0
Pension liabilities	5	8	3
Accrued expenses	10	10	0
Other liabilities	10	12	2
Total liabilities	1005	1010	5
P&L			
Equity	−5	−147	−142
Total liabilities and equity	1000	863	−137

Source: Authors

To determine the full amount of possible losses and the amount of equity potentially necessary to recapitalize the Bank, an independent valuer would be appointed by the Resolution Authority to determine the economic value of the assets and the liabilities. The valuer would request access to the IT systems of the Bank as well as to the numbers to conduct a fair valuation of the balance sheet of the Bank.

The analysis conducted by the independent valuer (Valuation 2) will bring certain valuation adjustments. The adjustments would be justified by the valuer, in our case, by the unrecognized losses of the loan portfolio, market value on property and equipment, disposal value for the investments and other valuation adjustments, based on conservative assumptions for the other assets (Table 3).

The Resolution Authority would need at this stage to choose a resolution tool based on the valuation adjustments provided by the independent valuer. We assume that the Resolution Authority adopts a bail-in strategy. Give the valuation adjustments and the losses incurred by the Bank, some subordinated liabilities and other liabilities that are part of the loss absorption capacity of the Bank would need to be written off to cover the losses not covered by the equity while the rest would be converted into equity (Table 4).

Following the restructuring decision to sell assets or restructure loan portfolio, a new business plan would be set up by the independent valuer to evaluate the new bank. This exercise (called equity valuation) would start with a new opening balance sheet that considers the actions stated above. In particular, the restructuring actions

Table 4 Balance sheet post bail-in

	Economic value	Bail-in
Assets (EUR m)		
Cash	100	100
Securities	40	40
Loans	500	500
Investments	70	70
Real estate & equipment	90	90
Intangibles	45	45
Other assets	18	18
Total assets	863	863
Liabilities		
Deposits	700	700
Subordinated debt	280	
Pension liabilities	8	3
Accrued expenses	10	
Other liabilities	12	
Total liabilities	1010	703
P&L		
Equity	−147	160
Total liabilities and equity	863	863

Source: Authors

Table 5 Equity valuation

Balance sheet after restructuring and bail-in	Opening balance business plan	E FY1	E FY2	E FY3	E FY4	E FY5
Cash	300	315	331	347	365	383
Securities	50	51	52	53	54	55
Loans	185	204	224	246	271	298
Real estate & equipment	50	50	50	50	50	50
Total assets	585	620	657	697	740	786
Liabilities						
Deposits	335	355	376	399	423	448
Total liabilities	335	355	376	399	423	448
P&L						
Equity	50	264	280	298	317	338
Total liabilities and equity	385	620	657	697	740	786

Source: Authors

and the bail-in resolution tool. During this exercise, the economic value would need to be rebased to accounting value allowing the independent valuer to assess the compensation due to the bailed-in creditors. The objective of the equity valuation is to calculate the value that would be redistributed to the creditors affected by the bail-in (Table 5).

This exercise would be compared with the value that would have been distributed in case of insolvency (Valuation 3-non-creditor worse off). This exercise would be informative for the Resolution Authority. It would estimate the amount of the potential compensation risk. The insolvency case (liquidation value) is derived by the valuer assuming a hypothetical scenario to estimate recoveries for creditors and shareholders in a counterfactual (hypothetical) insolvency scenario, i.e. liquidation (Table 6).

Bridge Bank Operationalization

One of the most used resolution plans focused on the transfer of assets and liabilities of the failing institution to a new bridge entity with a new full license to banking activities and investment services. The new Bank would likely be financed in terms of capital and liquidity by a Resolution fund or a Deposit Insurance. The asset separation would likely be operated by the resolution authority through an Asset Management Vehicle (AMV). The AMV is usually owned and controlled by the resolution fund and designed to take over non-performing assets from banks in resolution. The AMV would be capitalized by the resolution fund.

Resolution authority, during this process, would need to carry out a provisional valuation with a view to 1) determining the resolution measures, 2) revealing further losses that could lead to a capital shortfall and, in case subordinated debt would be insufficient to cover losses then 3) recapitalize the bank. In our example below, we see the potential operationalization of the Bridge bank. We assume:

- Overall balance sheet losses quantified in E15.04mn of which E12.25 from loan.
- Subordinated liabilities eligible for write-down were fully written down.
- The remaining subordinated debt of EUR 1.8 million left in this residual entity to be liquidated.
- In return for making up the negative net value of the bridge bank, the resolution fund will retain a senior claim of EUR ten million in the residual entity, as illustrated in the tables below.

The resolution authority proceeds to transfer the NPL to the AMV. In return for those NPL, the bridge bank received debt securities from the AMV. The resolution fund guaranteed these debt securities in the full amount. Finally, a further EUR ten million was injected by the resolution fund into the bridge bank in cash resulting in own funds of EUR 10.4 for the bridge bank, corresponding to a CET1 value of around 9% of RWA (Table 7).

Table 6 Summary valuation pre- and post-resolution

	Opening value	Valuation 1-FOLTF	Economic value	Valuation adjustment	Bail-in	Insolvency (Ins) valuation	Realization in Ins.
Assets (EUR m)							
Cash	100	100	100	0	100	100	100%
Securities	50	50	40	−10	40	30	60%
Loans	650	500	500	−100	500	139	75%
Investments	90	80	70	−10	70	60	75%
Real estate & equipment	110	100	90	−10	90	50	50%
Intangibles	50	50	45	−5	45	0	0%
Other assets	25	20	18	−2	18	10	50%
Total assets	1075	1000	863	−137	863	389	
Liabilities							
Deposits	700	700	700	0	700	335	100%
Subordinated debt	280	280	280	0		44	16%
Pension liabilities	5	5	8	3	3	5	100%
Accrued expenses	11	10	10	0		2.5	25%
Other liabilities	10	10	12	2		2.5	25%
Total liabilities	1006	1005	1010	5	703	389	
P&L							
Equity	69	−5	−147	−142	160	0	
Total liabilities and equity	1075	1000	863	−137	863	389	

Source: Authors

Table 7 Operationalization of a Bridge bank

Assets	E mln	Liabilities	E mln
Bank pre clean up			
Loans	106.2	Deposits	135.3
		Sub.	4.3
Other assets	40.9	Own funds	0.1
Total	147.1	Total	147.1
Residual entity (to be liquidated)			
		Loans from RA	10.1
		Sub debt	1.8
		Own funds	−11.8
Bridge bank (at set up)			
Assets transferred	123.5	Deposits	142.7
NPL	9.2		
Cash RF	10.1	Own funds	0.0
Total	142.7	Total	142.7
Bridge bank (after)			
Assets transferred	123.5	Deposits	142.7
AMV debt securities	9.2		
Cash from resolution fund	10.1		
Other assets	10.4	Own funds	10.4
Total	153.1	Total	153.1

Source: Authors

Purchasing and Assumption Operationalization

A P&A is the most utilized instrument to solve banking crisis under liquidation and resolution regime. It involves a transaction in which good, going concern bank, acquires some or all the assets of a failed bank and assumes the insured deposits. The main characteristics of the P&A transaction involves the acceleration of key growth opportunities for the acquirer, the lowest cost and most efficient option for the Deposit Insurance Fund, branches and offices of the failed bank would open as normal the day after the announcement. Usually, the acquirer would implement a comprehensive due diligence plan to support transaction assumptions and integration plan. This would include the possibility that the Deposit Insurance Fund would guarantee up-front a part of the asset acquired. The transferred price of part of assets and insured deposits of the failed bank are at book value. However, the deal usually is structured in a way to facilitate the acquirer from the liquidity and capital point of view. In an example below, taken from a real case an acquirer Bank (we will call it Bank A) purchased a failed bank (Bank F) with the support intervention of the local Deposit Insurance Fund (DIF). In particular: Bank A acquires substantial majority of assets and assumed certain liabilities of Bank F from the DIF

Table 8 P&A example. Bank failed balance sheet pre- and post-P&A

US$	Pre-P&A	Post-P&A	Comments
Assets			
Cash	0	0	
Securities	29.6	29.6	
Loan	172.9	150.3	Loss share agreements
Intangibles	0	1.1	From Core deposits
Other assets	5	4.8	
Total	207.5	185.8	
Liabilities			
Deposits	92.4	87.4	Eliminate Bank A dep on consolidation
DIFA	28.1	28.1	
Term financing	0	50	DIF provides 5-year financing
Other liabilities	1.4	2.3	
Total	121.9	167.8	

Source: Author elaboration of Auditor Y data

- $173B of loans and $30B of securities
- Approximately $92B of deposits and $28B of Deposit Insurance Advances.
- Bank A did not assume Bank F's corporate debt or preferred stock.

Bank A makes a payment of $10.6B to the DIF while DIF will provide loss share agreements with respect to most acquired loans. In particular:

- Single family residential mortgages: 80% loss coverage for 7 years
- Commercial loans, including CRE: 80% loss coverage for 5 years

DIF provides a new $50B 5-year fixed-rate term financing, the funding would be need to the buyer to fund receivership operations until asset liquidation generates enough working capital (Table 8).

Valuation in Liquidation

Abstract

Liquidation value is the value of the assets that remain if the company goes from a going concern to a gone concern status. In the liquidation process, creditors, lenders, and shareholders based on seniority of their claims will receive the proceeds from the liquidation of the company's assets. Banking assets included in liquidation value usually include loans, financial assets, tangible assets like real estate, equipment, investment. Liquidation procedures differentiates from resolution measures and presents several valuation methodologies peculiarities.

Liquidation Versus Resolution

Normal insolvency or liquidation proceedings are the default procedure when a bank is failing. However, as BRRD art 45 states, "*liquidation under normal insolvency proceedings might jeopardize financial stability, interrupt the provision of critical functions, and affect the protection of depositors. In such a case it is highly likely that there would be a public interest in placing the institution under resolution and applying resolution tools rather than resorting to normal insolvency proceedings*".

The practical difference between resolution and liquidation is that in resolution a resolution strategy will be chose and executed by the Resolution Authority trying to minimize the costs and protect the financial stability. On the other side, as of 14 Article 2(47) of BRRD states, "'normal insolvency proceedings' means collective insolvency proceedings which entail the partial or total divestment of a debtor and the appointment of a liquidator or an administrator normally applicable to institutions under national law and either specific to those institutions".

Another key difference between the two procedures is that under Resolution scheme "*no creditor shall incur greater loss than would have been incurred if the entity had been wound up (...) under normal insolvency proceedings (no creditor worse off principle)*" (Article 74 of the BRRD and Article 15(1) of the SRMR).

A. Santoni, F. Salerno, *How to Value a Bank*, Springer Texts in Business and Economics, https://doi.org/10.1007/978-3-031-43872-1_6

Historical Liquidation Value

One of the most complete databases to analyse liquidation value is the Federal Deposit Insurance Corporation (FDIC) website. This is a US insurance deposit agency with the mandate to maintain stability and public confidence in the nation's financial system.

An analysis provided by Bennet and Hulan[1] in 2014 assessed the average value in liquidation for USA Banks which failed in the period 1986–2007. Their sample was significant and constitutes 1213 of the 1244 banks that failed during the 1986–2007 period.

Their results revealed that:

1. The average discounted loss on the disposition of assets as a percent of total assets was 23.38%.
2. The mean ratio of discounted receivership expenses to assets was 12.02%, of which 3.49% is the average ratio of discounted direct receivership expenses to assets.
3. As a result, the average discounted total resolution cost to asset ratio was 33.18%.

The main conclusion was that an average failed bank during 1986–2007 has a negative market value of equity that is about one-third of the book value of its assets.

Another interesting result highlighted by their research was that the average time of bankruptcy process lasted 5 years, which was roughly twice that of a typical non-financial bankruptcy.

High-Level Liquidation Valuation

In this example, we provide a back of the envelope simulation of what there could be the main hypothesis underlining a liquidation valuation. A liquidator will be appointed following the failure of the Institution and Resolution Authority deciding that the Bank will be liquidated. Among the first tasks, there will be the assessment of the company's financial position. During this period, depositors will not have access to their deposits and no distributions to them will be made prior to the completion of a full assessment of the Bank's assets and liabilities. Below we provide an overview of the main liquidation valuation drivers based on a real-case liquidation report[2] and industry practices.

---

[1] Bennet and Unal (2014).

[2] SRB. Deloitte Valuation of difference in treatment. Banco Popular. https://www.srb.europa.eu/system/files/media/document/valuation_report_non_confidential_version_en.pdf.

A Practical Example

In this paragraph, we will describe the Resolution scheme adopted for a real case in 2017. We will use fictitious names to identify the parties involved. Banco X normal valuation proceeding was opened in June 2017. Auditor Y was hired by the Resolution Authority to provide a valuation on the difference between the Resolution and the Liquidation scenario. Auditor Y conclusion was that, under the most optimist scenario, the total assets realization would have been equals to 82% of the balance sheet of the day before the failing declaration. The total estimated losses would have been at best 18% of assets up to a maximum 27% in the worst-case scenario. In the following paragraph, we provide the key metrics utilized by Auditor Y to perform the analysis, most of them are common industry practices.

The valuator founded the analysis assessing three different timing scenarios. An unrealistic scenario of an 18-month recovery horizon that was the maximum prescribed by the National Insolvency Act. A 3-year scenario, with 50% of the portfolio maturing, and a more conservative scenario of 7 years where 75% of the portfolio maturing. Below is a table that summarizes the key findings of the three scenarios (Table 1).

The main assumptions utilized by the valuator were:

- Financial assets: Fair value methodology has been used as a proxy. The main differences between the book value and the amount estimated in liquidation was mainly due to Level 3 different valuation methodology. In the case of the worst-case scenarios, it has been included 0.1% adjustments of the financial assets to reflect the uncertainty in the market price. This uncertainty is linked to the valuation model uncertainty for the asset classified as Level 3 while for the other asset class level the uncertainty derived from future administrative costs and operational risk associated to the management.
- Joint ventures, associates: A mix of valuation methodologies has been utilized, as for example: market multiples comparable, adjusted book value, market capitalization.
- Real estate and foreclosed assets: Liquidator applied a valuation methodology that considered the latest appraisal value adjusted for some regional macroeconomic indicators such as unemployment, consumer price index (CPI). The value obtained at the expected time of sale was discounted considering maintenance costs (0.3% of the value per annum) and sale cost (approximal 5%).
- Deferred tax assets (DTAs): Only protected tax (tax for which recoverability does not depend on future profits) was considered.
- Intangible assets: Goodwill, computers software, trademarks were all considered having zero value following the license withdrawal of the Bank.
- Loans: One of the main assumptions applied in the valuation was the discount rate applied to the rump. The rump is the remaining part of the portfolio at the end of the considered period (e.g.: after 18 months for the first scenario). The remaining part of the portfolio is assumed to be discounted by a range of 7.9% in the worst case for the 18 months scenario up to 5.1% for the best case of 7 years scenario

Table 1 Banco X liquidation realization scenario

Assets	NBV E mln 6th June 2017	% Realization in different scenarios 18 months scenario Best case	Worst case	3 years scenario Best case	Worst case	7 years scenario Best case	Worst case
% realization							
Equity, fixed income, and derivatives portfolio	21,543	94.7%	94.7%	94.7%	94.7%	94.7%	94.7%
Loans and receivables	83,330	79.8%	76.1%	82.2%	78.8%	85.3%	82.3%
Joint ventures, associates, and subsidiaries	9908	84.6%	75.7%	84.6%	75.7%	84.6%	75.7%
Real estate assets	3728	67.4%	60.4%	76.0%	70.4%	79.0%	74.0%
Intangible assets	1198	0.0%	0.0%	0.0%	0.0%	0.0%	0.0%
Tax assets	5692	41.0%	41.0%	41.0%	41.0%	41.0%	41.0%
Other assets	1045	15.9%	15.9%	15.9%	15.9%	15.9%	15.9%
Total assets	126,444	79.3%	76.0%	81.2%	78.0%	83.3%	80.5%
Total insolvency realization							
Liquidation costs % Total assets		0.78%	0.78%	0.85%	0.85%	0.94%	0.94%
Total realization for shareholders and creditors		78.6%	73.6%	80.3%	77.2%	82.3%	79.5%

Source: Author elaboration of Auditor Y data

the different discount derives from the evolving mix of the different mix of asset class with the evolution of the portfolio.

Loans and receivables contained a significant portfolio of NPL in the case of Banco X (EUR 11.7 bn GBV). Important assumptions have been made on the recovery process and the NPL pricing. In particular:

- Recovery costs
  - Legal costs: an average 7%–10% of the assets
  - Asset disposal costs (agency fee, taxation): an average 5%
  - Asset management costs: depending on the year assumed. On average 10 bps.
- NPL pricing:
  - An average pricing in line with the market pricing at the time. A range between 20% and 29.8% P/GBV (Gross Book value).

For the liquidation costs representing up to 0.94% of total assets, the main components were:

- Remuneration costs (18% of total): it is represented by
  - The liquidator remuneration paid an amount equal to the lower between 4% of the assets and an absolute EUR 1.5 mln.
  - The lawyers (based on the scale of the Bar Association) and the court (up to EUR 300 K).
- Employee-related costs (47% of total): unpaid salaries for the last 30 days plus 20 days compensation for each year worked up to maximum 12 months' salary.
- Termination costs (23% of total): Mainly related to rental, IT termination contracts, and others.
- Operating costs (12% of total): Costs related to the operation costs of the liquidation process as IT costs, branches to maintain operational activities, headquarters.

Overall, we can say that the key drivers for calculating the impact in the liquidation scenario are:

- The haircut to the loan portfolio comes from the discount estimated needed to dispose NPL and the discount assumption on the performing rump sale
- Reduction in value for the assets as equity, intangibles, real estates, and tax assets.
- The costs associated with the liquidation procedure

Reference

Bennet, R., & Unal, H. (2014). *Understanding the components of bank failure resolution costs*. FDIC.

Loan Valuation

Abstract

During a period of recession or following period of economic financial turbulences, one of the main tasks to proper value a Bank is calculating the possible hidden losses arising from non-performing or illiquid assets. The two most important asset categories on Banks' balance sheet that will be hit by severe losses during crisis are loans and financial assets. In this chapter, we will concentrate on the methodology used to value loans while in the next chapter we will focus on financial assets.

Literature Review

There is plenty of academic literature available on the topics of bank valuation and accounting discrepancy and the market implication for this. We analysed the most relevant academic articles on the topics of banks valuation during severe economic downturn period. We found evidence that banks' balance sheets in some cases offer a distorted view of the financial health of the banks. Harry Huizinga and Luc Laeven in *Bank valuation and accounting discretion during a financial crisis*,[1] analysing quarterly accounting data on US bank holding companies for the period 2001–2008 found that the stock market implicitly discounted a 17% on banks' real estate loans.

Robert C. Pozen for the Harvard Business Review,[2] tried to answer to the dilemma if it is fair to blame Fair Value Accounting for the Financial Crisis. A lack of market-to-market valuation for loans and financial assets is usually treated with prudence since it could bring, in period of financial distress, high earnings and capital volatility. On the other side, investors incorrectly assume that most bank assets are valued at market prices, as bond prices were nose-diving. Other investors

[1] https://www.sciencedirect.com/science/article/pii/S0304405X12001341.

[2] https://hbr.org/2009/11/is-it-fair-to-blame-fair-value-accounting-for-the-financial-crisis.

A. Santoni, F. Salerno, *How to Value a Bank*, Springer Texts in Business and Economics, https://doi.org/10.1007/978-3-031-43872-1_7

failed to realize that notwithstanding the sharp markdowns of bonds available for sale would not put banks in violation of regulatory capital requirements. The dilemma is between transparency and correct asset valuation and too much volatility and risk of self-fulfillment prophecy. A suggestion put forward by Pozen is to have a double reporting that could match both sides. On the one side, a reporting earnings per share (EPS) and a CET 1 and on the other side an adjustement for the unrealized losses in the market value of its bond portfolio.

Yi Zheng and Da Wu studied the impact of opacity on bank valuation during the global financial crisis.[3] They demonstrated the relationship between in-transparency and bank value discount during a global recession. The most interesting aspect of their research was the analysis of the channel of transmission through which opacity negatively affects bank valuation. From one side, they found that the relationship between opacity and cost of equity during the financial crisis is positive and statistically significant. On the other side, they found that opacity is negatively associated with bank valuation during the financial crisis.

Loans Valuation: Cash Flow Adjustment Techniques

Differently from financial assets, most loans are not traded as such it is difficult to value them at fair value. An additional difficulty derives from the fact that loans can have very specific conditions mostly depending on borrower's credit quality. Lacking an objective market price, it is mostly important to leverage on a valuation methodology that would take into consideration the different loans specificality. The most used valuation methodology to value a loan is the cash flow adjustment technique that can be divided into two approaches: (a) a bottom-up approach and (b) a top-down approach. The main difference is that the first approach is very granular and analyses individual position while the second approach is most useful for wider portfolios.

Bottom-Up Approach Loan Valuation

The bottom-up approach is mainly used for simple structure loans. The initial estimate envisages the calculation of the unadjusted cash flows assuming all payments are made as per contractual terms. Once the projected unadjusted cash flow has been calculated, then we should calculate the potential adjustments.

For the second phase, the most important considerations would be:

Risk profile of the loan portfolio: concentration risks per product and geography and credit risk collateralized or not are the most important aspects to consider.

Time frame of the portfolio: the maturity horizon of the portfolio and the cash flows expectations.

[3] https://www.sciencedirect.com/science/article/abs/pii/S1057521923000960.

Table 1 Bottom-up approach loan valuation

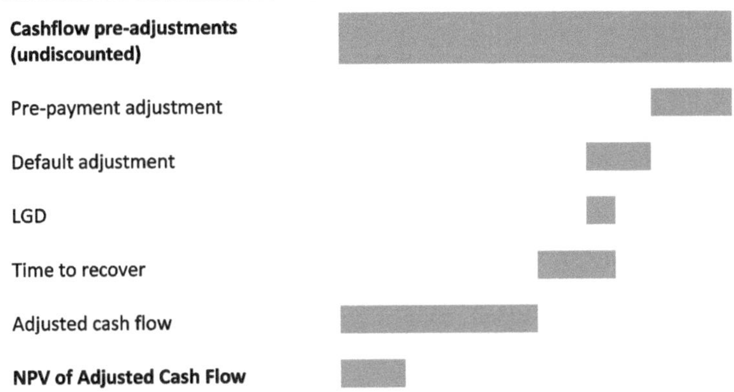

Source: Author elaboration

Current market interest rate and expected future ones.

The next steps would be to segment the loan portfolio between performing, sub-performing, and non-performing. On the Performing part, the second steps would be to highlight 1) the amortizing part, 2) the interest only part, and 3) the loan to value versus specific triggers (e.g.: <80%, 80%<>100%, >100%). The purpose of segmenting the loans is to derive key pricing assumptions for specific segments. After this segmentation process, the loans, groups can be finally valued collectively.

Once segmented then the loans can be finally valued using specific inputs that are part of the portfolio inputs and market inputs. The most important market inputs are the probability of default, the loss given default, the discount rate (dependent from the current interest rate), and the pre-payment rates (Table 1).

The calculation of the real economic value (REV) of the loans is so obtained from the bottom-up estimation of the future cash flows related to the loans. All future cash flows (net of workout costs) should be discounted using a sufficiently prudent credit spread ("a risk premium") over and above the risk-free rate:

$$\text{REV} = \sum_i E(CF_i)/(1 + R_{\text{risk}-\text{free},i} + \text{Credit Spread}_i)$$

The applicable credit spread premia depend on the default probability of the asset and its maturity. Losses on loans are linked to PDs, LGDs, etc., like loan portfolio valuation methodology. Macro trends are crucial to calculate losses on loans. In many cases, it is possible to find a statistical correlation between macro indicators (e.g. unemployment, GDP) and the potential losses.

The Expected Losses can be calculated by the following formula:

$$EL = EAD^{*}PD^{*}LGD$$

EAD is the exposure at default.

PD (probability of default) can be estimated using:

- Historical portfolio performance: This can be calculated relying on the performances of similar portfolios in similar situations in the past. Considering the vast statistical data available to the Bank, this exercise can be very significant and precise.
- Market data: There are several market sources (e.g. Rating Agencies, Central Banks, Central Bureau) that publish data that can be used to get the historical PD of a similar portfolio or forecast performance of portfolios (e.g. RMBS-Residential mortgage-backed security) or markets.

A simplified way to calculate the PD is using the Altman Z-score.[4] The original formula was based on five basic financial ratios to help determine the financial health of a company. It is a probabilistic model to screen for bankruptcy risk of a company. According to studies, the model showed an accuracy of 72% in predicting bankruptcy 2 years before it occurred, and it returned a false positive of 6%.

The original formula was created for publicly traded manufacturing companies:

$$Z - \text{score} = 1.2(A) + 1.4(B) + 3.3(C) + 0.6(D) + 1.0(E)$$

Where:

A = Working Capital (Current Assets – Current Liabilities)/Total Assets (Measures liquidity of firm)
B = Retained Earnings / Total Assets (measures accumulated profits compared to assets)
C = Earnings Before Interest & Taxes (EBIT)/Total Assets (measures how much profit the firm's assets are producing)
D = Market Value of Equity (Mkt. Cap. + Preferred Stock)/Total Liabilities (compares the company's value versus its liabilities)
E = Sales/Total Assets (efficiency ratio – measures how much the company's assets are producing in sales).

Z-score results:

Z-score of <1.81 represents a company in distress
Z-score between 1.81 and 2.99 represents the "caution" zone
Z-score of over 3.0 represents a company with a safe balance sheet

[4] Altman, *Corporate Financial Distress: A Complete Guide to Predicting, Avoiding, and Dealing with Bankruptcy* (1983).

The latest updated version of the Z-score was released in 2012. It was mainly used to evaluate US and non-US public and private companies, manufacturing, and non-manufacturing companies. From a Risk Management point of view, one of the main targets of the Altman Z-score is to evaluate corporate credit risk and probability of default.

There are several ways to estimate LGD (loss given default). A Bank that has been approved by its supervisor to use its own internal model for the calculation of capital ratios will estimate its own LGD. This estimate will be based on a different set of product type and collateral type. Among the main inputs utilized by Banks, we can highlight:

Changes in house price: Declining properties prices is usually the primary driver of losses. The value of collateral would need to be changed accordingly. Several sources are available to track the price movements, as for example Central Banks forecasts, analyst reports covering real estate investment trusts (REITs) and real estate companies.

Enforcement period: When considering the likely recovery of a bad loan the insolvency process is very important. The length of time that it can take between repossessing (e.g. court rulings) can severely impact the present value of the asset.

Costs associated with repossession: During the enforcement period, there are several costs associated with the process. These include maintenance costs, estate agents, legal, and sale costs.

A generally used formula to calculate loss given default is represented by:

$$\text{LGD} = \text{LGL} \times (1\text{–Cure rate})$$

LGL = Loss given default is equal to the exposure at default minus the collateral recovery value (Rval).

The recovery value (Rval) is derived by the following calculation:

$$\text{Rval} = \text{Vcoll} \times (1 - \text{HC}_t - \text{Cexp}_t - \text{Cmnt}_t - \text{Csc}_t)$$

Vcoll = Appraisal value of the collateral
HC_t = Expected repossession cost for the collateral
Cexp_t = Expected additional costs that may come from repossession of the collateral
Cmnt_t = Expected maintenance cost for the collateral from repossession
Csc_t = Expected additional costs that may come from the sale of the collateral

The cure rates are defined as the probability of the default loan returning performing. The parameter to utilize to calculate cure rate should reflect the credit characteristics of each loan and based on specific financial/economic conditions as macroeconomic variables (e.g. GDP, unemployment, property prices), changes in relevant laws (e.g. enforcement laws and bankruptcy code), changes in business conditions.

The recovery rate is very specific for the different assets and country. A recent paper of Bank of Italy[5] analysed the average recovery rate for bad loans in Italy between 2006 and 2015 equivalent to nearly two million positions and closed for a gross value of about €88 billion. During the period, the bad debt recovery rate averaged 43%, oscillating between a low of 34% and a high of 49%. The secured by collateral loan recovery rate was 54% (ranging from 45–61%), mainly coming from the portion sold by Banks to specialized providers. For unsecured positions, the average recovery rate was 35% (ranging from 26% to 41%); in this case too, the increase occurred on bad loans sold to third parties.

Top-Down Discounted Cash Flow Approach

The top-down approach can be used to support a quick valuation of a loan portfolio. The key advantage is that it relies on a single variable (discount rate analysis) that drives the discount cash flow valuation. The limitation is that the valuation is not precise, it cannot be considered a granular valuation as such do not adapt for valuation in resolution purposes for example.

The key criterion for a top-down valuation is based on obtaining the most suitable discount rate for each asset class. For example, in the case of corporate loans the discount rate can be obtained from the yield data on the corporate bonds. For mortgages and consumer loan the current market rates could be a valuable criterion to discount the cash flows. Credit default swaps (CDS) and external ratings could be a valuable market reference rate for stress-testing valuation.

Options for Addressing NPLs

Banks have different options for addressing NPL issues. They range from the internal workout to direct sale. In alternative, risk sharing agreements often with the intervention of the State, as in the case of asset management company, or third party as in the case of Asset Protection scheme or Securitization (Table 2).

The choice depends on several aspects as:

(a) Level reached in the balance sheet versus the total amount of loans. Usually, a limited level of NPL still suggests possible internal workout by the bank. This option would be characterized by different potential restructuring options. They usually would depend on the collection capacity of the Bank, the macro context, and an efficient legislative framework.
A high NPL level usually suggests the potential intervention by a third party or alternative options. For example:

[5] https://www.bancaditalia.it/pubblicazioni/note-stabilita/2017-0007/en_Note_di_stabilita_finanziaria_e_vigilanza_N._7.PDF?language_id=1.

Table 2 NPL strategies option

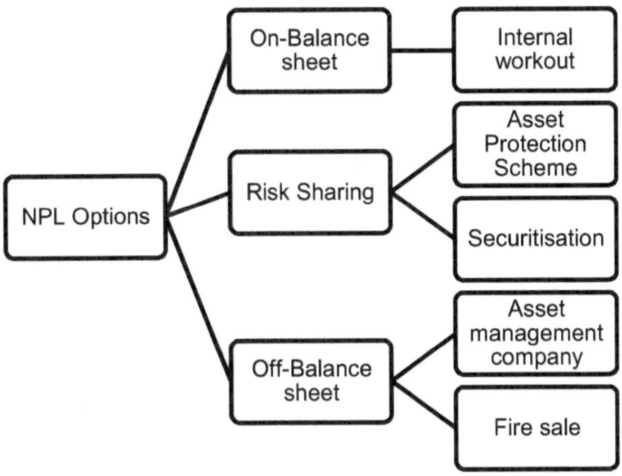

Source: Authors

(b) Presence of impediments for NPLs sale. This could derive from legislative "unfriendly" legal framework to lack of data from the Bank to cherry-picking attitude by the investors that would limit Banks willingness to sell. A State intervention through legislative reforms could open the market for NPL sale. The State in this case could intervene through an Asset Protection Scheme that would represent a short-term bridge to gain time between the need of Banks to sell and the closure of bid-ask spread on NPL price.

(c) Liquid secondary market and availability of third parties. Under these circumstances, the disposal of NPL portfolio to specialized investors via Securitization could represent one of the fastest ways to clean Banks balance sheet. Recently, to speed up the process, the State intervened as guarantor. The key steps to securitize NPL would foresee:
The first step would see the Bank selecting a pool of non performing exposures (NPEs) to sell to an Special Purpose vehicle (SPV). Second SPV would securitize the NPEs and issue three tranches of notes (senior, mezzanine, and junior). While senior notes are usually kept by the banks, junior and mezzanine notes are sold to investors. Proceeds from the notes will be used to pay off banks. This process usually sees the Bank carving out and selling their NPE servicers to investors, which will work out the NPEs as part of a service-level agreement (SLA). As a last step, the State provides guarantees on senior notes and in exchange the Bank will pay the state a certain fee for the guarantee. The guarantee would be activated only under certain conditions depending on the amount of junior or mezzanine notes sold.

(d) In a situation of systemic crisis that foresees a State intervention for a few Banks, usually an asset management company is evaluated. This solution has the benefit of being quicker than other solutions, less costly for the Banks, and could increase the liquidity of the NPL market for the country. Asset management company (AMC) usually prevents fire sales from several Banks in the same country that would further dampen NPL depressed prices.

(e) In a situation of Banks' capital available to sustain a possible fire sale of NPL and the implied losses, Banks could decide to direct sale on the market. This could bring the Bank to accept a sale below accounting value in exchange of improving its risk profile with possible positive. Sale of NPL in certain cases would even result in capital accretion (see below).

NPL Fire Sale: Valuation Considerations

However, in a situation of severe macroeconomic stress, Banks may decide to evaluate a possible fire sale of NPL. Valuation consideration in period of distress and forced disposal are different from normal going concern situation.

In this scenario, the calculation would start from the Collateral Value analysis to evaluate the clean PV of the stock of NPL. The difference versus the current value will be deducted from the current Book Value of the bank (shareholder equity).

The Haircuts to the value of the loans should reflect any discount on the collateral value originated from the new macro conditions and should be deducted from the collateral value. The haircut range is usually based on the decision taken by an independent valuer. To the reduced collateral value should be added possible residual interest payments due by the debtor.

Valuation should also be taken into consideration if senior claims are present. In the case of sale of NPL, a detailed assessment of whether there is other lieu on the collateral and the position of these lieu should be made.

During the process of NPL sale, it should also be considered costs associated with the sale of the collateral. Usually, they are related to attorney fees, bankruptcy and courts authority costs, accountant fees, notification fees, asset maintenance, and other costs related to the procedure.

The present value of the loans following a wind-down or a fire sale would be:

$$\text{PV}[\text{Collateral value} + \text{Residual cash flows} - \text{Haircut} - \text{Senior Claims} \\ - \text{Costs of procedure}]$$

To final transaction value would then be equal to

$$\text{PV} - \text{Transaction costs (buffer)}$$

Transaction costs would be the one-off costs related to the sale process that should reflect a potential discount link to conservative buffer that the buyer would assume to discount any execution risk.

Table 3 NPL value. A different perspective

Assumptions

Secured portfolio	
Gross Book Value	100
Recovery	60%
WAL of recovery	4 years

	Seller point of view	Buyer point of view		
Recovery	60	60		
Internal work-out costs	Not Included	-3	5%	of Net Recovery
External work-out costs	Not Included	-0.9	1.50%	of Net Recovery
Discount rate	5%	12%		IRR target
Recovery time value	-11	-24		
Net book value	49	36		Price Target
% GBV	49%	36%		

Source: Authors

Non-performing Loans Sale: The Difference Between the Seller and the Buyer Approach

One of the interesting aspects related to the NPL sale is the different point of view of Banks (usually the sellers) and the buyers (usually funds specialized in NPL management). The bid/offer gap is usually extremely wide and this is not only the result of a normal buyer/seller interaction process but it is also the effect of two different points of view. In particular, Banks in their accounting valuation of NPL tend to understimate the Legal and Maintanance costs of the assets that are important aspects for the buyer. The most important difference however depends on the expected IRR (internal rate of return) that the buyer would require in order to accept the risk.

We can summarize the difference using this table below (Table 3):

The price differences, summarized by the 13% gap between Net Book value as % GBV are mainly explained by the internal and external work-out costs not usually computed by the Seller (usually the Bank) and the assumptions utilized by the buyer. The internal rate of return expected by the buyer is usually one of the most important assumptions of the model.

The Regulator Approach on Gone Concern Scenario

ECB in 2017 published the Guidance on NPL.[6] The objective was to provide a supervisory tool with the aim of clarifying the supervisory expectations regarding NPL identification and management, valuation methodologies among others. In the document, the Authority gave an overview of the valuation methodology for the NPL.

ECB stressed the expectations regarding how the collateral behind the NPL should be valued in a gone concern scenario (that is where the collateral is exercised because debtor cash flows cannot be used anymore to repay the debt). In this situation, according to the Authority the collateral should be valued based on market value. This is the estimated amount for which an asset or liability should be exchanged on the valuation date between a willing buyer and a willing seller in an arm's length transaction after proper marketing and where the parties had each acted knowledgeably, prudently, and without compulsion.

The market value of the collateral should be adjusted, according to the ECB, considering the appropriate liquidation costs and market price discount to the open market value (OMV). Liquidation costs are defined by ECB as the additional costs that the seller would incur during collateral execution and the sales process. The liquidation costs should incorporate according to ECB:[7]

All applicable legal costs

Selling costs, taxes, and other expenses

Any additional maintenance costs to be incurred by the bank in relation to the repossession and disposal of the collateral

Any cash inflows up to the date of liquidation

In addition to the above liquidation costs, a market price discount, if appropriate, should be applied to the updated valuation as outlined below. The value of the discount would depend on the liquidity of the collateral type. It could range from a value close to zero for highly liquid and non-distressed collateral types, which are not affected by any significant correlation risks. However, Regulator expects that a minimum discount of 10% should be applied if the collateral is sold by auction. A

[6] https://www.bankingsupervision.europa.eu/ecb/pub/pdf/guidance_on_npl.en.pdf.

[7] https://www.bankingsupervision.europa.eu/ecb/pub/pdf/guidance_on_npl.en.pdf.

Table 4 NPL valuation methodology ECB (Guidance on NPL)

Example	*Assumption*	T = 0	T = 1	T = 2	T = 3	T = 4	T = 5
Time to liquidation/ sale	5 years						
Gross value of loan		E300					
Market value (MV) of collateral							E200
Discount to MV	10%						−E20
Selling costs (including taxes and other expenses)	10%						−E18
Maintenance costs	5%		−E10	−E10	−E10	−E10	−E10
Expected future cash flows (i) (effective interest rate)	5%		−E10	−E10	−E10	−E10	E152
Present value of collateral		E84					
Amount of impairment		E216					

Source: Authors, ECB (Guidance on NPL)

practical example, as illustrated below, has been provided by the Regulator to clarify the concept (Table 4).

NPL Sale: Capital Tied Up Sensitivity

A decision to dispose NPL for a Bank usually can be seen as an analysis of the possible alternative usage of the capital tied up as called opportunity cost.

A Bank is required to put capital aside for the NPL on the balance sheet. The level of capital is calculated by the required regulatory capital requirement (denoted as the ratio *reg. CAR%*) and the risk-weighted assets assigned to NPL.

The required capital is usually set at 10% of RWA.

NPLs are applied with a risk weight usually higher than performing loan. The rule is to set a 150% risk weight when specific provisions are less than 20% of the outstanding amount of the loan, and a 100% risk weight when specific provisions are above 20% of the outstanding amount of the loan.[8]

The calculation of possible capital relief coming from the sale of NPL would depend on the tied-up capital minus the haircut (if any). As such the first stage in the calculation would envisage computing the capital tied up by NPLs.

The second step would see the potential complication coming from two alternative scenarios with (i) no haircut; (ii) a haircut derived from the discount applied by the seller.

[8]Basel Committee on Banking Supervision (2019).

Table 5 From NPL capital relief to new assets

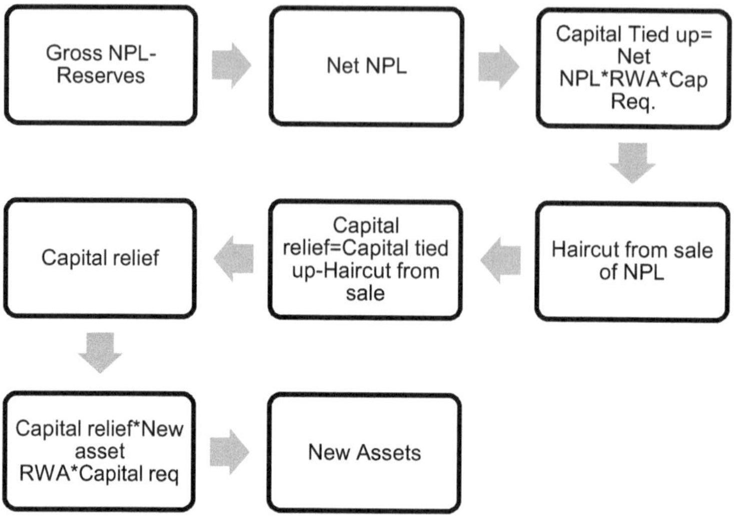

Source: Authors

- **Scenario1**: If there is *no haircut* (that is, no capital loss), then all capital is released and can unlock new lending.

$$\text{Capital relief} = \text{Tied} - \text{up capital}$$

- **Scenario 2**: With a *haircut scenario*, the capital released is computed as:

$$\text{Capital relief} = \text{Tied} - \text{up capital–Haircut required}^{*}\ \text{Gross NPL sold}$$

For a given NPL reduction, the sensitivity analysis would see two main inputs. From one side, the capital released by NPL disposal, which results from the combination of the lower capital charge and the possible haircut; and from the other the alternative use of the capital as new loans or additional financial assets that can be generated/bought from this capital relief.

As a result, the Bank choice will be driven by the result of this analysis with benefits in terms of capital generation from the new assets exceeding the losses generated by the sale of NPL (Table 5).

Reference

Basel Committee on Banking Supervision. (2019, December). *Calculation of RWA for credit risk. Standardized approach: individual exposures*. Basel Committee on Banking Supervision.

Financial Assets Valuation

Abstract

Accounting valuation is the process of valuing a company's assets and liabilities for financial reporting purposes. Several accounting-valuation methods may be used while preparing financial statements to value financial instruments (and non-financial). Accounting International Financial Reporting Standards (IFRS) require market transactions being recorded at fair value. To obtain a prudent value of the assets, regulation requires the fair value to account for additional valuation adjustments (AVA). In this chapter, we analyse prudent valuation methodologies for financial assets.

Prudent Valuation

The CRR (Art. 34 and 105) requires a prudent valuation to all fair-valued positions regardless of whether they are held in the trading book or banking book.

$$\mathrm{PV} = \mathrm{FVA} + \mathrm{AVA}$$

Valuation adjustments can be of two types:

- Fair value adjustments (FVA): Adjustments introduced to reach the fair value of a certain position.
- Additional valuation adjustments (AVA): Adjustments required on top of the accounting fair value to reach a prudent value associated with a given confidence level.

In general:

A. Santoni, F. Salerno, *How to Value a Bank*, Springer Texts in Business and Economics, https://doi.org/10.1007/978-3-031-43872-1_8

Table 1 Fair value and prudent value

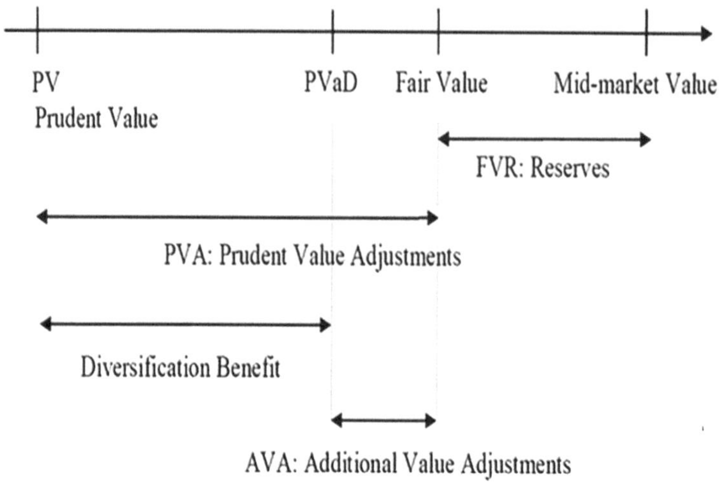

Source: Authors

$$\text{Prudent value (assets)} \leq \text{Accounting value (assets)}$$

$$\text{Prudent value (liabilities)} \geq \text{Accounting value (liabilities)}$$

To account for this risk, regulations require AVA to be deducted from the Common Equity Tier One (CET1) Capital (Table 1).

Fair Value

The starting point for the prudential valuation analysis is to start from the fair value. Accounting International Financial Reporting Standards (IFRS) require market transactions being recorded at fair value.

According to the IFRS:

- "The IFRS defines fair value as the price that would be received to sell an asset or paid to transfer a liability in an orderly transaction between market participants at the measurement date" (IFRS 13.9).
- "The best evidence of the fair value of a financial instrument at initial recognition is normally the transaction price (i.e. the fair value of the consideration given or received, see also IFRS 13)" (IFRS 9.B 5.1.2.A).

Fair value is the price that would be received or paid (. . .)

- At the measurement date under current market conditions (exit price) regardless of whether that price is directly observable or estimated
- Valuation techniques used shall maximize the use of relevant observable inputs and minimize the use of unobservable inputs (fair value hierarchy)

Fair valuation adjustments (FVA) need to be considered on top of native valuation to achieve fair value.

$$\text{Fair Value} = \text{EV} + \text{IPV} + \text{FVA}$$

- Model value (EV): Model or expected value.
- Independent price verification (IPV): Uncertainty about mid-price.
- Fair value adjustments (FVA): Institutions apply model and parameter uncertainty around mid-price as well as bid/ask (b/a) spreads to achieve exit price.

The uncertainty around these two components is addressed under Prudent Valuation by the Market Price Uncertainty and Close-Out-Costs adjustments. Counterparty valuation adjustments (CVA and DVA) are part of these adjustments.

Fair Value Calculation Techniques

When determining fair value, an entity shall use valuation techniques that are appropriate in the circumstances, for which sufficient data are available to measure fair value. Companies should maximize the use of observable inputs and minimize the use of unobservable ones.

IFRS 13 describes three techniques: market approach, cost approach, and income approach that we summarized below in the Table 2.

Fair Value Hierarchy

IFRS 13 establishes a Fair Value Hierarchy, which categorizes the inputs used in valuation techniques into three levels:

Table 2 Valuation techniques fair value

Market approach	Cost approach	Income approach
Market multiples	Current replacement costs plus obsolescence	Present value
Matrix pricing (compare with benchmarking securities)		Option pricing models
		Multi period excess earnings method

Source: Authors based on IFRS box

- Level 1 inputs are quoted (unadjusted) prices in active markets for identical assets or liabilities that the entity can access at the measurement date. (IFRS 13.76–79)
- Level 2 inputs are inputs other than quoted prices included within Level 1 that are observable for the asset or liability, either directly or indirectly. (IFRS 13.81–85)
- Level 2 inputs must be observable for substantially the full term of the asset or liability if these have a specified (contractual) term and include:
  - Quoted prices for similar instruments in active markets
  - Quoted prices for identical or similar instruments in markets that are not active
  - Other than quoted prices that are observable for example: interest rates and yield curves observable at commonly quoted intervals; implied volatilities; and credit spreads
  - Market-corroborated inputs: Inputs that are derived principally from or corroborated by observable market data by correlation or other means
- Level 2 inputs are inputs other than quoted prices included within Level 1 that are observable for the asset or liability, either directly or indirectly
  - Significant adjustments to Level 2 input may result in lower hierarchy
  - Level 2 assets are very heterogeneous
  - Neither accounting nor prudential rules require banks to report sub-categories within the Level 2 class
- Level 3 inputs are unobservable inputs for the asset or liability
  - Unobservable inputs shall be used to the extent that relevant observable inputs are not available (e.g. little, if any, market activity)
  - However, the fair value (FV) measurement objective remains: i.e. an exit price
  - Level 3 inputs are unobservable inputs for the asset or liability. (IFRS 13.86–90)

Day One Profit—DOP

Day one profit or loss or "D1 P&L" is the difference at the initial recognition of the instrument between the transaction price and the fair value. It can be a profit or a loss.

- Immediate recognition of D1 P&L is only permitted under IFRS (IFRS9. B5.1.2A) if the fair value is evidenced by:
- A quoted price in an active market for an identical financial asset or liability (Level 1)
- Based on a valuation technique whose variables include only data from observable markets. The valuation technique must also have been tested against market transactions (Level 2)
- Otherwise, i.e., in case unobservable data are involved, the D1 P&L must be deferred and recognized only when the parameters become observable, or when it is possible to justify the valuation of the instrument with similar transactions in an active market

- Note that although not mentioned in IFRS 13, it is prudent not to defer D1 losses but only D1 profits.

To avoid any double counting of the FVA, Net DOP should be calculated according to the following definitions:

(a) Gross DOP (1) = Mark to model – Transaction price
Mark to Model is the native valuation of the portfolio. This value is risk-managed by the front office and feeds automatically the P&L.
(b) FVA (2) encompasses the different FVAs (bid-ask, market price uncertainties, model risk, CVA/DVA…)
(c) P&L adj. (3) performed by IPV.
Note that DOP is calculated at trade level whereas FVA can be calculated at desk or portfolio level.

Additional Valuation Adjustments (AVA)

To identify the prudent valuation, an Institution should calculate the additional valuation adjustments necessary to adjust the fair value.

Regulation identifies two possible approaches to calculate the AVA. The simplified approach obtains the AVA by multiplying the aggregate absolute value of fair value positions by 0.1%. This methodology is mainly used when fair-valued assets plus fair value liabilities are less than EUR 15 bn.

Under Core Approach, [CDR PV, Art 9–17] defines the following nine risk categories that must be considered by Banks when calculating AVAs (Table 3):

The core approach is obligatory for institutions that are above the threshold of the simplified approach (EUR 15 bn B/S).

- Key features of the core approach:
  - AVA required for each risk sub-categories introduced in the regulatory texts
  - Every AVA calculated as the excess of valuation adjustment required to achieve the identified prudent value over any fair value adjustments addressing the same source of valuation uncertainty as the AVA
  - The prudent value of a position is linked to a range of plausible values and a specified target level of certainty (90%)
  - AVAs shall always be positive

[CDR PV, Art 7.2] specifies that where calculation of AVAs according to Art. 9–17 is not possible for certain positions, institutions should use the "fall-back" approach:

$$\text{AVA Fall Back} = 100\%\text{NUP} + 10\%\text{NLDeriv} + 25\%(\text{FV}–\text{NUP})\ \text{Non} - \text{Deriv.}$$

Where:

Table 3 Risk categories for the calculation of AVAs

Category	Objective
Market price uncertainty (MPU)	Uncertainty around mid-price
Close-out-cost (COC)	Uncertainty around exit price
Model risk (MR)	Uncertainty valuation model
Concentrated positions (CP)	Uncertainty on the exit value of a position that cannot be sold in less than 10 days
Unearned credit spread (UCS)	Uncertainty around CVA measures
Investing & funding cost (IFC)	Uncertainty around funding costs
Early termination (ET)	Uncertainty from losses coming from non-contractual early termination costs
Future administrative costs (FAC)	Uncertainty around future administrative and hedging costs
Operational risks (OR)	Uncertainty resulting from operational risk related to valuation processes

Source: Authors

- NUP = Net Unrealized Profit: Change, where positive, in FV since trade inception determined on a first-in-first-out basis.
- NLDeriv = Sum of notional of derivatives in absolute value.
- FV = Sum of FV for non-derivatives.

In summary, total AVA Core Approach can be calculated as:

$$\begin{aligned}\text{Total AVA} &= \text{AVA(9 CAT)} + \text{AVA Fall Back}\\ &= \{[0.5^{*}(\text{MPU} + \text{COC} + \text{MR}) + \text{UCS} + \text{IFC} + \text{ET} + \text{TAC} + \text{OR}]\\ &\quad +(100\%\text{NUP} + 10\%\text{NLDeriv} + 25\%(\text{FV–NUP})\ \text{Non Deriv.}\ \}\end{aligned}$$

Prudent Valuation Requirements

The CRR defines several requirements to ensure a prudent valuation. It also mandates the European Banking Authority (EBA) to develop Regulatory Technical Standards (RTS) specifying the conditions for implementing these requirements ([CDR PV]):

- Senior management shall be aware of the elements subject to mark to model and shall understand the materiality of the uncertainty introduced
- The model shall be assessed and challenged by qualified independent parties periodically, including validation of the mathematics, assumptions, and software implementation

- Formal model change procedures shall be in place
- Risk management shall be aware of the weaknesses of the models
- The model shall be developed or approved independently of the trading desk
  - Institutions shall revalue trading book positions at least daily (CRR Art 105(3))
  - Institutions are required to perform independent price verification (IPV) in addition to daily marking. IPV shall be performed at least monthly (CRR Art. 105(8))
  - IPV is the process of comparing front office or portfolio manager marks and valuations to a set of independently verifiable external prices to establish an accurate price to revalue positions
  - Pricing variances can arise due to genuine differences of opinions regarding the fair value of a position, and they can also arise due to deliberate mismarking by a trader
  - IPV on marked-to-model transactions evaluates the appropriateness of marking of the mid-accounting valuation
  - For most marked-to-model instruments, IPV is performed at the parameter level, rather than product/instrument level because the price of the derivatives is not available, only the values of the underlying assets are. IPV is in fact independent verification of implied market data. These data are mostly the Level 2 parameters
  - IPV on marked-to-model transactions often relies on consensus levels. The consensus levels are average mid-prices among the different contributors who pass the contribution test. The consensus levels do not represent exit prices as requested for the fair value estimation under IFRS
  - Valuation adjustment shall be considered if pricing sources are subjective or not available. A list of adjustments to be considered is provided in the CRR (Art. 105(10) to (13)).

Solvent Wind-Down

Abstract

In this chapter, we focus on some of the valuation and operational aspects of winding down a bank's trading book portfolio and the potential hidden exit costs. It provides a deep dive on valuation principles and exit strategies currently considered by industry practitioners when designing a solvent wind-down plan. It also provides the reader with an overview of key underpinning valuation or pricing concepts (i.e. "Fair value", "realisable value", and "Solvent Wind Down value"). Ultimately it shows that the cost to wind down a trading portfolio beyond the usual accounting carrying value might be mostly driven by wind-down operating costs (incl. Liquidity and funding costs) and two main pricing components, namely the Capital Valuation Adjustment and Margining Valuation Adjustment.

Wind-Down of Bank's Trading Book

The starting point to consider when designing a wind-down strategy is the portfolio segmentation together with the scenario analysis. It consists in segmenting the trading book and the potential hidden exit costs into different sub-portfolio depending on the trade complexity and counterparty. Following the full segmentation, the bank should clarify the specific target for each trading portfolio, and this encompasses several approaches that could be used in isolation or in combination.

This paper is an extract of the ECB Occasional Paper Wind down bank's trading book Operational aspects and the hidden costs. Alessandro Santoni, Ghislain Rossignol, Richard Akhouen.

It should not be reported as representing the views of the European Central Bank (ECB). The views expressed are those of the authors and do not necessarily reflect those of the ECB.

A. Santoni, F. Salerno, *How to Value a Bank*, Springer Texts in Business and Economics, https://doi.org/10.1007/978-3-031-43872-1_9

Additional steps would then be to aim at exit cash trades or derivatives maturing within the exit period. Particular attention should be placed on assets that are required for hedging purposes. Packaging and reselling to other dealers or Bank would be a key exit strategy that would be facilitated from liquidity perspective of the trades.

However, SWD execution clashes with several obstacles, among which the exit price determination is important. By design a solvent wind-down analysis is a quite extraordinary exercise that request unusual calculations and information compared to the regular accounting and risk reporting. Key elements of this intelligence are the pricing of market practices and a deep understanding on assumptions or overlays to be applied on the balance sheet carrying value to estimate the Solvent Wind-Down Value of a portfolio.

A sound SWD analysis should rely on realistic market evidence (when possible) and economically plausible and conservative assumptions. With that aim, the estimation of the overall exit cost should achieve a higher level of certainty compared to the one normally assumed under the accounting framework that for example does not integrate the risk premium for concentrated positions.

In this chapter, we investigated if valuation of the positions, under solvent wind-down conditions, could be considered as a Realizable Value. Despite the prudential framework requiring banks to measure in their capital the cost of exiting exposure under 90% of confidence, this prudent concept might not be sufficient to achieve realizable value, and consequently SWD value because of its procyclical effect, the lack of relevant information available on the market to measure the prudent exit price of L3 exposure with material non-observable input, and the non-financial risk not sufficiently factored in the prudent exit price (for example, legal risk, reputational risk). All in all, those facts do not question compliance of financial institution with their applicable accounting or regulatory framework, or means that bank's reported financial information is distorted, but it implies that under a SWD scenario different assumptions or overlays might have to be considered.

Key differentiating assumptions between SWD value and realizable value relate to expected operational costs, costs of future financial resources, concentration risk costs, and administrative costs. Additional premium to capture these costs should be considered by financial institutions when conducting a SWD analysis. Several other costs could influence SWD of derivatives and trading book portfolios such as the rump costs or the Day one Profit (DOP) deferral. Rump portfolio consists of the residual "sticky" positions, i.e. positions which will remain on the balance sheet of the institution after performing the wind-down exercise. Day one Profit deferral or day one losses deferral arise at the trade recognition when the transaction price and the fair value price using non-observable input are different. The exit of related transactions will generate capital effect, either capital benefit or capital drag because of the early exiting of the positions due to SWD exercise.

Ceteris paribus, the net capital impact of a SWD should be equal to the difference between the net expected cash flows impact resulting from a SWD and the adjustments already accounted for in the regulatory capital (i.e. excess cash flows over fair value, balance sheet correction—i.e. DOP/DOL deferral—AVA, net liquidity, and funding adjustments).

Derivative trading activities generate number of additional adjustments, known as X-Value Adjustments (XVAs). The global financial crisis saw a structural shift in the operation of the global banking sector. A significant change has been the introduction of new regulation to ensure that the banks are adequately capitalized. New regulation requires derivative instruments that are not cleared through central clearinghouses to be collateralized as required for exchange-traded instruments such as futures contracts. The costs of margining, and associated liquidity volatility, represent additional cost in trading derivatives. As a result, various derivative valuation adjustments are required, essentially to reflect the additional "costs" in holding derivative contracts. Overall, the industry has reached a good degree of maturity as regard the usage of the XVA metrics in pricing or accounting. That said the XVA remains an evolving landscape where significant degree of diversity can still be observed across the different metrics. From an accounting and regulatory perspective, provisions are mostly set aside for credit or debit valuation adjustment (CVA and DVA), Funding Valuation Adjustment (FuVA), and to a lesser extent for the costs of funding collateral also known as Collateral Valuation Adjustment (COLVA), while the margin valuation adjustment (MVA) and the capital costs of a derivative (i.e. the Capital Valuation Adjustment, KVA) are widely not yet considered for fair value or prudent value measurement purposes. Hence, when conducting a SWD analysis, Banks Recovery and Resolution plans should properly factor in such additional costs.

Orderly Versus Disorderly Wind-Down

A Bank being able to effectively implement a wind-down strategy, following a period of financial stress or experiencing the cancellation of its license, would enable to cease its regulated activities with minimal adverse financial impact and limited effect on its banking franchise customers, counterparties, or the wider financial markets. Wind-down of a trading portfolio would be characterized by an orderly execution when it is part of recovery option for a Bank under stress or part of a resolution strategy. Alternatively, or a disorderly wind-down when it is a result of a fire sale.

A disorder wind-down could be very costly for the Bank triggering significant value disruption of the trading book and potentially impacting banks' soundness and resolvability. In addition, it could create significant risks to financial systems particularly if this involves a G-SIBs with large portfolio and high interconnection. A key distinction between an orderly and a disorderly wind-down is the preparation phase. Planning a wind-down consists in creating the conditions to properly manage the exit of a Bank from a regulated business when the firm is no longer viable, or it has deliberately decided to exit from a regulated business.

Among various considerations, the most important aspects of a wind-down plan are the scenarios that create the condition for the company to exit the business, a plan to orderly execute the exit, an assessment of financial and non-financial resources needed to face the scenario described, and a detailed analysis of the potential risks that could represent an obstacle to a proper wind-down.

Ongoing Policy Discussion Overview

The topic related to orderly wind-down of trading portfolio recently tracked particular attention. The intense policy discussion on the topic originates from the negatives results of wind-down experiences of the massive Lehman Brothers trading portfolio. According to Michael J. Fleming and A. Sarkar,[1] Lehman creditors' recovery rate of the *over-the-counter (OTC)* derivative transactions was below historical averages of comparable failed firms. The settlement of OTC derivatives was a long and complex process with some of the losses borne by Lehman investors could have been avoided in a more orderly process. One of the key weaknesses of the Lehman bankruptcy process was the poor preparation and planning[2] of the wind-down procedure (preparation was initiated only 1 week before the bankruptcy).

Following the failure of Lehman wind-down experience, since 2015, Prudential Regulation Authority (PRA) asked overseas banks to conduct Solvent Wind-Down (SWD) exercises with the aim to support resolution authorities in understanding the cost of winding down trading books. Notwithstanding the preparation and the learning by doing process, the SWD exercise conducted by PRA from 2015 to 2021 highlighted several deficiencies from the UK Banks regarding their capacity to carry out an orderly wind-down of their trading activities in recovery and post-resolution restructuring. In October 2021, PRA published a consultation paper proposing that firms engaging in derivatives and trading activities *should be able to carry out a full or partial orderly wind-down of their trading activities in recovery and post-resolution restructuring*. The PRA proposals deriving from the approval of the consultation paper would result among other things in a new Supervisory Statement on trading activity wind-down, and amendments to Recovery Planning.

Trying to build up upon the US and the UK experience, the Single Resolution Board (SRB) made SWD of trading book its 2021–2023 priority topics to support IRTs in the drafting of resolution plans.[3] At global level, FSB in August 2019 published a discussion paper focused on solvent wind-down of the derivatives and trading book portfolio of G-SIBs. The FSB[4] Discussion Paper for Public Consultation highlighted consideration on Bank's capacity to execute a wind-down strategy of a derivative and trading book, on the resources (Capital and Liquidity) necessary to properly implement a wind-down, and the potential impact on the viability of the rest of the group.

---

[1] https://www.newyorkfed.org/medialibrary/media/research/epr/2014/1412flem.pdf.

[2] Valukas (2010).

[3] https://storage.polit-x.de/media/EU_SRB_EN/pdf/2020-11/ca4a9426d1f6be519fbc76e988b91849.pdf.

[4] https://www.fsb.org/wp-content/uploads/P030619-1.pdf.

Following the public consultation, FSB has decided not to develop further guidance. Member highlighted[5] that *regulatory divergence and differing business models could inadvertently lead FSB to develop fragmented requirements that are inconsistent across jurisdictions and therefore difficult for G-SIBs to implement.* Global financial markets association (GFMA)/institute of international finance (IIF)/ international swaps and derivatives association (ISDA) response[6] underlines substantial different SWD requirements among jurisdictions that makes a common guidance very challenging. Just as an example while the UK SWD primary objective is capital preservation, the US SWD objective is liquidity generation. FSB accepted these reservations while reinforcing his commitment to promote solvent wind-down planning as part of overall resolution planning.

Following the work initiated by FSB, SRB published[7] a guidance on how to demonstrate resolvability in relation to sizeable and complex trading books. The guidance sets out the scope and minimum expectations for SWD planning and potential execution. Banks are expected to properly plan an exit strategy and the potential financial implications for the different segments of its trading activities; this would need to be associated to a continuous monitoring and capacity to update the plan in a timely manner relying on internal capabilities to execute the wind-down. An essential element of the SWD would be a playbook focusing on governance as escalation points, parties involved in the decision-making, as human resources departments and communication with relevant stakeholders.

Operationalization of Wind-Down

In this chapter, we develop some operational aspects of a SWD, starting with some general considerations on solvent wind-down strategies and operational preconditions for a successful SWD. Among the most important operational factors to consider when designing a wind-down strategies are the scenario analysis consisting in segmenting the trading book into different sub-portfolio depending on variables such as products' characteristics or trade's counterparty. Following the full segmentation, the Bank should clarify the specific target for each trading portfolio, and this encompasses several approaches that could be used in isolation or in combination. Next step would then be aiming to start exiting maturing cash trades or derivatives maturing within the exit period. Particular attention would need assets required for hedging purposes. Packaging and reselling to other dealers or Bank would be a key exit strategy that would be facilitated from liquidity and transparency of the underlying assets.

[5] https://www.fsb.org/wp-content/uploads/P201219-1.pdf.

[6] https://www.iif.com/Portals/0/Files/content/Regulatory/08012019_joint_gfma-iif-isda_response_solvent_wind-down.pdf.

[7] https://www.srb.europa.eu/system/files/media/document/2021-12-01_Solvent-wind-down-guidance-for-banks.pdf.

Scenario Analysis

The starting point of a wind-down strategy, from an operational point of view, deals with identifying within a range of plausible scenario the SWD scenario that minimizes the impact on the franchise, counterparties, and the wider financial markets considering multiple internal/external dependencies (market conditions, expected return of buyers, legal constraints...). Because of its multidimensional nature, a SWD exercise remains data intensive and consequently very demanding in terms of IT infrastructure and modelling capabilities. It also requires strong involvement of Front Office, and support from other bank's functions (mainly Risk and Finance).

Scenario analysis might take several forms but practically consists in segmenting the trading book into different sub-portfolio depending on variables such as products' characteristics or trade's counterparty. This would lead to a segmented disposal strategy for each sub-portfolio highlighting the financial impact from a capital and liquidity point of view. The analysis should clarify all the assumptions and include rump identification and corresponding rump maintenance costs.

Wind-Down Strategies

The most important factors to consider when designing a wind-down strategies are the portfolio and product characteristics, including complexity (possibly apprehended through the fair value classification: i.e. Level 1, Level 2, or Level 3 instruments) and maturity, the existence of collateral agreements with the counterparts, the liquidity and depth of underlying primary (interbank market) or alternative financial market (unregulated financial counterparties such as hedge funds). In designing the wind-down strategy, a key element is to define the time horizon to execute the wind-down and within this timeframe designed a specific approach.

Once defined the strategies and the time frame to execute them, a detailed product and business segmentation would be required to price and wind-down the portfolio. This is represented by a detailed segmentation of the portfolio highlighting the governing jurisdiction, product type, OTC, booking entity, structural balance sheet hedge, differentiating derivatives books highlighting the one that is part of the banking book, Liquidity, fair value hierarchy level. The presence of Level 1 High-Quality Liquid Assets (HQLAs) would facilitate the task with no need of significant segmentation since expected to be easily liquidated.

Following the full segmentation, the Bank should clarify the specific target for each trading portfolio. Banks could in fact decide to sell some portfolio for explicit imminent liquidity needs or instead it decides to simply wait for the maturity of the trade. On the other side, the target could be to minimize costs because its counterparties are expected to exercise early termination rights. From capital point of view, a quick wind-down could release capital in the medium long term via reducing risk and RWA. In the short term, the capital relief benefit will be more than

Table 1 Different wind-down strategies approaches

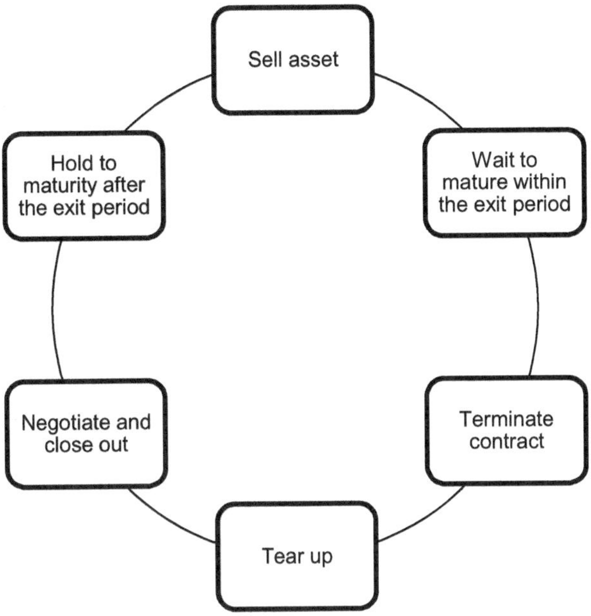

Source: Authors

offset by the costs, as selling quickly into a stressed market would likely imply high discounts. Finally, the Bank could simply classify some assets as illiquid with limited wind-down possibilities waiting for better market conditions.

The wind-down strategy should encompass several approaches that could be used in isolation or in combination (Table 1):

- Roll-off of position with finite contractual timespans (i.e. positions that are not replaced or rolled over as their maturity nears).
- Unwind of positions prior maturity date (i.e. close-out or termination of a financial instrument prior maturity though payment of a termination fee by the cancelling counterparty).
- Disinvestment of stand-alone financial instruments or portfolios through sale (primarily for cash instruments) or novation (through tripartite agreement by which both rights and obligations arising from derivative instruments are transferred to a third party).
- Compression or bilateral tear-up (post-trade balance sheet reduction technique whereby two or more counterparties replace multiple offsetting derivatives contracts with fewer contracts of the same net risk exposure but reduced portfolio notional value).

Next step would then be aiming to start exiting maturing cash trades or derivatives maturing within the exit period. Particular attention would need assets required for hedging purposes. Strategy could envisage also re-hedging activities to manage market risk during the wind-down activity. During this period, any position where the contract could be terminated or cancelled could be closed. Termination rights could be invoked. Exit costs of winding down the derivative and trading portfolios need to be factored including compensation paid to dealers or counterparties to incentivize them to close-out open positions.

Packaging and reselling to other dealers or Bank would be a key exit strategy that would be facilitated from liquidity and transparency of the underlying assets. Identification of the right buyers, essential tool to execute a rapid wind-down, would be based on selecting buyers that could be natural owners with a low weighted average cost of capital. Alternatively, by looking at the potential buyers' funding situation and capacity.

Operational Aspects to be Mitigated During the Wind-Down Strategy

An important element that would facilitate the derivatives portfolio wind-down in a resolution scenario is the concentration of the positions in few subsidiaries, all of which with material operating activities. This would facilitate liquidity support from the parent company. On the other side, entities could be sub-investment grade and unlikely re-establish investment grade status during the wind-down process. This will limit access to the bilateral OTC market forcing the entities to use exchange-traded and cleared instruments for hedging purposes.

An aspect that would need to be mitigated is the increased interconnectivity resulting from intercompany derivative transactions. Affiliates engaging in external activities should be subject to a certain cross-default and early termination rights. Derivatives contracts should envisage mutually agreed "tear-ups" and risk-free netting. A centralized clearing arrangement between material operating entities should be set to eliminate "interconnectedness risk". As common to other resolution process, a good cooperation between home/host authority is essential to facilitate the Bank's activity in the wind-down process.

Valuation Assumptions in the Trading Book Wind-Down

SWD execution clashes with several obstacles among which price determination, fair value, and costs to wind down. Estimating the cost to wind down a portfolio requires developing market intelligence on possible exit strategies. Key elements of this intelligence are pricing market practices and a deep understanding of assumptions or overlays to be applied on the balance sheet carrying value to estimate the Solvent Wind-Down Value of a portfolio. In this chapter, we propose an overall digest of valuation challenges, valuation principles, and exit strategies currently considered by industry practitioners when designing a solvent wind-down plan.

By design a solvent wind-down analysis is a quite extraordinary exercise that requests unusual calculations and information compared to the regular accounting and risk reporting. Quantifying the overall cost to exit a trading book implies:

- To develop a market intelligence on how market participants would price balance sheet-intensive transactions or portfolios (i.e. understand the factors or mechanisms at play behind the costs associated with winding down the in-scope portfolio)
- To understand the institution's ability to clear and compress derivative exposures, the market absorbing capacity (liquidity and breadth), and to capture the associated costs
- To recognize the issue of client consent to novate transactions
- To analyse the risk of additional margin requirement due to market instability and possible cascade effects on credit ratings, external factors as idiosyncratic reactions by counterparty or local authorities
- To identify the residual portfolio (rump) that is expected to remain on the balance sheet after the SWD period.

Fair Value: A Partial View of a Solvent Wind-Down (SWD) Value

A good SWD analysis should rest on realistic market evidence (when possible), economically plausible and conservative assumptions. In that aim, the estimation of the overall exit cost should achieve a higher level of certainty than the one normally assumed under the accounting framework. As mentioned in previous chapter, according to IFRS 13, the fair value of an instrument is defined as the exit price from the perspective of a market participant. This exit price should in principle include an adjustment for risk that market participants would include when pricing the asset or liability.

Is Fair Value reflective of Realizable Value under solvent wind-down conditions? This question of the accounting fair value as being a substantial and accurate representation of Realizable Value came under increased scrutiny by regulators who consequently sought to address measurement uncertainties by publishing prudent valuation guidance[8] and laying out prudent valuation requirements for regulatory purposes[9] (though not foreseen in all jurisdictions).

Despite the resulting favourable alignment across the accounting and prudential framework, fair value and prudent value might still deviate from realizable value, and consequently SWD value. All in all, those facts do not question compliance of financial institution with their applicable accounting or regulatory framework, or means that bank's reported financial information is distorted, but it implies that

[8] Basel Committee on Banking Supervision—CAP50 Prudent valuation guidance.

[9] European Banking Authority—Regulatory Technical Standards on prudent valuation.

Table 2 SWD financial assets

SWD exit cost components	Accounting adjustments	Prudent valuation adjustments
Market price uncertainty adjustment. Bid-offer costs/ model risk adjustment	Yes	Yes
Credit valuation adjustment, funding valuation adjustments, collateral valuation adjustment	Yes	Yes
Concentrated risk costs	No	Yes
Future administrative costs	No	Yes
Capital valuation adjustment	No	No
Other costs	No	No
Ticket-based costs	No	No

Source: Authors

under a SWD scenario different assumptions or overlays might have to be considered.

Accounting and Capital Reserve

The evaluation of the exit and operational costs associated with winding down a portfolio should be based on a combination of common, alternative, and additional assumptions and arrangements to those assumed under the accounting or prudential framework. In turn, the existing valuation reserves recognized under the different frameworks will ultimately generate capital benefits upon the exit of portfolio. Assume an outright Equity Vega exposure for which the institution has booked an accounting B/O adjustment, the bid-ask spread charged by the counterparty to flatten the risk will be compensated by the release of the accounting B/O adjustment.

All things being equal, the net capital impact of a SWD is to be the difference between the net expected cash flows impact resulting from a SWD (i.e. excess cash flows over fair value, balance sheet correction—i.e. DOP deferral-, net liquidity, and funding adjustments) and the adjustments already capitalized to regulatory capital. As an overview, the table below provides a list of the main derivative pricing drivers (non-exhaustive list) that might be considered by market practitioners when running a SWD analysis with an indication as whether those elements are commonly captured under the accounting or prudential frameworks (Table 2).

Each component is further detailed later in this section.

Fair Value and Market to Model Instruments

As we highlighted in the previous chapter, when Fair Value cannot be determined based on observed prices in active market, the Fair Value is estimated using valuation techniques. Those valuation techniques include statistical models and model parameters often derived from mid-market prices of instruments traded in active market. Where no observable information is available, the valuation technique can be based on proxy or historical approach.

When marked-to-model instruments are valued using mid-market prices, several adjustments are required to adjust the mid-model value:

- Market risk adjustments
  - Market price uncertainty adjustment (MPU) aims at covering the uncertainty of market parameters used to determine the mid-model value of a financial instrument
  - Close-out-costs adjustment (CoC): It is required to adjust the model mid-value to the appropriate bid/offer market value
  - Model risk adjustments (MoRi): This additional adjustment might be required when uncertainty arise from the model used in the valuation technique (e.g. risks not reflected in the valuation technique, specific market premium to account for the price difference between the cost-to-hedge value (hedging cost of the risk components) and the market price of the instrument
- Credit risk adjustments also commonly referred to as Credit Valuation Adjustment (CVA)
  Usually not included in the mid-model values or bid/offer prices, this adjustment reflects the expected credit losses arising from a derivative credit exposure considering the effect of collateral agreement, the counterparty credit risk and the loss given default of the counterparty
- Funding adjustments
  Funding adjustments might be required to discount derivative exposures at an appropriate market funding level. Funding Valuation Adjustment (FVA) is used for uncollateralized OTC derivatives to reflect the term cost of funding, and Collateral Valuation Adjustment (COLVA) accounts for the valuation difference between interbank offer rates discounting and the overnight rates discounting.

Those three costs categories usually known globally as "Fair Value adjustments" or "Fair Value reserves" are commonly embedded in the bank's valuation processes when measuring fair value and correspond to specific risk adjustments over valuation model mid-values.

Those adjustments might in several cases rest on assumptions or accounting choices that could be challenged under a SWD scenario, regarding for instance:

- *The level at which fair value adjustments are calculated*

Under the discretion allowed by accounting standards, institutions can decide the level at which offsetting positions and risks can be aggregated when measuring fair value (also known as portfolio exception). However, the netting level used for accounting purposes might not necessarily reconcile with how an institution would economically unwind its portfolio or risk under a SWD (e.g. group level vs. region/business/products level). Consequently, the price that would be received to sell a package of products might not necessary sum up with the accounting exit price estimated based on net position calculated at a different node level. As a result, additional hedging costs could possibly arise from replacement trades to replace or re-hedge the risk exposure upon close-out when using different netting assumptions.

– *The choice of the pricing convention*

Accounting standards do not preclude the use of one specific pricing convention but require institutions to use judgment to decide which price within the bid-ask spread that is most representative of fair value. In that respect, among a range of market practices, institutions providing market-making services might, for instance, consider that mid-prices are most representative of fair value. Conversely a bank acting as a broker could elect for a policy requiring using bid prices for long positions (assets) and ask prices for short positions (liabilities). In a SWD context, it is of major importance to understand the effect of the SWD exit strategy on the institution's capability to exit its portfolio according to the elected accounting pricing convention. All in all, additional bid-offer costs, in particular due to the loss of market maker exemption, should be regarded.

From Fair Value to SWD Value

As described above, Realizable Value and Fair Value tend to differ with Fair Value typically not embedding all pricing components.

$$\text{Realizable Value} \neq \text{Fair Value}$$

With:

$$\text{Realisable Value} = \text{Fair Value} + \text{Additional adjustments}$$

- Fair Value = "Risk – free" Model mid – value + MPU + CoC + MoRi + CVA + Funding adjustments
- Additional adjustments = Ticket based cost + Concentrated risk cost + Future Administrative cost + MVA + KVA + Other

At the opposite, it is expected that SWD value would be more aligned with Realizable Value, though not fully identical as both regimes follow different objectives.

$$\text{SWD Value} \approx \text{Realisable Value}$$

Key differentiating assumptions between SWD Value and Realizable Value relate to:

– Expected operational costs

The Realizable Value may incorporate the "likely to be incurred" additional costs (future hedging costs and administrative costs) of the investor to risk manage the portfolio until maturity (see 7.3), whereas SWD value may include not only the "likely to be incurred" additional costs (for position planned to be early terminated), but also all direct and indirect costs of the institution holding the position during the execution of the SWD plan, including re-hedging costs that could be faced during the close-out horizon.

– Costs of future financial resources

The costs of the anticipated capital and leverage resources may be estimated from the perspective of a market participant for the Realizable Value while the SWD Value may incorporate an overlay to account for the institution's own future cost of capital for holding a position until its derecognition from the institution's balance sheet.

– Market risk and liquidity costs

The evaluation of the Realizable Value may incorporate anticipated hedging and transaction costs that may not be relevant to consider in the SWD Value for rump positions.

The additional adjustments to Fair Value required to reflect the amount that could actually be realized through exiting a position and estimate ultimately the SWD value are presented in the following sections.

Ticket/Portfolio-Based Costs

When exiting cash position (ex. Structured loans) or derivative position, counterparts might face one-off costs that are likely to be factored into the bidding price (e.g. legal, advisory, tax, or additional administrative costs).

Concentrated Risk Costs

Concentration risk has been traditionally analysed in relation to credit activities. However, concentration risk does not only arise from credit activities but also from market activities. For this paper, the definition of concentration risk is aligned with the Joint Forum's working definition of concentration risks,[10] i.e. the risk arising

[10]BCBS publication on Cross-sectoral review of group-wide identification and management of risk concentrations.

from sizeable exposures, relative to the market liquidity and market volume, that may arise within or across different risk categories throughout an institution and can be difficult to unwind or hedge without significantly influencing the market price because of inadequate market depth.

Concentrated position can have potential to produce large losses and threaten institution's financial and capital situation, during periods of market stress. This explains why Supervisors promoted the development of holistic concentration risk management capabilities, notably in terms of recognition of concentration risk in pricing practices. However, unlike the incorporation of concentration risk in bank's pricing practices or capital prudent valuation adjustment for banks subject to prudent valuation requirements, accounting setters do not permit to reflect in fair value measurement premiums or discounts for concentrated position. As a general principle, the unit of account for financial instrument in the accounting standards is usually the individual financial instrument, expect when portfolio exception is used. In the latter case, the fair value of a portfolio is then conducted based on net risk position.

To conclude, additional premium for concentrated position should be considered by financial institutions when conducting a SWD analysis.

Future Administrative Costs

With maybe a few exceptions, market participants do not usually consider Future Administrative Costs as a risk and therefore do not integrate it in their valuation techniques when measuring fair value. As an example, when the portfolio is measured on a "cost-to-hedge" basis, e.g., using dynamic self-financing replication strategy, there is often an assumption of complete and frictionless market such as a financial product can be perfectly replicated using costless continuous rebalanced trading strategy.

However, in real world, a perfect replication strategy might be impossible. Therefore, even an optimal hedging strategy (i.e. limited hedging activities within risk appetite) is likely to generate transaction costs when readjusting the hedge, and administrative costs to risk manage the portfolio. In this respect, it could be reasonable to assume that complex products, like products sensitive to stochastic volatility and non-constant correlation risks, might attract higher costs due to more intensive risk management activities, more frequent rebalancing of the hedging portfolio, and incremental ad-hoc back-office treatments.

Additional premium should be considered by financial institutions when conducting a SWD analysis considering:

- The "likely to be incurred" additional costs that might be charged by a willing buyer to risk manage the portfolio until maturity, namely future hedging and administrative costs.
- The direct and indirect costs of running the business during the execution of the SWD plan, including the re-hedging costs that could be faced by the Institution during the close-out horizon.

Direct and indirect costs should be as comprehensive as possible and ideally cover among other elements the staff costs (including severance packages), IT costs, costs for facilities, costs from professional services, for infrastructure or legal services. Associated costs should also reflect the different operating models and intragroup dependencies.

Derivative Trading Wind-Down: XVA

Derivative trading activities generate a number of different costs, all together known as XVAs, that were progressively factored by the industry in derivative pricing practices; the credit or debit valuation adjustment of an uncollateralized derivative (CVA and DVA), the Funding Valuation Adjustment (FuVA), the costs of funding collateral also known as Collateral Valuation Adjustment (COLVA), the margin valuation adjustment (MVA), and the capital costs of a derivative (i.e. the Capital Valuation Adjustment, KVA). However, except for the more traditional XVAs related credit and funding adjustments (CVA/DVA, the FuVA, and COLVA), MVA and KVA are not considered yet for fair value or prudent value measurement purposes.

Initial Valuation Adjustment: MVA

MVA is a relatively new concept into the XVA family. It aims at capturing the additional costs resulting from the introduction of mandatory bilateral Initial Margin rules for non-centrally cleared derivatives (Initial margin protects derivative counterparts from gap risk, i.e. market movements experienced during the liquidation period of portfolio of the defaulting counterparty, also referred to as margin period of risk).

It is likely that for trades requiring mandatory initial margin (IM) posting a buyer would expect a discount to cover the corresponding funding costs. Hence, those costs should be factored in the exit price of a derivative portfolio under a SWD exercise.

Capital Valuation Adjustment (KVA)

Learning from the global financial crises, Regulators with the introduction of Basel III led banks to strengthen their capital requirement for holding derivative, including for market risk-free portfolio. Consider an unsecured 10–25 years swaption traded with a corporate client together with its hedging transaction. Assuming that during a wind-down period banks need to continue complying with regulatory requirements, both the trade and the hedging transaction create regulatory leverage, individually market risk (unless perfectly hedged), counterparty credit risk and CVA capital requirements.

Capital Valuation Adjustment, also referred to as KVA, reflects the cost of this additional derivative regulatory capital requirement, and consistency affected how market participants were used to price derivatives. Conceptually KVA requires to simulate future capital usage throughout the life of the trade. It represents a complex quantitative and computational challenge which arises from multiple elements:

– *The absence of market consensus on how to convert future RWA and regulatory leverage into future capital requirements*

KVA computation has not reached a mature state yet. Under a SWD KVA should reflect the KVA costs of the most punitive regime (most conservative approach).

– *How and as whether upcoming regulatory changes such as incoming fundamental review of the trading book (FRTB) should be considered*

The observation of the evolution of the XVA topic in the banking practices does not provide clear answer to this question. However, under a SWD it would be conservative to consider that a market participant would likely include the capital impact of incoming regulatory changes, notably for long dated and capital-intensive derivatives (transactions likely to be the most affected).

– *The choice of the underlying assumptions regarding the potential primary buyers and other related KVA inputs*

The KVA amount charges by a market participant, i.e. its cost to capital, might be very sensitive to heterogenous constraints such as its regulatory regime,[11] its type and activities, the regulatory effect of netting agreements.

Therefore, the KVA computation should not only be economically sound and reflect market generally accepted assumptions but also be consistent with the exit strategy and idiosyncratic constraints of identified primary buyers. For instance, primary buyers with high RWA constraints might offer less competitive prices compared to low RWA constrained institutions. Compare the effect of regulatory CVA impact of a portfolio made of transactions with non-financial counterparties, European banks exempted from CVA requirements for Non-Financial Counterparties would charge no or limited capital costs while US banks not benefiting from this exemption will. Similarly banks subject to capital prudent valuation requirements would be at competitive disadvantage compared to counterparts evolving under different regulatory regime.

[11] For example: Minimum leverage ratio target of 3% for European banks under the revised Capital Requirements Regulation as opposed to a supplementary leverage ratio (SLR) set around 5–6% for US G-SIBs.

Other Additional Costs to be Considered

1. Liquidity and Funding Costs

The wind-down of trading book activities is likely to impact the overall funding profile of an institution and might require treasury measures such as new issuances (in case of deficit of liquidity) or buy-back (in case of excess of liquidity). While exiting cash position might be immediately accretive to liquidity, on the other side excess of liquidity might increase negative carry, hence causing a significant drain on the bank's net interest earnings and additional SWD costs for remaining businesses.

2. Climate Risk Related Adjustments

Since recently, there is a growing attention from Supervisors on climate and environmental risks and on associated risk management practices, including in terms of pricing practices. For now, this topic has not translated to a change of derivatives practices. Several challenges need to be overcome first to adapt with the required level of sophistication the risk identification and assessment processes for climate and environmental risks. However, derivative pricing being an evolving framework, it may not be surprising seeing emerging climate change valuation adjustment soon.

DOP Deferral: A Specific Balance Sheet Correction

One of the principal drivers of the trading revenues for corporate and investment banks is what is commonly known as the revenue of the first day, also referred to as Day One P&L (DOP hereafter). DOP arises at the trade recognition from the difference between the transaction price and the exit price. Assume an exotic note sold in the over-the-counter market where the bank includes a spread when selling the product to cover its risks and generate a profit, the DOP would correspond to the difference between the price offered to the client (i.e. transaction price) and the cost to exit the corresponding market risk on the wholesale market (exit price).

According to IFRS 9 B5.1.2.1 A, the best evidence of the fair value of a financial instrument at initial recognition is normally the transaction price. If the fair value at initial recognition differs from the transaction price, the difference between the fair value and the transaction price as a gain or loss shall be recognized upfront to Profit and Loss account (P&L hereafter) if the fair value is observable.[12] In other cases, the difference between the fair value at initial recognition shall be deferred, then recognized through P&L during the remaining life of the trade (to the extent that

[12] i.e. evidenced by a quoted price in an active market for an identical asset or liability (i.e. a Level 1 input) or based on a valuation technique that uses only data from observable markets.

the difference arises from a change in a factor—including time—that market participants would consider when pricing the asset or liability).

From an economic perspective, this DOP deferral corresponds to a stock of uncertain unrealized P&L. Therefore, when a position is exited, the corresponding residual DOP deferral has to be recognized through P&L. Consider the sale of a transaction at 800€, with a fair value of 1000€ and a residual DOP deferral of 100€, the net P&L effect would be 100€ (+800€-1000€+100€). To put this into SWD perspective, the recognition of stock of DOP deferral upon the exit of related transaction will generate capital effect, either capital benefit when DOP deferral corresponds to unrealized gain, or capital drag when DOP corresponds to an unrealized loss.

Estimating Rump Costs

According to the FSB discussion paper on SWD of derivatives and trading book portfolios, the rump portfolio consists of the residual "sticky" positions, i.e. positions which will remain on the balance sheet of the institution after performing the wind-down exercise. Rump portfolio can be decomposed further between:

- Discretionary rump. Positions where costs to exit is greater than the hold-to-maturity value (expected or contractual maturities).
- Non-discretionary rump. Positions hard to liquidate despite reasonable efforts because of their complexity, illiquidity, lack of attractiveness, or the existence of any dependencies blocking the possibility for winding down (e.g. legal or regulatory hurdles).

The primary objective of the estimation of the all-in maintaining cost of a rump portfolio is to take a reasoned decision on the exit strategy that should be applied (trade-off between the cost of holding the trade and selling it). It consists mainly in estimating the future cash flows of the rump portfolio until its maturity. This estimation should consider a wide range of factors and reflect:

- The expected resources consumption of the rump portfolio over its expected lifetime, in terms of both financial resources (Capital, Leverage, Funding) and future operational costs.
- The expected credit losses (i.e. future impairments) of the asset until its maturity (irrespective of the existing impairment status of the asset at the SWD reference date).

Given the high degree of judgement necessary in this exercise, the estimation should be based on prudent assumptions.

Reference

Valukas, A. (2010, March 11). Report of the examiner in the Chapter 11 Proceedings of Lehman Brothers Holdings Inc. Retrieved from http://jenner.com/lehman/

Glossary

AVA Additional valuation adjustment
Basis point One hundredth of one percentage point (1%/100)
BRP Business reorganization plan
BRRD Bank Recovery and Resolution Directive
BU Banking Union
CBC Counterbalancing capacity
CET 1 Common equity Tier One capital
CF Critical function
CSA Credit support annex
DCF Discount cash flow
DDM Dividend discount model
FMI Financial market infrastructure
FOLTF Failing-or-likely-to-fail
FSB Financial Stability Board
G-SIBs Global Systemically Important Banks
IRT Internal Resolution Team
ISDA MA International Swaps and Derivatives Association Master Agreement
LLP Loan loss provisions
LLR Loan loss rate
LTD Loan to deposit ratio
MIS Management information system
NPL Non-performing loan
NRA National Resolution Authority
OFI Other Financial Institutions
PE ratio Price Earnings ratio
PEG PE divided by a growth rate
RWA Risk-weighted assets
SPV Special Purpose Vehicle
SWD Solvent wind-down

A. Santoni, F. Salerno, *How to Value a Bank*, Springer Texts in Business and Economics, https://doi.org/10.1007/978-3-031-43872-1